REFRIGERATION AND AIR CONDITIONING

FOURTH EDITION

About the Author

C P Arora was formerly Professor, in the Department of Mechanical Engineering, Indian Institute of Technology, Delhi. He did his MS from the University of Illinois, USA, under the TCM programme and was the first to obtain a PhD in engineering from the Indian Institute of Technology, Delhi. He has guided 11 students in completing their PhD theses. He has over 38 years of teaching experience and has been a Visiting Faculty at the University of Leeds, UK, and Visiting Professor at the University of Basrah, Iraq and California State University Sacramento, USA.

Professor Arora is a life member and was the President (1979–80) of the Indian Society of Mechanical Engineers. He was Chairman, NCST Panel of Refrigeration and Air Conditioning (1974–86); Chairman, Organizing Committee, Fourth National Symposium on Refrigeration and Air Conditioning (1975) and Editor of the *Journal of Thermal Engineering*. He has also published a number of research papers.

REFRIGERATION AND AIR CONDITIONING

FOURTH EDITION

C P Arora

Former Professor
Department of Mechanical Engineering
Indian Institute of Technology, Delhi

McGraw Hill Education (India) Private Limited

Published by McGraw Hill Education (India) Private Limited,
Registered Office: Anjana Complex No: 5/90 A, Butt Road, St. Thomas Mount, Chennai – 600016, Tamil Nadu, India.

Refrigeration and Air Conditioning, 4e

Fifth reprint 2026

Print Edition
ISBN -13: 978-93-90385-84-3
ISBN -10: 93-90385-84-9

E-book Edition
ISBN -13: 978-93-90385-92-8
ISBN -10: 93-90385-92-X

06 07 08 09 10 RAJ 30 29 28 27 26

Typeset at TVH Scan, WZ 391, Madipur, New Delhi 110 063, and printed and bound in India at Rajkamal Electric Press, Plot No.2, Phase-IV, Kundli, Haryana

Cover Designer: APS Compugraphics
Cover Image Source: Zaretska Olga / Shutterstock

Visit us at: www.mheducation.co.in
Toll free in India: 1800 103 5875
Write to us at: support.india@mheducation.com
CIN: U80302TN2010PTC111532

To

My Beloved Family

Sarla
Amitabh
Shubhra
Smita
and
Sangeeta

Contents

Preface

The need for a modern textbook in the field of refrigeration and air conditioning has been felt for a long time. This book presents a basic as well as applied thermodynamic treatment of the subject in a very comprehensive manner based on years of teaching and learning effort at the Indian Institutes of Technology, Mumbai and Delhi, and interaction with the industry.

The book is intended to serve as a textbook for undergraduate and postgraduate students of engineering. It should also serve as a useful reference for practising engineers. A conscious effort has been made in this book to maintain a reasonable level of rigour, but at the same time to employ simple techniques for solving fairly complex problems. Throughout the book, emphasis has been laid on physical understanding while at the same time relying on simple analytical treatment. A sound physical basis has also been laid for obtaining fairly precise estimates of refrigeration and air-conditioning equipment.

The presentation of the subject follows the classical line of separately treating the topics in refrigeration and air conditioning, the two being linked via the medium of the refrigerant evaporator. Accordingly, Chapters 1 to 13 are devoted to refrigeration and Chapters 14 to 22 to air conditioning. Chapters 23 and 24 deal with motors and controls and applications of refrigeration and air-conditioning process in food preservation.

The text and illustrative examples are in SI units throughout the book. Charts and tables, such as pressure-enthalpy diagrams for refrigerant 11 and carbon dioxide, enthalpy-composition diagrams for ammonia-water and lithium bromide-water systems, tables for solar radiation heat gain through glass, equivalent temperature differentials for walls and roofs, etc., have been adapted in SI units and are provided along with others, such as pressure-enthalpy diagram for refrigerant 12, psychrometric chart, etc.

Any claim to originality that may be advanced for the material presented here in refrigeration is with respect to (i) Ewing's construction to find the suction state for maximum COP, (ii) a comparison of refrigerants based on normal boiling points thus introducing the concept of thermodynamic similarity, (iii) a study of azeotropes, (iv) class of service of compressors, (v) illustrative examples on both air-cooled and water-cooled condensers, (vi) the sizing of the capillary tube according to Fanno-line flow, (vii) the influence of a refrigerant on the augmentation of boiling heat transfer, (viii) heat-transfer analysis of both dry and flooded evaporators, (ix) the simulation of the vapour compression system, and (x) the analysis and calculations for mixtures in the vapour-absorption system using enthalpy-composition diagrams.

The approach to the subject of air conditioning is both fundamental and practice-oriented. A basic calculation procedure is given for the preparation of psychrometric charts. Lucid explanations, expressions and diagrams are given to develop the understanding of sensible, latent and total heat processes and loads. A separate chapter is devoted to solar radiation, leading not only to the study of solar-heat gains and cutting-solar load, but also to provide to the reader the basic knowledge to enable him to design systems for solar-energy utilization. The chapter on air-conditioning equipment design makes use of the concept of enthalpy potential involving simultaneous heat and mass transfer. Examples on air transmission include the static regain method of duct designing which leads to a balanced air-distribution system.

Chapter 23 adequately fills the need to provide essential information on the electrical aspects of the control of refrigeration and air-conditioning equipment. It also gives methods for the control of room conditions at partial loads. Finally, Chapter 24 takes up typical applications of refrigeration and air-conditioning to food preservation. These include chilling, freezing, freeze-drying and heat-drying.

The twentieth century saw large scale development in commercial refrigeration and air conditioning, particularly after du Pont introduced a family of chloro-fluorocarbons, the so-called CFCs with the trade name of Freons. Now, as the new century begins, another revolution is taking place in the industry for replacing these very CFCs with alternatives on account of the ozone-depletion-potential of these refrigerants. These developments have been included in the revised edition of the book.

Many research and postgraduate students are interested in evaluating thermodynamic properties of new refrigerants and refrigerant mixtures. The basic procedure to evaluate the thermodynamic properties of *pure refrigerants* is, therefore, given in Chapter 1, and the same for *ideal and non-ideal mixtures* and particularly *Propane/Isobutane mixtures* in Chapter 4. Chapter 4 on refrigerants contains an exhaustive treatment of the topics *substitutes for CFC Refrigerants*, particularly CFC 12, and *Non-isothermal Refrigeration* using non-azeotropic mixtures of refrigerants. In addition, empirical relations for thermophysical properties of refrigerants, and *supercritical vapour compression cycle* for CO_2 as refrigerant with a potential to substitute for CFCs are also given in this chapter.

Chapter 9 on Evaporators includes many illustrative examples for *simulation and design of flooded and direct-expansion chillers* which include pressure drop calculations and use of *Slipcevic correlations* for tubes with roughened surfaces.

Since water-lithium bromide system has recently gained some popularity with the use of waste heat for refrigeration, the representation of vapour absorption cycle on *lnp versus 1/T* diagram and practical *single-effect* and *double-effect water-lithium bromide vapour absorption cycles* have been described in Chapter 12 on Vapour Absorption System.

In Chapter 20 on Design of A/C Apparatus the treatment of the topic has been greatly extended to include determination of *air-side heat transfer coefficient* and *cooling tower selection*. Examples include those on induced-draft counterflow and crossflow atmospheric cooling towers.

Prominent features added in the second edition were

(i) *Standard rating cycle for domestic refrigerators and second law efficiency* in Chapter 3

(ii) *Calorimetric method of determining refrigerating capacity of hermetic compressors* in Chapter 6, *R 22 centrifugal compressors* in Chapter 6 also due to the present trend of their use as substitutes for R 11 chillers

(iii) *Linde–Hampson process for liquefaction of gases* in Chapter 11; also, *reversed stirling cycle* in this chapter due to the application of this cycle in a big way in Philips Liquefier

(iv) *Clean spaces* in Chapter 16 and processing and transmission of air in clean rooms in Chapter 21

(v) *Flat-plate solar collector* in Chapter 17 as an extension of the topic of solar radiation

(vi) *Water vapour transmission* and use of vapour barriers in Chapter 18

(vii) *Building design features* and measures for conservation of energy in Chapter 19

(viii) *Static regain method* of *duct design* in Chapter 21

(ix) Example on *conversion of split-phase motor into capacitor-start motor* to increase starting torque which may help using compressor of one refrigerant with another refrigerant in Chapter 23

(x) *Freeze-drying of Yoghurt* in Chapter 24

Further, a major contribution to this edition is in the form of a detailed Appendix which is now presented in three parts as follows:

A. *Correlations* on thermodynamic properties of refrigerants R 12, R 152a, R 22, R 290 and R 600a

B. *Tables* on thermodynamic properties of the above and other refrigerants, R 290/R 600a mixtures, etc.

C. *Charts*

When the *second edition* was published in 2000, the refrigeration and air-conditioning industry was embarking on to an era of new refrigerants. Due to the problem of the depletion of the ozone layer, CFC refrigerants R 11, R 12, R 113, R 114, and R 502 were to be phased out on 31.12.2000, and alternative HFC and HCFC refrigerants were to be used from 1.1.2001.

The second edition did provide a study of the alternative refrigerants which were planned. But since 2000, certain new refrigerants have taken their place as substitutes. They have come to be accepted by the industry, and plants working on them have been designed and installed. For example, HFC 134a now occupies place of pride as a substitute for CFC R 12. However, HCFC R 22 continues to be used and loved by the industry, although an HFC blend R 410A is also favoured by some. At the same time, there is a newfound enthusiasm for ammonia. Further, HCFC R 123 has now replaced CFC R 11. Both the HCFCs, R 22 and R 123, are permitted for use till 2030.

In the *third edition* which was published in 2009, there was great emphasis on HCFC R 123 as centrifugal chiller refrigerant and HFC R 134a as long term alternative for many applications. Emphasis on HCFC R 22 and ammonia remained the same. Detailed comparisons were made, however, between HCFC R 22 and its HFC alternatives, and between HCFC R 123 and its alternatives.

After 2009, much development has taken place and certain refrigerants have been widely accepted. Hence, it has become absolutely necessary to include these developments in the fourth edition of this book. The following are the highlights of this edition:

- Discussion on HFC R 600a (Isobutane) as an alternative refrigerant to CFC R 12, which is used in refrigerators
- Covers HFC R 32 (Difluoromethane) as an alternative to HCFC R 22 that is used in air conditioners and its comparison with R 22 and R 410a
- Includes a new-found miracle HFC R 134a which is widely used in refrigerators and air conditioners and its comparison with HCFC R 123. Though HCFC R 123 continues to be in use, designers and manufacturers have also adapted the highly improbable HFC R 134a for use in centrifugal chillers in spite of it being a much lower boiling than HCFC R 123, high pressure and smaller volumes refrigerant.
- New section on Inverter Technology for capacity control of compressors by varying the speed

Thus, this edition is enriched with the latest information on refrigerants that are being phased out and new refrigerants that have been adopted by the industry.

To aid the instructors, the book also offers online supplements which are available at https://connect.mheducation.com*.

Acknowledgments

I bow with gratitude before the Divine Father, Mother, Friend, and Beloved, the source of all knowledge, Who made me an instrument to write this book.

At this juncture, I remember my father's words: "My investment is in my children". Truly speaking, the benefits of this book flow from the investment made by my father.

I want to express my heartfelt gratitude to the Divine for the Love, Kindness and Affection bestowed on me through my children and their spouses: Sangeeta–Vivek, Smita–Rajat, Shubhra–Hemant, and Amitabh–Shailaja and grandchildren Himali, Ishika, Vaibhav, Aakriti, Shreya, Atyant, and two new and loving grandchildren, Anisha and Rishi, born since the publication of the last edition.

I am indebted to my numerous students whose stimulating interest inspired this work. I am extremely grateful to my many friends and colleagues for appreciating the value of such a book and for urging me on to its completion. They include Prof. A K De and Prof. B B Parulekar of IIT Bombay, Prof. C P Gupta of the University of Roorkee, Prof. R D Garg, Prof. H B Mathur, Prof. S M Yahya, Prof. Prem Vrat, Prof. O P Chawla, Dr R S Agarwal, Dr P L Dhar and Dr M S Das of IIT Delhi, Prof. YVSR Sastry of the Delhi College of Engineering, Dr N J Dembi of Regional

* To know how to access the online material, please see the book cover.

Engineering College, Srinagar, Dr S N Saluja of Hull College of Higher Education, Mrs L I Trifonova of Higher Technological Institute, Sofia, Mr R S Mital of Voltas Limited, Mumbai, and Mr S K Mehta of Bhabha Atomic Research Centre, Mumbai.

I would like to express my grateful thanks to Cambridge University Press for granting permission to include some tables and charts in the Appendix from "Thermodynamic Tables in SI Units" by Haywood, and also to E.I. du Pont de Nemours and Co. for similar permission to include the thermodynamic properties of R 22.

I would also like to thank the Indian Institute of Technology, Delhi, for providing partial financial support in the preparation of the manuscript.

I am grateful to my research student Dr T P Ashok Babu of REC, Surathkal, for his timely and excellent piece of work on substitutes for CFC 12 and for permitting me to include data and tables from his thesis in this book.

I would also like to thank the reviewers of all the previous editions of this book for taking out time for providing their valuable feedback for enhancing it further.

May this wonderful subject of Refrigeration and Air Conditioning, and this book inspire teachers, students, and practicing engineers to explore new vistas in the field. Please feel free to send in your feedback at the book's website.

C P Arora

Publisher's Note:

McGraw Hill Education (India) invites suggestions and comments from you, all of which can be sent to *info.india@mheducation.com* (kindly mention the title and author name in the subject line).

Piracy-related issues may also be reported.

List of Principal Symbols

Capital letters

A Area
A_F Face area
C Velocity, thermal conductance, concentration (in mass transfer), clearance factor, heat capacity/specific heat
D Diameter, diffusion coefficient, mass of vapour distilled from generator
E Emissive power
F Force, genometric factor, rich solution circulation
G Mass velocity
H Enthalpy, head
I Solar radiation intensity
I_D Intensity of direct solar radiation
I_d Identity of diffuse solar radiation
K Dynamic loss coefficient
L Fin width, length, air mass
M Molecular weight, stability criterion in finite difference approximation
N Number of tubes
P Perimeter, power requirement
Q Heat transfer
Q_L Latent heat transfer
Q_s Sensible heat transfer
$\dot{Q}_v$ Volume flow rate of air
R Gas constant, thermal resistance
S Entropy
T Absolute temperature
U Internal energy, overall heat transfer coefficient
V Volume
V_p Piston displacement
W Work, moisture content of material
X Bypass factor

Small letters

a Velocity of sound, absorptivity
c Specific heat

c_p Specific heat at constant pressure
c_v specific heat at constant volume
d Solar declination angle
f Heat transfer coefficient, friction factor, specific rich solution circulation
g Acceleration due to gravity
h Specific enthalpy, heat transfer coefficient, hour angle
h_M Mass transfer coefficient
k Thermal conductivity
k_d Diffusion coefficient
k_ω Diffusion coefficient based on specific humidity
l Fin height, tube length
m Mass, polytropic index of expansion
n Polytropic index of compression, number of moles, recirculation number
p Pressure
Δ_p Pressure loss
p_s Static pressure
p_T Total pressure
p_v Velocity pressure
q Heat flux, heat transfer per unit mass
r Radius, compression ratio, reflectivity
s Specific entropy
t Celsius temperature
t_e Sol-air temperature
Δt_E Effective temperature difference
u Specific internal energy, tangential velocity
v Specific volume
w Specific work, moisture removal
x Distance, dryness fraction, liquid phase mole fraction
y Vapour phase mole fraction
z Height above datum

Greek letters

α Thermal diffusivity, wall solar azimuth angle
β Coefficient of thermal expansion, solar altitude
γ Adiabatic index, solar azimuth angle
δ Joule Thomson coefficient
ξ Coefficient of performance
ε Emissivity, heat exchanger effectiveness
λ Decrement factor
η Efficiency
η_p Polytropic efficiency
ϕ Flow coefficient, relative humidity, time lag
σ Stefan-Boltzmann constant, surface tension
μ Dynamic viscosity, head coefficient, degree of saturation
ν Kinematic viscosity
ρ Density

ψ Zenith angle
ξ Concentration by weight
τ Time, transmissivity
θ Angle of incident, excess temperature
ω Specific humidity, angular velocity
$\mathscr{H}_u$ Lockhart-Martinelli parameter for two phase turbulent flow

Dimensionless numbers
Bi Biot number
Bo Boiling number
Co Condensation number
Fo Fourier number
Gr Grashof number
$\mathbf{K_f}$ Load factor in boiling
Le Lewis number
Nu Nusselt number
Pr Prandtl number
Re Reynolds number
Sc Schmidt number
Sh Sherwood number
St Stanton number
θ Trouton number

Subscripts
A Absorber
C Convective
I Infiltration
R Radiative
S Apparatus dew point, wetted surface
TP two phase
a Ambient, poor solution, dry air
b Black body
c Cold, clearance, condensate, critical
d Dynamic loss, diffusion, vapour from generator, dew point, discharge
e Entrainment
f Friction, saturated liquid, fin, fouling, fluid
fg Vaporization
g Glass, saturated vapour, air-side
h Generator, hot
i Inside, initial
is Isentropic
k Heat rejection
m Log mean
max Maximum
min Minimum
n Nozzle, normal to surface

o	Outside, heat absorption or refrigeration, molar, stagnation
r	Radial, refrigerant-side, rich solution, reduced property
rel	Relative
s	Suction, at normal boiling point, saturation, saturated solid
sd	Shading
sg	Sublimation
t	Total, based on extended surface side area
u	Tangential
v	Vapour, volumetric
w	Wall, water
x	x-direction
∞	Free stream

Superscripts

*	Per ton refrigeration, thermodynamic wet bulb
′	Pure substance
L	Saturated liquid mixture, wet bulb
V	Saturated vapour mixture

Visual Preview

Concept Presentation

The student is first introduced to the theories and concepts related to the working of a refrigerator and an air conditioner.

Chapter
1
Introduction

1.1 A BRIEF HISTORY OF REFRIGERATION

The methods of production of cold by mechanical processes are quite recent. Long back in 1748, William Coolen of Glasgow University produced refrigeration by creating partial vacuum over *ethyl ether*. But, he could not implement his experience in practice. The first development took place in 1834 when Perkins proposed a hand-operated compressor machine working on ether. Then in 1851 came Gorrie's air refrigeration machine, and in 1856 Linde developed a machine working on ammonia.

Example 5.1 *A two-stage ammonia food-freezing plant—with a desired capacity of 528,000 kJ/h at – 40°C evaporating temperature and 35°C condensing temperature—has a flash intercooling system with a liquid subcooler. The vapour leaving the evaporator is at –30°C and entering the first-stage compressor is at –15°C. The vapour leaving the flash chamber is superheated by 10°C in the suction line to the second-stage compressor. Water intercooling is done to cool the vapour to 45°C. Adiabatic efficiencies of both compressors are 0.75. The volumetric efficiencies of first- and second-stage compressors are 0.65 and 0.77 respectively. Find the piston displacements, discharge temperatures and power requirements of the two compressors.*

Solution See Fig. 5.5. Reference state is –40°C at which $h_f = 0$, $s_f = 0$.

$$p_k = 1.35 \text{ MN/m}^2 \text{ (at 35°C)}$$

$$p_0 = 0.0718 \text{ MN/m}^2 \text{ (at – 40°C)}$$

$$p_i = \sqrt{1.35 \times 0.0718 \times \left(\frac{273+35}{273-40}\right)} = 0.355 \text{ MN/m}^2$$

The corresponding saturation temperature is

$$t_i = -5°C$$

Solved Examples

Solved Examples are provided in sufficient number in each chapter and at appropriate locations to aid in understanding of the concepts.

Practice Problems

Over 150 Practice Problems are given to provide hands-on practice to students in problem-solving.

Revision Exercises

3.1 A 15 TR R 22 vapour compression system operates between a condenser temperature of 40°C and an evaporator temperature of 5°C.
   (a) Determine the compressor discharge temperature:
      (i) Using the *p-h* diagram for R 22.
      (ii) Using saturation properties of R 22 and assuming the specific heat of its vapour as 0.8 kJ/kg. K.
      (iii) Using superheat tables for R 22.
   (b) Calculate the theoretical piston displacement and power consumption of the compressor per ton of refrigeration.

3.2 A simple saturation ammonia compression system has a high pressure of 1.35 MN/m^2 and a low pressure of 0.19 MN/m^2. Find per 400,000 kJ/h of refrigerating capacity, the power consumption of the compressor and COP of the cycle.

3.3 (a) A R 22 refrigerating machine operates between a condenser temperature of 40°C and an evaporator temperature of 5°C. Calculate the increase (per cent) in the theoretical piston displacement and the power consumption of the cycle:

Table 3.1 Categorization of refrigerants for suction state for maximum COP

Refrigerant	ε_s	t_{2s}, °C	$\dfrac{T_0}{T_{2s} - T_0}$	*Case*
NH_3	4.76	99	2.27	I
C_4H_{10} (Isobutane)	4.26	30	5.74	II
$C_2H_2F_4$ (Refrigerant 134a)	4.6	36.4	5.02	II
C_2H_6	2.41	50	3.97	II
CCl_3F(R 11)	5.03	44	4.38	I
CCl_2F_2 (R 12)	4.7	38.3	4.84	II
$CClHF_2$ (R 22)	4.66	53.3	3.68	I
C_3H_8 (Propane)	4.47	37.61	4.91	II

Tables

Refrigerants in use and alternative refrigerants have been compiled in a list and are compared, and comparison of CFC 11 alternatives is summarized for centrifugal compressors.

Figures

Normal boiling point of refrigerants is emphasized as an important performance criterion.

Apart from numerous self-explanatory figures, an innovative new pH diagram has been developed to estimate the properties of mixed refrigerants, as the need of the day is to find new refrigerant blends as alternatives. This is for PG and research students.

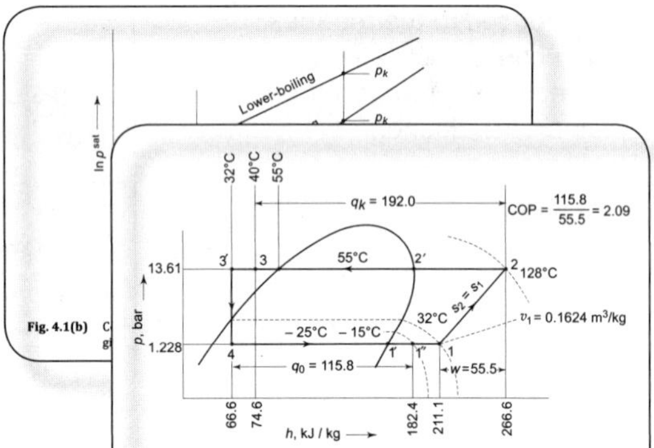

Fig. 4.3 R 12 cycle for refrigerator in Example 4.1

Example 9.4 ***Estimation of D-X Chiller Capacity (Simulation)***

The following specifications are given for an R 22 D-X Chiller.

Condensing temperature, t_k	*43°C*
Saturated suction temperature	*2°C*
Number of passes, n	*8*
Tubes in each pass	*12, 16, 20, 24, 30, 32, 32, 34*
Evaporator superheat	*5°C*
Inlet water temperature, t_{w1}	*11.1°C*
Outlet water temperature, t_{w2}	*7.2°C*
Refrigerant pressure drop in evaporator 0.14 bar (assumed)	
Shell diameter, D_s	*0.406 m*
Tube length between tube sheets, l	*2.213 m*
Tube ID, D_i	*0.0158 m*
Tube OD, D_0	*0.0191 m*

Simulation Problems

For advanced students, procedures to develop computer methods for design, simulation and optimization of refrigeration systems are given.

Chapter 1

Introduction

1.1 A BRIEF HISTORY OF REFRIGERATION

The methods of production of cold by mechanical processes are quite recent. Long back in 1748, William Coolen of Glasgow University produced refrigeration by creating partial vacuum over *ethyl ether*. But, he could not implement his experience in practice. The first development took place in 1834 when Perkins proposed a hand-operated compressor machine working on ether. Then in 1851 came Gorrie's air refrigeration machine, and in 1856 Linde developed a machine working on ammonia.

The pace of development was slow in the beginning when steam engines were the only prime movers known to run the compressors. With the advent of electric motors and consequent higher speeds of the compressors, the scope of applications of refrigeration widened. The pace of development was considerably quickened in the 1920 decade when du Pont put in the market a family of new working substances, the fluoro-chloro derivatives of methane, ethane, etc.—popularly known as chloro fluorocarbons or CFCs—under the name of *R*. Recent developments involve finding alternatives or substitutes for Freons, since it has been found that chlorine atoms in Freons are responsible for the depletion of ozone layer in the upper atmosphere. Another noteworthy development was that of the ammonia-water vapour absorption machine by Carre. These developments account for the major commercial and industrial applications in the field of refrigeration.

A phenomenon called *Peltier* effect was discovered in 1834 which is still not commercialized. Advances in *cryogenics*, a field of very low temperature refrigeration, were registered with the liquefaction of oxygen by Pictet in 1877. Dewar made the famous Dewar flask in 1898 to store liquids at cryogenic temperatures. Then followed the liquefaction of other permanent gases including helium in 1908 by Onnes which led to the discovery of the phenomenon of *superconductivity*. Finally in 1926, Giaque and Debye independently proposed adiabatic demagnetization of a paramagnetic salt to reach temperatures near absolute zero.

Two of the most common refrigeration applications, viz., a window-type room air conditioner and a domestic refrigerator, have been described in the following pages.

1.1.1 Room Air Conditioner

Figure 1.1 shows the schematic diagram of a typical window-type room air conditioner, which works according to the principle described below:

Consider that a room is maintained at constant temperature of 25°C. In the air conditioner, the air from the room is drawn by a fan and is made to pass over a *cooling coil*, the surface of which is maintained, say, at a temperature of 10°C. After passing over the coil, the air is cooled (for example, to 15°C) before being supplied to the room. After picking up the room heat, the air is again returned from the room to the cooling coil at 25°C.

Now, in the cooling coil, a liquid working substance called a *refrigerant*, such as $CHClF_2$ (monochloro-difluoro methane), also called *R 22* by trade name, or simply *Refrigerant 22* (R 22), enters at a temperature of, say, 5°C and evaporates, thus absorbing its latent heat of vaporization from the room air. This equipment in which the refrigerant evaporates is called an *evaporator*.

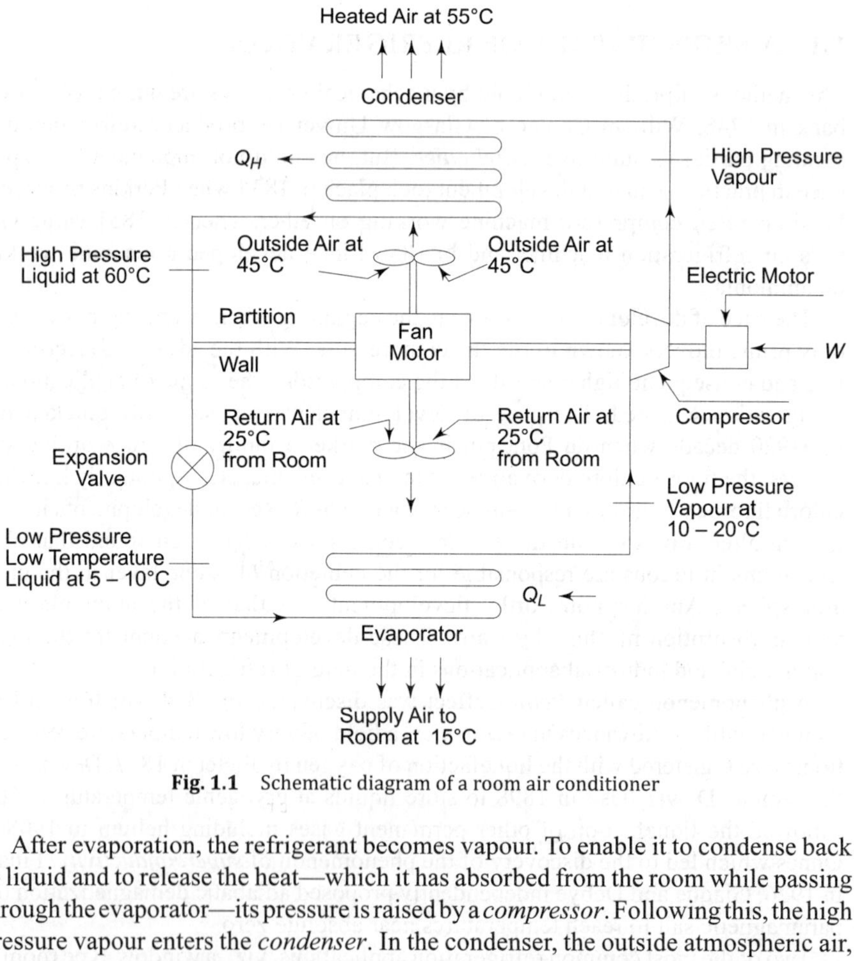

Fig. 1.1 Schematic diagram of a room air conditioner

After evaporation, the refrigerant becomes vapour. To enable it to condense back to liquid and to release the heat—which it has absorbed from the room while passing through the evaporator—its pressure is raised by a *compressor*. Following this, the high pressure vapour enters the *condenser*. In the condenser, the outside atmospheric air, say, at a temperature of 45°C in summer, is circulated by a fan. After picking up the latent heat of condensation from the condensing refrigerant, the air is let out into the environment, say, at a temperature of 55°C. The condensation of refrigerant may occur, for example, at a temperature of 60°C, 5°C above leaving air temperature of 55°C.

After condensation, the high pressure liquid refrigerant is reduced to the low pressure of the evaporator by passing it through a pressure reducing device called the *expansion device*, and thus the cycle of operation is completed. A partition wall separates the high temperature side of the condenser from the low temperature side of the evaporator.

The principle of working of large air conditioning plants is also the same, except that the condenser is *water cooled* instead of being *air cooled.*

1.1.2 Domestic Refrigerator

The working principle of a domestic refrigerator is exactly the same as that of an air conditioner. A schematic diagram of the refrigerator is shown in Fig. 1.2. Like the air conditioner, it also consists of the following four basic components:

(i) Evaporator; (ii) Compressor; (iii) Condenser; (iv) Expansion device.

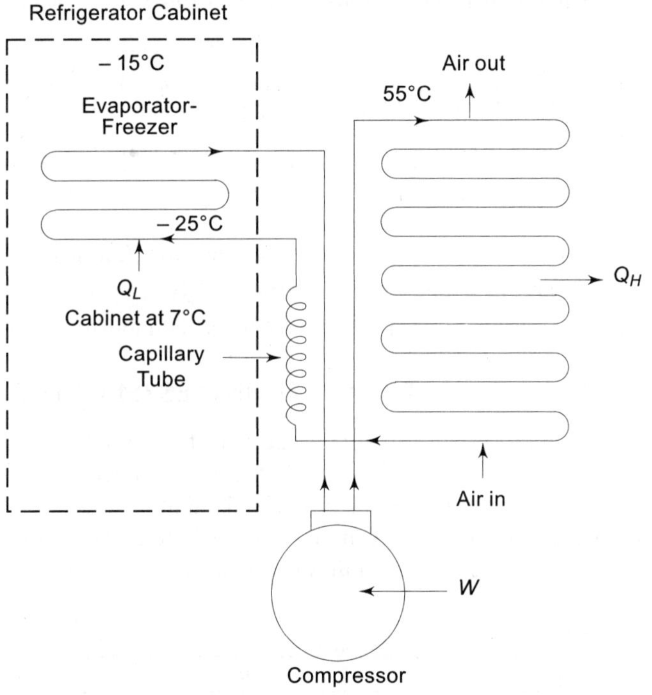

Fig. 1.2 Schematic diagram of a domestic refrigerator

But there are some design features which are typical of a refrigerator. For example, the evaporator is located in the *freezer* compartment of the refrigerator. The freezer forms the coldest part of the cabinet with a temperature of about –15°C, while the refrigerant evaporates inside the evaporator tubes at –25°C. Just below the freezer, there is a *chiller tray*. Further below are compartments with progressively higher temperatures. The bottom-most compartment which is meant for vegetables is the least cold one. The cold air being heavier flows down from the freezer to the bottom of the refrigerator. The warm air being lighter rises from the vegetable compartment to the freezer, gets cooled and flows down again. Thus, a *natural convection current* is set up which maintains a temperature gradient between the top and the bottom of the refrigerator. The temperature maintained in the freezer is about –15°C, whereas the mean inside temperature of the cabinet is 7°C.

The design of the condenser is also a little different. It is usually a *wire and tube* or *plate and tube* type mounted on the body of the refrigerator. There is no fan. The refrigerant vapour is condensed with the help of surrounding air which rises above by natural convection as it gets heated after receiving the latent heat of condensation from the refrigerant. The standard condensing temperature is 55°C.

Note *In both the room air conditioner as well as the refrigerator a long narrow bore tube, called the capillary tube, is employed as the expansion device.*

In the modern no-frost refrigerators, the evaporator is located outside the freezer compartment. The cold air is made to flow by forced convection by a fan.

Working Substances in Refrigerating Machines The working substances being used in air conditioners are R 22, R 32, R 134a and R 410a. In refrigerators R 12 has been used before the year 2000. But R 12 is a CFC (chlorofluorocarbon). Because of the ozone-layer depletion problem, alternatives such as the following are being used in place of R 12.

(i) Refrigerant 290 or R 290, viz., Propane (C_3H_8).

(ii) Refrigerant 134a or R 134a, viz., Tetrafluoroethane ($C_2H_2F_4$)

(iii) Refrigerant 600a or R 600a, viz., Isobutane (C_4H_{10}).

Presently R 600a, Isobutane, is being favoured by the industry for refrigerators.

1.2 SYSTEME INTERNATIONAL D'UNITES (SI UNITS)

SI or the International System of Units is the purest form and an extension and refinement of the traditional metric system. In SI, the main departure from the traditional metric system is in the use of *Newton* as the unit of force.

There are six basic SI units as given in Table 1.1. The units of other thermodynamic quantities may be derived from these basic units.

Table 1.1 Basic SI units

Quantity	*Unit*	*Symbol*
Length	metre	m
Mass	kilogram	kg
Time	second	s
Temperature	kelvin	K
Electric current	ampere	A
Luminous intensity	candela	cd

The unit of temperature is kelvin which measures the absolute temperature given by

$$T = t + 273.15$$

where t is the Celsius temperature in °C.

1.2.1 Unit of Force

Force F is proportional to mass m and acceleration a, so that

$$F = C(m)(a) \tag{1.1}$$

where C is a proportionality constant. The SI unit of force, viz., *Newton* denoted by the symbol N is derived from unit values taking the proportionality constant as unity. Thus, one newton is

$$1\text{N} = (1\text{ kg})\left(1\frac{\text{m}}{\text{s}^2}\right) = 1\ \frac{\text{kg}\cdot\text{m}}{\text{s}^2}$$

The MKS unit of force, kgf, defined by Eq. (1.1) is

$$1\text{ kgf} = \frac{1}{9.80665}\ (1\text{ kg})\left(9.80665\frac{\text{m}}{\text{s}^2}\right) = 1\text{ kgf}$$

which represents a unit weight or the gravitational force on one kilogram mass. In the above definition, the value of the constant C is taken as equal to the reciprocal of the standard gravitational acceleration so that one kilogram mass has one kilogram weight.

It can be seen that

$$1\text{ kgf} = (1\text{ kg})\left(9.80665\frac{\text{m}}{\text{s}^2}\right) = 9.80665\text{ N}$$

Also, we known that

$$1\text{ lbf} = 0.453592\text{ kgf}$$

1.2.2 Unit of Pressure

The SI unit of pressure p can also be derived from its definition as force per unit area. Thus

$$[p] = \frac{[F]}{[A]} = \text{N/m}^2$$

The unit is also called *Pascal* and is denoted by the symbol Pa.

Another common SI unit of pressure is bar which is equivalent to a pressure of 10^5 N/m^2 or 0.1 MN/m^2 or 100 kN/m^2. Its conversion to MKS and FPS units is as follows

$$1\text{ bar} = \frac{10^5/9.80665\,\text{kgf}}{10^4\text{ cm}^2} = 1.0197\text{ kgf/cm}^2\text{ or ata}$$

$$= \frac{1.02(2.54)^2}{0.453592} = 14.5\text{ lbf/in}^2$$

It can be seen that one *standard atmosphere* is given by

$$1\text{ atm} = 1.033\text{ kgf/cm}^2 = 14.696\text{ lbf/in}^2$$

$$= \frac{1.033}{1.0197} = 1.01325\text{ bar}$$

$$= 760\text{ mm Hg or }760\text{ torr}$$

Accordingly,

$$1 \text{ torr} = 1 \text{ mm Hg} = \frac{1}{760} \text{ atm} = 133 \text{ N/m}^2$$

The conversion of one *technical atmosphere*, i.e. ata is obtained as:

$$\begin{aligned} 1 \text{ ata} &= 1 \text{ kgf/cm}^2 = (9.80665)(10^4) = 980665 \text{ N/m}^2 \\ &= 0.980665 \text{ bar} \\ &= (0.980665)(14.5) = 14.22 \text{ lbf/in}^2 \\ &= \frac{98066.5}{133} = 736 \text{ torr or mm Hg} \end{aligned}$$

The conversion of other units of pressure are

$$1 \text{ cm } H_2O = \left(\frac{10^4 \times 1}{1000}\right) \text{kg} \left(9.80665 \frac{\text{m}}{\text{s}^2}\right) = 98.1 \text{ N/m}^2$$

$$1 \text{ in Hg} = (25.4 \text{ mm}) \left(133 \frac{\text{N/m}^2}{\text{mm Hg}}\right) = 3390 \text{ N/m}^2$$

1.2.3 Unit of Energy (Work and Heat)

The unit of work or energy is obtained from the product of force and distance moved. The SI unit of work is *Newton metre* denoted by Nm or *Joule* denoted by J. Thus

$$1 \text{ Nm} = 1 \text{ J} = (1 \text{ kg m/s}^2)(1 \text{ m}) = 1 \text{ kg m}^2/\text{s}^2$$

Since both heat and work are energy, the SI unit of heat is the same as the unit of work, viz., joule. The conversion of the MKS unit of heat, viz., kcal, is obtained from its mechanical equivalent of heat which is 427 kcal/kgfm. Thus:

$$\begin{aligned} 1 \text{ kcal} &= 427 \text{ kgf m} = (427)(9.80665 \text{ N})\text{m} \\ &= 4186.8 \text{ Nm or J} = 4.1868 \text{ kJ} \end{aligned}$$

Also

$$\begin{aligned} 1 \text{ kcal} &= (1 \text{ kg of water})(1°\text{C}) \\ &= \frac{1}{0.453} \times \frac{9}{5} \text{ 1b °F} = 3.968 \text{ Btu} \end{aligned}$$

Hence

$$\begin{aligned} 1 \text{ kcal} &= 4.1868 \text{ kJ} = 3.968 \text{ Btu} \\ 1 \text{ kJ} &= 0.948 \text{ Btu} = 0.239 \text{ kcal} \\ 1 \text{ Btu} &= 0.252 \text{ kcal} = 1.055 \text{ kJ} \end{aligned}$$

1.2.4 Unit of Power

The SI unit of power is *watt*, denoted by the symbol W. It is defined as the rate of doing 1 Nm of work per second. Thus

$$1 \text{ W} = 1 \text{ J/S} = 1 \text{ Nm/s}$$

It may also be noted that watt also represents the electrical unit of work defined by

$$1 \text{ W} = 1(\text{volt}) \times 1 (\text{ampere}) = 1 \text{ J/s}$$

The conversion of the horsepower unit can also be obtained

$$1 \text{ hp (imperial)} = 550 \frac{\text{ft.1bf}}{\text{s}} = \frac{(550 \times 0.3048 \text{ m})\,(0.453592 \times 9.80665 \text{ N})}{\text{s}}$$

$$= 746 \text{ Nm/s or J/s or W}$$

$$1 \text{ hp(metric)} = 75 \frac{\text{kgf.m}}{\text{s}} = (75 \times 9.80665 \text{ N}) \frac{\text{m}}{\text{s}}$$

$$= 736 \text{ Nm/s or J/s or W}$$

Further, the units of energy can be derived from those of power. Thus

$$1 \text{ J} = 1 \text{ Ws}$$

$$1 \text{ kWH} = 3{,}600{,}000 \text{ J} = 3{,}600 \text{ kJ} = 860 \text{ kcal} = 3{,}410 \text{ Btu}$$

$$1 \text{ hp/hr (imperial)} = 746 \times 3{,}600 \text{ J} = 2{,}680 \text{ kJ} = 641 \text{ kcal} = 2{,}540 \text{ Btu}$$

$$1 \text{ hp/hr (metric)} = 736 \times 3{,}600 \text{ J} = 2{,}650 \text{ kJ} = 632 \text{ kcal} = 2{,}510 \text{ Btu}$$

1.2.5 Unit of Enthalpy

The interconversion of units of enthalpy are as follows

$$1 \text{ kJ/kg} = 0.239 \text{ kcal/kg} = 0.42 \text{ Btu/lb}$$

$$1 \text{ kcal/kg} = 4.19 \text{ kJ/kg} = 1.8 \text{ Btu/lb}$$

$$1 \text{ Btu/lb} = 0.556 \text{ kcal/kg} = 2.33 \text{ kJ/kg}$$

Note *The definition of enthalpy (H) (and specific enthalpy (h)) is obtained by the application of the First Law of Thermodynamics to a thermodynamic process.*

1.2.6 Units of Entropy and Specific Heat

These are expressed as

$$1 \text{ kJ/kg.K} = 0.239 \text{ kcal/kg°C or Btu/lb°F}$$

$$1 \text{ kal/kg°C} = 1 \text{ Btu/lb°F} = 4.1868 \text{ kJ/kg.K}$$

Note *The definition of entropy (S) (and specific entropy (s)) is obtained by the application of the Second Law of Thermodynamics to a thermodynamic process.*

1.2.7 Unit of Refrigerating Capacity

The standard unit of refrigeration in vogue is *ton refrigeration* or simply ton denoted by the symbol TR. It is equivalent to the production of cold at the rate at which heat is to be removed from one US tonne of water at 32°F to freeze it to ice at 32°F in one day or 24 hours. Thus

$$1 \text{ TR} = \frac{1 \times 2{,}000 \text{ lb} \times 144 \text{ Btu/lb}}{24 \text{ hr}}$$

$$= 12{,}000 \text{ Btu/hr} = 200 \text{ Btu/min}$$

where the latent heat of fusion of ice has been taken as 144 Btu/lb. The term one ton refrigeration is a carry over from the time ice was used for cooling. In general 1

TR always means 12,000 Btu of heat removal per hour, irrespective of the working substance used and the operating conditions, viz., temperatures of refrigeration and heat rejection. This unit of refrigeration is currently in use in the USA, the UK and India. In many countries, the standard MKS unit of kcal/hr is used.

It can be seen that

$$1 \text{ TR} = 12{,}000 \text{ Btu/hr}$$

$$= \frac{12{,}000}{3.968} = 3{,}024.2 \text{ kcal/hr}$$

$$= 50.4 \text{ kcal/min} \approx 50 \text{ kcal/min}$$

Also, since 1 Btu = 1.055 kJ, the conversion of ton into equivalent SI unit is:

$$1 \text{ TR} = 12{,}000 \times 1.055 = 12{,}660 \text{ kJ/hour}$$

$$= 211 \text{ kJ/min} = 3.5167 \text{ kW}$$

Example 1.1 *The performance test of an air conditioning unit rated as 140.7 kW (40 TR) seems to be indicating poor cooling. The test on heat rejection to atmosphere in its condenser shows the following:*

Cooling water flow rate: 4 L/s

Water temperatures: In 30°C: Out 40°C

Power input to motor: 48 kW (95% efficiency)

Calculate the actual refrigerating capacity of the unit.

Solution Heat rejected in condenser

$$\dot{Q}_{\text{condenser}} = \dot{m}_w C_w \Delta t_w$$

$$= 4\,(4.1868)\,(40 - 30) = 167.5 \text{ kW}$$

Work input

$$\dot{W} = 48\,(0.95) = 45.6 \text{ kW}$$

Refrigeration capacity (by energy balance)

$$\dot{Q}_{\text{refrigeration}} = \dot{Q}_{\text{condenser}} - \dot{W}$$

$$= 167.5 - 45.6 = 121.9 \text{ kW (34.7 TR)}$$

The unit is operating below its rated capacity of 40 TR.

1.3 THERMODYNAMIC SYSTEMS, STATE, PROPERTIES, PROCESSES, HEAT AND WORK

Thermodynamic systems are of two types. They are either *closed* or *open* as illustrated in Fig. 1.3. A closed system is one across whose *boundary* only heat Q and work W flow. In an open system the working fluid also crosses the *control surface* drawn around the system. Everything outside the system is *surroundings*. The system plus surroundings combine to make the *universe*.

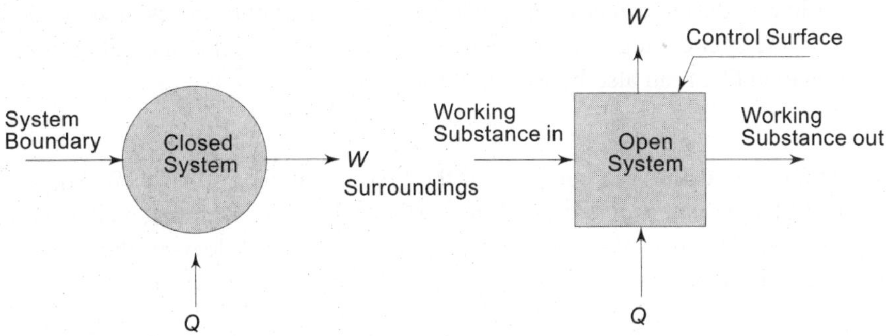

Fig. 1.3 Closed and open systems

The state of a thermodynamic system is characterised by its *properties*. The change of state of the working substance represents a *thermodynamic process*. Thermodynamic processes occurring in a closed system are called *non-flow processes*. Likewise, thermodynamic processes occurring in an open system are called *flow processes*.

Further, the processes that can be reversed such that the system and environment, both, can be restored to the initial state are called *reversible processes*. The processes which, when reversed, will not restore *both* the system and environment to the initial state are called *irreversible processes*.

The properties are either *intensive or extensive*. Intensive properties do not depend on the size of the system. These are, e.g., pressure p and temperature T. The extensive properties depend on the size of the system, e.g., volume V, internal energy U, enthalpy H, entropy S, etc. Their numerical values per unit mass of the working substance are called the *specific properties* denoted by lower case symbols, viz., v, u, h, s, etc. The specific properties are intensive properties.

A *thermodynamic process* is accompanied with *heat* and *work* interactions between the system and the surroundings. The heat added to the system is considered as positive, and that rejected by the system as negative. The sign convention for work is the opposite. The work done by the system is positive and the work done *on* the system is negative.

The heat and work interactions per unit mass of the *working substance* in the system are denoted as q and w.

Note *The work done in a reversible process in a simple compressible system is given by*

$$W = \int p \, dV$$

Note that in an irreversible process, the work is not given by $\int p \, dV$.

1.4 FIRST LAW OF THERMODYNAMICS

The first law of thermodynamics is mathematically stated as follows:

$$\oint \delta Q = \oint \delta W \tag{1.2}$$

Accordingly, during a *thermodynamic cycle*, viz., a cyclic process the system undergoes, the cyclic integral of heat added is equal to the cyclic integral of work done. Equation (1.2) can also be written for a cycle as

$$\oint(\delta Q - \delta W) = 0$$

Equation (1.3) below is a corollary of the first law. It shows that there exists a property U, named *internal energy* of the system/substance, such that a change in its value is equal to the difference in heat entering and work leaving the system. Accordingly, for a process in a closed system, the first law can be written as:

$$\delta Q = \delta U + \delta W \tag{1.3}$$

For the change of state of a system from initial state 1 to final state 2, this becomes

$$Q = U_2 - U_1 + W$$

Another property named *enthalpy* H can also be defined now as a combination of properties U, p and V,

$$H = U + pV, \; h = u + pv$$

For a reversible process, since $\delta W = p\mathrm{d}V$, the first law can also be written as

$$\delta Q = \mathrm{d}U + p\mathrm{d}V, \; \delta q = \mathrm{d}u + p\mathrm{d}v \tag{1.4a}$$

$$\delta Q = \mathrm{d}H - V\mathrm{d}p, \; \delta q = \mathrm{d}h - v\,\mathrm{d}p \tag{1.4b}$$

The first law can be applied to a process in an open system. Figure 1.4 represents an open system undergoing a *steady-state steady-flow* (SSSF) *process*. For the process, the first law takes the form of a *steady-flow energy equation* as in Eq. (1.5)

$$\dot{Q} = \dot{m}[(u_2 - u_1) + (p_2 v_2 - p_1 v_1) + \frac{1}{2}(C_2^2 - C_1^2) + g(z_2 - z_1)] + \dot{W}$$

$$= \dot{m}[(h_2 - h_1) + \frac{1}{2}(C_2^2 - C_1^2) + g(z_2 - z_1)] + \dot{W} \tag{1.5}$$

Here, in addition to change in internal energy, changes in kinetic and potential energies are also considered since these are significant. In addition, work, equal to $(p_2 v_2 - p_1 v_1)$, to make the fluid enter and leave the system called the *flow work* is also considered.

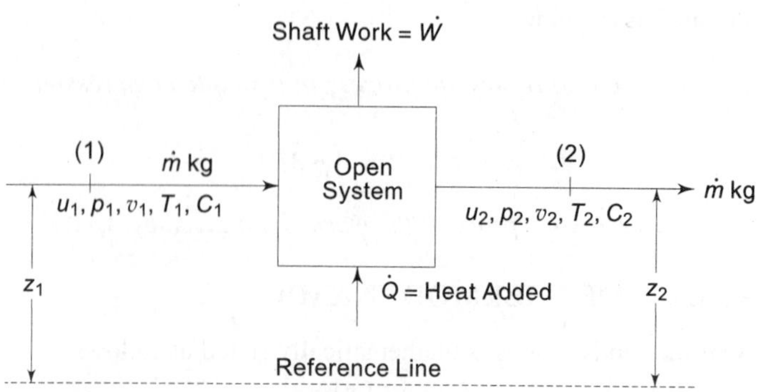

Fig. 1.4 Representation of a steady-state steady-flow process

Writing Eq. (1.5) on the basis of a unit mass entering and leaving the system, we have Eq. (1.6)

$$q + h_1 + \frac{C_1^2}{2} + gz_1 = h_2 + \frac{C_2^2}{2} + gz_2 + w \tag{1.6}$$

1.5 SECOND LAW OF THERMODYNAMICS

The second law of thermodynamics can be mathematically state for a thermodynamic cycle in the form of *Clausius Inequality* as given in Eq. (1.7)

$$\oint \frac{\delta Q}{T} \leq 0 \tag{1.7}$$

The equality holds for a reversible cycle, and the inequality for an irreversible cycle.

Just as the application of first law to a thermodynamic process led to the establishment of a new property, named internal energy (U), the application of the second law to a process leads to the establishment of another new property named *entropy* (S), defined as follows in Eq. (1.8)

$$\mathrm{d}S \equiv \left(\frac{\delta Q}{T}\right)_{\text{rev}} \tag{1.8}$$

Thus, for a reversible process, between two given states, from initial state 1 to final state 2 in a closed system, or inlet state 1 to exist state 2 in an open system, the change in entropy is given by

$$S_2 - S_1 = \int_1^2 \left(\frac{\delta Q}{T}\right)_{\text{rev}}, \; s_2 - s_1 = \int_1^2 \left(\frac{\delta q}{T}\right)_{\text{rev}}$$

It is found by applying Clausius inequality that for an irreversible process

$$S_2 - S_1 > \int_1^2 \left(\frac{\delta Q}{T}\right), \; s_2 - s_1 > \int_1^2 \left(\frac{\delta q}{T}\right)$$

Note *For a reversible process in a compressible system work done* $W = \int_1^2 p\mathrm{d}V$. *Hence, the area under the curve on P-V diagram gives work done in the process. Similarly, for a reversible process, heat transfer* $Q = \int_1^2 T\mathrm{d}S$. *Hence, the area under the curve on T-S diagram gives heat transfer during the process.*

1.6 NON-FLOW PROCESSES

Processes in a closed system are referred to as *non-flow processes*. Since the velocities are small, and hence dissipation due to friction is negligible, most non-flow processes are considered as reversible.

In a *reversible constant volume process*, $W = \int p\mathrm{d}V = 0$.

Hence, from first law, $Q = U_2 - U_1$.

In a *reversible constant pressure process*, $W = \int p\mathrm{d}V = p(V_2 - V_1)$, and from first law, $Q = (U_2 - U_1) + p(V_2 - V_1) = H_2 - H_1$.

Also, from second law and property relation, $Q = \int T\mathrm{d}S$

$$= T(S_2 - S_1) = H_2 - H_1.$$

In an *isothermal process*, $Q = T(S_2 - S_1) = (U_2 - U_1) + W$.

In an *adiabatic process*, $Q = 0$ and $W = -(U_2 - U_1)$. In a *reversible adiabatic process*, in addition, we have from second law, $Q = \int T\mathrm{d}S = 0$. Hence, $S_2 = S_1$. A reversible adiabatic process is, therefore, an *isentropic process*.

A general process can be represented by the *polytropic relation*

$$p_1 V_1^n = p_2 V_2^n = pV^n = \text{Constant}$$

in which n is the *polytropic index*. For the polytropic process, we have Eq. (1.9) for work

$$W = \int p\mathrm{d}V = \frac{n}{1-n}(p_2V_2 - p_1V_1)$$

$$= \frac{n}{1-n} p_1V_1 \left[\left(\frac{p_2}{p_1}\right)^{\frac{n-1}{n}} - 1\right] \tag{1.9}$$

1.7 STEADY-FLOW PROCESSES

The steady-flow energy equation is applicable to flow processes, viz. processes in an open system. In most flow processes, the irreversibility due to viscous friction cannot be neglected on account of significant velocities encountered. It is, therefore, necessary to calculate the unknown quantity by first assuming the process as reversible, and then multiplying or dividing the result by a *process efficiency*.

Boiling and Condensation

$$q = -h_2 - h_1 + \frac{1}{2}(C_2^2 - C_1^2) \cong h_2 - h_1$$

as the velocities are small, and the change in kinetic energy can be neglected. If there is a significant pressure drop as in *direct-expansion evaporators* in refrigeration, the process is not completely reversible.

Isothermal Process Neglecting kinetic and potential energies

$$q - w = h_2 - h_1$$

Adiabatic Process In nozzles and diffusers, there is no heat transfer or work done. In these devices, there is interconversion between kinetic energy and enthalpy. Thus we have

$$\frac{1}{2}(C_2^2 - C_1^2) = h_1 - h_2$$

In turbines and compressor, assuming no heat transfer and neglecting changes in kinetic and potential energies, we obtain for work

$$w = h_1 - h_2$$

To account for irreversibility in processes, the efficiencies are defined as follows:

Nozzle and Turbine Efficiency

$$\eta = \frac{h_1 - h_2}{h_1 - h_{2s}}$$

Diffuser and Compressor Efficiency

$$\eta = \frac{h_1 - h_{2s}}{h_1 - h_2}$$

In these equations, h_2 represents the enthalpy at the end in the actual process and h_{2s} represents the enthalpy if the process is executed reversibly, viz., isentropically between the pressure limits p_1 and p_2.

Throttling Process It is an irreversible adiabatic process. It is employed to reduce the pressure of a fluid by introducing a restriction in the flow passage as illustrated in Fig. 1.5.

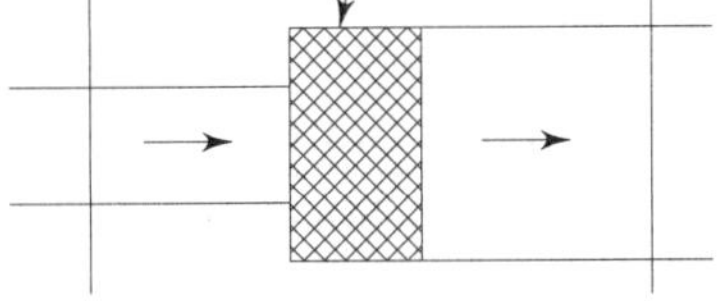

Fig. 1.5 Representation of a throttling process

As the process is adiabatic and no external work is done, we have

$$h_1 + \frac{C_1^2}{2} = h_2 + \frac{C_2^2}{2}$$

Since $p_2 < p_1$, $V_2 > V_1$. Hence if $A_1 = A_2$, $C_2 > C_1$. Making $A_2 > A_1$ such that $C_2 = C_1$, or since kinetic energy change is negligible, we find that in a throttling process

$$h_1 = h_2$$

Accordingly, throttling is also referred to as an *isenthalpic-expansion process*.

1.8 THERMODYNAMIC STATE OF A PURE SUBSTANCE

The working substance used in refrigerating machines is called a *refrigerant*. A refrigerant is usually a pure substance, though research is on to use mixtures also.

A knowledge of two independent properties is required to determine the thermodynamic state of a pure substance. In the case of equilibrium between two phases of a pure substance, only one independent property is required to define the state.

The equilibrium between phases, and the state of a pure substance are best described by the *phase diagrams* such as the $p - v$ and $T - s$ diagrams shown in Figs 1.7(a) and 1.7(b). The continuous lines on these diagrams represent the locii of equilibrium/saturation states in which phase change occurs, e.g., s and f_1 between solid and liquid, and f_2 and g between liquid and vapour. The temperature for change from solid state s to liquid state f_1 or vice versa, is the *melting/freezing/fusion temperature*. The temperature for change from liquid state f_2 to vapour state g, and vice versa, is the *boiling/condensation temperature*. Similarly the temperature for change between solid state s and vapour state g is the *sublimation/ablimation temperature*.

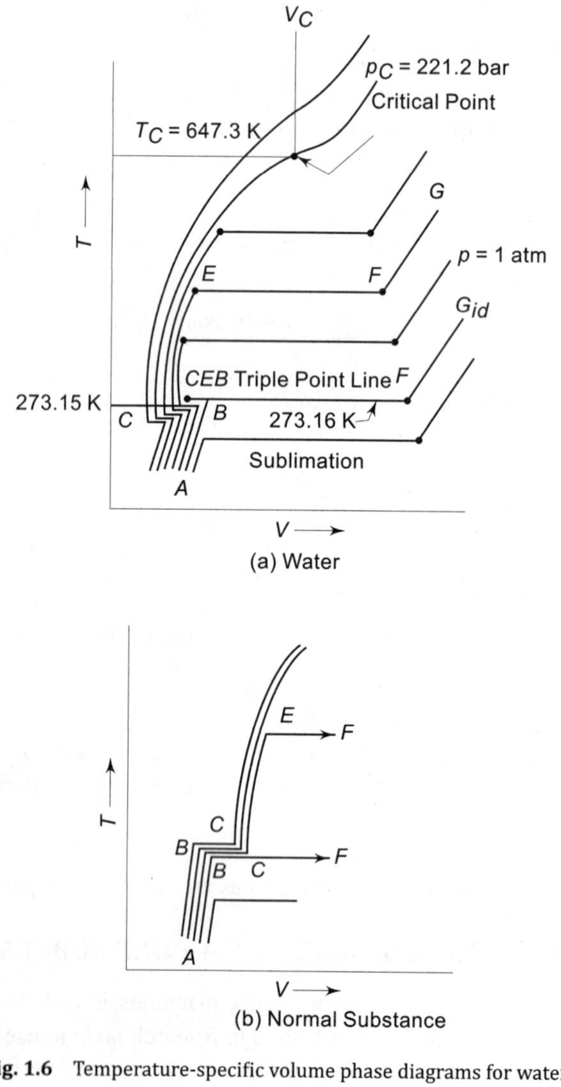

Fig. 1.6 Temperature-specific volume phase diagrams for water and normal substances

Consider a unit mass of solid at S below its melting point at 1 atm pressure. Let heat be supplied to it, and let us follow the events that occur at constant pressure as shown in Fig. 1.6(a) for water and Fig. 1.6(b) for a normal substance, Figs. 1.7(a) and 1.7(b) are for a normal substance. We observe the following from Fig. 1.7.

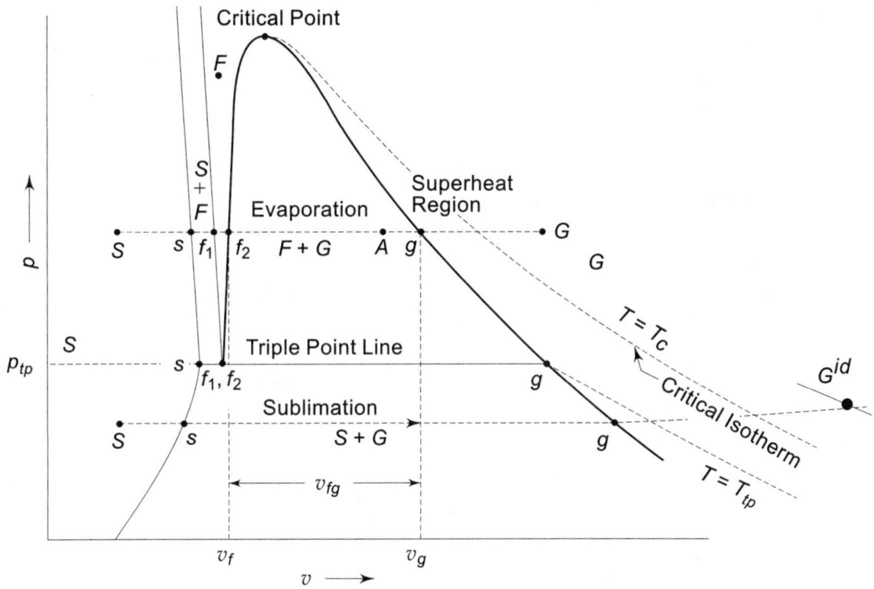

Fig. 1.7(a) *p-v* phase Diagram for a Normal Substance

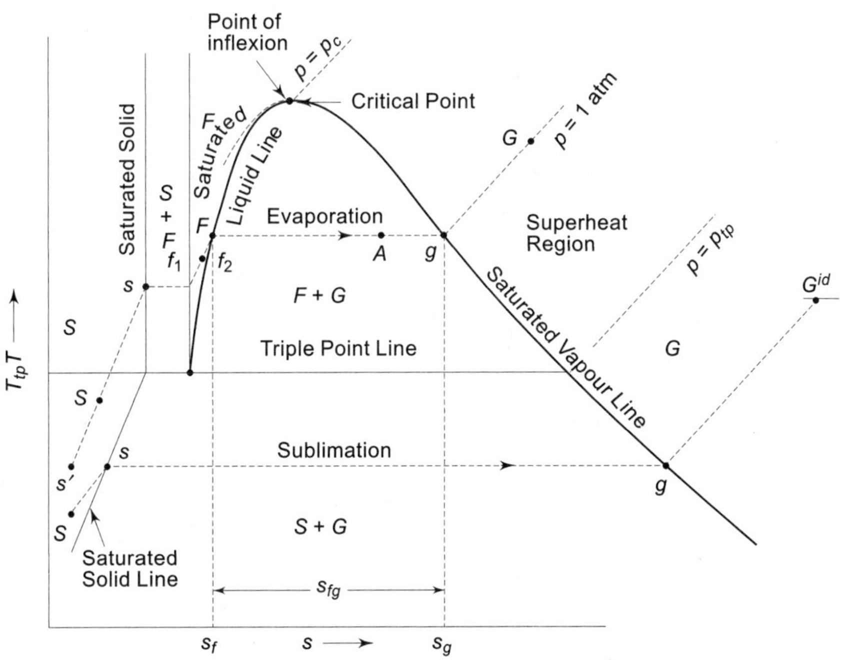

Fig. 1.7(b) *T*-s phase diagram for a normal substance

(i) Temperature rises until point *s* is reached.

(ii) Further heating results in melting of solid at constant (melting point) temperature until point f_1 is reached.

(iii) Change of phase from solid to liquid is complete at f_1. Heating of liquid now results in rise in temperature until point f_2 is reached.
(iv) Further heating at f_2 results in vaporization of liquid at constant (boiling point) temperature until point g is reached.
(v) Change of phase from liquid to vapour is complete at g. Heating of vapour/gas at g results in rise in temperature again until, say, point G above the boiling point temperature is reached.

States, f_1, f_2 and g are saturation states. f_1 represents saturated liquid state in equilibrium with saturated solid state s, and f_2 represents saturated liquid state in equilibrium will saturated vapour stage g. Now onwards, the subscripts 1 and 2 from f_1 and f_2 will be dropped, and the context will tell which one we are referring to.

Note that there is a large change in volume equal to $v_{fg} = v_g - v_f$ during vaporization from f_2 to g. Similarly, we have latent heat of vaporization $h_{fg} = h_g - h_f$, and entropy of vaporization $s_{fg} = s_g - s_f$.

State S below the melting point temperature $T_s = T_{f1}$ is a *subcooled solid state.* State G above the boiling point temperature $T_{f2} = T_g$ is a *superheated vapour state.* Any liquid state F between f_1 and f_2 below the boiling point temperature is a *subcooled liquid state.* Any state A in the liquid plus vapour $(F + G)$ region represents a mixture of vapour at g and liquid at f_2. The position of A is governed by the quality or dryness fraction x of vapour and $(1 - x)$ of liquid.

Consider now the reverse process of cooling of superheated vapour at G. The processes followed will be *desuperheating* from G to g, *condensation* from g to f_2, *subcooling* from f_2 to f_1, *freezing* from f_1 to s, and subcooling from s to S.

Consider now that the heating of solid is carried out at lower pressures. As the pressure is lowered, there is a marked decrease in boiling point temperature, and an increase in volume and enthalpy accompanying vaporization. This continues until *triple point pressure* p_{tp} and *temperature* T_{tp} are reached. On the *triple point line*, all the saturation states s, f_1, f_2 and g lie. Thus, all the three phases exist in equilibrium at the triple point. The triple point for water is at 273.16 K (0.01°C) temperature, and 0.006112 bar pressure.

Below the triple point pressure, saturated solid at s on heating directly changes into saturated vapour at g. This is called *sublimation*. Similarly, saturated vapour at g on cooling directly changes to saturated solid at s. This is called *ablimation* or *freeze-condensation*.

Consider now the heating at pressures above atmospheric. In general, at higher pressures, the boiling point increases, and there is a marked decrease in v_{fg}, h_{fg} and s_{fg}. At a certain high pressure, v_{fg}, h_{fg} and s_{fg} become zero. This is referred to as the *critical point*. The properties at this point are denoted as *critical pressure* p_c, *critical temperature* T_c and *critical volume* v_c. The same for water are:

$$p_c = 221.2 \text{ bar}, \; T_c = 647.3 \text{ K}, \; v_c = 0.00317 \text{ m}^3/\text{kg}$$

At pressures above critical, there is no definite transition between liquid and vapour phases.

Note that the gaseous state G^{id} at low pressures (tending to zero) and high temperatures represents the *ideal* or *perfect gas* state following the equation $pv = RT$.

1.9 HEAT EXCHANGE PROCESSES

Heat is normally absorbed or rejected by a working substance at a constant pressure. It has been seen that when a working substance exchanges heat, then either

(i) the temperature of the substance changes and the substance remains in a single phase, or

(ii) the temperature of the substance remains constant but a phase change occurs.

The heat transferred without a phase change results in a temperature rise and is called *sensible heat*, and that transferred resulting in a change of phase at constant temperature is called *latent heat*.

Sensible heat is measured by the expression

$$Q_S = mC_p \Delta T$$

where m is the mass, C_p is the specific heat at constant pressure, and ΔT is the temperature change of the working substance. In general, chilled water or salt solutions of either NaCl or $CaCl_2$ called *brines* are used as carriers of refrigeration for the absorption of heat. Chilled water is used for air conditioning in central air-conditioning plants. Brines have freezing points lower than 0°C and are, therefore, used as coolants in applications below 0°C refrigeration temperature, such as in cold storages, ice plants, skating rinks, etc.

Latent heat exchange processes correspond to those of melting, evaporation and sublimation, and vice versa, viz., fusion, condensation and desublimation and the heat transferred is measured by

$$Q_L = m\Delta h$$

where Δh is the latent heat for the corresponding process, e.g., h_{fg} for latent heat of vaporization, h_{sg} for sublimation and h_{sf} for fusion. The subscripts s, f and g denote, solid, liquid and gas respectively. Evaporation is the most commonly used method in refrigeration for the absorption of heat. Sublimation is used in a process called *freeze drying*. In this, the product is first frozen and cooled until it reaches below its triple point temperature. It is then placed in a chamber which is evacuated to a pressure sufficiently below the triple point pressure. Heat is then supplied to the product and in a *freeze-condenser* the sublimated vapours are condensed at a temperature of about –40°C. The food thus dried generally retains its original flavour and value and can be reconstituted into its original condition by the addition of water. Freeze drying is used in the manufacture of certain high-value food products and medicines.

Wide use is made of the melting of ice in homes, stores, transports, etc., for the preservation of products at refrigerated temperatures. The main drawbacks of ice are the temperature limitation of 0°C and the spoilage caused by the melted water. These drawbacks can be avoided by the use of solid carbon dioxide, also called *dry ice*, which sublimates at atmospheric pressure. Dry ice is rather widely used in refrigerated air transport. In recent times, liquid nitrogen has also become quite popular in transport refrigeration.

Example 1.2 *100 kg of ice at –5°C is placed in a bunker to cool some vegetables. 24 hours later, the ice has melted into water at 10°C. What is the average rate of cooling in kJ/h and TR provided by the ice? Given:*

Specific heat of ice, C_{ps} *= 1.94 kJ/kg.K*
Specific heat of water, C_{pf} *= 4.1868 kJ/kg.K*
Latent heat of fusion of ice at 0°C, h_{sf} *= 335 kJ/kg.*

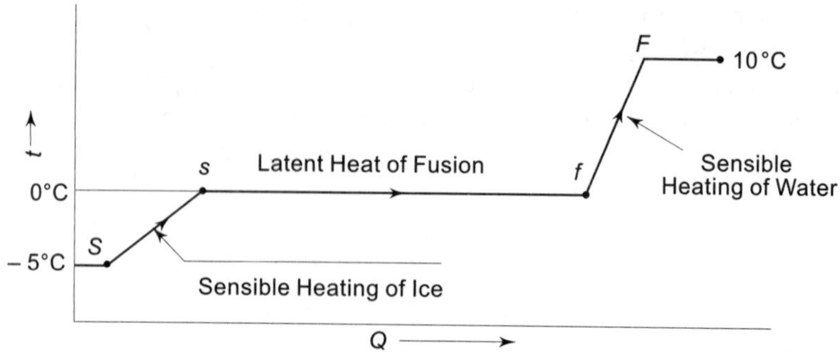

Fig. 1.8 Figure for Example 1.2

Solution The change of temperature with heat supplied is shown in Fig. 1.8.

$$Q = m\,[C_{ps}\,(t_s - t_S) + h_{sf} + C_{pf}\,(t_F - t_f)]$$

$$= 100\,[1.94\,(0 + 5) + 355 + 4.1868\,(10 - 0)] = 38{,}660 \text{ kJ}$$

$$\text{Capacity} = \frac{38{,}660}{24} = 1611 \text{ kJ/h}$$

$$= \frac{1611}{12{,}660} = 0.127 \text{ TR}$$

1.10 PRODUCTION OF LOW TEMPERATURES

The various principles and processes involved in the production of low temperatures are as follows:

(i) Throttling expansion of a liquid with flashing.
(ii) Reversible adiabatic expansion of a gas.
(iii) Irreversible adiabatic expansion (throttling) of a real gas.
(iv) Thermoelectric cooling.
(v) Adiabatic demagnetization.

1.10.1 Expansion of a Liquid with Flashing

Consider the throttling of a saturated liquid initially at 1 at pressure p_1 as shown in Fig. 1.9. The state after expansion is at 2. The process is accompanied by an increase in entropy along with a drop in pressure due to which the volume increases and a part of the liquid is vaporized, thus cooling the remaining liquid. The expansion, with the vaporization of a part of the liquid causing the lowering of its temperature, is called *flashing*. This method is most commonly used for obtaining low temperatures. It will be seen that for the process

$$h_1 = h_2 = h_{f2} + x_2\,h_{fg2} \qquad v_2 = v_{f2} + x_2 v_{fg2}$$

It may be noted that the temperature will similarly drop from T_1 to T_2 if the liquid had isentropically expanded from 1 to 2′. But the isentropic expansion of a liquid is not employed in refrigeration for reasons explained in Sec. 3.1.

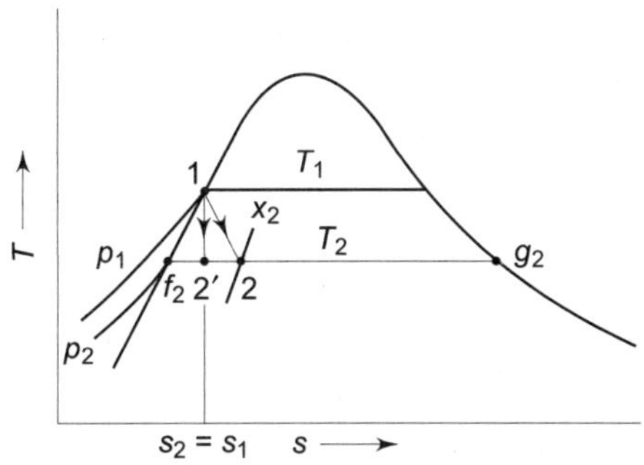

Fig. 1.9 Reversible and irreversible adiabatic expansion of a liquid

1.10.2 Reversible Adiabatic Expansion of a Perfect/Ideal Gas

This method is used with permanent gases such as air. The gas is initially compressed and cooled and then expanded reversibly in an adiabatic reciprocating or turbo-type expander, thus doing external work. The reversible adiabatic expansion of an ideal gas from p_1 to p_2 follows the path $pV^{\gamma} = C$. The work done is given by the shaded area shown in Fig. 1.10. The expression for work is obtained by adding the flow work components to the non-flow work. Thus, we have for specific work.

$$w = \oint p\mathrm{d}v = \int_a^1 p\mathrm{d}v + \int_1^2 p\mathrm{d}v + \int_2^b p\mathrm{d}v + \int_b^a p\mathrm{d}v = -\oint_1^2 v\mathrm{d}p$$

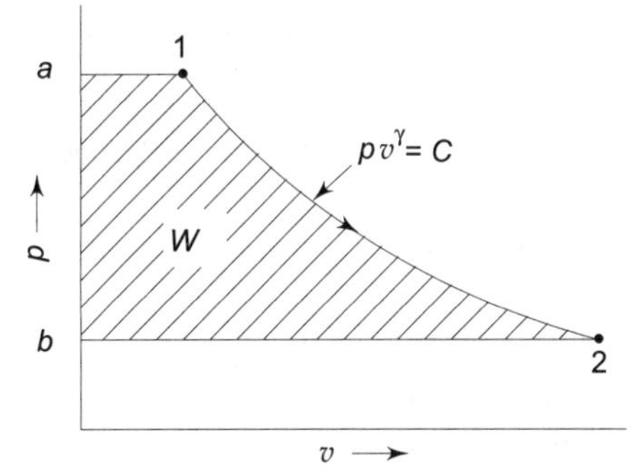

Fig. 1.10 Work done in a reversible adiabatic-expansion process

which gives on integration for a perfect gas

$$w = -\int v_1 \left(\frac{p_1}{p}\right)^{1/\gamma} \mathrm{d}p = \frac{\gamma}{\gamma - 1}\,(p_1 v_1 - p_2 v_2)$$

Also since $\quad q = 0$

we obtain from the steady-state steady-flow energy equation

$$h_1 - h_2 = w = -\int v\mathrm{d}p = \frac{\gamma}{\gamma - 1}(p_1 v_1 - p_2 v_2)$$

which for an ideal gas, on putting $\dfrac{\gamma R}{\gamma - 1} = C_p$, becomes

$$w = h_1 - h_2 = C_p\,(T_1 - T_2)$$

The temperature after expansion is given by

$$\frac{T_2}{T_1} = \left(\frac{p_2}{p_1}\right)^{\frac{\gamma-1}{\gamma}}$$

1.10.3 Irreversible Adiabatic Expansion (Throttling) of a Real Gas

Herein again, the initially compressed and cooled gas is expanded, but irreversibly, through a throttle device with restriction to flow. It has been shown that for this process

$$h_1 = h_2$$

and for a perfect gas $T_1 = T_2$, since enthalpy is a function of temperature only.

Real gases, however, show a departure from ideal gas behaviour and usually produce a substantial decrease in temperature under certain conditions, viz., initial low temperature and high pressure of the gas. The decrease in temperature dT corresponding to a drop in pressure dp, defined by

$$\left(\frac{\partial T}{\partial p}\right)_h = \delta \quad \text{or} \quad \mu_J$$

is known as the *Joule-Thomson coefficient*. For an ideal gas this coefficient is zero. For a real gas, we have:

For cooling	δ is positive	Initial state of gas at high pressure and low temperature
For no change in temperature	$\delta = 0$	Ideal gas behaviour
For heating	δ is negative	Initial state of gas at high temperature.

1.10.4 Thermoelectric Cooling

In 1834, Jean Peltier, a French watchmaker and an amateur scientist, discovered that cooling is produced at one junction of two dissimilar metals, if a current is passed through them. Simultaneously, heat is produced at the other junction (Fig. 1.11). The Peltier heats absorbed at the cold end and rejected at the hot end are given by

$$Q_c = \Pi_c I$$
$$Q_h = \Pi_h I$$

where Π_c and Π_h are the *Peltier coefficients* which are functions of temperatures T_c and T_h of the cold and hot ends respectively. It may be seen that if T_h is maintained at ambient temperature, T_c will be lower than the ambient temperature. It is also to be noted that which of the junctions or ends will become cold or hot depends on the direction of flow of the current.

The phenomenon is called the *Peltier effect*. The actual cooling produced at the cold end would, however, be reduced due to irreversible processes of Joule and Fourier heat transfers. The energy balance of the system is

$$\dot{Q}_k - \dot{Q}_o = EI$$

where I is the current, E is the emf applied and $\dot{Q}_o$ and $\dot{Q}_k$ are the actual amounts of heat flows at the cold and hot ends respectively.

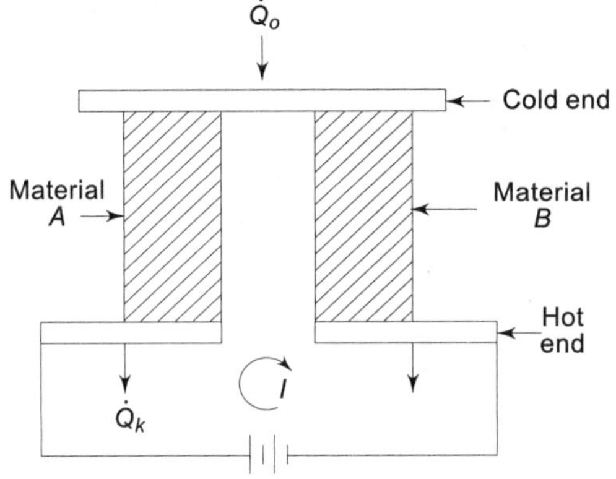

Fig. 1.11 Peltier effect

1.10.5 Adiabatic Demagnetization

In 1926, Giaque and Deby independently proposed that cooling could be produced by the adiabatic demagnetization of a paramagnetic salt. Figure 1.12 provides a schematic representation of such a process.

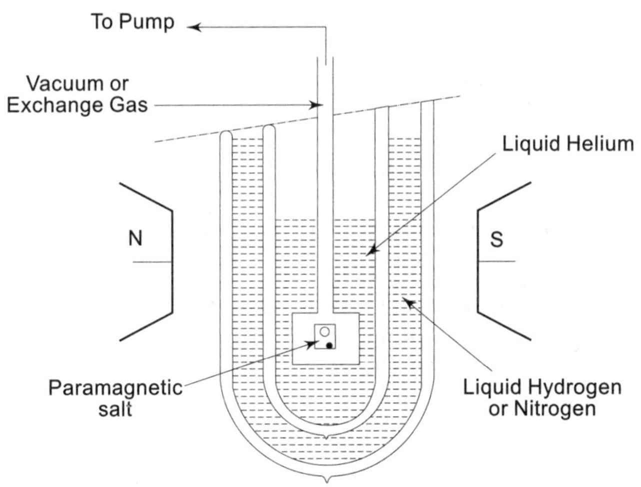

Fig. 1.12 Arrangement for adiabatic demagnetization of a paramagnetic salt

The paramagnetic salt is suspended in a tube containing low pressure gaseous helium as exchange gas to provide thermal communication with the surrounding bath of liquid helium. The liquid helium bath and the salt are first cooled to about 1 K by pumping helium to the lowest practical pressure. A magnetic field is applied, causing magnetization of the salt. The heat produced is removed by the helium bath such that the temperature of the bath again approaches 1 K. Next, the exchange gas is removed by pumping and the magnetic field is turned off. The temperature of the salt then decreases as a result of *adiabatic demagnetization*.

1.11 SATURATION PRESSURE VERSUS SATURATION TEMPERATURE RELATIONSHIP

The whole matter of phase equilibrium is best summarised in phase diagram in Fig. 1.13 between saturation pressure p^{sat} and saturation temperature T^{sat} of a substance. Any point on a line in this diagram represents two phases in equilibrium state such as s and f_1 on *fusion line*, f_2 and g on *vaporization line* and s and g on *sublimation line*. The triple point is one unique point where all saturation states s, f_1, f_2 and g conjoin. The vaporization line begins at the triple point, and ends at the critical point. It is called the *vapour pressure curve*.

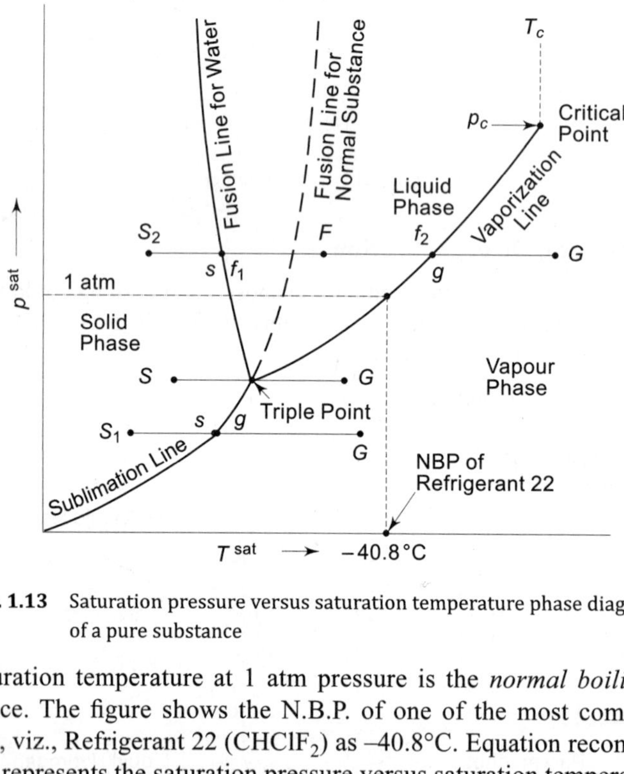

Fig. 1.13 Saturation pressure versus saturation temperature phase diagram of a pure substance

The saturation temperature at 1 atm pressure is the *normal boiling point* of the substance. The figure shows the N.B.P. of one of the most commonly used refrigerants, viz., Refrigerant 22 ($CHClF_2$) as –40.8°C. Equation recommended by Reynolds[14] represents the saturation pressure versus saturation temperature data of Refrigerant 22 very accurately.

$$\ln p^{sat} = 71.55415 - \frac{4,818.96}{T^{sat}} - 7.861 \ln T^{sat} + 9.0807 \times 10^{-3}\, T^{sat}$$

$$+ 0.445747 \frac{\left(381.17 - T^{sat}\right)}{T^{sat}} \ln\left(381.17 - T^{sat}\right) \tag{1.10}$$

Herein, the pressure is in Pa.

Example 1.3 *Find saturation pressure of Refrigerant 22 at 40°C (313 K).*

Solution Substituting values in Eq. (1.10), we obtain

$$\ln p^{sat} = 71.55415 - 15.396 - 45.1711 + 2.8423 + 0.4099 = 14.236$$

$$\Rightarrow \qquad p^{sat} = 1{,}527{,}660 \text{ Pa}$$

Note *ln p^{sat} versus $1/T^{sat}$ relationship can be simplified; and can be expressed in the form of Antonie equation as follows in Eq. (1.11)*

$$\ln p^{sat} = A - \frac{B}{T^{sat}} \tag{1.11}$$

which is a straight line, and is quite accurate.

Example 1.4 *Using the vapour pressure data of R 22 and propane at 40°C and –30°C given below, find the values of the constants of the Antonie equation for the two refrigerants, and verify their validity for pressure at 5°C.*

t^{sat}	T^{sat}	p^{sat}	
		R 22	*Propane*
°C	K	Pa	
40	313.15	1,533,500	1,366,400
5	278.15	584,000	547,750
–30	243.15	163,500	166,400

Solution Substituting values and solving, we get the Antonie equations:

For R 22

$$\ln p^{sat} = 22.01864 - \frac{2.43492 \times 10^3}{T^{sat}}$$

Putting $T^{sat} = 278.15$ K, we get

$$p^{sat} = 5.765 \times 10^5 \text{ Pa}$$

The error in comparison to actual value of 5.84×10^5 Pa is 1.28% only.

For Propane

$$\ln p^{sat} = 21.50251 - \frac{2.309424 \times 10^3}{T^{sat}}$$

At $\qquad T^{sat} = 278.15 \text{ K}, p^{sat} = 5.402 \times 10^5 \text{ Pa}$

The error in comparison to actual value of 547,750 Pa is 1.38% only.

Note *If a number of data points are used, and least squares method is employed for finding A and B, the accuracy would still be better.*

1.12 THE GASEOUS PHASE: EQUATION OF STATE

Pressure p, specific volume v and temperature T are the three measurable properties of a substance. The equation expressing their relationship in the gaseous phase is called the *equation of state*. This functional relationship between p-v-T of a gas can be either theoretical, or generalized or an empirical equation fitted from experimental data.

The simplest theoretical equation of state is the *ideal* or *perfect gas* equation representing behaviour of a gas at low pressures (*tending to zero*) and high temperatures such as at point G^{id} in Figs 1.6 and 1.7. This equation is

$$v^{id} = \frac{RT}{p}$$

where $R = \bar{R}/M$ is the *gas constant* for the particular gas, $\bar{R}$ is the *universal gas constant* having the value 8.3143 kJ/kmol K, and M is the molecular mass of the substance.

In addition, the internal energy and enthalpy of an ideal gas are functions of temperature only. These are given by,

$$du = C_{vo}\,dT\,,\ u_2 - u_1 = C_{vo}\,(T_2 - T_1)$$
$$dh = C_{po}\,dT,\ h_2 - h_1 = C_{po}\,(T_2 - T_1)$$

where C_{vo} and C_{po} are *zero-pressure constant volume* and *zero-pressure constant pressure* specific heats. It can be shown that

$$C_{po} - C_{vo} = R$$

and the ratio of specific heats C_{po}/C_{vo} is denoted as γ. Their values for air as a perfect gas are:

$$C_{po} = 1.005 \text{ kJ/kg. K},\ C_{vo} = 0.718 \text{ kJ/kg.K},\ R = 0.287 \text{ kJ/kg. K}$$
$$\gamma = 1.4,\ M = 28.966$$

For a *real gas*, the actual volume can be expressed by the general relation

$$v = v^{\text{real}} = z\frac{RT}{p}$$

where $z = v^{\text{real}}/v^{id} = pv/RT$ is called the *compressibility* of the gas.

Many different equations of state have been proposed to represent the real volume of gases. In 1949, Redlich-Kwong[12] (R-K) proposed the equation as follows in Eq. (1.12)

$$p = \frac{\bar{R}T}{\bar{v} - b} - \frac{a}{\sqrt{T}\,(\bar{v}^2 + \bar{v}\,b)} \tag{1.12}$$

in which v is the *molar volume* (volume of M kg of gas). Its constant a and b found from conditions of critical isotherm at the critical point are expressed in terms of critical constants T_c and p_c as follows:

$$a = 0.42748\,\frac{R^2\,T_c^{5/2}}{p_c},\ b = 0.08664\,\frac{RT_c}{p_c}$$

Often constants a and b are replaced by constants A and B written as:

$$A^2 = \frac{a}{R^2\,T^{2.5}},\ B = \frac{b}{RT},\ \text{and } h = \frac{b}{v} = \frac{b}{z\,RT/p} = \frac{Bp}{z}$$

so that the R-K equation can be written in terms of z in place of v as follows in Eq. (1.13)

$$z = \frac{1}{1-h} - \frac{a}{bRT^{1.5}}\left(\frac{h}{1+h}\right)$$

$$= \frac{1}{1-h} - \frac{A^2}{B}\left(\frac{h}{1+h}\right) = \frac{pv}{RT} \quad (1.13)$$

Note that h here is a function equal to b/v, and not the enthalpy.

A modification of the *R-K* equation is Peng-Robinson[11] (*P-R*) equation as given in Eq. (1.14)

$$p = \frac{RT}{v-b} - \frac{a}{v^2 + 2bv - v^2} \quad (1.14)$$

where $$a = \frac{0.45724\, R^2 T_c^2}{p_c}\left[1 + f(\omega)\left\{1 - \left(\frac{T}{T_c}\right)^{1/2}\right\}\right]^2$$

$$b = \frac{0.0778\, RT_c}{p_c}$$

$$f(\omega) = 0.37464 + 1.54226\,\omega - 0.26992\,\omega^2$$

$$\omega = Accentric\ factor = -\,1.0 - \log\,(p_r^{sat})T_{r=0.7}$$

$$T_r = \text{Reduced temperature} = \frac{T}{T_c}$$

$$p_r = \text{Reduced pressure} = \frac{p}{p_c}$$

A very popular empirical equation is the *Martin-Hou*[9] *equation*. A modified Martin-Hou equation, as recommended by Reynolds[14], which applies with great accuracy to one of the most commonly used refrigerants, viz., Refrigerant 22 ($CHClF_2$), is given in Eq. (1.15) in simplified form:

$$p = \frac{RT}{v-b} + \sum_{i=2}^{5} \frac{A_i + B_i T + C_i e^{-4.2T/T_c}}{(v-b)^i} \quad (1.15)$$

where R is the gas constant for the refrigerant equal to 96.15 J/kg.K, p is in Pa, v is in m^3/kg, and the constants are:

$b = 1.248556 \times 10^{-4}$

$A_2 = -\,1.169819 \times 10^2$	$B_2 = 1.164312 \times 10^{-1}$	$C_2 = -\,1.1841 \times 10^3$
$A_3 = -\,2.929526 \times 10^{-2}$	$B_3 = 2.303194 \times 10^{-4}$	$C_3 = 2.48896$
$A_4 = 2.419193 \times 10^{-4}$	$B_4 = -6.796677 \times 10^{-7}$	$C_4 = 0.0$
$A_5 = -2.434584 \times 10^{-7}$	$B_5 = 6.302018 \times 10^{-10}$	$C_5 = -1.2062 \times 10^{-6}$

Example 1.5 ***Solution Procedure for Martin-Hou Equation***

Using Eq. (1.15), find the specific volume of saturated vapour of Refrigerant 22 at 40°C (313 K) temperature. Critical temperature of the refrigerant is 369 K.

Solution The saturation pressure of Refrigerant 22 at 40°C is 1,533,500 Pa. At the pressure and 313 K temperature, the ideal gas volume is

$$v^{id} = \frac{RT}{p} = \frac{96.15(313)}{1,533,500} = 0.01962 \text{ m}^3/\text{kg}$$

Using a value close to this as *first approximation*, we find, by substituting values in Eq. (1.15) by successive approximations, that at $v = 0.0151$ m^3/kg, the right hand side is

$$\text{RHS} = 2{,}009{,}60 - 508{,}900 + 33{,}760 + 580 - 100$$

$$= 1{,}534{,}940 \text{ Pa} \cong \text{LHS}$$

The error is only 0.01%. Hence, the required volume is 0.0151 m^3/kg.

1.13 CLAPEYRON EQUATION

The *Clapeyron equation* relates the enthalphy of vaporization of a substance with the slope dp^{sat}/dT^{sat} of its vapour pressure curve. It is written as given in Eq. (1.16)

$$\left(\frac{dp}{dT}\right)^{sat} = \frac{h_{fg}}{T(v_g - v_f)} \tag{1.16}$$

The example below illustrates the application of this equation to find the latent heat of vaporization of a refrigerant.

Example 1.6 ***Calculation of Latent Heat of Vaporization of R 22.***

From the following handbook data for R 22,

t °C	p bar	v_f L/kg	v_g m^3/kg
–4	4.358	0.77	0.0536
0	4.976	0.778	0.0471
+4	5.657	0.787	0.0416

Calculate the latent heat of vaporization of R 22 of 0°C.

Solution

At 0°C, $$\left(\frac{\Delta p}{\Delta T}\right)^{sat} = \frac{129.9 \text{ kPa}}{8°\text{C}} = 16.237 \text{ kPa/°C}$$

$$(h_{fg})0°\text{C} = \left(\frac{\Delta p}{\Delta T}\right)^{sat} T(v_g - v_f)$$

$$= 16.237\ (273.16)\ (0.0471 - 0.00077)$$

$$= 205.48 \text{ kJ/kg}$$

Note *The value of* dp^{sat}/dT^{sat} *at 0°C can also be found by differentiation from Eq. (1.10).*

1.14 PROPERTY RELATIONS

$\delta Q = \mathrm{d}U + \delta W$ is the first law equation which is applicable to *all process* in a closed system. For the particular case of reversible processes in a closed system, since $\delta W = p\mathrm{d}V$ and $\delta Q = T\mathrm{d}S$, this equation takes the form

$$T\mathrm{d}S = \mathrm{d}U + p\mathrm{d}V \tag{1.17a}$$

$$T\mathrm{d}S = \mathrm{d}H - V\mathrm{d}p \tag{1.17b}$$

By dividing both sides by mass m, these can also be written on the basis of a unit mass of the substance as given below in Eqs. (1.18a) and (1.18b)

$$T\mathrm{d}s = \mathrm{d}u + p\mathrm{d}v \tag{1.18a}$$

$$T\mathrm{d}s = \mathrm{d}h - v\mathrm{d}p \tag{1.18b}$$

These are the two well-known T–ds equations. Although derived for a reversible process in a closed system, these are actually in terms of properties of a system/ substance. Hence, these are applicable to all processes whether reversible or irreversible, and whether in a closed system or in an open system. These equations are referred to as *property relations*, and are used to evaluate the changes in entropy in terms of changes in other properties employing the relationships as illustrated in Eqs. (1.19a) and (1.19b)

$$\mathrm{d}s = \frac{1}{T}\mathrm{d}u + \frac{p}{T}\mathrm{d}v \tag{1.19a}$$

$$\mathrm{d}s = \frac{1}{T}\mathrm{d}h - \frac{v}{T}\mathrm{d}p \tag{1.19b}$$

1.15 THERMODYNAMIC PROPERTIES OF REFRIGERANTS[2]

For establishing the thermodynamic properties of refrigerants, and for that matter, for any pure substance, the following minimum experimental data/correlations are required:

(i) p^{sat} versus T^{sat}, or T^{sat} versus p^{sat}.

(ii) p-v-T data or equation of state for gaseous phase.

(iii) Liquid density ρ or specific volume v_f.

(iv) Liquid specific heat C_f.

(v) Zero-pressure (ideal gas) constant pressure specific heat C_{po} or constant volume specific heat C_{vo} (= $C_{po} - R$) of the gaseous phase.

Then the latent heat is calculated from the Clapeyron equation. The only other properties that need to be calculated are internal energy, enthalpy and entropy. For the purpose, the relations given in Eqs (1.20), (1.21), (1.22), and (1.23) for changes in u, h and s at constant temperature in the gaseous phase are used:

$$(u_2 - u_1)_T = \int_1^2 \left[T\left(\frac{\partial p}{\partial T}\right)_v - p\right]\mathrm{d}v \tag{1.20}$$

$$(h_2 - h_1)_T = \int_1^2 \left[v - T\left(\frac{\partial v}{\partial T}\right)_p\right]\mathrm{d}p \tag{1.21}$$

$$(s_2 - s_1)_T = -\int_1^2 \left(\frac{\partial v}{\partial T}\right)_v dp \tag{1.22}$$

$$(s_2 - s_1)_T = \int_1^2 \left(\frac{\partial p}{\partial T}\right)_v dv \tag{1.23}$$

Note that Eq. (1.20) for internal energy change requires a p-explicit equation of state, where as Eq. (1.21) for enthalpy change requires a v-explicit equation of state. If only a p-explicit equation of state is available, then enthalpy change can be found from internal energy change using the relationship:

$$(h_2 - h_1)_T = (u_2 - u_1)_T + (p_2 v_2 - p_1 v_1)_T \tag{1.24}$$

1.15.1 Enthalpy Calculations

Figure 1.14 illustrates the method of calculation of enthalpy with the help of pressure-enthalpy ($p - h$) diagram. Since there is no absolute value of enthalpy, and only differences in enthalpies of state points are required in calculations, a *reference state* has to be chosen for the purpose to which an arbitrary value of enthalpy is assigned. In Fig. 1.14, the reference state chosen is that of saturated liquid at point 1. In the case of water, point 1, usually, is the saturated liquid state at its triple point (0.01°C) to which a value of $h_1 = h_{f1} = 0$ kJ/kg is assigned. In the case of refrigerants, the reference state chosen is that of saturated liquid at 0°C. And since the refrigerants work at temperatures below 0°C also, in order to avoid negative values of enthalpies in calculations, the value of enthalpy assigned at the reference temperature is, usually, $h_1 = h_{f1} = 200$ kJ/kg. Note that the pressure at 1 is $(p^{sat})_{0°C}$.

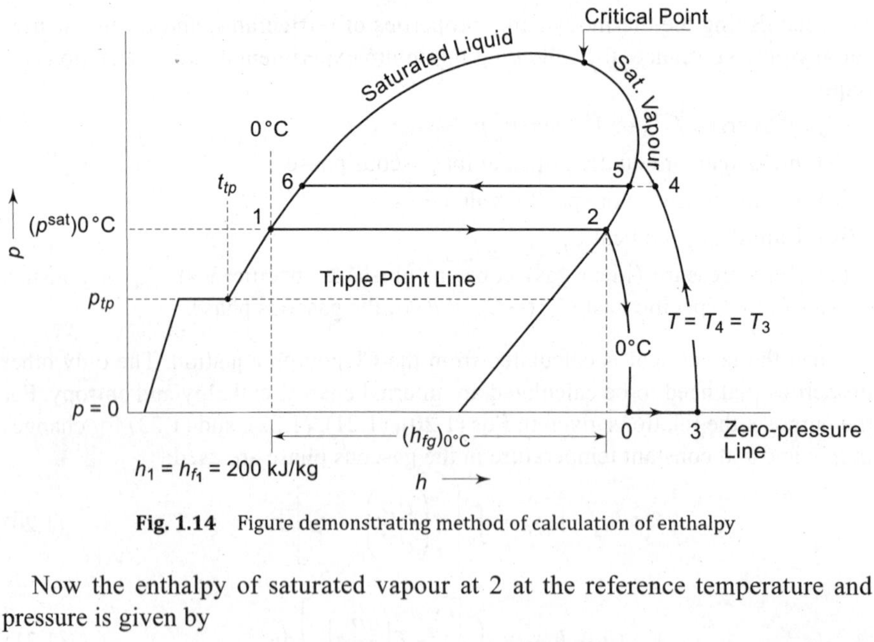

Fig. 1.14 Figure demonstrating method of calculation of enthalpy

Now the enthalpy of saturated vapour at 2 at the reference temperature and pressure is given by

$$h_2 = h_1 + (h_{fg})_{0°C} = 200 + (h_{fg})_{0°C}$$

Then, for the calculation of the enthalpy of vapour at any state 4 at temperature T_4 and pressure p_4, the path followed is from 2 to 0, 0 to 3 and 3 to 4. State 0 is at the same temperature as 2 but at zero pressure. Similarly, state 3 is at the same temperature as 4 but at zero pressure. Both states 0 and 3 are, therefore, ideal gas states at temperatures 0°C and T respectively. It is seen that

$$h_4 = h_2 + (h_0 - h_2) + (h_3 - h_0) + (h_4 - h_3)$$

where

$$h_3 - h_0 = \int_{T_3}^{T_4} C_{po}\, \mathrm{d}T$$

Now, $(h_4 - h_3)$ represents the change in enthalpy at constant temperature $T_4 = T_3$ as a result of change in pressure from 0 to $p_4 = p$. Similarly, $(h_2 - h_0)$ represents the change in enthalpy at 0°C as a result of change in pressure from 0 to p_2. These represent differences in enthalpies of real gas and ideal gas at the same temperature. The difference $h^R = (h - h^{id})_T$ is termed *residual enthalpy*, and can be found either from Eq. (1.21) or Eq. (1.24) by integrating between pressure limits of $p = 0$ to p, or between volume limits if $v = \infty$ (at $p = 0$) to v.

The enthalpy of saturated vapour at any point 5 can be similarly calculated. Then, the enthalpy of saturated liquid at 6 can be found from

$$h_6 = h_5 - (h_{fg})_T$$

or from

$$h_6 = h_1 + \int_{T_1}^{T_6} c_f\, \mathrm{d}T$$

Example 1.7 ***Calculation of Enthalpy of Vapour Using R-K Equation of State***

Using Redlich-Kwong equation of state, calculate the enthalpy of superheated vapour of propane at 2 MPa and 350 K. Take $h_f = 0$ at 200 K ($p^{sat} = 19.97$ kPa) as the reference state. Critical data for propane are:

$$T_c = 369.8 \text{ K}, p_c = 4.236 \text{ MPa}, v_c = 0.005066 \text{ m}^3\text{/kg}$$

Assume zero-pressure constant pressure specific heat of propane as constant as

$$C_{po} = 1.6794 \text{ kJ/kg.K}$$

Solution Refer to Fig. 1.14. Reference state 1 is at 200 K, and not 0°C. From Clapeyron equation, we find:

$$(h_{fg})_{200\text{ K}} = 456.24 \text{ kJ/kg}$$

Hence,
$$h_2 = h_1 + (h_{fg})_{200\text{ K}} = 456.24 \text{ kJ/kg}$$

Gas constant for propane

$$R = \frac{\bar{R}}{M} = \frac{8.3143}{44} = 0.18855 \text{ kJ/kg.K}$$

Constants of Redlich Kwong equation

$$a = 0.42748\, R^2 \frac{T_c^{5/2}}{p_c}$$

$$= 0.42748 \frac{(188.55)^2\,(369.8)^{5/2}}{4.236 \times 10^6} = 9416$$

$$b = 0.08664 \frac{RT_c}{p_c}$$

$$= 0.08664 \frac{(188.55)\,(369.8)}{4.236 \times 10^6} = 1.426 \times 10^{-3}$$

Substituting values in R-K equation, we get

$$v_2 = 1.865 \text{ m}^3/\text{kg}$$

$$v_4 = 0.02534 \text{ m}^3/\text{kg}$$

$$z_2 = \frac{p_2\, v_2}{RT_2} = 0.9876$$

$$z_4 = \frac{p_4\, v_4}{RT_4} = 0.7677$$

Redlich-Kwong[12] derived the following expression for residual enthalpy

$$\frac{h^R}{RT} = \frac{h - h^{id}}{RT} = -\frac{3}{2}\frac{A^2}{B} \ln\left[1 + \frac{Bp}{z}\right] + (z - 1) \qquad (1.25)$$

The constants at points 2 and 4 are:

$$A_2^2 = \frac{a}{R^2\, T_2^{2.5}} = 4.692 \times 10^{-7}$$

$$A_4{}^2 = \frac{a}{R^2\, T_2^{2.5}} = 1.1582 \times 10^{-7}$$

$$B_2 = \frac{b}{RT_2} = 3.782 \times 10^{-8}$$

$$B_4 = \frac{b}{RT_5} = 2.16 \times 10^{-8}$$

Residual enthalpies at points 2 and 4 are:

$$\left(\frac{h - h^{id}}{RT}\right)_2 = -\frac{3}{2}\cdot\frac{4.692 \times 10^{-7}}{3.782 \times 10^{-8}} \ln\left[1 + \frac{3.782 \times 10^{-8} \times 19.97 \times 10^3}{0.9876}\right] + (0.9876 - 1) = -\,0.02663$$

$$\Rightarrow \quad (h - h^{id})_2 = -\,(0.18855)\,(200)\,(0.02663) = -\,1.0 \text{ kJ/kg}$$

$$\left(\frac{h - h^{id}}{RT}\right)_4 = -\frac{3}{2}\cdot\frac{4.692 \times 10^{-7}}{3.782 \times 10^{-8}} \ln\left[1 + \frac{3.782 \times 10^{-8} \times 2 \times 10^6}{0.7677}\right] + (0.7677 - 1) = -\,0.6723$$

$$\Rightarrow \qquad (h - h^{id})_4 = -(0.18855)(350)(0.6723) = -44.35 \text{ kJ/kg}$$

Change in enthalpy at zero pressure

$$h_3 - h_0 = \int_{200}^{350} C_{po}\, \mathrm{d}T = 1.6794\,(350 - 200) = 251.91 \text{ kJ/kg}$$

Enthalpy of superheated propane vapour at 4

$$\begin{aligned} h_4 &= h_2 + (h_0 - h_2) + (h_3 - h_0) + (h_4 - h_3) \\ &= h_2 - (h - h^{id})_2 + (h_3 - h_0) + (h - h^{id})_4 \\ &= 456.24 + 1.0 + 251.91 - 44.35 = 664.8 \text{ kJ/kg} \end{aligned}$$

Note *The value from table is 641.6 kJ/kg. The error is due to the following reasons:*

(i) Empirical Redlich-Kwong equation has been used in place of an equation fitting actual p-v-T data.

(ii) A constant and approximate value of C_{p_o} has been used.

1.15.2 Enthalpy from Residual Internal Energy

If only a p-explicit equation is available, the enthalpy change at constant temperature between points 2 and 0, and 4 and 3 can be calculated by using Eq. (1.24). For the purpose, we have for *residual internal energy* from Eq. (1.20).

$$\begin{aligned} u^R = (u - u^{id})_T &= \int_{v=\infty}^{v} \left(T\left(\frac{\partial p}{\partial T}\right)_v - p \right) \mathrm{d}v_T \\ &= \int_{v}^{v=\infty} \left[p - T\left(\frac{\partial p}{\partial T}\right)_v \right] \mathrm{d}v_T \end{aligned}$$

Using this expression, one can find u_0 and then h_0 from Eq. (1.24) at $t_0 = 0°\text{C}$ ($T_0 = 273.15$ K) and $p_0 = 0$. Then, one can find the enthalpy at 4 using the relationship

$$\begin{aligned} h_4 &= h_0 + (h_3 - h_0) + (h_4 - h_3) \\ &= h_0 + \int_{T_o}^{T=T_3} C_{p0}\, \mathrm{d}T + \left|\left(u - u^{id}\right)_T\right|_{v=\infty}^{v=v_4} + p_4 v_4 - p_3 v_3 \end{aligned} \qquad (1.25a)$$

Putting $C_{po} = C_{vo} + R$, and $p_3 v_3 = RT_3$ (ideal gas state at T_3), we have

$$h_4 = h_0 - RT_0 + p_4 v_4 + \int_{T_o}^{T=T_3} C_{v_o}\, \mathrm{d}T + \left|\left(u - u^{id}\right)_T\right|_{v=\infty}^{v=v_4} \qquad (1.25b)$$

Note *Often, state 0 with enthalpy h_0 is employed as the new reference state for further calculation of enthalpies, and entropies in the entire vapour region.*

Example 1.8 ***Residual Internal Energy from Modified Martin-Hou Equation***

Derive an expression for $(u - u^{id})_T$ from Eq. (1.15).

Solution We have Eqs. (1.26), (1.27), (1.28) by differentiating partially with respect to T

$$\left(\frac{\partial p}{\partial T}\right)_v = \frac{R}{v-b} + \sum_{i=2}^{5} \frac{1}{(v-b)^i}\left[B_i - C_i \frac{k}{T_c} e^{-kT/T_c}\right] \tag{1.26}$$

$$\left[p - \left(\frac{\partial p}{\partial T}\right)_v T\right] = \frac{RT}{v-b} + \sum_{i=2}^{5} \frac{A_i + B_i T + C_i e^{-kT/T_c}}{(v-b)^i}$$

$$- \frac{RT}{v-b} + \sum_{i=2}^{5} \frac{B_i T - \frac{C_i kT}{T_c} e^{-kT/T_c}}{(v-b)^i} \tag{1.27}$$

$$\int_v^{v=\infty} \left[p - \left(\frac{\partial p}{\partial T}\right)_v T\right] \mathrm{d}v = \int_v^{v=\infty} \sum_{i=2}^{5} \frac{A_i + C_i\, e^{-kT/T_c}\left(1 + \frac{kT}{T_c}\right)}{(v-b)^i}$$

$$= \sum_{i=2}^{5} - \left[\frac{A_i + C_i\, e^{-kT/T_c}\left(1 + \frac{kT}{T_c}\right)}{(-i+1)\,(v-b)^{i-1}}\right] \tag{1.28}$$

Example 1.9 ***Reference State Enthalpy of R 22 Vapour***

Find the value of ideal gas enthalpy h_0 at $T_0 = 273.15$ K and $p_0 = 0$ of R 22 vapour using the modified Martin-Hou equation. T_c for R 22 is 369 K.

Solution Refer to Fig. 1.14.

First we find $p_2 = (p^{sat})_{273.15\,K} = 4.976 \times 10^5$ Pa from Eq. (1.10)

Next we find $v_2 = 0.0471$ m³/kg at p_2 and T_2 from Eq. (1.15)

Then,
$$h_2 = h_{f1} + (h_{fg})_{273.15\,K} = 200 + 205.48 = 405.48 \text{ kJ/kg}$$

$$u_2 = h_2 - p_2 v_2$$
$$= 405.48 - 497.6\,(0.0471) = 382.04 \text{ kJ/kg}$$

Now, substituting values of A_i, C_i, k, T_c, and b in Eq. (1.28), we have for $T_2 = 273.15$ K and $v_2 = 0.0471$ m³/kg,

$$u_2 - u_2^{id} = u_2 - u_0 = -\,7020 \text{ J/kg} = -\,7.02 \text{ kJ/kg}$$

$$\Rightarrow \quad u_0 = u_2 + 7.02 = 389.06 \text{ kJ/kg}$$

$$\Rightarrow \quad h_0 = u_0 + R\,T_0 = 389.06 + 0.09615\,(273.15)$$
$$= 415.32 \text{ kJ/kg}$$

Example 1.10 ***Enthalpy of R 22 Vapour***

Taking the value of reference state enthalpy h_0 as found in Example 1.9, calculate the enthalpy of R 22 superheated vapour at 1650 kPa and 95°C (368.15 K).

Solution From Eq. (1.15), $v = 0.0187$ m^3/kg at 1650 kPa and 368.15 K. Substituting values of constants, and v and T, we get:

$$\left|\left(u - u^{id}\right)_{368.15\,\mathrm{K}}\right|_{v=\infty}^{v=0.0187} = -\,27{,}440 \text{ J/kg} = -\,27.44 \text{ kJ/kg}$$

The expression for C_{vo} for R 22 from Appendix is as follows in Eq. (1.29)

$$C_{vo} = C_1 + C_2\,T + C_3\,T^2 - \frac{C_4}{T} \text{ J/kg.K} \tag{1.29}$$

where $C_1 = 117.767818$, $C_2 = 1.6997296$

$C_3 = -\,8.83043292 \times 10^{-4}$, $C_4 = 3.32541759 \times 10^5$

$$\int_{273.15}^{368.15} C_{vo}\,\mathrm{d}T = 70{,}750 \text{ J/kg} = 70.75 \text{ kJ/kg}$$

$$h = h_0 - RT_0 + pv + \int_{T_o}^{T} C_{vo}\,\mathrm{d}T + \left|\left(u - u^{id}\right)_T\right|_{v=\infty}^{v} \tag{1.30}$$

$$= 415.32 - 0.09615(273.15) + 1650(0.0187) + 70.75 - 27.44$$

$$= 463.22 \text{ kJ/kg}$$

1.15.3 Entropy Calculations

The method of calculation of entropy is similarly illustrated in Fig. 1.15. The value of entropy assigned to the reference state of saturated liquid at 0°C is usually $s_1 = s_{f1} = 1.0$ kJ/kg. K. Then, the entropy at 2 is

$$s_2 = s_1 + (s_{fg})_{0°\mathrm{C}} = 1.0 + \frac{(h_{fg})_{0°\mathrm{C}}}{273.15} \text{ kJ/kg. K}$$

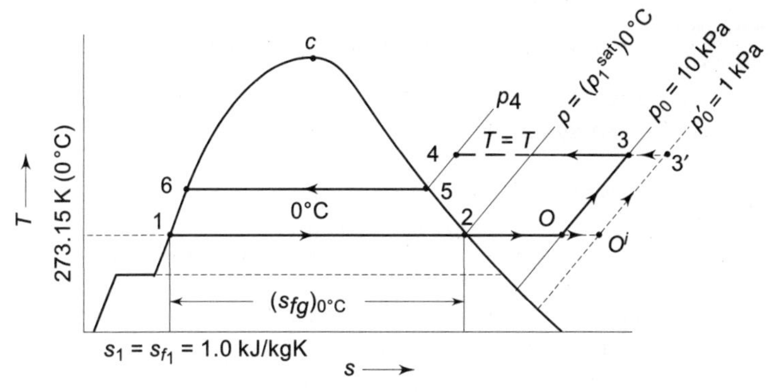

Fig. 1.15 Figure demonstrating method of calculation of entropy

And the entropy of vapour at any point 4 is given by

$$s_4 = s_2 + (s_0 - s_2) + (s_3 - s_0) + (s_4 - s_0)$$

$$= s_2 - \left|(s - s^{id})_{0°C}\right|_{p_o}^{p_2} + \int_{T_o}^{T_3} C_{p_o} \frac{dT}{T} + \left|(s - s^{id})_{T_4}\right|_{p_o}^{p_4}$$

We note that in the calculation of residual entropy $s^R = (s - s^{id})$, the lower limit of integration is not $p_o = 0$ but p_o equal to some very low pressure, say 10 kPa, at which the gas behaves as an ideal gas. Correspondingly, the volume at 0 is $v_0^{id} = RT_0/p_0$, and the volume at 3 is $v_3^{id} = RT_3/p_0$. This is done because of a mathematical anamoly since $s \to \infty$ as $p \to 0$. The path followed for entropy calculation is shown in Fig. 1.15. In case a lower reference pressure is chosen such as $p_0' = 1, 0.1, 0.01$, etc., kPa, this will not affect the values at 4 since the new path followed will now be $2 - 0'$, $0' - 3'$, and $3' - 4$.

Example 1.11 ***Residual Entropy from Modified Martin-Hou Equation***

Derive an expression for residual entropy from Eq. (1.15).

Solution $$s^R = (s - s^{id})_T = \int_{v^{id}}^{v} \left(\frac{\partial P}{\partial T}\right)_v dv$$

$$= \int_{v^{id}}^{v} \frac{R}{v-b}\, dv + \int_{v^{id}}^{v} \sum_{i=2}^{5} \frac{B_i - C_i \dfrac{k}{T_c} e^{-kT/T_c}}{(v-b)^i}$$

$$= R \ln \frac{p_o v}{RT} + \sum_{i=2}^{5} \frac{B_i - C_i \dfrac{k}{T_c} e^{-kT/T_c}}{(-i+1)(v-b)^{i-1}} \tag{1.31}$$

Note *In the integration of the first term $(v - b)$ has been approximated to v. The term containing $1/(v^{id} - b)^{i-1}$, obtained after the integration of the second term, is approximated to zero as $v^{id} >> v$.*

Example 1.12 ***Reference State Entropy of R 22 Vapour***

Find the reference state entropy s_0 of R 22 vapour at 273.15 K and 10 kPa using modified Martin-Hou equation.

Solution Ideal gas volume at 273.15 K

$$v_0^{id} = \frac{RT_0}{p_0} = \frac{0.09615\,(273.15)}{10} = 2.62625 \text{ m}^3/\text{kg}$$

Specific volume of saturated R 22 vapour at 273.15 K

$$v_2 = 0.0471 \text{ m}^3/\text{kg, from Eq. (1.15)}$$

Substituting values in Eq. (1.31), we get

$$(s - s^{id})_{0°C} = s_2 - s_0 = -\,0.4016 \text{ kJ/kg}$$

Now, $$s_2 = s_1 + \frac{(h_{fg})_{0°C}}{273.15} = 1 + \frac{205}{273.15} = 1.7518 \text{ kJ/kg.K}$$

$\Rightarrow$ $$s_0 = s_2 + 0.4016 = 2.1534 \text{ kJ/kg.K}$$

Note *It will be interesting to see how the value of s_0 will change if $p_0{}' = 1$ kPa is taken. It will be seen that $s_0{}' > s_0$. But, it will not effect the value of s_4 since $s_3{}' > s_3$. See Fig. 1.15.*

Example 1.13 ***Entropy of R 22 Vapour***

Find the entropy of superheated R 22 vapour at 1650 kPa and 95°C (368.15 K).

Solution Ideal gas volume at 368.15 K

$$v_4^{id} = \frac{RT}{p_0} = \frac{0.09615(368.15)}{10} = 3.5398 \text{ m}^3/\text{kg}$$

Specific volume of superheated vapour at 1650 kPa and 368.15 K

$$v_4 = 0.0187 \text{ m}^3/\text{kg, from Eq. (1.15)}$$

Using Eq. (1.29)

$$s_3 - s_0 = \int_{T_0}^{T} C_{p_o} \frac{dT}{T} = \int_{T_0}^{T} C_{v_o} \frac{dT}{T} + R \ln \frac{T}{T_0}$$

$$= R \ln \frac{T}{T_0} + \left[C_1 \ln \frac{T}{T_0} + C_2 (T - T_0) + \frac{C_3}{2}\left(T^2 - T_0^2\right) - 2C_4\left(\frac{1}{T} - \frac{1}{T_0}\right)\right] 10^{-3}$$

$$= 0.827 \text{ kJ/kg.K}$$

$$s_4 - s_3 = \left| \left(s - s^{id}\right)_{368.15\text{K}} \right|_{v^{id} = 3.5398}^{v = 0.0187} = -1.1488 \text{ kJ/kg.K}$$

Hence, $$s_4 = s_0 + (s_3 - s_0) + (s_4 - s_3)$$

$$= 2.1534 + 0.827 - 1.1488 = 1.8316 \text{ kJ/kg.K}$$

1.16 MODES OF HEAT TRANSFER

The difference in temperature provides the necessary potential for heat transfer. There are three *modes of heat transfer*. They are *conduction, convection* and *radiation*.

Essentially heat is transferred within a *stationary* medium by conduction, viz., from particle to particle, whether it be solid, liquid or gas. In convection, there must be a bulk flow of the fluid. Heat is carried away from the wall surface by the flowing fluid. Convection, however, takes place in two ways, viz., *forced convec-*

tion and *natural* or *free convection*. In forced convection, the flow of the fluid is produced by an external source such as a pump or a fan. Examples are the shell and tube condenser of a refrigeration plant in which the flow of water is maintained by a pump, and the air-cooled condenser of an air conditioner in which the flow of air is maintained by a fan. In natural or free convection, the flow of the fluid is produced by the difference in density due to temperature difference. The higher temperature fluid, being lighter, rises up and the lower temperature fluid, being heavier, settles down. Thus a natural convection current is set up in the fluid. One example is the air-cooled condenser of a domestic refrigerator. In radiation, heat is transferred in the form of electromagnetic waves. For radiative heat transfer, therefore, the presence of a medium is not necessary.

1.17 LAWS OF HEAT TRANSFER

The *heat flux* q is the heat transfer rate $\dot{Q}$ per unit area A normal to the direction of flow of heat. The various laws relate the heat transfer rate or heat flux to temperature difference ΔT. The unit of ΔT is degrees kelvin (K) or degrees celsius (°C). Both have the same numerical value.

1.17.1 Fourier Law of Heat Conduction

The heat flux by conduction is proportional to the *temperature gradient* within a body. Thus at any point P in a body, if the temperature gradient is $\frac{\partial T}{\partial x}$ as shown in Fig. 1.16, then the heat flux is given by

$$q = \frac{\dot{Q}}{A} \propto \frac{\partial T}{\partial x}$$

or
$$\dot{Q} = -kA\frac{\partial T}{\partial x} \tag{1.32}$$

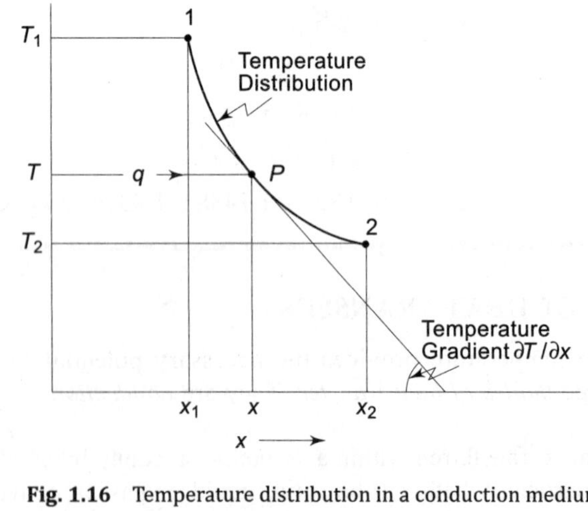

Fig. 1.16 Temperature distribution in a conduction medium

where k is the constant of proportionality, called *thermal conductivity* which is a property of the material of the body. The minus sign is inserted to make $\dot{Q}$ positive since $\frac{\partial T}{\partial x}$ is negative. Equation (1.32) is called the *Fourier law of heat conduction.* The SI unit of thermal conductivity can be derived as below

$$[k] = \frac{[\dot{Q}][\Delta x]}{[A][\Delta T]}$$

$$= \frac{(\text{J/s})\ \text{m}}{\text{m}^2\ \text{K}} = \text{Wm}^{-1}\ \text{K}^{-1}$$

Physically, thermal conductivity represents the amount of heat that will flow per unit time, per unit area normal to the direction of flow of heat, through a unit thickness of the material and when the temperature difference across the material is unity. Experimental measurements of q and $\Delta T/\Delta x$ can be made to determine the thermal conductivity of materials. It is seen that the order of decreasing thermal conductivity is as follows:

Metals; non-metals; liquids; insulating materials; gases.

Metals

Common metals used in heat exchangers are copper, aluminium and iron. Their thermal conductivities in $\text{Wm}^{-1}\ \text{K}^{-1}$ units are as follows:

Copper	387
Aluminium	203
Iron	73

It is seen that copper has the highest thermal conductivity. It is, therefore, used in condensers, evaporators, etc., in Freon refrigeration systems. However, ammonia attacks copper. Hence, iron is used in ammonia heat exchangers. As aluminium is cheap, and widely available, attempts are being made to use it in R 22 systems.

Thermal conductivities of other important substances (average values) are:

Liquids

Water	0.556
Ammonia	0.54
Freon 22	0.093

Gases

Air	0.024
Water vapour	0.0206

Low temperature insulating materials

Expanded polysterene (Thermocole)	0.037
Polyurethane foam(PUF)	0.0173

The low values of thermal conductivities of these insulating materials are essentially due to the low conductivity of gas/air trapped inside, as these are manufactured by blowing agents like air/R 134a into the melted polyesterene and polyurethane respectively.

1.17.2 Newton's Law of Cooling for Convection

When a fluid flows over a wall which is at a different temperature than the fluid, heat will flow from the wall to the fluid or from the fluid to the wall depending on the direction of the temperature gradient. Although this mode of heat transfer is named convection, the physical mechanism of heat transfer at the wall is a conduction process.

Consider a fluid flowing along a wall maintained a temperature T_w as shown in Fig. 1.17. The free stream temperature of the fluid is T_∞. Then a temperature field varying from T_w to T_∞ will establish itself in the fluid near the wall. Let δ_T be the distance from the wall of the point in the fluid at which the temperature of the fluid just becomes equal to the free stream temperature T_∞. The distance δ_T is termed the *thermal boundary layer*. The distance δ from the wall in which the fluid velocity becomes equal to the free stream velocity is called the *hydrodynamic boundary layer*.

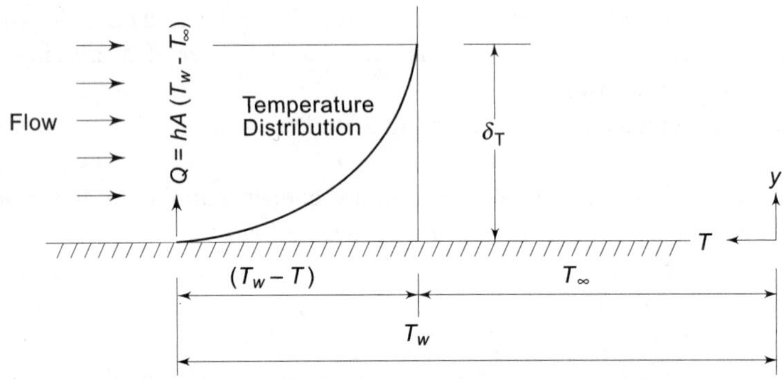

Fig. 1.17 Temperature distribution in thermal boundary layer in convection

The concept of the *heat transfer coefficient* or *film coefficient* or *surface conductance*, denoted by the symbols h or f, was introduced by Newton. He recommended the following equation to evaluate the heat transfer rate by convection.

$$\dot{Q} = hA\,(T_w - T_\infty) = hA\,\Delta T \tag{1.33}$$

where A is the wall surface area and $\Delta T = (T_w - T_\infty)$.

It must be noted that this h or f is not a property of the fluid. However, it depends on the *thermophysical* or *heat transport* properties of the fluid such as thermal conductivity k, dynamic viscosity μ, density ρ and specific heat C_p. In addition, it also depends on the hydrodynamic or flow parameters such as velocity of flow and characteristic dimensions.

The SI unit of the heat transfer coefficient can now be derived as follows:

$$[h] = \frac{[\dot{Q}]}{[A][\Delta T]}$$

$$= \frac{\text{J/s}}{\text{m}^2\,\text{K}} = \text{Wm}^{-2}\,\text{K}^{-1}$$

Typical values of h are given in Table 1.2.

Table 1.2 Typical values of convection heat transfer coefficients

Mode and Medium	*h*, W m^{-1} K^{-1}
Free convection, air	5 – 25
Forced convection, air	10 – 100
Forced convection, water	5,000 – 10,000
Boiling refrigerant	500 – 2,000
Condensing refrigerant	1,500 – 2,500

1.17.3 Thermal Radiation

The thermal radiation emitted by a body per unit area of its surface is called *emissive power* which is proportional to the fourth power of its absolute temperature.

For an *ideal radiator* or *black body*, the emissive power E_b is given by the *Stefan-Boltzmann law* that follows in Eq. (1.34)

$$E_b = \sigma T^4 \tag{1.34}$$

where σ is the constant of proportionality called the *Stefan-Boltzmann constant* having a numerical value of 5.669×10^{-8} W m^{-2} K^{-4}.

The emissive power of an actual radiator is expressed by the relation given in Eq. (1.35)

$$E = \varepsilon E_b = \varepsilon \sigma T^4 \tag{1.35}$$

where $\varepsilon = \dfrac{E}{E_b}$ is the emissivity of the actual surface.

Thermal radiations emitted by two bodies of areas A_1 and A_2, at temperatures T_1 and T_2 will then be given by

$$\dot{Q}_1 = \varepsilon_1 A_1 \sigma T_1^4 \text{ and } \dot{Q}_2 = \varepsilon_2 A_2 \sigma T_2^4$$

However, the heat exchange by thermal radiation between two bodies will also depend on the extent to which the two bodies "see" each other geometrically. The expression for such a heat exchange may be expressed as given in Eq. (1.36)

$$\dot{Q} = \sigma A_1 F_{12} (T_1^4 - T_2^4) = \sigma A_2 F_{21} (T_1^4 - T_2^4) \tag{1.36}$$

where $A_1 F_{12} = A_2 F_{21}$ is the *reciprocity relation*, and F_{12} and F_{21} are *geometric factors* which depend on the emissivities ε_1 and ε_2, and the geometry and orientation of the bodies.

In the case of a small body of area A_1 surrounded by a large body of area A_2, completely seeing each other, we have

$$F_{12} = \varepsilon_1 \tag{1.36}$$

so that

$$\dot{Q} = \varepsilon_1 A_1 \sigma (T_1^4 - T_2^4) \tag{1.37}$$

Often, the heat exchange by radiation $\dot{Q}_R$ between two surfaces is expressed in terms of a *radiation coefficient* h_R defined by

$$\dot{Q}_R = h_R A_1 (T_1 - T_2) = A_1 F_{12} \sigma (T_1^4 - T_2^4)$$

so that

$$h_R = \frac{F_{12} \sigma (T_1^4 - T_2^4)}{(T_1 - T_2)}$$

1.18 ELECTRICAL ANALOGY[1]

It is found convenient to handle complicated heat transfer problems involving composite materials and multi-modes by unifying the concept of heat transfer with that of the flow of electric current. Comparing the flow of electrical energy and the flow of heat, it is found that the following similarities hold.

Quantity	*Electrical Energy*	*Heat*
Driving potential	Voltage, V	Temperature difference, ΔT
Flow	Current, I	Heat transfer rate, $\dot{Q}$

The governing law for the transfer of electrical energy is Ohm's law

$$I = \frac{V}{R_e}$$

where R_e is the electrical resistance. One can similarly express the heat transfer rate by Eq. (1.38)

$$\dot{Q} = \frac{\Delta T}{R} = C\,\Delta T \tag{1.38}$$

where R is the *thermal resistance* and $C = 1/R$ is the *thermal conductance*. The units of the two quantities are W^{-1} K and WK^{-1} respectively.

Comparring Eq. (1.38) with Eqs (1.32), (1.33) and (1.36) we obtain expressions for conductive, convective and radiative resistances respectively as follows:

$$R_{COND} = \frac{\Delta x}{kA}, \text{ for plane wall of thickness } \Delta x \tag{1.39}$$

$$R_{CONV} = \frac{1}{hA} \tag{1.40}$$

$$R_{RAD} = \frac{\Delta T}{\sigma A_1 F_{12} (T_1^4 - T_2^4)} \tag{1.41}$$

1.18.1 Resistances in Series

When resistance are in series as shown in Fig. 1.18 (a), it implies that the heat transfer rate through all the resistances is the same. We can then write: as given in Eq. (1.42)

$$\dot{Q} = \frac{T_1 - T_2}{R_1} = \frac{T_2 - T_3}{R_2} = \frac{T_3 - T_4}{R_3} = \frac{\Delta T}{R} \tag{1.42}$$

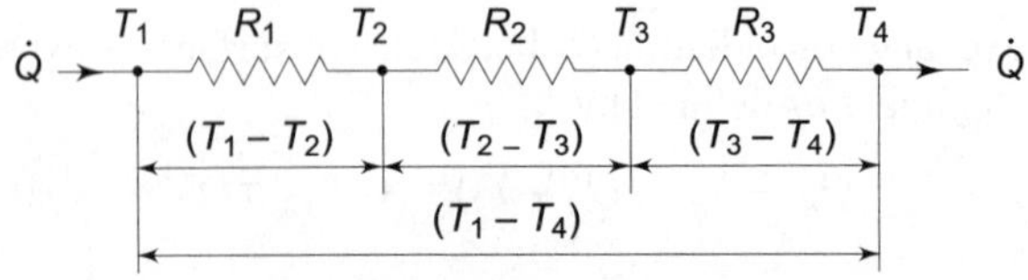

(a) Resistances in Series

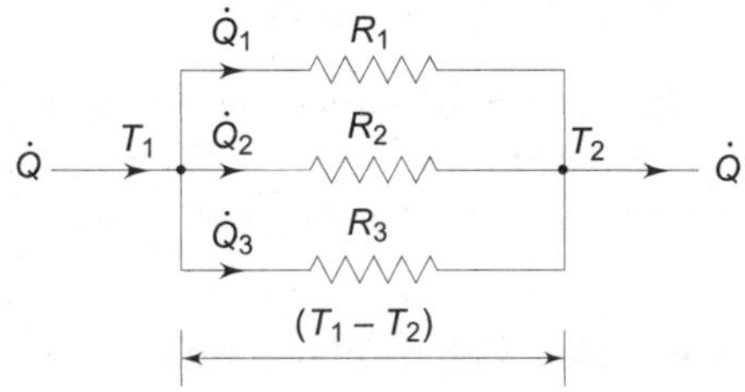

(b) Resistances in Parallel

Fig. 1.18 Thermal resistances in series and in parallel

where $\Delta T = (T_1 - T_4)$ is the overall temperature difference, and R is the overall thermal resistance. From Eq. (1.42), the individual temperature drops are:

$$T_1 - T_2 = \dot{Q}R_1$$

$$T_2 - T_3 = \dot{Q}R_2$$

$$T_3 - T_4 = \dot{Q}R_3$$

Adding, $$T_1 - T_4 = \dot{Q}\,(R_1 + R_2 + R_3)$$

$$\Rightarrow \qquad \dot{Q} = \frac{T_1 - T_4}{R_1 + R_2 + R_3} = \frac{\Delta T}{R}$$

so that the overall thermal resistance R is given by Eq. (1.43)

$$R = R_1 + R_2 + R_3 + \cdots + \tag{1.43}$$

1.18.2 Resistances in Parallel

When resistances are in parallel, as shown in Fig. 1.18(b), the net heat transfer rate is equal to the sum of the heat transfer rates through all sections. At the same time, the temperature drop ΔT across each resistance is the same. Hence,

$$\dot{Q}_1 = \frac{\Delta T}{R_1}$$

$$\dot{Q}_2 = \frac{\Delta T}{R_2}$$

$$\dot{Q}_3 = \frac{\Delta T}{R_3}$$

Adding $$\dot{Q} = \dot{Q}_1 + \dot{Q}_2 + \dot{Q}_3$$

$$= \Delta T\left[\frac{1}{R_1} + \frac{1}{R_2} + \frac{1}{R_3}\right] = \frac{\Delta T}{R}$$

whence we get for the overall thermal resistance Eq. (1.44)

$$\frac{1}{R} = \frac{1}{R_1} + \frac{1}{R_2} + \frac{1}{R_3} + \cdots + \cdots \tag{1.44}$$

1.19 STEADY-STATE CONDUCTION

In general, the temperature distribution throughout a body may vary with location and time. Under *steady-state* conditions, however, the temperature does not change with time. We shall now examine some steady-state heat conduction problems in one dimension.

1.19.1 Heat Flow Through a Slab or Plane Wall

Let there be a slab of thickness Δx, the two faces of which are maintained at temperature T_1 and T_2 as shown in Fig. 1.19. Consider a section at a distance x from one end. Let the temperature gradient at this section be $\mathrm{d}T/\mathrm{d}x$. Then heat entering the wall, per unit time, at this section is

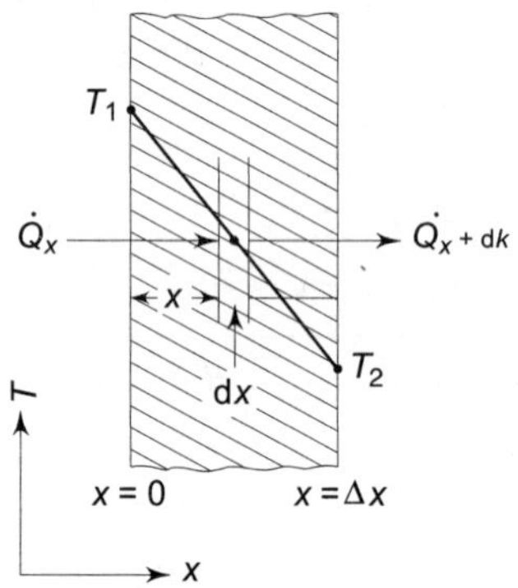

Fig. 1.19 Heat conduction through a plane wall

$$\dot{Q}_x = -kA\frac{\mathrm{d}T}{\mathrm{d}x} \tag{1.45}$$

Similarly, heat leaving the wall at $(x + \mathrm{d}x)$ is

$$\dot{Q}_{x+\mathrm{d}x} = \dot{Q}_x + \frac{\mathrm{d}\dot{Q}_x}{\mathrm{d}x}\mathrm{d}x \tag{1.46}$$

Under steady-state conditions

$$\dot{Q}_x = \dot{Q}_{x+\mathrm{d}x}$$

so that from Eqs (1.45) and (1.46) we have Eq. (1.47)

$$\frac{\mathrm{d}^2T}{\mathrm{d}x^2} = 0 \tag{1.47}$$

Using the boundary conditions

(i) $T = T_1$ at $x = 0$, (ii) $T = T_2$ at $x = \Delta x$

and solving the differential Eq. (1.47), we obtain for the temperature distribution in the wall

$$\frac{T - T_1}{T_2 - T_1} = \frac{x}{\Delta x} \tag{1.48}$$

which is linear with respect to x. The temperature gradient is

$$\frac{\mathrm{d}T}{\mathrm{d}x} = \frac{T_2 - T_1}{\Delta x} \tag{1.49}$$

which is constant. The heat transfer rate is given by the Fourier law in Eq. (1.50)

$$\dot{Q} = -kA\,\frac{\mathrm{d}T}{\mathrm{d}x} = \frac{kA\,(T_1 - T_2)}{\Delta x} = \frac{T_1 - T_2}{\Delta x / kA} \tag{1.50}$$

It is seen that the thermal resistance of the wall is

$$R = \frac{\Delta x}{kA}$$

1.19.2 Heat Flow Through a Cylinder

Consider a cylindrical shell of thickness dr at a distance r from the axis of a cylindrical tube or pipe of length L and outer and inner radii of r_0 and r_1 respectively, as shown in Fig. 1.20. The area of the inside surface of the shell will be $2\pi rL$. Then the heat flows in the radial direction at r and r + dr are:

$$\dot{Q}_r = -k\,(2\pi rL)\,\frac{\mathrm{d}T}{\mathrm{d}r}, \qquad \dot{Q}_{r+\mathrm{d}r} = \dot{Q}_r + \frac{\mathrm{d}\dot{Q}_r}{\mathrm{d}r}\cdot \mathrm{d}r$$

Under steady-state conditions, equating the two we get

$$\frac{\mathrm{d}\dot{Q}_r}{\mathrm{d}r} = 0$$

$$\Rightarrow \qquad \frac{\mathrm{d}}{\mathrm{d}r}\left(r\,\frac{\mathrm{dT}}{\mathrm{d}r}\right) = 0 \tag{1.51}$$

Using the boundary conditions

(i) $T = T_i$ at $r = r_1$

(ii) $T = T_o$ at $r = r_0$

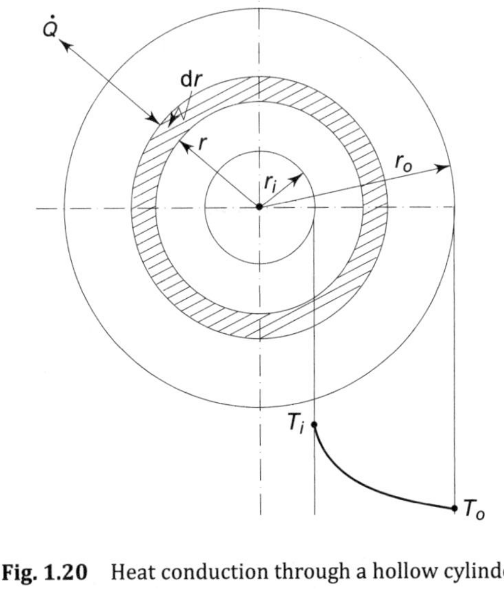

Fig. 1.20 Heat conduction through a hollow cylinder

and solving Eq. (1.51) we get for the temperature distribution

$$\frac{T - T_o}{T_i - T_o} = \frac{\ln\left(\frac{r}{r_o}\right)}{\ln\left(\frac{r_i}{r_o}\right)} \tag{1.51a}$$

and for heat flow we have Eq. (1.52)

$$\dot{Q} = -kA\frac{dT}{dr} = -k(2\pi rL)\frac{dT}{dr}$$

$$= \frac{T_i - T_o}{\left(\frac{1}{2\pi kL}\right)\ln\left(\frac{r_o}{r_i}\right)} \tag{1.52}$$

The thermal resistance of the cylindrical shell is then as follows in Eq. (1.53)

$$R = \frac{\ln\left(\frac{r_o}{r_i}\right)}{2\pi kL} \tag{1.53}$$

1.19.3 Heat Flow Through a Composite Wall with Convection Boundaries

Consider a wall comprising of more than one material and convection at the two surfaces as shown in Fig. 1.21. It is seen that all the thermal resistances are in series. An equivalent electrical analogue of the wall is also shown in the figure. The overall thermal resistance is given by Eq. (1.54).

$$R = \frac{1}{h_o A} + \frac{\Delta x_A}{k_A A} + \frac{\Delta x_B}{k_B A} + \frac{\Delta x_C}{k_C A} + \frac{1}{h_i A} \tag{1.54}$$

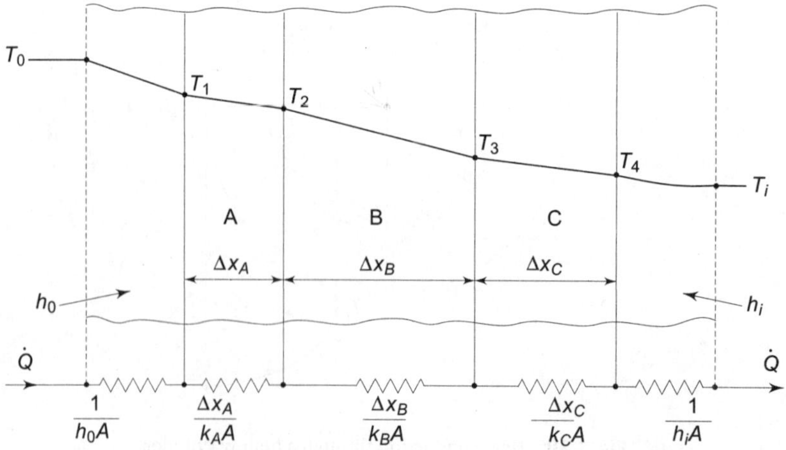

Fig. 1.21 Composite plane wall with resistances in series

where subscripts A, B and C refer to the three materials of the wall and h_i and h_o are the convective heat transfer coefficients between the inside and outside wall surfaces and surrounding air.

Similarly, if some of the resistances are in parallel while some are in series as shown in Fig. 1.22, then the overall thermal resistance is given by

$$R = \frac{1}{h_o A} + R_A + \left[\frac{1}{\frac{1}{R_B} + \frac{1}{R_C} + \frac{1}{R_D}}\right] + R_E + \frac{1}{h_i A} \quad (1.55)$$

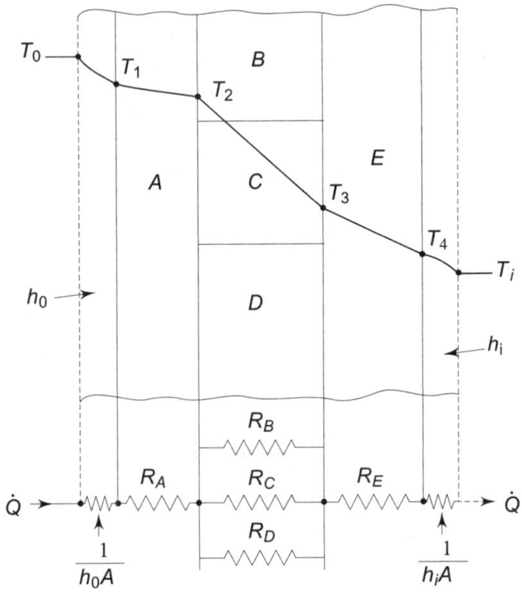

Fig. 1.22 Composite plane wall with resistances in series as well as in parallel

In both cases, if T_o and T_i are the outside and inside air temperatures respectively, then the heat flow rate is given by

$$\dot{Q} = \frac{T_o - T_i}{R}$$

Example 1.14 *An exterior wall of a house consists of 10.2 cm brick and 3.8 cm gypsum plaster. What thickness of loosely-packed rockwool insulation should be added to reduce the heat transfer through the wall by 80 per cent? The thermal conductivities of brick, gypsum plaster and rockwool are 0.7, 0.48 and 0.065 W.m^{-1} K^{-1} respectively.*

Solution Since ΔT is the same in both cases, we have $(\dot{Q}\,R)' = 0.2\,(\dot{Q}\,R) = \Delta T$

$$\Rightarrow \qquad \frac{\dot{Q}' \text{ with insulation}}{\dot{Q} \text{ without insulation}} = 0.2 = \frac{R \text{ without insulation}}{R' \text{ with insulation}}$$

The resistances of brick and plaster are (per unit area of wall)

$$R_{\text{brick}} = \frac{0.102}{0.7} = 0.145 \text{ m}^2.\text{K.W}^{-1}$$

$$R_{\text{plaster}} = \frac{0.038}{0.48} = 0.079 \text{ m}^2.\text{K.W}^{-1}$$

Then the resistance without insulation is

$$R = R_{\text{brick}} + R_{\text{plaster}}$$

$$= 0.145 + 0.079 = 0.224 \text{ m}^2.\text{K.W}^{-1}$$

and the resistance with insulation is

$$R' = \frac{R}{0.2} = \frac{0.224}{0.2} = 1.122 \text{ m}^2.\text{K.W}^{-1}$$

Hence, the resistance of rockwool is

$$R_{\text{rockwool}} = R' - R$$

$$= 1.122 - 0.224 = 0.898 \text{ m}^2.\text{K.W}^{-1}$$

The required thickness of rockwool insulation is, therefore,

$$\Delta x_{\text{rockwool}} = (k\,AR)_{\text{rockwool}}$$

$$= (0.065)\ 1\ (0.898) = 0.0585 \text{ m } (\approx 6 \text{ cm})$$

1.19.4 Heat Flow Through a Composite Cylinder

Consider a composite tube of three materials with fluids flowing inside as well as outside the tube as shown in Fig. 1.23. Summing up all the resistances which are in series, we obtain Eq. (1.56)

$$R = \frac{1}{h_i A_i} + \frac{\ln \frac{r_2}{r_1}}{2\pi k_A L} + \frac{\ln \frac{r_3}{r_2}}{2\pi k_B L} + \frac{\ln \frac{r_4}{r_1}}{2\pi k_C L} + \frac{1}{h_o A_o} \tag{1.56}$$

For the case of a single tube as in heat exchangers, this becomes

$$R = \frac{1}{h_i A_i} + \frac{\ln \frac{r_o}{r_i}}{2\pi k L} + \frac{1}{h_o A_o} \tag{1.57}$$

where A_o and A_i are the outside and inside tube surface areas respectively. Then the heat flow rate can be determined using the expression in Eq. (1.57).

Example 1.15 *A 3 cm OD pipe is to be covered with two layers of insulation, each having a thickness of 2.5 cm. The average thermal conductivity of one insulation is five times that of the other. Determine the percentage decrease in heat transfer if the better insulating material is next to the pipe than if it is the outer layer.*

Assume that the outside and inside surface temperatures of the composite insulation are fixed.

Solution Let the thermal conductivity of the poorer insulation be $k_1 = 5k$, and that of the better insulation be $k_2 = k$.

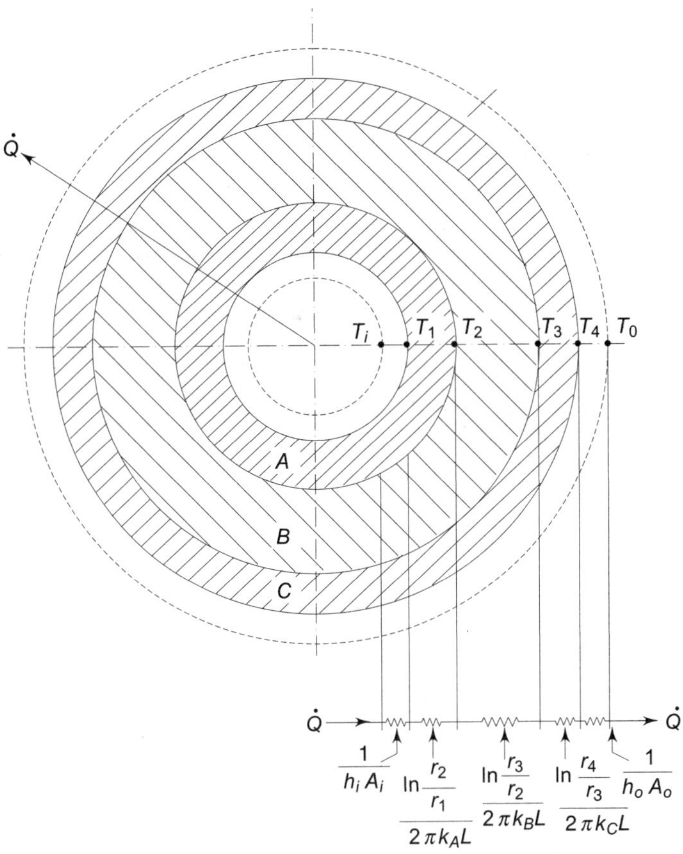

Fig. 1.23 Heat flow through a composite cylinder with convection boundaries

Case I: Better insulating material next to pipe

$$\dot{Q}_{\mathrm{I}} = \frac{\Delta T}{\dfrac{1}{2\pi k_1 L}\ln\dfrac{r_2}{r_1} + \dfrac{1}{2\pi k_2 L}\ln\dfrac{r_3}{r_2}}$$

$$= \frac{2\pi L \Delta T}{\dfrac{1}{k}\ln\dfrac{4}{2.5} + \dfrac{1}{5k}\ln\dfrac{6.5}{4}} = \frac{2\pi k L \Delta T}{1.0774}$$

Case II: Poorer insulating material next to pipe

$$\dot{Q}_{II} = \frac{2\pi L\Delta T}{\dfrac{1}{5k}\ln\dfrac{4}{2.5} + \dfrac{1}{k}\ln\dfrac{6.5}{4}} = \frac{2\pi kL\Delta T}{0.683}$$

Percentage reduction in heat flow

$$\frac{\dot{Q}_{II} - \dot{Q}_{I}}{\dot{Q}_{II}}\times 100 = \frac{\dfrac{1}{0.683} - \dfrac{1}{1.0774}}{\dfrac{1}{0.683}}\times 100 = 36.6\%$$

1.19.5 Overall Heat Transfer Coefficient

In order to calculate the rate of heat flow through a combination of resistances, the concept of *overall heat transfer coefficient* is introduced. Denoting it by the symbol U we express it by the relation

$$\dot{Q} = UA\,\Delta T = \frac{\Delta T}{R}$$

so that

$$\frac{1}{UA} = R$$

For a plane composite wall with all resistances in series as in Fig. 1.21, we find by comparison with Eq. (1.54) that

$$\frac{1}{UA} = \frac{1}{h_i A} + \frac{\Delta x_A}{k_A A} + \frac{\Delta x_B}{k_B A} + \cdots + \frac{1}{h_o A}$$

$$\Rightarrow \quad \frac{1}{U} = \frac{1}{h_i} + \frac{\Delta x_A}{k_A} + \frac{\Delta x_B}{k_B} + \cdots + \frac{1}{h_o} \qquad (1.58)$$

For a composite cylinder, the overall heat transfer coefficient can be based on either the outside or the inside tube surface area. Thus, we have

$$\dot{Q} = U_o A_o \Delta T = U_i A_i \Delta T = \frac{\Delta T}{R}$$

Then, by comparison with Eq. (1.56), we obtain

$$\frac{1}{U_o A_o} = \frac{1}{U_i A_i} = \frac{1}{h_i A_i} + \frac{\ln\dfrac{r_2}{r_1}}{2\pi k_A L} + \frac{\ln\dfrac{r_3}{r_2}}{2\pi k_B L} + \cdots + \frac{1}{h_o A_o} \qquad (1.59)$$

Equation (1.59) may be used to determine U_o or U_i. We see that $U_o A_o = U_i A_i$.

Example 1.16 *(a) Find the overall heat transfer coefficient of a flat built-up roof having the construction shown in Fig. 1.24.*

(b) Find the value of U if rigid roof deck insulation of resistance $R = 0.76$ K.W^{-1} is added to this construction.

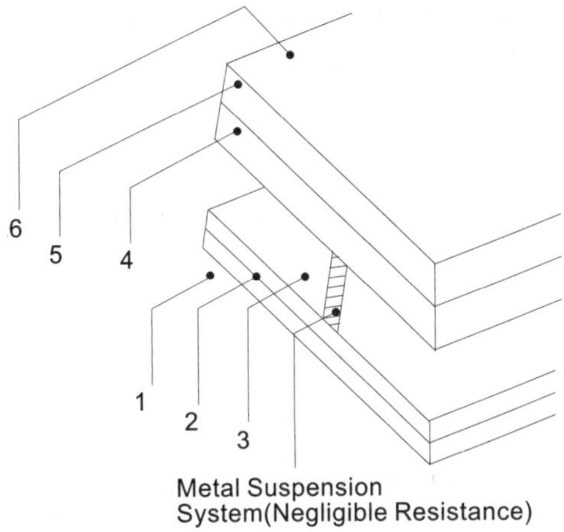

Fig. 1.24 Built-up roof construction for Example 1.16

Solution

(a) The values of resistances per unit area as found from ASHRAE Handbook[3] are as follows:

Resistance No.	*Construction material*	*Unit resistance, $\Delta x/k$ or $1/h$ m^2.K.W^{-1}*
1.	Inside surface (still air)	0.107
2.	Metal lath and 0.75 inch plaster	0.083
3.	Air space (greater than 9 cm width)	0.164
4.	Concrete slab, 5 cm	0.391
5.	Built-up roofing, 9.5 cm	0.058
6.	Outside surface (20 kmph wind velocity)	0.03
	Total thermal resistance, R_a	0.833

Overall heat transfer coefficient for construction '*a*'

$$U_a = \frac{1}{R_a} = \frac{1}{0.833} = 1.2 \text{ W.m}^{-2}\text{. K}^{-1}$$

(b) Total thermal resistance for construction '*b*'

$$R_b = R_a + 0.76 = 0.833 + 0.76 = 1.593 \text{ m}^2\text{. K.W}^{-1}$$

Overall heat transfer coefficient for construction '*b*'

$$U_b = \frac{1}{R_b} = \frac{1}{1.593} = 0.68 \text{ W.m}^{-2}\text{. K}^{-1}$$

1.20 HEAT TRANSFER FROM EXTENDED SURFACE

A *heat exchanger* is an apparatus which affects the transfer of heat from one fluid to another. The overall heat transfer coefficient of a heat exchanger surface is determined principally by the greatest single resistance. As an illustration, neglecting the thermal resistance of the metal wall of a heat exchanger, the overall heat transfer coefficient between the two fluids is given by

$$\frac{1}{UA} = \frac{1}{h_i A_i} + \frac{1}{h_o A_o}$$

Taking $A_i = A_o = A$, and values of h_i and h_o as 1000 and 10 W. m^{-2}. K^{-1} respectively, we see that

$$\frac{1}{U} = \frac{1}{1000} + \frac{1}{10} = 0.101$$

$$\Rightarrow \qquad U = 9.9 \text{ W.m}^{-2}.\text{K}^{-1}$$

so that the value of U is less than the value of the lower of the two heat transfer coefficients, viz., $h_o = 10$, in this case. The lower coefficient is, therefore, the *controlling coefficient*. The influence of the higher coefficient is only marginal. The size of the heat exchanger will thus be uneconomically large. There are two methods by which the heat exchange can be improved.

One is to augment the lower heat transfer coefficient.

The other is to employ an extended surface on the side of the lower coefficient.

Both methods decrease the thermal resistance, one by increasing h, and the other by increasing A.

Consider an extended surface in the form of a thin rod protruding from a surface into a surrounding fluid as shown in Fig. 1.25. The *root* or *base* of the rod is at temperature T_1 while the fluid temperature is T_∞. The length, cross-sectional area and perimeter of the rod are l, A and P respectively. The temperature distribution along the rod is also shown in Fig. 1.25.

Consider an element dx at a distance x from the base of rod. The heat flows by conduction within the rod, and by convection from its surface to the surrounding fluid. The energy balance over the element dx under steady-state gives

$$\dot{Q}_x - \dot{Q}_{x+\mathrm{d}x} = \dot{Q}_{\mathrm{conv}}$$

$$\left(-kA\frac{\mathrm{d}T}{\mathrm{d}x}\right)_x - \left(-kA\frac{\mathrm{d}T}{\mathrm{d}x}\right)_{x+\mathrm{d}x} = h(P\mathrm{d}x)\,(T - T_\infty) \qquad (1.60)$$

where T is the temperature of the rod at x, k is the thermal conductivity of the rod, and h is the heat transfer coefficient from the surface to the fluid. Equation (1.60) can be simplified to

$$\frac{\mathrm{d}^2\theta}{\mathrm{d}x^2} = m^2\theta = 0 \qquad (1.61)$$

by putting θ = excess temperature = $(T - T_\infty)$,

and $$m = \sqrt{\frac{hP}{kA}}$$

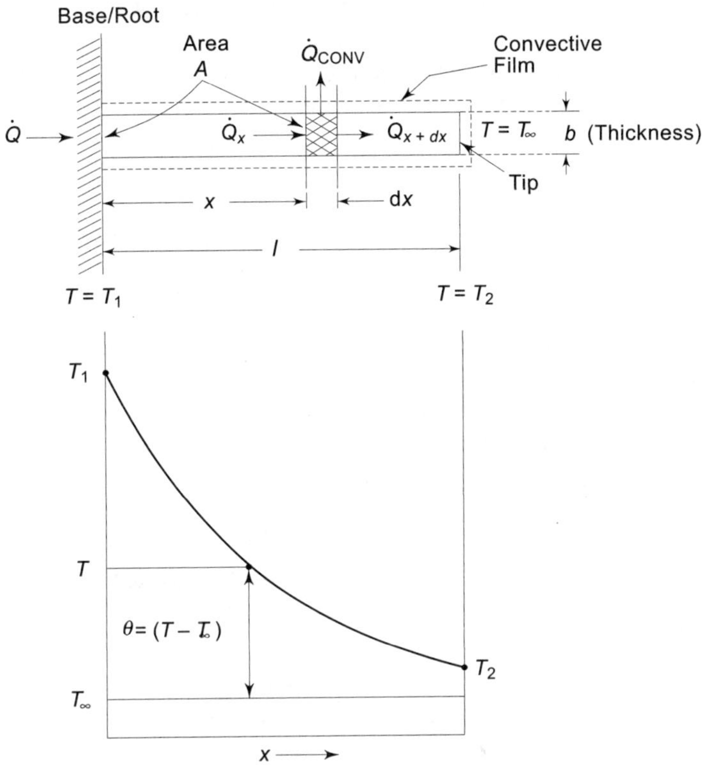

Fig. 1.25 Heat flow and temperature distribution along the length of a rod protruding from a surface

Using the boundary conditions

(i) $\theta = T_1 - T_\infty = \theta_1$ at $x = 0$, and

(ii) $\dfrac{d\theta}{dx} = 0$ at $x = l$ (no heat conduction at *tip*)

we obtain the solution of Eq. (1.61) for temperature distribution as

$$\frac{\theta}{\theta_1} = \frac{\cosh m(l - x)}{\cos ml} \tag{1.62}$$

Then the heat flow from the base of the rod (at $x = 0$) is given by Eq. (1.63)

$$\dot{Q}_1 = -kA \left|\frac{d\theta}{dx}\right|_{x=0}$$

$$= mkA\theta_1 \tanh ml$$

$$= \sqrt{hPkA}\ \theta_1 \tanh ml \tag{1.63}$$

The term $\sqrt{hPkA}\theta_1 = mkA\theta_1$ is constant. The value of tanh ml increases with ml and tends to unity. It becomes 0.964 at $ml = 2$ and 0.9951 at $ml = 3$ as seen from

Table 1.3. It is, therefore, apparent that increasing the length of an extended surface beyond $ml = 2$ or 3 does not improve the heat flow.

For a *rectangular fin* of height l, width L and thickness b, in place of a rod, as shown in Fig. 1.25 the cross-sectional area and perimeter are

$$A = bL, P = 2L + 2b = 2(L + b) \approx 2L$$

so that the parameter m is

$$m = \sqrt{\frac{hP}{kA}} = \sqrt{\frac{2h(L+b)}{kbL}} \cong \sqrt{\frac{2h}{kb}}$$

The entire fin is not at temperature T_1. It drops from T_1 at the root to T_2 at the tip. Hence, the whole fin surface is not equally effective.

The fin efficiency η_f is now defined as the ratio of the actual heat transferred by a fin to that which would be transferred if the entire fin surface were assumed to be at the base temperature. Thus

$$\eta_f = \frac{\dot{Q}_1 \text{ with fin}}{\dot{Q} \text{ with fin surface at base temperature}}$$

For a rectangular fin, therefore,

$$\eta_f = \frac{mk\,A\theta_1 \tanh ml}{h(Pl)\,\theta_1} = \frac{\tanh ml}{ml}$$

Table 1.3 gives the values of tanh ml and η_f as a functions of ml. It is seen that tanh ml, and hence $\dot{Q}_1$ increase rapidly at first as l increases. The increase, then, slows down finally reaching an asymptotic value at $ml \approx 3$. The fin efficiency is quite high upto $ml \approx 0.5$, but decreases rapidly as ml increases.

Table 1.3 Numerical values of tanh ml and η_f

ml	0	0.5	1.0	1.5	2	3	4	5	6
tanh ml	1	0.4621	0.7616	0.9052	0.964	0.995	0.9993	0.9999	1
η_f		0.924	0.7616	0.603	0.482	0.332	0.2498	0.20	0.167

1.20.1 Efficiency of Circular Fins and Finned Tube Arrays

A largely empirical method, developed by Schmidt[15] for finding the efficiency of a circular fin the configuration of which is shown in Fig. 1.26, is summarized as follow in Eqs. (1.64) and (1.65).

$$\eta_f = \frac{\tanh(mr\phi)}{(mr\phi)} \tag{1.64}$$

where

$$\phi = \left(\frac{R}{r} - 1\right)\left[1 + 0.35 \ln\left(\frac{R}{r}\right)\right] \tag{1.65}$$

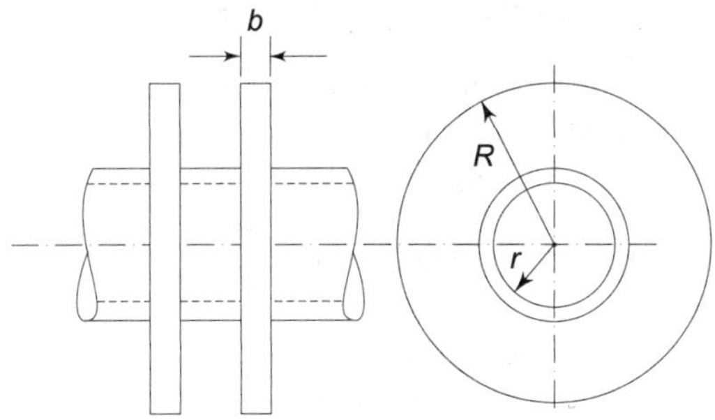

Fig. 1.26 Circular fins

However, in air conditioning, continuous plate fins are used in finned tube cooling coils. The two array configurations, viz., the rectangular tube array and the angular tube array (hexangular fin) are shown in Figs 1.27 and 1.28 respectively.

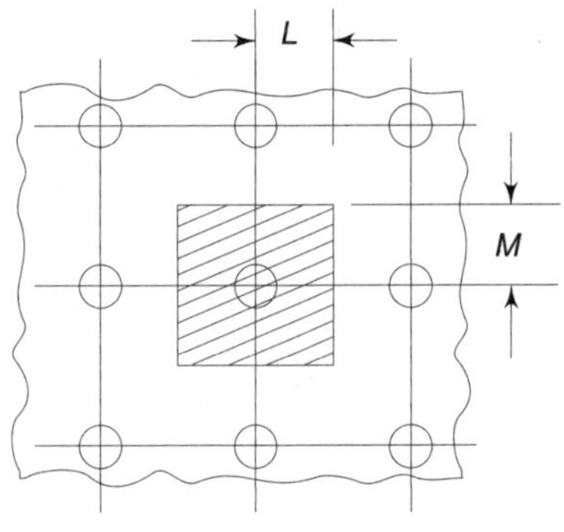

Fig. 1.27 Rectangular tube array fin

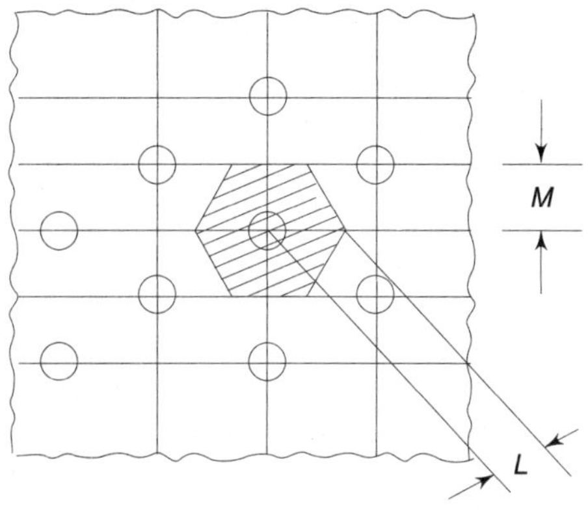

Fig. 1.28 Triangular tube array hexangular fin

Schmidt, again, provides an empirical method for calculating the efficiency of these fins. The method is based on selecting a circular fin with an equivalent radius R_e that has the same fin efficiency as the rectangular fin as follows in Eqs. (1.66) and (1.67).

For the Rectangular Tube Array Fin

$$\frac{R_e}{r} = 1.28\ \psi(\beta - 0.2)^{1/2} \tag{1.66}$$

For the Triangular Tube Array Hexangular Fin

$$\frac{R_e}{r} = 1.27\ \psi(\beta - 0.3)^{1/2} \tag{1.67}$$

In both the above expressions, from Figs. (1.27) and (1.28)

$$\psi = \frac{M}{r} \text{ and } \beta = \frac{L}{M}$$

in which L is always selected to be greater than or equal to M.

1.21 UNSTEADY-STATE CONDUCTION

In unsteady-state conduction, the temperature of the body changes with time. This change in temperature is represented by $\partial T/\partial \tau$ where τ stands for time. Thus, in the case of a plane wall element dx (Fig. 1.19), we have

$$\dot{Q}_x - \dot{Q}_{x+dx} = C_p (A\ dx) \frac{\partial T}{\partial \tau} \tag{1.68}$$

where ρ is the density, and C_p is the specific heat of the wall.

The right hand side in Eq. (1.68) represents the increase in stored energy of the element. Substituting for $\dot{Q}_x$ and $\dot{Q}_{x+dx}$ from Eqs (1.45) and (1.46) we obtain the unsteady-state heat conduction equation in one dimension as

$$\frac{\partial^2 T}{\partial x^2} = \frac{1}{\alpha} \frac{\partial T}{\partial \tau} \tag{1.69}$$

where $\alpha = k/\rho C_p$ is called the *thermal diffusivity* of the wall material. It will be seen that the units of thermal diffusivity are m^2 s^{-1}.

Various analytical methods have been used to solve Eq. (1.69). These include, among others, *the methods of separation of variables* and *the method of transformation of coordinates*. In all cases, the solutions are of the following dimensionless form

$$\frac{\theta}{\theta_i} = f\left(\frac{x}{L}, \text{Bi}, \text{Fo}\right)$$

where

θ = excess temperature at anytime τ at $x = t - t_{\text{reference}}$

θ_i = initial excess temperature at x

x/L = dimensionless distance

$$\text{Bi} = \textit{Biot number} = \frac{hL}{k}$$

$$\text{Fo} = \textit{Fourier number} = \frac{\alpha\tau}{L^2}$$

Here h represents the heat transfer coefficient at the surface and L is the distance between the centre line of the material and the surface. Many charts, such as those of Gurnie-Lurie and Heisler[5], present the solution of Eq. (1.69) in graphical form. In addition to analytical methods, there are also numerical methods employed to solve this equations. One such method using *finite difference approach* has been employed in Chapter 18.

1.22 FORCED CONVECTION CORRELATIONS

Correlations for the heat transfer coefficient h in forced convection are expressed in the following form

$$\text{Nu} = f(\text{Re}, \text{Pr})$$

where Nu, Re and Pr are dimensionless numbers expressed as follows:

$$\text{Nu} = \textit{Nusselt number} = \frac{hL}{k}$$

$$\text{Re} = \textit{Reynolds number} = \frac{Lu\rho}{\mu} \text{ or } \frac{Du\rho}{\mu}$$

$$\text{Pr} = \textit{Prandtl number} = \frac{C_p\,\mu}{k}$$

Here L is any characteristic length in the flow geometry. In the case of pipes, it is equal to the diameter D. Also u is the mean flow velocity and k, ρ, C_p and μ are the thermal conductivity, density, specific heat and dynamic viscosity of the fluid respectively.

The Reynolds number represents the ratio of inertia forces to viscous forces. It is a flow criterion. Generally, for a pipe, the flow is streamlined or *laminar* when $\text{Re} < 2100$, and *turbulent* when $\text{Re} > 2300$. For intermediate values of Re, the flow is in the transition region.

For turbulent flow in smooth tubes, the following correlation is commonly used.

Dittus-Boelter equation

$$\text{Nu} = 0.023\,(\text{Re})^{0.8}\,(\text{Pr})^n \tag{1.70}$$

where

$$n = 0.4 \text{ for heating}$$
$$= 0.3 \text{ for cooling}$$

For turbulent flow of fluids perpendicular to tubes such as water in the shell the following correlation (see Kern[7]) is used

$$\text{Nu} = 0.36\,(\text{Re})^{0.55}\,\text{Pr}^{1/3}\left(\frac{\mu}{\mu_w}\right)^{0.14} \tag{1.71}$$

For determining Nu and Re, the outside diameter of the tube is taken as the characteristic length. The velocity is measured in the free stream. The properties of the fluid are taken at the bulk mean temperature, whereas μ_w is at the wall temperature.

1.23 FREE CONVECTION CORRELATIONS

In free or natural convection, the flow velocity is developed as a result of the temperature difference causing a buoyancy force to act on the fluid. The flow criterion in the case of free convection is, therefore, the *Grashof number* which is defined by

$$\text{Gr} = \frac{g\beta\rho^2\,\Delta T L^3}{\mu^2}$$

in which β is the *coefficient of thermal expansion* of the fluid defined by

$$\beta = \frac{1}{v}\left(\frac{\partial v}{\partial T}\right)_p$$

For an ideal gas

$$v = \frac{RT}{p}, \left(\frac{\partial v}{\partial T}\right)_p = \frac{R}{p}, \text{ so that } \beta = \frac{1}{T}$$

Also, $\Delta T = (T_w - T_\infty)$ is the temperature difference between the wall and the surrounding fluid.

Free convection correlations are found to have the following form

$$\text{Nu} = f(\text{Gr}, \text{Pr})$$

In many cases, the relationship simplifies to

$$\text{Nu} = C\,(\text{Gr}\cdot\text{Pr})^m = C\,R_a^m$$

The product (Gr · Pr) is named Rayleigh number (Ra). It is a criterion of transition from laminar to turbulent flow in free convection.

For laminar flow, $10^4 < \text{Gr}\cdot\text{Pr} < 10^9$

For turbulent flow, $\text{Gr}\cdot\text{Pr} > 10^9$

For the case of air, Eq. (1.72) is simplified as written below in Eqs (1.73a), (1.73b), (1.74a) and (1.74b).

Vertical plate or cylinder to air

$$h = 1.42\left(\frac{\Delta T}{L}\right)^{1/4} \text{ laminar} \tag{1.73a}$$

$$h = 0.95\,(\Delta T)^{1/3} \text{ turbulent} \tag{1.73b}$$

Horizontal cylinder to air

$$h = 1.32\left(\frac{\Delta T}{D}\right)^{1/4} \text{ laminar} \tag{1.74a}$$

$$h = 1.24\left(\frac{\Delta T}{D}\right)^{1/3} \text{ turbulent} \tag{1.74b}$$

1.24 DESIGN OF HEAT EXCHANGERS

The thermal design of a heat exchanger is primarily concerned with the determination of the heat transfer area required to transfer a specified amount of heat between two fluids with specified flow rates and inlet temperatures.

In a heat exchanger, usually, the temperatures of fluids on both sides of the heat transfer surface vary as a result of heat exchange except in the case of the fluid undergoing change of phase. This variation has to be taken into account in any design procedure.

There are three basic types of heat exchangers. They are:

(i) Parallel flow.
(ii) Counter flow.
(iii) Cross flow.

In parallel flow, both the fluids flow past the heat transfer surface in the same direction as shown in Fig. 1.29. In counter flow, the two fluids flow in opposite directions as shown in Fig. 1.30. In cross flow, the two fluids flow at right angles to each other.

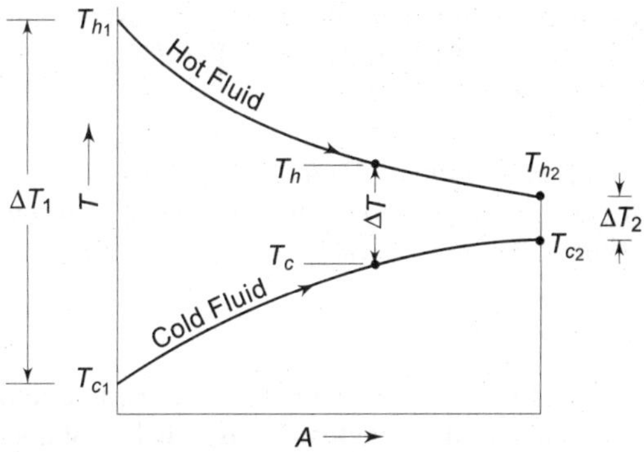

Fig. 1.29 Temperature variation in parallel flow heat exchanger

1.24.1 Log Mean Temperature Difference (LMTD) Method

It is proposed to calculate the required heat transfer surface area by the equation

$$\dot{Q} = UA\,\Delta T_m$$

where ΔT_m is the *log mean temperature difference* (LMTD) given by

$$\Delta T_m = \frac{\Delta T_1 - \Delta T_2}{\ln \dfrac{\Delta T_1}{\Delta T_2}} \tag{1.75}$$

where ΔT_1 and ΔT_2 are the temperature differentials between the two fluids at the two ends of the heat exchanger.

There are two great advantages of counter flow heat exchangers.

(i) A counterflow heat exchanger has a higher value of LMTD, for the same *end temperatures* as obtained from Eq. (1.75) and seen from Fig. 1.30. Hence, it requires smaller heat transfer surface area A than parallel flow heat exchanger for a specified heat transfer rate.

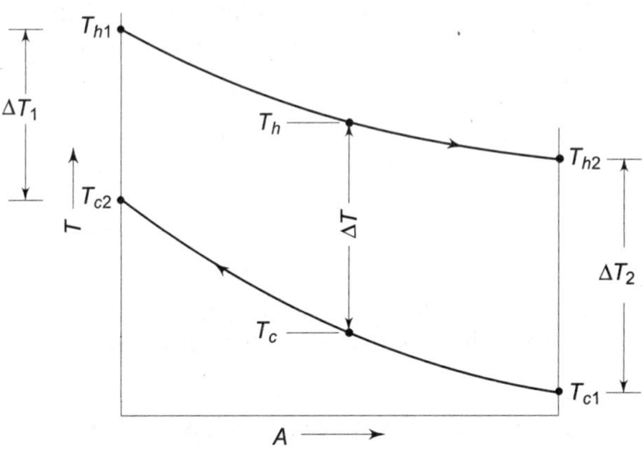

Fig. 1.30 Temperature variation in counter flow heat exchanger

(ii) It can also be seen from Fig. 1.30 that in the case of counterflow heat exchanger, the hot fluid can be cooled to the inlet temperature of the cold fluid (from t_{h1} to t_{c1}), and similarly, cold fluid can be heated to the inlet temperature of the hot fluid (from t_{c1} to t_{h1}) at least theoretically in the limit, if large heat transfer surface area is provided. This is not possible in parallel flow heat exchanger.

Note *In the case of phase change of a fluid, the temperature on one side is constant as in condensers and evaporators. In these cases, both types have the same LMTD, and there is no difference between counterflow and parallel flow.*

1.25 MASS TRANSFER

In *mass transfer*, we deal with the movement of species in a multicomponent system. The driving potentials for the transfer of mass can be obtained in various ways. But here we shall confine our discussion to *molecular diffusion* and *convective mass transfer*.

1.25.1 Molecular Diffusion: Fick's Law

Mass transfer by molecular diffusion of a specifies through another stationary medium is analogous to heat transfer by conduction. *Fick's Law* relates the *diffusion rate* or *mass flow* $\dot{m}_A$ of a species A to its driving potential which is the *concentration gradient* $\partial C_A/\partial x$ as follows in Eq. (1.76).

$$\frac{\dot{m}_A}{A} = -D\frac{\partial C_A}{\partial x} \tag{1.76}$$

where A is the normal area and D is called the *diffusion coefficient* or *mass transfer diffusivity*. The unit of diffusion coefficient is found to be $m^2\ s^{-1}$. It has the dimensions of kinematic viscosity $\nu = \mu/\rho$ or thermal diffusivity $\alpha = k/\rho C$.

1.25.2 Convective Mass Transfer

In molecular diffusion, bulk velocities are insignificant. Most physical applications of mass transfer, however, involve the bulk motion of fluids. This gives rise to convective mass transfer which is similar to convective heat transfer. Accordingly, we define the *mass transfer coefficient* h_M by the relation

$$\frac{\dot{m}_A}{A} = h_M(C_w - C_\infty) = \frac{k_\omega}{\rho}(\omega_w - \omega_\infty) \tag{1.77}$$

where C_w and C_∞ denote the concentrations of the species A at the wall and free stream. We see that h_M has the units of $m\,s^{-1}$. The expression $\rho\ h_M$ is replaced by *diffusion coefficient* k_ω, and C by specific humidity ω in the case of diffusion of water vapour into air. Its use is illustrated in Chapters 14 and 20. It is seen that $k_\omega = \rho h_M$ if $\Delta C = \Delta\omega$, which is a satisfactory assumption (see Chap. 14).

1.26 ANALOGY BETWEEN MOMENTUM, HEAT AND MASS TRANSFER

D has the same significance in mass transfer as ν in moment um transfer and α in heat transfer. Dimensionless numbers can be formed from the ratio of any two of these properties.

Prandtl Number (as defined earlier for convective heat transfer)

$$\mathbf{Pr} = \frac{\nu}{\alpha} = \frac{C_p \mu}{k}$$

Lewis Number

$$\mathbf{Le} = \frac{\alpha}{D} = \frac{k}{\rho C_p D}$$

Schmidt number

$$\mathbf{Sc} = \frac{\nu}{D} = \frac{\mu}{\rho D} = \frac{\text{Dynamic viscosity}}{\text{Mass diffusivity}}$$

where the product ρD is termed as *mass diffusivity*. It has the units of kgm^{-1} s^{-1}.

1.26.1 Analogy between Momentum and Heat Transfer

Then if $\nu = \alpha$ or Pr = 1, the velocity and temperature distributions in the flow will be the same. In that case, the heat transfer coefficient h can be determined from the knowledge of the friction factor f by the similarity relation called *Reynolds analogy*, viz.,

$$\text{St} = \frac{f}{2} \tag{1.78}$$

where St denotes the *Stanton number* defined by

$$\text{St} = \frac{\text{Nu}}{\text{Re}\cdot\text{Pr}} = \frac{h}{\rho u C_p} = \frac{h}{G C_p}$$

Here $G = \rho u$ is the *mass velocity.*
To account for some variation in Pr, Eq. (1.78) is modified to

$$j_H = \text{St}\cdot\text{Pr}^{2/3} = \frac{f}{2}\left(= \frac{h}{G C_p}\,\text{Pr}^{2/3} \right) \tag{1.79}$$

which is called the *Colburn analogy*, and j_H is the *Colburn j-factor* for heat transfer.

1.26.2 Analogy between Momentum and Mass Transfer

Similarly, the velocity and concentration profiles will have the same shape if $\nu = D$, or Sc = 1. Thus, the Schmidt number plays the same role in mass transfer as does the Prandtl number in heat transfer. An equivalent of the Nusselt number in mass transfer is the *Sherwood number* defined by

$$\text{Sh} = \frac{h_M L}{D} = \frac{k_\omega L}{\rho D}$$

where L is any characteristic length. The convective mass transfer correlations are, therefore, expressed in the form

$$\text{Sh} = f(\text{Re, Sc})$$

similar to forced convection heat transfer correlations which are usually expressed in the form Nu $= f(\text{Re}\cdot\text{Pr})$.

The Stanton number can then be replaced by the number

$$\frac{\mathrm{Sh}}{\mathrm{Re}\cdot\mathrm{Sc}} = \frac{h_M}{u} = \frac{\rho h_M}{G} \tag{1.80}$$

in mass transfer correlations, as given in Eq. (1.80). And similar to Eq. (1.78), we can write for mass transfer factor j_M as

$$j_M = \left(\frac{\mathrm{Sh}}{\mathrm{Re}\cdot\mathrm{Sc}}\right)\mathrm{Sc}^{2/3} = \left(\frac{\rho h_M}{G}\right)\mathrm{Sc}^{2/3} = \frac{f}{2} \tag{1.81}$$

The numerical value of the Schmidt number for the diffusion of water vapour into air at 25°C and 1 atm pressure is 0.6.

1.26.3 Analogy between Heat and Mass Transfer

Combining Eqs. (1.78) and (1.80) to eliminate $f/2$, we obtain a similarity relation between heat transfer and mass transfer coefficients as follows in Eq. (1.82)

$$\begin{aligned}\frac{h}{h_M} &= \rho C_p\left(\frac{\mathrm{Sc}}{\mathrm{Pr}}\right)^{2/3}\\ &= \rho C_p\left(\frac{\alpha}{D}\right)^{2/3}\\ &= \rho C_p\,\mathrm{Le}^{2/3}\end{aligned} \tag{1.82}$$

where Le = α/D is the *Lewis number.*

If Le = 1, the temperature and concentration profiles are the same and then $j_M = j_H$. Incidentally for air and water vapour mixtures at atmospheric pressure, the value of the Lewis number is approximately equal to unity. This simplifies the design of air-conditioning equipment. Writing h as f_g for heat transfer coefficient of air, and h_M as k_ω/ρ for diffusion of water vapour in air, we have for Le = 1

$$k_\omega = \frac{f_g}{C_p}$$

as used in Chapters 14 and 16.

Empirical relations for Sh and Nu for air over water droplets can be expressed as follows:

$$\mathrm{Sh} = \frac{h_M L}{D} = \frac{k_\omega L}{\rho D} = 2 + 0.6\,\mathrm{Re}^{0.5}\,\mathrm{Sc}^{0.33} \tag{1.83a}$$

$$\mathrm{Nu} = \frac{hL}{k} = 2 + 0.6\,\mathrm{Re}^{0.5}\,\mathrm{Pr}^{0.33} \tag{1.83b}$$

The average value of diffusivity D of water vapour in air is 2.495×10^{-5} m^2 s^{-1}.

Example 1.17 ***Mass Transfer Coefficient of Water Vapour in Air***

Air at 35°C and 1 atm flows at a mean velocity of 30 m/s over a flat plate 0.5 m long. Calculate the mass transfer coefficient of water vapour from the plate into air.

Assume the concentration of vapour in air as very small. The diffusion coefficient of water vapour into air is 0.256 × 10^{-4} m^2 s^{-1}. The Colburn j-factor for heat transfer coefficient is given by $j_H = 0.0296\ \mathbf{Re}^{-0.2}$.

Solution Properties of air at 35°C, 1 atm

$$\rho = 1.146 \text{ kg. m}^{-3}$$

$$C_p = 1.006 \text{ kJ. kg}^{-1} \text{ K}^{-1}$$

$$\mu = 2 \times 10^{-5} \text{ kg. m}^{-1} \text{ s}^{-1}$$

$$\text{Pr} = 0.706$$

$$\text{Sc} = \frac{\mu}{\rho D} = \frac{2 \times 10^{-5}}{(1.146)\,(0.256 \times 10^{-4})} = 0.682$$

$$\text{Re} = \frac{Lu\rho}{\mu} = \frac{(0.5)\,(30)\,(1.146)}{2 \times 10^{-5}} = 859{,}600$$

$$j_M = j_H = 0.0296 \text{ Re}^{-0.2} \tag{1.84}$$

$$= 0.0296\,(859{,}600)^{-0.2} = 1.925 \times 10^{-3}$$

$$h_M = j_M u/(\text{Sc})^{2/3}$$

$$= \frac{(0.001925)\,(30)}{(0.682)^{2/3}} = 0.075 \text{ m. s}^{-1}$$

References

1. Arora C P, *Heat and Mass Transfer*, Khanna Publishers, Delhi, 1979.
2. Arora C P, *Thermodynamics*, Tata McGraw-Hill, New Delhi, 1998.
3. Ashok Babu T P, A theoretical and experimental investigation of alternatives to CFC 12 in refrigerators, Ph. D. Thesis, IIT, Delhi, 1997.
4. ASHRAE, *Handbook of Fundamentals*, 1993.
5. Heisler M P, 'Temperature charts for induction and contact temperature heating', *Transactions ASME*, Vol. 69, pp. 227–236, 1947.
6. Holman, J P, *Heat Transfer*, McGraw-Hill, New York, 1968.
7. Kern D Q, *Process Heat Transfer*, McGraw-Hill, p. 137, 1950.
8. Martin J J, 'Correlations and equations used in calculating the thermodynamic properties of Freon refrigerants', ASME symposium, *Thermodynamic and Transport Properties of Gases, Liquids and Solids*, McGraw-Hill, Feb. 1959.
9. Martin J J and Y C Hou, 'Derivation of an equation of state for gases, *AIChE J.*, Vol. 1, pp. 142–151, 1955.
10. O'Leary, R A, *Refrigerating Engineering*, Vol. 42, No. 5, Nov. 1941.
11. Peng P V and D B Robinson, 'A new two constant equation of state, *Ind. Eng. Chem. Fundamentals,* Vol. 15, No. 1, pp. 59–64, 1976.

12. Redlich Otto and J N S Kwong, 'On the thermodynamics of solutions. An equation of state. Fugacities of gaseous solutions'. *Chem. Rev.*, Vol. 44, pp. 232–244, 1949.
13. Reid R C and T K Sherwood, *The Properties of Gases and Liquids*, McGraw-Hill, 4th Ed., 1988.
14. Reynolds C W, 'Thermodynamic properties in S.I. Graphs, tables and computational equations for forty substances, Stanford University, Stanford, CA94305, USA, 1979.
15. Schmidt T E, 'La production calorifique des surfaces munies d'ailettes', *Annexe Du Bull.* De 'L' Institut Int. Du Froid, Annexe G-5, 1945–46.

Revision Exercises

1.1 (a) A creamery must cool 11,350 kg of milk received each day from 30°C to 3°C in 3 hours. What must be the capacity of the refrigerating machine? Take the specific heat of milk as 3.92 kJ/kg.K.
(b) If the compressor in the above case operates 8 hours per day with surplus refrigeration accumulated in brine storage tanks, what must the capacity be?

1.2 Cold salt brine at – 2°C is used in a packing plant to chill food slabs from 40°C to 3.4°C in 18 hours. Determine the weight of brine required to cool 100 food slabs of 225 kg each if the final temperature of brine is 1.7°C, Given:

Specific gravity of brine = 1.05
Specific heat of brine = 3.77 kJ/kg.K
Specific heat of food slabs = 3.15 kJ/kg.K

1.3 A cold-storage room has inside dimensions of 12 m × 7.5 m × 3 m. The overall coefficient of heat transfer through the walls and ceiling is 0.55 W. m^{-2} K^{-1} and for the floor is 2.2 W.m^{-2}. K^{-1}. The inside temperature is to be maintained at –5°C. If the outside air is at 45°C and the air adjacent to the floor is at 30°C, determine the capacity of the refrigerating unit for 70 per cent running time.

1.4 10 kg of air at 65°C and 3.5 bar expands reversibly and polytropically to 1.5 bar. The index of expansion is 1.25. Find the final temperature, work done, heat transferred and change of entropy. Assuming the process to be adiabatic, find the final temperature and work done.

1.5 Air at 3.5 bar and 30°C flows at the rate of 0.5 kg/s through an insulated turbine. If the air delivers 11.5 hp to the turbine blades and the change in kinetic energy is negligible, at what temperature does the air leave the turbine?

1.6 A fluid expands from 3 bar and 150°C to 1 bar in a nozzle. The initial velocity is 90 m/s. The isentropic efficiency is likely to be 0.95. Estimate the final velocity assuming the fluid to be steam and air.

1.7 A real gas expands isenthalpically from 135 bar and –75°C to 7 bar. If the Joule-Thomson coefficient in the range is 0.2°C/bar, what is the final temperature of the gas?

1.8 R 22 saturated liquid at 35°C is throttled to a pressure corresponding to a temperature of 5°C. Determine the extent of flashing of the liquid into vapour.

1.9 The exterior wall of a house is made of 10 cm of common brick ($k = 0.7$ W. m^{-1} K^{-1}) followed by a 3.79 cm layer of gypsum plaster ($k = 0.48$ W.m^{-1}. K^{-1}). What thickness of thermocole insulation ($k = 0.037$ W.m^{-1}. K^{-1}) should be added to reduce the heat loss or gain through the wall to 25%?

1.10 Twelve thin brass fins, 0.75 mm thick, are placed axially on a 5 cm diameter × 5 cm long cylinder which stands vertically and is surrounded by air at 38°C. The fins extend 2.5 cm from the cylinder surface which is at a temperature of 150°C. The thermal conductivity of brass is 78 W.m^{-1} .$K^{-1}$and the heat transfer coefficient for air is 23 W.m^{-2}.K^{-1}. What is the rate of heat transfer to air from the cylinder?

1.11 Find the overall heat transfer coefficient between water and oil if the water flows through a copper pipe 1.8 cm 1D and 2.1 cm OD while the oil flows through the annulus between the copper pipe and a steel pipe. The water and oil side film heat transfer coefficients are 4600 and 1250 W.m^{-2}. K^{-1} respectively. The fouling factors on the water and oil sides may be taken as 0.0004 and 0.001 m^2 K.W^{-1} respectively. The thermal conductivity of the tube wall is 330 W. m^{-1}. K^{-1}.

1.12 In a R 134a condenser of a domestic refrigerator, having wire-and-tube construction, $A_o/A_i = 12$. Heat transfer coefficient of refrigerant condensing inside tubes is $h_i = 1950$ W.m^{-2}.K^{-1}. Heat transfer coefficient by natural convection of air outside tubes is $h_o = 19.5$ W. m^{-2}. K^{-1}. Air enters at 35°C and leaves at 43°C. Heat to be rejected in the condenser is 180 W. Determine the outside tube surface area.

1.13 Plot ln p^{sat} versus $1/T^{sat}$ diagrams for CO_2, R 22 ($CHCl\ F_2$), NH_3, C_3H_8, R 12 ($CCl_2\ F_2$), R 134a ($C_2\ H_2\ F_4$), R 152 a ($C_2\ H_4\ F_2$), Iso – $C_4\ H_{10}$ and R 11 ($C\ Cl_3F$). Find values of constants of Antonie equation.

1.14 Calculate the latent heat of vaporization of ammonia at 40°C using the following data from the saturation table of ammonia.

t °C	p bar	v_f L/kg	v_g m^3/kg
30	11.67	1.68	0.11
35	13.5	1.7	0.095
40	15.54	1.73	0.083
45	17.82	1.75	0.073
50	20.33	1.78	0.063

1.15 Determine the specific volume of saturated vapour by Redlich-Kwong equation for the following, and compare results with those given in the tables of properties.

(a) Propane at –25°C where $p^{sat} = 2.018$ bar.
(b) Propane at 50°C where $p^{sat} = 17.119$ bar.
(c) Isobutane at –25°C where $p^{sat} = 0.5865$ bar.
(d) Isobutane at 50°C where $p^{sat} = 6.92$ bar.

Chapter

2

Refrigerating Machine and Reversed Carnot Cycle

2.1 REFRIGERATING MACHINES

There are essentially two categories of thermal plants. These are:

(i) Thermal power plants or *work producing plants*.

(ii) Refrigeration/heat pump plants or *work consuming plants*.

The work producing plants or *heat engines* lead to the conversion of heat to work. The work consuming plants, viz., *refrigerators/heat pumps*, are not those which are in any way related to the conversion of work into heat. No ingenuity at all is required for the conversion of work into heat. In fact, all work (mechanical/electrical energy) that is consumed in machinery is ultimately dissipated as heat to the environment. The objective of work consuming plants, actually, is to lead to the flow of heat from a low temperature body to a high temperature body. The work is consumed to achieve this.

Examples of common work consuming plants, viz., refrigerators are the following:

Cold storages. Central air conditioning plants. Domestic refrigerators. Room air conditioners. Ice Plants. Food freezing and freeze-drying plants. Air liquefaction plants. etc.

Heat pumps are heating plants. But they operate in the same way as refrigerators.

Refrigeration equipment, in general, is relatively smaller in size as compared to work producing plants. The capacity of a power plant is in MW, whereas the capacity of a refrigeration system is in kW or even less. A very large super cold storage or a central air conditioning plant for a multistoreyed building may consume power in the range of 2000 to 5000 kW. A window-type room air conditioner may consume only 2.5 kW of power, and a domestic refrigerator just 100 to 250 W only.

2.2 A REFRIGERATING MACHINE—THE SECOND LAW INTERPRETATION

A *refrigerating machine* is a device which will either cool or maintain a body at a temperature below that of the surroundings. Hence, heat must be made to flow from a body at low temperature to the surroundings at high temperature.

However, this is not possible on its own. We see in nature that heat *spontaneously* flows from a high temperature body to a low temperature body.

The reverse process to complete the thermodynamic cycle, in which heat Q will flow back from the low temperature body to the high temperature body, is not possible. Thus, we see that a thermodynamic cycle involving heat transfer alone as shown in Fig. 2.1 cannot be devised. The logical conclusion is that there must be a process in which some work is done.

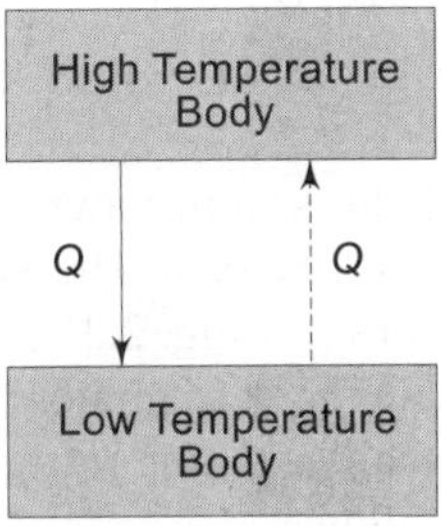

Fig. 2.1 A thermodynamic cycle involving heat transfer alone: Not possible

The second law of thermodynamics, like the first law, is based on the observations of actually existing processes and devices in nature. Figure 2.2 shows the schematic diagram of an actual refrigeration system which works on the well-known *vapour compression cycle*. Most refrigeration devices/plants, including air conditioners and refrigerators such as the ones illustrated in Chap. 1, work on this cycle only.

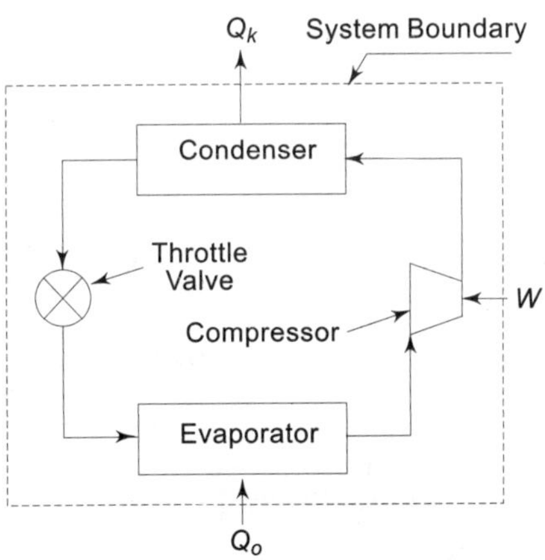

Fig. 2.2 Illustration of an actual refrigerator/heat pump: The simple vapour compression system

The heat and work interactions of the processes of the cycle are as follows:

(i) Heat Q_0 is absorbed in the evaporator by the evaporation of a liquid refrigerant at a low pressure p_0, and corresponding low saturation temperature T_0.

(ii) The evaporated refrigerant vapour is compressed to a high pressure p_k in the compressor consuming work W. The pressure after compression is such that the corresponding saturation temperature T_k is higher than the temperature of the surroundings.

(iii) Heat Q_k is then rejected in the condenser to the surroundings at high temperature T_k.

The application of the first law, $\oint \delta Q = \oint \delta W$, to the cycle gives:

$$-Q_k + Q_o = -W$$

$$Q_k - Q_o = W$$

This, also, represents an *energy balance* of the system in Fig. 2.2 obtained by drawing a boundary around it.

There are two statements of the second law of thermodynamics, the *Kelvin-Planck statement*, and the *Clausius statement*. The Kelvin-Planck statement pertains to heat engines such as E represented in Fig. 2.3.

The Clausius statement pertains to refrigerators/heat pumps R. The above observation from illustration of actually existing refrigerators/heat pumps leads to the Clausius statement which is as follows:

"*It is impossible to construct a device which will operate in a cycle and produce no effect other than the transfer of heat from a low temperature body to a high temperature body*".

The statement implies that a refrigerator R of the type shown in Fig. 2.4 which will absorb heat Q_o from a low temperature body and transfer it to a high temperature body is impossible.

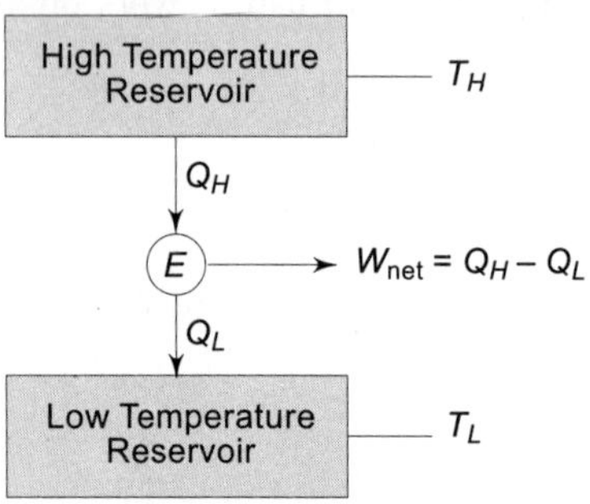

Fig. 2.3 Schematic representation of a heat engine

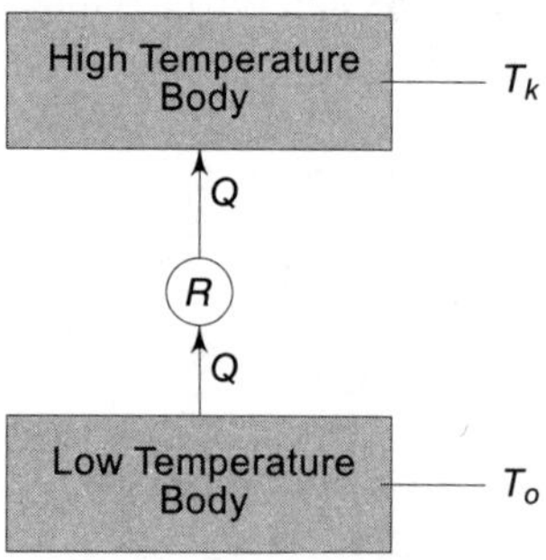

Fig. 2.4 Refrigerator without any Work Input: Impossible

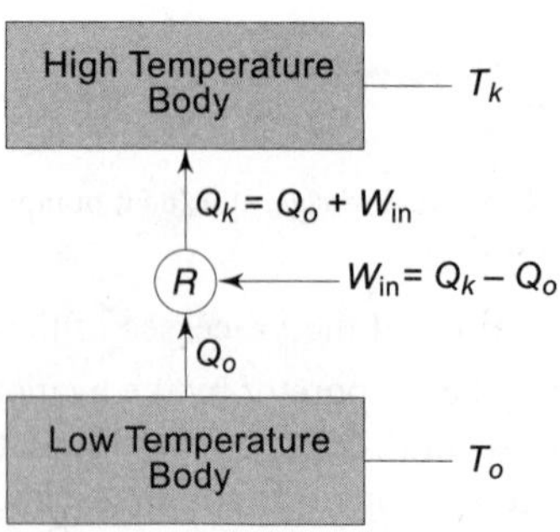

Fig. 2.5 Schematic representation of a refrigerator/heat pump

The only alternative is that there must be some work input W_{in}. Accordingly, we obtain a schematic representation of a refrigerating machine/heat pump as shown in Fig. 2.5, and from the first law $W_{in} = Q_k - Q_o$. Accordingly, heat transferred Q_k is more than heat absorbed Q_o by the amount of work input W_{in}.

Now, the stress on the words 'operating in a cycle' is significant. For, in a single process, it may be possible to obtain removal of heat from a low temperature body without doing external work, e.g., by the evaporation of a refrigerant, for instance liquid R 22 after throttling expansion to, say, 1 bar corresponding to a saturation temperature of –40°C as shown in Fig. 2.6. However, this could not happen continuously. This process could continue only if one had an infinite supply of high pressure R 22 liquid in the cylinder. But that is not possible in nature. To obtain refrigeration continuously, the refrigerant vapour after evaporation at low pressure will have to be brought back to the initial state of high pressure liquid again. That will mean forming a complete thermodynamic cycle.

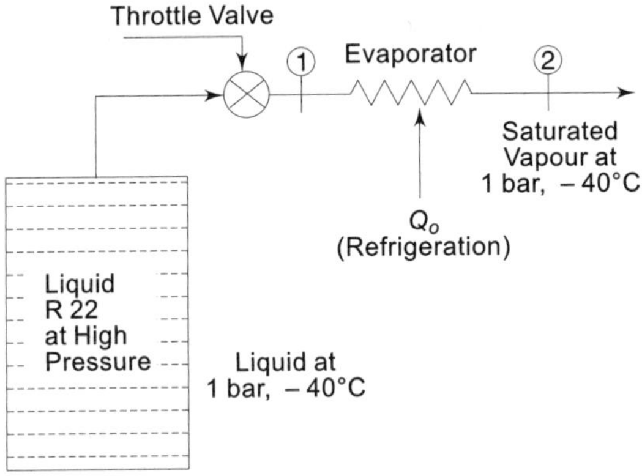

Fig. 2.6 A process: Producing refrigeration while $W = 0$

The Clausius statement eliminates the possibility of obtaining refrigeration without doing work. The statement necessitates a further clarification regarding heat-operated refrigerating machines such as the vapour absorption type or ejector type, using heat directly to produce refrigeration. Such systems may be considered as a combination of a heat engine and a refrigerating machine. The heat engine part of the system utilizes heat from a body at a higher temperature than the surroundings and delivers the required mechanical work, *within the system*, which is directly used by the refrigerating machine part. Thus the usual process of the conversion of thermal energy, first into work (or electrical energy) and then its utilization in a refrigerating machine, is replaced by a combined process.

2.3 HEAT ENGINE, HEAT PUMP AND REFRIGERATING MACHINE

It may be concluded from the preceding discussion that a reversible heat engine may be converted into a refrigerating machine by running it in the reversed direction. Schematically, therefore, a refrigerating machine is a reversed heat engine which can be seen by comparing Figs. 2.5 and 2.3.

As for the *heat pump*, there is no difference in the cycle of operation between a refrigerating machine and a heat pump. The same machine can be utilized either

(i) to absorb heat from a cold body (a cooled space) at temperature T_o and reject it to the surroundings at temperature $T_k \geq T_a$, or

(ii) to absorb heat from the surroundings at temperature $T_o \leq T_a$ and reject it to a hot body (a heated space) at temperature T_k,

where T_a is the temperature of the surroundings.

Figure 2.7 illustrates the manner of application of heat engine E, heat pump H and refrigerating machine R. It implies that the same machine can be used either for cooling or for heating. When used for cooling, it is called a refrigerating machine and when used for heating it is called a heat pump.

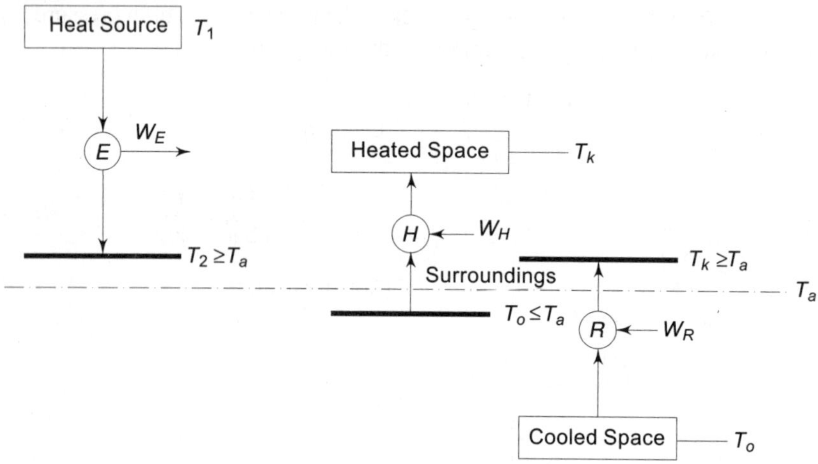

Fig. 2.7 Comparison of heat engine, heat pump and refrigerating machine

The main difference between the two is in their operating temperatures. A refrigerating machine operates between the ambient temperature $T_a \approx T_k$ and a low temperature T_o. A heat pump operates between the ambient temperature $T_a \approx T_o$ and a high temperature T_k. The heat engine operates between the heat source temperature T_1 and ambient temperature $T_a \approx T_2$.

Another essential difference is in their useful function. In a refrigerating machine, the heat exchanger that absorbs heat is connected to the conditioned space. In a heat pump, instead, the heat exchanger that rejects heat is connected to the conditioned space. The other heat exchanger in each case is connected to the surroundings. Thus if a refrigerating machine, that is used for cooling in summer, is to be used as a heat pump for heating in winter, it will be necessary, either

(i) to rotate the machine by 180° to interchange the positions of the two heat exchangers between the space and the surroundings, or

(ii) to exchange the functions of the two heat exchangers by the operation of valves, e.g., a four-way valve in a window-type air conditioner.

Such an operation of a refrigerating machine is termed as *reversed cycle heating*.

2.3.1 Energy Ratios or Coefficients of Performance

The performance of a heat engine is described by its thermal efficiency. The performance of a refrigerating machine or a heat pump is expressed by the ratio of useful heat to work, called the *energy ratio* or *coefficient of performance* (COP). Thus we have for a refrigerating machine:

Cooling energy ratio, or COP for cooling

$$\mathcal{E}_c = \frac{Q_o}{W} = \frac{Q_o}{Q_k - Q_o} = \frac{1}{\dfrac{Q_k}{Q_o} - 1} \tag{2.1}$$

And we have for a heat pump:

Heating energy ratio, or COP for heating

$$\mathcal{E}_h = \frac{Q_k}{W} = \frac{Q_k}{Q_k - Q_o} = \frac{1}{1 - \dfrac{Q_o}{Q_k}} \tag{2.2}$$

An idea about the approximate magnitude of the numerical values of these coefficients can be had from the following approximate calculations. The thermal efficiency of a heat engine is of the order of 30 per cent (say) so that

$$\eta_{\text{th}} = \frac{Q_k - Q_o}{Q_k} = 0.3$$

Then, if the engine is reversed in operation to work as a refrigerator or a heat pump with operating conditions unchanged (although in actual practice the operating temperatures will be different), we should have

$$\mathcal{E}_c = \frac{Q_o}{Q_k - Q_o} = \frac{1 - \eta_{\text{th}}}{\eta_{\text{th}}} = 2.33$$

and

$$\mathcal{E}_h = \frac{Q_k}{Q_k - Q_o} = \frac{1}{\eta_{\text{th}}} = 3.33$$

For vapour compression systems, $\mathcal{E}_c$ is of the order of 3 for water-cooled and 2 for air-cooled air-conditioning applications and 1 for domestic refrigerators.

For air cycle refrigeration systems, $\mathcal{E}_c \approx 1$ and for vapour absorption systems, it is well below unity. Steam ejector machines have still lower values.

However, the latter two are heat-operated refrigerating machines, and the definition of their coefficients of performance is altogether different. Hence, no comparison need be made at this stage.

2.3.2 Power Consumption of a Refrigerating Machine

Power consumption $\dot{W}$ of a refrigerating machine is determined in terms of kW. However, the power consumption of the motors is sometimes rated in horsepower (HP). We have

$$\text{HP} = \frac{\dot{W}}{0.746}$$

where $\dot{W}$ is in kW.

A quantity which is frequently used for comparison is horsepower per ton refrigeration. It is determined as follows:

$$\text{HP/kW refrigeration} = \frac{\dot{W}}{0.746\,\dot{Q}_o}$$

where $\dot{Q}_o$ is the refrigerating capacity in kW. Then, since 1 TR = 3.5167 kW,

$$\text{HP/TR} = \frac{\dot{W}}{0.746\,\dot{Q}_o}\,(3.5167)$$

$$= \frac{4.71\,\dot{W}}{\dot{Q}_o}$$

$$= \frac{4.71}{\mathcal{E}_c} \quad (2.3)$$

Thus HP/TR is inversely proportional to COP for cooling. In the above derivation, imperial horsepower has been used. If the horsepower is metric, then

$$\text{HP} = \frac{\dot{W}}{0.736}$$

$$\text{HP/TR} = \frac{4.78}{\mathcal{E}_0}$$

2.3.3 Heat Pump vs. Electric Resistance Heater

It may be seen from the simple transposition of Eqs. (2.1) and (2.2) that

$$\frac{Q_k}{W} = 1 + \frac{Q_o}{W}$$

$$\mathcal{E}_h = 1 + \mathcal{E}_c \quad (2.4)$$

The above relationship expresses a very interesting feature of a heat pump. According to Eq. (2.4), COP for heating is always greater than unity. It is so since Q_k is always greater then Q_o by the amount W.

Thus for the purpose of heating, it is far more economical to use a heat pump rather than an electric-resistance heater. For example, if W is the energy consumption of an electric resistance heater, the heat released to the space will be W only. But if this electrical energy W is utilized in a heat pump, the heat pumped to the space will be

$$Q_k = \mathcal{E}_h\,W = (1 + \mathcal{E}_c)\,W$$

Therefore, whatever is the value of $\mathcal{E}_c$ (even zero), Q_k will always be greater than or equal to W. The value of $\mathcal{E}_h$ for air-conditioning applications is of the order of 3. Then the heat pumped will be 3 W in a heat pump unit while the power consumption is only W. The heat pump, therefore, is a definite advancement over the simple electric-resistance heater. Only the cost of the heat pump (which is a refrigerating machine also) is prohibitive. But when an air-conditioning plant is already installed for cooling in summer, it would always be prudent to use it for heating as well in winter, operating as a heat pump.

Suppose a room loses $\dot{Q}_k$ kW of heat during winter. Then, it requires an equal amount of heat addition to maintain it at a desired temperature. Figure 2.8 illustrates the two methods for heating the room:

(i) Using electric resistance heater. The power consumption is

$$\dot{W} = \dot{Q}_k$$

(ii) Using heat pump. The power consumption is

$$\dot{W} = \frac{\dot{Q}_k}{\mathcal{E}_h} \cong \frac{1}{3}\dot{Q}_k$$

if $\mathcal{E}_h = 3$. Thus, power consumption of the heat pump is very much lower.

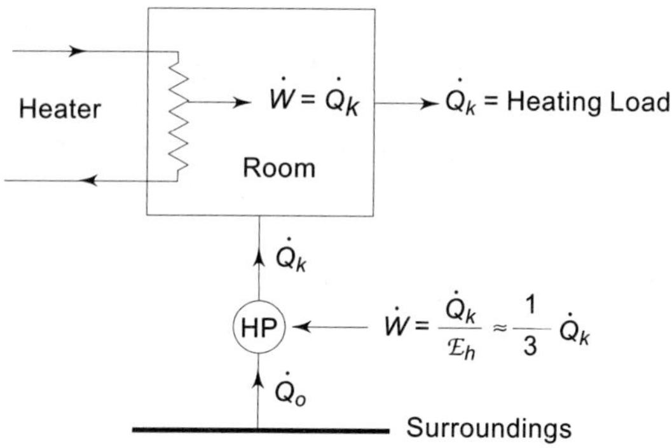

Fig. 2.8 Heating of a room by electric heater and heat pump

2.4 BEST REFRIGERATION CYCLE: THE CARNOT PRINCIPLE

It is possible to show that the cooling energy ratio of a refrigeration cycle working between two temperatures will be maximum when the cycle is a reversible one. For the purpose, consider a reversible refrigerating machine R and another irreversible refrigerating machine I, both working between two heat reservoirs at temperatures T_k and T_o, and absorbing the same quantity of heat Q_o from the cold reservoir at T_o as shown in Fig. 2.9 (a).

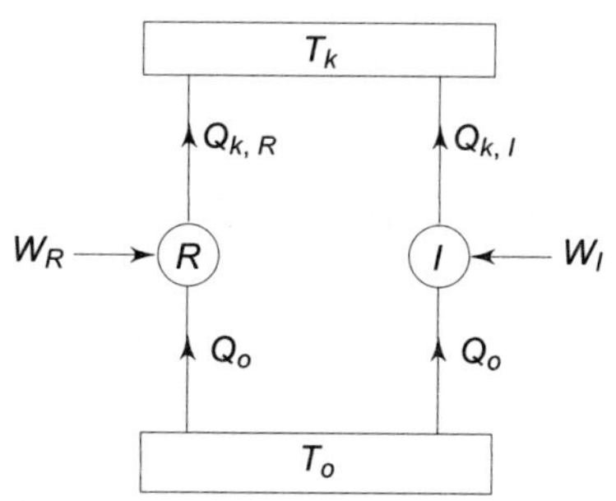

Fig. 2.9(a) Reversible and irreversible refrigerating machines

Now, to prove the contrary, let us assume that the COP of the irreversible machine is higher than that of the reversible machine, viz., $\mathcal{E}_I > \mathcal{E}_R$. Hence

$$\frac{\dot{Q}_o}{W_I} > \frac{\dot{Q}_o}{W_R}$$

$$W_R > W_1$$

And since
$$Q_{k,R} = Q_o + W_R$$
$$Q_{k,I} = Q_o + W_I$$

we have
$$Q_{k,R} > Q_{k,I}$$

and
$$Q_{k,R} - Q_{k,I} = W_R - W_I = W_{\text{net}}$$

If now, the reversible refrigerating machine is made to work as a heat engine and the irreversible refrigerating machine continues to work as a refrigerating machine, as shown in Fig. 2.9(b), the resultant combined system will work as a perpetual motion machine of the second kind taking heat equal to $Q_{k,R} - Q_{k,I}$ from the hot reservoir and converting it completely into work, thus violating the Kelvin-Planck statement of the Second Law applicable to heat engines as shown in Fig. 2.10.

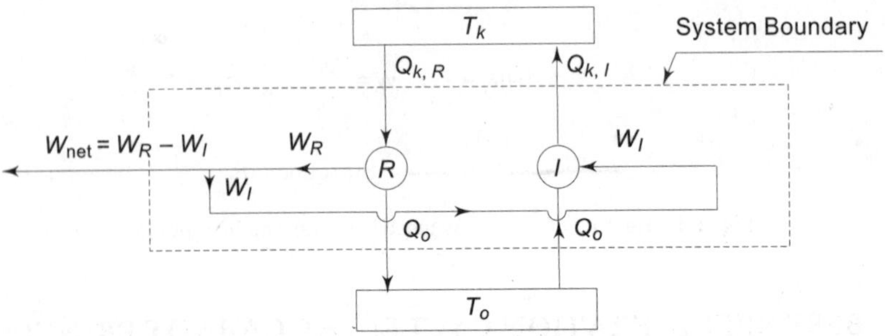

Fig. 2.9(b) Reversible refrigerating machine working as a heat engine in combination with on irreversible refrigerating machine

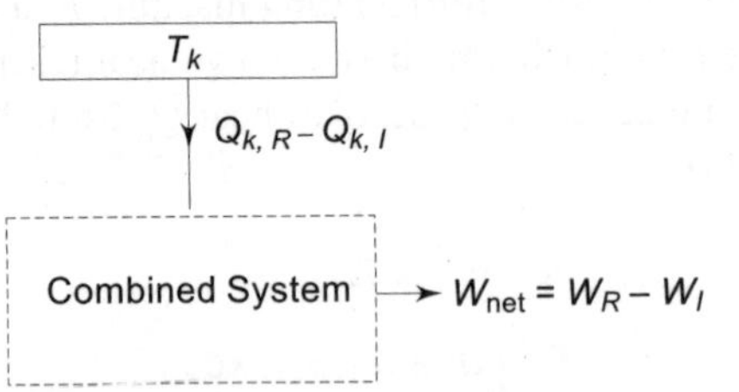

Fig. 2.10 Combined system resulting in a perpetual motion machine of the second kind thus violating the Second Law

It is, therefore, concluded that a refrigeration cycle operating reversibly between two heat reservoirs has the highest coefficient of performance. Likewise, it can also be shown that all reversible refrigeration cycles have the same COP. These are two corollaries of Second Law comprising the *Carnot Principle.*

2.4.1 Reversed Carnot Cycle

We now know that a reversible refrigeration cycle has the maximum COP. We know further that a reversible heat engine can be reversed in operation to work as a refrigerating machine.

Sadi Carnot, in 1824, proposed a reversible heat-engine cycle as a measure of maximum possible conversion of heat into work. A reversed Carnot cycle can, therefore, be employed as a reversible refrigeration cycle, which would be a measure of maximum possible COP of a refrigerating machine operating between two temperatures T_k of heat rejection and T_o of refrigeration.

A reversed Carnot cycle, for a unit mass of the working substance, is shown in Fig. 2.11 on a T–s diagram. The cycle consists of two isothermals and two isentropics as follows:

Process 1—2 Isentropic compression, $s_1 = s_2$
Process 2—3 Isothermal heat rejection to the hot reservoir at T_k = const.
Process 3—4 Isentropic expansion, $s_3 = s_4$
Process 4—1 Isothermal heat absorption from the cold reservoir at T_o = const.

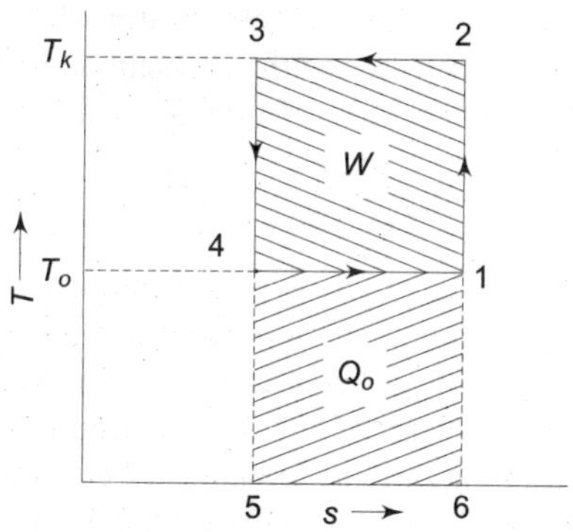

Fig. 2.11 Reversed Carnot Cycle

The areas on the T-s diagram, representing $\int Tds$ give the heat transfers, and work done in the cycle as follows:

Heat absorbed from cold body, $Q_o = T_o\,\Delta s$ = area 1-4-5-6
Heat rejected to hot body, $Q_k = T_k\,\Delta s$ = area 2-3-5-6
Work done, $W = Q_k - Q_o = (T_k - T_o)\,\Delta s$ = area 1-2-3-4

Hence we obtain Carnot values of COP for cooling and heating as

$$\mathcal{E}_{c,\,\text{Carnot}} = \frac{\dot{Q}_o}{W} = \frac{T_o}{T_k - T_o} = \frac{I}{\dfrac{T_k}{T_o} - I}$$

$$\mathcal{E}_{h,\,\text{Carnot}} = \frac{\dot{Q}_k}{W} = \frac{T_k}{T_k - T_o} = \frac{I}{I - T_o/T_k}$$

Effect of Operating Temperatures We, thus, see that the Carnot COP depends on the operating temperatures T_k and T_o only. It does not depend on the working substance (refrigerant) used.

For cooling, T_o is the refrigeration temperature and T_k is the temperature of heat rejection to the surroundings. The lowest possible refrigeration temperature is

$T_o = 0$ (absolute zero) at which $\mathcal{E}_c = 0$. The highest possible refrigeration temperature is $T_o = T_k$, i.e., when the refrigeration temperature is equal to the temperature of the surroundings (ambient) at which $\mathcal{E}_c = \infty$. Thus, Carnot COP for cooling varies between 0 and ∞.

For heating, T_o is the temperature of heat absorption from the surroundings and T_k is the heating temperature. Theoretically, the COP for heating varies between 1 and ∞.

It may, therefore, be noted that to obtain maximum possible COP in any application,

(i) the cold body temperature T_o should be as high as possible, and

(ii) the hot body temperature T_k should be as low as possible.

The lower the refrigeration temperature required, and higher the temperature of heat rejection to the surroundings, the larger is the power consumption of the refrigerating machine. Also, the lower is the refrigeration temperature required, the lower is the refrigerating capacity obtained.

Consider now, for example, a domestic refrigerator which produces refrigeration at –25°C (248 K). Let heat be rejected to the ambient at 60°C (333 K).

The maximum possible COP of this refrigerator would be

$$(\text{COP}_c)_{max} = \frac{248}{333 - 248} = \frac{248}{85} = 2.9$$

Thus, the refrigerator would produce a maximum of 290 W of refrigeration per 100 W of power consumption. The most popular size, 165 L internal volume refrigerators of most manufacturers produce 89 W of refrigeration, and the power of the electric motors running these refrigerators is around 110 W. Considering that the refrigerators have a running time of 75% only, the average power consumption would come to 82.5 W. Thus, the actual COP of these machines would be only 1.08.

Compare this with the COP of a room air conditioner which produces refrigeration at a higher temperature of about 5°C (278 K). Assuming the temperature of heat rejection the same as 60°C (333 K), the maximum possible COP would be

$$(\text{COP}_c)_{max} = \frac{278}{333 - 278} = \frac{278}{55} = 5$$

It would, thus, produce 5 kW of refrigeration per kW of power consumption. A 1.5 TR (5.3 kW) air conditioner will have a minimum power consumption of 1.05 kW. The actual air conditioner will, however, have power consumption of the order of 2.0 kW or more considering 75% running time. The actual COP is thus 2.6 or even less.

The above two examples show that the COP of a refrigeration system decreases, and power consumption increases as we go to lower and lower refrigeration temperatures.

2.4.2 Selection of Operating Temperatures

The selection of temperature T_0 depends on the particular application of refrigeration. Consider, for example, the simple summer air-conditioning system shown in Fig. 2.12(a). The room is maintained at temperature t_i equal to 25°C.

To offset the heat entering the room, the air must be supplied at a temperature lower than 25°C, at t_s equal to 15°C.

The air conditioner will, therefore, cool the room air from 25°C to 15°C and then supply it back to the room.

Accordingly, the refrigerant temperature t_o must be less than 15°C to absorb heat Q_o from the air maintaining a finite temperature difference Δt across the heat exchanger as shown in Fig. 2.12(b). If the temperature difference is zero, the *area requirement* of the heat exchanger will be infinite. Thus for air conditioning in summer, the temperature t_o is of the order of 0 to 10°C usually about 5°C. In a similar manner, the approximate refrigeration temperature requirements can be found out for various other applications as given in Table 2.1.

Table 2.1 Refrigeration temperature requirements of common applications

No.	*Application*	*Refrigeration Temperature, t_o, °C*
1.	Air conditioning in summer	0 to 10
2.	Cold storages	–10 to 2
3.	Domestic refrigerators	–25
4.	Frozen foods	–35
5.	Freeze drying and IQF (Instant Quick Freezing)	–35 to –45

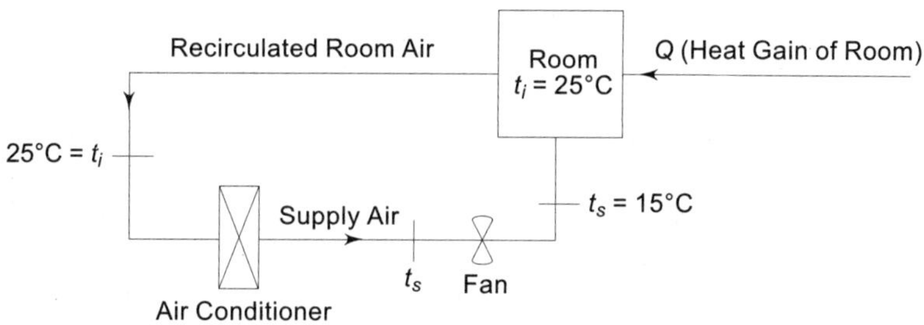

Fig. 2.12(a) Simple summer air-conditioning system

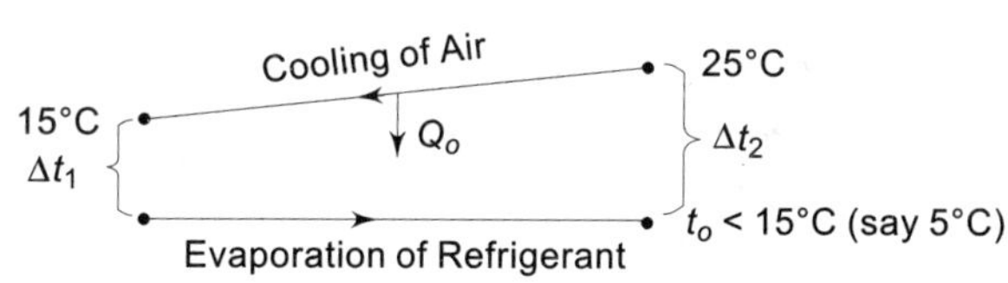

Fig. 2.12(b) Temperatures in a cooling coil

The selection of temperature T_k depends on the surrounding medium used for heat rejection. There are three possible media in the surroundings to which heat Q_k may be rejected, viz., air, water and ground. The units that use air as a cooling medium are called the *air-cooled units* and those using water are called the *water-cooled units*.

Consider an air-cooled unit. Let the surrounding air temperature (in summer) be 45°C. Also, let the rise in temperature of air, after absorbing heat Q_k, be 10°C, as shown in Fig. 2.13(a). Hence the temperature of heat rejection t_k has to be greater than 55°C, say 65°C, so that the temperature differentials Δt_1 and Δt_2, across the heat exchanger (HE) are 20°C and 10°C respectively, and the arithmetic mean temperature difference is (20 + 10)/2 = 15°C.

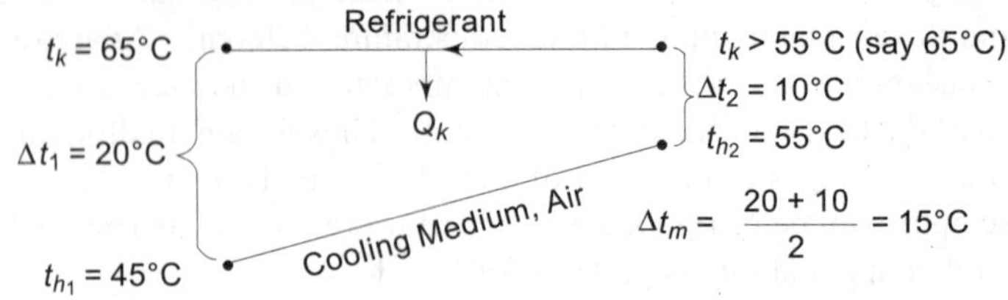

Fig. 2.13(a) Temperatures during heat rejection in an air-cooled refrigerating machine

Air is widely used as a cooling medium in small refrigerating machines such as refrigerators, water coolers, window-type air conditioners and small-package units.

Water as a cooling medium is preferable to air as it affords a lower value of T_k because of the following reasons:

(i) It is available at a temperature lower than that of air. Its temperature approaches the wet bulb temperature of the surrounding air. This is the limiting temperature to which heated water can be cooled in a cooling tower or a spray pond.

(ii) The specific heat of water is about four times that of air. Thus, for the same heat rejection Q_k and the same mass flow, the temperature rise of water is one-fourth that of air and correspondingly T_k is lower.

(iii) Water has a higher heat transfer coefficient than air mainly because of its high thermal conductivity, say, 5000 as against 100 in forced convection and 10 in free convection for air in $Wm^{-2} K^{-1}$. Thus, for the same heat rejected Q_k and the same area of the heat exchanger, the temperature difference Δt required across the heat exchanger is less which also results in a lower value of T_k.

Note *The lower value of heat transfer coefficient of air is made up by providing extended surface on air-side in actual equipment.*

Consider, for example, a water-cooled unit, in place of the air-cooled unit of the preceding example. Let the wet bulb temperature of the surrounding air be 28°C (even though the dry bulb temperature may be 45°C). Then the temperature of water from the cooling tower may be taken to be about 30°C. With the same mass flow for air and water, the temperature rise of water as a result of its higher specific heat will be 10/4 = 2.5°C only, though taken as 5°C in design as shown in Fig. 2.13 (b). The temperature of water leaving the heat exchanger is then 35°C.

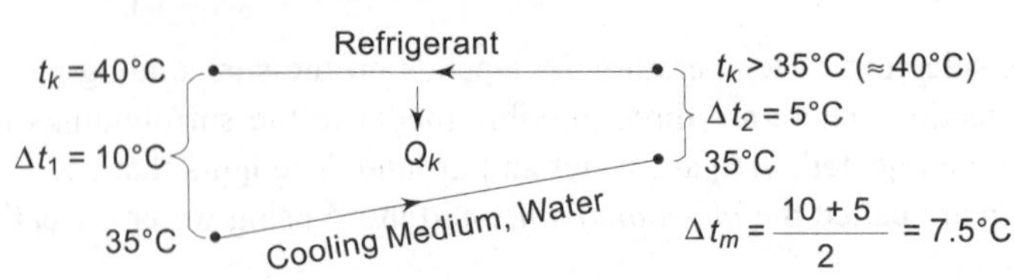

Fig. 2.13(b) Temperatures during heat rejection in a water-cooled refrigerating machine

Let the mean temperature difference required with water be 7.5°C instead of 15°C for air as a consequence of the higher heat transfer coefficient. Then the temperature of heat rejection required need only be 40°C for water as compared to 65°C for air. Normal design temperature T_k in Delhi is 43°C .

Thus, the use of water as a cooling medium results in a lower T_k, higher COP and lower power consumption in a refrigeration plant. At the same time, we have much more compact and smaller condenser even with a smaller value of Δt_m (7.5°C with water as against 15°C with air in the above illustration).

Large refrigeration plants, including central air-conditioning plants, therefore, are invariably water-cooled as the saving in the costs of power consumption exceeds the added cost of a water-cooling plant (cooling tower, pump, piping, etc.). Examples 2.1 and 2.2 amply illustrate the advantages of water over air, as a cooling medium. Although, the calculations are based on the reversed Carnot cycle for the working substance, yet finite temperature differentials have been assumed in the heat exchangers as external irreversibilities.

The ground is also sometimes used as a heat sink, specially in locations where extreme high and low temperatures are reached and the ground temperatures are lower in summer and higher in winter as compared to the surrounding temperatures.

Example 2.1

(a) *The ambient air temperatures during summer and winter in a particular locality are 45°C and 15°C respectively. Find the values of Carnot COP for an air conditioner for cooling and heating, corresponding to refrigeration temperatures of 5°C for summer and heating temperature of 55°C for winter. Assume suitable temperature differences in the exchanger that exchanges heat with the surroundings.*

(b) *If water from the cooling tower at 30°C is used as a cooling medium with 3°C temperature differential for air-conditioning in summer, what will be the Carnot COP for cooling?*

(c) *Also, find the theoretical power consumption per ton of refrigeration in each case. Assume no increase in the temperature of the surrounding air or water.*

Solution

(a) *Air as a cooling medium*

For summer: $t_k = 45 + 15 = 60°\text{C}$ (Assuming 15°C temperature difference for air-cooled)

$t_o = 5°\text{C}$

$$\mathcal{E}_c,\ \text{Carnot} = \frac{273+5}{60-5} = \frac{278}{55} = 5.05$$

For winter: $t_k = 55°\text{C}$

$t_o = 15 - 5 = 10°\text{C}$ (Assuming 5°C temp. difference)

$$\mathcal{E}_h,\ \text{Carnot} = \frac{273+55}{55-10} = \frac{328}{45} = 7.3$$

(b) *Water as a cooling medium*

$t_k = 30 + 10 = 40°\text{C}$ (Assuming 10°C temp. difference for water-cooled)

$t_o = 5°\text{C}$

$$\mathcal{E}_c, \text{ Carnot} = \frac{273+5}{40-5} = \frac{278}{35} = 7.94$$

(c) *Power consumption*

$$\dot{W} = \dot{Q}_o / \mathcal{E}_c$$

For the air-cooled unit, $\dot{W} = 211/5.05 = 41.8 \text{ kJ/min} = 0.7 \text{ kW}$

For the water-cooled unit, $\dot{W} = 211/7.94 = 26.6 \text{ kJ/min} = 0.44 \text{ kW}$

Note *Saving in power with water-cooled is 37%. In actual practice, it is much more.*

Example 2.2

(a) A reversed Carnot cycle air conditioner of 1 TR capacity operates with cooling coil temperature t_o = 5°C. The surrounding air at 43°C is used as a cooling medium rising to a temperature of 53°C. The temperature of heat rejection is t_k = 55°C. The overall heat transfer coefficient of the heat exchanger between the working substance and the surrounding air is U = 250 W. m^{-2} K^{-1}. Determine the mass flow rate of the surrounding air entering the heat exchanger, area of the heat exchanger, COP and power consumption of the air conditioner.

(b) If water at 30°C is used as a cooling medium in the same heat exchanger, will its area be adequate or inadequate for necessary heat rejection? Assume that the mass flow rate of water remains the same as that of air. Also for water, U = 2500 W. m^{-2} K^{-1}.

(c) If the area is reduced to 1 m^2, what will be the temperature of heat rejection t_k, temperature rise of water, COP, power consumption, percentage saving in power consumption and percentage reduction in the heat-exchanger area as compared with air as a cooling medium.

Solution

(a) *Air as a cooling medium* Referring to Fig. 2.13 (a), the various heat exchanger temperatures in this case are as follows:

$$t_{h1} = 43°\text{C}, \ t_{h2} = 53°\text{C}, \ t_k = 55°\text{C}$$

$$\text{COP} = \frac{273+5}{55-5} = \frac{278}{50} = 5.56$$

Power consumption

$$\dot{W} = \frac{\dot{Q}_o}{\mathcal{E}_c} = \frac{3.5167}{5.56} = 0.63 \text{ kW}$$

Heat rejected

$$\dot{Q}_k = \dot{Q}_o + \dot{W} = 3.5167 + 0.63 = 4.15 \text{ kW}$$

Mass flow rate of cooling air

$$\dot{m}_a = \frac{\dot{Q}_k}{C_p \, \Delta t_h} = \frac{4.15}{1.005(53-43)} = 0.413 \text{ kg/s}$$

LMTD in the heat exchanger

$$\Delta t_m = \frac{\Delta t_1 - \Delta t_2}{\ln \frac{\Delta t_1}{\Delta t_2}} = \frac{(55-43)-(55-53)}{\ln\left(\frac{55-43}{55-53}\right)} = \frac{10}{\ln 6} = 5.58°\text{C}$$

Area of the heat exchanger

$$A = \frac{\dot{Q}_k}{U\,\Delta t_m} = \frac{4.15 \times 10^3}{250(5.58)} = 2.975 \text{ m}^2$$

(b) *Water as a cooling medium*
First approximation: Let $t_k = 35°\text{C}$

$$\text{Then COP} = \frac{273+5}{35-5} = 9.27$$

$$\dot{Q}_k = \dot{Q}_o + \dot{W} = \dot{Q}_o + \frac{\dot{Q}_o}{\mathcal{E}_c} = \dot{Q}_o\left(1 + \frac{1}{\mathcal{E}_c}\right)$$

$$= 3.5167\left(1 + \frac{1}{9.27}\right) = 3.9 \text{ kW}$$

$$\Delta t_h = \frac{\dot{Q}_k}{\dot{m}_w C_p} = \frac{3.9}{0.413\,(4.1868)} = 2.26°\text{C}$$

$$t_{h2} = 30 + 2.26 = 32.26°\text{C}$$

LMTD in the heat exchanger

$$\Delta t_m = \frac{(35-30)-(35-32.26)}{\ln \frac{35-30}{35-32.26}} = 3.76°\text{C}$$

$$A = \frac{\dot{Q}_k}{U\Delta t_m} = \frac{3.9}{2.5\,(3.76)} = 0.41 \text{ m}^2$$

Checking with the actual area of 2.975 m^2, it is seen that temperature t_k could be much lower.

Second approximation of $t_k = 33°\text{C}$ gives $A = 0.95$ m^2.

The area required is still one-fourth of the available area of 2.975 m^2. Thus, even if t_k is taken equal to $t_{h2} = 32.24$ so that $\Delta t_2 = 0$, then $\Delta t_1 = 32.24 - 30 = 2.24°\text{C}$, and $\Delta t_m = 1.12°\text{C}$, it gives an area requirement of only (1.63/1/12) (0.95) = 1.38 m^2 which is still less than 2.975 m^2.

Hence the area of the heat exchanger is more than adequate with water as a cooling medium and can be reduced to at least one-third.

(c) *Water as a cooling medium* ($A = 1$ m^2)
By iteration, we find the solution, viz.,

$$t_k = 32.94°\text{C}$$

$$\text{COP} = \frac{273+5}{32.94-5} = 9.95$$

$$\dot{Q}_k = 3.5167\left(1 + \frac{1}{9.76}\right) = 3.87 \text{ kW}$$

$$\Delta t_h = \frac{3.87}{0.413(4.1868)} = 2.24°\text{C}$$

$$t_{h2} = 32.24°\text{C}$$

$$\Delta t_m = \frac{(32.9-30)-(32.9-32.24)}{\ln\dfrac{32.9-30}{32.9-32.24}} = 1.5°\text{C}$$

$$A = \frac{3.87}{2.5(1.5)} = 1.0 \text{ m}^2$$

$$\text{Power consumption, } \dot{W} = \frac{3.5167}{9.95} = 0.35 \text{ kW}$$

$$\text{Saving in power consumption} = \frac{0.63-0.35}{0.63} \times 100 = 44.4\%$$

Saving in area of the heat exchanger

$$= \frac{2.975-1}{2.975} \times 100 = 66\%$$

Note *Indiscriminate use of air-cooled window-type air conditioners is wasteful of energy and equipment. Insted of using large number of window units in a big building, it is much more desirable to install a central air conditioning plant which is always water-cooled. This will save power and reduce cost. It will also minimise thermal pollution of the environment by diminishing $\dot{W}$ and $\dot{Q}_k$.*

2.5 VAPOUR AS A REFRIGERANT IN REVERSED CARNOT CYCLE

The reversed Carnot cycle can be made almost completely practical by operating in the liquid-vapour region of a pure substance as shown in Fig. 2.14.

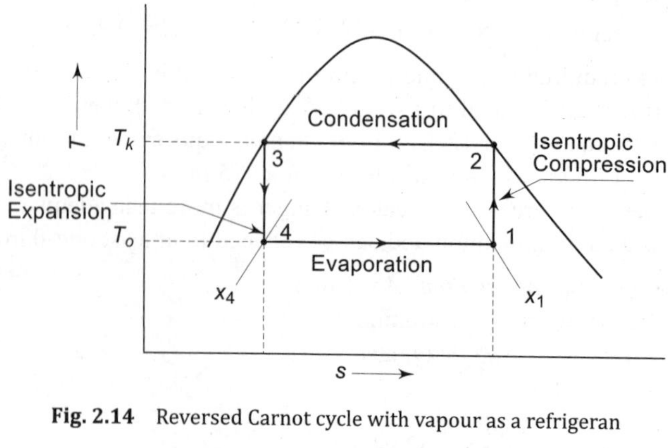

Fig. 2.14 Reversed Carnot cycle with vapour as a refrigeran

The isothermal processes of heat rejection (2-3) and heat absorption (4-1) of the Carnot cycle are achieved by making use of the phenomena of condensation and evaporation of a pure substance at constant pressure and temperature. This alternate condensation and evaporation of a working substance is accompanied by alternate isentropic compression (1-2) and expansion (3-4) processes.

It may be noted that the vapour during compression is wet although it is dry-saturated at the end of the process. Such a compression is called *wet compression.*

It may also be seen that the isentropic expansion of the liquid from 3 to 4 results in flashing of the refrigerant with consequent temperature drop from T_k to T_o although such expansion of a liquid with partial vaporization is practically difficult to achieve in a fast-moving piston and cylinder mechanism.

The thermodynamic analysis per unit mass of the refrigerant, for the four flow processes of the cycle, using steady-state steady-flow energy equation is as follows:

Refrigerating effect, $q_o = h_1 - h_4$

Heat rejected, $|q_k| = h_2 - h_3 = (h_{fg})_k$

Compressor work, $|w_C| = h_2 - h_1$

Expander work, $w_E = h_3 - h_4$

Net work,
$$|w| = |w_C| - w_E = (h_2 - h_1) - (h_3 - h_4)$$
$$= |q_k| - q_o = (h_2 - h_3) - (h_1 - h_4)$$

COP for cooling,
$$\mathcal{E}_c = \frac{q_o}{|w|} = \frac{h_1 - h_4}{(h_2 - h_3) - (h_1 - h_4)}$$
$$= \frac{1}{\left(\dfrac{h_2 - h_3}{h_1 - h_4}\right) - 1} \tag{2.5}$$

Note *q_k and w_C are negative with respect to the system (refrigerating machine). From here onwards we will consider their positive values only in our calculations.*

Example 2.3

(a) *A Carnot refrigerator has working temperatures of –30°C and 35°C. If it operates with R 12 as a working substance, calculate the work of isentropic compression and that of isentropic expansion, and refrigerating effect, heat rejected per kg of the refrigerant, and COP of the cycle.*

(b) *If the actual refrigerator operating on the same temperatures has a COP of 0.75 of the maximum, calculate the power consumption and heat rejected to the surroundings per ton of refrigeration.*

Solution Referring to Fig. 2.14, from the table of properties of R 12, we have:

$s_1 = s_2 = 0.6839$ kJ/(kg. K)	$s_{f4} = s_{f1} = 0.0371$ kJ/(kg. K)
$s_3 = s_4 = 0.2559$ kJ/(kg. K)	$s_{g4} = s_{g1} = 0.7171$ kJ/(kg. K)
$h_2 = 201.5$ kJ/kg	$h_{f4} = h_{f1} = 8.9$ kJ/kg
$h_3 = 69.5$ kJ/kg	$h_{g4} = h_{g1} = 174.2$ kJ/kg

Hence, by putting $s_1 = s_4$ and $s_4 = s_3$, we get the dryness fractions at 1 and 4 as

$$x_1 = \frac{0.6839 - 0.0371}{0.7171 - 0.0371} = 0.951$$

$$x_4 = \frac{0.2559 - 0.0371}{0.7171 - 0.0371} = 0.322$$

Enthalpies at 1 and 4

$$h_1 = 8.9 + 0.951\,(174.2 - 8.9) = 166 \text{ kJ/kg}$$

$$h_4 = 8.9 + 0.322\,(174.2 - 8.9) = 62.1 \text{ kJ/kg}$$

Work of compression, $w_C = h_2 - h_1 = 201.5 - 166 = 35.5$ kJ/kg

Work of expansion, $w_E = h_3 - h_4 = 69.5 - 62.1 = 7.4$ kJ/kg

Refrigerating effect, $q_o = h_1 - h_4 = 166 - 62.1 = 103.9$ kJ/kg

Heat rejected, $q_k = h_2 - h_3 = 201.5 - 69.5 = 132$ kJ/kg

Net work, $w = w_C - w_E = 35.5 - 7.4 = 28.1$ kJ/kg

$= q_k - q_o = 132 - 103.9 = 28.1$ kJ/kg

COP of the cycle, $\mathcal{E}_c = \dfrac{q_o}{w} = \dfrac{103.9}{28.1} = 3.74$

Alternatively, we have for Carnot COP

$$(\mathcal{E}_c)\text{ Carnot} = \frac{T_o}{T_k - T_o} = \frac{273 - 30}{35 - (-30)} = 3.74$$

(b) Actual COP, $\mathcal{E}_c = 0.75 \times 3.74 = 2.8$

Power consumption per ton, $\dot{W} = \dfrac{\dot{Q}_o}{\mathcal{E}_c} = \dfrac{3.5167}{2.8} = 1.256$ kW

Heat rejected per ton, $\dot{Q}_k = \dot{Q}_o + \dot{W} = 3.516 + 1.256 = 4.722$ kW

2.6 GAS AS A REFRIGERANT IN REVERSED CARNOT CYCLE

Figure 2.15 shows the Carnot cycle 1-2-3-4 with gas as a refrigerant, illustrated on *T-s* and *p-v* diagrams.

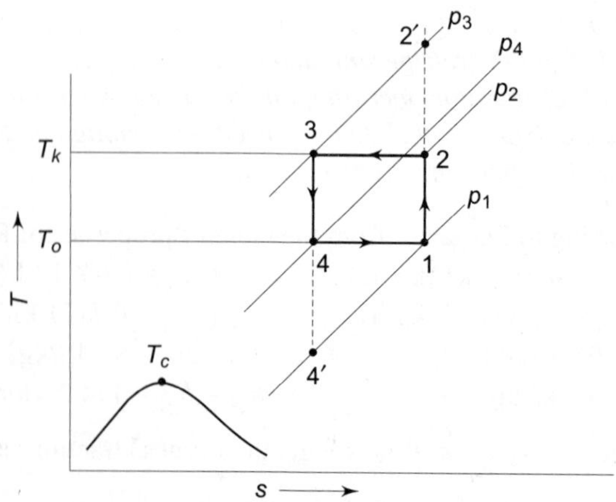

Fig. 2.15(a) Reversed Carnot cycle with gas as a refrigerant on *T-s* diagram

The four processes of the cycle are analysed as non-flow processes as follows in Eqs. (2.6), (2.7), (2.8) and (2.9):

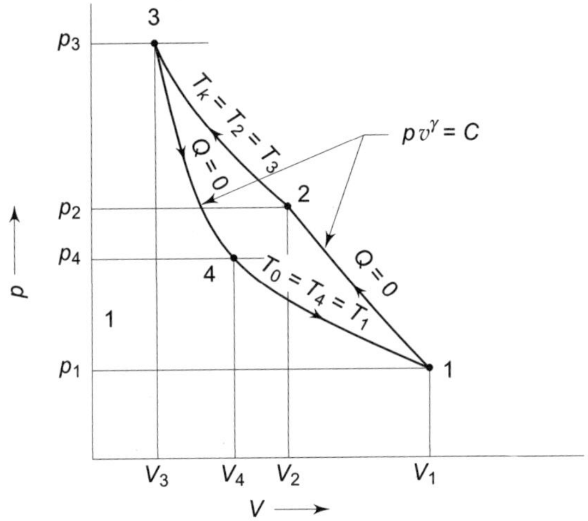

Fig. 2.15(b) Reversed Carnot cycle with gas as a refrigerant on *p-v* diagram

1–2 Isentropic compression: $Q = 0$

Pressure increases from p_1 to p_2

Specific volume reduces from v_1 to v_2

Temperature increases from $T_o = T_1$ to $T_k = T_2$

$$\text{Work done, } |w_{1-2}| = \frac{p_2 v_2 - p_1 v_1}{\gamma - 1} = \frac{R(T_k - T_o)}{\gamma - 1} \tag{2.6}$$

2–3 Isothermal compression and heat rejection: $T_2 = T_3 = T_k$

Pressure increases from p_2 to p_3

Specific volume reduces from v_2 to v_3

$$\text{Work done, } |w_{2-3}| = p_2 v_2 \ln \frac{v_2}{v_3} = RT_k \ln \frac{v_2}{v_3} \tag{2.7}$$

Heat rejected, $q_k = q_{2-3} = w_{2-3}$ (for a perfect gas in an isothermal process)

3–4 Isentropic expansion: $Q = 0$

Pressure falls from p_3 to p_4

Specific volume increases from v_3 to v_4

Temperature decreases from $T_k = T_3$ to $T_o = T_4$

$$\text{Work done, } w_{3-4} = \frac{p_3 v_3 - p_4 v_4}{\gamma - 1} = \frac{R(T_k - T_o)}{\gamma - 1} \tag{2.8}$$

4–1 Isothermal expansion and heat absorption: $T_4 = T_1 = T_o$

Pressure falls from p_4 to p_1

Specific volume increases from v_4 to v_1

$$\text{Work done, } w_{4-1} = p_4 v_4 \ln \frac{v_1}{v_4} = RT_o \ln \frac{v_1}{v_4} \tag{2.9}$$

$$\text{Refrigerating effect, } q_o = q_{4-1} = w_{4-1} \text{ (for a perfect gas)}$$

Network of the cycle.

$$w = |w_{2-3}| - w_{4-1} = RT_k \ln \frac{v_2}{v_3} - RT_o \ln \frac{v_1}{v_4} \tag{2.10}$$

$$\text{Refrigerating effect, } q_o = q_{4-1} = RT_o \ln \frac{v_1}{v_4} \tag{2.11}$$

Now for the isentropic processes 1-2 and 3-4

$$\frac{T_k}{T_o} = \left(\frac{p_2}{p_1}\right)^{\frac{\gamma-1}{\gamma}} = \left(\frac{p_3}{p_4}\right)^{\frac{\gamma-1}{\gamma}}$$

and

$$\frac{T_k}{T_o} = \left(\frac{v_1}{v_2}\right)^{\gamma-1} = \left(\frac{v_4}{v_3}\right)^{\gamma-1} = r^{\gamma-1}$$

where r is the compression ratio for the isentropic processes.

Hence $\quad \dfrac{p_2}{p_1} = \dfrac{p_3}{p_4}, \quad \text{and} \quad r = \dfrac{v_1}{v_2} = \dfrac{v_4}{v_3}$

or $\quad \dfrac{v_2}{v_3} = \dfrac{v_1}{v_4} = \dfrac{p_3}{p_2} = \dfrac{p_4}{p_1}$

since $\quad p_2 v_2 = p_3 v_3$ and $p_4 v_4 = p_1 v_1$

We have, for the COP for cooling

$$\mathcal{E}_c = \frac{q_o}{w} = \frac{RT_o \ln \frac{v_1}{v_4}}{RT_k \ln \frac{v_2}{v_3} - RT_o \ln \frac{v_1}{v_4}}$$

$$= \frac{T_o}{T_k - T_o}$$

$$= \frac{1}{\frac{T_k}{T_o} - 1} = \frac{1}{r^{\gamma-1} - 1} \tag{2.12}$$

Thus COP is a function of compression ratio only.

2.7 LIMITATIONS OF REVERSED CARNOT CYCLE

It is found that serious practical difficulties are encountered in the application of Carnot cycle.

In the reversed Carnot cycle with vapour as refrigerant, the isothermal processes of condensation and evaporation are internally reversible processes, and they are easily achievable in practice although there may be some problem in having only partial evaporation. However, isentropic compression and expansion processes have some limitations which are discussed in Chap. 3. In brief, it is difficult to design an expander to handle a mixture of largely liquid and partly vapour for the process 3-4. Also, because of the internal irreversibilities in the compressor and the expander, the actual COP of the Carnot cycle is very low, though the ideal cycle COP is the maximum. A cycle which is closest to the reversed Carnot vapour cycle is the vapour compression cycle described in Chap. 3.

There are two drawbacks of reversed Carnot cycle with gas as a refrigerant:

(i) Firstly, it is not possible to devise, in practice, isothermal processes of heat absorption and rejection, 4-1 and 2-3 in Fig. 2.15 with gas as the working substance. These are impractical as these will be infinitely slow.

(ii) Secondly, the cycle on p-v diagram is very narrow since the volume is changing both during the reversible isothermal and reversible adiabatic processes. Drawn correctly to scale, the Carnot p-v diagram is much thinner than the diagram illustrated in Fig. 2.15. As a result, the stroke volume of the cylinder is very large. The cycle, therefore, suffers from poor actual COP as a result of irreversibilities of the compressor and expander.

A gas refrigeration cycle, which is closest to reversed Carnot cycle with gas as a refrigerant, is described in Chap. 11.

2.8 ACTUAL REFRIGERATION SYSTEMS

Although the Carnot cycle is theoretically the most efficient cycle between given temperatures T_k and T_o, it has limitations for practical use. It is, therefore, found useful only as a criterion of perfection of cycle. In an actual cycle, the COPs, $\mathcal{E}_c$ and $\mathcal{E}_h$, will be less than their Carnot values. For the purpose of comparison between the actual and Carnot values, we define the *second law efficiency* or *exergetic efficiency* for cooling and heating, $(\eta_{II})_c$ and $(\eta_{II})_h$ as below:

$$(\eta_{II})_c = \frac{\mathcal{E}_c}{\mathcal{E}_{c,\text{Carnot}}}$$

$$(\eta_{II})_h = \frac{\mathcal{E}_h}{\mathcal{E}_{h,\text{Carnot}}}$$

Note that $\mathcal{E}_c$ and $\mathcal{E}_h$ are the first law COPs.

The conventional refrigeration systems work on the *vapour compression cycle* which is closest to the Carnot vapour cycle and has a high COP. *Gas cycle refrigeration* is used in aircraft refrigeration. Among the less conventional ones are the heat-operated refrigerating machines working on the *vapour absorption cycle* and *steam ejector cycle*.

There are also the low temperature refrigeration or cryogenic cycles, e.g., Linde cycle, Claude cycle, etc., used for the liquefaction of gases. Also, we have Philips liquefier which employs a cycle approaching the reversible Stirling cycle.

A recent development is the thermoelectric refrigeration as described in Fig. 1.11. But its COP is so poor that it cannot be exploited commercially. Temperatures approaching absolute zero have been obtained by adiabatic demagnetization, as described in Fig. 1.12, on a limited scale in laboratories.

Revision Exercises

2.1 (a) A refrigerator has working temperatures in the evaporator and condenser coils of –30 and 35°C respectively. What is the maximum possible COP of the refrigerator?

(b) If the actual refrigerator has a refrigerating efficiency of 0.75, calculate the refrigerating effect in kW and TR per kW of power input.

2.2 A reversed Carnot cycle has a COP for cooling of 4. Determine the temperature ratio T_k/T_o.

If the power consumption of the cycle is 7.5 kW, determine the refrigerating capacity of the machine in *TR*.

If the cycle is used as a heat pump with the same ratio of temperatures, determine its COP for heating and the quantity of heat pumped.

2.3 A Carnot refrigerator operates with Refrigerant 134a as a refrigerant condensing at 50°C and evaporating at –15°C.

Find its COP using the Carnot expression as well as the properties of R 134a. Also determine the power consumption per ton of refrigeration.

2.4 The overall volume compression ratio of a reversed Carnot cycle working with air as a refrigerant is 10. The temperature limits of the cycle are 40°C and 0°C.

Determine:

(i) the pressure, volume and temperature at each point of the cycle,

(ii) the work done in the cycle,

(iii) the refrigerating effect, and

(iv) the COP of the cycle.

2.5 An air conditioning system is operating in an ambient of 45°C. The room temperature is maintained at 25°C. Determine the power consumption of the system per ton of refrigeration if it is,

(a) air-cooled as in a window-type air conditioner;

(b) water-cooled as in a central air conditioning plant.

The cooling water from cooling tower is available at 30°C. Assume suitable operating temperatures. Actual COP of the system is only 50% of the COP of the reversible cycle.

2.6 Determine the power consumption of a domestic refrigerator if its refrigerating capacity is $\frac{1}{8}$TR. It is operating in an ambient of 40°C. Temperature in the freezer must be maintained at –15°C. COP of the system is half the Carnot COP. Assume suitable condensing and evaporating temperatures.

Chapter

3

Vapour Compression System

3.1 MODIFICATIONS IN REVERSED CARNOT CYCLE WITH VAPOUR AS A REFRIGERANT

The reversed Carnot cycle with vapour as a refrigerant can be used as a practical cycle with minor modifications. The isothermal processes of heat rejection and heat absorption, accompanying condensation and evaporation respectively, are nearly perfect processes and easily achievable in practice. The isentropic compression and expansion processes, however, have certain limitations which are discussed below and are, therefore, suitably modified.

3.1.1 Dry Versus Wet Compression

The compression process as shown in Fig. 3.1 involves the compression of wet-refrigerant vapour at 1′ to dry-saturated vapour at 2′. It is called *wet compression.* With a reciprocating compressor, wet compression is not found suitable due to the following reasons:

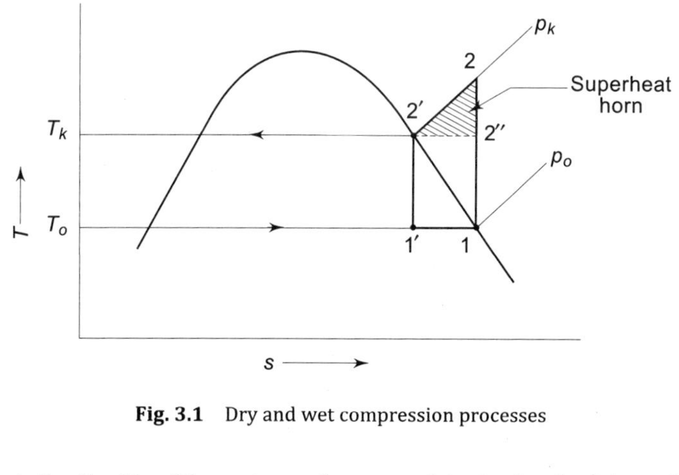

Fig. 3.1 Dry and wet compression processes

(i) First, the liquid refrigerant may be trapped in the head of the cylinder and may damage the compressor valves and the cylinder itself. Even though the state of vapour at the end of wet compression is theoretically dry-saturated, it is normal to expect some liquid droplets to remain suspended in the gas, as the time taken by the compression process is quite small compared to the time needed for evaporation of droplets. For example, in a modern high-speed compressor, say, running at 2800 rpm, the time available in one revolution is only 0.021 second.

(ii) Secondly, liquid-refrigerant droplets may wash away the lubricating oil from the walls of the compressor cylinder, thus increasing wear.

It is, therefore, desirable to have compression with vapour initially dry saturated at 1 as shown in Fig. 3.1, or even slightly superheated if a reciprocating compressor is used. Such compression is known as *dry compression*. The state of the vapour at the end of compression will, therefore, have to be at 2, at pressure p_k which is the saturation pressure of the refrigerant corresponding to the condensing temperature t_k, instead of being at 2″, which would be the state point if the Carnot cycle were to be executed. It results in the discharge temperature t_2 being higher than the condensing temperature t_k. Consequently, the refrigerant leaves the compressor superheated. The increased work of the cycle due to the substitution of wet compression by dry compression appears as the area 2–2′ – 2″, generally known as *superheat horn*.

It must, however, be stated here that wet compression in some cases is indeed desirable, and also practicable with the use of a continuous flow machine like a centrifugal or a screw compressor with no valves in place which are an essential feature of a reciprocating compressor. The improvement in COP with wet compression is amply illustrated in the case of ammonia in Example 3.4 in which it is shown that the power consumption per ton refrigeration with wet compression is less by 10 percent as compared with that of dry compression.

3.1.2 Throttling Versus Isentropic Expansion

Refrigerating machines are usually much smaller devices compared to power plants. Thus the net work required by refrigeration systems is quite small compared to the work done in power-generating plants.

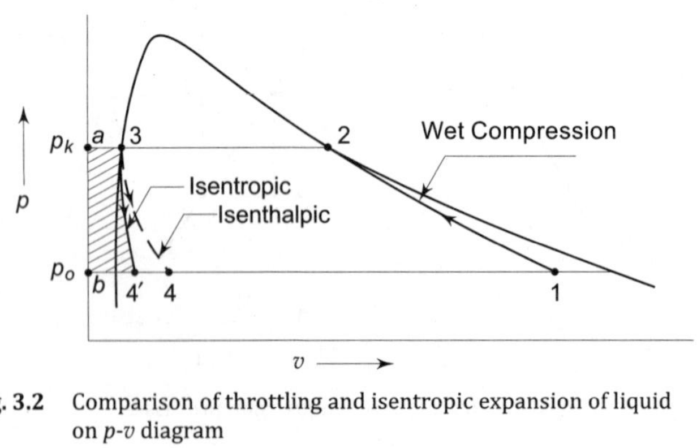

Fig. 3.2 Comparison of throttling and isentropic expansion of liquid on *p-v* diagram

Further, the positive work of the cycle, recovered during the isentropic expansion process, as shown by area 3-a-b-4′ in Fig. 3.2, is even smaller, as compared to the negative work of the cycle consumed during the isentropic compression process, shown by area 1-2-a-b. This is evident from the expression for work, viz., $-\int v dp$. Thus for the same pressure difference dp, the work depends on the volume v of the working substance. In the expander, the refrigerant is in the liquid state, whereas, in the compressor, it is in the gaseous state. The volume of the vapour is very large compared to the volume of the liquid ($v_g >> v_f$). Hence, the positive work of isentropic expansion is seldom large enough to justify the cost of an expander. On the other hand, the thermodynamic and friction losses of an expander, if employed, may even exceed the gain in work. Moreover, there are practical difficulties in smoothly expanding a liquid of a highly wet vapour in an expander.

Accordingly, the isentropic expansion process of the Carnot cycle may be replaced by a simple throttling process or an isenthalpic process by the use of an expansion device such as a throttle valve or a capillary tube.

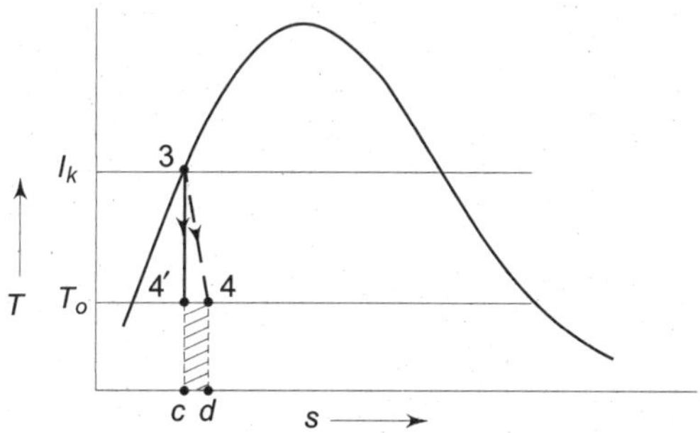

Fig. 3.3 Comparison of throttling and Isentropic expansion of liquid on *T-s* diagram

The process is an irreversible one and is accompanied by increase of entropy as shown by line 3-4 on the *T-s* diagram in Fig. 3.3. Thus, the substitution of the isentropic-expansion process 3-4′ by the isenthalpic/throttling process 3-4 would, theoretically, result in a loss of work represented by area 3-a-b-4′ on the *p-v* diagram and a decrease in the refrigerating effect represented by area 4′-c-d-4 on the *T-s* diagram. It can be shown that both these areas are equal.

3.2 VAPOUR COMPRESSION CYCLE

The cycle with the above two modifications is named as the *vapour compression cycle* and because of its high index of performance or efficiency, it is most widely used in commercial refrigeration systems. A complete vapour compression cycle is shown on the *T-s* diagram in Fig. 3.4 and on the *p-v* diagram in Fig. 3.5. Figure 3.4 also presents a comparison of the vapour compression cycle 1-2-3-4 with the reversed Carnot cycle 1-2″-3-4′, or 1-2″-3′-4, both operating between the same temperature limits of T_k and T_o.

In the vapour compression cycle:

Refrigerating effect, q_o = area 1-4-d-e

Heat rejected, q_k = area 2-2′-3-c-e

Work done, $w = q_k - q_o$ = area 1-2-2′-3-c-d-4-1

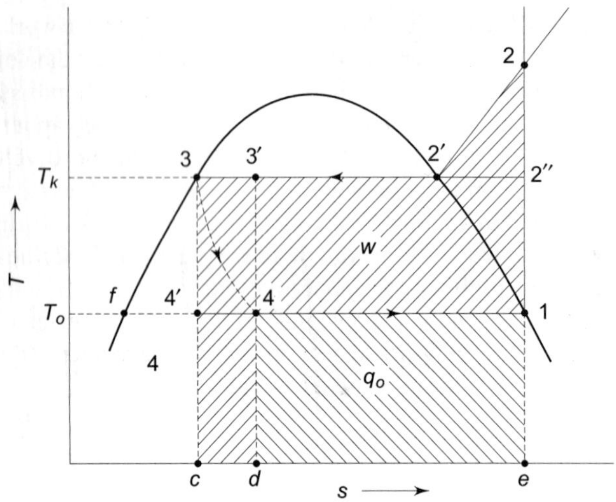

Fig. 3.4 Vapour compression cycle on *T-s* diagram

It may maybe seen that the vapour comptession cycle presents three deviations from the reversed Carnot cycle, as indicated below:

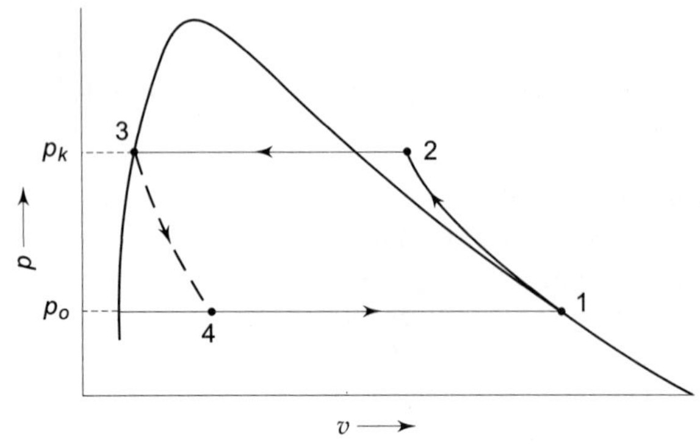

Fig. 3.5 Vapour compression cycle on *p-v* diagram

(i) Area 4-4′-c-d, representing a loss of the refrigerating effect, Δq_o, as a result of throttling.

(ii) Area 4-4′-c-d, also representing a loss of positive work, Δw_o, resulting from the failure to recover expansion work. It can be shown that areas 4-4′-c-d and 3-f-4′ are the same.

(iii) Area 2-2′-2″ of superheat horn, representing an increase of negative work, Δw_k, as a result of dry compression.

Consequently, the theoretical COP of the vapour compression cycle is lower than that of the reversed Carnot cycle. Nevertheless, it is closest to the Carnot cycle as compared to other cycles and its COP approaches nearest to the Carnot value.

3.3 VAPOUR COMPRESSION SYSTEM CALCULATIONS

A schematic vapour compression system is shown in Fig. 3.6. It consists of a *compressor*, a *condenser*, an *expansion device* for throttling and an *evaporator*. The compressor-delivery head, discharge line, condenser and liquid line form the high-pressure side of the system. The compressor and matching condenser together are also available commercially as one unit called the *condensing unit*. The expansion line, evaporator, suction line and compressor-suction head form the low-pressure side of the system. It may be pointed out here that, in actual systems unlike in Fig. 3.6, the expansion device is located as close to the evaporator as possible in order to minimise the heat gain in the low temperature expansion line.

In plants with a large amount of *refrigerant charge,* a *receiver* is installed in the liquid line. Normally, a *drier* is also installed in the liquid line particularly in flurocarbon systems. The drier contains silica gel and absorbs traces of moisture present in the liquid refrigerant so that it does not enter the narrow cross-section of the expansion device causing *moisture choking* by freezing. The thermodynamic processes are as follows:

Process 1-2 Isentropic compression: $s_2 = s_1$. $Q = 0$

Work done, $$w = -\int v\,\mathrm{d}p = -\int \mathrm{d}h = -(h_2 - h_1) \tag{3.1}$$

Process 2-3 Desuperheating and condensation: p_k = const.

Heat rejected, $q_k = h_2 - h_3$ (3.2)

Process 3-4 Isenthalpic expansion: $h_3 = h_4 = h_{f4} + x_4 (h_1 - h_{f4})$

$$\Rightarrow \quad x = \frac{h_3 - h_{f_4}}{h_1 - h_{f_4}} \tag{3.3}$$

Process 4-1 Evaporation: p_o = const.

Refrigerating effect, $q_o = h_1 - h_4$ (3.4)

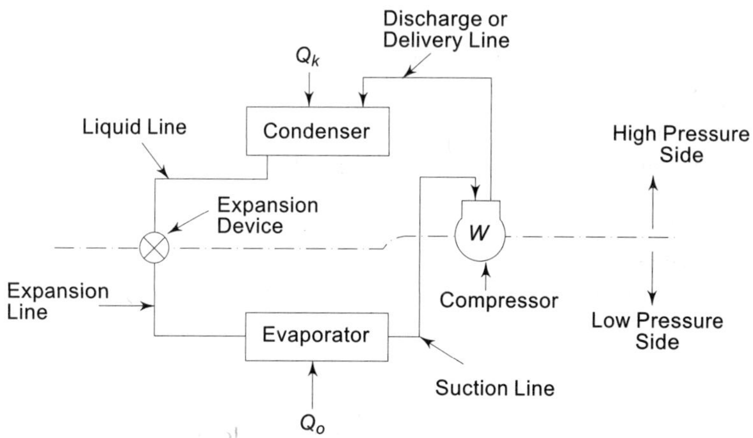

Fig. 3.6 Vapour compression system

3.3.1 Representation of Vapour Compression Cycle on Pressure-Enthalpy Diagram

It is worth noting that two of the processes are at constant pressure and one is at constant enthalpy. It is, therefore, found convenient to represent the vapour compression cycle on a pressure-enthalpy (*p-h*) diagram as shown in Fig. 3.7. Therefore, even though the fourth process is an isentropic one, the *p-h* diagram is still found convenient as the work done is given by the increase in enthalpy.

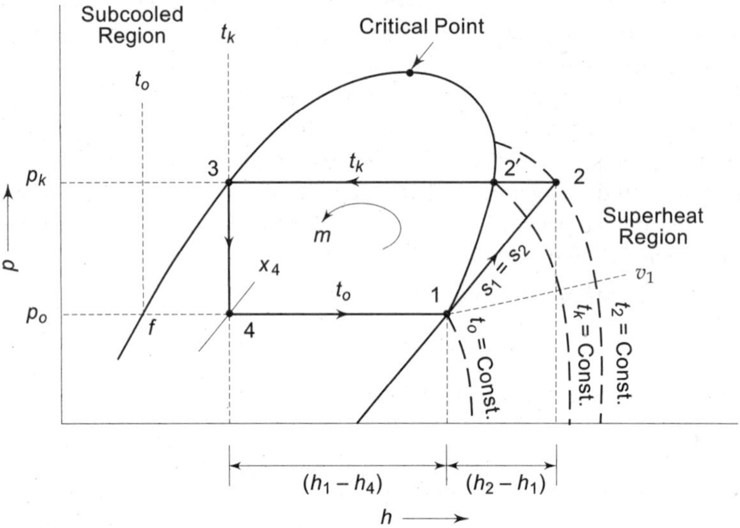

Fig. 3.7 Vapour compression cycle on *p-h* diagram

The cycle described and shown in Figs. 3.4, 3.5 and 3.7 is a *simple saturation cycle* implying that both the states, of liquid after condensation and vapour after evaporation, are saturated and lie on the saturated liquid and saturated vapour curves respectively. The condensation temperature t_k and evaporator temperature t_o, corresponding to the respective saturation pressures p_k and p_o, are also called *saturated discharge temperature* and *saturated suction temperature* respectively. However, the actual discharge temperature from the compressor is t_2.

Figure 3.7 also shows constant temperature lines in the subcooled and superheat regions along with constant volume lines. It may be noted that constant temperature lines in the subcooled liquid and low pressure vapour regions are vertical as the enthalpy of the liquid and the ideal gas are functions of temperature only and do not depend on pressure.

Further calculations of the cycle can be done as follows:

Heat rejected, $$q_k = q_o + w = h_2 - h_3$$

COP for cooling, $$\mathcal{E}_c = \frac{h_1 - h_4}{h_2 - h_1} \tag{3.5}$$

COP for heating, $$\mathcal{E}_h = \frac{h_2 - h_3}{h_2 - h_1} \tag{3.6}$$

$$\text{Refrigerant circulation rate, } \dot{m} = \frac{\text{refrigerating capacity}}{\text{refrigerating effect per unit mass}} = \frac{\dot{Q}_o}{q_o} \tag{3.7}$$

Specific volume of the vapour at suction = v_1

Theoretical piston displacement of the compressor or volume of the suction vapour,

$$\dot{V} = \dot{m} v_1 \tag{3.8}$$

Actual piston displacement of the compressor,

$$\dot{V}_p = \frac{\dot{m} v_1}{\eta_v}$$

where η_v is the volumetric efficiency.

$$\text{Power consumption, } \dot{W} = \dot{m} w = \dot{m}(h_2 - h_1) \tag{3.9}$$

$$\text{Heat rejected in the condenser, } \dot{Q}_k = \dot{m} q_k = \dot{m}(h_2 - h_3) \tag{3.10}$$

Expressing the power consumption per ton of refrigeration as *unit power consumption*, denoted by $\overset{*}{W}$, we have for mass flow rate and power consumption per ton refrigeration,

$$\overset{*}{m} = \frac{3.5167}{q_o} \text{ kg/(s)·(TR)} \tag{3.11}$$

$$\overset{*}{W} = \dot{m} w = 3.5167 \left(\frac{h_2 - h_1}{h_1 - h_4} \right) \text{ kW/TR} \tag{3.12}$$

It will be seen that

$$\overset{*}{W} \propto \frac{1}{\mathcal{E}_c}$$

Similarly, the suction volume requirement per ton is

$$\overset{*}{V} = \overset{*}{m} v_1 = \frac{211}{q_0} v_1 \text{ m}^3\text{/(min) (TR)} \tag{3.13}$$

The *isentropic discharge temperature* t_2 may be found by the following three methods.

(i) Graphically from *p-h* diagram by drawing the isentropic from point 1 to p_k = constant line, or by iteration, finding t_2 corresponding to $s_2 = s_1$.

(ii) Using saturation properties and the specific heat of vapour

$$s_1 = s_2 = s'_2 + C_p \ln \frac{T_2}{T'_2} \tag{3.14}$$

where $s'_2 = s_{g2}$ and $T'_2 = T_k$

(iii) Using superheat tables and interpolating for the degree of superheat $(T_2 - T'_2)$ corresponding to the entropy difference $(s_2 - s'_2)$ which is known.

The following examples illustrate the procedures.

Refrigerant Charge Refrigerant Charged in the system depends on the volume of the system, with liquid in some parts, and vapour in others. Note that mass flow rate is the mass of refrigerant circulated. It is not equal to the mass charged in the system.

Example 3.1 ***R 12 System***

A R 12 vapour compression system operating at a condenser temperature of 40°C and an evaporator temperature of 0°C develops 15 tons of refrigeration. Using the p-h diagram for R 12, determine.

(a) the discharge temperature and mass flow rate of the refrigerant circulated,
(b) the theoretical piston displacement of the compressor and piston displacement per ton of refrigeration,
(c) the theoretical horsepower of the compressor and horsepower per ton of refrigeration,
(d) the heat rejected in the condenser, and
(e) the Carnot COP and actual COP of the cycle.

Solution From the *p-h* diagram (table) of R 12

$$h_1 = 187.5 \text{ kJ/kg} \quad v_1 = 0.055 \text{ m}^3\text{/kg}$$
$$h_2 = 213.96 \text{ kJ/kg}$$
$$h_3 = h_4 = 74.6 \text{ kJ/kg}$$

(a) Discharge temperature $s_2 = s_1 = 0.6966$ gives

$$t_2 = 54.06°\text{C}$$

Refrigerating effect $q_0 = h_1 - h_4 = 187.5 - 74.6 = 112.8$ kJ/kg

Refrigerant circulated

$$\dot{m} = \frac{\dot{Q}_0}{q_0} = \frac{15 \times 211}{112.8} = 28.0 \text{ kg/min}$$

(b) Theoretical piston displacement of compressor

$$\dot{V} = \dot{m}v_1 = 28.0\,(0.055) = 1.514 \text{ m}^3\text{/min}$$

Piston displacement per ton

$$\overset{*}{V} = \frac{1.514}{15} = 0.101 \text{ m}^3\text{/min (TR)}$$

(c) Power consumption

$$\dot{W} = \dot{m}\,(h_2 - h_1)$$
$$= \frac{28.0\,(213.96 - 187.5)}{60} = 12.36 \text{ kW}$$

Theoretical horsepower of the compressor

$$\text{HP} = \frac{12.36 \times 10^3}{746} = 16.6$$

$$\text{HP/TR} = \frac{16.6}{15} = 1.1$$

(d) Heat rejected

$$\dot{Q}_k = \dot{m}(h_2 - h_3)$$
$$= \frac{28.0(213.96 - 74.6)}{60} = 65.1 \text{ kW}$$

(e) Carnot COP

$$\mathcal{E}_{max} = \frac{273}{40-(0)} = 6.8$$

COP of the cycle

$$\mathcal{E} = \frac{h_1 - h_4}{h_2 - h_1} = \frac{112.8}{213.96 - 187.5} = 4.3$$

Example 3.2 ***R 134a System***

Chlorine in the R 12 (CCl_2F_2) molecule depletes the ozone layer in the earth's upper atmosphere. R 12 has now been replaced by the ozone-friendly R 134 a ($C_2H_2F_4$)

For the conditions of Example 3.1, do calculations for R 134a, and compare results.

Solution From the table of properties of R 134a, we have

$p_0 = 0.2958$ MPa $\quad p_k = 1.0166$ MPa $\quad h_4 = 256.41$ kJ/kg

$h_1 = 398.6$ kJ/kg $\quad v_1 = 0.06931$ m^3/kg

$s_1 = 1.7541$ kJ/kgK $\quad s_2' = 1.7111$ kJ/kg K, $\quad h'_2 = 419.43$ kJ/kg

C_p at $p_k = 1.145$ kJ/kg K $\quad T_2' = 313$ K

(a) For isentropic compression,

$$s_2 = s_1 = 1.7541 + C_p \ln \frac{T_2}{313}$$

$$\Rightarrow \quad T_2 = 317.6 \text{ K } (44.4°\text{C})$$

$$h_2 = h_2' + C_p(T_2 - T_2')$$

$$= 419.43 + 1.145(4.4) = 424.5 \text{ kJ/kg}$$

$$w = h_2 - h_1 = 424.5 - 398.6 = 25.9 \text{ kJ/kg}$$

$$q_o = h_1 - h_4 = 398.6 - 256.41 = 142.19 \text{ kJ/kg}$$

$$\dot{m} = \frac{\dot{Q}_o}{q_o} = \frac{15 \times 211}{142.19} = 22.26 \text{ kg/min}$$

(b) $\dot{V} = \dot{m}v_1 = 22.26(0.06931) = 1.543$ m^3/min

$$\dot{V}^* = \frac{\dot{V}}{TR} = \frac{1.543}{15} = 0.1029 \text{ m}^3\text{/min (TR)}$$

(c) $\dot{W} = \dot{m}(h_2 - h_1) = \dfrac{22.26(424.5 - 398.6)}{60} = 9.61$ kW

$$\text{HP} = \frac{9.61 \times 10^3}{746} = 12.88$$

$$\text{HP/TR} = \frac{12.88}{15} = 0.86$$

(d) $\dot{Q}_k = \dot{m}(h_2 - h_3) = \dfrac{22.26}{60}(424.5 - 256.41) = 62.36$ kW

(e) $\mathcal{E} = \dfrac{q_o}{w} = \dfrac{142.19}{25.9} = 5.49$

Note *R 134a requires 2% more piston displacement, 22% less power consumption, and gives higher COP. Additionally, it has 10°C lower discharge temperature.*

Example 3.3 *Ammonia Ice Plant*

An ammonia ice plant operates between a condenser temperature of 35°C and an evaporator temperature of –15°C. It produces 10 tons of ice per day from water at 30°C to ice at –5°C. Assuming simple saturation cycle, using only tables of properties for ammonia, determine;

(a) the capacity of the refrigeration plant,
(b) the mass flow rate of refrigerant,
(c) the discharge temperature,
(d) the compressor cylinder diameter and stroke if its volumetric efficiency is $\eta_v = 0.65$, rpm N = 1200 and stroke/bore ratio L/D = 1.2,
(e) the horsepower of the compressor motor if the adiabatic efficiency of the compressor $\eta_a = 0.85$ and mechanical efficiency $\eta_m = 0.95$, and
(f) the theoretical and actual COP.

Solution

(a) *Refrigerating capacity*

$$\dot{Q}_0 = \frac{10 \times 1000}{24}[4.1868\,(30 - 0) + 335 + 1.94\,(0.5)]$$
$$= 195{,}960 \text{ kJ/h}$$
$$= \frac{195{,}960}{3600} = 54.43 \text{ kW}$$

Refer Fig. 3.8.

(b) From the table for ammonia,

$h_1 = h_{g\,(-15°C)} = 1443.9$ kJ/kg and $s_1 = s_{g(-15°C)} = 5.8223$ kJ/(kg) (K)

$h_4 = h_3 = h_{f\,(35°C)} = 366.1$ kJ/kg

$q_0 = h_1 - h_4 = 1443.9 - 366.1 = 1077.8$ kJ/kg

Mass flow rate

$$\dot{m} = \frac{195{,}960}{1077.8} = 181.7 \text{ kg/h}$$

(c) From the table, entropies and enthalpies of vapour saturated at 35°C and superheated by 50 K and 100 K are, respectively,

$s_{g\,(35°C)} = 5.2086,\quad s_{(50\,K)} = 5.6466\quad s_{(100K)} = 5.9806$

$h_{g\,(35°C)} = 1488.6\quad h_{(50\,K)} = 1633.6\quad h_{(100K)} = 1703$

Now $s_2 = s_1 = 5.8223 > 5.6466$

Hence the vapour is superheated by more than 50 K.
Interpolating for discharge temperature,

$$t_2 = 35 + 50 + \frac{5.8223 - 5.6466}{5.9806 - 5.6466}(50)$$
$$= 35 + 50 + 27.1 = 112.1°\text{C}$$

(d) Piston displacement of compressor

$$\frac{\pi D^2}{4} LN = \dot{V}_p = \frac{\dot{m}v_1}{\eta_v}$$

$$\Rightarrow \qquad \frac{\pi}{4}D^2 (1.2)\, D\, (1200)\, (60) = \frac{181.7(0.509)}{0.65}$$

whence $D = 0.128$ m

$L = 0.154$ m

(e) Enthalpy at discharge, by interpolation as for entropy

$$h_2 = 1633.6 + \frac{27.1}{50}(1761.6 - 1633.6) = 1633.6 + 69.4 = 1703 \text{ kJ/kg}$$

Here, 1633.6 kJ/kg is the enthalpy of vapour superheated by 50 K.
Power consumption

$$\dot{W} = \frac{\dot{m}(h_2 - h_1)}{\eta_a \eta_m} = \frac{181.7(1703 - 1443.9)}{3600 \times 0.85 \times 0.95} = 16.21 \text{ kW}$$

Horsepower $\text{HP} = \dfrac{16.21 \times 10^3}{746} = 21.73$ hp

(f) Theoretical COP $= \dfrac{h_1 - h_4}{h_2 - h_1} = \dfrac{1077.8}{1703 - 1443.9} = 4.16$

Actual COP $= \dfrac{\dot{Q}_0}{\dot{W}} = \dfrac{54.43}{16.21} = 3.36$

Example 3.4 ***Checking if Dry or Wet Compression is Desirable in Ammonia Systems***

An ammonia refrigerating machine has working temperatures of 35°C in the condenser and –15°C in the evaporator. Assume two cases;

(a) dry compression, and
(b) wet compression

Calculate for each, the following;

(i) the theoretical piston displacement per ton refrigeration,
(ii) the theoretical horsepower per ton refrigeration, and
(iii) the coefficient of performance.

Solution (a) *Dry compression* The cycle is shown as 1-2-3-4 in Fig. 3.8. Points 2″ and 2‴ correspond to 50°C and 100°C superheat respectively at 35°C condensing temperature. By interpolation for state 2 between states 2″ and 2‴ we already have values of entropy s_2 and enthalpy h_2 for dry compression from Example 3.3.

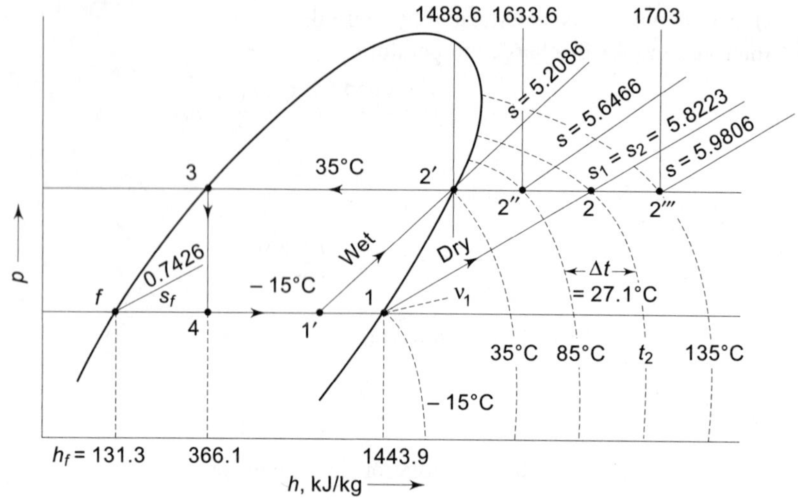

Fig. 3.8 Figure for Examples 3.3 and 3.4

Refrigerant circulation rate per ton refrigeration

$$m^* = \frac{211}{1077.8} = 0.1956 \text{ kg/(min) (TR)}$$

(i) Theoretical piston displacement per ton

$$\dot{V}^* = m^* v_1 \times 60 = 0.1956 \times 0.509 \times 60 = 5.975 \text{ m}^3\text{/h}$$

(ii) Theoretical horsepower per ton

$$\text{HP}^* = \frac{m^* w}{90 \times 0.736} = \frac{0.1956 \times 259.4}{60 \times 0.736} = 1.15 \text{ hp/TR}$$

(iii) $\text{COP} = \dfrac{q_0}{w} = \dfrac{1078.5}{259.1} = 4.16$

(b) *Wet compression* The cycle is shown as 1′-2′-3-4 in Fig. 3.8. The dryness at suction at 1′ can be calculated by interpolation for entropy.

$$x_1' = \frac{5.2086 - 0.7426}{5.8223 - 0.7426} = 0.88$$

$$h_1' = 131.3 + 0.88\,(1443.9 - 131.3) = 1286.1 \text{ kJ/kg}$$

$$v_1' = v_f + x\,(v_g - v_f) = 0.00152 + 0.88\,(0.509 - 0.00152)$$

$$= 0.448 \text{ m}^3\text{/kg}$$

$$q_0 = h'_1 - h_4 = 1286.1 - 366.1 = 920 \text{ kJ/kg}$$

$$w = h_2' - h'_1 = 1488.6 - 1286.1 = 202.5 \text{ kJ/kg}$$

Refrigerant circulation rate per ton

$$m^* = \frac{211}{920} = 0.229 \text{ kg/(min) (TR)}$$

(i) Theoretical piston displacement per ton

$$V^* = 0.229 \times 0.448 \times 60 = 6.159 \text{ m}^3\text{/h}$$

(ii) Theoretical horsepower per ton

$$\text{HP}^* = \frac{0.229 \times 202.5}{60 \times 0.736} = 1.05 \text{ hp/TR}$$

(iii) $\text{COP} = \dfrac{920}{202.5} = 4.55$

Note *It is seen that higher COP and lower power consumption is obtained with wet compression in the case of ammonia. However, compressor piston displacement is increased by (6.159–5.975)/5.975, viz., 3% though.*

3.4 EWING'S CONSTRUCTION

The theoretical COP of the vapour compression cycle with any *refrigerant* working between given evaporation and condenser temperatures is found to depend on the state of the suction vapour which may be wet, dry saturated or superheated. For example, in the limiting case, when the vapour is compressed immediately after throttling from 4 itself to 5 as shown in Fig. 3.9, the refrigerating effect as well as the COP are both zero. As the suction state is moved to the right, the refrigerating effect and COP go on increasing until point 1 is reached. After this point, the compression partly takes place in the superheat region and the heat is not rejected at constant temperature in the condenser. A trend of increasing COP *may continue* until the suction state 1 m for maximum COP is reached.

For some refrigerants and certain operating conditions, the maximum COP occurs with the suction state 1 m in the two phase region and for some others in the superheat region.

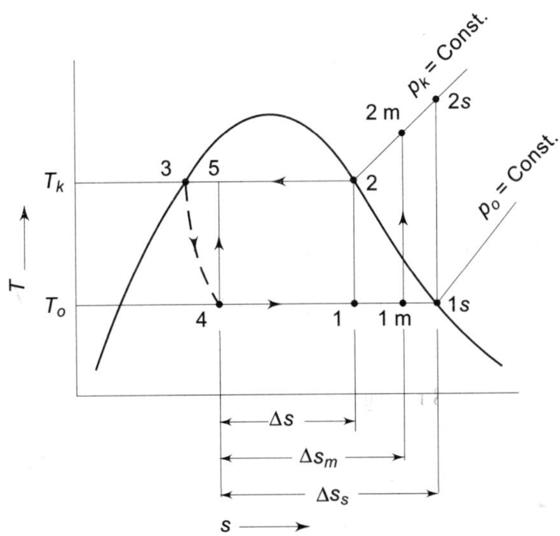

Fig. 3.9 Effect of variable suction state

Ewing[1] has shown that when this maximum occurs with state 1 m in the two-phase region, the COP is equal to that of a reversed Carnot cycle operating between the evaporator temperature T_0 and the temperature of superheated discharge vapour T_{2m}. To show this, consider the expression for COP

$$\mathcal{E} = \frac{h_1 - h_4}{h_2 - h_1} = \frac{(h_1 - h_4)/\Delta s}{(h_2 - h_4)/\Delta s - (h_1 - h_4)/\Delta s} \tag{3.15}$$

Now, an enthalpy-entropy diagram may be employed for Ewing's construction as shown in Fig. 3.10. It may be seen that the term $(h_1 - h_4)/\Delta s$ is the gradient of the evaporator pressure p_0 = const. line and is equal to T_0, as long as point 1 is in the two-phase region. The expression for COP, with the suction state in the two-phase region, is then as given in Eq. (3.16).

$$\mathcal{E} = \frac{T_0}{(h_2 - h_4)/\Delta s - T_0} \tag{3.16}$$

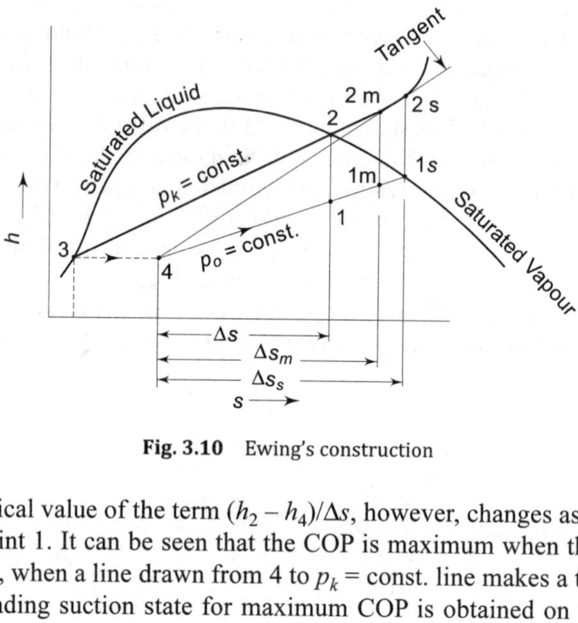

Fig. 3.10 Ewing's construction

The numerical value of the term $(h_2 - h_4)/\Delta s$, however, changes as point 2 shifts along with point 1. It can be seen that the COP is maximum when this gradient is minimum, i.e., when a line drawn from 4 to p_k = const. line makes a tangent at 2m. The corresponding suction state for maximum COP is obtained on the isentropic line at 1m. Using the thermodynamic relation.

$$\left(\frac{\Delta h}{\Delta s}\right)_p = T$$

we find that

$$\left[\frac{h_{2m} - h_4}{\Delta s_m}\right]_{p_k = \text{const.}} = T_{2m} \tag{3.17}$$

so that the slope of the tangent $\Delta h/\Delta s$ from point 4 to the p_k = const. line is equal to the discharge temperature T_{2m}. The expression for maximum COP is then as follows in Eq. (3.18).

$$\mathcal{E}_{max} = \frac{T_0}{\left[\dfrac{h_2 - h_4}{\Delta s}\right]_{min} - T_0} = \frac{T_0}{T_{2m} - T_0} \tag{3.18}$$

It must, however, be remembered that the above expression for $\mathcal{E}_{max}$ is valid only if the maximum COP occurs with the suction state 1 in the two-phase region.

3.4.1 Suction State for Maximum COP

Gosney[2] has given a method of finding whether the maximum COP occurs with the suction state in the two-phase region or in the superheat region. Consider Case I first, in which maximum COP occurs with the suction state in the two-phase region. Consider also a simple saturation cycle with suction state at 1s and the corresponding discharge state at 2s, the discharge temperature being T_{2s}, as shown in Figs. 3.9 and 3.10. Then from Eq. (3.16), the COP of the simple saturation cycle $\mathcal{E}_s$ is

$$\mathcal{E}_s = \frac{T_0}{\dfrac{h_{2s} - h_4}{\Delta s_s} - T_0} \tag{3.19}$$

Now, it can be seen from Fig. 3.10 that the slope of tangent at 2m < slope of line 4-2*s* < slope of tangent at 2s. It implies that

$$T_{2m} < \frac{h_{2s} - h_4}{\Delta s_s} < T_{2s} \tag{3.20}$$

Comparing Eqs. (3.18), (3.19) and (3.20), we see that

$$\mathcal{E}_{max} > \mathcal{E}_s > \frac{T_0}{T_{2s} - T_0} \quad \text{for Case I} \tag{3.21}$$

In a similar manner, it can be shown that for Case II, viz., for maximum COP to occur with the suction state in the superheat region,

$$\mathcal{E}_s < \frac{T_0}{T_{2s} - T_0} \quad \text{for Case II} \tag{3.22}$$

In this manner it can be found out whether the wet or superheated suction state is suitable for maximum COP for a particular refrigerant. Table 3.1 shows calculations for a number of common refrigerants for $t_0 = -15°C$ and $t_k = 30°C$ and classifies them either as Case I or as Case II.

Table 3.1 Categorization of refrigerants for suction state for maximum COP

Refrigerant	$\mathcal{E}_s$	t_{2s}, °C	$\dfrac{T_0}{T_{2s} - T_0}$	*Case*
NH_3	4.76	99	2.27	I
C_4H_{10} (Isobutane)	4.26	30	5.74	II
$C_2H_2F_4$ (Refrigerant 134a)	4.6	36.4	5.02	II
C_2H_6	2.41	50	3.97	II
CCl_3F(R 11)	5.03	44	4.38	I
CCl_2F_2 (R 12)	4.7	38.3	4.84	II
$CClHF_2$ (R 22)	4.66	53.3	3.68	I
C_3H_8 (Propane)	4.47	37.61	4.91	II

It can be readily seen that suction vapour superheat increases the COP in propane, isobutane, tetrafluoroethane (Refrigerant 134a) and Freon 12 systems, whereas, in the case of R 11 and R 22 and ammonia systems, maximum COP is obtained with wet suction state.

Example 3.5 *Suction State for Maximum COP in Isobutane Refrigerator*

Show that, for $t_0 = -15°C$ and $t_k = 30°C$, for isobutane the suction state for maximum COP lies in the superheat region.

Solution From the table of properties of isobutane for the saturated vapour state at 1 at –15°C, we have $p_1 = 0.895$ bar, $h_1 = 536.19$ kJ/kg

$$s_1 = 2.3073 \text{ kJ/kg. K}$$

Also, for the saturated vapour state 2′ at 30°C, we have.

$$p_2 = 4.08 \text{ bar}$$
$$s_2' = 2.3198 \text{ kJ/kg. K}$$

Since $s_2 = s_1 = 2.3073 < s'_2 = 2.3198$, the state 2 after compression is wet. Isobutane, therefore, has typical *T-s* diagram as shown in Fig. 3.11.

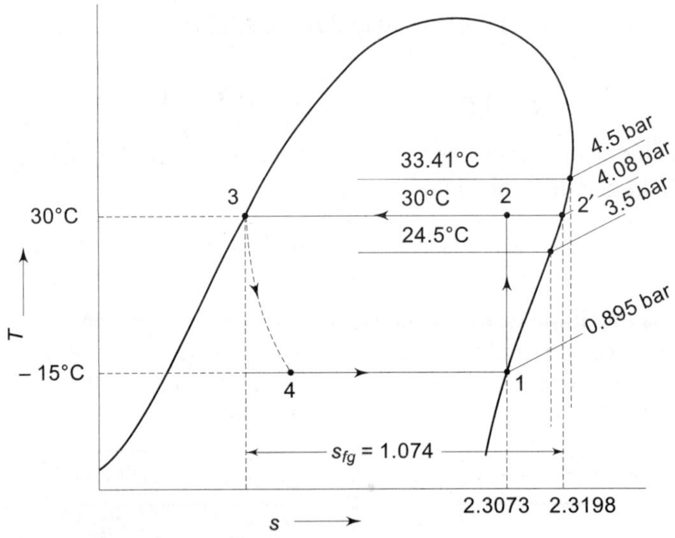

Fig. 3.11 Simple saturation cycle for isobutane for Example 3.5

From isobutane table, at 30°C

$$s_f = 1.2458 \text{ kJ/kg. K} = s_3$$
$$s_{fg} = 1.074 \text{ kJ/kg. K} = s_2' - s_3$$
$$h_f = 272.37 \text{ kJ/kg} = h_3 = h_4$$
$$h_{fg} = 325.59 \text{ kJ/kg} = h_2' - h_3$$

Hence, we have

$$s_1 = 2.3073 = 1.2458 + x_2\,(1.074) = s_2$$

$$\Rightarrow \qquad x_2 = 0.988 \text{ kJ/kg}$$

$$h_2 = h_3 + x_2\,h_{fg} = 272.37 + (0.988)\,(325.59)$$

$$= 598.12 \text{ kJ/kg}$$

$$\mathcal{E}_s = \frac{h_1 - h_4}{h_2 - h_1} = \frac{536.19 - 272.37}{598.12 - 536.19} = \frac{263.8}{61.9}$$

$$= 4.26$$

$$\frac{T_0}{T_{2s} - T_0} = \frac{258.15}{30 - (-15)} = \frac{258.15}{45}$$

$$= 5.74 < \mathcal{E}_s$$

Isobutane, therefore, belongs to Case II like R 12 and R 134a. The suction state for maximum COP lies in the superheat region.

3.5 STANDARD RATING CYCLE AND EFFECT OF OPERATING CONDITIONS

For the purpose of comparison, a case is made for the use of standard operating conditions. According to the current practice, the testing of single-stage compressors for air conditioning applications is carried out at $t_0 = 5°\text{C}$, $t_k = 40°\text{C}$ and a suction temperature of 20°C.

The following analysis shows the effect of change in operating conditions on the performance of the vapour compression cycle.

3.5.1 Effect of Evaporator Pressure

Consider a simple saturation cycle 1-2-3-4 with R 134a as the refrigerant as shown in Fig. 3.12 for operating conditions of $t_k = 40°\text{C}$ and $t_0 = 0°\text{C}$. It has been seen in Example 3.2 that for this cycle:

$$\text{Volume of suction vapour, } \dot{V}^* = 0.103 \text{ m}^3\text{/(min) (TR)}$$

$$\text{Unit power consumption, } \dot{W}^* = \frac{9.61}{15} = 0.64 \text{ kW/TR}$$

Now consider a change in the evaporator pressure corresponding to a decrease in the evaporator temperature to –10°C. The changed cycle is shown as 1′-2′-3-4′ in Fig. 3.12. Similar calculations for the changed conditions give:

Volume of suction vapour, $\dot{V}^* = 0.154 \text{ m}^3\text{/(min) (TR)}$

Unit power consumption, $\dot{W}^* = 0.87$ kW/TR

It is, therefore, seen that a drop in evaporator pressure corresponding to a drop of 10°C in saturated suction temperature increases the volume of suction vapour and hence decreases the capacity of a reciprocating compressor by

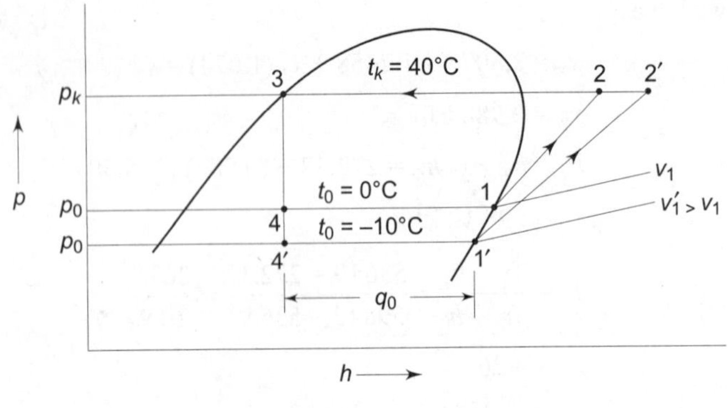

Fig. 3.12 Effect of evaporator pressure

$$\left(1 - \frac{V^*}{V^{*\prime}}\right) \times 100 = 33.3\%$$

and increases the power consumption per unit refrigeration by

$$\left(\frac{W^{*\prime}}{W^*} - 1\right) \times 100 = 36.4\%$$

It is observed that a decrease in evaporator temperature results in:

(i) Decrease in refrigerating effect from $(h_1 - h_4)$ to $(h_1' - h_4')$

(ii) Increase in the specific volume of suction vapour from v_1 to v_1'

(iii) Decrease in volumetric efficiency, due to increase in the pressure ratio, from η_v to η_v'.

(iv) Increase in compressor work from $(h_2 - h_1)$ to $(h_2' - h'_1)$ due to increase in the pressure ratio as well as change from *steeper* isentropic 1-2 to *flatter* isentropic 1′ – 2′ as discussed in Sec. 3.5.3.

Since $$\dot{Q}_0 = \dot{m} q_0 = \frac{\eta_v \dot{V}_p}{v_1} q_0$$

and $$\dot{W}^* = m^* w \tag{3.23}$$

expressions for the dependence of capacity and unit power consumption may now be written as follows:

$$\dot{Q}_0 \propto q_0 = (h_1 - h_4)$$

$$\propto \frac{1}{v_1}$$

$$\propto \eta_v$$

and $$W^* \propto m^* \propto \frac{1}{q_0} = \frac{1}{h_1 - h_4}$$

$$\propto w = h_2 - h_1$$

Note *The term volumetric efficiency* η_v *is only relevant in the case of positive displacement compressors.*

Hence, denoting the capacity and unit power consumption of the changed cycle (t_0 = – 10°C) by Q'_0 and $W^{*\prime}$, we have expression as follows in Eqs. (3.24) and (3.25).

$$\frac{Q'_0}{Q_0} = \frac{h'_1 - h_4}{h_1 - h_4} \cdot \frac{v_1}{v'_1} \cdot \frac{\eta'_v}{\eta_v} \tag{3.24}$$

$$\frac{\dot{W}^{*\prime}}{W^*} = \frac{\mathcal{E}_c}{\mathcal{E}'_c} = \frac{h_1 - h_4}{h'_1 - h_4} \cdot \frac{h'_2 - h'_1}{h'_2 - h_1} \tag{3.25}$$

Example 3.6 ***Variation in Capacity of Condensing Unit with Refrigeration Temperature***

A R 22 condensing unit is specified to give 40 TR capacity for ***air conditioning*** *under standard operating conditions of 40°C condensing and 5°C evaporating temperatures. What would be its capacity in TR for* ***food freezing*** *for which the evaporator temperature is –35°C?*

Solution Refer to Fig. 3.13.

(i) At t_0 = 5°C

$$\dot{Q}_0 = 40 \text{ TR} = 40\ (3.5167) = 140.67 \text{ kW}$$

$$q_0 = h_1 - h_4 = 407.15 - 249.7 = 157.45 \text{ kJ/kg}$$

$$\dot{m} = \frac{\dot{Q}_0}{q_0} = \frac{140.67}{157.45} = 0.8934 \text{ kg/s}$$

$$\dot{V}_p = \dot{m} v_1 = 0.8934\ (0.04035) = 0.036 \text{ m}^3\text{/s}$$

$$w = h_2 - h_1 = 26.85 \text{ kJ/kg}$$

$$\dot{W} = \dot{m} w = 24 \text{ kW}$$

$$\dot{q}_k = h_2 - h_3 = 434 - 249.7 = 184.3 \text{ kJ/kg}$$

$$\dot{Q}_k = \dot{m} q_k = 0.8934\ (184.3) = 164.7 \text{ kW}$$

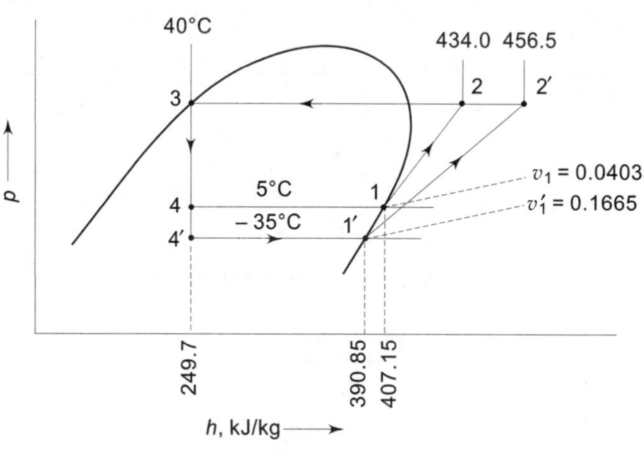

Fig. 3.13 *p-h* diagram for Example 3.6

(ii) At $t_0 = -35°C$, since $\dot{V}_p$ remains the same

$$\dot{m}' = \frac{\dot{V}_p}{v_1'} = \frac{0.036}{0.1665} = 0.2164 \text{ kg/s}$$

$$q'_0 = h'_1 - h'_4 = 390.85 - 249.7 = 141.15 \text{ kJ/kg}$$

$$\dot{Q}'_0 = \dot{m}' q'_0 = 0.2164\,(141.15) = 23.15 \text{ kW (6.6 TR)}$$

Thus, the compressor capacity would be reduced to 6.6 TR only from 40 TR. The compressor motor would become very much undersize as its power requirement would be

$$\dot{W} = \dot{m}' w' = 0.2164\,(456.5 - 390.85)$$
$$= 142.1 \text{ kW}$$

as against the existing motor of 24 kW only.

On the other hand, the condenser would be very much oversize as the heat rejected would be reduced to only

$$\dot{Q}'_k = \dot{m}'\, q'_k = 0.2164\,(456.5 - 249.7)$$
$$= 44.8 \text{ kW}$$

3.5.2 Effect of Condenser Pressure

An increase in condenser pressure, similarly results in a decrease in the refrigerating capacity and an increase in power consumption, as seen from the changed cycle 1-2′-3′-4′ for $t'_k = 45°C$ in Fig. 3.14. The decrease in refrigerating capacity is due to a decrease in the refrigerating effect and volumetric efficiency. The increase in power consumption is due to increased mass flow (due to decreased refrigerating effect) and an increase in specific work (due to increased pressure ratio), although the isentropic line remains unchanged. Accordingly, one can write for the ratios.

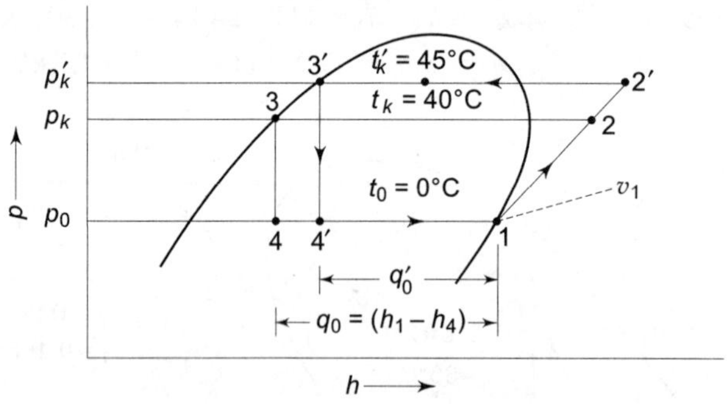

Fig. 3.14 Effect of condenser pressure

$$\frac{Q'_0}{Q_0} = \frac{h_1 - h'_4}{h_1 - h_4} \cdot \frac{\eta'_v}{\eta_v} \tag{3.26}$$

$$\frac{W^{*\prime}}{W^*} = \frac{\mathcal{E}'_c}{\mathcal{E}_c} = \frac{h_1 - h_4}{h_1 - h'_4} \cdot \frac{h'_2 - h_1}{h_2 - h_1} \tag{3.27}$$

Note *It is obvious that COP decreases both with decreasing evaporator and increasing condenser pressures.*

It may, however, be noted that the effect of increase in condenser pressure is not as severe, on the refrigerating capacity and power consumption per ton of refrigeration, as that of the decrease in evaporator pressure.

3.5.3 Effect of Suction Vapour Superheat

Superheating of the suction vapour is advisable in practice because it ensures complete vaporization of the liquid in the evaporator before it enters the compressor. Also, in most refrigeration and air-conditioning systems, the degree of superheat serves as a means of actuating and modulating the capacity of the expansion valve. It has also been seen that for some refrigerants such as R 134a, Isobutane, etc., maximum COP is obtained with superheating of the suction vapour.

It can be seen from Fig. 3.15, that the effect of superheating of the vapour from $t_1 = t_0$ to t'_1 is as follows:

(i) Increase in specific volume of suction vapour from v_1 to v'_1
(ii) Increase in refrigerating effect from $(h_1 - h_4)$ to $(h'_1 - h_4)$
(iii) Increase in specific work from $(h_2 - h_1)$ to $(h'_2 - h'_1)$.

It is to be noted that $(h'_2 - h'_1)$ is greater than $(h_2 - h_1)$. This is because, although the pressure ratio is the same for both lines, the initial temperature t'_1 is greater than t_1 and the work, given for example by the ideal gas expression,

$$w = \frac{\gamma RT_1}{\gamma - 1}\left[(p_2/p_1)^{\frac{\gamma-1}{\gamma}} - 1\right] = f\left(T_1, \frac{p_2}{p_1}, \gamma\right)$$

is a function of initial temperature T_1, pressure ratio p_2/p_1 and exponent γ. Thus, we see that it increases with the initial temperature T_1. That is why isentropic lines on the *p-h* diagram become flatter at higher temperatures as they move away more and more from the saturated vapour curve.

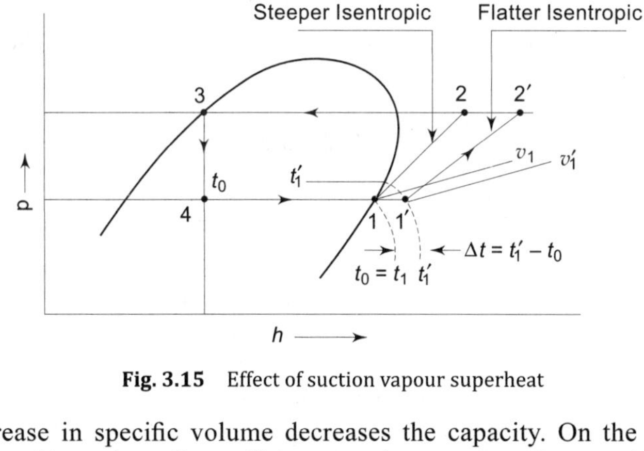

Fig. 3.15 Effect of suction vapour superheat

An increase in specific volume decreases the capacity. On the contrary, an increase in refrigerating effect will increase the capacity. The net effect of superheating is to theoretically reduce the capacity in ammonia systems and to increase it in R 12 systems. The ratio of the capacities with superheating Q'_0 and without superheating Q_0, can be written as

$$\frac{Q_0'}{Q_0} = \frac{h_1' - h_4}{h_1 - h_4} \cdot \frac{v_1}{v_1'} \tag{3.28}$$

Similarly, for the work done per unit of refrigeration, it may be seen that there are two contradictory influences, viz., an increase in the refrigerating effect decreasing the mass flow requirement and hence work and an increase in the specific work itself due to an increase in the suction temperature. The resulting work per unit refrigeration may, therefore, increase or decrease depending on the refrigerant and operating temperatures.

The effect on work or power consumption per unit refrigeration is given by the ratio given in Eq. (3.29).

$$\frac{W^{*\prime}}{W^*} = \frac{h_1 - h_4}{h_1' - h_4} \cdot \frac{h_2' - h_1'}{h_2 - h_1} \tag{3.29}$$

The COP of the cycle with superheat is given by

$$\begin{aligned} \mathcal{E}_c' &= \frac{h_1' - h_4}{h_2' - h_1'} \\ &= \frac{(h_1 - h_4) + (h_1' - h_1)}{(h_2 - h_1) + \left[(h_2' - h_1') - (h_2 - h_1)\right]} \end{aligned} \tag{3.30}$$

As both the numerator and the denominator increase, the numerical value of COP may increase or decrease or remain the same. It has been shown that in R 12 systems, superheating increases the COP whereas in R 22 and ammonia systems, it decreases it. In general, however, the effect of slight superheat on the volumetric efficiency of the reciprocating compressor and the COP is beneficial as it ensures complete vaporization of liquid refrigerant droplets in suspension in the suction vapour.

It may be noted that superheating outside the evaporator or cold space results in a loss.

3.5.4 Effect of Liquid Subcooling

It is possible to reduce the temperature of the liquid refrigerant to within a few degrees of the temperature of the water entering the condenser in some condenser designs by installing a *subcooler* between the condenser and the expansion valve. The effect of subcooling of the liquid from $t_3 = t_k$ to t_3' is shown in Fig. 3.16. It will be seen that subcooling reduces flashing of the liquid during expansion and increases the refrigerating effect. Consequently, the piston displacement and horsepower per ton are reduced for all refrigerants. The per cent gain is less pronounced in the case of ammonia because of its larger latent heat of vaporization as compared to liquid specific heat.

Normally, cooling water first passes through the subcooler and then through the condenser. Thus, the coolest water comes in contact with the liquid being subcooled. But this results in a warmer water entering the condenser and hence a higher condensing temperature and pressure. Thus, the advantage of subcooling is offset by the increased work of compression.

This can be avoided by installing parallel cooling water inlets to the subcooler and condenser. In that case, however, the degree of subcooling will be small and the added cost of the subcooler and pump work may not be worthwhile.

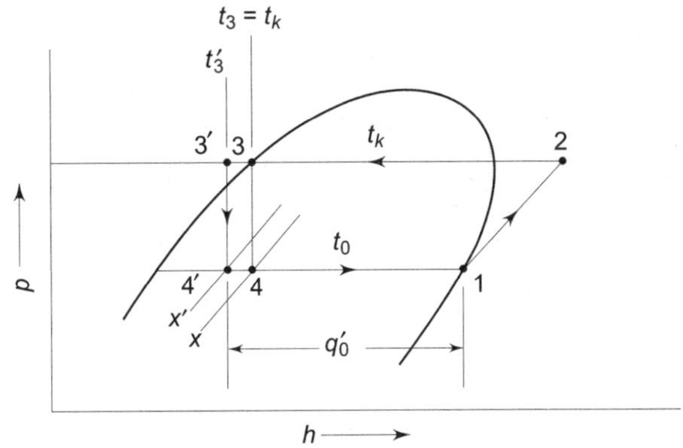

Fig 3.16 Effect of liquid subcooling

It may be more desirable to use the cooling water effectively in the condenser itself to keep the condensing temperature as near to the temperature of the cooling water inlet as possible.

In general, the functions of the condenser as well as the subcooler can be combined in the condenser itself by slightly oversizing the condenser.

3.5.5 Using Liquid–Vapour Regenerative Heat Exchanger

If we combine superheating of vapour with liquid subcooling, we have a *liquid-vapour regenerative heat exchanger.*

A liquid–vapour heat exchanger may be installed as shown in Fig. 3.17. In this, the refrigerant vapour from the evaporator is superheated in the regenerative heat exchanger with consequent subcooling of the liquid from the condenser. The effect on the thermodynamic cycle is shown in Fig. 3.18. Since the mass flow rate of the liquid and vapour is the same, we have from energy balance of the heat exchanger

$$q_N = h'_1 - h_1 = h_3 - h'_3 \tag{3.31}$$

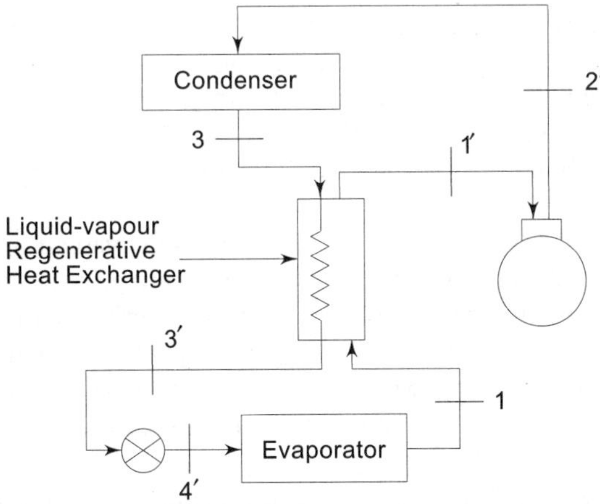

Fig. 3.17 Vapour compression system with liquid-vapour regenerative heat exchanger

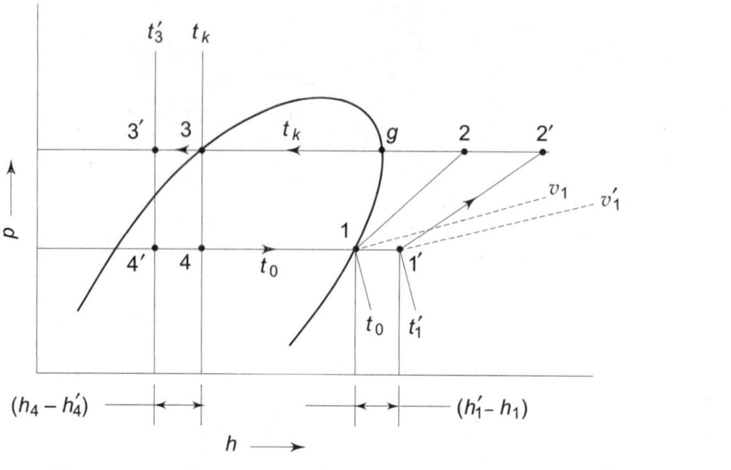

Fig. 3.18 Vapour compression cycle with liquid-vapour regenerative heat exchanger

The degree of superheat $(t'_1 - t_0)$ and the degree of subcooling $(t_k - t'_3)$ need not be the same as the specific heats of the vapour and liquid phases are different. The effect on the capacity, power requirement per unit refrigeration and COP is expressed as follows:

$$\frac{\dot{Q}'_0}{\dot{Q}_0} = \frac{h_1 - h'_4}{h_1 - h_4} \cdot \frac{v_1}{v'_1} \tag{3.32}$$

$$\frac{W^{*\prime}}{W^*} = \frac{h_1 - h_4}{h_1 - h'_4} \cdot \frac{h'_2 - h'_1}{h_2 - h_1} \tag{3.33}$$

$$\mathcal{E}'_c = \frac{(h_1 - h_4) + (h'_1 - h_1)}{(h_2 - h_1) + \left[(h'_2 - h'_1) - (h_2 - h_1)\right]} \tag{3.34}$$

In all the above expressions, both numerators and denominators increase. The net effect, whether positive, negative or zero, depends on the refrigerant used and the operating conditions. In practice, the suction volume per ton and horsepower per ton are reduced for R 12 and R 134a. Calculations show a slight increase in the suction volume and horsepower per ton for R 22 and ammonia. Experiments show, however, that the volumetric efficiency of most reciprocating compressors improves with superheat.

In particular, it must be stated that superheating of the vapour in a liquid-vapour regenerative heat exchanger is preferable to superheating in the evaporator itself, as the increased refrigerating effect $(h'_1 - h_1)$ due to superheating taking place from temperature t_0 to t'_1 is transferred as $(h_4 - h'_4)$ at temperature t_0, which lowers mean refrigeration temperature. Thus, we obtain the same increase in refrigerating effect at a lower temperature by the use of a liquid-vapour regenerative heat exchanger.

Example 3.7 ***Using Liquid-Vapour Regenerative Heat Exchanger in R 12 Systems***

(a) A R 12 simple saturation cycle operates at temperatures of 35°C and –15°C for the condenser and evaporator respectively. Determine the COP and HP/TR of the system.

(b) If a liquid-vapour heat exchanger is installed in the system, with the temperature of the vapour leaving the heat exchanger at 15°C, what will be the change in the COP and HP/TR?

Solution (a) *Simple saturation cycle* Refer to Fig. 3.18. We have from tables of properties of R 12

				Superheated			
				20 K		*40* K	
t	h_f	h_g	s_g	h	s	h	s
°C	kJ/kg	kJ/kg	kJ/(kg.K)	kJ/kg	kJ/(kg.K)	kJ/kg	kJ/(kg.K)
35	69.5	201.5	0.6839	216.4	0.731	231.0	0.7741
–15		181.0	0.7052	193.2	0.751	205.7	0.7942

Hence
$$h_3 = h_4 = 69.5 \text{ kJ/kg}$$
$$h_1 = 181 \text{ kJ/kg}$$
$$s_1 = 0.7052 \text{ kJ/(kg)} \cdot \text{(K)} = s_2$$

By interpolation for the degree of superheat at discharge
$$\Delta t = \frac{0.7052 - 0.6839}{0.731 - 0.6839}(20) = 9.04°\text{C}$$

Hence
$$h_2 = 201.5 + \frac{10.2}{20}(216.4 - 201.5) = 208.2 \text{ kJ/kg}$$
$$q_0 = 181 - 69.5 = 111.5 \text{ kJ/kg}$$
$$w = 208.2 - 181 = 27.2 \text{ kJ/kg}$$
$$\text{COP} = \frac{111.5}{27.2} = 4.09$$
$$\text{HP/TR} = \frac{4.761}{4.09} = 1.16$$

(b) *Liquid-vapour heat exchanger cycle*

Degree of superheat at suction
$$t_1' - t_1 = 15 - (-15) = 30°\text{C}$$

By interpolation for superheated vapour
$$h_1' = 193.2 + \frac{10}{20}(205.7 - 193.2) = 199.45 \text{ kJ/kg}$$
$$s_1' = 0.751 + \frac{10}{20}(0.7942 - 0.751) - 0.7726 \text{ kJ/(kg)} \cdot \text{(K)} = s'_2$$

Now, $h_1' - h_1 = 199.45 - 181 = 18.45 = h_3 - h'_3$

Hence, $h_3' = 69.5 - 18.45 = 51.05$ kJ/kg

By interpolation for the degree of superheat
$$\Delta t = 20 + \frac{0.7726 - 0.731}{0.7741 - 0.731}(20) = 20 + 19.3 = 39.3°\text{C}$$

Hence $h'_2 = 216.4 + \frac{19.2}{20}(231 - 216.4) = 230.5$ kJ/kg

$$q'_0 = h_1 - h'_3 = 181 - 51.05 = 129.95 \text{ kJ/kg}$$

$$w = 230.4 - 199.45 = 30.95$$

$$\text{COP} = \frac{129.95}{30.95} = 4.199$$

Increase in $\text{COP} = \frac{4.199 - 4.09}{4.09}(100) = 2.56\%$

$$\text{HP/TR} = \frac{4.761}{4.199} = 1.134$$

$$\text{Decrease in HP/TR} = \frac{1.16 - 1.134}{1.2}(100) = 2.5\%$$

Note *Theoretically, the increase in COP is not very large for R 12. And for R 22 there is, in fact, a decrease in COP. However, superheat improves the performance by ensuring complete vaporization of liquid.*

In refrigerators and air conditioners, the capillary tube is joined to the suction line, thus forming a regenerative heat exchanger.

Example 3.8 ***Using Liquid-Vapour Regenerative Heat Exchanger in R 134a Systems***

(a) *An R 134a simple saturation cycle refrigerator operates at 40°C condenser and –16°C evaporator temperatures. Determine COP and HP/TR.*

(b) *If a liquid–vapour regenerative heat exchanger is installed in the system, with the suction vapour at 15°C, what will be the effect on COP and HP/TR?*

Solution Refer Fig. 3.18

At $t_k = 40°\text{C}$ — $p_k = 1.0166$ MPa

$h_3 = 256.41$ kJ/kg — $h_g = 419.43$ kJ/kg

$C_p = 1.145$ kJ/kg K — $s_g = 1.711$ kJ/kg

At $t_0 = -16°\text{C}$ — $p_0 = 0.15728$ MPa

$h_1 = 389.02$ kJ/kg — $s_1 = 1.7379$ kJ/kg K

$C_p = 0.831$ kJ/kg K

(a) *Simple saturation cycle* Calculations for isentropic compression give for discharge temperature

$$s_2 = s_g + C_p \ln \frac{T_2}{T_g} = s_1$$

$$1.7111 + 1.145 \ln \frac{T_2}{313} = 1.7379$$

Discharge temperature and enthalpy

$$T_2 = 320.5 \text{ K } (47.4°\text{C})$$

$$h_2 = h_g + C_p(t_2 - t_g)$$
$$= 419.43 + 1.145(47.4 - 40) = 427.92 \text{ kJ/kg}$$

Refrigerating effect and specific work

$$q_0 = h_1 - h_3 = 389.02 - 256.41 = 132.61 \text{ kJ/kg}$$
$$w = h_2 - h_1 = 427.92 - 389.02 = 38.9 \text{ kJ/kg}$$

COP and HP/TR

$$\mathcal{E}_c = \frac{q_0}{w} = \frac{132.61}{38.9} = 3.41$$

$$\text{HP/TR} = \frac{4.761}{\mathcal{E}} = \frac{4.761}{3.41} = 1.396$$

(b) *Liquid–vapour regenerative heat exchanger cycle* Enthalpy and entropy of suction vapour

$$h_1' = h_1 + C_p(t_1' - t_1)$$
$$= 389.02 + 0.831(31) = 414.78 \text{ kJ/kg}$$

$$s_1' = s_1 + C_p \ln \frac{T_1'}{T_1}$$

$$= 1.7379 + 0.831 \ln \frac{273+15}{273-16} = 1.8325 \text{ kJ/kg K}$$

Insentropic compression gives

$$s_2' = s_1' = 1.8325 = s_g + C_p \ln \frac{T_2'}{T_g} = 1.7111 + 1.145 \ln \frac{T_2'}{313}$$

Discharge temperature and enthalpy

$$\Rightarrow \quad T_2' = 348 \text{ K } (75°\text{C})$$
$$h_2' = h_g + C_p(t_2' - t_g)$$
$$= 419.43 + 1.145(75 - 40) = 459.5 \text{ kJ/kg}$$

Energy balance of regenerative heat exchanger gives

$$q = h_1' - h_1 = 414.78 - 389.02 = 25.76 \text{ kJ/kg} = h_3 - h_3'$$

Refregerating effect and specific work

$$q_0' = q_0 + q = 132.61 + 25.76 = 158.37 \text{ kJ/kg}$$
$$w' = h_2' - h_1' = 459.5 - 414.78 = 44.72 \text{ kJ/kg}$$

COP and HP/TR

$$\mathcal{E}_c = \frac{q_0'}{w'} = \frac{158.37}{44.72} = 3.54$$

$$\text{HP/TR} = \frac{4.761}{\mathcal{E}_c} = \frac{4.761}{3.54} = 1.345$$

There is an increase of 3.8% in COP, and decrease in HP/TR by the same percentage nearly.

3.6 ACTUAL VAPOUR COMPRESSION CYCLE

Due to the flow of the refrigerant through the condenser, evaporator and piping, there will be drops in pressure. In addition, there will be heat losses or gains depending on the temperature difference between the refrigerant and the surroundings. Further, compression will be polytropic with friction and heat transfer instead of isentropic. The actual vapour compression cycle may have some or all of the items of departure from the simple saturation cycle as enumerated below and shown on the *p-h* and *T-s* diagrams in Figs. 3.19 and 3.20.

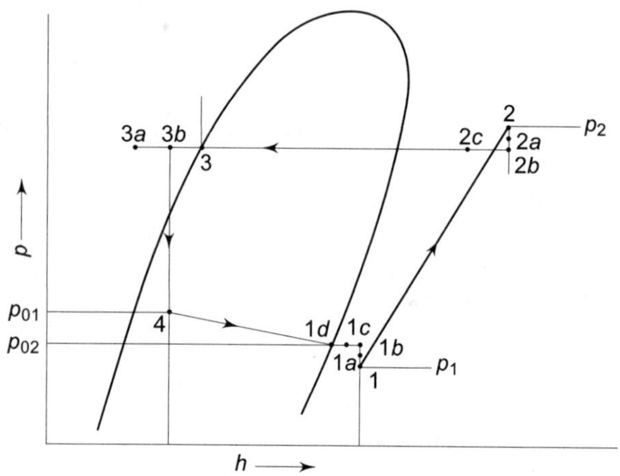

Fig. 3.19 Actual vapour compression cycle on *p-h* diagram

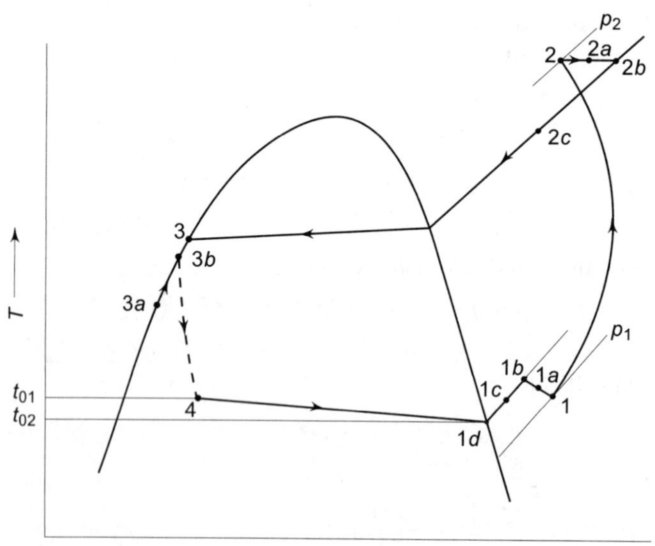

Fig. 3.20 Actual vapour compression cycle on *T-s* diagram

(i) Superheating of the vapour in the evaporator, 1d–1c.

(ii) Heat gain and superheating of the vapour in the suction line, 1c–1b.

(iii) Pressure drop in the suction line, 1b-1a.
(iv) Pressure drop due to *wire drawing* at the compressor-suction valve, 1a–1.
(v) Polytropic compression 1–2 with friction and heat transfer to the surroundings instead of isentropic compression.
(vi) Pressure drop at the compressor-discharge valve, 2–2a.
(vii) Pressure drop in the delivery line, 2a–2b.
(viii) Heat loss and desuperheating of the vapour in the delivery line, 2b–2c.
(ix) Pressure drop in the condenser, 2b–3.
(x) Subcooling of the liquid in the condenser or subcooler, 3–3a.
(xi) Heat gain in the liquid line, 3a–3b. The lines 3–3a and 3a–3b are along the saturated liquid line on the *T-s* diagram as the constant pressure lines in liquid region run close to it.
(xii) Pressure drop in the evaporator, 4–1d.

It may be noted that the pressure drop in the evaporator is large. This is due to the cumulative effect of two factors. Firstly, the pressure drops in the evaporator due to friction. This is called the *frictional pressure drop*. Secondly, as evaporation proceeds, the volume increases, and hence the velocity must also increase. The increase in kinetic energy comes from a decrease in enthalpy and, therefore, from a further pressure drop. This pressure drop is called the *momentum pressure drop*.

In the condenser, the pressure drop is not significant, since the frictional pressure drop is positive and the momentum pressure drop is negative.

It is, therefore, to be noted that pressure drop in the condenser is not very critical. But, it is very much so in the design of evaporators as it would increase power consumption greatly or alternatively reduce refrigerating capacity (see Sec. 9.5). The condensers, in any case, are usually oversized to keep the condenser pressure low and also to let condensers function as receivers. However, in the case of evaporators, both undersizing and oversizing are harmful.

It may also be noted that due to the pressure drop in the evaporator from p_{01} to p_{02}, the temperature in the evaporator does not remain constant. It correspondingly changes from t_{01} to t_{02}.

Further, due to various pressure drops, the capacity of the plant is decreased and the unit power consumption (per unit of refrigeration) is increased. Correspondingly, the COP of the actual cycle is reduced.

The compressor cylinder is hotter than the surroundings. Thus it loses heat to the refrigerant. The cooling of the compressor reduces the work of compression. Freon compressor are, therefore, provided with parabolic fins built into the body of the compressor cylinders on the outside during casting. The compressors are thus cooled by air by natural convection. Ammonia and even R 22 compressors are *water-jacketed* for the same purpose. Friction, however, increases the work of compression. The actual work of compression can be determined by knowing the initial and the final states and finding out n, the polytropic index of the compression curve. The heat rejected in the polytropic compression process can be obtained by applying the SSSF energy equation

$$q = (h_2 - h_1) + w$$

$$\Rightarrow \quad -w = (h_2 - h_1) - q \tag{3.35}$$

Equation (3.35) represents the energy balance of the compressor. It means:

Work of compression = Increase in enthalpy of gas + Heat lost in cooling

Example 3.9 *The following data were obtained from a test on a twin cylinder, single acting 15 cm × 20 cm, 320 rpm compressor ammonia refrigeration plant.*

Temperatures of refrigerant:	
After expansion valve, entering brine cooler	*–25°C*
Leaving brine cooler	*–18°C*
Entering compressor	*–8°C*
Leaving compressor	*140°C*
Entering condenser	*130°C*
Leaving condenser	*30°C*
Entering expansion valve	*32°C*
Pressures of refrigerant:	
Compressor discharge and condenser	*13.5 bar*
Compressor suction	*1.324 bar*
Brine circulation rate	*102 kg/min*
Temperature drop of brine in cooler	*7°C*
Specific heat of brine	*3.14 kJ/(kg) (K)*
Input to motor	*18.8 kW*
Motor efficiency at this load	*92%*
Compressor jacket cooling water	*5 kg/min*
Temperature rise of jacket water	*8.9°C*

Show the thermodynamic states at various points on p-h and T-s diagrams and calculate:

(a) *Refrigerating capacity in TR assuming 2 per cent loss of useful refrigeration by heat gain from room in brine cooler.*
(b) *Ammonia circulated.*
(c) *Compressor IHP and mechanical efficiency.*
(d) *Compressor volumetric efficiency.*
(e) *COP of the cycle.*

Solution Referring to Fig. 3.19, the various state points are given below:

State 3b: $t = 32°C$	$p = 13.5$ bar	$h = 351.5$ kJ/kg (at 32°C)
State 4: $t = -25°C$	$p = 1.516$ bar	$h = 351.5$ kJ/kg (saturation)
State 1d: $t = -28°C$ (saturation)		$p = 1.324$ bar
State 1c: $t = -18°C$	$p = 1.324$ bar	$h = 1451.3$ kJ/kg
State 1b: $t = -8°C$	$p = 1.324$ bar	$h = 1473.6$
State 1: $t = -8°C$		$h = 1473.6$ kJ/kg ($= h_{1b}$)
States 2, 2a and 2b:		
$t = 140°C$		$h = 1777$ kJ/kg
State 2c: $t = 130°C$	$p = 13.5$ bar	$h = 1751$ kJ/kg
State 3: $t = 35°C$ (saturation at 13.5 bar)		$p = 13.5$ bar
State 3a: $t = 30°C$	$p = 13.5$ bar	$h = 341.8$ kJ/kg

(a) Refrigerating capacity (from brine)

$$\dot{Q}_0 = 102 \times 3.14 \times 7 \times 1.02 = 2287 \text{ kJ/min}$$

$$= \frac{2287}{211} = 10.84 \text{ TR}$$

(b) Refrigerating effect

$$q_0 = h_{1c} - h_4 = 1451.3 - 351.5 = 1099.8 \text{ kJ/kg}$$

Ammonia circulated

$$\dot{m} = \frac{2287}{1099.9} = 2.08 \text{ kg/min}$$

(c) Note: From 2 to 2b, it is an isenthalpic process. It has been assumed here that enthalpy is a function of temperature only and hence $t_{2b} = t_2$.

Enthalpy increase during compression

$$h_2 - h_1 = 1777 - 1473.6 = 303.4 \text{ kJ/kg}$$

Total enthalpy increase of ammonia

$$H_2 - H_1 = \dot{m}\ (h_2 - h_1) = 2.08\ (303.4) = 631.1 \text{ kJ/min}$$

$$\text{Heat to jacket water, } \dot{Q}_j = 5 \times 4.1868 \times 8.9 = 186.3 \text{ kJ/min}$$

$$\begin{aligned}\text{Compressor work, } |\dot{W}| &= (H_2 - H_1) + |\dot{Q}_j| \\ &= 631.1 + 186.3 = 817.4 \text{ kJ/min} \\ &= 13.62 \text{ kW}\end{aligned}$$

$$\text{Compressor IHP} = \frac{13.62 \times 10^3}{746} = 18.26$$

$$\begin{aligned}\text{Compressor input} &= \text{Power consumption of motor} \times \text{motor efficiency} \\ &= 18.8 \times 0.92 = 17.3 \text{ kW}\end{aligned}$$

$$\text{Compressor BHP} = \frac{17.3 \times 10^3}{746} = 23.19$$

$$\text{Mechanical efficiency, } \eta_m = \frac{\text{IHP}}{\text{BHP}} = 0.787$$

(d) Gas enters compressor at 1b

$$v_{1b} = v_g \cdot \frac{273 + (-8)}{273 + (-28)}$$

But $v_g = 0.88$ m^3/kg (saturated vapour at –28°C)

Hence

$$v_{1b} = 0.88 \left(\frac{265}{245}\right) = 0.952 \text{ m}^3\text{/kg}$$ (It can also be obtained directly from the *p-h* diagram)

Actual volume flow rate of refrigerant

$$\dot{V} = \dot{m}v_{1b} = 2.08 \times 0.952 = 1.99 \text{ m}^3\text{/min}$$

Compressor piston displacement

$$\dot{V}_p = \frac{\pi}{4} D^2 LN\,(2) = \frac{\pi}{4}\,(0.15)^2\,(0.2)\,(320)\,(2) = 2.62 \text{ m}^3\text{/min}$$

Volumetric efficiency of the compressor

$$\eta_v = \frac{\dot{V}}{\dot{V}_p} = \frac{1.99}{2.262} = 0.882 \text{ (or 88.2\%)}$$

(e) $\text{COP} = \dfrac{\dot{Q}_0}{\dot{W}} = \dfrac{2287}{60 \times 13.62} = 2.797$

3.7 STANDARD RATING CYCLE FOR DOMESTIC REFRIGERATORS[3]

Figure 3.21 shows a standard *ten-state-points cycle*. This cycle approximates the design and operating conditions of a domestic refrigerator. The ten states points correspond to the following conditions:

(i) Vapour in cylinder before compression begins at $p_1 = p_{10} - \Delta p_s$, $t_1 = t_{10}$ where Δp_s is pressure drop at suction valve.

(ii) Vapour in cylinder after compression ends, 2*s* after isentropic, and 2*n* after actual polytropic compression at $p_2 = p_k + \Delta p_d$, where Δp_d is pressure drop at delivery valve.

(iii) Vapour at compressor shell outlet/condenser inlet at $p_3 = p_k$, $t_3 = t_2$.

(iv) Saturated vapour state in condenser at $t_4 = t_k = 55°\text{C}$ condensing temperature.

(v) Saturated liquid state in condenser at $t_5 = t_k = 55°\text{C}$.

(vi) Subcooled liquid leaving condenser at $t_6 = 43°\text{C}$.

(vii) Subcooled liquid leaving regenerator/entering capillary at $t_7 = 32°\text{C}$.

(viii) Liquid-vapour mixture exiting capillary/entering evaporator at $t_8 = t_0 = -25°\text{C}$ evaporator temperature.

(ix) Saturated vapour leaving evaporator at $p_9 = p_8 - \Delta p_E$ where Δp_E is pressure drop in evaporator.

(x) Superheated vapour leaving regenerator/entering compressor at $t_{10} = 32°\text{C}$.

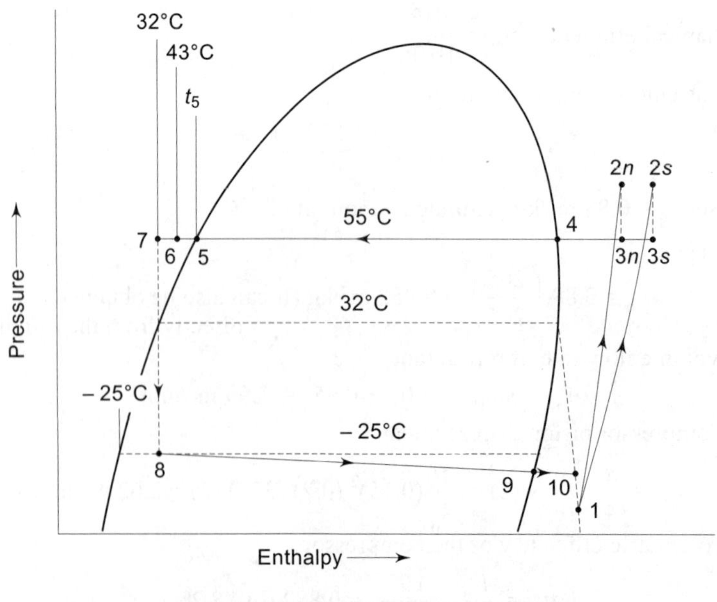

Fig. 3.21 Standard ten state point cycle

Note that this cycle corresponds to an ambient temperature of 43°C. Subcooling of liquid from $t_6 = 43°C$ to $t_7 = 32°C$ takes place in the regenerator. Superheating of vapour to $t_{10} = 32°C$ takes place in the suction line, regenerator and the compressor. For R 12, the following pressure drops are assumed:

$$\Delta p_E = 0.1 \text{ bar}, \Delta p_s = 0.1 \text{ bar}, \Delta p_d = 0.25 \text{ bar}$$

Example 3.10 ***Thermodynamic Calculations for 165 L R 12 Domestic Refrigerator***

An 89 W refrigerating capacity 165 L R 12 domestic refrigerator operates on the standard cycle (Fig. 3.21). Determine:

(i) *Isentropic discharge temperature.*
(ii) *Actual discharge temperature if experimental value of polytropic index n is found to be 1.032.*
(iii) *Motor watts (isentropic).*
(iv) *Heat rejected in the condenser.*
(v) *Volumetric efficiency of the compressor if its cylinder volume is 4.33 cc, and rpm of its motor is 2800.*

Solution The pressures are:

$$p_k = p_3 = p_4 = p_5 = p_6 = p_7 = (p^{sat})_{55°C} = 13.61 \text{ bar}$$
$$p_0 = p_8 = (p^{sat})_{-25°C} = 1.24 \text{ bar}$$
$$p_9 = p_{10} = p_8 - \Delta p_E = 1.24 - 0.1 = 1.14 \text{ bar}$$
$$t_9 = (t^{sat})_{1.14 \text{ bar}} = -27°C$$
$$p_1 = p_{10} - \Delta p_s = 1.14 - 0.1 = 1.04 \text{ bar}$$
$$t_1 = t_{10} = 32°C$$
$$p_2 = p_3 + \Delta p_d = 13.61 + 0.25 = 13.86 \text{ bar}$$

From the table of properties of R 12, we have:

$$h_5 = (h_f)_{55°C} = 90.3 \text{ kJ/kg}$$
$$h_6 = (h_f)_{43°C} = 77.65 \text{ kJ/kg}$$
$$h_7 = (h_f)_{32°C} = 66.56 \text{ kJ/kg} = h_8$$
$$h_9 = (h_g)_{-27°C} = 175.6 \text{ kJ/kg}$$
$$h_{10} = 211 \text{ kJ/kg} = h_1$$
$$s_1 = 0.85 \text{ kJ/kg. K}$$
$$v_{10} = 0.18 \text{ m}^3\text{/kg}$$

(i) Isentropic discharge temperature,

$s_{2s} = s_1 = 0.85$ kJ/kg. K at $p_2 = 13.86$ bar gives

$$t_{2s} = 138°C$$

$$\Rightarrow \quad h_{2s} = 267 \text{ kJ/kg} = h_{3s}$$

(ii) Actual discharge temperature,

$$T_{2n} = T_1 \left(\frac{p_2}{p_1}\right)^{\frac{\gamma-1}{\gamma}} = 305.15 \left(\frac{13.86}{1.04}\right)^{\frac{0.032}{1.032}} = 330.7 \text{ K}$$

$$t_{2n} = 57.5°\text{C}$$

$$\Rightarrow \quad h_{2n} = 234 \text{ kJ/kg} = h_{3n}$$

(iii) Refrigerating effect, mass flow rate and motor watts

$$q_0 = h_9 - h_8 = 175.6 - 66.56 = 109.1 \text{ kJ/kg}$$

$$\dot{m} = \frac{\dot{Q}_0}{q_0} = \frac{89 \times 10^{-3}}{109.1} = 0.817 \times 10^{-3} \text{ kg/s}$$

$$\dot{W} = \dot{m}\,w = \dot{m}\,(h_{2s} - h_1) = 0.817 \times 10^{-3}\,(267 - 211)$$

$$= 0.046 \text{ kW (46 W)}$$

Note *The actual power of motor in the hermetic unit of 165 L refrigerator is 110 W. This takes care of compressor and motor efficiencies.*

(iv) Condenser heat rejected

$$\dot{Q}_k = \dot{m}\,q_k = \dot{m}\,(h_{3n} - h_6) = 0.817 \times 10^{-3}\,(234 - 77.65)$$

$$= 0.128 \text{ kW (128 W)}$$

(v) Suction vapour volume

$$V_s = \frac{60\dot{m}}{N}\,v_{10} = \frac{0.817 \times 10^{-3}\,(0.18)}{2800/60}$$

$$= 3.1513 \times 10^{-6} \text{ m}^3\text{/rev} = 3.1513 \text{ cc/rev}$$

Volumetric efficiency of compressor

$$\eta_v = \frac{V_s}{V_p} = \frac{3.1513}{4.33} = 0.728\ (72.8\%)$$

Note *Results for alternatives to CFC 12 in refrigerators are discussed in Chap. 4.*

3.7.1 Pull-Down Characteristic and Ice-Making Time of Refrigerators[3]

According to the Indian Standards Institution (ISI) specification, for the testing of refrigerators, the environment temperature is maintained at 43°C. The *no-load test* is performed by adjusting the thermostat position corresponding to an average cabinet temperature of 7°C. The purpose of this test is to find the *pull-down period*, the no-load power consumption, and the percentage running time as per ISI code.

The pull-down period is the time required to reach the specified temperatures inside the cabinent after switching on the unit. The temperature at the geometric centre of the evaporator shall not exceed –5°C at pull-down.

Another test performed on the refrigerators is the *ice-making time* test. For this test, after stable operating conditions are obtained following pull-down period, a quantity of 0.5 kg of water at 30 ± 1°C in two standard ice trays is kept in the freezer, and the time for each 'ON' and 'OFF' is noted. After specified time of 3 hours, the ice trays are examined for the formation of ice, and the *running time* of the unit during the 3 hours of operation is obtained, say, from the data acquisition system monitoring 'ON-OFF' cycle. This running time of the unit is the ice-making time. If ice is not formed during 3 hours of operation, the system operation is to be continued.

For the case of R 12 refrigerator of Example 3.8, the pull down period is 90 minutes, and the ice making time is 125 minutes.

Purpose of these tests is to compare the values for different refrigerators, and for different refrigerants. Increase in pull-down period, running time, and ice making time implies inadequate capacity.

3.8 HEAT PUMP

In mild winter climate, a heat pump output of 2 to 3 times the compressor power input may be realized. If a condensing temperature of 46°C is assumed, while the room is maintained at 24°C, for the heat-pump cycle the following values of COP for heating versus evaporator temperature may be obtained:

$$4.5°\text{C evaporator, } COP_H = 6.5$$

$$-18°\text{C evaporator, } COP_H = 4$$

Note that evaporator temperature depends on outdoor air temperature.

These values are theoretical. These do not include losses due to pressure drops, *coil frosting* and *defrosting*, etc. Inspite of the lowering of COP with the outdoor air temperature and hence the evaporator temperature specially at the time of maximum heating load, these *air-to-air heat pumps* are finding increasing applications in mild climates.

However, despite overall actual COP ranging from 1.5 to 3, the operating cost of air-to-air heat pumps exceeds the cost with conventional *fossil-fuel-equipment*. But, *water-to-air* and *water-to-water* heat pumps which use relatively higher temperatures can compete with *fossil-fuel-equipment*. Warm water is thus a better heat source than outside air. In *solar-assisted heat pumps*, solar panels can provide warm water in the 20–35°C temperature range even in extreme cold climates.

Example 3.11 *A simple Refrigerant 134a (Tetrafluoroethane) heat pump for space heating operates between temperature limits of 15 and 50°C. The heat required to be pumped is 100 MJ/h.*

Determine:

(a) the dryness fraction of refrigerant entering the evaporator,
(b) the discharge temperature assuming the specific heat of vapour as 0.996 kJ/(kg) (K),
(c) the theoretical piston displacement of the compressor,
(d) the theoretical horsepower of the compressor, and
(e) the COP.

The specific volume of Refrigerant 134a saturated vapour at 15°C is 0.04185 m³/kg. The other relevant properties of R 134a are given below.

Saturation temperature	*Pressure*	*Specific enthalpy*		*Specific entropy*	
		h_f	h_g	s_f	s_g
°C	MN/m²	kJ/kg		kJ/(kg) (K)	
15	0.4887	220.48	417.1	1.0725	1.72
50	1.318	271.62	423.4	1.24	1.7072

Solution Refer to Fig. 3.7.

(a) Dryness fraction of refrigerant entering evaporator

$$x = \frac{h_3 - h_f}{h_1 - h_f} = \frac{271.62 - 220.48}{407.1 - 220.48} = 0.274$$

(b) Discharge temperature

$$s_1 = s_2 = s'_2 + C_p \ln \frac{T_2}{T'_2}$$

$$1.72 = 1.7072 + 1.246 \ln \frac{T_2}{273 + 50}$$

$$\Rightarrow \qquad T_2 = 327.15 \text{ K } (53°\text{C})$$

Enthalpy at discharge

$$h_2 = h'_2 + C_p (t_2 - t'_2)$$
$$= 423.4 + 1.246 (53 - 50) = 427.2 \text{ kJ/kg}$$

(c) Mass flow rate of refrigerant

$$\dot{m} = \frac{\dot{Q}_k}{q_k} = \frac{\dot{Q}_k}{h_2 - h_3} = \frac{100,000/3600}{427.2 - 271.62} = 0.179 \text{ kg/s}$$

Theoretical piston displacement of compressor

$$\dot{V} = \dot{m} v_1 = 0.179 (0.04212) = 7.539 \times 10^{-3} \text{ m}^3\text{/s}$$

(d) Power consumption

$$\dot{W} = \dot{m} (h_2 - h_1) = 0.179 (427.2 - 407.1) = 3.6 \text{ kW}$$

Theoretical horsepower of compressor

$$\text{HP} = \frac{3.6 \times 10^3}{746} = 4.82 \text{ hp}$$

Heating capacity

$$\dot{Q}_k = \dot{m}(h_2 - h_3) = 0.179(427.2 - 271.6) = 27.9 \text{ kW}$$

(e) Theoretical COP

$$\text{COP}_\text{H} = \frac{h_2 - h_3}{h_2 - h_1} = \frac{427.2 - 271.62}{427.2 - 407.1} = 7.75$$

Alternatively,

$$\text{COP} = \frac{\dot{Q}_k}{\dot{W}} = \frac{27.9}{3.6} = 7.75$$

Note that only 3.6 kW of power is consumed, and 27.9 kW of heat is delivered to conditioned space for heating. A simple electric heater would deliver only 3.6 kW.

3.9 SECOND LAW EFFICIENCY OF VAPOUR COMPRESSION CYCLE

The second law efficiency of a cycle is defined by the ratio

$$\eta_\text{II} = \frac{\text{Minimum available energy required for the cycle}}{\text{Actual available energy consumed in the cycle}}$$

Now, consider a vapour compression cycle refrigerator working between a surroundings temperature of T_k and a refrigeration temperature of T_0. The actual cycle is 1-2-3-4 as shown in Fig. 3.4. The network done in the cycle is w which is equal to the actual available energy consumed in the cycle

$$(AE)_{\text{actual}} = w = \frac{q_0}{\text{COP}_\text{I}} = \text{Area 1-2-3-c-d-4-1}$$

where COP_I is the actual first law COP of the cycle.

And the minimum available energy required for the same refrigerating effect q_0 is the work of a Carnot cycle such as 1-2″-3′-4 operating between the same temperatures T_0 and T_k. Thus

$$(AE)_{\text{minimum}} = w_{\text{Carnot}} = \frac{q_0}{\dfrac{T_0}{T_k - T_0}} = \text{Area 1-2″-3′-4-1}$$

The second law efficiency of the refrigerator is, therefore,

$$\eta_{\text{II}} = \frac{(AE)_{\text{minimum}}}{(AE)_{\text{actual}}} = \frac{w_{\text{Carnot}}}{w} = \frac{q_0 / w}{q_0 / w_{\text{Carnot}}} = \frac{\text{COP}_\text{I}}{\text{COP}_{\text{Carnot}}}$$

Thus, for the case of wet compression of NH_3 in Example 3.4 for $t_k = 35°\text{C}$ and $t_0 = -15°\text{C}$, we have:

$$\text{COP} = 4.55$$

$$\text{Carnot COP} = \frac{T_0}{T_k - T_0} = \frac{258}{35 + 15} = 5.15$$

Hence, the second law efficiency of the cycle is

$$\eta_{\text{II}} = \frac{4.55}{5.15} = 0.833\ (83.3\%)$$

Thus, the second law efficiency of vapour compression cycle is quite high. Its COP is close to Carnot cycle COP.

The ideal Carnot cycle is completely reversible. Hence, it does not involve any irreversibilities. The vapour compression cycle involves internal irreversibilities due to the throttling process and also due to the superheat horn. Hence, the value of $\eta_{\text{II}} < 1$.

The actual vapour compression cycle would also have the external irreversibilities of condensation and evaporation processes due to finite temperature differences required for heat transfer. η_{II}, therefore, is further reduced.

References

1. Ewing A, *Report of the Refrigeration Research Committee,* The Institution of Mechanical Engineers, UK, 1914.
2. Gosney W B, *The Maximum Coefficient of Performance of a Refrigerant,* Paper 2.74, XIIth International Congress of Refrigeration, Madrid, 1967.
3. ISI Code for Testing Refigerators No. 1476–1979.

Revision Exercises

3.1 A 15 TR R 22 vapour compression system operates between a condenser temperature of 40°C and an evaporator temperature of 5°C.

(a) Determine the compressor discharge temperature:

(i) Using the *p-h* diagram for R 22.

(ii) Using saturation properties of R 22 and assuming the specific heat of its vapour as 0.8 kJ/kg. K.

(iii) Using superheat tables for R 22.

(b) Calculate the theoretical piston displacement and power consumption of the compressor per ton of refrigeration.

3.2 A simple saturation ammonia compression system has a high pressure of 1.35 MN/m^2 and a low pressure of 0.19 MN/m^2. Find per 400,000 kJ/h of refrigerating capacity, the power consumption of the compressor and COP of the cycle.

3.3 (a) A R 22 refrigerating machine operates between a condenser temperature of 40°C and an evaporator temperature of 5°C. Calculate the increase (per cent) in the theoretical piston displacement and the power consumption of the cycle:

(i) If the evaporator temperature is reduced to 0°C.

(ii) If the condenser temperature is increased to 45°C.

(b) Why is the performance of a vapour compression machine more sensitive to change in evaporator temperature than to an equal change in the condenser temperature?

3.4 In a vapour compression cycle saturated liquid Refrigerant 22 leaving the condenser at 40°C is required to expand to the evaporator temperature of 0°C in a cold storage plant.

(a) Determine the percentage saving in net work of the cycle per kg of the refrigerant if an isentropic expander could be used to expand the refrigerant in place of the throttling device.

(b) Also determine the percentage increase in refrigerating effect per kg of refrigerant as a result of use of the expander. Assume that compression is isentropic from saturated vapour state at 0°C to the condenser pressure.

3.5 An ammonia refrigeration system operates between saturated suction temperature of –20°C, and saturated discharge temperature of +40°C. Compare the COP of the cycle using wet compression with that of the cycle using dry compression.

Assume that the vapour leaving the compressor is saturated in the case of wet compression, and the vapour entering the compressor is saturated in the case of dry compression. The refrigerant leaves the condenser as saturated liquid.

3.6 A standard vapour compression cycle using R 22 operates on simple saturation cycle at the following conditions:

Refrigerating capacity 15 TR
Condensing temperature 40°C
Evaporating temperature 5°C

Calculate:

(a) Refrigerant circulation rate in kg/s.
(b) Power required by the compressor in kW.
(c) Coefficient of performance.
(d) Volume flow rate of the refrigerant at compressor suction.
(e) Compressor discharge temperature.
(f) Suction vapour volume and power consumption per ton of refrigeration.

Refrigeration engineers assume that if this R 22 compressor is used with R 134a, its capacity would fall by about 40%. Examine this assumption by doing cycle analysis for R 134a with the same compressor.

3.7 An ammonia refrigeration plant operates between a condensing temperature of 40°C and an evaporating temperature of –10°C. The vapour is dry at the end of compression. Only the following property values are given:

t	h_f	h_g	s_f
°C	kJ/kg	kJ/kg	kJ/kg.K
40°C	371.5	1473	1.36
–10°C	135.4	1433	0.544

The specific heat of NH_3 vapour is 2.1897 kJ/kg.K. Calculate the theoretical coefficient of performance of the cycle.

3.8 A R 134a machine operates at –15°C evaporator and 35°C condenser temperatures. Assuming a simple-saturation cycle, calculate the volume of the suction vapour and power consumption per ton of refrigeration and COP of the cycle.

Calculate the same if the system has a regenerative heat exchanger with the suction vapour leaving at 20°C from the heat exchanger.

3.9 (a) What would be the necessary bore and stroke of a single acting four-cylinder, 350 rpm ammonia compressor working on simple saturation cycle between 35°C condenser and –15°C evaporator temperatures and developing 15 tons refrigeration. Given for the compressor:
Ratio of stroke to bore = 1
Volumetric efficiency = 0.7

(b) If the index of compression is 1.15, find the error introduced in calculating the work of compression by assuming the process to be isentropic. Also, find the heat rejected to the compressor-jacket cooling water.

3.10 A simple saturation cycle using R 22 is designed for a load of 100 TR. The saturated suction and discharge temperatures are 5°C and 40°C respectively. Calculate:

(a) The mass flow rate of refrigerant.
(b) The COP and isentropic horsepower.
(c) The heat rejected in the condenser.

Use the following data:

t	p	h_f	h_g	s_f	s_g	v_g
°C	bar	kJ/kg	kJ/kg	kJ/(kg.K)	kJ/(kg.K)	m³/kg
5	5.836	205.9	407.1	1.02115	1.7447	0.0404
40	15.331	249.53	416.4	1.16659	1.69953	

Specific heat of vapour is 0.65 kJ/(kg.K).

3.11 A commercial refrigerator using Isobutane operates on the simple saturation cycle with saturated suction and discharge temperature of –25°C and 55°C respectively.

(a) Calculate the COP and power required to run the compressor per ton of refrigeration.

(b) If the liquid is subcooled by 10°C in the condenser, calculate the COP and the power required per ton of refrigeration.

(c) If the liquid is further subcooled in a regenerative heat exchanger with superheating of the vapour by 30°C, what is the increase in COP and decrease in power required?

3.12 An ammonia ice plant operates on simple saturation cycle at the following temperatures.

Condensing temperature 40°C
Evaporating temperature –15°C

It produces 10 tons of ice per day at –5°C from water at 30°C. Determine:

(a) Capacity of the refrigeration plant.

(b) Mass flow rate of refrigerant.

(c) Isentropic discharge temperature.

(d) Compressor dimensions (bore and stroke) if its volumetric efficiency is assumed as 65%. The compressor is to run at 1400 rpm. Take stroke/bore ratio (L/D) as 1.2.

(e) Horsepower of the compressor if its adiabatic efficiency is taken as 85% and mechanical efficiency as 95%.

(f) Theoretical and actual COP.

3.13 A Refrigerant 22 vapour compression system meant for food freezing operates at 40°C condensing temperature and –35°C evaporating temperature. Its compressor is capable of pumping 30 L/s of vapour at suction.

(a) Calculate the COP of the system and its refrigerating capacity,

(b) If a regenerative heat exchanger is installed which allows suction vapour to be heated by 30°C with liquid from the condenser at 40°C to be cooled correspondingly, what is the new COP and refrigerating capacity?

3.14 A domestic refrigerator uses R 134a as refrigerant and operates between evaporator and condenser temperatures of –25°C and 55°C in an ambient of 43°C. The liquid leaves the condenser at 43°C. It is further subcooled to 32°C in a regenerative heat exchanger in which the vapour from the evaporator is heated to 32°C. The compressor displacement volume is 4.33 cm³. Calculate the refrigerating capacity and power consumption of the refrigerator if the volumetric efficiency of the compressor can be taken as 0.68, and its adiabatic efficiency 0.85. Also find the compressor discharge temperature. The compressor is directly coupled to a motor running at 2900 rpm.

Chapter

4

Refrigerants

4.1 A SURVEY OF REFRIGERANTS

The first refrigerant used was ether, employed by Perkins in his hand-operated vapour-compression machine. In the earlier days, ethyl chloride (C_2H_5Cl) was used as a refrigerant which soon gave way to ammonia as early as in 1875. At about the same time, sulphur dioxide (SO_2) in 1874, methyl chloride (CH_3Cl) in 1878 and carbon dioxide (CO_2) in 1881, found application as refrigerants. During 1910–30 many new refrigerants, such as N_2O_3, CH_4, C_2H_6, C_2H_4, C_3H_8, were employed for medium and low-temperature refrigeration. Hydrocarbons were, however, found extremely inflammable. Dichloromethane (CH_2Cl_2), dichloroethylene ($C_2H_2Cl_2$) and monobromomethane (CH_3Br) were also used as refrigerants for centrifugal machines.

A great breakthrough occurred in the field of refrigeration with the development of R (a trade name) in 1930s by E.I. du Pont de Nemours and Co.[32] R are a series of fluorinated hydrocarbons, generally known as *fluorocarbons*, derived from methane, ethane, etc., as bases.

With essentially fluorine, chlorine and sometimes bromine in their molecule, these form a series of refrigerants with a wide range of *normal boiling points* (boiling points or saturation temperatures at one atmosphere pressures) to satisfy the varied requirements of different refrigerating machines. The presence of fluorine in the molecule makes the compound non-toxic and imparts other desirable physical and physiological characteristics.

Plank[31] has given individual treatment to some 50 inorganic and organic refrigerants and many more have been listed. Among the most common inorganic refrigerants are:

Ammonia (NH_3)	Used with reciprocating and screw compressors, in cold storages, ice plants, food refrigeration, etc.
Water (H_2O)	Used in water-lithium bromide absorption system and steam-ejector system only for air conditioning
Carbon dioxide (CO_2)	Used as solid carbon dioxide or *dry ice* in frozen-food transport refrigeration.

Presently, the most commonly used organic refrigerants are the chloro-fluoro derivatives of CH_4 and C_2H_6. The fully halogenated ones with chlorine in their molecule are *chloro-fluorocarbons*, referred to as *CFCs*. Those containing H atom/s in the molecule along with Cl and F atoms are referred to as *hydrochlorofluorocarbons* or *HCFCs*. And those having no chorine atom/s in the molecule are *hydrofluorocarbons* or *HFCs*. Simple *hydrocarbons* are *HCs*, Thus, we have HCs, HFCs, HCFCs and CFCs.

CFCs and HCFCs were identified as high COP refrigerants, with many favourable properties, e.g., non-flammability, low toxicity, and material compatibility that led to their widespread use as refrigerants.

The most common organic refrigerants that have been employed in recent times are:

Refrigerant 11 or CFC 11 (CCl_3F)	Used with centrifugal compressors in large capacity central air-conditioning plants
Refrigerant 12 or CFC12 (CCl_2F_2)	Used with reciprocating compressors in small units, specially domestic refrigerators, water coolers, etc.
Refrigerant 22 or HCFC22 ($CHClF_2$)	Used with reciprocating compressors in window-type air conditioners and large units such as package units and central air conditioning plants. It is also used for low temperature refrigeration applications, cold storages, food freezing and storage, etc., with reciprocating and often with screw compressors.

Among the less common refrigerants were:

Refrigerant 113 or CFC 113 ($C_2Cl_3F_3$)	With centrifugal compressors for air conditioning
Refrigerant 114 or CFC 114 ($C_2Cl_2F_4$)	With rotary compressors
Refrigerant 142b or HCFC 142b ($C_2H_3ClF_2$)	For heat pump and high condensing temperature applications
Refrigerant 502	For large supermarket frozen food cabinets involving high pressure ratio applications

The F-atom in the molecule of these substances makes the substances physiologically more favourable. The Cl atom in the molecule is considered responsible for the depletion of ozone layer in the upper atmosphere (stratosphere), thus, allowing harmful ultra-violet radiation to penetrate through the atmosphere and reach the earth's surface. The H atom/s in the molecule impart/s a degree of flammability to the substance depending upon the number of these atoms.

4.2 DESIGNATION OF REFRIGERANTS

The international designation of refrigerants uses Refrigerant or R (or alternatively CFC, HCFC, HFC and HC as the case may be) as the designation followed by certain numerals. Thus, for a compound derived from a saturated hydrocarbon denoted by the chemical formula

$$C_mH_nF_pCl_q$$

in which $(n + p + q) = 2m + 2$, the complete designation is

$$R(m - 1)\ (n + 1)\ (p)$$

Hence, for dichloro-tetrafluoro-ethane in which there are two carbon atoms ($m = 2$), no hydrogen atom ($n = 0$) and four fluorine atoms ($p = 4$), the designation is R (2 – 1) (0 + 1) (4), viz., R 114 or CFC 114. In this manner, CCl_3F is R 11, CCl_2 F_2 is R 12, $CHClF_2$ is R 22, $C_2Cl_3F_3$ is R 113, CH_4 is R 50, C_2H_6 is R 170, C_3 H_8 is R 290, and so on.

The brominated refrigerants are denoted by putting an additional B and a number to denote as to how many chlorine atoms are replaced by bromine atoms. Thus R 13B1 is derived from R 13 with the replacement of one chlorine atom by a bromine atom. Its chemical formula is, therefore, CF_3Br.

In the case of *isomers*, i.e., compounds with the same chemical formula but different molecular structure, subscripts, a, b, etc., are used after the designations.

In this manner, there are 15 fluorochloro derivatives of methane (Table 4.1), 55 derivatives of ethane (Table 4.2), 332 of propane, and so on. There are also *azeotropes*, which are mixtures of refrigerants but which behave like pure substances. They are given arbitrary designations, e.g., R 502 for a mixture of 48.8 per cent R22 or HCFC22, and 51.2 per cent R 115 or CFC115.

Table 4.1 Derivatives of methane with normal boiling points (in°C) and designations

	No. of H atoms				
No. of F atoms	4-H	3-H	2-H	1-H	0-H
0-F	CH_4 – 164 R 50	CH_3Cl – 23.74 R 40	CH_2Cl_2 40 R 30	$CHCl_3$ 61.2 R 20	CCl_4 76.7 R 10
1-F		CH_3F –78 R 41	CH_2ClF –9 R 31	$CHCl_2F$ 8.9 R 21	CCl_3F 23.7 R 11
2-F			CH_2F_2 – 51.7 R 32	$CHClF_2$ – 40.8 R 22	CCl_2F_2 – 29.8 R 12
3-F				CHF_3 – 82.2 R 23	$CClF_3$ – 81.5 R 13
4-F					CF_4 –127.8 R 14

Unsaturated compounds, for which $(n + p + q) = 2m$, are distinguished by putting the digit 1 before $(m – 1)$. Thus ethylene is R 1150.

In the case of common inorganic refrigerants, numerical designations have been given according to their molecular weight added to 700. Thus ammonia, whose molecular weight is 17, is designated as R 717. Similarly, water is designated as R 718 and carbon dioxide as R 744.

Also, problem arises in the case of butane, C_4 H_{10} and higher hydrocarbons. In the case of butane, n = 10, a double digit figure. Hence, the prescribed method of designation cannot be used. Accordingly, *n*-butane and isobutane are assigned the designations arbitrarily as R 600 and R 600a respectively.

Table 4.2 Derivatives of ethane with normal boiling points (°C) and designations

No. of F atoms	***No. of H atoms*** 6-H	5-H	4-H	3-H	2-H	1-H	0-H
0–F	C_2H_6 – 88.6 R 170	C_2H_5Cl 12 R 160	$CH_2Cl–CH_2Cl$ 84 $CH_3–CHCl_2$ 57 R 150	$CH_2Cl–CHCl_2$ 113 $CH_3–CCl_3$ 75 R 140	$CHCl_2–CHCl_2$ 145 $CH_2Cl–CCl_3$ 128 R 130	$CHCl_2–CCl_3$ 162 R 120	C_2Cl_6 185 R 110
1-F		C_2H_5F R 161	$CH_3–CHClF$ 4 $CH_2Cl–CH_2F$ – 37.7 R 151	$CH_2Cl–CHClF$ 65 $CH_3–CCl_2F$ 42 $CHCl_2–CH_2F$ 32.1 R 141	$CHCl_2–CHClF$ 102 $CCl_3–CH_2F$ 90 $CH_2Cl–CCl_2F$ 86 R 131	$CHCl_2–CCl_2F$ 115.7 $CCl_3–CHClF$ 117 R 121	$CCl_3–CCl_2F$ R 111
2–F			$CH_2F–CH_2F$ – 24.7 $CH_3–CHF_2$ – 24.02 R 152	$CH_3Cl–CHF_3$ 35 $CH_2F–CHClF$ 27 $CH_3–CClF_2$ – 9.25 R 142	$CHClF–CHClF$ 66 $CH_2F–CCl_2F$ 62 $CHCl_2–CHF_3$ 60 $CH_2Cl–CClF_2$ 49 R 132	$CHClF–CCl_2F$ 85 $CCl_3–CHF_2$ 77 $CHCl_2–CClF_2$ 72 R 122	$CCl_2F–CCl_2F$ 92 $CCl_3–CClF_2$ 91.5 R 112

(*Contd*)

Table 4.2 (*Contd*)

No. of F atoms	*No. of H atoms* 6-H	5-H	4-H	3-H	2-H	1-H	0-H
3–F				(b) $CH_2F–CHF_2$ –35 (a) $CH_3–CF_3$ – 47.24 R 143	$CHClF–CHF_2$ 17 $CH_2F–CClF_2$ 7 $CH_2Cl–CF_3$ 8 R 133	$CHF_3–CCl_2F$ 38 $CHClF–CClF_2$ 32 $CHCl_2–CF_3$ 27.82 R 123	$CClF_3–CCl_2F$ 47.68 $CCl_3–CF_3$ 45.9 R 113
4-F					(a) $CH_2F–CF_3$ – 26.1 (b) $CHF_2–CHF_2$ – 20 R 134	$CHClF–CF_3$ –12 $CHF_2–CClF_2$ – 16 R 124	$CCl_2F–CF_3$ –12 $CClF_2–CClF_2$ 3.6 R 114
5-F						$CHF_2–CF_3$ – 48.1 R 125	$CClF_2–CF_3$ – 38 R 115
6–F							C_2F_6 –78.3 R 116

4.3 COMPARATIVE STUDY OF METHANE DERIVATIVES IN USE BEFORE THE YEAR 2000

A study of fluoro chloro derivatives of methane, along with their normal boiling points and designations as given in Table 4.1, makes an interesting reading. A comparative study of refrigerants in use before 2000 AD may be begun with the well-known refrigerant CFC 12 or R 12, viz., $CCl_2\,F_2$, with its normal boiling point (N.B.P.) of –29.8°C. The refrigerant has commonly been used in small-capacity unitary equipment with reciprocating compressors, such as domestic refrigerators, water coolers, car air conditioners, etc., for refrigeration from 0 to – 25°C. It is evident that R 12 will maintain positive pressure in the evaporator at – 25°C. Further, because of low value of γ, R 12 has low discharge temperature. It was, therefore, found very suitable for hermetically sealed units of refrigerators on which the manufacturers gave 5 to 10 years warrantee. Chances of burn-out of motor windings in hermetically sealed units are few if discharge temperature and hence winding temperature is low.

Just above R 12, along the column with zero H atoms, we have CFC 11 or R 11, viz., CCl_3F, with an N.B.P. of 23.7°C. This is a *higher boiling* refrigerant. Its boiling point is 53.5°C higher than that of R 12, and hence it is a low pressure substance. It has vacuum in the evaporator even in air conditioning applications at t_0 = 5°C. It, therefore, has a large specific volume of the suction vapour at any evaporator temperature and was, therefore, found suitable for use in large capacity centrifugal compressors (300 TR and above) for water chillers (water at about 7°C) for cooling air in central air-conditioning plants. R 11 could not be considered suitable for use with reciprocating compressors.

Above R 11, we have R 10, i.e., CCl_4 or carbon tetrachloride. It has a very high boiling point and was not found suitable even for centrifugal compressors.

Below R 12, in the same column, we have CFC 13 or R 13 having N.B.P. of – 81.5°C, and CFC 14 or R 14 with N.B.P. of – 127.8°C. Both are *lower boiling*, and hence high pressure substances. Their critical temperatures are also below the normal ambient temperatures. These, therefore, cannot be condensed at normal ambient temperatures. Because of these reasons, these substances were not used in common refrigeration and air conditioning applications.

To the left of R 12, in the row with one H atom, we have another well-known refrigerant HCFC 22 or R 22, viz., $CHClF_2$. It has an N.B.P. of – 40.8°C which is about 10°C lower than that of R 12. It is, therefore, a comparatively higher pressure refrigerant. Consequently, it has a smaller specific volume of the suction vapour and gives about 40 per cent more capacity in a positive displacement compressor as compared to R 12. Previously, R 22 was employed for air conditioning in large capacity plants and package units above 5 TR only. But now, because of the capacity bonus, it is used even in one ton window-type air conditioners. However, its value of γ is higher than that of R 12. Therefore, it has higher compressor discharge and winding temperatures. It is, therefore, necessary to affect greater cooling of motor windings in R 22 hermetically sealed units with the help of incoming suction vapours, and also by adequate heat transfer to surrounding air, even by forced convection, from its body. In any case, R 22 units have more frequent burn-out problems than R 12. Warrantees are given only for 1–2 years unlike 5–10 years for R 12.

Because of its lower boiling point, R 22 also finds application in food freezing, freeze drying, etc., maintaining positive pressure in evaporators with temperatures as low as – 40°C.

Above R 22, in the 1–H column, we have R 21. It is, however, a little chemically unstable. Further above is R 20 which is a very high boiling point substance unsuitable as a refrigerant. Below R 22, we have R 23 with an N.B.P. of – 82.2°C. It is similar to R 13.

Other substances in Table 4.1, are either chemically unstable or highly inflammable due to the pressence of H atoms in the molecule. Methane is the most flammable of all. R 40 or methyl chloride (CH_3Cl) with an N.B.P. of –23.74°C, used to be a very popular refrigerant in early 20th century. It is, however, poisonous and is odourless at the same time. Its leakage, therefore, gives no warning and may lead to death. Its use is, therefore, banned.

Note *Here, we notice that substances with normal boiling points in the range – 50 to + 50°C only find application as refrigerants in commercial refrigeration and air conditioning.*

4.4 COMPARATIVE STUDY OF ETHANE DERIVATIVES IN USE BEFORE THE YEAR 2000

A similar study can be made of ethane derivatives given in Table 4.2. These are mostly higher molecular weight and hence higher boiling substances. A few lower down in the table have their N.B.P. in the range – 50 to 50°C. One of them is R 113 or $C_2Cl_3F_3$. It has two isomers. One has an N.B.P. of 45.9°C. The N.B.P. of the other is 47.6°C. It was commonly used as a refrigerant for centrigual machines for air conditioning. It has higher boiling point than R 11. Hence it has lower pressure and larger volumes of the suction vapour as compared to R 11. It has the advantage that centrifugal units of smaller capacity, i.e., of the order of 150 TR, can be designed as against 300–500 TR of R 11. However, with R 113, there is vacuum, in both, evaporator and condenser.

Below R 113, we have R 114 with N.B.P. of 3.6°C. It is neither suitable for reciprocating nor for centrigual compressors. It was generally employed in rotary or low temperature centrifugal compressors.

Another known refrigerant is R 115. Having N.B.P. of – 38°C. It is similar to R 22 in its thermodynamic characteristics. It is used as a component of the azeotropic mixture R 502. The other component in the mixture is R 22. The azeotrope has a lower pressure ratio as compared to R 22. It was therefore, used in high condensing and low evaporating temperature applications, such as heat pumps, air-cooled frozen food cabinets, high ambient temperature applications, etc., so that the pressure ratios and, consequently, the discharge temperatures are low. This is a great advantage in hermetically sealed compressor units as the temperature of the compressor body and, therefore, the temperature of the enclosed motor windings depends on the discharge temperature of the gas. A lower winding temperature ensures a lower *burn-out* rate and hence a longer life of the sealed unit.

Then we have R 116 with N.B.P. of –78.3°C which is too low. Other derivatives of ethane have not been used as refrigerants *in the past*.

Note *Because of the problem of ozone layer depletion R 11, R 12, R 113, R 114, R 115 and R 502, all CFCs have been phased out.*

4.5 REFRIGERANTS IN USE AFTER THE YEAR 2000

After the finding that CFCs, and to a lesser extent HCFCs deplete the ozone layer, over 100 countries adopted Montreal Protocal (MP) of 1987 to phase out CFCs in the year 2000, and HCFCs by the year 2030.

HFCs and HCs do not deplete the ozone layer. They can be used even after 2030. Their production and use is not regualated by the MP. These refrigerants include R 32, R 125, R 134a, R 143a, R 236 fa, and R 245 fa, and their azeotropic or near azeotropic blends R 404A, R 407C, and R 410A.

HFCs have some global warming potential though some governments may consider regulating their use too.

Out of the refrigerants being used prior to 2000 AD, all CFCs R 11, R 12, R 113, R 114, and the azeotropic mixture R 502 have already been phased out.

Replacements for all important R 12, R 11, R 22, and R 502 to date to be used till 2030 are as follows:

4.5.1 Replacement for R 12

The replacement for R 12 in unitary equipment has been the easiest. Its place, for all practical purposes, has been taken over by R 134a.

Compare the NBP of –26.1°C of R 134a, which is very close to –29.8°C, the NBP of R 12.

As R 134a is an HFC, its use will continue well beyond 2030 provided its global warming potential is not considered a disqualification, and provided another more suitable refrigerant is not found.

4.5.2 Replacement for R 11

The replacement of R 11 in large capacity centrifugal machines for air conditioning has posed problems. So far, R 123 with its NBP of 27.82°C, closest to 23.71°C, the NBP of R 11, has been chosen as the most efficient option for R 11 in centrifugal chillers.

Another advantage of R 123 is its atmospheric life-time which is the shortest among the refrigerants.

Moreover, its benefit in reducing global warming is significant. It has very very low global warming potential.

Nevertheless, R 123 is an HCFC. Its use is slated for phase-out by 2030 despite its unique qualities. As an exception, R 123 may coutinue to be used well beyond 2030.

R 11 has also been replaced by HFC R 134a. But since NBP of R 134a (–26.1°C) is very very low compared to NBP of R 11 (+23.71°C), it is a much higher pressure refrigerant and it has very much smaller specific volumes of suction vapour. Hence, tonnage of a single unit of R 134a centrifugal compressor is much greater than 300–350 TR or so of a single R 11 centrifugal compressor.

Further, most R 134a centrifugal chillers have lower COPs, and they are few in number. R 245fa, an HFC, however, is being seriously considered for use in place of R 11. The NBP of R 245fa is 14.9°C. It is not far from 23.71°C, the NBP of R 11.

R 245fa offers potenial to approach R 123 efficiencies. Its commercialization, however, is uncertain. Its cost is high at present. And it has global warming potential and flammability concerns also.

4.5.3 Replacement for R 22

As R 22 is an HCFC, its use is permitted till 2030. Also, since it is currently the most favoured refrigerant in package units and chillers, its use continues. 70% of commercial refrigeration systems still use R 22.

Only in food refrigeration, its use is not being favoured. Before 2000 AD, the trend was to replace ammonia with R 22. Now, this trend has been reversed. Ammonia has come back in a big way as it is environment friendly. Ammonia has even higher COP.

Ammonia enjoys wide use today in industrial applications especially with reciprocating compressors in small capacity, and screw compressors in large capacity. Food industry is preparing for R 22 phase-out, and its replacement by ammonia.

However, in car air conditioners, R 134a has replaced R 22. R 134a has lower pressures.

But HCFC R 22 still dominates in air conditioning systems using positive displacement reciprocating, scroll, and screw compressors in medium capacity package units, and large capacity screw chillers. Some industries are using 50/50 percent R 32/R 125 blend, viz., R 410A also as a replacement for R 22.

4.5.4 Replacement for R 502

Presently, R 404A has found application in place of R 502. The NBP of R 404A is –46.22°C as against –45.6°C, the NBP of R 502. A 50/50 per cent R 125/R 143a blend, R 507a, is also an attractive alternative to R 502.

Note *Industries worldwide have turned to HCFCs R 22 and R 123. While HCFCs have a lower ozone depletion potential than CFCs, they still damage the ozone layer.*

4.6 SELECTION OF A REFRIGERANT

Refrigerants have to be physiologically non-toxic and-non flammable. Therodynamically, there is no working substance which could be called an ideal refrigerant. Different substances seem to satisfy different requirements and those also sometimes only partially. A refrigerant which is ideally suited in a particular application may be a complete failure in the other. In general, a refrigerant may be required to satisfy requirements which may be classified as thermodynamic, chemical and physical, as discussed in the following sections. The selection of a refrigerant for a particular application, therefore, depends on satisfying its essential requirements.

The choice of a refrigerant for a given application is governed mainly by the refrigerating capacity (very small, small, medium or large), and refrigeration temperature required, such as for air conditioning (5°C), cold-storage (–10 to 2°C), refrigerator (– 25°C), food freezing (– 40°C), etc.

4.7 THERMODYNAMIC REQUIREMENTS

The method of evaluation of thermodynamic properties is described in Chap. 1 in brief. Detailed thermodynamic properties tables and equations for various refrigerants are given in the appendix.

Table 4.3 gives the important thermodynamic properties of a number of refrigerants. The most important property of a refrigerant is its normal boiling point. Most of the other thermodynamic characteristics very much depend on it.

The thermodynamic requirements of refrigerants pertain to the condensing and evaporating pressures, critical temperature and pressure, freezing point, volume of the suction vapour per ton, COP, power consumption per ton, etc.

4.7.1 Significance of Normal Boiling Point

Pressure–temperature relationship of a pure substance is built into the *Clapeyron equation* which is expressed as

$$\frac{\mathrm{d}p^{\text{sat}}}{\mathrm{d}T^{\text{sat}}} = \frac{h_{fg}}{T^{\text{sat}}\,(v_g - v_f)} \tag{4.1}$$

Two simplifications may be introduced in the above equation. These are the assumptions as follows:

(i) The specific volume of the liquid is very small compared to that of the vapour, so that we may put

$$v_f = 0$$

(ii) The specific volume of the vapour is given by the ideal gas equation of state, so that

$$v_g = \frac{RT^{\text{sat}}}{p^{\text{sat}}}$$

Substituting these values in Eq. (4.1) we get the Clausius–Clapeyron equation

$$\frac{\mathrm{d}p^{\text{sat}}}{\mathrm{d}T^{\text{sat}}} = \frac{p^{\text{sat}} h_{fg}}{R\left(T^{\text{sat}}\right)^2} \tag{4.2}$$

Rearranging, we obtain

$$\frac{\mathrm{d}\,(\ln p^{\text{sat}})}{\mathrm{d}\left(\dfrac{1}{T^{\text{sat}}}\right)} = -\frac{h_{fg}}{R} = -b \tag{4.3}$$

where $b = h_{fg}/R$ represents the slope of $\ln p^{\text{sat}}$ versus $1/T^{\text{sat}}$ line. In the small range of condenser and evaporator pressures, the latent heat of vaporization may be assumed as constant, so that the value of b may also be taken as constant. Thus $\ln p^{\text{sat}}–1/T^{\text{sat}}$

Table 4.3 General data of refrigerants

Refrigerant	*Chemical Formula*	*Designation*	*M (Molecular Weight)*	t_s *(N.B.P.)* °C	t_c *(Critical Temperature)*°C	p_c *(Critical Pressure)* bar	v_c*(Critical Volume)* L/kg	t_f *(Freezing Point)* °C
Inorganic Refrigerants								
Water	H_2O	R 718	18.016	100.0	374.15	221.3	3.26	0.0
Ammonia	NH_3	R 717	17.031	– 33.313	133.0	112.97	4.13	–77.7
Carbon Dioxide	CO_2	R 744	44.01	–78.4	31.1	73.72	2.135	–56.6
Organic Refrigerants								
Monofluoro-Tetrachloro Methane	$CFCl_3$	R 11	137.39	23.7	197.78	43.7	1.805	–111
Difluoro-Dichloro Methane	CF_2Cl_2	R 12	120.92	– 29.75	112.04	41.15	1.793	–136
Trifluoro-Monochloro Methane	CF_3Cl	R 13	104.47	– 81.5	28.78	38.8	1.721	–180
Trifluoro-Monobromo Methane	CF_3Br	R 13 B1	148.9	– 58.7	67.8	40.5	—	–142
Tetrafluoro Methane	CF_4	R 14	88.01	– 128.0	– 45.5	37.5	1.50	– 194
Monofluoro-Dichloro Methane	$CHFCl_2$	R 21	102.92	8.9	173.5	51.6	1.915	–135
Difluoro-Monochloro Methane	CHF_2Cl	R 22	86.48	– 40.81	96.02	49.88	1.949	– 160
Trifluoro Methane	CHF_3	R 23	70.01	– 82.0	25.83	48.2	1.748	– 155
Difluoro Methane	CH_2F_2	R 32	52.024	– 51.7	78.41	58.3	2.326	

(*Contd*)

Table 4.3 (*Contd*)

Refrigerant	*Chemical Formula*	*Designation*	*M (Molecular Weight)*	t_s *(N.B.P.)* °C	t_c *(Critical Temperature)*°C	p_c *(Critical Pressure)* bar	v_c*(Critical Volume)* L/kg	t_f*(Freezing Point)* °C
Pentafluoro Ethane	CF_3CHF_2	R 125	120.02	– 48.1	66.25	36.31	1.748	
Tetrafluoro Ethane	CF_3CH_2F	R 134a	102.03	– 26.07	101.06	40.56	1.942	– 96.6
Monochloro-Tetra-fluoro Ethane	CF_3CHClF	R 124	136.475	– 12.05	122.5	36.34	1.786	
Dichloro-Trifluoro Ethane	CF_3CHCl_2	R 123	152.93	27.82	183.79	36.74	1.818	
- Do -	$CHClFCClF_2$	R 123a	152.93	29.9	189.62	38.89		
Tetrafluoro Ethane	CHF_2CHF_2	R 134	102.03	– 19.8	118.9			
Monofluoro-Dichloro Ethane	CH_3CCl_2F	R 141b	116.9	32.1	208.0	43.39		
Trifluoro-Trichloro Ethane	$CF_2Cl–CFCl_2$	R 113	187.89	47.68	214.1	34.15	1.735	–36.6
Tetrafluoro-Dichloro Ethane	$CF_2Cl–CF_2Cl$	R 114	170.91	3.6	145.8	32.75	1.715	–94
Pentafluoro-Monochloro Ethane	CF_2ClCF_3	R 115	154.48	–38.0	80.0	32.36	1.68	–106
Difluoro-Monochloro Ethane	$CH_3–CF_2Cl$	R 142b	100.495	–9.25	137.1	42.46	2.3	–131.0
Trifluoro Ethane	$CH_3–CF_3$	R 143a	84.04	–47.24	73.1	38.11	2.305	– 111.3
Difluoro Ethane	$CH_3–CHF_2$	R 152a	66.05	–24.02	113.3	45.2	2.717	–117.0
Ethyl Chloride	C_2H_5Cl	R 160	64.52	12.0	187.2	52.47	3.03	–138.7
n-perfluoro Butane	C_4F_{10}	—	238.04	– 2.0	113.2	23.3	1.588	—
Difluoro Ethane	$CH_2F–CH_2F$	R 152	66.05	30.7				

(*Contd*)

Table 4.3 *(Contd)*

Refrigerant	***Chemical Formula***	***Designation***	***M (Molecular Weight)***	***t_s (N.B.P.) °C***	***t_c (Critical Temperature) °C***	***p_c (Critical Pressure) bar***	***v_c (Critical Volume) L/kg***	***t_f (Freezing Point) °C***
Hydrocarbons								
Ethane	C_2H_4	R 170	30.06	−88.6	32.1	49.3	4.7	−183.2
Propane	C_3H_8	R 290	44.1	−42.1	96.8	42.56	4.545	−187.1
n-butane	C_4H_{10}	R 600	58.1	−0.56	153.0	35.3	4.29	−135
Isobutane	$(CH_3)_3CH$	R 600a	58.13	−11.67	135.0	36.45	4.526	−159.6
Unsaturated Hydrocarbons								
Propylene	C_2H_4	—	42.08	−47.7	94.4	46.0	4.2	−185
Dichloroethylene	$C_2H_2Cl_2$	—	96.9	90.9	243	54.9		−56.6
Aliphatic Amines								
Methylamine	CH_3NH_3	—	31.06	−6.7	156.9	74.5		−92.5
Ethylamine	$C_2H_4NH_3$	—	45.08	7.0	164.6	54.7		−93

relationship of refrigerants (pure substances) is nearly a straight line. The equation of the straight line is obtained by integrating Eq. (4.3)

$$\ln p^{\text{sat}} = a - \frac{b}{T^{\text{sat}}} \tag{4.5}$$

Exact equations, however, have some higher order terms as in Eq. (1.10).

Further, from the *Badylkes theory of thermodynamic similarity*[12] it is found that for substances belonging to the same family, the dimensionless *Trouton number* θ is constant. The number is defined by the expression

$$\theta = \frac{M(h_{fg})_s}{T_s} \tag{4.6}$$

where M = molecular weight

T_s = normal boiling point

$(h_{fg})_s$ = latent heat of vaporization at normal boiling point.

Thus, from Eqs. (4.4) and (4.5), we have

$$b = -\frac{h_{fg}}{R} = -\frac{M\,h_{fg}}{\bar{R}} = -\frac{\theta T_s}{\bar{R}} = f(T_s)$$

Accordingly, the slope of the $\ln p^{\text{sat}}$–$1/T^{\text{sat}}$ line is a function of the normal boiling point. Also from Eq. (4.5), at atmospheric pressure

$$a = \frac{b}{T_s} = f'\,(T_s)$$

Thus, both the constants in Eq. (4.5) are functions of the normal boiling point. The higher the boiling point, the steeper is the slope of $\ln p^{\text{sat}} - 1/T^{\text{sat}}$ line.

Also, the higher the boiling point, the further is the line from the origin. The $\ln p^{\text{sat}}$–$1/T^{\text{sat}}$ data for a number of refrigerants is plotted in Fig. 4.1 (a).

Figure 4.1(a) is plotted for $\ln p^{\text{sat}}$ along the ordinate against $1/T^{\text{sat}}$ along the abscissa for a number of refrigerants. Since the slope of the line is negative, the plot is usually made with $-(1/T^{\text{sat}})$ along the abscissa as shown qualitatively in Fig. 4.1(b).

Figure 4.1 (b) compares the pressures of lower-boiling and higher-boiling refrigerants at given evaporator and condenser temperatures. The following points emerge from this figure:

(i) At given T_0 and T_k, p_0 and p_k are lower for higher-boiling refrigerant. Hence, higher boiling substances are in general low pressure refrigerants. On the contrary, p_0 and p_k are higher for lower-boiling refrigerant. These are, therefore, high pressure refrigerants.

(ii) Because of the steeper characteristic, the high-boiling refrigerants have higher-pressure ratios p_k/p_0. And because of the flatter characteristic, the lower-boiling refrigerants have lower pressure ratios.

(iii) In addition, since slope $b = -h_{fg}/R$, the high-boiling having steeper slope have high latent heat of vaporization. And the lower boiling having flatter slope have low latent heat of vaporization.

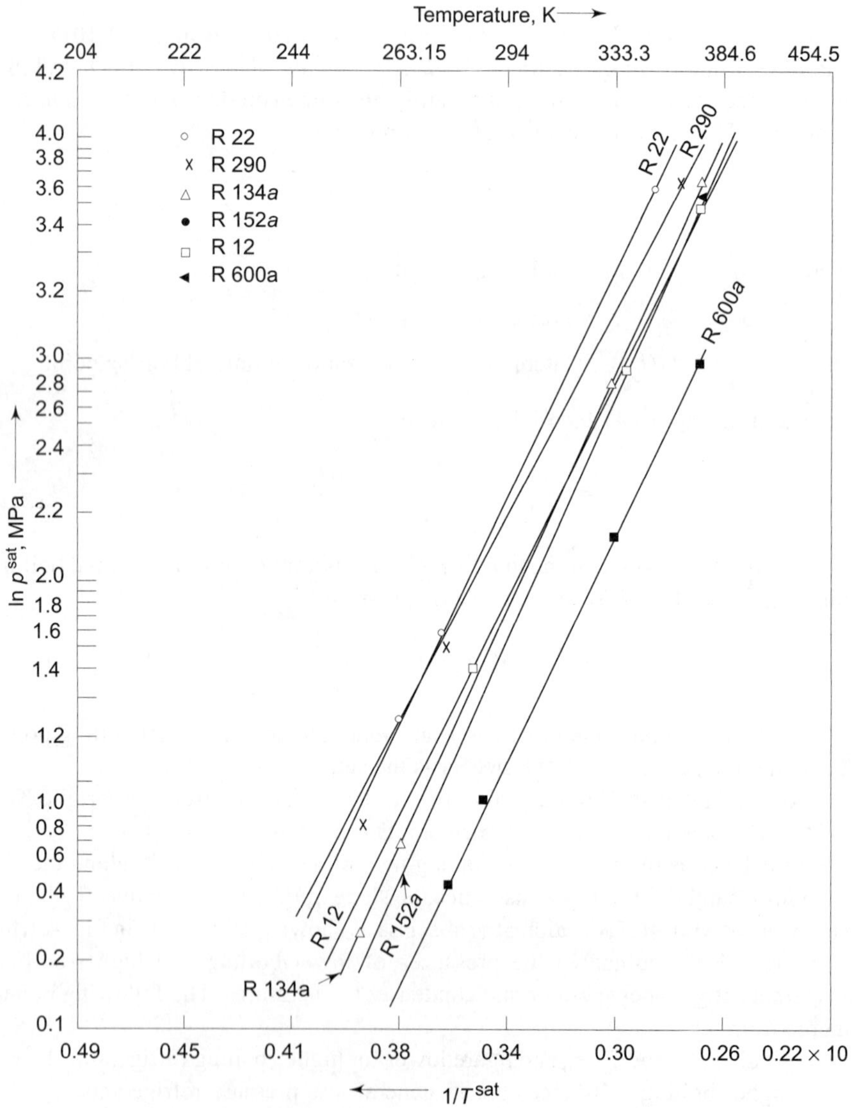

Fig. 4.1(a) ln p^{sat} versus $1/T^{sat}$ diagram of refrigerants

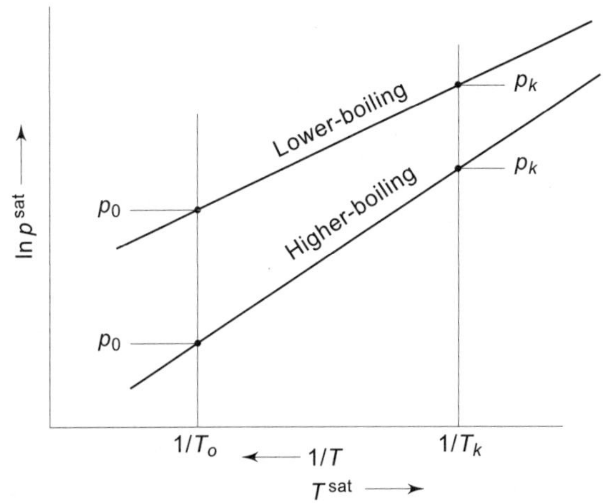

Fig. 4.1(b) Comparison of pressures of lower-boiling and higher boiling refrigerants at given evaporator and condenser temperatures

4.7.2 Standing and Condensing and Evaporating Pressures

When the refrigerant in the gas cylinder or in the system is allowed to remain standing for a period, and come to equilibrium with the ambient temperature, the pressure reeorded is named as the 'standing pressure'.

Low-boiling refrigerants have higher standing pressures. High-boiling refrigerants have lower standing pressures. For example, at 30°C ambient temperature, the standing pressure for R 134a will be 7.702 bar, and that for R 22 will be 11.919 bar, while the same for high-boiling R 123 will be just 1.0958 bar, viz., only slightly above atmospheric. As against standing pressure, condenser and evaporator pressures are the 'operating pressures.'

The evaporating pressure p_0 should be positive and as near atmospheric as possible. If it is too low, it would result in a large volume of the suction vapour. If it is too high, overall higher pressures, including condenser pressure p_k, would result necessitating heavier construction and consequently, higher cost of equipment.

A positive pressure is required in order to eliminate the possibility of the entry of air and moisture into the system. The normal boiling point of the refrigerant should, therefore, be preferably lower than the refrigeration temperature.

Table 4.4 gives the N.B.Ps, evaporating and condensing pressures, and their differentials and pressure ratios for some of the refrigerants for the standard rating cycle. The table shows that the N.B.P. determines working pressures. High pressure refrigerants are lower down in the table.

Thus, R 502, propane, R 22, R 404A, R 407C, R 410A, and ammonia are high-pressure refrigerants as compared to R 12. On the contrary, R 114, R 11, R 123, and R 113 are low-pressure refrigerants. And R 134a, R 152a and to an extent isobutane have pressures close to those of R 12. Water has extremely low pressures. R 11, R 113 and water operate at pressures below atmospheric. The condensers of these systems are, therefore, provided with purge valves at the top for R 11 or purge units for R 113 and water to remove air which will tend to collect in the condenser as uncondensed gas.

Table 4.4 also shows pressure ratios which are generally found to decrease as the boiling point decreases due to flatter ln p^{sat} versus $1/T^{sat}$ line.

Propane has the lowest pressure ratio among the high pressure, low-boiling group of refrigerants. R 152a has the lowest pressure ratio overall.

Table 4.4 Operating pressures of refrigerants

Refrigerant	t_s(N B P) °C	p_o at 5°C bar	p_k at 40°C bar	(p_k-p_v) bar	p_k/p_0
Water	100	0.00874	0.0738	0.06506	8.45
R 113	47.6	0.1903	0.7809	0.5906	4.10
R 123	27.82	0.4086	1.545	1.1362	3.78
R 11	23.71	0.4967	1.748	1.2513	3.52
R 114	3.6	1.069	3.454	2.385	3.23
R 600a (Isobutane)	– 11.67	1.88	5.361	3.481	2.85
R 152a	– 24.02	3.149	9.092	5.943	1.53
R 134a	– 26.074	3.5	10.167	6.667	2.545
R 12	– 29.8	3.62	9.6	5.98	2.65
Ammonia	– 33.33	5.16	15.54	10.38	3.01
R 22	– 40.81	5.836	15.331	9.495	2.63
R 290 (Propane)	– 42.1	5.478	13.664	8.186	2.49
R 407C	– 43.63	6.66	17.5	10.84	2.63
R 502	– 45.17	6.652	16.638	9.986	2.50
R 404A	– 46.22	7.13	18.3	11.17	2.57
R 410A	– 51.44	9.25	24.6	15.01	2.66
R 32	– 51.7	9.26	24.7	15.44	2.66

4.7.3 Critical Temperature and Pressure

For high COP, in general, the critical temperature should be very high so that the condenser temperature line on the *p-h* diagram is far removed from the critical point. This ensures reasonable refrigerating effect which becomes very small if the state of the liquid before expansion is near the critical point. Also, the critical pressure should be low so as to result in low condensing pressure.

Table 4.3 shows that except for carbon dioxide for which the critical temperature is 31°C for most of the common refrigerants, critical temperature is much above the normal condensing temperature.

4.7.4 Freezing Point

Table 4.3 also gives the freezing points of various refrigerants. As the refrigerant must operate in the cycle above its freezing point, it is evident that the same for the refrigerant must be lower than system temperatures. It will be seen that except in the case of water for which the freezing point is 0°C, other refrigerants have reasonably low values. Water can, therefore, be used only in air-conditioning applications, viz., above 0°C.

4.7.5 Volume of Suction Vapour

The volume of the suction vapour required per unit, say per ton, of refrigeration is an indication of the size of the compressor. Reciprocating compressors are used with refrigerants with high pressures and small volumes of the suction vapour. Centrifugal or turbo compressors are used with refrigerants with low pressures and large volumes of the suction vapour. A high volume flow rate for a given capacity is required in the case of centrifugal compressors to permit flow passages of sufficient width to minimise drag and obtain high efficiency.

Table 4.5 gives the suction volume requirements in m^3 per hour per ton of refrigeration of some refrigerants along with their normal boiling points. Calculations have been done for 30°C condensing and – 15°C evaporating temperatures except for water for which the evaporating temperature is 5°C, and carbon dioxide for which the condensing temperature is 25°C.

Table 4.5 Suction vapour volumes of refrigerants for t_k = 30°C and t_0 = –15°C

Refrigerant	t_s °C	q_0 kJ/kg	v m^3/kg	V^* m^3/h/TR
Water	100	2342.5(t_0 = 5°C)	147.2	825.6
R 113	47.6	122.6	1.649	186.9
R 123	27.82	142.3	0.8914	79.3
R 11	23.71	155.9	0.772	65.9
R 114	3.6	88.6	0.263	37.6
R 600a (Isobutane)	– 11.67	263.9	0.4073	21.24
R 152a	– 24.02	226.5	0.207	11.572
R 134a	– 26.074	148.0	0.1214	10.867
R 12	– 29.8	117	0.092	10.857
Ammonia	– 33.33	1103.1	0.512	6.124
R 22	– 40.81	162.7	0.0775	6.668
R 290 (Propane)	– 42.1	277.9	0.15425	7.737
R 407C	– 43.63	163.3	0.0805	6.241
R 404A	– 46.22	114.1	0.0537	5.96
R 410A	– 51.44	167.9	0.0545	4.11
R 32	– 51.7	—	—	—
Carbon Dioxide	– 78.52	158.7 (t_k = 25°C)	0.0166	1.33

It is seen from Table 4.5 that the required volume of the suction vapour increases with increasing normal boiling point. Inspite of the large latent heat of vaporization (and hence large refrigerating effect q_0) for ammonia, because of its large specific volume, its suction volume requirement is nearly the same as that of R 22. Both have nearly equal N.B.Ps. Thus, the normal boiling point is once again established as the single factor which determines the refrigerating capacity per unit volume of the suction vapour.

One can, therefore, conclude that high-boiling refrigerants, such as water, R 123 and R 11 are low-pressure refrigerants and require large suction volumes. They are suited for use with centrifugal compressors in large-capacity refrigerating machines. Conversely, low-boiling refrigerants, such as ammonia, R 22, propane, carbon dioxide, etc., which are high-pressure refrigerants require small suction volumes and are suited for use with reciprocating compressors in medium and smaller capacity refrigerating machines or with screw compressors in large capacity ones. In the middle, we have isobutane, R 152a, R 134a and R 12 mostly used with reciprocating compressors and now rotary compressors in very small capacity machines such as domestic refrigerators, car air conditioners, water coolers, etc.

Note that R 404A, 407C, and R 410A are mixture blends. They are not pure substances. The values for t_s (N.B.P.) in table represent their temperatures at which they begin to boil at 1 atm pressure.

4.7.6 Isentropic Discharge Temperatures

Table 4.6 gives isentropic discharge temperatures of refrigerants for operating conditions of $t_k = 30°C$ and $t_0 = -15°C$.

It is evident that isobutane has the lowest discharge temperature. It is thus best suited for domestic refrigerators. Sealed units of isobutane refrigerators can be expected to give a very long life.

R 134a has a much lower discharge temperature than R 22. In fact, R 22 has the highest discharge temperature. Accordingly, R 134a is now doing well in hermetically sealed compressors of car air conditioners. R 134a has lower pressures also.

Also comparing R 123 with R 11, we see that large tonnage semi-hermetic centrifugal compressor units with R 123 are better than R 11 units because of much lower discharge temperature.

Table 4.6 Pressure ratios, discharge temperatures, power consumptions and COPs of refrigerants $t_k = 30°C$, $t_0 = -15°C$.

Refrigerant	*NBP*	*Pressure Ratio*	*Discharge Temp.*	*Power Consumption*	*COP*
	°C		*°C*	*kW/TR*	
R 113	47.59	7.71	30	0.703	4.81
R 123	27.82	6.81	33	0.717	4.9
R 11	23.71	6.25	43	0.693	5.02
R 600a	– 11.67	4.58	**30**	0.756	4.62
R 152a	– 24.02	4.64	46.7	0.72	4.78
R 134a	– 26.074	4.71	37	0.76	4.6
R 12	– 29.75	4.1	38	0.746	4.7
Ammonia	– 33.33	4.76	39	0.739	4.76
R 22	– 40.81	4.66	**53**	0.753	4.66
R 290	– 42.09	4.5	36	0.767	4.5
R 407C	– 43.63	4.5	48	0.781	4.5
R 502	– 45.174	4.4	38	0.802	4.38
R 404A	– 46.22	4.21	36	0.833	4.21
R 410A	–51.44	4.41	51	0.781	4.41
R 32	– 51.7	—	—	—	—

4.7.7 Coefficient of Performance and Horsepower per Ton

Table 4.6 gives values of theoretical COP and power per ton for 30°C condensing and – 15°C evaporating temperatures for refrigerants. The Carnot value of COP for the same operating conditions is 5.73 and HP/TR is 0.82. It is seen that R 11 has its COP closest to the Carnot value. Other refrigerants, with the exception of carbon dioxide, also have quite high values of COP. From the point of view of COP and power consumption R 22 is better than ammonia. Practically, all common refrigerants have approximately the same COP and horsepower requirement.

The low value of the COP of carbon dioxide is due to the fact that its critical temperature is too low and the condensing temperature is very close to it.

One of the peculiarities of fluorocarbons is their high density coupled with a low-boiling point. Since these are high molecular weight and high density substances, their latent heats are not large on a mass basis as the Trouton number is observed to be constant for a group of similar substances. However, low latent heat does not affect the COP of the cycle as the isentropic work is also small due to

(i) low values of the ratio of specific heats, γ, as seen from Table 4.7, and

(ii) small volumes of the suction vapour as seen from Table 4.5.

The very large value of the latent heat of vaporization of ammonia does not help much in this respect as the ratio of its specific heats is also high.

Table 4.7 Average specific heats ratios and latent heats of refrigerants at their N.B.P.s

Refrigerant	*NBP* °C	γ	$(h_{fg})_s$ kJ/kg
Water	100	1.33	2257.9
R 113	47.6	1.09	147.2
R 11	23.7	1.11	182.5
R 600a (Isobutane)	– 11.7	1.086	365.5
R 152a	– 24.0	1.134	329.9
R 134a	– 26.1	1.102	217.0
R 12	– 29.8	1.126	165.8
Ammonia	– 33.3	1.31	1368.9
R 22	– 40.8	1.166	233.7
R 290 (Propane)	– 42.1	1.126	425.8
R 32	– 51.7	—	382.0

4.8 CHEMICAL REQUIREMENTS

The chemical requirements of refrigerants pertain to flammability, toxicity, reaction with water, oil and construction materials and damage to refrigerated products.

4.8.1 Flammability

Hydrocarbons, such as methane, ethane, propane and butane are highly explosive and flammable. Ammonia is also explosive in a mixture with air in concentrations of 16 to 25 per cent by volume of ammonia. None of the CFCs is explosive or flammable. Some of the HFCs which are being considered as alternatives to CFCs because of the ozone layer depletions problem, such as R 152, R 152a, R 143, etc., are flammable. However, R 134a is not flammable.

4.8.2 Toxicity

From the consideration of comparative hazard to life from gases and vapours, compounds have been divided into six groups by Underwriters Laboratories.[10] Group six contains compounds with a very low degree of toxicity. It includes CFCs. Group one, at the other end of the scale, includes the most toxic substances such as sulphur dioxide.

Because of flammability and toxicity, ammonia is not used in domestic refrigeration and comfort air-conditioning.

4.8.3 Action of Refrigerant with Water

Among refrigerants, ammonia is highly soluble in water. It forms a non-ideal mixture with water and possesses more than *Raoult's law solubility*. For this reason, a wet cloth is put at the point of leak to avoid harm to human beings in ammonia refrigeration plants.

The solubility of water in fluorocarbons is rather low. It ranges from poor for highly fluorinated compounds such as R 134a to fairly good for those containing comparatively less fluorine.

The presence of moisture is very critical in refrigeration systems operating below 0°C. If more water is present than can be dissolved by the refrigerant, then there is danger of ice formation and consequent choking in the expansion valve or capillary tube used for throttling in the system. This is called *moisture choking*. This is avoided by proper dehydration of the unit before charging (sealed systems are baked at 130 to 150°C before charging and sealing) and by the use of a silica gel drier in the liquid line.

The solvent power of R 22 for water is almost 23 times that of R 12. In R 22 systems, therefore, ice formation generally does not occur. Thus, in addition to small, medium and large capacity air conditioning applications, R 22 is also used in low temperature food refrigeration plants so that moisture choking would not occur even at freezing temperatures.

In general, moisture should not be allowed to enter refrigeration systems.

4.8.4 Action with Oil

The solubility behaviour of a refrigerant and oil and consequent changes in the viscosity of oil determine the steps which must be taken to provide good lubrication.

In compressors, some oil is carried by the high temperature refrigerant vapour to the condenser and ultimately to the expansion valve and evaporator. In the evaporator, as the refrigerant evaporates, a distillation process occurs and the oil separates from the refrigerant. A build-up of oil in the evaporator will result in a reduced heat transfer coefficient, *oil choking* in the evaporator due to restriction caused to refrigerant flow and even blockage and ultimately to oil starvation in the compressor.

The solubility behaviour of a refrigerant in mineral oil may be classified as in Table 4.8 as follows:

(i) Immiscible.

(ii) Miscible.

(iii) Partially miscible.

Refrigerants that are not miscible with oil, such as ammonia or carbon dioxide, do not present any problems. In such a case, an oil separator is installed a little away from the compressor in the discharge line and the separated oil is continuously returned to the crank-case of the compressor.

Table 4.8 Solubility of refrigerants and mineral oils

Completely Miscible	*Partially Miscible*			*Immiscible*
	High Miscibility	*Intermediate Miscibility*	*Low Miscibility*	
R 11	R 13 B1	R 22	R 13	NH_3
R 600a				
R 12	R 501	R 114	R 14	CO_2
R 290				
R 21			R 115	R 134a
R 152a				
R 500			R 502	

R 134a, which is already in use as an alternative to R 12 in domestic refrigerators, car air conditioners, etc., is also not miscible in mineral oil. But, since we cannot install an oil separator in hermetically sealed systems, we have to use an oil which is miscible. Accordingly, a synthetic oil, polyolester (POE), is used in R 134a system.

Refrigerants that are completely miscible with oil, such as R 12, R 152a, R 290, R 600a, etc., also do not present problems.

For R 123 naphthenic mineral oil is used. Oil which reaches the evaporator, is returned to the compressor along with the refrigerant as it is the refrigerant-oil mixture which boils off in the evaporator. Thus, the miscibility of the refrigerant with lubricating oil ensures *oil-return* to the compressor. Any oil that gets separated in the evaporator can be returned to the compressor either by gravity or by entrainment by the high velocity suction vapours. For the purpose, the evaporatory exit should be above the suction line. The diameters of the suction lines are so designed that the velocity of the returning gas is sufficient enough to carry away the oil sticking to the walls of the tubing. A high velocity, however, increases the pressure drop which is undesirable.

In systems in which a refrigerant is only partially miscible with oil, the return of the oil to the compressor creates problems.

Most fluorocarbons are miscible with oil in all concentrations and at all temperatures. However, with R 22, there is partial miscibility. Refrigerant and oil are miscible at the condenser temperature, but separation takes place at the evaporator temperature. Two liquid phases are formed at low temperature, one predominantly consisting of the refrigerant and the other, oil, thus resulting in oil separation. The temperature at which liquid separation occurs depends on the nature of the oil and its concentration. Thus, a solution of R 22 with 10 per cent oil will separate into two layers at –5°C, but with 1 per cent oil, separation does not occur until –51°C. With 18 per cent oil, separation will occur even at 0.5°C. No matter how little oil goes into the evaporator, as the evaporation of refrigerant proceeds, the composition of oil in the liquid solution increases and it is bound to pass through the *critical composition* for separation which usually lies between 15 and 20 per cent of oil in the liquid phase.

The return of oil in these compressors, therefore, presents a problem. At low refrigeration temperatures, it is all the more acute. One solution of the problem is to install an efficient oil separator. Another is to use synthetic oils instead of mineral oils which are completely miscible with the refrigerant at temperatures as low as – 80°C. Among the synthetic oils, a polybutyl silicate and alkyl benzenes have better miscibility with R 22.

Also, in *direct expansion evaporators*, oil-refrigerant emulsion can be easily carried to the compressor by high velocity suction vapours. These evaporators are, therefore, preferred over flooded evaporators, and are particularly used for low temperature refrigeration and with refrigerants such as R 22.

In *flooded evaporators*, the separated oil-rich layer, being lighter, floats on top of the boiling liquid, and because of the extremely low velocity of the suction vapour, it cannot be carried to the compressor. Thus, when flooded evaporators are used with R 22, a connection must be provided for the overflow of oil from the evaporator to the compressor crank-case.

In the case of ammonia, however, the oil being heavier than the refrigerant, it collects at the bottom of the evaporator and can be drained out, if necessary.

4.8.5 Action with Materials of Construction

The choice of a particular refrigerant has also a bearing on the selection of the materials of construction. Ammonia, for example, attacks copper and copper-bearing materials. Iron and steel are, however, found suitable for use with ammonia. Hence, piping, fittings and valves are manufactured from iron and steel. Joints in ammonia systems are, therefore, of flange type with gaskets, or threaded or welded.

Most fluorocarbons are outstanding in chemical stability but, nevertheless, undergo a variety of chemical reactions which are slow except at high temperatures. The recommended material for use for piping with halocarbons is copper. It is soft and amenable to bending, fitting, etc. The fittings are of *flare* type. *Brazing* is also used for joining.

In the case of aluminium-halocarbon reactions, the heat of evolution increases with the fluorine content, indicating chemical instability with aluminium with increasing number of fluorine atoms. But the reverse is true for other common metals, viz., iron, copper, zinc, nickel, chromium, etc. A more vigorous reaction can be expected between aluminium and R 134a than between aluminium and R 22. With regard to aluminium, it must be noted that it is generally protected by a tenacious oxide film which makes it non-reactive. Because of its cheapness and availability, aluminium is used in place of copper in fluorocarbon refrigeration equipment. However, the thermal conductivity of Al is half that of Cu. Hence, Cu continues to be preferred for construction of heat exchangers in Freon systems.

4.9 PHYSICAL REQUIREMENTS

As physical requirements, we may include properties such as dielectric strength, thermophysical properties such as thermal conductivity, viscosity, etc., and leak tendency and detection.

4.9.1 Dielectric Strength

The dielectric strength of a refrigerant is primarily important in hermetically sealed units in which the electric motor is exposed to the refrigerant as the windings are cooled by the suction vapour.

As compared to nitrogen, the relative dielectric strength of R 22 is 1.31. This shows that fluorocarbons are good electrical insulators. The values for carbon dioxide and ammonia are 0.88 and 0.82 respectively.

4.9.2 Thermal Conductivity

A high thermal conductivity is desirable for a high heat transfer coefficient. It is found that R 22 has better heat transfer characteristics than R 12. The thermal conductivities of R 22, k_f and k_g in liquid and vapour phases respectively, in W/mK are expressed by the relations:

$$(k_f)_{R\,22} = 2.17783 \times 10^{-4} - 4.36214 \times 10^{-7}\, T - 9.3985 \times 10^{-12}\, T^2 \tag{4.7}$$

$$(k_g)_{R\,22} = -\,4.6143 \times 10^{-7} + 1.96 \times 10^{-8}\, T + 5.7143 \times 10^{-11}\, T^2 \tag{4.8}$$

The thermal conductivity of *liquid* R 134a and other new refrigerants can be estimated by the following Sato-Reidel[33] *similarity relation*.

$$k_f = \frac{\dfrac{1.11}{\sqrt{M}}\left[3 + 20\,(1 - T_r)^{2/3}\right]}{\left[3 + 20\,(1 - T_{r_s})^{2/3}\right]}, \text{ W/mk} \tag{4.9}$$

where T_{rs} is the reduced temperature at the normal boiling point.

The thermal conductivity of *vapour* R 134a is given by Eq. (4.10)

$$(k_g)_{R\,134a} = 29.742 - 0.17962\, T + 0.4265 \times 10^{-3}\, T^2, \text{ W/mk} \tag{4.10}$$

The expressions for thermal conductivities of R 600a in W/m.K are

$$(k_f)_{R\,600a} = 0.21225 - 0.487 \times 10^{-3}\, T + 0.2746 \times 10^{-6}\, T^2 - 0.421 \times 10^{-9}\, T^3 \tag{4.11}$$

$$(k_g)_{R\,600a} = 0.99013 \times 10^{-2} - 0.87055 \times 10^{-4}\, T + 0.474 \times 10^{-6}\, T^2 - 0.3715 \times 10^{-9}\, T^3 \tag{4.12}$$

4.9.3 Viscosity

Again, low viscosity is desirable for a high heat transfer coefficient. The viscosity of fluorocarbons in vapour phase is expressed by the empirical relation[27].

$$\mu_g = (A\,\sqrt{T} - B)10^{-4}, \text{ cP} \tag{4.13}$$

The constants for R 22 are:

$$A = 575.67 \text{ and } B = 3.619$$

Correlations for viscosity can be written in the form

$$\log \mu = A + BT + CT^2 + DT^3 + \cdots \tag{4.14}$$

The following relations give viscosities of R 22 in Ns/m^2 more precisely.

$$(\log \mu_f)_{R\,22} = 13.8457 - 0.3283\text{T} + 2.5965 \times 10^{-3}\, T^2 - 1.046 \times 10^{-5}\, T^3 + 2.109 \times 10^{-8}\, T^4 - 1.691 \times 10^{-11}\, T^5 \tag{4.15}$$

$$(\mu_g)_{R\,22} = -1.00493 \times 10^{-3} + 1.918 \times 10^{-5}\,T - 1.4446 \times 10^{-7}\,T^2 + 5.4084 \times 10^{-10}\,T^3 - 1.005 \times 10^{-12}\,T^4 + 7.422 \times 10^{-16}\,T^5 \quad (4.16)$$

The expression for R 600a for liquid in m.Pa.s, and for vapour in μ.Pa.s are as follows:

$$(\ln \mu_f)_{R\,600a} = 0.002086 + \frac{0.21739 \times 10^{-4}}{T} + 0.0636\,T - 0.80755 \times 10^{-4}\,T^2 \quad (4.17)$$

$$(\ln \mu_g)_{R\,600a} = 0.01497 - \frac{0.88035 \times 10^{-2}}{T} + 0.00331\,T - 0.1854 \times 10^{-5}\,T^2 \quad (4.18)$$

The viscosity of liquid R 134a in Ns/m^2 can be found by the relation given by Shankland et. al.[36], and that of vapour can be estimated by using the similarity relation given by Falkovskii[15]:

$$(\ln \mu_f)_{R\,134a} = -3.3528 + \frac{714.25}{T} - 0.19969 \times 10^{-2}\,T \quad (4.19)$$

$$\mu_g = 1.286 \times 10^{-4}\sqrt{M}\,p_c^{2/3}\,T_r \quad (4.20)$$

4.9.4 Heat Capacity

Heat capacities of liquid and vapour R 22 in kJ/kg.K are given by the expressions:

$$(C_{pf})_{R\,22} = 1.11857 + 1.3508 \times 10^{-4}\,T - 8.0852 \times 10^{-6}\,T^2 + 3.042 \times 10^{-8}\,T^3 \quad (4.21)$$

$$(C_{pg})_{R\,22} = -0.318712 + 6.90956 \times 10^{-3}\,T - 1.7844 \times 10^{-5}\,T^2 + 1.8026 \times 10^{-8}\,T^3 \quad (4.22)$$

The specific heat of liquid R 134a is approximately given by:

$$(C_{pf})_{R\,134a} = 1.1978 + 0.00415T - 1 \times 10^{-6}\,T^2 + 1 \times 10^{-7}\,T^3 + 2.5 \times 10^{-9}\,T^4 \quad (4.23)$$

4.9.5 Surface Tension

The value of surface tension is required in boiling heat transfer coefficient calculations. The empirical relation is

$$\sigma = \sigma_0\,T_R^{\,a}\,(1 + b\,T_R^{\,c}) \quad (4.24)$$

where T_R is equal to $(1 - T/T_c)$. It is *not* equal to the reduced temperature which is $T_r = T/T_c$. The constants for R 22 are $\sigma_0 = 0.06993$ N/m, $a = 1.285$, $b = -0.154$, $c = 0.87$.

The equations used for estimation of σ by method of corresponding states are as follows:

$$\sigma = p_c^{2/3}\,T_c^{1/3}\,Q\,(1 - T_r)^{11/9} \quad (4.25)$$

$$Q = 0.1196\left[1 + \frac{T_{r_s}\ln(p_c/1.01325)}{1 - T_r}\right] - 0.279 \quad (4.26)$$

4.9.6 Leak Tendency

The leak tendency of refrigerants should be nil. Also, the detection of a leak should be easy. The greatest drawback of fluorocarbons is the fact that they are odourless. This, at times, results in a complete loss of costly gas from leaks without being detected. An ammonia leak can be very easily detected by its pungent odour. Leaks in ammonia plants are very common due to the use of glands and lead gaskets in joints and due to corrosion.

Several methods are available for the detection of leaks. The most common is the *soap-bubble method.* The other is the *halide torch method* used with fluorocarbons. In this, a methyl alcohol or hydrocarbon flame is used which is light blue in colour, but which turns bluish green in the presence of halocarbon vapours. For ammonia leaks, a burning sulphur taper is used which in the presence of ammonia forms white fumes of ammonium sulphite.

An *electronic leak detector* is most sensitive and is used in the manufacture and assembly of refrigeration equipment. The operation of the instrument depends on the flow of current due to the ionization of the leaking gas between two opposite-charged platinum electrodes.

4.9.7 Cost of Refrigerant

The consideration of the cost of a refrigerant has for many years commended the use of ammonia as a refrigerant in large industrial plants, such as cold storages and ice plants. The cost factor, however, is only relevant to the extent of the price of the initial charge of the refrigerant which is very small compared to the total cost of the plant and its installation. The cost of losses due to leakage is also important. In small-capacity units requiring only a small charge of the refrigerant, the cost of refrigerant is immaterial.

Ammonia is the cheapest refrigerant. Also it is environment-friendly. The future may see the return of ammonia in a big way.

4.10 OZONE DEPLETION POTENTIAL AND GLOBAL WARMING POTENTIAL OF CFC REFRIGERANTS

The earth's *ozone layer* in the upper atmosphere (stratosphere) is needed for the absorption of harmful ultraviolet rays from the sun. In 1985, the world was shocked to find a gaping hole above Antartic in the ozone layer that protects the earth from ultra-violet rays. These rays can cause skin cancer. CFCs have been linked to the depletion of this ozone layer. They have varying degrees of *ozone depletion potential* (*ODP*). In addition, they act as *greenhouse gases*. Hence, they have *global warming-potential* (*GWP*) as well. According to an international agreement (*Montreal Protocol*), the use of fully halogenated CFCs (no hydrogen at all in the molecule), that are considered to have high *ODP*, viz., the commonly used refrigerants R 11, R 12, R 113, R 114 and R 502, have been phased out from the year 2000 AD. HCFCs have much lower ODP. They have some GWP. They have to be phased out by 2030. But until 2030, they can be used. R 22, is an HCFC. Its *ODP* is only 5% of that of R 12. It will be phased out by the year 2030 AD. But R 22 continues to remain very popular as a refrigerant.

Even with the measures taken so far, as late as 2008, a 2.7 million square kilometers ozone layer hole was detected above Antarctic.

In 1974, two scientists including a radiochemist S. Rowland postulated that CFCs, because they are so stable, have a long life in the lower atmosphere. And, inspite of CFCs being heavier than N_2 and O_2, these slowly migrate into the upper atmosphere by molecular diffusion caused by partial pressure difference. It was hypothesized that the chlorine atoms from the molecule would be split off by the action of sunlight, and the free chlorine will react with ozone in the stratosphere, according to the following reactions:

$$CCl_2F_2 \xrightarrow{\text{sunlight}} CClF_2 + Cl$$

$$O_3 + Cl \xrightarrow{\text{sunlight}} ClO + O_2$$

Thus, O_3 will be depleted to O_2. The problem with CFC is that of chain reaction. A single atom of Cl released from CFC reacts taking out 100,000 O_3 molecules. That is why, even a small concentration of CFC also becomes inportant.

4.11 SUBSTITUTES FOR CFC REFRIGERANTS

CFC substitutes fall into four categories: HFCs and HCFCs, HCs, and inert gases.

Hydrocarbons (HCs) and hydrofluorocarbons (HFCs) provide an alternative to fully halogenated CFC refrigerants. They contain no chlorine atom at all and, therefore, have zero *ODP*. Even hydrochlorofluorocarbons (HCFCs) like R 22 and R 123 which do contain chlorine atom/s, but in association with H-atoms, have much reduced *ODP*. The association of one or more H-atoms allows them to dissociate (break apart) faster in the lower atmosphere of the earth. Chlorine, thus released, gets absorbed by rain water, etc., like the chlorine used in the chlorination of water. So fewer Cl-atoms reach the ozone layer in the upper atmosphere. However, HCFCs have a level of ODP in addition to GWP. Hence, these also have to be phased out ultimately. Till then, they can be used as *transitional refrigerants*.

The HFCs, on the other hand, because of their H-content may be flammable to some extent. The degree of flammability depends on the number of H-atoms in the molecule. Pure HCs are, of course, flammable to highly flammable.

The two most common refrigerants which have very high *ODP* are the highly chlorinated CFCs R 11 and R 12. The others are R 113 and R 502. So, R 11, R 12, R 113, R 114, and R 502 have been replaced by substitutes.

R 22, which is an HCFC, has 1/20th the *ODP* of R 11 and R 12. Hence, it can continue to be used for quite sometime. However, all HCFCs have *GWP*. Because of this, even R 22 will have to be ultimately phased out by 2030 AD.

Based on knowledge as at present, the list of refrigerants currently being used, and those which are potential candidates for future is as follows:

Methane Series

HCFC R 22 ($CHClF_2$), Monochlorodifluoromethane
HFC R 32 (CH_2F_2), Difluoromethane

Ethane Series

HCFC R 123 ($CHCl_2 - CF_3$), Dichlorotrifluroethane
HFC R 125 ($CHF_2 - CF_3$), Pentafluoroethane

HFC R 134a ($CH_2F - CF_3$), Tetrafluoroethane
HFC R 143a ($CH_3 - CF_3$), Trifluoroethane
HFC R 152a ($CH_3 - CHF_2$), Difluoroethane

Propane Series
HFC R 245fa ($C_3H_3F_3$), Pentafluoropropane
HC R 290 (C_3H_8), Propane

Butane Series
HC R 600a (C_4H_{10}), Isobutane

Zeotropic Blends
HFC R 404A [R 125/143a/134a (44/52/4)]
HFC R 407C [R 32/125/134a (23/25/52)]
HFC R 410A [R 32/125 (50/50)]

Azeotropic Blends
HFC R 507A [R 125/143a (50/50)]

Inorganic Refrigerants
R 717 (NH_3), Ammonia
R 718 (H_2O), Water
R 744 (CO_2), Carbon dioxide

The only refrigerant that is currently being used from methane series is R 22. The refrigerants being currently used from ethane series are R 134a and R 123. R 152a is also being used in Europe. Further R 600a, isobutane is also being used in domestic refrigerators in Europe.

We have seen that N.B.P. is the single most important characteristic of a substance to be used as a refrigerant. It also governs the type of application, the equipment, and the refrigerating capacity for which a particular refrigerant is to be used. From this point of view, substances with their N.B.Ps in the range – 50°C to + 50°C are considered suitable for use as refrigerants.

Devotta[17, 18], Mclinden[30] and others have conducted surveys of CFC alternatives. Table 4.9 gives the historic, current, and future candidate refrigerents as possible alternatives arranged in order of their decreasing N.B.Ps. The flammability or non-flammability of these substances is also mentioned. Note that flammability may not be hazardous in small units such as refrigerators, water coolers, etc., in which the amount of charge is small.

On the basis of N.B.Ps HCFC R 123 is a potential alternative to R 11; HFC R 134a is potential alternative to R 12; HCs R 600a and R 290 can also be used in place of R 12.

Note *It is seen that most of the substitutes are ethane series compounds. Note that R 404A, R 407C, and R 507A are near azeotropic blends. The first figure under their NBP column indicates dew temperature at atmospheric pressure. The second figure indicates bubble temperature. Further note that difference between dew and bubble temperatures for R 410A is only 0.08°C. It is like a pure substance. The difference for R 407C is large (7°C).*

Table 4.9 Common CFCs and possible alternatives with normal boiling points

Designation	*Category*	*Chemical Formula*	*N.B.P., °C*	*Flammability*
R 113	CFC	$C_2 Cl_3 F_3$	47.68	Non-flammable
R 141b	HCFC	$CH_3 CCl_2 F$	32.1	Slightly flammable
R 152	HFC	$CH_2 F CH_2 F$	30.7	Flammable
R 123	HCFC	$C H Cl_2 CF_3$	27.82	Non-flammable
R 11	CFC	$CCl_3 F$	23.7	Non-flammable
R 245 fa	HFC	–	14.9	Flammable
R 600a (Isobutane)	HC	$(CH_3)_3 CH$	– 11.67	Flammable
R 134	HFC	$CHF_2 CHF_2$	– 19.8	Non-flammable
R 152a	HFC	$CH_3 CHF_2$	– 24.02	Slightly flammable
R 134a	HFC	$CF_3 CH_2F$	– 26.07	Non-flammable
R 12	CFC	$CCl_2 F_2$	– 29.8	Non-flammable
R 717 (Ammonia)		NH_3	– 33.3	Flammable
R 22	HCFC	$CHClF_2$	– 40.8	Non-flammable
R 290 (Propane)	HC	$C_3 H_8$	– 42.1	Flammable
R 407 C	HFC	–	– 43.63/–36.63	–
R 502	CFC	–	– 45.4	Non-flammable
R 404 A	HFC	–	– 46.22/– 45.47	–
R 507 A	HFC	–	– 46.74	–
R 143a	HFC	$CH_3 CF_3$	– 47.35	Slightly flammable
R 125	HFC	$CHF_2 CF_3$	– 48.55	Non-flammable
R 410 A	HFC	–	– 51.44/– 51.36	–
R 32	HFC	CH_2F_2	– 51.7	Slightly flammable

It may be noted that R 123 has 98% less *ODP* and *GWP* than R 11. R 152a and R 134a have no Cl atoms. Hence, these have zero *ODP*. They have 74% less *GWP* as compared to R 12. Hydrocarbons have zero *ODP*.

HCFC 123 has 4.3°C higher boiling point than CFC 11. It is, therefore, a lower pressure replacement for CFC 11, thus having larger specific volume of suction vapour. Hence, its use results in 10–15% reduction in capacity if used in existing CFC 11 centrifugal compressors.

The boiling point of HFC 134a is again 3.6°C higher than that of CFC 12. It has lower suction pressure and larger suction vapour volume. For use in existing CFC 12 reciprocating compressors, it would require either an average increase in compressor speed of 5–8%, or an equivalent increase in cylinder volume.

HCFC 22 is, nowadays, gaining greater importance even as a transition refrigerant.

It may be noted that HCFC 22 is already being used with screw compressors in very large capacity, and with scroll compressors in package units.

Further, in addition to pure substances, binary mixtures of substances with *higher* and *lower boiling* points mixed in proportion to give an effective boiling point nearly equal to that of the refrigerant to be replaced (as illustrated in Fig. 4.2) can also be used.

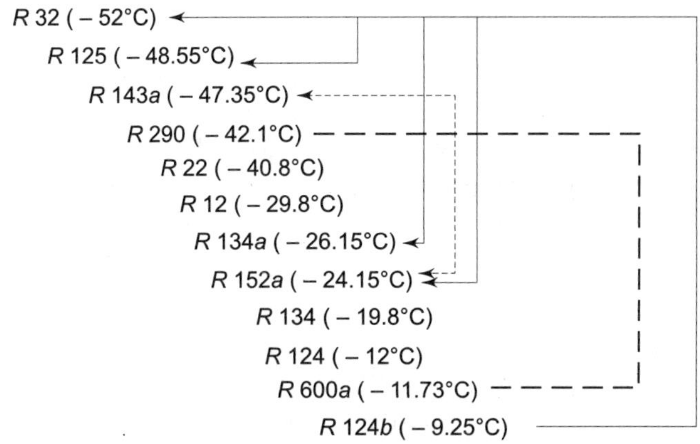

Fig 4.2 Forming binary mixtures of higher and lower boiling refrigerants

Figure 4.2 illustrates how this can be done to find sustitutes for R 12 and R 22.

An example is R 290/R 600a, viz., propane/isobutane mixture in 50-50% composition by mass. This mixture has been found suitable to replace R 12 in refrigerators.

The other example is R 410A formed by 50% R 32/50% R 125. It is being used as a substitute for R 22.

It may be pointed out that mixtures boil and condense through a range of temperature. This *temperature glide* might be an advantage in applications in which *non-isothermal refrigeration* is required (See Sec. 4.23).

Ternary mixtures can also be used. Rail cars of a company in USA use du Pont's SUVA HP 62 HFC R 404A (R 125 + R 143a + R 134a) blend.

To sum up, Table 4.10 mentions the common refrigerants used before the year 2000, viz., CFCs 11, R 12, and 502, and HCFC 22, and it lists the substitutes as follows:

(i) Short-term substitutes which are currently being used.
(ii) Long-term substitutes which might be used even after 2030.

Table 4.10 Short-and long-term substitute refrigerants

Refrigerants used Before 2000	*Substitute Refrigerent*	
	Shot Term	*Long Term*
CFC 11	HCFC 123 HFC 134a	HCFC 123, HFC 134a, HFC 245fa.
CFC 12		HFC 134a, HC 600a
HCFC 22	HCFC 22	HFC 32, HFC 134a, R 407C, R 410A. Other blends of HFC 32, HFC 134a, and others
R 502	HCFC 22	HFC 125, R 507A. Other blends of HFC 32, HFC 125.

4.12 SUBSTITUTES FOR CFC R 12

CFC R 12 was the most commonly used refrigerant in small hermetically sealed systems. This was because of its high stability, excellent thermodynamic properties, low index of compression making it suitable for use at extreme pressure ratios, etc., and good motor winding cooling characteristics.

Presently, HFC R 134a (Tetrafluoroethane) is the most preferred substitute for R 12. Its N.B.P. of – 26.67°C is closest to R 12's N.B.P. of –29.8°C. However, it should be noted that R 134a has relatively high GWP. The use of oil in R 134a system requires a very stringent quality control. R 134a is not soluble in mineral oil. The polyester-based synthetic oil that is used with it should be *totally dry*. This is difficult considering the fact that the synthetic ester oils are 100 times more *hygroscopic* than mineral oils.

With vibrations present, and car A/C not being a completely sealed system, there would be chances of frequent gas leaks. So R 134a is being used in place of R 12. It has lower pressures as well.

Why not hydrocarbons? Hydrocarbons have zero *ODP* and negligible *GWP*. Earlier researches for alternatives had excluded hydrocarbons because of their flammability. However, hydrocarbons are readily available, much cheaper and thermodynamically very suitable. In any case, the amount of refrigerant charge required in domestic refrigerators is only a few grams. So, R 290 (Propane) can be considered for use at least in domestic refrigerators. In fact, it is already in use in some refrigerators in Europe.

Another attractive HC refrigerant from the list in Table 4.9 for use in domestic refrigerators is R 600a (Isobutane). Whereas R 290 is a lower boiling, higher pressure and smaller suction vapour volume substitute, R 600a is higher boiling, lower pressure and larger suction vapour volume refrigerant.

But because of their widely different N.B.Ps, neither R 290 nor R 600a can be used as drop-in substitutes in place of R 12. However, by mass, a 50% R 290 + 50% R 600a mixture has exactly the same pressures as R 12. Its volume refrigerating capacity is also the same. Hence, this mixture can be considered as a drop-in substitute.

One more attractive alternative is R 152a (Difluoroethane). Its *GWP* is one order of magnitude less than that of R 134a.

4.12.1 Comparative Study of Alternatives to CFC R 12 in Domestic Refrigerators

In domestic refrigerators, the inside air is cooled through natural convection by the evaporating refrigerant, and the refrigerant is also condensed through natural convection by the outside air. In frost-free refrigerators, however, a fan is used to circulate air between freezer compartment and the evaporator coil. Both in the evaporator and the condenser, one side fluid is very low thermal conductivity air. This results in very low heat transfer coefficients. Hence, air-side coefficient becomes the *controlling coefficient*. The overall heat transfer coefficient is lower than this controlling coefficient. Accordingly, whichever be the refrigerant used, the overall coefficient would not change. Hence, if an alternative refrigerant is used in a refrigerator in place of CFC R 12, it would not be necessary to change the evaporator and the condenser. Only the compressor, its motor and the capillary tube would have to be changed.

Ashok Babu[9] has carried out an exhaustive theoretical analysis of five pure refrigerants R 290, R 22, R 134a, R 152a and R600a, and the binary mixture 50% R 290 + 50% R 600a for finding out the suitability of alternate refrigerants for R 12 in refrigerators, and for investigating the required modifications.

Analysis was done by Ashok Babu on a 165L, 89W refrigerating capacity, 4.33 cm^3 piston displacement refrigerator operating on the standard rating cycle as described in Fig. 3.21. The pressure drops Δp_s and Δp_d at suction and discharge valves were assumed as follows:

For R 290 and R 22	$\Delta p_s = 0.2$ bar,	$\Delta p_d = 0.4$ bar
For R 12, R 134a and R 152a	$\Delta p_s = 0.1$ bar,	$\Delta p_d = 0.25$ bar
For R 600a	$\Delta p_s = 0.03$ bar,	$\Delta p_d = 0.05$ bar

The operating parameters are given in Table 4.11. Two sets of calculations were done for performance parameters:

(i) One using the same R 12 compressor with alternate refrigerants considered as drop-in substitutes.

(ii) Another with an appropriate sized compressor to obtain the same refrigerating capacity of 89W with different refrigerants.

The results of calculations are given in Tables 4.12 and 4.13 respectively. From Table 4.12 it is clear that the higher boiling R 600a, R 152a and R 134a give decreased capacity and the lower boiling R 22 and R 290 give increased capacity.

Further comments about the respective alternatives are as follows:

R 600a (Isobutane) It has very large volume of the suction vapour $\dot{V}^*/\eta_v$, about 1.8 times that of R 12. Hence, it requires a much larger sized compressor. Pressure drop across the capillary is the lowest. Hence, R 600a requires a slightly shorter capillary than 3m of R 12. Further, R 600a has lowest value of γ. Hence, it has the lowest discharge temperature and, therefore, lowest winding temperature inspite of it having the highest pressure ratio. The compressor therefore runs very cool ensuring a long life for guarantee of 5 years as required. However, it has vacuum in the evaporator. Experiments show that it requires 1.75 times the size of R 12 compressor.

Additional advantage is its high COP equal to that of R 12.

R 152a (Difluoroethane) and R 134a (Tetrafluoroethane) Both these higher boiling substances have similar characteristics except that R 152a has slight vacuum in the evaporator at –25°C, and that the discharge temperature of R 152a is much higher because of its higher value of γ. And R 134a has the highest COP, even greater than that of R 12. Otherwise, both refrigerants require larger displacement compressors, viz., 1.35 times for R 134a and 1.43 times for R 152a.

Further, pressure drop in R 152a capillary is the same as in the case of R 12. R 134a has larger pressure drop, 13.85 bar, as against 12.37 bar for R 12. However, the exact capillary size needs to be calculated in each case according to $\dot{m}$, $p_k - p_o$ and thermophysical properties.

From Table 4.12, it can be seen that the refrigerating capacity is 61 W with R 134a as against 89 W for R 12. To obtain a capacity of 89 W with R 134a as well, the following modifications can be considered:

Table 4.11 Properties data of R 12 and new proposed refrigerants for $t_0 = -25°C$ and $t_k = 55°C$.

Refrigerants	*NBP* °C	p_0 bar	p_k bar	p_2/p_1	p_k-p_0 bar	v_1 m^3/kg	h_{fg} kJ/kg	γ
R 290	– 42.07	2.02	19.07	10.71	17.05	0.2913	404.99	1.126
R 22	– 40.76	2.01	21.74	12.23	19.74	0.1496	223.72	1.166
R 12	– 29.79	1.24	13.61	13.47	12.37	0.1803	163.34	1.126
R 134a	– 26.2	1.07	14.92	17.64	13.85	0.2592	221.83	1.102
R 152a	– 25	0.98	13.32	17.67	12.34	0.4314	328.25	1.134
R 600a	– 11.73	0.59	7.82	21.25	7.23	0.8876	379.49	1.086
R 290/R 600a		1.4	14.22	13.29	12.82	0.4142	390.13	1.104

Table 4.12 Calculated performance parameters of R 12 and new proposed refrigerants using same compressor as in R 12 refrigerator

Refrigerants	$\dot{m} \times 10^3$ kg/s	η_v	t_{2n} °C	t_{2s} °C	$\dot{Q}_o$ W	$\dot{Q}_k$ W	$\dot{Q}_v$ kJ/m^3	$\dot{W}_{is}$ W	$\dot{W}_a$ W	$V^* \times 10^3$ m^3/s/TR	W^* W/TR	$COP = \dot{Q}_o/\dot{W}_{is}$	τ_{stc} Nm
R 290	0.5896	0.936		125	155	282	808	88.4		4.35	2080	1.76	3.84
R 22	1.135	0.925		170	178	307	936	101.9		4.76	1995	1.75	4.42
R 12	0.8022	0.772	56.8	135	89	154	555	47.5	44.47	6.39	1786	1.84	2.77
R 134a	0.4437	0.608	60.5	125	65	113	518	33.86	32.46	6.79	1761	1.91	3.10
R 152a	0.2617	0.621	66.8	155	61	106	478	34.92	29.91	7.35	1723	1.75	2.77
R 600a	0.1941	0.894	53.7	110	48	86	260	25.85	24.68	13.54	1819	1.85	1.60
R 290/ R 600a	0.3039	0.612	70.8	116	77	129	609	35.0	28.1	5.78	2066	1.70	2.93

Table 4.13 Calculated performance parameters of R 12 and new proposed refrigerants for same refrigerating capacity of 89 W

Refrigerants	$\dot{m}\times10^3$ **kg/s**	$\dot{W}_{is}$ **W**	$\dot{Q}_k$ **W**	$\dot{Q}_k/(\dot{Q}_k)_{R\,12}$	V_p **c c**	$V_p/(V_p)_{R\,12}$	τ_{stQ} **Nm**
R 290	0.331	49.65	158.5	1.033	2.431	0.561	2.158
R 22	0.5564	49.93	150.4	0.980	2.122	0.49	2.168
R 12	0.80	47.93	153.4	1.	4.33	1.	2.77
R 134a	0.5973	45.58	151.9	0.990	5.828	1.346	4.167
R 152a	0.374	47.4	151.3	0.986	6.187	1.429	3.957
R 600a	0.355	47.27	156.4	1.02	7.917	1.83	2.933
R 290/R 600a	0.3462	51.3	147.2	0.96	4.933	1.139	3.333

(i) Using a next higher size compressor available with the manufacturer, having a displacement of 5.52 cm^3 in place of the existing 4.33 cm^3 of R 12 compressor.

(ii) Increasing evaporator temperature from – 25°C to –23°C.

(iii) Increasing liquid subcooling to 20 kJ/kg.

R 22 and R 290 (Propane) Both these are lower boiling and higher pressure refrigerants. Hence, they require much smaller displacement compressors, coupled to slightly larger motors. As seen from Table 4.13, R 22 requires only 0.49 times the size of R 12 compressor, and R 290 requires about 0.56 times. Both require longer capillaries due to larger pressure drops. Both have lower COPs.

Note that R 22 has the highest discharge temperature due to a very high value of γ, hence not suitable for domestic refrigerators. In any case, it is an HCFC. However, propane has low discharge temperature as compared to R 12. James and Missenden[23] have done studies on the use of propane in domestic refrigerators.

50% R 290/50% R 600a Mixture The binary mixture of equal proportions of R 290 and R 600a has nearly the same pressures as R 12. It also has the same pressure drop across the capillary as R 12. Table 4.13 shows 13.9% larger piston displacement required for the mixture compressor. However, experiments show that the mixture gives only about 8% less capacity using R 12 compressor. Again, the anamoly is due to assumption of pressure drops and heat transfer characteristics in the compressor. It is thus found that the equal mass R 290/R 600a mixture can be used as a *drop-in substitute* for R 12.

A point to be kept in mind, however, is the fact that the mixture undergoes temperature glide during boiling and condensation. If the composition is slightly affected, it may cause instable operation in the refrigerator. As such pure substances, propane and isobutane are preferable.

4.12.2 What if R 22 or R 290 is Used as a Drop-in Substitute in R 12 Refrigerator?

If R 22 is charged in a R 12 refrigerator, it will result in instant burn-out of the hermetically sealed unit:

(i) Due to the very high value of γ for R 22, its discharge temperature being the highest.

(ii) Coupled with higher mass flow rate of R 22, basically on account of higher pressure and smaller specific volume of suction vapour, it will cause increased load on the compressor motor.

On the other hand, things are not so bad with R 290 (Propane). Both on account of lower p_k/p_0 and lower γ, the discharge temperature of propane is much lower. However, the other two problems remain with R 290 viz.,

(i) smaller specific volume of suction vapour in compressor, and

(ii) higher evaporator pressure because of using same undersize capillary.

Hence, the motor would be overloaded, resulting in eventual burn-out.

The advantage of R 290 is its relatively lower discharge temperature.

Note *Accordingly, if compressor cylinder size is reduced by 44% keeping the size of the motor same, and capillary length is suitably increased, there is no reason why R 290 could not be used in place of R 12. In fact, R 290 compressor runs cooler than R 12; of course, R 600a compressor runs cooler than even R 290.*

4.12.3 Starting Torques of Motors

Starting torques (τ) of compressor motors are calculated on the basis of difference of discharge pressure and suction pressure and displacement volume of compressor. Thus,

$$\tau = \frac{1}{2}(p_k - p_0)V_p$$

The inertia forces have not been taken into account in this expression. Also, viscosity of refrigerants has not been considered. Oil effects on starting torques are also involved.

In Table 4.13, the starting torque requirements for appropriate sized compressors to produce the same refrigerating capacity of 89 W for different refrigerants—as calculated from the above expression—are listed. It is seen that R 134a (4.17 Nm), R 152a (3.96 Nm) and R 600a (2.93 Nm) demand higher starting torque motors than R 12 (2.77 Nm). Therefore, if available R 12 compressors of required larger displacement volumes are used with these substitutes, the starting torques of the motors could be inadequate, and starting problems might result. The motor could trip constantly.

In fact, the problem does exist in the cases of R 134a and R 152a. The solution to the problem is either to change the motor to a higher torque one, or to increase the starting torque of the existing motor by putting a capacitor in series with the starting winding.

The problem does not occur in the case of R 600a as the difference $(p_k - p_0)$ is small. But, it requires a larger V_p, and hence a higher starting torque of 2.93 Nm as against 2.77 Nm for R 12. But a larger displacement R 12 compressor would have a motor which would have a starting torque greater than 2.93 Nm, even if it is designed for R 12, because its $p_k - p_0$ is much more than that of R 600a.

4.12.4 Refrigerant Charge

The quantities of refrigerants charged by Ashok Babu in his experiments on 165L refrigerator are given in Table 4.14.

Table 4.14 Quantities of refrigerants charged

Refrigerant	*Quantity Charged in Grams*
R 12	120
R 134a	135
R 152a	66
R 600a	60
50% R 290/50%/R 600a	40

In the experiment, these quantities are about 33% more than otherwise required due to the instruments, connecting lines, etc. The R 12 charge specified by the manufacturer was only 90 g.

Note that refrigerant change with hydrocarbons R 290 and R 600a is very small. This is essentially due to their very high latent heat of vaporization, and hence very low mass flow rate. Inspite of these being hydrocarbons, the small charge favours their use in refrigerators ignoring flammability.

4.12.5 Summary of Factors Favouring/Disfavouring Use of Alternatives in Domestic Refrigerators/Water Coolers

Table 4.15 gives a summary of factors favouring and/or disfavouring use of each of the alternatives. In this summary, R 22 has been simply ruled out because of its very high discharge and winding temperatures, and also because it is an HCFC.

Table 4.15 Favourable and unfavourable factors of alternatives for use in domestic refrigerators

Refrigerant	*Favourable Factors*	*Unfavourable Factors*
R 290	1. Zero *GWP*. 2. Compatible with mineral oil. 3. Positive pressure in evaporator. 4. Low discharge and winding temperatures. 5. Small refrigerant charge required. 6. Easily available.	1. Flammable but charge is small. 2. Not very suitable at high ambient conditions of 43°C. 3. Very small piston displacement compressor required. May be suitable for large size refrigerators
R 134a	1. Non-flammable. 2. Zero ODP 3. Discharge and winding temperatures lower than R 12. 4. Highest COP, hence energy saving.	1. Significantly high *GWP*. 2. Synthetic oil used highly hygroscopic causing moisture and oil choking problems. 3. Stringent oil quality and perfect workmanship required.
R 152a	1. Zero *GWP*. 2. Compatible with mineral oil.	1. Slightly flammable. 2. Slight vacuum in evaporator at –25°C 3. High discharge/winding temperatures.
R 600a	1. Zero *GWP*. 2. Compatible with mineral oil. 3. Lowest system pressures. Very light construction.	1. Flammable (but quantity of charge very small in refrigerators).

(Contd)

Table 4.15 *(Contd)*

Refrigerant	*Favourable Factors*	*Unfavourable Factors*
	4. Lowest discharge/winding temperatures. Compressor runs very cool. 5. High COP. Low energy consumption 6. Lower starting torque motor required in comparison to R 134a and R 152a. 7. Quantity of charge very small. 8. Easily available.	
50% R 290 + 50% R 600a Mixture	1. Zero *GWP* 2. Compatible with mineral oil. 3. Pressures same as in R 12 system. Almost like a drop-in substitute. 4. Low discharge/winding temperatures. 5. Quantity of charge very small. 6. Easily available.	1. Flammable 2. Temperature glide during evaporation and condensation. 3. Leaks and composition imbalance affects functioning. 4. Low COP.

It is thus seen that both the hydrocarbons, R 290 and R 600a, have many factors which make them suitable for use in domestic refrigerators in place of R 12.

Note *Neither hydrocarbons because of their flammability, nor R 22 because of high pressures and temperatures, would be suitable for use in car air conditioners. R 134a has been chosen-for automotive air conditioning.*

In the third edition which came in 2009, there was great emphasis on HCFC R 123 as centrifugal chillers refrigerant and HFC R 134a as long term alternative for many applications. Emphasis on HCFC R 22 and ammonia remained as such. Detailed comparisons were made, however, between HCFC R 22 and its HFC alternatives, and between HCFC R 123 and its alternatives.

After 2009, much development has taken place and certain refrigerants have been accepted widely. For example, in place of CFC R 12, HFC R 600a (Iso-butane) and HFC R 134a are in wide use in refrigeraters. And though HCFC R 123 continues to be used, designers and manufactures have adopted the highly unlikely HFC R 134a for use in centrifugal chillers in spite of it being a much lower boiling than HCFC R 123, high pressure and smaller volumes refrigerant. As for HCFC R 22, HC R 32 is a new-found miracle. In addition, HFC blend R 410a and HFC R 134a are in use.

Further, inverter technology has become for common for capacity control.

Hence, it had become absolutely necessary and our due responsibility to include these developments in this revised edition.

4.12.6 Refrigerants Used in Most Refrigerators

R 600a (Isobutane) and R 134a have been the most popular refrigerants used in refrigerators and other small unitary equipment in recent times.

It is seen that R 600a has the most factors favourable for use in refrigerators. It has both zero OPD and zero GWP. It is compatible with mineral oil. It has lowest pressures. Hence lighter and cheaper construction equipment is required. It has lowest discharge and hence winding temperature. Therefore, the compressor runs the coolest leading to long life of the unit. The most important is the fact that the quantity of charge required in refrigerators is very small. Hence it can be used in spite of its flammability considerations.

R 134a is also preferred because it is a zero ODP non-flammable refrigerant. It also has the highest COP leading to energy savings in operation. Its discharge temperature is lower than that of R 12. Hence, the compressor runs cooler and has longer life.

Example 4.1 *Implications of Charging Propane in R 12 Refrigerator*

Figure 4.3 shows the p-h diagram of R 12 cycle in an 89 W domestic refrigerator. State 1″ leaving evaporator is at – 15°C.

State 1 after regenerator and at the end of suction line is at 32°C. Process 3–3′ represents subcooling in regenerative heat exchanger between vapour and liquid. State 3 leaving condenser is at 40°C.

What will be the results of charging this refrigerator with propane without changing hermetic compressor? Find to the first approximation.

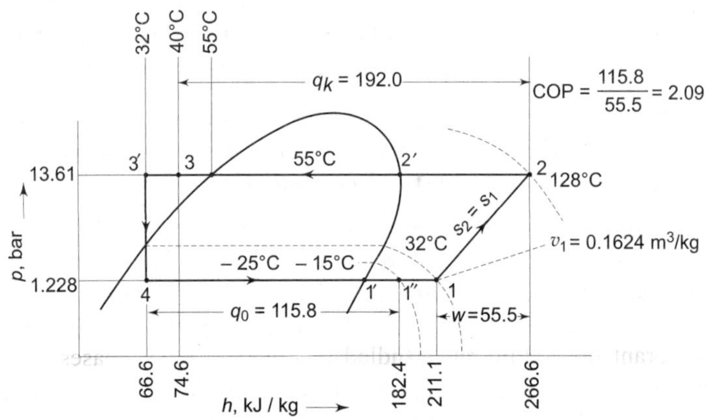

Fig. 4.3 R 12 cycle for refrigerator in Example 4.1

Solution Figure 4.4 shows the standard refrigerator cycles with propane.

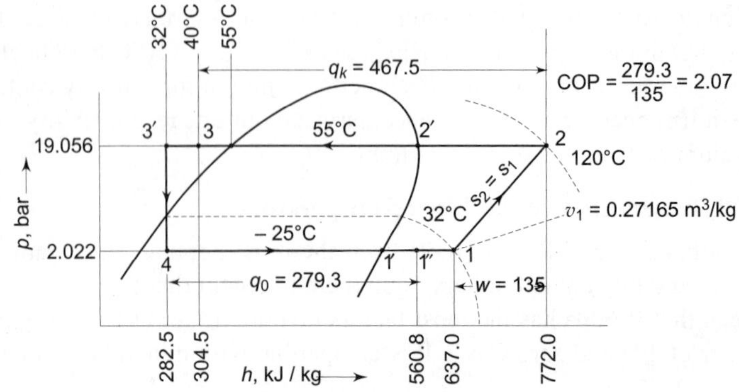

Fig. 4.4 Propane cycle for refrigerator in Example 4.1

Mass flow rate of refrigerant, piston displacement, motor watts and heat rejected in condenser for R 12 refrigerator are as follows:

$$(\dot{m})_{R\,12} = \frac{\dot{Q}_0}{q_0} = \frac{89 \times 10^{-3}}{115.8} = 7.68 \times 10^{-4}\,\text{kg/s}$$

$$\dot{V}_p = (\dot{m}v_1)_{R\,12} = 7.68 \times 10^{-4}\,(0.1624) = 1.247 \times 10^{-4}\,\text{m}^3\text{/s}$$

$$(\dot{W})_{R\,12} = (\dot{m}w)_{R\,12} = 7.68 \times 10^{-4}\,(55.5 \times 10^3) = 43\,\text{W}$$

$$(\dot{Q}_k)_{R\,12} = (\dot{m}q_k)_{R\,12} = 7.68 \times 10^{-4}\,(192 \times 10^3) = 147\,\text{W}$$

With the same compressor, and hence for the same piston displacement, we have with propane

$$(\dot{m})_{R\,290} = \frac{\dot{V}_p}{v_1} = \frac{1.247 \times 10^{-4}}{0.27165} = 4.59 \times 10^{-4}\,\text{kg/s}$$

$$(\dot{Q}_0)_{R\,290} = (\dot{m}q_0)_{R\,290} = 4.59 \times 10^{-4}\,(279.3 \times 10^3) = 128\,\text{W}$$

$$(\dot{W})_{R\,290} = (\dot{m}w)_{R\,290} = 4.59 \times 10^{-4}\,(135 \times 10^3) = 62\,\text{W}$$

$$(\dot{Q}_k)_{R\,290} = (\dot{m}q_k)_{R\,290} = 4.59 \times 10^{-4}\,(467.5 \times 10^3) = 215\,\text{W}$$

We clearly notice the following advantages in using the hydrocarbon HC 290 in place of CFC 12:

(i) Refrigerating capacity is increased from 43 W to 62 W, viz., by 44%.
(ii) COP *nearly* remains the same.
(iii) Discharge temperature is lowered from 128°C to 120°C.

However, there are implications in charging R 290 as a 'drop-in' substitute in a R 12 refrigerator without changing compressor, which are as follows:

(i) Refrigerant mass flow rate requirement decreases, due to higher latent heat and refrigerating effect of R 290

$$279.3 = (q_0)_{R\,290} > (q_0)_{R\,12} = 115.8\,\text{kJ/kg}$$

(ii) Refrigerant mass flow rate handled by compressor decreases due to large specific volume of R 290 suction vapour

$$0.27165 = (v_1)_{R\,290} > (v_1)_{R\,12} = 0.1624\,\text{m}^3\text{/kg}$$

This reduces mass flow handled by compressor as below:

$$4.59 \times 10^{-4} = (\dot{m})_{R\,290} < (\dot{m})_{R\,12} = 7.68 \times 10^{-4}\,\text{kg/s}$$

(iii) Conditions (i) and (ii) are complimentary. However, the effect of higher latent heat is greater. As a result, in spite of the larger specific volume and hence smaller mass flow rate handled, the same compressor gives more capacity, and correspondingly increases heat rejected in the condenser for R 290 because of its higher latent heat of vaporization. Thus

$$128\,\text{W} = (\dot{Q}_0)_{R\,290} > (\dot{Q}_0)_{R\,12} = 89\,\text{W}$$

$$215\,\text{W} = (\dot{Q}_k)_{R\,290} > (\dot{Q}_k)_{R\,290} = 147\,\text{W}$$

(iv) Motor watts increase for R 290 along with higher refrigerating capacity.

$$62\,\text{W} = (\dot{W})_{R\,290} > (\dot{W})_{R\,12} = 43\,\text{W}$$

This will result in considerable overloading and overheating of motor. Consequently, it may burn out inspite of the lower discharge temperature of R 290.

(v) Due to the enormous amount of heat to be rejected, and use of the same condenser, the condensing temperature and pressure and hence discharge temperature would rise. In case, the mean cooling medium temperature is assumed to be 38°C, the ΔT across R 12 condenser would be $(\Delta T_k)_{\text{R 12}} = 55 - 38 = 17°\text{C}$, and that across R 290 condenser would be

$$(\Delta T_k)_{\text{R 290}} = (\Delta T_k)_{\text{R 12}} \frac{(\dot{Q}_k)_{\text{R 290}}}{(\dot{Q}_R)_{\text{R12}}}$$

$$= 17 \left(\frac{215}{147}\right) = 25°\text{C}$$

Hence, the condensing temperature and pressure with R 290 would be

$$(t_k)_{\text{R 290}} = 38 + 25 = 63°\text{C}$$

$$(p_k)_{\text{R 290}} = 22.582 \text{ bar}$$

And the isentropic discharge temperature would be close to 145–150°C (much above 120°C).

(vi) Further, if the effect of using the same capillary is also considered, we see that the pressure drops required are as follows:

	R 12	R 290
Condenser pressure	13.61	19.056
Evaporator pressure	1.228	2.022
Pressure drop in capillary	12.382 bar	17.034 bar

Then, assuming same pressure drop as in R 12 capillary, viz., 12.382 bar, and neglecting the difference in thermophysical properties, we see that the pressure of propane at exist from the capillary and entering the evaporator would be

$$(p_0)_{\text{R 290}} = 22.582 - 12.382 = 10.2 \text{ bar}$$

This will correspond to an enormously high evaporation temperature of 28°C instead of – 25°C. It will also start a vicious cycle of a very high density of suction vapour, extremely large mass flow rate, capacity, motor watts, heat rejected in condenser, condenser temperature and pressure, discharge temperature, etc.

Example 4.2 *Make suggestions regarding changes to be made if R 12 in the refrigerator in Example 4.1 is to be replaced with propane.*

Solution Mass flow rate of propane required for the same refrigerating capacity of 89 W

$$\dot{m} = \frac{0.089}{279.3} = 3.18 \times 10^{-4} \text{ kg/s}$$

Compressor

Theoretical piston displacement required

$$\dot{V}_p = 3.18 \times 10^{-4} (0.27165) = 8.64 \times 10^{-5} \text{ m}^{3/s}$$

The piston displacement must, therefore, be reduced to

$$\frac{(\dot{V}_p)_{R290}}{(\dot{V}_p)_{R12}} = \frac{8.64 \times 10^{-5}}{1.247 \times 10^{-4}} = 0.53$$

viz., to 53% of R 12.
Motor Watts

$$\dot{W} = 3.18 \times 10^{-4} (135 \times 10^3) = 43 \text{ W}$$

Note *Hence, we need a new hermetic compressor with 47% less piston displacement but with the same size motor.*

Condenser $\dot{W} = 3.18 \times 10^{-4} (467.5 \times 10^3) = 149 \text{ W}$

Heat transfer required is the same as in R 12 refrigerator.
Evaporator
Heat transfer required is same as 89 W.

Note *The sizes of condenser and evaporator remain the same. Any change would depend upon the difference in thermophysical properties.*

Capillary Tube
Pressure drop required is 17.034 bar. Existing capillary gives 12.382 bar with R 12. Hence, a capillary of longer length is required with propane giving the required mass flow rate of 3.18×10^{-4} kg/s of R 290.

4.13 SUBSTITUTES FOR CFC R 11

R 11 Chillers represented one of the most efficient application of electrical energy in air conditioning.

The substitute for R 11 must have high molecular mass and N.B.P close to that of R 11 of 23.7°C. The substance having N.B.P. close to this is R 123 as seen from Table 4.6. HCFC R 123, therefore, has been found to be the most favoured transitional alternative to CFC R 11. It is close to a *drop-in* substitute, although it gives about 10–15% reduction in capacity. Toxicity of R 123 is relatively high compared to R 11. Due to the toxicity of R 123 and because of it being HCFC, it is very critical to identify a long term substitute for R 11.

Since R 134a is a HFC with zero ODP, it is being used in centrifugal chillers. However, it is a much lower boiling substance than R 11. It has higher pressures and smaller volumes. Beyerlien *et al.*[14] have suggested some fluorinated propanes and butanes as CFC and HCFC substitutes. Adcock *et al*[1] have suggested some fluorinated ethers, HFEs, as a new series of CFC substitutes. A comparative study of CFC11 and these substances was made by Devotta *et al.*[18] HFC 245 fa (Pentafluoropropane) and HFE 143 ether have their properties and performance parameters very close to those of R 11. Comparison of all four is given in Table 4.16.

Note that the ether is flammable, and flammability index of pentafluoropropane is uncertain.

Table 4.16 Comparative Data for HFC 134a, CFC 11, HCFC 123, HFC 245fa (Pentafluoropropane) and Ether for t_k = 40°C and t_0 = 0°C

	HFC 134a	*CFC11*	*HCFC 123*	*HFC 245 fa*	*HFE 143*
Formula	$C_2H_3F_4$	CCl_3F	$CHCl_2–CF_3$	$C_3 H_3 F_5$	CH_2 FO CHF_2
Mol. Mass	102.0	137.4	152.9	135.0	100.0
t_c, °C	101.1	197.8	183.68	154.1	186.83
p_c, MPa	4.06	4.37	3.662	3.64	4.141
N.B.P., °C	–26.07	23.7	27.82	14.9	30.06
p_0, MPa	0.293	0.04	0.03265	0.054	0.029
p_k, MPa	1.017	0.175	0.15447	0.252	0.145
q_0, kJ/kg	142.2	154.9	141.0	151.7	231.91
V^*, m^3/MJ	0.49	2.604	3.164	2.816	3.312
COP	5.49	6.06	6.1	5.861	6.023

4.13.1 Refrigerants Currently Used in Centrifugal Chillers

In recent times, the two refrigerants being used in centrifugal chillers are HCFC R 123 and HFC R 134a. R 123 chillers have a clear advantage when the highest COP is sought. It is equal to that of R 11 (Table 4.16). Hence, it has high energy efficiency. Also, it has lower pressures, as seen from Table 4.16, which helps in reducing the equipment cost and leakage.

However, it is an HCFC with two chlorine atoms scheduled for phase-out despite having low ODP and extremely low GWP.

R 134a is also widely used in place of R 11. It has preference for long term use as it is an HFC with zero ODP though its GWP is significant as seen from table below:

Refrigerant	*HCFC R 123*	*HFC R 134a*
Chemical formula	$CHCl_2F_3$	CH_2F_4
ODP	0.012	0
GWP	76	1320
Energy efficiency	13.5% More efficient than R 134a	–

As seen from Table 4.6, R 134a is a much lower boiling refrigerant. Its pressures are high. It requires multi-stage compression. And its volumes are small. Hence, it can be used only for a large capacity. When all environmental factors are balanced, both R 123 and R 134a offer advantages.

However, we have to choose R 134a over R 123 because of its zero ODP inspite of higher GWP and lower COP. HFC R 134 is considered as the long term alternative to R 11.

R 134a at the same times continues to be used as an alternative to R 11 centrifugal chillers replaced by R 134a screw chillers.

Presently (as in 2020), both are being used in centrifugal chillers. Note that R 134a is also being used in small and large capacity positive displacement compressors including screw compressors.

From the calculations in Examples 6.7 and 6.9, for the same operating conditions of condenser temperature (40°C) and evaporator temperature (6°C), the comparative values for suction vapour volume and COP of R 123 and R 134a are given in the following table:

Refrigerant	*Suction vapour volume, $m^3/s(TR)$*	*COP*
R 123	8.462×10^{-3}	7.5
R 134a	1.362×10^{-3}	6.8

It is seen that R 123 has suction vapour volume of 8.462×10^{-3} as against 1.362×10^{-3} of R 134a. R 134a requires only 1/6 the volume of R 123. This means that for the same width of passage, design capacity of R 134a compressor will be 6 times greater.

For example, if R 123 centrifugal compressor is designed for 100 TR minimum capacity, for nearly the same dimensions R 134a compressor will have to be designed for a minimum capacity of 600 TR.

While R 123 compressors are available in 100-500 TR range, R 134a compressors are in the range of 500–1000 TR each.

Another point to note here is that R 134 has a lower COP. Hence, R 123 has about 13.5% higher energy efficiency. R 134a is non-corrosive to metals like, copper, iron, and aluminium.

It is non-flammable and non-toxic also. But on coming in contact with open fire or high temperature by electric heaters, etc., it will decompose into toxic pungent compound of hydrogen fluoride. So, open fire is forbidden onsite using R 134a specially in large plants such as centrifugal chillers.

Also note that R 134a is 4.3 times heavier than air. If discharged in atmosphere from the unit while dissembling, it will accumulate at ground level. It can cause serious harm similar as with anasthesia or even worse. It is a very serious matter. So, in large capacity plants where the amount of charge is very high, empty R 134a from the plant (chiller) before dissembling the unit thereby preventing the refrigerant to be released in air. Also, measures should be taken to provide proper ventilation onsite.

4.14 SUBSTITUTES FOR HCFC R 22

It is seen that short-term substitute for HCFC R 22 remains HCFC R 22 by itself. HFC R 32, and HFC R 134a, both may as well remain the long-term substitutes. While R 32 is a higher pressure lower boiling refrigerant, R 134a is a lower pressure higher boiling refrigerant.

However, the average of bubble and dew temperatures of HFC R 407C, a mixture of R 32, R 125, and R 134a, is closest to –40.8°C the N.B.P. of HCFC 22. And R 410A is a near azeotropic mixture of HFCs R 32 and R 125. At present, these two blends are serious contenders to replace HCFC 22.

Another HFC blend is R 404A (see Table 4.9). Other blends of HFC 32, HFC 134a etc., may also be found till the deadline of 2030.

Table 4.17 compares the performance parameters of HCFC R 22 with the blend R 407C of R 32/R 125/R 134a (23/25/52%), and azeotropic mixture R 410A of R32/R 125 (50/50%).

Table 4.17 Comparison of Performance Parameters of R 407C and R 410A with HCFC R 22 for t_k = 40°C and t_0 = 0°C

	HCFC R 22		*R 407C*			*R 410A*	
t_c, °C	96.15		86.03			71.36	
p_c, MPa	4.99		4.63			4.903	
N.B.P.,°C	– 40.81	Bubble		Dew	Bubble		Dew
		– 43.63		– 36.63	– 51.44		– 51.36
p_o, MPa	0.498		0.5			0.8	
t_o, °C	0	– 3.85		+ 2.36	– 0.03		+ 0.08
p_k, MPa	1.5336		1.7			2.4	
t_k, °C	40	43.78		38.84	39.56		39.68
q_0, kJ/kg	155.4		152.1			155.8	
$\dot{m}$, kg/MJ	6.435		6.5733			6.55	
v_1, m^3/kg	0.0471		0.04687			0.03262	
V^*, m^3/MJ	0.303		0.308			0.2137	
w, kJ/kg	25.0		20.4			28.7	
COP	5.8		5.1			6.2	

Note that R 407C boils through –3.85°C to +2.36°C to give a mean refrigeration temperature of 0°C at 0.5 MPa. It condenses through 43.78°C to 38.84°C to give a mean condensation temperature of 40°C. So it boils and condenses over a range of temperatures of 6.2 and 4.9°C respectively. Hence, evaporators and condensers have to have counter flow designs. Issues relating to separation of refrigerant and oil in the evaporators have also to be examined.

Further, R 407C has much lower COP. But R 407C does have closely similar operating characteristics to R 22.

On the contrary, R 410A is nearly an azeotropic mixture. Its boiling and condensation ranges are negligible. It is like a pure substance. However, it is a much lower-boiling refrigerant than HCFC R 22. Hence, it has high pressures.

But it is shown that R 410A has higher COP.

And higher evaporator heat transfer coefficients of R 410A have facilitated sysyem designs with more, not less, dehumidification (refrigeration) capability as compared to R 22.

However, as the critical temperature of R 410A is lower, there is a recognized performance degradation at elevated ambient temperatures. This has effect on dehumidification capability.

On balance, performances of R 410A and R 22 systemes are comparable. Hence, it is being used as a substitute to R 22.

For the values presented in Table 4.17, see Example 4.3 for calculations.

Note that HFC R 134a, a medium pressure refrigerant is being used in Europe in conjunction with variable speed screw compressors that are competitive to R 22.

HFC R 32 has become very popular for its use in window-type and split unit air conditioners.

4.14.1 HFCs R 134a, R 32 and R 410a Presently in Use along with HCFC R 22

Currently, HCFC R 22 continues to be used because of its very high COP. But the three HFCs being simultaneously used are R 134a, R 32, and R 410a. Their NBPs are close to the NBP of R 22 as shown in comparison in the table below.

Refrigerant	*HCFC R 22*	*HFC R 410a*	*HFC R 32*	*HFC R 134a*
NBP, °C	–40.8	–51.44	–52.02	–24.02

It is seen that both R 410a and R 32 are lower boiling and higher pressure refrigerants. They have nearly same NBPs and vapour pressures at operating temperatures.

R 134a, on the contrary, is a higher boiling and lower pressure refrigerant. It has less leakage problem. Therefore, it can be used in car air conditioners in place of R 22.

An issue with R 410a is that the oil used in R 410a system absorbs moisture. Hence, proper evacuation must be done while installing the R 410a system.

Further, R 410a requires elimination of flooded evaporators to avoid composition change as it is a blend of two components.

R 32 is environment friendly. It is the refrigerant currently receiving attention. Compared to R 22, it has $1/3^{rd}$ GWP as seen from the table below.

Refrigerant	*CO_2*	*R 32*	*R 134a*	*R 22*	*R 410a*	*R 12*
GWP	1	675	1320	1810	2090	10,900
OPD		0	0	0.053	0	1

Thus, R 32 has an advantage over R 410a in respect of GWP. Considering the other factor of ODP, both R 32 and R 410a have zero ODP.

4.14.2 HFC R 32 (Difluromethane) Popular in Air Conditioners

R 32 has been a difficult refrigerant to handle because of inflammability consideration. Therefore, it was not used in practical till recently. But companies like DAIKIN, etc. have come out with R 32 split units along with inverter technology which are very popular.

Though R 32 is a hydrocarbon but its inflammability is very low. Moreover, since the amount of charge required in air conditioners is very small, about 70% of R 22, it is being considered safe for ACs. Still safety regulations must be observed while using R 32. It may not be used in larger capacity units. In large capacity positive displacement compressors, since the amount of charge is large, R 410a, remains the preferred choice.

Why R 32 is prefered in ACs? This is because it has zero ODP and very low GWP. With number of ACs multiplying and the world warming up, GWP has become a very important factor. Further, it has higher COP than R 410a.

The table below gives comparison of performance of R 32, R 22, and R 410a for simple saturation cycle with condensation temperature of 50°C and evaporator temperature of 0°C.

Refrigerant	*HFC 32*	*HFC R 410a*	*HCFC R 22*
Discharge Temperature, °C	106	83	88
COP	2.75	2.57	2.9
Volumetric Refrigerating Capacity, kJ/m^3	4812	4150	3010

It is seen that volumetric refrigerating capacity for R 32 is higher than that of R410a. So, it requires a smaller size compressor. More important than that is its COP which is much higher and closer to that of R 22. Hence, R 32 is found to be a more efficient substitute for R 22.

Further, there is a temperature glide of 0.11°C with R 410a in evaporator. Note that R 410a is not a pure substance. Hence, there is possibility of separation of components. R 32, on the other hand, is a single compound refrigerant. It is a pure substance which makes it easy to recycle.

4.14.3 HFC R 134a – A Wide Range Substitute for CFCs R 11 and R 12 and HCFC R 22

It is interesting to see how widely R 134a is being used presently from refrigerators, air conditioners, package units to large capacity chiller using both positive displacement and centrifugal compressors.

It has been already shown in Sec. 4.13.1 how R 134a is used with centrifugal chillers plants ranging 500–1000 TR replacing R 11.

It is also being used with large capacity screw compressors. R 134a screw chillers are used in place of R 11 centrifugal chillers. Screw compressors are known for efficiency and consistent part-load operations. Further, two-stage compression with economiser (refer to Fig. 5.4) is done to obtain maximum efficiency.

Multistaging is achieved within the single compressor only by providing an additional suction port which makes it possible to suck the refrigerant vapour at intermediate pressure.

We have already seen in Sec. 4.12.6 how R 134a is also used in small units such a refrigerators as substitute for CFC R 12.

However, note that GWP of R 134a is not very small. It is of the order of 1320 in comparison to unity of CO_2. But it is smaller than 1810 of R 22, 2090 of R 410a, and much smaller than 10,900 of R 12.

4.15 SUBSTITUTES FOR CFC R 502

R 502 is an azeotropic mixture of HCFCs R 22 and R 115 in 50/50% proportions by mass. Its N.B.P. is – 45.6°C.

A new azeotropic mixture R 507A formed by blending HFC 134a and HFC 125 in 50–50% proportions by mass, and near azeotropic mixture R 404A formed by HFCs R 125/R 143/R 134a in 44/52/4% proportions make attractive alternetives.

Where as R 507A is an azeotrope having N.B.P. of – 46.74°C which is very close to the N.B.P. of R 502, R 404A boils through a range from – 46.22°C to – 45.47°C at 1 atm pressure of 0.10132 MPa.

Note *Small deviations from azeotropic behaviour occur for R 507A.*

4.16 A TMOSPHERIC GASES AS SUBSTITUTES FOR CFC REFRIGERANTS

Concern for protecting the environment is on the rise. It is, therefore, suggested that the existing atmospheric gases themselves be used as refrigerants. These are air, water vapour, carbon dioxide, hydrogen, ammonia, etc.

While ammonia is already a very popular refrigerant in industrial applications such as cold storages and ice plants, its application can grow further. This has already put a stop to the replacement of ammonia by HCFC R 22.

It is, however, a challenge to thermal engineers to adapt water and other inert atmospheric gases in varied applications and high performance refrigeration systems.

Air cycle refrigeration (Chapter 11) is already used in aircraft air conditioning. Normalair-Garrett of UK, in cooperation with Hagenuk Faiveley of Germany, have devised the system that uses air cycle technology for car air conditioners. Environmental concerns have breathed life into natural refrigerant technologies including air cycle.

Hydrogen is also used as a working substance in Stirling-cycle in a Philips gas liquifier (Chapter 11). With modifications, it can be adapted for other common applications.

Water vapour refrigeration is used in steam ejector system (Chap. 13) driven by motive steam in very large capacity air conditioning plants.

Carbon dioxide was used as a refrigerant toward the end of nineteenth century. Since its critical temperature is low (31°C only), it cannot be condensed in high ambient conditions. But it can be used in a *supercritical vapour compression cycle*. Improving its COP, however, offers a challenge to thermodynamics engineers.

4.16.1 Supercritical Vapour Compression Cycle with CO_2 as Refrigerant

Figures 4.5 and 4.6 show the supercritical vapour compression cycle with CO_2 (R 744) as refrigerant on *T-s* and *p-h* diagrams respectively. The following operating conditions have been chosen for air conditioning cycle:

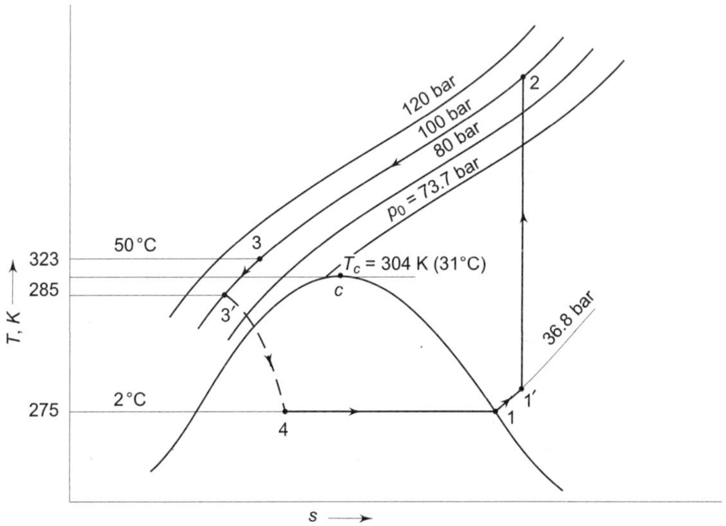

Fig. 4.5 Supercritical vapour compression cycle with CO_2 on *T-s* diagram

Evaporation temperature 2°C(p_0 = 36.8 bar)
Compressor discharge/Cooler pressure 100 bar
Temperature t_3 after cooler coil 50°C (323 K)

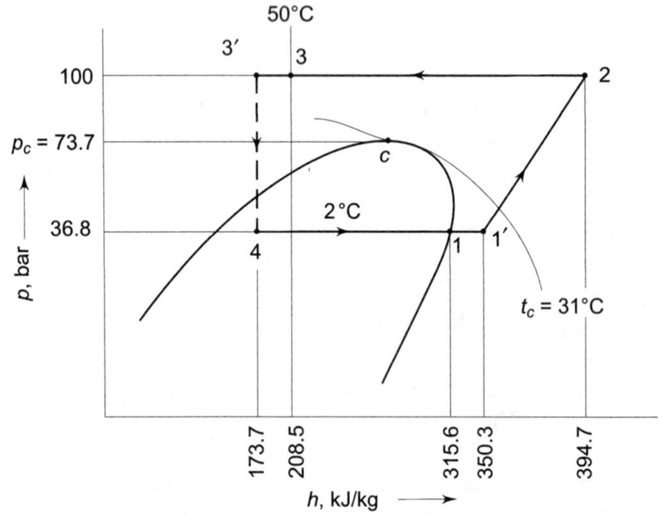

Fig. 4.6 Supercritical vapour compression cycle with CO_2 on *p-h* diagram

Assuming 20°C superheating of vapour in regenerative heat exchanger, the property values and calculations are as follows:

$$h_1 = 315.6 \text{ kJ/kg}, \quad s_1 = 1.157 \text{ kJ/kg K} \quad h_1' = 350.3 \text{ kJ/kg}$$

$$h_2 = 394.7 \text{ kJ/kg}, \quad v_1' = 0.014917 \text{ m}^3\text{/kg}$$

$$h_3 = 208.5 \text{ kJ/kg}, \quad h_3' = 173.7 \text{ kJ/kg}$$

$$\text{COP}_c = \frac{h_1 - h_3'}{h_2 - h_1} = \frac{142.2}{79.1} = 1.82$$

$$W^* = \frac{3.5164}{142.2}(79.1) = 1.95 \text{ kW/TR}$$

$$V^* = \frac{3.5164}{142.2}(0.014917) = 3.7 \times 10^{-4} \text{ m}^3\text{/s (TR)}$$

Results for discharge pressures of 80, 100 and 120 bar, and varying degrees of superheats are given in Table 4.18

It is observed that high-side pressure of 100 bar is optimum, and superheating of vapour by 30°C gives higher COP. This corresponds to a maximum possible suction temperature of 32°C.

Table 4.18 Performance parameters of CO_2 at $t_0 = 2°C$

Discharge pressure bar	*Superheat* °C	COP_c	W^* kW/TR	$V^* \times 10^4$ m³/s (TR)
80	10	1.17	3.0	7.0
	20	1.28	2.74	6.8
100	10	1.78	2.01	3.5
	20	1.82	1.95	3.7
	30	1.88	1.87	
120	10	1.51	2.33	3.4
	20	1.55	2.28	3.6

4.17 USING MIXED REFRIGERANTS

The diversification of the refrigeration and air-conditioning industry has led to a continuous search for newer working substances adapted to specific applications. The commonly used fluorocarbons are limited in number and there are gaps in their characteristics owing to their chemistry which need to be filled. The introduction of mixtures in compression refrigerating machines is a step in that direction.

Mcharness and Chapman[29], and Downing[20] had made experimental studies on a number of refrigerants and their mixtures. Varying the capacity of a refrigerating machine using varying proportions of different refrigerants has now become a well-known possibility.

The most important property of a mixture is the range of temperature at which it boils or condenses. This enables it to achieve non-isothermal refrigeration, and when used in applications involving cooling at a range of temperatures, it results in a higher COP and consequent power saving.[3,4]

Further, some mixtures form azeotropes which have interesting possibilities.

4.18 BINARY MIXTURES

Two properties are required to define the thermodynamic state of a pure substance. If, however, the state is saturated, only one property needs to be known.

A binary mixture consists of a *higher boiling* and a *lower boiling* component. In the case of a binary system, three independent properties are required to define the state, and two properties when the state is in an equilibrium saturated liquid state or saturated vapour state. One of these properties is invariably the composition.

4.18.1 Measures of Composition in Mixtures

Consider a liquid and/or vapour mixture of two substances. The number of moles of components in the mixture are n_1 and n_2. Then, the sum

$$n = n_1 + n_2$$

holds for each phase separately. Let x_1 and x_2 denote the mole fractions of the two components in the liquid phase. Then,

$$x_1 = \frac{n_1}{n}, \quad x_2 = \frac{n_2}{n}, \quad x_1 + x_2 = 1$$

Similarly, for the mole fractions y_1 and y_2 in the vapour phase, we have

$$y_1 + y_2 = 1$$

The relations between masses m_1 and m_2, moles n_1 and n_2 and molecular masses M_1 and M_2 are

$$m_1 = n_1 M_1, \quad m_2 = n_2 M_2$$

$$\Rightarrow \qquad m = m_1 + m_2 = n_1 M_1 + n_2 M_2 = nM$$

where M represents the effective molecular mass of the mixture. Thus,

$$M = \frac{m}{n} = \frac{1}{n}(n_1 M_1 + n_2 M_2)$$

$$= x_1 M_1 + x_2 M_2 \text{ for liquid mixture}$$

$$= y_1 M_1 + y_2 M_2 \text{ for vapour mixture}$$

If ξ_1^L and ξ_2^L represent mass fractions in liquid phase, then

$$\xi_1^L = \frac{M_1 x_1}{M_1 x_1 + M_2 x_2} = 1 - \xi_2^L, \quad \xi_2^L = \frac{M_2 x_2}{M_1 x_1 + M_2 x_2} = 1 - \xi_1^L \tag{4.27}$$

Similarly, for mass fractions ξ_1^V and ξ_2^V in vapour phase, we have

$$\xi_1^V = \frac{M_1 y_1}{M_1 y_1 + M_2 y_2} = 1 - \xi_2^V, \quad \xi_2^V = \frac{M_2 y_2}{M_1 y_1 + M_2 y_2} = 1 - \xi_1^V \tag{4.28}$$

The relation between mass fraction and mole fraction is expressed as follows:

$$x_1 \text{ or } y_1 = \frac{\xi_1 M_2}{\xi_1 M_2 + \xi_2 M_1} \tag{4.29}$$

4.18.2 Temperature-Composition Diagram

The nature of this diagram is shown in Fig. 4.7 for any given pressure p. The saturation temperatures of the two components at pressure p are t_1^{sat} and t_2^{sat}, the superscript sat denoting the saturation state of the pure substance. The temperature t_B, at which a liquid mixture begins to boil is called the *bubble point temperature*. A curve can be drawn passing through all such points for various liquid compositions $\xi = \xi^L$, as shown in Fig. 4.7. This is called the *bubble point curve*. It is the locus of saturated liquid states for the mixture.

Similarly, the temperature t_D at which a vapour mixture begins to condense is called the *dew point temperature*. A curve passing through all such points is called the *dew point curve*. It is the locus of saturated vapour states for the mixture. The region below the bubble point curve is the subcooled region, and that above the dew point curve is the superheated region. The region between the two curves is the liquid plus vapour region. The states on these curves are the saturation states. Any state A that falls in the liquid plus vapour region comprises of equilibrium states L of saturated liquid, and V of saturated vapour. The compositions of the two are ξ^L and ξ^V and the overall composition ξ at A is given in terms of the fraction z of vapour and $(1 - z)$ of liquid as given in Eq. (4.30)

$$\xi = z\xi^V + (1 - z)\xi^L \tag{4.30}$$

It may be noted that there will be a different loop like the one in Fig. 4.7 for each pressure.

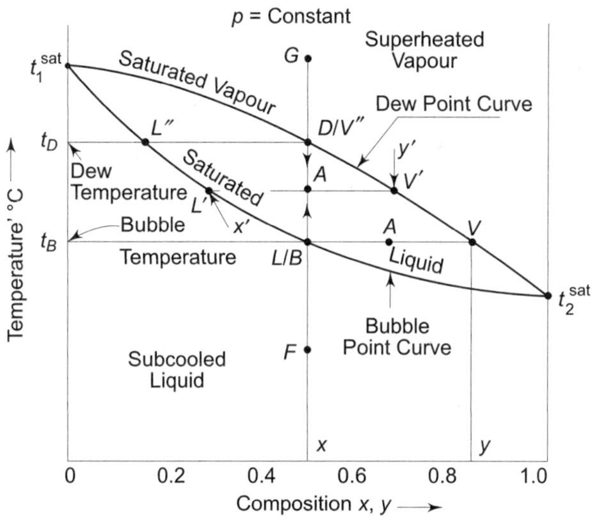

Fig. 4.7 *t-x-y* diagram of binary mixtures

Thus, when a subcooled liquid of composition x at F in Fig. 4.7 is heated it will first rise in temperature till it reaches the *bubble temperature* t_B at B/L when it begins to boil. The first vapour bubble that is formed is at V having the composition y. As the boiling proceeds, the temperature rises. Subsequent states during boiling are along the vertical such as point A at the same overall composition with liquid at L' and vapour at V' in equilibrium. When boiling is complete, the vapour is at D/V'' at *dew temperature* t_D. The last drop of liquid to vaporize is at L''.

Further heating will result in superheating of vapour to G. When superheated vapour at G is cooled, the reverse processes of desuperheating to D, condensation to B, and subcooling to F will take place.

It is observed that the *temperature glide/range* during evaporation and condensation is equal to the difference in dew and bubble temperatures ($t_D - t_B$) for the composition.

The temperature-composition diagram represents vapour-liquid equilibrium of mixtures at constant pressure. A counterpart to this diagram is the pressure composition *p-x-y* or $p - \xi$ diagram at constant temperature as shown in Fig. 4.8.

At a certain temperature for a given composition of the mixture, the pressure at which it begins to boil is called the *bubble pressure* p_B, and the pressure at which it begins to condense is the dew pressure p_D. Both these pressures are shown in Fig. 4.8. It is seen that dew pressure is lower than bubble pressure unlike dew temperature which is higher than bubble temperature. It merely states the fact that the higher the boiling point the lower the boiling pressure.

It may be pointed out that instead of the mole fractions x and y, compositions can be expressed in terms of mass fractions ξ^L and ξ^V also.

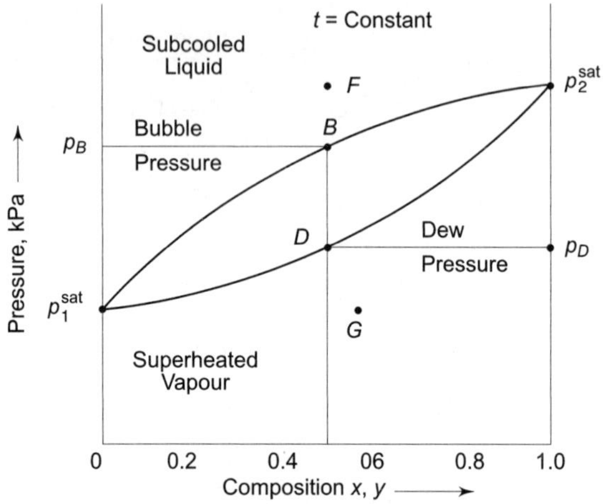

Fig. 4.8 *p-x-y* Diagram of Binary Mixtures

4.18.3 Enthalpy-Composition Diagram

Figure 4.9 shows the nature of the enthalpy-composition diagram. Points h_{f1} and h_{f2} represent the enthalpies of the saturated liquids of the two components, and h_{g1} and h_{g2}, those of the saturated vapours, at pressure p. At any composition ξ at a point A in the liquid plus vapour region for the mixture, the enthalpies of equilibrium saturated liquid and saturated vapour states L and V are h^L and h^V respectively. Then, the enthalpy at A is given by Eq. (4.31).

$$h = zh^V + (1 - z)h^L \tag{4.31}$$

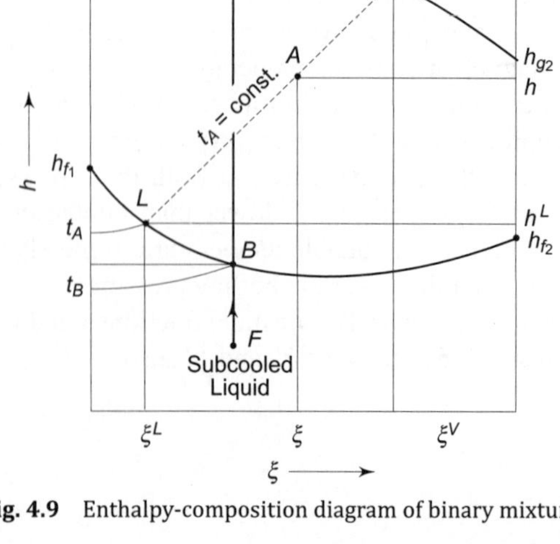

Fig. 4.9 Enthalpy-composition diagram of binary mixtures

Figure 4.9 also shows the constant temperature lines in the subcooled region. The line joining the equilibrium states such as L and V represent constant temperature lines in the two-phase region, and are called *tie lines*. Again, there will be a pair of curves for saturated liquid and saturated vapour mixtures for each pressure.

4.18.4 Pressure–Enthalpy Diagram of a Mixture

Fig. 4.10 shows the pressure enthalpy diagram of a mixture.

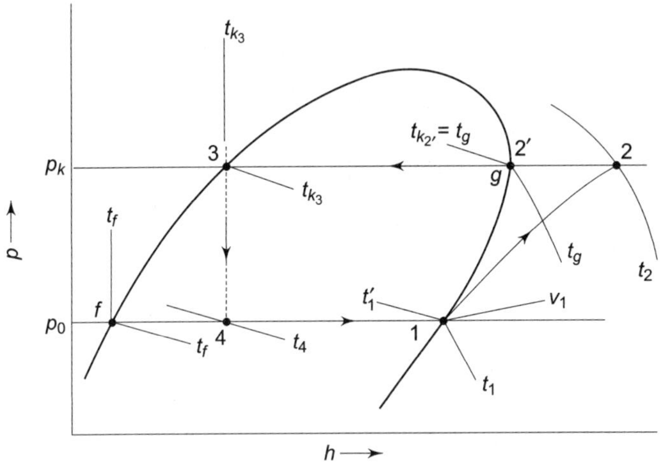

Fig. 4.10 Temperatures on pressure–enthalpy diagram of a mixture

The condensation and evaportion liness, 2 – 3 and 4 – 1, are horizontal at constant pressure, the throttling process line 3 – 4 is vertical at constant enthalpy, and the compression process line 1– 2 is isentropic as in the case of a pure substance. The only difference between pure substance and the mixture are the changes in temperature during condensation and evaporation. Thus

$$t_2' = t_g \neq t_3$$

and
$$t_1 \neq t_4 \neq t_f$$

The temperature at 2′ is the dew temperature t_g, and temperature at 3 is the bubble temperature at pressure p_k.

Similarly, temperature at 1 is the dew temperature at $p_{0,}$ and temperature at 4 is in between the bubble temperature t_f and dew temperature t_1 at p_0.

Thus, condensation temperature is not constant at t_k. It varies from $t_{k_{2'}} = t_g$ to $t_{k_3} = t_f$ at p_k.

Similarly, evaporation temperature is not constant at t_0. It varies from t_{o_4} to $t_{o_1} = t_g$ at p_0. Both condensation and evaporation processes are non-isothermal. Subscripts f and g refer to saturated liquid (bubble temperature) and saturated vapour (dew temperature) states respectively.

Further, average of $t_{k_{2'}}$ and t_{k_3} may be made equal to t_k, and average of t_{o_4} and t_{o_1} may be made equal to t_o.

Example 4.3 ***Comparison of HFC Blends R 407 and R 410A with HCFC 22***

Compare p_0, p_k, q_0, w, COP, v_1, m, and V^* of R 407C and R 410A with those of HCFC 22 for t_c = 0°C and t_k = 40°C.

Solution Refer Fig. 4.10

(a) *For HCFC 22* From the table of properties of HCFC 22 for t_o = 0°C, t_k = 40°C p_o = 0.498 MPa p_k = 1.5336 MPa $v_1 = 0.0471$ m³/kg Calculations give

$$q_0 = 156.3 \text{ kJ/kg}$$
$$w = 26.9 \text{ kJ/kg}$$
$$\text{COP} = 5.81$$
$$\dot{m} = 6.435 \text{ kg/MJ}$$
$$V^* = 0.303 \text{ m}^3\text{/MJ}$$

(b) *For R 407C* Choose p_0 = 0.5 MPa to get an average $t_0 \approx 0$°C

Bubble temperature t_f = –3.85°C

Dew temperature t_1 = 2.36°C

Choose p_k = 1.7 MPa to get average $t_k \approx 40$°C

Bubble temperature t_3 = 38.84°C

Dew temperature t_2' = 43.78°C

From the table of properties of R 407C, we further have

h_1 = 410.64 kJ/kg h_3 = 258.51 kJ/kg h_2' = 425.25 kJ/kg
s_1 = 1.7735 kJ/kgK s_2' = 1.7269 kJ/kgK
C_p at 1.7 MPa = 1.361 kJ/kgK

Isentropic compression gives

$$s_2 = s_2' + C_p \ln \frac{T_2}{T_2'} = s_1$$

$$1.7269 + 1.361 \ln \frac{T_2}{313} = 1.7735$$

$$\Rightarrow \quad T_2 = 328 \text{ K (54.9°C)}$$

Hence

$$h_2 = h_2' + C_p (T_2 - T_2')$$
$$= 425.25 + 1.361 (54.9 - 40) = 440.38 \text{ kJ/kg}$$
$$q_0 = h_1 - h_3 = 410.64 - 258.51 = 152.1 \text{ kJ/kg}$$
$$w = h_2 - h_1 = 440.38 - 410.64 = 29.74 \text{ kJ/kg}$$
$$\text{COP} = \frac{152.1}{29.74} = 5.11$$
$$v_1 \text{ at 0.5 MPa} = 0.04687 \text{ m}^3\text{/kg}$$
$$\dot{m} = \frac{Q_0}{q_0} = \frac{1000}{152.1} = 6.5733 \text{ kg/MJ}$$
$$V^* = \dot{m} v_1 = 29.74 (0.04687) = 0.3081 \text{ m}^3\text{/MJ}$$

(c) *For R 410A* Similarly, Choose $p_0 = 0.8$ MPa and $p_k = 2.4$ MPa
From the table of properties of R 410A

$$V^* = mv_1$$

$t_f = -0.03°C$ $t_1 = 0.08°C$

$t_3 = 39.56°C$ $t_2' = 39.68°C$

$h_1 = 421.33$ kJ/kg $h_3 = 265.52$ kJ/kg $h_2' = 425.33$ kJ/kg

$s_1 = 1.8102$ kJ/kg K $s_2' = 1.7294$ kJ/kgK

C_p at 2.4 MPa = 1.811 kJ/kgK

v_1 at 0.8 MPa = 0.03262 m^3/kg

Calculations as in (b) give

$T_2 = 328.5$ K (55.3°C)

$q_0 = 155.81$ kJ/kg

$w = 25.3$ kJ/kg

COP = 6.2

$\dot{m} = 6.55$ kg/MJ

$V^* = 0.214$ m^3/MJ

Note *We see that R 410A has higher COP than HCFC 22. The temperature range during evaporation is only 0.11°C. The temperature range during condensation is only 0.12°C.*

4.19 CLASSIFICATION OF MIXTURES

Mixtures can be classified broadly as *miscible and immiscible*. Miscible mixtures can be either *ideal* or *non-ideal*. Non-ideal mixtures may have *positive deviation* or *negative deviation from Raoult's law* as described in the following sections. Some non-ideal mixtures may also form *azeotropes*. An azeotrope is a constant temperature boiling mixture. Azeotropes may be *minimum boiling* or *maximum boiling* depending on whether the deviation from Raoult's law is positive or negative.

4.19.1 Ideal Mixtures

The laws governing ideal mixtures are Raoult's law for the liquid phase and Dalton's law for the vapour phase. According to Raoult's law, the *partial pressures* of the two components in a liquid mixture at temperature t are given by

$$p_1 = x_1 p_1^{sat} \tag{4.32a}$$

$$p_2 = x_2 p_2^{sat} \tag{4.32b}$$

where p_1^{sat} and p_2^{sat} are the saturation pressures of the two components at temperature t of the mixture, and x_1 and x_2 are the mole fractions of the components in the liquid phase.

According to Dalton's law for the vapour phase, the partial pressures are given by

$$p_1 = y_1 p, \tag{4.33 a}$$

$$p_2 = y_2 p \tag{4.33 b}$$

where y_1 and y_2 are the mole fractions of the components in the vapour phase, and p is the total pressure, which is also given by Eqs. (4.32a) and (4.32b) as

$$p = x_1 p_1^{sat} + x_2 p_2^{sat} \tag{4.34}$$

This is named as *Raoult's law pressure.*

Equation (4.34) can be used to find bubble temperature by iteration for a pressure p and liquid composition x_1, x_2 of the mixture. Another relation can be derived as follows in Eq. (4.35)

$$x_1 + x_2 = 1 = \frac{p_1}{p_1^{sat}} + \frac{p_2}{p_2^{sat}} = \frac{y_1 p}{p_1^{sat}} + \frac{y_2 p}{p_2^{sat}}$$

$$\Rightarrow \qquad p = \frac{p_1^{sat}\, p_2^{sat}}{p_1^{sat} - y_1\left(p_1^{sat} - p_2^{sat}\right)} \tag{4.35}$$

Equation (4.35) can be used to find dew temperature by iteration for a given pressure p and vapour composition y_1, y_2 of the mixture.

And if the temperature of the mixture is known, Eqn. (4.34) can be used to find bubble pressure for given x, and Eqn. (4.35) to find dew pressure for given y.

These set of equations can be used to draw the temperature-composition diagram for an ideal mixture. Equation (4.34) describes the variation of the total pressure of a mixture with composition. It is seen that it varies linearly with molar composition as shown in Fig. 4.11.

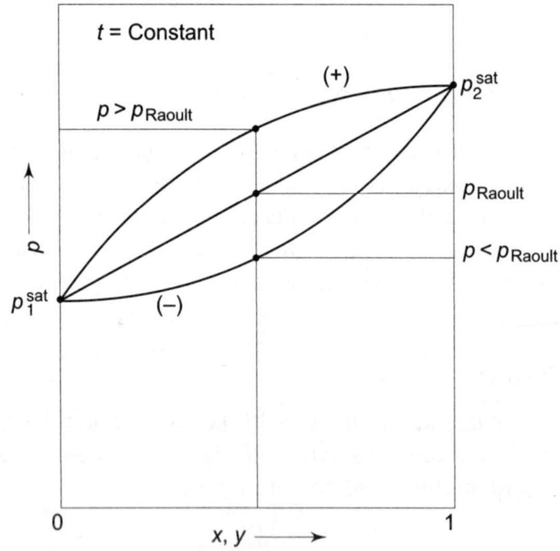

Fig. 4.11 Bubble point (saturated liquid) curves on *p-x-y* diagram showing deviations from Raoult's law

Another criterion of an ideal mixture is that the mixing of the constituents should cause no change in the average intermolecular forces of attraction. This amounts to saying that there is no change on mixing in the internal energy or volume. Accordingly, there is no evolution or absorption of heat on mixing. Thus, the enthalpy of an ideal mixture can be found by adding the enthalpies of the two components at the temperature and pressure of the mixture in proportion to the composition, i.e.,

$$h = \xi_1 h_1 + \xi_2 h_2.$$

Thus, from Figs. 4.7 and 4.9 the enthalpies at L and V are as given in Eqs. (4.36) and (4.37)

$$h^L = \xi_1^{\,L} [h_{f1} - C_{f1} (t_1^{\text{sat}} - t_A)] + \xi_2^{\,L} [h_{f2} + C_{f2} (t_A - t_2^{\text{sat}})] \tag{4.36}$$

$$h^V = \xi_1^{\,V} [h_{g1} - C_{p1} (t_1^{\text{sat}} - t_A)] + \xi [h_{g2} + C_{p2} (t_A - t_2^{\text{sat}})] \tag{4.37}$$

where C_f and C_p are the specific heats of liquid and vapour respectively at constant pressure. The expressions in brackets represent pure component enthalpies h_1^L and h_2^L for the liquid phase, and h_1^V and h_2^V for the vapour phase.

Example 4.4 ***Bubble and Dew Temperatures of Propane/Isobutane Mixtures (Illustrating Calculation Procedure)***

Determine the bubble and dew temperatures of, by mass, (1) 50% R 290 + (2) 50% R 600a mixtures at 14 bar and 1.4 bar.

Solution The molecular masses of R 290 and R 600a are $M_1 = 44$ and, $M_2 = 58$.

Given $\xi_1 = 0.5$ and $\xi_2 = 0.5$. From Eqn. (4.29),

$$x_1 \text{ or } y_1 = \frac{M_2\,\xi_2}{M_1\,\xi_1 + M_2\,\xi_2} = \frac{58(0.5)}{44(0.5) + 58(0.5)} = 0.5686$$

Solving Eqn. (4.34)

$$p = x_1\,p_1^{\text{sat}} + x_2\,p_2^{\text{sat}}$$

by iteration for $x_1 = 0.5686$ and $x_2 = 1 - x_1 = 0.4314$ for $p = 14$ bar, and 1.4 bar, we find that

$$(t_B)_{14\text{ bar}} = 54.3°\text{C}$$
$$(t_B)_{1.4\text{ bar}} = -25°\text{C}$$

Solving Eqn. (4.35)

$$p = \frac{p_1^{\text{sat}}\,p_2^{\text{sat}}}{p_1^{\text{sat}} - y_1\left(p_1^{\text{sat}} - p_2^{\text{sat}}\right)}$$

by iteration for $y_1 = 0.5686$ and $y_2 = 0.4314$, we get

$$(t_D)_{14\text{ bar}} = 62.7°\text{C}$$
$$(t_D)_{1.4\text{ bar}} = -16.3°\text{C}$$

4.19.2 Non-ideal Mixtures

Non-ideal mixtures are distinguished from the ideal ones in two respects:

(i) They show a marked change in volume on mixing in the vapour phase.

(ii) There is either absorption or evolution of heat on mixing in the liquid phase.

These two types of departures from ideal behaviour are characterised by deviations from (i) ideal gas equation of state and Dalton's law in vapour phase, and (ii) Raoult's law in liquid phase.

A mixture with positive deviation from Raoult's law exerts a higher vapour pressure than that given by the law at a particular temperature and composition as shown by the curve (+) for the saturated liquid states on the p-x diagram in Fig. 4.11.

The same is also shown by a curve (+) on the *t-x* diagram in Fig. 4.12 which shows that the bubble temperatures of such mixtures are lower at a particular pressure and composition.

Mixing of components forming such mixtures is accompanied by the absorption of heat. Such a *heat of mixing* Δh_m is considered to be *positive*. The enthalpy of the liquid mixture is, therefore, higher than the enthalpy obtained by the summation of the liquid enthalpies of the two components in proportion to their compositions, viz.

$$h^L = \xi_1 h_1^L + \xi_2 h_2^L + \Delta h_m \tag{4.38a}$$

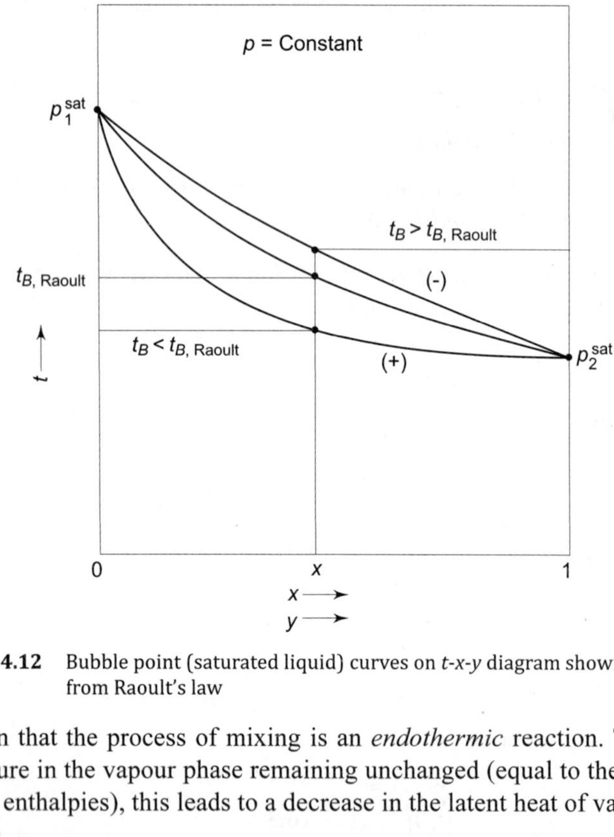

Fig. 4.12 Bubble point (saturated liquid) curves on *t-x-y* diagram showing deviations from Raoult's law

It is seen that the process of mixing is an *endothermic* reaction. The enthalpy of the mixture in the vapour phase remaining unchanged (equal to the sum of pure component enthalpies), this leads to a decrease in the latent heat of vaporization of the mixture.

In contrast to the above, a mixture with negative deviation from Raoult's law will have the following characteristics.

(i) It will exert a lower vapour pressure than given by Raoult's law at a particular composition and temperature as shown by curve (–) in Fig. 4.11.

(ii) The bubble temperature of the mixture will be higher at a particular pressure and composition as shown by curve (–) in Fig. 4.12.

(iii) Mixing of components results in the evolution of heat. The heat of mixing is thus negative. The enthalpy of the liquid mixture is, therefore, decreased resulting in an increased latent heat of vaporization. The process of mixing of liquids is *exothermic*, and the liquid enthalpy is given by Eq. (4.38 b)

$$h^L = \xi_1 h_1^L + \xi_2 h_2^L - \Delta h_m \tag{4.38 b}$$

An example of a mixture of the positive deviation type is a mixture of fluorocarbons, whereas the $NH_3 + H_2O$ system is an example of a mixture of the negative deviation type.

Example 4.5 ***Temperature-Composition and Enthalpy-Composition Diagrams for R 152a/R 22 Ideal Mixture***

Kim[25] suggested the use of R 152a/R 22 mixture in refrigerators. However, a refrigeration engineer proposes to use R152a/R22 mixture in place of pure R 22 in air conditioners. Draw $t - \xi$ and $h - \xi$ diagrams for the binary system, for 15 and 5 bar pressures, assuming ideal mixture behaviour (see Example 4.7 and 4.8 also).

Solution ***Temperature-composition diagrams***

Molecular masses $(M)_{R\ 152a} = 66.05$, $(M)_{R\ 22} = 86.48$

Let subscripts 1 and 2 denote R 152a and R 22 respectively.

At 15 bar

$$t_1^{sat} = 60°C,\ t_2^{sat} = 39.1°C$$

The bubble temperature will vary between 39.1°C and 60°C. Do calculations for 43, 47, 51, 55 and 57°C bubble temperatures.

Sample Calculations (Steps)

(i) At $t_B = 43°C$, $p_1^{sat} = 9.877$ bar and $p_2^{sat} = 1.642$ bar

(ii) Find equilibrium vapour composition:

$$y_1 = \frac{p_1^{sat} - p_1^{sat}\ p_2^{sat}/p}{p_1^{sat} - p_2^{sat}} = 0.15$$

(iii) Find equilibrium liquid composition:

$$x_1 = \frac{p_1}{p_1^{sat}} = \frac{y_1 p}{p_1^{sat}} = 0.77$$

(iv) Find equivalent mass fractions:

$$\xi_1^L = \frac{x_1 M_1}{x_1 M_1 + x_2 M_2} = 0.18$$

$$\xi_1^V = \frac{y_1 M_1}{y_1 M_1 + y_2 M_2} = 0.12$$

At 5 bar

Do calculations as above for 2, 4, 6, 8, 12 and 16°C, since

$$t_1^{sat} = 19.2°C, \quad t_2^{sat} = 0.2°C$$

Values obtained are given in Tables 4.19 and 4.20 respectively for 15 bar and 5 bar pressures. The same are plotted in Fig. 4.13.

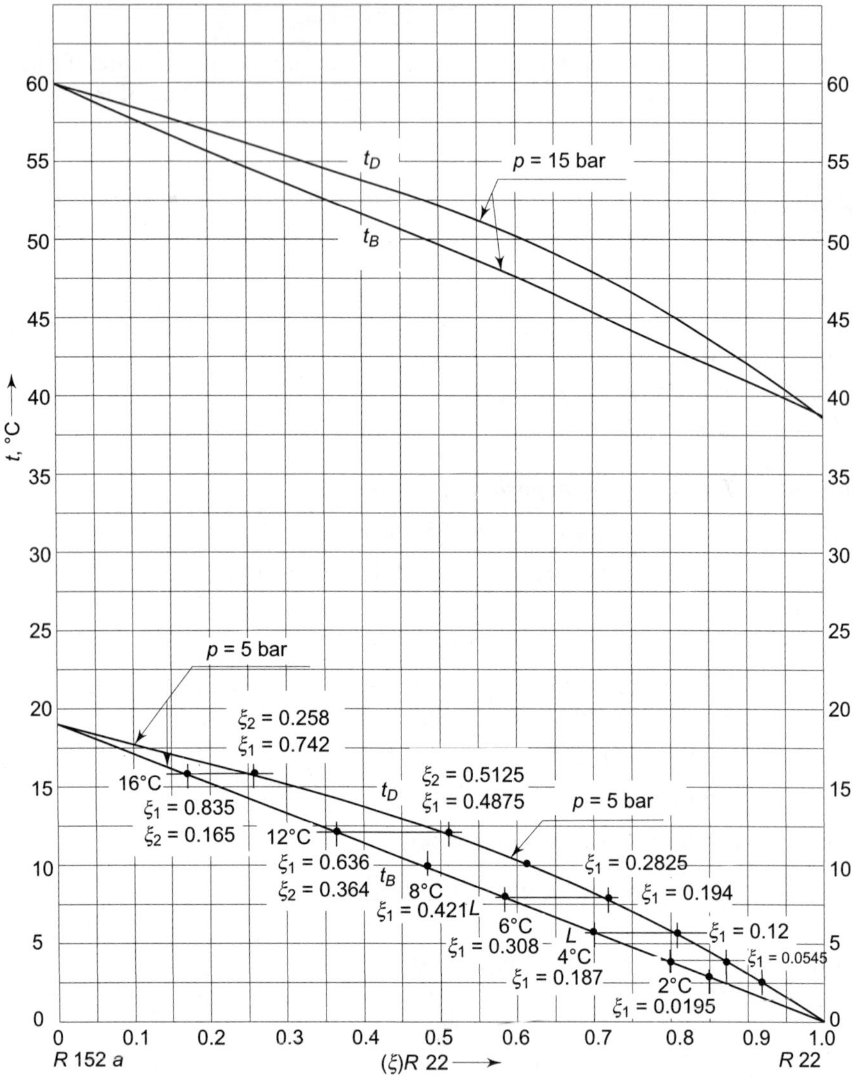

Fig. 4.13 Temperature-composition diagram of R 152/R 22 mixtures at 15 and 5 bar pressures

Table 4.19 Vapour-liquid equilibrium of (1) R 152a and (2) R 22 mixtures at 15 bar pressure

t °C	p_1^{sat} bar	p_2^{sat} bar	x_1	y_1	ξ_1^L	ξ_1^V
39.1		15	0.0	0.0	0.0	0.0
43	9.88	16.42	0.23	0.15	0.18	0.12
47	10.96	18.05	0.43	0.32	0.36	0.26
51	12.125	19.8	0.62	0.5	0.55	0.43
55	13.38	21.67	0.81	0.72	0.76	0.66
57	14.05	22.67	0.89	0.83	0.86	0.79
60	15		1.0	1.0	1.0	1.0

Table 4.20 Vapour-liquid equilibrium of (1) R 152a and (2) R 22 mixtures at 5 bar Pressure

t °C	p_1^{sat} bar	p_2^{sat} bar	x_1	y_1	ξ_1^L	ξ_1^V
0.2		5	0.0	0.0	0.0	0.0
2	2.835	5.308	0.117	0.07	0.0915	0.0545
4	3.04	5.657	0.268	0.163	0.1869	0.12
6	3.257	6.023	0.368	0.24	0.308	0.194
8	3.486	6.406	0.487	0.34	0.421	0.2825
12	3.98	7.226	0.696	0.555	0.636	0.4875
16	4.526	8.123	0.87	0.79	0.835	0.742
19.2	5		1.0	1.0	1.0	1.0

Enthalpy-composition Diagrams From the tables of properties we have the specific heats and saturated liquid and vapour enthalpies of pure refrigerants as follows:

p bar	t_1^{sat} °C	t_2^{sat} °C	h_{f1} kJ/kg	h_{f2} kJ/kg	h_{g1} kJ/kg	h_{g2} kJ/kg	C_{p1} kJ/kg.K	C_{p2} kJ/kg.K
15	60	39.1	311.9	247.9	540.5	415.8	1.61	0.91
5	19.2	0.2	233.0	200.0	520.0	406.0	0.864	0.89

Sample Calculations for Liquid and Vapour Enthalpies at 15 Bar At t = 43°C, $h^L = \xi_1^L h_1^L + \xi_2^L h_2^L$ = 0.18 (277.7) + 0.92 (253) = 258 kJ/kg.

Note *The values of h_1^L and h_2^L have been taken equal to saturation enthalpies h_{f1} and h_{f2} of pure refrigerants*

$$h^V = \xi_1^V [h_{g1} - C_{p1} (t_1^{sat} - t)] + \xi_2^V [h_{g2} + C_{p2} (t - t_2^{sat})]$$

$$= 0.12 [540.5 - 1.61 (60 - 43)] + 0.88 [415.8 + 0.91 (43 - 39.1)]$$

$$= 431.9 \text{ kJ/kg}$$

The results are presented in Tables 4.21 and 4.22 for 15 bar and 5 bar pressures, and in Fig. 4.14.

Table 4.21 Saturated liquid and vapour enthalpies of (1) R 152a and (2) R 22 mixtures at 15 bar

t °C	ξ_1^L	ξ_1^V	$h_1^L = h_{f1}$ kJ/kg	$h_2^L = h_{f2}$ kJ/kg	h^L kJ/kg	h^V kJ/kg
39.1	0.0	0.0		247.9	247.9	415.8 = h_{g2}
43	0.18	0.12	277.7	253.0	258.0	431.9
47	0.36	0.26	285.5	258.3	268.3	450.4
51	0.55	0.43	293.5	263.6	280.2	472.0
55	0.76	0.66	301.7	269.1	293.1	500.0
57	0.86	0.79	305.7	271.8	300.5	515.5
60	1.0	1.0	311.9		311.9	540.5 = h_{g1}

Table 4.22 Saturated liquid and vapour enthalpies of (1) R 152a and (2) R 22 mixtures at 5 bar

t °C	ξ_1^L	ξ_1^V	$h_1^L = h_{f1}$ kJ/kg	$h_2^L = h_{f2}$ kJ/kg	h^L kJ/kg	h^V kJ/kg
0.2	0.0	0.0		200.0	200.0	$406.0 = h_{g2}$
2	0.0915	0.0545	203.4	202.3	202.4	412.9
4	0.187	0.12	206.7	204.7	205.0	422.2
6	0.308	0.194	210.1	207.1	208.0	430.1
8	0.421	0.2825	213.5	209.4	211.0	440.3
12	0.636	0.4875	220.4	214.2	218.0	464.0
16	0.835	0.742	227.4	219.0	226.0	492.2
19.2	1.0	1.0	233.0			$520.0 = h_{g1}$

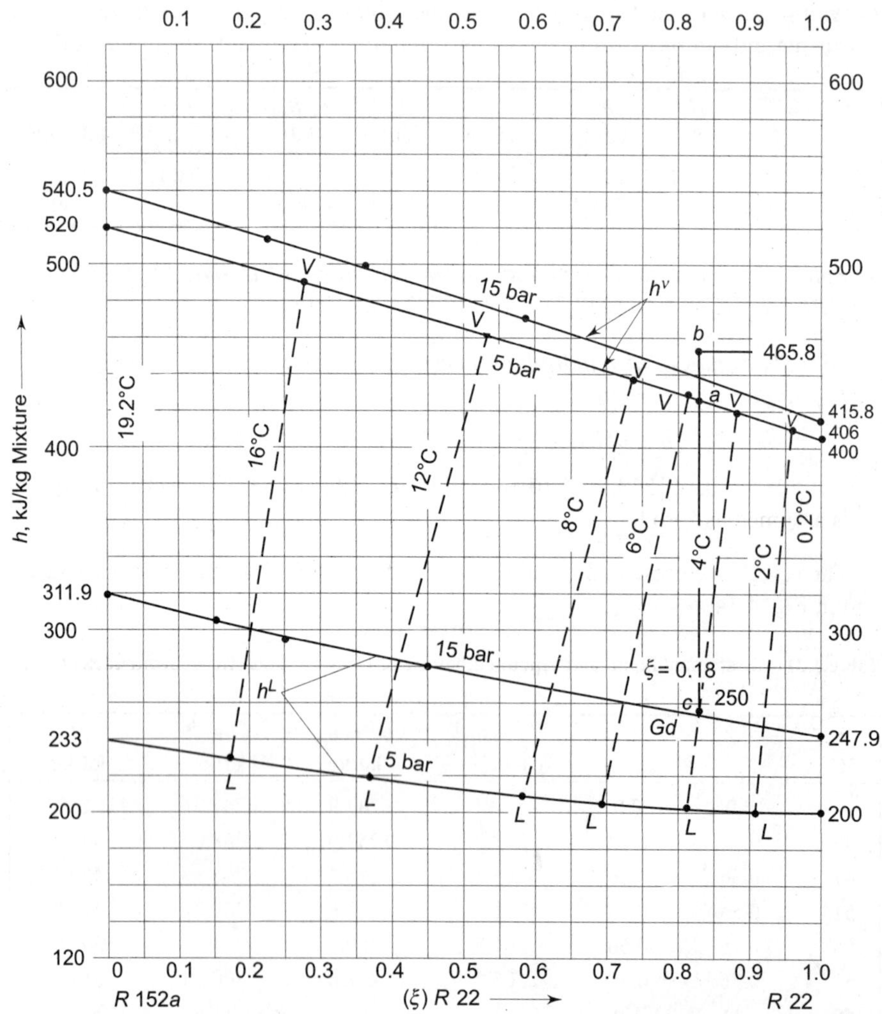

Fig. 4.14 Enthalpy-composition diagrams of R 152a/R 22 mixtures at 15 and 5 bar pressures

4.20 EVALUATION OF THERMODYNAMIC PROPERTIES OF R 290/R 600a MIXTURES

A more rigorous method of evaluating thermodynamic properties of mixtures has been described in the following pages. The assumptions that have been made in developing the thermodynamic properties of mixtures of R 290 and R 600a are as follows:

(i) There is no heat of mixing in liquid phase. So it obeys Raoult's law.

(ii) There is no change in volume due to mixing in the vapour phase. Ideal mixture behaviour has, therefore, been assumed and Dalton's law of partial pressures has been applied. However, real gas behaviour has been taken into account by employing the actual equations of state for the individual gases and the vapour mixture.

Note *Entropy of mixing has to be considered for ideal mixtures also, since mixing, in any case, is an irreversible process.*

4.20.1 Calculations of Specific Volume, Enthalpy and Entropy of Mixture in Liquid Phase[8]

The specific volume of liquid mixture v^L is calculated from Eq. (4.39)

$$Mv^L = \bar{v}^L = x_1 \bar{v}_1^L + x_2 \bar{v}_2^L \tag{4.39}$$

where x_1 and x_2 are the mole fractions and $\bar{v}_1^L$ and $\bar{v}_2^L$ are the molar volumes of its species.

The specific enthalpy of liquid mixture h^L is similarly calculated using the relations as given in Eq. (4.40)

$$Mh^L = \bar{h}^L = x_1 \bar{h}_1^L + x_2 \bar{h}_2^L \tag{4.40}$$

where $\bar{h}_1^L$ and $\bar{h}_2^L$ are the molar enthalpies of its species. The value of heat of mixing h_m is assumed as zero here.

The specific entropy of liquid mixture s_m^L is calculated from the relation given in Eq. (4.41)

$$Ms^L = \bar{s}^L = x_1 \bar{s}_1^L + x_2 \bar{s}_2^L - \bar{R}\ (x_1 \ln x_1 + x_2 \ln x_2) \tag{4.41}$$

where $\bar{s}_1^L$ and $\bar{s}_2^L$ are liquid entropies of pure species and the second term is the entropy of mixing.

4.20.2 Calculation of Specific Volume of Mixture in Vapour Phase

For the purpose, we need an equation of state for the components and the mixture. The constants for the mixture can be evaluated from the constants of the components by simple additive rules. So, Peng-Robinson[37] Equation (1.14) has been chosen in the following method. The pseudocritical temperature, pressure, and accentric factor for the mixture for use in the equation are given by:

$$T_c = y_1 T_{c1} + y_2 T_{c2}$$
$$p_c = y_1 p_{c1} + y_2 p_{c2}$$
$$\omega = y_1 \omega_1 + y_2 \omega_2$$

The constants for R 290 and R 600a are given in Table 4.23.

Table 4.23 Critical constants, molecular masses and accentric factors for R 290 and R 600a

Refrigerant	*M*	p_c, *MPa*	T_c, *K*	ω	$f(\omega)$
R 290	44	4.236	369.8	1.1442	0.60429
R 600a	58	3.685	409.1	1.1424	0.64783

4.20.3 Calculation of Enthalpy of Mixture in Vapour Phase

A method developed by Agarwal and Arora[2] will now be described.

Figure 4.15 shows the vapour-liquid domes of pure components 1 and 2, and their binary mixture of certain composition on a *p-h* diagram.

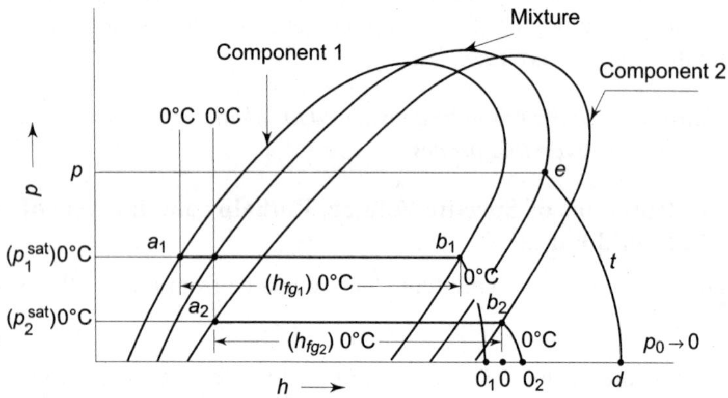

Fig. 4.15 Proposed method for vapour mixture enthalpy calculation

The figure illustrates how the enthalpies of saturated liquid and saturated vapour may be calculated. The proposed method assumes values for reference state enthalpies for saturated liquid states of both the pure components 1 and 2 at a_1 and a_2 at 0°C.

Consider now the saturated vapour state *e* on the dome for the mixture as shown in Fig. 4.15. Let the pressure and temperature of the mixture at *e* be *p* and *T* respectively.

Now, the state at the same temperature *T* but at zero pressure is shown by point *d* in the figure. Then, the enthalpy at *e* is related to ideal gas state enthalpy at *d* at temperature *T* by the residual enthalpy term. Thus

$$h_e = h_d + |h - h^{id}|_{p0 \to 0}^{p}$$

The enthalpy h_d, is related to the ideal gas state enthalpy h_0 at point 0 at $t_0 = 0$°C ($T_0 = 273.15$ K) and $p_0 \to 0$ by the relation

$$h_d - h_0 = \int_{T_0}^{T} C_{p0}\, dT$$

where C_{p0} is the zero-pressure constant pressure specific heat of the mixture. Now, h_0 is found from

$$h_0 = \xi_1\, h_{01} + \xi_2\, h_{02}$$

where h_{01} and h_{02} are ideal gas state enthalpies of components at point o_1 and o_2. These, in turn are related to real gas saturated vapour state enthalpies h_{b1} and h_{b2} by residual enthalpy terms, while h_{b1} and h_{b2} are given by

$$h_{b1} = h_{a1} + (h_{fg1})_{0°C},\ h_{b2} = h_{a2} + (h_{fg2})_{0°C}$$

Reference state enthalpies h_{a1} and h_{a2} have assigned values, say, equal to 200 kJ/kg. Note that the pressures at b_1 and b_2 are $(p_1^{sat})_{0°C}$ and $(p_2^{sat})_{0°C}$ respectively.

Now, to find h_e from h_d for the mixture, and h_{o1} and h_{o2} from h_{b1} and h_{b2} for the components, we have to evaluate residual enthalpy terms:

$$\left|h - h^{id}\right|_0^p \quad \text{for the mixture}$$

$$\left|h - h^{id}\right|_0^{p_1^{sat}} \quad \text{and} \quad \left|h - h^{id}\right|_0^{p_2^{sat}} \quad \text{for the components}$$

Since Peng-Robinson equation is a p-explicit equation, we find $|h - h^{id}|$ from $|u - u^{id}|$ using the procedure described in Sec. 1.15.2. For the purpose, we find the differential $(\partial p/\partial T)_v$ from Eq. (1.14), as given below in Eq. (4.42).

$$\left(\frac{\partial p}{\partial T}\right) = \frac{R}{v-b} + \frac{kf(\omega)\left[1 + f(\omega)\left(1 - T_T^{\frac{1}{2}}\right)\right]}{\left[(v+b)^2 - 2b^2\right]\sqrt{T\,T_c}} \tag{4.42}$$

in which
$$k = \frac{0.45724\, R^2\, T_c^2}{p_c}$$

Combining the relations as above, we find the expression for vapour phase enthalpy as written below in Eq. 4.43.

$$h = h_o + pv - RT_o + \int_{T_o}^{T} (C_{p_o} - R)\, dT - \frac{d}{2\sqrt{2}b} \ln v \left[\frac{v + b - \sqrt{2}b}{v + b + \sqrt{2}b}\right] \tag{4.43}$$

in which
$$d = a + kf(\omega)\sqrt{T_r}\left[1 + f(\omega)\left(1 - \sqrt{T_r}\right)\right]$$

C_{po} for mixture is obtained from Gibbs law

$$\bar{C}_{p_o} = y_1\, \bar{C}_{p_{o1}} + y_2\, \bar{C}_{p_{o2}}$$

For 50% R 290 and 50% R 600a mixture with $h_{a1} = h_{a2} = 200$ kJ/kg as the reference state, it is found that

$$h_o = 568 \text{ kJ/kg}$$

at $t_0 = 0°C$ and $p_o = 0$ for the mixture.

4.20.4 Calculation of Entropy of Mixture in Vapour Phase

Vapour phase entropy of the mixture is calculated using the relation similarly derived by Smith and Van Ness[37] as given below:

$$s = s_o - R\,(y_1 \ln y_1 + y_2 \ln y_2) + R \ln\left[\frac{RT(v-b)}{p_o}\right]$$
$$+ \int_{T_o}^{T} \left(C_{p_o} - R\right)\frac{dT}{T} + X\left[\frac{1}{2\sqrt{2}b} \log\left(\frac{v + b - \sqrt{2}b}{v + b + \sqrt{2}b}\right)\right] \tag{4.44}$$

$$\text{where} \qquad X = \frac{k\, f(\omega)\left[1 + f(\omega)\left(1 - \frac{T}{T_c}\right)^{\frac{1}{2}}\right]}{\sqrt{TT_c}}$$

The second term on right hand side in Eq. (4.44) represents entropy of mixing, and

$$s_o = \xi_1^V s_{o1} + \xi_2^V s_{o2}$$

For 50% R 290 and 50% R 600a mixture with $s_{a1} = s_{a2} = 1$ kJ/kg. K as the reference state entropics, it is found that

$$s_o = 2.6593 \text{ kJ/kg.K}$$

at $t_o = 0°C$ and $p_o = 1.01$ bar. Note that ideal gas state is taken, not at $p_o = 0$ but at $p_o = 1.01$ bar for calculations of residual entropy in order to avoid the mathematical anamoly of $s \to \infty$ as $p \to 0$ as explained in Chap. 1.

4.21 AZEOTROPIC MIXTURES

As stated earlier, azeotropes are essentially a class of non-ideal mixtures having bubble point temperature equal to dew point temperature. Hence, they boil and condense at constant temperature like pure substances. Azeotropes are generally formed when the difference in the boiling points of the two components is not very large and when the deviations from ideal behaviour are large enough. Hence, there are azeotropes with positive deviation from Raoult's law as well as those with negative deviation from it as described by Figs. 4.16 and 4.17. An example of an azeotrope which has positive deviation from Raoult's law is R 22/R 12 azeotrope. Such an azeotrope has a bubble or dew point which is lower than the boiling point of either of the components as shown in Fig. 4.16 and is, therefore, called a *minimum-boiling azeotrope*. On the other hand, an azeotrope with negative deviation from Raoult's law has a bubble point which is higher than the boiling point of either of the components as shown in Fig. 4.17, and is called a *maximum boiling azeotrope*.

It is to be noted from Figs. 4.16 and 4.17 that there is *an azeotropic composition* ξ_{azeo} for the mixture at the given pressure and temperature. This composition changes with the variation in pressure and temperature. However, near the azeotropic composition the bubble and dew point curves become flat. Thus in the range of usual condensation and evaporation temperatures in refrigerating machines, the azeotropic concentration more or less remains the same. Therefore, a system charged with an azeotrope may be considered as working at all sections without any change in composition.

It is also to be noted that a maximum boiling azeotrope, on account of the negative heat of mixing, will have a higher latent heat of vaporization compared to the molal average latent heat. On the contrary, a minimum boiling azeotrope will have a lower latent heat.

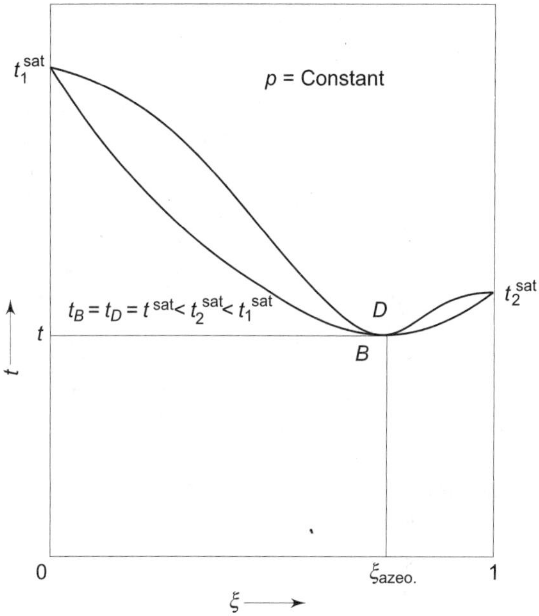

Fig. 4.16 *t-ξ* diagram of a minimum boiling azeotrope

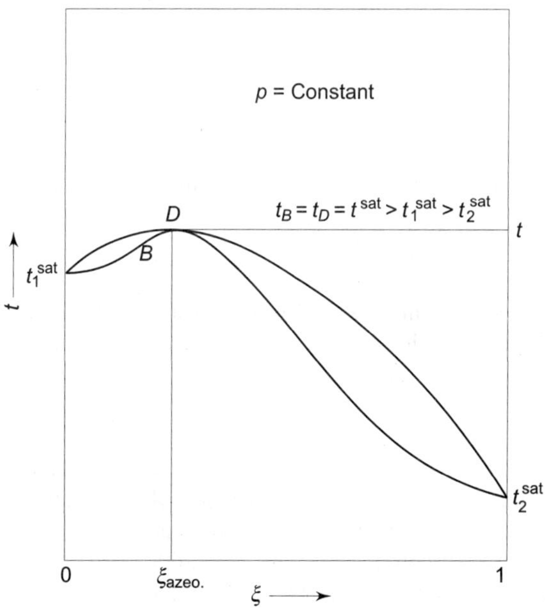

Fig. 4.17 *t-ξ* diagram of a maximum boiling azeotrope

4.22 USE OF MINIMUM AND MAXIMUM BOILING AZEOTROPES

Eiseman[21] has done a study of a large number of azeotropes. Most fluorocarbon mixtures have positive deviations from Raoult's law. Thus, they form minimum boiling azeotropes.

The main property of a minimum boiling azeotrope is its boiling point which is lower than the N.B.P. of even the lower boiling component. It enables the system with an azeotrope to have a higher evaporator pressure as compared to the lower boiling component. This, in general, results in an increased density of the suction vapour and, therefore, higher capacity.

Further, due to the depression of the boiling point of the azeotrope, slope '*b*' of the vapour pressure line of the azeotrope is flatter than the slope of the line of the lower boiling component. Thus, below a certain temperature, say point *A*, in Fig. 4.18, the vapour pressure of the azeotrope is higher, and above this temperature, it is lower than the vapour pressure of the lower boiling component. Accordingly, for the azeotrope, the evaporator pressure p_0 at T_0 is higher, and the condenser pressure p_k at T_k is lower than the values for the lower boiling component. Thus the compression ratio and discharge temperature are considerably reduced.

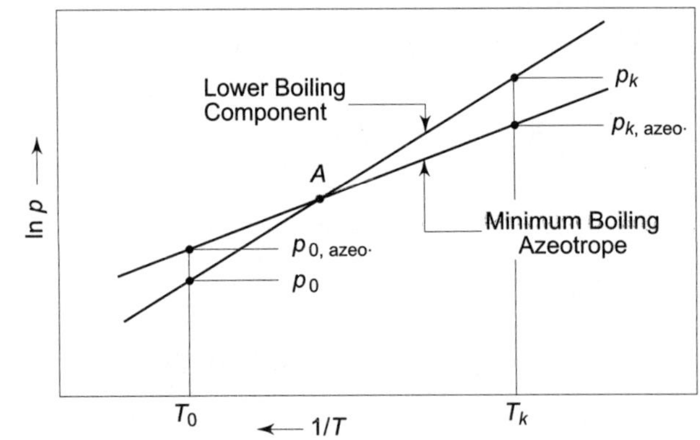

Fig. 4.18 Comparison of ln p-$1/T$ diagrams of a minimum boiling azeotrope and lower boiling component

It is not intended here to describe all the known azeotropes. However, Table 4.24 gives examples of such common minimum boiling azeotropes.

Refrigerant 500 R 500 was discovered in 1950 as a special refrigerant of the Carrier Corporation under the trade name of Carrene 7. This azeotrope consists of 73.8 per cent R 12 and 26.2 per cent R 152, difluoroethane.

The N.B.P. of R 500 is about 3.5°C lower than that of R 12. The azeotrope has a refrigerating effect per unit of swept volume about 18 per cent more than that of pure R 12. In the past this azeotrope was used to replace R 12 in 60-cycle units when they were operated on a 50-cycle current. This, essentially, kept the capacity of the unit same even with decreased speed.

Table 4.24 Data of minimum boiling azeotropes

Azeotrope	*Refrigerants*	*Weight %*	*N.B.P.*	*N.B.P. of Lower Boiling Component, °C.*
R 500	R 12/R 152	73.8/26.2	– 33.3	– 29.8 (R 12)
R 501	R 22/R 12	75/25	< – 40.8	– 40.8 (R 22)
R 502	R 22/R 115	48.8/51.2	– 45.6	– 40.8 (R 22)
R 507A	R 125/R 143a	50/50	– 46.74	– 48.09 (R 125)
R 404 A	R 125/R 143a/R 134a	44/52/4	– 46.22 /– 45.47	– 48.09 (R 125)
R 410 A	R 32 / R 125	50/50	– 51.44 /– 51.36	– 51.65 (R 32)

Refrigerant 501 Spauchus[39] has published vapour pressure data of R 22 + R 12 mixtures. Using his values, pressure ratios are determined for $t_k = 50°C$ and $t_0 = 0°C$ and are given in Table 4.25.

Table 4.25 Pressure ratios at various compositions of R 22 in R 22 + R 12 mixtures

Mole *Fraction*, x_{22}	1	0.974	0.95	0.9	0.85	—	0
Pressure ratio, p_k/p_0	3.906	3.897	3.897	3.841	3.85	—	3.947

Table 4.25 shows that for the azeotropic mixture the pressure ratio is lower than that for pure R 22. It exhibits a minimum in the neighbourhood of 85 mole per cent of R 22. Lower pressure ratios—coupled with a lower value of the compression index for the mixture as compared to R 22—yield lower discharge temperatures and hence lower winding temperatures in hermetically-sealed units. A higher density of the suction vapour results in increased cooling and further lowering of the temperature of windings.

Refrigerant 502 The behaviour of R 502 is similar to that of the R 22/R 12 azeotrope. It boils at a temperature of about 4.8°C lower than R 22. Thus, the higher capacity and lower compression ratio of the azeotrope, in relation to R 22, result in significantly lower discharge temperatures and lower winding temperatures. The use of R 502 had, therefore, proven very advantageous in supermarket air-cooled and low temperature frozen-food cabinets, and heat pump applications.

Note *All the three R 500, R 501 and R 502 contain CFCs. Hence, these have been phased out.*

Refrigerant 507A Döring *et al.*[19] have shown that a blend containing, 50–50% by mass, R 143a with an N.B.P. of – 47.24°C and R 125 with an N.B.P. of – 48.09°C shows azeotropic behaviour, making it an attractive alternative to replace R 502, and even R 22, in certain cases, since both R 125 and R 143a are HFCs (no chlorine atom/s in the molecules). In a test conducted with an open-type compressor, it was found that R 507 has compressor discharge temperature approximately 8°C below that of even R 502. It gives approximately 5-6% higher capacity than R 502.

Refrigerant 404A This *near-azeotrope refrigerant mixture* (NARM) is a *ternary* blend. It is also considered to be a zero-ODP replacement for R 502. Snelson *et al.*[38] have conducted experiments on this mixture using an open-type compressor. The lubricating oil used has to be polyol ester oil because of the presence of R 134a in the mixture. The results show that the volumetric refrigerating capacity of R 404A and R 502 are nearly the same. R 404A has about 2–4% higher pressure ratio than R 502. Inspite of this, it has 5–6°C lower compressor discharge temperature which would make it a suitable substitute for R 502 in high compression ratio applications. However, its COP is lower.

Refrigerant 410A R 410A is another near azeotropic mixture of 50/50% HFCs R 32 and R 125. It is already being used as a substitute for R 22.

Interesting possibilities exist for the use of the maximum boiling azeotropes which are a result of negative deviations from Raoult's law. They have a higher latent heat of vaporization and hence a higher refrigerating effect. Thus, an increased coefficient of performance can be obtained.

Unfortunately, maximum boiling azeotropes are rare in nature. A known example is that of an azeotrope formed by R 22 and dimethyl ether, $(CH_3)_2O$, in about 50/50 per cent composition by weight. The azeotrope boils at – 20.3°C as compared to – 40.8°C for R 22, and – 23.7°C for dimethyl ether.

4.23 NON-ISOTHERMAL REFRIGERATION

There are two kinds of duties a refrigerating machine may be required to perform:

(a) *Isothermal refrigeration:* Pumping of heat from a constant low temperature.

(b) *Non-isothermal refrigeration:* Removal of heat at varying temperature.

The former is generally termed as *space refrigeration* and the latter as *process refrigeration.*

In space refrigeration, the cold body is at a constant low temperature, such as in domestic refrigerators, cold storages, etc. The vapour compression cycle used for the purpose suffers from the irreversibility of the heat transfer processes during evaporation and condensation. This is due to the finite temperature difference between the refrigerant, the cold body and the coolant. This finite temperature difference could, however, be made small or even made to approach zero by the use of infinite heat transfer surfaces. This, therefore, does not present a thermodynamic obstacle to remove this source of inefficiency.

In process refrigeration, a body is cooled through a range of temperature. This also has numerous applications, such as in food freezing, beer chilling, most chemical processes and even cooling of air for air conditioning. Here again, the conventional vapour compression system is used. The evaporation temperature t_o of the refrigerant has to be lower than the lowest cold body temperature t_c, and the condensation temperature t_k, similarly, higher than the highest coolant temperature t_h. This is shown in Fig. 4.19. Thus, $t_0 < t_{c1} < t_{c2}$ and $t_k > t_{h1} > t_{h2}$. The system, as for space refrigeration, suffers from the inherent irreversibility of the heat transfer processes during evaporation and condensation, and in the case of process refrigeration, this cannot be made to approach zero even with infinite surfaces. For even if it were possible to make temperature differentials—$(t_{c2} - t_o)$ and $(t_k - t_{h2})$ at one end of the evaporator and condenser respectively—approach zero by the use of infinitely large heat transfer surfaces, there would still be existing finite temperature differentials—$(t_{c1} - t_o)$ and $(t_k - t_{h1})$—at the other ends.

The scope of using non-azeotropic mixtures is described in detail by Tchaikovski and Arora.[41] The most important property of a mixture is its ability to boil through a range of temperature. If a non-azeotropic mixture of two or more refrigerants is used, a definite boiling range can be obtained for the evaporation as well as condensation processes. The actual range will depend upon the proportion of individual components and the temperature and pressure of the mixture. Thus in Fig. 4.19, the evaporation could be made to proceed along path t_o to t_o' and condensation along t_k to t_k'. In this manner, the excessive temperature differences at the other two ends could also be made small. The average refrigeration temperature could thus be increased. And, the average heat rejection temperature could be decreased at the same time. Since the Carnot COP of a refrigerating machine is given by.

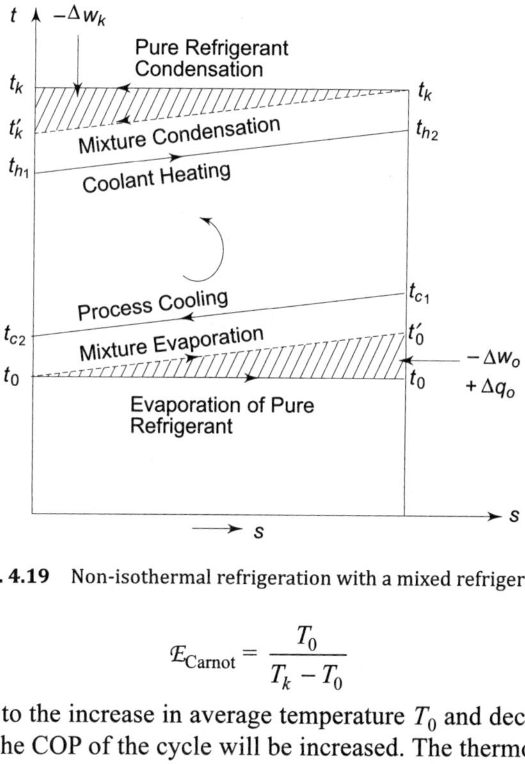

Fig. 4.19 Non-isothermal refrigeration with a mixed refrigerant

$$\mathcal{E}_{\text{Carnot}} = \frac{T_0}{T_k - T_0}$$

we find that due to the increase in average temperature T_0 and decrease in average temperature T_k, the COP of the cycle will be increased. The thermodynamic obstacle to the elimination of irreversibility of the heat transfer processes can thus be removed by a judicious selection of a mixture of refrigerants and its composition.

It can be seen from Fig. 4.19 that the use of mixed refrigerant could decrease the work by

area Δw_0 due to non-isothermal evaporation and
area Δw_k due to non-isothermal condensation.

The area $\Delta w_0 = \Delta q_0$ also represents increase in refrigerating effect. It would, therefore, result in increase in COP and hence power saving.

Consider for example a case in which air is cooled from $t_{c1} = 25°C$ to $t_{c2} = 15°C$, and temperature of cooling air in condenser rises from $t_{h1} = 45°C$ to $t_{h2} = 55°C$. These are typical values for a window-type air conditioner. Then if pure refrigerant is used, the practical evaporator and condenser temperatures could be $t_0 = 0°C$ and $t_k = 65°C$. The corresponding value of Carnot COP would be 4.55.

Now, if a mixture of refrigerants is used for evaporating from $t_0 = 0°C$ to $t_0' = 10°C$, and condensing from $t_k = 60°C$ to $t_k' = 50°C$, then the Carnot COP would be 5.55. The saving in power would be 18%.

Note *It is important to observe that this effect can be utilized only if there is counterflow of the two fluids which is possible only if the mixture is boiling and condensing* ***inside tubes*** *as in window-type air conditioner. Otherwise, if for example, the condensation is outside tubes as in shell and tube condensers, the condensation temperature will rise above the outlet temperature of the coolant ($t_k' \geq 55°C$ and $t_k > t_k'$). This will cause increase in average temperature of heat rejection, and hence in condenser pressure, work, and power consumption.*

Example 4.6 ***Vapour Compression Cycle of a Hydrocarbon Mixture Refrigerator on p-h Diagram***

A hydrocarbon refrigerator operating on equal proportions of propane and isobutane by mass works on simple saturation cycle. The condenser and evaporator pressures are 14 bar and 1.4 bar. Show the cycle on p-h diagram and determine its discharge temperature and theoretical COP.

Solution The values of bubble and dew temperatures at these pressures have already been determined in Example 4.4. Accordingly, the *p-h* diagram is drawn as shown in Fig. 4.20.

From tables of properties for 50–50% R 290/R 600a mixture we have,

$$h_1 = 545.0 \text{ kJ/kg}, h_3 = 341.15 \text{ kJ/kg} = h_4$$
$$s_1 = 2.47 \text{ kJ/kg.K} = s_2$$

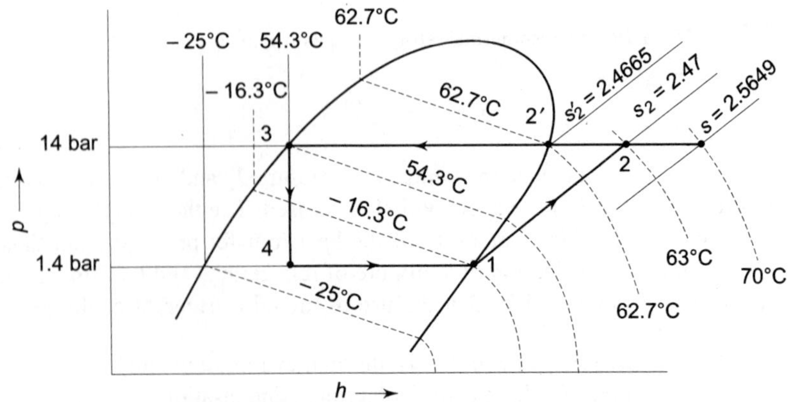

Fig. 4.20 Vapour compression cycle for a refrigerator with 50–50% propane/isobutane mixture

From superheat table for the mixture, for known p_2 and s_2, we have:
Discharge temperature

$$t_2 = 63°C$$

$$\Rightarrow \quad h_2 = 661.0 \text{ kJ/kg}$$

Hence
$$\text{COP} = \frac{q_0}{w} = \frac{h_1 - h_4}{h_2 - h_1} = \frac{545 - 34.15}{661.0 - 545} = 4.7$$

Example 4.7 ***Using Binary Mixture R 152a/R 22 for Non-Isothermal Refrigeration in Air Conditioners.***

It is proposed to substitute R 22 by a mixture of 18% R 152a and 82% R 22 by mass in an air conditioner operating on simple saturation cycle. The condenser and evaporator pressures are 15 bar and 5 bar. Locate all the four state points of the cycle on t-ξ, h-ξ and p-h diagrams.

Also, find the discharge temperature, temperature after throttling, and theoretical COP of the cycle.

Solution The cycle *abcd* with the mixture is shown on *p-h* diagram in Fig. 4.21.

The parameters of states as obtained from *t-ξ* and *h-ξ* diagrams in Figs. 4.13 and 4.14, and plotted in Figs. 4.22 (a) and (b) are as follows:

State a Saturated vapour

$$p = 5 \text{ bar}, t_a = 5.75°C, h_a = 428 \text{ kJ/kg}$$

State b Superheated vapour

$$p = 15 \text{ bar}$$

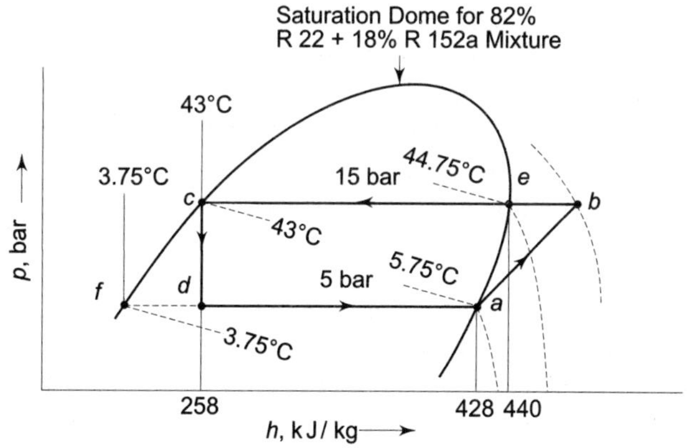

Fig. 4.21 Pressure-enthalpy diagram for R 22/R 152a mixture for Example 4.7

To find its temperature from $s_b = s_a$

State c Saturated liquid

$$p = 15 \text{ bar}, t_c = 43°C, h_c = 258 \text{ kJ/kg}$$

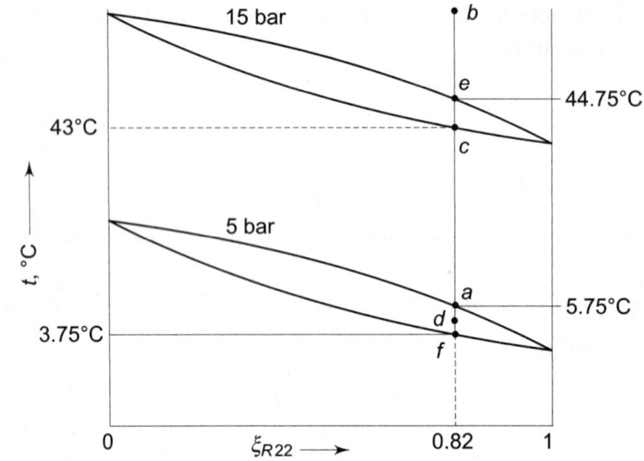

(a) Temperature-composition diagram

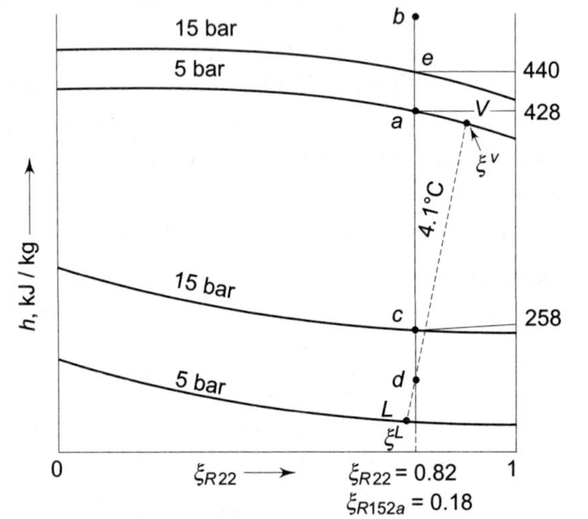

(b) Enthalpy composition diagram

Fig. 4.22 Figure for Example 4.7

State d Liquid-vapour mixture after throttling
$p = 5$ bar, $h_d = 258$ kJ/kg $= h_c$
Temperature at d lies between 3.75°C at f and 5.75°C at a.

State e Saturated vapour mixture at 15 bar is at 44.75°C, and its enthalpy is 440 kJ/kg

Entropy at a

$$s_a = (\xi\, s)_{\text{R 152a}} + (\xi\, s)_{\text{R 22}} + \Delta s_m$$

Values of specific entropies of R 152a and R 22 are found from their superheat tables at 5 bar and 5.75°C. And Δs_m is the entropy of mixing. Thus

$$s_a = 0.18\ (2.1183) + 0.82\ (1.7518) + \Delta s_m$$
$$= 1.8178 + \Delta s_m \text{ kJ/kg.K}$$

Entropy at e is similarly found at 15 bar and 44.75°C (317.75 K)

$$s_e = 0.18\ (2.0838) + 0.82\ (1.6996) + \Delta s_m$$
$$= 1.7686 + \Delta s_m \text{ kJ/kg.K}$$

Discharge temperature at b

The equation to solve is

$$s_b = s_e + \xi_1 C_{p1} \ln \frac{T_b}{317.75} + \xi_1 C_{p2} \ln \frac{T_b}{317.75} = s_a$$

$$1.7686 + (0.18 \times 1.61 + 0.82 \times 0.91) \ln \frac{T_b}{317.75} = 1.8178$$

$$\Rightarrow \quad T_b = 324 \text{ K } (t_b = 51°\text{C})$$

Enthalpy at b

$$h_b = h_e + \xi_1 C_{p1} (t_b - t_e) + \xi_1 C_{p2} (t_b - t_e)$$
$$= 440 + (0.18 \times 1.61 + 0.82 \times 0.91)(51 - 44.75) = 455.8 \text{ kJ/kg}$$

Temperature at d

Overall composition and enthalpy during throttling process remain same. Hence

Composition balance equation

$$(\xi_1)_d = \left[\xi_1^L (1 - z) + \xi_1^V z\right]_d = (\xi_1)_c = 0.18$$

Enthalpy Balance equation

$$h_d = (h^L)_f (1 - z) + (h^V)_a z = 202\ (1 - z) + 428z = h_c = 258$$

The two equations can be solved for t_d by iteration. The equilibrium L-V states at 5 bar from Tables 4.19 and 4.21 are plotted on t-ξ and h-ξ diagrams in Figs. 4.22 (a) and 4.22(b). It is seen that state d is at a temperature slightly above 4°C. The method of solution by iteration is as follows:

(i) Assume t_d.
(ii) Find ξ_1^L and ξ_1^V.
(iii) Find z from the composition-balance equation.
(iv) Verify if enthalpy-balance equation is satisfied.
(v) If not, assume another value of t_d.

The solution is

$$t_d = 4.1°\text{C}$$

Theoretical COP of the cycle

$$\varepsilon = \frac{h_a - h_d}{h_b - h_a} = \frac{428 - 258}{455.8 - 428} = \frac{170}{27.8} = 6.1$$

Note *Ghosh and Sarkar[22] obtained a saving of 2.7 to 4.3% in power consumption with 9.87% R 152a in R 22/R 152a mixture.*

Example 4.8 ***Estimating Power Saving in Air Conditioners with R 152a/R 22 Mixture.***

Find the theoretical COP of R 22 cycle equivalent to the mixture cycle in Example 4.7. Estimate the power saving, and discuss the other effects.

Solution In order to maintain the condenser and evaporator sizes same, we have to design for same LMTD in the two heat exchangers. This means, we have to maintain mean condensation and evaporation temperatures as same. Hence, for the equivalent R 22 cycle,

$$t_k = \frac{t_e + t_c}{2} = \frac{44.75 + 43}{2} = 43.9°\text{C}$$

$$t_0 = \frac{t_a + t_d}{2} = \frac{5.75 + 4.1}{2} = 4.9°\text{C}$$

The corresponding R 22 cycle is shown in Fig. 4.23.

COP of equivalent R 22 cycle

$$\mathcal{E}_{\text{R 22}} = \frac{h_1 - h_4}{h_2 - h_1} = \frac{407.5 - 254.1}{435.5 - 407.5} = 5.5$$

Power saving by using mixture for same refrigerating capacity

$$\frac{W_{\text{R 22}} - W_{\text{mix}}}{W_{\text{R 22}}} = \frac{1/\mathcal{E}_{\text{R 22}} - 1/\mathcal{E}_{\text{mix}}}{1/\mathcal{E}_{\text{R 22}}} = \frac{1/5.5 - 1/6.1}{1/5.5} = 0.1\ (10\%)$$

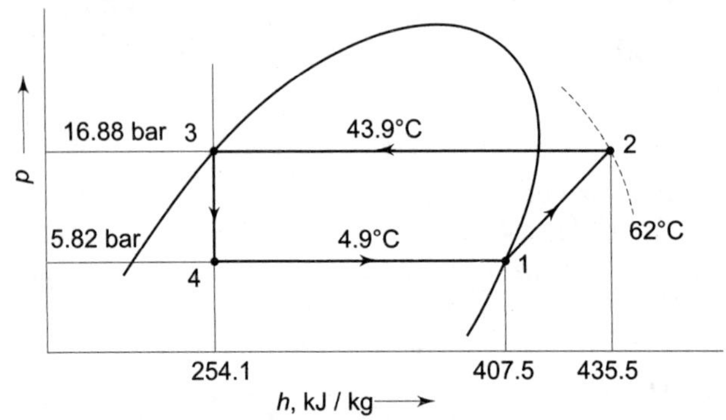

Fig. 4.23 R 22 cycle for Example 4.8

Other Effects:

(i) Discharge temperature is lowered from 62°C for R 22 to 51°C for mixture.

(ii) Compressor displacement volume will increase for mixture due to larger volume of R 152a which is a higher boiling substance.

(iii) R 152a improves the miscibility of R 22 in mineral oil.

4.24 REFRIGERANT PIPING AND DESIGN

A well-designed refrigeration plant is one with a well-sized and well-laid refrigerant piping. Some salient features of this are discussed below. The material used for fluorocarbon refrigerant piping is either seamless copper tubing or iron, whereas for ammonia only iron pipes are used. Their sizes are given in terms of the outside diameter (OD) for copper tubing and normal iron pipe size (IPS) for iron pipes.

4.24.1 Location and Arrangement of Piping

The following factors are important:

(i) To minimize tubing and refrigerant charge requirement and pressure drops, refrigerant lines should be as short and direct as possible.

(ii) Piping is to be planned for a minimum number of joints using as few elbows and fittings as possible, although providing for sufficient flexibility to absorb compressor vibrations.

(iii) It is to be so arranged that normal inspection and servicing of the compressor and equipment is not hindered.

(iv) It should be so run that it does not interfere with the removal of compressor-cylinder heads, access plates or any other internal part such as the stator in hermetically-sealed units.

(v) Sufficient clearance is to be provided between the piping and wall or between pipes for insulation.

(vi) Locations where copper tubing will be exposed to mechanical injury, should be avoided.

(vii) In the case of iron pipes, *hangers* should be provided close to vertical risers to and from compressors to keep the piping weight off the compressors.

(viii) Hangers should not be placed more than 2 to 3 m apart and should be within 0.6 m of the change in direction of piping.

(ix) Valves should be located in inlet and outlet lines to condensers, receivers, evaporators and long lengths of pipes to permit isolation and to facilitate *pumping down* and *pumping out*. Valve stems should be horizontal so that there is less chance of dirt or scale accumulating on the valve seat.

4.24.2 Vibration and Noise in Piping

Vibration and noise can be eliminated by proper design and support of the piping. The undesirable effects of vibration are:

(a) Breaking of brazed or soldered joint and consequent loss of charge.

(b) Transmission of noise through piping and building.

The vibration of piping is caused by the rigid connection of the refrigerant piping to a reciprocating compressor. It is impossible to eliminate it completely. But steps such as the following can be taken to mitigate its effects:

(i) Run the suction and discharge lines at least 15 pipe diameters in each of two or three directions before securing the first hanger. In this manner, piping can absorb vibrations without being overstressed.

(ii) A flexible metal hose is often used to absorb vibration transmitted along small rises of pipes. For great effectiveness, it should be installed at right angle to the direction of vibration. Most compressor vibrations have the greatest amplitude in the horizontal plane. Therefore, the flexible hose should be installed in the vertical lines near the compressor. Two isolators, one vertical and one horizontal, may be required to do an efficient job.

(iii) Vibration and noise are also caused by gas pulsations and turbulence in gas which is increased at high velocities. It is usually more important in discharge lines. Under some conditions, these pulsations may occur at such a frequency as to cause resonance, sufficiently strong to break piping connections. In such a case, change the size or length of the resonating line or install a properly sized hot *gas muffler* immediately after the compressor.

(iv) When the noise is due to turbulence, as in centrifugal systems, and isolating the line is not effective enough, the installation of a larger line to reduce the gas velocity is often effective.

4.24.3 Flow Rates

The volume flow rate in the suction line can be easily calculated from its capacity. The volume flow rate in the discharge line, however, cannot be calculated accurately as the discharge temperature is very much a function of the compressor design and efficiency. The discharge volume can be approximated by the formula

$$V_d = V_s \times \frac{p_s}{p_d} \times 1.2$$

where subscripts s and d refer to suction and discharge states respectively.

4.24.4 Pressure Drop in Refrigerant Piping

It is customary to evaluate the pressure drop in the refrigerant piping in terms of the number of degrees change in the saturation temperature as a result of this loss. The refrigerant piping is designed mainly on the basis of practical line sizes without excessive pressure drops.

The pressure losses without a change of phase can be calculated using Darcy's formula

$$\frac{\Delta p}{\rho} = \frac{fLC^2}{2gD}$$

where Δp is the pressure drop, ρ the density, L the length, D the diameter, C the velocity and f the friction factor. For friction factor f in suction and discharge lines, a value of 0.003 may be taken, whereas for liquid lines, it may be taken as 0.004.

4.24.5 Liquid Lines

The liquid line presents fewest problems. It is desirable to have a slightly subcooled liquid reach the liquid-feed device to prevent the formation of flash gas. The flash gas in the liquid line causes an increase in the pressure drop and further flashing, reduction in the capacity of the expansion device, noise and erratic control of the liquid refrigerant entering the evaporator. The pressure drop in turn is due to friction as well as decrease in the static head due to the elevation of the evaporator above the condenser.

Generally, liquid always leaves the condenser in a subcooled state. An additional refrigerant charge will increase subcooling. Shell-and-tube condensers are designed to maintain a liquid level in the shell which can provide for subcooling. The construction of the aircooled condensers is also such that the liquid is in contact with the cooling air surface until it gets subcooled and leaves. In the case of water-cooled condensers, even liquid lines can be provided with insulation.

Where long vertical lines are used, it is almost impossible to prevent flashing caused by static pressure drop. It is simple to determine the degree of subcooling required to offset friction and static pressure drop. The static pressure drop associated with an increase in elevation is: 0.1275 bar for every 1.11 m lift for R 22.

When a system is equipped with a receiver to maintain a seal of the liquid refrigerant in the control devices, the liquid line entering the receiver is usually designed generously. A velocity of 0.5 m/s is typical to ensure pressure equalization between the condenser and receiver and to prevent vapour locking. With this velocity a vapour-equalising line should be provided from the top of the condenser to the top of the receiver. Otherwise, a lower velocity can be used.

For liquid line between the receiver and expansion device, it is recommended to design it on the basis of pressure loss of 0.05 bar per 100 m of length.

In general, the pressure drop should not be greater than that corresponding to 0.9°C change in the saturation temperature for ammonia, and 0.9 to 1.8°C for fluorocarbons. A change of 0.9°C in the saturation temperature at 38°C condenser comes to a pressure drop of approximately 0.227 bar for ammonia, and 0.2 bar for R 22.

4.24.6 Suction Lines

Suction lines are the most critical from the viewpoints of design and construction. The considerations involved are the following:

(i) Correct size for practical pressure drop.

(ii) Capability to return oil to the compressor by entrainment by the suction vapour under minimum loading conditions, especially in the case of fluorocarbons.

(iii) If there are suction risers, gas velocities to ensure oil return have to be increased. Thus a minimum tonnage is prescribed depending on the size of the suction line for oil entrainment in suction risers as given in Table 4.26 for R 22.

Table 4.26 Tonnage of suction piping up suction risers

Refrigerant	t_0, °C	***Pipe OD*, cm**			
		1.27	1.59	1.905	2.86
R 22	– 40	0.09	0.16	0.27	0.79
	–10	0.13	0.24	0.39	1.2
	5	0.18	0.33	0.54	1.6

(iv) Double-suction risers may be used for full-load operation and single risers for part-load operation as shown in Fig. 4.24. Thus, excessive pressure drop at full load is avoided, and oil return at part load is ensured. When the load reduces, the oil cannot be entrained in the beginning which collects in the U-bend and forms a seal. Afterwards, the gas flows only through one riser.

(v) Prevention of drainage of oil from an active evaporator into an idle evaporator. For this purpose, arrangements are made as shown in Figs 4.25 to 4.27. The common-suction line should either be horizontal or pitched down towards the compressor.

It is customary to design suction lines so that the total loss in pressure is equivalent to drops of about 1.8 and 0.9°C in saturation temperatures for fluorocarbons and ammonia respectively. It is also desirable to provide for less pressure drop in low-temperature installations because of increased penalties on the compressor size and performance. Since the pressure loss is also a function of the length of the line and number of fittings, it is obviously desirable to have the shortest possible runs of suction lines.

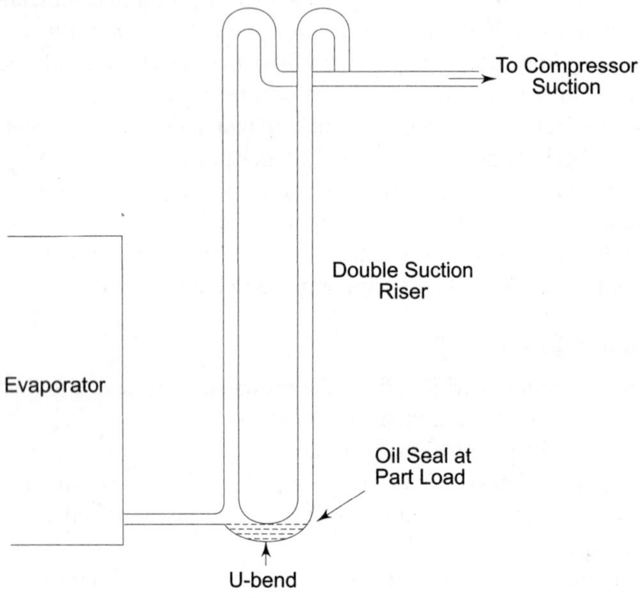

Fig. 4.24 Double-suction riser

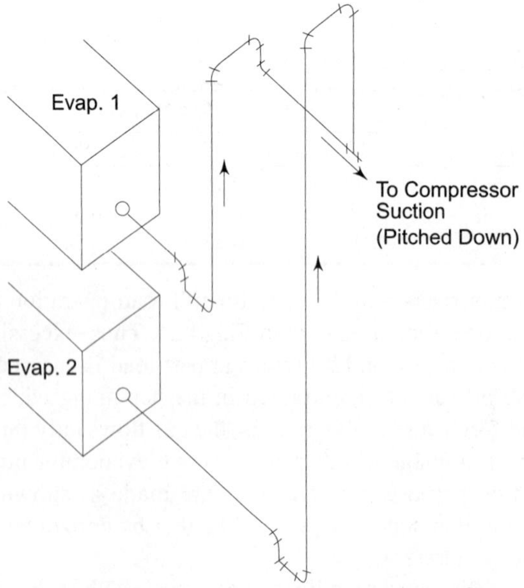

Fig. 4.25 Multiple evaporators at different levels with compressor above

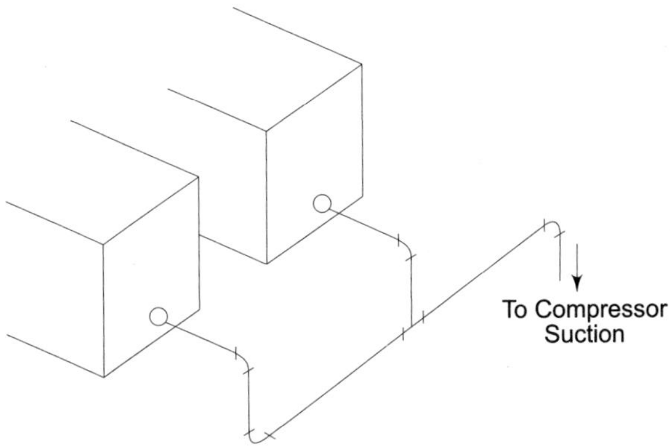

Fig. 4.26 Multiple evaporators at same level with compressor below

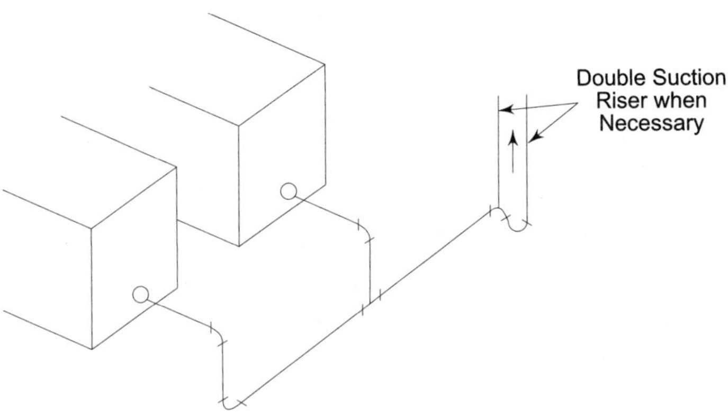

Fig. 4.27 Multiple evaporators at same level with compressor above

4.24.7 Discharge Lines

Even though a low pressure drop is desired, the discharge lines should not be oversized to the extent that the gas velocities are reduced so much that the refrigerant will not be able to carry along the entrained oil. They should be so designed and constructed as to prevent the refrigerant and oil in the line from condensing and draining back to the compressor, especially during shutdown or operation at low ambient where long outdoor discharge lines are required as in evaporative condensers.

Whenever the condenser is located above the compressor, the discharge line should loop to the floor before rising to the compressor as shown in Fig. 4.28. It is not desirable to exceed a pressure drop equivalent to 1 or 2°C in the saturation temperature because of the penalty on volumetric efficiency and hence the capacity of the compressor. Discharge lines should be selected on the basis of 0.45 to 0.67 bar per 100 m pressure drop.

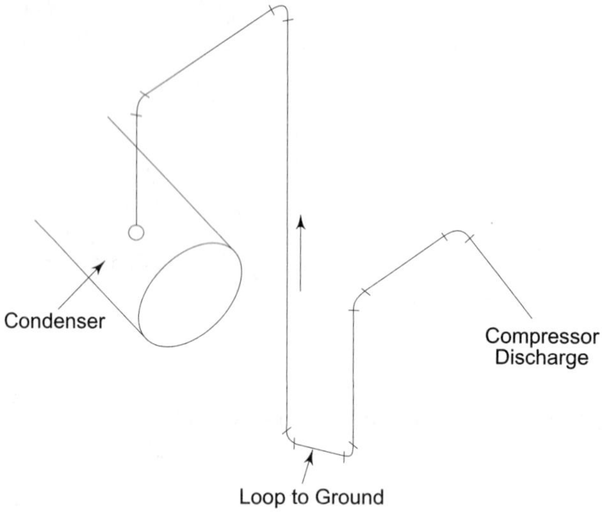

Fig. 4.28 Hot-gas loop

4.24.8 Piping at Multiple Compressors

The piping of compressors operating in parallel must be carefully done to ensure proper operation.

There should be a common suction header. This will ensure equal oil return to each compressor and that all compressors run at the same suction pressure. The suction header should be run above the level of the compressor suction inlet so that all oil can drain into the compressor by gravity. If the header is below the compressor-suction inlets, it can become an oil trap. Further, branch-suction lines should be taken from the side of the header to ensure equal distribution of oil and to prevent the accumulation of liquid refrigerant in an idle compressor in the case of slop over. The horizontal takeoffs to the various compressors should be of the same size as the suction header. No reduction is to be made in the branch suction lines until the vertical drop is reached.

It is always desirable, even with two or more compressors, to use a single condenser (parallel operation).

4.25 LUBRICANTS IN REFRIGERATION SYSTEMS

In refrigeration systems, oil must perform certain functions other than minimizing friction, such as sealing the gas between suction and discharge ports, acting as a coolant to transfer heat from the crank-case to the compressor shell and to dampen the noise generated by moving parts. For hermetic units, it must also have a high dielectric strength. It must have adequate fluidity at low temperatures so that it may easily return to the compressor from the evaporator. For good heat transfer in the evaporator, oil should remain miscible with the refrigerant at evaporator temperatures. Further, it should not contain any suspended matter, wax or moisture which might choke the expansion device. In hermetic units, oil is charged only once for the life-time of the unit (minimum five years). It should, therefore, be chemically stable in the presence of the refrigerant, metals, winding insulation and extraneous contaminants.

The oils of mineral origin contain paraffins, napthenes, aromatics and non-hydrocarbons. Paraffinic and napthemic oils are saturates. They have excellent chemical stability but poor solubility for refrigerants, such as R 22, and are also poor lubricants. Aromatics have good solubility and lubrication properties. Non-hydrocarbons are the most reactive but are good for lubrication. The resultant properties depend on the proportional composition of the four constituents.

Synthetic oils, such as alkylbenzenes and phosphate esters, have desirable properties for use with R 22. Polyalkylene glycols (PAGs), modified PAGs, and esters are primary lubricants that are being tested for use with R 134a.

Normally, one should choose an oil with the lowest viscosity which will, at the same time, give the necessary sealing properties with the refrigerant used. Thus, the oil which gives the maximum volumetric efficiency is the best. The prescribed oil kinematic viscosities for reciprocating compressors for use with ammonia and R 22 are 150-300 SSU (Saybolt Universal Units) at 38°C.

4.26 SECONDARY REFRIGERANTS

In large refrigeration plants, secondary refrigerants or coolants such as water, brines, glycols and sometimes even halocarbons are used for carrying refrigeration from the plant room to the space where it is usefully applied, instead of directly obtaining it by the evaporating refrigerant at the place of application. This is done in order to reduce the quantity of the refrigerant charge in the system and to reduce pressure losses in lines. The desirable properties of secondary coolants are low freezing point, low viscosity, non-flammability, good stability and low vapour pressure. Chilled water is used as a secondary refrigerant in air-conditioning applications. For low-temperature applications, brines, glycols and hydrocarbons are used.

4.26.1 Brines

Brines are formed by dissolving salt in water. The phase diagram for a brine solution is shown in Fig. 4.29.

The function of salt in water is to depress its freezing point. The temperature at which the freezing point is lowest is called the *eutectic temperature*, and the composition at this temperature is called the *eutective composition*.

If a brine solution has less than eutectic composition, such as at A, its crystallization temperature or freezing point will be lowered to t_B. If this solution at A is cooled, ice crystals will begin to form at B. As a result, the solution will become richer in salt content. At any point C, the mixture will consist of ice at C_2 and solution at C_1. At point D, the solution will have the composition corresponding to D_1. On further cooling, the solution will freeze as a whole without any separation of ice crystals.

Similarly, if a brine solution of more than eutectic composition such as at E is cooled, it will first separate into salt and solution, until it reaches point D_1 again, where after the solution will freeze as a whole.

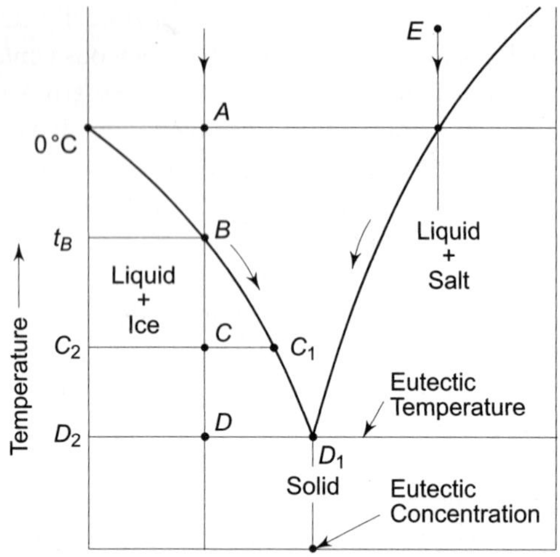

Fig. 4.29 Phase diagram for brine

Brines are mainly used in industrial ice plants, cold storages, skating rinks, etc. Common brines are water solutions of calcium chloride ($CaCl_2$) and sodium chloride (NaCl). Their relevant properties are given in Tables 4.27 and 4.28.

Table 4.27 Properties of $CaCl_2$ brine

Pure **$CaCl_2$** *% Weight*	*Crystallization Temperature* **°C**	*Specific Gravity at* **15°C**	*Specific Heat* **kJ/(kg. K)**
0	0	1.0	4.1868
5	–1.8	1.044	3.869
10	–5.4	1.087	3.58
15	–10.3	1.133	3.32
20	–18.0	1.182	3.086
25	–28.3	1.233	2.88
29.87 (Eutectic)	–55.0	1.29	2.74

Table 4.28 Properties of NaCl brine

Pure **NaCl** *% Weight*	*Crystallization Temperature* **°C**	*Specific Gravity at* **15°C**	*Specific Heat* **kJ/(kg. K)**
0	0	1.0	4.1868
5	–2.9	1.035	3.93
10	–6.6	1.072	3.72
15	–10.9	1.111	3.55
20	–16.5	1.15	3.4
25	–8.8	1.191	3.29
25.2 (Eutectic)	0	1.2	

As seen from the tables, $CaCl_2$ brine has lower crystallization temperatures at higher concentrations. It is, therefore, more widely used except when it may attack a product by direct contact, such as fish, in which case NaCl brine is used. Corrosion is, however, a serious problem in the use of $CaCl_2$ brine. To reduce corrosion, excessive contact of air with brine should be avoided. *Corrosion inhibitors* such as sodium dichromate are added to keep alkaline conditions, i.e., a pH value of 7 to 8.5. The recommended dichromate concentration is 2 kg/m^3 of brine. Caustic soda may be added to correct for acidity, i.e., a pH below 7. Dichromate can be added to correct for excessive alkalinity, i.e., a pH above 8.5.

4.26.2 Inhibited Glycols

Ethylene glycol and propylene glycol are also used as freezing point depressants and heat transfer media in solution with water. Freezing points of ethylene glycol solutions are lower than those of propylene glycol. It is more commonly used except when, because of its toxicity, it is found to attack a food product by direct contact, in which case propylene glycol is used.

Glycol solutions have lower corrosivity when properly inhibited. The specific gravities of ethylene glycol and propylene glycol are 1.1155 and 1.0381 respectively. Their specific heats at 15°C are 9.82 and 10.38 kJ/(kg. K) respectively.

References

1. Adcock J L, et al., 'Fluorinated ethers a new series of CFC substitutes', *Proc. Int. CFC and Halon Conference*, Baltimore, MD, USA, 1991.
2. Agarwal R S and C P Arora, 'Calculation of thermodynamic properties of binary mixtures of refrigerants', *J. Thermal Engineering*, Vol.1, No.1, pp. 9–15, 1980.
3. Arora C P, *Thermodynamics*, Tata McGraw-Hill, New Delhi, 1997.
4. Arora C P, *'Power Savings and Low Temperatures in Refrigerating Machines Using Mixed Refrigerants'*, Ph D Thesis, IIT Delhi, 1968.
5. Arora C P, 'Power savings in refrigerating machines using mixed refrigerants', *Proc. XIIth International Congress of Refrigeration*, Madrid, 1967.
6. Arora C P, 'Low temperatures in refrigerating machines using R 12 and R 13 mixtures', *Proc, XIIIth International Congress of Refrigeration*, Washington, 1971.
7. Arora C P, *'An Investigation into the Use of a Mixture of R 12 and R 22 in Window Type Air Conditioners'*, Paper presented to Roorkee University for Khosla Research Prize, 1974.
8. Arora C P, B K Bhalla and Addai Gassab, 'A study on the performance of window-type air conditioners using R 22/R 12 azeotrope', *Proc. XVth International Congress of Refrigeration*, Venice, 1979, Paper No. B2–24.
9. Ashok Babu T P, 'A theoretical and experimental investigation of alternatives to CFC 12 in refrigerators', Ph D Thesis, IIT Delhi, 1997.
10. ASHRAE, *Handbook of Fundamentals*, 1972.
11. ASHRAE, *Handbook of Systems*, 1973.
12. Badylkes, IS, *Working Substances and Processes in Refrigerating Machines (Russian)*, Gostorgizdat, 1962.

13. Badylkes, I S, 'Thermodynamic properties of azeotropic mixtures of R 22 and Freon 115,' *I.I.R. Bulletin*, Annex 1965–4, pp.195–202.

14. Beyerlien A L, et al., Physical property data on fluorinated propanes and butanes as CFC and HCFC substitutes', *Proc. Int. CFC and Halon Conference*, Baltimore, MD, USA, 1991.

15. Bretsznajder S, *Prediction of Transport and Other Physical Properties of Fluids*, Pergamon Press, 1971.

16. Chhaya K and A Tandon, 'Propane as an alternative to R 12 in domestic refrigerator,' B.Tech. Thesis, IIT Delhi, 1994.

17. Devotta S and S Gopichand, 'Comparative assessment of HFC 134a and some refrigerants as alternatives to CFC 12,' *Int. J. Refrigeration*, Vol. 15, No. 2, 1992.

18. Devotta S, *et al.*, Comparative assessment of some HCFCs, HFCs and HFEs as alternatives to CFC 11", *Int. J. Refrigeration*, Vol. 17, No. 1, 1994.

19. Döring R, H Buchwald and J Hellmann, 'Results of experimental theoretical studies of the azeotropic refrigerant R 507', *Int. J. Refrigeration*, Vol. 20, No. 2, 1997.

20. Downing R C 'Mixed refrigerants', *Service Manual RT-38*, E.II. du Pont de Nemours Co. Wilmington, Delaware, U.S.A.

21. Eiseman, B J, 'The azeotrope of monochlorodifluoromethane and dichlorodifluoromethane', *J. American Chemical Society*, Vol. 79, No. 4, 1957, p. 6086.

22. Ghosh S and B Sarkar, 'Using binary mixture of refrigerants in the window type air conditioner', B. Tech. Thesis, IIT Delhi, 1995.

23. James R W, and J F Missenden, 'The use of propane in domestic refrigerators', *Int., J. Refrigeration*, Vol. 15, No. 2, 1992.

24. Jung D S and R Radermacher, 'Performance simulation of single evaporator domestic refrigerators charged with pure and mixed refrigerants', *Int. J. Refrigeration*, Vol. 14, July 1991.

25. Kim K, *et al.*, 'R 22/R 152a mixtures and cyclopropane as substitutes for R 12 in single evaporator refrigerators,' *ASHRAE Trans.*, Vol. 99, Part 1, pp. 1439–1446, 1993.

26. Kriebel M, 'Phasenglechgewichte zwicshen Flussigkeit and Damf im binaren system R 22-R 12', *Kaltetechnik Klimatisierung*, Vol. 19, No. 1, 1967.

27. Latto B and Al Saloum, A.J.M. 'Extensive viscosity correlations for refrigerants in liquid and vapour', *ASHRAE Trans.*, Vol. 76, Part I, 1970, pp. 64-80.

28. Löffler H J, *Kaltetechnik*, Vol. 9, No. 5, 1957, p. 135.

29. Mcharness R C and Chapman D D, 'Refrigerating capacity and performance data', *ASHRAE J.*, Jan. 1962, pp. 49–58.

30. McLinden M O, 'Thermodynamic properties of CFC alternatives: A survey of available data', *Int. J. Refrigeration*, Vol. 13, May 1990.

31. Plank R, *Handbuch der Kaltetechnik*, Vol. IV, Springer Verlag, Berlin/Gottingen, 1959.

32. 'Properties and applications of freon fluorocarbons', *Technical Bulletin B-2*, Freon Products Division, E.I., du Pont de Nemours, Wilmington, Delaware, U.S.A.

33. Reid R C, et al., *The Properties of Gases and Liquids*, McGraw-Hill, 1988.
34. Sand J R, et al., 'Experimental performance of ozone-safe alternative refrigerants', *ASHRAE Trans.*, part 2, Vol. 96, pp. 173–182, 1990.
35. Sand J R and S K Fischer, 'Modelled performance of non-chlorinated substitutes for CFC 11 and CFC 12 in centrifugal chillers," *Int. J. Refrigeration*, Vol. 17, No. 1, 1994.
36. Shankland I R, et al., 'Thermal conductivity and viscosity of a new stratospherically safe refrigerant tetrafluoroethane (R 134a)', *ASHRAE Trans.*, Vol. 94, pp. 305–313, 1988.
37. Smith J M and H C Van Ness, *Introduction to Chemical Engineering Thermodynamics*, 4th Ed., McGraw-Hill, p. 493.
38. Snelson W K and J W Linton, 'System drop-in tests of refrigerant blend R 125/R 143/R 134a (44%/52%/4%) compared to R 502, *ASHRAE Trans. Res.*, No. 3834, pp. 17–24.
39. Spauchus H O, 'Vapour pressures of mixtures of refrigerants R 12 and R 22', *ASHRAE J.*, Vol. 4, No. 9, Sept. 1962, pp. 49–51, 123.
40. Spauchus H O, 'HFC 134a as a substitute refrigerant for CFC 12,' *Proc. 11R Commissions B1, B2*, Purdue University, USA, pp. 397–400, 1988.
41. Tchaikovski V F and C P Arora, 'Using mixtures of refrigerants, *J. Inst. Engrs.* (India), Vol. XLIV, No. 5, pp. 128–134, 1964.
42. Wijaya H and M W Spatz, 'Two-phase flow heat transfer and pressure drop characteristics of R 22 and R 32/R 125,' *ASHRAE Trans*: *Symposia*, No. CH-95-14-2, pp. 1020–1026, 1995.

Revision Exercises

4.1 Explain in brief which refrigerant/s would you choose for each of the following applications and why?
   (i) A cold storage of 100 TR capacity using reciprocating compressor.
   (ii) An 800 TR air conditioning plant using centrifugal compressor/s.
   (iii) A small capacity frozen food cabinet to maintain –30°C temperature.

4.2 Refrigeration engineers usually presume that if a R 12 car air conditioner compressor is operated with R 134a, its 'cooling capacity' would fall by about 10%. Examine this assumption by a realistic vapour compression cycle analysis.

4.3 Calculate the latent heat of vaporization of R 134a at – 25°C and + 50°C. The p^{sat} versus T^{sat} relationship of R 134a is

$$\ln p^{sat} = 24.8 - \frac{3980}{T^{sat}} - 0.024\, T^{sat}$$

where p^{sat} is in kPa and T^{sat} in K. Use the specific volumes data from the table of properties of R 134a.

Also, find the constants a and n if the h_{fg} versus T^{sat} relationship can be expressed by the equation

$$h_{fg} = a\left(1 - \frac{T^{\text{sat}}}{T_c}\right)^n$$

Critical temperature for R 134a is $T_c = 374.25$ K

4.4 What are azeotropic and non-azeotropic mixtures? Explain, in brief, their advantages giving examples.

4.5 Calculate and compare the specific volume of suction vapour, refrigerating effect, mass flow rate, discharge pressure and temperature, piston displacement, power consumption and COP for a 150 W refrigerating capacity domestic refrigerator operating on simple saturation cycle with –25°C evaporator and 55°C condenser temperatures for R 290, R 134a, R 152a and R 600a. Also compare their pressure drops across capillary.

4.6 If R 134a refrigerator in Prob. 4.5 is charged with propane without changing the hermetic compressor, what would be the motor wattage, isentropic discharge temperature, and heat required to be rejected in condenser? Assume the same operating conditions, and ignore the effect on capillary. At what refrigerating capacity, would the refrigerator be operating with propane? What are the implications of charging R 134a refrigerator with propane?

4.7 (a) A propane refrigerator has to operate at $t_k = 55$°C and $t_0 = -25$°C. For maximum COP, find if the suction state should be in wet or superheat region.

(b) Also find if it is advantageous to use liquid vapour regenerative heat exchanger.

4.8 Draw t-ξ and h-ξ diagrams for R 134a/R 22 mixtures for 2, 4, 6, 10, 15, 20 bar pressures. Do calculations for bubble and dew temperatures and enthalpies assuming ideal mixtures for 10, 20, 30 and 50 per cent R 134a.

4.9 Find the bubble and dew pressures of R 134a/R 22 mixtures at 0, 5, 10, 45, 55 and 60°C temperatures and 10, 20, 30 and 50 per cent R 134a compositions.

4.10 (a) An R 32 1.5TR air conditioner operates on simple saturation cycle. The operating conditions are:

Condensing temperature	60°C
Evaporation temperature	5°C
Temperatures of conditioned air	25°C In, 15°C Out
Cooling air temperatures	45°C In, 55°C Out

Find the piston displacement of compressor, motor watts and COP of the cycle.

(b) Find the operating conditions of an equivalent R 134a/R 32 cycle which will maintain the same LMTD in condenser and evaporator.

(c) Find the piston displacement of compressor required, motor watts and COP of the cycle for the conditions in (b).

Chapter

5

Multipressure Systems

5.1 INTRODUCTION

The simple vapour compression system is a two-pressure system. Systems with more than two pressures may arise either due to *multistage,* viz., *compound compression* to minimize work, or due to feeding of the refrigerant to a *multi evaporator system*. *Cascade systems* which employ more than one refrigerant also have multipressures although each refrigerant circuit operates on a two-pressure system only.

5.2 MULTISTAGE OR COMPOUND COMPRESSION

It has been shown in Sec. 3.5 that the slope of the constant entropy lines on the *p-h* diagram decreases for the isentropics away from the saturated vapour line. Multistage or compound compression with interstage cooling is one effective method of reducing work of compression by working on isentropics closer to the saturation curve.

It is, however, desirable to employ compound compression only when the pressure ratio between the condenser and evaporator is greater than 4 or 5. This will happen either as a result of a very high condensing temperature, and/or a very low evaporator temperature.

As shown in Secs. 6.3 and 6.6, with increasing pressure ratio, the volumetric efficiency and hence the refrigerating capacity of a reciprocating compressor tends to zero. Thus multistaging is necessary to reduce the power consumption and also to increase the refrigerating capacity in high condensing temperature and/or low evaporator temperature applications.

The two methods employed for cooling between stages are *water intercooling* and *flash intercooling* with flash gas removal.

5.2.1 Flash Gas Removal

In compound compression the throttling expansion of the liquid may also be done in stages as shown in Figs. 5.1(a) and 5.1(b). Thus the liquid from the condenser at

6 first expands into a *flash chamber* to 7 at the intermediate pressure p_i, and then the liquid from the flash chamber at 8 enters the evaporator through another expansion valve and expands to 9.

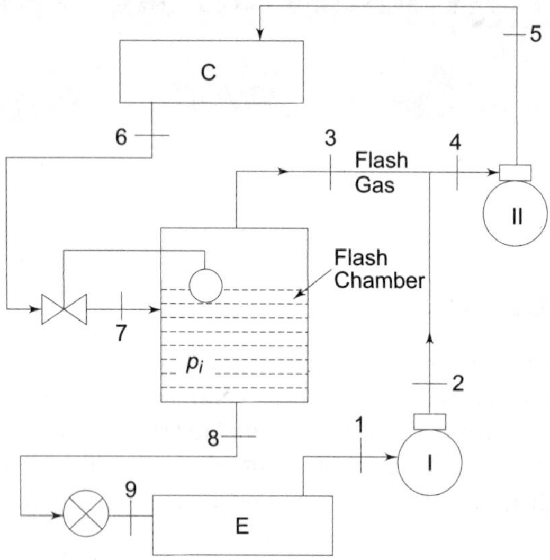

Fig. 5.1(a) Schematic diagram of the system with flash gas removal

In a system without a flash chamber, the liquid from the condenser expands straight to the evaporator pressure as shown by the process line 6–10 in Fig. 5.1(b). This is wasteful of energy as the vapour flashed at the intermediate pressure at 3 is also throttled to 11 at the evaporator pressure and is, therefore, required to be again recompressed to the intermediate pressure. A system with a flash chamber, thus, eliminates the undesirable throttling of the vapour generated at the intermediate pressure.

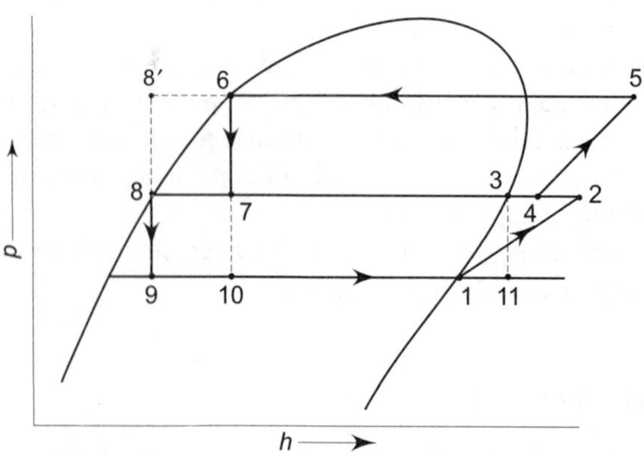

Fig. 5.1(b) Thermodynamic cycle for the system of Fig. 5.1(a)

Flash gas removal with multistage compression, therefore, results in power economy, and is always desirable whichever be the refrigerant used.

Another method of obtaining the same result as that of flash gas removal is to employ the flash chamber as a *liquid subcooler* as shown in Fig. 5.2. The liquid subcooler subcools the liquid by the evaporation of the liquid refrigerant in the flash chamber. Figure 5.2 also gives the mass balance of the liquid subcooler. Herein, m_1 is the mass flow rate through first-stage compressor, and m_2 is the mass flow rate through second-stage compressor. The thermodynamic states correspond to those of Fig. 5.1 (b) except that the state of liquid entering the evaporator expansion valve shifts to 8′ from 8.

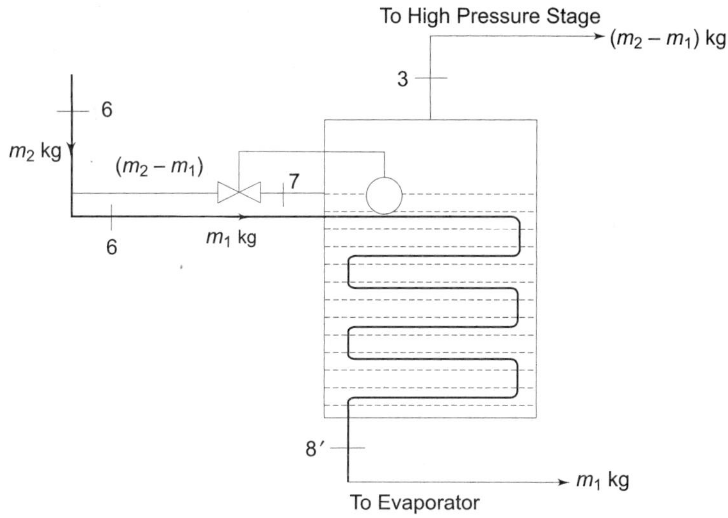

Fig. 5.2 Flash chamber as a liquid subcooler

5.2.2 Flash Intercooling

For flash intercooling, the compressed vapours from the lower stage are led and bubbled through the liquid in the flash chamber as shown in Fig. 5.3(a). The vapours are thus cooled to the saturation temperature at the pressure of the flash chamber and a part of the liquid evaporates which goes to the higher stage along with the vapours from the lower stage. Flash intercooling thus enables the higher stage compression to take place along the steeper isentropic, nearer the saturated vapour line.

Figures 5.3 (a) and (b) show that the discharge vapours from the lower stage at 2 are cooled to 3 by the evaporation of a part of the liquid refrigerant from the flash chamber at 6. Thus, the vapours entering the high-stage compressor m_2 comprise these vapours in addition to the vapours m_1 from the low-stage.

Thus, although the specific work is reduced in the high stage because of working along the steeper isentropic 3-4 instead of the isentropic 2-2′, the increase of the actual mass flow through the higher stage may increase the work of the higher stage.

It is found that in the case of ammonia, the mass of the liquid evaporated for flash intercooling is extremely small because of its high latent heat of vaporization and the isentropics become very flat at higher temperatures. Hence flash intercooling will decrease the power requirement. Flash intercoolers are, therefore, commonly used in multistage ammonia plants.

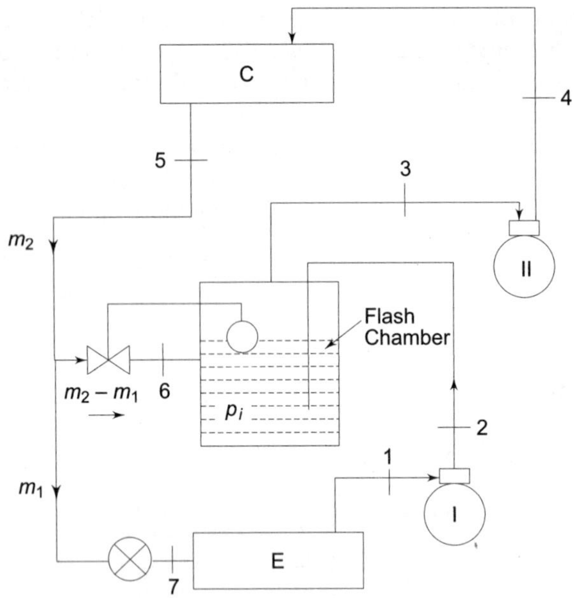

Fig. 5.3(a) Schematic diagram of a two-stage compression system with flash inter cooling

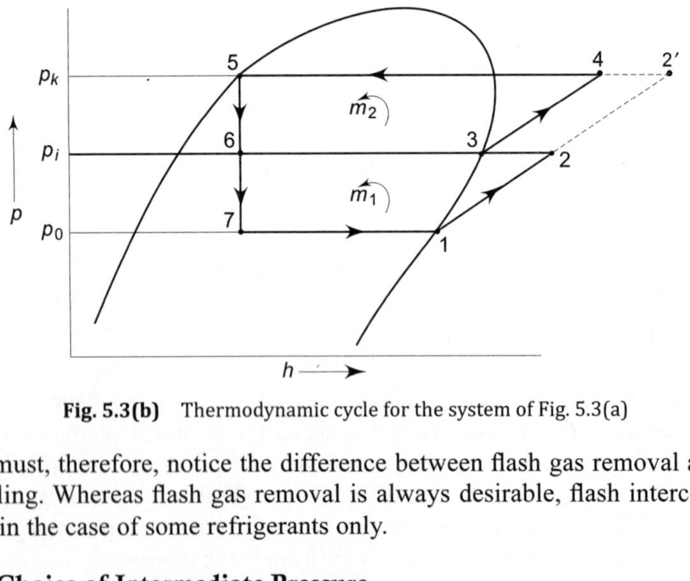

Fig. 5.3(b) Thermodynamic cycle for the system of Fig. 5.3(a)

One must, therefore, notice the difference between flash gas removal and flash intercooling. Whereas flash gas removal is always desirable, flash intercooling is suitable in the case of some refrigerants only.

5.2.3 Choice of Intermediate Pressure

It is well known that for minimum total work, the intermediate pressures are decided by the stage pressure ratio which should be

$$r = \left(\frac{p_k}{p_0}\right)^{1/n}$$

where n is the number of stages. In a two-stage system, this gives a geometric mean value for the intermediate pressure as given in Eq. (5.1).

$$p_i = \sqrt{p_k\, p_0} \tag{5.1}$$

However, this condition is true for *complete intercooling* to the initial temperature. In refrigeration systems, complete intercooling is not possible as it is done by ambient water or air. Also, it is seen that the discharge temperature of the low-stage is much lower than the discharge temperature of the high-stage. To reduce the discharge temperature of the higher stage, therefore, the pressure ratio of the lower stage can be increased, with a corresponding reduction in the pressure ratio of the higher stage. The expression in Eq. (5.2) is recommended for determining the intermediate pressure in refrigeration systems

$$p_i = \sqrt{p_k\, p_0\, T_k/T_0} \tag{5.2}$$

It is, however, realized that the choice of intermediate pressure is not very critical. If it is different from the optimum, then the work of one stage will be increased while that of another stage will be decreased. In the final analysis, the total work will not be much different from the minimum.

5.2.4 Complete Multistage Compression System

A complete two-stage compression system with flash-gas removal and water and flash intercooling is shown in Fig. 5.4(a) and its *p-h* diagram in Fig. 5.4(b). Example 5.1 illustrates the calculation procedure for an equivalent system with a liquid subcooler.

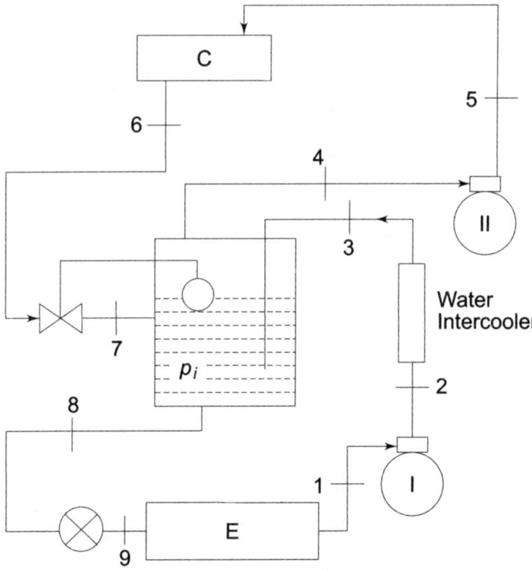

Fig. 5.4(a) Schematic diagram of a two-stage compression system with flash gas removal, and water and flash intercooling

It may be noted that a part of the heat of low-stage compression can be removed by water intercooling, in case the discharge temperature from the stage is substantially higher than the cooler water temperature.

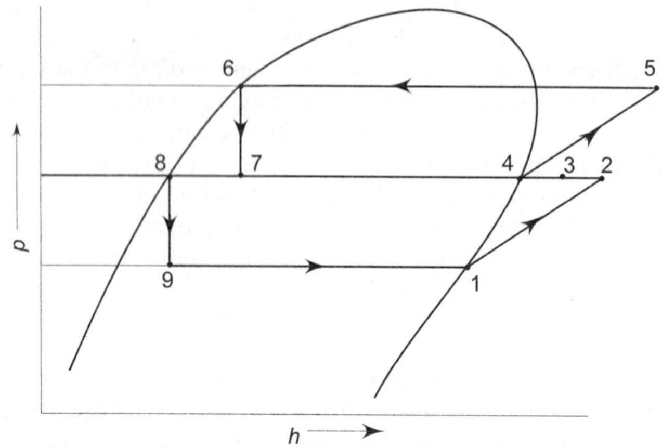

Fig. 5.4(b) Thermodynamic cycle for the system of Fig. 5.4(a)

Example 5.1 *A two-stage ammonia food-freezing plant—with a desired capacity of 528,000 kJ/h at – 40°C evaporating temperature and 35°C condensing temperature—has a flash intercooling system with a liquid subcooler. The vapour leaving the evaporator is at –30°C and entering the first-stage compressor is at –15°C. The vapour leaving the flash chamber is superheated by 10°C in the suction line to the second-stage compressor. Water intercooling is done to cool the vapour to 45°C. Adiabatic efficiencies of both compressors are 0.75. The volumetric efficiencies of first- and second-stage compressors are 0.65 and 0.77 respectively. Find the piston displacements, discharge temperatures and power requirements of the two compressors.*

Solution See Fig. 5.5. Reference state is –40°C at which $h_f = 0$, $s_f = 0$.

$$p_k = 1.35 \text{ MN/m}^2 \text{ (at 35°C)}$$

$$p_0 = 0.0718 \text{ MN/m}^2 \text{ (at – 40°C)}$$

$$p_i = \sqrt{1.35 \times 0.0718 \times \left(\frac{273 + 35}{273 - 40}\right)} = 0.355 \text{ MN/m}^2$$

The corresponding saturation temperature is

$$t_i = -5°\text{C}$$

First Stage

Enthalpy of vapour leaving the evaporator

$$h_1' = 1390 + \{-30 - (-40)\}\left(\frac{1499 - 1300}{50}\right)$$

$$= 1411.8 \text{ kJ/kg}$$

Enthalpy of vapour entering the compressor

$$h_1 = 1390 + \{-15 - (40)\}\left(\frac{1499 - 1390}{50}\right)$$

$$= 1444.5 \text{ kJ/kg}$$

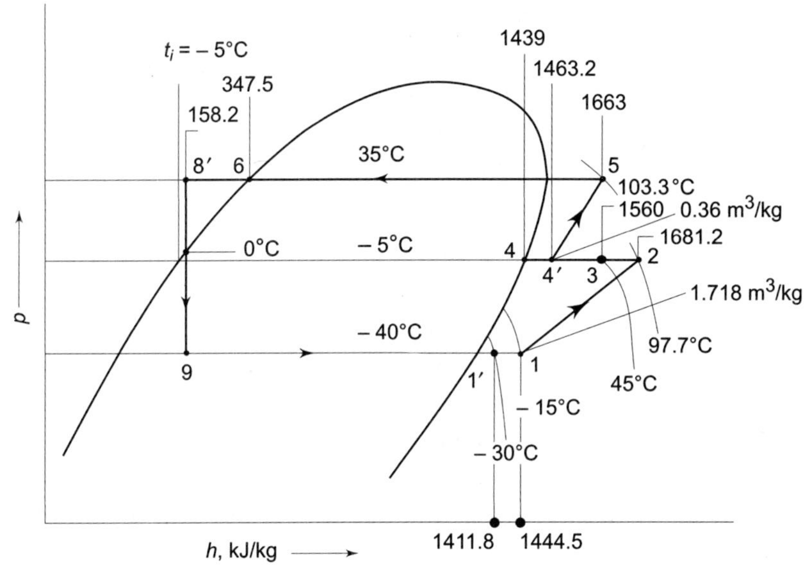

Fig. 5.5 Figure for Example 5.1

Enthalpy of the liquid

$$h_9 = h'_8 = 158.2 \text{ kJ/kg}$$

Mass flow rate of the refrigerant through the first-stage compressor

$$\dot{m}_1 = \frac{528{,}000}{1411.8 - 158.2} = 420 \text{ kg/h}$$

$$v_1 = v_{g1}\left(\frac{273-15}{273-40}\right) = 1.552\left(\frac{258}{233}\right) = 1.718 \text{ m}^3\text{/kg}$$

Piston displacement of the first-stage compressor

$$\dot{V}_1 = \frac{(420)(1.718)}{0.65} = 1110 \text{ m}^3\text{/h}$$

Entropy of vapour during isentropic compression from 1 to 2

$$s_1 = s_2 = 5.963 + \{-15 - (-40)\}\left(\frac{6.387 - 5.693}{50}\right)$$
$$= 6.175 \text{ kJ/kg·K}$$

Degree of superheat of vapour after isentropic compression

$$= 100 + \frac{6.175 - 6.157}{6.157 - 5.822}(50) = 102.7°\text{C}$$

Discharge temperature

$$t_2 = -5 + 102.7 = 97.7°\text{C}$$

$$h_{2,\text{ isen}} = 1675 + \frac{2.7}{50}(1675 - 1560) = 1681.2 \text{ kJ/kg}$$

Isentropic work

$$w_{\text{isen}} = 1681.2 - 1444.5 = 236.7 \text{ kJ/kg}$$

Actual work $w = \dfrac{236.7}{0.75} = 316$ kJ/kg

Power requirement of first-stage compressor

$$\dot{W}_1 = \frac{(420)(316)}{2650} = 50 \text{ hp}$$

Second Stage

Enthalpy of vapour entering the flash chamber

$$h_3 = 1560 \text{ kJ/kg}$$

and $h_4 = 1439$ kJ/kg

Liquid enthalpy $h_7 = h_6 = 347.5$ kJ/kg

Energy balance of the evaporator gives

$$\dot{m}_4 = \frac{1560 - 158.2}{1439 - 347.5}(420) = 540 \text{ kg/h}$$

$$t_4' = 5°\text{C}$$

Specific volume of vapour entering the second-stage compressor

$$v_4' = 0.347 \frac{(273 + 5)}{(273 - 5)} = 0.36 \text{ m}^3\text{/kg}$$

Piston displacement of the second-stage compressor

$$\dot{V}_4 = \frac{(540)(0.36)}{0.77} = 252 \text{ m}^3\text{/h}$$

Entropy of vapour during isentropic compression from 4′ to 5

$$s_4' = s_5 = 5.407 + \frac{10}{50}(5.822 - 5.407) = 5.490 \text{ kJ/kg.K}$$

Degree of superheat of vapour after isentropic compression

$$= 50 + \frac{5.490 - 5.368}{5.702 - 5.368}(50) = 68.3°\text{C}$$

Discharge temperature

$$t_5 = 35 + 68.3 = 103.3°\text{C}$$

$$h_{5,\,\text{isen}} = 1616 + \frac{18.3}{50}(1744 - 1616) = 1663 \text{ kJ/kg}$$

Enthalpy of vapour at compressor suction

$$h_4' = 1439 + \frac{10}{50}(1560 - 1439) = 1463.2 \text{ kJ/kg}$$

Isentropic work $w_{\text{isen}} = 1663 - 1463.2 = 199.8$ kJ/kg

Actual work

$$w = \frac{199.8}{0.75} = 266.5 \text{ kJ/kg}$$

Power requirement of second-stage compressor

$$\dot{W}_2 = \frac{(540)(266.5)}{2650} = 54.3 \text{ hp}$$

5.3 MULTI-EVAPORATOR SYSTEMS

Very often, a situation arises when varied types of cooling loads are connected to the same refrigeration system. Each load may require an evaporator working at a different refrigeration temperature. The whole system may, therefore, be operated either at a suction pressure equal to the lowest evaporator pressure leading to a single-compressor system, or at various suction pressures with individual compressors for each evaporator, leading to a multi-compressor system.

5.3.1 Single Compressor-Individual Expansion Valves

A two-evaporator single-compressor system with *individual expansion valves* for each evaporator and one compressor is shown in Fig. 5.6(a). Operation under these conditions means the dropping of pressure from high pressure evaporators through back pressure valves as can be seen in Fig. 5.6(b). This necessitates the compression of the vapour from the higher temperature evaporators through a pressure ratio greater than necessary, and hence at the expense of operating economy.

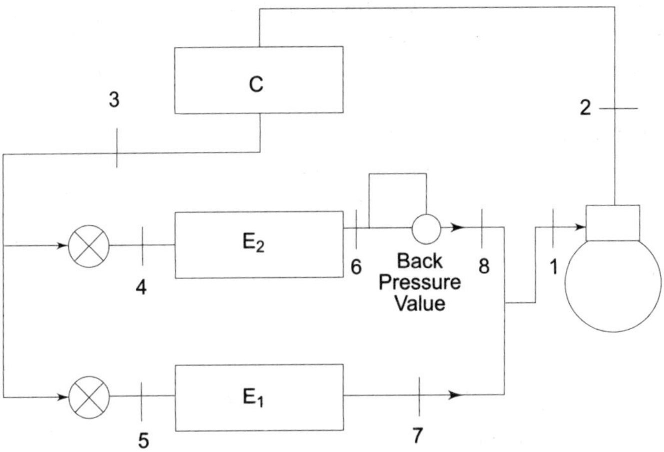

Fig. 5.6(a) System with two evaporators and single compressor, with individual expansion valves

The mass flow rates through evaporators 1 and 2 are respectively

$$m_1 = \frac{Q_{0_1}}{q_{0_1}} = \frac{Q_{0_1}}{h_7 - h_5}$$

$$m_2 = \frac{Q_{0_2}}{q_{0_2}} = \frac{Q_{0_2}}{h_6 - h_4}$$

The enthalpy of the vapour mixture entering the compressor is

$$h_1 = \frac{m_1 h_7 + m_2 h_6}{m_1 + m_2} \tag{5.3}$$

and the net work done is given by

$$W = (m_1 + m_2)(h_2 - h_1) \tag{5.4}$$

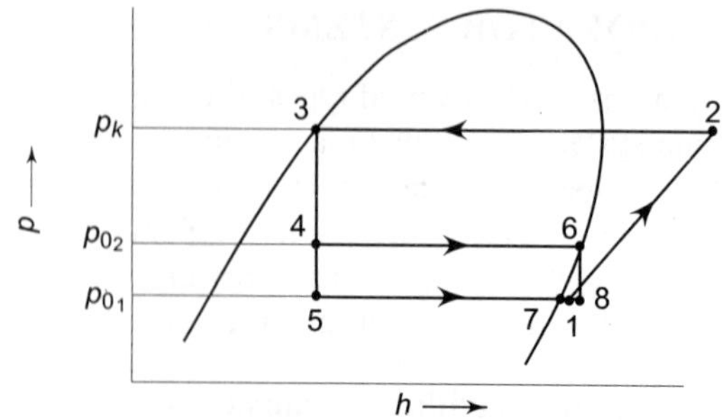

Fig. 5.6(b) Thermodynamic cycle for the system of Fig. 5.6(a)

5.3.2 Single Compressor—Multiple Expansion Valves

The operation of a two-evaporator single-compressor system with *multiple arrangement* of expansion valves is shown in Figs 5.7(a) and (b). The only advantage of the arrangement is that the flashed vapour at the pressure of the high temperature evaporator is not allowed to go to the lower temperature evaporator, thus improving its efficiency. To gain thermodynamic advantage from this, it will be necessary to use individual compressors for each evaporator, thus eliminating the throttle losses of the back pressure valves.

The refrigerating effects for the evaporators are

$$q_{0_1} = h_8 - h_6$$
$$q_{0_2} = h_7 - h_4$$

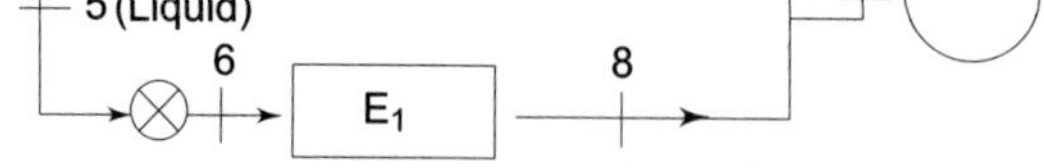

Fig. 5.7(a) System with two evaporators and single compressor with multiple expansion valves

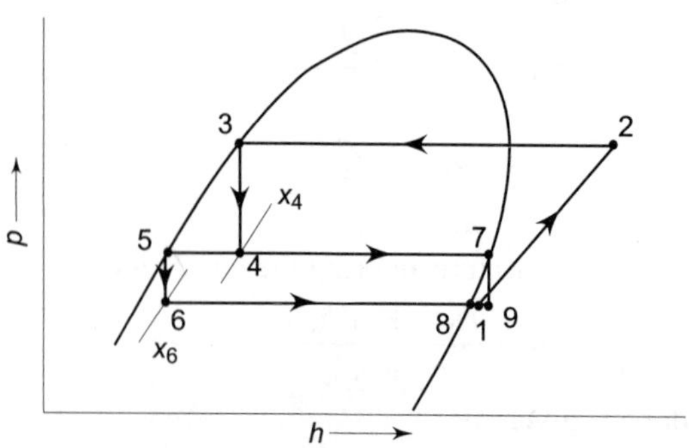

Fig. 5.7(b) Thermodynamic cycle for the system of Fig. 5.7(a)

The mass flow rates of the refrigerant comprising the liquid and vapour fractions, are:

Evaporator 1 $$m_1 = \frac{Q_{0_1}}{q_{0_1}}$$

Evaporator 2 $$m_2 = \frac{Q_{0_1}}{q_{0_1}} + m_1\left(\frac{x_4}{1 - x_4}\right)$$

where the second term on R.H.S. represents the mass of vapour flashed at 4 corresponding to the mass of liquid going to the second evaporator.

5.3.3 Individual Compressors—Multiple Expansion Valves

The total power requirement can be reduced by the use of an individual compressor for each evaporator and by the multiple arrangement of expansion valves as shown in Figs 5.8(a) and (b). This amounts to parallel operation of evaporators and is called *sectionalizing*. There may be a separate condenser for each compressor or a common condenser for the whole plant.

The calculations for mass flow rates are similar to those of the preceding case.

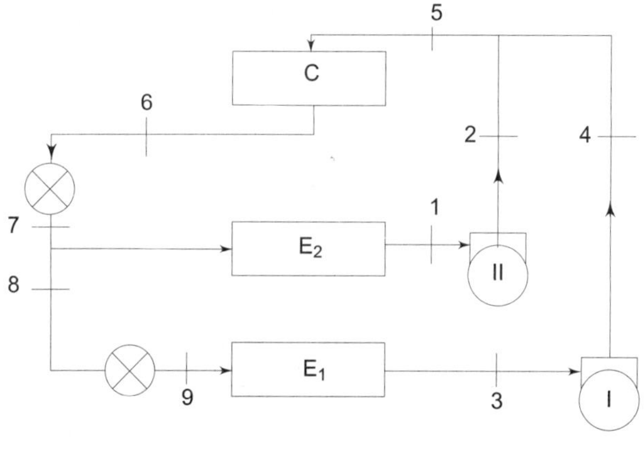

Fig. 5.8(a) System with two evaporators, individual compressors and multiple expansion valves

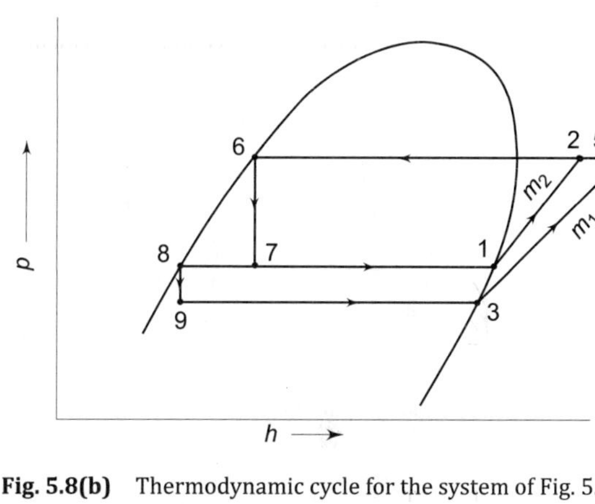

Fig. 5.8(b) Thermodynamic cycle for the system of Fig. 5.8(a)

5.3.4 Individual Compressors with Compound Compression and Flash Intercooling

When one or more evaporators are in operation at very low temperatures, and individual compressors are installed, then compound compression can be used to effect power saving as shown in Figs 5.9(a) and (b) with multiple arrangement of expansion valves and flash intercooling. The flash chamber is maintained at the pressure of the high temperature evaporator.

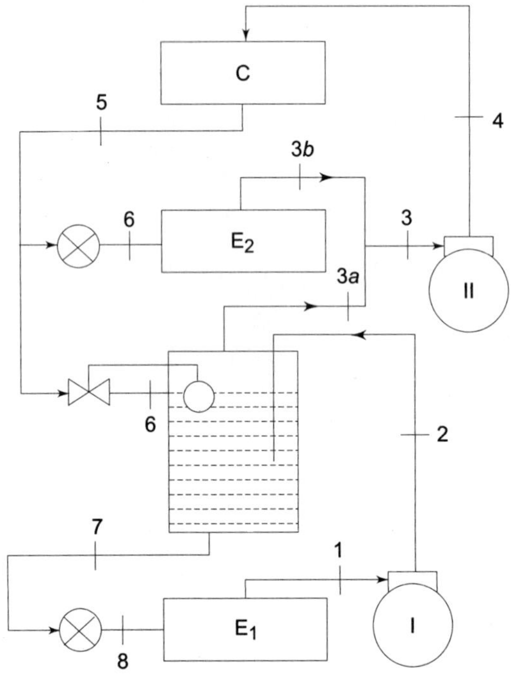

Fig. 5.9(a) System with two evaporators, compound compression and flash intercooling

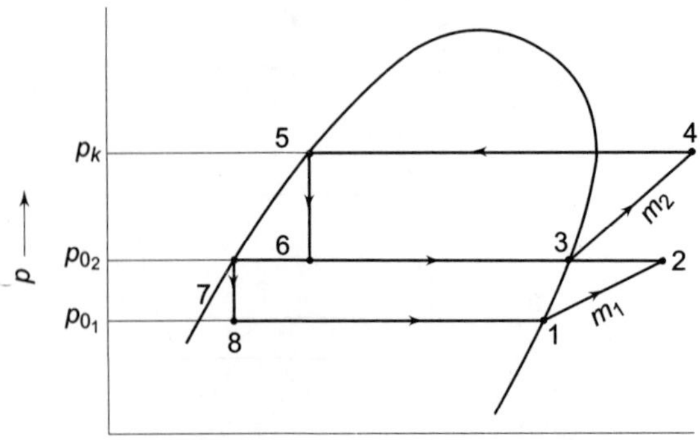

Fig. 5.9(b) Thermodynamic cycle for the system of Fig. 5.9(a)

5.4 CASCADE SYSTEMS

The use of a single refrigerant in a simple vapour compression cycle for the production of low temperatures is limited by the following reasons.

(i) Solidification temperature of the refrigerant.

(ii) Extremely low pressures in the evaporator and large suction volumes if a high-boiling refrigerant is selected.

(iii) Extremely high pressures in the condenser if a low-boiling refrigerant is selected.

(iv) Very high pressure ratio p_k/p_0 and, therefore, a low coefficient of performance.

(v) Difficulties encountered in the operation of any mechanical equipment at very low temperatures.

We know that multistage compression is employed when low evaporator temperatures are required and when the pressure ratio p_k/p_0 is high. Refrigerant 22 is used in a two-stage system up to –50°C and in a three-stage system up to about –65°C.

If vapour compression systems are to be used for the production of low temperatures, the common alternative to stage compression is the *cascade system* in which a series of refrigerants, with progressively lower boiling points, are used in a series of single-stage units. The system provides a solution to all the problems mentioned above except the last one.

The cascade system combines two or more vapour compression units as shown in Fig. 5.10. The high temperature cascade produces refrigeration at a certain low temperature t_{0_2}. The low temperature cascade produces refrigeration at a still further low temperature t_{0_1}, using the refrigerating effect of high temperature cascade at temperature t_{0_2} for rejecting heat in its condenser at temperature t_{k_1}, which in the limit is equal to t_{0_2}. In practice, however, there is a certain overlap between these temperatures, i.e. the temperature t_{k_1} is about 5°C higher than the temperature t_{0_2}.

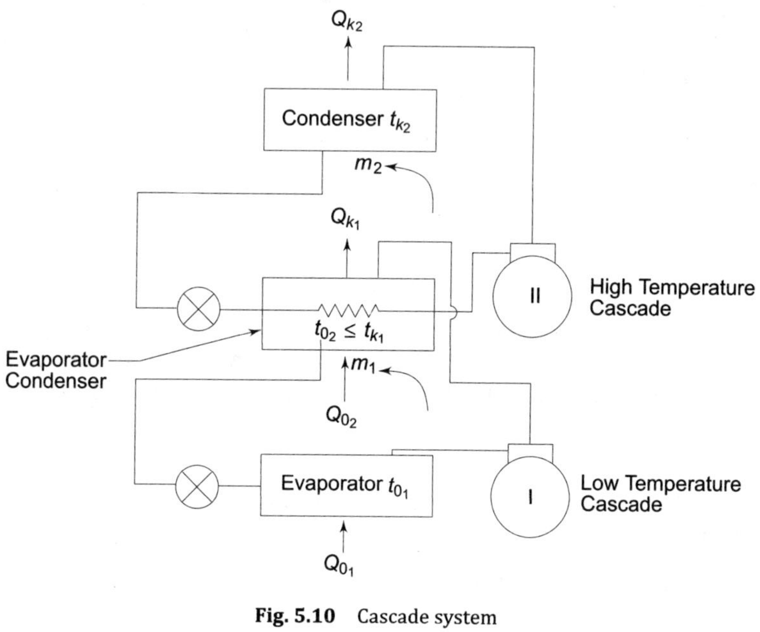

Fig. 5.10 Cascade system

It is also to be noted that each cascade works on a separate refrigerant. Each refrigerant can be chosen in such a way that it operates best within the required comparatively narrow temperature limits. The high temperature cascade uses a high-boiling refrigerant such as NH_3 or R 22, whereas the low temperature cascade uses a low-boiling refrigerant such as CO_2, ethylene, methane, etc., depending on the requirements. The use of a low boiling and, therefore, a high pressure refrigerant ensures a smaller compressor displacement in the low temperature cascade and a higher coefficient of performance.

The cascade system was first used by Pictet in 1877 for the liquefaction of oxygen employing SO_2 and CO_2 as intermediate refrigerants. Another set of refrigerants commonly used for the liquefaction of gases in a three-stage cascade system is ammonia, ethylene and methane. Ammonia is also used in the high temperature cascade for the manufacture of solid carbon dioxide.

An additional advantage of the cascade system is that oil from one compressor cannot wander to the other compressors as it generally happens in multistage systems.

In the system shown in Fig. 5.10, the high and low temperature systems have to be balanced with each other. This means that the heat absorbed in the high temperature cascade evaporator must be equal to the heat rejected in the low temperature cascade condenser. Thus, the two systems are designed such that

$$Q_{0_2} = Q_{k_1}$$

In operation, however, the balancing problem creates difficulty since during the pull-down period, the high temperature cascade system is inadequate. Hence, it has to be slightly oversized.

5.4.1 Optimum Coupling Temperature between Cascade Circuits

The intermediate temperature between the two cascade circuits, such as $t_{0_2} \le t_{k_1}$, may be called the *coupling temperature*. For optimum sizing of the two circuits, Schmidt[5] has shown that the optimum coupling temperature can be approximately represented by the square-root of the condensing temperature in warm, and evaporating temperature in cold circuit. This is based on the assumption that the Carnot COP of the two circuits is the same, as shown in Eq. (5.5).

$$\frac{T_{0_1}}{T_{k_1} - T_{0_1}} = \frac{T_{0_2}}{T_{k_2} - T_{0_2}} \tag{5.5}$$

so that when $T_{0_2} = T_{k_1}$, we have

$$T_{0_2} = \sqrt{T_{k_2} \times T_{0_1}} = T_{k_1} \tag{5.6}$$

Schmidt derives a more accurate relation in terms of the coefficients of the vapour pressure curves of the two refrigerants. This derivation is based on the condition that the pressure ratios of the compressors in each circuit are the same. Thus

$$\frac{p_{k_1}}{p_{0_1}} = \frac{p_{k_2}}{p_{0_2}} \tag{5.7}$$

Then if the pressure-temperature relationships of the two substances are governed by the relation.

$$\log p = a - \frac{b}{T}$$

we have

$$\log p_{k_1} - \log p_{0_1} = b_1 \left(\frac{1}{T_{k_1}} - \frac{1}{T_{0_1}} \right) \tag{5.8}$$

$$\log p_{k_2} - \log p_{0_2} = b_2 \left(\frac{1}{T_{k_2}} - \frac{1}{T_{0_2}} \right) \tag{5.9}$$

and from Eq. (5.7) we obtain

$$\frac{b_1}{b_2} \left(\frac{1}{T_{k_1}} - \frac{1}{T_{0_1}} \right) = \frac{1}{T_{k_2}} - \frac{1}{T_{0_2}} \tag{5.10}$$

which for $T_{0_2} = T_{k_1}$ becomes

$$T_{0_2} = \frac{b_1 + b_2}{b_2/T_{k_2} + b_1/T_{0_1}} = T_{k_1} \tag{5.11}$$

A more appropriate semi-empirical relation is suggested in Eq. (5.12)

$$T_{0_2} = \frac{b_1 + 1.05\, b_2}{b_2/T_{k_2} + b_1/T_{0_1}} - 10 \tag{5.12}$$

5.5 SOLID CARBON DIOXIDE—DRY ICE

Figure 5.11 shows the relevant thermodynamic characteristics of carbon dioxide which make it suitable to be used in the solid state at atmospheric pressure, commonly known as *dry ice*. The name is derived from the fact that solid carbon dioxide sublimates into vapour at atmospheric pressure and temperature. The sublimation of solid carbon dioxide into vapour at normal atmospheric pressure is made possible because its triple point pressure is 5.18 bar, i.e., greater than one atmosphere. Also, the triple point temperature is – 56.6°C, which is much below the normal ambient temperature. Hence, heat can flow from the surroundings to solid carbon dioxide, exposed to atmosphere, to sublimate it. Thus the heat of sublimation of carbon dioxide in the solid state at atmospheric pressure can be conveniently utilized to provide refrigeration specially for the preservation of foods in transport.

The triple point pressure for water is 0.00611 bar, viz., much below one atmosphere. Thus, at atmospheric pressure water-ice first melts and then evaporates. Also, the triple point temperature of water is only 0.01°C and its fusion temperature is 0°C. Thus water-ice, even with melting, can be used for refrigeration only above 0°C. The normal boiling point of water is 100°C.

In the case of carbon dioxide, the fusion temperature and normal boiling point have no meaning. We have, however, the normal sublimation temperature which is – 78.52°C.

5.6 MANUFACTURE OF SOLID CARBON DIOXIDE[1,6]

The critical temperature of carbon dioxide is very low, viz., 31°C whereas its critical pressure is quite high, viz., 73.8 bar. The use of carbon dioxide in a simple vapour compression cycle is possible only if condensation is achieved at a temperature below 31°C. Even then the condensing pressure will be very high. Such a cycle is shown in Fig. 5.11. For manufacture of solid CO_2, the gaseous carbon dioxide is compressed from 1 to 2, condensed from 2 to 3 at 28°C and 69 bar and then expanded to atmospheric pressure at 4 in a *snow chamber*. The solid carbon dioxide or dry ice can be removed at 5; the vapour at 6, after mixing with the make-up gas at 7, is led to the compressor.

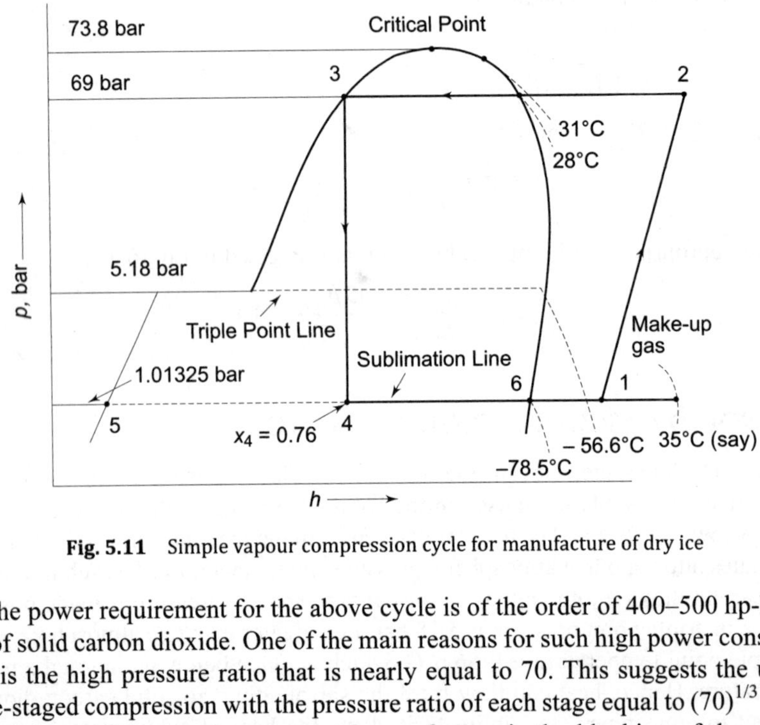

Fig. 5.11 Simple vapour compression cycle for manufacture of dry ice

The power requirement for the above cycle is of the order of 400–500 hp-hr per ton of solid carbon dioxide. One of the main reasons for such high power consumption is the high pressure ratio that is nearly equal to 70. This suggests the use of three-staged compression with the pressure ratio of each stage equal to $(70)^{1/3} = 4.1$.

Another problem in the process of manufacture is the blocking of the expansion device by the formation of dry ice. This is eliminated by first producing liquid carbon dioxide at a pressure slightly above the triple point pressure and then reducing its pressure to one atmosphere in a snow chamber. This makes it necessary to have two snow chambers, one receiving liquid carbon dioxide at a pressure higher than the triple point pressure, and the other reducing pressure to form solid carbon dioxide and vapour, both working alternately. Such a method is known as the *pressure snow chamber method*.

Thus if 6 bar is chosen as the pressure of liquid formation in the snow chamber, then the various stage pressures would be nearly equal to 69, 20, 6 and 1.01325 bar respectively.

Further, the condensing temperature of 28°C is impractical when the cooling water temperature itself is 30°C or more in summer. Also, even with 28°C, the COP of the system would be very low as it is close to the critical temperature. The high pressure stage of the carbon-dioxide cycle is, therefore, replaced with a separate ammonia circuit in cascade with the carbon dioxide circuit.

The schematic diagram of a system using the pressure snow chamber method and ammonia in the cascade circuit is shown in Fig. 5.12. The condensing pressure of carbon dioxide is generally kept at 15 to 20 bar. The *p-h* diagram of the carbon dioxide circuit is shown in Fig. 5.13. In the position shown in Fig. 5.12, high pressure liquid carbon dioxide is being prepared in snow chamber II at 6 bar, whereas snow chamber I is shown to be reducing in pressure from 6 bar to 1.01325 bar. Dry ice is formed at 12 and the vapour leaves at 13. During this cycle of operation, expansion valve A is open and B is closed. Also, shut-off valves C and F are open and D and E are closed.

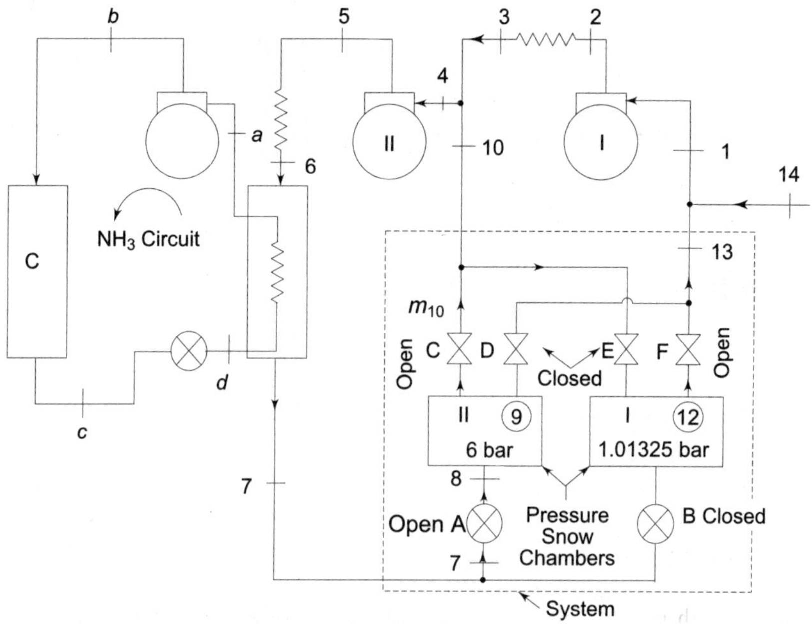

Fig. 5.12 Pressure snow chamber method with ammonia in the cascade circuit for manufacture of dry ice

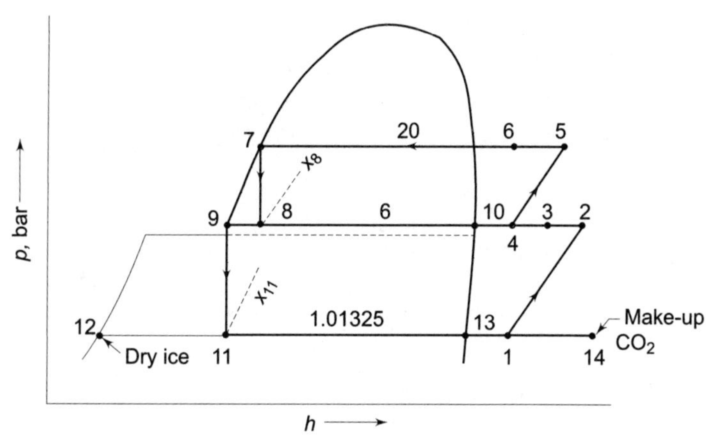

Fig. 5.13 *p-h* diagram of CO_2 circuit for pressure snow chamber method

For the analysis of the cycle, the section enclosed within the broken lines, shown in Fig. 5.12, may be considered as a system. Then assuming steady-state conditions, writing the mass balance for the system, we obtain

$$m_7 = m_8 = m_9 + m_{10}$$

and

$$m_9 = m_{11} = m_{12} + m_{13}$$

Thus, from the above two relations

$$m_7 = m_{10} + m_{12} + m_{13} \tag{5.13}$$

Also, by energy balance

$$m_7 h_7 = m_{10} h_{10} + m_{12} h_{12} + m_{13} h_{13} \tag{5.14}$$

In Eqs. (5.13) and (5.14), m_{12} represents the yield of dry ice which can be considered as specified. The mass of make-up CO_2, m_{14} is also equal to m_{12}. The three unknowns m_7, m_{10} and m_{13} in the two equation can be found by solving these equations together with the relation for dryness at 11, viz.,

$$\frac{m_{13}}{m_{12}} = \frac{x_{11}}{1 - x_{11}} \tag{5.15}$$

Example 5.2 *Calculate the power required to produce 500 kg of dry ice per hour using the pressure snow chamber method in conjunction with ammonia in the cascade circuit. The pressures and temperatures may be assumed as follows:*

Temperature in snow chamber after liquid throttling = – 50°C
Condensing temperature of CO_2 = – 5°C
Temperature of make-up gas = 30°C
Condensing temperature of NH_3 = 35°C
Evaporating temperature of NH_3 = – 10°C
Temperature after water-intercooling = 35°C
Temperature after second-stage cooling = 40°C

Solution The various enthalpies are as follows:

For CO_2 Circuit

From chart

$$h_{13} = 312 \text{ kJ/kg} \qquad h_7 = 72 \text{ kJ/kg}$$
$$h_{14} = 433 \text{ kJ/kg} \qquad h_9 = -18 \text{ kJ/kg}$$
$$h_{10} = 320 \text{ kJ/kg} \qquad h_{12} = -250 \text{ kJ/kg}$$

Dryness fractions

$$x_8 = \frac{h_7 - h_9}{h_{10} - h_9} = \frac{72 + 18}{320 + 18} = 0.266$$

$$x_{11} = \frac{h_9 - h_{12}}{h_{13} - h_{12}} = \frac{-18 + 250}{312 + 250} = 0.413$$

Consider yield = $m_{12} = m_{14} = 1$ kg. Then

$$m_{13} = \frac{x_{11}}{1 - x_{11}}, m_{12} = \frac{0.413}{0.587}(1) = 0.704 \text{ kg}$$

$$m_9 = m_{12} + m_{13} = 1.704 \text{ kg}$$

$$m_{10} = \frac{x_8}{1-x_8} m_9 = \frac{0.266}{0.734}(1.704) = 0.618 \text{ kg}$$

$$m_7 = m_8 = m_9 + m_{10} = 1.704 + 0.618 = 2.232 \text{ kg}$$

$$h_1 = \frac{m_{13}h_{13} + m_{14}h_{14}}{m_{13} + m_{14}} = \frac{(0.704)(312) + (1)(433)}{1.704} = 383 \text{ kJ/kg}$$

From chart

$$h_2 = 435 \text{ kJ/kg}$$

$$t_2 = 75°\text{C}$$

Water intercooling is, therefore, essential. Then

$$t_3 = 35°\text{C}$$

$$h_3 = 396 \text{ kJ/kg (From chart)}$$

$$h_4 = \frac{m_{10}h_{10} + m_3h_3}{m_{10} + m_3} = \frac{(0.618)(320) + (1.604)(396)}{2.322} = 376 \text{ kJ/kg}$$

From chart

$$t_4 = 11°\text{C}$$

$$h_5 = 471 \text{ kJ/kg}$$

$$t_5 = 120°\text{C}$$

After water inter-cooling

$$t_6 = 40°\text{C}$$

$$h_6 = 380 \text{ kJ/kg}$$

Power requirement of the first-stage CO_2 compressor

$$\dot{W}_1 = \dot{m}_1(h_2 - h_1)$$

$$= \frac{500}{3600}(1.704)(435 - 383)$$

$$= 12.31 \text{ kW}$$

Power requirement of the second-stage CO_2 compressor

$$\dot{W}_2 = \dot{m}_4(h_5 - h_4)$$

$$= \frac{500}{3600}(2.322)(471 - 376)$$

$$= 30.6 \text{ kW}$$

For NH_3 Circuit

$$h_a = 1450 \text{ kJ/kg}$$

$$h_b = 1680 \text{ kJ/kg}$$

$$h_c = 366 = h_d$$

Energy balance of the CO_2 condenser and NH_3 evaporator gives

$$m_{NH_3} = \frac{h_6 - h_7}{h_a - h_d} m_6$$

$$= \frac{380 - 72}{1450 - 366} (2.322) = 0.66 \text{ kg}$$

Power requirement of the NH_3 compressor

$$\dot{W}_{NH_3} = \dot{m}_{NH_3} (h_b - h_a)$$

$$= \frac{500}{3600} (0.66)(1680 - 1450)$$

$$= 21.08 \text{ kW}$$

Total power requirement $= 12.31 + 30.6 + 21.08 = 64$ kW

5.7 SYSTEM PRACTICES FOR MULTI-STAGE SYSTEMS

Rotary vane or centrifugal compressors are commonly used for the booster stage of a multistage system where large gas volumes are handled. There are also some applications in which oil-free and dry-cylinder compressors are used where the effects of oil in the refrigerant are of great consideration.

When LP stage temperatures are below –70°C, the pull-down load may be three times the normal. Compressor motors are, therefore, selected for about 150 per cent above the normal loading.

Some thought must also be given for sizing the condenser for the maximum amount of heat rejection that is expected during the pull-down period.

One problem in low temperature evaporators is the return of oil to the compressor. Another is the pressure drop through the evaporator. A D-X (direct-expansion) type evaporator is the most common because of its improved ability to return oil to the compressor, as well as the smaller charge of the refrigerant required with it. Sometimes an oil separator is used in low-temperature applications such as freeze-drying. The effect of pressure drop in evaporators at very low pressures is very serious because of its large magnitude which is a result of the large increase in volume of the refrigerant during vaporization. Hence, pumped systems such as a *flash-cooler type* evaporator can be used, in which the refrigerant is used like brine under pressure by a pump from a flash chamber so that vaporization does not take place inside the evaporator, but only on return to the flash cooler. Or a *recirculation type* evaporator may be used.

Receivers should not be used as the temperature of the surroundings may make it difficult for the liquid to enter into it.

It may be necessary to raise the temperature of the low-stage suction gas for maintaining the lubricating ability of the oil in the compressor. In cascade systems, this can be best done through a heat exchange between this gas and the high pressure stage liquid.

It is desirable to charge the system with a *fade-out charge* such that on shut down, all liquid in the system will evaporate into gas without excessively increasing the system pressure. In practice, however, the system may not have enough volume to permit this. In that case, an expansion tank may be provided and connected to the evaporating side of the system.

References

1. Anon, 'Cut costs in making dry ice', *Chemical Engineering*, Vol. 63, No. 8, Aug. 1956, p. 114.
2. Arora C P and P L Dhar, 'Optimization of multistage refrigeran compressors', *Proc. XIIIth International Congress of Refrigeration*, Washington, 1971, Paper No. 326, pp 693–700.
3. Dhar P L and C P Arora, 'Optimum interstate temperature for cascade system' *Proc. Second National Symposium on Refrigeration and Air Conditioning*, University of Roorkee, March 1973, pp. 211–215.
4. Missimer D J, 'Cascade refrigeration systems for ultra low temperatures', *Refrigerating Engineering*, Vol. 64, No. 2, Feb. 1956, p. 37.
5. Schmidt H, 'Die bemessung von kältekompressoren in kaskadenschaltung', *Kältetechnik*, Vol. 17, No. 5, May 1965, pp. 151–155.
6. Stickney A P, 'The thermodynamics of CO_2 cycles', *Refrigerating Engineering*, Vol. 24, No. 6. Dec. 1932, p. 334.

Revision Exercises

5.1 An R 22 refrigerating plant with back pressure control has a capacity of 30,000 kJ/h at –30°C and 45,000 kJ/h at –25°C. The refrigerant flow is controlled by thermostatic expansion valves with 8°C superheat. The condensing temperature is 35°C. There is no heat exchanger. The vapours are superheated by 15°C in the suction line. The compressor is single-acting and has four cylinders. It runs at 930 rpm and its bore/stroke ratio is 1.25. Volumetric efficiency may be assumed as 70 per cent. Calculate the dimensions of the cylinders. Also calculate the power consumption and COP.

5.2 An ammonia refrigerating plant is working at an evaporating temperature of – 30°C and a condensing temperature of 37°C. There is no subcooling of the liquid refrigerant, and the vapour is in the dry-saturated condition at the inlets to the compressors. The capacity is 150 kW refrigeration. Estimate the power consumption

(i) when one-stage is used,

(ii) when two-stage compression with flash intercooling is used, and

(iii) when two-stage compression with flash chamber and liquid subcooler is used. Assume suitable intermediate pressure.

Explain why it is not advisable to use multistage compression to produce refrigeration temperatures of the order of, say, – 90°C.

5.3 A two-stage 10 tons NH_3 refrigeration plant with shell and coil type flash intercooler as shown in Fig. 5.14 is operating at the following conditions:

Condensation temperature	35°C
Evaporator temperature	– 40°C
Interstage pressure	2.91 bar
Subcooling of liquid in flash intercooler by	30°C
Thermostatic expansion valve setting	5°C superheat

Assuming that the vapours leaving the LP compressor are intercooled to 40°C by water before entering the flash chamber, determine:

(a) Refrigerant mass flow rate in evaporator.
(b) Refrigerant mass flow rate through HP compressor.
(c) Total power consumption.
(d) Total power consumed in a single-stage NH_3 plant for the same duty.

State what are the advantages of using flash intercooler in comparison to single stage system.

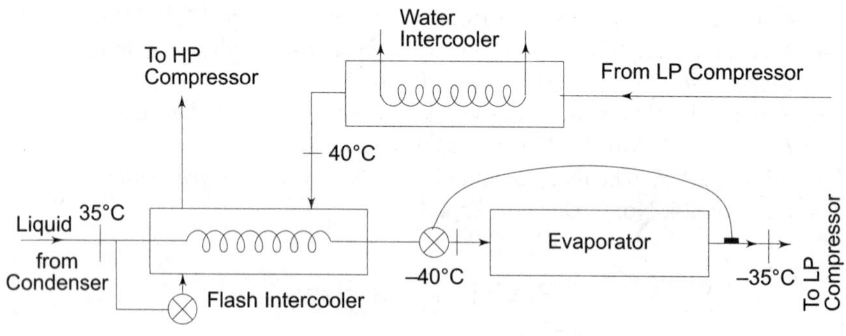

Fig. 5.14 Figure for Prob. 5.3

5.4 A two-stage R 22 plant with flash intercooler for food freezing has two 45 mm bore and 40 mm stroke compressors as follows:

LP Compressor: No. of cylinders 6
Rpm 1000
Volumetric efficiency 75%

HP Compressor: No. of cylinder 4
Rpm 800
Volumetric efficiency 69%

Find the refrigerating capacity of the plant when operating at a condenser temperature of 40°C and an evaporator temperature of – 40°C.

Also, find the interstage pressure.

Chapter

6 Refrigerant Compressors

6.1 TYPES OF COMPRESSORS

The compression of the suction vapour from the evaporator to the condenser pressure can be achieved by mechanical compression, ejector compression or by a process combination of absorption of vapour, pumping and desorption. The latter two come under the category of heat-operated refrigerating machines discussed in Chaps. 12 and 13. For the mechanical compressor, fundamentally, there are two types of machines:

(i) *Positive displacement machines*, viz., reciprocating, rotary, scroll and screw compressors.

(ii) *Non-positive displacement machines*, viz., centrifugal compressors.

Positive displacement machines ensure positive admission and delivery preventing undesired *reversal of flow* within the machine as achieved by the use of valves in the case of reciprocating compressors. They have intermittent operation, subjecting the fluid to non-flow processes, and work is transferred by virtue of a hydrostatic force on the moving boundary.

Non-positive displacement machines, viz., centrifugal compressors have no means to prevent the reversal of flow. The fluid is subject to flow processes and the work is transferred by virtue of the change of momentum of a stream of fluid flowing at a high speed over blades or vanes attached to a rotor.

However, positive displacement machines can also be regarded as open systems being steadily supplied with the working fluid although the internal processes are intermittent as in a closed system. The net work of compression including the work of discharge and suction strokes is the same as the work done in a flow process in an open system shown as follows.

6.1.1 Work in Reciprocating Compressor[1]

The *p-v* diagram for the *machine cycle* of a reciprocating compressor is shown in Fig. 6.1 along with the skeleton diagram of the cylinder and piston mechanism. When the piston is in the extreme left position of the *inner dead centre* (IDC), the volume occupied by the gas is $V_c = V_3$ called the *clearance volume*, i.e., the volume between the IDC position of the piston and the cylinder head. As the piston moves outward, the clearance gas expands to 4, where the pressure inside the cylinder is equal to the pressure at the *suction flange* of the compressor. As the piston moves further, the *suction valve S* opens and the vapour from the evaporator is sucked in till the extreme right position of the *outer dead centre* (ODC) is reached. At this point the volume occupied by the gas is V_1. The *stroke* or *swept volume* or *piston displacement* is

$$V_p = (V_1 - V_3) = \frac{\pi D^2}{4} L$$

where D is the *bore* or diameter and L is the *stroke*, i.e., the distance travelled by the piston between I.D.C. and O.D.C. of the cylinder. At 1, the suction valve closes as the piston moves inwards and the compression begins. At 2, the pressure in the cylinder is equal to the pressure at the *discharge flange* of the compressor. A further inward movement of the piston results in the pressure in the cylinder exceeding the condenser pressure. This opens the discharge valve D and the vapour from the cylinder flows into the condenser till the piston again reaches the IDC position. Gas equal to the clearance volume V_c remains in the cylinder and the cycle is repeated.

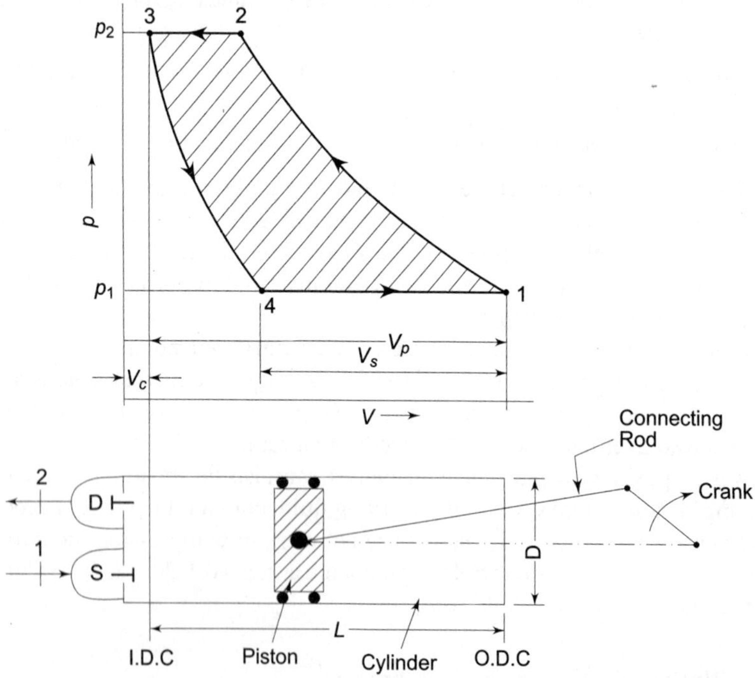

Fig. 6.1 Cylinder and piston mechanism and *p-V* diagram of a reciprocating compressor

The work done for compression for the machine cycle is given by the cyclic integral of $p dV$. Hence

$$W = \oint p dV = \int_1^2 p dV + \int_2^3 p dV + \int_3^4 p dV + \int_4^1 p dV$$

$$= \int_1^2 p\mathrm{d}V + p_2(V_3 - V_2) + \int_3^4 p\mathrm{d}V + p_1(V_1 - V_4)$$

$$= \text{Area 1-2-3-4}$$

It will be seen that this area is also expressed by the term $-\oint V\mathrm{d}p$. Hence

$$W = \oint_1^2 p\mathrm{d}V = -\oint_1^2 V\mathrm{d}p = m\oint_1^2 p\mathrm{d}v = -m\oint_1^2 v\mathrm{d}p$$

where m is the mass of the suction vapour. Thus, the specific work in a reciprocating compressor is given by

$$w = \oint_1^2 p\mathrm{d}v = -\oint_1^2 v\mathrm{d}p \tag{6.1}$$

where 1 and 2 are the limits of integration from suction state 1 to the discharge state 2 as indicated in Fig. 6.1.

6.1.2 Work in Centrifugal Compressor[1]

In a steady-flow process, the gas enters the centrifugal compressor, passes over the blades in a centrifugal field and is subjected to momentum change, leaving finally, through a diffuser at the discharge pressure. From the steady-flow energy equation

$$q = (h_2 - h_1) + w \tag{6.2}$$

and from the combined First and Second Laws for reversible process,

$$q = \int_1^2 T\mathrm{d}s = \int_1^2 (\mathrm{d}h - v\mathrm{d}p) = (h_2 - h_1) - \int_1^2 v\mathrm{d}p \tag{6.3}$$

Comparing the two expressions, we have for work

$$w = -\int_1^2 v\mathrm{d}p \tag{6.4}$$

It is thus seen that the work of compression is the same for both reciprocating and centrifugal compressors and is given by the expression $-\int_1^2 v\mathrm{d}p$, integrated between the suction and discharge states. Equation (6.3), therefore, represents the energy equation for both compressors, viz.,

$$q = (h_2 - h_1) - \oint_1^2 v\mathrm{d}p$$

For an adiabatic compression process, in which $q = 0$, it gives

$$w = -\oint_1^2 v\mathrm{d}p = (h_2 - h_1) \tag{6.5}$$

Note *The expression for work done is the same, viz.,* $w = -\int_1^2 v\mathrm{d}p$ *whether it is a reciprocating compressor or a centrifugal compressor.*

6.2 THERMODYNAMIC PROCESSES DURING COMPRESSION

Here is a comparison of the theoretical and actual thermodynamic processes during compression. Theoretical compression processes considered here are those of constant entropy and constant temperature. The actual compression process is, however, close to polytropic.

6.2.1 Isentropic Compression

For the isentropic compression process

$$pv^{\gamma} = p_1 v_1^{\gamma} = p_2 v_2^{\gamma}$$

so that

$$w = -\int v\,dp = -\int_1^2 v_1 (p_1/p)^{1/\gamma}\,dp$$

$$= -\frac{\gamma}{\gamma - 1} p_1 v_1 \left[\left(\frac{p_2}{p_1} \right)^{\frac{\gamma - 1}{\gamma}} - 1 \right] \tag{6.6}$$

For a perfect gas, this becomes

$$w = -C_p (T_2 - T_1) \tag{6.7}$$

since

$$p_1 v_1 = RT_1,\ C_p = \frac{\gamma R}{\gamma - 1} \text{ and } \left(\frac{p_2}{p_1} \right)^{\frac{\gamma-1}{\gamma}} = \frac{T_2}{T_1}$$

6.2.2 Isothermal Compression

The initial and final states are known to have pressure and temperature as p_1, T and p_2, T. Hence h_1, s_1 and h_2, s_2 can be found from the table of properties and work can be evaluated using the steady-flow energy equation, viz.,

$$q = T(s_2 - s_1) = (h_2 - h_1) + w \tag{6.8}$$

For a perfect gas

$$h_1 = h_2, \text{ hence } q = w.$$

Also, $pv = p_1 v_1 = p_2 v_2 =$ const. Hence

$$q = w = -\int_1^2 v\,dp = -p_1 v_1 \ln \frac{p_2}{p_1} = -RT_1 \ln \frac{p_2}{p_1}$$

$$= T(s_2 - s_1) \tag{6.9}$$

Adiabatic and isothermal processes are shown as $1 - 2t$ and $1 - 2s$ respectively on the p-v and T-s diagrams in Figs. 6.2 and 6.3. The shaded area on the p-v diagram represents the difference in work. For the same pressure ratio, isothermal work is seen to be less than the isentropic work. The shaded area on the T-s diagram in Fig. 6.3 represents the heat transfer in the isothermal process.

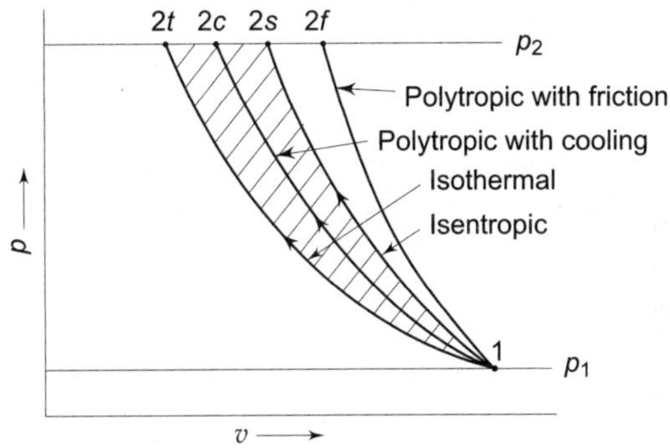

Fig. 6.2 Isothermal, isentropic and polytropic compression processes on *p-v* diagram

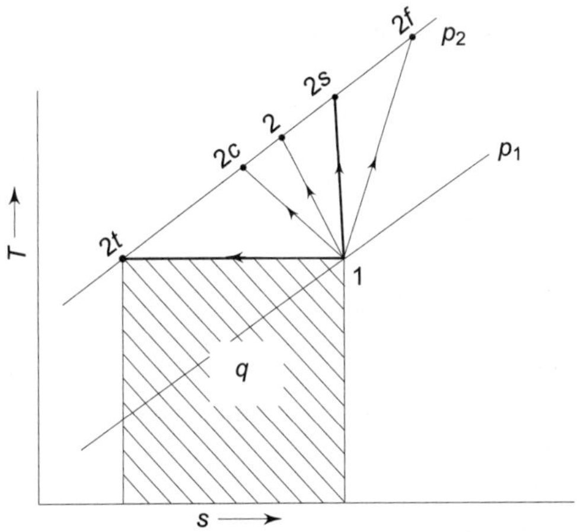

Fig. 6.3 Isothermal, isentropic and polytropic compression processes on *T-s* diagram

6.2.3 Polytropic Compression

Any general process can be expressed by the *polytropic law*

$$p_1 v_1^n = p_2 v_2^n = p v^n = \text{const.}$$

in which the polytropic *index* of compression n is such that $1 < n < \gamma$.

The work done in a polytropic compression process is thus

$$w = -\int v\mathrm{d}p = -\int v_1 \left(\frac{p_1}{p}\right)^{1/n} \mathrm{d}p = -\frac{n}{n-1}\, p_1 v_1 \left[\left(\frac{p_2}{p_1}\right)^{\frac{n-1}{n}} - 1\right] \tag{6.10}$$

For a perfect gas, since $p_1\, v_1 = RT_1$, this becomes

$$w = -\frac{n}{n-1}\, RT_1 \left[\left(\frac{p_2}{p_1}\right)^{\frac{n-1}{n}} - 1\right] = -\frac{n}{n-1}\cdot\frac{\gamma-1}{\gamma}\, C_p\,(T_2 - T_1) \tag{6.11}$$

Equation (6.11) shows that work is a function of initial temperature, pressure ratio and polytropic index n, viz.,

$$w = f(T_1, p_2/p_1, n)$$

For isentropic compression, $n = \gamma$, and for isothermal compression, $n = 1$. Higher the value of γ for a substance, more the work required, and higher is the discharge temperature. To reduce the work of compression and also to lower the discharge temperature, it is necessary to resort to cooling during compression, specially, in the case of substances with high value of γ.

The term polytropic compression is used in different senses. It may mean either (a) reversible but non-adiabatic compression, in which heat is removed during the process or, (b) irreversible but adiabatic compression, in which there is friction but no heat transfer, or (c) both.

Reciprocating compressors may approach case (a) with the cooling of the cylinder, provided the velocities are small. Thus n will be less than γ.

Centrifugal compressors approach case (b), i.e., friction effects are considerable but the flow is nearly adiabatic and n will be greater than γ.

We know that the isothermal compression process is the best but it would be extremely slow and is not possible to achieve in practice. Actual compression processes are nearly adiabatic. We can reduce the work of compression to some extent by cooling the compressor cylinder and achieve a process of the type (a) viz., *polytropic with cooling* such as $1 - 2c$ in Figs 6.2 and 6.3. Such a process will have work less than that of an adiabatic process and more than that of an isothermal process. If the value of γ for a gas is very high, such as 1.4 for air and 1.3 for ammonia, and/or the pressure ratio is also high, the compressor cylinders are cooled by water-jacketing. If the value of γ is not so high as in the case of fluorocarbons, cooling by air through natural convection is found satisfactory. To augment the heat transfer, the cylinder bodies of these compressors are cast with *fins* on the external surface.

In centrifugal compressors, however, the process is nearly adiabatic. But an adiabatic compression process (no cooling) will normally be accompained with friction. Such a process can also be represented by the polytropic law and will be of the type (b), viz., *polytropic with friction* as represented by the line $1 - 2f$ in Figs 6.2 and 6.3.

An actual compression process in reciprocating compresors will be accompanied with both cooling and friction. Such a process can also be represented by the polytropic law with an appropriate value of the index of compression n. The discharge state after compression may be either to the left or to the right of the point $2s$, depending on the degree of cooling and friction.

To compare the performance of a compressor, we define the following efficiencies

$$\textit{Isothermal efficiency, } \eta_T = \frac{\text{Isothermal work}}{\text{Actual work}}$$

$$\textit{Adiabatic efficiency, } \eta_a = \frac{\text{Isentropic work}}{\text{Actual work}}$$

Adiabatic efficiency is the most commonly used term.

The minimum work of compression, with cooling, is isothermal work. Often isothermal efficiency is, therefore, used to express the performance of reciprocating compressors which are invariably cooled.

The minimum work of compression, without cooling, is isentropic work. Adiabatic efficiency is, therefore, used to express the performance of centrifugal compressors in which it is not possible to arrange cooling during compression.

6.3 VOLUMETRIC EFFICIENCY OF RECIPROCATING COMPRESSORS

Volumetric efficiency η_v is the term defined in the case of positive displacement compressors to account for the difference in the displacement or swept volume V_p in-built in the compressor and volume V_s of the suction vapour sucked and pumped. It is expressed by the ratio

$$\eta_v = \frac{V_s}{V_p}$$

6.3.1 Clearance Volumetric Efficiency

The clearance or gap between the IDC position of the piston and cylinder head is necessary in reciprocating compressors to provide for thermal expansion and machining tolerances. A clearance of $(0.005\ L + 0.5)$ mm is normally provided. This space, together with the volume of the dead space between the cylinder head and valves, forms the clearance volume. The ratio of the clearance volume V_c to the swept volume V_p is called the *clearance factor* C, i.e.,

$$C = \frac{V_c}{V_p}$$

This factor is normally ≤ 5 per cent.

The effect of clearance in reciprocating compressors is to reduce the volume of the sucked vapour, as can be seen from Fig. 6.1. The gas trapped in the clearance space expands from the discharge pressure to the suction pressure and thus fills a part of the cylinder space before suction begins. Considering only the effect of clearance on volumetric efficiency, we have from Fig. 6.1, for *clearance volumetric efficiency*

$$\eta_{Cv} = \frac{V_1 - V_4}{V_p} = \frac{(V_p + V_c) - V_4}{V_p}$$

The volume occupied by the expanded clearance gases before suction begins, is

$$V_4 = V_c \left(\frac{p_2}{p_1}\right)^{1/\gamma} = CV_p \left(\frac{p_2}{p_1}\right)^{1/\gamma}$$

so that

$$\eta_{Cv} = \frac{V_p + CV_p - CV_p\,(p_2/p_1)^{1/\gamma}}{V_p}$$

$$= 1 + C - C\left(\frac{p_2}{p_1}\right)^{1/\gamma} \tag{6.12}$$

It is seen that lower the value of γ, lower the η_v, and higher the value of γ, higher the η_v. The expression for volumetric efficiency can also be written in the form,

$$\eta_v = 1 + C - \frac{V_4}{V_p} = 1 + C - C\frac{V_4}{V_c} = 1 + C - C\frac{v_{\text{suction}}}{v_{\text{discharge}}} \tag{6.12a}$$

6.3.2 Variation of Volumetric Efficiency with Suction Pressure

Figure 6.4 shows the nature of variation of the *p-V* diagram of a reciprocating compressor with decrease in suction pressure for constant discharge pressure. It is seen that with decreasing suction pressure, or increasing pressure ratio, the suction volume V_s and hence volumetric efficiency η_V decrease until both become zero at a certain low pressure p_1''. Thus the refrigerating capacity of a reciprocating compressor tends to zero with decreasing evaporator pressure.

It can be observed from Eq. (6.12) that the clearance volumetric efficiency will be zero for a pressure ratio given by

$$\frac{p_2}{p_{1_{\min}}} = \left(\frac{1}{C}+1\right)^{\gamma} \tag{6.13}$$

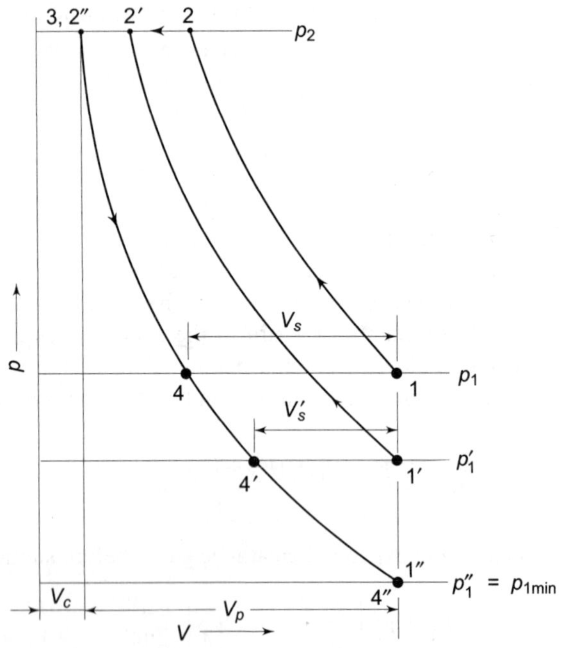

Fig. 6.4 Decrease in suction volume of a reciprocating compressor with decreasing evaporator pressure

For a given discharge pressure p_2, the above expression gives the value of $p_{1_{\min}}$, the lowest pressure possible for obtaining any capacity from a given compressor.

6.3.3 Exponent of the Re-expansion Curve

The clearance gases are expanded and compressed in every cycle. The curve 3-4 is, therefore, called the *re-expansion curve*.

To reduce the discharge temperature and the work of compression, the compressor cylinder is normally cooled either by *water-jacketing* as in the case of ammonia, or simply by the surrounding air by natural convection as in the case of R 134a. The small volume of the hot-clearance gas at 3, therefore, comes in contact with a larger and cooler surface of the cylinder walls of the compressor, thus losing heat at the beginning of expansion. Initially, the re-expansion curve is, therefore, steeper than the adiabatic curve. During, the latter part of the expansion process, the gas is cooler than the cylinder walls, receives heat in turn, and the curve becomes flatter. This shows that *the exponent of the re-expansion curve, m, is less than γ for the greater part of the process*.

The effect of heat gain from the walls is to increase the volume V_4 occupied by the expanded clearance gases, and thus to decrease the suction volume. Replacing the adiabatic exponent γ by the polytropic exponent m, we have for clearance volumetric efficiency

$$\eta_{C_v} = 1 + C - C\left(\frac{p_2}{p_1}\right)^{1/m} \tag{6.13a}$$

In the case of ammonia, the adiabatic exponent is high, i.e., $\gamma = 1.31$. Its compression, therefore, results in high discharge and cylinder-wall temperatures resulting in heat transfer to the gas during the re-expansion process. The effective value of m is, therefore, less than γ, giving a low value of volumetric efficiency. Cooling by water-jacketing of ammonia compressors, therefore, results in lower discharge and cylinder-wall temperatures and hence, less heat transfer to the gas and higher volumetric efficiency. The value of the exponent m with cooling approaches the value of the adiabatic exponent.

The value of γ for fluorocarbons is quite low. As a result, the volume V_4 is quite large and the volumetric efficiency is low. It is, therefore, necessary to provide minimum possible clearance (less than 5%) in these compressors. Also, because of the low value of γ, the compressors need to be air cooled only.

A clearance of 6–7% is considered permissible with refrigerants with high value of γ such as ammonia. In addition, they are water-cooled.

In high speed compressors (1500 rpm or more), the value of re-expansion coefficient m approaches the adiabatic exponent γ.

6.3.4 Effect of Valve Pressure Drops[3]

For the flow of any fluid, the pressure must drop in the direction of flow. Both suction and discharge valves will open only when there is a pressure drop across them. The effect of these pressure drops on the indicator diagram of the compressor is shown in Fig. 6.5. It is seen that as a result of throttling or pressure drop on the suction side the pressure inside the cylinder at the end of the suction stroke is p_s while the pressure at the suction flange is p_1. The pressure in the cylinder rises to the suction flange pressure p_1 only after the piston has travelled a certain distance inward during which the volume of the fluid has decreased from $(V_p + V_c)$ to V_1. Assuming the compression index to be n instead of γ, as the compression process is also polytropic due to heat exchange with cylinder walls and friction, we have

$$V_1 = (V_p + V_c)\left(\frac{p_s}{p_1}\right)^{1/n}$$

The expression for volumetric efficiency becomes

$$\eta_v = \frac{V_1 - V_4}{V_p} = \frac{(V_p + V_c)\left(\frac{p_s}{p_1}\right)^{1/n} - V_c\left(\frac{p_2}{p_1}\right)^{1/m}}{V_p}$$

$$= (1 + C)\left(\frac{p_s}{p_1}\right)^{1/n} - C\left(\frac{p_2}{p_1}\right)^{1/m} \quad (6.14)$$

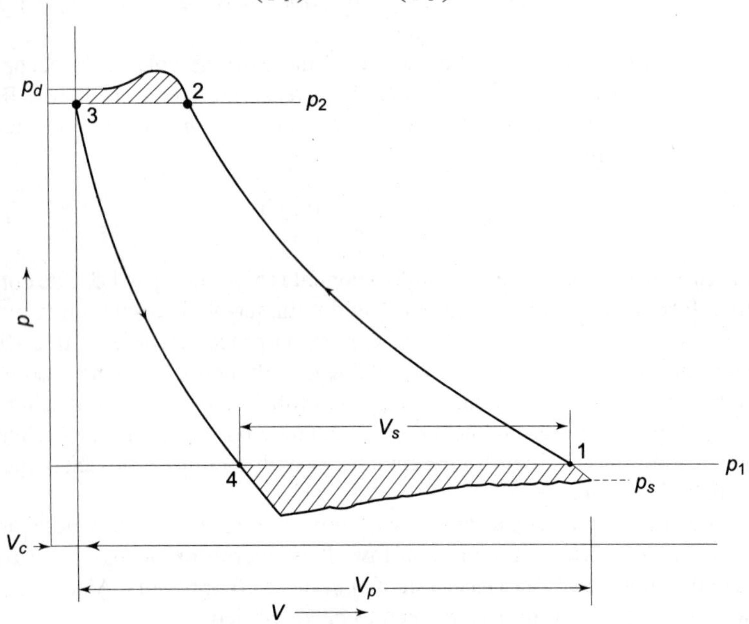

Fig. 6.5 Effect of valve pressure drops

Considering the effect of pressure drop at the discharge valve as well, it can be shown that the expression for volumetric efficiency is

$$\eta_v = (1 + C)\left(\frac{p_s}{p_1}\right)^{1/n} - C\left(\frac{p_d}{p_1}\right)^{1/m} \quad (6.15)$$

6.3.5 Leakage Loss

The effect of leakage past piston rings and under the suction valve elements is normally accounted for by allowing 1.5 per cent leakage per unit of the compression ratio r, which is equal to p_2/p_1.

Old worn out compressors tend to have more leakage and hence they lose their cooling capacity.

6.3.6 Overall Volumetric Efficiency

Considering the effect of wire-drawing at the valves, polytropic compression, re-expansion, and leakage, we may write the expression for the overall or total volumetric efficiency as follows

$$\eta_v = (1 + C)\left(\frac{p_s}{p_1}\right)^{1/n} - C\left(\frac{p_d}{p_1}\right)^{1/m} - 0.015\, r \tag{6.16}$$

Some authors give slightly different expressions for the total volumetric efficiency. The methods of improving the volumetric efficiency include the following:

(i) Providing clearance as small as possible.

(ii) Maintaining low pressure ratio.

(iii) Cooling during compression.

(iv) Reducing pressure drops at the valves by designing a light-weight valve mechanism, minimizing valve overlaps and choosing suitable lubricating oils.

6.4 EFFECT OF CLEARANCE ON WORK

The effect of the clearance volume on the work of compression is mainly due to the different values of the exponents of the compression and expansion processes. If the exponents are different, the net work is given by

$$W = -\int_1^2 V\mathrm{d}p + \int_4^3 V\mathrm{d}p$$

$$= -\frac{n}{n-1} p_1 V_1 \left[\left(\frac{p_2}{p_1}\right)^{\frac{n-1}{n}} - 1\right]$$

$$+\frac{m}{m-1} p_1 V_4 \left[\left(\frac{p_2}{p_1}\right)^{\frac{m-1}{m}} - 1\right] \tag{6.17}$$

When the two exponents are equal, i.e., $m = n$

$$W = p_1 V_s \left[\left(\frac{p_2}{p_1}\right)^{\frac{n-1}{n}} - 1\right] \tag{6.18}$$

where $V_s = V_1 - V_4$ = volume of the vapour sucked. Thus the work is only proportional to the suction volume. The clearance gas merely acts like a spring, alternately expanding and contracting. In practice, however, a large clearance volume results in a low volumetric efficiency and hence large cylinder dimensions, increased contact area between the piston and cylinder and so, increased friction and work. Shaded areas in Fig. 6.5 represent additional work due to valve pressure drops.

Example 6.1 *1 m^3 of a gas is compressed adiabatically (γ = 1.4) from 1 bar to 5 bar in a reciprocating compressor with 8 per cent clearance. If the exponent of the re-expansion curve is 1.1 instead of 1.4, find the percentage increase in the work of compression.*

Solution *Adiabatic compression and expansion*

$$W = \frac{\gamma}{\gamma-1} p_1 V_s \left[\left(\frac{p_2}{p_1}\right)^{\frac{\gamma-1}{1}} -1\right]$$

$$= \left(\frac{1-4}{1.4.-1}\right) \times 10^5 \times 1\,[(5)^{\frac{1.4-1}{1.4}} - 1] = 204{,}050 \text{ Nm}$$

Adiabatic compression and polytropic expansion (Fig. 6.1).

$$V_4 = 0.08\left(\frac{5}{1}\right)^{1/1.1} = 0.3456 \text{ m}^3$$

$$V_1 = 0.3456 + 1.0 = 1.3456 \text{ m}^3$$

$$W = \frac{n}{n-1} p_1 V_1 \left[\left(\frac{p_2}{p_1}\right)^{\frac{n-1}{n}} - 1\right] - \frac{m}{m-1} p_1 V_4$$

$$\left[\left(\frac{p_2}{p_1}\right)^{\frac{m-1}{m}} - 1\right]$$

$$= \frac{1.4}{1.4-1} \times 10^5 \times 1.3456\ [(5)^{\frac{1.4-1}{1.4}} - 1]$$

$$- \frac{1.1}{1.1-1} \times 10^5 \times 0.3456\ [(5)^{\frac{1.1-1}{1.1}} - 1] = 218{,}870 \text{ Nm}$$

$$\text{Increase in work} = \frac{218{,}870 - 204{,}050}{204{,}050} \times 100 = 7.26\%$$

6.5 PRINCIPAL DIMENSIONS OF A RECIPROCATING COMPRESSOR

The principal dimensions of a reciprocating compressor are the bore D and stroke L. These are to be decided in conjunction with the rpm N or mean piston speed $C_m = \left(\frac{2LN}{60}\right)$. Thus, there are three parameters, D, L and N or C_m to be selected. Stroke to bore ratio $\theta = L/D$ is a very important consideration in compressor design. Only one equation is, however, available for design, i.e., the equation for the suction volume

$$\dot{V}_s = \frac{\dot{Q}_0}{q_0} \cdot v_1 = \eta_v \dot{V}_p$$

where

$$\dot{V}_p = \frac{\pi D^2}{4} \cdot LN \times 60 = 47.1\, D^2\, LN \text{ m}^3\text{/h}$$

so that $\dot{V}_s = 47.1\, D^3\, \theta\, N\, \eta_v = 47.1\, D^2\, (30\, C_m)\, \eta_v$ (6.19)

whence

$$D = 0.277 \sqrt[3]{\frac{\dot{V}_s}{\theta\, N\, \eta_v}} = 0.0266 \sqrt{\frac{\dot{V}_s}{C_m\, \eta_v}} \tag{6.20}$$

Assumptions are, therefore, necessary for two parameters. To minimize the friction and wear, the mean piston speed is selected in the range of 1.5 to 5 m/s. At the same time the stroke to bore ratio, viz., $\theta = L/D$, is chosen on the basis of the delivery pressure p_d. The piston force defined by

$$F_p = \frac{\pi D^2}{4} \cdot p_d$$

depends on the delivery pressure and cylinder diameter. To minimize the piston force and hence the inertia force, a large value of θ is chosen in high-pressure compressors so that the diameter is smaller. The following values are normally adopted in practice:

Vacuum pumps and high-speed air compressors	$\theta \le 0.5$
Fluorocarbon compressors	$\theta \approx 0.8$
Ammonia compressors	$\theta \approx 1.0$
High-pressure compressors	$\theta = 4.6$

6.6 PERFORMANCE CHARACTERISTICS OF RECIPROCATING COMPRESSORS[5]

Catalogue data of refrigerant compressors is normally in the form of tables and charts giving the refrigerating capacity and brake horsepower as a function of condenser and evaporator temperatures, which are the two principal design parameters. The effects of these temperatures or corresponding pressures on the capacity and power consumption of reciprocating compressors are analysed below.

The effect of suction pressure, keeping the discharge pressure constant, is shown in Fig. 6.6. The mass flow rate of the refrigerant in a reciprocating compressor is given by

$$\dot{m} = \frac{\dot{V}_p}{v_1} \eta_v \tag{6.21}$$

As suction pressure drops, the specific volume of the suction vapour v_1 increases and the volumetric efficiency η_v decreases. As a result the mass flow rate decreases. The volumetric efficiency and hence the mass flow become zero at a certain low value of suction pressure (Sec. 6.3.2).

The refrigerating capacity is now given by

$$\dot{Q}_0 = \dot{m}\, q_0$$

$$= \eta_v \frac{\dot{V}_p}{v_1} (h_1 - h_4) \tag{6.22}$$

The refrigerating effect varies very little with the suction pressure. The capacity, therefore, varies almost according to the mass flow rate of the refrigerant which in turn is proportional to η_v/v_1. The capacity decreases as p_0 decreases not only because of increase in v_1, but also because of decrease in η_v. The capacity $\dot{Q}_0 \to 0$ as $p_0 \to p_{0\min}$. Thereafter it increases with p_0 as shown in Fig. 6.6.

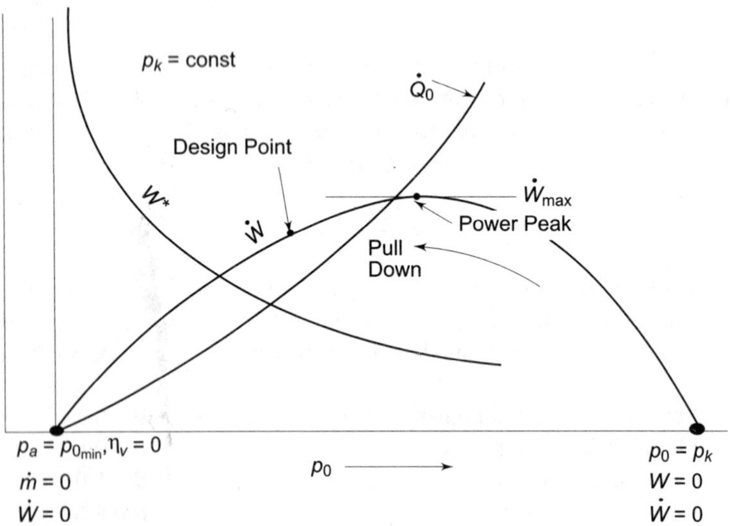

Fig. 6.6 Performance characteristics of a reciprocating compressor as a function of evaporator pressure

It is seen that the piston displacement per ton is given by

$$V^* = \frac{211}{q_0} \cdot \frac{v_1}{\eta_v} \text{ m}^3/\text{min (TR)}$$

Hence

$$\dot{Q}_0 = \left(\frac{\dot{V}_p}{\dot{V}^*}\right) \propto \frac{1}{V^*} \tag{6.23}$$

Thus, the capacity is inversely proportional to the piston displacement per ton. It was shown in Sec. 3.5.1 that a drop of 5°C in the evaporator temperature increases V^* from 0.103 to 0.154 m^3/min/TR in an R 134a system. The corresponding decrease in capacity was found to be 33.3 per cent.

The power consumption of the compressor is given by

$$\dot{W} = \dot{m}\, w$$

$$= \eta_V \frac{\dot{V}_p}{v_1} \cdot \left(\frac{h_2 - h_1}{\eta_m \eta_a}\right) \tag{6.24}$$

The mechanical efficiency η_m is generally of the order of 0.95. The adiabatic efficiency η_a depends on the compression ratio and many other factors, such as the size and disposition of valves, cylinder dimensions, heat transfer, etc.

It may be noted that w is zero when the suction pressure is equal to the discharge pressure usually at starting and $\dot{m}$ is zero at a certain low value of the suction pressure. Hence $\dot{W}$ is zero at two values of the suction pressure, viz.,

$$\dot{W} = 0 \text{ when } \dot{m} = 0 \text{ (At } p_0 = p_{0min})$$
$$\dot{W} = 0 \text{ when } w = 0 \text{ (At starting)}$$

Therefore, the power consumption curve passes through the maximum, a peak, as shown in Fig. 6.6.

Generally refrigeration systems operate on the left-hand side of this curve. But just after starting, the compressor passes through the power peak. The compressor motors are, therefore, oversized to enable them to take the peak load during *pull-down*. The starting current is more than the running current.

It has been shown in Sec. 3.5 that the power per ton $\dot{W}^*$ decreases with increasing suction pressure.

The effect of the discharge pressure can similarly be analysed. At constant suction pressure, an increase in the discharge pressure will cause a reduction in the volumetric efficiency due to higher compression ratio. The mass of refrigerant circulated will thus be reduced. At the same time the specific work will increase. But there is a continuous increase in the power consumption and power per ton. The capacity will be decreased due to decrease in the mass flow and slight decrease in the refrigerating effect.

Example 6.2 *(a) An R 22 hermetic (directly-coupled motor) reciprocating compressor with 4 per cent clearance is to be designed for 7.5 TR capacity at 4°C evaporating and 40°C condensing temperatures. The compression index may be taken as 1.15. The number of cylinders may be selected as two and the mean piston speed C_m as 3 m/s approximately. The stroke to bore ratio for fluorocarbons may be taken as 0.8. Pressure drops at suction and discharge valves may be assumed as 0.2 and 0.4 bar respectively. Determine:*

(i) Power consumption of the compressor and COP of the cycle.
(ii) Volumetric efficiency of the compressor.
(iii) Bore and stroke of the compressor. Choose motor with 1400 rpm.
(b) If the evaporator temperature drops to –2°C with the condenser temperature and other conditions remaining same, what will be the capacity and power consumption of this compressor and COP of the system?

Solution Refer to Figs 3.6 and 3.7.

(a) At 4°C evaporator temperature and 40 °C condensing temperature

$$p_0 = 5.657 \text{ bar}, p_k = 15.335 \text{ bar}$$

Suction and discharge pressures are

$$p_s = 5.657 - 0.2 = 5.457 \text{ bar}$$
$$p_d = 15.335 + 0.4 = 15.735 \text{ bar}$$

(i) Enthalpy of the suction vapour

$$h_1 = 406.8 \text{ (At 4°C. Saturated vapour)}$$

Specific volume of suction vapour

$$v_1 = 0.0431 \text{ m}^3\text{/kg}$$

Enthalpy of the liquid from the condenser

$$h_3 = h_4 = 249.1 \text{ (At 40°C. Saturated liquid)}$$

Refrigerating effect

$$q_0 = 406.8 - 249.1 = 157.7 \text{ kJ/kg}$$

Specific work

$$w = \frac{1.15}{1.15-1}(5.457 \times 10^5 \times 0.0431)\left[\left(\frac{157.735}{5.437}\right)^{\frac{1.15-1}{1.15}} - 1\right]\frac{1}{10^3}$$

$$= 26.6 \text{ kJ/kg}$$

Mass flow rate

$$\dot{m} = \frac{7.5 \times 3.5167}{157.7} = 0.1673 \text{ kg/s}$$

Power consumption

$$\dot{W} = 0.1673 \times 26.6 = 4.45 \text{ kW}$$

$$\text{COP} = \frac{157.7}{26.6} = 5.9$$

(ii) Volumetric efficiency

$$\eta_v = (1 + C)\left(\frac{p_s}{p_1}\right)^{1/n} - C\left(\frac{p_d}{p_1}\right)^{1/n} - 0.015\, r$$

$$= (1 + 0.04)\left(\frac{5.457}{5.657}\right)^{1/1.15} - 0.04\left(\frac{15.735}{5.657}\right)^{1/1.15} - 0.015\left(\frac{15.735}{5.457}\right)$$

$$= 1.0065 - 0.0474 - 0.029 = 0.8585$$

(iii) Piston displacement per cylinder

$$\dot{V}_p = \frac{1}{2}\frac{\dot{m}v_1}{\eta_v} = \frac{0.1673 \times 0.0431 \times 60}{2 \times 0.8585} = 0.252 \text{ m}^3\text{/min}$$

$$= \frac{\pi D^2}{4} LN = \frac{\pi D^2}{4} L\left(\frac{30C_m}{L}\right)$$

whence

$$D = \sqrt{\frac{4\dot{V}_p}{\pi\, 30\, C_m}} = \sqrt{\frac{4 \times 0.252}{\pi(30)(3)}} = 0.06 \text{ m or } 60 \text{ mm}$$

$$L = 0.8 \times 0.06 = 0.048 \text{ m or } 48 \text{ mm}$$

$$N = \frac{30C_m}{L} = \frac{30 \times 3}{0.048} = 1875 \text{ rpm}$$

As the motor is of 1400 rpm, L, D and C_m will have to be altered for same $\dot{V}_p$. Keeping C_m and D the same, one gets

$$L = \frac{30C_m}{N} = \frac{30 \times 3}{1400} = 0.064 \text{ m or } 64 \text{ mm}$$

$$\theta = \frac{L}{D} = \frac{64}{60} = 1.07$$

which is very high. It will result in increased frictional losses. So, we have to reduce L. The compromise solution is:

$$L = 56 \text{ mm}, D = 64 \text{ mm}, \theta = 0.875, C_m = 2.61 \text{ m/s}$$

(b) At -2°C evaporator temperature

$$p_0' = 4.659 \text{ bar}$$

Also $p_s' = 4.659 - 0.2 = 4.459$ bar

$$h_1' = 404.6 \text{ kJ/kg}$$

$$v_1' = 0.0525 \text{ m}^3\text{/kg}$$

$$q_0' = 404.6 - 249.1 = 155.5 \text{ kJ/kg}$$

$$w' = \frac{1.15}{1.15-1}(4.459 \times 10^5 \times 0.0525)\left[\left(\frac{15.735}{4.459}\right)^{\frac{1.15-1}{1}} - 1\right]\frac{1}{10^3}$$

$$= 32.0 \text{ kJ/kg}$$

$$\eta_v = (1 + 0.04)\left(\frac{4.459}{4.659}\right)^{1/1.15} - 0.04\left(\frac{15.735}{4.459}\right)^{1/1.15} - 0.015\left(\frac{15.735}{4.459}\right)$$

$$= 0.82$$

$$\dot{m}' = \frac{2 \times 0.252 \times 0.82}{60 \times 0.0525} = 0.1312 \text{ kg/s}$$

Capacity

$$\dot{Q}_0 = 0.1312 \times 155.5 = 20.4 \text{ kW}$$

Power consumption

$$\dot{W}' = 0.1312 \times 32 = 4.2 \text{ kW}$$

$$\text{COP}' = \frac{155.5}{32} = 4.86$$

Note *With a decrease in the evaporator temperature and pressure, although the specific work increases, the total power consumption of compressor does not increase since the mass flow rate decreases. The power consumption at 4°C evaporator temperature is 4.45 kW (Capacity 7.5 TR). The power consumption at –2°C evaporator temperature is 4.2 kW only (Capacity 5.8 TR).*

6.6.1 Calorimetric Method of Measuring Refrigerating Capacity of Small Compressors

Figure 6.7 illustrates the construction and schematic arrangement of a calorimeter used to measure the refrigerating capacity of small compressors.

The calorimeter encloses the evaporator in the form of a coil, and a set of electric heaters dipped below the level of the liquid column of a secondary refrigerant. When the heaters are on, vapours of secondary refrigerant rise above. They get condensed by coming in contact with the cold surface of the evaporator coil. The condensate returns back to the sump of liquid below.

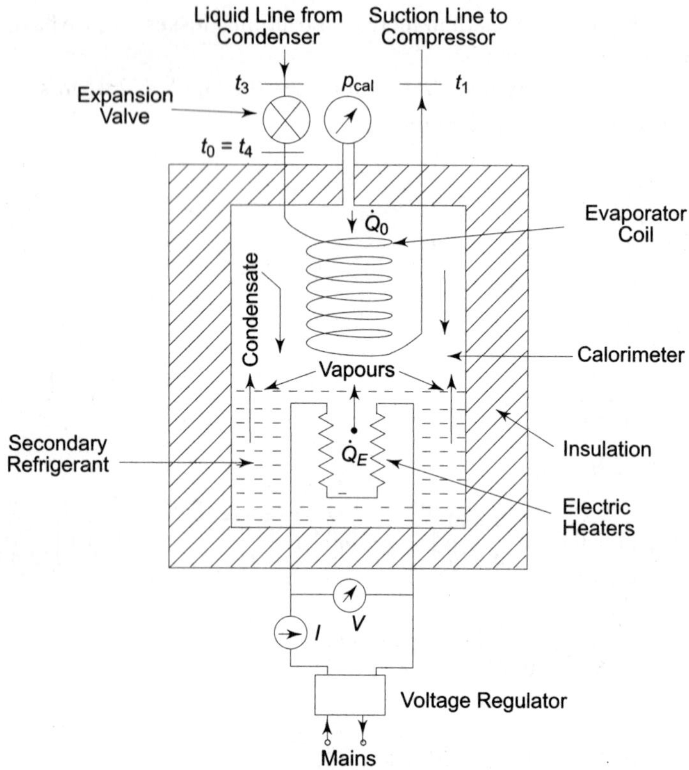

Fig. 6.7 Calorimeter for measuring refrigerating capacity of small compressors

If heat added by electric heaters $\dot{Q}_E$ is more than the refrigerating capacity $\dot{Q}_0$, then more vapours are produced, and less condensate returns. As a result, the calorimeter pressure p_{cal} rises. On the contrary, when $\dot{Q}_E < \dot{Q}_0$, the calorimeter pressure drops. If $\dot{Q}_E = \dot{Q}_0$, p_{cal} and t_1 become constant.

Now, the procedure is different depending on the type of expansion valve used.

(a) If an *automatic expansion valve* is used which maintains constant temperature $t_0 = t_4$ and pressure $p_0 = p_4$, then electric load $\dot{Q}_E$ is adjusted with the help of voltage regulator until we get constant p_{cal} and t_1. This value of $\dot{Q}_E = \dot{Q}_0$ for the adjusted t_0.

(b) If a *thermostatic expansion valve* is used which maintains a constant degree of superheat $(t_1 - t_4)$, then different electric loads $\dot{Q}_E$ are given. On reaching steady state (equilibrium), we note down the temperature $t_0 = t_4$ for each $\dot{Q}_E$ which is equal to $\dot{Q}_0$.

Thus, we get a plot between $\dot{Q}_0$ and t_0.

6.7 CAPACITY CONTROL OF RECIPROCATING AND ROTARY COMPRESSORS

Refrigerating capacity control with reciprocating compressors running at constant speed consists in controlling the quantity of the gas delivered to match the fluctuating load. In an efficient control system, the power consumption of the compressor should be proportional to the amount of gas delivered. Various control methods have been discussed below.

6.7.1 On and Off Control

Simple on-off or start-stop control in conjunction with a thermostat is used to advantage in small unitary equipment, such as refrigerators, air conditioners, water coolers, etc. It is particularly suited where there is a sudden large demand followed by periods of small or no demand on capacity. Switching off is preferable to running at partial load because of the low part-load efficiency of squirrel-cage induction motors employed in refrigerant compressors.

6.7.2 Holding the Valves Open

This is the most commonly used method of unloading in multi-cylinder V/W compressors. It is accomplished by the lifting of suction valves, usually of two cylinders together, by means of *depressors* or *fingers*. The gas drawn in the cylinder during the suction stroke is displaced back into the suction line during the discharge stroke. No work is involved except the frictional work during such *idling*. The depressor roads are operated by a cam which is turned either manually or by means of a thermostat or a pressure-stat (for suction pressure). This is the best method of capacity control when the condition is of intermittent unloading that requires idling for short periods only.

More and more cylinders are unloaded, as the suction pressure or evaporator temperature continues to drop, by means of stepped control.

6.7.3 Hot Gas Bypass

In this method, a part of the hot compressed gas is bypassed back to the suction line, through a constant-pressure throttle valve. The valve admits hot gas in the suction line as the evaporator pressure tends to drop, thus maintaining constant suction pressure. This method is not efficient as the suction temperature tends to increase as more and more hot gas is bypassed. The compressor work also tends to increase, thus resulting in the loading of the compressor instead of unloading.

6.7.4 Using Multiple Units

The use of multiple units in reciprocating compressors is very common. This is resorted to when unloading is required for long periods. Some units may serve the purpose of stand-by equipment as well, thus allowing partial load operation in the case of breakdown/routine maintenance of other units.

6.7.5 Varying Compressor Speed by Using Inverter Technology

Let us first understand what is inverter technology. An inverter is a device for converting frequency. Inverter air conditioners vary their capacity by varying the speed of the compressor which is done by adjusting the power supply frequency as explained in the diagram below:

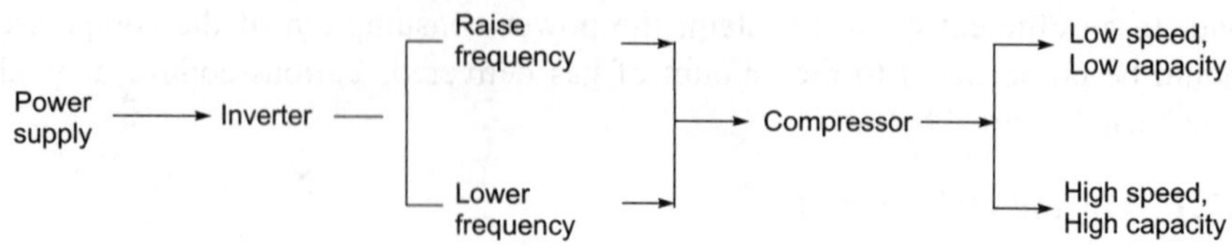

An inverter adjusts the speed of the compressor to control refrigerant vapour flow rate.

A compressor without an inverter starts and stops. The current and power consumption goes down when it stops. But it goes up sharply when it starts. As a result, it has an overall higher power consumption. The air conditioners with inverters are more energy-saving.

Further, an inverter AC will maintain room temperature close to the set value without fluctuation. But a non-inverter AC will experience wide fluctuations from the set value. For example, a non-inverter AC with temperature set at 24°C maintains room temperature ranging from 23°C to 25°C, thus resulting in high temperature fluctuations. Therefore, the inverter ACs are widely used.

Example 6.3

(a) *A $1\frac{1}{2}$-ton R 22 air conditioner operates on a simple-saturation cycle between an evaporator temperature of 4°C and a condenser temperature of 54.5°C. Find the theoretical COP for cooling, piston displacement and horsepower of the compressor motor.*

(b) *If the same air conditioner is used as a heat pump for winter heating with evaporator and condenser temperatures of –5°C and 60.4°C respectively, find the theoretical COP for heating and the capacity of the compressor for heating in kW.*

(c) *What should be the horsepower of the motor so that it is adequate for both summer and winter air conditioning?*

Solution (a) *Refrigeration Cycle* (Fig. 3.7)

$$h_1 = 406.8 \text{ kJ/kg}, \ s_1 = 1.746 \text{ kJ/kg.K} = s_2,$$

$$h_2 = 439.3 \text{ kJ/kg}, \ t_2 = 74°\text{C}$$

$$h_3 = h_4 = 268.3 \text{ kJ/kg}$$

$$q_0 = 406.8 - 268.3 = 138.5 \text{ kJ/kg}$$

$$w = 439.8 - 406.8 = 33 \text{ kJ/kg}$$

COP,

$$\mathcal{E}_c = \frac{138.5}{33} = 4.2$$

$$\dot{m} = \frac{1.5 \times 211}{138.5} = 2.2852 \text{ kg/min}$$

$$\dot{V} = 2.2852 \times 0.0416 = 0.095 \text{ m}^3\text{/min}$$

$$\dot{W} = \frac{2.2852 \times 33}{60} = 1.27 \text{ kW}$$

$$\text{HP} = \frac{1.27 \times 10^3}{746} = 1.7$$

(b) *Heat pump cycle*

$$h_1 = 403.5 \text{ kJ/kg} \qquad h_2 = 448.0 \text{ kJ/kg} \qquad t_2 = 82°\text{C}$$

$$h_3 = h_4 = 276.6 \text{ kJ/kg} \qquad v_1 = 0.0554 \text{ m}^3\text{/kg}$$

$$q_0 = 403.5 - 276.6 = 126.9 \text{ kJ/kg}$$

$$w = 448.0 - 403.5 = 44.5 \text{ kJ/kg}$$

$$q_k = 448.0 - 276.6 = 171.4 \text{ kJ/kg}$$

COP, $$\mathcal{E}_h = \frac{171.4}{44.5} = 3.85$$

Mass flow in the compressor (piston displacement = 0.095 m^3/min)

$$\dot{m} = \frac{\dot{V}}{v_1} = \frac{0.095}{0.0554} = 1.715 \text{ kg/min}$$

Capacity for heating of the compressor

$$\dot{Q}_k = \dot{m}\, q_k = \frac{1.715\,(171.4)}{60} = 4.9 \text{ kW}$$

(c) *Horsepower requirement*

$$\text{HP} = \frac{1.715\,(44.5)}{60} \times \frac{10^3}{746} = 1.71$$

Thus the power requirement for the heat pump is nearly the same as for the refrigerating machine. It would, however, be necessary to use a larger size motor if the air conditioner is also to be used for winter heating in case the ambient temperature drops further such that the evaporator temperature falls below – 5°C.

6.8 CONSTRUCTION FEATURES OF RECIPROCATING COMPRESSORS

The reciprocating compressors are classified according to their enclosing pattern as *hermetically sealed, semi-sealed* and *open-type*. The hermetically sealed and semi-hermetic compressors have their motor enclosed along with the cylinder and crankcase inside a *dome*. The motor windings are cooled by incoming suction vapours. These have the advantage of no leakage, less noise and compactness.

The crankcase is usually of cast iron. Aluminium is also used in small, open and welded hermetic compressors. The crankcase encloses the shaft and oil sump. The cylinder can be integral with the crankcase or can be in a separate cylinder block. The valves are located in the cylinder head.

Crank-shafts are of forged steel with hardened bearing surfaces, and dynamically balanced. Pistons are usually made of cast iron or aluminium. Cast iron pistons with a clearance of 0.0004 cm/cm of the cylinder diameter would provide for adequate sealing without piston rings. Piston rings are essential with aluminium pistons.

The most crucial parts in reciprocating compressors are suction and discharge valves. Their design should provide for long life and low pressure drops. To minimize the pressure drop, their weight should be small. Reed valves with varying clamping arrangements are commonly employed in refrigeration.

The lubrication systems in refrigerant compressors vary. Both *splash* and *forced feed* systems are employed.

Shaft seals are used in open-type compressors. Normally *stationary* seals with *bellows* are employed. Recently, *rotary* synthetic seals, tightly fitted to the shaft in a carbon nose, have gained popularity. They are much less expensive.

6.9 ROTARY COMPRESSORS

Rotary compressors are positive displacement, direct-drive machines. There are essentially two designs of this compressor:

(i) Rolling piston type. (ii) Rotating vane type.

In the rolling piston type, shown in Fig. 6.8(a) the roller is mounted on an eccentric shaft with a single blade, which is always in contact with the roller by means of a spring. The theoretical piston displacement is

$$V_p = \frac{\pi H(A^2 - B^2)}{4} \tag{6.25}$$

where A and B are respectively the diameters of the cylinder and rolling piston and H is the height of the cylinder.

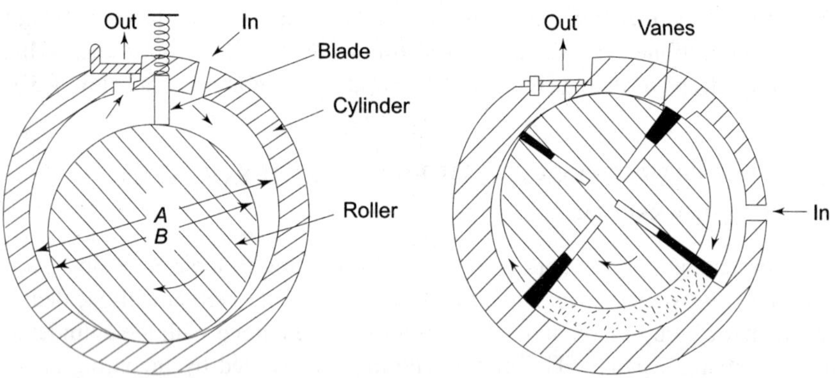

(a) Rolling Piston Type (b) Rotating Vane Type

Fig. 6.8 Rotary compressors

In the rotating vane type, as shown in Fig. 6.8 (b) with four vanes, the rotor is concentric with the shaft. The vanes slide within the rotor but keep contact with the cylinder. The assembly of rotor and the vanes is off-centre with respect to the cylinder.

In both designs, the whole assembly is enclosed in a housing (not shown in the figures), filled with oil and remains submerged in oil. An *oil film* forms the seal between the high-pressure and the low-pressure sides. When the compressor stops, this seal is lost and the pressure equalizes.

Rotary compressors have high volumetric efficiencies due to negligible clearance. They are normally used in a single stage up to a capacity of 5 TR.

6.10 SCREW COMPRESSORS

Rotary screw compressors also belong to the category of positive displacement compressors. The machine essentially consists of two helically-grooved rotors as illustrated in Fig. 6.9 which rotate in a housing.

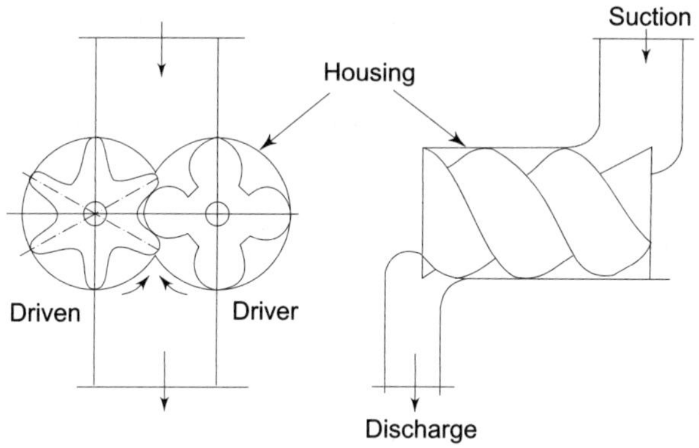

Fig. 6.9 Sectional and side views of a screw compressor

The male rotor consists of *lobes* and is normally the driving rotor. The female rotor has *gullies* and is normally the driven rotor. A four-lobe male rotor will drive a six-gully female rotor at two-thirds of its speed. At 3600 rpm, the number of compressed gas discharges of a four-lobe rotor will be 4 × 3600 = 14,400 per minute.

As in the case of other positive displacement machines, there are three basic continuous phases of the working cycle, viz., suction, compression and discharge. When the male rotor turns clockwise, an interlobe space between a pair and housing, nearest to the suction end, opens and is filled with the gas. There are four such pairs to be filled during one revolution in a four-lobe rotor and the suction periods overlap one another.

When remeshing starts, the volume decreases and the pressure rises. The charge is moved helically and compressed until the trapped volume reaches the discharge end. The compression ratio is thus fixed.

Further rotation simply empties the rotors of the high pressure gas until the last traces of the gas are squeezed out, irrespective of the pressure in the condenser.

On completion of the discharge phase, there is no residual gas remaining in the rotors. As a result, there is no expansion of clearance gases. The compressor has no suction and discharge valves.

There are leakage paths in a screw compressor mainly across the line of mesh between the rotors and across the clearance between the rotors and the housing. To eliminate leakage, oil is injected in a number of small jets directed towards the mesh. Oil injection also serves the purpose of cooling and lubricating along with that of sealing the leakage paths.

The rotor profile is patented, the patent rights being held by Svenska Rotor Maskiner AB, Sweden.

A slide valve, closely following the shape of the rotors is used for capacity control. At full load the valve is closed. At part load, the valve opens enabling a return flow passage to be formed so that a part of the gas drawn into the interlobe spaces can flow back to the suction side.

The screw compressor combines many advantageous features of both centrifugal and reciprocating compressors, along with some of its own. As it is a positive displacement machine, high pressure refrigerants as in reciprocating compressors, such as R 22 and ammonia are used in it. As it is a high speed rotary machine, a large volume, as in centrifugal compressors, can be handled by it. It is, therefore, found extremely suitable for large capacity low temperature applications such as in food refrigeration, and also in large capacity central air conditioning plants, with R 134a and R 22, as alternative to R 11 (CFC 11) centrifugal compressors.

Unlike centrifugal compressors, it has no surging problems. Like reciprocating compressors, it has small pipe dimensions and positive pressures due to the use of high pressure refrigerants. Like centrifugal compressors, it has high compression efficiency, continuous capacity control, unloaded starting and no balancing problems. Also, the compressor is suitable for large capacity installations.

6.11 SCROLL COMPRESSORS

Scroll compressors are valve-less positive displacement machines like rotary and screw compressors. Because of their simplicity, they have become very popular with the industry in recent years.

In scroll compressors, compression is achieved by two interfitting, spiral-shaped 'scroll members' one of which is a 'fixed scroll' and the other an 'orbiting scroll' as indicated in the cross-sectional views in Fig. 6.10.

Scroll compressors are currently preferred in residential and commercial refrigeration air-conditioning, heat pump, and automotive air-conditioning applications.

Capacities of a single compressor range from 1 to 14 TR only. For larger capacity, multiple units can be used.

The compressor requires machining of scroll members which have close tolerances. Such machining has become possible only recently with the developments in precision manufacturing processes. Gas sealing is an important mechanical feature of scroll compressors.

The compressor gives high efficiency, and has low noise level. As such, it has come to replace reciprocating and even rotary machines in these capacities.

6.11.1 Scroll Compression Process

The fixed and orbiting scrolls have geometrically matching surfaces. The two scrolls are fitted to form pocktes of gas.

Suction gas is sealed in pockets of a given volume at the outer periphery of the scrolls. Compression is achived by progressively reducing the size of these pockets, as the motion of the orbiting scroll, relative to the fixed scroll, moves the pockets 'inwards' towards the discharge port.

As is the case with screw compressors, the scroll compressor has a built-in volume ratio, defined by the geometry of the scrolls, and the location of the discharge port.

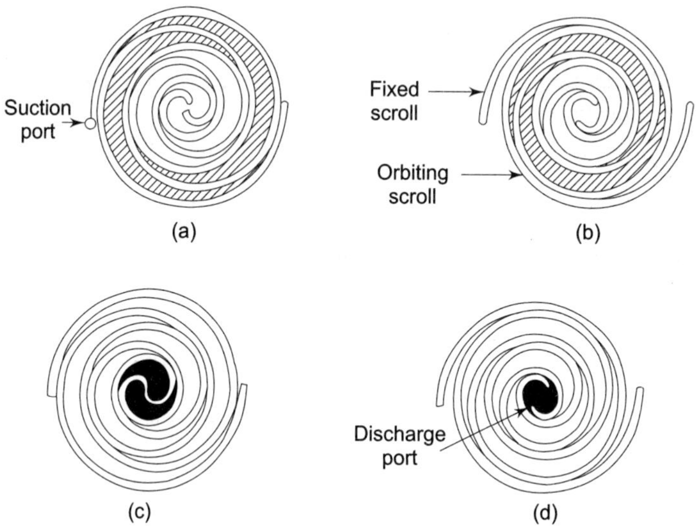

Fig. 6.10 Scroll compression process

Figure 6.10 shows the scroll compression process in a sequence of suction, compression and discharge.

Fig. 6.10(a)

The outermost pockets are sealed off. Suction gas is trapped just before compression begins.

Fig. 6.10(b)

At stage (b), orbiting motion moves the gas to the centre. Pocket volume progressively reduces, and pressure rises, as shown in (c).

Fig. 6.10(d)

At (d), the gas reaches the discharge port in the centre, and discharge takes place. Simultaneously, suction begins at the outer periphery.

6.11.2 Capacity Control of Scroll Compressors

There are two methods

(i) *Variable Speed Control* Variable speed scroll compressor uses an inverter drive to convert a fixed frequency AC current into one with adjustable voltage and frequency which permits varying the speed of the induction motor.

(ii) *Variable Displacement Scroll Compressor* The control mechanism connects or disconnects the compression chamber to suction side by closing and opening the porting holes.

When all porting holes are closed, the compressor runs at full capcity.
When all porting holes are open to suction side, we get the smallest capacity.

6.12 CENTRIFUGAL COMPRESSORS

A single-stage centrifugal compressor mainly consists of the following four components as shown in Fig. 6.11.

(i) An *inlet casing* to accelerate the fluid to the impeller inlet.

(ii) An *impeller* to transfer energy to the fluid in the form of increased static pressure (enthalpy) and kinetic energy.

(iii) A *diffuser* to convert the kinetic energy at the impeller outlet into enthalpy resulting in pressure rise.

(iv) A *volute casing* to collect the fluid and to further convert the remaining kinetic energy into enthalpy resulting in further pressure rise.

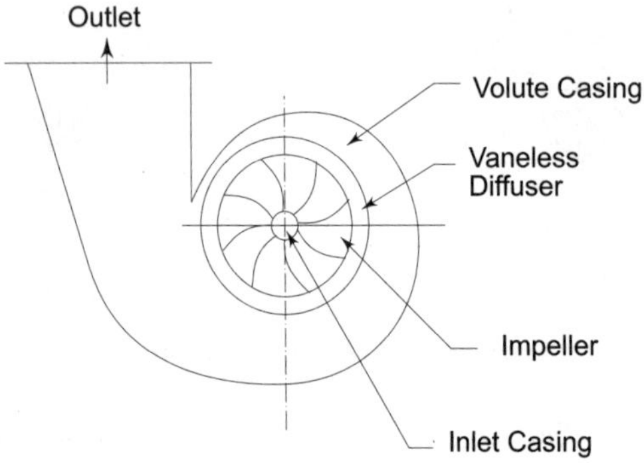

Fig. 6.11 Elements of a centrifugal compressor

Besides these, there are intercoolers in a multistage compressor that are generally integrated with the casing. The casing is usually made of cast iron and the impeller, of high speed (chrome-nickel) steels. The maximum stress is developed at the root of the blades.

The diffuser is normally of the vaneless type as it permits more efficient part load operation which is quite usual in any air-conditioning plant. A vaned diffuser will certainly cause shock losses if the compressor has to run at reduced capacity and flow (part-load).

6.12.1 Application of Steady-Flow Energy Equation to a Centrifugal Stage

Compression in a centrifugal compressor is achieved by the self-compression of the refrigerant by centrifugal force as well as by the conversion of kinetic energy of the high-velocity vapours into static enthalpy. Applying the steady-flow energy equation to processes from the entrance to the discharge for one stage of compression as shown on the *h-s* diagram in Fig. 6.12, we obtain the following relations:

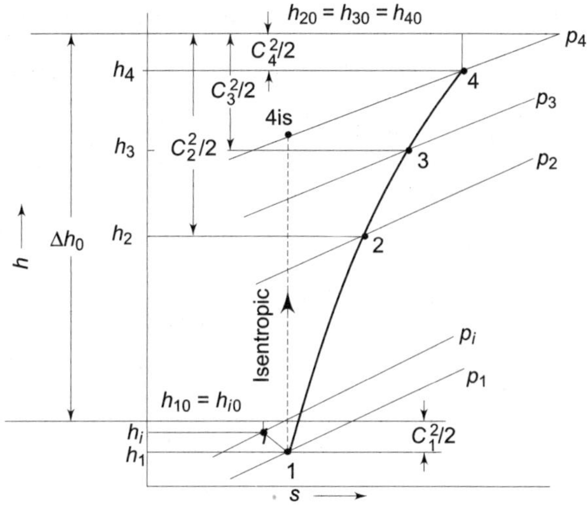

Fig. 6.12 Mollier diagram of centrifugal stage

Flow through Inlet Casing (Process i-1) In this process the fluid is accelerated, but there is no energy transfer. Hence, the stagnation enthalpy remains constant but there is a drop in the static pressure and enthalpy.

$$h_{io} = h_i + \frac{C_i^2}{2} = h_1 + \frac{C_1^2}{2} = h_{10} \tag{6.26}$$

(Subscript 0 refers to stagnation state.)

Flow through Impeller (Process 1-2) In this, work is done by the impeller. Energy is transferred to the fluid and the velocity, pressure and enthalpy are increased. The *energy equation* for the process is

$$w = (h_2 - h_1) + \frac{C_2^2 - C_1^2}{2} = h_{20} - h_{10} \tag{6.27}$$

Flow through Diffuser and Volute Casing (Processes 2–3 and 3–4) There is no energy transfer, but kinetic energy is converted into static enthalpy in both processes.

Diffuser: $$h_{20} = h_2 + \frac{C_2^2}{2} = h_3 + \frac{C_3^2}{2} = h_{30} \tag{6.28}$$

Volute casing: $$h_{30} = h_3 + \frac{C_3^2}{2} = h_4 + \frac{C_4^2}{2} = h_{40} \tag{6.29}$$

Combining the two equations for the process from 2 to 4

$$h_{20} = h_2 + \frac{C_2^2}{2} = h_4 + \frac{C_4^2}{2} = h_{40} \tag{6.30}$$

and further combining with Eq. (6.27), we obtain

$$w = h_{20} - h_{10} = h_{40} - h_{10} = \Delta h_0 = (h_4 - h_1) + \frac{C_4^2 - C_1^2}{2} \tag{6.31}$$

which represents the overall energy balance for a centrifugal compressor stage.

6.12.2 Application of Momentum Equation to a Centrifugal Stage

Figure 6.13 represents the impeller of a stage of a centrifugal compressor and Fig. 6.14 shows the velocity triangles at the impeller inlet and outlet. In Fig. 6.13, C is the fluid velocity, C_{rel} is the relative velocity of the fluid with respect to the impeller and u is the impeller tip *speed.* C_u and C_r are the tangential and radial components of the fluid velocity as shown in Fig. 6.14.

Now the torque I is given by the rate of change of angular momentum of the fluid. Thus, per unit mass of the fluid, we have

$$I = C_{u_2} r_2 - C_{u_1} r_1 \tag{6.32}$$

where r_2 and r_1 are the radii of the impeller at the outlet and inlet respectively.

The work done by the impeller rotating at an angular velocity ω is then given by

$$\begin{aligned} w &= I\omega = \omega (C_{u_2} r_2 - C_{u_1} r_1) \\ &= C_{u_2} u_2 - C_{u_1} u_1 \end{aligned} \tag{6.33}$$

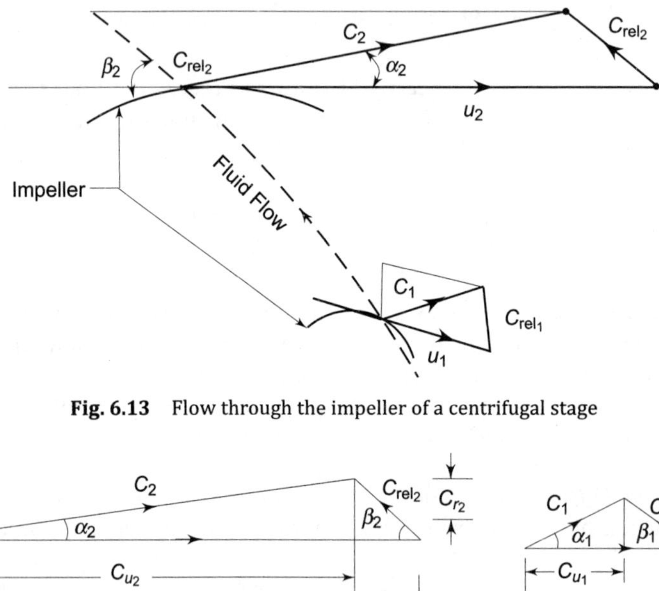

Fig. 6.13 Flow through the impeller of a centrifugal stage

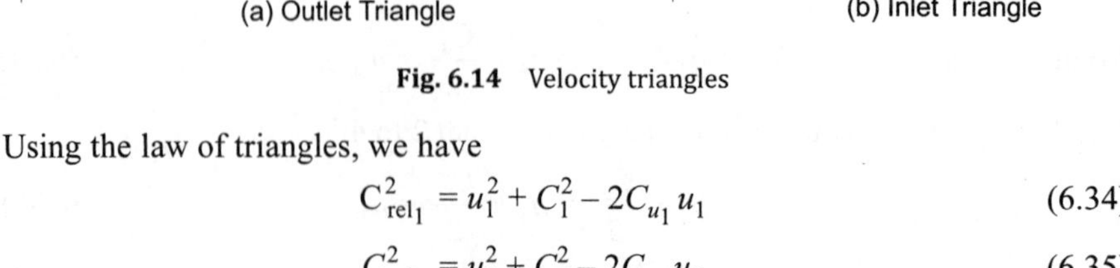

Fig. 6.14 Velocity triangles

Using the law of triangles, we have

$$C_{rel_1}^2 = u_1^2 + C_1^2 - 2C_{u_1} u_1 \tag{6.34}$$

$$C_{rel_2}^2 = u_2^2 + C_2^2 - 2C_{u_2} u_2 \tag{6.35}$$

Substituting for $C_{u_2}u_2$ and $C_{u_1}u_1$ from these equations in Eq. (6.33), we obtain

$$w = C_{u_2}u_2 - C_{u_1}u_1 = \frac{u_2^2 - u_1^2}{2} + \frac{C_{\text{rel}_1}^2 - C_{\text{rel}_2}^2}{2} + \frac{C_2^2 - C_1^2}{2} \tag{6.36}$$

Equations (6.33) and (6.36) are the two forms of the *fundamental equation of turbomachinery*.

6.12.3 Combined Energy and Momentum Equations

Combining Eqs (6.27) and (6.36) we obtain

$$\begin{aligned} w = h_{40} - h_{10} = h_{20} - h_{10} &= (h_2 - h_1) + \frac{C_2^2 - C_1^2}{2} \\ &= \frac{u_2^2 - u_1^2}{2} + \frac{C_{\text{rel}_1}^2 - C_{\text{rel}_2}^2}{2} + \frac{C_2^2 - C_1^2}{2} \end{aligned} \tag{6.37}$$

It is seen that the first two terms together represent the *static head*, corresponding to the pressure rise in the impeller, viz., the static enthalpy increase given by

$$h_2 - h_1 = w - \frac{C_2^2 - C_1^2}{2} = \frac{u_2^2 - u_1^2}{2} + \frac{C_{\text{rel}_1}^2 - C_{\text{rel}_2}^2}{2} \tag{6.38}$$

and the term $\left(\frac{C_2^2 - C_1^2}{2}\right)$ represents the *dynamic head* denoting the kinetic energy increase in the impeller. The two parts of the static head are respectively the *centrifugal head*, $\left(\frac{u_2^2 - u_1^2}{2}\right)$ developed due to the application of the centrifugal force, and *relative head*, $\left\{\frac{C_{\text{rel}_1}^2 - C_{\text{rel}_2}^2}{2}\right\}$ developed due to the change in cross-sectioned area of flow. The relative head is generally very small and is normally made zero by making the flow areas at inlet and outlet the same by reducing the width of the shrouds from b_1 at inlet to b_2 at outlet such that

$$\pi D_1 b_1 = \pi D_2 b_2$$

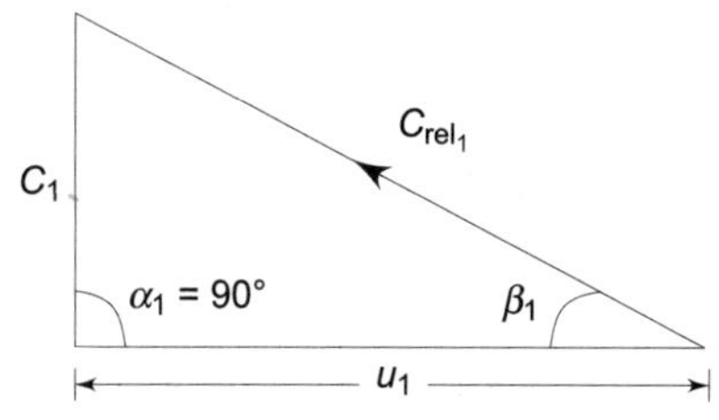

Fig. 6.15 Inlet velocity triangle for no prewhirl

Equation (6.33) shows that the head developed in the impeller is reduced because of the tangential *prewhirl* component of the velocity C_{u1} in the direction of rotation of the impeller at the inlet. With the selection of an appropriate inlet blade angle β_1 at the design point, the velocity C_{u1} can be reduced to zero as shown in Fig. 6.15 so that the head developed with no prewhirl becomes

$$w = \Delta h_0 = h_{40} - h_{10} = C_{u_2} u_2 \tag{6.39}$$

In such a case $\alpha_1 = 90°$, and the absolute velocity C_1 becomes radial.

6.12.4 Polytropic Efficiency

Figure 6.16 shows the performance characteristics at various percentages of the rated speed. It also shows a family of curves for polytropic efficiencies superimposed on the curves for various speeds.

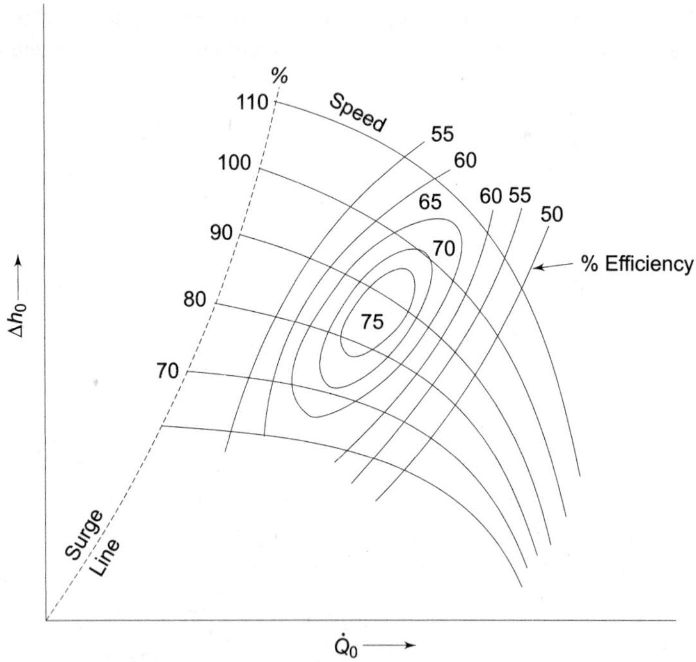

Fig. 6.16 Efficiency curves of a centrifugal compressor at different percentages of rated speed and capacity

An overall polytropic efficiency η_P is defined as the ratio of the isentropic work $\Delta h_{0\text{is}}$ to the actual work Δh_0

$$\eta_P = \frac{\Delta h_{0\text{is}}}{\Delta h_0} \tag{6.40}$$

From Fig. 6.17, the actual work done is

$$\Delta h_0 = h_{40} - h_{10} = (h_4 + C_4^2/2) - h_{10} \tag{6.41}$$

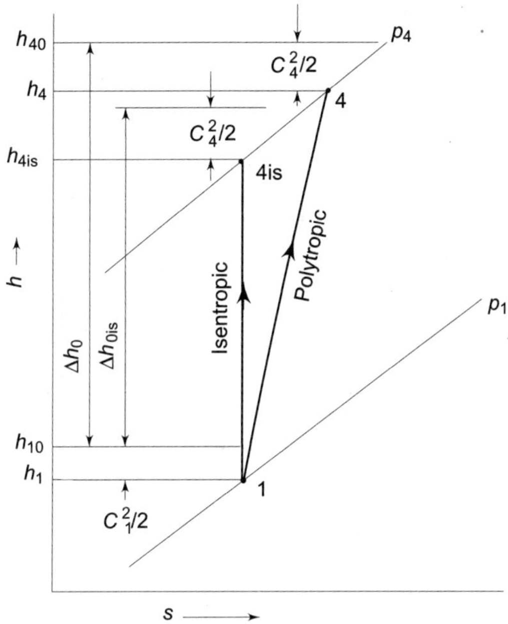

Fig. 6.17 Comparison of isentropic and polytropic compression processes on a Mollier diagram

Also, $4is$ is the point after isentorpic compression to the same pressure p_4. Considering same kinetic energies at $4is$ and 4, we have for work after isentropic compression

$$\Delta h_{0\text{is}} = (h_{4\text{is}} + C_4^2/2) - h_{10} \tag{6.42}$$

Enthalpy $h_{4\text{is}}$ can be found from the isentropic relationslip.

6.12.5 Pressure Ratio Developed in a Centrifugal Stage

Combining steady-flow energy Eq. (6.31) and momentum Eq. (6.33) we have

$$w = \Delta h_0 = C_{u_2} u_2 - C_{u_1} u_1 = (h_4 - h_1) + \frac{C_4^2 - C_1^2}{2} \tag{6.43}$$

whence

$$h_4 - h_1 = w - \frac{C_4^2 - C_1^2}{2}$$

$$= (C_{u_2} u_2 - C_{u_1} u_1) - \frac{C_4^2 - C_1^2}{2} \tag{6.44a}$$

For the isentropic process, since $\mathrm{d}h = v\mathrm{d}p$, we can write for the enthalpy change in terms of the pressure ratio p_4/p_1 developed in the compressor stage

$$h_4 - h_1 = \int_1^4 v\mathrm{d}p = \frac{\gamma}{\gamma - 1}\, p_1 v_1 \left[\left(p_4/p_1\right)^{\frac{\gamma-1}{\gamma}} - 1 \right] \tag{6.44b}$$

Combining Eqs (6.44a) and (6.44b), we have

$$h_4 - h_1 = \frac{\gamma}{\gamma - 1} p_1 v_1 \left[\left(\frac{p_4}{p_1} \right)^{\frac{\gamma-1}{\gamma}} - 1 \right]$$

$$= (C_{u_2} u_2 - C_{u_1} u_1) - \frac{C_4^2 - C_1^2}{2} \tag{6.45}$$

whence we get for the pressure ratio developed

$$r = \frac{p_4}{p_1} = \left\{ \frac{\gamma - 1}{\gamma} \frac{1}{p_1 v_1} \left[(C_{u_2} u_2 - C_{u_1} u_1) - \frac{C_4^2 - C_1^2}{2} \right] + 1 \right\}^{\frac{\gamma}{\gamma - 1}} \tag{6.46}$$

The pressure ratios developed separately in the impeller and diffuser (including volute casing) respectively are similarly given by,

$$h_2 - h_1 = \frac{\gamma}{\gamma - 1} p_1 v_1 \left[\left(\frac{p_2}{p_1} \right)^{\frac{\gamma-1}{\gamma}} - 1 \right]$$

$$= \frac{u_2^2 - u_1^2}{2} + \frac{C_{\text{rel}_1}^2 - C_{\text{rel}_2}^2}{2} \tag{6.47}$$

$$h_4 - h_2 = \frac{\gamma}{\gamma - 1} p_2 v_2 \left[\left(\frac{p_4}{p_2} \right)^{\frac{\gamma-1}{\gamma}} - 1 \right] = \frac{C_2^2 - C_4^2}{2} \tag{6.48}$$

6.12.6 Compressor with Radial Blades

A radial blade compressor is simple in construction. In this, the outlet blade angle β_2 is 90°. It can be seen from the outlet velocity triangle in Fig. 6.18 that $C_{u_2} = u_2$ for a radial blade compressor. Then, from Eqs. (6.43) and (6.44a),

$$w = \Delta h_0 = u_2^2 - C_{u_1} u_1 = (h_4 - h_1) + \frac{C_4^2 - C_1^2}{2} \tag{6.49}$$

Under assumptions of no prewhirl and negligible kinetic energies at the inlet and outlet, we obtain

$$w = \Delta h_0 = h_4 - h_1 = \frac{\gamma}{\gamma - 1} p_1 v_1 \left[\left(\frac{p_4}{p_1} \right)^{\frac{\gamma-1}{\gamma}} - 1 \right] = C_{u_2} u_2 = u_2^2 \tag{6.50}$$

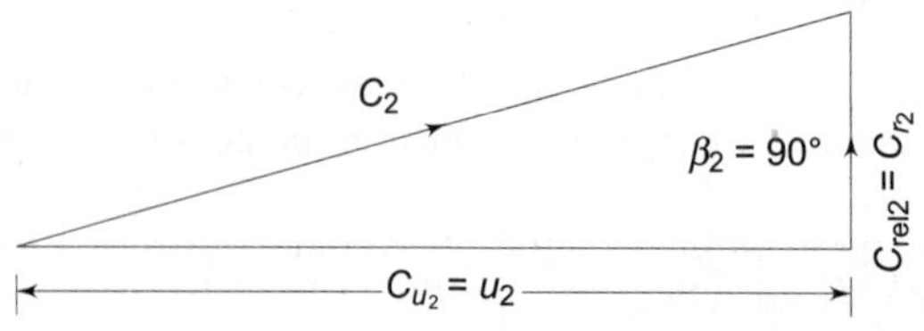

Fig. 6.18 Outlet velocity triangle for radial blade compressor

and for the pressure ratio developed

$$r = \frac{p_4}{p_1} = \left[\frac{\gamma - 1}{\gamma}\frac{1}{p_1 v_1}u_2^2 + 1\right]^{\frac{\gamma}{\gamma - 1}} \tag{6.51}$$

6.12.7 Class of Service of Centrifugal Compressors

In refrigeration, a compressor has to develop the required pressure ratio corresponding to the condenser and evaporator pressures. Whether a compressor would develop the required ratio in a single stage or would require more stages can be verified from Eq. (6.46). However, to analyse the results, a few approximate assumptions can be made.

(i) No prewhirl, i.e., radial inlet, so that

$$C_{u_1} = 0$$

(ii) No significant kinetic energy at inlet and outlet, i.e.,

$$C_1 = 0 \text{ and } C_4 = 0$$

(iii) Relative velocity extremely small compared to the tip speed of the impeller, i.e.,

$$C_{\text{rel}_2} << u_2$$

so that

$$C_2 = C_{u_2} = u_2$$

Thus we may write for the pressure ratio from Eq. (6.46)

$$r = \frac{p_4}{p_1} = \left[\frac{\gamma - 1}{\gamma}\frac{1}{p_1 v_1}u_2^2 + 1\right]^{\frac{\gamma}{\gamma - 1}} \tag{6.52}$$

Comparing Eqs (6.51) and (6.52), it is seen that the governing relations for a compressor with radial blades and for backward-curved blades with the assumption that $C_{u_2} \approx u_2$, are the same.

Equation (6.52) shows that the pressure ratio developed in a centrifugal stage depends on:

(i) Density $\rho_1 = \frac{1}{v_1}$ and pressure p_1 or the acoustic velocity

$$a = \sqrt{\gamma R T_1} = \sqrt{\gamma\, p_1 v_1}$$

(ii) Adiabatic exponent γ

(iii) Tip speed u_2.

Thus the pressure ratio increases with the density of the suction vapour. It also increases as γ decreases. Therefore, high molecular weight substances with a simultaneously low value of γ such as R 11 and R 113 are more suitable for centrifugal compressors. However, now, both R 11 and R 113 have been phased out.

Also, at lower suction pressure p_1, the pressure ratio required to be developed is more, although the actual pressure rise in a single stage will be still less. The effect of both the suction pressure p_1 and density ρ_1 is accounted in the suction temperature T_1 or acoustic velocity a. The acoustic velocity of most fluorocarbons is of the order of 100 m/s only. Some refrigerants which are normally not used in centrifugal machines become quite suitable for them at low evaporator temperatures because of a lower value of acoustic velocity and also a larger specific volume of vapour at the suction condition. Their use may require multistaging of the compressor.

With the further assumption of $u_1 \approx 0$, it is seen from Eq. (6.47) that

$$h_2 - h_1 = \frac{u_2^2}{2}$$

and from Eq. (6.48)

$$h_4 - h_2 = \frac{u_2^2}{2}$$

Thus, under the assumptions made, half of the head is developed in the impeller and half in the diffuser.

Also, the net work done is, by addition

$$w = \Delta h_0 = u_2^2$$

Or, directly from Eq. (6.45), with no prewhirl ($C_{u1} = 0$), negligible C_4 and C_1 and $\beta_2 = 90°$ ($C_{u_2} = u_2$), we get the same expression.

6.13 PERFORMANCE CHARACTERISTICS OF A CENTRIFUGAL COMPRESSOR[5]

The principal performance curve of a centrifugal machine is the *head-flow characteristic*. We may write for the tangential velocity at the exit

$$C_{u_2} = u_2 - C_{r2} \cot \beta_2 \tag{6.53}$$

and substituting this expression in Eq. (6.39), we have for the head developed

$$\begin{aligned} w &= \Delta h_0 = u_2 (u_2 - C_{r_2} \cot \beta_2) \\ &= u_2^2 - u_2 C_{r_2} \cot \beta_2 \\ &= (\omega r_2)^2 - (\omega r_2) C_{r_2} \cot \beta_2 \end{aligned} \tag{6.54}$$

Thus we find that for a given compressor, with r_2 and β_2 fixed, and running at a certain speed ω, the head developed is a straight line function of the radial velocity C_{r_2}. The flow rate, in turn, is proportional to C_{r_2}. The limiting head is, of course, u_2^2 which is developed at $C_{r_2} = 0$, i.e., at zero flow rate. This occurs when the impeller is simply rotating in a mass of the fluid with the delivery valve closed.

It is seen that the nature of the characteristic depends on the *outlet blade angle* β_2 as follows:

For backward-curved blades, $\beta_2 < 90°$, head decreases with C_{r_2} (flow) and hence with $\dot{Q}_v$.

For radial blades, $\beta_2 = 90°$, head $= u_2^2 =$ const.

For forward-curved blades, $\beta_2 > 90°$, head increases with C_{r_2} (flow) and hence with $\dot{Q}_v$.

From the point of view of optimal design, an outlet blade angle of 32° is normally preferred.[2] A simple design will, however, have radial blades.

Figure 6.19 shows the theoretical head-flow characteristic for the three cases of angle β_2. For the case of backward-curved blades, it is a *drooping characteristic*. The actual characteristic can, however, be obtained by considering the following losses as shown in Fig. 6.19.

(i) Leakage loss L_1 proportional to the head.
(ii) Friction loss L_2 proportional to $C_{rel}^2/2$ and hence $\dot{Q}_v^2$.
(iii) Entrance loss L_3 due to turning of the fluid to enter the impeller, being zero at the *design point*, which also corresponds to maximum efficiency.

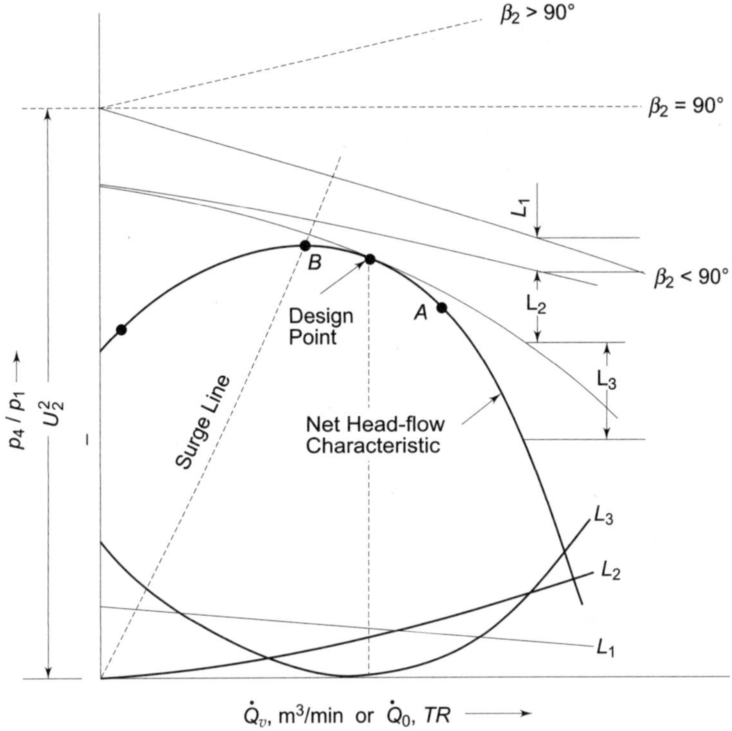

Fig. 6.19 Performance characteristic and losses of a centrifugal compressor

Expressing the radial velocity and head developed in dimensionless form we have

Flow coefficient, $$\phi = \frac{C_r}{u}$$

Head coefficient, $$\mu = \frac{\Delta h_0}{u_2^2} = \frac{C_{u_2}}{u_2}$$

and from Eq. (6.54)

$$\mu = 1 - \phi_2 \cot \beta_2 \tag{6.55}$$

This relation also shows that, for radial blades and with no pre-whirl, μ is equal to unity, i.e., the head developed is equal to u_2^2.

6.13.1 Surging

Consider *A* in Fig. 6.19 as the point of operation at full load. When the refrigeration load decreases, the point of operation shifts to the left until point *B* of maximum head is reached. If the load continues to decrease to the left of *B*, say to *C*, the pressure ratio developed by the compressor becomes less than the ratio required between the condenser and evaporator pressure, viz.,

$$\frac{p_4}{p_1} < \frac{p_k}{p_0}$$

Hence some gas flows back from the condenser to the evaporator, thus increasing the evaporator pressure and decreasing p_k/p_0. The point of operation suddenly shifts to A. As the refrigeration load is still less, the cycle will repeat itself. This phenomenon of reversal of flow in centrifugal compressors is called *surging*. It occurs when the load decreases to below 35 per cent of the rated capacity and causes severe stress conditions in the compressor as a result of *hunting*. Figures 6.16 and 6.19 show the *surge line* drawn through the points of maximum head at different speeds.

6.13.2 Capacity Control of Centrifugal Compressors

Centrifugal compressors require high tip speeds u_2 to develop the necessary pressure ratio. The high tip speed is achieved by employing either a large diameter impeller or high rpm or both. Because of large u_2, the velocities in general including the flow velocity C_r are high. Also, there must be a reasonable width of the shrouds to minimize friction and achieve high efficiency. Thus, because of the sufficiently large flow area (diameter D and width of shrouds b) required and large flow velocity, the satisfactory volume that can be handled by a centrifugal compressor is about 30–60 cubic metres per minute. A single centrifugal compressor, therefore, can be designed for a minimum capacity approximately of the order of 250 TR with R 11 and 150 TR with R 113 for the purpose of air conditioning. These were the two common refrigerants which were being used with centrifugal compressors in the past.

The centrifugal compressors are, therefore, used in large capacity installations in which the load varies through a wide range and hence, capacity control in them, although simple, is very important.

One of the common methods is by varying the compressor speed through a speed-reduction gear. The decrease in speed results in an operation on a lower head-flow characteristic giving a lower volume flow rate corresponding to the same pressure ratio.

The use of variable inlet whirl vanes is another method of capacity control frequently employed with a constant-speed drive. The capacity is varied by changing the angle at which the gas enters the impeller. The gas then enters with prerotation and this results in a decrease in flow.

Both methods are efficient and the work done remains proportional to the volume handled.

6.13.3 Relations for a Perfect Gas

For a centrifugal compressor with no prewhirl and negligible kinetic energies at inlet and outlet, we have

$$\frac{p_4}{p_1} = \left[\frac{\gamma-1}{\gamma}\frac{1}{p_1 v_1}u_2^2 + 1\right]^{\frac{\gamma}{\gamma-1}}$$

This equation is valid for both radial blade compressors and backward-curved blade compressors, but with extremely small relative velocities so that $C_{u_2} \approx u_2$.

Now for a perfect gas, $p_1 v_1 = RT_1$ and $\dfrac{\gamma R}{\gamma - 1} = C_p$, so that for isentropic compression, we have

$$\frac{p_4}{p_1} = \left[\frac{u_2^2}{C_p T_1} + 1\right]^{\frac{\gamma}{\gamma-1}} \tag{6.56}$$

and

$$w_{is} = \Delta h_{0is} = C_p T_1 \left[\left(\frac{p_4}{p_1}\right)^{\frac{\gamma}{\gamma-1}} - 1\right] \tag{6.57}$$

For the case of actual compression

$$w = \Delta h_0 = u_2^2$$

Hence, we have for the adiabatic efficiency, also referred to as polytropic efficiency,

$$\eta_P = \frac{\Delta h_{0_{is}}}{\Delta h_0} = \frac{C_p T_1 \left[\left(\dfrac{p_4}{p_1}\right)^{\frac{\gamma}{\gamma-1}} - 1\right]}{u_2^2}$$

whence the required tip speed of the impeller is obtained as

$$u_2 = \sqrt{\frac{C_p T_1 \left[\left(\dfrac{p_4}{p_1}\right)^{\frac{(\gamma-1)}{\gamma}} - 1\right]}{\eta_P}}$$

Example 6.4

(a) *A centrifugal compressor with an impeller diameter of 45 cm is running at 4200 rpm. Refrigerant 12 is used and the suction condition is dry saturated at 5°C and 3.6255 bar. Determine the maximum pressure ratio developed by the compressor. Is it adequate if condensing temperature is 40°C?*

(b) *If the outlet blade angle β_2 is 32° and the pressure ratio is 1.5, what will be the flow velocity? Assume no prewhirl and negligible kinetic energies at inlet and outlet.*

Solution

(a) Maximum pressure ratio is developed when flow is nil, i.e., at zero flow-coefficient. Then

$$\phi_2 = \frac{C_{r_2}}{u_2} = 0 \text{ or } C_{r_2} = 0 \text{ and } C_{u_2} = u_2$$

The tip speed of the impeller is

$$u_2 = \frac{\pi D_2 N}{60} = \frac{\pi (0.45)(4200)}{60} = 98.96 \text{ m/s}$$

Then from Eq. (6.52), assuming no prewhirl and zero kinetic energy at inlet and outlet, we have for the maximum pressure ratio

$$\frac{p_4}{p_1} = \left[\frac{\gamma - 1}{\gamma} \frac{1}{p_1 v_1} u_2^2 + 1\right]^{\frac{\gamma}{\gamma - 1}}$$

$$= \left[\frac{1.15 - 1}{1.15} \frac{1}{3.6255 \times 10^5 \times 0.0475} (98.96)^2 + 1\right]^{\frac{1.15}{1.15 - 1}} = 1.56$$

At 0°C, $p_0 = 3.626$ bar
At 40°C, $p_k = 9.607$ bar

Required pressure ratio $\frac{p_k}{p_0} = \frac{9.607}{3.266} = 2.65$

Note *It is obvious that the pressure ratio developed is grossly inadequate. So what should be done? Increase the tip speed (diameter or rpm).*

(b) Again from Eq. (6.46)

$$\frac{p_4}{p_1} = \left[\frac{\gamma - 1}{\gamma} \frac{1}{p_1 v_1} C_{u_2} u_2 + 1\right]^{\frac{\gamma}{\gamma - 1}}$$

$$1.5 = \left[\frac{1.15 - 1}{1.15} \frac{1}{3.6255 \times 10^5 \times 0.0475} C_{u_2} (98.96) + 1\right]^{\frac{1.15}{1.15 - 1}}$$

$$C_{u_2} = 72.5 \text{ m/s}$$

Now $C_{u_2} = u_2 - C_{r_2} \cot \beta_2$

Hence $C_{r_2} = \frac{u_2 - C_{u_2}}{\cot \beta_2} = \frac{98.96 - 72.5}{\cot 32°} = 42.5$ m/s

Example 6.5 *R 11 Compressor*

(a) R 11 saturated vapour at 7°C is to be compressed to a saturated discharge temperature of 36°C. The adiabatic efficiency may be taken as 77.3 per cent.

The outlet blade angle β_2 is 32° and the flow coefficient ϕ_2, for maximum efficiency, is given by the expression

$$\phi_2 = (1 - \sin \beta_2) \tan \beta_2$$

If the compression is to be achieved in a single stage, what will be the diameter of the impeller running at 3600 rpm?

(b) If the blades are radial, what will be the required diameter of the impeller? The outlet and inlet velocities may be considered as negligible. The ratio of the specific heats for R 11 is 1.13.

(c) Determine the ideal COP, and capacity of one compressor if volume flow rate is 100 m^3/min.

Solution From tables of properties for R 11, the pressures are

$$p_1 = p_0 = 0.5373 \text{ bar}, \quad p_k = 1.529 \text{ bar}, \quad \frac{p_k}{p_0} = 2.85$$

The specific volume and enthalpy at suction are

$$v_1 = 0.308 \text{ m}^3/\text{kg}$$

$$h_1 = 391.4 \text{ kJ/kg}$$

From the *p-h* diagram for R 11, the enthalpy after isentropic compression is

$$h_4 = 410.9 \text{ kJ/kg}$$

Isentropic work

$$w_{is} = h_4 - h_1 = 411.9 - 392.5 = 19.4 \text{ kJ/kg}$$

Also, approximately, based on γ value

$$w_{is} = \Delta h_{0is} = \frac{\gamma}{\gamma - 1} p_1 v_1 \left[\left(\frac{p_4}{p_1} \right)^{\frac{\gamma-1}{\gamma}} - 1 \right]$$

$$= \frac{1.13}{0.13} (0.5373 \times 10^5) \times 0.308 \left[(2.85)^{0.13/1.13} - 1 \right] \frac{1}{10^3} = 18.4 \text{ kJ/kg}$$

For actual compression

$$w = \Delta h_0 = \frac{\Delta h_{0_{isen}}}{\eta_P} = \frac{18.4}{0.773} = 24.1 \text{ kJ/kg}$$

(a) *For backward curved blades*

$$\phi_2 = (1 - \sin \beta_2) \tan \beta_2 = (1 - \sin 32°) \tan 32° = 0.294$$

$$\mu = (1 - \phi_2 \cot \beta_2) = \frac{\Delta h_0}{u_2^2} = (1 - 0.294 \times \cot 32°) = 0.528$$

Hence

$$u_2^2 = \frac{\Delta h_0}{\mu} = \frac{24.1 \times 10^3}{0.528} = 4.564 \times 10^4 \text{ m}^2/\text{s}^2$$

$$u_2 = 213.2 \text{ m/s}$$

The required diameter of the impeller is

$$D_2 = \frac{60 u_2}{\pi N} = \frac{(60)(213.2)}{(3600)} = 1.131 \text{ m}$$

(b) *For radial blades*

$$\beta_2 = 90°, C_{u2} = u_2, \Delta h_0 = C_{u2} u_2 = u_2^2, \mu = 1$$

Hence

$$u_2 = \sqrt{\Delta h_0} = \sqrt{24.1 \times 10^3} = 155.2 \text{ m/s}$$

$$D_2 = \frac{60 \, (155.2)}{3600 \, \pi} = 0.824 \text{ m}$$

(c) *Enthalpy of liquid from condenser at 36°C*

$$h_f = 231.4 \text{ kJ/kg}$$

Refrigerating effect

$$q_0 = 391.4 - 234.9 = 156.5 \text{ kJ/kg}$$

Minimum mass flow rate

$$\dot{m} = \frac{100/60}{0.308} = 5.41 \text{ kg/s}$$

$$\text{COP}_c = \frac{q_0}{w_{is}} = \frac{156.5}{19.4} = 7.9$$

Minimum capacity

$$\dot{Q}_0 = \dot{m}\, q_0 = 5.41\,(156.5) = 871.8 \text{ kJ/s} = 248 \text{ TR}$$

6.14 ALTERNATIVES TO R 11 (CFC 11)

Table 6.1 below shows that R 123 (HCFC 123), presently being used in place of R 11, having normal boiling point closest to the N.B.P. of R 11, is the best substitute for CFC 11 in centrifugal chillers. It is non-flammable. However, it has not been cleared as far as toxicity is concerned. Moreover, it is an HCFC. Hence, the search for better alternative/s continues.

Table 6.1 N.B.Ps of alternatives to CFC11

Refrigerant	*N.B.P. °C*
CFC 11	23.71°C
HCFC 123	27.82°C
HFC 245fa	14.9

A future alternative is HFC 245fa. It is a comparatively lower boiling, higher pressure refrigerant. It can be considered for use after HCFC 123 is also phased out. R 134a is another possibility.

6.14.1 Comparison of Performance Parameters of CFC R 11 and Alternatives R 123, R 245fa and R 134a in Centrifugal Chillers

Consider centrifugal chillers operating on air conditioning simple saturation cycle with 40°C condenser and 6°C evaporator/chiller temperatures. The performance parameters for historically most commonly used refrigerant in centrifugal plants, viz., R 11, and the refrigerant being used presently R 123 with those of the other alternatives R 134a and R 245fa are given in Table 6.2.

Table 6.2 Comparison of performance parameters of centrifugal compressors for air conditioning simple saturation cycle $t_k = 40°C$, $t_0 = 6°C$

Refrigerant	*N.B.P*	v_1	q_0	$w_s = \Delta h_{0_{isen}}$	COP	$\dot{V}_s^* \times 10^3$	$\dfrac{\dot{Q}_{v_r}}{(\dot{Q}_{v_r})_{R11}}$	u_2	$\dot{Q}_0$
	°C	m³/kg	kJ/kg	kJ/kg		m³/s(TR)		m/s	TR
R 11	23.7	0.33186	156.5	22.0	7.9	7.457	1	213.2	248
R 123	27.8	0.3624	143.9	20.8	7.5	8.462	0.88	218.0	197
R 245fa	14.9	0.2394	156.2	22.1	7.1	5.39	1.38	232.6	309
R 134a	–26.1	0.05644	145.7	21.44	6.8	1.362	5.5	216.5	1224

Calculations have been done for COP assuming adiabatic efficiency of the compressor as $\eta_C = 0.773$. Thus

$$\text{COP} = \frac{q_0}{w_a} = \frac{q_0}{w_s/0.773} = \frac{q_0}{\Delta h_{0_{\text{iscn}}}/0.773}$$

Similarly, calculations have been done for tip speed u_2 by assuming $\eta_C = 0.773$, and head coefficient $\mu = 0.528$ or 0.59 as used by Atwood and Murphy[2]. Thus

$$u_2 = \sqrt{\frac{\Delta h_0}{\mu}} = \sqrt{\frac{\Delta h_{0_{\text{iscn}}}/0.85}{\mu}}$$

Suction vapour volume calculations have been done for one TR. Table 6.2 also gives ratios of volumetric refrigerating capacities $\dot{Q}_{v_r}$ with respect to the same for R 11, viz., $\left(\dot{Q}_{v_r}\right)_{\text{R11}}$. We see that

$$\frac{\dot{Q}_{v_r}}{\left(\dot{Q}_{v_r}\right)_{\text{R11}}} = \frac{\left(\dot{V}_s\right)_{\text{R11}}}{\dot{V}_s}$$

Values in Table 6.2 are taken from Examples 6.7, 6.8 and 6.9 that follow.

It can be noticed from Table 6.2 that as far as COP is concerned R 134a has the lowest value. Carnot COP is 8.2. These is not much difference between R 11 and R 123.

R 11 centrifugal compressors can be substituted by R 134a compressors with about 3.4% increase in tip speed. However, the compressors would give 5.5 times the refrigerating capacity of R 11. Thus, a 248 TR R 11 centrifugal compressor would give 1224 TR with R 134a otherwise suction vapour volume will be so small that the width of impeller shrouds will be less than 1 mm.

Example 6.6 *R 22 Centrifugal Compressor*

Find the tip speed, diameter and width of shrouds at impeller outlet for a 600 TR R 22 centrifugal compressor for air conditioning simple saturation cycle. Condenser and evaporator temperatures are 40°C and 5°C. Design speed is 2950 rpm. Assume adiabatic efficiency of compressor as 0.85, and head coefficient as 0.59. Check mach number at outlet.

Solution For the cycle (Fig. 3.7), for isentropic compression

$$h_1 = 407.1 \text{ kJ/kg}, \qquad v_1 = 0.040356 \text{ m}^3\text{/kg}$$
$$h_{2s} = 430.4 \text{ kJ/kg}, \qquad t_{2s} = 57.5°\text{C}$$
$$h_3 = 249.7 \text{ kJ/kg} = h_4$$

From calculations in Table 6.2

Refrigerating effect, $q_0 = h_1 - h_4 = 157.5$ kJ/kg

Isentropic work, $w_s = h_{2s} - h_1 = 23.3 \text{ kJ/kg} = (\Delta h_0)_{\text{isen}}$

Actual work, $$w_a = \frac{w_s}{\eta_a} = \frac{23.3}{0.85} = 27.4 \text{ kJ/kg} = \Delta h_0$$

Actual discharge condition

$$h_2 = h_1 + w_a = 434.5 \text{ kJ/kg}$$
$$t_2 = 73°\text{C}, v_2 = 0.0168 \text{ m}^3\text{/kg}$$

Tip speed of impeller

$$u_2 = \sqrt{\Delta h_0 / \mu} = \sqrt{\frac{27.4 \times 10^3}{0.59}} = 215.6 \text{ m/s}$$

Taking outlet blade angle $\beta_2 = 32°$, we have for flow coefficient

$$\phi_2 = (1 - \sin \beta_2) \tan \beta_2 = 0.294$$

Radial flow velocity

$$C_{r2} = \phi_2 u_2 = 0.294 \, (215.6) = 63.39 \text{ m/s}$$

Impeller diameter

$$D_2 = \frac{60 \, u_2}{\pi N} = \frac{60 \, (215.6)}{\pi \, (2950)} = 1.4 \text{ m}$$

Width of shrouds at outlet

$$b_2 = \frac{\dot{m} \, v_2}{\pi D_2 C_{r_2}} = \frac{\dot{Q}_0}{q_0} \frac{v_2}{\pi D_2 C_{r_2}} = \frac{600 \times 3.5167 \, (0.0168)}{(157.5) \pi \, (1.4) \, (63.39)}$$

$$= 0.81 \times 10^{-3} \text{ m } (0.81 \text{ mm})$$

From outlet velocity triangle, absolute velocity $C_2 = 141.8$ m/s

Accoustic velocity and Mach number at outlet

$$a_2 = \sqrt{\gamma R T_2} = \sqrt{1.16 \, (96.13) \, (273 + 73)} = 184.7 \text{ m/s}$$

$$M_2 = \frac{C_2}{a_2} = \frac{141.8}{184.7} = 0.77$$

Note *Width of shrouds of 0.81 mm is extremely small. The efficiency of compressor will be very poor. It will be interesting to see what the dimensions will be with R 134a. R 134a compressor, though is already being used for much larger capacities.*

Example 6.7 ***R 123 Compressor***

(a) *R 123 saturated vapour at 6°C is compressed to saturated discharge temperature of 40°C.*
Outlet blade angle is 32°. Compressor efficiency is 0.773. Use expression for flow coefficient for maximum efficiency. Dertermine the tip speed and diameter of the impeller if it runs at 3600 rpm, and compression is achieved in a single stage.

(b) *Determine the COP of the system, suction vapour volume rate per ton of refrigeration, and volumetric refrigerating capacity relative to R 11.*

(c) *Determine the capcity if minimum volume flow rate has to be 100 m^3/min for best efficiency*

(d) *Compare R 123 with R 11.*

Solution From the table of properties of R 123

At $t_0 = 6°$C

$p_1 = p_0 = 0.04264$ MPa $\qquad v_1 = 0.34759$ m^3/kg

$h_1 = 385.05$ kJ/kg $\qquad s_1 = 1.6631$ kJ/kg

At $t_k = 40°$C

$p_k = 0.15447$ MPa

$(h_f)_{40°C} = 240.59$ kJ/kg $\quad$ $(s_f)_{40°C} = 1.1383$ kJ/kg K

$(h_g)_{40°C} = 405.54$ kJ/kg $\quad$ $(s_g)_{40°C} = 1.6651$ kJ/kg K

Required pressure ratio

$$\frac{p_k}{p_0} = \frac{0.15447}{0.04264} = 3.62$$

Since $(s_g)_{40°C} > s_1$, the state after compression is wet. For dryness fraction after compression, we have

$$s_1 = s_2 = s_f + x\,(s_g - s_f)$$

$$1.6631 = 1.1383 + x\,(1.6651 - 1.1383)$$

$$\Rightarrow \qquad x = 0.9934$$

Enthalpy after compression

$$h_2 = h_f + x\,(h_g - h_f)$$
$$= 240.59 + 0.9934\,(405.54 - 240.59) = 404.45 \text{ kJ/kg}$$

Isentropic work

$$w_{is} = (\Delta h_0)_{is} = h_2 - h_1 = 404.45 - 385.05 = 19.4 \text{ kJ/kg}$$

Actual work

$$w = \Delta h_0 = \frac{w_{is}}{\eta_p} = \frac{19.4}{0.773} = 25.1 \text{ kJ/kg}$$

For maximum efficiency

$$\phi_2 = 0.294 \qquad \mu = 0.528 = \frac{\Delta h_0}{u_2^2}$$

Tip speed

$$u_2^2 = \frac{25.1\times10^3}{0.528} = 47.54 \times 10^3$$

$$u_2 = 218 \text{ m/s}$$

$$D_2 = \frac{60u_2}{3600\pi} = \frac{60(218)}{3600\pi} = 1.157 \text{ m}$$

(b) Refrigerating effect

$$q_0 = 385.05 - 240.59 = 144.46 \text{ kJ/kg}$$

Ideal COP of system

$$\mathcal{E}_c = \frac{q_0}{w_{is}} = \frac{144.46}{19.4} = 7.5$$

Mass flow rate of refrigerant per ton

$$\dot{m} = \frac{\dot{Q}_0}{q_0} = \frac{3.5167}{144.46} = 0.02434 \text{ kg/s}$$

Suction vapour volume rate per ton

$$\dot{V}_s^* = \dot{m}v_1 = 0.02434\,(0.34759) = 8.462 \times 10^{-3} \text{ m}^3\text{/s}$$

Volumetric refrigerating capacity relative to R 11

$$\dot{Q}_{v_r} = \frac{(\dot{Q}_v)_{R11}}{(\dot{Q}_v)_{R123}} = \frac{7.457\times10^{-3}}{8.462\times10^{-3}} = 0.881$$

Suction vapour volume for R 11 is 7.457×10^{-3} m^3/s/TR.

(c) Suction vapour volume rate in given compressor $\dot{V}_s = \frac{100}{60} \text{m}^3\text{/s}$
Refrigerating capacity

$$\dot{Q}_0 = \frac{\dot{V}_s}{\dot{V}_s^*} = \frac{100/60}{8.462 \times 10^{-3}} = 197 \text{ TR}$$

(d) Comparison of R 123 with R 11
We note the following
(i) COPs of R 123 and R 11 are nearly the same.
(ii) Suction vapour volume is more for R 123. Hence, a smaller capacity centrifugal compressor design in possible, e.g., 197 TR for R 123 against 248 TR for R 11 for the same operating conditions, and nearly same dimensions.
(iii) The tip speeds required for both R 123 and R 11 are the same.

Thus R 123 is an ideal replacement for R 11. Though on HCFC, its characteristics are so favourable that its use might continue well beyond 2030.

Example 6.8 ***R 245 fa Compressor***

For the design and operating conditions of Example 6.7, dertermine the performance parameters for R 245fa. Compare the result with R 123.

Solution Form the table of properties of R 245fa

$$\begin{aligned} p_1 &= p_0 = 0.06995 \text{ MPa} & p_4 &= p_k = 0.25179 \text{ MPa} & \frac{p_k}{p_0} &= 3.6 \\ h_1 &= 409.44 \text{ kJ/kg} & s_1 &= 1.7505 \text{ kJ/kg K} \\ (h_f)_{40°\text{C}} &= 253.24 \text{ kJ/kg} & (s_f)_{40°\text{C}} &= 1.1813 \text{ kJ/kg K} \\ (h_g)_{40°\text{C}} &= 434.97 \text{ kJ/kg} & (s_g)_{40°\text{C}} &= 1.7616 \text{ kJ/kg K} \end{aligned}$$

$s_g > s_1$, hence state after compession is wet. Dryness fraction after cmpression

$$s_2 = s_f + x\,(s_g - h_f) = s_1$$

$$1.1813 + x\,(1.7616 - 1.1813) = 1.7505 \text{ kJ/kg K}$$

$$\Rightarrow \quad x = 0.981$$

Enthalpy after isentropic compression

$$\begin{aligned} h_2 &= h_f + x\,(h_g - h_f) \\ &= 253.24 + 0.981(434.97 - 253.24) \\ &= 431.52 \text{ kJ/kg} \end{aligned}$$

Isentropic work

$$w_{\text{is}} = (\Delta h_0)_{\text{is}} = h_2 - h_1 = 431.52 - 409.44 = 22.08 \text{ kJ/kg}$$

Actual work

$$w = \frac{w_{\text{is}}}{\eta_p} = \frac{22.08}{0.773} = 28.56 \text{ KJ/kg}$$

Tip speed

$$u_2 = \sqrt{\frac{28.56 \times 10^3}{0.528}} = 232.6 \text{ m/s}$$

Impeller diameter = $D_2 = \frac{60u_2}{3600\pi} = \frac{60(232.6)}{3600\pi} = 1.235 \text{ m}$

Refrigerating effect

$$q_0 = h_1 - (h_f)_{40°C} = 409.44 - 253.24 = 156.2 \text{ kJ/Kg}$$

$$\text{Ideal COP} = \frac{156.2}{22.08} = 7.1$$

Mass flow rate per TR

$$\dot{m} = \frac{3.5167}{156.2} = 0.0225 \text{ kg/s}$$

Suction vapour volume per TR

$$\dot{V}_s^* = \dot{m}\, v_1 = 0.0225\ (0.23939) = 5.39 \times 10^{-3} \text{ m}^3\text{/s}$$

Volumetric refrigerating capacity relative to R 11

$$\dot{Q}_{v_r} = \frac{7.457\times10^{-3}}{5.39\times10^{-3}} = 1.38$$

Refrigerating capacity for the given volume flow rate

$$\dot{Q}_0 = \frac{100/60}{5.39\times10^{-3}} = 309 \text{ TR}$$

Comparison of Results with R 123

(i) Pressure ratios for both R 245fa and R 123 are nearly the same.

(ii) As the critical temperature for R 245fa (154.1°C) is lower, its condensation temperature is closer to critical temperature. Hence, its COP is lower than R 123 COP. Critical temperature for R 123 is 183.7°C.

(iii) As R 245fa is a lower N. B. P, and higher pressure refrigerant, it gives higher refrigerating capacity for the same volume flow rate as compared to both R 123 and R 11.

Example 6.9 ***R 134a Compressor***

For the design and operating conditions of Example 6.7, determine the performance parameters for R 134a. Compare result with R 11.

Solution From the table of properties of R 134 a,

$p_0 = 0.36198$ MPa $\quad p_k = 1.0166$ MPa

$v_1 = 0.05644$ m^3/kg

$h_1 = 402.06$ kJ/kg $\quad s_1 = 1.724$ kJ/kg K

$(h_f)_{40°C} = 256.41$ kJ/kg $\quad (h_g)_{40°}\text{C} = 419.43$ kJ/kg K

$(C_b)_{40°C} = 1.145$ kJ/kgK $\quad (s_g)_{40°C} = 1.711$ kJ/kgK

Since $(s_g)_{40°C} < s_1$, state after compression is superheated. For discharge temperature after isentropic compression.

$$s_2 = (s_g)_{40°C} + C_p \ln \frac{T_2}{T_1} = s_1$$

$$1.711 + 1.145 \ln \frac{T_2}{(273+6)} = 1.724$$

$$\Rightarrow \quad T_2 = 316.5 \text{ K } (43.5°\text{C})$$

Entropy after isentropic compression

$$h_2 = (s_g)_{40°C} + C_p(T_2 - T_g)$$

$$= 419.43 + 1.145\ (3.5) = 423.5\ \text{kJ/kg}$$

Isentropic and actual work

$$w_{is} = 423.5 - 402.06 = 21.44\ \text{kJ/kg}$$

$$w_a = \frac{w_{is}}{\eta_p} = \frac{21.44}{0.773} = 27.7\ \text{kJ/kg}$$

Taking $\mu = 0.59$, we have for tip speed

$$u_2 = \sqrt{\frac{27.7 \times 10^{-3}}{0.59}} = 216.5\ \text{m/s}$$

Impeller diameter $$D_2 = \frac{60u_2}{3600\pi} = \frac{60(216.7)}{3600\pi} = 1.15\ \text{m}$$

Refrigerating effect

$$q_0 = 402.06 - 256.41 = 145.7\ \text{kJ/kg}$$

$$\text{Ideal COP} = \frac{q_0}{w_{is}} = \frac{145.7}{21.44} = 6.8$$

Mass flow rate per ton

$$\dot{m} = \frac{3.5167}{145.7} = 0.0241\ \text{kg/s}$$

Suction vapour volume per ton

$$\dot{V}_s^* = \dot{m}v_1 = 0.0241(0.05644) = 1.362 \times 10^{-3}\text{m}^3/3$$

Volumetric refrigerating capacity relative to R 11

$$\dot{Q}_{v_r} = \frac{7.457 \times 10^{-3}}{1.362 \times 10^{-3}} = 5.5$$

Refrigerating capacity for the given volume flow rate

$$\dot{Q}_0 = \frac{100/60}{1.362 \times 10^{-3}} = 1224\ \text{TR}$$

Comparison of results with R 11

(i) The COP of R 134a is the lowest.

(ii) Its relative volumetric refrigerating capacity is 5.5. That means, for nearly same dimension, R 134a centrifugal compressor will have 1224 TR capacity as against 248 TR of R 11. Thus, R 134a centrifugal compressors are feasible and practical in 500–1000 TR capacity.

A smaller capacity R 134a compressor is not possible. If it is intended that a smaller capacity is needed, width of shrouds will become very small. And the compressor will have poor efficiency.

6.15 COMPARISON OF PERFORMANCE OF RECIPROCATING AND CENTRIFUGAL COMPRESSORS

The advantages of the centrifugal compressor over the reciprocating compressor are high efficiency over a large range of load and a large volume of the suction vapour and hence a larger capacity for its size.

The centrifugal compressor has also many other advantageous features. The most important is the *flat head-capacity characteristic* as compared to that of a reciprocating compressor as shown in Fig. 6.20(a) for typical compressors at a constant condensing temperature of 38°C and constant rpm. It is seen that the variation in the evaporator temperature is only 2 to 7.5°C for a load variation of 100 to 240 TR for a centrifugal compressor, whereas it is – 11 to 6°C for the same load variation for a reciprocating compressor.

Another advantageous feature is the *non-overloading characteristic* as shown in Fig. 6.20(b). It is seen that for a centrifugal machine, there is a decrease in the power requirement with an increase in the condensing temperature. This is due to the fact that the flow rate (refrigerating capacity) decreases as the head required increases (Fig. 6.19), while the power consumption represents the product of the two quantities. Thus there is no overloading of the motor with increasing condenser temperature. This feature is accompanied by a rapid fall in the flow rate and hence capacity as shown in Fig. 6.20(c). The power requirement decreases although the horsepower per ton increases. For reciprocating compressors, there is a small increase in the power requirement with an increase in the condensing temperature with consequent overloading of the motor. This is accompanied with a very small decrease in the refrigerating capacity. This latter aspect of a reciprocating compressor is, however, an advantage so that the capacity of a reciprocating machine is not affected much by an increase in the condensing temperature followed by adverse ambient conditions.

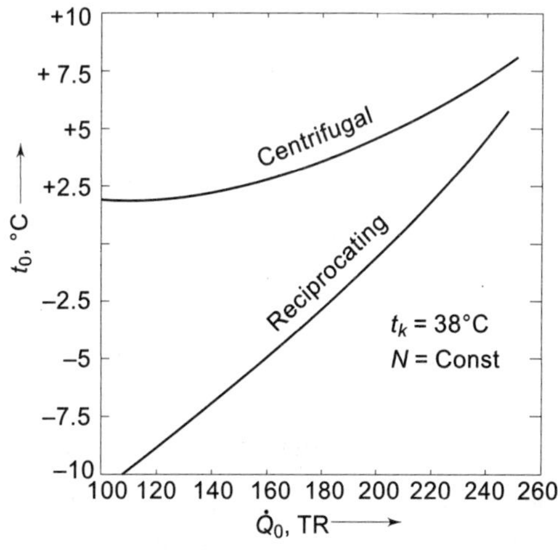

Fig. 6.20(a) Nature of capacity variation with evaporator temperature t_0 for centrifugal and reciprocating compressors

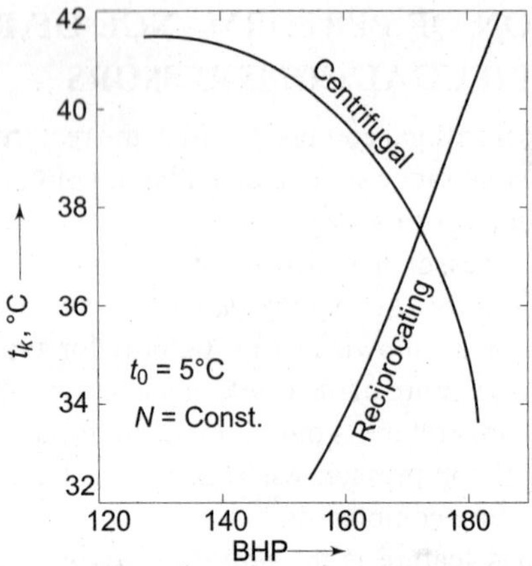

Fig. 6.20(b) Non-overloading characteristic of centrifugal compressor as compared to that of reciprocating compressor

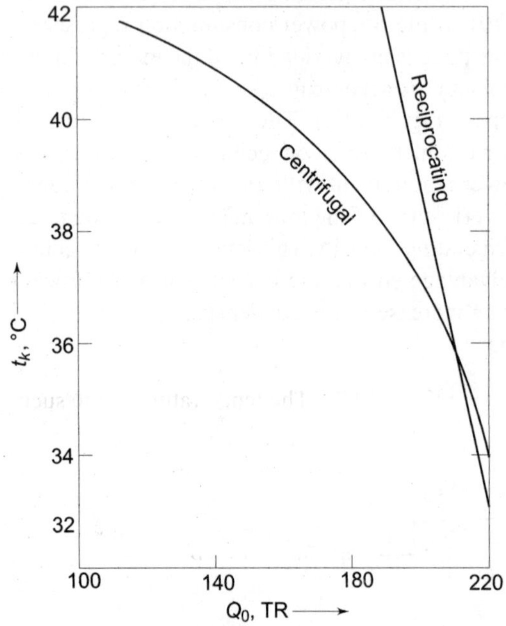

Fig. 6.20(c) Nature of capacity variation with condenser temperature for centrifugal and reciprocating compressors

References

1. Arora C P, *Thermodynamics*, Tata McGraw-Hill Pub. Co., New Delhi, 1998.
2. Atwood T and K P Murphy, 'An investigation of refrigerants for single stage centrifugal water chillers', *ASHRAE Trans.*, Vol. 76, pp. 81–95, 1970.
3. Chlumsky V, *Reciprocating and Rotary Compressors*, E & FN Spon, London, 1965.

4. Ferguson T B, *The Centrifugal Compressor Stage*, Butterworths, London, 1963.
5. Stoecker W F, *Refrigeration and Air Conditioning*, McGraw-Hill, New York, 1958, p. 174.
6. Vincent, E T, *The Theory and Design of Gas Turbines and Jet Engines*, McGraw-Hill, New York, 1950, p. 306.

Revision Exercises

6.1 (a) Calculate the clearance volumetric efficiency of an ammonia compressor ($\gamma = 1.31$) operating between a condenser temperature of 35°C and an evaporator temperature of – 15°C. The clearance factor is 0.07.

(b) If the throttling losses are 0.15 bar in the suction manifold and valve and 0.25 bar in the discharge valve, calculate the volumetric efficiency.

(c) The measurement of the output of compressor showed an overall volumetric efficiency of 65.2 per cent. The temperature at the suction flange is 10°C but that in the beginning of compression is 30°C. What were the leakage losses in the compressor?

6.2 Catalogue data of 4.8% clearance R 134a compressor with piston displacement of 2 m^3/min shows the capacity to be 12.7 TR with a bhp of 13.5 when the suction conditions are 18°C and 3.19 bar and the condensing temperature is 35°C. The refrigerant leaves the condenser as saturated liquid. Calculate the actual and clearance volumetric and adiabatic compression efficiencies of the compressor at these conditions.

6.3 An R 22 compressor is working at –18°C evaporating and 40°C condensing temperatures. The superheat at the bulb of the thermostatic expansion valve is 7°C and the temperature of the suction vapour is increased by 22°C in a heat exchanger by simultaneous cooling of the liquid refrigerant. The efficiency of the heat exchanger is 0.75. The temperature of the suction vapour is 20°C at the compressor inlet. The liquid refrigerant is subcooled by water to 35°C. Estimate the dimensions of the cylinders of the compressor.

Given:		
	Bore/stroke ratio	1
	Compressor speed	1420 rpm
	Compressor capacity	15 TR
	Number of cylinders	4

Assume a pressure drop of 5 per cent of the value of the pressure at the compressor valves. Also assume for the compression process, $n = 1.13$ and for the expansion process, $m = \gamma = 1.148$.

6.4 A single-stage centrifugal compressor, operating at 5000 rpm, receives saturated R 134a at 6°C and discharges it at a pressure of 1.0166 MPa. If the conditions of the suction gas remain unchanged, to what pressure can the compressor discharge if the speed is increased to 5500 rpm?

6.5 A single-wheel centrifugal compressor has an impeller with a diameter of 60 cm. The speed of the impeller is 5000 rpm. The working substances are R 11, R 123, R 245fa and R 134a. If the suction vapour is at 5°C saturation, determine the saturated discharge temperatures in each case.

6.6 A single-stage R 123 centrifugal compressor is running at 5000 rpm. The suction condition is 5°C dry saturated. The condenser temperature is 35°C. The outlet blade angle is 32°. The flow coefficient may be selected as 0.3 at which the polytropic efficiency is 0.8. Determine the diameter of the impeller. Assume no prewhirl and negligible velocities at inlet and outlet.

6.7 The variation of volumetric efficiency with pressure ratio for a R 134a compressor of an automotive air conditioner is as follows:

p_k/p_0	2	3	4	5	6	7
η_V	80	79.5	72.5	66.5	61	56.7

Somehow, the air flow over the condenser of the air conditioner gets completely blocked. Estimate the maximum pressure at discharge if the evaporation temperature is 5°C.

6.8 (a) Given for an R 134a centrifugal compressor:
Suction state 8°C saturated vapour
Condensing temperature 40°C
Estimate the minimum tip-speed required for a single stage compressor, and the impeller diameter and width of shrouds at discharge for a refrigerating capacity of 500 TR. Take rpm as 2800.

(b) If instead a 2-stage compressor is used running at 2800 rpm, and if both wheels are to be of the same diameter, estimate the new impeller diameter and the widths of shrouds at discharge for the two stages for 500 TR capacity.

6.9 (a) Determine the evaporator temperature at which a single-stage R 22 plant will cease to produce any refrigerating effect in Delhi where the condenser temperature is 42°C. The compressor has 5% clearance, and the index of compression is 1.15.

(b) For a particular condenser pressure, how does the power requirement of a reciprocating compressor vary with change in evaporator pressure? Explain giving reasons.

6.10 The particulars of a 20 TR R 22 food freezer are as follows:

Evaporator temperature	– 30°C
Condenser temperature	40°C
Temperature of vapour leaving evaporator	– 25°C
Subcooling of liquid with vapour in heat exchanger	5°C
Polytropic index of compressor	1.15
Superheating of suction vapour after passing through suction valve	15°C
Compressor clearance factor	0.05
Pressure drop in valves	0.2 bar suction 0.5 bar discharge

The compressor has 6 cylinders with bore equal to stroke. Its rpm is 1500, and mechanical efficiency is 0.75. Determine

(i) Mass flow rate of refrigerant.
(ii) Piston displacement of compressor.
(iii) Cylinder diameter.
(iv) Motor horse power.
(v) Cooling requirement of compressor in kW.

6.11 Typical operating conditions of a 1/8 TR domestic refrigerator are as follows:

Condenser temperature	55°C
Evaporator temperature	– 25°C
Temperature of vapour at evaporator exit	– 15°C
Heat gained by suction vapour in cooling motor windings	15% of motor power input

The compressor bore/stroke ratio is 0.8. It runs at 2800 rpm. Pressure drops in suction and discharge reeds are 10% of absolute pressure. Isentropic efficiency of compressor is 85%, and clearance is 5%. Making appropriate assumptions, if necessary, estimate if the refrigerant is (a) R 134a, (b) R 600a;
(i) Power drawn from mains.
(ii) Bore, stroke and piston speed of the compressor.

6.12 Catalogue data of a compressor is as follows:

No. of cylinders 2	Bore 45 mm	Rpm 4000
Displacement 121 cc	Stroke 38 mm	Refrigerant R 134a

Evaporation Temperature	*Condensing Temperature*			
	40°C		60°C	
	Capacity TR	Power kW	Capacity TR	Power kW
+ 10°C	4.68	3.68	(a)	(b)
– 20°C	(c)	(d)	(e)	(f)

Fill in the blanks in the above table. Make appropriate assumptions. Compare calculated values with those cited in the catalogue, viz., (a) 3.68 (b) 6.0 (c) 1.33 (d) 2.4 (e) 0.78 (f) 3.6.

Chapter

7

Condensers

7.1 HEAT REJECTION RATIO

It is ultimately in the condenser that heat is rejected in a vapour-compression refrigeration machine. The vapour at discharge from the compressor is superheated. Desuperheating of the vapour takes place in the discharge line and in the first few coils of the condenser. It is followed by the condensation of the vapour at the saturated discharge temperature or condensing temperature t_k. In some condensers, subcooling may also take place near the bottom where there is only liquid. However, the sensible heat of the desuperheating and subcooling processes is very small as compared to the latent heat of the condensation process.

The loading on the condenser per unit of refrigeration is called the *heat rejection ratio*.

Since

$$Q_k = Q_0 + W$$

we have for the heat rejection ratio

$$\frac{Q_k}{Q_0} = 1 + \frac{1}{\mathcal{E}_c} \tag{7.1}$$

Thus the ratio depends on the COP which in turn depends on the condenser and evaporator temperatures. In actual air-conditioning equipment for R 22, operating at $t_k = 40°C$ and $t_0 = 5°C$, the heat rejection ratio is approximately 1.25 to 1.35.

7.2 TYPES OF CONDENSERS

The type of a condenser is generally characterized by the cooling medium used. Thus there are three types of condensers:

(i) Air-cooled condensers.
(ii) Water-cooled condensers.
(iii) Evaporative condensers.

There is also a fourth type dependent on *ground-cooling* which is not used commonly.

7.2.1 Air-cooled Condensers

In air-cooled condensers, heat is removed by air using either natural or forced circulation. The condensers are made of steel, copper or aluminium tubing provided with fins to improve air-side heat transfer coefficient. The refrigerant flows inside the tubes and the air flows outside.

Air-cooled condensers are used only in small capacity machines, such as refrigerators and small water coolers which use vertical *wire and tube* or *plate and tube* construction with natural circulation, and window-type and package air conditioners which have tubes with 5–7 fins per cm and use forced circulation of air. The current practice in the forced-convection type is to use 10-15 m^2 of the total surface area per ton of refrigeration based on 2–5 m/s *face velocity* of air over the coil.

Air-cooled condensers are seldom made in sizes over 5 TR because of high head pressure, excessive power consumption and objectionable fan noise.

7.2.2 Water-cooled Condensers

Water-cooled condensers can be of three types, viz., *shell and tube, shell and coil* and *double tube*. The shell-and-tube type, with water flowing through passes inside tubes and the refrigerant condensing in the shell is the most commonly used condenser. Figure 7.1 shows the arrangement for a *two-pass* condenser. A shell-and-tube condenser also serves the purpose of a receiver, specially for *pumping down* the refrigerant, because there is sufficient space in the shell. The bottom portion of the condenser also serves the purpose of a sub-cooler as the condensed liquid comes in contact with the entering water at a lower temperature. Thus, we see that shell and tube condensers, and, for that matter, all types of condensers are usually overdesigned. This also keeps the head pressure low, and saves power. Further, the outside surface of the shell is made available for heat transfer. The shell is made of steel. Copper tubes are used for fluorocarbons and steel tubes for ammonia.

The shell-and-coil condenser consists of an electrically-welded closed shell containing a water coil sometimes of finned tubing.

In the double-tube arrangement, the refrigerant condenses in the outer tube and water flows through the inner tube in the opposite direction.

Water-cooled condensers are invariably used in conjunction with *cooling towers*, *spray ponds*, etc. Heated water from the condenser is led to the cooling tower where it is cooled by *self-evaporation* into a stream of air. After cooling, the water is pumped back to the condenser.

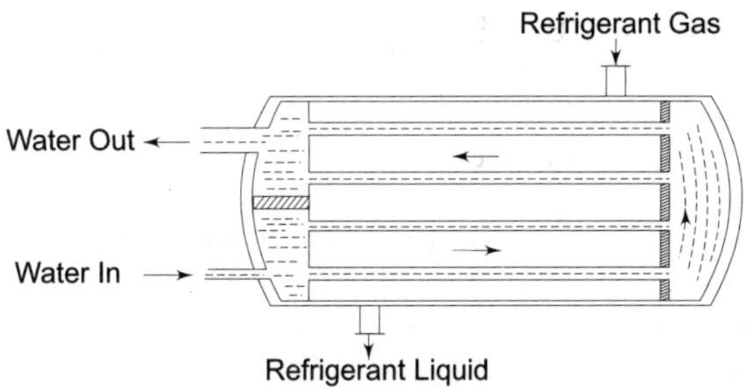

Fig. 7.1 Schematic representation of a two-pass water-cooled shell and tube condenser

Cooling towers are of two types, *natural draft* with natural convection of air, and either *induced* or *forced draft* with forced convection of air by a fan.

7.2.3 Evaporative Condensers

Figure 7.2 shows the schematic diagram of an evaporative condenser. The refrigerant first rejects its heat to water and then water rejects its heat to air, mainly in the form of evaporated water. Air leaves with high humidity as in a cooling tower. Thus an evaporative condenser combines the functions of condenser and cooling tower.

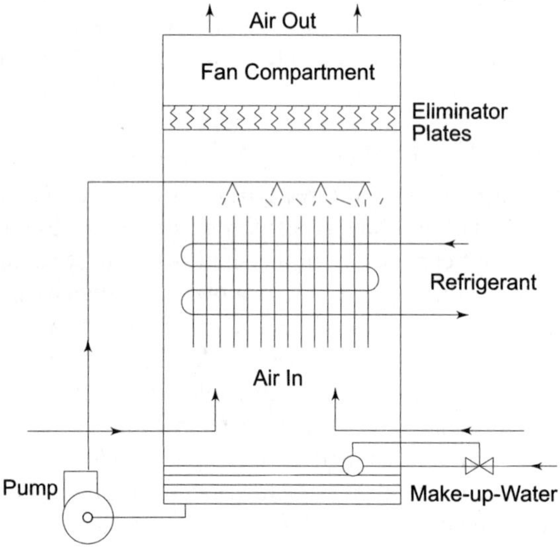

Fig. 7.2 Evaporative condenser

Evaporative condensers are commonly used on large ammonia plants as they are found to be cheaper. Such condensers require a larger amount of the refrigerant charge due to the longer length of the refrigerant piping. But in the case of ammonia systems this is immaterial since the refrigerant is quite cheap.

7.3 HEAT TRANSFER IN CONDENSERS

The heat transfer in a water-cooled condenser is given by Eq. (7.2)

$$\dot{Q}_k = UA\,\Delta t = \frac{\Delta t}{R} \tag{7.2}$$

where U is the overall heat transfer coefficient based on the surface area A of the condenser and Δt is the overall temperature difference. Figure 7.3 shows the components of the heat-transfer resistance in a water-cooled condenser, viz., the outside refrigerant film, metal wall, scaling on water-side surface and inside-water film. The overall resistance is obtained by adding all the resistances which are in series. Thus,

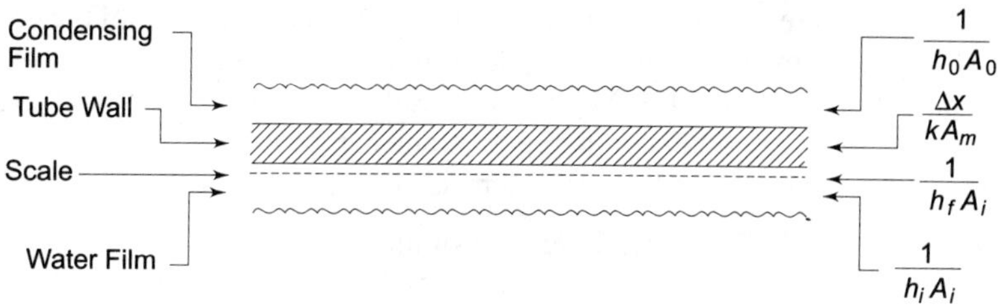

Fig. 7.3 Thermal resistance in a water-cooled condenser

$$R = \frac{1}{U_0 A_0} = \frac{1}{h_0 A_0} + \frac{\Delta x}{k A_m} + \frac{1}{h_f A_i} + \frac{1}{h_i A_i} \tag{7.3}$$

where

U_0 = Overall heat-transfer coefficient based on the outside surface area
h_0 = Condensing film coefficient of heat transfer
A_0 = Outside or refrigerant-side area
Δx = Thickness of the metal-tube wall
k = Thermal conductivity of the tube material
A_m = Mean tube surface area
h_f = Coefficient of heat transfer through the scale
A_i = Inside or water-side area
h_i = Water-side coefficient of heat transfer.

The overall heat-transfer coefficient can be determined from Eq. (7.3) after estimating the individual resistances.

7.3.1 Desuperheating

In condenser design, normally no separate calculations are made for the desuperheating section. Though the heat transfer coefficient is lower for this section, the temperature differential is greater as compared to that of the condensing section. The two factors tend to cancel the effects of each other. Thus, the practice is to take the refrigerant-side temperature equal to the condensing temperature and the heat-transfer coefficient equal to the condensing coefficient for the whole length of the condenser.

7.3.2 Condensing Heat Transfer Coefficient

Refrigerants are usually good wetting agents. In condensation, therefore, they follow a *film-wise* pattern. Nusselt, using his laminar liquid film theory, derived the following expression for the overall coefficient of heat transfer for condensation on a vertical surface of height x

$$h = 0.943 \left[\frac{k_f^3 \rho_f (\rho_f - \rho_g) g h_{fg}}{\mu_f \, x \, \Delta t} \right]^{1/4} \tag{7.4}$$

where k_f, ρ_f and μ_f are the thermal conductivity, density and viscosity of the condensate film. These are evaluated at the average film temperature $t_f = \frac{1}{2}(t_k + t_w)$. And Δt is the temperature differential $(t_k - t_w)$ where t_w is the tube-wall temperature. Density ρ_g of the vapour is very small compared to ρ_f and can be neglected.

Note that the condensate film acts as a resistance to heat transfer.

Condensation Outside Horizontal Tubes A similar expression for the average coefficient of heat transfer for vapour condensing on the outside of horizontal tubes of diameter D_0, as in shell and tube condensers, is

$$h_0 = 0.725\left[\frac{k_f^3\,\rho_f^2\,g h_{fg}}{N D_0 \mu_f\,\Delta t}\right]^{1/4} \tag{7.5}$$

where D_0 is the outside diameter of tubes and N is the number of tubes in a vertical row.

Equations (7.4) and (7.5) show that the heat-transfer coefficient increases with thermal conductivity. A large density results in a thinner growth of the condensate film. The heat-transfer coefficient, therefore, increases with the density. Similarly, a high latent heat implies a lower condensation rate and hence a thinner film and higher heat-transfer coefficient. On the other hand, a higher viscosity results in a thicker film and hence a lower heat transfer coefficient. Similarly, a higher temperature differential results in higher condensation rate, a thicker film and a lower heat-transfer coefficient.

Further, a large vertical tube length x or a horizontal tube diameter D_0 cause a thicker growth of film before it drains of from the tube and hence a lower average heat-transfer coefficient. Finally, the condensate film formed on the lower tubes in a vertical bank is thicker, giving a lower average value of the coefficient.

Equation (7.5) of Nusselt has been arranged by Kirkbride[7] in the following form

$$\text{Co} = 1.514\,(\text{Re}_f)^{-1/3} \text{ for } \text{Re}_f < 1800 \tag{7.6}$$

where Co is the *condensation number* expressed by

$$\text{Co} = h_0\left[\frac{\mu_f^2}{k_f^3\,\rho_f\,(\rho_f - \rho_g)\,g}\right]^{1/3} \simeq \frac{h_0}{k}\left(\frac{v_f^2}{g}\right)^{1/3} \quad \text{since } \rho_g << \rho_f \tag{7.7}$$

and Re_f is the film Reynolds number and $v_f = \dfrac{\mu_f}{\rho_f}$ is the kinematic viscosity of the liquid. If D_e is equivalent dia, and A_f is flow area

$D_e = 4A_f/P$

$$\text{Re}_f = \frac{D_e u \rho_f}{\mu_f} = \frac{4A_f}{P}\cdot\frac{u\rho_f}{\mu_f} = \frac{4\dot{m}}{P\mu_f} = \frac{4\Gamma}{\mu_f} = \frac{4h_0 A\Delta t}{P\mu_f h_{fg}} \tag{7.8}$$

where $\Gamma = \dfrac{\dot{m}}{P}$ = Loading per tube per unit depth

$\dot{m}$ = Mass flow rate per tube or condensation rate = $\rho_f u A_f = \dfrac{h_0\,A\Delta T}{h_{fg}}$

P = Shear perimeter of tube = πD_0

A = Heat transfer area of one tube.

These expressions are applicable to laminar films only considered to be at film Reynolds numbers below 1800. Figure 7.4 shows the nature of variation of Co with Re_f in the case of vertical tubes.

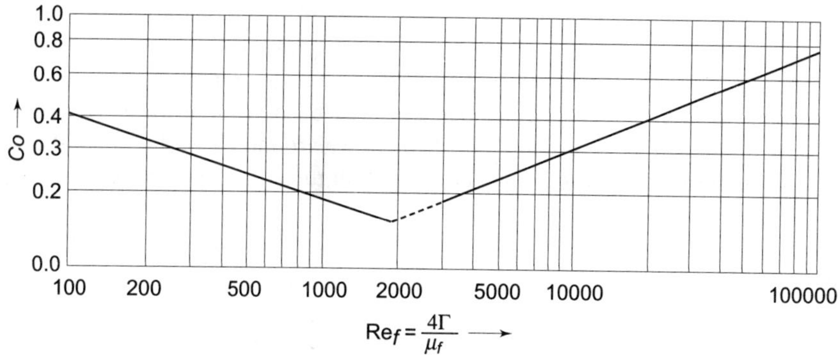

Fig. 7.4 Semi-empirical condensation curve for vertical tubes

For laminar flow Co decreases with Re_f. After transition to turbulent flow, Co increases with Re_f. Similarly, corresponding to Eq. (7.6) for laminar film, correlation by Kirkbride for the turbulent condensing film outside horizontal tubes is

$$Co = 0.0077\ Re_f^{0.4} \text{ for } Re_f > 1800 \tag{7.9}$$

Condensation Inside Horizontal Tubes The phenomenon of condensation inside horizontal tubes is quite complicated and not amenable to analytical treatment. The flow rate of the vapour greatly influences the heat-transfer coefficient in this forced convection-condensation situation. The rate of liquid accumulation also influences the coefficient. Akers, Deans and Crossef[1] give the following relation for the condensing heat-transfer coefficient inside tubes when the condensation rate or the length of tube is large,

$$\frac{D_i}{k_f} = 0.026\ Pr_f^{1/3}\ Re_m^{0.8} \tag{7.10}$$

where Pr_f is the Prandtl number of the liquid equal to $(C_p\ \mu/k)_f$ and Re_m is the mixture Reynolds number defined by

$$Re_m = \frac{D_i}{\mu_f}\ [\rho_f u_f + \rho_g u_g (\rho_f/\rho_g)^{1/2}] \tag{7.11}$$

in which u_f and u_g are the liquid and vapour velocities.

Most refrigeration condensers such as in refrigerators and air conditioners operate in this manner with condensation inside long horizontal tubes. Correlation in Eq. (7.10) is valid for $Re_g > 20{,}000$ and $Re_f > 5000$. However, in this turbulent flow regime (flow is annular), the calculations using Eq. (7.10) show that heat transfer coefficient increases with liquid flow rate, and it is independent of temperature difference. The experimental data does not agree with both these facts. The equation presented below is, therefore, recommended

$$h_i = h_{fi} \left\{ 1 + x \left[\frac{\rho_f}{\rho_g} - 1 \right] \right\}^{1/2} \tag{7.12}$$

where h_{fi} is the sensible heat transfer coefficient calculated from Dittus-Bolter equation assuming that total fluid is flowing as liquid with condensate properties. Eq. (7.12) enables one to calculate local heat transfer coefficient at any section where the quality of vapour is x. A detailed method of calculation of h_i for high velocity annular flow condensation is, however, given by Rohsenow[9].

At low condensation rates or in short tubes, as in evaporative condensers, when the vapour velocity is low, the condensate flow becomes stratified. The vapour condensed in the upper part of the tube flows down around the periphery to the bottom. The layer of liquid thus formed at the bottom runs off in an open-channel type of flow.

At these low vapour velocities ($Re_f < 35{,}000$), the experimental results agree very well with Chato's[2] correlation

$$h_i = 0.555\left[\frac{g\rho_f(\rho_f - \rho_g)k_f^3 h'_{fg}}{\mu_f \Delta T_i D_i}\right]^{1/4} \tag{7.13}$$

where the modified latent heat is calculated from

$$h'_{fg} = h_{fg} + \frac{3}{8}\, C_f \Delta T_i \tag{7.14}$$

Chato also established the effect of inclining the tubes to the horizontal. He found that the heat transfer coefficient increases by 10 to 20 per cent at inclination angles of 10 to 20° as a result of faster drainage of the condensate.

7.3.3 Water-side Coefficient

The water-side coefficient is expressed by the well-known *Dittus-Boelter equation*[2] which relates the Nusselt number with Reynolds and Prandtl numbers for forced convection for turbulent flow inside tubes, viz.,

$$\frac{h_i D}{k} = 0.023\left(\frac{Du\rho}{\mu}\right)^{0.8}\left(\frac{C_p\,\mu}{k}\right)^{0.4} \tag{7.15}$$

All the fluid properties are evaluated at the bulk mean temperature of the fluid.

7.3.4 Fouling Factor

Condensers should always be designed considering the formation of scales, i.e., *fouling* inside the tube. The following *fouling factors*[10] are recommended.

$$\frac{1}{h_f} = 0.00009\ \text{J}^{-1}\ \text{sm}^2\ \text{K for copper tubes and R 22 condensers}$$
$$= 0.0009\ \text{J}^{-1}\text{sm}^2\ \text{K for steel tubes and ammonia condensers.}$$

For hard water conditions as in Delhi, the recommended value is 0.001 to 0.0025 $m^2\,K/W$.

7.3.5 Air-side Coefficient

In refrigerant condensers with forced convection, air flows outside the tubes in cross-flow to the refrigerant. Average heat-transfer coefficients for the forced convection of air across a tube may be calculated from the Grimson's's[5] equation

$$\frac{hD}{k} = C\left(\frac{Du_\infty\rho}{\mu}\right)^n \text{Pr}^{1/3} \tag{7.16}$$

where constants C and n are tabulated as functions of Reynolds number in Table 7.1.

Table 7.1 Constants for Grimson's equation

Re	*C*	*n*
0.4–4	0.989	.33
4–40	0.911	.385
40–4000	0.683	.466
4000–40,000	0.193	.618
40,000–400,000	0.266	.805

Many heat-exchanger arrangements involve multiple rows of tubes. The tubes may be *staggered* or *in-line* as shown in Fig. 7.5.

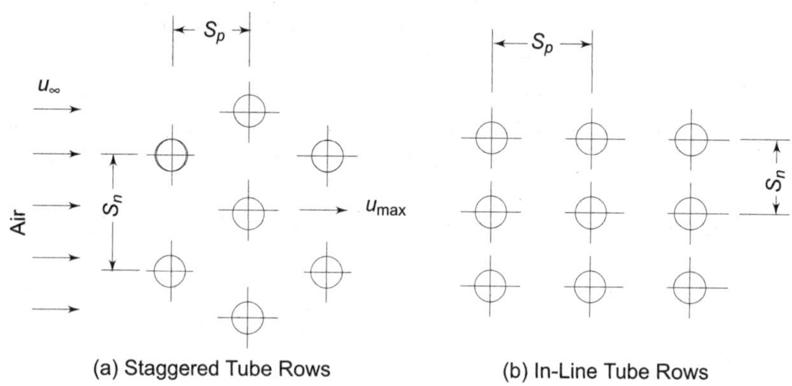

Fig. 7.5 Tube arrangements

In the case of tube banks, u_∞ is the face velocity (FV) measured at a distance from the tube

$$u_\infty = \frac{\text{Volume flow rate}}{\text{Face area}} = \frac{\dot{Q}_v}{(FA)} = (FV)$$

The Reynolds number is, however, based on the maximum velocity occurring in the tube bank which is

$$u_{max} = u_\infty \frac{S_n}{S_n - D}$$

For the case of forced convection of air over finned tubes, Table 20.1 gives the values of h_{air} or f_g as a function of (FV). These values can be correlated by the relation

$$h_{air} = 38\,(FV)^{0.5} \tag{7.17}$$

For the case of natural convection of air on horizontal tubes, the following simplified relations may be used.

For laminar flow, $10^4 < Gr_f Pr_f < 10^9$

$$h_{air} = 1.32\left(\frac{\Delta T}{D_0}\right)^{1/4} \tag{7.18}$$

For turbulent flow, $Gr_f Pr_f > 10^9$

$$h_{air} = 1.24\,(\Delta T)^{1/3} \tag{7.19}$$

In these equations Gr_f is the Grashoff's number, a characteristic of natural convection, expressed by

$$Gr_f = \frac{g \beta \Delta T D^3}{\nu^2} \tag{7.20}$$

where $\beta = \frac{1}{T_f}$. T_f is the mean temperature of the film and ν is the kinematic viscosity.

7.3.6 Augmentation of Condensing Heat-Transfer Coefficient

It is well understood that the overall heat-transfer coefficient such as U_0, always, has a value lower than the lower of the two film coefficients, viz., h_0 and h_i, howsoever large the other film coefficient may be. Thus, it is advantageous to decrease the major heat-transfer resistance. A familiar way to decrease this resistance is to provide an *extended surface* or *fins* on the side of the lower heat-transfer coefficient.

In shell-and-tube condensers the water-side coefficient is of the order of 6000 $Wm^{-2} K^{-1}$. Whereas a typical value of the condensing coefficient for ammonia is 8000 $Wm^{-2} K^{-1}$, the same for R 22 is of the order of 1500 $Wm^{-2} K^{-1}$ only.

Thus in R 22 condensers, it is desirable to use fins on the refrigerant side.

Of late, the use of *integrally-finned tubes*, such as the one shown in Fig. 7.6, has become quite popular in R 22 condensers. These fins are a part of the tube itself and are not mounted separately. Acme Industries, USA[8] have perfected a configuration with one fin per mm having a height of approximately 1.5 mm, obtained by cold compression of the tube.

In air-cooled condensers, the air-side coefficient is much lower than the refrigerant side coefficient. Hence, fins must be provided on the air-side.

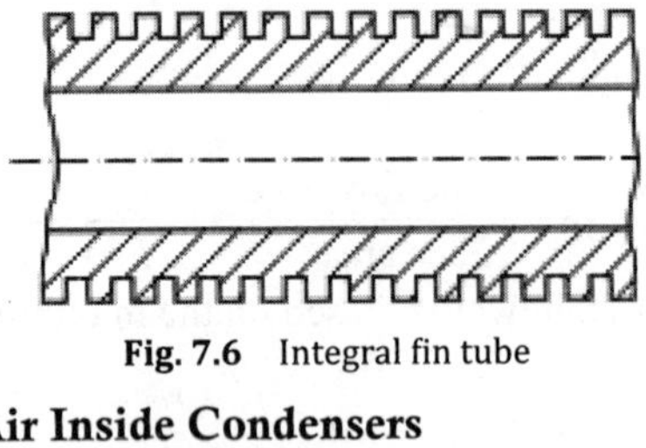

Fig. 7.6 Integral fin tube

7.3.7 Influence of Air Inside Condensers

Air, which has extremely low normal boiling and critical points, remains as a *non-condensable gas*, if present in refrigerant condensers. The presence of even one per cent of air by volume can reduce the condensing coefficient by as much as 50 per cent. This is because air tends to cling to the surface and because of its very low thermal conductivity, it introduces a large thermal resistance. Another detrimental effect of air is the increase of head pressure. Firstly, it is because of a decrease in the condensing coefficient and consequent operation at a higher temperature differential Δt and hence a higher pressure p_k in the condenser. Secondly, the head pressure is further increased over and above the saturation pressure p_k of the refrigerant by an amount equal to the partial pressure exerted by air or non-condensables present.

If any part of a refrigeration system, such as in R 123, operates under vacuum, there is a possibility of air leakage into the system. This requires a provision for its removal by purging from the top of the condenser where it will normally collect. A purge valve is also provided in the condensers of the ammonia system because these systems suffer from leakage. Pressure vessels of lithium bromide-water vapour absorption refrigeration systems which operate under high vacuum are provided with regular purge units with a vacuum pump.

Example 7.1 ***Design of Shell and Tube Condenser***

(a) *Determine the length of tubes in a 3-pass, shell-and-tube R 22 condenser with 108 tubes for 40 TR chiller. The condensing temperature is 43°C. The heat rejection ratio is 1.25. Water is cooled to 30°C in the cooling tower. The temperature rise of water may be taken as 4.8°C. Use integral fin copper tubes with an O.D. of 1.59 cm, an I.D. of 1.37 cm with 748 fins/m length of tube. Fins are 1 mm thick and 1 mm high over tubes.*

(b) *Determine the length for plain tube as well. Allow for 10 bottom tubes for subcooling.*

Solution Let the tube arrangement be as shown in Fig. 7.7 with nine vertical rows so that the average number of tubes per row is

$$N = \frac{108}{12} = 9$$

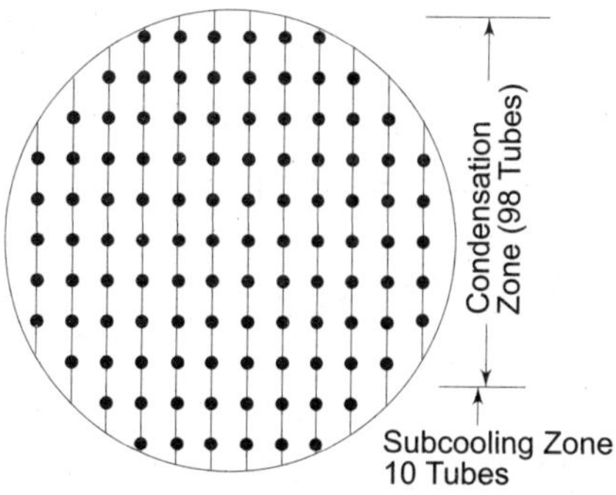

Fig. 7.7a Tube arrangement for Example 7.1

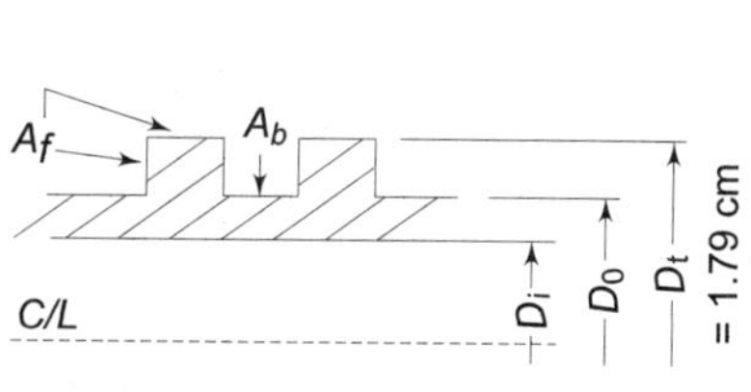

Fig. 7.7b Fin construction

(a) *Integral fin tubes* Assume a temperature drop of 5°C through the condensing film. The mean temperature of the film is

$$t_f = 43 - \frac{5}{2} = 40.5°\text{C}$$

At this temperature, the properties of liquid R 22 are

$$k_f = 0.0315 \text{ Wm}^{-1} \text{ K}^{-1}$$

$$\rho_f = 1129 \text{ kg. m}^{-3}$$

$$\mu_f = 0.221 \; cP = 2.21 \times 10^{-4} \text{ kg. m}^{-1} \text{ s}^{-1}$$

Also $$h_{fg} = 163.3 \times 10^3 \text{ J.kg}^{-1} \text{ at } 43°\text{C}$$

Condensing coefficient

$$h_0 = 0.725 \left[\frac{k_f^3 \, \rho_f^2 \, g h_{fg}}{ND \, \mu_f \, \Delta t} \right]^{1/4}$$

$$= 0.725 \left[\frac{(0.815)^3 \, (1129)^2 \, (9.81) \, (163.3 \times 10^3)}{9(0.0159) \, (2.21 \times 10^{-4}) \, (5)} \right]^{1/4} = 1180 \text{ Wm}^{-2} \text{ K}^{-1}$$

Heat rejected in the condenser

$$\dot{Q}_k = 40 \times 3.5167 \times 1.25 = 175.83 \text{ kW}$$

Water flow rate

$$\dot{m}_w = \frac{\dot{Q}_k}{C\Delta t} = \frac{175.83}{(4.187)\,(4.8)} = 8.74 \text{ kg. s}^{-1}$$

Velocity of water in tubes

$$u_w = \frac{\dot{m}_w / \rho_w}{(108/3)\left(\dfrac{\pi D_i^2}{4}\right)} = \frac{8.74 \times 10^{-3}}{36\,(\pi/4)\,(0.0137)^2} = 1.644 \text{ ms}^{-1}$$

Bulk mean temperature of water

$$t_w = 30 + \frac{4.8}{2} = 32.4°\text{C}$$

For water at 32.4°C

$$k = 0.623 \text{ Wm}^{-1} \text{ K}^{-1}$$

$$\mu = 7.65 \times 10^{-4} \text{ kg.m}^{-1} \text{ s}^{-1}$$

$$\text{Pr} = \frac{C\mu}{k} = \frac{(4.1868 \times 10^3)\,(7.65 \times 10^{-4})}{0.623} = 5.14$$

$$\text{Re} = \frac{Du_w\rho}{\mu} = \frac{(0.0137)\,(1.644)\,(1000)}{7.65 \times 10^{-4}} = 24{,}070$$

Water-side coefficient

$$\text{Nu} = \frac{h_i D}{k} = 0.023\,(\text{Re})^{0.8}\,(\text{Pr})^{0.4} = 141.4$$

$$h_i = 141.4 \times \frac{0.623}{0.01} = 6432 \text{ Wm}^{-2} \text{ K}^{-1}$$

Area Calculations Consider 1 m length of tube.

Fin surface area

$$A_f = 748\,\pi \left[\frac{2(0.0179^2 - 0.0159^2)}{4} + (0.0179)\,0.001 \right] = 0.12144 \text{ m}^2$$

Bare tube surface area

$$A_b = \pi\,(0.0159)\,(1 - 0.748) = 0.01258 \text{ m}^2$$

Total tube surface area

$$A_t = A_f + A_b = 0.0134 \text{ m}^2$$

Outside and inside tube surface areas

$$A_0 = \pi D_0 = 0.04993 \text{ m}^2$$

$$A_i = \pi D_i = 0.04302 \text{ m}^2$$

Outside to inside surface area ratio of the tube

$$\frac{A_0}{A_i} = \frac{D_0}{D_i} = \frac{1.59}{1.37} = 1.1607$$

Extended surface to outside and inside tube surface area ratios

$$A_t/A_0 = \frac{0.0134}{0.04993} = 2.684$$

$$A_t/A_i = \frac{0.0134}{0.04302} = 3.046$$

Fin efficiency as calculated (circular fins)

$$\eta_f = 0.992$$

Overall heat transfer coefficient based on total extended surface side area A_t

$$\frac{1}{U_t} = \frac{1}{h_0}\frac{A_t}{(\eta_f A_f + A_b)} + \frac{1}{h_f}\frac{A_t}{A_i} + \frac{1}{h_i}\frac{A_t}{A_i}$$

$$= \frac{1}{1180\ (0.993)} + 0.00009\ (3.046) + \frac{1}{6432}\ (3.046),$$

since $\dfrac{\eta_f A_f + A_b}{A_t} = 0.993$

$\Rightarrow$ $U_t = 624\ \text{Wm}^{-2}\ \text{K}^{-1}$

Note *It can be seen that the condensing coefficient is very small compared to the water-side coefficient and hence the use of integral fins is beneficial.*

Log-mean temperature difference

R 22 temperatures: In 43°C; Out 43°C

Water temperatures: In 30°C; Out 34.8°C.

$$\Delta t_m = \frac{(43 - 30) - (43 - 34.8)}{\ln\ (13/8.2)} = 10.43°\text{C}$$

Extended surface area

$$A_t = \frac{\dot{Q}_k}{U_t\ \Delta t_m} = \frac{175.83 \times 10^3}{624 \times 10.43} = 27.02\ \text{m}^2$$

Outside or bare tube surface area

$$A_0 = \frac{27.02}{2.684} = 10.07\ \text{m}^2$$

Tube length (Number of tubes for condensation = 108 – 10 = 98)

$$L = \frac{A_0}{98\pi D_0} = \frac{10.07}{98\pi\ (0.0159)} = 2.06\ \text{m}$$

Check for condensing film temperature drop

$$\Delta t = \frac{\dot{Q}_k}{h_0(\eta_f A_f + A_b)} = \frac{\dot{Q}_k}{h_0(0.993 A_t)} = \frac{175.83 \times 10^3}{1180(0.993 \times 27.02)} = 5.55°\text{C}$$

It is close to the assumed value of 5°C. Calculations may be repeated with Δt value of 5.55°C. The solution converges in the second iteration giving.

$h_0 = 1150$ W/m^2.K, $U_t = 617$ W/m^2. K, $A_t = 27.32$ m^2, $A_0 = 10.2$ m^2, $L = 2.08$ m, $\Delta t = 5.6$°C,

(b) *Plain tubes* The procedure is the same as for integral fin tubes. However, the lower condensing coefficient becomes the controlling coefficient. The thermal resistance on the condensing-side will increase. Hence, (i) the total heat transfer surface area will have to be increased (ii) or, alternatively, condensing temperature will have to be raised. Assuming the same condensing temperature, we have:

Overall heat-transfer coefficient based on bare tube surface area

$$\frac{1}{U_0} = \frac{1}{h_0} + \frac{1}{h_f}\frac{A_0}{A_i} + \frac{1}{h_i}\left(\frac{A_0}{A_i}\right)$$

$$= \frac{1}{1150} + 0.00009\,(1.1607) + \frac{1}{6432}\,(1.1607)$$

$$= 11.54 \times 10^{-4}$$

$$U_0 = 866 \text{ Wm}^{-2}\text{ K}^{-1}$$

Bare tube surface area

$$A_0 = \frac{175.83 \times 10^3}{866 \times 10.43} = 19.47 \text{ m}^2$$

Note *The bare tube surface area is increased from 10.07 to 19.47 m^2 in the absence of integral fins.*

Tube length $$L = \frac{19.47}{98\Pi(0.0159)} = 3.98 \text{ m}$$

Note *Since the tube length is very large, it will be necessary to increase the number of passes. Check for the assumption of refrigerant-side temperature drop.*

$$\Delta t = \frac{175.83 \times 10^3}{1150 \times 19.47} = 7.85°\text{C}$$

It is more than the assumed value of 5.6°C. Further approximation is necessary. To reduce the size of the condenser, a higher condensing temperature of 45–46°C may be considered. But, it will increase the running cost.

Example 7.2 *Design of Forced Convection Air-cooled Condenser*

An air-cooled condenser for a package air conditioning unit is to be designed to transfer 22.2 kW of heat from R 22 condensing inside 1.27 cm OD and 1.12 cm ID tubes at 55°C. The mass flow rate of the refrigerant is 0.162 kg/s. The refrigerant-side heat-transfer coefficient is given by

$$\mathbf{Nu} = 0.026\,(\mathbf{Pr})^{1/3}(\mathbf{Re}_m)^{0.8}$$

Air circulated is 120 cmm (cubic metres per minute) entering at 40°C. The air-side heat-transfer coefficient is approximated by

$$\mathbf{Nu} = 0.193\,(\mathbf{Pr})^{1/3}(\mathbf{Re})^{0.618}$$

The Prandtl number of air can be taken as 0.71. The face velocity of air may be taken as 6 m/s. The air side has 6 fins/cm so that finned surface to outside bare tube surface area ratio is 20. Determine the finned surface area of the condenser. Neglect the resistance of the metal wall. Assume efficiency of finned surface $\eta_f = 0.9$.

Solution *Refrigerant side heat-transfer coefficient*

Inside area of the tube, $$A_i = \frac{\pi D_i^2}{4} = \frac{\pi (0.0112)^2}{4} = 9.852 \times 10^{-5}\ \text{m}^2$$

Mass velocity of refrigerant, $$\rho u = \frac{\dot{m}}{A_i}$$

Assume a refrigerant-side temperature drop of 5°C. So the mean film temperature is (55 – 5/2) = 52.5°C. At this temperature, R 22 liquid film properties are

$$k_f = 0.08\ \text{Wm}^{-1}\ \text{K}^{-1}$$
$$\rho_f = 1072\ \text{kg.m}^{-3}$$
$$\mu_f = 2.12 \times 10^{-4}\ \text{kg. m}^{-1}\ \text{s}^{-1}$$
$$C_p = 1.52\ \text{kJ. kg}^{-1}\ \text{K}^{-1}$$

Hence $$\mathbf{Re}_m = \frac{D_i}{\mu_f}\frac{\dot{m}}{A_i} = \frac{0.0112\ (0.162)}{2.12 \times 10^{-4}\ (9.852 \times 10^{-5})} = 86{,}870$$

$$\mathbf{Pr}_f = \left(\frac{C_p \mu}{k}\right)_f = \frac{(1.52 \times 10^3)\ (2.12 \times 10^{-4})}{0.08} = 4.03$$

$$\mathbf{Nu} = \frac{h_i D_i}{k_f} = 0.026\ (4.03)^{1/3}\ (86{,}870)^{0.8} = 370$$

$$h_i = \frac{0.08 \times 370}{0.0112} = 2643\ \text{Wm}^{-2}\ \text{K}^{-1}$$

Air side heat transfer coefficient

Face area, $$FA = \frac{\dot{Q}_v}{u} = \frac{120/60}{6} = \frac{1}{3}\ \text{m}^2$$

Equivalent diameter, $$D_E = \sqrt{\frac{4}{\pi}(FA)} = \sqrt{\frac{4}{\pi}\frac{1}{3}} = 0.65\ \text{m}$$

Note *See Kern for more accurate calculations.*

Mean temperature of air, $$t_a = 40 + \frac{\Delta t_a}{2} = 40 + 4.63 = 44.63°\text{C}$$

Note *See calculations for* Δt_a *below.*

At this temperature for air

$$\mu = 2 \times 10^{-5}\ \text{kg. m}^{-1}\ \text{s}^{-1}$$
$$k = 0.03\ \text{Wm}^{-1}\ \text{K}^{-1}$$

$$\mathbf{Re} = \frac{D_E \rho u}{\mu} = \frac{0.65 \times 1.2 \times 6}{2 \times 10^{-5}} = 234{,}000$$

$$\mathbf{Nu} = 0.193\,(0.71)^{1/3}\,(234{,}000)^{0.618} = 358$$

$$h_0 = \frac{k}{D_E}\,\mathbf{Nu} = \frac{0.03}{0.65}\,(358) = 16.4\ \text{Wm}^{-2}\,\text{K}^{-1}$$

Overall heat transfer coefficient Neglecting the metal-wall resistance, the overall heat-transfer coefficient based on the total fin-side surface area is given by

$$\frac{1}{U_t A_t} = \frac{1}{h_i A_i} + \frac{1}{h_0 A_t \eta_f}$$

$$\frac{1}{U_t} = \frac{1}{h_i}\frac{A_t}{A_0}\frac{A_0}{A_i} + \frac{1}{h_0}\frac{1}{\eta_f}$$

$$= \frac{1}{2643}(20)\left(\frac{1.27}{1.12}\right) + \frac{1}{16.4\,(0.9)}$$

$$\Rightarrow \qquad U_t = 13.1\ \text{Wm}^{-2}\,\text{K}^{-1}$$

Finned surface area Temperature rise of air

$$\Delta t_a = \frac{\dot{Q}_k}{\dot{Q}_v \rho C_p} = \frac{22.2 \times 60}{120 \times 1.2 \times 1.005} = 9.25°\text{C}$$

Air leaving temperature

$$t_{a_2} = 40 + 9.25 = 49.25°\text{C}$$

Log mean temperature difference

$$\Delta t_m = \frac{(55-40)-(55-49.25)}{\ln \dfrac{55-40}{55-49.25}} = 9.62°\text{C}$$

Finned surface area

$$A_t = \frac{\dot{Q}_k}{U_t \Delta t_m} = \frac{22.2 \times 10^3}{(13.1)\,(9.62)} = 176\ \text{m}^2$$

Note *Temperature drop through films may be checked.*

7.4 WILSON'S PLOT

Wilson's plot[7] for condensers, as shown in Fig. 7.8, is obtained by plotting a number of experimental values of $\dfrac{1}{U_0}$ against $\dfrac{A_0}{A_i}\left(\dfrac{1}{h_i} + \dfrac{1}{h_f}\right)$ which varies with the flow rate of water. It is seen that the plot, if extended to the abscissa equal to zero, gives the intercept on the ordinate, corresponding to zero water-side resistance (very high water velocity) so that

$$\frac{1}{U_0} = \frac{1}{h_0} + \frac{\Delta x}{k}\frac{A_0}{A_m}$$

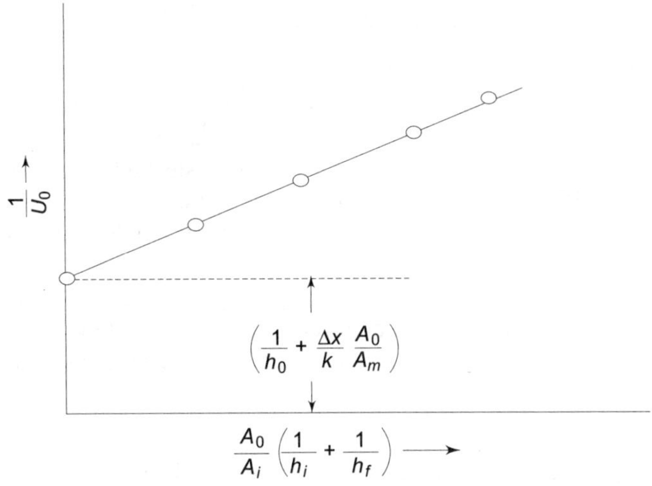

Fig. 7.8 Wilson's plot

This result may be used to determine the condensing coefficient h_o under actual operating conditions for any condenser.

References

1. Akers W W, O K Crosser and H A Deans, Proc. 2nd Nat. Heat Transfer Conf., *ASME/AIChE*, Aug. 1958.
2. Chato J C, *ASHRAE J.* Feb. 1962, p. 52.
3. Colburn A P, *Trans. A.I. Ch. E.*, Vol. 30, 1934, pp. 187–193.
4. Dittus F W and Boelter, LMK, Univ. Calif. (Berkeley) Pub. Eng., Vol. 2, 1930, p. 443.
5. Grimson E D, Correlation and utilisation of new data on flow resistance and heat transfer for cross-flow of gases over tube banks', *Trans. ASME*, Vol. 59, 1937, pp. 583–594.
6. Kern, D Q, *Process Heat Transfer*, McGraw-Hill, New York, 1950, p. 556.
7. Kirkbride C G, 'Heat transfer by condensing vapours on vertical tubes', *Trans. A.I. Ch. E.*, Vol. 30, 1934, pp. 170–186.
8. Kratz, A P, H J Macintire and R E Gould, *Heat Transfer in Ammonia Condensers*, Pt. III, Univ. Illinois Exp. Stn. Bull. 209, June 17, 1930.
9. Rohsenow W M and J P Hartnett, *Handbook of Heat Transfer*, McGraw-Hill Book Co., 1973, pp. 12–20 to 26.
10. Stoecker W F, *Refrigeration and Air Conditioning*, McGraw-Hill, New York, 1958.
11. Young E H and D J Ward, *Refining Engineer*, Vol. 29, No. 11, Oct. 1957, pp. C7-11 and No. 12, Nov. 1957, pp. C32–36.

Revision Exercises

7.1 A 10 TR ammonia ice plant has a two-pass shell-and-tube water-cooled condenser. The water enters at 30°C. The temperature rise of water may be taken as 2.5°C. The refrigerant condensing temperature is 35°C. The heat rejection ratio of the plant is 1.35. The condenser has 24 steel tubes of 21 mm OD and 13.65 mm ID. Determine the length of the tubes.

7.2 A $1\frac{1}{2}$ ton R 22 air conditioner, operating at 58°C condensing and 5°C evaporating temperatures, has an air-cooled condenser with 12.7 mm OD and 11.2 mm ID copper tubes. The air-side surface has fins with the finned surface to bare tube surface area ratio of 12. A fan blows 30 cm of air at 44°C with a face velocity of 4.5 m/s. The air-side heat-transfer coefficient is approximated by

$$\mathbf{Nu} = 0.24\ (\mathbf{Re})^{0.6}$$

The refrigerant-side heat-transfer coefficient is given by

$$\mathbf{Nu} = 0.026\ (\mathbf{Pr})^{1/3}\ (\mathbf{Re})^{0.8}$$

Determine the finned-surface area of the condenser and the length of the tubing.

7.3 In an experiment[8] on condensation of ammonia, the following values were measured for the overall heat-transfer coefficient U_0 and water velocity

U_0 (Wm^{-2} K^{-1})	2300	2070	1930	1760	1360	1130	865
u (ms^{-1})	1.22	0.975	0.853	0.731	0.488	0.366	0.244

The condenser tubes were 51 mm OD and 46 mm ID. The thermal conductivity of the material of the tubes is 60 Wm^{-1} K^{-1}. Using the method of Wilson's plot, determine the condensing heat-transfer coefficient. The water-side coefficient can be expressed as

$$h_i = 3967.3\ u^{0.8}$$

7.4 Do the reverse calculations in Example 7.1. Assume all data is known as given, and calculate. Estimate the capacity of the condenser. Show that it is 176 kW (Simulation).

7.5 Catalogue data of a water-cooled condenser of a manufacturer gives the following details:

Condensing temperature	48.9°C
Water inlet temperature	37.8°C
Water flow rate	20.694 kg/s
Capacity	145 tons

Estimate the capacity of this condenser with the same water flow rate but with an inlet temperature of 30°C and a condensation temperature of 42°C. The evaporation temperature may be assumed to be constant at 2.2°C.

Chapter

8

Expansion Devices

8.1 TYPES OF EXPANSION DEVICES

An *expansion device* in a refrigeration system normally serves two purposes. One is the thermodynamic function of expanding the liquid refrigerant from the condenser pressure to the evaporator pressure. The other is the control function which may involve the supply of the liquid to the evaporator at the rate at which it is evaporated. The latter has an important role and determines the efficiency with which the evaporator surface is utilized.

An expansion device is essentially a restriction offering resistance to flow so that the pressure drops, resulting in a throttling process. Basically there are two types of expansion devices. These are:

(i) Variable-restriction type.
(ii) Constant-restriction type.

In the variable-restriction type, the extent of opening or area of flow keeps on changing depending on the type of control. There are two common types of such control devices, viz., the *automatic expansion valve* and the *thermostatic expansion valve*.

In addition, there are the float-valves which are also variable restriction type devices. The float valves again are of two types: The *high-side float* and the *low-side float*.

The high-side float maintains the liquid at a constant level in the condenser and the low-side float maintains the liquid at a constant level in the evaporator.

The constant-restriction type device is the *capillary tube* which is merely a long tube with a narrow diameter bore.

8.2 AUTOMATIC OR CONSTANT-PRESSURE EXPANSION VALVE[1]

The basic function of an automatic expansion valve is to maintain constant pressure in the evaporator.

A typical design of an automatic-expansion valve is shown in Fig. 8.1. The liquid refrigerant enters the inlet (1) and flows through the strainer (2) to the orifice (3), expanding into the outlet (10). The needle (9) carried by the element (13) fits into the orifice in a position determined by the forces acting. The position of the needle (9) and the assembly (13) with respect to the orifice (3) determines the extent of opening or closing of the orifice.

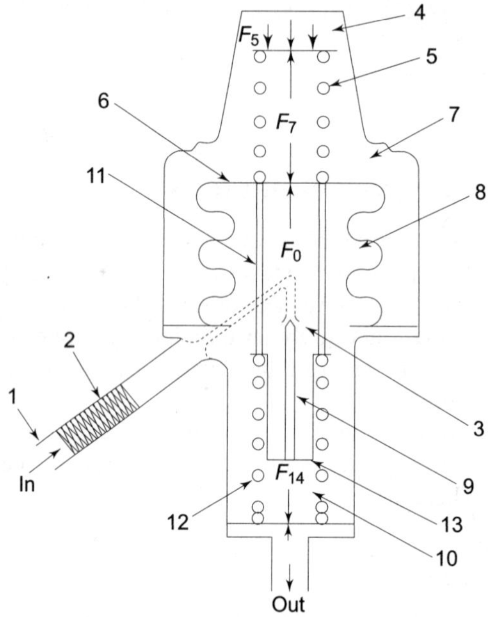

Fig. 8.1 Automatic-expansion valve

The forces acting on the bellows (8) are transferred to the needle through the push pins (11). The forces that tend to move the needle down and hence to open the orifice are:

(i) Force F_7 on the bellows (8) corresponding to the atmospheric pressure in the chamber (7).

(ii) Force F_5 of the regulating spring (5) adjusted by means of the knob (4).

The forces acting on the bellows and element (13) that tend to move the needle up and hence to close the orifice are:

(i) Force F_0 on the bellows corresponding to the evaporator pressure p_0.

(ii) Force F_{14} of the follow-up spring (12).

During *off-cycle*, the valve is closed so that

$$F_0 + F_{14} > F_7 + F_5$$

When the compressor starts, F_0 drops as the evaporator pressure decreases. The valve starts opening and the liquid refrigerant begins to enter the evaporator. Force F_{14} is increased due to the compression of the follow-up spring whereas force F_5 is decreased due to the elongation of the regulating spring. When the equilibrium position is reached, we have

$$F_0 + F_{14} = F_7 + F_5$$

in which F_0 corresponds to a constant-evaporator pressure p_0. The adjustment of the tension in the regulating spring controls the force F_5 and hence the equilibrium pressure p_0 maintained in the evaporator.

Consider the effect of variation in the cooling load on the working of this valve. When the load decreases, vapour formation diminishes, and p_0 tends to drop as the compressor continues to suck vapour. As a result, the valve opens more and allows more refrigerant to enter the evaporator, thereby maintaining constant pressure. The evaporator becomes flooded and a decreased superheat results.

On the contrary, when the load increases, evaporation increases, and p_0 tends to rise, the valve tends to close, allowing less refrigerant to enter the evaporator and an increased superheat results.

Thus we see that an automatic-expansion valve is not suitable for a varying-load requirement. It increases the flow of the liquid refrigerant into the evaporator when the load decreases, and decreases the flow when an increase in load occurs. It is, however, used in applications such as milk coolers where precise control of the evaporator temperature is needed and where the cooling load is more or less constant.

8.3 THERMOSTATIC-EXPANSION VALVE[1]

A thermostatic-expansion or T-X valve maintains a constant degree of superheat in the evaporator.

A typical design of a thermostatic-expansion valve is shown in Fig. 8.2. It is only slightly different in construction from the automatic-expansion valve. In this case, the downward force on the bellows $(F_7 + F_5)$ which tends to open the valve is replaced by a force F_0' exerted on the bellows through the capillary tube (1) which is connected to the bulb (2). The bulb in turn is fixed at the outlet end of the evaporator as shown in Fig. 8.3.

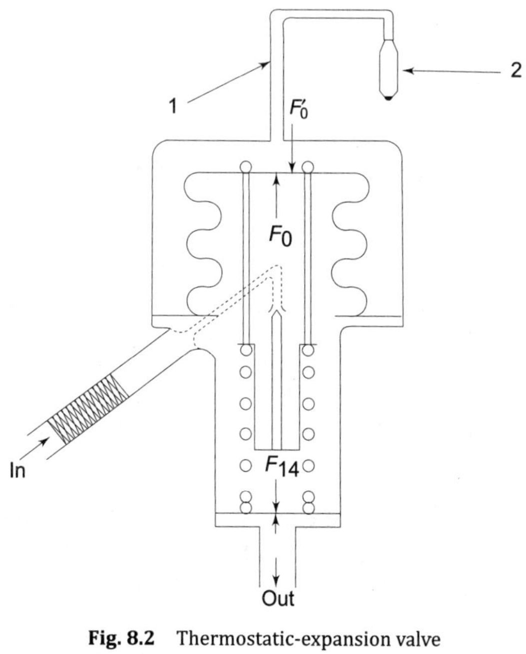

Fig. 8.2 Thermostatic-expansion valve

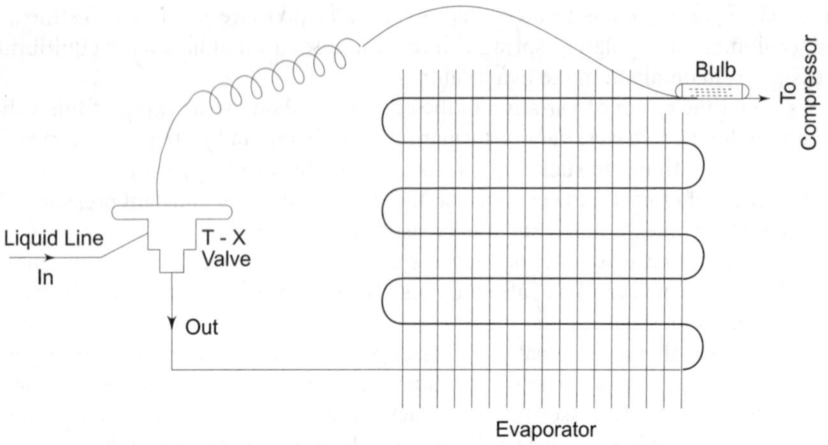

Fig. 8.3 Arrangement showing installation of thermostatic expansion valve and its thermal bulb

The forces acting upwards on the bellows which tend to close the valve are F_{14} of the follow-up spring and F_0 corresponding to the evaporator pressure p_0 as in the case of the automatic-expansion valve.

The capillary and bulb are generally filled with the same liquid refrigerant (power fluid) as in the refrigeration system. Force F_0' corresponds to the saturation pressure p_0' of the refrigerant at temperature t_0' at the exit end of the evaporator as shown in Fig. 8.4.

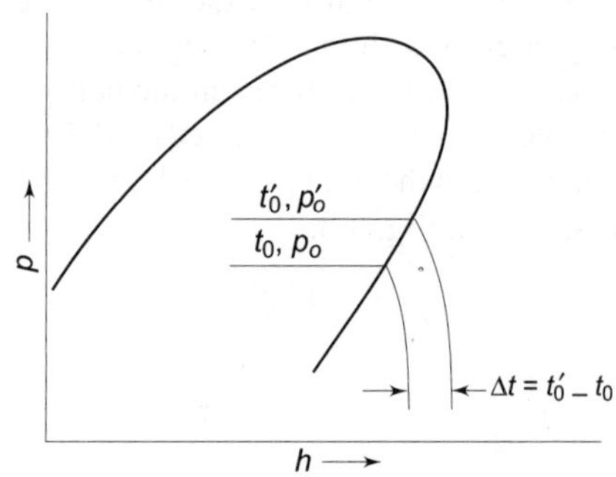

Fig. 8.4 Degree of superheat with thermostatic-expansion valve

During off-cycle, the temperatures at the evaporator inlet and outlet are the same so that $(F_0' - F_0)$ is zero. Since the follow-up spring is adjusted with some initial compression, the valve remains closed as

$$F_0 + F_{14} > F_0'$$

or,
$$F_{14} > (F_0' - F_0) = f(t_0' - t_0)$$

When the compressor starts, F_0 drops. A temperature difference $(t_0' - t_0)$ is created between the outlet and inlet of the evaporator, i.e., a degree of superheat results. The corresponding pressure difference $(p_0' - p_0)$ between the outside and inside of the bellow causes a downward force $(F_0' - F_0)$ on the bellow and the valve opens. Due to compression, F_{14} also increases. The opening of the valve is determined by the equilibrium position at which

$$(F_0' - F_0) = F_{14}$$

By adjusting the initial compression of the follow-up spring and hence the force F_{14}, the force $(F_0' - F_0)$ and hence the degree of superheat can be controlled. Normally, the setting of the spring is done in the factory to allow the valve to start opening at a superheat of 5°C.

8.3.1 Application of Thermostatic-Expansion Valves

The performance characteristic of the thermostatic-expansion valves is most suitable for application in air conditioning and refrigerant plants.

When the cooling load increases, the refrigerant evaporates at a faster rate in the evaporator than the compressor can suck. As a result, the pressure p_0 and the degree of superheat in the evaporator increase. The increase in superheat causes the valve to open more and to allow more refrigerant to enter the evaporator. At the same time, the increase in suction pressure p_0 also enables the compressor to deliver increased refrigerating capacity.

When the cooling load decreases, the refrigerant evaporates at a slower rate than the compressor is able to suck. As a result, the evaporator pressure drops and the degree of superheat decreases. The valve tends to close and the compressor delivers less refrigerating capacity at a decreased suction pressure.

Thus the thermostatic-expansion valve, as opposed to the automatic-expansion valve, is capable of meeting varying load requirements. The evaporator pressure and hence the evaporator temperature, however, do not remain constant. Another advantageous feature of the T-X valve is that by maintaining a constant degree of superheat of the suction vapour, it keeps the evaporator always full of the refrigerant, regardless of the changes in the cooling load. This ensures the efficient utilization of the evaporator surface even under extreme loading conditions and the safety of the compressor (by not allowing the liquid to enter it) under part-load conditions. These characteristics have enabled T-X valves to have universal application in air conditioning and many refrigeration systems.

8.3.2 External Equilizer

It has so far been assumed that pressure p_0 in the evaporator is throughout constant. In actual practice, however, there is always a significant pressure drop in the long evaporator tubing.

The pressure at the evaporator outlet is, therefore, very much less than the pressure at the evaporator inlet. If the pressure at the evaporator in let itself is exerted on the inside of the bellows of the thermostatic-expansion valve, then for a given compression F_{14} of the follow-up spring, the resulting superheat of the vapour will be higher than the adjusted superheat. Assuming that the pressure in the evaporator drops from p_{01} at the inlet to p_{02} at the outlet, the corresponding saturation temperatures being t_{01} and t_{02} respectively, it is seen from Fig. 8.5, that the compression F_{14} of the spring adjusted for a superheat of $\Delta t_1 = (t_0' - t_{01})$ results in an actual superheat of $\Delta t_2 = (t_0' - t_{02})$.

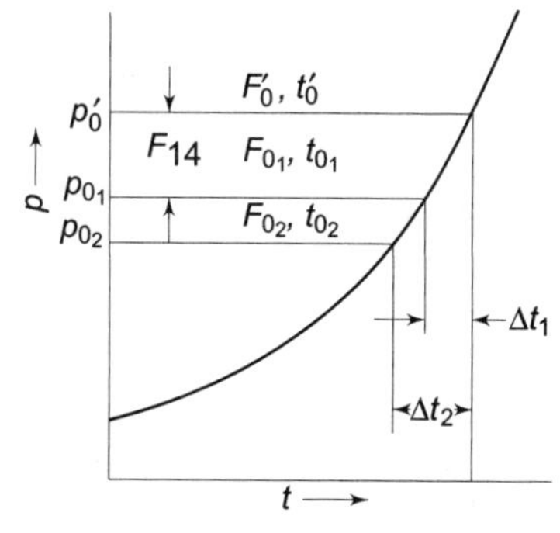

Fig. 8.5 Effect of evaporator pressure drop on degree of superheat

Further, it may be noted that under partial-load conditions, the superheat tends to increase in a normal-expansion valve. And if the pressure drop in the evaporator is large, it will result in a large amount of superheat of the vapour.

The remedy is to use an external-equalizer connection. An external-equalizer connection, as shown in Fig. 8.6, transmits the pressure at the outlet of the evaporator to the inside of the bellows of the expansion valve. At the same time, there is no connection between the evaporator inlet and inside of the bellows. Thus the pressure acting inside the bellows is always equal to the pressure at the evaporator outlet, irrespective of the extent of pressure drop in the evaporator tubing.

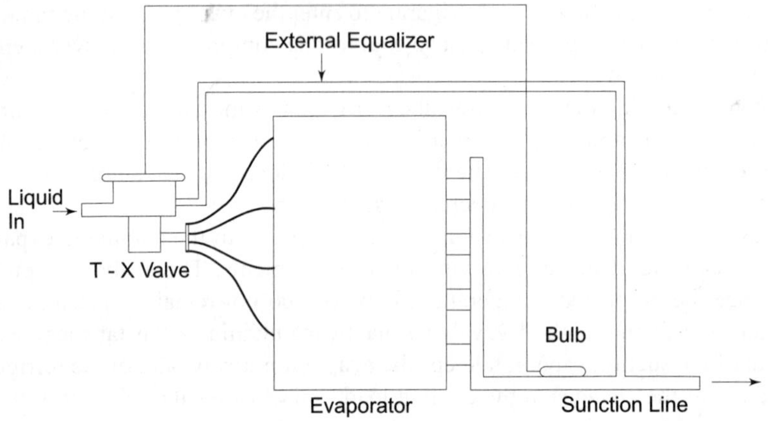

Fig. 8.6 External-equalizer connection

Example 8.1 *An R 134a Thermostatic-expansion valve, not equipped with an external equalizer, has a superheat setting of 7°C while supplying the refrigerant to the evaporator at 0°C. The power fluid is the same as the refrigerant.*

(a) Determine the difference in pressure on opposite sides of the diaphragm or bellows required to open the valve.

(b) If the temperature at the evaporator inlet is –5°C and the pressure drop through the coil is 0.3 bar, what is the degree of superheat of the suction gas leaving the evaporator?

Solution (a) Referring to Fig. 8.5, for the evaporator at 0°C

$$p_{0_1} \text{ (at 0°C)} = 2.928 \text{ bar}$$

$$p_0' \text{ (at 7°C)} = 3.748 \text{ bar}$$

The required pressure difference or initial compression of the follow-up spring per unit area of diaphragm is

$$f_{14} = p_0 - p_{0_1} = 3.748 - 2.928 = 0.82 \text{ bar}$$

(b) If the temperature at the evaporator inlet is – 5°C, then the pressure at the evaporator inlet is

$$p_{0_1} \text{ (at – 5°C)} = 2.435 \text{ bar}$$

For the same adjustment of the follow-up spring compression, the equilibrium pressure of the power fluid is

$$p_0' = p_{0_1} + f_{14} = 2.435 + 0.82 = 3.255 \text{ bar}$$

The corresponding saturation temperature of the power fluid in the bulb is

t'_0 (at 3.255 bar) = 2.47°C

Pressure at evaporator outlet

$$p_{0_2} = p_{0_1} - \Delta p = 2.435 - 0.3 = 2.135 \text{ bar}$$

Saturation temperature at the evaporator outlet

t_{0_2} (at 2.135 bar) = – 7.9°C

Effective superheat of the suction gas

$$\Delta t_2 = t'_0 - t_{0_2} = 2.47 - (-7.9) = 10.37°\text{C}$$

8.3.3 Cross-Charged Expansion Valves

It can be observed from the nature of the vapour-pressure curve of a refrigerant, as shown in Fig. 8.7, that the degree of superheat once adjusted by the tension F_{14} of the follow-up spring at a certain evaporator temperature, increases at lower evaporator temperatures. In a similar manner, it can be shown that a decreased superheat will result at a higher evaporator temperature.

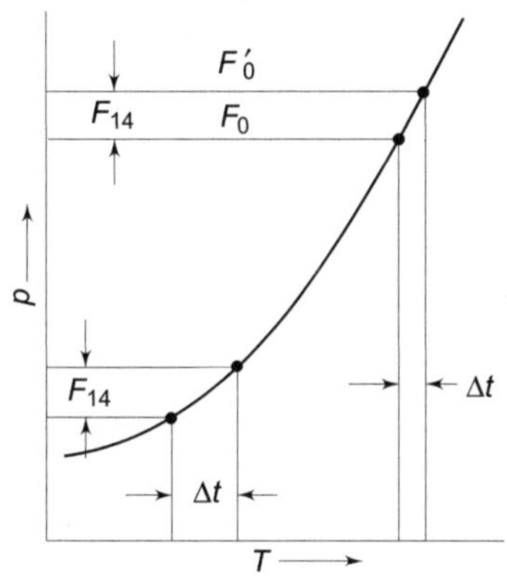

Fig. 8.7 Increase in superheat at lower evaporator temperature

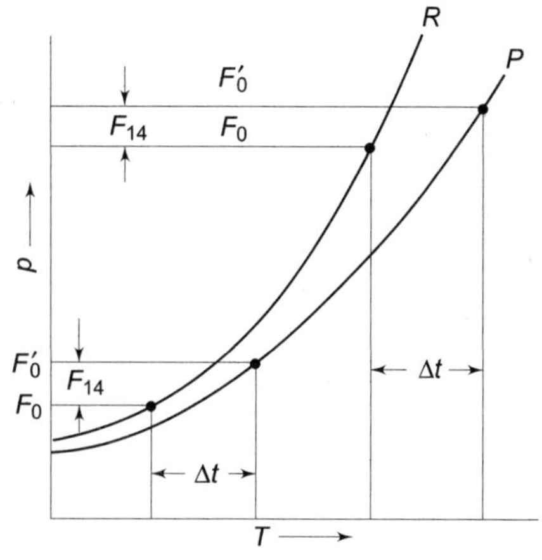

Fig. 8.8 Maintaining uniform superheat by the use of cross-charged expansion valve

The method by which the superheat may remain more or less unaffected by a change in the evaporator temperature employs a *power fluid* in the bulb as different from the refrigerant. The power fluid has a flatter *p-T* curve. Figure 8.8 shows curve *R* for the refrigerant and a flatter curve *P* for the power fluid. The power fluid is a slightly higher boiling substance. It has lower pressures. It is seen that the degree of superheat Δt almost remains constant at different evaporator temperatures. Such a thermostatic expansion valve is called *cross-charged.*

8.3.4 Thermostatic Charge and Fade-Out Point

The charge of the power fluid contained inside the bulb and the capillary tube of a thermostatic-expansion valve limits the maximum suction pressure up to which the expansion valve modulates.

Consider the thermodynamic change of state of the power fluid as its temperature varies. The process take place along a constant-volume line as shown in

Figs 8.9 and 8.10. Initially, at some stage, the fluid may exist at state A with the liquid and vapour in equilibrium at temperature t_A. With increase in temperature, the pressure increases according to the p-T relationship of the fluid until point B is reached. After B, at which all the liquid has turned into vapour, any further increase in temperature results in an insignificant increase in the pressure (according to the constant-volume process for a gas) inside the bulb and capillary of the expansion valve. As shown in Fig. 8.9, the increase in temperature equal to $t_C - t_B$ increases the pressure only by $p_D - p_B$. Thus at B, the pressure of the power fluid almost reaches a limiting value. The valve, therefore, does not open any wider with increase in the temperature, and the suction pressure stays constant at its value at B. The point B is called the *fade-out point*.

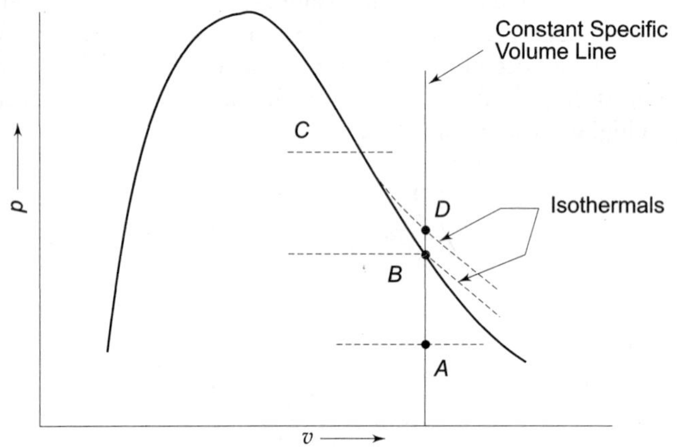

Fig. 8.9 Fade out point on p-v diagram

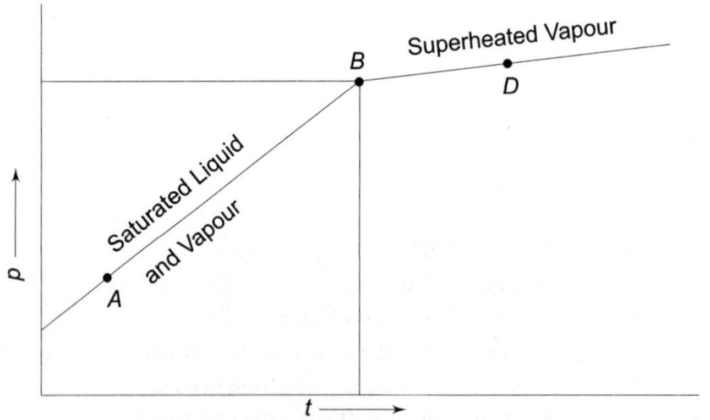

Fig. 8.10 Fade-out point on p-t diagram

It can be seen that the fade-out point depends on the specific volume of the power fluid, i.e., the mass of the refrigerant charged in the power element, as the total volume of the element is fixed. Thus, the limiting value of the suction pressure can be increased by decreasing the specific volume, i.e., increasing the charge of the power fluid. Similarly, the limiting suction pressure can be lowered by decreasing the charge. It can thus provide a built-in motor overload protection such that the refrigerating machine always operates on the left-hand side of the peak of the compressor power versus suction pressure characteristic (see Fig. 6.6).

Such an expansion valve with a limited liquid charge of the refrigerant is normally referred to as *gas-charged.* Gas-charged valves must be carefully applied in order to avoid a loss of control from the bulb. If the diaphragm or the bellow chamber becomes colder than the bulb, the small amount of charge in the element may condense and the bulb may contain only gas. Thus the valve will throttle or close.

A large mass of the refrigerant charged in the element will make it *liquid-charged.* In this case, the bulb contains liquid under all temperature conditions. Thus the bulb always controls the valve operation even with a colder diaphragm or bellow.

The charging of the refrigerant itself in the power element results in an increase in operating superheat as the evaporator temperature decreases. This limits its use for moderately high evaporator temperatures. For evaporator temperatures which are substantially below 0°C, *liquid cross-charged* expansion valves may be used. They have superheat characteristics which remain fairly constant throughout the evaporator temperature range.

8.4 CAPILLARY TUBE AND ITS SIZING

The capillary tube is a fixed restriction-type device. It is a long and narrow tube connecting the condenser directly to the evaporator. The pressure drop through the capillary tube is due to the following two factors:

(i) Friction, due to fluid viscosity, resulting in *frictional pressure drop.*

(ii) Acceleration, due to the flashing of the liquid refrigerant into vapour, resulting in *momentum pressure drop.*

The cumulative pressure drop must be equal to the difference in pressure at the two ends of the tube. The mass flow through the capillary tube will, therefore, adjust so that the pressure drop through the tube just equals the difference in pressure between the condenser and the evaporator. For a given state of the refrigerant, the pressure drop is directly proportional to the length and inversely proportional to the bore diameter of the tube.

A number of combinations of length and bore are possible for a capillary tube to obtain the desired flow and pressure drop. However, once a capillary tube has been selected, it will be suitable only for the design pressure drop and flow. It cannot satisfy the flow requirements with changing condenser and evaporator pressures. Even then, the capillary tube is the most commonly used expansion device in small refrigeration units, such as domestic refrigerators, window-type and split air conditioners, water coolers, etc. The advantages of a capillary tube are its simplicity, low cost and the absence of any moving parts. Also, it is found most advantageous with on-off control because of its unloading characteristics. Thus, when the compressor stops, it allows high and low pressures to equalize, thereby enabling the compressor motor to re-start on no load. Accordingly, smaller low-starting torque motors can be used.

The sizing of a capillary tube implies the selection of bore and length to provide the desired flow for the design condenser and evaporator pressures. The method employed by manufacturers is usually that of *cut and try*. The principle of design based on methods proposed by Stocker,[6] and Hopkins[4] and Cooper *et al.*[2] is presented here.

A capillary of a particular bore dia D and cross-sectional flow area A is first selected. Step decrements in pressure are then assumed and the corresponding required increments of length calculated. These increments can be totalled to give the complete length of the tubing for a given pressure drop.

Consider that the state of the entering refrigerant is saturated liquid. The mass flow rate $\dot{m}$ is known. The condenser and evaporator temperatures are t_k and t_0, and corresponding pressures are p_k and p_0 respectively. Divide this temperature drop into a number of parts. Let the corresponding pressure drops be Δp_1, Δp_2 and Δp_3, $\cdots$, etc., as shown in Fig. 8.11. Now there are two approaches to design.

(i) Isenthalpic expansion, as shown by line *k-a*.

(ii) Adiabatic or *Fanno-line* expansion, as shown by line *k-b*.

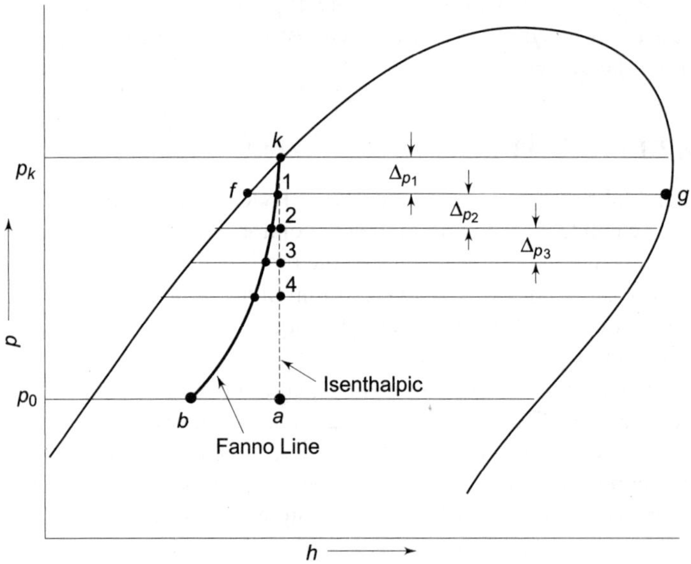

Fig. 8.11 Incremental pressure drops in a capillary tube

Isenthalpic expansion is the common thermodynamic assumption. In actual practice, however, expansion takes place adiabatically, viz., according to Fanno-line flow. Thus enthalpy does not remain constant since, with pressure drop, the volume increases and an increase in kinetic energy is obtained from a decrease in enthalpy. Nevertheless, it may be noted from Fig. 8.11 that in the first few steps of pressure drop, there is not much difference between isenthalpic and Fanno-line flow.

The steps of calculations to be followed in both cases are the same and are as follows for the first element.

(i) Determine the quality at the end of the decrement assuming isenthalpic flow. Then at point 1 at pressure p_1

$$x_1 = \frac{h_k - h_{f_1}}{h_{fg_1}} \tag{8.1}$$

(ii) Determine the specific volume

$$v_1 = v_{f1} + x_1(v_{g1} - v_{f1}) \tag{8.2}$$

(iii) Calculate the velocities from the continuity equation at both the ends of the element

$$u_k = \frac{\dot{m}v_k}{A} \text{ and } u_1 = \frac{\dot{m}v_1}{A}$$

or

$$\frac{u}{v} = \frac{\dot{m}}{A} = G = \text{Constant} \tag{8.3}$$

where G is mass velocity.

(iv) For Fanno-line flow, an iteration procedure is necessary. This is done by applying the correction to enthalpy since $h_1 \neq h_k$. Thus

$$h_1 = h_k - \frac{u_1^2}{2} \tag{8.4}$$

The calculations for quality, specific volume, velocity and enthalpy may be repeated until the final value of h_1 is equal to its value in the preceding iteration.

Note *The procedure appears time-consuming but, fortunately, it converges very fast.*

(v) Determine the pressure drop due to the acceleration, Δp_A, from the momentum equation

$$A\mathrm{d}p = -\dot{m}\,\mathrm{d}u$$

whence

$$\Delta p_A = \frac{\dot{m}}{A}(u_k - u_1) = G(u_k - u_1) \tag{8.5}$$

(vi) Determine the pressure drop due to the friction, Δp_F, from

$$\Delta p_F = \Delta p - \Delta p_A \tag{8.6}$$

(vii) Equate the required frictional pressure drop to

$$\Delta p_F = \frac{\rho f \Delta L u^2}{2D} \tag{8.7}$$

where

$$\rho = \frac{1}{v}$$

$$\Delta L = \text{length of the element}$$

Substituting

$$\dot{m} = \rho u A,\ \rho u = \frac{\dot{m}}{A} = G, \text{ we have} \tag{8.8}$$

$$\Delta p_F = \frac{G}{2D} fu\,\Delta L = Yfu\,\Delta L \text{ (say)} \tag{8.9}$$

where

$$Y = \frac{G}{2D} \tag{8.10}$$

from which the length ΔL may be calculated. For this purpose, the mean values of u and f for the liquid and vapour phases present may be taken for the section. The friction factor is a function of Reynolds number which in turn is expressed as

$$\textbf{Re} = \frac{Du\rho}{\mu} = \frac{DG}{\mu} = \frac{Z}{\mu} \text{ (say)} \tag{8.11}$$

where $$Z = DG \tag{8.12}$$

Niaz and Davis[5] have proposed the following correlation for evaluating the friction factor:

$$f = \frac{0.324}{\text{Re}^{0.25}}$$

They mention that the length obtained by using this correlation is about 10% greater than the experimental length. Based on available analytical and experimental data, the expression for friction factor in terms of a straight capillary given by Gorasia *et al*[3] is:

$$f_f = \frac{17.24}{\text{Re}^{0.62}} \quad \text{and} \quad f_g = \frac{17.24}{\text{Re}^{0.62}}$$

where

$$\text{Re}_f = \frac{GD}{\mu_f} \quad \text{and} \quad \text{Re}_g = \frac{GD}{\mu_g}$$

where subscripts f and g refer to liquid and gas phases respectively. The friction factor for the liquid-vapour mixture flowing in the capillary is found by taking into account the percentage weightage of each phase. Thus

$$f = f_f(1 - x) + f_g x$$

The examples that follow illustrate the method of calculation as it can be applied for the design of capillaries for air conditioners and refrigerators.

Example 8.2 ***Design of 1 TR R 22 Air Conditioner Capillary***

A capillary tube in a one-ton R 22 air conditioner has a bore of 2.3 mm. Saturated liquid from the condenser enters at a temperature of 48°C and flows adiabatically through the tube until its temperature is 5°C. Determine its length. The friction factor is given by

$$f = \frac{0.32}{\text{Re}^{0.25}}$$

Assume intermediate sections at 40, 30, 20 and 10°C.

Solution From the simple saturation cycle, the mass flow rate is

$$\dot{m} = 0.02417 \text{ kg/s}$$

The cross-sectional area of the capillary tube is

$$A = \frac{\pi}{4}(0.0023)^2 = 4.15 \times 10^{-6} \text{ m}^2$$

Let $$G = \frac{\dot{m}}{A} = \frac{0.02417}{4.15 \times 10^{-6}} = 5.83 \times 10^3 \text{ kgs}^{-1} \text{ m}^{-2}$$

$$= \frac{u}{v}$$

$$Y = \frac{G}{2D} = \frac{5.83 \times 10^3}{2(0.0023)} = 1.2674 \times 10^6 \text{ kgs}^{-1} \text{ m}^{-3}$$

$$Z = DG = (0.0023)\ (5.83 \times 10^3) = 13.41 \text{ kgs}^{-1} \text{ m}^{-1}$$

In actual practice, the enthalpy does not remain constant in flow through a capillary. Assuming isenthalpic flow, however, the properties and velocities at various sections are found and are given in Table 8.1. The last column in the table gives the actual decrease in enthalpy due to increase in kinetic energy, e.g., at the first point

$$\Delta h = h_k - h_1 = -\Delta(KE) = \frac{10.49^2 - 5.32^2}{2} = 41 \text{ J/kg} = 0.041 \text{ kJ/kg}$$

Table 8.1 Calculations for Example 8.2—Isenthalpic flow but $\Delta h = -\Delta(KE)$

Section	t	p	x	$10^3 v$	$u = Gv$	Δh
	°C	*bar*		m³/kg	m/s	kJ/kg
k	48	18.548	0	0.9137	5.32	0
1	40	15.335	0.07	1.804	10.49	0.041
2	30	11.819	0.144	3.375	19.63	0.179
3	20	9.099	0.209	5.704	33.18	0.536
4	10	6.807	0.267	9.167	53.33	1.408
5	5	5.84	0.292	11.572	67.03	2.232

Calculations can now be done for actual Fanno-line flow, starting from $h_k = 260.314$ kJ/kg.

Point 1 At 40°C, from the table of properties for R 22

$$h_f = 2497 \text{ kJ/kg} \qquad v_f = 0.0884 \times 10^{-3} \text{ m}^3/\text{kg}$$

$$h_{fg} = 166.9 \text{ kJ/kg} \qquad v_g = 15.1 \times 10^{-3} \text{ m}^3/\text{kg}$$

Then,
$$h_1 = h_k - \Delta h$$
$$= 260.314 - 0.041 = 260.273 \text{ kJ/kg}$$

Hence,
$$x_1 = \frac{260.273 - 249.7}{166.9} = 0.0633$$
$$v_1 = 0.884 \times 10^{-3} + 0.0633\ (15.1 - 0.884)10^{-3}$$
$$= 1.78 \times 10^{-3} \text{ m}^3/\text{kg}$$
$$u_1 = Gv_1 = 5.83 \times 10^3 \times 1.78 \times 10^{-3} = 10.38 \text{ m/s}$$

Recheck,
$$\Delta h_1 = \frac{10.38^2 - 5.32^2}{2} = 40 \text{ J/kg} = 0.04 \text{ kJ/kg}$$

Note *Further iteration can be done for greater accuracy. Proceeding in this manner, iterated values are obtained at various points which are given in Table 8.2.*

Table 8.2 Calculations for Example 8.2-Fanno-line flow

Section	*x*	*(h)iterated* kJ/kg	$10^3 \times (v)$ *iterated* m^3/kg	*(u)iterated* m/s	*(Δh) iterated* kJ/kg
k	0.0	260.5	0.91	5.317	0.0
1	0.064	260.46	1.8	10.474	0.04
2	0.133	260.324	3.356	19.525	0.18
3	0.191	259.977	5.634	32.773	0.52
4	0.240	259.163	8.937	51.99	1.34
5	0.261	258.424	11.114	64.655	2.08

Friction Factor Calculations: Point 1 (40°C)

Viscosities,
$$\mu_f = 0.221 \text{ cP}$$
$$\mu_g = 0.0134 \text{ cP}$$
$$\mu_1 = (1 - x_1)\,\mu_f + x_1\,\mu_g$$
$$= (1 - 0.064)0.221 + (0.064)(0.0134) = 0.2071 \text{ cP}$$

$$\mathbf{Re}_1 = \frac{Z}{\mu_1} = \frac{13.41}{0.2071 \times 10^{-3}} = 64{,}780$$

$$f_1 = \frac{0.32}{(64{,}780)^{0.25}} = 0.02$$

The calculations for various points are given in Table 8.3.

Table 8.3 Friction factor calculations for Example 8.2

Point	*t* °C		μ_f cP	μ_g cP	μ cP	**Re**	*f*
k	48	0	0.215	0.01368	0.215	62,400	0.020
1	40	0.064	0.221	0.0134	0.2071	64,780	0.02
2	30	0.133	0.229	0.0131	0.1979	66,800	0.02
3	20	0.191	0.239	0.0127	0.1917	68,400	0.0197
4	10	0.240	0.25	0.0124	0.1934	68,400	0.0197
5	5	0.261	0.257	0.0123	0.1973	69.300	0.0198

Length Calculations: Consider section *k*–1.

Total pressure drop

$${}_k\Delta p_1 = 18.543 - 15.331 = 3.212 \text{ bar}$$

Acceleration pressure drop

$$\Delta p_A = G\,\Delta u = (5.83 \times 10^3)(10.474 - 5.317) = 32.8 \times 10^3 \text{ N/m}^2$$

Friction pressure drop

$$\Delta p_F = \Delta p - \Delta p_A = 3.212 \times 10^5 - 0.328 \times 10^5 = 2.884 \times 10^5 \text{ N/m}^2$$

Mean friction factor

$$f = \frac{0.0202 + 0.02}{2} = 0.0201$$

Mean velocity

$$u = \frac{5.317 + 10.474}{2} = 7.82 \text{ m/s}$$

Incremental length $\quad {}_k\Delta L_1 = \dfrac{\Delta p_F}{Yfu} = \dfrac{2.884 \times 10^5}{(1.2674 \times 10^6)(0.0201)(7.82)} = 1.389 \text{ m}$

The calculations for various sections are given in Table 8.4.

Table 8.4 Capillary tube length calculations for Example 8.2

Sections	Δp	Δp_A	Δp_F	ΔL
	bar	***bar***	***bar***	***m***
k–1	3.212	0.328	2.884	1.389
1–2	3.416	0.574	2.842	0.706
2–3	2.818	0.857	1.961	0.276
3–4	2.292	1.283	1.009	0.087
4–5	0.969	0.844	0.125	0.008
		Total length required = $\Sigma\Delta L$ = 2.466 m		

References

1. Andersen, S A, *Automatic Refrigeration*, Maclaren and Sons Ltd. for Danfoss, Nordborg, Denmark, 1959.
2. Cooper L, C K Chu and W R Brisken, "Simple selection method for capillaries derived from physical flow conditions" *Refrigerating Engineering*, Vol. 65.
3. Gorasia, J N, N Dubey and K K, Jain, "Computer-aided design of capillaries of different configurations, *ASHRAE Transaction* 1991, pp. 132–138.
4. Hopkins N E, 'Rating the restrictor tube', *Refrigerating Engineering*, Vol. 58, No. 11, Nov. 1950, p. 1087.
5. Niaz, R H, and G, Davis, 'Adiabatic two-phase flow in a capillary tube', *Symp. Ser. of Can. Soc. for Chem. Eng.*, Vol. 1, 1969, pp. 259–269.
6. Stoecker, W F, *Refrigeration and Air Conditioning*, McGraw-Hill, New York 1958, pp. 111–129.

Revision Exercises

8.1 (a) Calculate the pressure of R 134a as it expands through a capillary tube having a bore of 1.05 mm if 4.028×10^{-2} kg/s of saturated liquid enters the capillary tube at a temperature of 45°C. Neglect heat transfer.

(b) Determine the temperature at which the condition of *choked flow* occurs as the expansion proceeds.

8.2 An R 134a thermostatic expansion valve uses R 12 itself as the power fluid set to fade out at a pressure of 0.35 MN/m^2. The internal volume of the power element is 15 cm^3. Calculate the mass of R 12 contained in the element.

8.3 (a) An R 22 expansion valve is factory set for a superheat of 7°C when supplying refrigerant to an evaporator at 5°C. If the evaporator is operating at – 15°C, what will be the effective superheat of the suction vapour?

(b) If the valve is cross-charged with R 134a as the power fluid, what will be the effective superheat?

8.4 An R 22 thermostatic expansion valve with R 22 itself as the power fluid is not equipped with an external equalizer. It supplies a coil in which there is pressure drop due to friction. The superheat setting made on the valve at the factory is 5°C with 0°C evaporator.

(i) What is the difference in pressure on opposite sides of the valve required to open the valve?

(ii) When the pressure at evaporator inlet is 4.22 bar, how many degrees is the suction gas superheated at 11.5°C evaporator exit temperature if the pressure drop through the coil is 0.53 bar?

8.5 How much would have been the superheat in Prob. 8.4 (ii) if the valve had been equipped with an external equalizer?

8.6 Design a capillary for a 165 L $\left(\dot{Q}_0 = 89W\right)$ refrigerator working on

(a) R 12 (b) R 134a

The R 12 compressor is 4.33 CC.

The R 134a compressor is 5.48 CC.

Use Fig. 3.21 for cycle analysis.

Chapter

9

Evaporators

9.1 TYPES OF EVAPORATORS

The evaporator is the component of a refrigeration system in which heat is removed from air, water or any other body required to be cooled by the evaporating refrigerant.

Evaporators are mainly classified as *flooded* or *direct-expansion*, viz., *dry*. In flooded evaporators, the liquid refrigerant covers the entire heat-transfer surface. In or dry evaporators, a part of the heat-transfer surface is used for superheating the vapour. A float valve is used for the expansion of the refrigerant in the case of a flooded evaporator, whereas a thermostatic-expansion valve (in the case of large units) or a capillary tube (in the case of small units) is used in conjunction with a dry evaporator.

A distinction can also be made on the basis of the flow of refrigerant inside tubes in dry evaporators and outside tubes in flooded evaporators.

An illustration of a flooded evaporator used as a water chiller with the refrigerant in the shell is shown in Fig. 9.1.

As far as direct-expansion (D-X) evaporators are concerned, there are a number of types of the same such as the following:

(i) Direct-expansion chiller.
(ii) Direct-expansion cooling coil for air with forced convection.
(iii) Direct-expansion coil for air-blast freezer.
(iv) Natural convection evaporator for freezers of domestic refrigerators.
(v) Evaporator coils submerged in brine tanks for ice plants.

In contrast to flooded chiller, Fig. 9.2 shows a D-X chiller in which water flows in the shell across a number of *baffles*, and refrigerant, from the liquid line, directly expands inside tubes and flows through a number of passes simultaneously evaporating and absorbing heat from recirculated water.

Another common type of dry evaporator is the direct-expansion (D-X) coil as shown in Fig. 9.3 with fins on the air side used in air-conditioning equipment. The coil shown in the figure has 4 rows of tubing, and only one tube-circuit. Large D-X evaporators have more than one refrigerant circuit otherwise the refrigerant pressure drop will become too large.

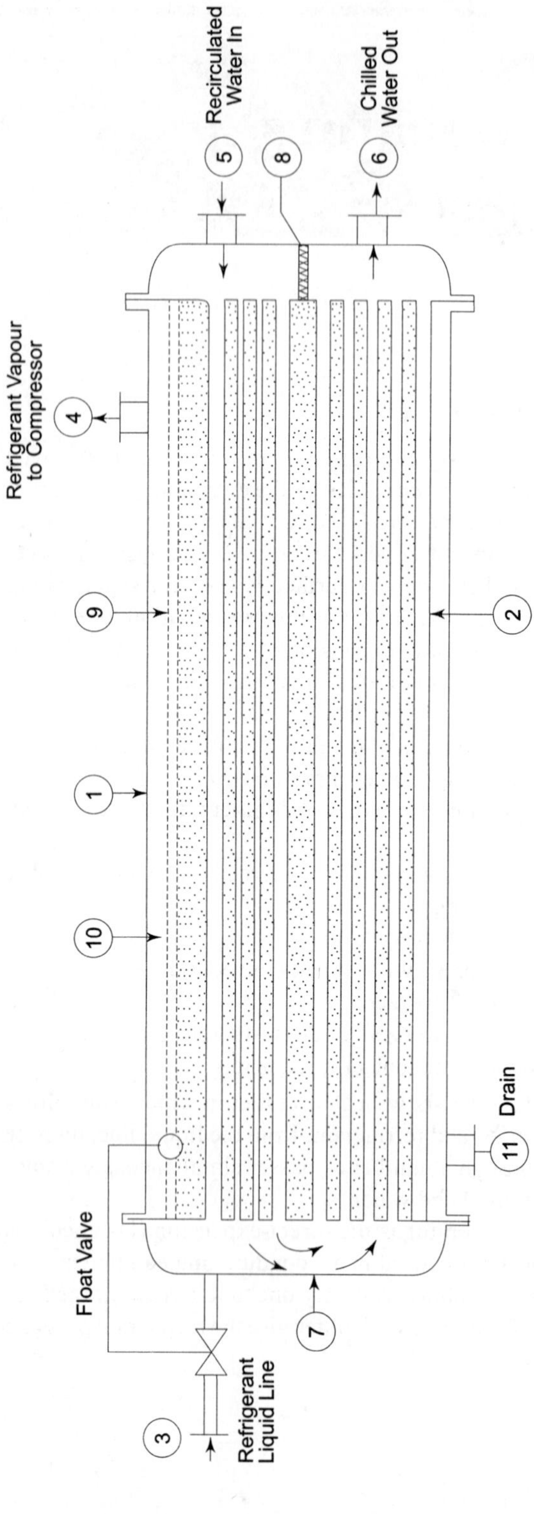

1. Shell
2. Tubes
3. Refrigerant Inlet
4. Refrigerant Outlet
5. Water Inlet
6. Water Outlet
7. Header
8. Partition
9. Liquid Refrigerant Level
10. Vapour Space
11. Drain

Fig. 9.1 Flooded chiller

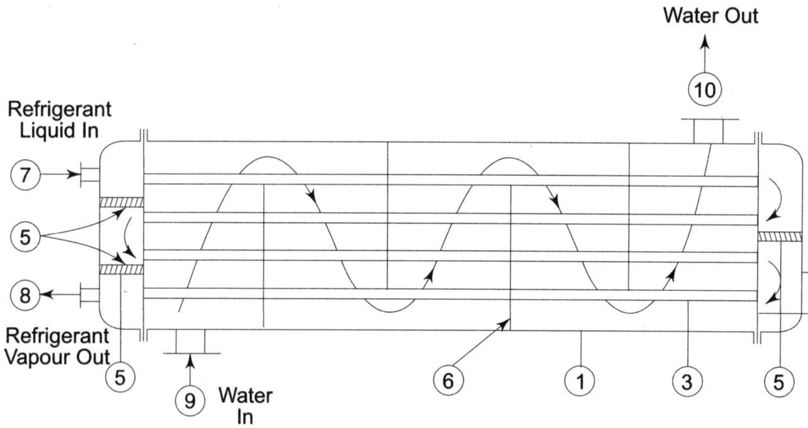

1 Shell
2 Headers
3 Tubes
4 Tube Sheet
5 Partitions
6 Baffles
7 Refrigerant Inlet (from Expansion Valve)
8 Refrigerant Outlet
9 Water Inlet
10 Water Outlet

Fig. 9.2 Direct-expansion chiller

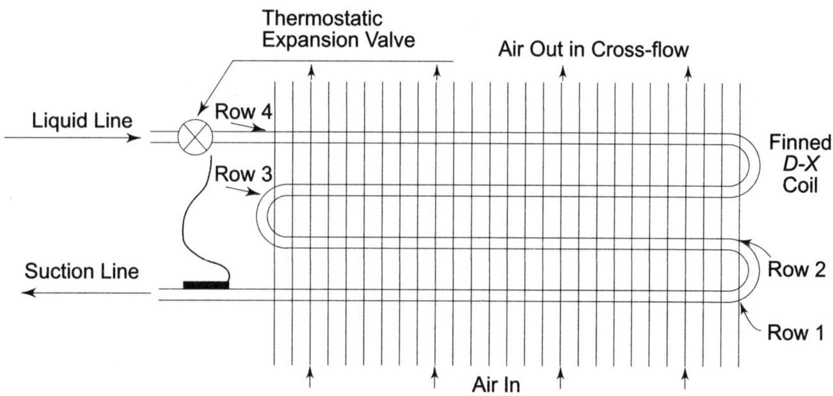

Fig. 9.3 Illustration of a dry evaporator, direct-expansion cooling coil with thermostatic-expansion valve

The name direct-expansion is derived from the fact that the refrigerant expands directly inside the tubing, and evaporates, thus cooling the medium outside. To facilitate the return of oil to the compressor, D-X evaporators are fed from the top by a thermostatic-expansion valve as shown in Figs 9.2 and 9.3.

The D-X coil for air-blast freezers is similar in construction except that the refrigerant temperature is much lower, viz., – 35 to – 40°C. As a result there is frost formation in between the fins. The flow of air could, therefore, get blocked. The frosting is more at the inlet end. Accordingly, larger fin-spacing is provided at the inlet end. The spacing progressively decreases as the air gets dehumidified.

Present-day frost-free refrigerators also have similar freezer coils. These are placed outside the freezer/refrigerator cabinet. Air flows by forced convection between the coil and the freezer compartment.

The conventional refrigerators use *bonded evaporators*. The refrigerant flows and evaporates inside channels formed by binding engrooved plates. Heat is absorbed from air on the surface of plates by natural convection.

The ice plants use ammonia evaporator coils submerged in agitated brine tanks. The amount of refrigerant charge in such coils is, no doubt, large. But, it does not matter since ammonia is not expensive.

9.2 HEAT TRANSFER IN EVAPORATORS

The three heat-transfer resistances in evaporators are:

(i) Refrigerant side for the transfer of heat from solid surface to the liquid refrigerant.
(ii) Metal wall.
(iii) Cooled-medium side which could be due to air, water, brine or any other fluid or a wetted surface on a cooling and dehumidifying coil.

The heat transfer from solid surface to the evaporating refrigerant is of primary interest here. However, the mechanism of boiling is so complex because of the influence of such factors as surface tension, saturation temperature, latent heat and nature of the solid surface, in addition to the usual transport properties, that it is very difficult to predict the heat-transfer coefficient analytically. Nevertheless, an attempt is made here to present correlations applicable to evaporating refrigerants screened through the large amount of published information available on the subject.

In commercial equipment, the boiling process occurs in two types of situations: one, of *pool boiling* as in flooded evaporators with refrigerant boiling on the shell-side, and the other, of *flow* or *forced convection boiling* as in direct-expansion evaporators with refrigerant on the tube-side.

9.2.1 Pool Boiling

Pool boiling occurs in flooded evaporators. Experiments on pool boiling show three distinct regimes as illustrated in Fig. 9.4.

Ordinarily, during the boiling of a liquid at its saturation temperature, such as in a pool, simple evaporation or *natural convection evaporation* occurs at the free surface without the formation of bubbles when the solid wall temperature t_w is only a few degrees above the saturation temperature of the evaporating substance t_s. It is shown by Regime 1 in Fig. 9.4. But if $\Delta t = (t_w - t_s)$ is increased, vapour bubbles form and agitate the liquid in the vicinity of the heating surface, that is, in Regime 2 in Fig. 9.4. With increasing Δt the bubbles rise and break through the free surface. In this Regime 3 heat transfer coefficients are higher. This kind of boiling with bubble growth (Regimes 2 and 3) is called *nucleate boiling*. Eventually, as $(t_w - t_s)$ is increased further, the heating surface gets blanketed with vapour bubbles resulting in a process called *film boiling*. Thus, there are three distinct regimes of pool boiling.

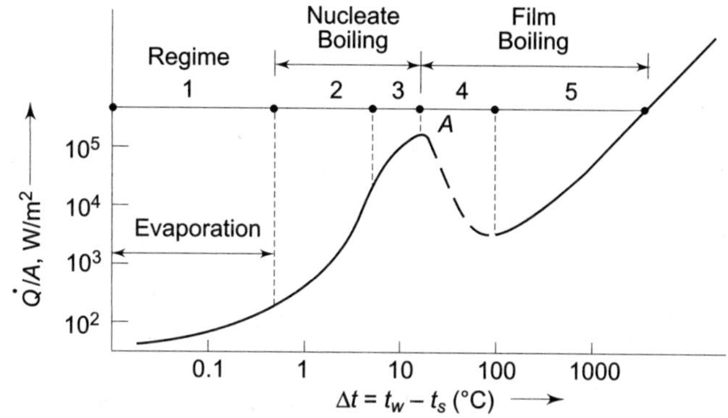

Fig. 9.4 Heat flux vs. temperature difference in pool boiling

Heat transfer in refrigerant evaporators corresponds to Regime 2 of the nucleate boiling in which the bubbles formed condense in the liquid before reaching the surface.

9.2.2 Heat Transfer Coefficient for Nucleate Pool Boiling

A method, based on a logical explanation of the mechanism of heat transfer associated with the boiling process, was presented by Rohsenow[14] for correlating heat-transfer data for nucleate boiling of liquids for the case of pool boiling. The suggested relation was

$$\frac{C_f \Delta t}{h_{fg}} = C_{sf}\left\{\frac{\dot{Q}/A}{\mu_f h_{fg}}\left[\frac{\sigma}{g(\rho_f - \rho_g)}\right]^{\frac{1}{2}}\right\}^r \left(\frac{C\mu}{k}\right)_f^s \tag{9.1}$$

where σ is the vapour-liquid surface tension of the fluid, and the various fluid properties are evaluated at the saturation temperature corresponding to the evaporation pressure. The constant C_{sf} is a function of the particular fluid-heating surface combination.

By plotting the experimental data of a known fluid (i.e., **Pr** = constant) for the first term on the RHS versus the LHS, Rohsenow obtained the value of the exponent r as 0.33. Similarly, a cross-plot of $C_f \Delta t/h_{fg}$ vs. **Pr** for constant values of the first term on the RHS gives s equal to 1.0 for water and 1.7 for R. Then, a minimum of one test point, i.e., a value of $\dot{Q}/A$ and its corresponding value of Δt, is all that is needed to evaluate C_{sf}. Piret and Isbin[12] have found its value for the CCl_4-copper combination as 0.013. The same value can also be used for copper and R 22 and other fluorocarbons.

Ratiani and Avaliani[13] recommended the following simple correlation for pool boiling of Freon refrigerants

$$h = 1.35\, q^{0.7}\, p^{665/T_s^{1.3}} \tag{9.2}$$

where h is in W/m^{2}K, q is the heat flux in W/m^2, p is the pressure in bar and T_s is the normal boiling point in K. Combining this relation with $q = h\,\Delta T$, we obtain

$$h = 2.72\, \Delta T^{2.33}\, p^{2214/T_s^{1.3}} \tag{9.3}$$

For the case of R 22, this reduces to

$$h = 2.72\, \Delta T^{2.33}\, p^{1.86} \tag{9.3a}$$

9.2.3 Flow or Forced-Convection Boiling

Boiling will take place when a liquid flows by forced convection over a surface maintained at a temperature higher than the saturation temperature of the fluid. Boiling under these conditions is called *flow* or *forced-convection boiling*. Flow boiling occurs in D-X evaporators. Heat-transfer rates in forced-convection boiling are substantially higher than those of pool boiling.

For forced-convection boiling, Rohsenow and Griffith suggested that the total heat flux be separated into two parts: One, a *boiling flux* $\dot{Q}_b/A$, and the other, a *convective flux* $\dot{Q}_c/A$. Thus the total flux can be expressed as

$$q = q_b + q_c$$

For computing q_b, the normal pool boiling correlation may be used. For computing q_c, e.g., forced-convection inside tubes, the Dittus-Boelter equation can be used. In this equation, however, it is recommended that the constant 0.023 be replaced by 0.019.

The forced-convection boiling processes taking place in dry evaporators are very complex. At the inlet end, the flow is of a low quality. As the evaporation proceeds, the quality increases, a critical condition (tube wall dry-out) appears and then the remainder of the tube may have superheated vapour with liquid droplets.

Hoogendooren discovered seven distinct possibilities for two-phase flow, five of which are given below.

(i) *Laminar* or *stratified flow* in which both phases flowing beside each other are separated by a plane surface.

(ii) *Wavy flow* in which the two phases are separated by a wavy surface.

(iii) *Slug flow* in which slugs of the liquid are enclosed in the gas stream.

(iv) *Annular flow* with the liquid flowing in an annular form and the gas flowing in the central core. The occurrence of this phase of boiling can be expected when the mixture velocity exceeds 2 m/s.

(v) *Foam* or *mist flow* with the flow patterns corresponding to definite mass flow rates or liquid velocities.

The transitions occur as the flow proceeds from the inlet end to the outlet end of the evaporator.

The major portion of the heat transfer occurs in the horizontal tube through the wetted part of the tube wall. The heat-transfer coefficient increases when the flow becomes annular, but when most of the liquid is evaporated and the dry-out point is reached, the coefficient falls off rapidly to a low value.

To a large degree, the complete length of a single evaporator tube can be divided into three convenient parts, viz., stratified flow at the inlet, first slug, and then annular flow in the middle, and mist flow at the outlet.

A significant feature of boiling heat transfer is the dependence of the heat-transfer coefficient on quality. The nature of variation is shown in Fig. 9.5. For water h is maximum at $x = 0.4$. For fluorocarbons the peak occurs at $0.4 < x < 0.8$. If this is true, a substantial improvement in the evaporator capacity is achieved with wet operation, represented by line CD in Fig. 9.5 as against dry operation with superheat as from A to B. The average value of h-when the quality changes from C to D—is much higher than the same when it changes from A to B as in normal evaporators without recirculation. This is achieved by recirculation (see Sec. 9.4.3).

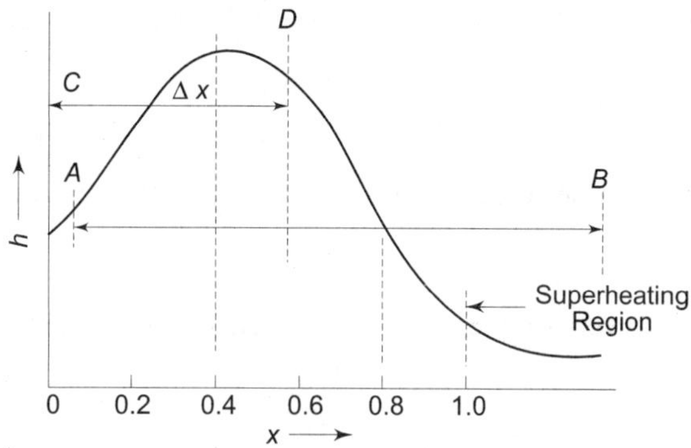

Fig. 9.5 Variation of flow-boiling heat transfer coefficient with dryness fraction of refrigerant

9.2.4 Forced Convection Boiling Correlations

Boiling of liquids has been of considerable interest to nuclear, chemical and mechanical engineers and, consequently, a lot of work has been done to obtain general correlations for predicting heat-transfer coefficients and pressure drops for flows with boiling. However, as verified by Andersen *et al.*, most of these do not satisfactorily predict the heat-transfer coefficients and pressure drops for evaporating refrigerants. Only Bo Pierre's[2] correlation predicts somewhat satisfactory values for the average heat-transfer coefficient but this does not give the local heat-transfer coefficient. His correlations are given below.

$$\text{Nu}_f = 0.0009\,(\text{Re}_f^2\, K_f)^{0.5} \tag{9.4}$$

for incomplete evaporation, viz., dryness fraction at exit, $x_0 \le 0.9$, and

$$\text{Nu}_f = 0.0082\,(\text{Re}_f^2\, K_f)^{0.5} \tag{9.5}$$

for complete evaporation, viz., exit condition 6°C superheated.

Here Nu_f and Re_f are the liquid Nusselt and Reynold numbers respectively, and K_f is the *load factor* defined by

$$K_f = \frac{\Delta x\, h_{fg}}{L}$$

where L is the length of the tube, and Δx is the change in dryness fraction.

The expression for incomplete evaporation has been found to conform to experimental data more precisely.

There are some expressions which involve the Lockhart-Martinelli parameter $\mathcal{H}$ relating the two-phase heat-transfer coefficient h_{TP} with the single liquid phase coefficient h_f.

Chaddock

$$h_{TP} = 1.85\, h_f \left[\text{Bo} \times 10^4 + \left(\frac{1}{\mathcal{H}_{tt}}\right)^{0.67}\right]^{0.6} \tag{9.6}$$

where $\textbf{Bo} = \text{Boiling number} = \dfrac{\dot{Q}/A}{(\dot{m}/A)h_{fg}} = \dfrac{q}{Gh_{fg}} = \dfrac{h\Delta T}{Gh_{fg}}$

$\mathcal{X}_{tt}$ is the Lockhart-Martinelli parameter for turbulent flow given by

$$\mathcal{X}_{tt} = \left(\frac{1-x}{x}\right)^{0.9}\left(\frac{\rho_g}{\rho_f}\right)^{0.5}\left(\frac{\mu_f}{\mu_g}\right)^{0.1}$$

and h_f is the liquid phase forced-convection heat-transfer coefficient for turbulent flow.

Bogdanov[3]

Bogdanov gives a very simple correlation for refrigerants boiling inside tubes, viz.,

$$h = z^2 \frac{G^{0.4}\Delta T}{D_i} \tag{9.7}$$

where G is mass velocity in kg/m^2. s and h is in W/m^2. K. z is a factor depending on the refrigerant and temperature as given in Table 9.1 for R 22 and R 142.

Table 9.1 Values of z in Equation (9.7)

Temp., °C	*R 22*	*R 142*
– 30	0.99	0.61
– 10	1.22	0.76
+ 10	1.54	0.94
+ 30	1.82	1.17

Lavin and Young[10]

$$h_{TP} = C\, h_f \left(\frac{1}{\mathcal{X}_{tt}}\right)^n \tag{9.8}$$

where $C = 3.5$ and $n = 0.75$ for R 12 as found by Murty[11].

All the above expressions are applicable to horizontal-tube evaporators.

There are other correlations which have also been proposed for the nucleate range. They yield a varying exponent of $\dot{Q}A$ vs. Δt increasing with Δt, in the range of 2 to 4. Thus

$$q = \dot{Q}A = C\,(\Delta t)^{2 \text{ to } 4} \tag{9.9}$$

Dividing both sides by Δt, we get the boiling heat-transfer coefficient

$$h = \frac{\dot{Q}}{A\Delta t} = C\,(\Delta t)^n$$

where n lies between 1 and 3. All boiling correlations can be reduced to this form.

This form of equation is very suitable for computer work as precise values of the constants C and n at different x can be given for different parts of the nucleate boiling range which can simultaneously occur in the case of refrigerant boiling inside a tube.

Of late, Chawla[4] and Dembi, Dhar and Arora[6] have obtained generalized semi-empirical correlations to satisfactorily predict the pressure drop and heat transfer of the refrigerants evaporating inside horizontal tubes. Dembi, Dhar and Arora's correlations have minimum RMS error and are as follows:

For forced convection vaporization (annular flow region)

$$\frac{hD}{k_f} = 0.115\,[x^4(1-x^2)]^{0.11}\left[\frac{G^2 h_{fg}}{g\sigma\rho_f}\right]^{0.44}[\mathrm{Pr}_f]^{0.7} \tag{9.10}$$

For nucleate flow boiling (stratified flow region)

$$\frac{hD}{k_f} = 23388.5\left[\frac{q}{\rho_g\, h_{fg}\, w''}\right]^{0.64}\left[\frac{gD}{h_{fg}}\right]^{0.27}\left[\frac{G^2 D}{\sigma \rho_f}\right]^{0.14} \tag{9.11}$$

where w'' is the *vapour bubble growth rate parameter* correlated as follows:

$$w'' = 0.36 \times 10^{-3}\left(\frac{p_c}{p}\right)^{1.4} \tag{9.12}$$

Equation (9.10) applies to very low Δt values while Eq. (9.11) applies to high Δt values, corresponding to similar regimes in pool boiling. In case of doubt, the one giving a higher numerical value of h may be used.

9.2.5 Horizontal vs. Vertical Tube

The boiling in vertical tube evaporators is characterised by the separation of the liquid from the vapour emerging from its surface. It, therefore, represents a case of pool boiling. Also, the hydrostatic pressure exerted by a column of liquid flooding the evaporator causes an increase in the saturation temperature. The heat-transfer coefficient as well as $t_w - t_s$ within vertical tube evaporator are, therefore, very small.

On the contrary, in horizontal tube coil evaporators, it is only at small velocities of flow inside the tube that the separation of phases occurs. As the length of a single coil is considerable and the vapour content of the refrigerant increases steadily, a vapour-liquid emulsion appears with an ever increasing volume flowing correspondingly faster resulting in a high value of the heat-transfer coefficient as in the case of forced-convection boiling. High refrigerant vapour velocities also ensure the return of oil to the compressor.

9.2.6 Effect of Oil in Refrigerant on Heat Transfer

It is generally assumed that the effect of oil is to decrease the boiling heat transfer coefficient because of its high viscosity. A 5 per cent solution of oil may have about 45 per cent greater viscosity than pure refrigerant. The corresponding effect on the heat-transfer coefficient would be to diminish it by 60 per cent. In this calculation the effect of slightly increased thermal conductivity and decreased specific heat due to oil has not been considered.

The experiments, however, show that the average heat-transfer coefficient first increases reaching a peak at about 4 per cent oil content and then decreases. The higher coefficient is attributed to the onset of annular flow with oil at low volume flow rates, i.e., in the inlet region of the tube. Also, the oil may act as a wetting agent for refrigerant so that the bubbles are swept away from the surface in the low-velocity region.

Notwithstanding this improvement in h at small oil concentrations, the value of h at 9 to 10 per cent oil would be about the same as that for an oil-free refrigerant.

Example 9.1 ***Basic Design of Direct-Expansion Chiller***

Some of the design details of a Refrigerant 22, 20 TR, D-X chiller are as follows:

Effective tube length	*221.5 cm*
Diameter of tubes	*1.905 cm OD, 1.704 cm ID*

Number of refrigerant passes	*8*
Entering water temperature	*11.1°C*
Leaving water temperature	*7.2°C*
Refrigerant temperature at inlet	*2.2°C*
Condensing temperature	*43.3°C*

The water-side heat-transfer coefficient h_0 may be taken as 4,650 W.m^{-2} °C^{-1}. The refrigerant-side coefficient may be approximated by

$$h_i = 230\ \Delta t\ \text{W.m}^{-2}\ {}^\circ\text{C}^{-1}$$

Find the number of tubes in the last pass. Assume equal enthalpy change in all passes. Neglect the thickness and thermal resistance of the tube wall.

Solution Heat transfer in each pass

$$\dot{Q} = \frac{(20)(3.5167)}{8} = 8.792 \text{ kW}$$

Temperature drop of water in each pass

$$\Delta t_w = \frac{11.1 - 7.2}{8} = \frac{3.9}{8} = 0.49°\text{C}$$

Mass flow rate of water

$$\dot{m}_w = \frac{(20)(2.5167)}{(4.1868)(3.9)} = 4.307 \text{ kg/s}$$

For Eighth Pass

Entering water temperature, $t_{w_1} = 11.1°\text{C}$

Leaving water temperature, $t_{w_2} = 11.1 - 0.49 = 10.61°\text{C}$

Mean temperature difference

$$\Delta t_m = \frac{0.49}{\ln (11.1 - 2.2)/(10.61 - 2.2)} = 8.85°\text{C}$$

Assume mean tube wall temperature $t_m = 8.8°\text{C}$. Then the refrigerant-side film coefficient is

$$h_i = 230\ (8.8 - 2.2) = 1518 \text{ W. m}^{-2}\ \text{K}^{-1}$$

Overall heat-transfer coefficient based on outside tube surface area

$$\frac{1}{U_0} = \frac{1}{h_0} + \frac{1}{h_i}\frac{A_0}{A_i} = \frac{1}{4650} + \frac{1}{1518}\left(\frac{1.905}{1.704}\right)$$

$$\Rightarrow \qquad U_0 = 1051 \text{ W.m}^{-2}\ \text{K}^{-1}$$

Heat transfer area of the pass

$$A_0 = \frac{\dot{Q}}{U_0 \Delta t_m} = \frac{8.792 \times 10^3}{(1051)(8.85)} = 0.945 \text{ m}^2$$

$$A_i = 0.945 \times \frac{1.704}{1.905} = 0.845 \text{ m}^2$$

Check for the assumed valve of t_m

$$t_m = 2.2 + \frac{\dot{Q}}{h_i A_i} = 2.2 + \frac{8.792 \times 10^3}{(1518)(0.845)} = 9.05°\text{C}$$

Further iteration converges at

$$t_m = 9°\text{C}, \qquad h_i = 1564 \text{ W.m}^{-2} \text{K}^{-1}.$$
$$U_0 = 1076 \text{ Wm}^{-2} \text{K}^{-1}, \qquad A_0 = 0.924 \text{ m}^2$$

Number of tubes in the pass

$$n = \frac{A}{\pi D_0 L} = -\frac{0.924}{\pi(0.01905)(2.215)} = 6.974$$

Take 7 tubes. Then the heat transfer in the pass is

$$\dot{Q} = \frac{7}{6.974}(8.792) = 8.82 \text{ kW}$$

Water temperature drop in the pass remains same as assumed, viz.,

$$\Delta t_w = \frac{7}{6.974}(0.49) = 0.49°\text{C}$$

Note *In the above calculations, a large number of simplifications have been made which include the following:*

(i) The water-side coefficient is constant.

(ii) The refrigerant-side coefficient is not dependent upon the dryness fraction and an extremely simplified expression is used.

(iii) Tube wall and fouling resistances (0.0005 m^2 k/W) have been neglected. Recommended fouling resistance for Chillers is 0.0005 m^2 k/W.

(iv) The refrigerant-side pressure drop is neglected. If considered, this will reduce the temperature potential in the chiller and hence increase the surface area and number of tubes.

(v) The effect of tube spacing, baffle pitch, etc., on the water-side coefficient is ignored.

(vi) LMTD will also be affected by baffle pitch, baffle cut, etc.

9.3 EXTENDED SURFACE EVAPORATORS

Extended surfaces are used in the evaporators both on the refrigerant-side as well as on the cooled medium side depending on which heat-transfer coefficient is lower.

Flooded evaporators are used in chillers for water or brine. The pool-boiling refrigerant-side heat-transfer coefficient in flooded evaporators is very low. The order of magnitude is 396 W/m^2.K as against 4170 W/m^2.K for water flowing inside tubes. Accordingly, integral fins on the outside of tubes as illustrated in Figs 7.6 and 9.6(a) are used on the refrigerant side.

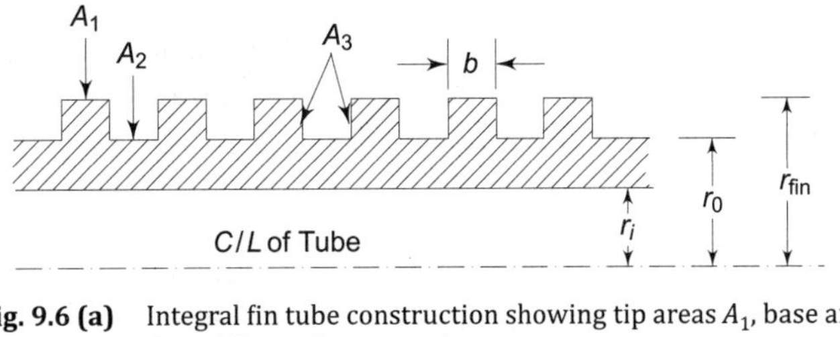

Fig. 9.6 (a) Integral fin tube construction showing tip areas A_1, base areas A_2 and fin surface areas A_3

Direct-expansion evaporators are used both for

(i) chilling water or brine, and

(ii) cooling air.

In direct-expansion chillers, the refrigerant-side heat-transfer coefficient with refrigerant boiling inside tubes is lower than water-side coefficient. Hence, either *inner fins* to increase the area, or *enhanced surfaces* to improve the heat-transfer coefficient, or both are used.

In direct-expansion cooling coils for air, there may be natural or forced convection on the air-side. In both cases, the heat-transfer coefficients on the air-side are very low.

In natural convection evaporators as used in domestic refrigerators, the air-side coefficient is very-very low. With the use of *bonded evaporators* that provide extended surface on the air-side, the heat transfer conductance is somewhat improved. Still the air-side coefficient remains the controlling coefficient in such evaporators. The overall coefficient is lower than the effective natural convection air-side coefficient.

The present *frost-free refrigerators*, however, use forced convection of air by a fan between freezer compartment of the refrigerator and the evaporator coil which is located outside the cabinet.

In forced convection evaporators also as used in air conditioners, the air-side coefficient is quite low.

Extended surface evaporators, as shown in Fig. 9.6(b) are extensively used for cooling of air since the air-side heat-transfer coefficient is much lower than the boiling coefficient. Coupled with forced convection, high overall heat-transfer coefficients can be achieved in D-X coil evaporators.

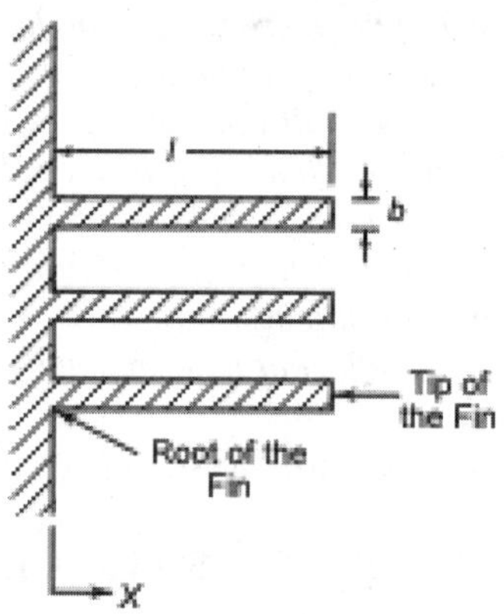

Fig. 9.6 (b) Configuration of rectangular fins

Generally, fins for cooling coils are made of thin aluminium (material with high thermal conductivity) sheets mounted on the copper tubes by applying oil pressure of the order of 150 bar inside the tubes so that perfect contact between the roots of the fins and the tubes is achieved. In all, fins are provided with a spacing corresponding to about 5 fins per cm. The thickness b and the height l of the fins [Fig. 9.6(b)] are of the order of 1 mm and 1.5 cm respectively. Air velocities of the order of 2.5 to 3 m/s are used.

If we assume the entire fin at the same temperature as the root of the fin and the outside total/finned surface to the inside surface area ratio as A_t/A_i, then the reduced air-side heat transfer resistance can be written as

$$R_{\text{air}} = \frac{A_i}{A_t f_g}$$

where f_g is the film coefficient on the air-side. The symbol f is used here instead of h for the heat-transfer coefficient so that it is not confused with enthalpy in the later chapters.

However, the temperature over the fin varies from the metal wall temperature at the root to approximately the air temperature at the tip. This will result in decreased heat flux. The ratio of the heat actually transferred to the heat which would be transferred if the entire fin were at the root temperature, is defined as the *fin efficiency* η_f and is given by the expression (see Chapter 1)

$$\eta_f = \frac{\tanh(ml)}{ml}$$

where $$m = \sqrt{\frac{2f_g}{kb}}$$

in which k is the thermal conductivity of the fin material.

Accordingly, the air-side resistance can be modified to

$$R_{\text{air}} = \frac{A_i}{(A_{\text{unfinned}} + \eta_f A_{\text{finned}}) f_g}$$

where A_{unfinned} and A_{finned} are the unfinned and finned areas of the surface.

9.3.1 Cooling and Dehumidifying Coils

Extended surface cooling and dehumidifying coils used in air conditioning can now be designed. Taking into account the extended surface, the equations for heat transfer now $\mathrm{d}Q_I$ through the air film only, and $\mathrm{d}Q_{II}$ through the remaining combined thermal resistance of the refrigerant, metal wall and condensed water layer are respectively (see Sec. 20.3 and Fig. 20.4)

$$\mathrm{d}\dot{Q}_I = \frac{\eta_f A_t f_g}{A_i C_p}(h - h_s)\,\mathrm{d}A_i \tag{9.13}$$

$$\mathrm{d}\dot{Q}_{II} = U_i (t_s - t_r)\,\mathrm{d}A_i \tag{9.14}$$

Combining Eqs (9.13) and (9.14), since $\mathrm{d}\dot{Q}_I = \mathrm{d}\,\dot{Q}_{II}$, we have

$$\frac{t_s - t_r}{h - h_s} = \frac{\eta_f\, f_g\, A_t}{U_i\, C_p\, A_i} = \text{const.} \tag{9.15}$$

where t_r = Refrigerant temperature

t_s = Temperature of the wetted surface

h = Enthalpy of air

h_s = Enthalpy of saturated air at the temperature of the wetted surface

U_i = Overall heat-transfer coefficient based on inside tube surface area accounting for the thermal resistance of refrigerant side film, metal wall and outside condensate layer

C_p = Specific heat of air.

A detailed discussion on cooling and dehumidifying coils is given in Chapter 20.

Example 9.2 ***Simulation of Flooded Chiller***

The word 'simulation' means recreation and validation of design and performance.

In an R 22 flooded chiller, the design conditions are as follows:

Evaporation temperature,	t_0 = 2°C
Chilled water inlet temperature,	t_{w_1} = 11.1°C
Chilled water circulation rate,	$\dot{m}_w$ = 24.332 kg/s

Construction details of the chiller are as follows:

Tube OD, $D_0 = 0.01704\ m$ *Tube ID,* $D_i = 0.01379\ m$

Tube length, $L = 3.69\ m$

Number of passes, $n = 2$

Number of tubes per pass $= 114$

As the refrigerant-side heat transfer coefficient in pool boiling is very much lower than water-side coefficient, integral fin tubes with the following specifications are used:

Number of fins $= 748$ *per metre*

Diameter over fins, $D_{fin} = 0.01905\ m$

Fin thickness, $b = 0.8\ mm$

Estimate the capacity of the chiller in kW, accurate to the first approximation.

Solution *First Approximation* Let the temperature drop across the outside refrigerant film be $\Delta T_0 = 5°C$, and let the water leaving temperature be $t_{w2} = 7°C$. Surface tension of R 22 at 2°C from Eq. (4.24) and constants given is

$$\sigma = 0.0115 \text{ N/m}$$

Thermodynamic and thermophysical properties of R 22 at 2°C are:

$h_{fg} = 203.7$ kJ/kg $\qquad C_f = 1.177$ kJ/kg.K

$\rho_f = 1279$ kg/m^3 $\qquad \rho_g = 22.573$ kg/m^3

$\mu_f = 0.231 \times 10^{-3}$ kg/ms $\qquad \mu_g = 1.2 \times 10^{-5}$ kg/m.s

$k_f = 0.1$ W/m.K $\qquad \text{Pr}_f = 3.07$

Refrigerant-side pool-boiling heat-transfer coefficient from Rohsenow correlation Eq. (9.1) written as below:

$$h_0\,\Delta T = \mu_f h_{fg}\left[\frac{g(\rho_f - \rho_g)}{\sigma}\right]^{1/2}\left[\frac{C_f \Delta T}{0.013 h_{fg}\,\text{Pr}_f^{1.7}}\right]^3$$

$$h_0\,(5) = (0.231 \times 10^{-3})\,(203.7 \times 10^3)$$

$$\left[\frac{9.81(1279 - 22.573)}{0.0115}\right]^{1/2}\left[\frac{1.177 \times 10^3 (5)}{0.013(203.7 \times 10^3)(3.07)^{1.7}}\right]^3$$

$$\Rightarrow \qquad h_0 = 396 \text{ W/m}^2\text{.K}$$

Mean temperature of water

$$t_w = \frac{t_{w_1} + t_{w_2}}{2} = \frac{11.1 + 7}{2} = 9.05°C$$

Thermophysical properties of water at 9.05°C

$k = 0.579$ W/m.K $\qquad \mu = 1.346 \times 10^{-3}$ kg/m.s

$\rho = 1001.1$ kg/m^3 $\qquad C = 4.1988$ kJ/kg.K

$\text{Pr} = 9.663$

Area of flow of water (114 tubes)

$$A = 114\,\frac{\pi}{4}\,(0.01379)^2 = 0.0175 \text{ m}^2$$

Mass velocity of water

$$G = \frac{\dot{m}_w}{A} = \frac{24.332}{0.0175} = 1430 \text{ kg/s.m}^2$$

Reynolds number of water

$$\text{Re} = \frac{GD_i}{\mu} = \frac{1430(0.01379)}{1.346\times 10^{-3}} = 14{,}651$$

Water-side forced-convection heat-transfer coefficient inside tubes using Dittus-Boelter equation

$$\frac{h_i D_i}{k} = \text{Nu} = 0.023\ (\text{Re})^{0.8}\ (\text{Pr})^{0.3}$$

$$\frac{h_i(0.01379)}{0.579} = 0.023\ (14{,}651)^{0.8}\ (9.663)^{0.3}$$

$$\Rightarrow \qquad h_i = 4170 \text{ W/m}^2\text{.K}$$

Area Calculations Figure 9.6(a) shows the construction of an integral fin tube. Now, per metre length of tube, we have:

$$A_i = \pi D_i L = \pi (0.01379) = 0.0433 \text{ m}^2$$

$$A_0 = \pi D_0 L = \pi (0.01704) = 0.0548 \text{ m}^2$$

$$A_1 = 748\ \pi D_{\text{fin}}\, b = 748\ \pi (0.01905)\ (0.8 \times 10^{-3}) = 0.0358 \text{ m}^2$$

$$A_2 = A_0 - A_1 = 0.0548 - 0.0358 - 0.019 \text{ m}^2$$

$$2\,A_3 = \frac{2\pi}{4}(\text{D}_{\text{fin}}^2 - \text{D}_0^2)\ 748 = \frac{\pi}{2}\ (0.01905^2 - 0.01704^2)\ 748$$

$$= 0.0852 \text{ m}^2$$

Fin area $\quad A_{\text{fin}} = A_1 + 2A_3 = 0.0358 + 0.0852 = 0.121 \text{ m}^2$

Base area $\quad A_{\text{base}} = A_2 = 0.019 \text{ m}^2$

Total effective heat transfer area

$$A_t = A_{\text{base}} + \eta_f A_{\text{fin}} = 0.019 + (1)\ (0.121) = 0.14 \text{ m}^2$$

Fin efficiency for these circular short fins is found to be $\eta_f = 1$ from Eq. (1.64).

Heat Transfer Calculations Neglecting metal wall resistance and fouling factor, we have for overall heat-transfer coefficient based on total effective heat transfer area A_t

$$\frac{1}{U_t A_t} = \frac{1}{h_i A_i} + \frac{1}{h_0 A_t} = R_i + R_0$$

$$= \frac{1}{4170\,(0.0433)} + \frac{1}{396\,(0.14)} = 5.58 \times 10^{-3} + 0.01804$$

$$\Rightarrow \qquad U_t A_t = 42.42 \text{ W/K per metre tube length}$$

Temperature differentials at inlet and outlet

$$\Delta T_1 = 11.1 - 2 = 9.1°\text{C},\ \Delta T_2 = 7 - 2 = 5°\text{C}$$

LMTD, $$\Delta T_m = \frac{9.1 - 2}{\ln(9.1/5)} = 6.85°\text{C}$$

Heat transfer rate $\dot{Q} = (U_t A_t)\,\Delta T_m = (42.42)\,(6.85) = 290$ W/m length
Total heat transfer rate: Estimated capacity of chiller

$$\dot{Q} = 290\,(2 \times 114 \times 3.69) = 244{,}000 \text{ W} = 244 \text{ kW}$$

Checking t_{w_2} Assumption

$$t_{w_2} = t_{w_1} - \frac{\dot{Q}}{\dot{m}_w C_w} = 11.1 - \frac{244}{(24.332)(4.1988)} = 8.706°\text{C}$$

Successive iterations give

$$t_{w_2} = 8.43°\text{C},\ \Delta T_m = 7.69°\text{C},\ \dot{Q} = 274 \text{ kW (Stop)}$$

Second Approximation It is seen that refrigerant-side film resistance is

$$R_0 = 0.01804 \text{ K/W per metre length of tube.}$$

Checking assumption and updating value of ΔT_0

$$\Delta T_0 = R_0\,\dot{Q} = (0.01804)\,(290) = 5.23°\text{C}$$

So, our assumption of $\Delta T_0 = 5°$C was quite good (only 4.4% away from 5.23°C). However, calculations may be repeated with the new value.

9.4 AUGMENTATION OF BOILING HEAT TRANSFER

As stated earlier, the overall resistance to heat transfer is greatly dependent upon the largest of the component resistances constituting it. In evaporators for conditioning air with bare tubes, the value of the heat-transfer coefficient on the refrigerant-side is several times higher than that of the air-side and the intensification of the boiling process has no practical influence on heat-transfer conditions. On the other hand, the necessity of intensification of the boiling process becomes fully evident in evaporators with finned tubes, viz., cooling coils for air, where the air-side effective coefficient reaches very high values. The same applies all the more to evaporators for cooling liquids, such as brine chillers for ammonia ice-plants and water chillers for air conditioning, in which the water/brine side coefficient reaches values of the order of 5000 W/m^2.K, or even higher.

As a typical example, we consider the case of a 1 TR evaporator, working with a nominal heat flux of 23.3 kW/m^2 and having a refrigerant flow rate of 62.0, and 11.5 kg/hr for R 22 and ammonia respectively, for $t_0 = -15°$C and $t_k = 30°$C. Assuming a single tube of 1 cm dia as the evaporator tube, we obtain the refrigerant side heat transfer coefficients as below:

R 22	R 717	
4036	3559	W m^{-2} °C^{-1}

On the other hand, the water side coefficient can be as high as 6000 W m^{-2} °C^{-1}. Thus, there is a large scope for the improvement of the boiling heat-transfer coefficient.

The modern trend towards compact, high heat-flux chillers, using small hydraulic dia tubes, thus necessitates the use of suitable augmentation techniques.

9.4.1 Influence of Refrigerant Vapour Pressure Characteristic

The effect of using an augmentation method is to increase the heat transfer rate. But, it increases the pressure drop as well. The effect of increased pressure drop is either

(i) to cause a lowering of the compressor suction pressure and thus to decrease the capacity and performance of the whole system, or

(ii) to cause a decrease in the temperature potential (at the same suction pressure) for heat transfer and thus nullify, to some extent, the increase in the heat transfer obtained. Figures 9.8 and 10.7 illustrate this case. Augmentation of heat-transfer coefficient decreases the size of the evaporator, whereas decrease in ΔT increases its size.

A proper optimal balance between the two is, therefore, necessary. It is obvious that a refrigerant which causes a small change in the saturation temperature for a given change in the saturation pressure, will have a smaller fall in the temperature potential due to increased pressure drop in the evaporator. A qualitative estimate of the suitability of various refrigerants for use with augmentation techniques is thus provided by Table 9.2, which gives the temperature change corresponding to a saturation pressure change of 0.01 bar. It can be clearly seen that R 22 is the most suitable refrigerant for augmentation since it has the least saturation temperature change.

Table 9.2 Temperature change of refrigerants for 0.01 bar pressure drop

Temperature **Range, °C**	**R 134a**	**R 21**	**R 22**	**R 717**
0 to 5	0.0876	0.312	0.0569	0.0585
– 10 to – 20	0.1475	5.65	0.0893	0.098
– 35 to – 40	0.335	14.9	1.792	2.26

It can be anticipated that with R 21, no substantial improvement in net heat transfer unit pressure drop may be possible while with R 134a, R 717 and R 22, significant improvements (in that order) may be obtained.

9.4.2 Augmentation Techniques

A large number of methods for the augmentation of heat transfer exist. These methods either increase the coefficient or increase the effective surface area, or do both. The most common of them are as follows:

(i) Turbulence promoters.
(ii) Swirl flow generators.
(iii) Extended surfaces.
(iv) Inner fins.
(v) Recirculation.
(vi) Roughening of surfaces.

9.4.3 Recirculation

This is one method of augmentation of boiling heat transfer. Evaporators with pumped circulation of the refrigerant are widely used in ammonia systems and are becoming popular in large refrigeration systems using R 22 as well. With recirculation the heat-transfer coefficient increases not only with an increased rate of circulation of the refrigerant (i.e. increased Reynolds number), but also due to a decrease in the average dryness fraction over the length of the evaporator tube and consequent flooded operation as explained in Fig. 9.5. Simultaneously, the pressure drop in the evaporator is also increased.

A recirculation number n is defined for such an evaporator as

$$n = \frac{\text{Refrigerant circulation rate}}{\text{Refrigerant evaporation rate}}$$

Where the inlet quality is zero, the change of vapour quality Δx is related to recirculation number n, according to the relation

$$\Delta x = \frac{1}{n}$$

An illustration of a coil evaporator with forced recirculation of the refrigerant is given in Fig. 9.7. The system requires an accumulator on the low-pressure side fed by a float valve. A disadvantage of this system is the necessity of a pump to circulate the refrigerant requiring pump, piping, and additional power consumption and liquid-line insulation.

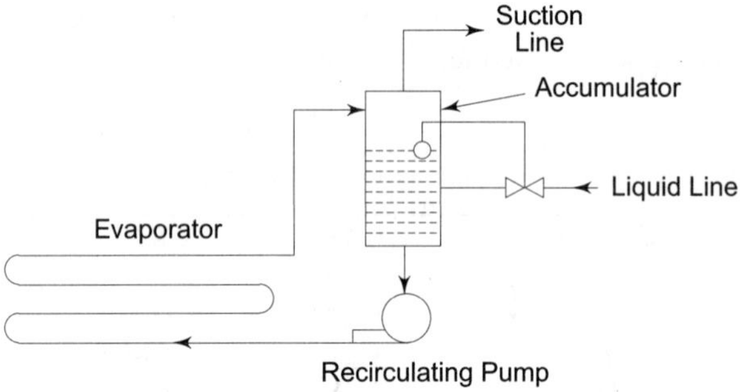

Fig. 9.7 Recirculation type evaporators

Overfeeding of the evaporator by recirculation assures full-flooded operation and in this regard an optimum heat-transfer rate is possible. For example, a certain amount of evaporator surface is required to superheat gas to actuate a thermostatic-expansion valve. This length of tubing is saved by obtaining flooded operation by the use of recirculation as in flooded evaporators. At the same time, a high heat transfer coefficient is attained corresponding to forced convection boiling at high Reynolds number instead of pool boiling.

In ammonia systems a recirculation number of $n = 4$ is generally employed. It means that only one-fourth of the feed is vaporized, i.e., the quality $x_0 = 0.25$ at the exit of the evaporator so that $\Delta x = 0.25 = 1/n$.

However, the pressure drop in the evaporator also increases with quality and reaches a maximum. With the increasing rate of recirculation, the pressure drop is increased. And as usual, a high pressure drop in the evaporator will offset the advantage of the increased heat-transfer coefficient by the decrease in the available temperature potential at the same suction pressure. An optimum recirculation number, therefore, depends on a number of factors, such as the cost of the tubing, pump, and insulation of the liquid line, pump work, etc. Due to the complexity of the problem, a simplified choice of an optimum recirculation number can be made on thermodynamic considerations alone. At a given evaporator duty, it is the one resulting in the highest possible overall heat-transfer coefficient. An even better choice is the recirculation number resulting in the minimum heat transfer surface of the evaporator corresponding to the design pressure in the suction line.

An added advantage of recirculation-type evaporators is the convenience of separation and drainage of excess oil from the accumulator. Also, both in ammonia and R 22 systems, a higher COP is obtained with flooded operation as the suction state is not superheated (see Chap. 3). Another advantage of liquid recirculation for refrigeration is the ease of control of the refrigerant, specially in multi-evaporator system, i.e., the elimination of such problems as precise setting of an expansion valve, hunting or flooding over.

The main disadvantage in recirculation by the use of a pump is the cavitation problem encountered, speciality since it works at low pressures and temperatures.

9.4.4 Use of Wire Screens in D-X Evaporators[4]

An appreciable increase in the heat-transfer coefficient of boiling R 22 is possible by fitting wire screens to the inside wall of the tube. The largest increase is obtained at low-quality values where they act as turbulence promoters. Also, a considerable decrease in the fluctuation of the wall temperature is achieved by using them in the *dry-out region* of high-quality vapour. The increase in pressure drop in the annular flow region may not justify their use throughout the evaporator.

9.4.5 Using Roughened Surfaces: Slipcevic Correlations[16,17,18]

As seen in Example 9.2, pool-boiling refrigerant-side heat-transfer coefficient in flooded evaporators is less than 1/10th the value of water-side coefficient. Two remedies are adopted to decrease the resistance and increase the conductance on refrigerant side:

(i) Increasing area by providing integral fins as in Example 9.2.

(ii) Improving heat-transfer coefficient by introducing roughness by any of the following methods:

 (a) One stage chemical etching with HNO_3.

 (b) Two stage chemical etching with HNO_3 and $(NH_4)\ S_2O_3$.

 (c) Rubbing with coarse grain emery.

Slipsevic gives the following correlations for evaluating refrigerant-side heat-transfer coefficient for flooded chillers with roughened integral fin tubes. The roughness parameter R_p is specified as the height of grains on roughened surface.

$$\mathrm{Nu} = \frac{hL}{k_f} = 230\ \phi_a^{0.1}\ \mathrm{Re}_r^{n1} \left(\frac{Ph}{P_{r_f}}\right)^{n_2} \varepsilon^{0.133} \tag{9.16}$$

where

$$n_1 = 0.75\ \phi_a^{-0.144}\ \phi_p^{0.088\phi_a^{-0.25}} \tag{9.17}$$

$$n_2 = 0.75\ \phi_a^{-0.13}\ \phi_p^{-0.28} \tag{9.18}$$

Area ratio,

$$\phi_a = \frac{A_t}{A_0} \tag{9.19}$$

Pressure ratio,

$$\phi_p = \frac{p}{p_c} \tag{9.20}$$

Roughness factor,

$$\varepsilon = \frac{R_p}{L} \tag{9.21}$$

Laplace constant, $$L = \left[\frac{\sigma}{g(\rho_f - \rho_g)}\right]^{0.5} \tag{9.22}$$

Capacity ratio, $$Ph = \frac{CT}{h_{fg}} \tag{9.23}$$

Refrigerant Reynolds No., $$\text{Re}_r = \frac{q_t L}{h_{fg}\,\mu_f} \tag{9.24}$$

where q_t is the heat flux based on outside total finned tube surface area. P_{r_f} is the liquid Prandtl number.

Note *Subscripts i, 0 and t refer to inside tube, outside tube and total outside finned tube surface areas respectively.*

Example 9.3 ***Design of Flooded Chiller with Roughened Surface***[15]

An R 22 flooded chiller is to be designed with the following specifications:

Refrigerating capacity	*422 kW*
Evaporator temperature	*2°C*
Condensing temperature	*42°C*
Chilled water temperatures	*11.1°C Inlet: 7°C Outlet*
Number of passes	*2*

ID, OD and length of tubes, diameter over integral fins, fin thickness and spacing, thermodynamic and thermophysical properties of R 22 and water are the same as in Example 9.2. Take into account the inside and outside fouling and metal tube wall resistances.

Given fouling resistances and thermal conductivity of tubes as:

$$(R_{fouling})_i = 0.00007\ m^2K/W$$

$$(R_{fouling})_0 = 0.00009\ m^2K/W$$

$$k_{tubes} = 0.386\ kW/m.K$$

Tubes are roughened with roughness parameter

$$R_p = 0.5\ \mu\text{m}$$

Determine the number of tubes required. Assume equal number of tubes in each pass. Use Slipcevic correlations. Assume fin efficiency $\eta_f = 1$.

Solution Evaporator pressure, $p_0 = 5.283$ bar (at 2°C)

Critical pressure of R 22, $p_c = 49.36$ bar

$$\phi_p = \frac{p_0}{p_c} = \frac{5.283}{49.36} = 0.107$$

From Example 9.2, per metre length of tube, we have

$$\phi_a = \frac{A_t}{A_0} = \frac{0.13998}{0.0548} = 2.5544$$

Hence,

$$n_1 = 0.75\,(2.5544)^{-0.144}\,(0.107)^{0.088\,(2.5544)^{-0.25}} = 0.54$$

$$n_2 = 0.75\,(2.5544)^{-0.13}\,(0.107)^{-0.28} = 1.235$$

$$L = \left[\frac{\sigma}{g(\rho_f - \rho_g)}\right]^{0.5} = \left[\frac{0.0115}{9.81(1279 - 22.573)}\right]^{0.5} = 9.656 \times 10^{-4}\ \text{m}$$

$$\varepsilon = \frac{R_p}{L} = \frac{0.5 \times 10^{-6}}{9.656 \times 10^{-4}} = 5.178 \times 10^{-4}$$

First Approximation

Assume number of tubes, $n = 228$

Length of tubes, $l = 3.69$ m

Area/m length of tube $= 0.14\ \text{m}^2$

$$q_t = \frac{\dot{Q}}{A_t n l} = \frac{422 \times 10^3}{0.14 \times 228(3.69)} = 3583\ \text{W/m}^2$$

$$\text{Re}_r = \frac{q_t\, L}{h_{fg}\,\mu_f} = \frac{(3583)(9.656 \times 10^{-4})}{(203.7 \times 10^{-3})(0.231 \times 10^{-3})} = 0.0735$$

$$Ph = \frac{C_f T}{h_{fg}} = \frac{1.77(273 + 2)}{203.7} = 1.589$$

$$P_{rf} = 2.798$$

$$\text{Nu} = \frac{h_0 L}{k_f} = 230\,(\phi_a)^{0.1}\,\text{Re}_r^{n1}\left(\frac{Ph}{P_{r_f}}\right)^{n_2} \varepsilon^{0.133}$$

$$= 230\,(2.5544)^{0.1}\,(0.0735)^{0.54}\left(\frac{1.589}{2.798}\right)^{1.235} (5.178 \times 10^{-4})^{0.133} = 11.2$$

$$h_0 = \frac{11.2\,(0.1)}{9.656 \times 10^{-4}} = 1127\ \text{W/m}^2.\text{K}$$

Area ratio $\dfrac{A_t}{A_i} = \dfrac{0.14}{0.0433} = 3.231$

Overall heat transfer coefficient

$$\frac{1}{U_t} = \frac{A_t}{h_i A_i} + (R_{\text{fouling}})_i \frac{A_t}{A_i} + (R_{\text{fouling}})_0 + \frac{\ln(D_0/D_i)}{2\pi k \ln} A_t + \frac{1}{h_0 \eta_{\text{fin}}}$$

$$= \frac{3.231}{4170} + 0.00007\,(3.231) + 0.00009$$

$$+ \frac{\ln(0.01704/0.01379)}{2\pi(386)(3.69)(228)} + \frac{1}{1127(1)}$$

$$\Rightarrow \qquad U_t = 495.5\ \text{W/m}^2.\text{K}$$

Log-mean temperature difference

$$\Delta T_m = \frac{(11.2-2)-(7-2)}{\ln(9.2/5)} = 6.89°\text{C}$$

Total finned tube surface area required

$$A_t = \frac{\dot{Q}_0}{U_t\,\Delta T_m} = \frac{422\times10^3}{495.5(6.89)} = 123.6 \text{ m}^2$$

Inside tube surface area required

$$A_i = \frac{A_t}{3.231} = \frac{123.6}{3.231} = 38.254 \text{ m}^2$$

Number of tubes required

$$n = \frac{A_i}{\pi D_i\, l} = \frac{38.254}{\pi(0.01379)(3.69)} = 240$$

Second Approximation Assume number of tubes now, $n = 240$, and repeat calculations. We find that there is no change in n. The solution has thus converged after the first approximation itself.

9.5 PRESSURE DROP IN EVAPORATORS

Pressure drop in evaporators is quite significant particularly in direct-expansion evaporators. Pressure drops in evaporator tubes not only due to friction but also due to increase in momentum as a result of increase in volume and hence velocity consequent upon vaporization of liquid into vapour. The effect of pressure drop is detrimental to the performance of the refrigerating machine as discussed in Sec. 10.5. It essentially decreases the suction pressure and hence refrigerating capacity.

The total pressure drop Δp_T in the evaporator has the following three components

(i) Friction pressure drop Δp_F.

(ii) Acceleration or momentum pressure drop Δp_M.

(iii) Pressure drop/gain due to change in level.

The static pressure drop gradient due to change in level is usually small. The other two pressure drops can be calculated using Chawla's[5] correlations as follows:

$$\left(\frac{\mathrm{d}p}{\mathrm{d}y}\right)_T = \left(\frac{\mathrm{d}p}{\mathrm{d}y}\right)_F + \left(\frac{\mathrm{d}p}{\mathrm{d}y}\right)_M \tag{9.25}$$

$$\left(\frac{\mathrm{d}p}{\mathrm{d}y}\right)_F = \frac{0.3164}{(GD_i/\mu_g)^{0.25}}\,\frac{G^2x^{7/4}}{2D_i\rho_g}\left[1+\frac{1-x}{x\varepsilon R_\rho}\right]^{19/8} \tag{9.26}$$

$$\left(\frac{\mathrm{d}p}{\mathrm{d}y}\right)_M = \frac{4Gqx}{D_i h_{fg}\rho_f}\left[\left(1-\frac{1}{\varepsilon R_\rho}\right)\left(1+\frac{1-x}{x}\varepsilon\right)+\left(1+\frac{1-x}{x\varepsilon R_\rho}\right)(1-\varepsilon)\right] \tag{9.27}$$

where the two-phase flow parameter ε is defined as

$$\varepsilon = 62\left(\frac{1-x}{xR_\mu}\right)[\text{Re}_f\ \text{Fr}_f]^{-1/6}\,[R_\rho]^{-0.9} \tag{9.28}$$

and $$\text{Re}_f = \frac{G(1-x)^2}{\mu_f} \tag{9.29}$$

$$\text{Fr}_f = G^2(1-x)^2/(\rho_f^2\, g\, D_i)$$

$$= \frac{G^2(1-x)^2}{(1279)^2(9.81)(0.0158)} = \frac{G^2(1-x)^2}{253550} \tag{9.30}$$

$$R_\rho = \rho_f/\rho_g = 56.93 \tag{9.31}$$

$$R_\mu = \mu_f/\mu_g = 18.92 \tag{9.32}$$

Calculated values in Eqs (9.30) to (9.32) are for R 22 in Example 9.5.

Example 9.4 ***Estimation of D-X Chiller Capacity (Simulation)***

The following specifications are given for an R 22 D-X Chiller.

Condensing temperature, t_k	*43°C*
Saturated suction temperature	*2°C*
Number of passes, n	*8*
Tubes in each pass	*12, 16, 20, 24, 30, 32, 32, 34*
Evaporator superheat	*5°C*
Inlet water temperature, t_{w1}	*11.1°C*
Outlet water temperature, t_{w2}	*7.2°C*
Refrigerant pressure drop in evaporator 0.14 bar (assumed)	
Shell diameter, D_s	*0.406 m*
Tube length between tube sheets, l	*2.213 m*
Tube ID, D_i	*0.0158 m*
Tube OD, D_0	*0.0191 m*
Tube pitch (triangular), P_T	*0.0222 m*
Number of baffles	*21*
Baffle pitch, P_B	*0.0762 m*
Baffle cut	*0.094 m*

Estimate the refrigerating capacity of the chiller. Take thermodynamic and thermophysical properties of R 22 at 2°C from Example 9.2. Use flow boiling correlation for first and second passes, forced convection boiling correlation for middle passes, and Dittus-Boelter equation for $x \geq 0.975$ for finding refrigerant-side heat-transfer coefficients.

Solution Figure 9.8 shows the *p-h* diagram. Suction temperature is 7°C with 5°C superheat. The temperature at evaporator inlet is 2.86°C because of 0.14 bar pressure drop. The quality at evaporator inlet is $x_4 = 0.2517$.

Refrigerant mass flow rate $$\dot{m}_r = \frac{\dot{Q}_0}{h_1 - h_4}$$

First Approximation Assume a refrigerating capacity of 140.7 kW. Then

$$\dot{m}_r = \frac{140.7}{409.7 - 253.7} = 0.9 \text{ kg/s}$$

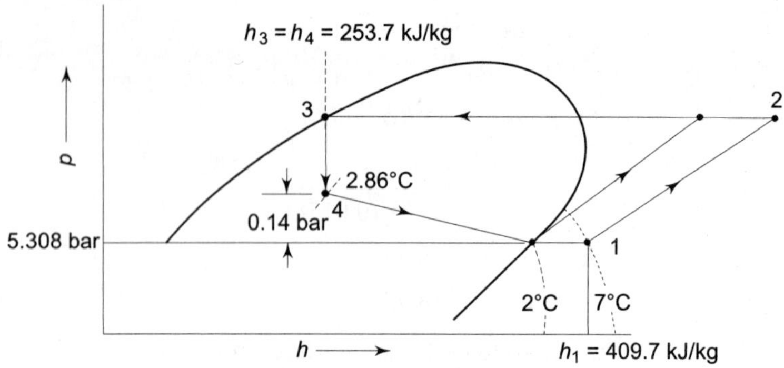

Fig. 9.8 *p-h* diagram for Example 9.4

Refrigerant flow area in a pass with n number of tubes

$$A = n\frac{\pi D_i^2}{4} = n\frac{\pi}{4}(0.0158)^2 = 1.96 \times 10^{-4}\, n \text{ m}^2$$

Refrigerant mass velocity $G = \dfrac{\dot{m}_r}{A} = \dfrac{0.9}{1.96\times 10^{-4} n} = \dfrac{4590}{n}$ kg/s.m^2

Shell-side heat-transfer coefficient (Kern's correlation[9])

$$\text{Nu} = \frac{hD_0}{k} = 0.36\left(\frac{DG_s}{\mu}\right)^{0.55}(\text{Pr})^{1/3}\left(\frac{\mu}{\mu_{\text{wall}}}\right)^{0.14} \tag{9.33}$$

Neglecting viscosity variation between bulk and wall fluid, we have

$$\text{Nu} = \frac{hD_0}{k} = 0.36\left(\frac{D_e G_s}{\mu}\right)^{0.55}(\text{Pr})^{1/3}$$

Now, tube pitch, $P_T = 0.0222$

Tube OD, $D_0 = 0.0191$ m

Equivalent diameter for triangular pitch

$$D_e = \frac{4}{\pi D_0}\left[0.86P_T^2 - \frac{\pi D_0^2}{4}\right] \tag{9.34}$$

$$= 9.15 \times 10^{-3} \text{ m}$$

Calculation of mass flow rate of water

$$\dot{Q}_0 = \dot{m}_w\, C_p\,(t_{w_1} - t_{w_2})$$

Assuming a net refrigerating capacity of 140.7 kW, we get

$$140.7 = \dot{m}_w\,(4.1868)\,(11.1 - 7.2)$$

$\Rightarrow$ $\dot{m}_w = 8.617$ kg/s

Cross-flow area S_s is defined as

$$S_s = D_s\,(P_T - D_0)\cdot\frac{P_B}{P_T} \tag{9.35}$$

We have shell dia, $D_s = 0.406$ m

Baffle pitch, $P_B = \dfrac{\text{Tube length}}{\text{No. of baffles}} = \dfrac{2.213}{22} = 0.1$ m

$$S_s = 0.406\ (0.0222 - 0.0191)\ \frac{0.1}{0.0222} = 5.67 \times 10^{-3}\ \text{m}^2$$

Shell-side mass velocity of water

$$G_s = \frac{\dot{m}_w}{S_s} = \frac{8.617}{5.67 \times 10^{-3}} = 1601\ \text{kg/s.m}^2$$

Bulk mean temperature of water

$$t_w = \frac{t_{w_1} + t_{w_2}}{2} = \frac{11.1 + 7.2}{2} = 9.15°\text{C}$$

Thermophysical properties of water at 9.15°C

$\rho = 1001.5\ \text{kg/m}^3$, $\mu = 1.304 \times 10^{-3}$ kg/m.s

$k = 0.586$ W/m.K, $\text{Pr} = 9.39$

Substituting values in Eq. (9.33), we get for the shell-side (water-side) heat-transfer coefficient

$$h_0 = 3939\ \text{W/m}^2\text{.K}$$

Heat Transfer Calculations for Passes I and II (nucleate Flow Boiling Regime)
First Pass (Number of tubes, $n_I = 12$)
Refrigerant mass velocity in Pass I

$$G_I = \frac{4590}{n_I} = \frac{4590}{12} = 382.5\ \text{kg/s.m}^2$$

Heat transfer area of Pass I

$$A_I = n_I\, \pi\, D_i\, l = 12\ \pi\, (0.0158)\ (2.213) = 1.3175\ \text{m}^2$$

Assume equal heat transfer in all passes. Then, for heat transfer in Pass I

$$\dot{Q}_\text{I} = \frac{\dot{Q}}{n} = \frac{140.7}{8} = 17.6\ \text{kW} = \dot{m}_r\, \Delta h = q_I\, A_I$$

$\Rightarrow$ Δh_I = Enthalpy rise of refrigerant in Pass I = 19.55 kJ/kg

$\Rightarrow$ q_I = Heat flux in Pass I = 13.34 kW/m^2

Mean temperature of refrigerant

$$t_r = \frac{2.86 + 2}{2} = 2.43°\text{C}$$

Vapour bubble growth rate parameter for calculation of inside (refrigerant-side) heat-transfer coefficient using nucleate flow boiling correlation

$$w'' = 0.36 \times 10^{-3} \left(\frac{p_c}{p}\right)^{1.4} = 0.36 \times 10^{-3} \left(\frac{49.78}{5.308}\right)^{1.4} = 0.008$$

Substituting values in nucleate flow boiling correlation Eq. (9.11), we get

$$\frac{h_i (0.0158)}{0.0971} = 23388.5 \left[\frac{q}{(22.57)(203.7)(0.0008)}\right]^{0.64}$$

$$\times \left[\frac{9.81(0.0158)}{203.7 \times 10^3}\right]^{0.27} \left[\frac{382.5^2 (0.0158)}{0.0115(1279)}\right]^{0.14}$$

$\Rightarrow$ $$h_i = 689\ q^{0.64}$$

First Approximation Assume $q = 13.34$ kW/m^2 as before. Then

$$h_i = 689\,(13.34)^{0.64} = 3617 \text{ W/m}^2\text{.K}$$

$$\frac{1}{U_i} = \left[\frac{1}{h_i} + \frac{1}{h_{\text{fouling}}} + \frac{D_i \ln(D_0/D_i)}{2k_{Cu}} + \frac{1}{h_0}\frac{D_i}{D_0}\right]$$

$$= \left[\frac{1}{h_i} + 0.00009 + \frac{0.0158 \ln(0.0191/0.0158)}{2(387)} + \frac{1}{3939}\left(\frac{0.0158}{0.0191}\right)\right]$$

$$= \left[\frac{1}{h_i} + 0.00031\right]$$

$$\Rightarrow \quad U_i = 1730 \text{ W/m}^2\text{.K} = 1.73 \text{ kW/m}^2\text{.K}$$

$$\dot{q}_I = U_i\,(T_{\text{water}} - T_{\text{refrigerant}}) = 1.73\,(9.15 - 2.43) = 11.63 \text{ kW/m}^2$$

which is less than assumed value of 13.34 kW/m^2. After a number of iterations, the solution converges at

$$h_i = 3144 \text{ W/m}^2\text{ K},\ U_i = 1592 \text{ W/m}^2\text{K},\ q_I = 10.7 \text{ kW/m}^2$$

Heat transfer in Pass I

$$\dot{Q}_I = \dot{q}_I\,A_I = 10.7\,(12)\,\pi\,(0.0158)\,(2.213) = 14.1 \text{ kW}$$

Second Pass (Number of tubes $n = 16$, $G = 286.9$ kg/s.m^2)
Similar calculations using nucleate flow boiling correlation give

$$h_i = 2858 \text{ W/m}^2\text{K},\ U_i = 1503 \text{ W/m}^2\text{K},\ \dot{q}_{II} = 10.1 \text{ kW/m}^2$$

$$\dot{Q}_{II} = q_{II}\,A_{II} = 10.1\,(16)\Pi\,(0.0158)\,(2.213) = 17.75 \text{ kW}$$

Heat Transfer Calculations for Pass III Onwards until $x \le 0.975$ (Forced Convection Vaporization Regime)
Third Pass ($n = 20$, $\dot{G} = 229.51$ kg/s.m^2)
Change in quality of refrigerant in Passes I and II

$$\Delta x = \frac{\dot{Q}}{\dot{m}_r\,h_{fg}} = \frac{14.1 + 17.75}{0.9(203.7)} = \frac{31.85}{183.3} = 0.1737$$

Quality of refrigerant entering pass III

$$x_{\text{in}} = 0.2517 + 0.1737 = 0.425$$

Take mean value of x in Pass III as 0.46.
Substituting values in forced convection vaporization correlation Eq. (9.10), we get

$$\frac{h_i(0.0158)}{0.0971} = 0.115\,[0.46^4\,(1 - 0.46^2)]^{0.11}$$

$$\left[\frac{229.51^2(203.7 \times 10^3)}{9.81(0.0115)(1279)}\right]^{0.44} \quad [3.07]^{0.7}$$

$$\Rightarrow \quad h_i = 2908 \text{ W/m}^2\text{.K}$$

$$U_i = 1517 \text{ W/m}^2\text{.K}$$

$$\dot{Q}_{\text{III}} = U_i A_i (t_{\text{water}} - t_{\text{refrigerant}})$$

$$= 1.517 (20) \pi (0.0158) (2.213) (9.15 - 2.43) = 22.4 \text{ kW}$$

$$\Delta h_{\text{III}} = \frac{22.4}{0.9} = 24.9 \text{ kJ/kg}$$

$$\Delta x_{\text{III}} = \frac{24.9}{203.7} = 0.122$$

$$x_{\text{mean}} = 0.425 + \frac{0.122}{2} = 0.48$$

Further iteration gives

$$h_i = 2920 \text{ W/m}^2\text{.K},\ U_i = 1520 \text{ W/m}^2\text{.K},\ \dot{Q}_{\text{III}} = 21.3 \text{ kW},$$

$$\dot{q}_{\text{III}} = 9.7 \text{ kW/m}^2,\ x_{\text{out}} = 0.52$$

Fourth Pass ($n = 24$, $G = 191.3$ kg/s.m^2), similarly

$$h_i = 2712 \text{ W/m}^2\text{. K},\ U_i = 1461 \text{ W/m}^2\text{.K},\ \dot{Q}_{\text{IV}} = 25.9 \text{ kW}$$

$$\dot{q}_{\text{IV}} = 9.833 \text{ kW/m}^2,\ x_{\text{out}} = 0.664$$

Fifth Pass ($n = 30$, $G = 153$ kg/s. m^2), similarly

$$h_i = 2363 \text{ W/m}^2\text{.K},\ U_i = 1353 \text{ W/m}^2\text{.K},\ \dot{Q}_{\text{V}} = 29.9 \text{ kW}$$

$$\dot{q}_{\text{V}} = 9.08 \text{ kW/m}^2,\ x_{\text{out}} = 0.827$$

Sixth Pass ($n = 32$, $G = 143.4$ kg/s.m^2), similarly

$$h_i = 2195 \text{ W/m}^2\text{. K},\ U_i = 1297 \text{ W/m}^2\text{.K},\ \dot{Q}_{\text{VI}} = 29.8 \text{ kW}$$

$$x_{\text{out}} = 0.99$$

Since, $x_{\text{out}} > 0.975$, calculations have to be modified for sixth pass. We see that sixth pass has two sections.

Section A Forced convection boiling section for which

$$(x)_{\text{mean}} = \frac{0.827 + 0.975}{2} = 0.901$$

$$h_i = 2210 \text{ W/m}^2\text{.K},\ U_i = 1302 \text{ W/m}^2\text{.K}$$

$$(\dot{Q}_{\text{VI}})_A = \dot{m}_r\ \Delta h = 0.9\ (0.975 - 0.827)\ (203.7) = 27.1 \text{ kW}$$

Length of forced convection boiling section

$$L_A = \frac{(\dot{Q}_{\text{VI}})_A}{U_i\, n \pi\, D_i \Delta T} = \frac{27.1 \times 10^3}{1302(32)\pi(0.0158)(9.15 - 2.43)} = 1.948 \text{ m}$$

Section B Forced convection Dittus-Boelter equation section

$$L_B = 2.213 - 1.948 = 0.265 \text{ m}$$

$$(x_B)_{\text{mean}} = \frac{0.975 + 1}{2} = 0.9875$$

assuming $x_{\text{out}} \cong 1$.

Properties of liquid-vapour mixture in B

$$\mu = x\, \mu_g + (1 - x)\, \mu_f$$

$$= 0.9875\ (1.2 \times 10^{-5}) + 0.0125\ (2.31 \times 10^{-4})$$

$$= 1.473 \times 10^{-5} \text{ kg/m.s}$$

Similarly, $C_p = 0.606$ kJ/kg.K

$k = 0.01035$ W/m.K

Heat transfer calculations using Dittus-Boelter equation for L_B

$$\text{Re} = \frac{GD_i}{\mu} = \frac{143.4(0.0158)}{1.473\times10^{-5}} = 154{,}000$$

$$\text{Pr} = \frac{C_p\,\mu}{k} = \frac{(0.606\times10^3)(1.473\times10^{-5})}{0.01035} = 0.862$$

$$h_i = 0.023\,(154{,}000)^{0.8}\,(0.862)^{0.4}\left(\frac{0.01035}{0.0158}\right) = 200.6\ \text{W/m}^2\text{.K}$$

$$U_i = 188.6\ \text{W/m}^2\text{.K}$$

$$(\dot{Q}_{\text{VI}})_B = 0.53\ \text{kW}$$

Total heat transfer in Pass VI

$$\dot{Q}_{\text{VI}} = (\dot{Q}_{\text{VI}})_A + (\dot{Q}_{\text{VI}})_B = 27.1 + 0.53 = 27.63\ \text{kW}$$

$$x_{\text{out}} = 0.827 + \frac{27.63}{(0.9)(203.7)} = 0.978 < 1\ \text{(assumed)}$$

Repeat calculations do not change the value of $(\dot{Q}_{\text{VI}})_B$.

Seventh Pass ($n = 32$, $G = 143.4$ kg/s.m^2), similarly

$$\dot{Q}_{\text{VII}} = 4.5\ \text{kW},\ x_{\text{out}} = 1.002\ \text{(superheated vapour)}$$

Eighth Pass ($n = 34$, $G = 135$ kg/s.m^2)

Taking properties of R 22 vapour at mean temperature of 4.5°C, we have

$$C_{pg} = 0.609\ \text{kJ/kg.K},\ \mu_g = 1.22\times10^{-5}\ \text{kg/m.s},\ k_g = 9.378\times10^{-3}\ \text{W/m.K}$$

$$\text{Pr} = 0.792$$

$$\text{Re} = \frac{135(0.0158)}{1.22\times10^{-5}} = 174{,}800$$

$$h_i = 0.023\,(174{,}800)^{0.8}\,(0.792)^{0.4}\left(\frac{9.378\times10^{-3}}{0.0158}\right) = 194.4\ \text{W/m}^2\text{.K}$$

$$U_i = 183.2\ \text{W/m}^2\text{.K}$$

$$\dot{Q}_{\text{VIII}} = 3.2\ \text{kW}$$

Total capacity of D-X chiller $\dot{Q} = \Sigma\,\dot{Q}_{\text{pass}} = 144.3$ kW

Note *Calculated capacity of 144.3 kW is greater than assumed value of 140.7 kW. Repeat calculations till convergence is reached.*

Example 9.5 ***Pressure Drop Calculations in D-X Evaporator***

In Example 9.4, verify if the pressure drop in the evaporator is 0.14 bar as assumed.

Solution The results of calculations using Chawla's correlations are given in Table 9.3.

Table 9.3 Pressure drop calculations for Example 9.5

Pass	x	G kg/s.m^2	Re_f	ε	q W/m^2	Δp_F Pa	Δp_M Pa	Δp_T Pa
I	0.28	382.5	550,616	0.0255	10,700	2901	1.7	2903
II	0.378	286.9	308,619	0.0227	10,100	1893	2.3	1895
III	0.48	229.5	172,322	0.01697	9,700	1568	2.3	1570
IV	0.63	191.3	72,723	0.01229	9,833	1250	2.1	1252
V	0.75	153.0	26,554	0.0086	9,080	754	2.0	756
							Total	8376

The pressure drops in passes VI, VII and VIII are essentially due to friction of vapour flow only. These are 1053, 1023 and 929 Pa respectively. The total pressure drop is, therefore, 11381 Pa (0.114 bar), a little lower than assumed.

Note *The number of tubes in passes are increased from pass I to pass VIII as more and more vapour is formed. The calculations show that* Δp_M *is kept small in this manner.*

References

1. Andersen S W, D G Rich and D F Geary, 'Evaporation of Refrigerant 22 in a Horizontal 3/4" OD Tube', *ASHRAE Trans.*, Vol. 72, Part I, 1966, pp. 28–42.
2. Bo Pierre, 'Flow Resistance with Boiling Refrigerants', *ASHRAE J*, Vol. 6, Nos. 9 and 10, 1964.
3. Bodganov S M, *ASHRAE J*, pp. 59-60, 1967.
4. Chawla J M, 'Correlations of Convective Heat Transfer Coefficient for Two-phase Liquid-Vapour Flow'. Paper B 5.7, Heat Transfer, Vol. V, 1970, *Proceedings of the IV International Conference on Heat Transfer*, Paris.
5. Chawla J M, 'Wärmeübergang and Drukabfall in Waagerchten Rohren beider Strömung Von Verdampfenden Kältemitteln,' *Kältetechnik-Klimatisierung*, Vol. 19, pp. 246–252, 1967.
6. Dembi N J, Dhar P L and C P Arora, 'An investigation into the use of wire screens in D-X-evaporators', Paper No. B 1 25, *XV International Congress of Refrigeration*, Venice, 1979.
7. Dhar P L, *Optimization of Refrigeration System*, Ph. D. Thesis, Indian Institute of Technology, Delhi, 1974.
8. Dembi N J, P L Dhar and C P Arora, 'Heat Transfer and Pressure Gradient for R 22 Boiling in a Horizontal Tube', Paper No. B 1/26, *XV International Congress of Refrigeration,* Venice. 1979.
9. Kern D Q, *Process Heat Transfer*, McGraw-Hill, pp. 136–139, 1950.
10. Lavin J G and E H, Young 'Heat transfer of evaporating refrigerants in two-phase flow', *A.I.Ch.E.J.*, Vol. 11, Nov. 1965, pp. 1124–32.
11. Murty S V R, *Pressure Drop and Heat Transfer Characteristics of R 12 Boiling Inside a Horizontal Tube*, M. Tech. Thesis, Indian Institute of Technology, Delhi, 1969.
12. Piret E L and H S Isbin, 'Two-phase heat transfer in natural circulation evaporators', *A.I.Ch.E. Heat Transfer Symposium*, St. Louis, Dec. 1953.

13. Ratiani C V and D I Avaliani, *Kholod. Tekh.*, Vol. 3, pp. 23–28, 1965.
14. Rohsenow W M, 'A method of correlating heat transfer data for surface boiling of liquids', *Trans. ASME*, Vol. 74, 1952.
15. Samuel C R, *Simulation and Optimization of Heat Transfer Equipment Used in Refrigeration Systems*, Ph. D. Thesis, Indian Institute of Technology, Delhi, 1983.
16. Slipsevic B, 'Ein Beitrag zum Wärmeübergang beim Blasensieden von Kältemitteln an einzelnen Rippenrohren, *Klima + Kälte Ingenieur*, pp. 69–76, 1974.
17. Slipsevic B, Über den Wärmeübergang beim Blasensieden von Kältemitteln an Rippenrohrbündeln, *Klima + Kälte* Ingenieur, pp. 127–134, 1975.
18. Slipsevic B, 'Wärmebergang bei der Blasenverdampfung von Kältemitteln an Glatten und Berippten Rohrbündeln, *Klima + Kälte Ingenieur*, pp. 279–286, 1978.

Revision Exercises

9.1 A direct-expansion cooling and dehumidifying coil for air has 1.27 cm OD and 1.12 cm ID serpentine copper tubing in each row. The fins are made of aluminium and are 1 mm thick. The spacing between the tubes is 1.27 cm and that between the fins corresponds to 5 fins per cm. The tube arrangement is staggered and the spacing between the rows is 1.27 cm. Determine.

(i) The ratio of the total finned surface area to the inside tube surface area.
(ii) The total finned surface area per unit face area of the cooling coil.
(iii) Fin efficiency.
(iv) Overall heat-transfer coefficient accounting for the thermal resistance of the condensate layer, tube wall and the refrigerant film. The refrigerant side heat-transfer coefficient is 1100 $Wm^{-2} K^{-1}$.

9.2 In a 112 TR R 22 flooded chiller, the water inlet temperature is 11.67°C. The mass flow rate of water to be chilled is 18.93 kg/s. The evaporating temperature of the refrigerant is 2.8°C. The copper tubes used are 1.27 cm OD and 1.12 cm ID. There are two parallel circuits and eight passes for water. The refrigerant-side heat-transfer coefficient is given by Rohsenow's correlation. Calculate the overall heat-transfer surface area of the chiller.

9.3 A direct-expansion chiller provides 20 tons of refrigeration for water entering at 20°C and flowing at the rate of 250 kg/min. The evaporator is fed through an automatic expansion valve. The evaporator temperature is 2°C.

It is now required to use the same chiller for cooling 100 kg/min of water at 30°C. Estimate the water outlet temperature.

Assume that the overall heat-transfer coefficient U in the evaporator is given by

$$U = C\sqrt{\dot{m}_w}$$

where $\dot{m}_w$ is the mass flow rate of water, and C is a constant.

9.4 In Example 9.2, investigate the effect of roughening of tubes on the size of flooded evaporator.

9.5 In Example 9.4, investigate the effect of keeping the number of tubes same as 24 in Passes VI to VIII.

Chapter

10

Complete Vapour Compression System

10.1 THE COMPLETE SYSTEM

The four components of the vapour-compression system, viz., the compressor, condenser, expansion device and evaporator, each have their individual characteristics which are the functions of evaporator and condenser temperatures. These have been discussed earlier in Chapters 6 to 9. Then, the performance of the complete system can be obtained by superimposition or simultaneous solution of the characteristics of the individual components. This can be done either graphically or analytically.

Often, the system may also include ancillary components as well, such as cooling tower, a boiler, ducting, etc. As the number of components increases, an analytical approach provides the only feasible answer. However, the graphical solution gives a better visual appreciation and comprehension about the behaviour of various components and the system.

Note *The performance of the complete system also depends upon the mass of total refrigerant charge in the system.*

10.2 GRAPHICAL METHOD[4]

The graphical method involves the use of tables and charts from the catalogue data of individual components and successive graphical analysis, taking two components at a time as illustrated by the following examples.

10.2.1 Performance Characteristics of the Condensing Unit

The performance characteristics of a *condensing unit* (combination of compressor and condenser) are obtained by superimposing the characteristics of the two as shown in Fig. 10.1.

The figure shows the characteristics of a typical reciprocating compressor running at a given speed superimposed on those of a typical shell- and tube-condenser with the inlet temperature and flow rate of cooling water fixed.

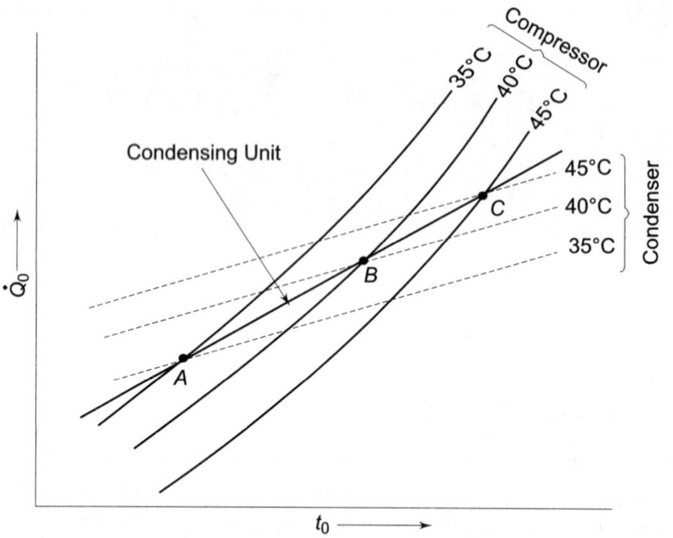

Fig. 10.1 Performance characteristic of a condensing unit by graphical method

It is seen that the compressor capacity increases with evaporator pressure, and decreases with condenser pressure. On the other hand, condenser capacity increases with condenser pressure (temperature) for the same water inlet temperature because of larger ΔT available for heat transfer. The increase in condenser capacity with evaporator pressure is small. This increase is due to the increase in mass flow rate of the refrigerant through the compressor, and consequent increase in heat-transfer coefficient in the condenser.

The characteristic of the condensing unit ABC is obtained by joining the points of intersection of the characteristics corresponding to the same condensing temperatures for the two components.

In this case it may be noted that the condensing temperature is not constant along this characteristic. It also decreases as the evaporator temperature decreases since a lower evaporator temperature results in a reduced refrigerating capacity and hence a decreased loading on the condenser. Simultaneously, the condensing temperature (and hence pressure) drops. The performance of the condensing unit will, however, change if the flow and/or inlet temperature of the cooling water are changed.

10.2.2 Performance Characteristics of Compressor-Capillary Tube

The mass flow rate through a capillary of a given bore and length depends on the pressure difference between the inlet and outlet. Thus, the flow rate increases with increasing condensing pressure and decreases with increasing evaporator pressure, as shown in Fig. 10.2.

Superimposed on the same figure are the compressor characteristics. Intersection points 1, 2 and 3 indicate the *balance points* between the compressor and the capillary tube. At the balance points, the compressor pumps out as much refrigerant from the evaporator as the expansion device feeds into it. Any unbalanced condition can only be temporary.

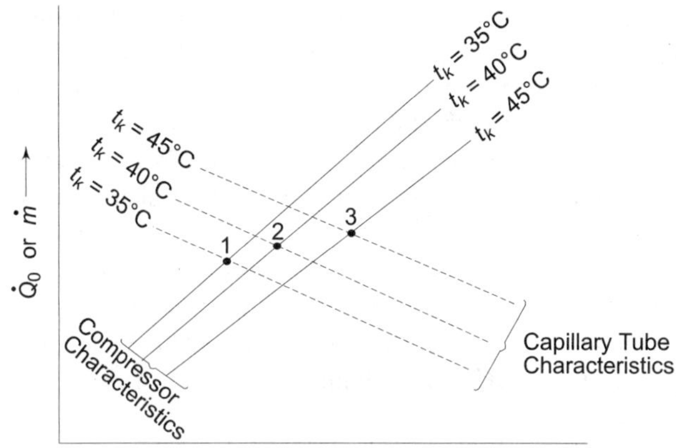

Fig. 10.2 Compressor-capillary tube balance points by graphical method

However, the compressor and the capillary tube cannot fix the balance point independently. The evaporator heat-transfer characteristics must also be satisfied. If they are not satisfied at the compressor-capillary tube balance point, an unbalanced condition can result which may cause either *starving* or *flooding* of the evaporator.

Figure 10.3 shows a balance point between the compressor and the capillary tube. Consider that the load on the evaporator coil increases to more than the refrigeration capacity at the balance point so that the evaporator temperature and hence the pressure rises from A to B. At B, the compressor pumps more refrigerant than the expansion device can feed, i.e., $\dot{m}_C > \dot{m}_E$. As a result, the evaporator will soon be starved of the refrigerant.

The corrective process should then start to work. As a result of the above, the liquid refrigerant will *back-up* into the condenser, thereby reducing the effective heat-transfer surface. The temperature and hence the condensing pressure will, therefore, rise. An increased pressure difference across the capillary will in turn increase the feeding rate of the capillary. At the same time, the compressor will give decreased refrigerating capacity, and thus a new balance point will emerge.

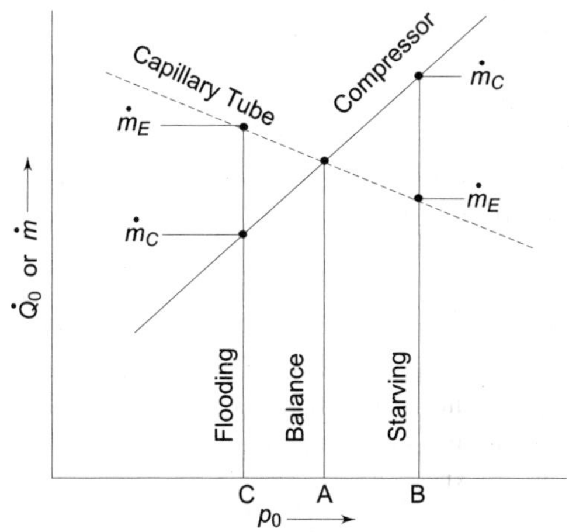

Fig. 10.3 Unbalanced conditions in compressor-capillary tube system

Consider now that the load on the cooling coil decreases. Consequently, the evaporator pressure will drop to point C. Now the compressor will be pumping out less refrigerant than the capillary can feed. It will, therefore, result in the flooding of the evaporator. Eventually, the liquid refrigerant may even enter the compressor and cause damage. In capillary-tube systems, therefore, the refrigerant charge is carefully measured, and only that much refrigerant is charged as would just fill the evaporator.

The corrective situation in this case occurs when some gas from the condenser starts flowing through the capillary tube, reducing its mass flow rate because of a high specific volume of the vapour. At the same time, due to decreased condenser pressure, compressor will start pumping more refrigerant from the evaporator, and again a new balance point will emerge.

10.3 ANALYTICAL METHOD

The performance of the complete system can also be obtained by a solution of the four simultaneous equations representing the individual characteristics. The method, though slightly difficult because of the nonlinearity of equations, is more general and amenable to computer analysis and optimization, thus providing economic designs for refrigeration and air conditioning systems.

The essential requirements of the method are the necessary governing equations representing the performance of the individual components. These can be established either by fitting an equation to the available experimental data, or by modelling a mathematical equation derived from the fundamental analysis of the performance of the equipment.

10.3.1 Equation Fitting

When catalogue data is available, an equation can be fitted to represent the performance characteristics of a component. The most common form of representation is a polynomial. For example, the enthalpy of saturated vapour can be expressed as a function of temperature by a relation of the type

$$h = a_0 + a_1\left(\frac{T}{100}\right) + a_2\left(\frac{T}{100}\right)^2 + \cdots + a_6\left(\frac{T}{100}\right)^6 \tag{10.1}$$

Likewise, property correlations can be established for specific heat, pressure-temperature relationship, etc. Another form of representation is the exponential form

$$y - b = ax^m \tag{10.2}$$

Yet another equation is like the ln p_{sat} versus $1/T_{sat}$ relationship.

When a variable is a function of two variables, e.g., the pressure ratio developed by a centrifugal compressor is a function of speed N and refrigerating capacity $\dot{Q}_0$ (see Fig. 6.18), then we may express as follows:

$$\frac{p_k}{p_0} = a + b\dot{Q}_0 + c\dot{Q}_0^2 \tag{10.3}$$

where

$$a = A_0 + A_1N + A_2N^2$$
$$b = B_0 + B_1N + B_2N^2$$
$$c = C_0 + C_1N + C_2N^2$$

This representation requires nine data points to find the constants. The *least square method* may be used to find the constants if a large number of data points are available.

10.3.2 Mathematical Modelling

In mathematical modelling no catalogue data are necessary. An example of mathematical modelling is provided by analytical expressions/equations which are derived to represent the performance of heat exchangers. Thus for a counter flow heat exchanger, as shown in Fig. 10.4 in which subscripts *c* and *h* refer to cold and hot fluids respectively, we have for the rate of heat transfer by LMTD method,

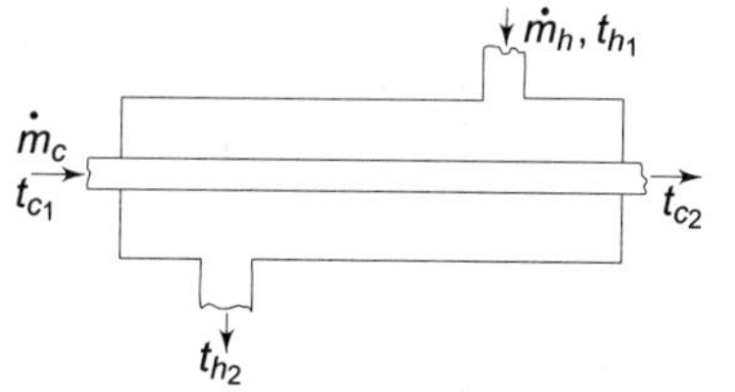

Fig. 10.4 Schematic representation of a simple heat exchanger

$$\dot{Q} = \dot{m}_h\, C_h\, (t_{h_2} - t_{h_1}) \tag{10.4}$$

$$= \dot{m}_c\, C_c\, (t_{c_1} - t_{c_2}) \tag{10.5}$$

$$= UA\, \frac{(t_{c_1} - t_{h_2}) - (t_{c_2} - t_{h_1})}{\ln(t_{c_1} - t_{h_2})/(t_{c_2} - t_{h_1})} \tag{10.6}$$

The three Eqs (10.4), (10.5) and (10.6) contain three unknowns, $\dot{Q}$, t_{h_2} and t_{c_2}. Eliminating $\dot{Q}$ and t_{h_2} and solving, we get

$$t_{c_2} = t_{c_1} - (t_{c_1} - t_{h_1}) \left[\frac{1 - e^D}{\left(\frac{\dot{m}_c C_c}{\dot{m}_h C_h}\right) - e^D} \right] \tag{10.7}$$

where
$$D = UA \left[\frac{1}{\dot{m}_c C_c} - \frac{1}{\dot{m}_h C_h} \right] \tag{10.8}$$

By substituting t_{c_2} in Eqs (10.5) and (10.4) we can obtain $\dot{Q}$ and t_{h_2} respectively.

Applying this to a condenser (Fig. 10.5), the heat-transfer equation becomes

$$\dot{Q}_k = (UA)_k\, \frac{(t_k - t_{w_1}) - (t_k - t_{w_2})}{\ln \dfrac{t_k - t_{w_1}}{t_k - t_{w_2}}} = \dot{m}_w\, C_w\, (t_{w_2} - t_{w_1}) \tag{10.9}$$

Equation (10.9) can be written as

$$\frac{(UA)_k}{\dot{m}_w C_w} = \ln \left(\frac{t_k - t_{w_1}}{t_k - t_{w_2}} \right) = -\ln \left(\frac{t_k - t_{w_2}}{t_k - t_{w_1}} \right)$$

or

$$e^{-(UA)_k/\dot{m}_w C_w} = \frac{t_k - t_{w_2}}{t_k - t_{w_1}} = \frac{t_k - t_{w_2} + t_{w_1} - t_{w_1}}{t_k - t_{w_1}} = 1 - \frac{t_{w_2} - t_{w_1}}{t_k - t_{w_1}}$$

whence

$$t_{w_2} = t_{w_1} + (t_k - t_{w_1})\left[1 - e^{(UA)_k/\dot{m}_w C_w}\right] \tag{10.10}$$

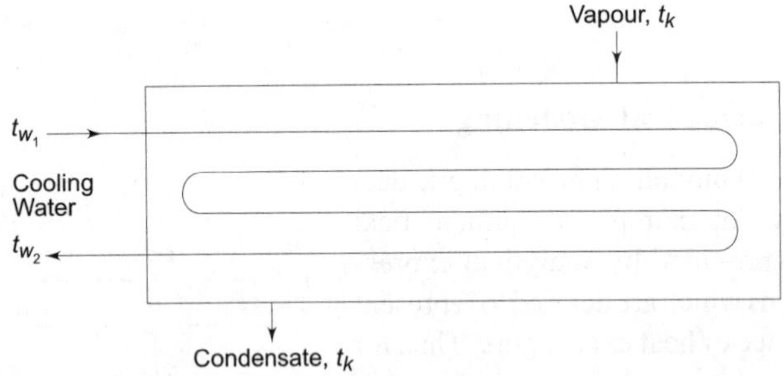

Fig. 10.5 Schematic representation of a 4 tube-pass, 1 shell-pass condenser

Applying this to an evaporator (Fig. 10.6) we similarly have

$$\dot{Q}_0 = (UA)_0 \frac{(t_{b_1} - t_0) - (t_{b_2} - t_0)}{\ln\left(\dfrac{t_{b_1} - t_0}{t_{b_2} - t_0}\right)} = \dot{m}_b C_b (t_{b_1} - t_{b_2})$$

and

$$t_{b_2} = t_{b_1} - (t_{b_1} - t_0)\left[1 - e^{-(UA)_0/\dot{m}_b C_b}\right] \tag{10.11}$$

where t_b is the brine or chilled water temperature.

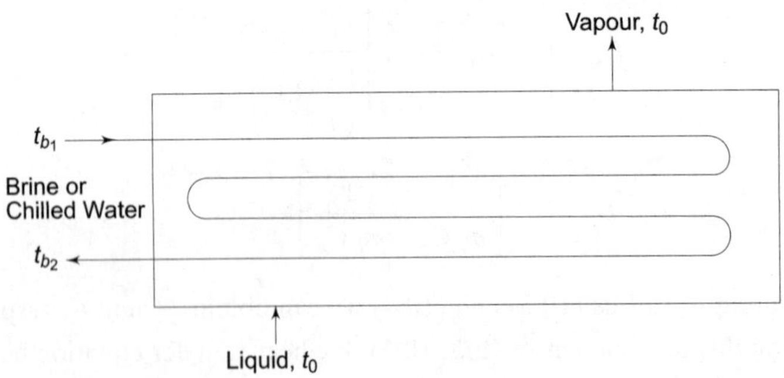

Fig. 10.6 Schematic representation of a 4-tube pass, 1 shell-pass flooded evaporator (Chiller)

10.3.3 System Simulation

The equation representing performance characteristics, energy and flow equations, properties, etc., can be used to simulate the system. The arrangement of the system at times permits a *sequential calculation* and at other times may need a *simultaneous calculation*. If the equations are linear, the well-known method of *triangularization* may be used. The calculations become difficult when the equations are non-linear. A common method used then is that of starting with *trial values* and following the method of successive iterations. Quite often it results in divergence and the method fails. A method which is devoid of this difficulty is the *Newton-Raphson method.*

10.4 NEWTON–RAPHSON METHOD

Let us consider that the following three non-linear equations are to be solved for the three unknown variables x, y, z

$$f_1(x, y, z) = 0 \tag{10.12}$$

$$f_2(x, y, z) = 0 \tag{10.13}$$

$$f_3(x, y, z) = 0 \tag{10.14}$$

Then the procedure for their solution involves the following steps:

Step 1 Assume trial values of the unknown variables, say x_t, y_t and z_t.

Step 2 Substituting the trial values in Eqs (10.12), (10.13) and (10.14), determine the values of f_1, f_2 and f_3. If the trial values are correct, then the three functions will be zero. If not, then the numerical values will represent some errors.

Step 3 Determine the partial differentials of the functions with respect to the unknown variables and equate the aggregate errors as follows:

$$\frac{\partial f_1}{\partial x} \cdot \Delta x + \frac{\partial f_1}{\partial y} \cdot \Delta y + \frac{\partial f_1}{\partial z} \cdot \Delta z = \Delta f_1 \tag{10.15}$$

$$\frac{\partial f_2}{\partial x} \cdot \Delta x + \frac{\partial f_2}{\partial y} \cdot \Delta y + \frac{\partial f_2}{\partial z} \cdot \Delta z = \Delta f_2 \tag{10.16}$$

$$\frac{\partial f_3}{\partial x} \cdot \Delta x + \frac{\partial f_3}{\partial y} \cdot \Delta y + \frac{\partial f_3}{\partial z} \cdot \Delta z = \Delta f_3 \tag{10.17}$$

where Δx, Δy and Δz represent the errors in the trial values of the unknown variables, e.g.,

$$\Delta x = x_t - x_c$$

where x_c represents the correct value of x.

Step 4 Equations (10.15), (10.16) and (10.17) can be solved simultaneously for the error values Δx, Δy and Δz and the new correct values are found as follows:

$$x_c = x_t - \Delta x$$

$$y_c = y_t - \Delta_y \quad \text{and} \quad z_c = z_t - \Delta z$$

The calculations are repeated with the new values until functions f_1, f_2, f_3, etc., and errors Δf_1, Δf_2, Δf_3, etc., are close to zero.

The following example is based on this approach of system simulation as made by Stoecker[5] and is applied here to a vapour compression refrigeration system. The equations used are extremely simplified versions of the actual performance equations. The aim here is to illustrate the procedure in applying the method which is normally done on a computer.

Example 10.1 *The performance data of a chilled water central air-conditioning plant in the range of operation is given below:*

Condenser	$(UA)_k$	$= 20{,}700\ W\ °C^{-1}$
Cooling water flow rate,	$\dot{m}_w$	$= 8.05\ kg/s$
Evaporator	$(UA)_0$	$= 23{,}840\ W°C^{-1}$
Chilled water flow rate,	$\dot{m}_b$	$= 7.2\ kg/s$
Compressor		
Refrigerating capacity,	$\dot{Q}_0$	$= 116{,}300 + 2910\ (t_0 - 5) - 1165\ (t_k - 43)\ W$
Power consumption,	$\dot{W}$	$= 13{,}490 + 1165\ t_0 + 350\ t_k\ W$

If the cooling water enters the condenser at 30°C and chilled water enters the evaporator at 11°C, find the temperatures of the cooling water leaving the condenser and the chilled water leaving the evaporator, and the refrigerating capacity and power consumption of the plant.

Solution The mathematical expression for power, cooling and chilled water temperatures, t_w and t_b and refrigerating capacity are:

$$\dot{W} = \dot{Q}_k - \dot{Q}_0 = (8.05)\,(4187)\,(t_w - 30) - 7.2\,(4187)\,(11 - t_b)$$

$$t_w = 30 + (t_k - 30)\,[1 - e^{-20{,}700/(8.05)\,(4187)}]$$

$$t_b = 11 - (11 - t_0)\,[1 - e^{-23{,}840/(7.2)\,(4187)}]$$

$$\dot{Q}_0 = (7.2)\,(4187)\,(11 - t_b)$$

The simplified expressions are

$$\dot{W} = 33{,}705\ t_w + 30{,}146\ t_b - 1{,}342{,}756$$

$$t_w = 0.459\ t_k + 16.23$$

$$t_b = 0.547\ t_0 + 4.99$$

$$\dot{Q}_0 = 331{,}610 - 30{,}146\ t_b$$

The functions can be written as

$$f_1 = 33{,}705\ t_w + 30{,}146\ t_b - 1{,}342{,}756 - \dot{W} = 0$$

$$f_2 = 0.459\ t_k + 16.23 - t_w = 0$$

$$f_3 = 0.547\ t_0 + 4.99 - t_b = 0$$

$$f_4 = 331{,}610 - 30{,}146\ t_b - \dot{Q}_0 = 0$$

where $\dot{W}$ and $\dot{Q}_0$ are separately defined in terms of t_k and t_0. The independent unknown variables are t_k, t_0, t_w and t_b

Assume trial values of variables as follows:

$$t_k = 40°C \qquad t_0 = 3°C$$

$$t_w = 33°C \qquad t_b = 6°C$$

Then

$$\dot{W} = 13{,}490 + 1165(3) + 350(40) = 30{,}985\ W$$

$$\dot{Q}_0 = 116{,}300 + 2910(3 - 5) - 1165\ (40 - 43) = 113{,}975\ W$$

Substituting these values, we obtain calculated values of functions and their aggregate errors:

$$f_1 = -80{,}600 \qquad \Delta f_1 = f_1 - 0 = -80{,}600$$
$$f_2 = 1.59 \qquad \Delta f_2 = f_2 - 0 = 1.59$$
$$f_3 = 0.631 \qquad \Delta f_3 = f_3 - 0 = 0.631$$
$$f_4 = 36{,}759 \qquad \Delta f_4 = f_4 - 0 = 36{,}759$$

Differentiating the functions, we get

$$\frac{\partial \dot{W}}{\partial t_k} = 350; \quad \frac{\partial \dot{W}}{\partial t_0} = 1165$$

$$\frac{\partial \dot{Q}_0}{\partial t_k} = -1165; \quad \frac{\partial \dot{Q}_0}{\partial t_0} = 2910$$

and

$$\frac{\partial f_1}{\partial t_w} = 33{,}705; \quad \frac{\partial f_1}{\partial t_b} = 30{,}146; \quad \frac{\partial f_1}{\partial t_k} = -350; \quad \frac{\partial f_1}{\partial t_0} = -1165$$

$$\frac{\partial f_2}{\partial t_w} = -1; \quad \frac{\partial f_2}{\partial t_b} = 0; \quad \frac{\partial f_2}{\partial t_k} = 0.459; \quad \frac{\partial f_2}{\partial t_0} = 0$$

$$\frac{\partial f_3}{\partial t_w} = 0; \quad \frac{\partial f_3}{\partial t_b} = -1; \quad \frac{\partial f_3}{\partial t_k} = 0; \quad \frac{\partial f_3}{\partial t_0} = 0.547$$

$$\frac{\partial f_4}{\partial t_w} = 0; \quad \frac{\partial f_4}{\partial t_b} = -30{,}146; \quad \frac{\partial f_4}{\partial t_k} = +1165; \quad \frac{\partial f_4}{\partial t_0} = -2910$$

Setting up expressions for aggregate errors, we have

$$\Delta f_1 = -33{,}705\, \Delta t_w + 30{,}146\, \Delta t_b - 350\, \Delta t_k - 1165\, \Delta t_0 = -80{,}600$$
$$\Delta f_2 = -\Delta t_w + 0.459\, \Delta t_k = 1.59$$
$$\Delta f_3 = -\Delta t_b + 0.547\, \Delta t_0 = 0.631$$
$$\Delta f_4 = -30{,}146\, \Delta t_b - 1165\, \Delta t_k - 2910\, \Delta t_0 = 36{,}759$$

Solving these four equations simultaneously, we obtain for errors in trial values, and correct values of variables as follows:

$$\Delta t_0 = -0.89°\text{C};\ t_0 = 3 + 0.89 = 3.89°\text{C}$$
$$\Delta t_k = +0.41°\text{C};\ t_k = 40 - 0.41 = 39.59°\text{C}$$
$$\Delta t_w = -1.4°\text{C};\ t_w = 33 + 1.4 = 34.4°\text{C}$$
$$\Delta t_b = -1.12°\text{C}\ t_b = 6 + 1.12 = 7.12°\text{C}$$

Also,

$$\dot{W} = 13{,}490 + 1165\,(3.89) + 350(39.59) = 27{,}794 \text{ W}$$

$$\dot{Q}_0 = 116{,}300 - 2910\,(3.89 - 5) - 1165(39.59 - 43) = 117{,}043 \text{ W}$$

The revised values of the functions become

$$f_1 = 3541,\ f_2 = 0,\ f_3 = 0,\ f_4 = -72$$

Subsequent iterations do not appreciably alter the values of the variables calculated above. Hence, these values of t_0, t_k, t_w and t_b represent the actual operating conditions. The refrigerating capacity obtained is 117 kW, and the power consumption of the compressor is 27.8 kW, giving a COP of 4.2. Thus, we see that convergence is very fast in Newton-Raphson method.

Note *Operating conditions for central A/C plants in Delhi:*

Evaporator Temperature 3 to 6°C

Condenser Temperature 43°C

Cooling Water Temperature 32.3°C In, 36.3°C Out (In Condenser)

Chiled Water Temperature 12°C In, 6.5°C Out (In Chiller)

10.5 OPTIMAL DESIGN OF EVAPORATOR

The basic objective of system simulation as discussed in this chapter and a very simple illustration of which is provided in Example 10.1, is to predict the performance of the system with varying operating parameters. Another objective is optimization of design by varying the design parameters and seeing their effect on performance.

Complete simulation and optimization of the four-component vapour compression system is extremely difficult. However, an attempt can be made to optimize the design of one component by keeping the specifications of the other components unaltered in the existing design. Here too, the problem becomes complicated as the change in the design of one affects the performance of the others.

The ultimate aim of every industrial optimisation problem is invariably to reduce costs—the fixed cost as well as the running cost. In the present case, the main components of the variable capital cost are the condenser and the evaporator due to the enormous amount of copper-tubing used. The capital cost of the compressor is not affected very much with slight variations in condenser and evaporator pressures. Further, the condenser design is not very sensitive to the refrigerant pressure drop (or rise) as the frictional and decelerational components are of opposite nature. The evaporator is extremely sensitive to design as the frictional and accelerational pressure drop components add to each other. Pressure drop causes a decrease in the temperature potential and hence an increase in heat-transfer surface area. On the other hand, a tubing design with a larger pressure drop may give a higher heat-transfer coefficient and hence a decreased heat-transfer surface area.

Dhar and Arora[1,2,3] have carried out the optimization of the design of a chiller subject to the constraints that the capacity of the system and operating conditions remain unaltered. In order to do so, the evaporator exit conditions have been constrained to be the same as that in the original design. This ensures that there are no changes in the performance of the compressor and condenser, the only change being in the expansion valve design which has to be suitably modified to take care of the different pressure drop in the optimized chiller design. Thus, as shown in Fig. 10.7, if the pressure drop in a chiller is more than that in the original design, the existing expansion valve has to be replaced by one having a slightly bigger orifice diameter. This is because this valve has to supply the same amount of refrigerant flow at a lesser pressure drop across the orifice, viz., $p_3 - p_4'$ instead of $p_3 - p_4$. Obviously, this will not affect the performance of the refrigeration system and will also not alter the cost of the other components, including the expansion valve and running cost. But it will decrease the temperature difference between the cooled medium and the evaporating referigerant thus increasing the size and cost of the evaporator.

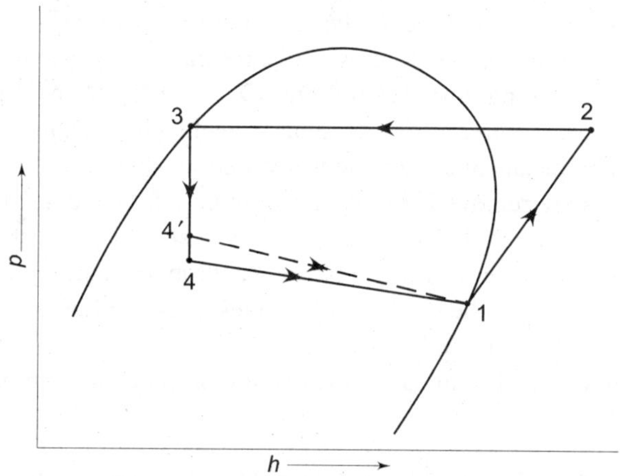

Fig. 10.7 Pressure drop in evaporator with constant suction state

10.6 INSTALLATION, SERVICE AND MAINTENANCE OF VAPOUR COMPRESSION SYSTEMS

Utmost precaution should be taken so that no component gets contaminated with foreign matter, dirt moisture etc. Refrigerant system is a closed one. Hence this care is essential.

While installing, it must be ensured that each component, viz., compressor, condenser, evaporator and expansion device, and even other items in the lines can be isolated from the rest of the system for repair and replacement. This can be done by making provision for valves at inlet and outlet of the component.

Further, each component should be made accesible for the same purpose.

Before proceeding to start the plant, it is first necessary to test it so that there are no leaks.

Pressure Test After the plant has been installed, all joints, fittings, etc., must be tested for leakage by a 'Pressure Test'.

For pressure testing, all valves should be wide open with exception of the charging line, drain, and purge valves.

The test is performed either with CO_2 or N_2 gas. If N_2 is used, then one must connect the nitrogen gas cylinder to the plant through a 'needle valve' and not the ordinary hand shut-off valve as the pressuree in the N_2-cylinder is very high, of the order of 2000 psi (150 atm.). If a needle valve is not used for slow opeining, there is a danger of serious accident happening.

Even with CO_2, better use a regular stop valve. After connecting the N_2 or CO_2 cylinder, the gas valve should be 'slightly' cracked open. The gas is thus allowed to enter the system slowly until the pressure is raised to 1.5 to 2 times the highest expected pressure in the system. This is usually 12–14 atmospheres in R 134 a and R 22 systems. If at any time it is intended to open or shut off the gas supply, this should be done by means of the valve on the gas cylinder, and not by means of the regulatory stop valve in the lime. If the regulatony stop valve is shut and the gas cylinder valve is left open, the full pressure of the gas, say N_2 about 150 atm, could be applied to the charging line. THIS WOULD BE VERY DANGEROUS.

After applying pressure, all joints, glands, welds, etc., should be thoroughly tested for leaks by means of a soap and water solution.

After leakages are removed, the plant should be left 'standing' at the test pressure of 12 to 14 atm for 24 hours. After 24 hours, it should be checked whether the pressure has dropped or not. If the pressure has dropped, apply pressure, and test the leakages again. Leave the plant standing at test pressure for 24 hours once again until satisfied that there are no leaks.

Note that minor changes in gas pressure could also be due to room temperature variation also.

Note *Never use O_2 for pressure testing. It may cause an explosion when mixed with oil in the compressor.*

Evacuating and Dehydrating the Plant After pressure testing, the whole system must be completely evacuated so as to make sure that no air or moisture is trapped inside before charging the refrigerant.

Air in the system raises the pressure in the system and increases load on the compressor.

Moisture in the system can cause moisture-choking in expansion device, if temperature goes below 0°C, and certain other problems like dilution of lubricating oil, oil separation, etc.

The evacuation must be continued until reaching a vacuum of 29 inches or 758-mm Hg. In order to do this, a vacuum pump (often a 2-stage one) and gauges are necessary. Sometimes, the compressor itself is used to evacuate the plant.

The process of evacuation not only removes air from the system but it also simultaneously causes moisture inside to be evaporated, and to be removed along with the air.

The underlying principle, in moisture evaporation, is the fact that water, like any other substance, has an exact p versus T relationship. For example, at normal atmospheric pressure water boils at 100°C, but at 1-mm Hg pressure or 759-mm Hg vacuum it boils at –17°C. Hence, by reducing the pressure in the system, moisture in the system can be made to evaporate, and the resultant vapour is drawn off by the vacuum pump along with air leaving the system quite dry.

We often heat the system also simultaneously to help in evaporation of any moisture trapped in the system while evacuating.

For evacuating the system, all line valves should necessarily be kept wide open as for pressure testing. Evacuation has to proceed until a pressure no more than 2 mm or better still, 1-mm Hg is obtained.

The pump should then be islolated by shutting off the stop valve connecting it to the system. The pump can then be stopped.

Note *The vacuum pump should never be switched off before isolating it from the system, otherwise oil from the pump will be drawn into the system, and vacuum will be lost.*

Air can also be removed by using the condensing unit compressor itself as a vacuum pump. The Compressor pumps the suction line first. Then the liquid line valve is opened allowing vactum to be produced in the liquid line also. Be careful that no oil is pumped.

Do not put a 'drier' in the liquid line until most of the air has been removed. During vacuumising, a short length of tubing may be used in place of the drier.

Using Compressor as Vacuum Pump To use compressor itself as vacuum pump, take steps as follows:

(i) Put a compound gauge on compressor suction service valve.

(ii) Remove the air by running compressor until a vacuum of 20 inch /500-mm Hg or more is reached.

(iii) Then instal a purge line, equipped with hand shut-off valve in to the gauge opening of discharge service valve.

(iv) Turn the discharge service valve stem shutting off condenser opening.

(v) Now turn the suction service valve fully, and pump a vacuum. The air being removed now will be discharged out of the purge line.

(vi) After creating as high a vacuum as possible with the compressor, 70–75% of air is removed.

The refrigerant cylinder valve on the suction side may be cracked slighty open, permitting very small quantity of refrigerant to enter the suction line.

(vii) When this has also been pumped out of the system, the amount of air left will be negligilble, say les than 5%.

(viii) A second refrigerant flush followed by evacuating will reduce air to, say 1%.

(ix) After the system has been purged of air, quickly instal a drier in the liquid line, purge and seal the connection.

(x) Proceed to test for leaks.

Charging the Refrigerant Charging of refrigerant in the system is done in two ways:

(i) Charging from suction line.

(ii) Charging from liquid line.

Suction line charging is done in the case of small units in which the amount of charge required is small.

For the purpose, the refrigerant cylinder is made to stand vertically on the ground so as to ensure that only vapour enters the compressor. The refrigerant cylinder is connected to the suction-side of the compressor either through the suction service valve or through a separate 'charging line' in hermetically sealed units.

After charging, the suction service valve is closed, and the gas cylinder is disconnected.

In the case of hermetically sealed units, the charging line is 'pinched' with a 'pinch-off tool', and then the pinch brazen or soldered.

To make sure that the correct weight of charge is made, many methods are used.

In unitary equipment, manufacturers charge a measured amount of refrigerant from experience.

In the field it is ensured that the required head pressure is reached.

In refrigerators, it is seen that frosting appears over the capillary tube. Then the refrigerator is tested for pull-down period, or ice-making time, and so on.

Charing from the liquid line is done in large plants where the amount of charge required is large. Charging refrigerant as vapour from suction line will take inordinately long time. For the purpose, the gas cylinder is often kept at a height in a tilted position so that the liquid enters the system. The pressure in the system in water- cooled units will be below pressure in gas cylinder to enable liquid to flow in.

Adequate charge is usually ensured by getting required operating pressures in the sysyem.

Liquid refrigrant level can also be checked in the condenser or receiver through a sight-glass.

Starting a System Before starting the system, first check the electrical system. Make sure that the phase is correct, voltage is correct, etc. Recording type volt and ammeters should be connected.

Make sure water or air circuit is turned on. Check the electrical meters, pressure guages, water flow, etc.

Shut down the unit at first sign of trouble. To start the compressor at full load, it can get overloaded. Hence, liquid line valve should be closed. Slowly open evaporator line, and then liquid line. After starting, check high and low side pressures.

Checking Oil-Level in Compressor When the plant starts running, a certain amount of oil will pass round the system, and then return to the compressor. A certain amount may remain in the system. Hence it may become necessary to add more oil.

Frequent inspection of oil level in the compressor has to be made when the plant is newly installed. When it is required to add oil, proceed as follows:

(i) Stop the machine.
(ii) Isolate the compressor by closing delivery and suction valves.
(iii) Slacken the oil-filling plug, and let the gas disperse gradually.
(iv) Remove the plug when pressure reaches atmospheric level.
(v) Charge the oil to reach the correct level which is when the oil is filled to the top of the oil-filling plug hole.
(vi) Replace the plug. Open the valves. And start the machine.

Note *It should not be necessary constantly to add oil as the oil does not deteriorate, and it is not lost.*

Off-design Gauge Pressures For gauge readings showing off-design pressures, any of the following reasons could be seen.

(i) *High Condenser Pressure* Shortage of condenser water, or water temperature from cooling tower high.
Overcharge of gas in the system.
Air in the system.

(ii) *Low Evaporator Pressure* Undercharge of gas.
Dirty liquid strainer.
Moisture or dirt in expansion device.

(iii) *Low Condenser Pressure* Shortage of gas/Under charge.

It may be seen that chocked strainers can cause most of these symptoms. Hence they should be checked first including those on the cooling tower.

In unitary equipment with hermetically sealed units, there are no gauges. The problems are diagnosed as follows:

(i) *Inadequate Cooling—Comperssor Continuously Running* This could be due to undercharge. Condenser pressure will be low. It will be indicated by condenser running cool. The refrigerator body temperature housing condenser will not be warm.

(ii) *No Cooling—Compressor Continuously Running* It means the gas has leaked out. The unit requires complete servicing including repair for leak, and recharging.

(iii) *In A/C Units, Ice Formation on Cooling Coil but no Cooling in Room While the Compressor is Running* The reason is simple. Air filters are choked. It requires cleaning of filters. By far, this is the primary cause of complaints of inadequate cooling in all A/C units. Filters need to be regularly cleaned.

Undercharge of Refrigerant Indications for undercharg are the following:

(i) Low condenser pressure.
(ii) Low condenser temperature. Condenser running cool in refrigerator and air conditioners.
(iii) Large gas bubbles appearing in liquid line sight-glass.
(iv) Inadequate cooling as there is no liquid to evaporate in evaporator while the compressor is sucking vapour.
(v) Body of expansion device without frost. The remedy is to charge more refrigerant till condenser and evaporater pressure gauges record set readings, or condenser runs warm enough, or till requird cooling and temperatures are obtained.

Adding Refrigerant with Compressor Running While the compressor is running, the refrigerant is added as vapour from suction side. For the purpose:

Connect gas cylinder to charging/suction line. Keep cylinder vertical. Make sure no liquid enters the compressor. Open the gas cylinder valve slightly, and purge the air out of charging line.

Tighten Charging Pipe Connection Open the compressor suction services valve/charging line valve slightly.

Open fully the gas cylinder valve. Allow refrigerant to pass into the system until required pressures are obtained, or all bubbles disappear in the liquid sight-glass, or condenser runs warm enough, or evaporator side is seen to give adequate cooling .

After charging, first close the valve on gas cylinder, and then close the suction service valve, etc.

Finally remove the gas cylinder, and seal it.

Note that operating unit with undercharge is very detrimental. It will result in continuous motor operation, and overloding of compressor and motor.

Lack of refrigerant is also indicated by an increase in liquid line and drier temperature.

Overcharge of Refrigerant Overcharge is indicated by high condenser pressure. And liquid sight-glass runs full of liquid. In unitary equipment, it will be indicated by high temperature of condenser as the condenser pressure rises, and there is no adequate surface area for vapour to condense.

For decresing charge, in small units, the refrigerant can be allowed to escape from the discharge line.

In large units, the refrigerant can be taken out from liquid line into a cylinder, which may be cooled with ice to maintain its pressure lower than condenser pressure. Otherwise, use an evacuated cylinder.

The operation should be performed very slowly in order to minimise the possibility of system becoming undercharged.

Pumping Down The process of pumping the whole of the refrigerant in the system into the condenser is referred to as 'pumping down'.

After pumping down, and closing the 'in' and 'out' valves of the condenser, the condenser gets isolated. Thus any system component can be removed for repair or replacement without loss of charge.

Pumping down is done in the following steps:

(i) Close liquid line valve/condenser 'out' valve.

(ii) Keep the cooling water running in condenser.

(iii) Run the machine until evaporater pressure drops to 1 psig. DO NOT PUMP IN VACUUM.

(iv) Stop the machine.

(v) Close the delivery valve, and condenser 'in' valve.

At the time of restarting, refrigerant should be allowed to enter evaporater slowly in order to prevent damage from too rapid cooling of the components.

Air in the System Air in the system is indicated by high condenser pressure. This usually occurs in small units in case no proper care is taken during evacuation and charging. In large plants, it could be due to various reasons, and also because of negative evaporator pressure in the case of some refrigerants.

To expel the air, the following procedure is adopted.

(i) Shut the liquid line valve. Pump down the refrigerant in the condenser.

(ii) Stop the machine, shut the condenser 'in' valve.

(iii) Allow the liquid in the condenser to be cooled by allowing the cooling water to continue to flow.

(iv) If no air is present, the condenser gauge reading should be corresponding to temperature of cooling water. If the condenser gauge reading is above that, it is definite that air is in the system.

Expel the air by slackening the 'purge valve' on the condenser until the condenser gauge reading is reduced corresponding to that of the cooling water temperature.

(v) Retighten the purge valve. Open the liquid line slowly.

(vi) Retighten the joints, and test for leaks so that no air again enters the system.
(vii) Open the delivery line. Start the machine.
(viii) Add refrigerant if necssary. Some refrigerant will have escaped with air during purging.

Moisture in the System Moisture in the system primarily creates problem in applications such as refrigerator, freezers, etc., in which the evaporator temperature is below 0°C. It will then cause 'moistrue choking' of expansion device due to freezing of moisture. The evaporator gauge, in such cases, will register a high vacuum.

If the evaporater operates above 0°C, ice formation does not occur. But there may be other problems such as dilution of oil.

In capillary systems, moisture choking will be indicated by a point on the capillary which will show large difference in temperature of upstream and downstream side. There may be frosting on the downstream side at the point of choking, while the rest of the capillary will be warm.

Again the cause of moisture in the system is the lack of care during evacuation and charging.To remove moisture, thaw any ice formed by warming the expansion device.

The procedure for evacuation, drying, and charging has to be repeated in small units.

In both small and large units, the solution is to instal a 'drier' in the liquid line.

Servicing Condenser Air-cooled condenser need 6-monthly cleaning and painting. Water-cooled condensers require regular/routine cleaning of tubes inside which water is flowing. Water in Delhi is particularly hard. Further, there may be muck coming from cooling tower. So the water-side of tubes will get corroded. This will reduce the water-side of heat transfer coefficient. Consequently, the plant will run at high head pressure.

Cleaning of tubes is done by circulating a 5% solution of HCl to dissolve the scales that are formed.

Note *Routine servicing of open-type compressor is also required for replacing bearings, grinding valve reeds, to repair compressor valve-plate, and compressor crankshaft seal, etc.*

References

1. Dhar P L, *Optimization in Refrigeration Systems*, Ph. D. Thesis. Indian Institute of Technology, Delhi, 1974.
2. Dhar P L and C P Arora, 'Design of direct expansion chillers', *Proc. Fourth National Symposium on Refrigeration and Air Conditioning*, Delhi, Sept. 11–12, 1975, pp. 230–239.
3. Dhar P L and C P Arora, 'Optimizing the use of fins in D-X chillers', *Proc. Fourth National Symposium on Refrigeration and Air Conditioning*, Delhi, Sept. 11–12, 1975, pp. 1–5.
4. Stoecker W F, *Refrigeration and Air Conditioning*, McGraw Hill, New York, 1968, pp. 158–171.
5. Stoecker W F, *Design of Thermal Systems*, McGraw Hill, Kogakusha, 1971.

Revision Exercises

10.1 For a R 134a compressor, the following performance data at a condensing temperature of 35°C are given:

Evaporator Temperature t_0(°C)	*Refrigerating Capacity* $\dot{Q}_0(W)$	*Power Consumption* $P(W)$
–20	2560	1800
–15	3325	2000
–10	4190	2200

Fit suitable equations for the refrigerating capacity and power consumption as a function of the evaporator temperature and constant condensing temperature of 35°C and find their numerical values at an evaporator temperature of 0°C.

10.2 A refrigeration plant serves as a water chiller. Data on the individual components are as follows:

Evaporator: $UA = 30595 \text{ W°C}^{-1}$
Chilled water flow rate, $\dot{m}_c = 6.795$ kg/s

Condenser: $UA = 26375 \text{ W°C}^{-1}$
Cooling water flow rate, $\dot{m}_w = 7.55$ kg/s

Compressor: The refrigerating capacity and power consumption as functions of evaporating and condensing temperatures are given by the equations

$$\dot{Q}_0 \text{ (kW)} = -768 + 5.1\,T_0 - 2T_k$$

$$P\text{(kW)} = -0.732\,T_0 + 0.001746T_0^2 + 0.31\,T_k$$

Determine the condensing and evaporating temperatures, refrigerating capacity and power consumption for inlet-chilled water temperature of 2°C and inlet cooling water temperature of 30°C.

Chapter

11

Gas Cycle Refrigeration

11.1 LIMITATIONS OF CARNOT CYCLE WITH GAS AS A REFRIGERANT

Although the Carnot cycle is the most efficient cycle between fixed temperature limits and is useful as a criterion of perfection, it has its inherent drawbacks when a gas is used as a refrigerant, as can be seen from its *p-v* diagram shown in Fig. 2.1b. These are:

(i) Extreme pressures and large volumes are developed since the pressure rise takes place both during isentropic compression as well as isothermal heat-rejection processes.

(ii) Isothermal heat-transfer processes with a gas, viz., a medium with a finite specific heat, are impossible to achieve in practice.

(iii) The *p-v* diagram of the cycle working with a gas is so narrow that even small irreversibilities of the individual processes cause a significant increase in the work done. This increase in work forms a large proportion of the net work of the cycle.

11.2 REVERSED BRAYTON OR JOULE OR BELL COLEMAN CYCLE

In the reversed Brayton or Joule or Bell Coleman cycle, as it is variously called, the two isothermal processes 2′-3 and 4′-1 of the reversed Carnot cycle are replaced by the more practical isobaric processes 2-3 and 4-1 respectively, as shown on the *T-s* diagram in Fig. 11.1. This involves, however, a loss in the refrigerating effect Δq_0 equivalent to the area 1-4-4′, and increases in the net work of the cycle, Δw_0 and Δw_k, corresponding to the areas 1-4-4′ and 3-2-2′ respectively.

A schematic diagram of an air-refrigeration machine working on a simple gas cycle is shown in Fig. 11.2. In this arrangement, the compressor and expander are shown coupled together since the expander work is utilised to provide a part of the compressor work. Point 4 in Fig. 11.1 represents the state of the refrigerated air which would absorb heat at a constant pressure p_0 until it attains the temperature corresponding to point 1. At 1, the air is isentropically compressed to 2, from p_0 to p_k pressure, after which it is cooled at constant pressure p_k to 3. The cooling medium is invariably the surrounding atmospheric air as the cycle is presently employed only in aircraft refrigeration. The air is finally expanded isentropically to 4 whereby it gets cooled to T_4.

Note *It may be noted that if ever the cycle is used for cooling on ground, water can be used as a cooling medium.*

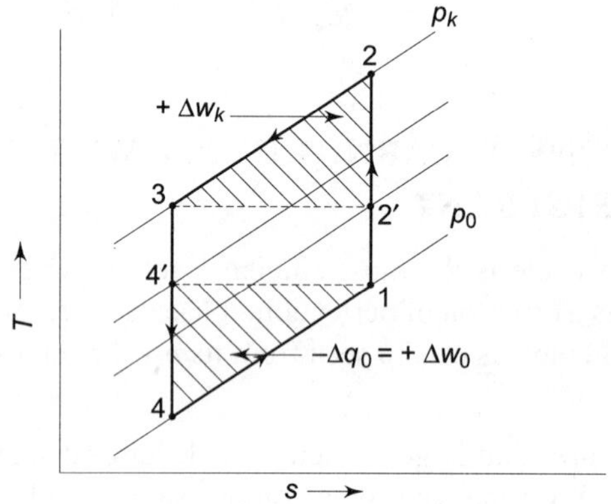

Fig. 11.1 Comparison of Joule or reversed Brayton cycle with reversed Carnot cycle

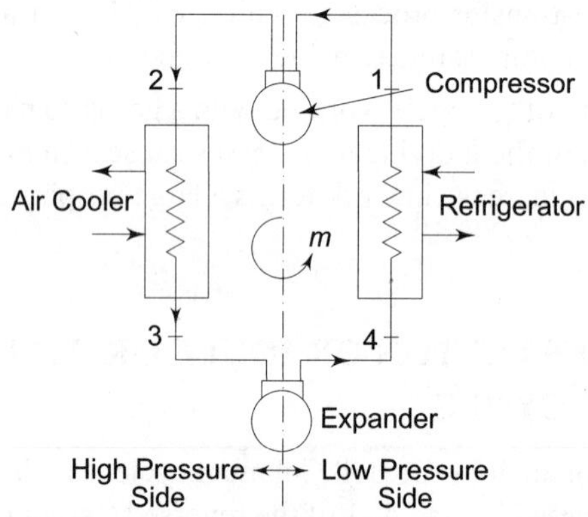

Fig. 11.2 Schematic diagram of simple gas cycle

11.2.1 Analysis of Gas Cycle

The relationship between the various temperatures of the cycle is given by the isentropic relation applied to both compression and expansion processes, viz.,

$$\frac{T_2}{T_1} = \frac{T_3}{T_4} = \left(\frac{p_k}{p_0}\right)^{\frac{\gamma-1}{\gamma}} \tag{11.1}$$

Also, assuming air to be a perfect gas, we have per unit mass of air circulated:

Refrigerating effect, $$q_0 = C_p\,(T_1 - T_4) \tag{11.2}$$

Heat rejected, $$|q_k| = C_p(T_2 - T_3) \quad (11.3)$$

Compressor work, $$|w_C| = \frac{\gamma}{\gamma - 1}(p_2 v_2 - p_1 v_1)$$
$$= C_p(T_2 - T_1) \quad (11.4)$$

Expander work, $$w_E = \frac{\gamma}{\gamma - 1}(p_3 v_3 - p_4 v_4)$$
$$= C_p(T_3 - T_4) \quad (11.5)$$

Network of the cycle, $$|w| = |w_C| - w_E = |q_k| - q_0$$
$$= C_p[(T_2 - T_3) - (T_1 - T_4)] \quad (11.6)$$

The coefficient of performance is, therefore,

$$\mathcal{E} = \frac{q_0}{|w|} = \frac{1}{\left(\dfrac{T_2 - T_3}{T_1 - T_4}\right) - 1} \quad (11.7)$$

which on substitution from Eq. (11.1) can be expressed as

$$\mathcal{E} = \frac{1}{\left(\dfrac{p_k}{p_0}\right)^{\frac{\gamma-1}{\gamma}} - 1} = \frac{1}{r^{\frac{\gamma-1}{\gamma}} - 1} \quad (11.8)$$

The COP of the gas cycle is, therefore, a function of the pressure ratio $r = p_k/p_0$ only. The lower the value of the pressure ratio, the higher is the COP. The variation of COP with pressure ratio is given in Table 11.1.

Table 11.1 Variation of COP of air cycle with pressure ratio

p_k/p_0	1	2	3	4	5	6
$\mathcal{E}$	∞	4.56	2.71	2.05	1.72	1.5

It may be noted that the pressures p_k and p_0, and hence the pressure ratio r, have limitations on account of the operating temperatures, viz.,

(i) T_1 as the highest refrigeration temperature, and
(ii) T_3 as the lowest ambient temperature.

Point 1 on the diagram is fixed by the temperature T_1 and also the pressure p_0 which is generally equal to the surrounding atmospheric air pressure. Point 3 is fixed because of the limitations of the ambient temperature T_3 to which the gas can be cooled. Pressure p_k can, however, be varied within wide limits, starting from $p_{k,\,min}$ onwards as shown in Fig. 11.3. With the compressor discharge pressure equal to $p_{k,\,min}$, the refrigerating capacity of the machine is zero. The air is alternately compressed and expanded between points 2_{min} and 1. The net work is also zero and hence the COP is indeterminate.

However, as the pressure p_k is increased, although the refrigerating effect (area under the curve 4-1) and hence the capacity of the refrigerating machine increases, the work of the cycle also increases. For example, when the discharge pressure is p_k, the refrigerating effect is 1-4-5-6 and the net work is 1-2-3-4. When the discharge pressure is increased to p_k', the increase in the refrigerating effect is 4-4′-5′-5 and that in the net work is 2-2′-3′-4′-4-3-2. If is evident that the increase in work is much more than the increase in the refrigerating effect. As a result, the COP decreases with increasing p_k. Looking at Table 11.1, we find that a compression ratio of 3 to 4 in a single stage is reasonable.

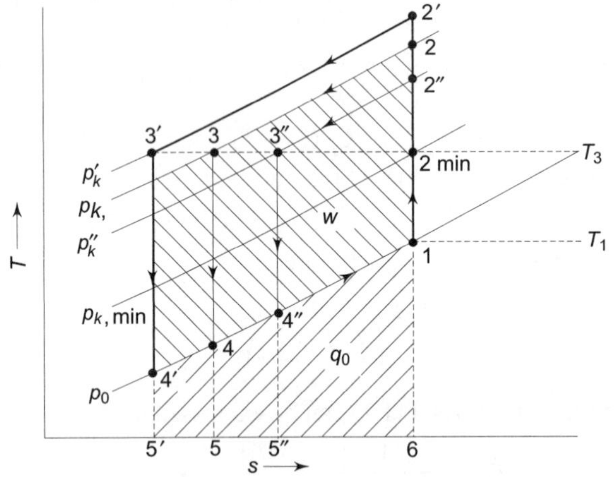

Fig. 11.3 Effect of discharge pressure p_k on the performance of the gas cycle

11.2.2 Polytropic and Multistage Compression

As discussed in Sec. 6.2.3, polytropic compression with cooling would reduce the net work of the cycle by reducing the average temperature of the compression process and the value of the compression index from γ to n. Then, the expression for compressor work becomes

$$|w_C| = \frac{n}{n-1}(p_2 v_2 - p_1 v_1)$$

$$= \frac{n}{n-1}\frac{\gamma-1}{\gamma} C_p (T_2 - T_1) \tag{11.9}$$

The network is

$$|w| = |w_C| - w_E$$

$$= C_p\left[\frac{n}{n-1}\cdot\frac{\gamma-1}{\gamma}(T_2 - T_1) - (T_3 - T_4)\right] \tag{11.10}$$

and the COP is

$$\mathcal{E} = \frac{T_1 - T_4}{\frac{n}{n-1}\frac{\gamma-1}{\gamma}(T_2 - T_1) - (T_3 - T_4)} \tag{11.11}$$

When the compression ratio becomes too high, it becomes necessary to use multistage compression with intercooling.

11.2.3 Actual Gas Cycle

The actual gas cycle will have significant pressure drops, Δp_k and Δp_0, in heat exchangers and irreversibilities of the compressor and expander as shown in Fig. 11.4. There are also other losses associated with heat gain from the surroundings and pressure drops at the compressor and expander.

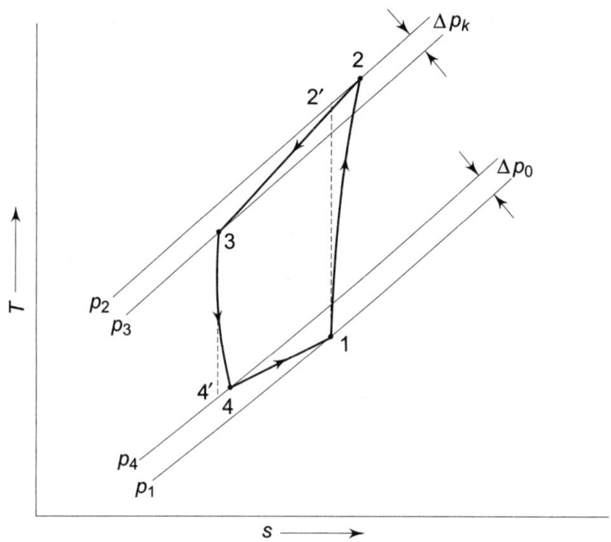

Fig. 11.4 Actual gas cycle

11.3 APPLICATION TO AIRCRAFT REFRIGERATION

The gas cycle, presently, is exclusively used in air-conditioning systems of military and commercial aircrafts, though refrigerated cargo aircrafts generally use dry ice. It is more appropriate to call it *air cycle refrigeration* since only air has found application as a working substance in this cycle. The coefficient of performance of this cycle is lower than that of the vapour compression cycle. Nevertheless, the air cycle continues to be favoured for aircraft refrigeration because of its many advantages. The present day jet aircrafts have very high cooling loads because of their large occupancy, electronic equipment and high velocity and consequent heat generation due to skin friction.

The air cycle can work as an open cycle or as a closed cycle system. A *closed air cycle system* or a *dense air machine* has many thermodynamic advantages. It can work at a suction pressure p_0 higher than the atmospheric. This reduces the volumes handled by the compressor and the expander. Also, the operating pressure ratio p_k/p_0 can be reduced, resulting in a higher coefficient of performance.

In an *open air-cycle system* the air after expansion is directly led to the conditioned space. It is, therefore, necessary to expand air to one atmosphere pressure. This requires larger volumes to be handled. Notwithstanding this, the sizes of the compressor and expander, nowadays, are not affected significantly since both of them are of the turbo-type. The rotational speed of the assembly is of the order of 100,000 rpm. The open cycle system, has another advantage over the closed cycle system, in respect that it does not require a heat exchanger for the refrigeration process. This saves the weight and cost of the equipment. It, however, has one disadvantage, viz., when the air drawn from the refrigerated space is humid, it might produce fog and ice at the end of the expansion process and clog the line. A drier in the circuit is required in such a case.

The main considerations involved in an aircraft application in order of importance are weight, space and operating power. These have been discussed in detail by Scofield[9]. Though the power per ton of refrigeration is considerably more for air-cycle refrigeration than for a vapour-compression system, the bulk and weight advantages of the air-cycle system, due to no heat exchanger at the cold end and a common turbo-compressor for both the engine and refrigeration unit, result in a greater overall power saving in the aircraft. The advantages of air cycle with regard to its application in aircraft refrigeration can be listed as follows:

(i) Small amounts of leakages are tolerable with air as the refrigerant.
(ii) Air cycle in its simplest form as an open system requires only one heat exchanger.
(iii) Availability of the refrigerant in mid air is also an important consideration.
(iv) Cabin pressurization and airconditioning can be combined into one operation.
(v) Initial compression of the air is obtained by the *ram effect*, viz., conversion of the high kinetic energy of the ambient air relative to the aircraft into enthalpy and hence pressure rise. The power for this, however, is derived from the engine of the aircraft since the process in the ram causes a drag on the engine. Refrigeration units for aircrafts have been built in various sizes weighing from 3 to 100 kg, and ranging in refrigerated air capacity from 3 to 70 kg per min.

11.3.1 Simple Aircraft Refrigeration Cycle with Ram Compression

The initial compression of the ambient air due to ram effect and a simple aircraft refrigeration cycle are shown in Fig. 11.5. The ram effect is shown by the line 1-2. Point 2′ denotes the state after isentropic diffusion to pressure p_2' and temperature T_2'.

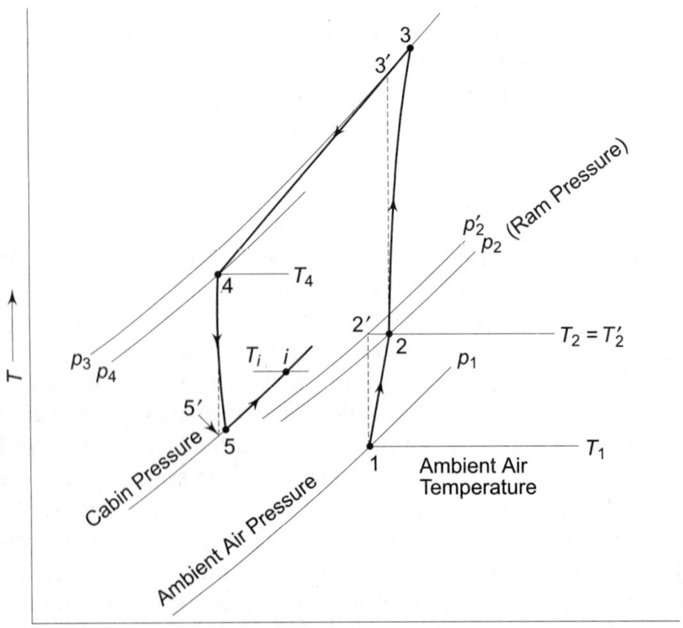

Fig. 11.5 Simple aircraft refrigeration cycle

The energy equation for the process of diffusion is

$$h_2 = h_2' = h_1 + \frac{C^2}{2}$$

where C is the velocity of the aircraft. We have from the above

$$C_p T_2 = C_p T_2' = C_p T_1 + \frac{C^2}{2}$$

$$T_2 = T_2' = T_1 + \frac{C^2}{2C_p} \tag{11.13}$$

The above relation can be simplified to

$$\frac{T_2'}{T_1} = 1 + \frac{C^2}{2C_p T_1} = 1 + \frac{C^2}{\frac{2\gamma R}{(\gamma-1)} T_1} = 1 + \frac{\gamma-1}{2}\frac{C^2}{a^2}$$

$$= 1 + \frac{\gamma-1}{2} M^2 \tag{11.14}$$

where $a = \sqrt{\gamma R T_1}$ is the acoustic velocity and M is the Mach number of the flight.

The temperature $T_2 = T_2'$ is the stagnation temperature of the ambient air. The stagnation pressure after isentropic diffusion, p_2', is given by the relation

$$\frac{p_2'}{p_1} = \left(\frac{T_2'}{T_1}\right)^{\frac{\gamma}{\gamma-1}} \tag{11.15}$$

The irreversible compression in the ram, however, results in air reaching point 2 instead of point 2′, i.e., at the same stagnation temperature but at a reduced stagnation pressure p_2 which is obtained from the knowledge of the ram efficiency η_R defined by

$$\eta_R = \frac{\text{Actual pressure recovery}}{\text{Ideal pressure recovery}}$$

$$= \frac{p_2 - p_1}{p_2' - p_1} \tag{11.16}$$

The rest of the cycle is also shown in Fig. 11.5. The temperature after the cooler (process 3-4), T_4 is shown to be higher than the stagnation temperature T_2 of the ambient air. It implies that the air cannot be cooled to a temperature below T_2 unless some method is devised to extract the kinetic energy of the ambient air. Another point to note is that the pressure after expansion p_5 is slightly above the cabin pressure. It is also higher than the ambient air static pressure existing at the altitude at which the aircraft is flying. The refrigerating effect produced is

$$q_0 = C_p (T_i - T_5) \tag{11.17}$$

where T_i is the room temperature maintained inside the cabin. The network of the cycle is the difference of work for process 2-3 of the compressor and process 4-5 of the expander plus the ram air work, $C_p (T_2 - T_1)$, which is derived directly from the engine.

The ambient air temperature varies with the altitude of the flight of the aircraft. Generally, the temperature drops by 0.64°C per 100 m of height from the sea level temperature.

Example 11.1 *An air-cooling system for a jet plane cockpit operates on the simple cycle. The cockpit is to be maintained at 25°C. The ambient air pressure and temperature are 0.35 bar and – 15°C respectively. The pressure ratio of the jet compressor is 3. The plane speed is 1000 kilometres per hour. The pressure drop through the cooler coil is 0.1 bar. The pressure of the air leaving the cooling turbine is 1.06 bar and that in the cockpit is 1.01325 bar. The cockpit cooling load is 58.05 kW. Calculate:*

(a) Stagnation temperature and pressure of the air entering the compressor.
(b) Mass flow rate of the air circulated.
(c) Volume handled by the compressor and expander.
(d) Net power delivered by the engine to the refrigeration unit.
(e) COP of the system.

Solution Refer to Fig. 11.5.

(a) Speed of aircraft $C = \dfrac{1000 \times 1000}{3600} = 277.8$ m/s

Stagnation temperature $T_2 = T_1 + \dfrac{C^2}{2C_p}$

$$= 258 + \frac{(277.8)^2}{2(1.005)10^3} = 258 + 38.5 = 296.5 \text{ K}$$

Stagnation pressure $p_2 = 0.35 \left(\dfrac{296.5}{258}\right)^{\frac{1.4}{1.4-1}} = 0.57$ bar

(b) Discharge pressure from the jet compressor $p_3 = 3(0.57) = 1.71$ bar
Discharge temperature from the jet compressor

$$T_3 = T_2 \left(\frac{p_3}{p_2}\right)^{\frac{\gamma-1}{\gamma}} = 296.5\ (3)^{0.286} = 406 \text{ K}$$

$$t_3 = 133°\text{C}$$

Other conditions are: $T_4 = T_2 = 296.5$ K

$$p_4 = 1.71 - 0.1 = 1.61 \text{ bar}$$

$$p_5 = 1.06 \text{ bar}$$

$$T_5 = T_4 \left(\frac{p_5}{p_4}\right)^{\frac{\gamma-1}{\gamma}} = \frac{296.5}{(1.71/.106)^{0.286}} = 263 \text{ K}$$

$$t_5 = -\ 10°\text{C}$$

Refrigerating effect and mass flow rate of air

$$q_0 = 1.005\ (25 + 10) = 35.18 \text{ kJ/kg}$$

$$\dot{m} = \frac{58.05 \times 3600}{35.18} = 5950 \text{ kg/h}$$

(c) Volume handled by the compressor

$$\dot{V}_C = \frac{mRT_2}{p_2} = \frac{5950(0.286 \times 10^3)(296.5)}{0.57 \times 10^5} = 9050 \text{ m}^3\text{/h}$$

Volume handled by the expander

$$\dot{V}_E = \frac{mRT_5}{p_5} = \frac{5950(0.286\times10^3)(263)}{1.06\times10^5} = 430 \text{ m}^3/\text{h}$$

(d) Ram work (done by the engine in overcoming the drag of ram diffusion)

$$|w_R| = C_p\,(T_2 - T_1) = 1.005\ (38.5) = 38.7 \text{ kJ/kg}$$

Compressor work $|w_C| = C_p\,(T_3 - T_2) = 1.005\ (406 - 296.5) = 111.0$ kJ/kg

Expander work $w_E = C_p\,(T_4 - T_5) = 1.005\ (296.5 - 263) = 33.7$ kJ/kg

Network $w = |(w_R + w_C)| - w_E = 38.7 + 111.0 - 33.7 = 116.0$ kJ/kg

Net horsepower $\text{HP} = \dfrac{5950(116)}{2650} = 260$

(e) $\text{COP} = \dfrac{35.18}{116} = 0.3$

Example 11.2 *(a) An aircraft flying at an altitude of 8000 m, where the ambient air is at 0.341 bar pressure and 263 K temperature, has a speed of 900 km/h. The pressure ratio of the air compressor is 5. The cabin pressure is 1.01325 bar and the temperature is 27°C. Determine the power requirement of the aircraft for pressurization (excluding the ram work), additional power required for refrigeration and refrigerating capacity on the basis of 1 kg/s flow of air.*

(b) Determine the same if the following are to be accounted

Compressor efficiency,	$\eta_C = 0.82$
Expander/turbine efficiency,	$\eta_T = 0.77$
Heat exchanger effectiveness,	$\varepsilon = 0.8$
Ram efficiency	$\eta_R = 0.84$

Solution Refer to Fig. 11.5

(a) Speed of aircraft $C = \dfrac{(900)(1000)}{3600} = 250$ m/s

Stagnation temperature $T_2 = T_1 + \dfrac{C^2}{2C_p}$

$$= 263 + \frac{(250)^2}{2(1.005)10^3} = 263 + 31.1 = 294.1 \text{ K}$$

Stagnation pressure $p_2' = p_1 \left(\dfrac{T_2}{T_1}\right)^{\frac{\gamma}{\gamma-1}} = 0.341 \left(\dfrac{294.1}{253}\right)^{\frac{1.4}{1.4-1}} = 0.504$ bar

Power requirement for pressurization (excluding ram work)

$$\dot{W}_1 = m\,C_p\,T_2 \left\{\left(\frac{p_{\text{cabin}}}{p_2'}\right)^{\frac{\gamma-1}{\gamma}} - 1\right\}$$

$$= (1)\,(1.005)\,(294.1) \left\{\left(\frac{1.01325}{0.504}\right)^{\frac{1.4-1}{1.4}} - 1\right\} = 65.34 \text{ kW}$$

Compressor discharge temperature $T_3 = T_2 \left(\frac{p_3}{p_2'}\right)^{\frac{\gamma-1}{\gamma}} = 294.1\ (5)^{0.286} = 466.0$ K

Expander exit temperature $T_5 = T_4 \left(\frac{p_4}{p_5}\right)^{\frac{\gamma-1}{\gamma}}$

$$= \frac{294.1}{\left(\frac{5 \times 0.504}{1.01325}\right)^{0.286}} = 226.6 \text{ K}$$

Power required for refrigeration (excluding ram work)

$$\dot{W}_2 = |\dot{W}_C| - \dot{W}_E = \dot{m}\, C_p\, (T_3 - T_2 - T_4 + T_5)$$
$$= 1\ (1.005)\ (466 - 294.1 - 294.1 + 226.6) = 104.4 \text{ kW}$$

Additional power for refrigeration $\dot{W}_1 = \dot{W}_2 - \dot{W}_1 = 104.4 - 65.34 = 39.06$ kW

Refrigerating capacity $\dot{Q}_0 = \dot{m}\, q_0 = \dot{m}\, C_p\, (T_{cabin} - T_5)$

$$= (1)\ (1.005)\ (300 - 226.6) = 73.77 \text{ kW}$$

(b) Ram pressure $p_2 = p_1 + \eta_R\, (p_2 - p_1)$

$$= 0.341 + 0.84\ (0.504 - 0.341) = 0.478 \text{ bar}$$

Power requirement for pressurization (excluding ram work)

$$\dot{W}_1 = \frac{(1)(1.005)(294.1)}{0.82}\left\{\left(\frac{1.01325}{0.478}\right)^{0.286} - 1\right\} = 86.4 \text{ kW}$$

Temperature after isentropic compression $T_3' = 466.0$ K

Temperature after actual compression $T_3 = T_2 + \frac{T_3' - T_2}{\eta_C}$

$$= 294.1 + \frac{466 - 294.1}{0.82} = 503.7 \text{ K}$$

Temperature after heat exchanger $T_4 = T_3 - \varepsilon\, (T_3 - T_2)$

$$= 503.7 - 0.8\ (503.7 - 294.1) = 336.0 \text{ K}$$

Temperature after isentropic expansion

$$T_5' = \frac{336}{\left(\frac{5 \times 0.504}{1.01325}\right)^{0.286}} = 258.9 \text{ K}$$

Actual temperature after expansion $T_5 = T_4 - \eta_T\, (T_4 - T_5')$

$$= 336 - 0.77\ (336 - 258.9) = 276.7 \text{ K}$$

Power required for refrigeration (excluding ram work)

$$\dot{W}_2 = \dot{m}\, C_p\, (T_3 - T_2 - T_4 + T_5)$$
$$= (1)\ (1.005)\ (503.7 - 294.1 - 336 + 276.7) = 151.1 \text{ kW}$$

Additional power for refrigeration $\dot{W} = 151.1 - 86.4 = 64.7$ kW

Note *(i) The difference between the power requirement for refrigeration and that for pressurization is not very large.*

(ii) The effect of irreversibilities is very large on the power requirement and refrigerating capacity.

11.3.2 Dry Air Rated Temperature

A comparison of the different types of air-refrigeration systems is made in terms of *dry air rated temperature* (DART). This is the temperature of the discharge air from the expander if there is no condensed moisture present and is calculated by the usual procedure, assuming the constant pressure specific heat of 1.005 kJ/kg for air. Thus the dry air rated discharge temperature will be essentially that resulting from a calculation which assumes that air, at a humidity so low that the expanded discharge is unsaturated, enters the system. The rating of the aircraft units is given in terms of kg of air per unit time at the design DART. Thus the capacity of the machine giving $\dot{m}$ kg/s of air at a DART of t_0 to maintain a cabin at temperature t_i is

$$\dot{Q}_0 = \dot{m}\, C_p\, (t_i - t_0)$$

Note *The refrigerating capacity of the air cycle unit in kW has no particular relevance. It is usually expressed in terms of the mass flow rate of air at DART. In a modern jet aircraft, typical values are 1.8 kg/s of air at 10°C DART just after take-off, and 1.4 kg/s of air at 12.8°C when the aircraft is cruising.*

11.3.3 Air-cycle Systems for Aircraft Refrigeration

In air-cycle systems for aircraft refrigeration, the air is bled from the compressor of the jet or gas turbine engine or is supplied from an auxiliary compressor source and cooled in an air to air heat exchanger having an effectiveness of approximately 90 per cent. Messinger[6] has described a few systems commonly employed for further processing of this compressed and cooled air before being supplied to the room. The common systems are: simple, Bootstrap, regenerative and reduced ambient types. These are discussed below. There are many other types which are variations or combinations of these.

Simple System In the simple system, shown in Fig. 11.6, the compressed air after cooling in air-cooler is passed through a cooling turbine. The work of this turbine is to drive a fan which draws cooling air through the heat exchanger. Air is discharged from the turbine at a pressure slightly above the cabin pressure. The fan is put on the downstream side thus avoiding the additional temperature rise of the cooling air. The thermodynamic cycle is shown in Fig. 11.5. This system is good for ground cooling since the fan driven by the turbine is a source of providing cooling air for the heat exchanger. The fan power overcomes the drag of the air-cooler. However, the turbine work is not available for the compressor.

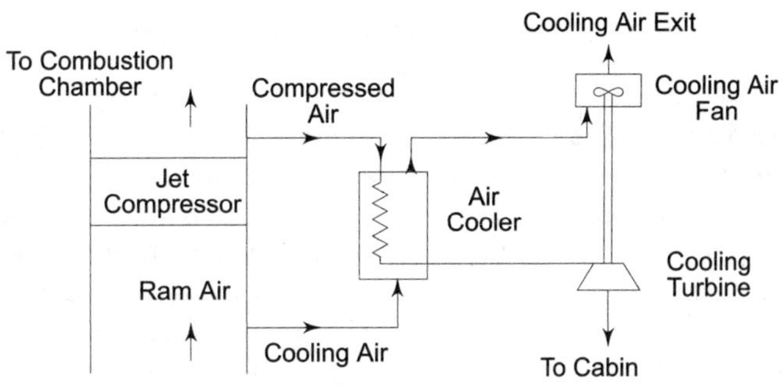

Fig. 11.6 Simple system

Bootstrap System The Bootstrap system, shown in Fig. 11.7, has two heat exchangers instead of one and the expansion turbine drives a compressor rather than a fan. Thus, it cannot be used for ground cooling. The thermodynamic cycle is shown in Fig. 11.8. The primary purpose of the Bootstrap system is to provide an additional cooling capacity when the primary source of air does not have a sufficiently high pressure to provide the amount of cooling required. The turbine drives the secondary compressor to raise the pressure of primary air before it enters the turbine. This is accomplished at the expense of drag in the cooling circuit since both the heat exchangers depend on the ram for the flow of cooling air.

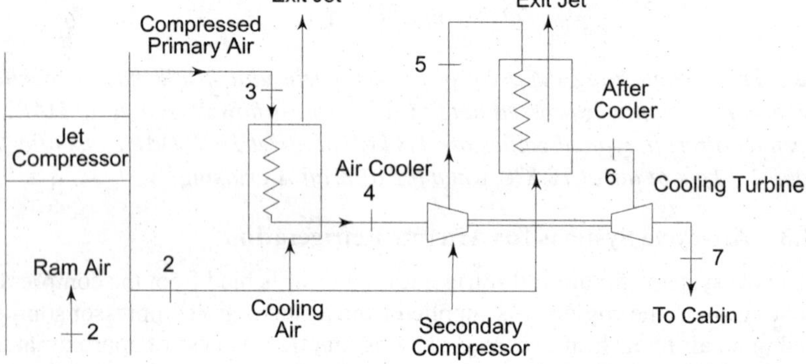

Fig. 11.7 Bootstrap system

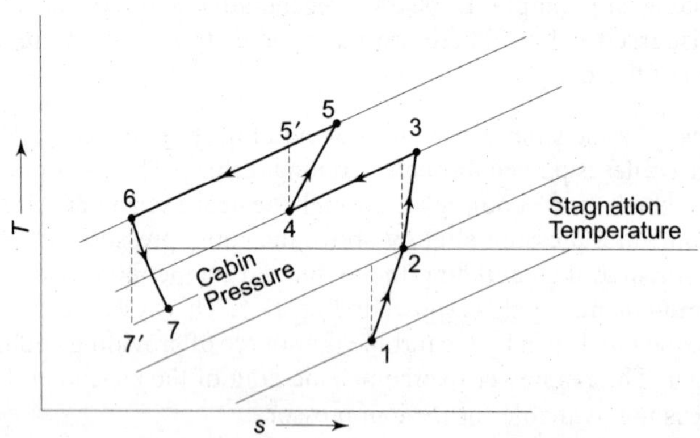

Fig. 11.8 Thermodynamic cycle for Bootstrap System

Regenerative System The regenerative system, shown in Fig. 11.9 also has two heat exchangers but does not require ram air for cooling the air in the second heat exchanger. It is a modification of the simple system (and is good for ground cooling) with the addition of a secondary heat exchanger in which the air from the primary heat exchanger is further cooled with a portion of the refrigerated air bled after expansion in the turbine as shown in Fig. 11.10. It provides lower turbine discharge temperatures but at the expense of some weight and complications.

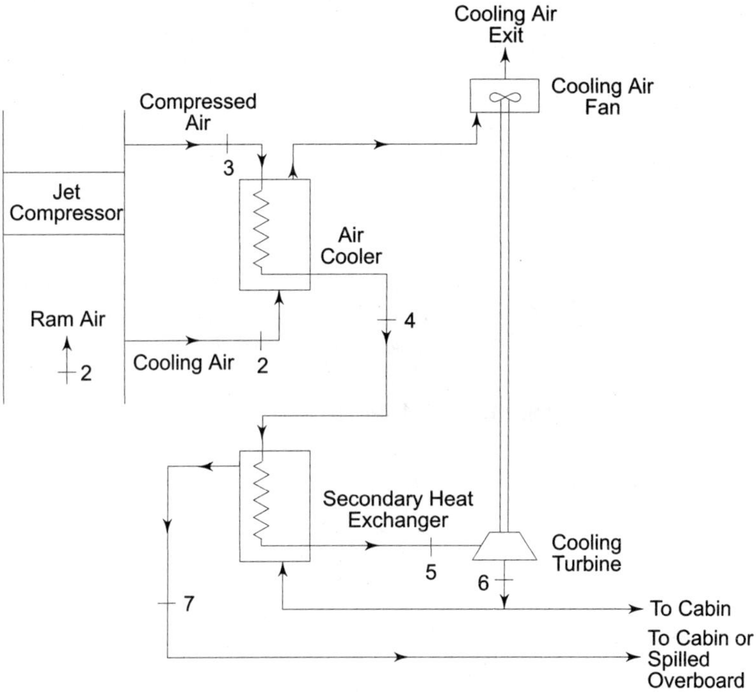

Fig. 11.9 Regenerative system

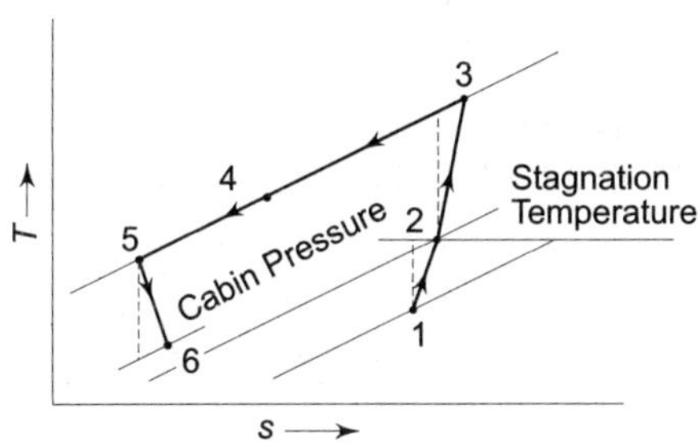

Fig. 11.10 Thermodynamic cycle for regenerative system

Reduced Ambient System In the reduced ambient system (Fig. 11.11) there are two expansion turbines—one in the cabin air stream and the other in the cooling air stream. Both turbines are connected to the shaft driving the fan which absorbs all the power. The turbine for the ram air operates from the pressure ratio made available by the ram air pressure. The thermodynamic cycle is shown in Fig. 11.12. The system shows promise for applications in an exceptionally high speed aircraft when the ram air temperature is too high. The cooling turbine reduces the temperature of cooling air to the level of static temperature of ambient air. Thus, primary compressed air can be cooled to, say, T_4 below the stagnation temperature T_2 and a little above the static temperature T_1.

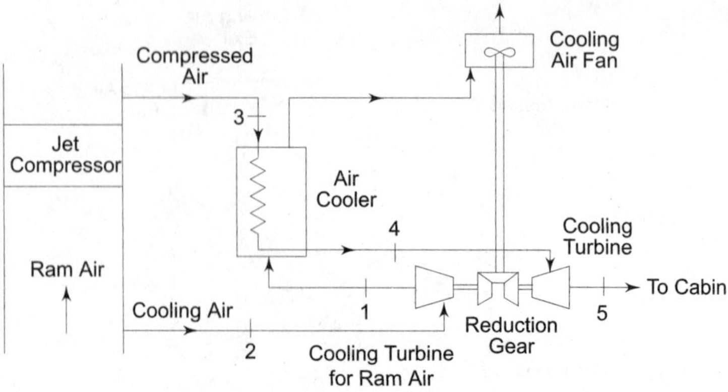

Fig. 11.11 Reduced ambient system

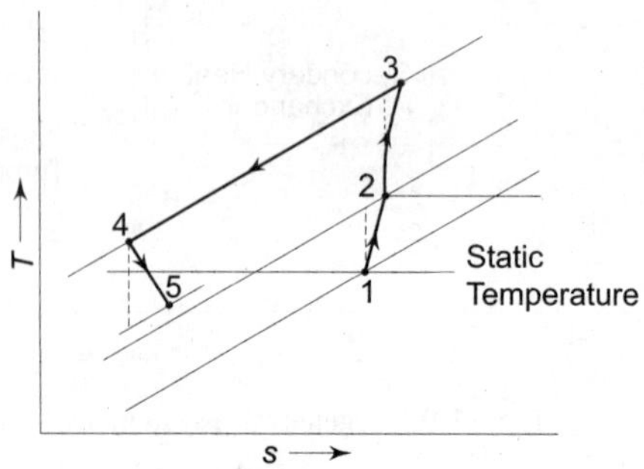

Fig. 11.12 Thermodynamic cycle for reduced ambient system

It is interesting to compare these systems with respect to the aircraft speed. From the results shown in Fig. 11.13, the following brief conclusions may be drawn:

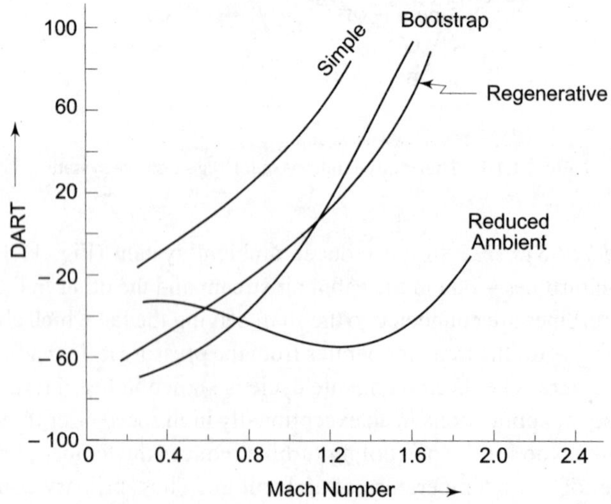

Fig. 11.13 Comparison of DART vs. Mach number variation for common aircraft refrigeration systems

(i) The simple system is preferable at low speeds.
(ii) Above a speed of 1000 kmph, the regenerative system is necessary. The Bootstrap system is also used.
(iii) Reduced ambient system may be useful for supersonic aircrafts and rockets.

Note *In actual aircraft air conditioning, the turbine may either drive the fan as in the simple cycle, or it may drive the secondary compressor as in the Bootstrap cycle, or it may do both in a cycle called compound cycle. The ground cooling in an aircraft is provided by an external A/C unit. It is not desirable to run the jet engine to provide power for ground cooling.*

Example 11.3 *The following performance is expected in a certain Bootstrap system:*

Turbine efficiency,	η_T	= 85%
Secondary compressor efficiency,	η_C	= 77%
Secondary heat exchanger effectiveness,	ε	= 0.9

The cabin pressure is maintained at one standard atmosphere. The cooling air temperature entering the secondary heat exchanger is 32°C. The compressed air leaves the primary heat exchanger at 64°C. The refrigerated air is required to enter the cabin at 4.5°C. Calculate:

(i) The temperature of air entering the cooling turbine.
(ii) The pressures of air at discharge from primary and secondary compressors.

Solution Refer to Figs 11.7 and 11.8.
Point 3 represents the state after the primary heat exchanger. It is given that

$$T_4 = 337 \text{ K}$$
$$T_7 = 277.5 \text{ K}$$
$$p_7 = 1.01325 \text{ bar}$$

The following equations can be written.

Secondary heat exchanger effectiveness $\varepsilon = \dfrac{T_5 - T_6}{T_5 - (273+32)} = 0.9$

$$\Rightarrow \quad T_6 = 274.5 + 0.1\, T_5 \qquad \text{(I)}$$

Secondary compressor efficiency $\eta_C = \dfrac{T_5' - T_4}{T_5 - T_4} = 0.77$

$$\Rightarrow \quad T_5' - 337 = 0.77\,(T_5 - 337) \qquad \text{(II)}$$

Cooling turbine efficiency $\eta_T = \dfrac{T_6 - T_7}{T_6 - T_7'} = 0.85$

$$\Rightarrow \quad 0.85\,(T_6 - T_7') = T_6 - 277.5 \qquad \text{(III)}$$

Compression process $\left(\dfrac{p_5}{p_4}\right)^{0.286} = \dfrac{T_5'}{337}$ (IV)

Expansion process $\left(\dfrac{p_5}{1.01325}\right)^{0.286} = \dfrac{T_6}{T_7'}$ (V)

An additional equation is obtained by equating the turbine work to secondary compressor work,

$$0.85\,(T_6 - T_7') = T_5 - 337 \tag{VI}$$

We have six unknowns, p_5, p_4, T_4, T_5, T_7' and T_5' and six equations. The solutions are:

$T_5 = 371.1$ K	(From I, II, III and VI)	$T_7' = 271.6$ K	(From III)
$T_5' = 363.3$ K	(From II)	$p_5 = 1.64$ bar	(From V)
$T_6 = 311.6$ K	(From III and VI)	$p_4 = 1.26$ bar	(From IV)

Example 11.4 *The air-conditioning unit of a pressurized jet aircraft receives its air from the compressor driven by the engine at a pressure of 1.22 bar. The pressure and temperature of the surrounding air at the height of the aircraft are 0.227 bar and 217 K respectively.*

The air-conditioning unit consists of a secondary compressor and a turbine mounted on the same shaft. The pressure and temperature of air leaving the turbine are 1 bar and 280 K. Calculate the pressure after the secondary compressor and temperature of air at exit from the cooler. Assume that all processes are reversible.

Solution Refer to Figs 11.14 and 11.15.

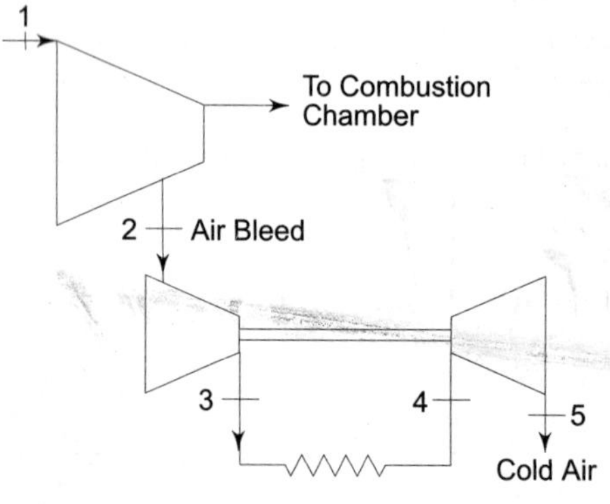

Fig. 11.14 System for Example 11.4

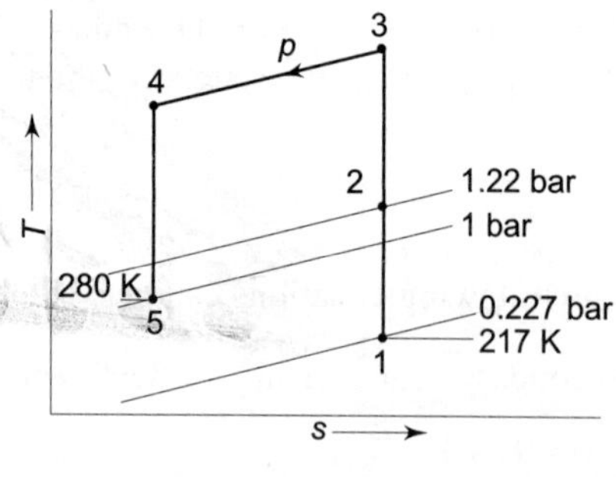

Fig. 11.15 Thermodynamic cycle for Example 11.4

Temperature of the air bled from the jet engine compressor

$$T_2 = T_1 \left(\frac{p_2}{p_1}\right)^{\frac{\gamma-1}{\gamma}} = 217 \left(\frac{1.22}{0.227}\right)^{0.286} = 351 \text{ K}$$

Work of the secondary compressor

$$w_{2\text{-}3} = C_p\,(T_3 - T_2) = C_p\,T_2 \left[\left(\frac{p}{1.22}\right)^{0.286} - 1\right]$$

Work of the turbine $\quad w_{4-5} = C_p\,(T_4 - T_5) = C_p\,T_5 \left[\left(\frac{p}{1}\right)^{0.286} - 1\right]$

Equating the two, we have $351 \left[\left(\frac{p}{1.22} \right)^{0.286} - 1 \right] = 280 \, (p^{0.286} - 1)$

Solving, we get $p = 3.06$ bar

Temperature of air at the exit from the cooler $T_4 = T_5 \left(\frac{p}{1} \right)^{0.286}$

$$= 280 \, (3.06)^{0.286} = 385 \text{ K}$$

11.4 RANQUE–HILSCH TUBE

This device, called the vortex tube or Ranque-Hilsch[1, 5, 7] tube, consists of a straight length of a tube with a concentric orifice located in a diaphragm near one end and a nozzle located tangentially near the outer radius adjacent to the orifice plate (Fig. 11.16). Compressed gas enters the tube tangentially through a nozzle forming a vortex kind of motion. The diaphragm prevents leftward motion of the vortex which, therefore, travels towards the righthand side of the tube called the hot end. A hot stream at temperature T_h which is above the temperature of supply, say, T_3 ejects from the hot end through the throttle valve, while the cold stream at temperature T_c below the temperature of supply is received at the cold end through the orifice. The throttle-valve opening controls the temperature and proportion of the cold stream with respect to the hot stream; the larger the throttle valve opening, the lower the temperature of the cold stream and the smaller its fraction and vice versa. The throttle valve is placed sufficiently distant from the nozzle and the diaphragm immediately close to it.

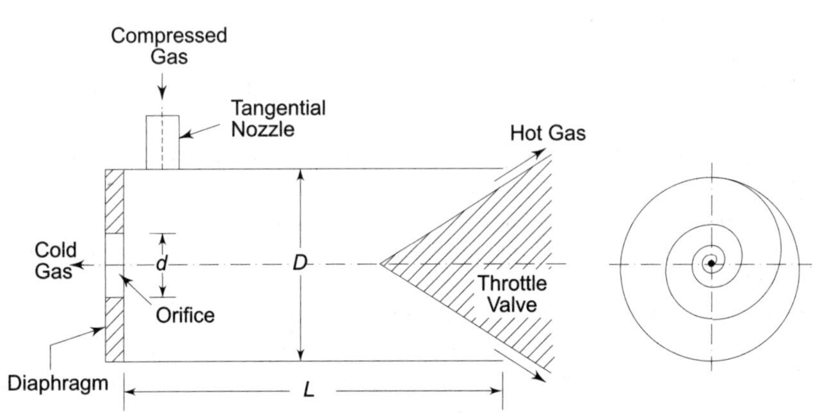

Fig. 11.16 Vortex tube

The geometry of the vortex tube can be described by the diameter of the vortex tube D, diameter of the orifice d, length of the vortex tube L, geometry and number of nozzles and design of the valve. The setting of the valve determines the ratio of the hot and cold mass flow fractions μ_h and μ_c respectively. The optimum diameter of the orifice is found to be half the tube diameter.

It may be seen from Fig. 11.16 that the vortex tube system is a modification of the open-type air-refrigeration system with the expander having been replaced by a vortex tube. In the Joule cycle, a temperature drop is obtained equal to the isentropic temperature drop, $\Delta T_s = T_3 - T_4$ (Fig. 11.17). The work of expansion is utilized to either run a cooling fan or a secondary compressor. The temperature drop obtained with the vortex tube, $T_3 - T_c$, is smaller than the isentropic drop. The air first expands to a state n in the nozzle reaching a temperature T_4 and velocity C given by

$$\frac{C^2}{2} = C_p\,(T_3 - T_4)$$

If all this kinetic energy could be removed, we would obtain cold air at temperature T_4. But without the separation of kinetic energy, it is as good as air at temperature T_3 which is the stagnation temperature of the gas.

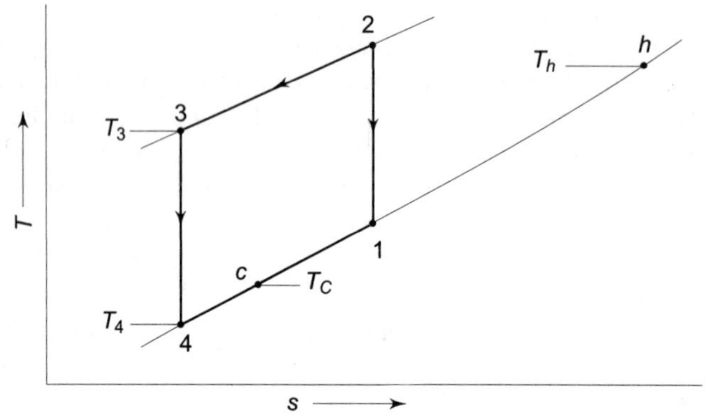

Fig. 11.17 Temperatures in vortex tube

From the nozzle, the high-velocity gas travels from the periphery of the tube to the axis during which the separation of kinetic energy occurs. The kinetic energy is retained by the outer layers due to which they are heated and emerge from the hot end of the tube at state h. The central core after having lost some kinetic energy emerges from the cold end at state c, i.e., at a temperature slightly above the static temperature of the expanded gas. The pressure of the cold-gas stream is usually lowered further due to expansion in the vortex chamber.

It can be seen that the temperature drop is less than the isentropic-temperature drop. Also, the quantity of cold air is only a fraction of the air supplied. Thus the COP of the system is much lower than that of the Joule cycle. The energy balance of the vortex tube gives

$$T_3 = \mu_c T_c + \mu_h T_h \tag{11.18}$$

where $\mu_c + \mu_h = 1$ and T_c and T_h are the stagnation temperatures of the two streams.

The energy efficiency η_e of the vortex tube can now be defined as the ratio of the coefficient of performance of the vortex tube to that of an ideal refrigerating machine in which the work done by the expander is not recovered, viz.,

$$\eta_e = \frac{\mu_c (T_3 - T_c)}{T_3 - T_4} \tag{11.19}$$

It is seen that if all the air goes through the throttle valve, there will be no energy separation. Also, if all the air leaves through the orifice, there will again be no energy separation. Thus, there will be optimum values of fractions μ_c and μ_h for the maximum temperature drop, refrigerating effect and COP.

11.5 THE JOULE–THOMSON COEFFICIENT AND INVERSION CURVE

The Joule–Thomson coefficient defined in Sec. 1.10.3 by the expression

$$\mu_J = \left(\frac{\partial T}{\partial p}\right)_h$$

is a thermodynamic property. It may be readily obtained for a gas by making it undergo a throttling process through an insulated expansion valve. The pressure p_1 and temperature T_1 upstream are maintained constant and the pressure p_2 downstream is varied by operating the valve manually. For each setting of the valve, both p_2 and T_2 are measured. These states 1, 2, 2′, 2″, 2‴, etc., are then plotted on a T versus p diagram as shown in Fig. 11.18.

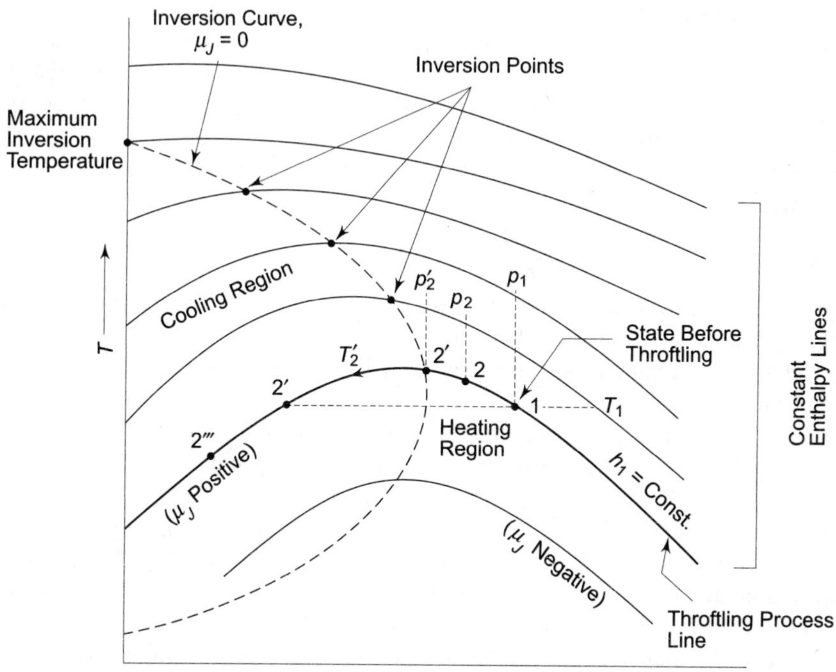

Fig. 11.18 Family of isenthalpic lines, and inversion curve

All these states have the same enthalpy. Hence, it represents an *isenthalpic line*. By changing the initial state p_1, T_1 and hence h_1, a family of such isenthalpic lines can be drawn as shown in Fig. 11.18. The locus of the maxima of these isenthalpics is called the *inversion curve*. The point at which this maximum occurs, i.e., the point at which the slope of *T-p* curve changes from positive to negative, is called the *inversion point*. The slope of these isenthalpics $(\partial T/\partial p)_h$ is equal to the Joule-Thomson coefficient.

It is seen that μ_J is positive for the region inside the inversion curve. Hence, it represents the cooling region. On the other hand μ_J is negative for the region outside the inversion curve. Hence, it represents the heating region. At the points of maxima, viz., at the inversion points, $\mu_J = 0$. Accordingly, the inversion curve represents the locus of initial states of a gas corresponding to which there is neither cooling nor heating on Joule–Thomson expansion.

Thus, if the gas at 1 in Fig. 11.18 is throttled, its temperature will increase or decrease depending upon the final pressure after throttling. The temperature will continue to increase upto a final pressure equal to p_2'. Thereafter, the decreasing trend will start. When the final pressure is p_2'', the change in temperature is zero. If the final pressure is less than p_2'', say, equal to p_2''', then only cooling will be obtained. It is evident that, to obtain cooling, the initial state should be below the inversion pressure p_2' and inversion temperature T_2' corresponding to a given enthalpy h_1.

It is evident that the throttling of the gas at 1 at pressure p_1 will result, initially, in rise in temperature. Hence, to obtain cooling, this gas must be precooled to a temperature below the inversion temperature T_2' at this pressure, say, to 2″ at temperature T_2'', before being throttled to 2‴ at pressure p_2'''. It is, thus, seen that for every pressure there is an inversion temperature. The inversion curve, precisely, represents this combination of inversion temperatures and pressures. The point of intersection of the inversion curve with the temperature axis ($p = 0$) gives the *maximum inversion temperature*.

11.5.1 Linde–Hampson Process

Figure 11.19 shows the Linde–Hampson process for the liquefaction of gases utilizing the Joule–Thomson expansion.

The gas at 1 is compressed, say, isothermally usually in two stages to 2. It may even be cooled by external refrigeration before and after 2. The lower the temperature before throttling, the greater the Joule–Thomson coefficient, and hence greater the *yield*. The yield *y* is the ratio of the mass of the liquid produced to the mass of the gas compressed. The compressed gas is, then, cooled to 3 in a counterflow regenerative heat exchanger with the help of cold vapour returning from the separator. At 3, the gas is throttled to 4. In the separator, the liquid is removed at *f*. The saturated cold vapour at g is heated to 5 in the heat exchanger. Make-up gas is added at 6, and the mixture of 5 and 6 enters the compressor at 1.

The energy balance of the combined heat exchanger and separator gives

$$\dot{m}h_2 = \dot{m}_f h_f + (\dot{m} - \dot{m}_f)h_5$$

whence we obtain for the yield

$$y = \frac{\dot{m}_f}{\dot{m}} = \frac{h_5 - h_2}{h_5 - h_f} \tag{11.20}$$

Note *No yield is possible unless* $h_2 < h_5$.

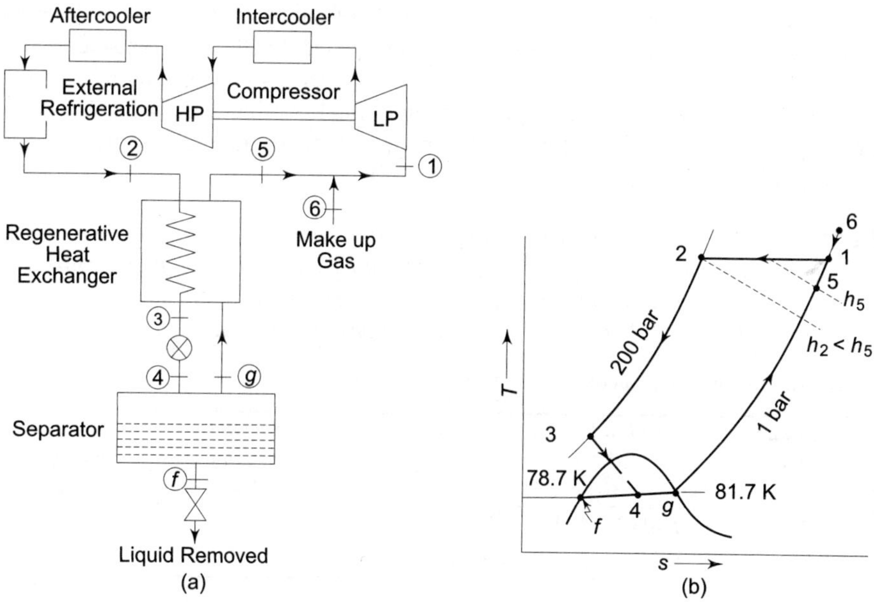

Fig. 11.19 Linde–Hampson process

Applying SSSF energy equation, we get for the work of compression per unit mass of the gas compressed

$$\frac{\dot{W}}{\dot{m}} = q + (h_1 - h_2) \tag{11.21}$$

For reversible isothermal compression, this becomes

$$\frac{\dot{W}}{\dot{m}} = T(s_2 - s_1) - (h_1 - h_2) \tag{11.22}$$

This expression represents the minimum work of compression. The work per unit mass of the yield is given by

$$W = \frac{\dot{W}}{\dot{m}_f} = \frac{\dot{W}}{\dot{m}y} = \frac{h_5 - h_f}{h_5 - h_2}\,[T(s_2 - s_1) + (h_1 - h_2)] \tag{11.23}$$

In Eq. (11.20), the value of h_f is set by the pressure p_1 which is 1 bar. Note that at 1 bar, $T_f = 78.7$ K while $T_g = 81.7$ K. As air is a mixture of O_2 and N_2, $T_f \neq T_g$. The value of h_5 is set by the ambient condition/design of regenerative heat exchanger. Consequently, the compressor delivery pressure p_2 is the main variable in the cycle controlling h_2 and the yield. The lower the value of h_2, the greater the yield.

A number of modifications are introduced in the basic Linde-Hampson cycle to improve the yield and COP. One of these is to employ external refrigeration as shown in Fig. 11.19 to precool, the gas. This auxiliary vapour compression refrigeration system is inherently more efficient in cooling the gas than the throttling process. The refrigeration system is less irreversible than a throttling process.

Another modification is to precool the gas by reversible adiabatic expansion through an expander/turbine. In the *Claude process* of gas liquefaction, the Linde-

Hampson process is modified so that about 10% of the compressed gas is diverted from the main stream, expanded reversibly and adiabatically in an expander, and reunited with the cold vapour returning from the separator before entering the heat exchanger. This diverted stream provides additional cooling and further lowering of temperature before throttling of the main stream. In addition, some work is recovered in the expander.

Note *(i) Linde-Hampson process is similar to the supercritical vapour compression cycle with the difference that the liquid from the mixture after throttling is separated instead of being sent to the evaporator.*

(ii) Tables 11.2, 11.3 and 11.4 give the properties of air.

Table 11.2 Saturation properties of air at 1 bar*

Pressure bar	*Temp.* K		*Specific volume* m³/kg		*Specific enthalpy* kJ/kg		*Specific entropy* kJ/kg.K	
p	*Bubble*	*Dew*	v_f	v_g	h_f	h_g	s_f	s_g
1	78.7	81.7	0.00114	0.224	zero	205.3	zero	2.559

Table 11.3 Enthalpy of air*

p **bar**	***h*** **kJ/kg**							
Temp., K →	120	130	140	150	160	200	250	300
100	70.1	98.1	128.0	159.5	190.3	276.7	347.6	408.5
150	70.4	95.8	121.5	147.5	172.8	257.5	335.1	400.5
200	72.1	95.8	119.4	142.7	165.5	246.0	325.9	394.3
250	75.5	98.2	120.6	142.3	163.4	239.8	319.7	389.8
300	80.0	102.2	123.6	144.3	164.1	236.8	315.9	386.8
350					166.8	236.2	314.1	385.2
400					170.5	237.1	313.7	384.9

Table 11.4 Entropy of air*

p **bar**	***s*** **kJ/kg.K**							
Temp., K →	120	130	140	150	160	200	250	300
100	0.625	0.849	1.070	1.288	1.486	1.972	2.291	2.513
150	0.572	0.775	0.966	1.145	1.308	1.782	2.130	2.369
200	0.532	0.723	0.898	1.059	1.206	1.657	2.015	2.265
250	0.507	0.689	0.855	1.005	1.142	1.569	1.926	2.182
300	0.489	0.667	0.826	0.968	1.095	1.502	1.856	2.115
350					1.064	1.451	1.800	2.060
400					1.040	1.412	1.754	2.013

* Extracted from Haywood R.W., Thermodynamic Tables in S.I. Units, Cambridge University Press, 1968.

Example 11.5 ***Minimum Work of Compression for Liquefaction of Air***

Determine the yield and minimum work of compression per unit mass of liquid air as a function of compressor discharge pressure. Given $T_1 = 300$ K.

Solution From Tables 11.2 to 11.4, at $p_1 = 1$ bar

$$h_f = 0,\ h_g = 205.3 \text{ kJ/kg}$$

$$h_1 = 427.8 \text{ kJ/kg},\ s_1 = 2.559$$

At $p_2 = 150$ bar, $T_2 = 300$ K, we have

$$h_2 = 400.5 \text{ kJ/kg},\ s_2 = 2.369 \text{ kJ/kg.K}$$

Assuming that state 5 coincides with state 1, we have:

Yield $$y = \frac{h_5 - h_2}{h_5 - h_f} = \frac{h_1 - h_2}{h_1} = 1 - \frac{h_2}{h_1} = 1 - \frac{400.5}{427.8} = 0.0638 \text{ kg/kg}$$

Minimum work $$W = \frac{1}{y}\,[T(s_2 - s_1) + (h_1 - h_2)]$$

$$= \frac{1}{0.0638}\,[300\,(2.369 - 2.559) + 427.8 - 400.5]$$

$$= -\,465.5 \text{ kJ/kg yield}$$

Similarly, we have for $p_2 = 200$ bar

$h_2 = 394.3$ kJ/kg, $s_2 = 2.265$ kJ/kg.K

$y = 0.0783$ kg/kg, $W = -\,689.6$ kJ/kg

And, for $p_2 = 300$ bar

$h_2 = 386.8$ kJ/kg, $s_2 = 2.115$ kJ/kg.K

$y = 0.0968$ kg/kg, $W = -\,962.4$ kJ/kg

Note *We, thus, see that by increasing compressor discharge pressure, the yield increases, and the work per unit yield also increases. Thus, the COP decreases.*

11.6 REVERSED STIRLING CYCLE

Figure 11.20 shows the ideal reversed Stirling cycle on *T-s* and *p-v* diagrams.

It comprises of two isothermal and two constant-volume procesess. Process 1–2 is isothermal compression with heat rejection Q_C to the surroundings at temperature $T_C = T_k$, and process 3–4 is isothermal expansion with absorption Q_E from the cold body at temperature $T_E = T_0$. Processes 2–3 and 4–1 are constant volume heat-transfer processes. It is seen that

$$_2Q_3 = {}_4Q_1$$

So, if a perfect regenerator is employed between processes of cooling 2–3, and heating 4–1, then, in the cycle, all cooling/refrigeration $Q_0 = Q_E$ takes place at a constant temperature T_0, and all heat rejection $Q_k = Q_C$ takes place at a constant temperature T_k. The COP of the ideal Stirling cycle, therefore, equals the Carnot COP, $T_0/(T_k - T_0)$.

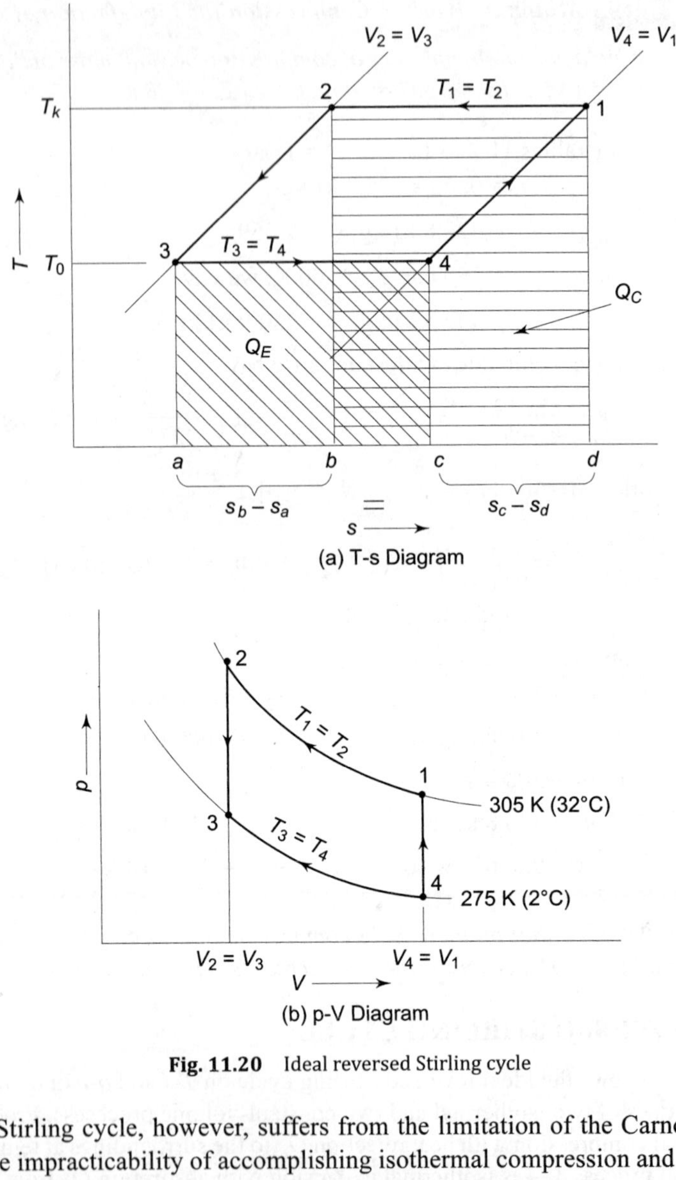

Fig. 11.20 Ideal reversed Stirling cycle

The Stirling cycle, however, suffers from the limitation of the Carnot cycle as far as the impracticability of accomplishing isothermal compression and expansion processes with a gas is concerned. But, it does not suffer from the other drawback of the Carnot cycle, viz., the narrow *p-v* diagram. The Stirling cycle is, therefore, not very much susceptible to the internal efficiencies of the compressor and the expander. On the other hand, it has the problem of executing constant volume heat-transfer processes 2–3 and 4–1 with change in pressure in the regenerator.

11.6.1 Practical Stirling Cycle

Figure 11.21 shows how the constant volume processes of the Stirling cycle can be approximated by a machine having two pistons.

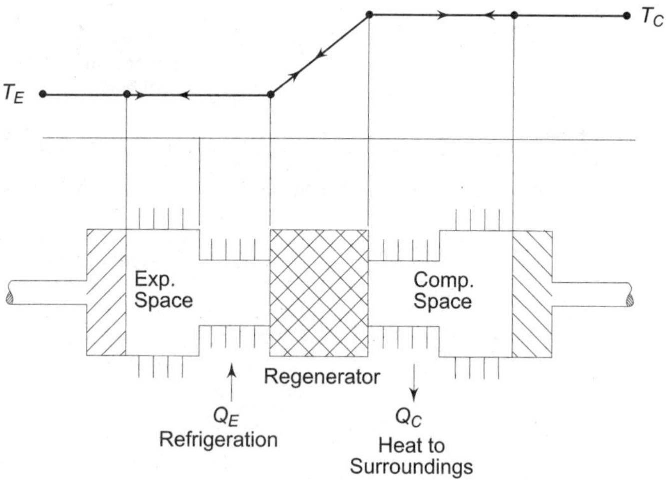

Fig. 11.21 Practical Stirling cycle

A regenerator acts as perfect storage of heat and cold, and divides the working space into two parts. One is the expander/cold space, and the other is the compressor/hot space. The two spaces are maintained at constant temperatures T_E and T_C by heat transfers Q_E and Q_C respectively. Two pistons move in the two spaces cyclically with a phase difference, maintaining the total volume of the two spaces together nearly constant, thus enabling the execution of constant volume heat transfer processes. Flow of the gas takes place in alternate directions between the compressor and the expander via the regenerator.

In practice, two arrangements that are possible are as follows:

1. α-machine, Fig. 11.22, with separate cylinders for compression and expansion spaces.

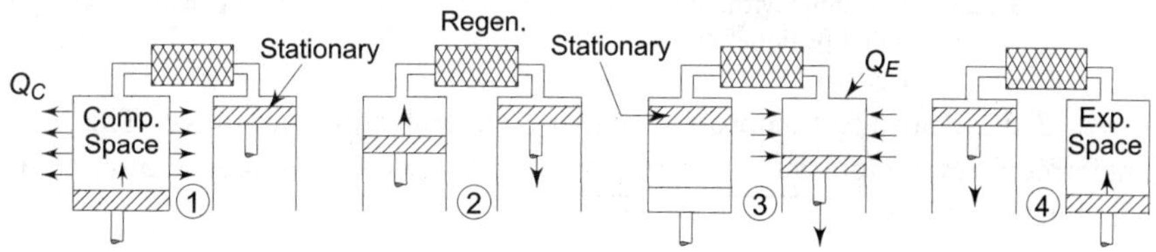

Fig. 11.22 α-machine with separate cylinders for compression and expansion spaces

2. β-machine, Fig. 11.23, with *integral cylinder* for compression and expansion spaces as developed by Philips for air liquefaction.

The figures also show the disposition of pistons at 4 terminal points which are as follows:

Terminal Point 1 Compression space is filled with gas. There is no gas in expansion space.

Process 1–2 Expander-piston is stationary. Compressor-piston moves. The gas is compressed rejecting heat Q_C.

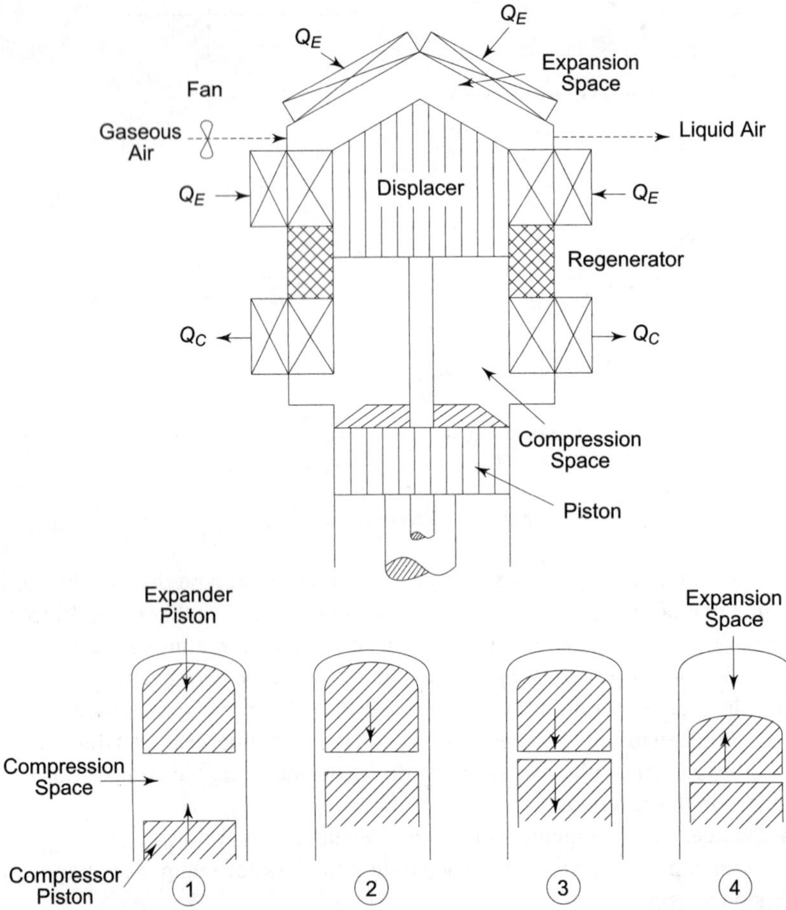

Fig. 11.23 β-machine with integral cylinder for compression and expansion spaces as in Philips liquefier

Process 2–3 Both the compressor and the expander pistons move. The gas is transferred from compression space to expansion space via the regenerator, getting cooled in the process.

Process 3–4 Compressor-piston is stationary. Expander-piston moves. The gas is expanded absorbing heat Q_E.

Process 4–1 Both the compressor and the expander pistons move. The gas is transferred back from expansion space to compression space via the regenerator, getting heated in the process.

In this manner, the cycle is repeated. Figure 11.24 illustrates this operation in the form of *cyclic piston displacement diagram* which holds good for both α and β machines. The upper line represents the expansion space volume. The lower line represents the compression space volume. The difference between the two represents the total volume of the gas between the two spaces.

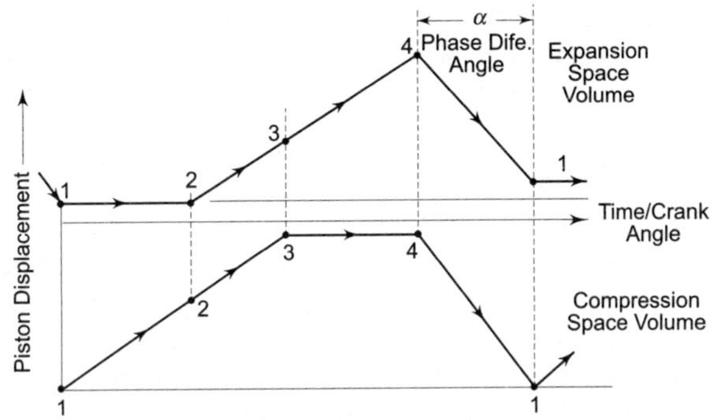

Fig. 11.24 Cyclic piston displacement diagram of α and β machines

Note *(i) The pressure at any instant of time is the same in both the spaces.*

(ii) The α-machine has the disadvantage of large dead volume. This is the volume that is not swept by any of the pistons.

(iii) The β-machine, on the other hand, has the disadvantage of heat transfer from the hot space to the cold space along the cylinder walls, thus reducing the refrigerating effect.

(iv) The maximum volumes of the compressor and expander spaces are V_C and V_E. The volumes occupied by the gas in the two spaces at any instant of time are V_c and V_e. The total volume of the gas is

$$V_t = V_c + V_e$$

Philips, Holand have developed an air liquefier which works on the β-machine. The air enters the top of the expansion space head; it gets cooled and condensed and leaves as liquid air from the other side of the space as shown by broken lines in Fig. 11.23.

Sharma[10] obtained temperatures between 0 to – 20°C with air as well as hydrogen using α-configuration.

11.6.2 Hydrogen as Working Substance in Stirling Cycle

Crucial factors for high COP in Stirling cycle are:

(i) High regenerator efficiency.
(ii) High heat-transfer coefficients in compressor and expander spaces.
(iii) Low pressure drop in system.
(iv) Low work of compression.

This means we require a working substance (gas) which has high thermal conductivity simultaneously with high thermal diffusivity. It should also have low viscosity, and low value of γ, the ratio of specific heats. Table 11.5 compares these values for three important gases, viz., air, hydrogen and helium.

Table 11.5 Values of thermal conductivity k, viscosity μ, thermal diffusivity α, and ratio of specific heats γ of air, hydrogen and helium at room temperature

Gas	k W/m.K	$\mu \times 10^5$ kg/m.s	$\alpha \times 10^4$ m^2/s	γ
Air	0.02624	1.983	0.2216	1.4
Hydrogen	0.182	0.8963	1.554	1.409
Helium	0.1491	2.012	1.8	1.667

It is seen that hydrogen has the highest thermal conductivity and lowest viscosity. Its thermal diffusivity is also quite high, and the value of ratio of specific heats is quite low. Hydrogen is, therefore, the preferred gas for use in Stirling cycle refrigeration.

11.6.3 Why the Stirling Cycle Refrigeration?

Stirling cycle is a viable alternative to vapour-compression system presently using CFCs. Further, hydrogen as working substance is environment-friendly. Simultaneously, with providing an alternative to CFCs, it has high thermodynamic performance. With regenerator, the COP approaches Carnot value.

Unprecedented opportunities, therefore, exist for applying it to commercial refrigeration and air conditioning applications. The limitation, however, is that it can be employed in small refrigerating capacity applications as the heat transfer surface available is the area of the expansion space only, and the heat-transfer coefficients with gas are very low.

11.6.4 Basic Design Parameters in Stirling Machine

There are three volumes in a Stirling machine, the maximum volumes of compression and expansion spaces V_C and V_E, and the volume of dead space V_D, viz., the volume of the working space which is not swept by any piston. Then, there is the phase difference angle α between the two pistons. And the temperatures maintained in the two spaces are T_C and T_E. The basic design parameters written in dimensionless form are:

Ratio of volumes $$\kappa = \frac{V_C}{V_E}$$

Reduced dead volume $$X = \frac{V_D}{V_E}$$

Ratio of temperatures $$\tau = \frac{T_C}{T_E}$$

Further, we have total of swept volumes

$$V_T = V_C + V_E = (\kappa + 1)\, V_E$$

11.6.5 Derived Parameters

Some parameters derived from basic parameters, which appear in theoretical analysis, are the following:

Angle $$\theta = \tan^{-1} \frac{\kappa \sin\alpha}{\tau + \kappa\cos\alpha}$$

$$S = \frac{2X\tau}{\tau + 1} = \text{Reduced dead volume}$$

$$A = \sqrt{\tau^2 + 2\tau\kappa\cos\alpha + \kappa^2}$$

$$B = \tau + \kappa + 2S$$

$$\delta = \frac{A}{B}$$

11.6.6 Theoretical Analysis of the Cycle[11, 12, 13]

Figure 11.24 represents a theoretical piston displacement versus crank angle diagram. Schmidt[8] presented his classical analysis of the cycle in which the motion of the pistons is simple harmonic as it has to be in a practical device. Schmidt made the following assumptions:

(i) Regenerator is perfect.
(ii) Instantaneous pressure is the same throughout the system.
(iii) Gas obeys ideal gas equation $pV = RT$.
(iv) Compressor and expander heat exchangers are perfect.
(v) Temperatures T_C and T_E are constant.

At any instant of time, the volumes of the expansion and compression spaces are given by

$$V_e = \frac{1}{2} V_E (1 + \cos \phi)$$

$$V_c = \frac{1}{2} V_C [1 + \cos (\phi - \alpha)] = \frac{1}{2} \kappa V_E [1 + \cos (\phi - \alpha)]$$

where ϕ is the crank angle of expansion space. Note that lower case subscripts e and c denote instantaneous values of T, p, V and mass, and upper case subscripts E and C denote maximum or constant values.

Now, equating the sum of masses of gas in the expansion, compression and dead spaces to total mass m_T, we have

$$\frac{p_e V_e}{RT_e} + \frac{p_c V_c}{RT_c} + \frac{p_d V_d}{RT_d} = \frac{1}{2} K \frac{V_E}{RT_c} = m_T \tag{11.24}$$

where K is a constant which depends on the total mass of gas charged in the system. Note, that this will also affect the system pressures.

Now, since compression and expansion are isothermal, hence $T_e = T_E$ and $T_c = T_C$. Then, substituting for volumes, putting $p_e = p_c = p_d = p$, eliminating R, and rearranging, we get

$$\frac{K}{p} = \frac{T_C}{T_E} (1 + \cos \phi) + \kappa [1 + \cos (\phi - \alpha)] + 2 \frac{V_D}{V_E} \frac{T_C}{T_D}$$

$$= \tau (1 + \cos \phi) + \kappa [1 + \cos (\phi - \alpha)] + 2 S \tag{11.25}$$

where $$T_D = \frac{T_E + T_C}{2} = \frac{T_C}{2}\left(1 + \frac{T_E}{T_C}\right) = \frac{T_C}{2}\left(\frac{\tau + 1}{\tau}\right)$$

Equation (11.25) can be written as

$$\frac{K}{p} = \sqrt{(T + \kappa \cos \alpha)^2 + (\kappa \sin \alpha)^2} \cos (\phi - \theta) + \tau + \kappa + 2S$$

$$= \sqrt{\tau^2 + 2\tau\kappa \cos \alpha + \kappa^2} \cos (\phi - \theta) + \tau + \kappa + 2S$$

$$\Rightarrow \quad p = \frac{K}{B[1 + \delta \cos(\phi - \theta)]} \tag{11.26}$$

The instantaneous pressure p is

(a) minimum when $\phi = \theta$, $\phi - \theta = 0$, and
(b) maximum when $\phi = \theta + \pi$, $(\phi - \theta) = \pi$.

Thus, we have $p_{\min} = \dfrac{K}{B(1+\delta)}, p_{\max} = \dfrac{K}{B(1-\delta)}, \dfrac{p_{\max}}{p_{\min}} = \dfrac{1+\delta}{1-\delta}$

Accordingly, from Eq. (11.26), $p = p_{\max}\dfrac{1-\delta}{1+\delta\cos(\phi-\theta)}$

$$= p_{\min}\frac{1+\delta}{1+\delta\cos(\phi-\theta)} \tag{11.27}$$

Mean Cycle Pressure

$$p_{\text{mean}} = \frac{\pi}{2}\int_0^{2\pi} p\mathrm{d}\,(\phi-\theta) = \frac{\pi}{2}\int_0^{2\pi} p_{\max}\,\frac{1-\delta}{1+\delta\cos(\phi-\theta)}\,\mathrm{d}(\phi-\theta)$$

$$= p_{\max}\sqrt{\frac{1-\delta}{1+\delta}} \tag{11.28}$$

Note *The values of p_{max}, p_{min} and p_{mean} depend on each other. But, ultimately, these depend on the charging pressure, i.e., on the mass of the gas charged.*

Heat Transferred and Work Done In the isothermal expansion and compression processes

$$Q = W = \int p\,\mathrm{d}\,V$$

Heat Transferred in Expansion Space/Refrigeration Produced Putting,

$$V_e = \frac{1}{2}\,V_E\,(1+\cos\phi)$$

$$\mathrm{d}\,V_e = -\frac{1}{2}\,V_E\sin\phi\,\mathrm{d}\,\phi$$

$$p \cong p_{\text{mean}}\,[1+\Delta\cos(\phi-\theta)]\text{ approximately}$$

$$\Delta = \frac{2\delta}{1+\sqrt{1-\delta^2}}$$

and integrating, we get

$$Q_E = \int_0^{2\pi} p\,\mathrm{d}\,V_e = -\frac{1}{2}\int_0^{2\pi} p_{\text{mean}}\,V_E\,[1-\Delta\cos(\phi-\theta)]\sin\phi\,\mathrm{d}\,\phi$$

$$= -\frac{1}{2}\,p_{\text{mean}}\,V_E\left\{-\cos\phi-\Delta\left[-\cos\theta\,\frac{1}{2}\cos 2\phi+\sin\theta\left(\frac{1}{2}\phi-\frac{1}{4}\sin^2\phi\right)\right]\right\}_0^{2\pi}$$

$$= \frac{\pi}{2}\,p_{\text{mean}}\,V_E\,\Delta\sin$$

$$\theta = \pi\,p_{\text{mean}}\,V_E\,\frac{\delta\sin\theta}{1+\sqrt{1-\delta^2}} \tag{11.29}$$

Heat Transferred in Compression Space/Heat Rejected (Similarly)

$$Q_C = \int_0^{\pi} p\,\mathrm{d}\,V_c$$

$$= \pi\,p_{\text{mean}}\,V_E\,\frac{\kappa\delta\sin(\theta-\alpha)}{1+\sqrt{1-\delta^2}} \tag{11.30}$$

Work Done and COP These have their respective Carnot values.

11.6.7 Optimal Design for Stirling Machine

In the case of internal combustion engines, the *mean effective pressure* (MEP) is used as a basis of comparing the power output. In a Stirling cycle engine or refrigerator, however, there are always two working spaces. In this case, therefore, total swept volume rather than the expander volume, and the maximum pressure p_{max} rather than mean pressure are used to make Eq. (11.29) non-dimensional so that the *specific refrigerating effect* becomes,

$$\frac{Q_E}{p_{max}\, V_T} = \frac{\pi\sqrt{1-\delta}\,\delta \sin\theta}{(1+\kappa)\sqrt{1+\delta}\left[1+\sqrt{1-\delta^2}\right]} \tag{11.31}$$

Replacing p_{max} in terms of p_{min}, Eq. (11.31) can also be written as

$$\frac{Q_E}{p_{min}\, V_T} = \frac{\pi\sqrt{1+\delta}\ \delta \sin\theta}{(1+\kappa)\sqrt{1-\delta}\left[1+\sqrt{1-\delta^2}\right]} \tag{11.32}$$

It is seen from Eq. (11.31) that

$$\begin{aligned} Q_E\ &\alpha\, N, \text{ Speed} \\ &\alpha\, p_{max}, \text{ viz., amount of gas charged} \\ &\alpha\, V_T \end{aligned}$$

Accordingly, to double the refrigerating capacity, one may just double any, or a combination of these parameters. In practice, however, it is not so. The increase in pressure and speed has an adverse effect also because of friction and *thermal saturation* of the *matrix* in the regenerator as well as the compressor and expander heat exchangers. After reaching an optimum, the increase in p and N in excess may actually decrease Q_E.

Other parameters to optimise are τ, κ, α and X. As in all refrigerating machines, Q_E increases as τ decreases, or as T_E increases and T_C decreases. Q_E also increases as $\kappa = V_C/V_E$ increases. It decreases with $X = V_D/V_E$. Hence dead volume should be as small as possible. Q_E is maximum at phase difference angle α around 90°. But it is remarkably insensitive to variation of α between 60° – 120°.

We see that there is an optimum value of volume ratio κ for a given temperature ratio τ. Thus, if

$$T_E = 278 \text{ K } (5°\text{C}),\ T_C = 318 \text{ K } (45°\text{C})$$

$$\tau = \frac{318}{278} = 1.144,\ X = 1,\ \alpha = 90°$$

then, optimum volume ratio is found to be

$$\kappa = V_C/V_E = 1.5$$

and specific refrigerating effect becomes

$$\left(\frac{Q_E}{p_{max}\, V_T}\right)_{opt} = 0.138$$

The analysis and optimization given above is for an ideal Stirling cycle refrigerating machine. But, there are various factors which affect the performance in an actual machine. The major ones are the following:

1. Compression and expansion are not isothermal but polytropic. After Schmidt, the next major contribution to the theory was made by Finkelstein[2, 3] who considered variation in the temperature of the two spaces.
2. Heat transfer processes in the compressor and expander are irreversible. There is a finite temperature difference between the working fluid and the medium on the other side.
3. Dead volume ratio may be large.
4. Regenerator efficiency is less than unity. Accordingly:
   Gas cooled from T_C to $T_E' > T_E$
   Gas heated from T_E to $T_C' < T_C$
5. Aerodynamic losses in the compressor, expander, heat exchangers, regenerator and connecting lines are present.
6. Heat gain of expansion space from surroundings and heat leakage from hot space to cold space will also offset refrigerating effect.

Note *A factor of safety of 3 is recommended to account for these.*

Example 11.6 ***Theoretical Stirling Cycle Calculations***

In a Stirling cycle refrigerating machine operating at T_C = 318 K and T_E = 278 K, the volumes of the cylinders are:

$$V_C = V_E = 6 \times 10^{-5}\ m^3$$

The dead volume ratio is X = 1.5. The phase difference angle is α = 90°. The speed of the machine is 720 rpm. The working fluid is charged so much as to give a maximum pressure of p_{max} = 20 bar.

Find the refrigerating capacity, power, consumption and mean and minimum pressures of the ideal cycle.

Solution The temperature ratio is $\tau = T_C/T_E = 1.144$, and swept volume ratio is $\kappa = V_C/V_E = 1$. Now

$$S = \frac{2X\tau}{1+\tau} = 1.6007$$

$$\delta = \frac{\sqrt{\tau^2 + \kappa^2 + 2\tau\kappa\cos\alpha}}{\tau + \kappa + 2S} = 0.2843$$

$$\theta = \tan^{-1}\left(\frac{\kappa \sin\alpha}{\tau + \kappa\cos\alpha}\right) = 61.6°$$

$$\sin\theta = 0.6581$$

$$Q_{max} = \frac{Q_E}{p_{max} V_T} = \frac{\pi}{1+\kappa}\sqrt{\frac{1-\delta}{1+\delta}}\frac{\delta\sin\theta}{1+\sqrt{1-\delta^2}} = 0.112$$

$$Q_E = 0.112\,(20 \times 10^5)\,(2 \times 6 \times 10^{-5}) = 26.9\ \text{J}$$

Refrigeration capacity, power consumption and pressures

$$\dot{Q}_E = Q_E N/60 = \frac{26.9(720)}{60} = 323\ \text{W}$$

$$\dot{W} = \dot{Q}_E/\text{COP} = (\tau - 1)\,\dot{Q}_E = 46.5\ \text{W (Actual)}$$

$$p_{\text{mean}} = p_{\max} \sqrt{\frac{1-\delta}{1+\delta}} = 14.93 \text{ bar}$$

$$p_{\min} = p_{\max} \left(\frac{1-\delta}{1+\delta}\right) = 11.14 \text{ bar}$$

11.6.8 Design Aspects of Regenerator

The regenerator is a critical component in Stirling machines. The regenerator efficiency should be close to 0.98 to 1. A 1% reduction in efficiency results in 21% reduction in Q_E at 75 K. At the same time, heat exchange in regenerator is very large, viz.,

$$Q_{\text{REG}} \cong 10 \text{ to } 50\ Q_E$$

Hence, the desirable properties of regenerator *matrix* material are:

(a) High heat capacity (high density and specific heat).

(b) Large thermal conductivity.

(c) Low pressure drop inspite of large heat-transfer surface of matrix.

Comparing a few materials, Table 11.6 clearly highlights that MS pipe can be used for the casing, while copper would be the best choice for the matrix.

Table 11.6 Comparison of regenerator materials

	MS	*Copper*	*Bronze*	*Brass*
C_p, kJ/kg.K	0.465	0.383	0.343	0.385
k, W/m.K	54	386	26	111

After choosing the material, it is also very important to select a *mesh number* of proper size. The mesh wire diameter d_m should be optimum.

(a) *If d_m is too small*, heat penetrates to the centre before the *blow time* expires. This means *storage volume* is insufficient.

(b) *If d_m is too large*, heat does not penetrate to the centre within the blow time. This means some volume does not contribute to storage.

Consequently, it is important to select a mesh size for given operating conditions, viz., working fluid, pressure, temperature, mass/volume flow rate, speed, etc.

References

1. Arora C P, 'Theoretical analysis of Ranque-Hilsch tube', *Proc. Seventh Congress on Theoretical and Applied Mechanics*, Bombay, 1961, pp. 241–48.
2. Finkelstein T, 'Generalized thermodynamic analysis of Stirling engines', *SAE Annual Meeting*, 1960.
3. Finkelstein T, 'Optimization of phase angle and volume ratio for Stirling engines,' *SAE Annual Meeting*, 1960.
4. 'Fundamentals and Equipment', *ASHRAE Guide and Data Book*, 1963.
5. Hilsch R, 'The use of the expansion of gases in a centrifugal field as a cooling process', *The Review of Scientific Instruments*, Vol. 18, No. 2, Feb. 1947, p. 108.
6. Messinger B L, 'Refrigeration for air conditioning pressurized transport aircraft', *Trans. SAE Journal*, Vol. 54, No. 3, March, 1946.

7. Ranque G J, 'Experience sur la de'tente giratoire avec productions simultane'es d'ume', chappement d' air chand et. d'un e chappement d' air froid', *Le Journal de Physique et le Radium*, Vol. 4, series 7.
8. Schmidt G, 'Theorie der Lehmann'schen, calorischen Maschine', *Z. Ver. Dtsch. Ing.*, Vol. 15, No. 1, 1871.
9. Scofield P C, 'Air cycle refrigeration', *Refrigerating Engineering*, Vol. 57, No. 6, June 1949, pp. 558–563, 61.
10. Sharma S K, 'Development of Stirling cycle refrigerator', M.Tech. Thesis, IIT Delhi, 1993.
11. Walker G, 'Operating cycle of the Stirling engine with particular reference to the function of the regenerator', *J. Mech. Engg. Sc.*, Vol. 3, No. 4, 1961.
12. Walker G, 'An optimization of the principal design parameters of Stirling cycle machines', *J. Mech. Engg. Sc.*, Vol. 4, No. 3, 1962.
13. Walker G, *Cryocoolers Part 1; Fundamentals*, Plenum Press, 1983.
14. Walker G, *Cryocoolers Part 2: Applications*, Plenum Press, 1983.

Revision Exercises

11.1 An open air refrigeration system operating between pressures of 16 bar and 1 bar is required to produce 33.5 kW refrigeration. The temperature of air leaving the refrigerated room is –5°C and that leaving the air cooler is 30°C. Assume no losses and clearance. Calculate for the theoretical cycle:

(i) Weight rate of air circulated per minute.
(ii) Piston displacement of compressor and expander.
(iii) Network
(iv) COP.

11.2 A dense air machine operates between 17 bar and 3.4 bar. The temperature of the air after the cooler is 15°C and after the refrigerating coils is 6°C. Determine:

(i) Temperature after expansion and compression.
(ii) Air circulated per minute/TR.
(iii) Work of the compressor and expander/TR.
(iv) Theoretical COP and hp/TR.

11.3 A dense air machine is to produce 10 tons of refrigeration with a compressor entrance pressure of 4.5 bar and a temperature of 0°C. The compressor discharge pressure is 20.5 bar. A pressure drop of 0.6 bar takes place in the air cooler and of 0.2 bar in the refrigerating coils. Assume the following pressure drops due to throttling in valves:

Compressor suction	0.2 bar
Compressor discharge	0.4 bar
Expander admission	0.3 bar
Expander exhaust	0.15 bar

The air temperature leaving the cooler is 38.5°C. Also assume the indices of the compression and expansion processes as 1.25 and 1.35 respectively. The clearance factor of the compressor is 2 per cent, and its mechanical efficiency is 0.8. Neglect the heat losses in piping. The compressor and expander are double-acting. Calculate:

(i) Refrigeration per kg of air and kg of air circulated per min.
(ii) Water flow required per min, the rise in temperature of water being 8°C.
(iii) Indicated work of compressor and expander.
(iv) Horsepower.
(v) Water required in the compressor jacket for 5.5°C rise.
(vi) Volumetric efficiency of the compressor.
(vii) Piston displacement.
(viii) Bore, stroke and rpm of the compressor, stroke/bore ratio = 1.3.

11.4 An air-cycle unit operating on the simple system is designed for the following conditions: 7.5 kg/min of conditioned air, air entering the turbine at 4.4 bar, cabin pressure 1 bar, dry air rated discharge temperature of –6°C and a turbine efficiency of 80 per cent.
(a) At what temperature does the air enter the turbine?
(b) How much power does the turbine supply to the fan?

11.5 (a) For an air cycle refrigeration system, the highest refrigeration temperature is 15°C and the exit temperature of the air from the cooling coils is 40°C. Find the minimum pressure ratio necessary for producing refrigeration at 1 atmosphere pressure.
(b) If the capacity of the plant is 33.3 kW and the pressure ratio is 4, calculate:
(i) The weight rate of air circulated.
(ii) Theoretical displacements of the compressor and expander.
(iii) The theoretical horsepower of the plant and its COP.

11.6 Dry air is supplied to the cabin superchargers of an aeroplane at 1 bar and is discharged at 2 bar and 135°C. Twenty-five kilograms per minute of this air is supplied to the primary intercooler of an air cycle refrigeration system. Determine:
(a) The air temperature leaving the intercooler if 25 kW heat is removed. From the intercooler, the air enters a secondary compressor where the pressure is increased to 3.2 bar. Determine:
(b) The discharge air temperature.
(c) The compressor horsepower (mechanical efficiency = 86%). The air then flows through a secondary intercooler and 38 kW heat is removed. Determine:
(d) The air temperature leaving the secondary intercooler. The air then enters an expansion turbine ($n = 1.3$), where the pressure is dropped to 1.02 bar. Determine:
(e) The final discharge air temperature.
(f) The work recovered from the expansion turbine (mechanical efficiency = 65%).
(g) The kW of refrigeration developed if air is allowed to be heated to 24°C.

11.7 (a) An aircraft is flying at a speed of 1000 km/h at a height where the surrounding air pressure and temperature are 0.35 bar and – 15°C. Calculate the limiting temperature to which air can be cooled after compression.
(b) A *reduced ambient cycle* is used in the above aircraft refrigeration system. The pressure ratio of the jet compressor is 3. Effectiveness of the air cooler is 0.75. Determine DART and COP of the cycle.

Chapter

12

Vapour-Absorption System

12.1 SIMPLE VAPOUR-ABSORPTION SYSTEM

The function of the compressor in the vapour-compression system is to continuously withdraw the refrigerant vapour from the evaporator and to raise its pressure and hence temperature, so that the heat absorbed in the evaporator, along with the work of compression, may be rejected in the condenser to the surroundings.

In the vapour-absorption system, the function of the compressor is accomplished in a three-step process by the use of the absorber, pump and generator or reboiler as follows:

(i) *Absorber*: Absorption of the refrigerant vapour by its weak or poor solution in a suitable absorbent or adsorbent, forming a strong or rich solution of the refrigerant in the absorbent/adsorbent.

(ii) *Pump*: Pumping of the rich solution raising its pressure to the condenser pressure.

(iii) *Generator or Desorber*: Distillation of the vapour from the rich solution leaving the poor solution for recycling.

A simple vapour-absorption system, therefore, consists of a condenser, an expansion device and an evaporator as in the vapour-compression system, and in addition, an absorber, a pump, a generator or desorber and a pressure-reducing valve to replace the compressor. The schematic representation of the system is shown in Fig. 12.1 in which various components of the system are arranged according to their pressures and temperatures. The refrigerating effect is shown as Q_0 at temperature T_0 and the heat rejected in the condenser as Q_c at temperature $T_c = T_k$ of the environment. The compressor work is replaced by the heat supplied in the generator Q_h plus pump work Q_P. Cooling must be done in the absorber to remove the latent heat of the refrigerant vapour as it changes into the liquid state by absorption by the weak solution. Let this heat rejected in the absorber be Q_A at absorber temperature $T_A = T_k$. Then the energy balance of the system is

$$Q_0 + Q_P + Q_h = Q_C + Q_A = Q_k \tag{12.1a}$$

where Q_k represents the net heat rejected to the environment.

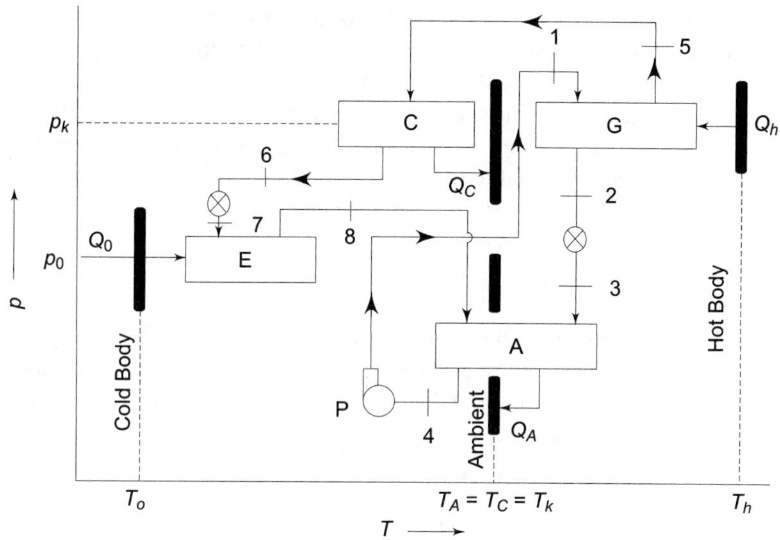

Fig. 12.1 Schematic representation of simple vapour absorption system

The pump work $Q_P = -\int v\,\mathrm{d}p$, is very small compared to the compressor work in the vapour compression system, as the specific volume v of the liquid is extremely small compared to that of the vapour ($v_f << v_g$). The energy consumption of the system is mainly in the generator in the form of heat supplied Q_h.

Generally, calculations for the vapour-absorption system are based on a unit mass of the refrigerant vapour distilled from the generator. All heat and mass flow quantities per unit mass of the vapour are then referred to as *specific quantities*. Thus, if D is the mass of the vapour distilled, then the various specific heat quantities are

$$q_0 = \frac{Q_0}{D},\ q_P = \frac{Q_P}{D}, \text{ and so on.}$$

The energy balance equation per unit mass of the vapour distilled is then

$$q_0 + q_P + q_h = q_C + Q_A = q_k \tag{12.1b}$$

The overall coefficient of performance may be denoted by ζ and is expressed as

$$\zeta = \frac{\text{Refrigerating effect}}{\text{Energy supplied}}$$

$$= \frac{q_0}{q_h + q_P} = \frac{q_0}{q_k - q_0} \tag{12.2}$$

$\cong q_0/q_h$, neglecting pump work which is small compared to q_h.

12.2 MAXIMUM COEFFICIENT OF PERFORMANCE OF A HEAT OPERATED REFRIGERATING MACHINE

The vapour-absorption system is a *heat-operated refrigerating machine*. It may be considered as a combination of a heat engine and a refrigerating machine as illustrated in Fig. 12.2. The energy supplied to the system is in the form of heat Q_h at temperature T_h. The thermodynamic cycle is considered to comprise of a heat-engine E cycle, operating between the heat source temperature T_h and the temperature of heat rejection T_k, and a refrigerator R cycle operating between the refrigeration temperature T_0 and temperature of heat rejection T_k. The work done in the heat engine part of the cycle is equal to the work requirement of the refrigeration part of the cycle. Thus, one may write for the COP of the cycle

$$\zeta = \frac{Q_0}{Q_h} = \frac{W}{Q_h} \times \frac{Q_0}{W} = \eta_{\text{th}} \times \mathcal{E}_c$$

which implies that the COP of a heat-operated refrigerating machine is equal to the product of the thermal efficiency of the heat engine part of the cycle and the coefficient of performance for cooling of the refrigeration part of the cycle. Accordingly, COP ζ of vapour absorption system expressed by Q_0/Q_h should not be compared with COP $\mathcal{E}$ of vapour compression system which is Q_0/W.

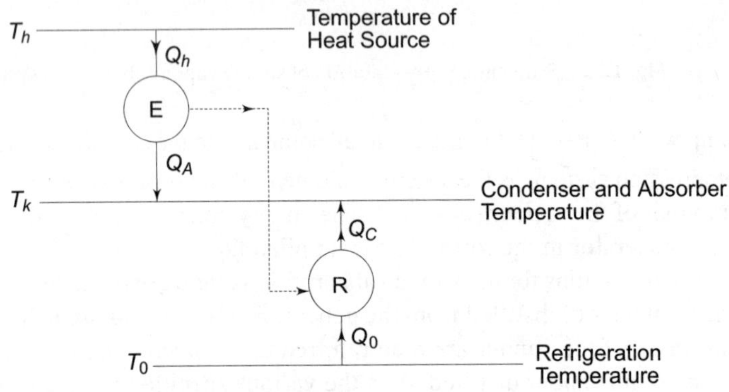

Fig. 12.2 Representation of a heat-operating machine as a combination of a heat engine and a refrigerator

The COP of a heat-operated refrigerating machine should, therefore, be maximum when each of the two terms has a maximum value which would be so when both are equal to their respective Carnot values. Thus

$$\zeta_{\max} = \eta_{\text{th}_{\text{Carnot}}} \times \mathcal{E}_{c_{\text{Carnot}}} = \left(\frac{T_h - T_k}{T_h}\right)\left(\frac{T_0}{T_k - T_0}\right)$$

$$= \left(1 - \frac{T_k}{T_h}\right)\left(\frac{1}{\dfrac{T_k}{T_0} - 1}\right) \tag{12.3}$$

It can be seen from the above expression that the COP depends on temperatures T_h, T_k and T_0 and in order for it to be high one should have:

(i) Temperature T_h of the heat source as *high* as possible,

(ii) Temperature T_k of the heat sink as *low* as possible, and

(iii) Temperature T_0 of refrigeration as *high* as possible.

Note *However, absorption system is not a reversible cycle. There is a degree of irreversibility due to mixing of refrigerant and absorbent. Its COP, first, increases with increase in generator temperature, it reaches an optimum value, and then it starts decreasing as a result of increase in irreversibility at high generator temperatures.*

In the vapour-absorption cycle, the system formed by the generator-valve-absorber-pump may be considered to represent the heat engine part of the cycle. This part in which only the refrigerant-absorbent solution circulates, is named as the *solution circuit*. On the other hand, the condenser-expansion device-evaporator form the usual refrigeration part of the cycle. This part is named as the *refrigerant circuit*.

It may be noted that, in case, the condenser and absorber temperatures are not the same and are equal to T_C and T_A respectively, then the maximum possible COP is given by

$$\zeta_{\max} = \frac{T_h - T_A}{T_h} \times \frac{T_0}{T_C - T_0} \tag{12.4}$$

Example 12.1 *In a vapour-absorption refrigeration system, the refrigeration temperature is – 15°C. The generator is operated by solar heat where the temperature reached is 110°C. The temperature of the heat sink is 55°C. What is the maximum possible COP of the system?*

Solution

$$T_0 = 273 - 15 = 258 \text{ K}$$
$$T_h = 273 + 110 = 383 \text{ K}$$
$$T_k = 273 + 55 = 328 \text{ K}$$
$$\zeta_{\max} = \frac{383-328}{383} \times \frac{258}{328-258} = 0.0914 \times 3.69 = 0.34$$

Note *Thus, the COP of 0.34 of the heat-operated vapour absorption system is equivalent to COP of 3.69 of vapour compression system in this case.*

12.3 COMMON REFRIGERANT-ABSORBENT SYSTEMS

Some of the desirable characteristics of a refrigerant-absorbent pair for an absorption system are low viscosity to minimize pump work, low freezing point and good chemical and thermal stability. Irreversible chemical reactions of all kinds, such as decomposition, polymerization, corrosion, etc., are to be avoided.

In addition to the above, two main thermodynamic requirements of the mixture are:

(i) *Solubility requirement*: The refrigerant should have more than Raoult's law solubility in the absorbent or adsorbent so that a strong solution, highly rich in the refrigerant, is formed in the absorber by the absorption of the refrigerant vapour.

(ii) *Boiling points requirement*: There should be a large difference in the normal boiling point of the two substances, at least 200°C, so that the absorbent exerts negligible vapour pressure at the generator temperature. Thus, almost absorbent-free refrigerant is boiled off from the generator and the absorbent alone returns to the absorber. If absorbent vapour goes with the refrigerant vapour to refrigerant circuit, the refrigeration produced will not be isothermal. In case a solid adsorbent is used, it does not exert any vapour pressure. Thus, pure refrigerant vapour only will go to the refrigerant circuit.

The two commonly used pairs are those of refrigerant NH_3 + absorbent H_2O, and refrigerant H_2O + adsorbent $LiBr_2$.

In the ammonia-water system, ammonia is the refrigerant and water is the absorbent. Ammonia forms a highly non-ideal solution in water. Hence, from the point of view of the solubility requirement, it is satisfactory. But the difference in their boiling points is only 138°C. Hence the vapour leaving the generator contains some amount of water.

Thus, the ammonia-water system is not suitable from the point of view of the boiling point requirement.

In the water-lithium bromide system, water is the refrigerant and lithium bromide is the adsorbent. Hence the mixture is used only in air-conditioning applications since water freezes at 0°C. The mixture is again non-ideal and is satisfactory from the point of view of the solubility requirement. Since lithium bromide is a salt, it exerts no vapour pressure. So the vapour leaving the generator is a pure refrigerant. The mixture, therefore satisfies the boiling point requirement also. However, it is corrosive and the plant works under high vacuum, both in condenser and evaporator. Hence, a purge unit is used.

The thermodynamic properties of the aqua-ammonia system are available in the form of Jennings and Shannon[7] tables. These tables give values of saturated liquid and vapour enthalpies, equilibrium temperatures and vapour compositions as a function of pressure and liquid composition. Kohloss and Scott[8] developed a diagram based on similar data by Scatchard.[16]

The enthalpy composition ($h - \xi$) diagram for the aqua-ammonia system is due to Merkel and Bosnjakovic.[12] An advantage of this diagram is that it makes it possible to geometrically represent the complete vapour-absorption cycle. There are auxiliary lines on this diagram which are used to locate the equilibrium vapour state, say A'', corresponding to a certain liquid state, say A', by drawing a tie line and vice versa, as shown in Fig. 4.19.

The properties of $H_2O - LiBr_2$ system were established by Rosenfeld and Karnaukh[15], Lower and McNeely.[11] An enthalpy-composition diagram is available for the liquid mixture. The vapour phase consists of pure water only. Some data in the form of vapour pressure, specific heat, specific weight and differential heat of mixing have been given by Niebergall.[13]

Research work is in progress on various other refrigerant absorbent combinations[1, 2, 5, 6, 10, 17, 18, 21].

12.4 MODIFICATIONS TO SIMPLE VAPOUR-ABSORPTION SYSTEM

There are two drawbacks in the simple vapour-absorption cycle. These are:

(i) Low temperature of the rich solution entering the generator and high temperature of the poor solution entering the absorber.

(ii) Presence of absorbent in the vapour leaving the generator and going to refrigerant circuit.

The two are separately discussed along with proposed modifications.

12.4.1 Using Liquid-Liquid Heat Exchanger in the Solution Circuit

Referring to Fig. 12.3, it is observed that in the simple system, saturated rich solution at 4 at refrigerant concentration ξ_r^L in the liquid phase must be heated from the absorber temperature t_A to the bubble temperature t_1 at the generator pressure, whereas the saturated poor solution at 2 from the generator at temperature t_h and refrigerant concentration ξ_a^L again in the liquid phase must be cooled to temperature t_3 in the absorber. This is inefficient since useful heat must be added in the generator and the same must be rejected in the absorber.

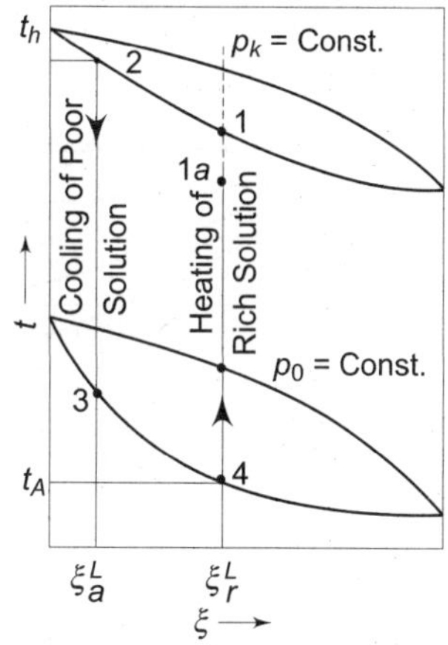

Fig. 12.3 Diagram showing cooling of poor solution and heating of rich solution

The system can be improved by incorporating a regenerative heat exchanger between the poor and rich solutions. This will reduce the amount of heat added in the generator and hence increases the COP, and decrease the amount of heat rejected in the absorber. The sizes of the generator and absorber will also be reduced. This heat exchanger is called a *liquid-liquid heat exchanger*. In such a case, the state after heating of the rich solution at 1a will be a subcooled state below the saturation temperature at 1 at pressure p_k.

12.4.2 Using Analyser—The Exhausting Column, and Dephlegmator—The Rectifying Column

In a system like that of ammonia-water, the vapours distilled from the generator contain a considerable amount of absorbent vapour which subsequently reaches the evaporator after condensation. As a result, evaporation would not be isothermal and the required low temperature would not be reached in the evaporator as illustrated in Fig. 12.4.

The temperature entering evaporator at the beginning of evaporation is t_{01}, and leaving evaporator is t_{0_2} which is equal to maximum permissible refrigeration temperature. The temperature after *complete evaporation* is very high, equal to the dew point temperature at evaporator pressure and composition.

Let p_0 be the required evaporator pressure with pure refrigerant corresponding to evaporation temperature t_0. With the mixed vapour at V with concentration ξ_a^V leaving the generator in equilibrium with the boiling poor solution at L with concentration ξ_a^L, the condensate is at point 6 at pressure p_k and temperature t_C of the condenser.

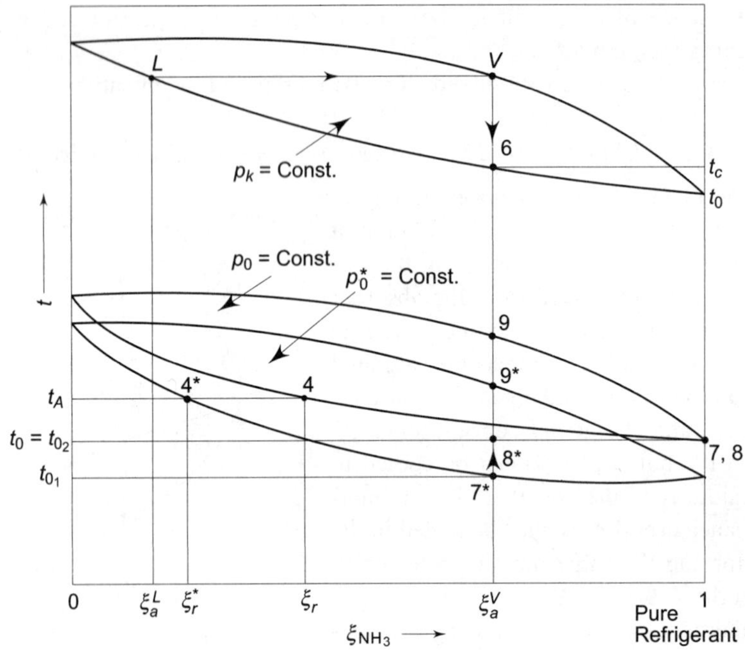

Fig. 12.4 Effect of presence of absorbent in evaporator

The temperature at the end of throttling to p_0 and complete evaporation is t_9 in the case of refrigerant mixed with absorbent instead of $t_7 = t_8 = t_{01} = t_{02}$ with the pure refrigerant at pressure p_0. In order to obtain maximum permissible refrigeration temperature $t_0 = t_{0_2}$, at the end of evaporation, it is necessary to have expansion to a lower evaporator pressure p_0^* instead of p_0. The liquid entering the evaporator at first evaporates somewhat like a pure substance at pressure p_0* corresponding to temperature $t_{0_1} = t_7$*. The liquid gradually becomes weaker in the refrigerant and at the end of partial or incomplete evaporation to 8*, it reaches temperature $t_{0_2} = t_8^*$. Thus, evaporation is incomplete.

Not only is the evaporator pressure reduced due to concentration ξ_a^V being less than unity, it can be seen that the rich solution concentration at the given absorber temperature t_A is also reduced, say, from ξ_r to ξ_r^*, because of a lower absorber pressure. This will ultimately result in a smaller difference in the rich and poor solution concentrations and hence a lower COP as explained in Sec. 12.5.3.

We thus see that there are three adverse effects of the absorbent vapour from the generator entering the refrigerant circuit. These are the following:

(i) Evaporator pressure has to be lowered from p_0 to p_0^* to attain the required refrigeration temperature of $t_0 = t_{0_2}$.

(ii) Rich solution refrigerant concentration decreases from ξ_r to ξ_r^*, thus increasing rich solution circulation f and decreasing COP.

(iii) Evaporation is incomplete. All the liquid does not evaporate. Part of the liquid has to go to the absorber without producing any refrigerating effect. Hence, COP decreases further.

Example 12.2 *With assumed numerical values, examine the effect of presence of water in vapour leaving generator in NH_3—H_2O system on solution circulation rates, refrigeration temperature and COP.*

Solution Consider following condenser and evaporator temperatures:

$$t_k = t_C = 50°C, \qquad t_0 = -17.5°C$$

Assuming pure ammonia vapours are evolved in generator, we have for pressures from the table of properties of ammonia:

$$p_k = (p_{sat})_{50°C} = 20.33 \text{ bar}$$
$$p_0 = (p_{sat})_{-17.5°C} = 2.25 \text{ bar}$$

Now, if generator temperature is $t_h = 156°C$, we have for poor solution concentration of NH_3

$$\xi_a^L = 0.2 \text{ (sat. liq. at 20.33 bar, 156°C)}$$

and if absorber temperature is $t_A = 40°C$, we have for rich solution concentration of NH_3.

$$\xi_r^L = 0.34 \text{ (sat. liq. at 2.25 bar, 40°C)}$$

The concentration of NH_3 in vapour leaving generator

$$\xi_d = \xi_a^V = 0.73 \text{ (sat. vap. at 20.33 bar, 156°C)}$$

Specific Solution Circulation Rates Then, from Eqs (12.7) and (12.8), we have per unit mass of vapour leaving generator, circulation rates of rich solution leaving absorber and poor solution leaving generator respectively as

$$f = \frac{\xi_d - \xi_a}{\xi_r - \xi_a} = \frac{0.73 - 0.2}{0.34 - 0.2} = \frac{0.53}{0.14} = 3.8 \text{ kg/kg vapour}$$

$$f - 1 = 2.8 \text{ kg/kg vapour}$$

But, the difference in N.B.P. of NH_3(–33°C) and that of H_2O (100°C) is small, only 133°C. The consequences of the same are the following:

(i) Considerable amount of absorbent water is present in the vapour leaving generator. This goes to the refrigerant circuit. It is equal to $(1 - \xi_d) = 1 - 0.73 = 0.27$ (27%) in this case.

(ii) As a result, temperature of – 17.5°C would not be attained after throttling to 2.25 bar.

(iii) To attain this temperature, an evaporator pressure lower than 2.25 bar would be required.

(iv) Let the throttling be done to an evaporator pressure of $p_0^* = 1$ bar. Even with this low pressure, the temperature attained after throttling would be $t_{0_1} = -19°C$ only.

(v) After complete evaporation, the temperature of this liquid-vapour NH_3/H_2O mixture would be as high as $t_9^* = 76°C$ (Fig. 12.4). As the maximum refrigeration temperature required is $t_{0_2} = -17.5°C$, it is seen that only a small fraction of the mixture could be allowed to evaporate. The rest would have to go to the absorber, along with vapour, without producing any refrigerating effect. Hence, Q_0 would be very much reduced.

(vi) Further, with the lowering of absorber pressure to 1 bar, the new rich solution concentration would be

$$\xi_r^* = 0.234 \text{ (sat. liq. at 1 bar, 40°C)}$$

Thus, $\xi_r^* - \xi_a = 0.235 - 0.2 = 0.035$ would become very small resulting in extremely large solution circulation rates f and $(f - 1)$, and hence large quantity of heat added in generator, as seen from Eq. (12.9), and very low COP.

To return the absorbent to the generator and to allow, as far as possible, only the refrigerant vapour to enter the condenser, two elements are added to the simple absorption system. These are:

(i) The *analyser* or the exhausting column.

(ii) The *dephlegmator* and *rectifier* or the rectifying column.

The analyser or the exhausting column is installed on top of the generator as shown in Fig. 12.5. The vapour, leaving the generator with refrigerant concentration ξ_a^V in equilibrium with the boiling poor solution having concentration ξ_a^L, enters the analyser at 2^V (Fig. 12.7). As it travels upwards, counterflow to the entering rich solution at 1 with concentration ξ_r^L, the vapour encounters heat and mass exchange with the falling rich solution ultimately leaving the analyser enriched in the refrigerant with vapour concentration $\xi_d = \xi_5 \leq \xi_r^V$ in equilibrium with the rich solution having concentration ξ_r^L.

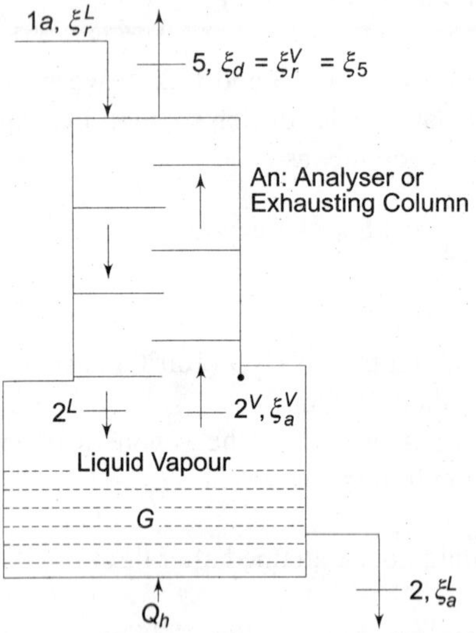

Fig. 12.5 Analyser or exhausting column

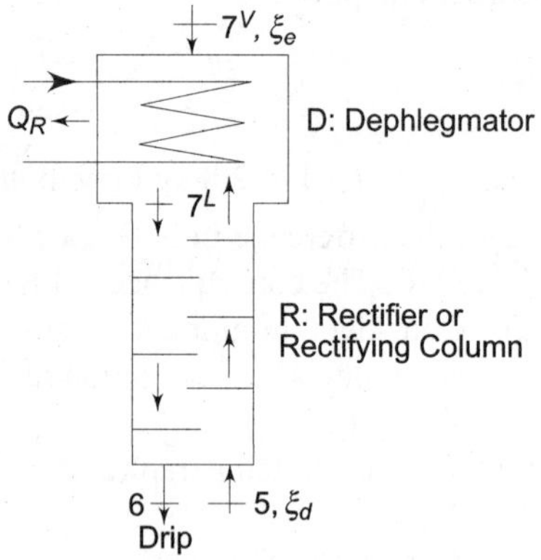

Fig. 12.6 Dephlegmator and column rectifying

We see from Fig. 12.7 that this method has the additional advantage of returning some heat from the vapour to the generator in the form of preheating of the rich solution from t_{1a} to t_1 to t_2 with simultaneous cooling of the vapour from t_2 to $t_5 = t_1$.

The enriched vapour from the generator-analyser now enters the dephlegmator or the rectifying column as shown in Fig. 12.6 wherein heat is removed from the vapour by the circulation of a cooling medium. The leaving state 7^V of the vapour is determined by the temperature $t_7 = t_R$ of the cooling medium (Fig. 12.7) at the end of the rectifying column. A part of the vapour is condensed at 7^L and is returned as *drip* 6 to the analyser. The leaving vapour having concentration ξ_e (higher than ξ_d) goes to the refrigerant circuit. In the rectifying column, heat Q_R is rejected to the cooling medium.

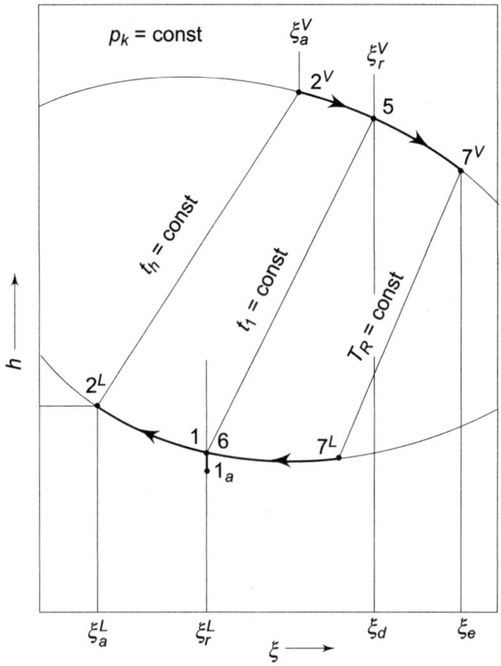

Fig. 12.7 Thermodynamic processes of analyser and exhausting and rectifying columns on h–ξ diagram

The latter method of increasing the refrigerant concentration of the vapour has a drawback in that it involves a loss of useful heat added in the generator which is rejected to the cooling medium in the dephlegmator. The drip returns to the generator and has to be evaporated again. The conditions under which the use of a dephlegmator would improve the COP would depend on the working pair being used and the operating conditions.

Note *It may be noted that the use of analyser and dephlegmator is not necessary in the case of systems such as lithium bromide-water in which case the absorbent does not exert any significant vapour pressure at all.*

12.5 ACTUAL VAPOUR-ABSORPTION CYCLE AND ITS REPRESENTATION ON ENTHALPY-COMPOSITION DIAGRAM[3,11]

Figure 12.8 shows the schematic arrangement of the actual vapour-absorption cycle and Fig. 12.9 presents its thermodynamic cycle on the h – ξ diagram. The system consists of generator G, analyser AN, dephlegmator D and condenser C on the high pressure side, and evaporator E and absorber A on the low pressure side. Pump P, expansion valve VI and pressure-reducing valve VII separate the two sides. In addition, liquid-vapour heat exchanger HE I and liquid-liquid heat exchanger HE II are also provided.

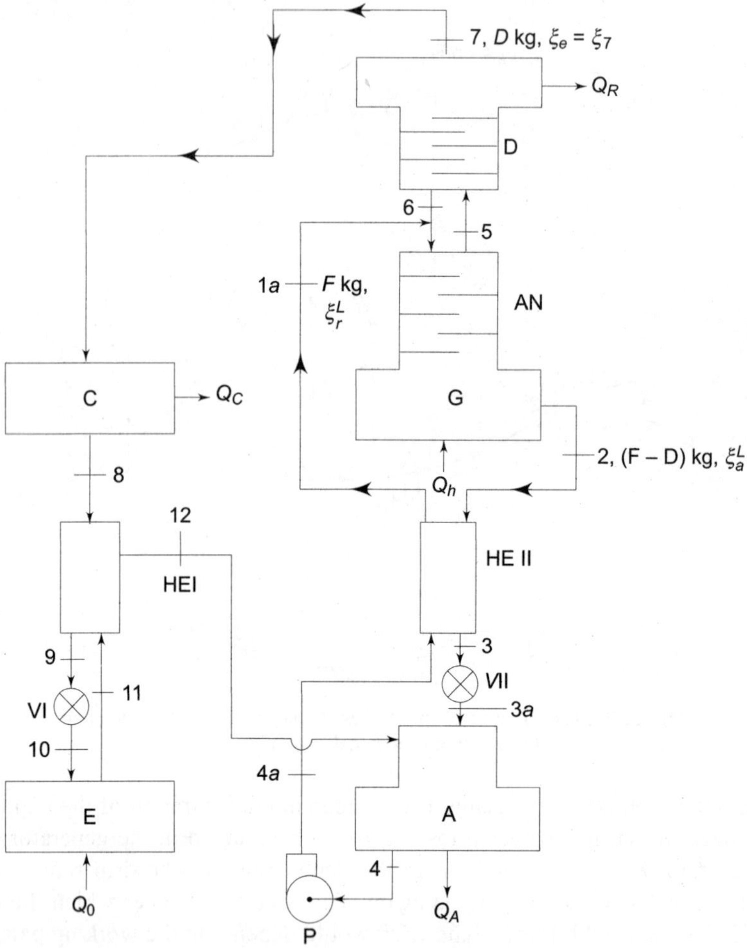

Fig. 12.8 Schematic diagram of actual vapour absorption system

The vapours at 5, distilled from the generator-analyser, enter the dephlegmator. The vapours from the dephlegmator with higher concentration of the refrigerant at 7 then enter the refrigerant circuit, whereas the drip at 6 returns to the generator-analyser. The vapours are condensed to 8 in the condenser, precooled to 9 in the liquid vapour regenerative heat exchanger and throttled to 10 before entering the evaporator. The state 10 is at the same point as state 9 on the h-ξ diagram as both enthalpy and composition remain the same before and after throttling. The refrigerant entering the evaporator at 10, leaving the evaporator at 11 and the liquid vapour heat exchanger at 12, comprises of a liquid plus vapour mixture. The refrigerant is finally absorbed by the poor solution at 2 returning from the generator after being cooled in the liquid-liquid heat exchanger to 3 and throttled to 3a, whereas the rich solution from the absorber at 4 is pumped to 4a and heated to 1a before entering the analyser.

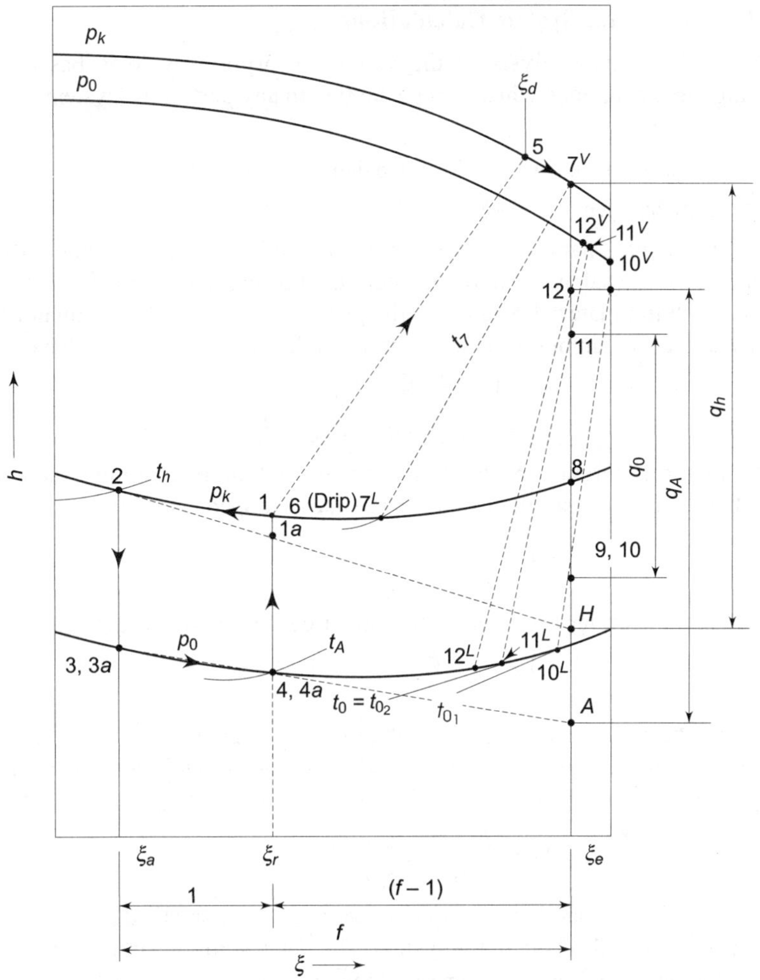

Fig. 12.9 Representation of vapour-absorption cycle on h-ξ diagram

The state points 1, 2, 3 and 4 can be located on the h-ξ diagram according to their temperatures and compositions as the enthalpy of liquid is independent of pressure. Also, point 3a lies at 3 only (isenthalpic process) and point 4a lies approximately at 4 itself as the pump work is very small. State point 5 of the vapour is along the isothermal tie line drawn from 1. State point 7 is on the tie line corresponding to the dephlegmator temperature t_7 and condenser pressure p_k. Point 8 is the saturation state at p_k and at the same composition as 7. Point 9 after subcooling of the liquid can be plotted according to the temperature and composition and point 10 is at 9 itself (isenthalpic process). Point 11 is on the tie line corresponding to the evaporator leaving temperature $t_{02} = t_0$ and pressure p_0. The composition is same at 7, 8, 9, 10,

11 and 12. Point 12 can be similarly located by knowing the temperature from the energy balance of the heat exchanger.

12.5.1 Absorption-System Calculations

The thermodynamic analysis of the vapour-absorption cycle is based on the following three equations which can be applied to any part of the system.

(i) Mass balance $\Sigma m = 0$

(ii) Material balance (or partial mass balance) $\Sigma m\xi = 0$

(iii) Energy balance $\Sigma Q + \Sigma mh = 0$

Thus, if the mass handled by the pump is F and the refrigerant vapour distilled is D, then by mass balance, the weak solution returning to the absorber is (F-D).

Also, applying material balance to the generator-analyser-dephlegmator taking the refrigerant, say ammonia, as the material under consideration, we have

$$F\xi_r = (F - D)\,\xi_a + D\xi_e \tag{12.5}$$

in which $\xi_r = \xi_{1a}$, $\xi_a = \xi_2$ and $\xi_e = \xi_7$

represent the refrigerant concentrations of rich solution, poor solution and vapour respectively. The superscripts L and V have been dropped. This gives.

$$f = \frac{F}{D} = \frac{\xi_e - \xi_a}{\xi_r - \xi_a} \tag{12.6}$$

Also, when the rectifier is not used, the vapour concentration is $\xi_d = \xi_5$,

$$f = \frac{\xi_d - \xi_a}{\xi_r - \xi_a} \tag{12.7}$$

Now, f represents the quantity of rich solution handled by the pump per unit mass of the vapour distilled. This quantity is termed as the *specific rich solution circulation*. The *specific weak solution circulation* is similarly given by

$$(f - 1) = \frac{F - D}{D} = \frac{\xi_e - \xi_r}{\xi_r - \xi_a} \tag{12.8}$$

It must be pointed out that these quantities are of great practical importance from the point of view of the COP of the system. The lower solution circulation rates will ensure a higher COP. It can be seen that f mainly depends on the value of $(\xi_r - \xi_a)$. The latter quantity is called *degasing range*. This quantity should be as large as possible. This means that ξ_r should be as high as possible, and ξ_a should be as low as possible. This will happen when refrigerant has more than Raoult's law solubility in absorbent at the absorber temperature (negative deviation from Raoult's law), and when the N.B.Ps of the refrigerant and absorbent are wide apart.

Accordingly, both the thermodynamic requirements as stated in Sec. 12.3 would be satisfied.

Further, we see that if the N.B.P. of the absorbent is very high, or if the absorbent which is used exerts no vapour pressure at all, then pure refrigerant vapour only will leave the generator, so that $\xi_e = \xi_d = 1$, and we have:

$$f = \frac{1 - \xi_a}{\xi_r - \xi_a} \quad \text{and} \quad f - 1 = \frac{1 - \xi_r}{\xi_r - \xi_a}$$

We shall now apply energy balance to the individual components of the system based on unit mass of the vapour distilled from the generator.

Solution Circuit Energy Balance *Generator without dephlegmator* States 5 and 7 are the same.

$$q_h + f h_{1a} = h_5 + (f - 1)\, h_2$$

whence

$$q_h = h_5 - h_2 + f(h_2 - h_{1a}) \tag{12.9}$$

Introducing an auxiliary quantity defined by

$$h_H = h_2 - f(h_2 - h_{1a}) \tag{12.10}$$

we get

$$q_h = h_5 - h_H \tag{12.11}$$

Equation (12.9) shows that the heat added in the generator depends upon specific solution circulation f.

Generator with dephlegmator $q_h + f h_{1a} = h_7 + (f - 1)\, h_2 + q_R$

whence

$$q_h = h_7 - h_2 + f(h_2 - h_{1a}), + q_R$$

$$= (h_7 - h_H) + q_R \tag{12.12}$$

Liquid-liquid heat exchanger II Points 3, 4a and 1a represent subcooled liquid states at pressure p_k. Let q_{II} represent the heat transfer in heat exchanger II. Then

$$q_{II} = (f - 1)\, C_a\, (t_h - t_3)$$

$$= f C_r\, (t_{1a} - t_A) \tag{12.13}$$

where t_h and t_A are generator and absorber temperatures, and C_a and C_r are the specific heats of the poor and rich solutions respectively. Thus

$$t_{1a} = t_A + \frac{f-1}{f}\frac{C_a}{C_r}\,(t_h - t_3) \tag{12.14}$$

Similarly

$$h_{1a} = h_4 + \frac{f-1}{f}\,(h_2 - h_3) \tag{12.15}$$

Since the specific heat of weak ammonia solution is smaller than that of the strong solution and also since $(f - 1)$ is less than f, it is seen that

$$(t_{1a} - t_A) < (t_h - t_3)$$

Absorber $q_A + f h_4 = h_{12} + (f - 1)\, h_3$

whence

$$q_A = h_{12} - h_3 + f(h_3 - h_4) \tag{12.16}$$

Introducing another auxiliary quantity for the absorber defined by

$$h_A = h_3 - f(h_3 - h_4) \tag{12.17}$$

we get

$$q_A = h_{12} - h_A \tag{12.18}$$

Since $h_{3a} = h_3$ and $h_4 \approx h_{4a}$, we can see from Eqs (12.10) and (12.17) that $h_A \approx h_H$.

Pump The pump work is given by

$$q_P = f v_4\, (p_k - p_0)/\eta_P \tag{12.19}$$

where η_p is the pump efficiency.

Refrigerant Circuit Energy Balance The mass flow rate as well as the composition is the same at all sections.

Condenser

$$q_C = h_7 - h_8 \tag{12.20}$$

Liquid-vapour heat exchanger I

$$q_I = h_8 - h_9 = h_{12} - h_{11} \tag{12.21}$$

Expansion valve V I

$$h_9 = h_{10} = (1 - z)\, h_{10}^L + z\, h_{10}^V \tag{12.22}$$

This along with the material balance equation

$$\xi_9 = \xi_{10} = (1 - z)\, \xi_{10}^L + z\, \xi_{10}^V \tag{12.23}$$

can be solved by trial and error to get the temperature $t_{10} = t_{01}$ after expansion and at inlet to evaporator. Symbol z is used here for vapour dryness fraction (instead of x) since x denotes liquid mole fraction in the case of mixtures.

Evaporator

$$q_0 = h_{11} - h_{10} \tag{12.24}$$

The overall energy balance gives

$$q_h + q_0 + q_P = q_C + q_A + q_R = q_k$$

12.5.2 Auxiliary Quantities

It can be seen from Fig. 12.9 that if points 2 and 1a are joined by a line and the straight line is extended to meet the vertical line corresponding to the leaving vapour concentration ξ_e at H, then the projections of 2-1a, 1a-H and 2-H on the abscissa are proportional to 1, $(f - 1)$ and f respectively. Also, the enthalpy at H is given by

$$h_H = h_2 - f(h_2 - h_{1a}) \tag{12.25}$$

which is the same as the auxiliary quantity for the generator as defined in Eq. (12.10).

Similarly, if one joins points 3 and 4 and extends the line to meet the same vertical at A, then we get the auxiliary quantity for the absorber

$$h_A = h_3 - f(h_3 - h_4) \tag{12.26}$$

as defined by Eq. (12.17).

These auxiliary quantities make it possible to geometrically represent the generator and absorber heat quantities on the h-ξ diagram as shown in Fig. 12.9.

12.5.3 Rich and Poor Solution Concentrations

As stated earlier, the specific solution circulation f is a very significant quantity. The larger the value of f, the more is the pump work. Higher solution circulation rates of both rich and poor solutions involve larger pressure drops in the system and hence still more pump work. In addition, as seen from Eq. (12.9), a higher value of f results in a larger heat input to the generator q_h and hence a lower COP. Also, the heat rejected in the absorber q_A is more. And hence high f implies larger sizes of both generator and absorber and accompanying losses in addition to a lower COP.

It is, therefore, desirable to have f as small as possible. It is seen from Eq. (12.6) that for f to be small, the degasing range $(\xi_r - \xi_a)$ should be as large as possible. In other words, the rich-solution concentration ξ_r should be as high, and the poor-solution concentration ξ_a should be as low as possible.

The absorber pressure in the absorption system is equal to the evaporator pressure p_0 and the generator pressure is similarly equal to the condenser pressure p_k. Hence at a given generator pressure, the poor solution concentration is determined by the heating temperature t_h, and at a given absorber pressure, the rich solution concentration is determined by the cooling temperature t_A. The higher the t_h, the lower is ξ_a. And lower the t_A, the higher is ξ_r. Hence the generator temperature should be as high as possible, whereas the absorber temperature should be as low as possible.

Example 12.3 *In an ammonia-absorption system with an analyser but without a dephlegmator the following data are given:*

Condenser pressure	*20.3 bar*
Evaporator pressure	*2.1 bar*
Generator temperature	*156°C*
Absorber temperature	*40°C*

Determine, per unit mass of the vapour distilled:

(a) *Specific solution circulation rates.*
(b) *Temperature t_{01} at inlet to evaporator if the liquid from the condenser is cooled by 13°C in the liquid-vapour heat exchanger.*
(c) *The refrigerating effect if the maximum refrigeration temperature is 5°C.*
(d) *The heat transfer in the liquid-liquid heat exchanger.*
(e) *The heat added in the generator.*
(f) *The pump work.*
(g) *The coefficient of performance.*
(h) *The heat rejected in the absorber and condenser.*
(i) *Energy balance of the system.*

Solution Refer to Figs 12.7 and 12.8.

At 20.3 bar and 156°C

Poor solution concentration, $\xi_a = 0.2$

At 2.1 bar and 40°C

Rich solution concentration, $\xi_r = 0.34$

Also, the concentration of vapour leaving the analyser, in equilibrium with the entering rich solution is

$$\xi_5 = \xi_r^V = 0.913$$

It is convenient to put the thermodynamic properties and flow rates at various sections in a tabular form as in Table 12.1.

(a) Specific rich solution circulation rate

$$f = \frac{\xi_5 - \xi_a}{\xi_r - \xi_a} = \frac{0.913 - 0.2}{0.34 - 0.2} = 5.1 \text{ kg/kg of vapour}$$

Specific poor solution circulation rate

$$f - 1 = 5.1 - 1 = 4.1 \text{ kg/kg of vapour}$$

(b) Temperature after condensation

$$t_8 = 53°\text{C (Saturated liquid at 0.913 composition and 20.3 bar).}$$

Temperature after subcooling

$$t_9 = 53 - 13 = 40°C$$

For the throttling-expansion process

$$(1 - z)\,\xi_{10}^L + z\,\xi_{10}^V = \xi_{10} = \xi_9 = 0.913$$

$$(1 - z)\,h_{10}^L + z\,h_{10}^V = h_{10} = h_9 = 444 \text{ kJ/kg}$$

Table 12.1 Data for Example 12.3

State Point	*Pressure* p	*Temperature* t	*Concentration* ξ	*Enthalpy* h	*Flow rate* $\dot{m}$
	bar	°C	kg NH_3/kg mixture	kJ/kg	kg/s
1	20.3	—	0.34	—	—
2	20.3	156	0.2	616	4.1
3	20.3	67	0.2	205	4.1
3a	2.1	67	0.2	205	4.1
4	2.1	40	0.34	63	5.1
4a	20.3	40	0.34	63	5.1
1a	20.3	—	0.34	—	5.1
5.7	20.3	—	0.913	1947	5.1
8	20.3	53	0.913	507	5.1
9	20.3	40	0.913	444	5.1
10	2.1	–16	0.913	444	5.1
11	2.1	5	0.913	1281	5.1
12	2.1	—	0.913	—	5.1

There are two variables, temperature t_{10} and vapour fraction z. A trial and error solution can be obtained by assuming t_{10}. But in such a solution with the h-ξ diagram, point 10 fluctuates greatly. A simpler and quite accurate method is to assume the vapour state 10^V at $\xi_{10}^V = 1$ and then join the point 10^V to 9 and extend the line to intersect the saturated liquid line for 2.1 bar at 10^L which gives the temperature after expansion and the temperature at inlet to evaporator as

$$t_{0_1} = t_{10} = -16°C$$

(c) Temperature after the evaporator

$$t_{0_2} = t_{11} = 5°C$$

Draw the isothermal tie line for 5°C and 2.1 bar. The intersection with $\xi_{11} = 0.913$ locates point 11.

Refrigerating effect

$$q_0 = h_{11} - h_{10} = 1281 - 444 = 837 \text{ kJ/kg}$$

(d) Heat transfer in the liquid-liquid heat exchanger

$$q = (f - 1)(h_2 - h_3) = 4.1\,(616 - 205) = 1685 \text{ kJ/kg}$$
$$= f(h_{1a} - h_4) = 5.1\,(h_{1a} - 63)$$

whence

$$h_{1a} = 393 \text{ kJ/kg}$$

(e) Heat added in the generator

$$q_h = h_5 - h_2 + f(h_2 - h_{1a}) = 1947 - 616 + 5.1\,(616 - 393)$$
$$= 2468 \text{ kJ/kg vapour}$$

(f) Specific volume of the solution at 40°C, at the inlet to the pump

$$v_4 = \xi_r v_{NH_3} + (1 - \xi_4)\, v_{H_2O} = 0.34\,(1.726 \times 10^{-3}) + 0.66\,(1.008 \times 10^{-3})$$
$$= 1.251 \times 10^{-3}\ \text{m}^3/\text{kg}$$

Pump work

$$q_P = f v_4 (p_k - p_0) = 5.1 \times 1.251 \times 10^{-3}\,(20.3 - 2.1) \times 10^5$$
$$= 11{,}600\ \text{J/kg} = 11.6\ \text{kJ/kg refrigerant}$$

(g) Coefficient of performance

$$\text{COP} = \frac{q_0}{q_h + q_P} = \frac{837}{2468 + 11.6} = 0.34$$

(h) Enthalpy of vapour entering absorber

$$h_{12} = h_{11} + (h_8 - h_9) = 1281 + (507 - 444) = 1344\ \text{kJ/kg}$$

Heat rejected in the absorber

$$q_A = h_{12} - h_3 + f(h_3 - h_4) = 1344 - 205 + 5.1\,(205 - 63)$$
$$= 1865\ \text{kJ/kg vapour}$$

Heat rejected in the condenser

$$q_C = h_5 - h_8 = 1947 - 507 = 1440\ \text{kJ/kg}$$

(i) Energy balance
Heat rejected

$$q_k = q_A + q_C = 1865 + 1440 = 3305\ \text{kJ/kg vapour}$$

Heat received

$$q_0 + q_h + q_P = 837 + 2468 + 11.6 = 3316.6\ \text{kJ/kg refrigerant}$$

Thus the energy balance checks very closely.

12.6 REPRESENTATION OF VAPOUR ABSORPTION CYCLE ON $\ln p - \frac{1}{T}$ DIAGRAM

For the representation of the processes of vapour absorption machines, it is necessary to develop state diagrams that contain composition. Enthalpy-concentration diagram, used for ammonia-water system, is already well-known in this respect. Similarly, $\ln p - 1/T$ diagram, with composition of the mixture as parameter, has been adopted for the representation of the vapour absorption cycle, particularly for water-lithium bromide system.

We know that the saturation pressure versus saturation temperature relationship of pure substances can be expressed in the form

$$\ln p = a - b\,\frac{h_{fg}}{T}$$

In the case of binary mixtures, also, the above relationship can be used for composition separately. And, in case there is any heat of mixing of the solution Δh_m, the above equation will assume the form

$$\ln p = a' - b' \frac{h_{fg} + \Delta h_m}{T} \text{ or } \ln p = A - \frac{B}{T}$$

It is to be noted that both h_{fg} and Δh_m have been assumed to be independent of temperature T. It has also been assumed that during boiling in the generator, no appreciable quantity of the absorbent is evaporated, meaning that the solvent exerts a negligible vapour pressure. This assumption is true for water-lithium bromide mixture.

Thus, $\ln p$ versus $\frac{1}{T}$ relation, to a great extent, follows a straight line for a pure substance as well as for a mixture. We have one such line for the refrigerant and another for the absorbent. We can obtain such straight lines for each composition of the mixture. The constant composition lines are parallel if $\Delta h_m = 0$. They diverge towards higher temperatures if Δh_m is positive, and converge if Δh_m is negative as shown in Fig. 12.10. The absorption system refrigerant-absorbent pairs have negative heat of mixing. Hence, these lines are convergent for absorption system.

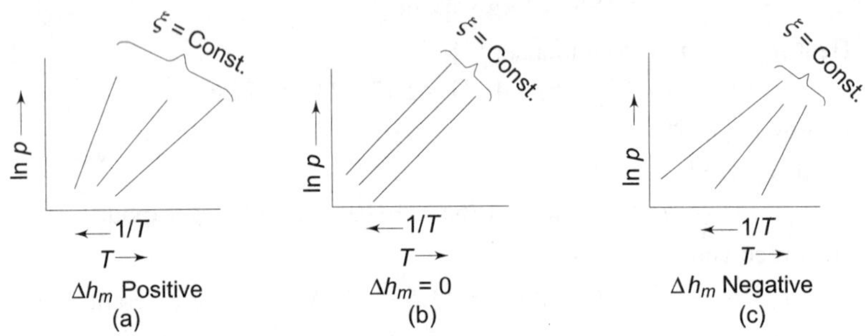

Fig. 12.10 Nature of $\ln p - 1/T$ lines for mixtures

Figure 12.11 shows a simple absorption system of Fig. 12.1 with the liquid-liquid heat exchanger incorporated in it in addition. The $\ln p - 1/T$ diagram of the cycle in Fig. 12.12 represents its working. It is a *single-effect* $H_2O - LiBr_2$ vapour absorption system since it has one-stage of generation of vapour. The *double-effect* is described in Sec. 12.8.

Line R on this diagram represents $\ln p^{sat}$ versus $1/T^{sat}$ relationship of the refrigerant ($\xi = 1$). Then, we draw the two horizontal lines corresponding to the condenser pressure p_k at condenser temperature T_c, and evaporator pressure p_0 at evaporator temperature T_0. The various state points are located as follows:

State 6 Saturated liquid refrigerant at condenser temperature T_C on $\xi = 1$ line.

States 7 & 8 Liquid plus vapour state 7 after throttling, and saturated vapour state 8 at T_0 on $\xi = 1$ line.

State 4 Saturated liquid mixture leaving absorber at p_0 and absorber temperature t_A. This gives the refrigerant rich solution concentration ξ_r.

State 4a It lies at 4 itself since its temperature and composition are the same. However, it is at pressure p_k. Hence, it represents a sub-cooled liquid state.

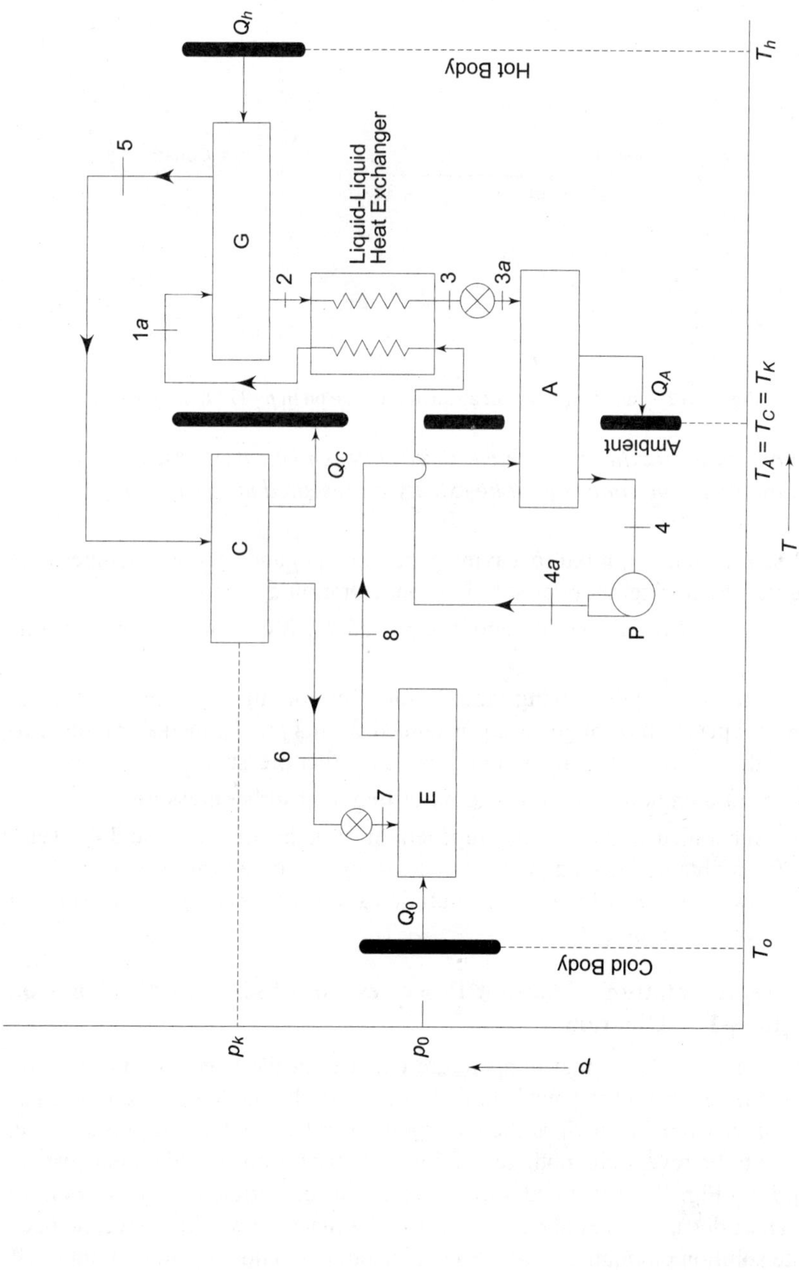

Fig. 12.11 Schematic representation of simple vapour absorption system with liquid-liquid regenerative heat exchanger

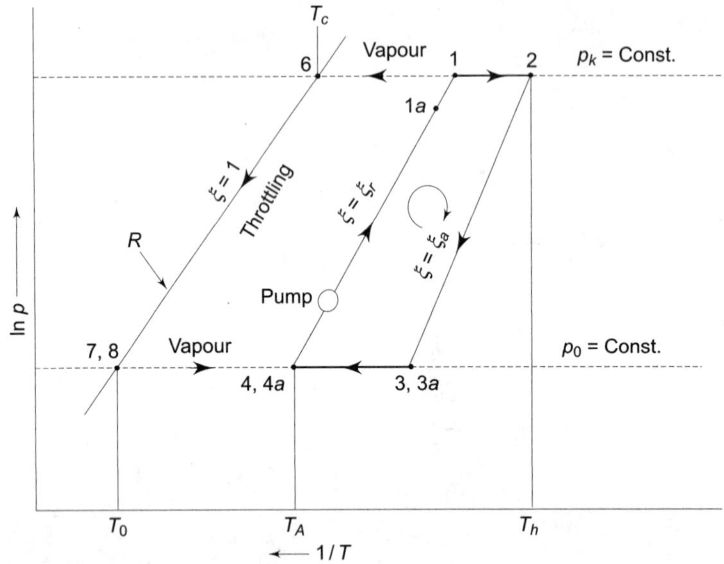

Fig. 12.12 Representation of absorption cycle on ln $p - 1/T$ diagram

Note *For binary mixtures, we have three independent properties. On a two-dimensional plane, the state cannot be exactly represented unless it is a saturation state.*

State 2 Saturated liquid mixture leaving generator at p_k and generator temperature t_h. This gives the refrigerant poor solution concentration ξ_a.

State 5 Superheated refrigerant vapour at p_k and T_h. It cannot be shown on this diagram.

State 3 Subcooled liquid leaving regenerator after cooling at p_k and ξ_a. It is assumed the hot solution from generator is cooled from T_2 to T_3, the saturation temperature of the solution of composition ξ_a and absorber pressure p_0.

State 3a Saturated liquid mixture after throttling to absorber pressure p_0.

State 1a Subcooled liquid leaving regenerator after heating at p_k and ξ_r. Temperature T_{1a} is obtained from energy balance of regenerative heat exchanger. Note that T_{1a} is lower than T_1 which is the saturation state at pressure p_k for mixture of refrigerant concentration ξ_r.

12.6.1 Representation of Carnot Processes for Absorption Machines on ln p–$1/T$ Diagram

It is seen from Fig. 12.12 that temperature changes in the generator from T_{1a} to $T_2 = T_h$ as the refrigerant vapour is boiled off. Similarly, the temperature changes in the absorber from T_{3a} to T_4 as the absorption of refrigerant vapour takes place. For process to be reversible, both generation and absorption should be isothermal. It is seen from Fig. 12.13 that under these conditions of Carnot processes, we have $\xi_r = \xi_a$. Thus, degasing breadth/range is zero. The machine would, therefore, need an infinite solution circulation rate *f*. Such a machine would have maximum COP, which is impossible to have.

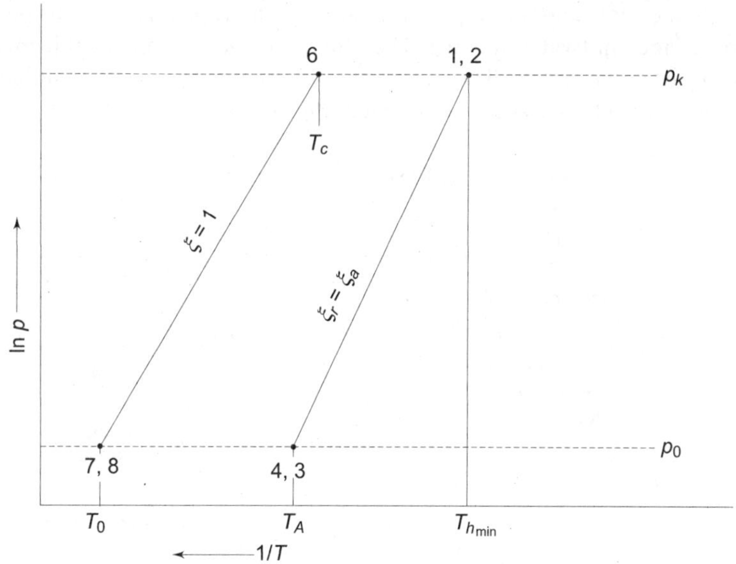

Fig. 12.13 Carnot processes for absorption machines on ln $p - 1/T$ diagram

12.6.2 Minimum Generator Temperature for Absorption Machines

Pressures p_k and p_0 are decided by condenser temperature T_C and evaporator temperature T_0. Then state 4 in Fig. 12.13 at absorber pressure p_0 is located from given absorber temperature T_A. This decides the rich solution concentration ξ_r, and state 1. In a Carnot cycle machine, $\xi_a = \xi_r$ and state 2 is at 1 itself. This T_2 is equal to T_h, the theoretically minimum generator temperature required for given T_C, T_0 and T_A. It is much above the ambient temperature. On account of other irreversibilities, the required minimum temperature in generator is still higher.

Note *It is thus seen that an absorption cycle cannot be devised simply from a heat source having temperature higher than the ambient temperature. To make the absorption system possible, its temperature should be above this* $T_{h_{min}}$.

12.7 PRACTICAL SINGLE-EFFECT WATER–LITHIUM BROMIDE ABSORPTION CHILLER

Several Inc., U.S.A. developed an absorption machine which uses solution of lithium bromide in water, water being the refrigerant and lithium bromide which is a highly hygroscopic salt-the adsorbent. Thus the solution leaving the absorber, being rich in *refrigerant water*, is a weak solution of salt in water. But for the sake of consistency, we shall continue to use ξ_r to denote the refrigerant concentration which, in this case, will mean the concentration of water in the solution leaving the absorber. Similarly, the solution returning from the generator is a strong solution of lithium bromide in water, but it being still poor in refrigerant water, ξ_a will, therefore, denote the concentration of water in the solution leaving the generator. The corresponding lithium bromide concentration will then be $(1 - \xi_r)$ and $(1 - \xi_a)$ respectively.

The salt does not exert any vapour pressure. So the vapour leaving the generator is a pure refrigerant (water vapour). Therefore, the analyser and dephlegmator do not form a part of the system. The equipment is normally designed for chilled-water applications with a flash system as shown in Fig. 12.14.

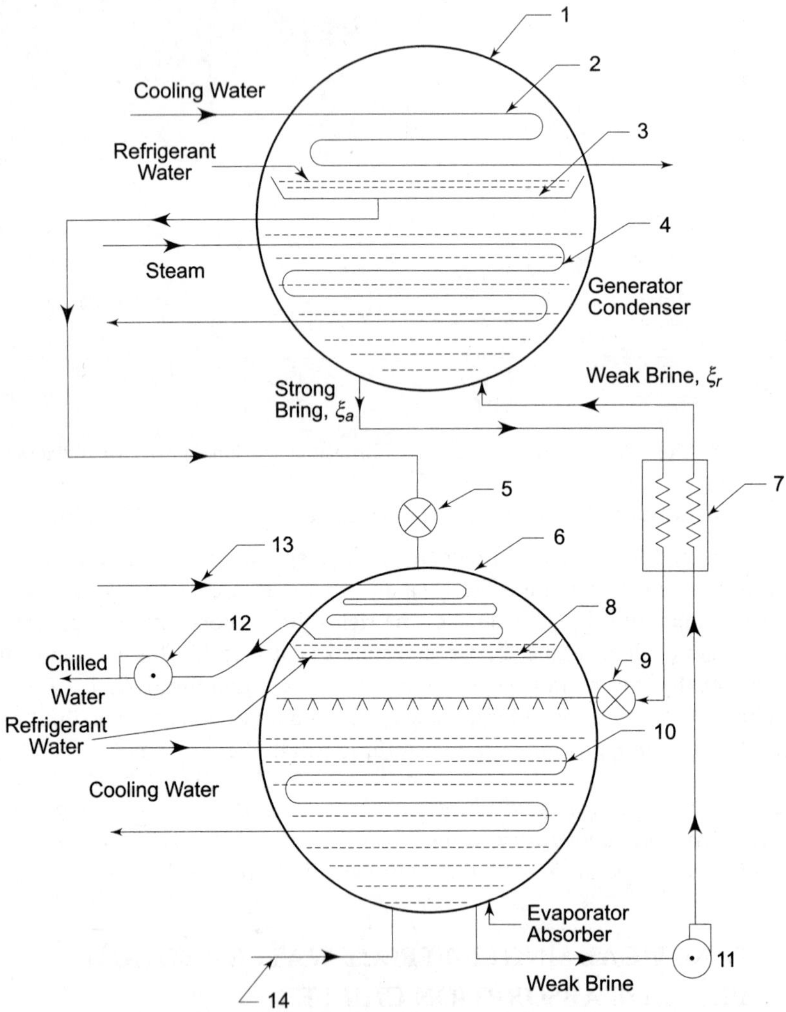

Fig. 12.14 Single-effect water-lithium bromide absorption chiller

The generator and condenser are housed in the single-cylindrical vessel 1 and the flash evaporator and absorber in another similar vessel 6. Water is boiled off from vessel 1 by the steam coils 4. The vapour is condensed over the condenser *cooling water* coils 2 and collected in the tray 3. The condensate is flashed through expansion valve 5 into the vessel 6. The refrigerant water is collected in the tray 8. Chilled water is circulated by the pump 12 and is returned to the system at 13.

The strong brine from vessel 1 flows by gravity through the heat exchanger 7 and the pressure reducing valve 9 to the vessel 6. The flashed water vapour filling the space in 6 is absorbed by this solution. The absorber heat is removed by the cooling coils 10. Again, there is separate cooling water line for the absorber. The weak salt solution leaving the absorber is then returned to the generator by the pump 11 through the liquid-liquid regenerative heat exchanger 7.

Thus, we see that there are three kinds of water flowing in the system:

(i) *Refrigerant water.*

(ii) *Chilled water* (secondary refrigerant).

(iii) *Cooling water.*

The chilled water, used as a secondary refrigerant, and refrigerant water are kept separate. If refrigerant water itself is used as chilled water, and it goes to the air conditioning plant and returns to the *sealed system*, it will bring in air with it. And thus it will break the vacuum.

Both the vessels 1 and 6 are maintained under high vacuum, vessel 1 corresponding to the condensing temperature (e.g., 55.3 mm Hg pressure at 40°C) and vessel[6] corresponding to the flashed refrigerant water temperature (e.g., 4.9 mm Hg pressure at 1°C). To remove air and other non-condensables that may enter the sealed system through pump glands, a two-stage *purge unit* is provided. To avoid corrosion, the temperature in the boiler should not be higher than 120°C. The overall COP of the system is reported to be approximately 0.7. The lithium bromide-water system is thus found to be more suitable in applications involving low heat-source temperatures such as are obtained with low-pressure (1 to 8 bar) or even exhaust (say 0.4 bar) steam, waste heat, solar energy, etc.

Absorption chillers are available in capacities from 100 TR upwards up to 7500 TR.

Example 12.4 *Calculations for Single-Effect Water-Lithium Bromide System*

The operating conditions for a water-lithium bromide chilled-water plant for air conditioning are as follows:

Generator temperature	*97°C*
Condenser temperature	*40°C*
Chilled-water temperature	*10°C*
Absorber temperature	*40°C*

Find temperature of the solution entering generator assuming hot solution is cooled to the saturation temperature at absorber pressure. Determine for one ton refrigeration capacity, the following:

(a) Thermodynamic conditions at all points.

(b) Coefficient of performance.

(c) Mass flow rates at all sections.

(d) Energy balance of complete system.

Solution (a) Refer to Figs 12.11 and 12.12. From table of water vapour pressures in the Appendix, we obtain the condenser and evaporator pressures corresponding to their respective temperatures.

Condenser and generator pressure

$$p_k = 55.32 \text{ mm Hg (At } 40°\text{C)}$$

Flash chamber and absorber pressure

$$p_0 = 9.21 \text{ mm Hg (At } 10°\text{C)}$$

Now, from ln p–$1/T$ diagram we get first the $LiBr_2$, and then the refrigerant water concentration in rich and poor solutions at states 4 and 2.

State 4 Saturated cold solution from absorber at

$$p = 9.21 \text{ mm Hg and } t = 40°\text{C}$$
$$\xi_{LiBr_2} = 0.55$$
$$h_4 = 93.5 \text{ kJ/kg (From } h - \xi \text{ diagram)}$$

Rich solution concentration of water (refrigerant)

$$\xi_r = 1 - \xi_{LiBr_2} = 1 - 0.55 = 0.45$$

State 2 Saturated hot solution from generator at

$$p = 55.32 \text{ mm Hg and } t = 97°\text{C}$$
$$\xi_{LiBr_2} = 0.65$$
$$h_2 = 248 \text{ kJ/kg (From } h - \xi \text{ diagram)}$$

Poor solution concentration of water (refrigerant)

$$\xi_a = 1 - 0.65 = 0.35$$

State 1 Saturated solution at condenser pressure and 0.55 $LiBr_2$ concentration

$$t_1 = 74°\text{C (From ln } p\text{–}1/T \text{ diagram)}$$
$$h_1 = 166 \text{ kJ/kg (From } h - \xi \text{ diagram)}$$

State 3 Saturated solution at evaporator pressure and 0.65 $LiBr_2$ concentration

$$t_3 = 60°\text{C}$$
$$h_3 = 180 \text{ kJ/kg}$$

State 3a has the same enthalpy, temperature and composition as state 3. But it is at the generator pressure. It represents a state subcooled from 2 to 3 at 55.32 mm Hg pressure

State 4a $t = 4°\text{C}$ and $\xi_{LiBr_2} = 0.55$

$$h_{4a} \cong h_4 = 93.5 \text{ kJ/kg (Neglecting pump work)}$$

Note *The enthalpy is read against temperature and composition for all solution states as it is independent of pressure. It may be noted that point 4a after pumping represents a subcooled state at 55.32 mm Hg pressure.*

Specific solution circulation rates

$$f = \frac{1-0.35}{0.45-0.35} = 6.5 \text{ kg/kg vapour}, f - 1 = 5.5 \text{ kg/kg vapour}$$

Heat available in hot solution for transfer

$$= (f - 1)(h_2 - h_3) = 5.5\,(248 - 180) = 374 \text{ kJ}$$

Heat required by cold solution for heating

$$= f(h_1 - h_4) = 6.5\,(166 - 93.5) = 471 \text{ kJ} > 374 \text{ k}$$

Hence, cold solution at 4a cannot be heated to 1. Let it be heated to 1a.

State 1a Energy balance of the liquid-liquid heat exchanger gives

$$f(h_{1a} - h_4) = (f - 1)(h_2 - h_3)$$

whence

$$h_{1a} = h_4 + \frac{(f-1)}{f}(h_2 - h_3) = 93.5 + \frac{5.5}{6.5}(248 - 180) = 151 \text{ kJ/kg}$$

State 5 It is that of water vapour at 55.32 mm Hg pressure and 97°C temperature. At these conditions it represents a superheated vapour state. The enthalpy of water vapour above the reference state of saturated water at 0°C can be found either from steam table having reference state $h_f = 0$ at $t = 0$°C or from the empirical relation

$$h = (2501 + 1.88\, t) \text{ kJ/kg}$$

Using the latter procedure

$$h_5 = 2501 + 1.88\,(97) = 2683 \text{ kJ/kg}$$

State 6 Saturated water at 40°C

$$h_6 = 4.1868\,(40) = 167.5 \text{ kJ/kg}$$

State 7 $p = 9.21$ mm Hg and $t = 10$°C (liquid + vapour)

$$h_7 = h_6 = 167.5 \text{ kJ/kg}$$

State 8 $p = 9.21$ mm Hg at $t = 10$°C (saturated vapour)

$$h_8 = 2501 + 1.88\,(10) = 2520 \text{ kJ/kg}$$

(b) Refrigerating effect

$$q_0 = h_8 - h_7 = 2520 - 167.5 = 2352.5 \text{ kJ/kg}$$

Heat added in the generator per unit mass of vapour distilled

$$q_h = h_5 - h_2 + f(h_2 - h_{1a})$$

$$= 2683 - 248 + 6.5\,(248 - 151) = 3066 \text{ kJ/kg vapour}$$

$$\text{Coefficient of performance COP} \approx = \frac{q_0}{q_h} = \frac{2352.5}{3066} = 0.77$$

(c) Water vapour distilled per ton refrigeration

$$D = \frac{211}{q_0} = \frac{211}{2352.5} = 0.09 \text{ kg/min}$$

Mass flow rate of cold solution from the absorber

$$F = fD = 6.5\,(0.09) = 0.585 \text{ kg/min}$$

Mass flow rate of hot solution from the generator

$$F - D = 0.585 - 0.09 = 0.495 \text{ kg/min}$$

(d) Head rejected in the condenser

$$Q_C = \frac{D}{60}(h_5 - h_6) = \frac{0.09}{60}(2683 - 167.5) = 3.77 \text{ kW}$$

Heat rejected in the absorber

$$Q_A = Dq_A = D[h_8 - h_3) + f(h_3 - h_4)]$$

$$= \frac{0.09}{60}[(2520 - 180) + 6.5\,(180 - 93.5)] = 4.35 \text{ kW}$$

Heat supplied in the generator

$$Q_h = Dq_h = \frac{0.09}{60}(3066) = 4.6 \text{ kW}$$

Heat absorbed by chilled water (per ton refrigeration)

$$Q_0 = 3.5167 \text{ kW}$$

Net heat received = $Q_h + Q_0 = 4.6 + 3.5167 = 8.12$ kW

Net heat rejected = $Q_C + Q_A = 3.77 + 4.35 = 8.12$ kW

Again, the energy balance checks very closely.

12.8 DOUBLE-EFFECT $H_2O - LiBr_2$ ABSORPTION SYSTEM

A single-stage like the single-effect absorption system is not suited to utilize a heat source at a temperature higher than a certain point unlike other heat-operated refrigerating machines that follow the Carnot trend, viz., the higher the temperature of the heat source, the higher the COP. In fact, the COP decreases as the heat source temperature increases beyond a point. This is because the absorption system is not a reversible refrigerating machine. Because of the mixing process of refrigerant and absorbent, a degree of irreversibility is involved. That is why, the COP of an absorption system levels with the increase in generator temperature, and then it starts decreasing.

It is found that the single-effect system gives best results upto a heat source temperature of 105°C. Above that temperature, it is worthwhile to switch over to double-effect system as illustrated in Figs 12.15 and 12.16.

Example 12.5 ***Calculations for Double-Effect H_2O-$LiBr_2$ System***

The high pressure generator of a double-effect H_2O-$LiBr_2$ vapour absorption system, shown in Fig. 12.15 operates on steam supplied at 8 bar pressure. The following conditions are specified:

Pressure in high-pressure generator	*100 kPa*
Temperature in high-pressure generator	*150°C*
Condenser temperature	*40°C*
Absorber temperature	*35°C*
Evaporator temperature	*10°C*

The water vapour from the high-pressure generator I is condensed in the low-pressure generator II. The temperature of solution leaving generator II is 95°C.

(i) Show all the thermodynamic states on $\ln p - 1/T$ diagram.
(ii) Find enthalpies at all state points.
(iii) Determine the heat added in generator I per kg of water in refrigerant circuit.
(iv) Determine COP.
(v) What would have happened if a single-effect system were to be used?

Solution (i) The thermodynamic states are shown on $\ln p - 1/T$ diagram in Fig. 12.16. All states are saturation states except 3*a* and 4*a*.

(ii) Condenser pressure at 40°C = 55.32 mm Hg
Evaporator pressure at 10°C = 9.27 mm Hg

Solution Circuit (Enthalpy is found from composition and temperature)

State 1 $p_1 = 9.27$ mm Hg, $t_1 = 35°C \Rightarrow \xi_{LiBr_2} = 0.525$, $\xi_{H_2O} = \xi_1 = 0.475$

$h_1 = 77.4$ kJ/kg $\cong h_2$

State 3 $p_3 = 55.32$ mm Hg, $\xi_{LiBr_2} = 0.525 \Rightarrow t_3 = 68°C$

$h_3 = 146.5$ kJ/kg

State 4 $p_4 = 100$ kPa, $\xi_{LiBr_2} = 0.525 \Rightarrow t_4 = 137°C$

$h_4 = 295.7$ kJ/kg

State 5 $p_5 = 100$ kPa, $t_5 = 150°C \Rightarrow \xi_{LiBr_2} = 0.578$, $\xi_{H_2O} = \xi_5 = 0.422$
$h_5 = 324.5$ kJ/kg

State 6 $p_6 = 55.32$ mm Hg, $\xi_{LiBr_2} = 0.578 \Rightarrow t_6 = 80°C$
$h_6 = 184$ kJ/kg $= h_{17}$

State 7 $p_7 = 55.32$ mm Hg, $t_7 = 95°C \Rightarrow \xi_{LiBr_2} = 0.642$, $\xi_{H_2O} = \xi_7 = 0.358$
$h_7 = 240$ kJ/kg

State 8 $p_8 = 9.27$ mm Hg, $\xi_{LiBr_2} = 0.642 \Rightarrow t_8 = 59°C$
$h_8 = 162.9$ kJ/kg

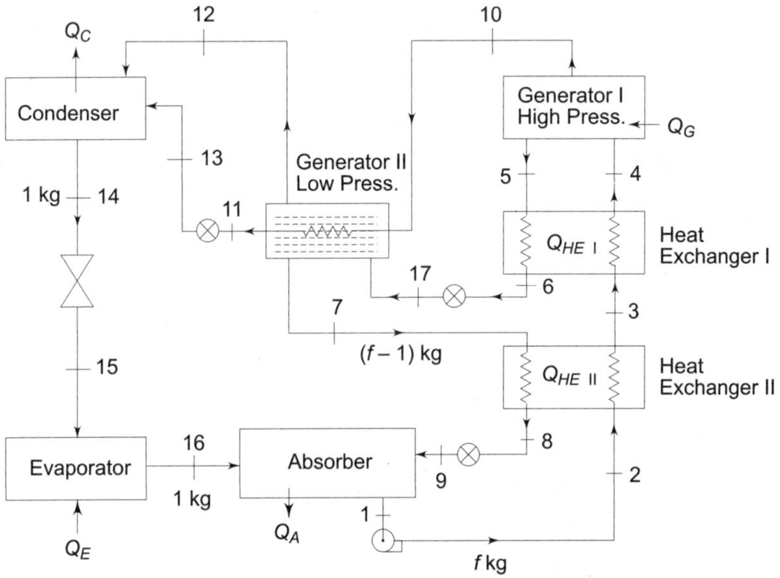

Fig. 12.15 Double-effect vapour absorption system

Refrigerant Circuit (Enthalpies of Water)

$h_{10} = 2776.4$ kJ/kg (Superheated vapour at 100 kPa, 150°C from steam table)

$h_{11} = (h_f)_{100\text{ kPa}} = 417.5$ kJ/kg $= h_{13}$ (Liquid-Vapour Mixture at 55.32 mm Hg)

$h_{12} = 2501 + 1.88\,(95) = 2680$ kJ/kg (Low pressure vapour)

$h_{14} = (h_f)$ 40°C $= 167.6$ kJ/kg $= h_{15}$

$h_{16} = (h_g)$ 10°C $= 2501 + 1.88\,(10) = 2519.8$ kJ/kg

(iii) *Solution Circulation Rates* Rich in refrigerant (weak in $LiBr_2$) solution circulation rate

$$f = m_1 = m_2 = m_3 = m_4$$

$$= \frac{1-\xi_a}{\xi_r - \xi_a} = \frac{1-\xi_8}{\xi_1 - \xi_9} = \frac{1-0.358}{0.475-0.358} = 5.487 \text{ kg/kg refrigerant}$$

Poor in refrigerant (strong in $LiBr_2$) solution circulation rate

$$f - 1 = m_9 = m_8 = m_7 = 5.487 - 1 = 4.487 \text{ kg/kg refrigerant}$$

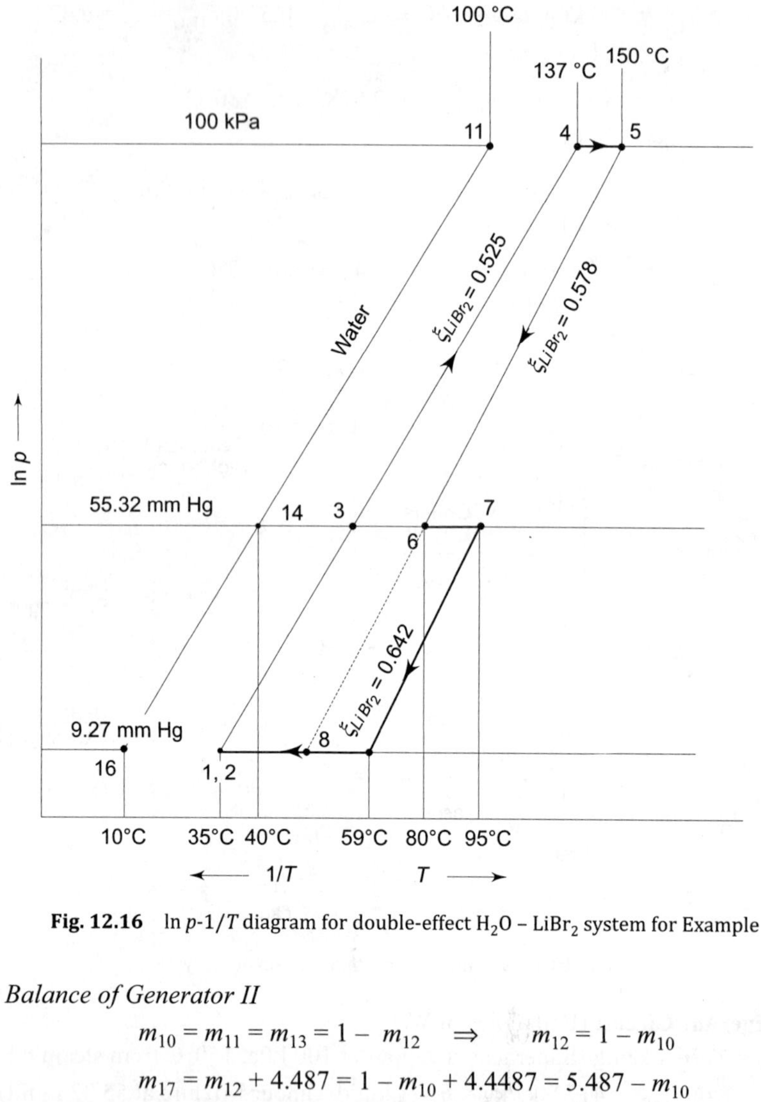

Fig. 12.16 ln p-1/T diagram for double-effect H_2O – $LiBr_2$ system for Example 12.3

Mass Balance of Generator II

$$m_{10} = m_{11} = m_{13} = 1 - m_{12} \quad \Rightarrow \quad m_{12} = 1 - m_{10}$$

$$m_{17} = m_{12} + 4.487 = 1 - m_{10} + 4.4487 = 5.487 - m_{10}$$

Enthalpy Balance of Generator II

$$m_{10}\, h_{10} + m_{17}\, h_{17} = m_{12}\, h_{12} + m_{13}\, h_{11} + m_7\, h_7$$

Substituting values and solving, we get

$$m_{10} = 0.774 \text{ kg}$$

$$m_{12} = 1 - m_{10} = 0.226 \text{ kg}$$

$$m_{17} = m_{12} + 4.487 = 4.713 \text{ kg}$$

Heat added in Generator I (From enthalpy balance)

$$(Q_G)_I = m_5\, h_5 + m_{10}\, h_{10} - m_4\, h_4$$

$$= 4.713\,(324.5) + 0.774\,(2776.4) - 5.487\,(295.7)$$

$$= 2065.2 \text{ kJ/kg refrigerant}$$

(iv) Refrigeration produced

$$Q_0 = h_{16} - h_{14} = 2519.8 - 167.6 = 2352.2 \text{ kJ/kg refrigerant}$$

$$\text{COP} = \frac{Q_0}{(Q_G)_I} = \frac{2352.2}{2065.2} = 1.14$$

(v) If a single-effect system were used, then we would find:

(a) Line 5-6 when extended to absorber pressure-since the solution is very strong in $LiBr_2$-would end up at point 8′.

(b) Further, ξ_a would be 0.422 instead of 0.358. Degasification breadth would be very small. Hence, COP would be very low.

12.9 ELECTROLUX REFRIGERATOR

Electrolux principle works on *3-fluid system*. There is no solution circulation pump. Total pressure is the same throughout the system. The third fluid remains mainly in the evaporator thus reducing partial pressure of refrigerant to enable it to evaporate at low pressure and hence low temperature.

The schematic diagram of the electrolux refrigerator working on NH_3–H_2O system with H_2 as the third fluid is shown in Fig. 12.17. Liquid NH_3 evaporates in the evaporator in the presence of H_2. Hydrogen is chosen as it is non-corrosive and insoluble in water.

A *thermosyphon bubble pump* is used to lift the weak aqua from the generator to the separator. The discharge tube from the generator is extended down below the liquid level in the generator. The bubbles rise and carry slugs of weak NH_3–H_2O solution into the separator.

Two *U-bends* are provided as *vapour-locks* to prevent H_2 from getting into the high side or solution circuit.

Partial pressure of H_2 provides the pressure difference of NH_3 between the condenser and the evaporator. Accordingly, we have:

In condenser Pure NH_3 vapour pressure = Total pressure

In evaporator NH_3 vapour pressure = Total pressure – Partial pressure of H_2

For example, consider the condenser temperature as 50°C, and evaporator temperature as – 15°C. The corresponding vapour pressures of NH_3 are:

$$\text{Condenser, } p_k = 20.33 \text{ bar}$$

$$\text{Evaporator outlet, } p_{0_2} = 2.36 \text{ bar}$$

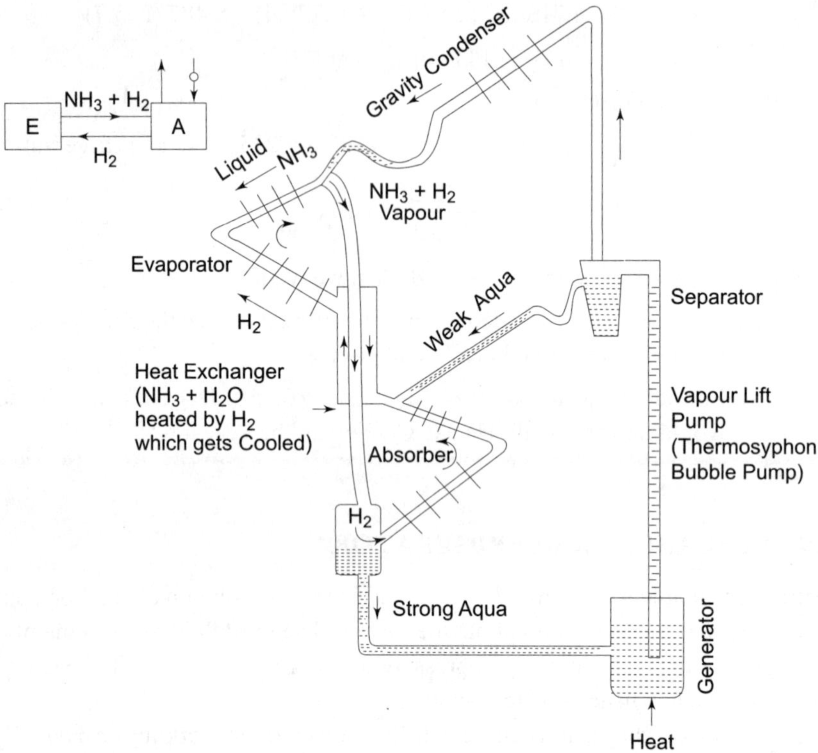

Fig. 12.17 Electrolux refrigerator

The approximate pressures in various parts of the system, then will be as given in Table 12.2.

Table 12.2 Partial pressures in electrolux refrigerator in bar

Section	NH_3	H_2O	H_2	Total
Condenser	20.33	0	0	20.33
Evaporator inlet	1.516	0	18.814	20.33
Evaporator exit	2.36	0	17.97	20.33
Generator top	15.54	4.79	0	20.33

It has been assumed that vapours leaving generator top are in equilibrium with entering rich solution at 40°C, at which temperature saturation pressure of NH_3 is 15.54 bar. It has also been assumed that the temperature at evaporator inlet is – 25°C at which temperature saturation pressure of NH_3 is 1.516 bar.

12.10 NEW MIXTURES FOR ABSORPTION SYSTEM

We know that NH_3 and H_2O do not form an ideal pair for absorption system because the difference in their N.B.Ps is not large enough. And, although, H_2O and $LiBr_2$

form a very good pair from the points of view of solubility and N.B.P. requirement, but the system suffers from the problems of corrosion, and of maintaining high vacuum both on the low pressure side as well as on the high pressure side. Hence, the efforts are on to find the most suitable pair for the system.

Zellhoeffer et al.[21] determined the solubility of R 21 and R 22 in a number of solvents such as ethers, esters, amides and amines. They found that dimethyl ether of tetraethylene glycol (DME–TEG) with chemical formula $CH_3O\ (CH_2CH_2O)_4CH_3$ is an extremely good solvent for these refrigerants. Mastrangelo made a further study of the solubility of various fluorocarbons in DME-TEG.

Ever since the pioneering work of Zellhoeffer et al., the investigators[2, 3, 5, 6, 10, 17, 18] have been trying to find suitable R 22 based refrigerant-absorbent mixtures to replace NH_3–H_2O system. For a study, Arora et al[1]. chose DME-TEG, isobutyl acetate, dimethyl formamide (DMF) with the chemical formula HCO. $N(CH_3)_2$ and diethyl formamide (DEF) with the chemical formula $HCON(C_2\ H_5)_2$ as absorbents for R 22. The basic ln p–$1/T$ data for the study was obtained for NH_3–H_2O and R 22 – Isobutyl acetate systems from Sellerio[17], for R 22-DME-TEG system from Mastrangelo and R 22-DMF and R 22-DEF systems from Thieme and Albright[18].

Table 12.3 Solution circulation rates

System	Δ(N.B.P.), °C	ξ_r	ξ_a	f
$NH_3 + H_2O$	133	0.5	0.22	3.78
R 22 + DME-TEG	316.6	0.496	0.264	3.18
R 22 + Isobutyl Acetate	158.8	0.52	0.34	3.67
R 22 + DMF	193.8	0.623	0.409	2.76
R 22 + DEF	218.3	0.563	0.433	4.36

Calculations were done for the operating conditions

$$t_h = 120°\text{C},\ t_C = t_A = t_k = 40°\text{C},\ t_0 = 5°\text{C}$$

It was assumed that only pure refrigerant circulates in the refrigerant circuit. The rectification losses were assumed negligibly small. Table 12.3 gives the differences between the N.B.Ps of refrigerant and absorbent, and the results of calculations for rich solution circulation rates. From the table, DMF with minimum solution circulation rates appears to be the best absorbent. Another suitable asbrobent is DME-TEG. Using Eisman method of calculation, it was also shown by Arora et al., that R 22-DMF system gives high COP. Recently, Bapat[2] and Das[5] have done detailed study of properties and analysis of R 22-DMF and R 22-DEF respectively.

Many studies have, however, been conducted on the use of solid adsorbents/ dessicants.

Ammonia as refrigerant in combination with solid adsorbents such as calcium chloride, sodium thiocyanate, lithium thiocynate, lithium nitrate, etc., is being tried in vapour absorption system. All these solid adsorbents are, however, highly corrossive in nature.

References

1. Arora C P, A K Mittal and A K Gupta, 'An analysis of the properties of mixtures for vapour absorption refrigeration'. Reprint No. 2.57, *Proc. XIII International Congress of Refrigeration*, Washington, 1971.
2. Bapat S L, *Thermodynamic Properties of Dimethyl Formamide and R 22 for Vapour Absorption Refrigeration System*, Ph. D. Thesis, IIT Delhi, 1982.
3. Berestneff A A, 'A new development in absorption refrigeration', *Refrigerating Engineering*, Vol. 57, No. 6, June, 1945, p. 553.
4. Bosnjakovic F, *Technical Thermodynamics*, Holt, Reinhert and Winston, 1965, pp. 255–279.
5. Das M S, *A Thermodynamic Study of Diethyl Formamide and R 22 Combination for Vapour Absorption Refrigeration System*, Ph D. Thesis, IIT Delhi, 1984.
6. Eisemen B J. Jr, 'Why R 22 should be favoured for absorption refrigeration', *ASHRAE J.*, Vol. 1, No. 12, Dec. 1959, pp. 45–50.
7. Jennings B H and F P Shannon, *ASRE Handbook*, 1951, p. 187.
8. Kohloss F H Jr and G L, Scott in 'Equilibrium properties of aqua-ammonia in chart from', *Refrigerating Engineering*, Vol. 58, No. 10, Oct. 1950, p. 970.
9. Lower H, 'Thermodynamic and physical properties of aquous lithium–bromide solution', *Report of Technical University*, Karlsruhe, Germany, 1961.
10. Mastrangelo S V R, 'Solubility of some chlorofluoro-hydrocarbons in tetra-ethylene glycol dimethyl ether', *ASHRAE J.*, Vol. 1, No. 10, Oct. 1959, pp. 64–68.
11. McNeely L A, 'Thermodynamic properties of aquous solutions of lithium bromide', *ASHRAE Trans*. 1979, Part One, pp. 413–434.
12. Merkel-Bosnjakovic, *Diagrams and Tables Relating to Absorption Refrigerators*, Springer, Berlin, 1929.
13. Neibergall, W, 'Absorption refrigerating machines', *Händbuch der Kältetechnik*, Vol. VII, Edited by R. Plank, Springer, Berlin, 1959.
14. Richter K H, 'Multistage absorption refrigeration systems', *Journal of Refrigeration*, Sept/Oct. 1962, pp. 105–111.
15. Rosenfeld L M and M C Karnaukh, *Kholodilnaya Tekhnika*, 1958, No. 1, pp. 37–42.
16. Scatchard G, *et al.*, 'Thermodynamic properties saturated liquid and vapour of ammonia-water mixture', *Refrigerating Engineering*, Vol. 53, No. 5, May, 1947, pp. 38.
17. Sellerio, U L, 'Impiego di idro carburi alognetied in particolarlara dell'r 22. nelle machine frigorifere and assorbimento a funzionamento continuous', *Estratto da Il CALORE Ressegna Technical Mensile Dell Associazione Naxionale per Il Controllo Della Combustion,* 1965, N-7.
18. Thieme A and L F Albright, 'Solubility of Refrigerants 11, 21, and 22 in Organic Solvents Containing a Nitrogen Atom and in Mixtures of Liquids,' *ASHRAE J.*, July 1961, p. 71.
20. Threlkeld J L and G F Zellhoefer, *Thermal Environmental Engineering*, Prentice Hall, 1962, pp. 96-104.

21. Zellhoefer G F, 'Solubility of halogenated hydrocarbon refrigerants in organic solvents', *Industrial and Engineering Chemistry*, Vol. 29, May 1937, pp. 548–51.

Revision Exercises

12.1 For a simple ammonia-absorption system, the following are given:

Condenser pressure	12.5 bar
Evaporator pressure	1.8 bar
Rich solution concentration	0.36
Poor solution concentration	0.25

Find on the basis of 1 ton refrigerating capacity:

(a) The temperature at the end of the evaporator if the vapour is assumed to be dry saturated.

(b) The mass rate of flow of the refrigerant absorbent mixture in the evaporator.

(c) The mass rate of circulation of the rich and poor solution.

(d) Generator and absorber temperatures.

(e) Heat added in the generator if the rich solution is assumed to be heated in the heat exchanger to 80°C.

12.2 (a) For an ammonia absorption system:

$p_k = 10$ bar, $t_h = 130°C$, $\xi_r = 0.38$

$t_{0_2} = -12°C$, $p_0 = 2.75$ bar

Find the temperature and concentration of the vapour leaving the generator-analyser. Assume the specific heats of the rich and poor solutions as same. Calculate the cooling energy ratio for a system in which there is no rectifier.

(b) Calculate the same if there is a rectifier which cools the vapours to 60°C and the evaporator pressure is raised to 3.4 bar.

12.3 In a lithium bromide-water system the condenser and evaporator temperatures are 35 and 8°C respectively. The generator temperature is 85°C and the absorber temperature is 30°C. Assume a pressure drop of 2.5 mm Hg between the generator and condenser and 1 mm Hg between the evaporator and absorber. Determine the heat rejection rates in the condenser and absorber per unit of refrigerating capacity. Also find the heat input to the generator, and COP.

12.4 Let an analyser be added in the system of Prob. 12.1, and do the calculations for (a) to (e).

12.5 For an NH_3 absorption cycle, we have

Condenser pressure	14 bar
Evaporator pressure	1.4 bar
Absorber temperature	36°C
Generator temperature	110°C

(a) Find out if such a cycle is possible or not.

(b) Change the evaporator pressure to 2.8 bar. Check if the cycle is possible now.

(c) Assume temperature leaving evaporator is $t_{0_2} = 5°C$. What is the refrigerating effect?

(d) Also find the COP.

(e) Add analyser in the system, and find COP.

(f) Add a dephlegmator which cools the vapour to 40°C. Determine the effect on COP.

12.6 (a) Find the temperature range through which liquid ammonia containing (a) 1% water, (b) 10% water will evaporate at a pressure of 2 bar.

(b) It is proposed to design a solar refrigeration NH_3–H_2O vapour absorption system operating under the following conditions:

Condenser pressure 16 bar

Evaporator pressure 2 bar

Absorber temperature 35°C

Hot water from flat plate solar collector can be obtained at a temperature of 80°C at the most. Examine the feasibility of operating the refrigeration system with this hot water.

(c) If not feasible, then suggest suitable measures to make it feasible.

12.7 (a) The operating conditions in a $H_2O – LiBr_2$ vapour absorption system are as follows:

Condenser and absorber temperatures 45°C

Evaporator temperature 5°C

Determine the minimum temperature of heat source required to produce any refrigeration at all.

(b) Saturated liquid mixture of 80% NH_3 plus 20% H_2O at 16 bar is throttled to 2 bar pressure. What is the temperature after throttling, and the temperature after complete evaporation of the mixture after throttling?

12.8 Introduce double-effect generation in Prob. 12.3, and find the new COP.

Chapter

13

Ejector-Compression System

13.1 WATER AS A REFRIGERANT

Ejector compression can be used as an alternative to centrifugal compression. In general, the thermodynamic characteristics of refrigerants employed in both are the same. The fact that so far only water has been used merely points out to the fact that steam is readily available as a source of motive power for driving the ejector. Some work has been done on the use of fluorocarbons in ejector compression system[1, 2, 3, 4]. The system using water as the refrigerant in a vapour-compression cycle in which compression is achieved by the principle of jet compression employing its own vapour as the motive vapour, is generally known as *water-vapour refrigeration* or *steam-ejector system.*

In a steam-ejector system, refrigeration is obtained by direct evaporation and subsequent self-cooling of water. This is possible because the latent heat of vaporization of water is high compared to its specific heat (530 kg of water can be cooled by 1°C by the evaporation of only 1 kg of itself). Also, there is no need to use a secondary refrigerant. Thus, irreversibility due to additional secondary refrigerant between primary refrigerant and cooled body is eliminated. Besides, water as a refrigerant possesses the most desirable physical and physiological properties.

Notwithstanding these, water has certain limitations also. These are:

(i) Enormous volume to be handled, e.g., at 8°C, 665 m^3/hr/TR are required, whereas for R 123 it is only 30.3 m^3/hr/TR. Centrifugal compressors could also be used with water vapour but they would be expensive. Ejector compression is the most economical to use with water as refrigerant.

(ii) High operating vacuum, e.g., at 8°C evaporator, the pressure is 0.01072 bar and the same at 35°C condenser, is 0.0595 bar.

(iii) High freezing point; it can be used for refrigeration above 0°C only, such as is required for air conditioning.

Accordingly, water is found suitable as a refrigerant with ejector compression, and not with centrifugal compressors for air conditioning.

Though the overall COP of the system is low, the steam-jet refrigeration system has proven its utility in applications such as air conditioning for comfort where steam is available and safety is the prime consideration. It is also suitable for applications where direct vaporization is used for the concentration or drying of heat-sensitive foods and chemicals. In such cases, the use of certain heat exchangers is eliminated.

13.2 STEAM EJECTOR SYSTEM

A simple water vapour refrigeration system and its thermodynamic cycle are shown in Figs 13.1 and 13.2. Water at 9 expands to 11 through throttle openings into the *flash chiller*. Due to vaporization (flashing) of a part of it, the remaining water is chilled to the required temperature T_0. The pressure in the flash chamber is maintained at p_0 which corresponds to the saturation pressure at T_0. The chilled water at 7 can be recirculated after taking up the load in the cooling coil of the air-conditioning equipment. Corresponding to the amount of water vaporized, the make-up water at 6 is added to the flash chamber through a throttle valve. Water vapour at 0 can be compressed to 4 by an ejector driven by steam (motive vapour) at 1. The processes taking place within the ejector are shown in Fig. 13.3. The compressed water vapour (steam) at 4 is finally condensed to 5 and is pumped to the boiler.

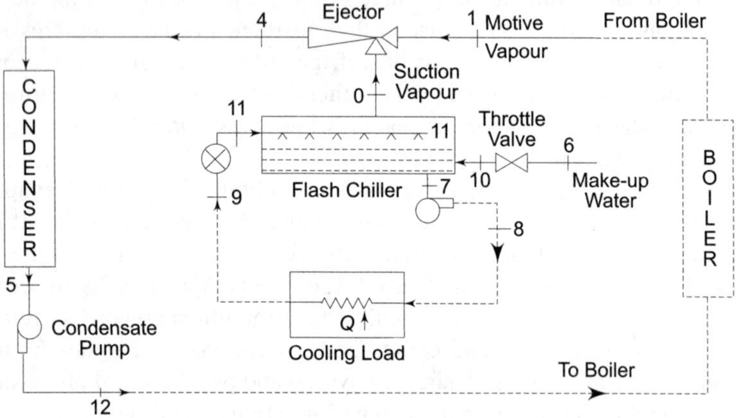

Fig. 13.1 Simple water-vapour refrigeration system

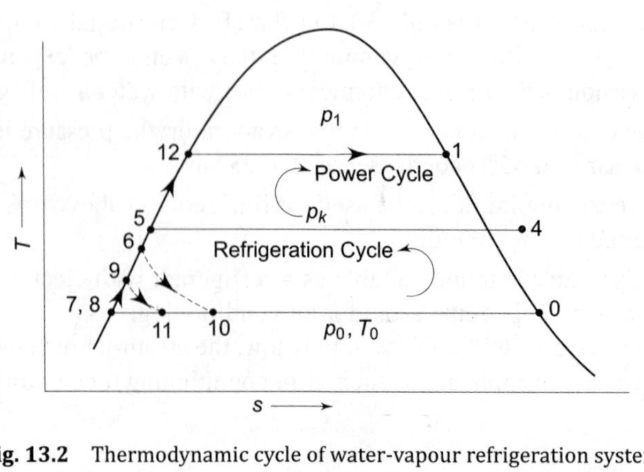

Fig. 13.2 Thermodynamic cycle of water-vapour refrigeration system

It is seen that steam ejector system is another kind of heat-operated refrigerating machine. The processes between boiler pressure p_1 and condenser pressure p_k form the heat engine part of the cycle. And the processes between evaporator pressure p_0 and p_k form the refrigeration part of the cycle.

Note *The control of capacity in the steam ejector system is possible either by throttling the steam inlet pressure, or by providing more than one nozzle operating in parallel, each driving a certain amount of vapour from the flash chamber.*

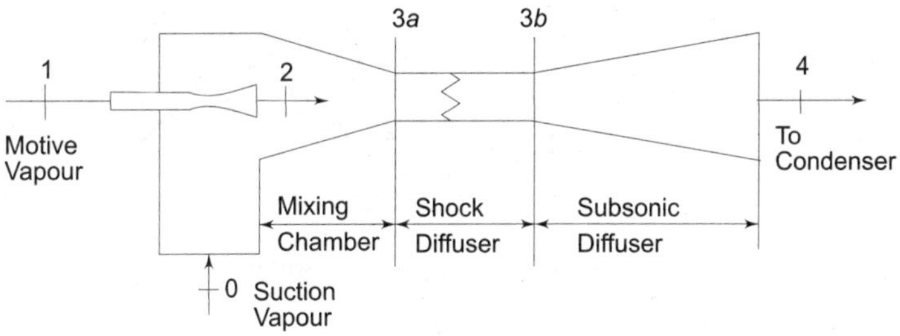

Fig. 13.3 Schematic diagram of steam ejector

13.3 THEORETICAL ANALYSIS OF THE STEAM EJECTOR

It can be seen that because of the extremely low pressure in the flash chamber, the suction vapour consists of a very large volume which makes even a centrifugal compressor uneconomical to use. The suction vapour, therefore, in water vapour refrigeration is invariably compressed by an ejector using steam as the motive vapour.

A schematic steam ejector is shown in Fig. 13.3 and pressure variation along its length in Fig. 13.4. The high-pressure motive vapour at 1 expands to a pressure slightly above the pressure of the suction vapour at 0. The high velocity jet at 2 entrains the suction vapour and mixing takes place at constant pressure. The state after mixing at 3*a* is still at a very high velocity (supersonic). The mixing chamber is followed by a constant area section where a normal shock may occur. After the shock at 3*b*, the fluid stream consisting of both the motive and suction vapours is compressed to the condenser pressure to state 4 in the diffuser section. The thermodynamic states at various points are shown on the temperature-entropy diagram in Fig. 13.5. The following analysis of the ejector is based on the work of Kalustian *et al.*[5]

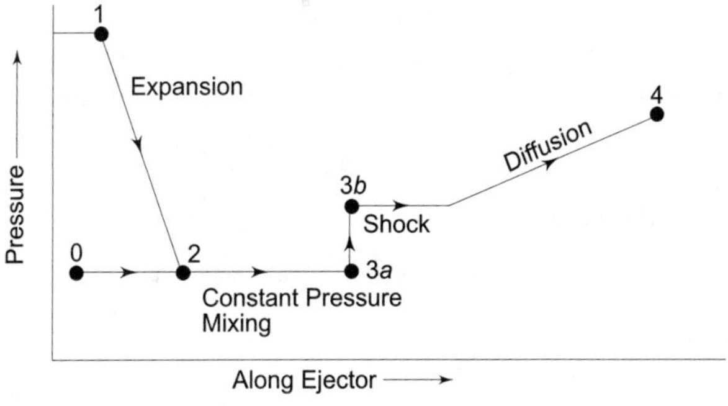

Fig. 13.4 Pressure variation along ejector

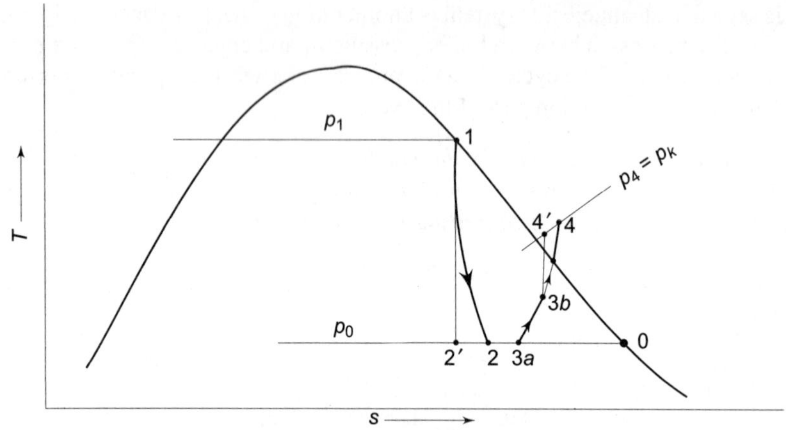

Fig. 13.5 Thermodynamic processes within ejector

Nozzle If η_n is the efficiency of the nozzle, then the jet velocity C_2 is given by

$$C_2 = \sqrt{2\eta_n(h_1 - h_2)} \tag{13.1}$$

Mixing Section Here m and C denote the mass flow rate and velocity. The continuity equation for the mixing section is

$$m_1 + m_0 = m_3 \tag{13.2}$$

The momentum equation is

$$(m_1 + m_0)\, C_{3a} - m_1 C_2 - m_0 C_0 = 0 \tag{13.3}$$

Since the process is at constant pressure, the force acting is zero (frictional force has been neglected in this equation). Considering velocity C_0 of the suction vapour as negligible, we have

$$m_3 C_{3a} = m_1 C_2 \tag{13.4}$$

This equation assumes no loss of momentum in the mixing process. Actual mixing process is accompanied with *entrainment loss* of mechanical energy as defined by Eq. (13.11).

Finally the energy equating for the mixing process is

$$m_1 h_1 + m_0 h_0 = m_3 \left(h_{3a} + \frac{C_{3a}^2}{2} \right) \tag{13.5}$$

Shock Diffuser Since the velocity at $3a$ is supersonic, it is necessary to design the ejector assuming the presence of a shock at this point. *A* constant-area section is also provided to stabilize the shock. We have the following equations for the shock:

Continuity: $$m_3 = \frac{C_{3a} A}{v_{3a}} = \frac{C_{3b} A}{v_{3b}} \tag{13.6}$$

where v is the specific volume

Momentum: $$(p_{3b} - p_{3a})A = (C_{3a} - C_{3b}) m_3 \tag{13.7}$$

Energy: $$h_{3a} + \frac{C_{3a}^2}{2} = h_{3b} + \frac{C_{3b}^2}{2} \tag{13.8}$$

Equation of state: Steam property tables may be used. (13.9)

The solution of the above four equations for the shock may be obtained by drawing two curves as shown in Fig. 13.6. The solution of Eqs (13.6), (13.8) and (13.9) is represented by a *Fanno line* and, the solution of Eqs (13.6), (13.7) and (13.9) is represented by a *Rayleigh line*. The intersection of the two lines, then, gives a solution of all the four equations. Consequently, we obtain state 3b after the shock. If the vapour before shock (at $3a$) is dry saturated or superheated, shock tables may be used and the value of the adiabatic index γ may be taken as 1.33 for steam.

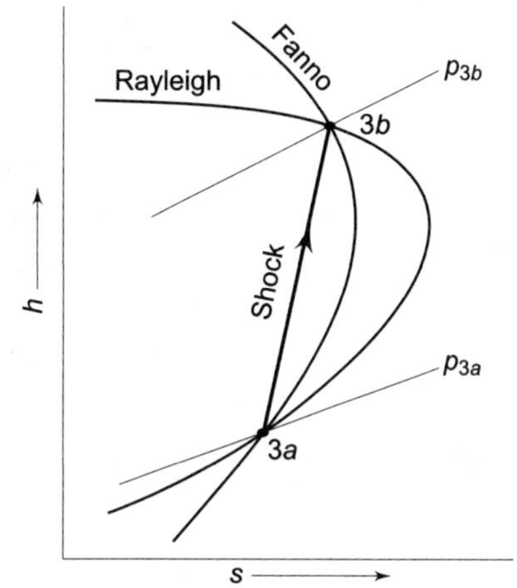

Fig. 13.6 Process in shock diffuser

Diffuser After the shock at $3b$, the velocity is subsonic. The kinetic energy at $3b$ is converted into enthalpy in the subsonic diffuser section. Neglecting the velocity at the end of the diffuser and taking η_d as the diffuser efficiency, we have

$$h_4 - h_{3a} = \frac{C_{3b}^2}{2} = \frac{h_4' - h_{3b}}{\eta_d} = \frac{\Delta h_{is}}{\eta_d} \tag{13.10}$$

where h_4' is the enthalpy after isentropic diffusion to p_4, and Δh_{is} is the isentropic enthalpy rise. The significant quantities in this analysis are the mass ratio of the motive vapour to suction vapour and the cross-sectional areas at various sections. The method of solution involves two iterations, one for the mass ratio and the other for the shock equations. The solution is checked by the pressure developed at 4 which should be equal to the required condenser pressure. Example (13.1) illustrates the calculation procedure.

When simplifications are made such that the shock is not considered (i.e., state $3b$ is the same as $3a$ or say 3) and the efficiency of the mixing process with entrainment efficiency η_e is considered, defined by a mechanical energy balance equation

$$\eta_e m_1 \frac{C_2^2}{2} = m_3 \frac{C_3^2}{2} \tag{13.11}$$

which replaces the momentum Eq. (13.3). Then the solution of Eqs (13.1), (13.11), (13.5) and (13.10) gives for the mass ratio

$$m = \frac{m_1}{m_0} = \frac{h_4' - h_3}{(h_1 - h_3)\eta_n \eta_e \eta_d - (h_4' - h_3)} \tag{13.12}$$

or defining the overall ejector efficiency by

$\eta = \eta_n \eta_e \eta_d$, we have

$$m = \frac{h_4' - h_3}{(h_1 - h_3)\eta - (h_4' - h_3)} \tag{13.13}$$

For the case of an ideal ejector, $\eta = 1$, so that

$$m = \frac{h_4 - h_3}{h_1 - h_4} \tag{13.14}$$

Thus, we find that to keep the mass ratio low, and to maintain the COP high, the enthalpy h_1 and hence the temperature and pressure of the motive vapour should be as high as possible. We note the following:

Refrigerating effect, $q_0 = h_0 - h_9$

Heat supplied, $q_h = m\,(h_1 - h_5)$

$$\text{COP} = \frac{q_0}{q_h} = \frac{h_0 - h_9}{m(h_1 - h_5)}$$

Example 13.1 *In a steam-jet refrigeration system, the motive vapour is saturated at 150.3°C, and the chilled water temperature is 6°C. The mass ratio of the motive vapour to refrigerant vapour is 2.5. Find the saturated discharge temperature of the ejector. Assume the nozzle, entrainment and diffuser efficiencies as 0.85, 0.65 and 0.8 respectively.*

Solution From the steam tables

$p_1 = 0.048$ bar at 150.3°C $\qquad p_0 = 0.00935$ bar at 6°C

$h_1 = 2745.7$ kJ/kg $\qquad s_1 = 6.833$ kJ/kg K

At 6°C,

$s_f = 0.091$ kJ/kg K $\qquad s_g = 9.001$ kJ/kg K

$h_f = 25.2$ kJ/kg K $\qquad h_g = 2512.6$ kJ/kg $= h_0$

Nozzle After isentropic expansion, the dryness fraction is

$$x_2' = \frac{6.833 - 0.091}{9.001 - 0.091} = 0.7567$$

Enthalpy after expansion

$$h_2' = 25.2 + 0.7567\,(2512.6 - 25.2) = 1907.4 \text{ kJ/kg}$$

Velocity of motive vapour leaving nozzle

$$C_2 = \sqrt{2\eta_n(h_1 - h_2)} = \sqrt{(0.85)(2)(2745.7 - 1907.4)10^3} = 1193.8 \text{ m/s}$$

Mixing Section The momentum equation gives

$$\eta_e m C_2^2 = (m + 1)C_{3a}^2$$

$$0.65(2.5)(1193.8^2) = (2.5 + 1)C_{3a}^2$$

$$\Rightarrow \qquad C_{3a} = 813.4 \text{ m/s}$$

The energy equation gives

$$mh_1 + h_0 = (m + 1)\left(h_{3a} + \frac{C_{3a}^2}{2}\right)$$

$$(2.5)\,(2745.7) + 2512.6 = (3.5)\left(h_{3a} + \frac{813.4^2}{2 \times 10^3}\right)$$

whence

$$h_{3a} = 2348.3 \text{ kJ/kg}$$

Dryness fraction at $3a$

$$x_{3a} = \frac{h_{3a} - h_f}{h_{fg}} = \frac{2348.3 - 29.2}{2487.4} = 0.932$$

The state $3a$ corresponds to a dryness fraction of 0.932 at 6°C saturation temperature. Its specific volume can be found

$$v_{3a} = xv_g + (1 - x)v_f = 0.932(137.8) + (1 - 0.932)\,(0.001) = 128.5 \text{ m}^3/\text{kg}$$

Shock Diffuser The continuity, momentum and energy equations are

$$\frac{\dot{m}_3}{A} = \frac{C_{3b}}{v_{3b}} = \frac{C_{3a}}{v_{3a}} = \frac{813.4}{128.5} = 6.33 \qquad \text{(I)}$$

$$p_{3b} = p_{3a} + (C_{3a} - C_{3b})\frac{\dot{m}_3}{A} = 0.00935 + \frac{6.33}{10^5}(813.4 - C_{3b}) \qquad \text{(II)}$$

$$h_{3b} = h_{3a} + \frac{C_{3a}^2}{2} - \frac{C_{3b}^2}{2} = 2348.3 + \frac{813.4^2}{2\times 10^3} - \frac{C_{3b}^2}{2\times 10^3} \qquad \text{(III)}$$

Equations (I), (II) and (III) are to be solved in conjunction with the steam tables for the four variables, p_{3b}, v_{3b}, h_{3b} and C_{3b}. The following iterative procedure may be adopted.

(i) Assume C_{3b}.
(ii) Calculate v_{3b} from Eq. (I).
(iii) Calculate p_{3b} from Eq. (II).
(iv) Find h_{3b} from the steam tables for the calculated values of p_{3b} and v_{3b}.
(v) Again calculate h_{3b} from Eq. (III).
(vi) Compare the two values and if necessary, repeat the procedure. The solution obtained is

$C_{3b} = 222$ m/s $\qquad v_{3b} = 35.07$ m³/s

$p_{3b} = 0.0468$ bar $\qquad h_{3b} = 2654.5$ kJ/kg

At 0.0468 bar pressure, the saturation properties are as follows:

$t_g = 31.7$°C $\qquad v_g = 30.056 \text{ m}^3/\text{kg} < 35.07 \text{ m}^3/\text{kg}$

$h_g = 2559.4$ kJ/klg $\qquad s_g = 8.42$ kJ/kg K

Hence, the vapour after the shock is superheated. The ratio of temperatures is

$$\frac{T_{3b}}{T_g} = \frac{v_{3b}}{v_g} = \frac{35.07}{30.056} = 1.167$$

Temperature after shock

$$T_{3b} = (273 + 31.7)\,(1.167) = 365.5 \text{ K or } 82.5°\text{C}$$

Degree of superheat

$$\Delta t = 82.5 - 31.7 = 50.8°\text{C}$$

Enthalpy after shock

$$h_{3b} = h_g + C_p\,\Delta t = 2559.4 + 1.885\,(50.8) = 2655.2 \text{ kJ/kg}$$

Note *The above value of enthalpy agrees well with the value of 2654.5 kJ/kg calculated from Eq. (III). Hence the assumption of velocity is correct.*

Entropy after shock

$$s_{3b} = s_g + C_p \ln\frac{T_{3b}}{T_g} = 8.42 + 1.885 \ln 1.167 = 8.711 \text{ kJ/kg.K}$$

Diffuser Enthalpy rise after actual diffusion to 4 (pressure $p_4 = p_k$)

$$\Delta h = h_4 - h_{3b} = \frac{C_{3b}^2}{2} = \left(\frac{222}{2}\right)\frac{1}{10^3} = 24.6 \text{ kJ/kg}$$

Enthalpy rise after isentropic diffusion to 4′ (same pressure)

$$\Delta h_{\text{is}} = h_4' - h_{3b} = \eta_d \, \Delta h = (0.8)\,(24.6) = 19.7 \text{ kJ/kg}$$

Enthalpy after isentropic diffusion to 4′

$$h_4' = h_{3b} + \Delta h_{\text{is}} = 2655.2 + 19.7 = 2674.9 \text{ kJ/kg}$$

Entropy after isentropic diffusion

$$s_4' = s_{3b} = 8.711 \text{ kJ/kg.K}$$

For the state after isentropic diffusion, we follow an iterative procedure. For the purpose we assume the saturated discharge temperature and check for entropy after diffusion. The calculations are given in Table 13.1.

Table 13.1 Iterations for saturation temperature and pressure after diffusion

Sat. Discharge/Condensing Temp. $t_k \rightarrow$	*32°C*	*33°C*	*33.8°C*	*34°C*
Saturation properties at t_k				
p, bar	0.0475	0.05035	—	0.0532
h_g, kJ/kg	2560.0	2561.8	—	2563.6
s_g, kJ/kg K	8.414	8.394	—	8.374
Superheat enthalpy				
$\Delta h = h_4' - h_g = 2674.9 - h_g$	114.9	113.1	—	111.3
Degree of superheat				
$\Delta T = \dfrac{\Delta h}{C_p} = \dfrac{\Delta h}{1.885}$	60.95	60.0	—	59.05
Superheat entropy				
$\Delta s = 1.885 \ln \dfrac{T_g + \Delta T}{T_g}$	0.3434	0.3375	—	0.3316
Entropy				
$s = s_g + \Delta s$	8.757	8.7315	8.711	8.7056

Since the entropy after isentropic diffusion s_4' is 8.711 kJ/kg.K, we find by interpolation that the saturated discharge temperature and hence the required condensing temperature is 33.8°C. This corresponds to a condensing pressure of 40.1 mm Hg (5.57 kN/m^2).

Note *The condensing temperature attained is too low for a cooling water temperature of 30°C. Mass ratio of motive vapour has to be increased.*

References

1. Arora C P, 'Prospects of using refrigerant ejectors', *J. of Refrigeration*, Vol. 7, No. 6, Nov./Dec. 1964, pp. 117–121.
2. Arora C P, 'Other refrigerants in ejector compression system', *Air Conditioning and Refrigeration in India*, Dec. 1966.
3. Badylkes I S, I.I.R Bulletin, Annexe 1958, *J. of Refrigeration*, Vol. 1, No. 7, Nov/Dec. 1958, pp. 168, 169 and 176.
4. Cavallini A, G Lovison and G Trappanese, 'Experimental research on a fluorinated hydrocarbon jet refrigeration plant', *Proc. XII International Congress of Refrigeration*, 1967.
5. Kalustian P, 'Analysis of the ejector cycle', *Refrigerating Engineering*, Vol. 38, No. 4, Oct. 1934, pp. 188–193, 208.
6. Martinovsky V S, 'Use of waste heat for refrigeration, *Kholodilnaya Tekhnika*, Vol. 30, No. 1, Jan./March, 1953, p. 60.

Revision Exercises

13.1 (a) In a steam jet refrigeration system, the evaporator temperature is 5°C and the condensing temperature is 35°C. The motive vapour is dry saturated steam at 120°C. Assuming ideal processes, find the mass ratio of the motive vapour to the refrigerant vapour.

(b) Find the same if the nozzle, entrainment and diffuser efficiencies are 0.85, 0.65 and 0.8 respectively. Assume no shocks.

13.2 Repeat the calculations in Problem 13.1 (b) above assuming a normal shock in the constant area section.

13.3 Find the dimensions of the various sections of the ejector in Problem 13.2 for a 5,000-ton refrigerating capacity.

Chapter

14

Properties of Moist Air

14.1 BRIEF HISTORY OF AIR CONDITIONING

The art of air conditioning developed only gradually from the predecessor arts of cooling, cleaning, heating and ventilating.

Leonardo da Vinci had built a ventilating fan by the end of the 15th century. Later, Boyle in 1659 and Dalton in 1800 discovered the laws that are very well known. The first text on heating and ventilating was written by Robertson Buchanan, a Glasgow civil engineer, in 1815. Fans, boilers and radiators had been invented by the middle of the 19th century. Refrigeration technology was soon to follow. In 1853, Professor Alexander Twining of New Haven produced 725 kg of ice a day using a double-acting vacuum and compression pump, employing the 1834 invention of the vapour compression cycle by Jacob Perkins. He used sulphuric ether as the refrigerant. While quick developments took place in the manufacture of ice, a few machines appeared which chilled air by blowing it over brine or direct-expansion pipe coils.

Towards the latter half of the 19th century, the developments in the art of humidifying air went along with the progress of textile industry in England. Devices for measuring pressure, temperature, humidity and flow of air were perfected during this period. To bring the various groups of engineers together, societies such as the American Society of Refrigerating Engineers (ASRE) in 1904 were formed.

It is worth mentioning here the name of A.R. Wolff who designed air-conditioning systems for as many as hundred buildings during his life-time. But it is W.H. Carrier (1876–1950) who is known as the 'Father of Air Conditioning'. While Working with Buffalo Forge Co., he developed formulae for optimizing the application of forced-draft fans, developed ratings of pipe-coil heaters and set up a research laboratory. He engineered and installed the first year-round air-conditioning system, providing for the four major functions of heating, cooling, humidifying and dehumidifying. He made use of air washers for controlling the dew point of air by heating or chilling recirculated water. In 1911, Carrier presented his remarkable paper 'Rational Psychrometric Formulae' in an ASME meeting. The paper related the dry bulb, wet bulb and dew point temperatures of air with its sensible, latent and total heats, and presented a theory of adiabatic saturation.These formulae, together with the accompanying psychrometric chart, laid the subject of air conditioning on a firm thermodynamic basis. Carrier also employed the centrifugal compressor for refrigeration in 1922.

After Carrier, there was a tremendous increase in the use of air conditioning in cotton, silk, rayon, tobacco, paper, pharmaceuticals, candy and printing industries. As far as air conditioning for comfort is concerned, it got off the ground in motion-picture theatres in 1920 in Chicago employing CO_2 machines and in 1922 in Los Angeles employing NH_3 compressors. In the last 30 years, the industry has grown in leaps and bounds all over the world, using various new refrigerants.

In the following chapters, attention will henceforth be focussed on the art and science of air conditioning which is the greatest single application of refrigeration, in addition to that of heating and ventilation. For this purpose it is necessary to study the properties of the working substance in air conditioning, viz., moist air.

14.2 WORKING SUBSTANCE IN AIR CONDITIONING

An important thing for the student of air conditioning is to appreciate that the working substance under study, viz., *moist air*, is a mixture of two gases. One of these is *dry air* which itself is a mixture of a number of gases and the other is *water vapour* which may exist in a saturated or superheated state.

One might ask whether moist air can be considered as a *pure substance*. But a pure substance is *homogeneous and invariable in chemical composition*. Thus, a homogeneous mixture of gases is a pure substance until its components do not change in phase. Dry air is a good example of such a kind of pure substance. Water vapour is certainly a pure substance. But moist air is not a pure substance in any process in which condensation or evaporation of moisture occurs. In such a case, regular charts have to be developed to describe the thermodynamic properties of the mixture under different conditions and compositions.

It is, thus, seen that moist air consists of two parts: one, comprising dry air, considered as the *fixed part*, and the other, solely of water vapour, considered as the *variable part*.

The dry air part is a mixture of a number of permanent gases with approximate compositions as given in Table 14.1.

Table 14.1 Composition of dry part in atmospheric air

Component	*Molecular Mass*	*Part by Volume*	*Part by Mass*
N_2	28.02	0.7803	0.7547
O_2	32.0	0.2099	0.2319
Ar	39.91	0.0094	0.0129
CO_2	44.0	0.0003	0.0005
H_2	2.02	0.0001	0.0000

Both dry air and water vapour can be considered as perfect gases since both exist in the atmosphere at low pressures. Hence, perfect gas laws can be applied to them individually. In addition, *Gibbs-Dalton laws* for non-reactive mixtures of gases can be applied to the dry air part only to obtain its properties as a single pure substance, before establishing the properties of moist air.

14.2.1 Dalton's Law of Partial Pressures

Consider a homogeneous mixture of non-reacting ideal gases 1, 2, etc., at temperature T, pressure p and occupying volume V as shown in Fig. 14.1 (a). Let the number of moles of individual gases be $n_1, n_2, \cdots$ etc., and their respective masses be $m_1, m_2, \cdots$ etc. Then we have for total number of moles n and total mass m

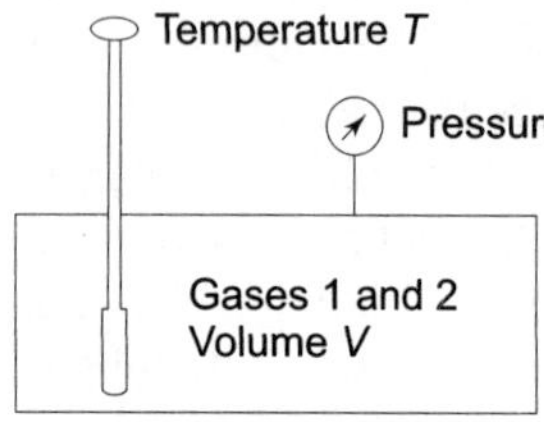

Fig. 14.1(a) Gas mixture

$$n = n_1 + n_2 + \cdots = \Sigma\, n_i \tag{14.1}$$

$$m = m_1 + m_2 + \cdots = \Sigma\, m_i \tag{14.2}$$

where i is the number of each gas.

In the Dalton's model, each gas is conceived of as existing *separately* at the temperature T and total volume V of the mixture as shown in Fig. 14.1 (b).

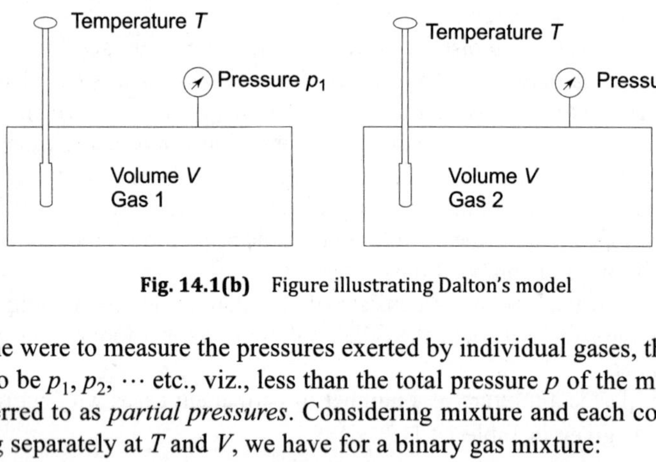

Fig. 14.1(b) Figure illustrating Dalton's model

If one were to measure the pressures exerted by individual gases, they would be found to be $p_1, p_2, \cdots$ etc., viz., less than the total pressure p of the mixture. These are referred to as *partial pressures*. Considering mixture and each component gas existing separately at T and V, we have for a binary gas mixture:

Mixture $\quad pV = n\,\bar{R}T \qquad n = \dfrac{pV}{\bar{R}T}$

Components $\quad p_1 V = n_1\,\bar{R}T \qquad n_1 = \dfrac{p_1 V}{\bar{R}T}$

$\quad p_2 V = n_2\,\bar{R}T \qquad n_2 = \dfrac{p_2 V}{\bar{R}T}$

Substitution in $n_1 + n_2 = n$ gives

$$p = p_1 + p_2 \text{ or } p = \Sigma_i\, p_i \tag{14.3}$$

Thus, for a mixture of ideal gases, the total pressure p is equal to the sum of the partial pressures. This is known as the *Dalton's law* of *partial pressures*.

14.2.2 Amagat Law of Partial Volumes

In the Amagat model, each component gas is considered as existing separately at the total pressure p and temperature T of the mixture as shown in Fig. 14.1 (c).

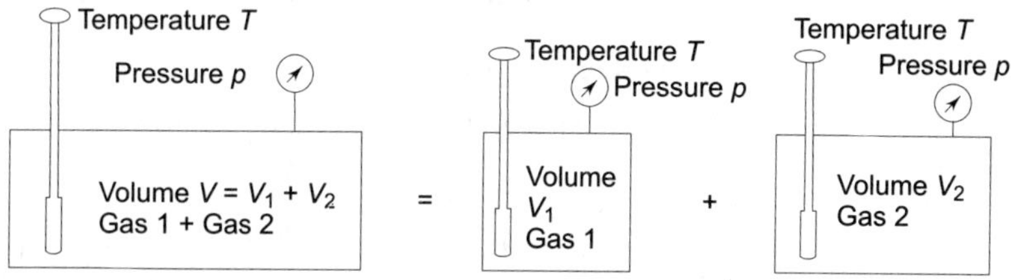

Fig. 14.1(c) Figure illustrating Amagat model

Let the volumes of individual gases under these conditions be V_1, V_2, ⋯ etc. These are referred to as *partial volumes*. Again, applying the ideal gas equation of state to mixture and components, we have:

Mixture $\quad pV = n\bar{R}T \qquad n = \dfrac{pV}{\bar{R}T}$

Components $\quad pV_1 = n_1\bar{R}T \qquad n_1 = \dfrac{pV_1}{\bar{R}T}$

$$pV_2 = n_2\bar{R}T \qquad n_2 = \frac{pV_2}{\bar{R}T}$$

Substitution in $n_1 + n_2 = n$ gives

$$V = V_1 + V_2 \text{ or } V = \Sigma_i V_i \tag{14.4}$$

Thus, we see that the total volume is equal to the sum of the partial volumes. This is known as the *Amagat law of partial volumes*. The ratios V_1/V, V_2/V, ⋯ etc., are referred to as volume fractions.

14.2.3 Mole Fractions of Component Gases

It is seen from the above section that

$$\frac{p_1}{p} = \frac{V_1}{V} = \frac{n_1}{n} = y_1 \tag{14.5}$$

Thus, the ratio of partial pressure to total pressure, and volume fraction are equal to the *mole fraction* y_i of the gas. It also shows that both Dalton's law and Amagat law are equivalent.

14.2.4 Molecular Mass of Mixture

Since $\quad m = m_1 + m_2$, and $m = Mn$, $m_1 = M_1 n_1$, $m_2 = M_2 n_2$, we have

$$Mn = M_1 n_1 + m_2 n_2$$

We have

$$M = y_1 M_1 + y_2 M_2 \qquad \text{or} \qquad M = \Sigma_i y_i M_i \tag{14.6}$$

where M represents the molecular mass of the mixture. Note that $n_1 = m_1/M_1$, $n_2 = m_2/M_2$, etc. Similarly, for the mixture, $n = m/M$.

14.2.5 Gibbs' Theorem

Gibbs' Theorem further enunciates that the internal energy of a mixture is equal to the sum of internal energies of the individual components, taken each at the temperature and volume of the mixture. Thus, we have for the internal energy of the mixture.

$$mu = m_1u_1 + m_2u_2 \tag{14.7}$$

It can also be shown that the enthalpy and specific heat of the mixture can, similarly, be written as

$$mh = m_1h_1 + m_2h_2 \tag{14.8}$$

$$mC = m_1C_1 + m_2C_2 \tag{14.9}$$

14.2.6 Molecular Masses and Gas Constants for Dry Air and Water Vapour

From the respective mole fractions and molecular masses of component gases, the molecular mass of the dry air part may be computed. For the purpose, it may be observed from Eq. (14.16) that a part by volume represents the mole fraction. Thus using the values for mole fractions from Table 14.1, we have

$$\begin{aligned} M_a &= \Sigma\, M y \qquad (14.10)\\ &= 28.02\ (0.7803) + 32\ (0.2099)\\ &\quad + 39.91\ (0.0094) + 44\ (0.0003) + 2.02\ (0.0001)\\ &= 28.966 \end{aligned}$$

where subscript a denotes dry air. Knowing that the value of the universal gas constant is 8.3143 kJ/kg mole K, the gas constants for the two parts of moist air are as follows:

Dry air $\quad M_a = 28.966$

$$R_a = \frac{8.3143}{28.966} = 0.2871 \text{ kJ/kg.K}$$

Water vapour $\quad M_v = 18.016$

$$R_v = \frac{8.3143}{18.016} = 0.461 \text{ kJ/kg.K}$$

where subscript v refers to water vapour.

Example 14.1 *One cubic metre of H_2 at 1 bar and 25°C is mixed with one cubic metre of N_2 at 1 bar and 25°C. For the mixture at the same conditions, find:*

(a) Mole fractions of components. *(b) Partial pressures of components.*
(c) Mass fractions of components. *(d) Molecular weight of the mixture.*
(e) Gas constant of the mixture. *(f) Volume of the mixture.*

Solution Molecular weights

$$M_1 = M_{H_2} = 4.003, \text{ and } M_2 = M_{N_2} = 28.02$$

Gas constants

$$R_1 = \frac{8.3143}{4.003} = 2.077 \text{ kJ/kg.K}$$

$$R_2 = \frac{8.3143}{28.02} = 0.2967 \text{ kJ/kg.K}$$

Masses

$$m_1 = \frac{p_1V_1}{R_1T_1} = \frac{(1\times10^5)(1)}{(2.077\times10^3)(298)} = 0.1616 \text{ kg}$$

$$m_2 = \frac{p_2V_2}{R_2T_2} = \frac{(1\times10^5)(1)}{(0.2967\times10^3)(298)} = 1.131 \text{ kg}$$

$$m = m_1 + m_2 = 0.1616 + 1.131 = 1.2926 \text{ kg}$$

Number of moles

$$n_1 = \frac{m_1}{M_1} = \frac{0.1616}{4.003} = 0.0404$$

$$n_2 = \frac{m_2}{M_2} = \frac{1.131}{28.02} = 0.0403$$

$$n = n_1 + n_2 = 0.0807$$

(a) Mole fractions

$$y_1 = \frac{n_1}{n} = \frac{0.0404}{0.0807} = 0.5$$

$$y_2 = \frac{n_2}{n} = \frac{0.0403}{0.0807} = 0.5$$

(b) Partial pressures

$$p_1 = y_1 p = 0.5\ (1) = 0.5 \text{ bar}$$

$$p_2 = y_2 p = 0.5\ (1)\ \ 0.5 \text{ bar}$$

(c) Mass fractions

$$\frac{m_1}{m} = \frac{0.1616}{1.2926} = 0.125$$

$$\frac{m_2}{m} = \frac{1.131}{1.2926} = 0.875$$

(d) Molecular weight of mixture

$$M = y_1 M_1 + y_2 M_2 = 0.5(4.003) + 0.5(28.02) = 16.01$$

(e) Gas constant of mixture

$$R = \bar{R}/M = \frac{8.3143}{16.01} = 0.5193 \text{ kJ/kg.K}$$

(f) Volume of mixture

$$V = \frac{mRT}{p} = \frac{1.2926(0.5193\times10^3)(298)}{1\times10^5} = 2 \text{ m}^3$$

Also, from Eq. (14.4)

$$V = V_1 + V_2 = 1 + 1 = 2 \text{ m}^3$$

14.3 PSYCHROMETRIC PROPERTIES

Dry air and water vapour form a binary mixture. A mixture of two substances requires three properties to completely define its thermodynamic state, unlike a pure substance which requires only two. One of the three properties can be the composition. The properties of moist air are called *psychrometric properties* and the subject which deals with the behaviour of moist air is known as *psychrometry*.

Water vapour is present in the atmosphere at a very low partial pressure. At this low pressure and atmospheric temperature, the water vapour behaves as a perfect gas. The partial pressure of dry air is also below one atmosphere which may also be considered to behave very much as a perfect gas. The Gibbs-Dalton laws of perfect gas mixtures can, therefore, be applied to the case of moist air.

In air-conditioning practice, all calculations are based on the dry air part since the water vapour part is continuously variable. For defining and calculating the relevant psychrometric properties, we may consider a certain volume V of moist air at pressure p and temperature T, containing m_a kg of dry air and m_v kg of water vapour as shown in Fig. 14.2. The actual temperature t of moist air is called the *dry bulb temperature* (DBT). The total pressure p which is equal to the *barometric pressure* is constant. The other relevant properties will now be discussed.

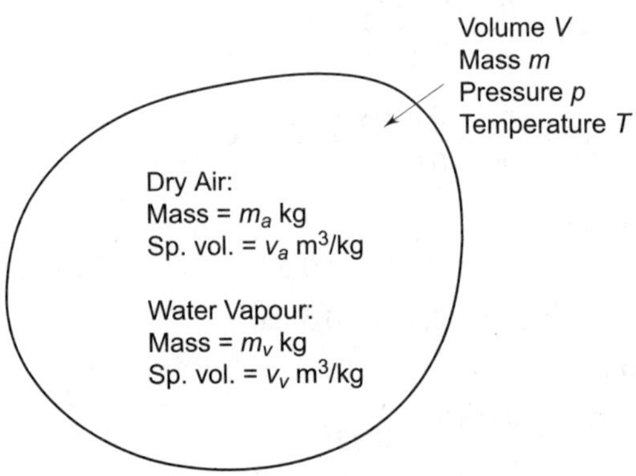

Fig. 14.2 A mixture of dry air and water vapour (moist air)

14.3.1 Specific Humidity or Humidity Ratio

Specific or *absolute humidity* or *humidity ratio* or *moisture content* as it is variously called denoted by the symbol ω is defined as the ratio of the mass of water vapour (w.v.) to the mass of dry air (d.a.) in a given volume of the mixture. Thus

$$\omega = \frac{m_v}{m_a} = \frac{V/v_v}{V/v_a} = \frac{v_a}{v_v} \tag{14.11}$$

where the subscripts a and v refer to dry air and water vapour respectively.

Now $\quad p_a v_a = \frac{\bar{R}}{M_a} T \qquad p_a V = m_a \frac{\bar{R}}{M_a} T \qquad (14.12)$

$$p_v v_v = \frac{\bar{R}}{M_v} T \qquad p_v V = m_v \frac{\bar{R}}{M_v} T \tag{14.13}$$

Substituting for m_v and m_a from these expressions in Eq. (14.11), we obtain

$$\omega = \frac{M_v p_v}{M_a p_a} = \frac{18.016}{28.966} \frac{p_v}{p_a} = 0.622 \frac{p_v}{p_a} \tag{14.14}$$

The units of ω are kg of water vapour per kg of dry air. If multiplied by 1000, it can be expressed as

$$\omega = 622 \frac{p_v}{p_a} \text{ g w.v./kg d.a.} \tag{14.15}$$

Also, since p denotes the actual total atmospheric pressure, then from Dalton's law

$$p = p_a + p_v \tag{14.16}$$

so that

$$\omega = 0.622 \frac{p_v}{p - p_v} \tag{14.17}$$

Considering that the total atmospheric pressure remains constant at a particular locality, we can see that

$$\omega = f(p_v)$$

viz., the specific humidity is a function of the partial pressure of water vapour only. Accordingly, if there is no change in specific humidity or the moisture content of air, the partial pressure of water vapour also does not change.

It may be noted that since p_v is very small and p, which is the barometric pressure, is constant, the denominator in Eqs (14.14) and (14.17) remains more or less constant, i.e.,

$$p_a = p - p_v \approx p$$

Hence ω is *approximately* a linear function of p_v.

The concept of specific humidity is that if we take 1 kg of dry air, viz.,

$$m_a = 1 \text{ kg}$$

then the mass of water vapour associated with this dry air, in the same volume is

$$m_v = \omega \text{ kg}$$

so that the total mass of this volume of moist air is

$$m = (1 + \omega) \text{ kg}$$

Thus specific humidity is *not* a mass fraction of water vapour which would have been $\omega/(1 + \omega)$, but a ratio of the mass of water vapour to that of dry air in a certain volume of the mixture.

14.3.2 Dew Point Temperature

Figure 14.3 shows the normal thermodynamic state 1 of water vapour in moist air. The water vapour existing at temperature T of the mixture and partial pressure p_v of the vapour in the mixture is normally in a superheated state. Moist air containing moisture in such a state is considered as *unsaturated air*.

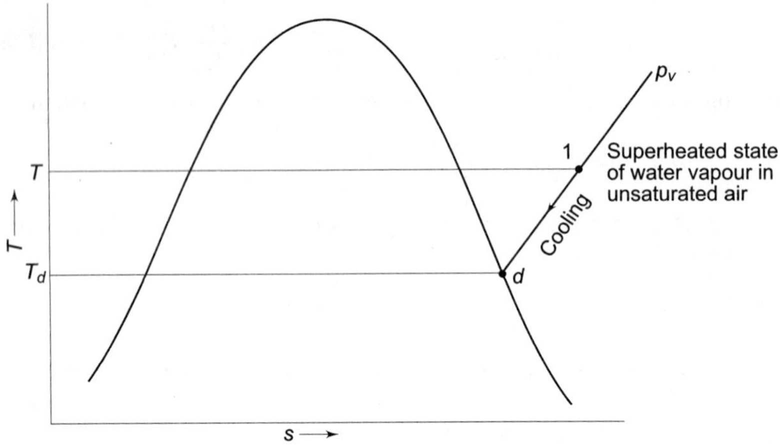

Fig. 14.3 Thermodynamic state of water vapour in moist air

If a sample of such unsaturated moist air containing superheated water vapour is cooled (at constant pressure), the mixture will eventually reach the saturation temperature t_d of water vapour corresponding to its partial pressure p_v, at which point the first drop of dew will be formed, i.e., the water vapour in the mixture will start condensing. This temperature t_d is called the *dew point temperature* (DPT). It is, therefore, the temperature to which moist air must be cooled at constant pressure before condensation of moisture takes place.

Moisture can be removed from humid air by bringing the air in contact with a *cold surface* or *cooling coil* whose temperature is below its dew point temperature. During the process of cooling, the partial pressure p_v of water vapour and specific humidity ω remain constant until the vapour starts condensing.

It is seen that the dew point temperature can be found by knowing, from the steam tables, the saturation temperature t_d at the partial pressure p_v of the water vapour.

Example 14.2 *In a dew point apparatus a metal beaker is cooled by gradually adding ice water to the water initially at room temperature. The moisture from the room air begins to condense on the beaker when its temperature is 12.8°C. If the room temperature is 21°C and the barometric pressure is 1.01325 bar, find the partial pressure of water vapour in the room air and parts by mass of water vapour in the room air.*

Solution Partial pressure of water vapour at DPT of 12.8°C

$$p_v = 1.479 \text{ kN/m}^2$$

Partial pressure of dry air

$$p_a = 101325 - 1479 = 99846 \text{ N/m}^2$$

Specific humidity

$$\omega = \frac{m_v}{m_a} = 0.622 \frac{p_v}{p_a} = 0.622 \times \frac{1479}{99846} = 0.009214 \frac{\text{kg w.v.}}{\text{kg d.a.}}$$

Parts by mass of water vapour

$$\frac{m_v}{m} = \frac{\omega}{1+\omega} = \frac{0.009214}{1.009214} = 0.00913 \frac{\text{kg w.v.}}{\text{kg mixture}}$$

14.3.3 Degree of Saturation

Figure 14.4 again shows the superheated thermodynamic state 1 of water vapour in unsaturated moist air representing the control volume V in Fig. 14.2. The water vapour exists at the dry bulb temperature T of the mixture and partial pressure p_v.

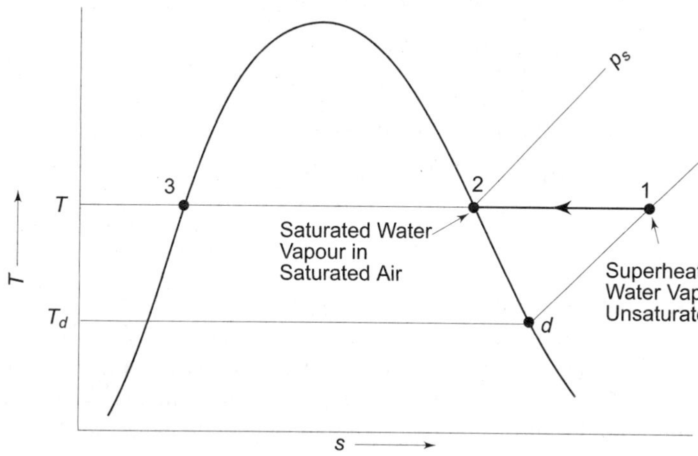

Fig. 14.4 An imaginary isothermal process representing change of state of water vapour in unsaturated air to that of saturated air at the same temperature

Consider now that more water vapour is added in this control volume V at temperature T itself. The partial pressure p_v will go on increasing with the addition of water vapour until it reaches a value p_s corresponding to state 2 in Fig. 14.4, after which it cannot increase further as p_s is the saturation pressure or maximum possible pressure of water at temperature T. The thermodynamic state of water vapour is now saturated at point 2. The air containing moisture in such a state is called *saturated air*. In this state the air is holding the maximum amount of water vapour (the specific humidity being ω_s, corresponding to the partial pressure p_s) at temperature T of the mixture. The maximum possible specific humidity, ω_s at temperature T is thus

$$\omega_s = 0.622 \frac{p_s}{p - p_s} \tag{14.18}$$

The ratio of the actual specific humidity ω to the specific humidity ω_s of saturated air at temperature T is termed as the *degree of saturation* denoted by the symbol μ. Thus

$$\mu = \frac{\omega}{\omega_s} = \frac{p_v}{p_s}\left[\frac{1 - p_s/p}{1 - p_v/p}\right] \tag{14.19}$$

We thus see that the degree of saturation is a measure of the capacity of air to absorb moisture.

14.3.4 Relative Humidity

Relative humidity denoted by the symbol ϕ or RH is defined as the ratio of the mass of water vapour m_v in a certain volume of moist air at a given temperature to the mass of water vapour m_{v_s} in the same volume of saturated air at the same temperature. Thus, referring to Fig. 14.4 again, if v_v and v_s are the specific volumes of water vapour in the actual moist air and saturated air respectively at temperature T and in volume V, viz., at points 1 and 2 respectively, we see that

$$\phi = \frac{m_v}{m_{v_s}} = \frac{p_v V/\bar{R}T}{p_s V/\bar{R}T} = \frac{p_v}{p_s} \tag{14.20}$$

Also,

$$\phi = \frac{V/v_v}{V/v_s} = \frac{v_s}{v_v} \tag{14.21}$$

Using the perfect-gas relationship between points 1 and 2, viz.,

$$p_1 v_1 = p_2 v_2 \text{ or } p_v v_v = p_s v_s \tag{14.22}$$

we have

$$\phi = \frac{p_v}{p_s} = \frac{v_s}{v_v} \tag{14.23}$$

Thus relative humidity turns out to be the ratio of partial pressures of water vapour in a certain unsaturated moist air at a given temperature T to the saturation pressure of water vapour (or partial pressure of water vapour in saturated air) at the same temperature T. It is usually measured in percentage. When p_v is equal to p_s, ϕ is equal to unity, and the air is saturated and is considered to have 100 per cent RH.

In general, since partial pressure is a direct measure of the moisture holding of dry air, relative humidity is considered to be a more commonly understood measure of the degree of saturation of air.

From Eqs (14.14) and (14.23), it can be shown that

$$\omega = 0.622\phi \frac{p_s}{p_a} \tag{14.24}$$

$$\phi = \frac{\omega}{0.622} \frac{p_a}{p_s} \tag{14.25}$$

Also, from Eqs (14.19) and (14.23), we get

$$\mu = \phi \left[\frac{1 - p_s/p}{1 - p_v/p}\right] \tag{14.26}$$

$$\phi = \frac{\mu}{1 - (1-\mu)p_s/p} \tag{14.27}$$

14.3.5 Enthalpy of Moist Air

According to Gibbs' law, the enthalpy of a mixture of perfect gases is obtained by the summation of the enthalpies of the constituents. Thus the enthalpy of moist air h is equal to the sum of the enthalpies of dry air and associated water vapour, i.e.,

$$h = h_a + \omega h_v \tag{14.28}$$

per kg of dry air, where h_a is the enthalpy of the dry air part and ωh_v is the enthalpy of the water vapour part.

Considering the change in enthalpy of a perfect gas as a function of temperature only, the enthalpy of the dry air part, above a datum of 0°C, is expressed as

$$h_a = C_{pa}\, t = 1.005\, t \text{ kJ/kg } (= 0.24\, t \text{ Btu/lbm where } t \text{ is in °F}) \qquad (14.29)$$

where C_{pa} = 1.005 kJ/(kg.K) is the specific heat of dry air, and t is the dry-bulb temperature of air in °C.

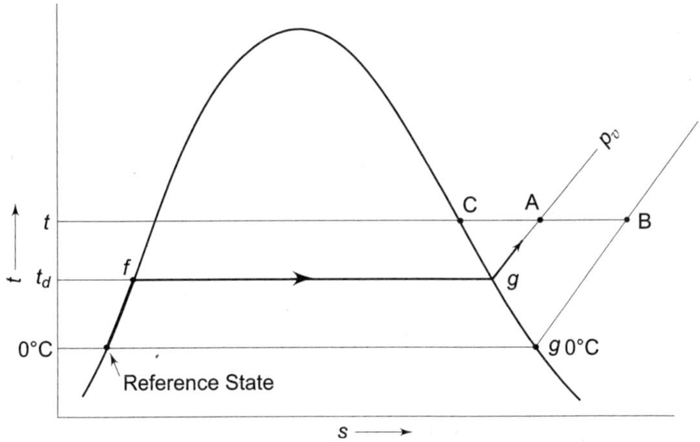

Fig. 14.5 Evaluation of enthalpy of water vapour part

Again, taking the reference state enthalpy as zero for saturated liquid at 0°C, the enthalpy of the water vapour part, viz., at point A in Fig. 14.5, is expressed as

$$h_v = h_A = C_{p_w}\, t_d + (h_{fg})_d + C_{pv}\,(t - t_d) \text{ kJ/kg} \qquad (14.30)$$

where

C_{p_w} = specific heat of liquid water
t_d = dew point temperature
$(h_{fg})_d$ = latent heat of vaporization at DPT
C_{p_v} = specific heat of superheated vapour

Taking the specific heat of liquid water as 4.1868 kJ/(kg.K) and that of water vapour as 1.88 kJ/(kg.K), in the range 0 to 60°C, we have

$$h_v = 4.1868\, t_d + (h_{fg})_d + 1.88\,(t - t_d) \qquad (14.31)$$

This expression for enthalpy is rather unwieldy for the purpose of calculations. It may, however, be pointed out that at low pressures for an ideal gas, the enthalpy is a function of temperature only. Thus in Fig. 14.5, the enthalpies at points B and C are also the same as the enthalpy at A. Accordingly, enthalpy of water vapour at A, at DPT of t_d and DBT of t, can be determined more conveniently by the following two methods:

(i) $$h_A = h_C = (h_g)_t \qquad (14.32)$$

(ii) $$h_A = h_B = (h_g)_{0°C} + C_{p_v}(t - 0) \qquad (14.33)$$

Thus, employing the second expression and taking the latent heat of vaporization of water at 0°C as 2501 kJ/kg, we obtain the following empirical expression for the enthalpy of the water vapour part

$$h_v = 2501 + 1.88t \text{ kJ/kg } (= 1061 + 0.444\, t \text{ Btu/lbm, where } t \text{ is in °F}) \tag{14.34}$$

and combining Eqs (14.29) and (14.34), we have for the enthalpy of moist air

$$h = 1.005t + \omega(2500 + 1.88\, t) \text{ kJ/kg d.a.} \tag{14.35}$$

$$(= 0.24t + \omega\,(1061 + 0.444t) \text{ Btu/lbm d.a. where } t \text{ is in °F})$$

14.3.6 Humid Specific Heat

Equation (14.35) for the enthalpy of moist air can also be written in the form

$$\begin{aligned} h &= (C_{p_a} + \omega\, C_{p_v})\, t + \omega\,(h_{fg})_{0°C} \\ &= C_p\, t + \omega\,(h_{fg})_{0°C} \end{aligned} \tag{14.36}$$

where

$$\begin{aligned} C_p &= C_{p_a} + \omega\, C_{p_v} \\ &= (1.005 + 1.88\omega)\, \frac{\text{kJ}}{(\text{kg d.a.})(\text{K})} \quad (= 0.245 \text{ Btu/lbm d.a.R}) \end{aligned} \tag{14.37}$$

is termed as the *humid specific heat*. It is the specific heat of moist air $(1 + \omega)$kg per kg of dry air. The term $C_p\, t$ governs the change in enthalpy of moist air with temperature at constant specific humidity, and the term $\omega\,(h_{fg})_{0°C}$ governs the change in enthalpy with the change in specific humidity, i.e., due to the addition or removal of water vapour in air.

Since the second term 1.88ω is very small compared to the first term 1.005, an approximated value of C_p of 1.0216 kJ/(kg d.a.) (K) may be taken for all practical purposes in air-conditioning calculations.

Example 14.3 *A mixture of dry air and water vapour is at a temperature of 21°C under a total pressure of 736 mm Hg. The dew-point temperature is 15°C. Find:*

(i) Partial pressure of water vapour.
(ii) Relative humidity.
(iii) Specific humidity.
(iv) Specific enthalpy of water vapour by the three methods of Fig. 14.5.
(v) Enthalpy of air per kg of dry air.
(vi) Specific volume of air per kg of dry air.

Solution (i) From steam tables, the partial pressure of water vapour at 15°C DPT is

$$p_v = 12.79 \text{ mm Hg} = 12.79\,(133.5) = 1707.5 \text{ N/m}^2.$$

(ii) Saturation pressure of water vapour at 21°C DBT

$$p_s = 18.65 \text{ mm Hg} = 18.65\,(133.5) = 2489.8 \text{ N/m}^2$$

Relative humidity

$$\phi = \frac{p_v}{p_s} \times 100 = \frac{12.79}{18.65} \times 100 = 68.58\%$$

(iii) Specific humidity

$$\omega = 0.622\, \frac{p_v}{p_a} = 0.622\, \frac{(12.79)}{(736 - 12.79)} = 0.622\, \frac{(12.79)}{723.21}$$

$$= 0.011 \text{ kg w.v./kg d.a.}$$

(iv) Latent heat of vaporization of water at dry bulb and dew-point temperatures of air

$$(h_{fg})_{21°C} = 2452 \text{ kJ/kg}$$
$$(h_{fg})_{15°C} = 2466.2 \text{ kJ/kg}$$

Specific enthalpy of water vapour from Fig. 14.5 by the three methods

$$h_C = (4.1868)\,(21) + 2452 = 2540 \text{ kJ/kg w.v.}$$
$$h_A = (4.1868)\,(15) + 2466.2 + 1.88\,(21 - 15)$$
$$= 2540.3 \text{ kJ/kg w.v.}$$
$$h_B = 2501 + 1.88\,(21) = 2540.5 \text{ kJ/kg w.v.}$$

Note *The three values are extremely close to each other.*

(v) Enthalpy of air using the value of specific enthalpy of water vapour from the empirical relation

$$h = h_a + \omega h_v$$
$$= 1.005\,(21) + 0.011\,(2540.5)$$
$$= 21.1 + 27.9 = 49.0 \text{ kJ/kg d.a.}$$

(vi) Specific volume of air is equal to the volume of 1 kg of dry air or 0.011 kg of water vapour. Based on the dry air part

$$v = v_a = \frac{R_a T}{p_a} = \frac{287.3(273+21)}{(723.21)(133.5)} = 0.875 \text{ m}^3\text{/kg d.a.}$$

14.4 WET BULB TEMPERATURE (WBT)[10]

A method of measuring humidity by a *psychrometer* may now be described. A psychrometer comprises of a dry bulb thermometer and a wet bulb thermometer.

The dry bulb thermometer is directly exposed to the air and measures the actual temperature of air. The bulb of the wet bulb thermometer is covered by a wick thoroughly wetted by water. The temperature which is measured by the wick-covered bulb of such a thermometer indicates the temperature of liquid-water in the wick and is called the *wet bulb temperature*. Let it be denoted by the symbol t'.

Referring to Fig. 14.6, when unsaturated moist air at A flows over water in the wick at f, the partial pressure p_v of the water vapour in the mixture is too low for equilibrium with the liquid which exerts the saturation pressure p_s. As a result the liquid evaporates into air, increasing the vapour pressure to p_{s_1}. The enthalpy of vaporization is derived from liquid water itself as well as air. Accordingly, the state of water drops to f_1, its vapour pressure to p_{s_1}, and the state of leaving water vapour (and also temperature of air) moves to 1.

The process is repeated until the equilibrium temperature t' is reached in the wick, when the water in the wick is at f' and the water vapour in the air after leaving the bulb is saturated at g'. The leaving temperature of air is also t'. At this equilibrium state, the enthalpy of vaporization is completely derived from air. Thus, we have at this point the heat required to evaporate water to saturate air equal to the heat surrendered by air.

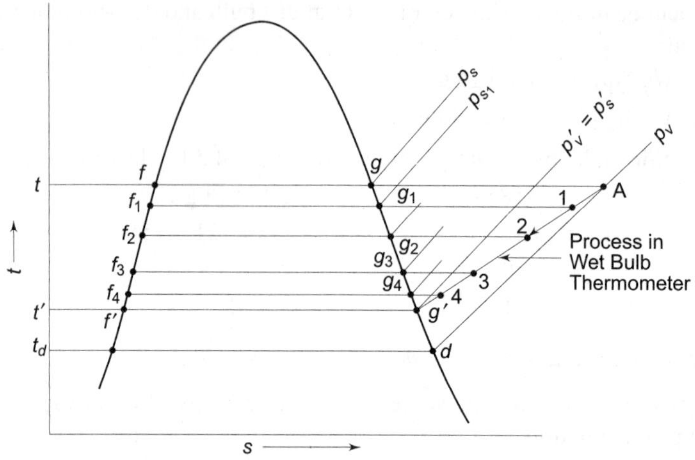

Fig. 14.6 Change of state of water vapour in air flowing over a wet bulb thermometer

The vapour pressure of water p_v' in air is now equal to the saturation pressure p_s' at the temperature t' of the wet bulb. There is only one possible equilibrium wet bulb temperature t' for the complete saturation of a given initial state A of the moist air. The change of state of water vapour is from A to g'.

The difference between the dry bulb and wet bulb temperatures is called *wet bulb depression* (WBD). Thus

$$\text{WBD} = (t - t')$$

If the ambient air is saturated, viz., the RH is 100 per cent, then there will be no evaporation of water on the bulb and hence WBT and DBT will be equal. The wet bulb depression will be zero. Thus WBT is an indirect measure of the dryness of air.

The wet bulb temperature is essentially not a thermodynamic property. It is the temperature of equilibrium reached by heat transfer from air to water in the wick due to the temperature difference $(t - t')$ causing the evaporation of water and the consequent diffusion of water vapour into air due to the partial pressure difference $(p_v' - p_v)$, where p_v' is the saturation water vapour pressure at temperature t'. Referring to Fig. 14.7, this equilibrium condition can be expressed by the energy balance equation

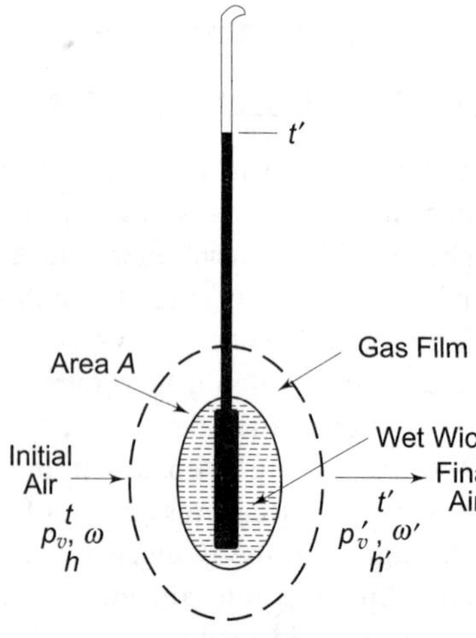

Fig. 14.7 Flow of air over the wick-covered bulb of a wet bulb thermometer

$$f_g A\,(t - t') = m_v\, h_{fg}' = k_d A\,(p_v' - p_v)\, h_{fg}' \tag{14.38}$$

where f_g = Heat-transfer coefficient of the air film around the wetted surface

A = Area of the wetted surface

m_v = Rate of evaporation and diffusion of water vapour

k_d = Mass transfer coefficient of water vapour through the film of air and water vapour mixture

h_{fg}' = Enthalpy of vaporization at the wet bulb temperature

Rearranging Eq. (14.38), we obtain for the wet bulb temperature

$$t' = t - \frac{k_d}{f_g} h_{fg}'(p_v' - p_v) \tag{14.39}$$

As the specific humidity is solely a function of vapour pressure, the vapour pressure difference in Eq. (14.39) can be replaced by the corresponding specific humidity difference ($\omega' - \omega$) and with suitable modification of the units, the mass-transfer coefficient k_d may be replaced by the mass transfer coefficient k_ω based on the specific humidity difference so that the expression for the wet bulb temperature becomes

$$t' = t - \frac{k_\omega}{f_g} h_{fg}'(\omega' - \omega) \tag{14.40}$$

It is thus seen that the equilibrium wet bulb temperature depends on the heat and mass-transfer coefficients f_g and k_ω respectively and the initial state of air. Further, the units of the mass-transfer coefficient k_ω can be obtained from Eq. (14.40) as

$$[k_\omega] = \frac{[t][f_g]}{[h_{fg}][\omega]} = \frac{[\text{K}][\text{Js}^{-1}\ \text{m}^{-2}\ \text{K}^{-1}]}{[\text{J kg}^{-1}][\text{kg kg}^{-1}]} = \frac{\text{kg}}{\text{s.m}^2}$$

Comparing Eqs (14.39) and (14.40), we see that

$$k_d(p_v' - p_v) = k_\omega(\omega' - \omega) = 0.622\, k_\omega \frac{(p_v' - p_v)}{p_a}$$

$$\Rightarrow \quad k_\omega = \frac{1}{0.622} p_a k_d = \frac{RT}{0.622}(\rho_a k_d) \tag{14.41}$$

14.5 THERMODYNAMIC WET BULB TEMPERATURE OR TEMPERATURE OF ADIABATIC SATURATION

For any state of unsaturated moist air, there exists a temperature t^* at which the air becomes adiabatically saturated by the evaporation of water into air, at exactly the same temperature t^*. Figure 14.8 is a schematic representation of this process, called the *adiabatic saturation process*. The leaving air is saturated at temperature t^*. The specific humidity is correspondingly increased to ω^*. The enthalpy is increased from a given initial value h to the value h^*. The weight of water added per kg of dry air is $\omega^* - \omega$ which adds energy to the moist air of amount equal to $(\omega^* - \omega)$ h_f^*, where h_f^* is the specific enthalpy of the injected water at t^*.

Therefore, since the process is strictly adiabatic, we have by energy balance

$$h + (\omega^* - \omega) h_f^* = h^* \tag{14.42}$$

Now $$h = C_{p_a} t + \omega h_v$$

and $$h^* = C_{p_a} t^* + \omega^* h_v^*$$

where $$h_v^* = h_f^* + h_{fg}^* \text{ (saturated vapour)}$$

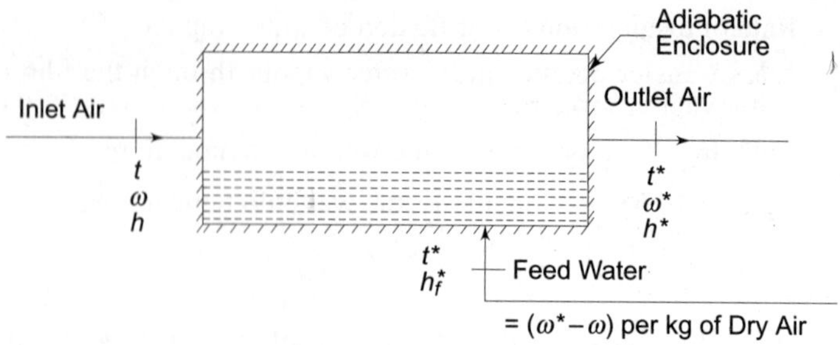

Fig. 14.8 Adiabatic saturator

and, h_v can be expressed in terms of h_v^*, such that

$$h_v = h_v^* + C_{p_v}(t - t^*) = h_f^* + h_{fg}^* + C_{p_v}(t - t^*)$$

Substituting these values in Eq. (14.42), we obtain

$$C_{p_a} t + \omega [h_f^* + h_{fg}^* + C_{p_v}(t - t^*)] + (\omega^* - \omega) h_f^* = C_{p_a} t^* + \omega^* [h_f^* + h_{fg}^*] \quad (14.43)$$

Simplifying Eq. (14.43) we get

$$(C_{p_a} + \omega C_{p_v}) t - (C_{p_a} + \omega^* C_{p_v}) t^* = h_{fg}^* (\omega^* - \omega)$$

or

$$C_p t - C_p^* t^* = h_{fg}^* (\omega^* - \omega) \quad (14.44)$$

Considering the humid specific heats C_p and C_p^* as approximately the same, we can simplify Eq. (14.44) to

$$C_p (t - t^*) = h_{fg}^* (\omega^* - \omega)$$

whence

$$t^* = t - \frac{h_{fg}^*}{C_p} (\omega^* - \omega) \quad (14.45)$$

It can be seen from Eq. (14.45) that there is only one solution for t^*, and it depends only on t and ω which are the thermodynamic properties representing the initial state of air. Thus, the temperature of adiabatic saturation is a thermodynamic property of moist air.

14.5.1 Difference between t' and t^*

Let us compare the expressions for the wet bulb temperature t' and the temperature of adiabatic saturation t^*, i.e.,

$$t' = t - \frac{k_\omega}{f_g} h'_{fg} (\omega' - \omega)$$

$$t^* = t - \frac{h_{fg}^*}{C_p} (\omega^* - \omega)$$

It follows that if

$$\frac{k_\omega}{f_g} = \frac{1}{C_p}$$

or
$$\frac{f_g}{k_\omega C_p} = Le = 1 = \left(\frac{\alpha}{D}\right)^{2/3} \tag{14.46}$$

then $t' = t^*$, i.e., the two temperatures are equal. The dimensionless quantity $f_g/k_\omega C_p$ is called the *Lewis number*.[5,6] Fortunately, for the air and water vapour mixture at low pressures, this number is approximately equal to unity ($Le = 0.945$).

Thus by an entirely fortuitous circumstance, the measurable wet bulb temperature is equal to the thermodynamic wet bulb temperature. There is no theoretical basis, however, to assume that the two temperatures are the same. For any other kind of gas and vapour mixture these would not be the same. In a way, it is also somewhat unfortunate that in the case of air and water vapour mixture, the two temperatures are exactly the same and they are confused as meaning one for the other.

14.5.2 Measurement of Psychrometric Properties

It may be seen that there is no convenient way of measuring ω, μ or ϕ. They are properties which have to be calculated. The measurable properties are dry bulb, wet bulb and dew point temperatures.

To define the complete thermodynamic state of the air and water vapour mixture, one needs to know three properties. Usually, two of these are the total pressure (barometer pressure) and dry bulb temperature. The third measurable property is either the dew point temperature or the wet bulb temperature.

The dew point temperature is measured by cooling a bulb in a stream of air until the first dew appears on the bulb. The wet bulb temperature is measured by rotating the wick-covered bulb of a thermometer at 160 to 660 rpm in air.

From the dew point temperature, the saturation pressure of water can be obtained from the steam table which in turn is equal to the actual partial pressure of water vapour in the air. It is generally difficult to accurately measure the dew point temperature.

The wet bulb temperature is easily measured with the help of a psychrometer. The wet bulb temperature, as stated earlier, is not a thermodynamic property. Therefore, no analytical expression can be derived to relate WBT with p_v or other thermodynamic properties. Nevertheless, since WBT happens to be equal to the temperature of adiabatic saturation, it can be considered as a thermodynamic property. It is thus an easily measurable property, and empirical relations exist to obtain the value of p_v in terms of t'. They are:

(i) *Modified Apjohn equation*
$$p_v = p_v' - \frac{1.8p(t-t')}{2700} \tag{14.47}$$

(ii) *Modified Ferrel equation*
$$p_v = p_v' - 0.00066p\,(t-t')\left[1+\frac{1.8t}{1571}\right] \tag{14.48}$$

(iii) *Carrier equation*
$$p_v = p_v' - \frac{(p-p_v')(t-t')(1.8)}{2800-1.3(1.8t+32)} \tag{14.49}$$

where all the temperatures are in °C, and p_v' is the saturation pressure at the wet bulb temperature t'. The pressures p, p_v' and p_v may be in any consistent units.

Example 14.4 *Calculate, (i) relative humidity, (ii) humidity ratio, (iii) dew point temperature, (iv) density and (v) enthalpy of atmospheric air when the DBT is 35°C, WBT is 23°C and the barometer reads 750 mm Hg.*

Solution (i) From the saturation properties of water, at a WBT of 23°C

$$p_v' = 21.06 \text{ mm hg}$$

Then from Eq. (14.49)

$$p_v = p_v' - \frac{(p - p_v')(t - t')(1.8)}{2800 - 1.3(1.8t + 32)}$$

$$= 21.06 - \frac{(750 - 21.06)(35 - 23)(1.8)}{2800 - 1.3(1.8 \times 35 + 32)}$$

$$= 21.06 - 5.88 = 15.18 \text{ mm Hg}$$

From saturation properties of water, at a DBT of 35°C

$$p_s = 42.4 \text{ mm Hg}$$

Relative humidity

$$\phi = \frac{p_v}{p_s} \times 100 = \frac{15.18}{42.2} \times 100 = 36\%$$

(ii) Humidity ratio

$$\omega = 0.622 \frac{p_v}{p - p_v} = 0.622 \frac{15.18}{(750 - 15.18)} = 0.01285 \text{ kg w.v./kg d.a.}$$

(iii) Dew point temperature

$$t_d = 17.7\text{°C (Saturation temperature at 15.18 mm Hg pressure)}$$

(iv) Partial densities

$$\rho_a = \frac{p_a}{R_a T} = \frac{(750 - 15.18)(133.5)}{(287.1)(273 + 35)} = 1.1086 \text{ kg/m}^3$$

$$\rho_v = \frac{p_v}{R_v T} = \frac{(15.18)(133.5)}{(461)(273 + 35)} = 0.0143 \text{ kg/m}^3$$

Density of moist air

$$\rho = \rho_a + \rho_v = 1.1086 + 0.0143 = 1.1229 \text{ kg/m}^3$$

(v) Enthalpy

$$h = 1.005t + \omega(2500 + 1.88t)$$

$$= 1.005(35) + 0.01285(2500 + 1.88 \times 35) = 68.15 \text{ kJ/kg d.a.}$$

14.6 PSYCHROMETRIC CHART[7,9]

All data essential for the complete thermodynamic and psychrometric analysis of air-conditioning processes can be summarised in a *psychrometric chart*. At present, many forms of psychrometric charts are in use. The chart which is most commonly used is the $\omega - t$ chart, i.e., a chart which has specific humidity or water vapour pressure along the ordinate and the dry bulb temperature along the abscissa. The chart is normally constructed for a standard atmospheric pressure of 760 mm Hg or 1.01325 bar, corresponding to the pressure at the mean sea level. A typical layout of this chart is shown in Fig. 14.9. The procedures for drawing various constant property lines on this chart will now be described.

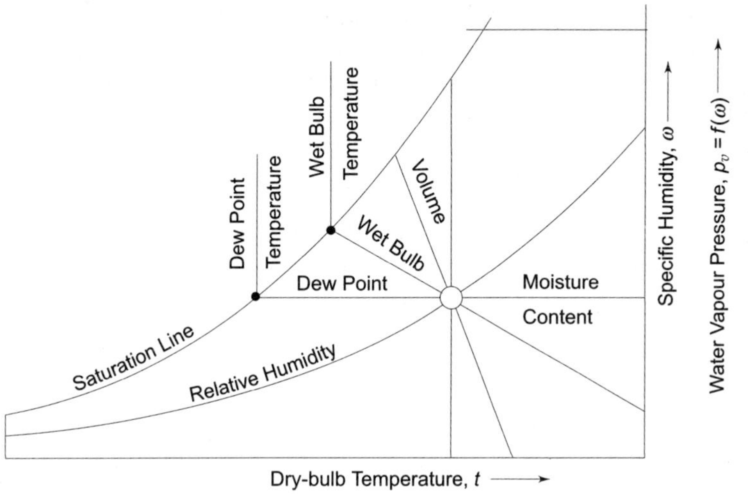

Fig. 14.9 Constant property lines on a psychrometric chart

14.6.1 Saturation Line

The saturation line represents the states of saturated air at different temperatures. As an example of fixing such a state on the chart, consider an atmosphere A at 20°C and saturation as shown in Fig. 14.10. From the steam tables at 20°C, water vapour pressure

$$p_s = p_v = 17.54 \text{ mm Hg} = 2342 \text{ N/m}^2$$

Partial pressure of dry air

$$p_a = p - p_v = 101325 - 2342 = 98983 \text{ N/m}^2$$

Specific humidity at 20°C saturation

$$\omega_s = \frac{0.622 p_v}{p_a} = \frac{0.622(2342)}{98983} = 0.01472 \text{ kg w.v./kg d.a.}$$

Knowing t and ω, point A can be plotted. In a similar manner, saturation states at other temperatures can also be plotted to draw the saturation line on the psychrometric chart.

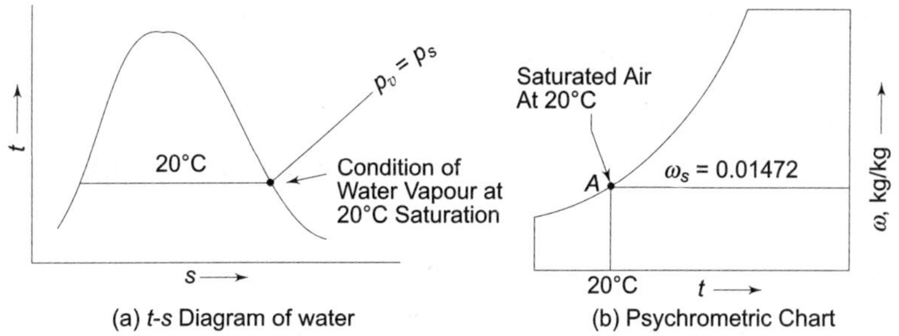

Fig. 14.10 Saturated air at 20°C

14.6.2 Relative Humidity Lines

Relative humidity is the ratio of the actual vapour pressure to the vapour pressure which would exist in a saturated mixture at the temperature of the air. The saturation line on the chart is, therefore, the line of 100 per cent RH since for all points on this line $p_v = p_s$. The lines on this chart for any other desired value of RH can be constructed as follows.

Taking 50 per cent RH as an example, the point on the 20°C line corresponding to this RH must be at the intersection C (Fig. 14.11) with the line representing a vapour pressure of

$$p_v = 0.5 \times 2342 = 1171 \text{ N/m}^2$$

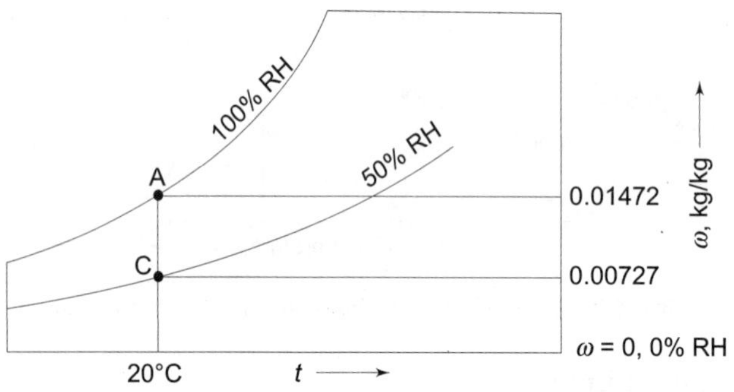

Fig. 14.11 Drawing 50 per cent RH line

At this point

$$p_a = 101325 - 1171 = 100154 \text{ N/m}^2$$

$$\omega = \frac{0.622(1171)}{100154} = 0.00727 \text{ kg w.v./kg d.a.}$$

Likewise, points for other temperatures can be plotted to construct the complete 50 per cent RH line. It may be noted that the $\omega = 0$ line also corresponds to zero per cent RH.

14.6.3 Constant Specific Volume Lines

Consider a line corresponding to a specific volume of, say, 0.85 m^3/kg d.a. as shown in Fig. 14.12. One point on this line is A which is on the saturation curve. To locate this point, the following trial and error procedure is to be followed:

(i) Assume t at A and find $p_s = p_v$.
(ii) Find p_a.
(iii) Calculate v and check.

In this case, an assumed value of $t = 20$°C is taken initially at which

$$p_s = p_v = 2342 \text{ N/m}^2$$

$$p_a = 101325 - 2342 = 98983 \text{ N/m}^2$$

$$v_a = R_a T/p_a = \frac{287.3(273+20)}{98983} = 0.85 \text{ m}^3\text{/kg d.a.}$$

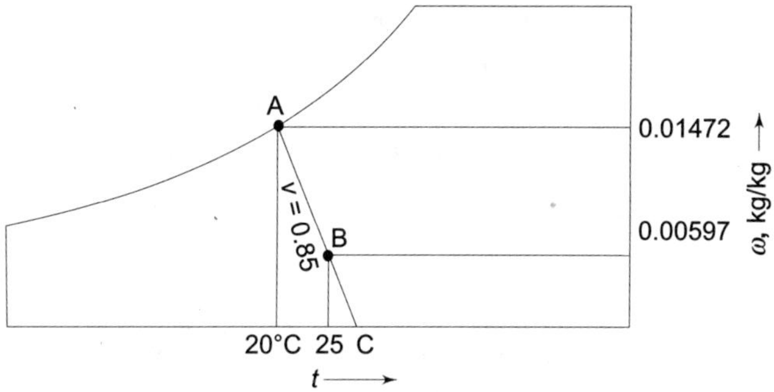

Fig. 14.12 Drawing 0.85 m^3/kg d.a. constant-volume line

It checks with the required value. If it does not agree with the required value, a new value of *t* may be assumed. Now to plot point A, we calculate

$$\omega = \frac{0.622(2342)}{98983} = 0.01472 \text{ kg w.v./kg d.a.}$$

Thus point A can be plotted. Consider now another point B on this line at a DBT of 25°C. For this point with the same specific volume

$$p_a = \frac{R_a T}{v_a} = \frac{287.3(273+25)}{0.85} = 100724 \text{ N/m}^2$$

$$p_v = 101325 - 100724 = 601 \text{ N/m}^2$$

$$\omega = \frac{0.622(601)}{100724} = 0.00597 \text{ kg w.v./kg d.a.}$$

Thus point B can also be plotted. Likewise, a number of other unsaturated states can be plotted up to $\omega = 0$, to give the complete constant-volume line. It may be noted that at any point on this line

$$v = v_a = \omega v_v$$

14.6.4 Constant Thermodynamic Wet Bulb Temperature Lines

Consider the energy balance equation (14.41) for the adiabatic saturation process, viz.,

$$h + (\omega^* - \omega)h_f^* = h^*$$

Rearranging, we have

$$h - \omega h_f^* = h^* - \omega^* h_f^* = \text{const.}$$

Now in Fig. 14.13, point A at 20°C DBT also has a WBT of 20°C. For any other point on the 20°C constant WBT line, we must have

$$\Sigma = h - \omega h_f^* = h^* - \omega^* h_f^* = \Sigma^* = \text{const.} \tag{14.50}$$

where Σ is called the *sigma heat function* which is constant along the constant WBT line. All states on this constant wet bulb temperature or constant sigma heat function line have a combination of *h* and *ω* which must satisfy Eq. (14.50). Now for point A

$$\omega^* = 0.01472 \text{ kg w.v./kg d.a.}$$

$$\begin{aligned} h^* &= (1.005 + 1.88\,\omega^*)\,t^* + 2501\,\omega^* \\ &= (1.005 + 1.88 \times 0.01472)20 + 2500\,(0.01472) \\ &= 57.45 \text{ kJ/kg d.a.} \end{aligned}$$

$$\begin{aligned} h_f^* &= 83.9 \text{ kJ/kg (at 20°C)} \\ \Sigma^* &= h^* - \omega^* h_f^* \\ &= 57.45 - 0.01472\,(83.9) \\ &= 57.45 - 1.24 = 56.21 \text{ kJ/kg d.a.} \end{aligned}$$

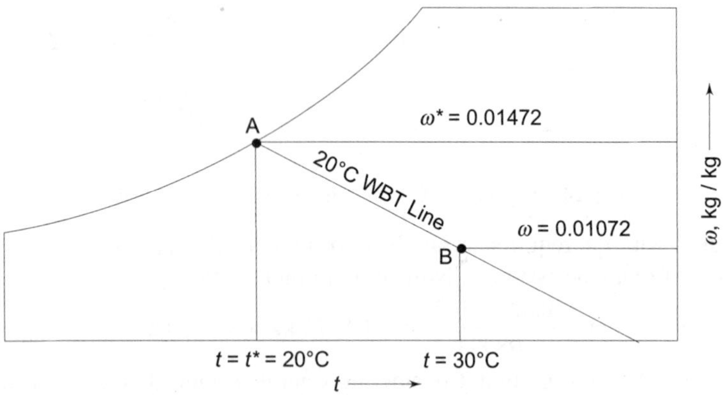

Fig. 14.13 Drawing 20°C constant WBT line

Then for point B, say at 30°C

$$\Sigma = h - \omega h_f^* = \Sigma^* = 56.21$$

$\Rightarrow$ $\quad (1.005 + 1.88\,\omega)\,30 + 2501\omega - 83.9\,\omega = 56.21$

whence $\quad \omega = 0.01072$ kJ/kg d.a.

and $\quad h = \Sigma + \omega\, h_f^* = 56.21 + 0.01072\,(83.9) = 57.11$ kJ/kg d.a.

Similarly, the points at the other dry bulb temperatures for a sigma heat function of 56.21 kJ/kg d.a. can be plotted to enable the construction of the 20°C WBT line as shown in Fig. 14.13.

14.6.5 Constant Enthalpy Lines

It may be observed from the preceding calculations that the difference of the enthalpies at the saturation point A and another point B on the constant WBT line is very small. This difference denoted by the symbol D is called *enthalpy deviation* and is given by

$$D = h - h^* = (\omega - \omega^*)\, h_f^*$$

In spite of the small enthalpy deviation, separate lines of constant wet bulb temperature and constant enthalpy can be shown on the psychrometric chart, if it is made sufficiently large. Many psychrometric charts follow this practice. Another method suggested by Palmatier and Wile[8] uses constant wet bulb lines and also shows constant enthalpy deviation lines as shown in Fig. 14.14. The enthalpy deviation at B is

$$D = h - h^* = 57.11 - 57.45 = -\,0.34 \text{ kJ/kg d.a.}$$

Thus the enthalpy scale only gives the values of enthalpies h^* for points on the saturation line (Fig. 14.14). To find the enthalpy at any other point we go along the constant WBT line, and determine the enthalpy by the expression

$$h = h^* + D$$

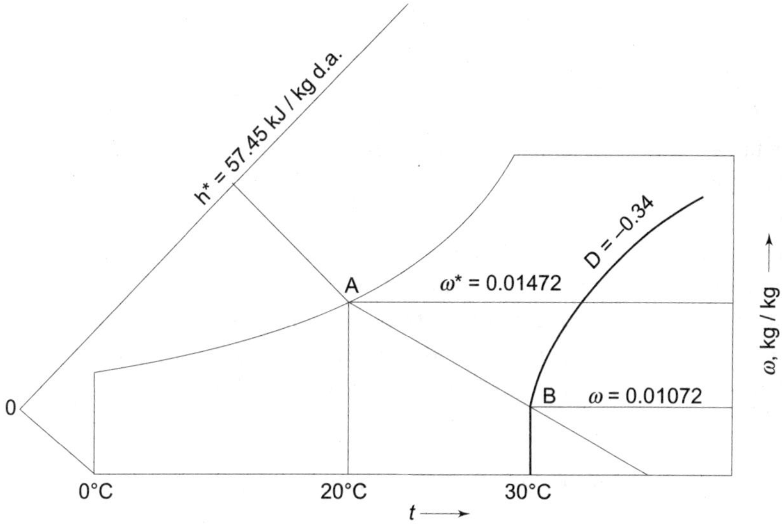

Fig. 14.14 Enthalpy deviation line

14.6.6 Psychrometric Charts for Low and High Temperatures and High Altitudes

Psychrometric charts are prepared for subzero temperatures for extreme winter-heating calculations, and for high temperatures for calculations of drying processes. In a subzero temperature chart, the water vapour pressure is that of sublimated vapour over ice since 0.01 °C is the triple point of water. Also, the latent heat used for enthalpy calculations is that of sublimation instead of vaporization.

Psychrometric charts are to be specially constructed for localities at higher altitudes and consequently lower barometric pressures. An idea of the variation of barometric pressure with altitude is obtained from values given in Table 14.2.

Table 14.2 Barometric pressure with altitude

Height above MSL m	*Barometric Pressure* mm Hg
Seal level	760
760	693
1520	633
2280	577

14.7 APPLICATION OF FIRST LAW TO A PSYCHROMETRIC PROCESS

The change of state of moist air is referred to as a *psychrometric process*. Such a process may involve any one or a combination of the following processes.

(i) Cooling.
(ii) Heating.
(iii) Dehumidifying.
(iv) Humidifying.

By far, a combination of cooling and dehumidifying is mostly used in *summer air conditioning*. On the contrary, *winter air conditioning* commonly involves *heating* and *humidifying* of air.

For the analysis of equipment involving psychrometric processes, the first law in the form of SSSF energy equation can be conveniently applied. For the purpose, the specific volume, enthalpy, etc., of moist air are either calculated from fundamental procedures as outlined in Example 14.5, or are obtained from the psychrometric chart. The following example illustrates the calculation procedure.

Example 14.5 *Cooling and Dehumidification of Moist Air*

Moist air at standard atmospheric pressure is passed over a cooling coil. The inlet and exit states are as follows:

Number	*State*	*DBT* °C	*ϕ or RH* %
1	Inlet	30	50
2	Exit	15	80

Show the process on a psychrometric chart. Determine the amount of heat and moisture removed per kg of dry air.

Solution Figure 14.15 shows the schematic diagram of the cooling coil, and Fig. 14.16 shows the cooling and dehumidification process on a psychrometric chart. Consider air around the cooling coil as forming the control volume in which moist air enters at 1 and leaves at 2. The condensed moisture leaves at 3. Applying continuity equation we have, by dry air and moisture balance

$$\dot{m}_{a_1} = \dot{m}_{a_2} = \dot{m}_a$$

$$\dot{m}_{w_3} = \dot{m}_a\,(\omega_1 - \omega_2)$$

Steady-state steady-flow (SSSF) energy equation:

$$\dot{Q} + \dot{m}_a\,h_1 = \dot{m}_a\,h_2 + \dot{m}_{w_3}\,h_{w_3}$$

Dividing the equation by $\dot{m}_a$ we get for 1 kg of dry air

$$q + h_1 = h_2 + (\omega_1 - \omega_2)\,h_{w_3}$$

$$q + h_{a_1} + \omega_1\,h_{v_1} = h_{a_2} + \omega_2\,h_{v_2} + (\omega_1 - \omega_2)\,h_{w_3}$$

Saturation pressures of water at 25°C and 15°C from steam tables are

$$p_{01} = (p_{sat})_{30°C} = 4.246 \text{ kPa}, \; p_{02} = (p_{sat})_{15°C} = 1.7051 \text{ kPa}$$

Partial pressures of water vapour in air at inlet and outlet

$$p_{v_1} = \phi_1\,p_{s_1} = 0.5\,(4.246) = 2.123 \text{ kPa}$$

$$p_{v_2} = \phi_2\,p_{s_2} = 0.8\,(1.7051) = 1.3641 \text{ kPa}$$

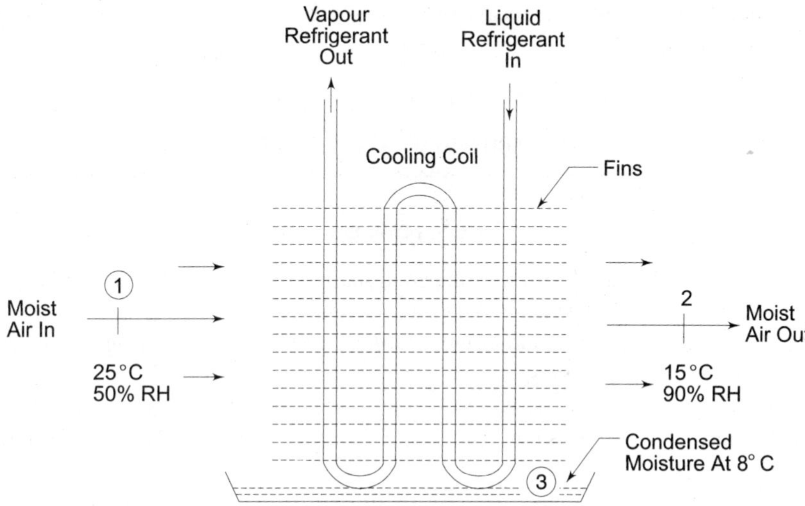

Fig. 14.15 Cooling and dehumidification of air for Example 14.5

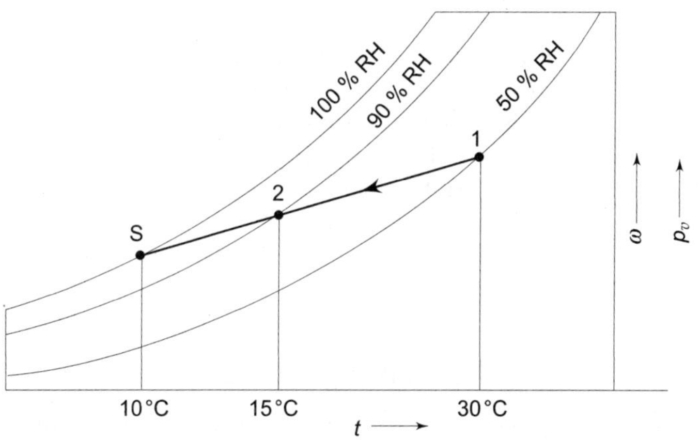

Fig. 14.16 Cooling and dehumidification process on psychrometric chart in Example 14.5

Specific humidities

$$\omega_1 = 0.622\ \frac{p_{v_1}}{p_{a_1}} = 0.622\ \frac{2.123}{101.325 - 2.123} = 0.01331 \text{ kg w.v./kg d.a.}$$

$$\omega_2 = 0.622\ \frac{p_{v_2}}{p_{a_2}} = 0.622\ \frac{1.3641}{101.325 - 1.3641} = 0.00849 \text{ kg w.v./kg d.a.}$$

Specific enthalpies of water vapour and condensed moisture are:

$$h_{v_1} = (h_g)_{30°\text{C}} = 2556.3 \text{ kJ/kg}$$

$$h_{v_2} = (h_g)_{15°\text{C}} = 2528.9 \text{ kJ/kg}$$

$$h_{w_3} = (h_f)_{15°\text{C}} = 63 \text{ kJ/kg}$$

Substituting values in SSSF energy equation, we get

$$q = (h_{a_2} - h_{a_1}) + \omega_2 h_{v_2} - \omega_1 h_{v_1} + (\omega_1 - \omega_2) h_{\omega_3}$$

$$= 1.005 (15 - 30) + 0.00849 (2528.9) - 0.01331 (2556.3)$$

$$+ (0.01331 - 0.00849) 63 = - 27.3 \text{ kJ/kg d.a.}$$

References

1. Goff J A, 'Standardisation of thermodynamic properties of moist air', *Trans. ASHVE*, Vol. 55, 1949, p. 463.
2. Haywood R W, *Thermodynamic Tables in SI Units*, Cambridge University Press Cambridge, 1968.
3. Keenan J H, *Thermodynamics*, Chapter XIII, John Wiley, New York, 1941.
4. Keenan J H and F G Keys, *Thermodynamic Properties of Steam*, John Wiley, New York, 1936.
5. Lewis W K, 'The evaporation of a liquid into a gas', Trans. ASME, Vol. 44, 1922, p. 325.
6. Lewis W K, 'The evaporation of a liquid into a gas—a correction', *Mechanical Engineering*, Vol. 55, Sept. 1933, p. 1567.
7. Palmatier E P, 'Construction of the normal temperature *ASHRAE* psychrometric Chart', *ASHRAE J.*, Vol. 5, No. 5, May 1963, p. 55.
8. Palmatier E P and D D Wile, 'A new psychrometric chart', *Refrigerating Engineering*, Vol. 52, No. 1, July, 1946, p. 31.
9. Raber B F and F W Hutchinson, *Refrigeration and Air Conditioning Engineering*, John Wiley, New York, 1949.
10. Threlkeld J L, *Thermal Environmental Engineering*, Prentice-Hall, New Jersey, 1962.

Revision Exercises

14.1 For a dry bulb temperature of 25°C and a relative humidity of 50 per cent, calculate the following for air, when the barometric pressure is 740 mm Hg.

(a) Partial pressures of water vapour and dry air.
(b) Dew point temperature.
(c) Specific humidity.
(d) Specific volume.
(e) Enthalpy.

14.2 A sample of moist air has a dry bulb temperature of 43°C and a wet bulb temperature of 29°C. Calculate the following without making use of the psychrometric chart:

(a) Partial pressure of water vapour.
(b) Specific humidity.
(c) Relative humidity.
(d) Dew point temperature.
(e) Humid specific heat.
(f) Enthalpy.
(g) Degree of saturation.
(h) Sigma heat function.

14.3 A sample of air has dry and wet bulb temperatures of 35°C and 25°C respectively. The barometric pressure is 760 mm Hg. Calculate:

(a) Humidity ratio, relative humidity and enthalpy of the sample.

(b) Humidity ratio, relative humidity and enthalpy, if the air were adiabatically saturated. The use of steam tables only is permitted.

14.4 Investigate the effect of humidity on the density of moist air by computing the vapour density for an air water vapour mixture at 26°C and relative humidities of 0, 50 and 100 per cent. Also, for each case, compare the values of the degree of saturation to the values of relative humidity.

14.5 Air at a condition of 30°C dry bulb, 17°C wet bulb and a barometric pressure of 1050 m bar enters an equipment where it undergoes a process of adiabatic saturation, the air leaving with a moisture content of 5 g/kg higher than what it was while entering. Calculate:

(i) Moisture content of air entering the equipment.

(ii) Dry bulb temperature and enthalpy of the air leaving the equipment.

14.6 (a) Moist air is at 25°C temperature. Its dew point is measured as 20°C. The barometric pressure is 755 mm Hg. What are the values of specific and relative humidities of the air?

(b) If this air is cooled to 15°C dry bulb temperature and 50% relative humidity, what will be the amount of total heat removed per unit mass of dry air? What will be the corresponding amount of moisture removed?

14.7 (a) The temperature of air entering an adiabatic saturator is 42°C, and the leaving air temperature is 30°C. Compute the humidity ratio and relative humidity of entering air.

(b) The conditions inside a room are 25°C and 50% degree of saturation. The inside surface temperature of the glass window is 10°C. Will the moisture condense from room air upon the window glass?

14.8 Using tables and applying first law of thermodynamics, estimate the heat transfer rates in the following two cases:

(a) Heating of 1.2 m^3/s of air at 15°C and 90% RH to 50°C without the addition of moisture.

(b) Cooling of 1.5 m^3/s of moist air at 30°C and 60% RH to 15°C and 80% RH. The condensate leaves at 20°C.

Chapter

15

Psychrometry of Air-Conditioning Processes

This chapter presents the psychrometry of common air-conditioning processes showing how the state of moist air is altered as the respective processes take place.

15.1 MIXING PROCESS

Let us consider the adiabatic mixing of different quantities of air in two different states at constant pressure. Let subscripts 1 and 2 refer to the two streams of air, and let m_a refer to the mass of dry air in the stream. Then by *moisture balance*, we have for the specific humidity of the mixture

$$m_{a_3}\,\omega_3 = m_{a_1}\,\omega_1 + m_{a_2}\,\omega_2$$

$$\Rightarrow \qquad \omega_3 = \frac{m_{a_1}\omega_1 + m_{a_2}\omega_2}{m_{a_3}} \tag{15.1}$$

where by *dry air mass balance*,

$$m_{a_3} = m_{a_1} + m_{a_2}$$

is the mass of dry air in the mixture.

Also, by *energy balance*, we similarly get the expression for the enthalpy of the mixture

$$h_3 = \frac{m_{a_1} h_1 + m_{a_2} h_2}{m_{a_3}} \tag{15.2}$$

Substituting expressions from Eq. (14.36) for the enthalpies in the above equation, we have

$$(C_p t_3 + h_{fg0}\omega_3) = \frac{m_{a_1}}{m_{a_3}}(C_p t_1 + h_{fg0}\,\omega_1) + \frac{m_{a_2}}{m_{a_3}}(C_p t_2 + h_{fg0}\,\omega_2)$$

Simplifying, we get an expression for the temperature of the mixture

$$t_3 = \frac{m_{a_1} t_1 + m_{a_2} t_2}{m_{a_3}} + \frac{h_{fg0}}{C_p}\left[\frac{m_{a_1}}{m_{a_3}}\omega_1 + \frac{m_{a_2}}{m_{a_3}}\omega_2 - \omega_3\right]$$

The second term in the above expression being negligible, we can write

$$t_3 \approx \frac{m_{a_1} t_1 + m_{a_2} t_2}{m_{a_3}} \tag{15.3}$$

The sign of approximation has been used since an assumption has been made that the humid specific heat C_p is the same for all three streams.

Thus, if the psychrometric chart had been plotted on a ω–h coordinate system, the state point for the mixture would lie on the straight line joining the two states. On the ω–t coordinate system (Fig. 15.1) it is only approximately so. The position of the mixture state is such that it divides the straight line joining states 1 and 2 in the inverse ratio of the masses m_{a_1} and m_{a_2} of the two dry air streams.

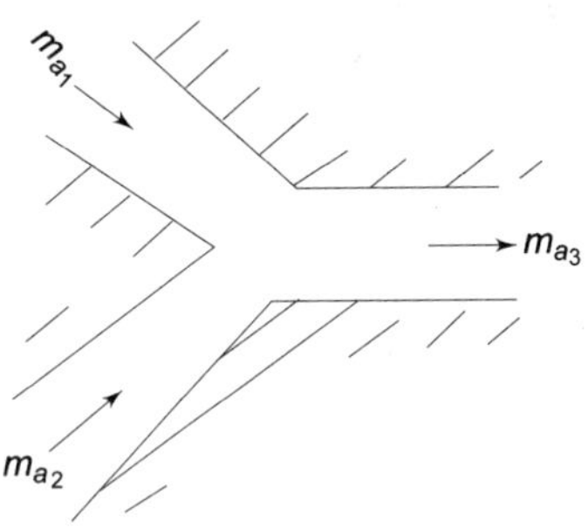

Fig. 15.1(a) Adiabatic mixing of air atreams

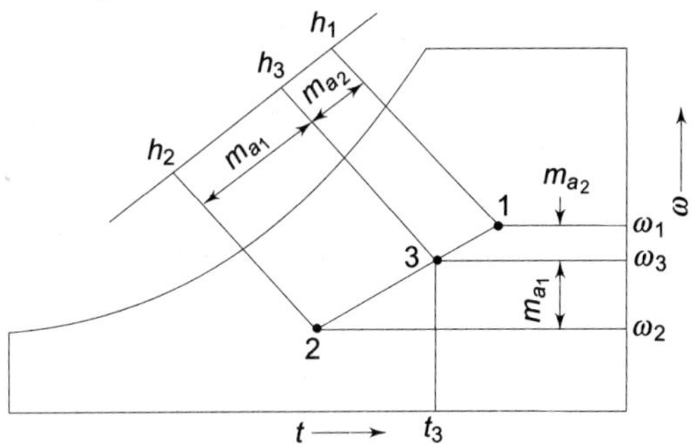

Fig. 15.1(b) Mixing process on psychrometric chart

Example 15.1 *30 m³/min of a stream of moist air at 15°C DBT and 13°C WBT are mixed with 12 m³/min of a second stream at 25°C DBT and 18°C WBT. Barometric pressure is one standard atmosphere. Determine the dry bulb and wet bulb temperatures of the resulting mixture.*

Solution From the psychrometric chart:

For the first stream

$t_1 = 15°\text{C}$, $t_1' = 13°\text{C}$, $v_1 = 0.827 \text{ m}^3/\text{kg d.a.}$,

$\omega_1 = 0.0084$ kg w.v./kg d.a., $h_1 = 36.85$ kJ/kg d.a.

For the second stream

$t_2 = 25°\text{C}$, $t_2' = 18°\text{C}$, $v_2 = 0.859 \text{ m}^3/\text{kg d.a.}$,

$\omega_2 = 0.01$ kg w.v./kg d.a., $h_2 = 51.1$ kJ/kg d.a.

Dry air mass flow rates

$$m_{a_1} = \frac{30}{0.827} = 36.2 \text{ kg d.a./min}$$

$$m_{a_2} = \frac{12}{0.859} = 13.9 \text{ kg d.a./min}$$

For the mixture

$$m_{a_3} = 36.2 + 13.9 = 50.1 \text{ kg d.a./min}$$

$$\omega_3 = \frac{36.2(0.0084)+13.9(0.01)}{50.01} = 0.00886 \text{ kg w.v./kg d.a.}$$

$$h_3 = \frac{36.2(36.85)+13.9(51.1)}{50.1} = 40.8 \text{ kJ kg./d.a.}$$

From the psychrometric chart for calculated values of ω_3 and h_3

$$\text{DBT of mixture, } t_3 = 17.5°\text{C}$$

$$\text{WBT of mixture, } t_3' = 14.5°\text{C}$$

Example 15.2 *A stream of moist air at 2°C dry bulb and 80 per cent relative humidity mixes with another stream of moist air at 30°C dry bulb and 10°C dew point in the-ratio by mass of one part of the first to two parts of the second. Calculate the temperature and specific humidity of the air after mixing.*

Solution *For the first stream*

$$\omega_1 = 0.0035 \text{ kg w.v./kg d.a.}$$

$$h_1 = 10.78 \text{ kJ/kg d.a.}$$

Mass of dry air per unit mass of moist air

$$m_{a_1} = \frac{1}{1+\omega_1} \text{ kg} = \frac{1}{1.0035} = 0.9955 \text{ kg}$$

For the second stream

$$\omega_2 = 0.00765 \text{ kg w.v./kg d.a.}$$

$$h_2 = 49.67 \text{ kJ/kg d.a.}$$

Mass of dry air per two-units mass of moist air

$$m_{a_2} = \frac{2}{1+\omega_2} \text{ kg} = \frac{2}{1.00765} = 1.9848 \text{ kg}$$

For the mixture

$$h = \frac{0.9965(10.78)+1.9848(49.67)}{0.9965+1.9848} = 36.68 \text{ kJ/kg d.a.}$$

$$\omega = \frac{0.9965(0.0035)+1.9848(0.00765)}{0.9965+1.9848} = 0.00627 \text{ kg w.v./kg d.a.}$$

15.1.1 Mixing with Condensation

When a large quantity of cold air mixes with a quantity of warmer air at a high relative humidity, there is a possibility of condensation of water vapour, as seen in Fig. 15.2, and the mixture will then consist of saturated air and the condensate.

If the DBT of the mixture falls below 0°C, the condensate may eventually freeze.

It may be noted that due to condensation, the specific humidity of the mixture ω_4, will be reduced to below ω_3 given by Eq. (15.1), as shown in Fig. 15.2 (b). Correspondingly, the temperature of the air would be increased to t_4 from t_3 due to the release of the latent heat of the condensate. Now, if ω_c represents the mass of the condensate per unit mass of the mixture, we have by moisture and energy balance

$$\omega_c = \omega_3 - \omega_4$$

or $$\omega_4 = \frac{m_{a_1}\omega_1 + m_{a_2}\omega_2}{m_{a_1} + m_{a_2}} - \omega_c \tag{15.4}$$

and $$m_{a_4} h_4 + m_{a_4}\omega_c h_{f_4} = m_{a_4} h_3$$

or $$h_4 = \frac{m_{a_1} h_1 + m_{a_2} h_2}{m_{a_1} + m_{a_2}} - \omega_c h_{f_4} \tag{15.5}$$

where h_{f4} is the enthalpy of the condensate at temperature t_4 of the mixture. The two variables to be solved from Eqs (15.4) and (15.5) are t_4 and ω_c. By assuming different values of t_4 and substituting for ω_4, h_4 and h_{f_4}, the two equations can be solved by trial and error to obtain the final state after mixing.

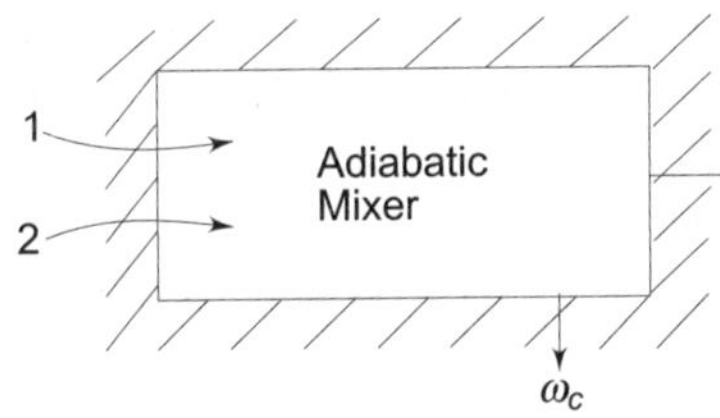

Fig. 15.2(a) Adiabatic mixer with condensation

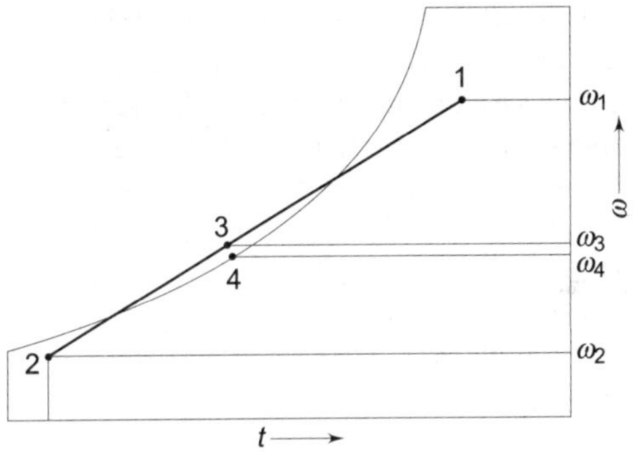

Fig. 15.2(b) Mixing process with condensation

Mixing with condensation rarely occurs in air-conditioning processes. However, during winter it is a common phenomenon in nature for fog or frost to form. It is due to the mixing of cold air near the earth's surface with the humid and warm air which develops towards the evening or after rains.

15.2 BASIC PROCESSES IN CONDITIONING OF AIR

Four basic thermodynamic processes and four combinations of processes by which the state of moist air can be altered are shown in Fig. 15.3. They are:

(i) Sensible heating – process OA
(ii) Sensible cooling – process OB
(iii) Humidifying – process OC
(iv) Dehumidifying – process OD
(v) Heating and humidifying – process OE
(vi) Cooling and dehumidifying – process OF
(vii) Cooling and humidifying – process OG
(viii) Heating and dehumidifying – process OH.

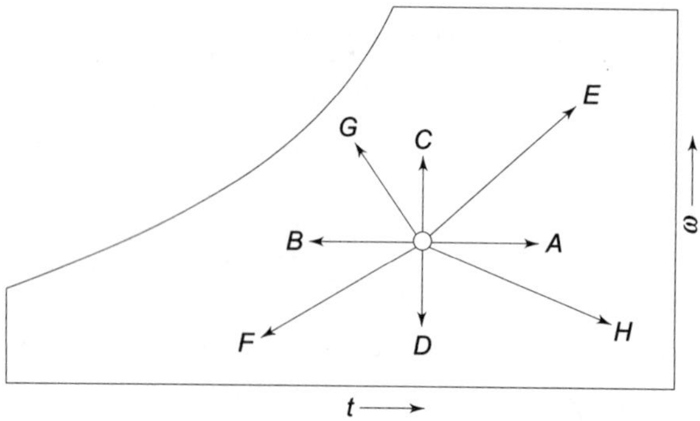

Fig. 15.3 Basic psychrometric processes

The first two processes, viz., sensible heating and cooling, involve only a change in the dry bulb temperature, whereas the processes of humidifying and dehumidifying involve a change in the specific humidity. Thus, when the state of the air moves from O to A or to B, there is no change in the moisture content of the air; if the state changes from O to C or to D, the DBT remains constant. However, most practical moisture-transfer processes involve a change in temperature as well. The last four fundamental processes listed above involve both changes in temperature as well as in humidity.

We shall, now, consider calculations for processes involving changes in temperature and humidity.

15.2.1 Sensible Heat Process-Heating or Cooling

When the state of moist air is altered along the ω = constant line such as AB in Fig. 15.4, the heat has to be transferred which goes to change the temperature of the air. The heat transfer, is given by

$$\begin{aligned} Q_S &= m_a (h_B - h_A) \\ &= m_a C_p (t_B - t_A) \\ &= m_a C_{p_a} (t_B - t_A) + m_a \omega C_{p_v} (t_B - t_A) \\ &= m_a (1.005 + 1.88\,\omega)(t_B - t_A) \quad (15.6) \end{aligned}$$

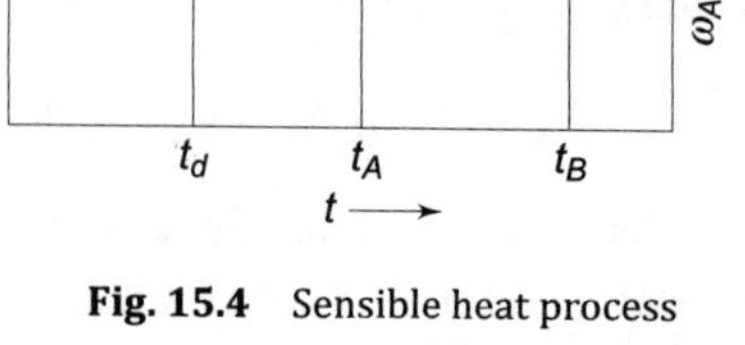

Fig. 15.4 Sensible heat process

where C_p is the humid specific heat. This heat, denoted by the subscript S, is called the *sensible heat*. If a building to be air conditioned receives or loses heat due to transmission or other reasons, it is supposed to have *sensible heat load*. Heat gain in buildings will require the conditioning of air to lower temperatures, causing a *cooling load* on the air-conditioning equipment. However, heat loss in buildings will require the heating of air causing a *heating load* on the equipment.

In Eq. (15.6), $\dot{m}_a$ denotes the mass flow rate of dry air. Generally the flow rate of air is measured in terms of cubic metres of air per minute (cmm). Then the mass flow rate of dry air can be calculated from

$$\dot{m}_a = \dot{Q}_v \rho \quad (15.7a)$$

where $\dot{Q}_v$ is the volume flow rate of air. Expressing this in cmm, we have

$$\dot{m}_a = \frac{(\text{cmm})\rho}{60} \text{ kg d.a./s} \tag{15.7b}$$

For the purpose of calculations, *standard air* is taken at 20°C and 50 per cent RH. The density of standard air is approximated to 1.2 kg/m^3 d.a. The value of humid specific heat is taken as 1.0216 kJ/(kg d.a.) (K). Substituting these in Eq. (15.6), we obtain

$$\dot{Q}_S = \frac{(\text{cmm})(1.2)(1.0216)}{60}\Delta t$$
$$= 0.0204 \text{ (cmm) } \Delta t, \text{ kW } (= 1.08 \text{ (cfm) } \Delta t \text{ Btu/h}) \tag{15.8}$$

In English units, Δt is in °F.

It may be noted that, whereas, simple heating of moist air can be done to any desired temperature, simple cooling can be done only up to the dew point temperature, viz., up to t_d in Fig. 15.4. Cooling below this temperature will result in the condensation of moisture.

15.2.2 Latent Heat Process-Humidification or Dehumidification

When the state of air is altered along the t = constant line, such as BC in Fig. 15.5, moisture in the form of vapour has to be transferred to change the humidity ratio of the air. This transfer of moisture is given by

$$G = m_a(\omega_C - \omega_B) \tag{15.9}$$

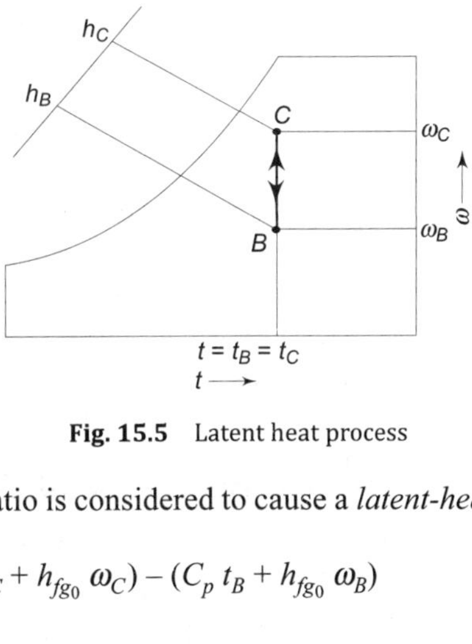

Fig. 15.5 Latent heat process

Because of this change in the humidity ratio, there is also a change in the specific enthalpy of the air given by $(h_C - h_B)$ as shown in Fig. 15.5. In air-conditioning practice, this change in enthalpy due to the change in the humidity ratio is considered to cause a *latent-heat transfer* given by

$$Q_L = m_a(h_C - h_B) = m_a(C_p t_C + h_{fg_0}\omega_C) - (C_p t_B + h_{fg_0}\omega_B)$$
$$= m_a h_{fg_0}(\omega_C - \omega_B)$$
$$= G h_{fg_0} = 2500\, G \tag{15.10}$$

In the above derivation C_p has been taken as the same for both B and C. it is evident that though the term latent heat appears to be a misnomer, the quantity Q_L is proportional to the latent heat of vaporization h_{fg_0}. It implies that if water is to be evaporated or condensed at the temperature of air, the required heat transfer would be Q_L.

Accordingly, if a building gains or loses moisture, it is supposed to have a *latent-heat load.* A gain of moisture will require the condensation of moisture for the dehumidification of air in the conditioning apparatus, and hence a cooling load. On the other hand, a loss of moisture will necessitate the evaporation of water for the humidification of air in the apparatus and hence a heating load. Substituting 2500 for h_{fg_0} in Eq. (15.10) and using the expression for $\dot{m}_a$ from Eq. (15.7b) we obtain

$$\dot{Q}_L = \frac{(\text{cmm})(1.2)(2501)}{60}\,\Delta\omega$$
$$= 50\,(\text{cmm})\,\Delta\omega,\ \text{kW}\ (= 0.68\,(\text{cfm})\,\Delta\omega\ \text{Btu/h}) \tag{15.11}$$

In English units, $\Delta\omega$ is in grains/Ibm d.a. Note that 7000 grains = 1 Ibm.

15.2.3 Total Heat Process

Consider now a change in the state of air along the path AC as shown in Fig. 15.6. This involves both a change in temperature as well as in the humidity ratio. The change in temperature causes a sensible heat load given by

$$Q_S = m_a\,(h_B - h_A)$$
$$= m_a\,C_p\,(t_C - t_A) \tag{15.12}$$

The change in the humidity ratio causes a moisture transfer given by

$$G = m_a\,(\omega_C - \omega_A)$$

and a latent heat load given by

$$Q_L = m_a\,(h_C - h_B)$$
$$= m_a\,h_{fg0}\,(\omega_C - \omega_A) \tag{15.13}$$

Adding Eqs (15.12) and (15.13) we obtain an expression for *total heat load* as

$$Q = Q_S + Q_L$$
$$= m_a\,(h_C - h_A) \tag{15.14a}$$
$$= m_a\,[C_p\,(t_C - t_A) + h_{fg0}\,(\omega_C - \omega_A)] \tag{15.14b}$$

Again, expressing the mass flow rate in cmm, we get

$$\dot{Q} = \frac{(\text{cmm})(1.2)}{60}\,\Delta h$$
$$= 0.02\,(\text{cmm})\,\Delta h, \tag{15.15a}$$

which is the same as

$$\dot{Q} = (\text{cmm})\,(0.0204\,\Delta t + 50\,\Delta\omega),\ \text{kW} \tag{15.15b}$$

15.2.4 Sensible Heat Factor (SHF)

The ratio of the sensible heat transfer to the total heat transfer is termed as the *sensible heat factor*. Thus

$$\text{SHF} = \frac{Q_S}{Q_S + Q_L} = \frac{Q_S}{Q}$$

Substituting from Eqs (15.8) and (15.15), we obtain

$$\text{SHF} = \frac{h_B - h_A}{(h_B - h_A) + (h_C - h_B)} = \frac{h_B - h_A}{h_C - h_A}$$
$$= \frac{0.0204\Delta t}{0.0204\Delta t + 50\Delta\omega} = \frac{0.0204\Delta t}{0.02\Delta h} \tag{15.16}$$

It may be observed from Fig. 15.6 that point B divides the total enthalpy change $(h_C - h_A)$ in the ratio of SHF and 1 – SHF. The sensible heat transfer taking place along AB is proportional to SHF and the latent heat transfer along BC is proportional to 1 – SHF. The process line AC is called the *sensible heat factor line* or *process* or *condition line*.

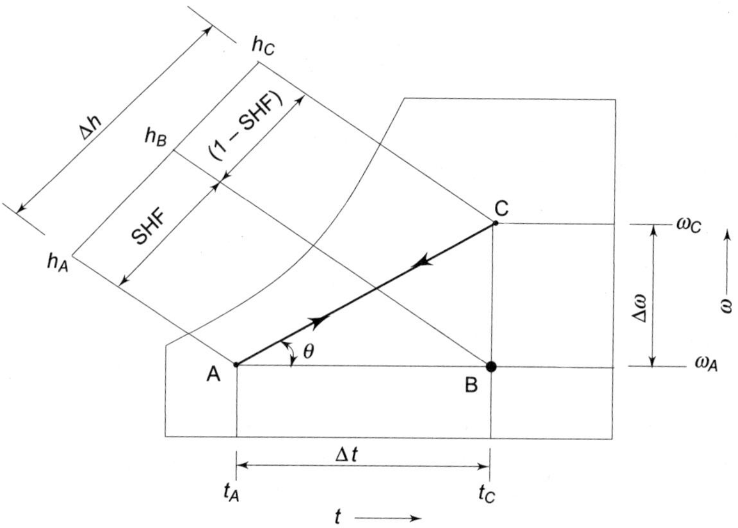

Fig. 15.6 Total heat process

It is obvious that a sensible heat factor of unity corresponds to no latent heat transfer and the SHF line is horizontal on the psychrometric chart. However, a zero SHF line is vertical on the psychrometric chart and implies no sensible heat transfer. An SHF of 0.75 to 0.8 is quite common in air-conditioning practice in a normal dry climate. A lower value of SHF, such as 0.65, implies a high latent heat load, which is quite common in a humid climate.

Simplifying Eq. (15.16), we have

$$\text{SHF} = \frac{1}{1 + 2451\dfrac{\Delta\omega}{\Delta t}} = \frac{1}{1 + \tan\theta} \tag{15.17}$$

where

$$\tan\theta = \frac{\Delta\omega}{\Delta t}$$

$$= \frac{1}{2451}\left(\frac{1}{\text{SHF}} - 1\right) \tag{15.18}$$

We see that θ is the slope of the SHF line AC on the psychrometric chart, which is purely a function of SHF.

Thus, when a process line is to be drawn on the psychrometric chart, the following two things have to be known:

(i) Initial state of air

(ii) Sensible heat factor

The following are the ways in which the SHF line can be drawn on the psychrometric chart with this data:

(i) In the first method, we can calculate tan θ. Then move vertically a certain distance $\Delta\omega$ from the initial state, and then horizontally a distance

$$\Delta t = (\Delta\omega)\,(\tan\theta)$$

Finally, join the point obtained to the initial state point. However, this method is prone to grave error since $\Delta\omega$ is numerically small, and tan θ tends to a value close to zero. The method is given here only to illustrate the principle involved. It is never used.

(ii) In the second method, move vertically a certain enthalpy change Δh_L. This is proportional to the latent heat change. Then move horizontally equivalent to the sensible heat change in terms of enthalpy given by

$$\Delta h_S = (\Delta h_L)\left(\frac{\text{SHF}}{1-\text{SHF}}\right)$$

Again, join the final point to the initial point.

(iii) The third method uses a nomographic method with some charts in which a scale is provided for SHF. There is also a reference point provided which is joined to the appropriate SHF on the scale. Then a line from the initial state point can be drawn parallel to the above line which will give the required SHF line.

(iv) The fourth and the best method is by calculation. In the case of cooling and dehumidification of air, we have from Eq. (15.16)

$$\frac{0.0204(t_C - t_{\text{ADP}})}{0.0204(t_C - t_{\text{ADP}}) + 50(\omega_C - \omega_{\text{ADP}})} = \text{SHF}$$

where, t_C, ω_C are the conditions at C, and t_{ADP} is the *apparatus dew point temperature* (see Chap. 19). It is the temperature at S where the SHF-line CA cuts the saturation curve when extended. This equation can be solved by iteration. Once $t_S = t_{\text{ADP}}$ is found, the point on saturation curve can be joined to the initial state point C to give the SHF line. Or, in general, take any value of t_D on the line and find ω_D by iteration.

15.3 PSYCHROMETRIC PROCESSES IN AIR-CONDITIONING EQUIPMENT

Eight basic psychrometric processes are listed in Sec. 15.2. However, all of them cannot be achieved in practice by the use of known air-conditioning equipment. Even if a certain process can be carried out in a particular range, it may not be possible to achieve it in the complete range. The limitations of practical psychrometric processes and the types of equipment used for them are discussed below. But we will discuss the concept of bypass factor first which is a vital para-meter signifying the performance of air-conditioning equipment.

15.3.1 Bypass Factor

Figure 15.7 shows the process that the moist air undergoes while flowing over a surface. The air enters at 1 and leaves at 2 when the surface is maintained at *S*. In the transfer of heat and water vapour, in any conditioning process, the particles of air stream come in contact with a surface. The state of the *contacted air* is that of saturated air at the temperature of the surface. There is thus the equivalent of perfect contact of a definite portion of the air particles with the surface or no contact or an equivalent bypass of the remaining particles. The *uncontacted air* remains at the entering state. The end state of the air is the same as that produced by the complex mixing of contacted and uncontacted particles, viz., 2 as shown in Fig. 15.7.

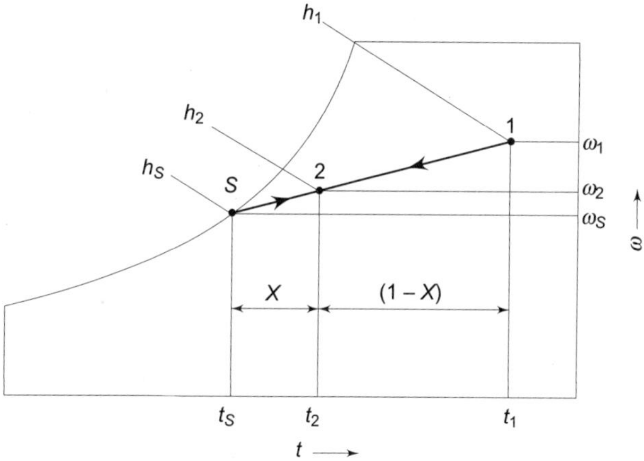

Fig. 15.7 Bypass factor and leaving air state

Thus one can define a *bypass factor* (BPF) of the apparatus representing the fraction of "uncontacted" air in terms of the states 1, 2 and S, as

$$X = \frac{t_2 - t_S}{t_1 - t_S} = \frac{\omega_2 - \omega_S}{\omega_1 - \omega_S} = \frac{h_2 - h_S}{h_1 - h_S}$$

Conversely, one can define a *contact factor* $(1 - X)$ representing a fraction of the contacted air. Thus the bypass factor can be defined in terms of temperature or specific humidity or enthalpy of air. In the absence of any specific data, values from all the three may be considered to be the same. It may be seen in Fig. 15.7 that the resulting state 2 divides the line joining 1 to S in the ratio of X and $1 - X$.

15.3.2 Cooling and Dehumidifying Coils and Apparatus Dew Point of Coil

Moist air can be made to flow over a battery of *cooling coils* which may be of the *direct-expansion type*, or may carry chilled-water or brine as the secondary refrigerant (Fig. 15.8).

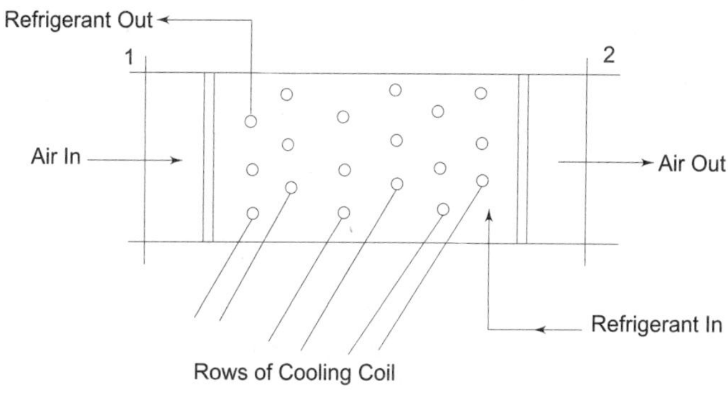

Fig. 15.8 Cooling and dehumidifying coil

Sensible or simple cooling of air takes place when it flows over dry cooling coil whose surface temperature t_S is lower than the dry bulb temperature of the air as shown in Fig. 15.9. The air is cooled along the constant DPT line. The leaving air state depends on the bypass factor of the coil. Thus in Fig. 15.9, the leaving air state is 2 for a bypass factor of X. The bypass factor can be decreased and the leaving air state can be made to approach the coil surface temperature by increasing the number of rows in the coil thus improving the contact between air and surface. There is a minimum limit to the coil temperature for simple cooling, viz., t_d which is equal to the dew point temperatures of entering air.

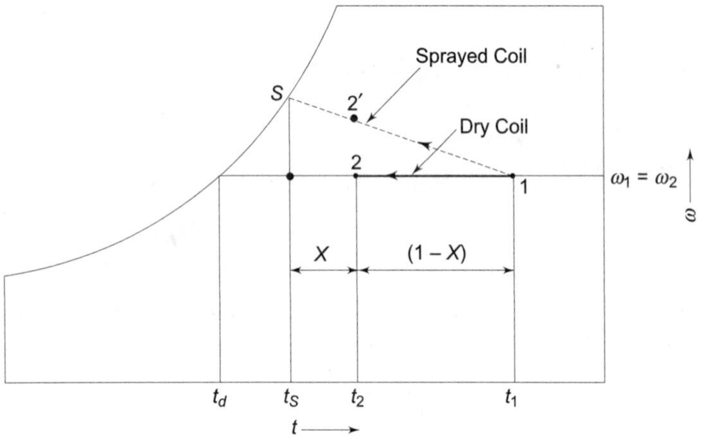

Fig. 15.9 Simple cooling, and sprayed coil processes

Figure 15.9 also shows a process along a broken line 1 – S, the leaving air state being 2′. Such a process would, however, occur if the coil were wet, say, it were sprayed with water at temperature t_S. It is seen that the cooling process is accompanied with humidification. Such coils are used, often, in the air conditioning of textile mills.

There is no method by which one can obtain simple dehumidification of air. Dehumidification processes are accompanied with either simultaneous cooling or heating of air.

Dehumidification will take place along with cooling if moist air flows over a cooling coil, whose mean surface temperature t_S is below the dew point temperature t_d of the entering air, as shown in Fig. 15.10.

This temperature t_S of the cold surface is named *apparatus dew point of the coil*, or simply as *coil ADP*. Between the air and the surface, both sensible and latent heat transfers will take place. For sensible heat transfer, the driving potential is the temperature differential $(t - t_S)$. For latent heat transfer, the driving potential is the partial pressure difference $(p_v - p_{v_S})$ or the corresponding specific humidity difference $(\omega - \omega_S)$, where p_{v_S} is the partial pressure of water vapour in the air in the immediate vicinity of the cold surface at temperature t_S. The actual path followed in the process will be a curve 1 – S depending on the heat and mass transfer coefficients. We shall assume this path to be a straight line 1 – S. The leaving air state will then be at 2 as a result of the bypass factor of the coil.

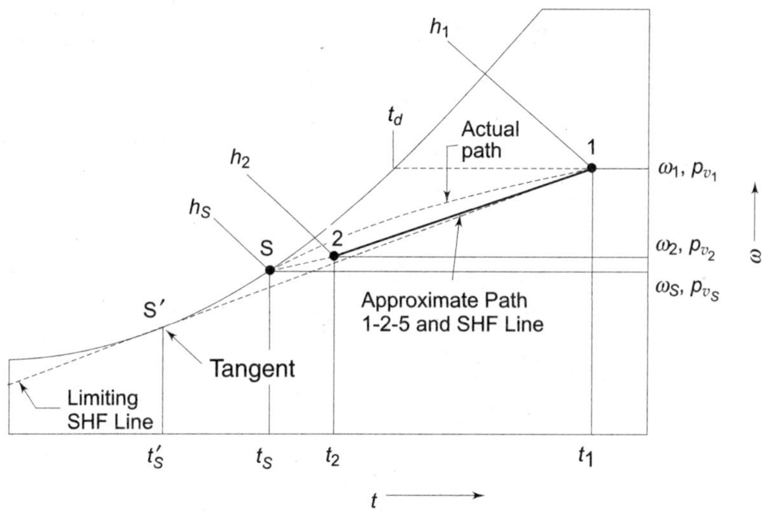

Fig. 15.10 Cooling and dehumidification

There is, however, a limitation to the practical limit of this process. This limit is up to the condition line $1 - S'$ in Fig. 15.10, where it becomes a tangent to the saturation line. A sensible heat factor lower than that of the line $1 - S'$ cannot be achieved in any conditioning process with the given entering air state at 1. Even for the process $1 - S'$, a very low value of the cooling surface temperature t'_S would be required, resulting in a low coefficient of performance of the refrigeration unit. Such a situation arises when the latent heat load is high and the SHF line is steep (very low value of SHF).

Example 15.3 *39.6 cmm of a mixture of recirculated room air and outdoor air enter a cooling coil at 31°C DB and 18.5°C WB temperatures. The effective surface temperature of the coil is 4.4°C. The surface area of the coil is such as would give 12.5 kW of refrigeration with the given entering air state. Determine the dry and wet bulb temperatures of the air leaving the coil and the coil bypass factor.*

Solution Refer to Fig. 15.10.

At the apparatus dew point $\omega_s = 5.25$ g/kg d.a.

$h_S = 17.7$ kJ/kg d.a.

State of entering air $\omega_1 = 8.2$ g/kg d.a.

$v_1 = 0.872$ m^3/kg d.a.

$h_1 = 52.5$ kJ/kg d.a.

Mass flow rate of dry air $\dot{m}_a = \dfrac{\dot{Q}_v}{v} = \dfrac{39.6}{0.872} = 44.41$ kg d.a./min

Cooling load per kg of dry air $h_1 - h_2 = \dfrac{\dot{Q}}{\dot{m}_a} = \dfrac{(12.5)(60)}{44.41} = 16.89$ kJ/kg d.a.

Enthalpy of air leaving the coil $h_2 = 52.5 - 16.89 = 35.61$ kJ/kg d.a.

Equation for the condition line $\dfrac{h_1 - h_2}{h_1 - h_S} = \dfrac{\omega_1 - \omega_2}{\omega_1 - \omega_S}$

$$\frac{52.5 - 35.61}{52.5 - 17.7} = \frac{8.2 - \omega_2}{8.2 - 5.25}$$

whence $\omega_2 = 6.77$ g w.v./kg d.a.

Dry and wet bulb temperatures of air leaving the coil for calculated values of h_2, ω_2 from psychrometric chart

$$t_2 = 18.6°\text{C}$$

$$t_2' = 12.5°\text{C}$$

Coil bypass factor $X = \dfrac{h_2 - h_S}{h_1 - h_S} = \dfrac{35.61 - 17.7}{52.5 - 17.7} = 0.515$ (very high)

15.3.3 Heating Coils

Sensible or simple heating of air takes place when it flows over a *heating coil* similar to the cooling and dehumidifying coil of Fig. 15.8, whose surface temperature t_S is higher than the dry bulb temperature of air (Fig. 15.11). There is no critical limit to the coil temperature for sensible heating. The heating medium flowing through the coil is usually steam, or hot gases from a furnace. In general, three methods are commonly employed for winter heating of air.

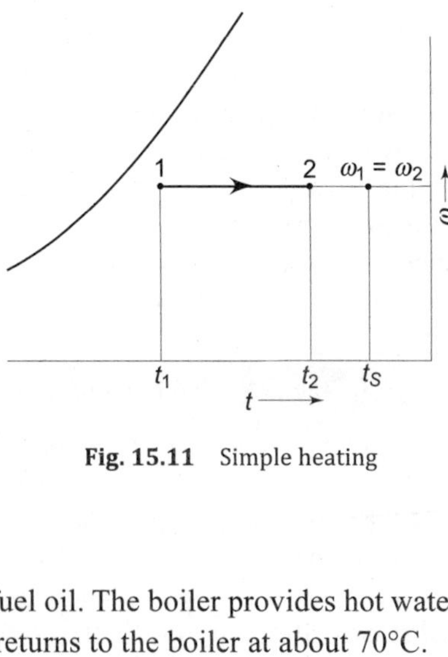

Fig. 15.11 Simple heating

(i) Hot water or steam coils
(ii) Direct-fired furnace gases coils
(iii) Finned electric strip heaters.

In hot water coils, boilers are run on light fuel oil. The boiler provides hot water at about 92°C to the heating coils. The water returns to the boiler at about 70°C.

Furnaces are either oil-fired (at 75% efficiency) or gas-fired (at 80–85% efficiency)

15.3.4 Air Washer

Figure 15.12 shows the schematic representation of an *air washer*. It involves the flow of air through a spray of water. During the course of flow, the air may be cooled or heated, humidified or dehumidified, or simply adiabatically saturated, depending on the mean surface temperature of water. The water is, accordingly, externally cooled or heated or simply recirculated by a pump. Make-up water is added for any loss in the case of humidification of air. Eliminator plates are provided to minimise the loss of water droplets.

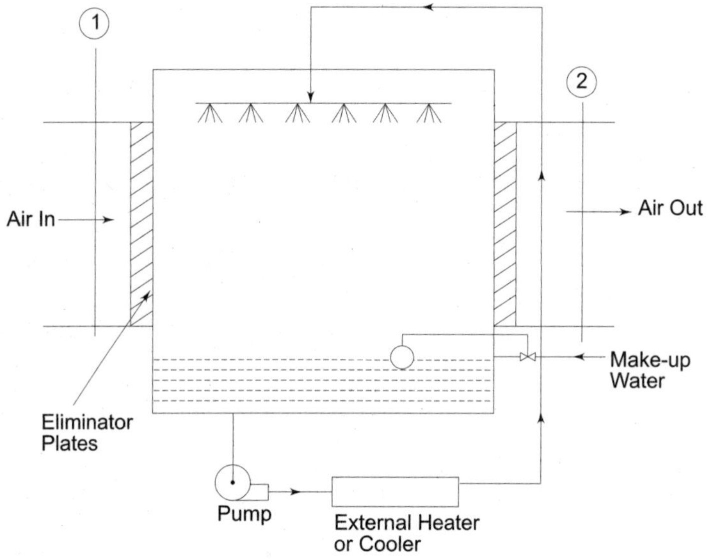

Fig. 15.12 Air washer

Figure 15.13 shows the thermodynamic changes of state of air along paths 1–2 in an air washer, depending on the mean surface temperature of water droplets t_S which is equal to the actual temperature of water t_w.

Thus, the droplets of water act as wetted surface, and both sensible and latent heat transfers take place. Their directions depend on the temperature and vapour pressure potentials. The following processes are possible:

Process 1-2 A: *Heating and humidification* ($t_S > t_1$)

The mean surface temperature of water is greater than the dry bulb temperature of air. The water is externally heated.

Process 1-2B: *Humidification* ($t_S = t_1$)

The mean surface temperature of water is equal to the dry bulb temperature of air. The enthalpy of air increases. Hence the water is required to be externally heated.

Process 1-2C: *Cooling and humidification* ($t_1' < t_S < t_1$)

The mean surface temperature of water is less than the dry bulb temperature of air but greater than the wet bulb temperature of air. Though the air is cooled, its enthalpy increases as a result of humidification. The water is, therefore, required to be externally heated.

Process 1-2D: *Adiabatic saturation* ($t_1' = t_S$)

This is the case of *pumped recirculation* of water without any external heating or cooling as discussed in Sec. 14.5. The recirculated water reaches the equilibrium temperatures which is equal to the thermodynamic wet bulb temperature of air.

Process 1-2E: *Cooling and humidification* ($t_d < t_S < t_1'$)

The process is similar to 1-2C with the difference that the enthalpy of air decreases in this case. Accordingly, water is required to be externally cooled.

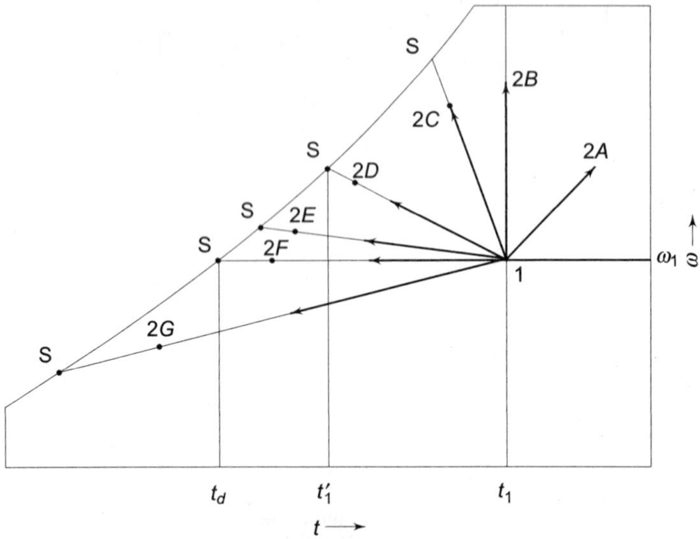

Fig. 15.13 Range of psychrometric processes with an air washer

Process 1-2F: *Cooling* ($t_S = t_d$)

The temperature of water is equal to the dew point temperature of air. Water is required to be cooled.

Process 1-2G: *Cooling and dehumidification* ($t_S < t_d$)

The mean water surface temperature is lower than the dew point temperature of air. Air is simultaneously cooled and dehumidified. The process is exactly similar to that of a cooling and dehumidifying coil. Again, the limiting process is along the condition line tangent to the saturation line drawn from initial state 1.

It is thus seen that the air washer affords means for an *year-round air-conditioning system*.

Consider the energy balance of an air washer, the mass balance of which is shown in Fig. 15.14. Let $\dot{m}_a$ and $\dot{m}_w$ be the mass flow rates of dry air and water respectively. The energy balance gives

$$\dot{m}_a\ (h_2 - h_1) = \dot{m}_w\ C_{p_w}\ t_{w_3} - [\ \dot{m}_w\ -\ \dot{m}_a\ (\omega_2 - \omega_1)]\ C_{p_w}\ t_{w_4}$$

or
$$\dot{m}_a\ (h_2 - h_1) = \dot{m}_w\ C_{p_w}\ (t_{w_3} - t_{w_4}) + \dot{m}_a\ (\omega_2 - \omega_1)\ C_{p_w}\ t_{w_4}$$

Neglecting the effect of the temperature of water in the last term, we obtain by combining it with the left-hand side term

$$\dot{m}_a\ \left(\Sigma_2 - \Sigma_1\right) = \dot{m}_w\ C_{pw}\ (t_{w_3} - t_{w_4}) \tag{15.19a}$$

Thus for any section of the air washer

$$\dot{m}_a\ \mathrm{d}\Sigma\ = -\ \dot{m}_w\ C_{p_w}\ \mathrm{d}t_w \tag{15.19b}$$

For the case of the adiabatic saturation process, $\mathrm{d}\Sigma = 0$.

Hence
$$\mathrm{d}t_w\ = 0 \text{ and } t_{w_3} = t_{w_4}.$$

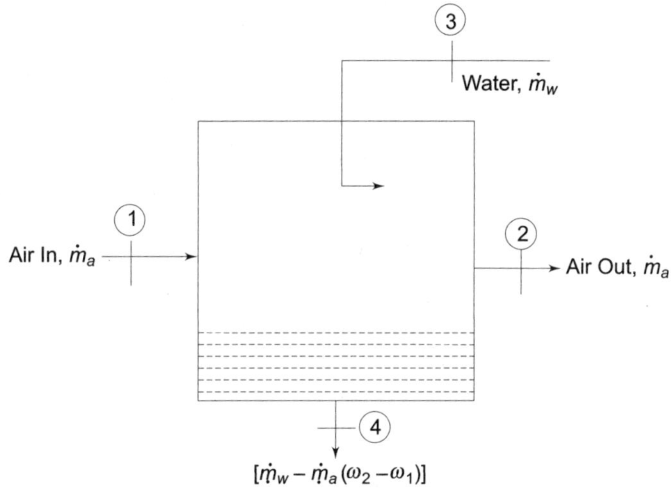

Fig. 15.14 Mass balance of an air washer

Thus we can make the following conclusions:

(i) If the spray water is heated external to the washer, the WBT of air increases.

(ii) If the spray water is cooled external to the washer, the WBT of air decreases.

(iii) If the spray water is neither heated nor cooled external to washer, the WBT of air is not changed. It is an adiabatic saturation process.

We define the humidifying efficiency of an air washer as

$$\eta_H = \frac{h_2 - h_1}{h_S - h_1} = \frac{\omega_2 - \omega_1}{\omega_S - \omega_1}$$

It can be seen that the bypass factor X can be expressed as

$$X = \frac{\omega_S - \omega_2}{\omega_S - \omega_1} = 1 - \frac{\omega_2 - \omega_1}{\omega_S - \omega_1} = 1 - \eta_H$$

Thus the humidifying efficiency is the same as the contact factor.

15.3.5 Use of Hygroscopic Solutions in Air Washer

Hygroscopic solutions, such as brines, glycols, etc., exert lower vapour pressures as compared to pure water at the same temperature as shown in Fig. 15.15. The saturation curve A on the psychrometric chart represents the vapour pressure of water p_{v_A}. Representative vapour-pressure curves for certain hygroscopic solutions are shown as B, C, D, E., etc. The vapour pressures p_{v_B}, p_{v_C}, etc., of the solutions are lower than the vapour pressure p_{v_A} of water at the same temperature t.

Now, if a hygroscopic solution such as C at temperature t is circulated in an air washer instead of water at the same temperature, the condition line will be 1-C instead of 1-A for the initial state 1 of air. Thus, a spray of the hygroscopic solution is more effective for dehumidification. The solution after absorbing moisture, however, becomes dilute and has to be regenerated again by heating and driving off the water vapour.

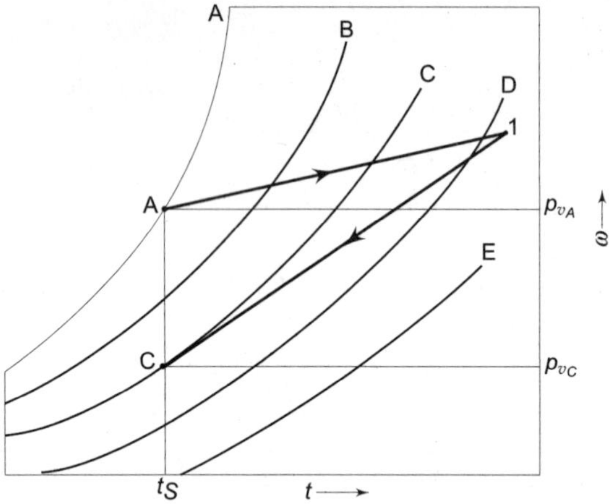

Fig. 15.15 Dehumidification of air by hygroscopic solutions

15.3.6 Adiabatic Dehumidifier

Adiabatic dehumidification is based on the principle of adsorption, viz., capillary action. The vapour which is condensed at the surface of the adsorbent is drawn into capillaries, thereby reducing the vapour pressure at the surface causing a pressure gradient, and hence a mass transfer from the passing air stream to the absorbing surface. As the capillaries get filled with water, the attraction decreases and the rate of dehumidification falls off.

Thermodynamically, an adsorption process is the reverse of the adiabatic saturation process as shown in Fig. 15.16. As the air passes over the adsorbing surface, water vapour flows to the surface through the air film, condenses and releases its latent heat which raises the adsorbent and air temperatures. Thus the heat of condensation supplies the sensible heat for the heating of air.

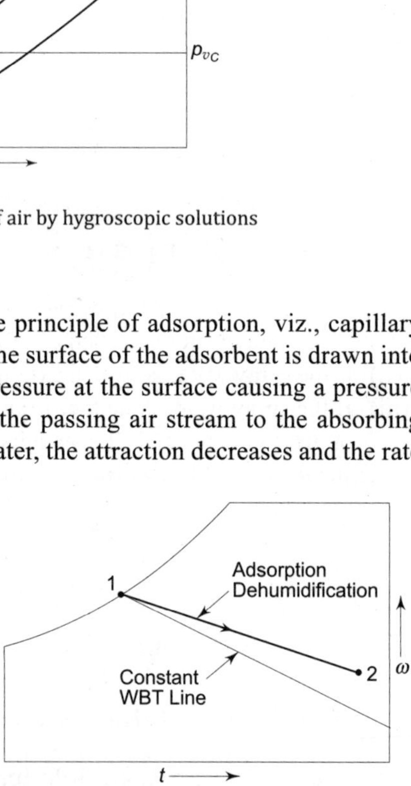

Fig. 15.16 Adsorption dehumidification process

In actual practice, however, the process is accompanied with a release of an additional heat called the *heat of adsorption*. This heat, with adsorbents, such as silica gel and activated alumina, is very large. Thus the sensible heat gain of air exceeds the loss of latent heat and the process line 1–2 lies above the constant WBT line, as shown in Fig. 15.16.

It is to be noted that after adsorption, the material becomes saturated and has to be reactivated by heating as in the case of hygroscopic solutions.

15.3.7 Water Injection

Let liquid water at temperature t_f be injected and sprayed into a flowing air stream with the help of nozzles. The condition of the air will change depending on the amount of water that evaporates. The enthalpy of vaporization will come from the enthalpy of the air.

Let us consider that the amount of water that has been evaporated m_v is exactly equal to the amount injected. The process line is as shown in Fig. 15.17. The air flow rate is m_a. The mass and enthalpy balances give

$$\omega_2 = \omega_1 + \frac{m_v}{m_a} \tag{15.20}$$

$$h_2 = h_1 + \frac{m_v}{m_a} h_f$$

$$= h_1 + (\omega_2 - \omega_1)h_f \tag{15.21}$$

where h_f is the enthalpy of liquid water. It is evident from Eq. (15.21) that if water is injected at the wet bulb temperature of the air, the sigma heat function is constant, and the process follows the constant WBT line 1–2b. Otherwise, the process follows line 1–2a or 1–2c, depending on whether the temperature of water is lower or higher than the WBT of air. Nevertheless since the term $(\omega_2 - \omega_1)\ h_t$ is extremely small compared to h_1 and h_2, lines 1–2a and 1–2c are very close to line 1–2b, irrespective of the temperature of the injected water.

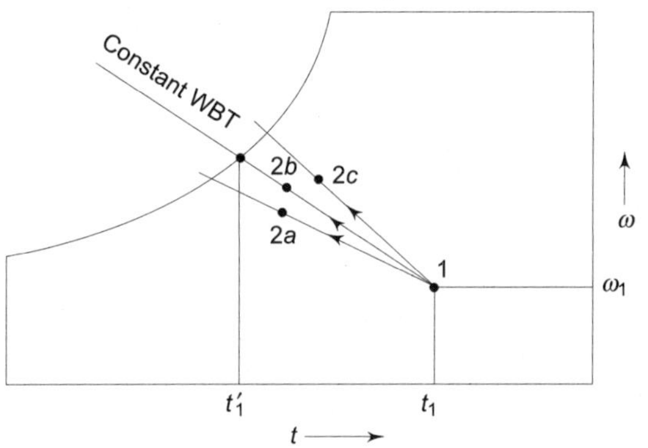

Fig. 15.17 Process with liquid water injection

Note *In recent years, this method of evaporative cooling has gained popularity specially in places where there is shortage of water since it eliminates at least 2–3% water loss which occurs in air washer equipment. Also, it provides for individual control of supply conditions for different spaces. And the equipment is simple and economical.*

15.3.8 Steam Injection

Steam is normally injected into fresh outdoor air which is then supplied for the conditioning of textile mills where high humidities have to be maintained. The process can be analysed by considering mass and energy balances. If m_v is the mass of steam supplied with enthalpy h_v and m_a the mass of dry air, then the leaving air state is given by

$$\omega_2 = \omega_1 + \frac{m_v}{m_a} \tag{15.22}$$

$$h_2 = h_1 + \frac{m_v}{m_a} h_v \tag{15.23}$$

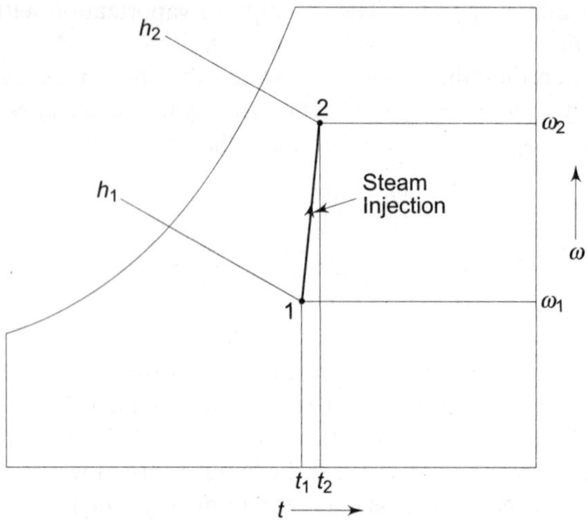

Fig. 15.18 Process with steam injection

The process is shown in Fig. 15.18. The dry bulb temperature of air changes very little during the process.

Note *The reference state for zero enthalpy for* h_1, h_2 *and* h_v *must be the same.*

Example 15.4 *Moist air enters a chamber at 5°C DBT and 2.5°C thermodynamic WBT at a rate of 90 cmm. The barometric pressure is 1.01325 bar. While passing through the chamber, the air absorbs sensible heat at the rate of 40.7 kW and picks up 40 kg/h of saturated steam at 110°C. Determine the dry and wet bulb temperatures of the leaving air.*

Solution This is a case of simple heating and humidification of air by the addition of steam as shown in Fig. 15.19. The air mass flow rate is

$$\dot{m}_a = \frac{(\text{cmm})60}{v} = \frac{90 \times 60}{0.792} = 6820 \text{ kg d.a./h}$$

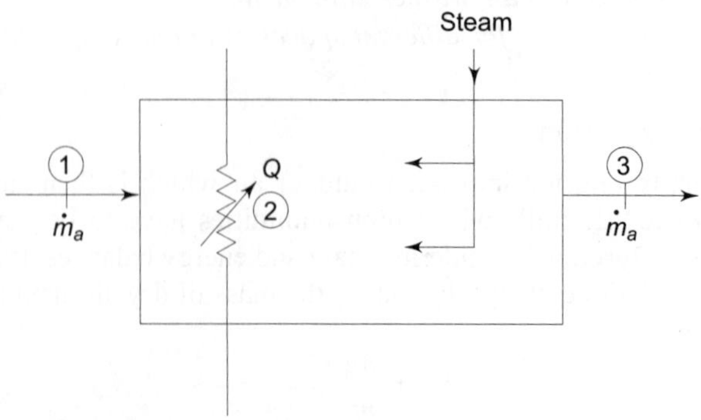

Fig. 15.19 (a) System for Example 15.4

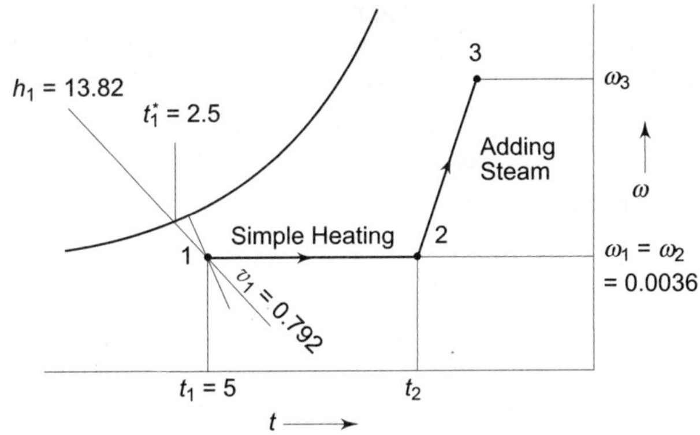

Fig. 15.19 (b) Psychrometric processes for Example 15.4

where 0.792 m^3/kg d.a. is the specific volumes of the entering air.

By moisture balance

$$\dot{m}_a\ (\omega_3 - \omega_1) = 40$$

$$\omega_3 = \omega_1 + \frac{40}{\dot{m}_a} = 0.0036 + \frac{40}{6820}$$

$$= 0.00947 \text{ kg w.v./kg d.a.}$$

By energy balance

$$\dot{m}_a\ (h_3 - h_1) = (40.7)\ (3600) + 40\ h_v$$

where h_v = 2691.3 kJ/kg is the enthalpy of saturated steam at 110°C.

Thus

$$h_3 = 13.82 + \frac{1}{6820}\ [146{,}540 + 40\ (2691.3)] = 51.1 \text{ kJ/kg d.a.}$$

From psychrometric chart, at 3

$$\text{DBT} = 26.5°\text{C}$$

$$\text{WBT} = 18.1°\text{C}$$

15.4 SIMPLE AIR-CONDITIONING SYSTEM AND STATE AND MASS RATE OF SUPPLY AIR

The problem of air-conditioning a space essentially reduces to the calculation of the state and mass rate of air to be supplied to the space-necessary to pick up its sensible and latent heat loads. For the simplest air-conditioning system, consider a space which is to be maintained at the room or inside conditions of, say, dry bulb temperature t_i and humidity ratio ω_i.

Let Q_S represent the sensible heat gain and G, the moisture gain of the room from internal and external sources. In the usual nomenclature we may denote the room sensible heat and the room latent heat as RSH and RLH respectively. Thus the room total heat, viz., RTH is

$$\text{RTH} = \text{RSH} + \text{RLH}$$

In the case of the air-conditioning apparatus, the return or recirculated air from the room is conditioned to a supply dry bulb temperature of t_s and a humidity ratio of ω_s. Air is then supplied to the space by a supply air fan. The schematic diagram of the whole system is shown in Fig. 15.20. Assuming steady-state conditions and the supply air flow rate as $(\text{cmm})_s$, we have the following two equations by sensible and latent heat balances.

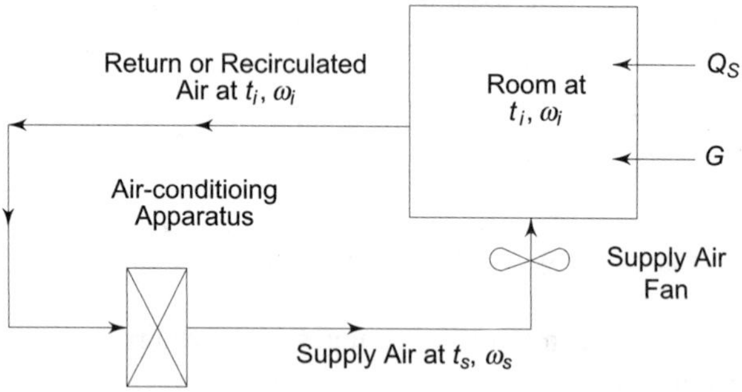

Fig. 15.20 Simple air-conditioning system

Sensible Heat Balance

$$\text{RSH} = \dot{Q}_S = \dot{m}_a C_p (t_i - t_s)$$
$$= 0.0204\,(\text{cmm})_s\,(t_i - t_s) \tag{15.24}$$

Moisture or Latent Heat Balance

$$\text{RLH} = \dot{Q}_L = \dot{m}_a (h_{fg})_0 (\omega_i - \omega_s)$$
$$= 2500\text{G} = 50\,(\text{cmm})_s\,(\omega_i - \omega_s) \tag{15.25}$$

We thus have only two equations and three independent variables, viz. t_s, ω_s and $(\text{cmm})_s$, to solve. Hence one of the three variables is to be known in advance or fixed from experience. Generally, it is either the supply air flow rate or the supply air temperature. The other two parameters can then be calculated.

Thus, we find that there will be one supply air state corresponding to each supply air rate. An innumerable number of such combinations are possible. But for all these states we have from Eqs (15.24) and (15.25)

$$\frac{0.0204(t_i - t_s)}{0.0204(t_i - t_s) + 50(\omega_i - \omega_s)} = \frac{Q_S}{Q_S + Q_L} \tag{15.26}$$

which is the same as Eq. (15.16) for the SHF line. Thus all the supply air states lie on a line, the slope of which is governed by Eq. (15.26). The locus of these states is the sensible heat factor or the condition line drawn from the room state. Figure 15.22 shows a number of such supply air states, s_1, s_2, $\cdots$, etc., in a summer air-conditioning plant, that satisfy the governing equations.

A similar treatment is possible for a winter air-conditioning system. The following example gives an illustration of the calculation procedure for an industrial problem in which an adsorption type of dehumidifier is used.

Example 15.5 *A room for process work is maintained at 20°C DBT and 25% RH. The outside air is at 40°C DB and 25°C WB temperatures. Twelve cmm of fresh air is mixed with a part of recirculated air and passed over the adsorption dehumidifier. It is then mixed with another part of recirculated air and is sensibly cooled in a cooler before being supplied to the room at 14°C.*

The room sensible and latent heat gains are 6 and 0.8 kW respectively. Calculate the volume flow rate of the air entering the dehumidifier and the amount of heat removed in the cooler. The performance of the adsorbent material is as follows:

Entering moisture content, g/kg d.a.	2.86	4.29	5.7	7.15	8.57	10.0	11.43	12.86	14.29
Leaving moisture content, g/kg d.a.	0.43	0.57	1.0	1.57	2.15	2.86	3.57	4.57	5.23

The heat of adsorption may be taken as 390 kJ/kg of the moisture adsorbed.

Solution The schematic arrangement of the plant is shown in Fig. 15.21 (a) and its processes in Fig. 15.21(b).

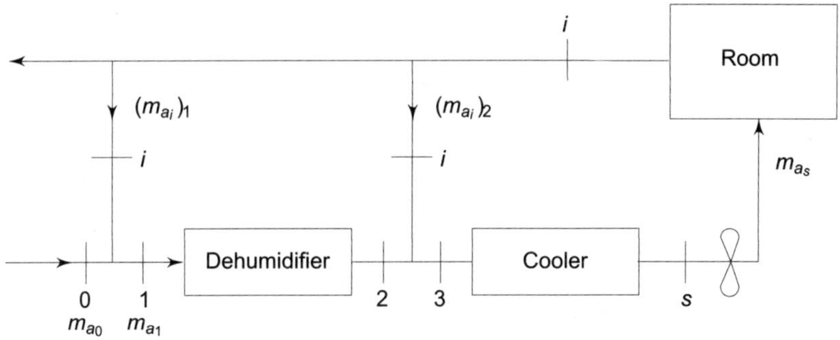

Fig. 15.21(a) System for Example 15.5

Supply air rate

$$\dot{m}_{as} = \frac{\dot{Q}_S}{C_p(t_i - t_s)} = \frac{6}{(1.0216)(20-14)} = 0.979 \text{ kg d.a./s}$$

Latent heat balance gives the value of ω_s which is the same as ω_3.

$$\dot{m}_{a_s}(\omega_i - \omega_s)\, 2500 = \dot{Q}_L$$

whence

$$\omega_s = \omega_1 - \frac{\dot{Q}_L}{m_{a_s}(2500)}$$

$$= 0.00365 - \frac{0.8}{(0.979)(2500)} = 0.0033 \text{ kg w.v./kg d.a.}$$

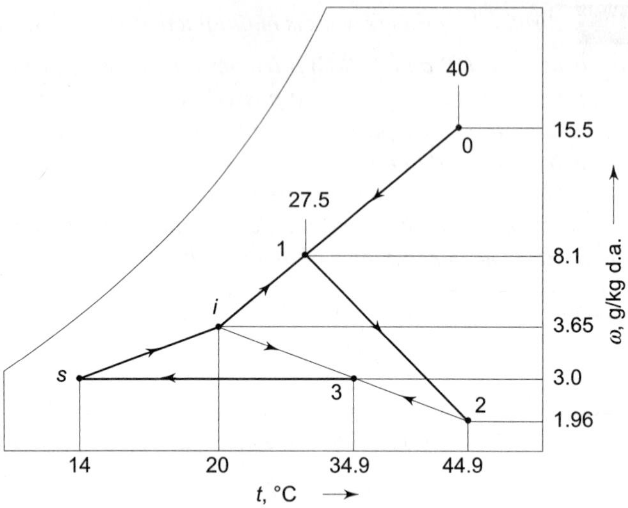

Fig. 15.21(b) Psychrometric processes for Example 15.5

Fresh air $\dot{m}_{a_0} = \dfrac{\dot{Q}_v}{v_0} = \dfrac{12}{(60)(0.91)} = 0.22$ kg d.a./s

Total recirculated air

$$\dot{m}_{a_i} = \dot{m}_{a_s} - \dot{m}_{a_0} = 0.979 - 0.22 = 0.759 \text{ kg d.a./s}$$

Of this $(m_{ai})_1$ is added before the dehumidifier, and $(m_{ai})_2$ after the dehumidifier.

Mixing before the dehumidifier

$$m_{a_0}\,\omega_0 + (m_{a_i}),\,\omega_i = [m_{a_0} + (m_{a_i})_1]\,\omega_1$$

$$0.22\,(15.5) + (m_{a_i})_1\,(3.65) = [0.22 + (m_{a_1})_1]\,\omega_1$$

$$\Rightarrow \qquad (m_{a_i})_1 = \frac{0.22\omega_1 - 3.41}{3.65 - \omega_1} \qquad \text{(I)}$$

Mixing after the dehumidifier

$$(\dot{m}_{a_1})\,\omega_i + \dot{m}_{a_2}\,\omega_2 = \dot{m}_{a_s}\,\omega_s$$

$$[0.759 - (\dot{m}_{a_i})_1]\,(3.65) + [0.22 + (\dot{m}_{a_i})_1]\,\omega_2 = 0.759\,(3.3)$$

or

$$(\dot{m}_{a_i})_1 = \left(\frac{0.22\omega_2 + 0.195}{3.65 - \omega_2}\right) \qquad \text{(II)}$$

Equating (I) and (II), we have

$$\frac{0.22\omega_1 - 3.41}{3.65 - \omega_1} = \frac{0.22\omega_2 + 0.195}{3.65 - \omega_2} \qquad \text{(III)}$$

The relationship between the entering moisture content ω_1 and the leaving moisture content ω_2 is given in terms of the performance data. Solving Eq. (III) simultaneously or graphically with the tabulated data, we get

$$\omega_1 = 8.1 \text{ g w.v./kg d.a.}$$
$$\omega_2 = 1.96 \text{ g w.v./kg d.a.}$$

From Eq. (I)

$$(\dot{m}_{a_i})_1 = \frac{0.22(8.1) - 3.41}{3.65 - 8.1} = 0.366 \text{ kg d.a./s}$$

and $$(\dot{m}_{a_i})_2 = m_{a_i} - (m_{a_i})_1 = 0.759 - 0.366 = 0.393 \text{ kg d.a./s}$$

Dry air flow through the dehumidifer

$$\dot{m}_{a_1} = m_{a_s} - (m_{a_1})_2 = 0.979 - 0.393 = 0.586 \text{ kg d.a./s}$$

Heat liberated due to the condensation of moisture

$$\dot{Q}_{\text{cond}} = \dot{m}_{a_1} (\omega_1 - \omega_2) h_{fg} = 0.586 \left(\frac{8.1 - 1.96}{1000}\right)(2500) = 9.0 \text{ kW}$$

Heat of adsorption

$$\dot{Q}_{\text{ads}} = \dot{m}_{a_1} (\omega_1 - \omega_2)(390) = 0.586 \left(\frac{8.1 - 1.96}{1000}\right)(390) = 1.5 \text{ kW}$$

Temperature rise of air in the dehumidifier

$$t_2 - t_1 = \frac{\dot{Q}}{\dot{m}_a C_p} = \frac{9 + 1.4}{(0.586)(1.0216)} = 17.4°\text{C}$$

Temperature of air entering the dehumidifier

$$t_1 = \frac{(\dot{m}_{a_1})_1 t_i + \dot{m}_{a_0} t_0}{\dot{m}_a} = \frac{(0.366)(20) + 0.22(40)}{0.586} = 27.5°\text{C}$$

Temperature of air leaving the dehumidifier

$$t_2 = t_1 + 17.4°\text{C} = 27.5 + 17.4 = 44.9°\text{C}$$

Specific volume of air through the dehumidifier from the psychrometric chart

$$v_1 = 0.86 \text{ m}^3\text{/kg d.a.}$$

Volume flow rate of air through the dehumidifier

$$\dot{Q}_{v_1} = \dot{m}_a v_1 = (0.586)(0.86)(60) = 30.2 \text{ cmm}$$

Temperature after mixing after the dehumidifier

$$t_3 = \frac{\dot{m}_{a_1} t_2 + (\dot{m}_{a_i})_2 t_i}{\dot{m}_{a_s}} = \frac{0.586(44.9) + 0.393(20)}{0.979} = 34.9°\text{C}$$

From the psychrometric chart

$$h_3 = 44.6 \text{ kJ/kg d.a.}$$

$$h_s = 22.4 \text{ kJ/kg d.a.}$$

Heat removed in the cooler

$$\dot{Q} = \dot{m}_{a_s} (h_3 - h_s) = 0.979 (44.6 - 22.4) = 21.7 \text{ kW}$$

15.5 SUMMER AIR CONDITIONING-APPARATUS DEW POINT

In summer, the outside air temperature and humidity are both high. The room, therefore, gains heat as well as moisture. It is thus required to cool and dehumidify the recirculated room air in the air-conditioning apparatus either by the use of a cooling coil or by an air washer in which chilled water is sprayed. The process follows the *room sensible heat factor* (RSHF) line. The room sensible heat factor is the ratio of the room sensible heat to the room total heat

$$\text{RSHF} = \frac{\text{RSH}}{\text{RSH} + \text{RLH}} = \frac{\text{RSH}}{\text{RTH}}$$

In a cooling and dehumidification process, the temperature at which the RSHF or condition line intersects the saturation curve is called the *room apparatus dew point* (*Room* ADP). Thus t_{ADP} in Fig. 15.22 denotes the effective surface temperature t_S. The condition line i-S represents the locus of all possible supply air states. One extremity of the condition line is *i* which would be the supply air state with an infinite quantity of supply air. The other extremity is *S* which is the supply state with the minimum supply air requirement corresponding to the given condition line. It is not possible to have any other supply air state with a DBT lower or higher than the ADP on the saturation curve which would satisfy the given condition line.

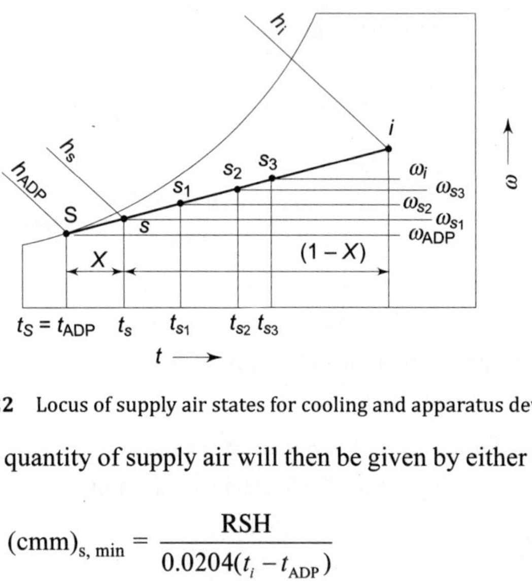

Fig. 15.22 Locus of supply air states for cooling and apparatus dew point

The minimum quantity of supply air will then be given by either of the following three equations:

$$(\text{cmm})_{s,\,\min} = \frac{\text{RSH}}{0.0204(t_i - t_{\text{ADP}})} \tag{15.27a}$$

$$= \frac{\text{RLH}}{50(\omega_i - \omega_{\text{ADP}})} \tag{15.27b}$$

$$= \frac{\text{RTH}}{0.02(h_i - h_{\text{ADP}})} \tag{15.27c}$$

In the case of an actual coil with a bypass factor of *X*, the leaving air state will be at *s*. It is seen that the effect of the bypass factor is to decrease the difference in temperature between the room air and supply air, and hence to increase the supply air quantity over its minimum value.

For any supply air state, the temperature difference $(t_i - t_s)$ available to counteract the room sensible heat load is called the *dehumidified rise* and the corresponding *dehumidified air quantity* $(\text{cmm})_d$ which is equal to the quantity of the supply air, is obtained by the equation for sensible heat balance, and considering the effect of bypass factor

$$(\text{cmm})_d = (\text{cmm})_s = \frac{\text{RSH}}{0.0204(t_i - t_s)} = \frac{\text{RSH}}{0.0204(t_i - t_{\text{ADP}})(1 - X)} \tag{15.28}$$

It can also be found from the equations of latent heat or total heat balances.

15.5.1 Summer Air-Conditioning System with Ventilation Air-Zero Bypass Factor

The introduction of fresh outside air for the ventilation of conditioned space is necessary to dilute the carbon dioxide and odours and other air contaminants for maintaining the purity of room air. Accordingly, the simple air-conditioning system of Fig. 15.20 is modified, so that the supply air to the room comprises fresh air and recirculated room air. An amount equivalent to the fresh air is ejected from the room. The schematic diagram of the system is shown in Fig. 15.23, and the processes for the cases of cooling and dehumidification are shown in Fig. 15.24.

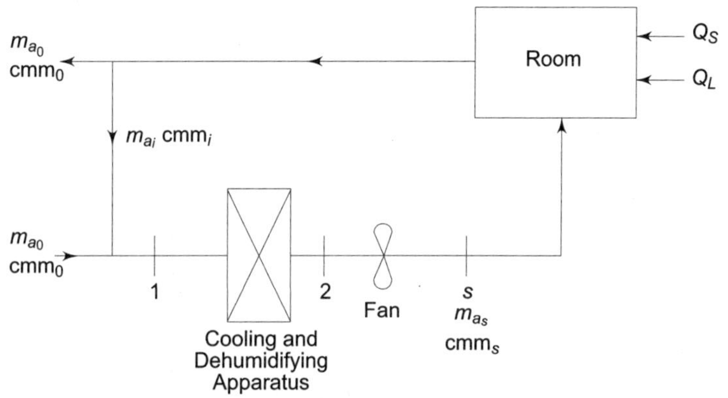

Fig. 15.23 Schematic diagram of system with ventilation air

In Fig. 15.24, 0 and *i* represent the outside and inside air states and 1 is the state of air after the mixing of recirculated room air with ventilation air. The mixture entering the conditioning apparatus comprises recirculated room air m_{ai} and ventilation air m_{a_0}. The room sensible heat factor (RSHF) line is drawn from the inside condition *i* to intersect the saturation curve at room ADP at 2. Point 2 represents the supply air state for a minimum rate of supply air. The line 1–2, therefore, represents the condition line for the apparatus and is called the *grand sensible heat factor* (GSHF) line. It is noted that the line i–2 is the condition line for the room or the RSHF line and line 1–2 is the condition line for the apparatus or the GSHF line intersecting the saturation curve at *coil apparatus dew point* (*Coil* ADP). Note that, in this case, the coil ADP and the room ADP are the same.

In the absence of ventilation air the load on the air-conditioning apparatus is that due to the room sensible heat and room latent heat. When ventilation air is used, there is an additional load on the apparatus named the *ventilation* load equivalent to the change of state of the ventilation air from the outside condition to inside condition. This becomes evident when we write for the total load on the air-conditioning apparatus in terms of the change of state of the mixture air m_{a_s} from 1 to 2. Thus

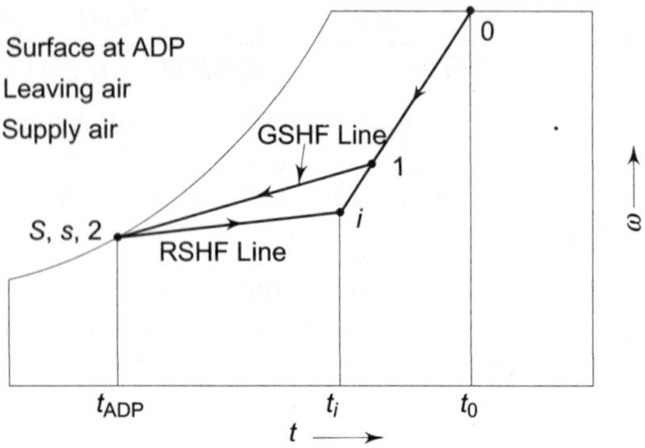

Fig. 15.24 Summer air-conditioning processes with ventilation air and zero bypass factor

$$
\begin{aligned}
\dot{Q} &= \dot{m}_{a_s}(h_1 - h_2) = (\dot{m}_{a_i} h_i + \dot{m}_{a_0} h_0) - \dot{m}_{a_s} h_2 \\
&= (\dot{m}_{a_s} - \dot{m}_{a_0})\, h_i + \dot{m}_{a_0} h_0 - \dot{m}_{a_s} h_2 \\
&= \dot{m}_{a_s}(h_i - h_2) + \dot{m}_{a_0}(h_0 - h_i) \qquad (15.29)
\end{aligned}
$$

The first term on the right-hand side in Eq. (15.29) represents the room load and the second term, the load due to the ventilation air as explained earlier. Accordingly, if $(\text{cmm})_0$ is the outside ventilation air volume flow rate, then the *outside air sensible heat* (OASH) and *outside air latent heat* (OALH) loads are

$$\text{OASH} = \dot{Q}_{S_0} = 0.0204\,(\text{cmm})_0\,(t_0 - t_i) \qquad (15.30)$$

$$\text{OALH} = \dot{Q}_{L_0} = 50\,(\text{cmm})_0\,(\omega_0 - \omega_i) \qquad (15.31)$$

Also for the *outside air total heat* (OATH), we have

$$\text{OATH} = \dot{Q}_0 = \text{OASH} + \text{OALH} = 0.02\,(\text{cmm})_0\,(h_0 - h_i) \qquad (15.32)$$

Note *These equations apply to winter air conditioning as well.*

The break-up of the load on the air conditioning apparatus is now as follows:

Room Load

Sensible	RSH	
Latent	RLH	
Total	RTH	= RSH + RLH

Ventilation Load

Sensible	OASH	
Latent	OALH	
Total	OATH	= OASH + OALH

Air-conditioning Equipment Load

Total sensible	TSH	=	RSH + OASH
Total latent	TLH	=	RLH + OALH
Grand total	GTH	=	TSH + TLH

In Fig. 15.24, the process line 1–2 represents the grand sensible heat factor line for the process in an air-conditioning apparatus. The grand sensible heat factor is the ratio of the total sensible heat to the grand total heat. Thus

$$\text{GSHF} = \frac{\text{TSH}}{\text{TSH+TLH}} = \frac{\text{TSH}}{\text{GTH}}$$

Example 15.6 *The air-handling unit of an air-conditioning plant supplies a total of 4500 cmm of dry air which comprises by weight 20 per cent fresh air at 40°C DBT and 27°C WBT, and 80 per cent recirculated air at 25°C DBT and 50 per cent RH. The air leaves the cooling coil at 13°C saturated state. Calculate the total cooling load, and room heat gain.*

Solution Refer to Fig. 15.24. From the psychrometric chart the following conditions are noted:

Condition	*DBT* °C	*WBT* °C	*RH* %	*Sp. Hu.* g w.v./kg d.a.	*Enthalpy* kJ/kg d.a.
Outside	40	27		17.2	85
Inside	25		50	10.0	50.8
ADP	13		100	9.4	37.0

Condition of air entering the cooling coil

$$\omega_1 = 0.2\,(17.2) + 0.8\,(10) = 11.44 \text{ g w.v./kg d.a.}$$
$$h_1 = 0.2\,(85) + 0.8\,(50.8) = 57.64 \text{ kJ/kg d.a.}$$
$$t_1 = 0.2\,(40) + 0.8\,(25) \quad = 28°\text{C}$$

Specific volume of air entering the cooling coil

$$v_1 = 0.869 \text{ m}^3\text{/kg d.a.}$$

Mass flow rate of air entering the cooling coil

$$\dot{m}_{a1} = \frac{4500}{(60)(0.869)} = 86.31 \text{ kg d.a./s}$$

Total cooling load

$$\dot{Q} = \dot{m}_{a_1}\,(h_1 - h_2) = 86.31\,(57.64 - 37) = 1781.4 \text{ kW}$$

Fresh air load $\dot{Q}_0 = \dot{m}_{a_0}\,(h_0 - h_i) = 0.2\,(86.31)\,(85 - 50.8) = 590.4 \text{ kW}$

Room heat gain

$$\text{RTH} = \dot{Q} - \dot{Q}_0 = 1781.4 - 590.4 = 1191 \text{ kW}$$

Note *A point to be noted here is that the fresh air load is a significant part of the cooling load. In the present case, it is 33 per cent. In some applications such as operation theatres in hospitals where 100 per cent fresh air is taken, it is of overriding importance. In cinema halls and theatres also, the predominant load is due to occupancy and fresh air. In such cases, the peak load occurs when the outside wet bulb temperature is maximum. This usually occurs between 2 and 5 p.m.*

Example 15.7 *An air-conditioned space is maintained at 27°C DBT and 50 per cent RH. The ambient conditions are 40°C DBT and 27°C WBT. The space has a sensible heat gain of 14 kW. Air is supplied to the space at 7°C saturated.*

Calculate:

(i) Mass of moist air supplied to the space in kg/h.

(ii) Latent heat gain of space in kW.

(iii) Cooling load of the air washer in kW if 30 per cent of the air supplied to the space is fresh, the remainder being recirculated.

Solution Refer to Fig. 15.24. From the psychrometric chart.

Condition	*DBT* °C	*WBT* °C	*RH* %	*Sp. Hu.* g w.v./kg d.a.	*Enthalpy* kJ/kg d.a.
Outside	40	27	—	17.2	85
Inside	27	—	50	11.2	56.1
Supply	7	—	100	6.2	23.0

(i) Mass of dry air supplied to space

$$\dot{m}_a = \frac{\dot{Q}_S}{C_p \Delta t} = \frac{14(3600)}{1.0216(27.7)} = 2467 \text{ kg d.a./h}$$

Ratio of moist air to dry air in supply air = $(1 + \omega_2) = 1.0062$

Mass of moist air supplied to space

$$\dot{m} = (1 + \omega_2)\, \dot{m}_a = 1.0062\ (2467) = 2482 \text{ kg/h}$$

(ii) Latent heat gain of space

$$\dot{Q}_L = \dot{m}_a (\omega_1 - \omega_2)\, h_{fg_0} = \frac{2467(11.2 - 6.2)}{(3600)(1000)}(2500) = 8.57 \text{ kW}$$

(iii) For point 1

$$t_1 = 0.7\ (27) + 0.3\ (40) = 30.9°\text{C}$$

On the line joining *o* to *i*, locate point 1 at 30.9°C. Then, from the psychrometric chart

$$h_1 = 64.9 \text{ kJ/kg d.a.}$$

Cooling load of the air washer

$$\dot{Q} = \dot{m}_a (h_1 - h_2) = \frac{2467}{3600}(64.9 - 23.0) = 28.71 \text{ kW}$$

15.5.2 Summer Air-conditioning System with Ventilation Air-Bypass Factor X

In the case when the bypass factor of cooling and dehumidifying apparatus is not zero, it is evident from Fig. 15.24 that if the surface temperature is t_S, viz., equal to room ADP, the leaving air state 2 will not be at *S*; it will be on the line joining 1 to *S*. Hence the supply air state will not lie on the room sensible heat factor line which is essential to satisfy the room sensible and latent heat load requirements. In such a case, which usually occurs in actual practice, it will be necessary to lower the apparatus dew point, i.e., the effective surface temperature of the air-conditioning apparatus, as shown in Fig. 15.25, in such a way that the leaving air state 2 lies on the RSHF line *i-S*, and also the new surface temperature t'_S or coil ADP is such that the following condition for the bypass factor *X* is satisfied

$$X = \frac{t_2 - t_S'}{t_1 - t_S'} = \frac{\omega_2 - \omega_S}{\omega_1 - \omega_S'} = \frac{h_2 - h_S'}{h_1 - h_S'}$$

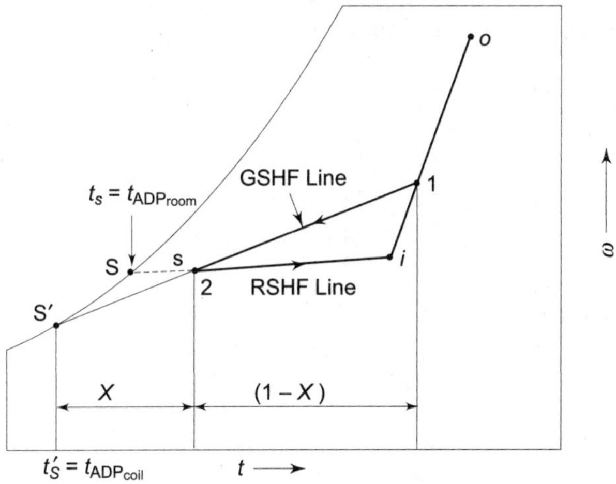

Fig. 15.25 Summer air-conditioning processes with ventilation air and finite bypass factor

It will be seen that the effect of the bypass factor is to lower the ADP of the surface, and hence to decrease the coefficient of performance of the refrigerating machine. It is also seen that the position of the grand sensible heat factor line is changed. This is explained in greater detail in Sec. 19.7. The supply air temperature t_s is now increased to the leaving air temperature t_2. Further, the dehumidified rise now is $t_i - t_2$. The dehumidified air quantity may be calculated accordingly, which will be found to be more than that of the apparatus with a zero bypass factor. The recirculated air quantity can then be calculated by the difference of the dehumidified (supply) air and ventilation air quantities. Since ventilation air quantity is fixed according to requirement, this leads to a variation in the recirculated air quantity. The greater the BPF, the more the recirculated air quantity. As a result, point 1 after mixing, shifts closer towards *i*.

Example 15.8 *A building has the following calculated cooling loads:*

$$RSH\ gain = 310\ kW$$
$$RLH\ gain = 100\ kW$$

The space is maintained at the following conditions:

$$Room\ DBT = 25°C$$
$$Room\ RH = 50\%$$

Outdoor air is at 28°C and 50% RH. And 10% by mass of air supplied to the building is outdoor air. If the air supplied to the space is not to be at a temperature lower than 18°C, find:

(a) *Minimum amount of air supplied to space in m^3/s.*
(b) *Volume flow rates of return (recirculated room) air, exhaust air, and outdoor air.*
(c) *State and volume flow rate of air entering the cooling coil.*
(d) *Capacity, ADP, BPF and SHF of the cooling coil.*

Solution Refer to Fig. 15.25. Room SHF is 0.756.

Draw room SHF line. Its intersection with $t = 18°C$ vertical gives supply air state point s which is the same as coil leaving air state point 2. From psychrometric chart

$$h_i = 50.5 \text{ kJ/ kg d.a.}, h_2 = h_s = 41.2 \text{ kJ/kg d.a.},$$
$$v_s = 0.836 \text{ m}^3\text{/kg d.a.}$$
$$h_0 = 92.0 \text{ kJ/kg d.a.}$$

(a) Supply air quantity and volume flow rate (minimum)

$$\dot{m}_{a_s} = \frac{\text{RTH}}{h_i - h_s} = \frac{410}{50.5 - 41.2} = 44.09 \text{ kg/s}$$

$$\dot{Q}_{v_s} = \dot{m}_{a_s} v_s = (44.09)(0.836) = 36.86 \text{ m}^3\text{/s}$$

(b) Quantity and volume flow rate of outdoor/exhaust air

$$\dot{m}_{a_0} = 0.1 \ \dot{m}_{a_s} = 0.1\ (44.09) = 4.41 \text{ kg/s}$$
$$\dot{Q}_{v_0} = \dot{m}_{a_0} v_0 = 4.41\ (0.91) = 4.01 \text{ m}^3\text{/s.}$$

Quantity and volume flow rate of return air

$$\dot{m}_{a_i} = \dot{m}_{a_s} - \dot{m}_{a_0} = 44.09 - 4.41 = 39.68 \text{ kg/s}$$
$$\dot{Q}_{v_i} = \dot{m}_{a_i} v_i = 39.68\ (0.86) = 34.05 \text{ m}^3\text{/s}$$

Here, $v_0 = 0.91$ and $v_i = 0.86 \text{ m}^3\text{/kg d.a.}$ are the specific volumes of outdoor and indoor air respectively.

(c) State of air entering cooling coil

$$t_1 = 0.9\ t_i + 0.1\ t_0 = 0.9\ (25) + 0.1\ (38) = 26.3°C$$
$$t'_1 = 19.2°C \text{ at } 26.3°C \text{ DBT on the line joining } i \text{ to } o.$$
$$\Rightarrow \quad v_1 = 0.865 \text{ from psychrometric chart}$$
$$h_1 = 54.6 \text{ kJ/kg d.a.}$$

Volume flow rate of air entering the cooling coil

$$\dot{Q}_{v_1} = \dot{m}_{a_s} v_1 = 44.09\ (0.865) = 38.14 \text{ m}^3\text{/s}$$

(d) Refrigerating capacity of the coil

$$\dot{Q}_{\text{coil}} = \text{GTH} = \dot{m}_{a_s}(h_1 - h_2) = 44.09\ (54.6 - 41.2) = 591 \text{ kW}$$

Coil ADP is obtained by the intersection of the line joining 1 to 2 with the saturation curve. Thus

$$t_{\text{ADP}} = 9°C$$

BPF and SHF of the coil

$$\text{BPF} = \frac{t_2 - t_{\text{ADP}}}{t_1 - t_{\text{ADP}}} = \frac{18 - 9}{26.3 - 9} = 0.52$$

$$\text{GSHF} = \frac{\text{TSH}}{\text{GTH}} = \frac{\dot{m}_{a_s} C_p (t_1 - t_2)}{591} = \frac{44.09(1.0216)(25 - 18)}{591} = 0.533$$

Note *Room SHF of 0.756 is normal. But GSHF of 0.533 is very low. This is due to high OALH; because of the outside air being very humid and hot, $\omega_0 = 21$ g/kg d.a. and $h_0 = 92.0$ kJ/kg d.a. Low GSHF and high BPF of the coil chosen, viz., 0.52 has resulted in low ADP of 9°C. This would result in low evaporator temperature and high power consumption of the refrigerating machine. It would be better to choose a coil of low BPF. That would require a higher coil ADP and evaporator temperature, and hence would give better performance. But, then the supply air temperature would be lower than 18°C, and quantity of supply air would be reduced.*

Example 15.9 *The conditioning plant of a room consists of a fresh air intake, a cooling coil-followed by a mixing chamber for the cooled fresh air and recirculated room air, and a supply fan as shown in Fig. 15.26. The cooling coil handles all fresh air and has a BPF of 0.1. (See also Sec. 23.7).*

The ratio of fresh air to recirculated air is determined by modulating dampers. The other data is as follows:

Inside conditions : DBT = 24°C, RH = 50%
Outside conditions : DBT = 30°C, WBT = 23.3°C
Heat gains : RSH = 14.7 kW, RLH = 3 kW
Supply air quantity : 191 cmm

Neglecting temperature changes in the fan and duct, determine:

(i) DBT and moisture content of supply air.
(ii) Mass flow rate of moist air supplied to room.
(iii) DBT and moisture content of air leaving cooling coil.
(iv) Load on the cooling coil.

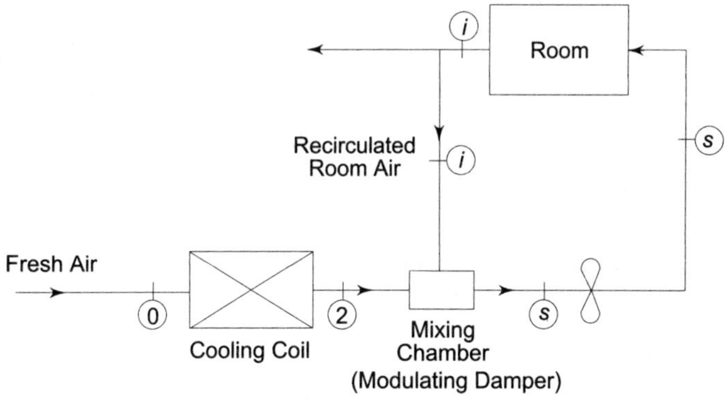

Fig. 15.26 Conditioning plant for Example 15.9

Solution (iii) The processes are shown in Fig. 15.27

$$\text{RSHF} = \frac{14.7}{14.7+3} = \frac{14.7}{17.7} = 0.831$$

This is the slope of line *i–s–*2. However, it is not possible to fix points s and 2 at this stage. But, we know that

$$\text{BPF} = \frac{\text{Line } 2-\text{S}}{\text{Line } 0-\text{S}} = 0.1$$

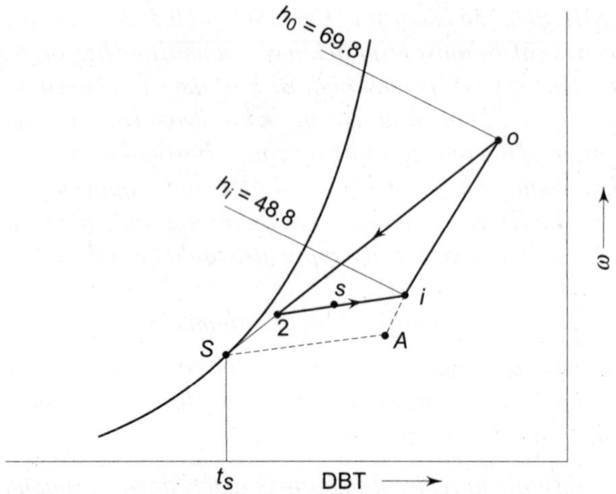

Fig. 15.27 Processes in Example 15.9

Accordingly, point 2 can be fixed by trial and error on the RSHF line. However, the following construction is simpler and more accurate. Draw line *o*–*i* and extend it to *A* such that:

$$\frac{\text{Line}\, i - A}{\text{Line}\, o - A} = 0.1$$

And then draw AS parallel to RSHF line $i - s$. Intersection with saturation curve gives coil ADP of 11.1°C at *S*. Join 0 – *S*. It cuts *i*–*s* extended at 2. Thus, we get condition of air leaving coil as:

$$t_2 = 12.2°\text{C},\ \omega_2 = 8.5 \text{ g/kg d.a.},\ h_2 = 33.8 \text{ kJ/kg d.a.}$$

(i) DBT and moisture content of supply air

$$\dot{m}_{a_s} = \frac{\dot{Q}_{v_s}}{v_s} = \frac{191/60}{v_s}$$

$$\text{RTH} = \dot{m}_{a_s}(h_i - h_s) = \frac{191/60}{v_s}(48.8 - h_s) = 17.7$$

This equation can be solved by trial and error. Thus, point *s* can be located on RSHF line. It is found that:

$$v_s = 0.856 \text{ m}^3\text{/kg d.a.},\ t_s = 20°\text{C},$$
$$h_s = 43.8 \text{ kJ/kg d.a.},\ \omega_s = 9.1 \text{ g/kg d.a.}$$

(ii) Mass flow rate of moist air supplied to room

$$\dot{m}_s = \frac{191/60}{0.856}(1 + 0.0091) = 3.753 \text{ kg/s}$$

(iv) Fresh air through coil

$$\dot{m}_{a_2} t_2 + (\dot{m}_{a_s} - \dot{m}_{a_2})\, t_i = \dot{m}_{a_s} t_1$$

$$\dot{m}_{a_2}(12.2) + \left(\frac{191}{0.856} - \dot{m}_{a_2}\right)(24) = \frac{191}{0.856}(20)$$

$$\Rightarrow \quad \dot{m}_{a_2} = 76.1 \text{ kg/min} = 1.268 \text{ kg/s} = \dot{m}_{a0}$$

Load on cooling coil

$$\text{GTH} = \dot{m}_{a_2}(h_0 - h_2) = 1.268\,(69.8 - 33.8) = 45.7 \text{ kW}$$

Example 5.10 *Given for a conditioned space:*

Room sensible heat gain = 20 kW
Room latent heat gain = 5 kW
Inside design conditions: 25°C DBT, 50% RH
Bypass factor of the cooling coil = 0.1

The return air from the space is mixed with the outside air before entering the cooling coil in the ratio of 4 : 1 by weight. Determine:

(i) Apparatus dew point.
(ii) Condition of air leaving cooling coil.
(iii) Dehumidified air quantity.
(iv) Ventilation air mass and volume flow rates.
(v) Total refrigeration load on the air conditioning plant.

Solution Refer to Fig. 15.25. From the psychrometric chart

Condition	*DBT* °C	*WBT* °C	*RH* %	*Sp. Hu*, g w.v./kg d.a.	*Enthalpy* kJ/kg d.a.	*Sp. Vol.* m^3/kg d.a.
Outside	43	27.5		17.0	87.5	0.922
Inside	25		50	10.0	50.8	

Condition of air entering the cooling coil

$$\omega_1 = 0.8\,\omega_i + 0.2\,\omega_0 = 0.8\,(10) + 0.2\,(17) = 11.4 \text{ g w.v./d.a.}$$
$$h_1 = 0.8\,(50.8) + 0.2\,(87.5) = 58.1 \text{ kJ/kg d.a.}$$
$$t_1 = 0.8\,(25) + 0.2\,(43) = 28.6°\text{C}$$

(i) and (ii). Both parts have to be worked out together. There are two methods. One method is to draw the RSHF line and then draw a line from 1 to S on the saturation curve so that (S – 2)/(1 – 2) are in the ratio of 1 : 9. Another method is to do the same thing using calculations as given below.

Ratio of room sensible and latent heats

$$\frac{\text{RSH}}{\text{RLH}} = \frac{0.0204(25 - t_2)}{50(0.01 - \omega_2)} = \frac{20}{5} \qquad \text{(I)}$$

Relations for bypass factor

$$\frac{t_2 - t_{\text{ADP}}}{t_1 - t_{\text{ADP}}} = \frac{t_2 - t_{\text{ADP}}}{28.6 - t_{\text{ADP}}} = 0.1 \qquad \text{(II)}$$

$$\frac{\omega_2 - \omega_{\text{ADP}}}{\omega_1 - \omega_{\text{ADP}}} = \frac{\omega_2 - \omega_{\text{ADP}}}{0.0114 - \omega_{\text{ADP}}} = 0.1 \qquad \text{(III)}$$

Solving Eqs (I), (II) and (III) by iteration for t_{ADP}, t_2 and ω_2, we obtain

$$t_{\text{ADP}} = 11.8°\text{C (Corresponding } \omega_{\text{ADP}} = 8.6 \text{ g w.v./kg d.a.)}$$
$$t_2 = 13.5°\text{C}$$
$$\omega_2 = 0.0089 \text{ kg w.v./kg d.a.}$$

(iii) Dehumidified air quantity

$$(\text{cmm})_d = \frac{\text{RSH}}{0.0204(t_i - t_2)} = \frac{20}{0.0204(25-13.5)} = 85.25$$

(iv) Specific volume of supply air

$$v_2 = 0.822 \text{ m}^3/\text{kg d.a.}$$

Mass flow rate of supply air

$$\dot{m}_{a_s} = \frac{(\text{cmm})_d}{(60)v_2} = \frac{85.25}{(60)(0.822)} = 1.729 \text{ kg d.a./s}$$

Mass flow rate of fresh air

$$\dot{m}_{a_0} = 0.2\ \dot{m}_{a_s} = 0.2\ (1.729) = 0.346 \text{ kg d.a./s}$$

Volume flow rate of fresh air

$$Q_{v0} = \dot{m}_{a_0}\ v_0(60) = (0.346)\ (0.922)\ (60) = 19.12 \text{ cmm}$$

(v) Outside air total heat

$$\text{OATH} = \dot{m}_{a_0}\ (h_i - h_0) = 0.346\ (87.5 - 50.8) = 12.7 \text{ kW}$$

Total refrigeration load on the air conditioning plant

$$\text{GTH} = \text{RTH} + \text{OATH} = (20 + 5) + 12.7 = 37.7 \text{ kW}$$

15.6 WINTER AIR CONDITIONING

In winter, the building sensible heat losses are partially compensated by the solar heat gains and the internal heat gains such as those from occupancy, lighting, etc. Similarly, the latent heat loss due to low outside air humidity is more or less offset by the latent heat gains from occupancy. Thus in winter, the heating load is likely to be less than the cooling load in summer. However, the actual situation both in summer and winter depends on the swing of the outside temperature and humidity with respect to the inside conditions.

Further, certain sensible heat gains (negative loads) such as the solar heat may not be present at the time of peak load, and hence they are not counted. On the other hand, latent heat gains from occupancy, etc., are always present and should be taken into account. As a result, the design heating load for winter air conditioning is predominantly sensible.

In general, the processes in the conditioning apparatus for winter air conditioning for comfort involve heating and humidifying. Two of the typical process combinations are:

(i) Preheating the air with steam or hot water in a coil followed by adiabatic saturation and reheat.

(ii) Heating and humidifying air in an air washer with pumped recirculation and external heating of water followed by reheat.

The processes for the two systems are shown in Fig. 15.28. The first system with preheating and adiabatic saturation follows processes 1–2 and 2–3 respectively. The second system replaces the two processes with heated water spray in the air washer and the process line is 1–3. The leaving air state 3 from the air washer may be affected by its saturation efficiency. The reheating process 3-s is common to both. The supply air states should lie on the room sensible heat factor line. It is, therefore, determined by the RSHF and by the choice of supply air rate which is usually known from summer air-conditioning calculations.

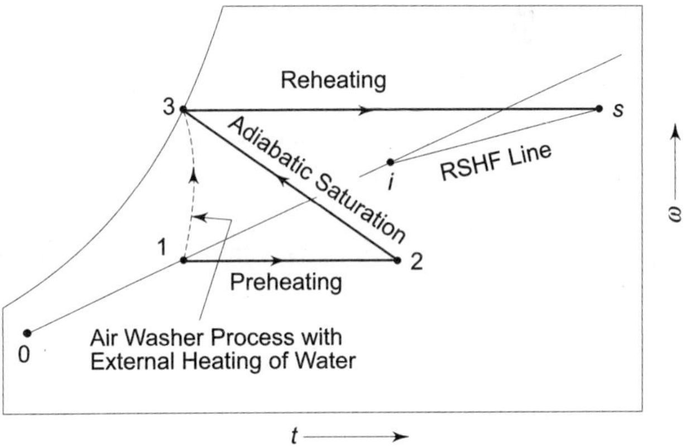

Fig. 15.28 Winter air-conditioning processes

Example 15.11 *In an industrial application for winter air conditioning, an air washer is used with heated water spray followed by a reheater. The room sensible heat factor may be taken as unity. The design conditions are:*

Outside: 0°C DBT and dry
Inside: 22°C DBT and 50% RH
Room heat loss : 703 kW
The following quantities are known from the summer design.

Ventilation air	*1600 cmm*
Supply air	*2800 cmm*
Spray water quantity	*500 kg/min*

The air washer saturation efficiency is 90 per cent. The make-up water is available at 20°C. Calculate:

(i) *The supply air condition to space.*
(ii) *The entering and leaving air conditions at the spray chamber.*
(iii) *The entering and leaving spray water temperatures.*
(iv) *The heat added to the spray water.*
(v) *The reheat, if necessary.*

Solution Refer to Fig. 15.29.

(i) Supply air temperature

$$t_s = \frac{\text{RSH}}{0.0204(\text{cmm})_s} + t_i = \frac{703}{0.0204(2800)} + 22 = 34.2°\text{C}$$

A RSHF = 1 line from the room condition *i* can be drawn as shown in Fig. 15.29. Intersection with the 34.2°C DBT line locates the supply air state. The supply air WBT is found to be 19.6°C, and its specific humidity ω_s to be 0.0082 kg/kg d.a.

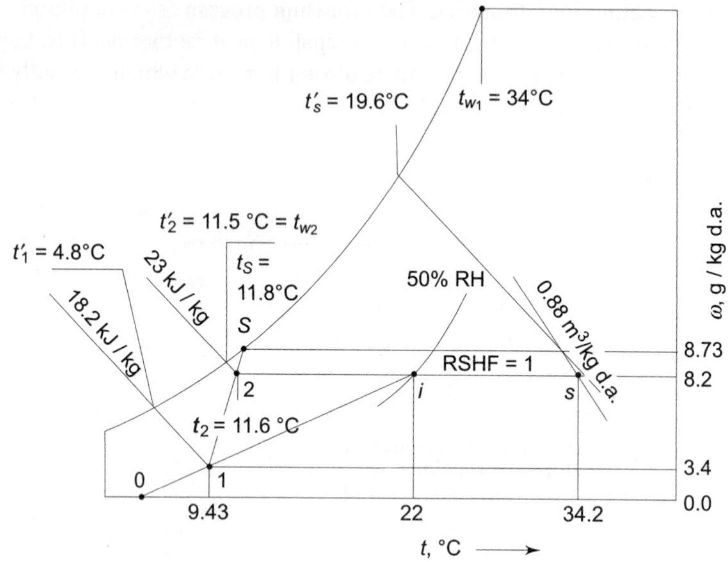

Fig. 15.29 Figure for Example 15.11

(ii) Entering air conditions (calculated from the volume flow rates).

$$t_1 = \frac{(\text{cmm})_0\, t_0 + (\text{cmm})_i\, t_i}{(\text{cmm})_i}$$

$$= \frac{(1600)(0)+(2800-1600)(22)}{2800} = 9.43°\text{C}$$

Similarly, $$\omega_1 = \frac{(1600)(0)+1200(0.0082)}{2800} = 0.0035 \text{ kg w.v./kg d.a.}$$

Wet bulb temperature of entering air (from the psychrometric chart)

$$t_1' = 4.8°\text{C}$$

Specific humidity of leaving air

$$\omega_2 = \omega_s = \omega_i = 0.0082 \text{ kg w.v./kg d.a.}$$

Expression for saturation or humidifying efficiency

$$\eta_H = \frac{\omega_2 - \omega_1}{\omega_S - \omega_1}$$

$$0.9 = \frac{0.0083 - 0.035}{\omega_S - 0.0035}$$

which gives the specific humidity at the wetted-surface temperature t_S as

$$\omega_S = 0.00873 \text{ kg w.v./kg d.a.}$$

From the psychrometric chart

$$t_S = 11.8°\text{C}$$

Dry bulb temperature of leaving air

$$t_2 = \eta_H (t_S - t_1) + t_1 = 0.9\,(11.8 - 9.43) + 9.43 = 11.6°\text{C}$$

Wet bulb temperature of leaving air (from the psychrometric chart)

$$t_2' = 11.5°C$$

(iii) The temperature of the leaving spray water and the wet bulb temperature of the leaving air may be taken to be the same. Hence, the leaving spray water temperature $t_{w2} = 11.5°C$.

Energy balance of the air washer

$$\frac{(\text{cmm})_s}{v_s}(h_2 - h_1) = \dot{m}_w C_{p_w}(t_{w_1} - t_{w_2})$$

$$\frac{2800}{0.88}(33.0 - 18.2) = (500)(4.187)(t_{w_1} - 11.5)$$

$$\Rightarrow \quad t_{w_1} = 34°C$$

(iv) Make-up water

$$\Delta \dot{m}_w = \frac{(\text{cmm})_s}{v_s}(\omega_2 - \omega_1) = \frac{2800}{0.88}(0.0082 - 0.0035) = 14.95 \text{ kg/min}$$

Heat added to make-up water

$$\dot{Q}_1 = \frac{14.95}{60}(4.187)(34 - 20) = 14.61 \text{ kW}$$

Heat added to raise temperature of spray water

$$\dot{Q}_2 = \frac{500}{60}(4.187)(34 - 11.5) = 785.1 \text{ kW}$$

Heat added to spray water

$$\dot{Q} = \dot{Q}_1 + \dot{Q}_2 = 14.61 + 785.1 = 799.71 \text{ kW}$$

(v) Reheat $= 0.0204\,(\text{cmm})_s\,(t_s - t_2) = 0.0204\,(2800)\,(34.2 - 11.6) = 1290.9$ kW

Revision Exercises

15.1 250 kg/h of air saturated at 2°C is mixed with 50 kg/h of air at 35°C and 80 per cent RH. Determine the final state of air.

15.2 Air at 20°C DBT and 19°C DPT enters a heating and humidifying apparatus, from which it leaves at 35°C DBT and 28°C DPT. Moisture is supplied as liquid water at 25°C to humidify the air. Find the quantity of heat that must be added per kg of dry air through the apparatus.

15.3 500 kg of air is supplied per minute to an auditorium maintained at 21°C and 40 per cent RH. The outside air at 5°C DBT and 60 per cent RH is first passed over heating coils and is heated until its WBT is equal to the room WBT. It is then passed through an adiabatic saturator and is finally heated to 45°C before being supplied to the room. Determine:

(i) The heat added in both the heating coils.

(ii) The mass of water evaporated in the air washer.

15.4 Given:

Room conditions: 26°C DBT, 19°C WBT

Outside conditions: 35°C DBT, 27°C WBT

Room heat gains:
Sensible heat = 11.1 kW
Latent heat = 3.9 kW.
The conditioned air supplied to the room is 50 cmm and contains 25 per cent fresh air and 75 per cent recirculated room air. Determine:
(a) The DBT and WBT of supply air.
(b) The DBT and WBT of mixed fresh and recirculated air before the cooling coil.
(c) The apparatus dew point and bypass factor of the coil.
(d) The refrigeration load on the cooling coil and the moisture removed by the coil.

15.5 In an auditorium which is to be maintained at a temperature not exceeding 24°C, and a relative humidity not exceeding 60%, a sensible heat load of 132 kW and 84 kg/h of moisture has to be removed. Air is supplied to the auditorium at 15°C.
(a) How many kg of air per hour must be supplied?
(b) What is the dew point temperature of supply air and what is its relative humidity?
(c) How much latent heat is picked up in the auditorium?
(d) What is the sensible heat factor?
(e) What is the ADP of the coil, and what is its BPF?

15.6 An air conditioned space is maintained at 25°C DBT and 50% RH. The outside conditions are 40°C DB and 25°C WB. The space has a sensible heat gain of 24.5 kW. Conditioned air is supplied to the space as saturated air at 10°C. The equipment consists of an air washer. The air entering the air washer comprises 25% outside air, the remainder being recirculated room air.
Calculate:
(i) Volume flow rate of air supplied to space.
(ii) Latent heat gain of space.
(iii) Cooling load of air washer.

15.7 Given for the air conditioning of a room
Room conditions: 26.5°C DBT and 50 per cent RH
Room sensible heat gain = 26.3 kW
Room sensible heat factor = 0.82
Find:
(i) The room latent heat gain.
(ii) The apparatus dew point.
(iii) The cmm of air if it is supplied to the room at the apparatus dew point.
(iv) The cmm and specific humidity of air if it is supplied to the room at 17°C.

15.8 (a) 28.5 cmm of room air at 25.5°C DBT and 50 per cent RH is mixed with 28.5 cmm of outside air at 38°C DBT and 27°C WBT. Find the ventilation load and the condition of air after mixing.
(b) The above mixture of air is passed through an air conditioning equipment. If the wet bulb temperature of air after the equipment is 14.5°C, determine the heat removed by the equipment.

15.9 An air washer cools and dehumidifies 18,200 kg of dry air per hour from 41°C DBT and 24°C WBT. Chilled water enters the washer at 7°C with a flow of 18,500 kg per hour. The washer is 88 per cent effective. What is the heat removed from the air in kW?

15.10 Saturated steam at standard atmospheric pressure is injected into a passing air stream in an amount sufficient to raise the absolute humidity from 0.0057 to 0.0143 kg w.v./kg d.a. If the air enters the humidifier at 21°C DBT, determine its leaving state.

15.11 300 cmm of outside air at 5°C and 60 per cent RH are heated and humidified to maintain room conditions at 21°C and 40 per cent RH. Devise a suitable system for the air conditioning and find the kW of various heaters and the moisture added. The supply air temperature is 45°C and the room sensible heat factor is 0.75.

15.12 Moist air at 31°C dry bulb, 22°C wet bulb and 1013.25 mbar barometric pressure flows over a cooler coil and leaves it at a state of 10°C dry bulb and 7.95 g w.v./kg d.a.

(a) If the air is required to offset a sensible heat gain of 2.5 kW and a latent heat gain of 0.35 kW in a room being air-conditioned, calculate the mass of dry air which must be supplied to the room in order to maintain a dry-bulb temperature of 23.5°C inside.

(b) What will be the relative humidity in the room?

(c) If the sensible heat gain is diminished by 1.75 kW but the latent heat gain remains unchanged, at what temperature and moisture content must the air be supplied to the room?

15.13 A summer air conditioning plant mixes 70 cmm of outside air at 35°C DB and 23°C WB with 210 cmm of return air at 24°C DB and 50% RH. The mixture passes over a cooling coil. Air off the coil is 90% RH. The room SHF is 0.7.

(i) Find the ADP, and air off the coil dew point and dry bulb temperatures.

(ii) How much cooling in kW is the unit doing?

(iii) How much of the total load is sensible, and how much is latent?

15.14 A conditioned room with partial recirculation of room air is to be maintained at 24°C DB, 17°C WB. The local outside environment conditions are 35°C DB, 26°C WB. The sensible heat load is 48 kW. The latent heat load from occupants and infiltration, but excluding ventilation, is 57 kW. Based on the occupancy, 36 cmm of ventilation air is required. Find:

(a) the temperature of air entering the room,

(b) the volume of air passing through the room,

(c) the state of air entering the conditioner,

(d) the required ADP, and

(e) the system BPF.

Chapter

16

Design Conditions

When one refers to design conditions, it is implied that these pertain to the room or inside, and the ambient meaning outside conditions. Before proceeding with the design, the inside and outside design conditions have to be specified. Also, in most cases, the supply air design conditions have to be fixed as well.

16.1 CHOICE OF INSIDE DESIGN CONDITIONS

The inside design conditions depend on the particular air-conditioning application. Among the many applications of air conditioning, the important ones can be identified as:

(i) Cold storage.
(ii) Industrial air conditioning.
(iii) Comfort air conditioning.

16.1.1 Cold Storage

Though cold storage is understood to be merely an application of refrigeration, it is in fact a complete air-conditioning system in which room air is cooled to much lower temperature over a cooling coil and supplied back to the storage space. The conditions maintained inside the storage space depend on the nature of the product stored. It is to be noted that in cold storages, often, strict control of both temperature and relative humidity is required. Also, the storage life depends a great deal on the temperature at which a product is stored. The required storage conditions for a number of important food products are given in Table 16.1. It is seen that in the case of bananas, there is no storage period. Instead, there is a period of ripening. Bananas cannot be stored after they have ripened. The best temperature for slow ripening is 14.5°C. Further, in the case of milk, the storage temperature is 0.5°C, whereas its highest freezing temperature is – 0.6°C. Thus, air is to be maintained within a close tolerance of 0°C so that milk does not freeze. The same is also true for the pasteurization process. Hence, it is necessary to use automatic-expansion valves in milk-chilling and storage plants so as to have constant temperature in the evaporator. The *ice-bank type* water chillers are also used which permit the use of a

Table 16.1 Storage conditions and properties of food products

Product	***Temperature*** °C	***Relative Humidity*** %	***Approximate Storage Life***	***Water Content %***	***Highest Freezing Point*** °C	***Sp. Heat Above Freezing*** kJ/kg.K	***Sp. Heat Below Freezing*** kJ/kg.K
Apples	–1 to 0	85–90		84.1	–1.5	3.643	1.884
Bananas	14.5	95	For ripening in 8–10 days	74.8	–0.8	3.35	1.76
Butter	0 to 4.4	80–85	2 months	15.5–16.5	—	1.382	—
	–18 to –23	80–85	1 year	15.5–16.5	—	—	—
Milk, Pasteurized	0.5	—	7 days	87	–0.6	3.77	1.93
Eggs	–1.5 to –0.5	80–85	6–9 months	66	–2.2	3.06	1.68
Fish, fresh	0.5 to 1.5	90–96	5–15 days	62–85	–2.2	2.93–3.6	—
frozen	–23.5 to –18	90–95	8–10 months	62–85	—	—	1.59–1.884
Grapes	–0.5	85–90	3–8 weeks (American)	81.9	–1.3	3.6	1.842
	—	—	3–6 months (European)	81.6	–2.2	3.6	1.842
Beef, fresh	0 to 1	88–92	1–6 weeks	62–77	–2.2	2.93–3.52	—
frozen	–23.5 to –18	90–95	9–12 months	62–77	—	—	1.59–1.8
Mangoes	10	85–90	2–3 weeks	81.4	–0.9	3.56	1.842
Potatoes, late crop	3 to 4.5	85–90	5–8 months	77.8	–0.6	3.433	1.8
Tomatoes, green	14 to 21	85–90	2–4 weeks	94.7	–0.6	3.98	2.01
ripe	7 to 10	85–90	2–7 days	94.1	–0.5	3.98	2.01

refrigeration system having a considerably less capacity, and to supply chilled water at a constant temperature of 0°C.

16.1.2 Industrial Air Conditioning

There are various categories of applications requiring varying standards of inside design conditions. One category comprises those where constancy of temperature is the prime consideration, such as metrology laboratories, precision-machine shops, computer centres, etc. In these a variation of ± 10 to 20 per cent in relative humidity will not have much effect. The other category may comprise paper and textile mills where the relative humidity is to be maintained constant at a high value, e.g., of the order of, say, 70 to 75 per cent in a textile mill. The temperature requirements of such spaces are not severe. There is still another category of applications where strict control of both temperature and relative humidity is required. These pertain to chemical and biological process industries.

16.1.3 Comfort Air Conditioning and Effective Temperature

Extensive tests have been conducted on the effects of temperature, humidity and air velocity. The results are not in complete agreement. Also, there is the problem of measuring comfort in terms of a single parameter which could include all three parameters governing comfort, namely, air temperature, humidity and air velocity in addition to air purity. Often, a single parameter called the *effect temperature* is used as an index of comfort.

Effective temperature (ET) is defined as that temperature of saturated air (100% RH) at which the subject would experience the same feeling of comfort as experienced in the actual unsaturated environment.

Correspondingly, another effective temperature ET^* can also be defined as the temperature at 50% RH at which the subject would experience exactly same feeling of comfort as at ET at 100% RH, and as experienced in the actual environment.

Because ET and ET^* depend on activity and clothing, it is not possible to generate a universal ET chart. Figure 16.1, however, shows pairs of constant ET lines (a) for activity with 180 W/m^2 of body surface heat generation, standard clothing, and air velocity < 0.2 m/s, and (b) with 3 times heat generation, and with clothing just half of (a). Figure 16.1 shows, for example, a line of ET = 22°C or ET^* = 25°C for activity and clothing as in (b).

At lower humidities, the DBTs of air can be higher for the same ET and for the same feeling of comfort. Thus at a higher DBT, the body would lose more heat in the form of latent heat, i.e., by the evaporation of perspiration. An increase in temperature can also be compensated by an increase in velocity. For example, an increase of 2 to 3°C in DBT can be compensated by increasing the air velocity from 0.1 to 0.3 m/s.

Note that ET or ET^* combine temperature and humidity into a single index. Accordingly, two environments at same ET^* evoke the same thermal response even though they may be having different temperatures and humidities provided they have the same air velocities.

The slope of a constant ET^* line depends on 'skin wettedness', and moisture permeability of clothing. Skin wettedness represents fraction of skin that is covered with water to account for rate of moisture loss by evaporation.

Note that activity determines metabolic rate, and hence skin wettedness.

Thus, effective temperture for a given temperature and humidity depends on activity and clothing. Figure 16.1 shows ET* lines 'a' that correspond to low activity levels. That means lower metabolic rates, and hence low skin wettedness.

At low activity levels, air humidity has little effect, and constant ET* lines are nearly vertical as at 'a'. As activity and hence metabolic rate and skin wettedness increase, the lines become more horizontal as at 'b'. The influence of humidiy is more pronounced.

Since ET* depends on activity and clothing, it is not possible to have a universal ET* chart.

In Fig. 16.1, ET* lines at 'a' correspond to metabolic rate '1' which is taken as 58.1 W/m^2 of naked body area. Lines 'a' also correspond to 'clothing factor' of '0.6'. That is equivalent to 60% of some standard clothing. Lines 'b' corresspond to more normal situations of metabolic rate of '3', viz., $3 \times 58.1 = 144.3$ W/m^2, and clothing factor of 50% of that in 'a'.

According to Malhotra[5], the effective temperatures and also the range of DBT at 50 per cent RH for comfort are as follows:

Climate	*ET, °C*	*Corresponding DBT at 50% RH, °C*
Hot and dry	21.1 to 26.7	23.9 to
Hot and humid	22 to 25.6	26.7

The general practice is to recommend the following optimum inside design conditions for comfort for summer air conditioning:

ET 22°C or ET* = 25°C

DBT 25 ± 1°C

RH 50 ± 5%

The corresponding room air velocity is 0.4 m/s. The points of this equal comfort are shown in Fig. 16.1.

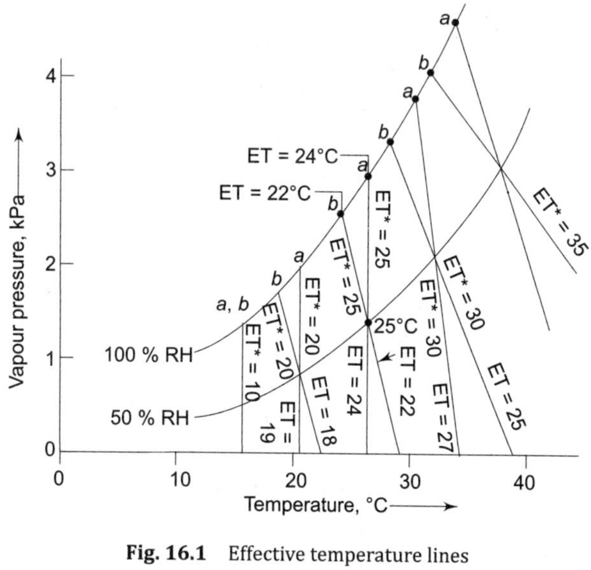

Fig. 16.1 Effective temperature lines

During winter, the body gets acclimatized to withstand lower temperatures. Consequently, a DBT of 21°C at 50 per cent RH and 0.15–0.2 m/s air velocity is quite comfortable.

In addition to the maintenance of temperature, humidity and air velocity, it is also important to maintain the purity of room air. Even if there are no sources of production of pollutants within the conditioned space, the carbon dioxide content of air increases because of the occupants. It is, therefore, necessary to introduce fresh air or *ventilation air* into the space. The requirement of *ventilation air* is much more when some occupants are smoking. In the case of auditoriums, because of very large occupancy, the ventilation air requirement is very large. Hence the need to prohibit smoking in auditoriums and assembly halls. Table 16.2 gives the ventilation air requirements for some applications.

Table 16.2 Ventilation air requirements

Application	*Smoking Status*	*Recommended cmm/person*	*Minimum*	
			cmm/person	*cmm/m floor area*
Apartments	Some	0.56	0.28	—
Offices and factories	Occasional-Some	0.28–0.6	0.21	—
Restaurants	Some	0.4	—	—
Board rooms	Very heavy	1.4	0.56	0.03
Department stores	None	0.21	0.14	0.0015
Theatres	None	0.21	0.14	—
Hotel rooms	Heavy	0.84	0.7	—
Hospital wards	None	0.84	—	—
Hospital operation theatres	None	All outdoor	—	—

16.1.4 Comfort Chart

Studies have shown correlation between comfort level, temperature, humidity, sex, length of exposure, etc. For example for exposure for a period of one hour, the values of index Y for comfort on ASHRAE thermal sensation scale for men, women and both sexes separately are expressed as follows:

Men $\quad Y = 0.22\,t + 0.233\,p_v - 5.673$

Women $\quad Y = 0.272\,t + 0.248\,p_v - 7.245$

Both $\quad Y = 0.245\,t + 0.248\,p_v - 6.475$

where t is dbt in °C, and p_v is vapour pressure in kPa. ASHRAE scale refers to value for feeling of comfort of Y as below:

+ 3	+ 2	+ 1	0	– 1	– 2	– 3
Hot	Warm	Slightly Warm	Neutral	Slightly Cool	Cool	Cold

It is seen that women are more sensitive to temperature, and less sensitive to humidity than men. In genral 3°C change in temperature or a 3 kPa change in water vapour pressure is necessary to change thermal sensation index from one to the next.

Based on the concept of ET, some 'comfort charts' have been developed. Fig. 16.2 shows ASHRAE summer and winter comfort zones in the form of comfort chart. This chart may be referred to when a compromise in the inside design conditions is to be achieved.

The chart in Fig. 16.2 is valid for typical summer and winter clothing during primarily sedentry activity. The warmer and cooler temperature borders coincide with lines of constant ET*.

In the middle of zone, a person wearing standard clothing will have comfort sensation index Y very near to netural zero.

Near the boundary of the warmer zone, a person would feel about Y = 0.5 warmer. Near the boundary of the cooler zone, one would feel about Y = – 0.5 cooler. The upper and and lower limits of humidity levels of the comfort zones in the chart are less precise.

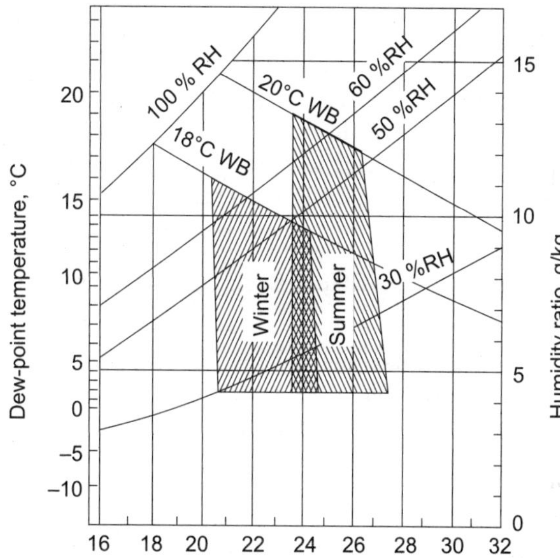

Fig. 16.2 ASHRAE summer and winter comfort zones

16.2 COMFORT[2]

The salient features governing human comfort are discussed below.

16.2.1 The Metabolic Rate

The rate at which body produces heat is called the *metabolic rate*. The heat produced by a normal healthy person while sleeping is called the basal metabolic rate which is of the order of 60 W. The maximum value may be 10 times as much as this for a person engaged in sustained hard work.

The temperature of the body remains comparatively constant at about 36.9°C (98.4°F) for tissues at the surface or the skin and about 37.2°C for the deep tissues or the core. It is found that the body temperature in the morning after sleep is about

0.5°C less than its temperature in the afternoon. A value of 40.5°C (104.9°F) is considered serious and 43.5°C (110°F) is certainly fatal.

Human comfort is influenced by physiological factors determined by the rate of heat generation within the body and the rate of heat dissipation to the environment.

16.2.2 Mechanism of Body Heat Loss

The body loses heat to the surroundings mainly by convection C, radiation R, and evaporation of moisture E. In addition, there is heat loss by respiration having sensible component C_{res} and latest component E_{res}. The total heat loss from the body is thus

$$Q = (C + R + C_{res}) + (E + E_{res})$$

There are two components of this heat loss: $C + R + C_{res}$ forms the sensible heat component Q_S, and $E + E_{res}$ forms the latent heat component Q_L. The sensible heat component depends on the temperature difference between the surface of the body and the surroundings, and the latent heat component, similarly, depends on the difference in the water vapour pressures.

In summer, the temperature difference available for sensible heat transfer is less. Thus the convective and radiative heat losses are reduced. To maintain thermal equilibrium, the body starts perspiring to increase the evaporative loss. On the other hand, in winter, the sensible heat transfer is increased; the evaporative losses thus tend toward zero.

16.2.3 Mathematical Model of Heat Exchange between Man and Environment

The heat exchange between man and his environment can be expressed by the following energy balance equation

$$M - W = Q + S \tag{16.2}$$

where

M = Metabolic rate

W = Work done by man

Q = Rate of convective, radiative, respirative and evaporative heat losses

S = Rate of heat storage.

In summer, the body temperature has a tendency to rise since the stored energy S is positive. The blood flow rate through the extremities increases, and the body starts perspiring. This is called the condition of *vasodilation*. In winter, the temperature tends to fall, the stored energy may be negative and the blood flow rate through the extremities becomes low. This leads to the condition of *vasoconstriction* resulting in shivering.

For the condition of equilibrium or *thermal neutrality* of the body, there should be no stored energy, and hence no change in body temperature. For a feeling of comfort, thermal neutrality is the required condition. Any variation in body temperature acts as a stress signal to the brain which ultimately results in either perspiration or shivering.

The net heat release rate of the body due to oxidation is

$$H = (M - W) = M(1 - \eta)$$

where η is the thermal efficiency of the body heat engine. Work done W is positive when the body performs work. Both M and η are governed by the activity of the man. The values of heat liberated depending on the activity are given in Table 19.1. The thermal efficiency in most cases is zero, except in cases of high activity, such as when playing outdoor games—it is of the order of 20 per cent.

16.3 OUTSIDE DESIGN CONDITIONS

It is observed that there is a kind of sinusoidal relationship between the air dry bulb temperature and the sun time. For example, in the month of June in a certain locality where the sunrise is at about 5 a.m. and the sunset at about 7 p.m., the time of minimum temperature falls at about 4 a.m. and that of maximum temperature at about 4 p.m., i.e., with a lapse of about 12 hours.

As regards relative humidity, it is seen that it reaches a minimum value in the afternoon. Since the mean daily maximum dry bulb temperature occurs between 1 p.m. and 4 p.m., it is reasonable to assume that the minimum relative humidity would occur during the same period.

Accordingly, meteorological data of the locality may be collected in the form of the mean daily or monthly maximum and minimum temperatures, in combination with corresponding relative humidities or wet bulb temperatures.

For the outside design conditions in summer, it is, therefore, recommended to use the mean monthly maximum dry bulb temperature and its corresponding wet bulb temperature. Very often the value of the wet bulb temperature cannot be ascertained from the data provided by metereological laboratories. The wet bulb temperatures fall usually at the time of maximum dry bulb temperature. However, it is essential to take the value of the wet bulb temperature in this manner, and not the value of the maximum wet bulb temperature as that would lead to an erroneously high cooling load. This is because the relative humidity is the lowest when the dry bulb temperature is the highest, and vice versa. This does not mean that the wet bulb temperature would also be the lowest when the dry bulb temperature is the highest during a single day. In fact, the wet bulb temperature remains more or less uniform on any particular day.

As for winter, the concept of degree-days is used. It is found that the fuel consumption in winter for the heating of buildings varies almost directly as the difference between the outside temperature and a reasonably comfortable inside temperature of 18.5°C (65°F). Thus the fuel consumption would be practically nil if the outside temperature is 18.5°C. On the other hand, the fuel consumption, would double if the outside temperature dropped from 13.5 to 8.5°C. A *degree-day* is obtained for every degree when the mean outside temperature is below 18.5°C during the 24-hour period. Accordingly, if in a given locality the outside temperature average of 30 days is 10°C, the degree days for the period would be

$$(18.5 - 10)(30) = 255$$

This concept enables the calculation of fuel consumption for a given period for a building. For example, the steam consumption S can be calculated by the following equation

$$S = \frac{Q_D \text{ (degree-days)(24)(3600)}}{1055} \tag{16.3}$$

where Q_D = Heat loss of the building per degree of temperature difference between inside and outside in kW, and

1055 = Approximate amount of heat released by steam in kJ/kg.

The value of Q_D can be calculated by knowing the total heat loss Q of the building for a particular temperature difference. Thus

$$Q_D = \frac{Q}{t_i - t_0} \tag{16.4}$$

For the calculation of the outside design temperature take, for example, a typical January month for a locality for which the number of degree-days is 677. Then the outside design temperature may be obtained as follows:

$$31\ (18.5 - t_0) = 677$$

$$\Rightarrow \qquad t_0 = -\ 3.3°\text{C}$$

16.4 CHOICE OF SUPPLY DESIGN CONDITIONS[3]

Consider a room that is completely sealed having an internal sensible heat gain of $\dot{Q}_S$. The inside dry bulb temperature t_i of the room will then rise above the outside temperature t_0 until the heat transferred to the outside is equal to the heat generated inside. Under this equilibrium condition

$$t_i = t_0 + \frac{\dot{Q}_S}{UA} \tag{16.5}$$

where U is the overall heat-transfer coefficient of the structure and A is its surface area. For example, if for a typical cubical room of side equal to 3.16 m

$$\dot{Q}_S = 2 \text{ kW},\ U = 1.2 \text{ Wm}^{-2}\text{ K}^{-1},\ A = 6(3.16^2) = 60 \text{ m}^2$$

then, if the outside temperature is 40°C, the inside temperature in the absence of air conditioning will become

$$t_i = 40 + \frac{2 \times 10^3}{(1.2)(60)} = 67.8°\text{C}$$

To improve the situation, it would help to ventilate the room. If the fresh outside air is supplied as ventilation air to the room, the temperature attained will be reduced. Let such air supplied be equal to n air changes per hour. One *air-change* is equivalent to the amount of air supplied per hour equal to the volume of the room. If V is the volume of the room, then the temperature attained in the room will be given by the energy balance equation

$$\dot{Q}_S = \frac{\rho C_p n V (t_i - t_0)}{3600} + UA\ (t_i - t_0)$$

whence

$$t_i = t_0 + \frac{3600 \dot{Q}_S}{\rho C_p n V + 3600 UA} \tag{16.6}$$

In the case of the example of the room taken earlier, if we take $n = 10$, then

$$t_i = 40 + \frac{(3600)(2)}{(1.2)(1.0216)(10)(3.16^3) + (3600)\left(\frac{1.2}{1000}\right)(60)} = 51.1°\text{C}$$

Thus, if the internal heat gain of the room is high, ventilation will help decrease the room temperature.

In practice, however, the situation is not so bad as it appears from the above calculations. Due to the effect of fall in the outside dry bulb temperature at night, and the storage of heat due to the heat capacity of the structure, the room temperature would not rise to the extent shown above, and t_i may even be below the instantaneous value of t_0 if internal heat gain is not much.

Nevertheless, to maintain conditions of comfort in the room, it is necessary to condition the air and supply it to the room, generally, at a lower temperature and moisture content in summer, and a higher temperature and moisture content in winter. It was shown in the preceding chapter that the supply air condition lies on the room sensible heat factor line. Thus, thermodynamically, there is a wide choice available for the supply air-condition. For each supply air condition, there is a supply air rate. Corresponding to the minimum weight of the supply air, the supply condition is at the apparatus dew point for summer, or slightly away from it in the case of a finite bypass factor of the conditioning apparatus. This minimum weight of the supply air may not be adequate from the point of view of good room air distribution aiming at equalizing both the temperature and humidity throughout the conditioned space, in addition to providing a certain air movement at the occupancy level. For good room air distribution, it is found necessary to supply the conditioned air equivalent to 8–12 air changes per hour. The normally acceptable value is 10 air changes per hour. A lower value will cause *stagnancy* whereas a higher value will cause *draft*, both leading to discomfort. Some consultants in India take supply air 15 air changes per hour as an insurance against dirty filters. This comes to about 2.5 cfm/ft^2 of floor area with a height of about 10 feet.

In case of evaporative cooling, this value is close to 20 air changes per hour.

Corresponding to the specified amount of the quantity of supply air, the supply air temperature for summer air conditioning would be 12.5 to 7.5°C below the room temperature. For a room temperature of 25°C, the supply air temperatures is close to 13° to 15°C.

In case, the minimum weight of the supply air does not meet the air change requirement, it is possible to increase the quantity of the supply air by mixing it with an additional amount of recirculated room air after the A/C apparatus as shown in Fig. 16.3.

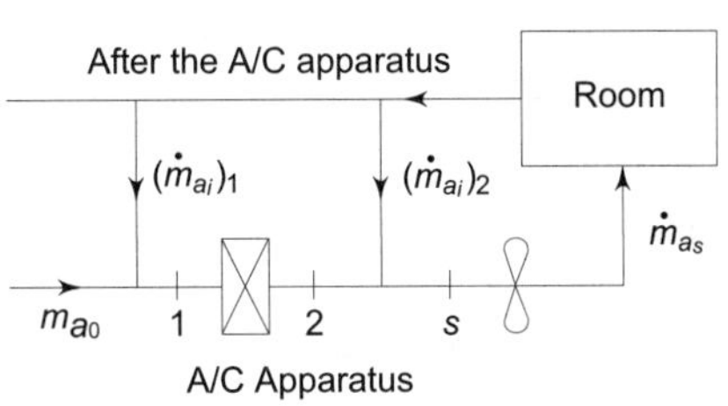

Fig. 16.3 Recirculated room air added after the A/C apparatus

The effect of such mixing is merely to increase the bulk of the supply air, and correspondingly change its temperature and moisture content, but in no way does it affect the capacity of this air to meet the room sensible and latent heat loads. In such a case, the conditioned air quantity is $[(\dot{m}_{a_i})_1 + \dot{m}_{a0}]$ but the supply air quantity becomes $[(\dot{m}_{a_i})_1 + \dot{m}_{a_0} + (\dot{m}_{a_i})_2]$. Accordingly, the supply air condition moves to s as shown in Fig. 16.4. The dehumidified air rise available is $t_i - t_s$.

Note *The supply air temperature for winter heating is kept at about 45–55°C.*

Example 16.1 *An air-conditioned space is maintained at 26°C DBT and 50 per cent RH when the outside conditions are 35°C DBT 28°C WBT.*

(a) If the space has a sensible heat gain of 17.6 kW and air is supplied to the room at a condition of 8°C saturated, calculate:
   (i) The mass and volume flow rates of air supplied to the room.
   (ii) The latent heat gain of the space.
   (iii) The cooling load of the refrigeration plant if 15 per cent of total weight of air supplied to the space is fresh air, and the remainder is recirculated air.

(b) If the supply air to the room is to be maintained at a level of at least 6 cmm per ton of the cooling load, find:
   (i) The new supply air temperature.
   (ii) The mass and volume flow rates of the supply air, and recirculated air added after the A/C apparatus.

Solution Refer to Figs 16.3 and 16.4.

(a) From the psychrometric chart

$$h_0 = 90 \text{ kJ/kg}$$
$$h_i = 53.5 \text{ kJ/kg}$$
$$h_2 = 25 \text{ kJ/kg (8°C saturated)}$$
$$h_A = 43.8 \text{ kJ/kg (8°C DPT, 26°C DBT)}$$

Point A is obtained by drawing the horizontal from point 2 at ADP, and a vertical from i.

(i) Mass rate of supply air

$$\dot{m}_{a_s} = \frac{\dot{Q}_S}{h_A - h_2} = \frac{(17.6)(3600)}{43.8 - 25} = 3370 \text{ kg/h}$$

Specific volume of air at 2

$$v = 0.805 \text{ m}^3\text{/kg}$$

Volume rate of supply air

$$\dot{Q}_{v_s} = \dot{m}_{a_s} v = \frac{(3370)(0.805)}{60} = 45.2 \text{ cmm}$$

(ii) Latent heat gain of space

$$\dot{Q}_L = \dot{m}_{a_s}(h_i - h_A) = \frac{3370}{3600}(53.5 - 43.8) = 9.1 \text{ kW}$$

(iii) Cooling load of the refrigeration plant

$$\dot{Q} = \dot{m}_{a_s}(h_1 - h_2) = \frac{3370}{3600}\left[\frac{3}{4}(90) + \frac{1}{4}(53.5) - 25\right] = 52.3 \text{ kW}$$

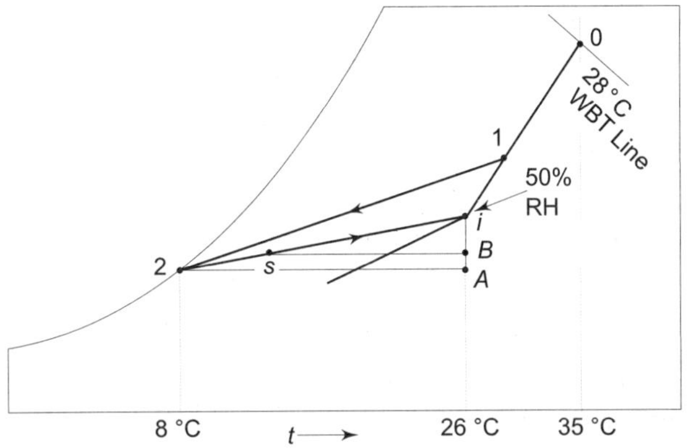

Fig. 16.4 Processes on psychrometric chart for the system of Fig. 16.3 and Example 16.1

(b) In this case, the supply air moves to point s from 2.

(i) Tons refrigeration

$$TR = \frac{52.3}{3.5167} = 14.9$$

Supply air volume flow rate

$$\dot{Q}_{v_s} = 6(14.9) = 89.4 \text{ cmm}$$

Supply air temperature is then given by

$$\dot{Q}_S = \dot{m}_{a_s} C_p(t_i - t_s)$$

$$17.6 = \frac{89.4}{60v_s}(1.0216)(26 - t_s)$$

Solving by trial and error

$$t_s = 17°\text{C}$$

$$v_s = 0.828 \text{ m}^3/\text{kg}$$

(ii) Mass flow rate of the supply air

$$\dot{m}_{a_s} = \frac{(89.4)(60)}{0.828} = 6478 \text{ kg/h}$$

Recirculated room air added after the conditioning apparatus

$$(\dot{m}_{a_i})_2 = 6478 - 3370 = 3108 \text{ kg/h}$$

Volume flow rate of this air

$$(Q_{v_i})_2 = (\dot{m}_{a_i})_2 v_i = \frac{3108}{60}(0.86) = 44.55 \text{ cmm}$$

16.4.1 Purposes of Ventilation

The fresh air or ventilation air is a must in any comfort air-conditioning system. Its purposes are the following:

(a) *To provide oxygen* The oxygen concentration is 21% by volume in atmospheric air. It should not be allowed to fall below 15% under any circumstances whatsoever. The metabolic oxygen requirement of individuals varies between 0.03 m^3/hr for very light work to 0.15 m^3/hr for very heavy work.

(b) *To remove carbon dioxide* The CO_2 concentration in atmospheric air is 0.03% by volume. It should not be allowed to rise above 5% under any circumstance. For CO_2 dilution, a minimum fresh air flow of 0.2 cmm per sedentary adult is recommended. In spaces of heavy activity and smoking, 0.7 to 1.1 cmm per person is required. 0.42 cmm of fresh air per person is used as a design standard.

(c) *To remove odours* 0.42 cmm of fresh air per person is required to remove body odours. The actual air requirement depends on room size and level of activity.

(d) *To remove heat and humidity* Removal of *body heat* and moisture addition by ventilation is the controlling factor. If this is accomplished, all other requirements will be met. For sedentary adults, the body heat generation is taken as 116 W per person. The ventilation requirement can be determined from the room sensible and latent heat generation rates using the following equation:

$$Q_S = Q_{v_0}\, \rho\, C_p\, \Delta t \tag{16.7}$$

$$Q_L = Q_{v_0}\, \rho\, h_{fg}\, \Delta \omega \tag{16.8}$$

where Δt and $\Delta\omega$ are the *allowable changes* in room *DBT* and humidity.

(e) *To dilute toxicity* This is required when toxic and hazardous fumes/particles are being generated in the space. A very detailed study is required for the purpose. However, the *dilution ventilation* is not very effective. In such cases, an outright removal by *exhaust ventilation* is recommended.

Note *During mild weather, e.g., March/April at the beginning of summer, and September/October at the beginning of winter, recourse to only ventilation calculated froms Eqs (16.7) and (16.8) may be a satisfactory substitute for air conditioning.*

16.5 CRITICAL LOADING CONDITIONS[6]

There are five critical loading conditions. Three are determined by the maximum values of the three loads, viz., sensible, latent and total, and the other two are determined by the maximum and minimum values of the room sensible heat factor. The condition lines for the five cases are drawn in Fig. 16.5 for the case of summer air conditioning involving cooling and dehumidifying of air.

It is seen that for a certain minimum quantity of supply air, RSH_{max} determines the minimum supply air temperature t_{smin}, and RLH_{max} determines the minimum supply air humidity ω_{smm}. Similarly, the RTH_{max} determines the minimum supply air enthalpy h_{smin}. A curve through these three supply air states *b*, *c*, *d* forms the locus of the supply air states to maintain the room at *i* under the three maximum load conditions, at the assigned constant minimum supply air rate which may be fixed in terms of the maximum sensible heat load.

The other two conditions of maximum and minimum room sensible heat factors often occur at *partial loads*. The supply conditions for these two cases can, similarly, be determined as *e* and *a* for the two cases respectively. Curve *a b c d e* forms the complete locus of supply air states at the minimum supply air rate. A similar curve *a′ b′ c′ d′ e′* can be drawn corresponding to the maximum supply air rate. Between these two curves fall an infinite number of loci corresponding to different supply air rates and states. A perfect air-conditioning system should be able to handle all such conditions.

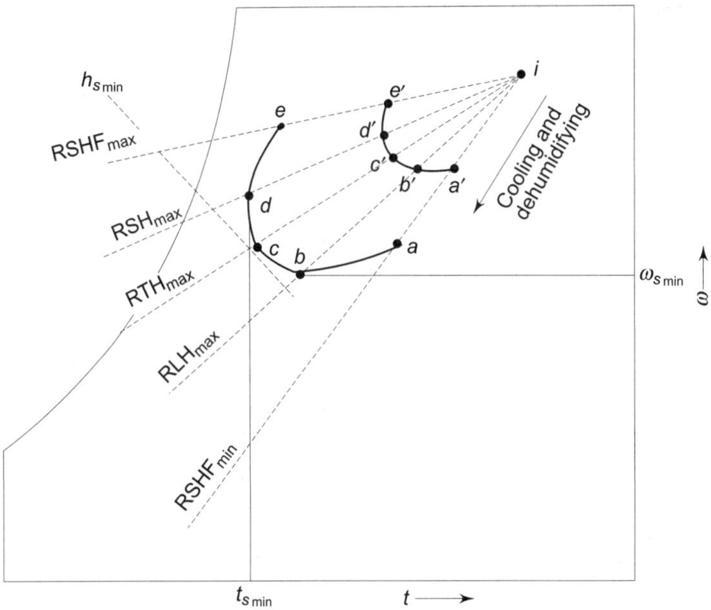

Fig. 16.5 Five critical loading conditions

The investigation of each of the above critical loading conditions is the most complex task that faces an air-conditioning engineer. In most applications, one or more of these critical condition lines will coincide on the psychrometric chart, thus simplifying the problem.

Figure 16.6 shows the four basic types of supply problems. Lobe *A* represents the region of the supply air states for comfort air conditioning for summer which is associated with the cooling and dehumidification of air. Lobe *B* represents the same for winter which is accompanied with heating and humidification. Lobes *C* and *D* are, respectively, for cooling and humidification, and heating and dehumidification. These process combinations do not normally arise in comfort air-conditioning, but often are of great importance in industrial air-conditioning systems.

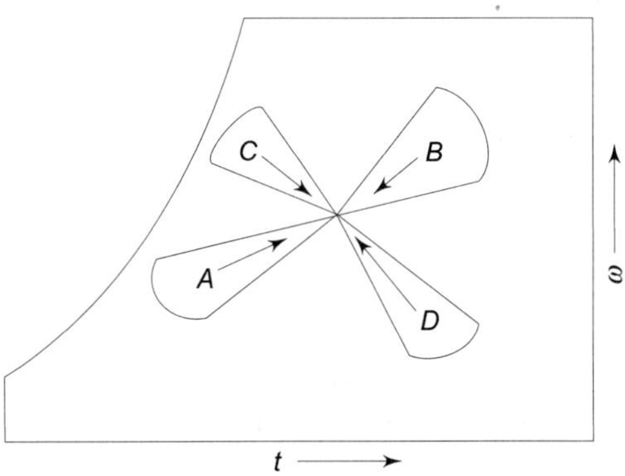

Fig. 16.6 Four types of supply problems

16.6 CLEAN SPACES

Apart from contamination of conditioned spaces from undesirable and obnoxious gases which is taken care of by ventilation, the application of *clean rooms/spaces* pertains, primarily, to the problem of *particulate contamination*. This contamination not only poses a threat to human health, but also to certain industrial processes involving assembly of microscopically small electronic devices, manufacturing of optical instruments, drugs, etc., and medical and biological applications. These processes are carried out in clean rooms/spaces wherein airborne particles are limited and air flow patterns are regulated. In addition, the temperature and humidity are controlled, the space is pressurized so that no outside air can infilterate into the space, and special materials of construction are used so as to eliminate generation of any particulate matter within the conditioned space.

The limiting of air borne particles is achieved by using a series of filters of various grades, and selecting air flow patterns as described in Chap. 21.

The design conditions are specified by particle size and number of particles per unit volume as follows:

Class 100,000 Particles not to exceed 3, 531,000/m^3 (100,000/ft^3) of size 0.5 μm and larger, or 24,700/m^3 of size 5 μm and larger.

Class 10,000 Particles not to exceed 353,000/m^3 (10,000/ft^3) of size 0.5 μm and larger, or 2,295/m^3 (65/ft^3) of size 5 μm and larger.

Class 100 Particles not to exceed 3,500/m^3 (100/ft^3) of size 0.5 μm and larger.

In these applications, *high efficiency particulate air* (*HEPA*) filters are used. These filters must have an efficiency of 99.97% for 0.3 μm particles as determined by *Dioctyl Phthalate* (DOP) test. Air recirculating through HEPA filters is called *primary air*. The portion of primary air circulated through A/C equipment is called *secondary air*. In addition, we have *make-up air* which is added through normal filters to secondary air for the purpose of ventilation and pressurization.

References

1. *ASHRAE Guide and Databook*, Application Volume, 1964.
2. Fanger P O, *Thermal Comfort*, McGraw-Hill, New York, 1972.
3. Jones W P, *Air-Conditioning Engineering*, Edward Arnold, London, 1973.
4. Kadambi V and F W Hutchinson, *Refrigeration, Air Conditioning and Environmental Control in India*, Prentice-Hall of India, New Delhi, 1968.
5. Malhotra H S, 'Environmental comfort zone in warm and humid atmosphere', *J. Scientific and Industrial Research*, 14A, pp. 469-473.
6. Raber B F, and F W Hutchinson, *Refrigeration and Air Conditioning Engineering*, John Wiley, New York, 1949.

Revision Exercises

16.1 (a) For an office building in a city having a calculated heating load of 940 kW, estimate the weight of steam required during the heating season for which the number of degree-days are 3413. The design outside and inside temperatures are –17 and 21°C respectively.

(b) Also estimate the consumption if gas, oil or coal is used instead of steam. The heating values of the three alternative fuels are 29,806 kJ/m^3, 4,315,910 kJ/m^3 and 29,810 kJ/kg respectively.

16.2 (a) A hermetically sealed room 4 m × 4 m × 4 m has an internal sensible heat gain of 3 kW. The overall heat-transfer coefficient of its walls, floor and ceiling is 1.1 W m^{-2} K^{-1}. The outside temperature is constant at 37°C. Calculate the steady-state temperature of the room.

(b) What will be the temperature maintained inside the room if the ventilation air supplied is equivalent to 10 air changes per hour?

(c) For the above room with other conditions same, calculate the steady outside temperature that will result in an inside temperature of 26°C.

16.3 (a) An air-conditioned building has a space volume of 1000 m^3 and a room sensible heat gain of 20 kW. The room is maintained at 25°C DBT and 50 per cent RH. The outside design conditions are 43°C DBT and 26.5°C WBT. The ventilation air is 20 per cent of the supply air. The apparatus dew point of the coil is 11°C and its bypass factor is 0.1. Determine:

(i) The state and mass flow rate of the supply air.
(ii) The latent heat gain of the room.
(iii) The cooling load of the refrigeration plant.

(b) If the supply air rate is to be maintained at 10 air changes per hour, determine the mass and volume flow rates of the recirculated room air added after the air conditioning apparatus.

Chapter

17

Solar Radiation

17.1 DISTRIBUTION OF SOLAR RADIATION

Solar radiation forms the greatest single factor of cooling load in buildings. It is, therefore, necessary to study the subject not only for the purpose of load calculation, but also from the point of view of load reduction. Further, the subject has acquired new dimensions in the present-day world in the context of solar energy utilization for heating as well as cooling.

For all practical purposes, the sun is the source of energy for the continuation of life on earth. It is a sphere of intensely hot gaseous matter. It is a fusion reactor—the most important of its reactions is the combination of hydrogen to form helium, the difference in mass being converted to energy. This fusion energy is produced in the interior of the solar sphere at a temperature of many millions of degrees. The energy is transferred to the surface of the sun by radiation and convection. The surface is opaque. For all practical purposes, it may be considered to be radiating energy as a black body at an effective temperature of 6000 K. The spectrum of the wavelength of radiation stretches from 0.29 to 4.75 μm. As a consequence of high temperature, the maximum radiation intensity is found to be at a wavelength of 0.5 μm.

The mass of the sun is about 332,000 times that of the earth and its diameter is about 1,392,400 km. The earth is about 12,710 km in diameter. It makes one rotation about its axis in 24 hours, and a revolution around the sun in a period of approximately $365\frac{1}{4}$ days.

The earth revolves round the sun in an elliptical orbit. The earth is closest to the sun on January 1, and remotest from it on July 1 (about 3.3 per cent farther away). The mean distance of the earth from the sun is 149,500,000 km. The intensity of solar radiation outside the earth's atmosphere varies inversely with the square of the distance between the centre of the earth and the centre of the sun. Accordingly, the earth receives 7 per cent more radiation in January than in July.

The earth's axis of rotation is, however, tilted 23.5° with respect to its orbit around the sun as shown in Fig. 17.1. This angle of tilt is essentially responsible for the distribution of solar radiation over the earth's surface and, consequently, the change of seasons.

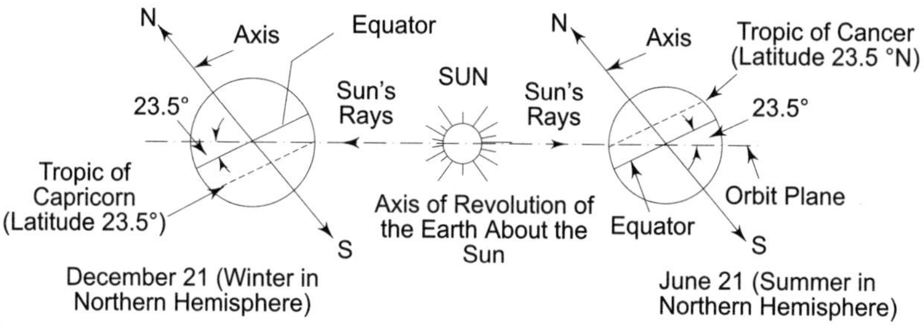

Fig. 17.1 Earth's position with respect to the sun during summer and winter

17.1.1 Solar Radiation Intensity Outside Earth's Atmosphere

The solar radiation intensity—normal to the sun's rays incident upon a plane surface situated in the outer limits of the earth's atmosphere—varies with the time of the year as the distance of the earth from the sun changes. Its value when the earth is at its mean distance from the sun is called the *solar constant*. The presently accepted standard value of the solar constant as determined by Thekaekara and Drummond[11] in 1971 is 1353 W/m^2. The variation of the normal solar radiation intensity outside the earth's atmosphere, denoted as I_{n0}, with the time of the year can be obtained by applying a correction factor to the solar constant as given in the Smithsoman Physical Tables and reproduced here in Table 17.1.[12]

Table 17.1 Solar constant correction factors[12]

Month	*Day of the Month*			
	1	*8*	*15*	*22*
January	1.0335	1.0325	1.0315	1.03
February	1.0288	1.0263	1.0235	1.0207
March	1.0173	1.0141	1.0103	1.0057
April	1.0009	0.9963	0.9913	0.9875
May	0.9841	0.9792	0.9757	0.9727
June	0.9714	0.9692	0.9680	0.967
July	0.9666	0.967	0.9680	0.9692
August	0.9709	0.9726	0.9757	0.9785
September	0.9828	0.9862	0.9898	0.9945
October	0.9995	1.0062	1.0087	1.0133
November	1.0164	1.0207	1.0238	1.0267
December	1.0288	1.0305	1.03	1.028

The spectral distribution of this extra-terrestrial radiation is given in Table 17.2.

Table 17.2 Spectral distribution of extra-terrestrial radiation[2]

Region	*Ultraviolet*	*Visible Region*	*Infra-red*
Wavelength Range	0 – 0.38 μm	0.38 – 0.78 μm	0.78 – ∞ μm
Percentage Radiation	7	47.3	45.7

17.1.2 Direct and Diffuse Solar Radiation

From outside the earth's atmosphere, the solar heat reaches any part of the earth's surface in two ways. A part of the sun's radiation travels through the atmosphere and reaches the earth's surface directly. This part is called *direct* or *beam radiation*. It is *specular* in nature and is incident on a surface at an angle which is determined by the line joining the centre of the sun to the centre of the earth. Thus if the orientation of the surface is changed, this radition can be increased or decreased. It is maximum when the surface is normal to the sun's rays, and zero when it is parallel to them.

A major part of the sun's radiation is scattered, reflected back into space and absorbed by the earth's atmosphere. A part of this radiation is re-radiated and reaches the earth's surface uniformly from all directions. It is called *diffuse* or *sky radiation*. It is diffuse in nature and does not normally change with the orientation of the surface.

The total solar radiation reaching a surface is equal to the sum of the direct and diffuse raditions.

The difference between the solar radition outside the earth's atmosphere and the total radiation reaching the earth's surface is governed by the distance travelled by the radiation through the atmosphere to reach the surface, and the amount of haze in the atmosphere.

The sky radiation is usually a very small part of the total radiation in a clear sky. But with a hazy or cloudy sky, the sky radiation increases while direct radiation is depleted considerably.

17.1.3 Depletion of Direct Solar Radiation by Earth's Atmosphere

After entering the earth's atmosphere, the solar radiation is scattered in all directions by air and water vapour molecules and dust particles. A part of this radiation is also absorbed, particularly by ozone, in the upper atmosphere and water vapour and carbon dioxide in the atmosphere near the earth. To some extent, it is also absorbed by oxygen as well. Thus, the depletion of the direct solar radiation is quite large even on clear days when most of the air-conditioning load requirement occurs.

In order to take this depletion into account, the concept of *air mass* has been introduced. Unit air mass corresponds to the condition of a clear sky and the sun at the zenith, i.e., when the sun is directly overhead or at an altitude angle of 90°. The *altitude angle* is defined as the angle in a vertical plane between the sun's rays and the projection of the sun's rays on a horizontal plane. Thus in Fig. 17.2, the angle APH is the altitude angle. However, when the sun is at *B*, then the altitude angle is equal to 90° and the distance BP corresponds to an air mass equal to unity. In this position, the sun's rays have to travel through the shortest distance through the atmosphere, and the depletion of the direct solar radiation is minimum.

The air mass can now be defined as the path length of solar radiation through the atmosphere, assuming the vertical path at the mean sea level as unity. Thus for any altitude angle β, we see from Fig. 17.2, that

$$\frac{\text{AP}}{\text{BP}} = \frac{\sin 90^\circ}{\sin \beta}$$

or considering BP as unity, we have for the air mass

$$L = \text{AP} = \frac{1}{\sin\beta} \tag{17.1}$$

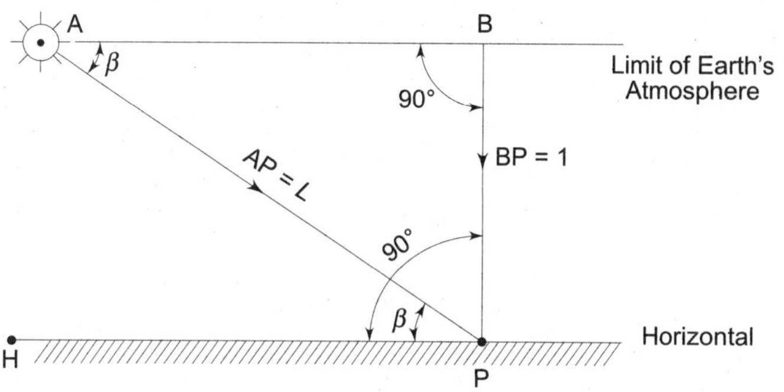

Fig. 17.2 Length of path of direct solar radiation through the atmosphere and altitude angle

Thus AP corresponds to the air mass L for an altitude angle of β. We see that the air mass is a measure of the length of the path the radiation has to travel in the earth's atmosphere. It can be readily seen that for $\beta = 0$, the air mass $L = \infty$. The zero value of the air mass corresponds to the solar-radiation intensity outside the earth's atmosphere.

Table 17.3 gives values of air mass and the corresponding solar altitude angles along with the values of the direct solar radiation, as given by Moon.[7]

Figure 17.3 shows a plot of these values of the direct normal solar radiation against the air mass. The variation can be expressed in the form of an exponential relationship.

$$I_n = I_{n_0} e^{-KL} \tag{17.2}$$

where I_n = Direct solar radiation normal to the sun's rays at a location such as P in Fig. 17.2

I_{n_0} = Direct solar radiation normal to the sun's rays at the outer limits of the atmosphere

K = Solar-extinction constant.

Table 17.3 Solar radiation Intensity vs. air mass

Air Mass	0	1.0	1.5	2.0	3.0	4.0	5.0	6.0	8.0
Solar Altitude		90°	41°48′	30°	19°3′	14°30′	11°32′	9°36′	7°14′
Radiation, W/m^2	1320	925	826	744	616	523	454	384	337

Equation (17.2) may also be written in the form of a straight-line relationship between log I_n and L, viz.,

$$\begin{aligned} \log_{10} I_n &= \log_{10} I_{n_0} - (K \log_{10} e)L \\ &= \log_{10} I_{n_0} - K_1 L \end{aligned} \tag{17.3}$$

where $K_1 = 0.434\,K$

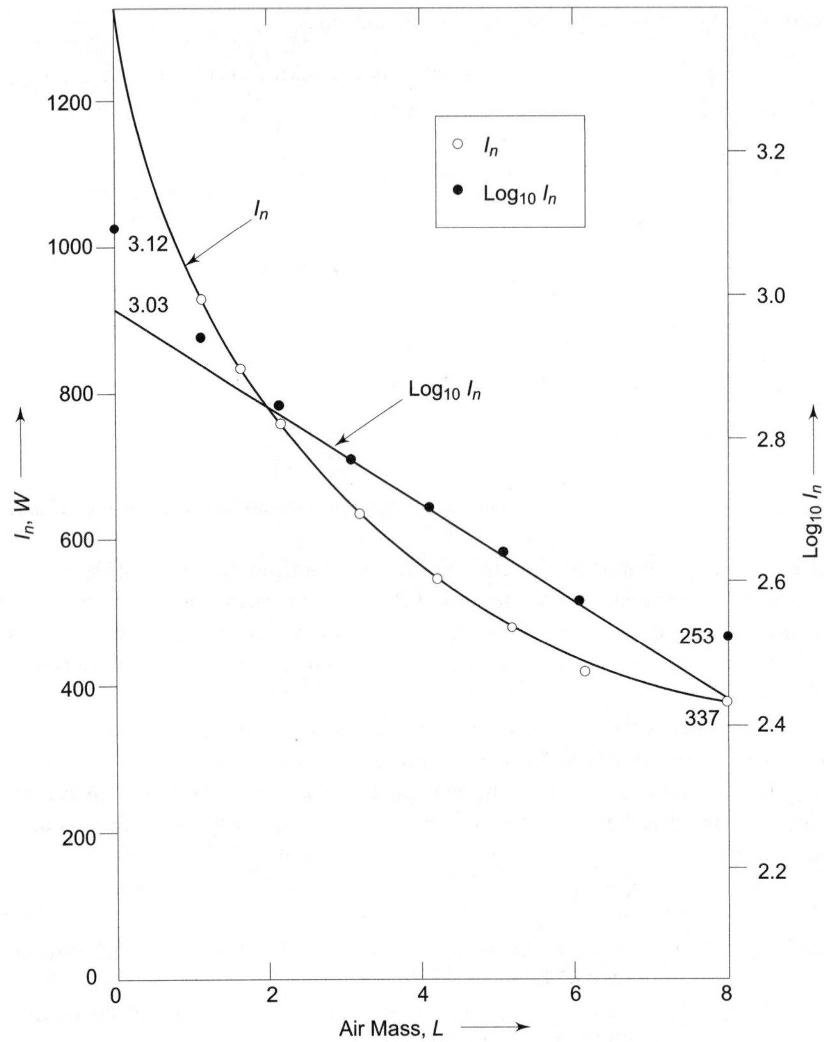

Fig. 17.3 I_n and $\log_{10} I_n$ vs. Air Mass

The points from Table 17.3 are plotted in Fig. 17.3. Most of these points satisfy the straight-line relationship of Eq. (17.3) if

$$\text{Log}_{10}\, I_{n0} = 3.03$$

This gives

$$I_{n0} = 1082 \text{ W/m}^2$$

Such an equation is, however, not satisfied by the extreme values corresponding to $L = 0$ and $L = 8$ in Fig. 17.3. Now by taking the values at any other point on the line, such as $L = 2$, $I_n = 744$ W/m^2, we can find the value of the solar extinction coefficient. Thus

$$K_1 = \frac{\log I_{n0} - \log I_n}{L} = \frac{\log 1082 - \log 744}{2} = 0.0789$$

and since $K_1 = 0.434\,K$

$\Rightarrow$ $K = (0.434)^{-1}\,(0.0789) = 0.182$

Accordingly, the equation for the depleted normal solar radiation becomes

$$I_n = 1082\,e^{-0.182L}\ \text{W/m}^2 \tag{17.2a}$$

This equation takes care of the absorption of the solar radiation by the atmosphere, the length of which is determined by the altitude angle. Besides the solar altitude, other factors that can be considered to affect the air mass are the atmospheric conditions.

17.1.4 Variation in Solar Intensity with Altitude of a Place

The above calculations of the air mass and solar intensity are applicable to the mean sea level. However, the intensity increases for places at higher altitudes. The increase in intensity with height for different solar altitude angles is given in Table 17.4.

Table 17.4 Percentage increase in solar intensity with height

Height Above Mean Sea Level, m	*Solar Altitude Angle, Degrees*									
	10	**20**	**25**	**30**	**35**	**40**	**50**	**60**	**70**	**80**
	Percentage Increase in Solar Intensity									
1000	17.5	14	13	12	10	10	9	8	8	8
1500		26	23	20	15	17	16	15	15	15
2500		40	35	32	30	28	26	24	23	22
3000		42	37	33	32	30	28	27	26	25

17.2 EARTH-SUN ANGLES AND THEIR RELATIONSHIPS[12]

At first we shall define the fundamental angles describing the orientation of the earth with respect to the sun. This will be followed by the definitions of certain derived angles and relationships between these angles.

17.2.1 Fundamental Earth-Sun Angles

The position of any point on the earth's surface, in relation to the sun's rays, is described at any instant by the *latitude* of the place *l*, *hour angle h* and sun's *declination d*. These angles are illustrated in Fig. 17.4 for a point P in the northern hemisphere. The point O in the figure represents the centre of the earth.

The angle which the line *OP* makes with its projection OA in the equatorial plane is the latitude *l* of the point *P*.

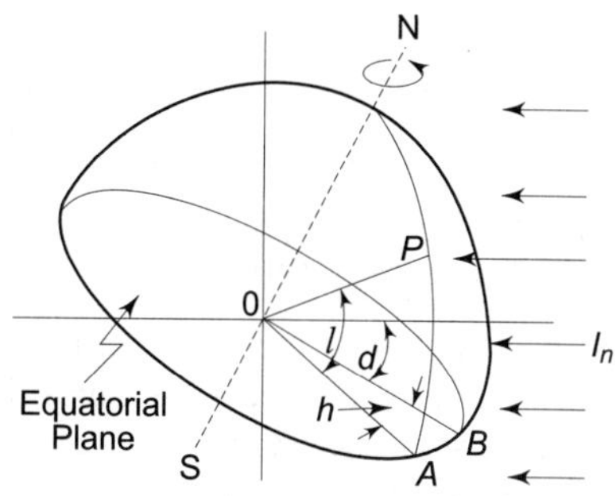

Fig. 17.4 Latitude angle, hour angle and sun's declination

The hour angle h is the angle between the projection OA and the projection OB of the line joining the centre of the earth to the centre of the sun. The hour angle is zero at solar noon at P. Thus OB represents the projection of the centre-to-centre line, at noon i.e., the line of the sun's rays. The hour angle is a measure of the time of the day with respect to solar noon. One hour of time corresponds to $360/24 = 15°$ of the hour angle. Thus at 3 p.m. solar time, the hour angle is 45°.

The declination d is the angle between the centre-to-centre line and its projection OB. It is, thus, the angle between the sun's rays and the equatorial plane.

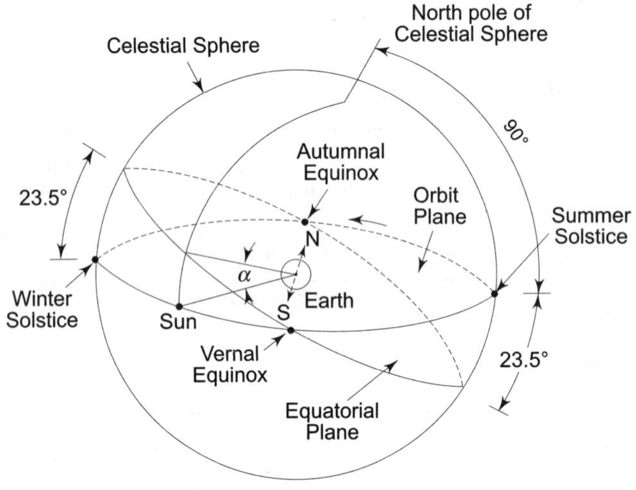

Fig. 17.5 Relative motion of the sun with respect to the earth

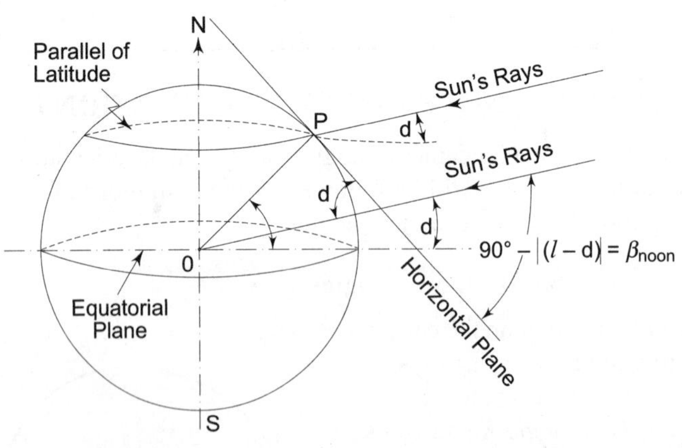

Fig. 17.6 Another representation of sun's declination as angle between the sun's rays and equatorial plane

If we take the earth as the centre of the universe, the sun would appear to orbit the earth as shown in Fig. 17.5. This orbit would be the same as the earth's own orbit around the sun. The sphere with a diameter equal to the distance between the winter and summer solstice positions of the sun is called *celestial sphere*. The declination is shown here as the angle between the orbit plane and the equatorial plane. Another representation of sun's declination is shown in Fig. 17.6 as the angle between the sun's rays and the equatorial plane. It will be seen that the sun's declination is maximum at the solstice positions and is equal to 23.5° south ($d = -23.5°$) of the equatorial plane at the time of the winter solstice on December 21 and 23.5° north ($d = +23.5°$) of the plane at the time of the summer solstice on June 21. At the autumnal and vernal equinoxes it is equal to zero. The weekly variation of d over the year is given in Table 17.5. During the year d can be approximated by a sinusoidal variation

$$d = 23.47 \sin \frac{360(284 + N)}{365} \tag{17.4}$$

where N is the day of the year counted from January 1.

17.2.2 Derived Solar Angles

Some other angles used in solar radiation calculations which can be derived in terms of the fundamental angles are the sun's *zenith angle* ψ, *altitude angle* β and *azimuth angle* γ. For a point P on the surface of the earth, these angles are shown in Fig. 17.7.

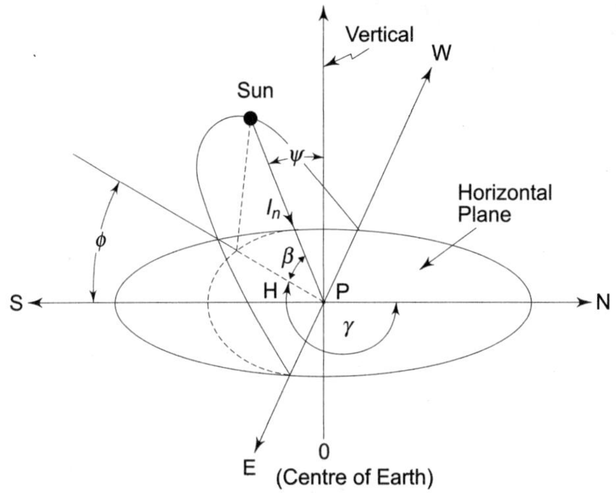

Fig. 17.7 Derived solar angles

The altitude angle β has already been defined in Sec. 17.1.3. It is the angle between the sun's rays and the horizontal plane. The zenith angle is the angle between the sun's rays and the vertical line at P. Thus

$$\psi = \frac{\pi}{2} - \beta.$$

Finally, the azimuth angle γ is the angle measured from the north direction to the projection of the sun's rays in the horizontal plane. Note that $\phi = \gamma - \pi$ is the azimuth angle measured from the south direction.

17.2.3 Relationship between Angles

The relationships between the derived and fundamental earth-sun angles are given below.

In Fig. 17.8, the plane x-y corresponds to the equatorial plane and the z-axis corresponds to the earth's axis. Vector I_n representing the sun's radiation is in the

Table 17.5 Weekly variation of sun's declination angle and equation of time[12]

Day	*1*		*8*		*15*		*22*	
	Declination	***Equation of Time***	***Declination***	***Equation of Time***	***Declination***	***Equation of Time***	***Declination***	***Equation of Time***
Month	***Deg:Min***	***Min:Sec***	***Deg:Min***	***Min:Sec***	***Deg:Min***	***Min:Sec***	***Deg:Min***	***Min:Sec***
January	–23:08	–3:16	–22:20	–6:26	–21:15	–9:12	–19:50	–11:27
February	–17:18	–13:34	–15:13	–14:14	–12:55	–14:15	–10:17	–13:41
March	–7:51	–12:36	–5:10	–11:04	–2.25	–9:14	0:21	–7:12
April	4:16	–4:11	6:56	–2:07	9:30	–0:15	11:57	1:19
May	14:51	2:50	16:53	3:31	18:41	3:44	20:14	3:30
June	21:57	2:25	22:47	1:15	23:17	–0.09	23:27	–1:40
July	23:10	–3:33	22:34	–4:48	21:39	–5:45	20:25	–6:19
August	18:12	–6:17	16:21	–5:40	14:17	-4:35	12:02	–3:04
September	8:33	–0:15	5:58	2:03	3:19	4:29	0:36	6:58
October	–2:54	10:02	–5:36	12:11	–8:15	13.59	–10:48	15:20
November	–14:12	16:20	–16:22	16:16	–18:18	15:29	–19:59	14:02
December	–21:41	11:14	–22:38	8:26	–23:14	5:13	–23:27	1:47

x-z plane. It coincides with the line joining the centre of the earth to the centre of the sun. O-X is then the projection of I_n on the equatorial plane. The angle d as shown, therefore, represents the sun's declination.

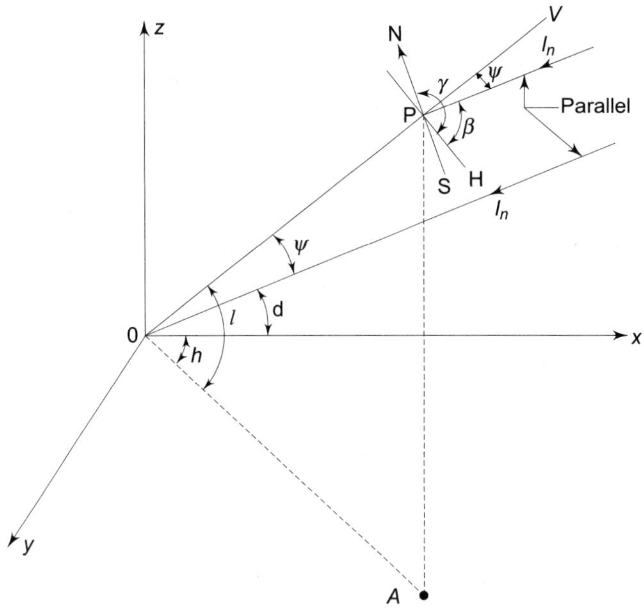

Fig. 17.8 Relationship between angles

The projection of P on the equatorial plane x-y is A. The angle POA is, therefore, the latitude l at P, and the angle XOA in the x-y plane is the hour angle h.

The line OP represents the vertical through P, and the line PH the horizontal. The line PN is pointing north. It is, therefore, perpendicular to OP as well as the z-axis.

Now, let a_1, b_1 and c_1 be the direction cosines of OP. Also let a_2, b_2 and c_2 be the direction cosines of I_n. Then

$$a_1 = \cos l \cos h; \qquad a_2 = \cos d$$
$$b_1 = \cos l \sin h; \qquad b_2 = 0$$
$$c_1 = \sin l; \qquad c_2 = \sin d$$

The zenith angle ψ is the angle between OP and I_n. Therefore, for the altitude angle β, we have

$$\sin \beta = \sin\left(\frac{\pi}{2} - \psi\right) = \cos \psi = a_1 a_2 + b_1 b_2 + c_1 c_2$$
$$= \cos l \cos h \cos d + \sin l \sin d \tag{17.5}$$

Similarly, it can be shown that the sun's azimuth angle is given by

$$\tan \phi = \tan (\gamma - \pi) = \frac{\sin h}{\sin l \cos h - \cos l \tan d} \tag{17.6}$$

At solar noon, $h = 0$, so that from Eq. (17.6)

$$\gamma = \pi \text{ if } l > d \quad \text{and} \quad \gamma = 0 \text{ if } l < d$$

In the use of Eq. (17.6), the convention followed are given below:

In the afternoon: The value of h is positive. Equation (17.6) may give either a positive or a negative value of $(\gamma - \pi)$.

(i) If $(\gamma - \pi)$ is positive, say ϕ, then ϕ is measured clockwise from S-direction.

(ii) If $(\gamma - \pi)$ is negative, say $-\phi$, then ϕ is measured anticlockwise from N-direction.

In the morning before noon: The value of h is negative. Again Eq. (17.6) may give positive or negative values of $(\gamma - \pi)$.

(i) If $(\gamma - \pi)$ is negative, say $-\phi$, then ϕ is measured anticlockwise from S-direction.

(ii) If $(\gamma - \pi)$ is positive, say ϕ, then ϕ is measured clockwise from N-direction.

Another relation for finding the sun's azimuth angle is the following:

$$\cos \phi = \frac{\sin\beta \sin l - \sin d}{\cos\beta \cos l}$$

It can be shown by putting $h = 0$ in Eq. (17.5) that the altitude angle at noon is given by

$$\sin \beta = \cos l \cos d + \sin l \sin d = \cos (l - d) \text{ or } \cos (d - l)$$

$$\Rightarrow \quad \beta_{noon} = \frac{\pi}{2} - |(1 - d)| \tag{17.7}$$

where $|(1 - d)|$ is the net positive value of $(1 - d)$. Equation (17.7) makes it possible to determine the maximum altitude angle at a particular place and a particular time of the year.

The sunset hour angle h_0 can be found by setting $\beta = 0$ in Eq. (17.5). Thus,

$$\cos h_0 = -\tan l \tan d \tag{17.8}$$

The hour angle at sunrise is also the same but will be negative of this value.

Similarly, from Eq. (17.6), the solar azimuth angle at sunset is given by

$$\tan (\gamma_0 - \pi) = \frac{\sin h_0}{\sin l \cos h_0 - \cos l \tan d} \tag{17.9}$$

Example 17.1

(a) Find out the sun's altitude and solar azimuth angle at 3 p.m. solar time on August 1 for a location at 30°N latitude.

(b) Also, find the maximum altitude angle, sunshine hours and solar azimuth angles at sunrise and sunset.

(c) Determine the solar radiation intensity normal to the sun's rays at 3 p.m.

Solution Given $l = 30°\text{N}$

From Table 17.5, on August 1,

$$d = 18°12'$$

(a) At 3 p.m., $h = 45°$

From Eq. (17.5), for the altitude angle

$$\sin \beta = \cos l \cos h \cos d + \sin l \sin d$$

$$= \cos 30° \cos 45° \cos 18°12' + \sin 30° \sin 18°12' = 0.7379$$

$$\beta = 47°33'$$

From Eq. (17.6), for the solar azimuth angle

$$\tan(\gamma - \pi) = \frac{\sin h}{\sin l \cos h - \cos l \tan d}$$

$$= \frac{\sin 45°}{\sin 30° \cos 45° - \cos 30° \tan 18°12'} = 10.28$$

$$\phi = \gamma - \pi = 84°24'$$

$$\gamma = 180° + 84°24' = 264°24'$$

Note *Since γ is measured from north, the direction of the sun's rays is at an angle of 84°24′ from south towards west. Thus on August 1, the sun is almost due west at 3 p.m.*

(b) Maximum altitude angle

$$\beta_{noon} = \frac{\pi}{2} - |(1 - d)| = 90° - (30° - 18°12') = 78°12$$

Note *The sun is almost at the zenith.*

Hour angle at sunset

$$\cos h_0 = -\tan l \tan d = -\tan 30° \tan 18°12' = -0.1898$$

$$h_0 = 100.9°$$

Time at sunset $\quad \tau = \dfrac{100.9°}{15} = 6.73$ hours after solar noon

Sunshine hours $\quad \Delta\tau = 6.73 + 6.73 = 13.46$ hrs

Solar azimuth angle at sunset

$$\tan \phi_0 = \tan(\gamma_0 - \pi) = \frac{\sin h_0}{\sin l \cos h_0 - \cos l \tan d}$$

$$= \frac{\sin 100.9°}{\sin 30° \cos 100.9'' - \cos 30° \tan 18°12'} = -2.5889$$

$$\phi_0 = \gamma_0 - \pi = -68.88° = -68°53'$$

The solar azimuth angle at sunset is 68°53′ measured anticlockwise from N-direction, i.e., 21° 7′ north of west.

The solar azimuth angle at sunrise can, similarly, be found as 21°7′ north of east.

(c) Normal solar radiation intensity at 3 p.m.

$$I_n = 1082\, e^{-0.182L} = 1082\, e^{-0.182/\sin 69.2°} = 890.6 \text{ W/m}^2$$

17.3 TIME

Solar radiation calculations are made in terms of the solar time. We will now relate the solar time to the local time in a particular locality.

However, first we introduce the *Greenwich mean time* which is the time at Greenwich meridian taken at zero longitude. Midnight at Greenwich is 0 h and noon is 12 h. Then we have the *local civil time* (LCT). On a particular longitude, LCT is more advanced towards the east, and less advanced towards the west. The difference amounts to four minutes for each degree of difference in longitude.

Local solar time (LST) is the time that would be shown by a sun dial. It may be pointed out here that whereas a *civil day* is exactly 24 hours, a *solar day* is not exactly so due to the irregularities of the earth's rotation, obliquity of the earth's orbit and other factors. The difference between the local solar time and local civil time is called the *equation of time*. Thus

$$\text{LST} = \text{LCT} + \text{Equation of time} \tag{17.10}$$

Table 17.5 gives the weekly values of the equation of time.

The actual official time in a locality may be different from the civil time. It may be based on the mean longitude of a country. This time is called the *central standard time* (CST). For example, the CST in India is based on 82° 30′ E longitude of Allahabad. To find the solar time at 3 p.m. on June 22 at Delhi where the longitude is 77°12′ E, we find the difference in LCT of Allahabad and Delhi which is 0 h 21 min 12 s. Then, at Delhi at 3 p.m. CST, LCT is 15 h-(0h 21 min 12 s) = 14 h 38 min 48 s. Equation of time on June 22 is – 1 min 40 s. Hence the solar time is 14 h 38 min 48 s – 0 h 1 min 40 s = 14 h 37 min 8 s.

Example 17.2 ***Determination of Local Solar Time***

(a) Determine the local solar time corresponding to 10:00 a.m. central standard time on February 8 for a location at 95°W longitude. The central standard time is equal to the local civil time at 90°W longitude.

(b) Determine the same on June 22 for a location at 82.5°W when the central standard time is equal to the local civil time at 77.1°W.

Solution (a) The local civil times at 90° and 95°W longitude

$$\text{LCT}_{90} = 4(90) = 360 \text{ min (6 h)}$$

$$\text{LCT}_{95} = 4(95) = 380 \text{ min (6 h 20 min)}$$

The difference in time at the two locations is 20 minutes. At 95°W longitude, therefore, LCT is 20 minutes less advanced than LCT at 90°W longitude. Thus at 95°W longitude

$$10:00 \text{ a.m. CST} = 9{:}40 \text{ a.m. LCT}$$

The equation of time, from Table 17.5 on February 8 = – 14 min 14 s ≈ – 14 min

Local solar time $\quad \text{LST} = \text{LCT} + \text{equation of time}$

$$9{:}40 - 0{:}14 = 9:26 \text{ a.m.}$$

(b) Difference in time at the two longitudes

$$\Delta\tau = 4(82.5 - 77.1) = 22 \text{ min}$$

Thus at 77.1°W

$$10:00 \text{ a.m. CST} = 10:00 - 0:22 = 9:38 \text{ a.m. LCT}$$

Equation of time on June 22 = – 1 min : 40 s ≈ 2 min

Local solar time $\quad \text{LST} = 9:38 - 0:02 = 9:36 \text{ a.m.}$

Note *There is a very small difference between LCT and LST. Should the difference between CST and LCT be kept to a minimum, the central standard time will be very close to the local solar time.*

17.4 WALL SOLAR AZIMUTH ANGLE AND ANGLE OF INCIDENCE

We shall now refer to angles which are in relation to the orientation of vertical walls of buildings.

Consider a building block as shown in Fig. 17.9. Also consider the wall facing SW direction. Accordingly, the altitude angle β and solar azimuth angle γ are also shown.

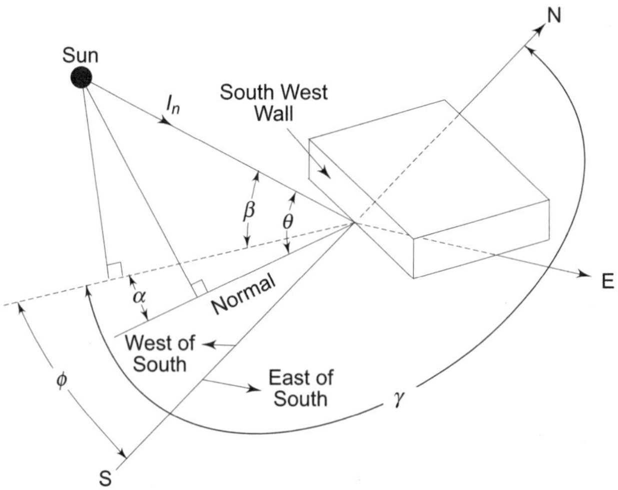

Fig. 17.9 Solar angles of a building wall, $\alpha = \phi \pm \psi$

Now the location of the sun with respect to the wall is defined by the *wall solar azimuth angle* α and *angle of incidence* θ. The wall solar azimuth angle α is the angle in the horizontal plane between the projection of the sun's rays and the normal to the surface. Also, the sun's angle of incidence θ on a wall is the angle between the sun's rays and the normal to the wall as shown in Fig. 17.9.

The angle α can be found from the solar azimuth angle γ and the orientation of the wall itself. Note that α is positive when the wall faces west of south and negative when it faces east of south. For the angle of incidence θ on a vertical wall, we may derive the relation

$$\cos \theta = \cos \beta \cos \alpha \tag{17.11}$$

17.5 DIRECT SOLAR RADIATION ON A SURFACE

We are now in a position to estimate the intensity of the direct solar radiation incident on a surface. It can be determined if the intensity I_n normal to the sun's rays and the angle of incidence θ at the point under consideration are known. The direct radiation I_D, then, is equal to the component of I_n perpendicular to the given surface. Thus, in general,

$$I_D = I_n \cos\theta \tag{17.12}$$

17.5.1 On a Vertical Surface

The direct solar radiation on a vertical surface is shown in Fig. 17.10. It is seen that the component of I_n in the horizontal plane is $I_n \cos\beta$. Its further resolution into a component normal to the vertical wall gives

$$I_{DV} = I_n \cos\beta \cos\alpha = I_n \cos\theta \tag{17.13}$$

so that $\cos\theta = \cos\beta \cos\alpha$ as already derived for a vertical wall in Eq. (17.11).

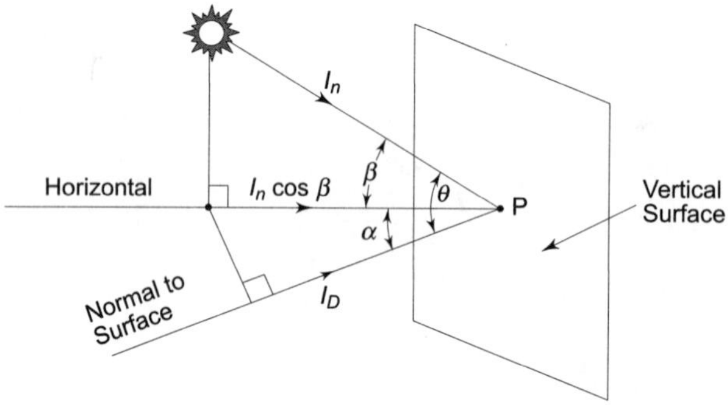

Fig. 17.10 Direct solar radiation on a vertical surface

17.5.2 On a Horizontal Surface

The direct solar radiation on a horizontal surface is shown in Fig. 17.11. The component of I_n normal to the horizontal surface is

$$I_{DH} = I_n \sin\beta = I_n \cos\theta \tag{17.14}$$

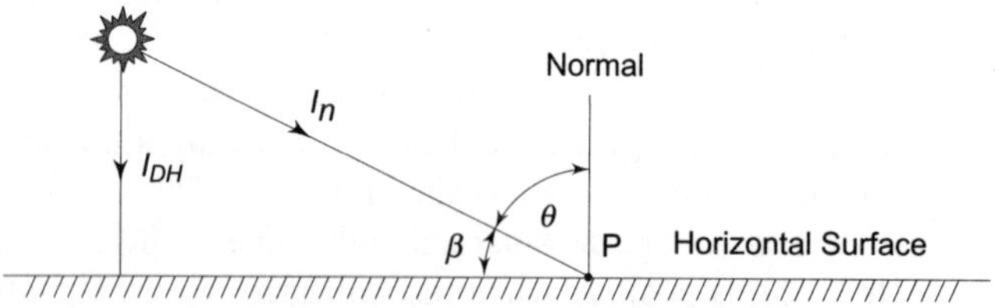

Fig. 17.11 Direct solar radiation on a horizontal surface

so that $\cos\theta = \sin\beta$. The angle of incidence θ in this case is equal to

$$\left(\frac{\pi}{2} - \beta\right)$$

17.5.3 On an Inclined Surface

Consider a surface inclined to the vertical at an angle ϕ, away from the sun as shown in Fig. 17.12. Let the solar azimuth angle of the surface, in case it is vertical, be α. Also let the solar altitude angle be β.

Then the horizontal and vertical components of I_n at point P on the surface are

$$BP = I_n \cos\beta \cos\alpha \quad \text{and} \quad AB = I_n \sin\beta$$

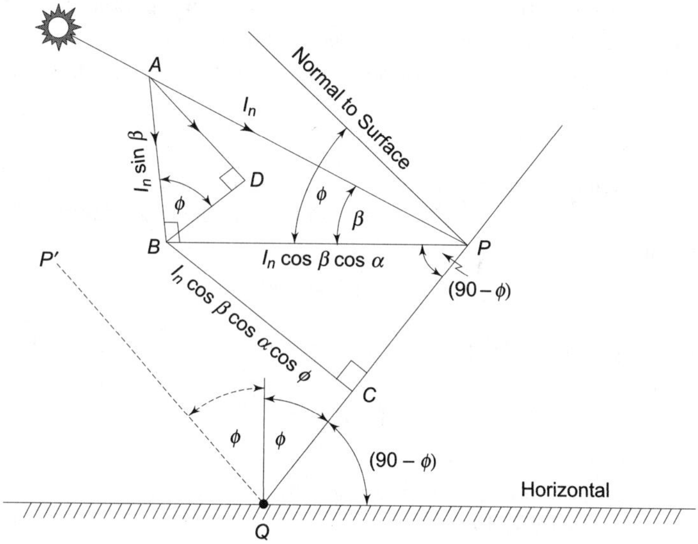

Fig. 17.12 Direct solar radiation on an inclined surface

Now, their respective components normal to the surface are

$$BC = I_n \cos\beta \cos\alpha \cos\phi$$

and

$$AD = I_n \sin\beta \sin\phi$$

Adding the two terms, we obtain for the direct solar radiation normal to the surface at P

$$I_D = I_n (\cos\beta \cos\alpha \cos\phi + \sin\beta \sin\phi)$$

$$= I_n \cos\theta \tag{17.15}$$

so that the angle of incidence for a surface inclined away from the sun is given by

$$\cos\theta = \cos\beta \cos\alpha \cos\gamma + \sin\beta \sin\alpha \tag{17.16}$$

It may be observed here that if the surface is inclined towards the sun as $P'Q$ in Fig. 17.12, then ϕ will be negative. Then the second term on the right-hand side of Eqs (17.15) and (17.16) will also be negative.

The ratio of direct radiation on a tilted surface to that on a horizontal surface R_D is known as the *geometric factor*. From Eqs (17.14) and (17.16), the geometric factor is

$$R_D = \sin\phi + \cos\alpha \cot\beta \cos\phi \tag{17.17}$$

17.6 DIFFUSE SKY RADIATION ON A SURFACE

As stated earlier, a large proportion of the direct solar radiation is either absorbed or scatterd by the constituents of the earth's atmosphere. A part of this radiation also reaches the earth in the form of diffuse radiation. The magnitude of this radiation is such that even on clear days it cannot be neglected even though it is quite small compared to the direct solar radiation. On cloudy days, the direct solar radiation is very much reduced and the diffuse sky radiation increases.

Diffuse sky radiation mostly consists of short-wavelength radiation as it is more readily scattered by the atmosphere. Also, as it is non-directional in nature, it is difficult to analyse it. The measurements show that diffuse sky radiation for clear days is given approximately by the equation

$$I_d = C I_n F_{ss} \tag{17.18}$$

where C is a dimensionless coefficient given in Table 17.6 and F_{ss} is the angle factor between the surface and the sky. For this purpose, the angle factor between the surface and the ground is given by

$$F_{sg} = \frac{1}{2}(1 - \sin \phi) \tag{17.19}$$

where ϕ is the angle of tilt of the surface with the vertical. Then, the factor between the sky and the surface is

$$F_{ss} = 1 - F_{sg} \tag{17.20}$$

Thus, for a vertical wall, $\phi = 0$, $\sin \phi = 0$, $F_{sg} = 0.5$ and $F_{ss} = 0.5$.

ASHRAE recommends the following empirical coefficients C derived by Stephenson, given in Table 17.6.

Table 17.6 Dimensionless coefficient for sky radiation

Date	*C*	*Date*	*C*
Jan. 21	0.058	July 21	0.136
Feb. 21	0.060	Aug. 21	0.122
March 21	0.071	Sept. 21	0.092
April 21	0.097	Oct. 21	0.073
May 21	0.121	Nov. 21	0.063
June 21	0.134	Dec. 21	0.057

Example 17.3 *Calculate the direct and diffuse sky radiation intensities on a roof with a 30° angle of tilt to the horizontal facing south-east at 3 p.m. solar time on July 21 at a location of 30°N latitude.*

Solution Given:

$$l = 30°$$
$$h = 45°$$
$$\phi = 60°$$

From Table 17.5, on July 21 $\quad d = 20°25'$

Altitude angle $\quad \sin \beta = \cos l \cos h \cos d + \sin l \sin d$

$$= \cos 30° \cos 45° \cos 20° 25' + \sin 30° \sin 20°25' = 0.7482$$

$$\beta = \sin^{-1} 0.7482 = 48.4°$$

Air mass $\quad L = \dfrac{1}{\sin \beta} = \dfrac{1}{\sin 48.4°} = 1.3365$

Normal solar radiation intensity

$$I_n = 1082\, e^{-0.182\, L} = 1082\, e^{-0.182\,(1.3365)} = 848 \text{ W/m}^2$$

Solar azimuth angle

$$\tan(\gamma - \pi) = \frac{\sin h}{\sin l \cos h - \cos l \tan d}$$

$$= \frac{\sin 45°}{\sin 30° \cos 45° - \cos 30° \tan 20°25'} = 22.4$$

$$\gamma - \pi = \tan^{-1} 22.4 = 87° \text{ (measured from south)}$$

Wall solar azimuth angle measured from normal to wall, viz., SE direction

$$\alpha = 87° + 45° = 132°$$

Angle of incidence

$$\cos\theta = \cos\beta \cos\alpha \cos\phi + \sin\beta \sin\phi$$

$$= \cos 48.4° \cos 132° \cos 60° + \sin 48.4° \sin 60° = 0.4258$$

$$\theta = \cos^{-1} 0.4258 = 64.8°$$

Direct solar radiation intensity

$$I_D = I_n \cos\theta = 848\,(0.4258) = 361 \text{ W/m}^2$$

Angle factor between the surface and the ground

$$F_{sg} = \frac{1}{2}(1 - \sin\phi) = \frac{1}{2}(1 - \sin 60°) = 0.067$$

Angle factor between the sky and the surface

$$F_{ss} = 1 - F_{sg} = 1 - 0.067 = 0.935$$

Coefficient for the sky radiation, from Table 17.6

$$C = 0.136$$

Diffuse sky radiation intensity

$$I_d = C I_n F_{ss} = (0.136)\,(848)\,(0.935) = 108 \text{ W/m}^2$$

17.7 HEAT GAIN THROUGH GLASS

Glass construction forms a significant part of modern building structures. It is, therefore, important to study how a space gains heat through glass.

Solar radiation—direct and diffuse—incident upon a glass surface is, in parts, transmitted, reflected and absorbed. This is shown in Fig. 17.13.

Thus if τ, r and a represent the respective fractions known as *transmissivity*, *reflectivity* and *absorptivity*, then

$$\tau + r + a = 1$$

The heat gain of a space through glass then comprises

(i) all the transmitted radiation,
(ii) a part of the absorbed radiation that travels to the room, and
(iii) the heat transmitted due to the difference between the outside and inside temperature.

The direct radiation enters the space only if the glass is receiving the direct rays of the sun. The diffuse radiation enters the space even when the glass is not facing the sun.

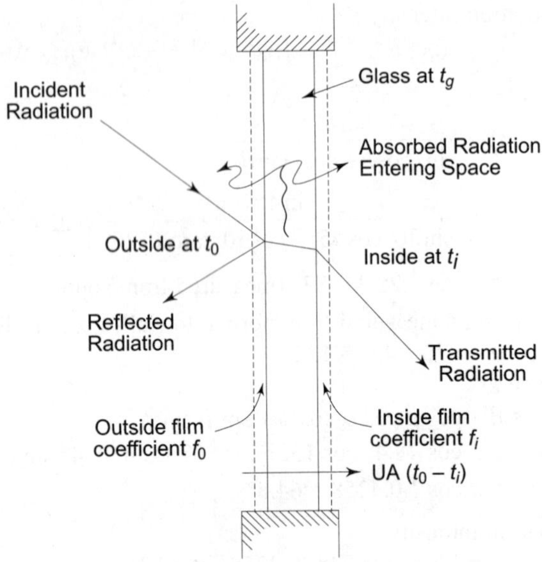

Fig. 17.13 Heat transfer through glass

The absorbed radiation raises the temperature of glass, and the glass then transmits this heat partly to the outside and partly to the inside. Thus from Fig. 17.13, if t_g represents the temperature of glass, then the heat gain of the space is given by

$$Q = A_{\text{sun}}\, \tau_D\, I_D + A\, \tau_d\, I_d + f_i\, A\, (t_g - t_i) \tag{17.21}$$

where f_i is the inside film-coefficient of heat transfer, subscripts D and d denote the terms for direct and diffuse radiations respectively, A_{sun} is the glass area directly exposed to the sun and A is the total glass area.

Writing a steady-state energy balance for the glass sheet itself, we have

$$A_{\text{sun}}\, a_D\, I_D + A\, a_d\, I_d = A[f_i\, (t_g - t_i) + f_0\, (t_g - t_0)] \tag{17.22}$$

where f_0 is the outside film-coefficient of heat transfer.

In Eq. (17.22), the left-hand side represents the heat gain of glass due to absorption, and the right-hand side represents the heat loss by convection and radiation from its two surfaces. In this equation, the thermal resistance of glass has been neglected. Eliminating t_g between Eqs (17.21) and (17.22), we get for the heat gain of space

$$\dot{Q} = (A_{\text{sun}}\, \tau_D\, I_D + A\tau_d I_d) + \frac{A_{\text{sun}}\, a_D\, I_D + A a_d\, I_d}{\left(1 + \dfrac{f_0}{f_i}\right)} + UA\,(t_0 - t_i) \tag{17.23}$$

where U is the overall coefficient of heat transfer given by

$$\frac{1}{U} = \frac{1}{f_i} + \frac{1}{f_0} \tag{17.24}$$

If the thermal resistance of glass is considered, we have for U

$$\frac{1}{U} = \frac{1}{f_i} + \frac{\Delta x}{k_g} + \frac{1}{f_0} \tag{17.25}$$

where Δx is the thickness of glass and k_g—its thermal conductivity.

It is seen from Eq. (17.23) that the first two terms represent the transmitted radiation, the third term represents the portion of the radiation absorbed by glass that travels to the room, and the last term represents the heat transmission by conduction, convection and radiation due to the temperature difference on the two sides of glass.

Now the common glass used in building construction is *ordinary glass*. It is the crystal glass of single thickness and single or double strength. The transmissivity and absorptivity of glass are functions of the angle of incidence. The variation of τ and a for direct radiation with θ for ordinary glass is given in Table 17.7. As the angle of incidence increases beyond 60°, the transmitted radiation rapidly falls down to zero.

Table 17.7 Transmissivity and absorptivity of ordinary glass[2]

Angle of Incidence	0°	20°	40°	50°	60°	70°	80°	90°
Transmissivity	0.87	0.87	0.86	0.84	0.79	0.67	0.42	0
Absorptivity	0.05	0.05	0.06	0.06	0.06	0.06	0.06	0

For diffuse radiation, the transmissivity and absorptivity are taken as 0.79 and 0.06 respectively, regardless of the angle of incidence.

Example 17.4 *Effect of Tilt of Glass on Solar Gain*

(a) For a 2.5 m wide and 3 m high unshaded glass window in a south-west wall, calculate the heat gain of space assuming a single vertical glass. The following data are given:

Direct radiation normal to sun's rays	*720 W/m²*
Diffuse radition	*95 W/m²*
Sun's altitude angle	*60°*
Solar azimuth angle	*15° west of south*
Outside temperature	*40°C*
Inside temperature	*21°C*
Outside surface heat-transfer coefficient	*17.5 W/m²K*
Inside surface heat-transfer coefficient	*11.5 W/m²K*

(b) Determine the per cent reduction in the solar heat gain if the glass is sloped inwards so that the bottom of the glass is recessed 0.5 m inside with respect to the top.

Solution (a) From Fig. 17.14, the wall solar azimuth angle

$$\alpha = 45 - 15 = 30°$$

Angle of incidence $\cos\theta = \cos\beta\cos\alpha = \cos 60° \cos 30° = 0.433$

$$\theta = \cos^{-1} 0.433 = 64.3°$$

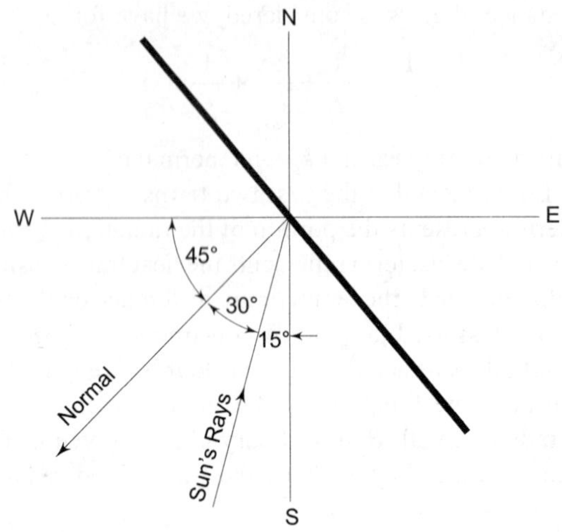

Fig. 17.14 Figure for Example 17.4(a)

Direct solar radiation normal to the glass

$$I_D = I_n \cos\theta = 720\ (0.433) = 312\ \text{W/m}^2$$

From Table 17.7, at 64.3° angle of incidence

$$\tau_D = 0.74,\ a_D = 0.06$$

Total transmitted radiation

$$\dot{Q}_1 = A(\tau_D I_D + \tau_d I_d) = (2.5 \times 3)\ (0.74 \times 312 + 0.79 \times 95) = 2294\ \text{W}$$

Energy balance for the glass gives

$$A\ (a_D I_D + a_d I_d) = A f_0\ (t_g - t_0) + A f_i\ (t_g - t_i)$$

whence

$$t_g = \frac{\sum aI + f_i t_i + f_0 t_0}{f_i + f_0}$$

$$= \frac{0.06(312+95)+11.5(21)+17.5(40)}{11.5+17.5} = 33.3°\text{C}$$

Heat transfer to the space from the inside surface of the glass

$$\dot{Q}_2 = A f_i\ (t_g - t_i) = (3 \times 2.5)\ (11.5)\ (33.3 - 21) = 1061\ \text{W}$$

Total heat gain of the space

$$\dot{Q} = \dot{Q}_1 + \dot{Q}_2 = 2294 + 1061 = 3355\ \text{W}$$

Note *The quantity $\dot{Q}_2$ includes the radiation entering the room, that is absorbed by the glass and the heat transmitted due to the temperature difference on the two sides of the glass.*

(b) The glass is sloping as shown in Fig. 17.15. The angle of inclination of the glass to the vertical is

$$\sin\phi = \frac{0.5}{3.04} = 0.1645$$

$$\Rightarrow \qquad \phi = \sin^{-1} 0.1645 = 9°28'$$

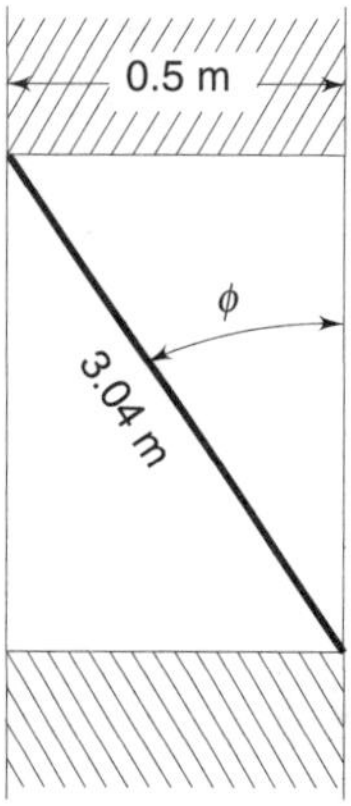

Fig. 17.15 Figure for Example 17.4(b)

Angle of incidence

$$\cos\theta = \cos\beta\cos\alpha\cos\phi - \sin\beta\sin\phi$$

$$= \cos 60^\circ \cos 30^\circ \cos 9^\circ 28' - \sin 60^\circ \sin 9^\circ 28'$$

$$= 0.2739$$

$$\theta = 74^\circ$$

Direct radiation normal to the glass

$$I_D = 720\ (0.2739) = 197\ \text{W/m}^2$$

Transmissivity for direct radiation

$$\tau_D = 0.57$$

Temperature of the glass

$$t_g = \frac{0.06(197+95)+11.5(21)+17.5(40)}{11.5+17.5}$$

$$= 33.1^\circ\text{C}$$

Heat gain of the space

$$\dot{Q} = A[\tau_D I_D + \tau_d I_d + f_i(t_g - t_i)]$$

$$= (3 \times 2.5)\ [0.57\ (197) + 0.79\ (95) + 11.5\ (33.1 - 21)] = 2449\ \text{W}$$

Per cent reduction in heat gain $= \left(\dfrac{3355-2449}{3355}\right) 100 = 27\%$

17.8 SHADING FROM REVEALS, OVERHANGS AND FINS[8]

Most glass areas in buildings are provided with reveals, overhangs and fins in the form of vertical and horizontal projections from walls. Thus some portions of glass are shaded to cut down the direct solar radiation. The shading, however, does not eliminate the diffuse radiation.

Consider, now, a vertical projection of depth D_x on a glass window of width X and height Y as shown in Fig. 17.16. Then the shading x on the width is given by the tangent of the wall solar azimuth angle

$$x = D_x \tan\alpha \tag{17.26}$$

Again, consider a horizontal projection of overhang D_y (it need not be equal to the depth of vertical projection) causing a shading y on the height. Then the shading y is given by

$$y = Q'Q'' = QQ' \tan\beta$$

But

$$QQ' = \frac{D_y}{\cos\alpha} = D_y \sec\alpha$$

Hence

$$y = D_y \sec\alpha\tan\beta \tag{17.27}$$

The sunlit area of the glass is

$$A_{\text{sun}} = (X - x)\,(Y - y)$$

$$= (X - D_x \tan\alpha)\,(X - D_y \sec\alpha\tan\beta) \tag{17.28}$$

Note *In Fig. 17.16, both vertical and horizontal projections, are shown equal.*

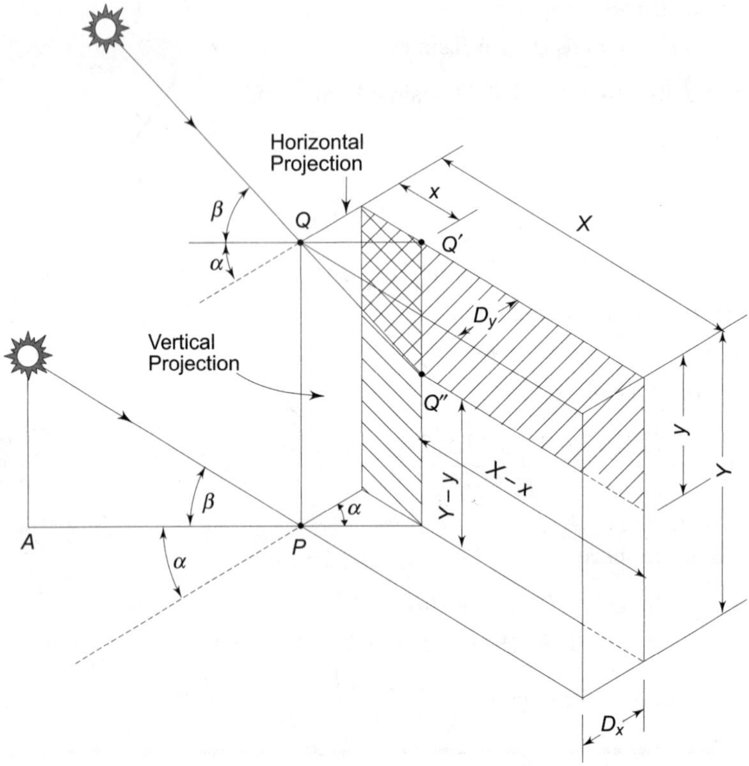

Fig. 17.16 Shading on glass due to horizontal and vertical projections

Example 17.5 *For a window opening in a south-west wall of a building, the following data are given*

Latitude	*40°N*
Outside air temperature	*40°C*
Inside air temperature	*25°C*
Time and day	*1 p.m. on 21st July*
Size of glass window	*4 m × 4 m*
Recess in window	*0.4 m*
Absorptivity of glass	*0.06*

The incident radiation normal to the sun's rays at 40°N latitude is given by the empirical relation

$$I_n = 1082\, e^{\frac{-0.182}{\sin\beta}}, \text{ W/m}^2$$

Calculate the total heat gain of the space through the glass. The outside and inside heat-transfer coefficients from the glass surface may be taken as 23 and 8 W/m²K respectively.

Solution For 21st July $\qquad d = 20°25'$

At 1 p.m. $h = 15°$

Solar altitude angle

$$\sin\beta = \cos l \cos h \cos d + \sin l \sin d$$

$$= \cos 40° \cos 15° \cos 20°25' + \sin 40° \sin 20°25' = 0.9165$$

$$\beta = \sin 0.9165 = 66°25'$$

Solar azimuth angle

$$\tan(\gamma - \pi) = \frac{\sin h}{\sin l \cos h - \cos l \tan d}$$

$$= \frac{\sin 15°}{\sin 40° \cos 15° - \cos 40° \tan 20°25'} = 0.768$$

$$(\gamma - \pi) = \tan^{-1} 0.768 = 37°31'$$

$$\gamma = 180° + 37°31'$$

The direction of the sun's rays is 37°31′ west of south as shown in Fig. 17.17.

Wall solar azimuth angle, from Fig. 17.17

$$\alpha = 45° - 37°31' = 7°29'$$

Sunlit area $x = D_x \tan\alpha = 0.4 \tan 7°29' = 0.0525$ m

$$y = D_y \sec\alpha \tan\beta = 0.4 \sec 7°29' \tan 66°25' = 0.924 \text{ m}$$

$$A_{sun} = (X - x)(Y - y) = (4 - 0.0525)(4 - 0.924) = 12.14 \text{ m}^2$$

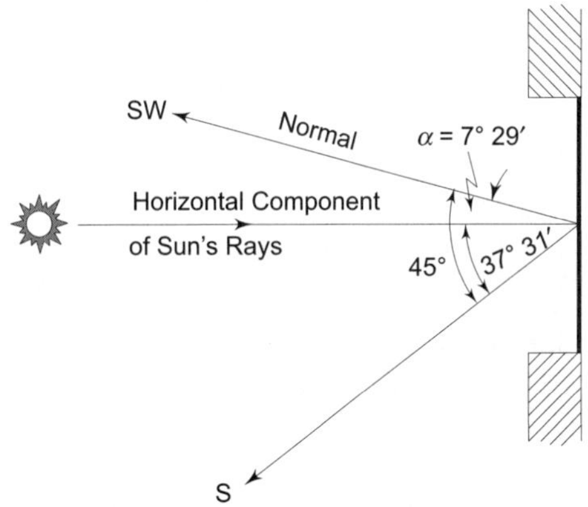

Fig. 17.17 Figure for Example 17.5

Total glass area $A = XY = 4 \times 4 = 16$ m^2

Radiation normal to the sun's rays

$$I_n = 1080\, e^{-0.182/\sin\beta} = 1080\, e^{-0.182/\sin 66°25'} = 886 \text{ W/m}^2$$

Direct radition normal to the window

$$I_D = I_n \cos\beta \cos\alpha = 884 \cos 66°25' \cos 7°29' = 350 \text{ W/m}^2$$

Coefficient for diffuse radiation on July 21

$$C = 0.136$$

Diffuse radiation

$$I_d = CI_n F_{ss} = 0.136\ (886)\ (0.5) = 60\ \text{W/m}^2$$

Angle of incidence

$$\cos\theta = \cos\beta\cos\alpha = \cos 66°25' \cos 7°29' = 0.3966$$

$$\theta = 66°38'$$

Transmissivity for direct radiation from Table 17.7

$$\tau_D = 0.71$$

Transmissivity for diffuse radiation may be taken as

$$\tau_d = 0.79$$

Total transmitted radiation

$$\dot{Q}_1 = A_{\text{sun}}\ \tau_D\ I_D + A\tau_d I_d$$

$$= 12.14\ (0.71)\ (350) + 16(0.79)\ (60) = 3775\ \text{W}$$

Overall heat-transfer coefficient

$$\frac{1}{U} = \frac{1}{f_0} + \frac{1}{f_i} = \frac{1}{23} + \frac{1}{8}$$

$$U = 5.94\ \text{W/m}^2\text{.K}$$

Note *This is the value of U through most single-pane glass windows with no storm sash, and no indoor shading. With no storm sash but with indoor shading, its value is 4.7. With storm sash having 25 mm air space, the value is 2.6. The value can be further reduced by employing low emittance coatings.*

Energy balance gives

$$aI = f_0\ (t_g - t_0) + f_i(t_g - t_i)$$

whence

$$t_g = \frac{aI + f_0 t_0 + f_i t_i}{f_0 + f_i}$$

$$= \frac{0.06(340+62)+23(40)+8(25)}{23+8} = 3.9°\text{C}$$

Total absorbed radiation entering the space

$$\dot{Q}_2 = \frac{A_{\text{sun}} a_D I_D + A a_d I_d}{1+\dfrac{f_0}{f_i}}$$

$$= \frac{12.14(0.06)(350)+16(0.06)(60)}{1+\dfrac{23}{8}} = 81\ \text{W}$$

Heat entering due to temperature difference

$$\dot{Q}_3 = UA\ (t_0 - t_i) = 5.94(16)\ (40 - 25) = 1426\ \text{W}$$

Total heat gain through the glass

$$\dot{Q} = \dot{Q}_1 + \dot{Q}_2 + \dot{Q}_3 = 3775 + 81 + 1426 = 5282\ \text{W}$$

Note *The heat gain due to the absorbed radiation is only 1.5 per cent of the total heat gain.*

17.9 EFFECT OF SHADING DEVICE[2]

The effect of the shading device is to further curtail the heat gain of the conditioned space. The effectiveness is more if the shading device is outside the space than when it is inside. This is because the inside shading devices dissipate all of their absorbed heat into the conditioned space. They must also reflect the solar heat back to the glass which absorbs some of it. On the other hand, the outside shading devices dissipate all their absorbed as well as the reflected radiation into the surroundings. Figure 17.18 illustrates the effect of an inside shading device such as venetian blinds.

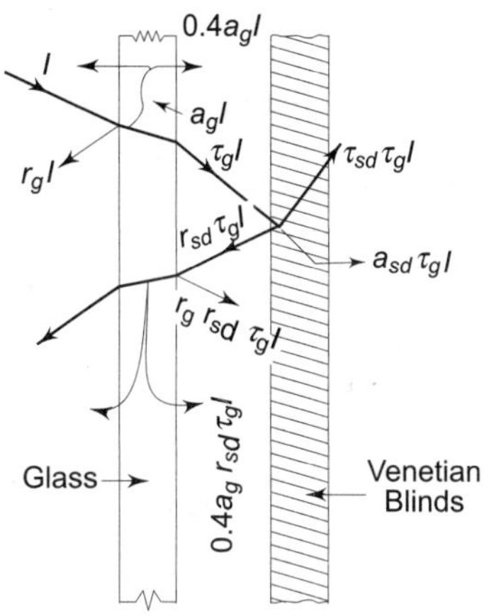

Fig. 17.18 Heat gain of space through glass with venetian blinds shading

Consider a radiation I incident on an ordinary glass surface at an angle of incidence of θ, say equal to 30° for which transmissivity is 0.86. A part of this radiation will be reflected, a part will be transmitted and a small part, of the order of 5 to 6 per cent, will be absorbed.

The heat absorbed by glass increases its temperature above that of the outside and inside air. A part of this heat, therefore, travels to the outside and a part to the inside. It was seen in Examples 17.4 and 17.5 that about 40 per cent of the absorbed radiation enters the space. The total heat gain of the space from the direct solar radiation is then equal to the sum of the transmitted radiation and about 40 per cent of the absorbed radiation. Thus, the radiation heat gain through a 6.35 mm thick, regular plate glass is

$$Q_g = \tau_g I + 0.4 a_g I = [0.86 + 0.4(0.06)]I = 0.884I$$

where $\tau_g = 0.86$ and $a_g = 0.06$ are the transmissivity and absorptivity of the glass.

Now, if an inside-shading device is used for which the transmissivity, reflectivity and absorptivity are τ_{sd}, r_{sd} and a_{sd} respectively, then from Fig. 17.18, the heat gain of the space to first approximation is

$$Q_{sd} = [0.4a_g + \tau_g(a_{sd} + \tau_{sd} + r_{sd}r_g + 0.4a_g r_{sd})]I \tag{17.29}$$

The values of τ, r and a for an angle of incidence of 30° for various types of glass and shading devices are given in Table 17.8. The values for venetian blinds are

$$\tau_{sd} = 0.12,\ r_{sd} = 0.51,\ a_{sd} = 0.37$$

Accordingly, the radiation heat gain of space through the glass with venetian blinds is

$$Q_{sd} = [0.4(0.15) + 0.86(0.37 + 0.12 + 0.51(1 - 0.86 - 0.15) + 0.4\,(0.15)\,(0.15))]I = 0.492I$$

The *solar heat-gain factor* represents the ratio of the radiation heat gain with the shading device to that of plain glass. Thus, in the present case, the solar heat-gain factor is

$$R = \frac{Q_{sd}}{Q_g} = \frac{0.492I}{0.884I} = 0.56.$$

Table 17.8 gives such factors for various categories of glass and shading devices.

Table 17.8 Radiation properties of glass and shading devices

Type of Glass or a Shading Device	*Absorptivity* a	*Reflectivity* r	*Transmissivity* τ	*Solar Factor* R	
Ordinary glass	0.06	0.08	0.86	1.00	
6.35 mm regular plate	0.15	0.08	0.77	0.94	
Heat absorbing glass	By mfg	0.05	(1 – 0.05 – a)	–	
Double pane ord. glass				0.9	
Double pane reg. plate glass				0.8	compared to ordinary glass
Venetian blind light	0.37	0.51	0.12	0.56	
medium	0.58	0.39	0.03	0.65	
dark	0.72	0.27	0.01	0.75	
Cotton cloth, beige	0.26	0.51	0.23	0.56	
Cotton cloth, dark green	0.02	0.28	0.7	0.76	
Dacron cloth, white	0.02	0.28	0.7	0.76	

17.10 TABLES FOR SOLAR HEAT GAIN THROUGH ORDINARY GLASS

The factors affecting the solar heat gain through ordinary glass are:

(i) Location of the point on the earth's surface given by the latitude of the place
(ii) Time of day
(iii) Time of year
(iv) Facing direction of window.

To simplify the air-conditioning calculations of the solar heat gain through ordinary glass, tables have been prepared by Carrier[2] Air Conditioning Co. Table 17.9(a) to (e) provide this data in SI units for latitudes from 0 to 50° from January to December and for each hour of the day. The heat gain includes the direct and diffuse solar radiation plus the portion of the heat absorbed by glass which enters the room. It is to be noted that the transfer of heat across the glass—because of the temperature difference between the outside and the room air—is not included. The data are based on the following conditions:

(i) Glass windows have a typical wood-sash construction in which the glass area is equal to 85 per cent of the sash area. Metal-sash windows may be considered all glass so that a correction factor of 1/0.85 = 1.17 has to be applied to them.
(ii) No haze in the air. For hazy conditions, a reduction up to a maximum of 15 per cent may be made.

(iii) Sea level elevation. For higher altitudes, an increase of 0.23 per cent per 100 m of height above the mean sea level may be made.

(iv) A sea level dew point temperature of 19.3°C. This corresponds to about 4 cm of *precipitable water vapour* which is the water vapour contained in a column of air from the mean sea level to the outer edge of the atmosphere. For local DPT above and below 19.3°C, substract and add respectively 6.3 per cent per 5°C difference in DPT.

(v) North latitude. For south latitude, for December and January, add 7 per cent.

It is seen that the values in Table 17.9 include the transmitted radiation and a part of the absorbed radiation that enters the space. This is verified in Example 17.5 for the case of the S-W wall without any shading.

Transmitted radiation

$$\dot{Q}_1 = \tau_D I_D + \tau_d I_d = 0.71(350) + 0.79(60) = 296 \text{ W/m}^2$$

Absorbed radiation entering the space

$$\dot{Q}_2 = \frac{a_D I_D + a_d I_d}{1 + f_0 / f_i}$$

$$= \frac{0.06(350 + 60)}{1 + 23/8} = 6 \text{ W/m}^2$$

Total radiation entering the space

$$\dot{Q} = \dot{Q}_1 + \dot{Q}_2 = 295 + 6 = 302 \text{ W/m}^2$$

And considering the glass area as 85 per cent of the actual sash area, the radiation heat gain of the glass is

$$\dot{Q}_{\text{corrected}} = \dot{Q}(0.85) = 302(0.85) = 257 \text{ W/m}^2$$

The numerical value from Table 17.9(e) for 40°N latitude, on July 23, at 1 p.m. on a S-W wall is 259 W/m^2. It is thus seen that the calculation procedure outlined in this chapter for solar radiation through glass is extremely precise.

Note *As the solar and transmission heat gain through glass forms a major component of the cooling load in buildings, it is desirable from the point of view of energy conservation to minimize the glass areas. It is recommended that the glass areas should not exceed 25 per cent of the floor or carpet areas in buildings. Further, a cursory glance at Tables 17.9(a) through (f) will show that the maximum heat gain is from the west-facing glass. Hence, providing glass on the walls facing west, south-west and north-west should be avoided, as far as possible.*

Example 17.6 ***Time of Peak Solar Heat Gain through Glass***

A building located at 30°N latitude has equal glass areas on the south and west walls. Find the time of the peak solar heat gain of the space through the glass areas in the afternoon.

Table 17.9 (a) Solar heat gain through ordinary glass (W/m²)

0° North Latitude		*a.m.*												*p.m.*	*0° South Latitude*	
Time of yr.	***Exposure***	**6**	**7**	**8**	**9**	**10**	**11**	**Noon**	**1**	**2**	**3**	**4**	**5**	**6**	***Exposure***	***Time of yr.***
	North	0	142	205	233	246	252	259	252	246	233	205	142	0	South	
	North east	0	375	492	486	426	300	167	63	44	41	35	19	0	South east	
	East	0	365	464	426	293	136	44	44	44	41	35	19	0	East	
	South east	0	117	132	85	47	44	44	44	44	41	35	19	0	North east	
June 21	South	0	19	35	41	44	44	44	44	44	41	35	19	0	North	December 22
	South west	0	19	35	41	44	44	44	44	47	85	132	117	0	North west	
	West	0	19	35	41	44	44	44	136	293	426	464	366	0	West	
	North west	0	19	35	41	44	63	167	300	420	486	492	375	0	South west	
	Horizontal	0	88	274	464	603	685	713	685	603	464	274	89	0	Horizontal	
	North	0	117	170	192	205	208	211	208	205	192	170	117	0	South	
	North east	0	372	483	473	391	271	136	50	44	41	35	35	0	South-East	
	East	0	382	479	438	303	136	44	44	44	41	35	35	0	East	
July 23	South east	0	145	164	114	57	44	44	44	44	41	35	19	0	North east	Jan. 21
&	South	0	19	35	41	44	44	44	44	44	41	35	19	0	North	&
May 21	South west	0	19	35	41	44	44	44	44	57	114	164	145	0	North west	Nov. 21
	West	0	19	35	41	44	44	44	136	303	438	479	382	0	West	
	North west	0	19	35	41	44	50	136	271	391	473	482	372	0	South west	
	Horizontal	0	91	287	476	615	703	735	703	615	476	287	91	0	Horizontal	
	North	0	54	88	98	104	107	107	107	104	98	88	54	0	South	
	North east	0	347	445	420	322	192	76	44	44	41	38	19	0	South east	
	East	0	407	514	467	325	145	44	44	44	41	38	19	0	East	
Aug. 24	South east	0	211	249	205	110	47	44	44	44	41	38	19	0	North east	Feb. 20
&	South	0	19	35	41	44	44	44	44	44	41	38	19	0	North	&
April 20	South west	0	19	35	41	44	44	44	47	110	205	249	211	0	North west	Oct. 23
	West	0	19	35	41	44	44	44	145	325	467	514	407	0	West	
	North west	0	19	35	41	44	44	16	192	322	420	444	347	0	South west	
	Horizontal	0	98	965	473	650	738	773	738	650	473	306	98	0	Horizontal	
	North	0	19	38	41	44	44	44	44	44	41	38	19	0	South	
	North east	0	300	372	319	215	97	44	44	44	41	38	19	0	South east	

(Contd)

Table 17.9a (*Contd*)

Sept. 22 & March 22	East	0	423	527	476	338	148	44	44	44	41	38	19	0	East	March 22 & Sept. 22
	South east	0	300	372	318	215	98	44	44	44	41	38	19	0	North east	
	South	0	19	35	41	44	44	44	44	44	41	38	19	0	North	
	South west	0	19	35	41	44	44	44	98	215	319	372	300	0	North west	
	West	0	19	35	41	44	44	44	148	338	476	527	426	0	West	
	North west	0	19	35	41	44	44	44	98	215	319	372	295	0	South west	
	Horizontal	0	101	315	514	662	757	789	757	662	514	315	101	0	Horizontal	
Oct. 23 & Feb. 20	North	0	19	38	41	44	44	44	44	44	41	38	19	0	South	April 20 & Aug. 24
	North east	0	211	249	205	110	47	44	44	44	41	38	19	0	South east	
	East	0	407	514	467	325	145	44	44	44	41	38	19	0	East	
	South east	0	347	445	420	322	192	76	44	44	41	38	19	0	North east	
	South	0	54	88	98	104	107	107	107	104	98	88	54	0	North	
	South west	0	19	38	41	44	44	76	192	322	420	444	347	0	North west	
	West	0	19	38	41	44	44	44	145	325	467	514	407	0	West	
	North west	0	19	38	41	44	44	44	47	110	205	249	211	0	South west	
	Horizontal	0	98	306	473	650	738	773	738	650	473	306	98	0	Horizontal	
Nov. 21 & Jan. 21	North	0	19	35	41	44	44	44	44	44	41	35	19	0	South	May 21 & July 23
	North east	0	145	164	114	57	44	44	44	44	41	35	19	0	South east	
	East	0	382	479	438	303	136	44	44	44	41	35	19	0	East	
	South east	0	372	483	473	391	271	136	50	44	41	35	19	0	North	
	South	0	117	170	192	205	208	211	208	205	192	170	117	0	North	
	South west	0	19	35	41	44	50	136	271	39	473	483	372	0	North west	
	West	0	19	35	41	44	44	44	136	303	438	479	382	0	West	
	North west	0	19	35	41	44	44	44	44	57	114	164	145	0	South west	
	Horizontal	0	91	287	476	615	703	735	703	615	476	287	91	0	Horizontal	
Dec. 22	North	0	19	35	41	44	44	44	44	44	41	35	19	0	South	June 21
	North east	0	117	132	85	47	44	44	44	44	41	35	19	0	South east	
	East	0	366	463	426	293	136	44	44	44	41	35	19	0	East	
	South east	0	375	492	486	420	306	167	63	44	41	35	19	0	North east	
	South	0	142	205	233	246	252	259	252	246	233	68	142	0	North	
	South west	0	19	35	41	44	63	167	300	420	486	292	375	0	North west	
	West	0	19	35	41	44	44	44	136	293	426	464	366	0	West	
	North west	0	19	35	41	44	44	44	44	47	85	132	117	0	South west	
	Horizontal	0	88	274	464	603	686	713	685	603	464	275	88	0	Horizontal	

Table 17.9 (b) Solar heat gain through ordinary glass (W/m^2)

10° North Latitude		*a.m.*												*p.m.*	*10° South Latitude*	
Time of yr.	*Exposure*	6	7	8	9	10	11	Noon	1	2	3	4	5	6	*Exposure*	*Time of yr.*
	North	60	139	160	142	139	136	129	136	139	142	158	139	6	South	
	North east	173	413	483	442	334	205	88	44	44	41	35	25	6	South east	
	East	170	423	489	438	309	129	44	44	44	41	35	25	6	East	
	South east	57	155	173	136	79	44	44	44	44	41	35	25	6	North east	
June 21	South	6	25	35	41	44	44	44	44	44	41	35	25	6	North	Dec. 22
	South west	6	25	35	41	44	44	44	44	70	135	173	154	57	North west	
	West	6	25	35	41	44	44	44	129	309	438	489	423	170	West	
	North west	6	25	35	41	44	57	80	205	334	442	483	413	173	South west	
	Horizontal	13	139	338	524	647	735	767	735	647	524	338	139	13	Horizontal	
	North	16	107	123	110	104	90	94	98	104	110	123	107	16	South	
	North east	132	401	467	420	34	177	69	44	44	41	35	22	3	South east	
	East	158	426	498	448	309	136	44	44	44	41	35	22	3	East	
July 23	South east	82	100	208	177	101	44	44	44	44	41	35	22	3	North east	Jan. 21
&	South	3	22	35	41	44	44	44	44	44	41	35	22	3	North	&
May 21	South west	3	22	35	41	44	44	44	44	101	177	208	180	82	North west	Nov. 21
	West	3	22	35	41	44	44	44	136	309	448	498	426	158	West	
	North west	3	22	35	41	44	44	69	177	344	420	466	401	132	South west	
	Horizontal	9	132	338	524	662	744	779	744	662	524	338	132	9	Horizontal	
	North	3	47	50	47	47	44	44	44	47	47	50	47	3	South	
	North east	54	356	410	350	252	107	44	44	44	41	35	22	3	South east	
	East	79	435	514	470	328	145	44	44	44	41	35	22		East	
Aug. 24	South east	57	249	297	268	189	85	44	44	44	41	35	22	3	North east	Feb. 20
&	South	3	22	35	41	44	44	44	44	44	41	35	22	3	North	&
April 20	South west	3	22	35	41	44	44	44	85	189	268	297	249	57	North west	Oct. 23
	West	3	22	35	41	44	44	44	145	252	470	514	435	79	West	
	North west	3	22	35	41	44	44	44	107	47	350	410	356	54	South west	
	Horizontal	6	120	331	527	672	763	788	763	672	527	331	120	6	Horizontal	
	North	3	19	35	41	44	44	44	44	44	41	35	19	3	South	
	North east	3	281	325	252	142	54	44	44	44	41	35	19	3	South east	

(Contd)

Table 17.9b (*Contd*)

Sept. 22 & March 22	East	3	410	517	476	334	148	44	44	44	41	35	19	3	East	March 2 & Sept 22
	South east	3	306	401	385	297	177	66	44	44	41	35	19	3	North east	
	South	3	19	35	50	76	85	88	85	76	60	41	19	3	North	
	South west	3	19	35	41	44	44	66	177	297	385	401	306	3	North west	
	West	3	19	35	41	44	44	44	148	334	476	517	410	3	West	
	North west	3	19	35	41	44	44	44	54	142	252	325	281	3	South west	
	Horizontal	3	98	306	505	653	741	779	741	653	505	54	98	3	Horizontal	
Oct. 23 & Feb. 20	North	0	16	32	41	44	44	44	44	44	41	32	15	0	South	April 20 & Aug. 24
	North east	0	183	208	454	88	44	44	44	44	41	32	15	0	South east	
	East	0	372	489	457	315	126	44	44	44	41	32	15	0	East	
	South east	0	325	464	470	388	256	145	57	44	41	32	15	0	North east	
	South	0	57	126	173	205	240	230	224	205	173	126	57	0	North	
	South west	0	16	32	41	44	57	245	256	388	470	464	325	0	North west	
	West	0	16	32	41	44	44	44	126	315	457	489	372	0	West	
	North west	0	16	32	41	44	44	44	44	88	139	208	183	0	South west	
	Horizontal	0	69	268	438	609	694	726	693	609	438	268	69	0	Horizontal	
Nov. 21 & Jan. 21	North	0	13	28	38	41	44	44	44	41	38	28	13	0	South	May 21 & July 23
	North east	0	85	117	54	41	44	44	44	41	38	28	13	0	South east	
	East	0	312	451	416	293	123	44	44	41	38	28	13	0	East	
	South east	0	312	483	508	461	328	221	98	54	38	28	13	0	North east	
	South	0	110	205	287	303	328	334	328	303	287	205	110	0	North	
	South west	0	13	28	38	54	98	221	344	461	508	483	312	0	North west	
	West	0	13	28	38	41	44	44	123	293	416	451	312	0	West	
	North west	0	13	28	38	41	44	44	44	41	54	117	85	0	South west	
	Horizontal	0	54	196	413	551	637	662	637	552	413	196	54	0	Horizontal	
Dec. 21	North	0	13	28	38	41	44	44	44	41	38	28	13	0	South	June 21
	North east	0	47	88	54	41	44	44	44	41	38	28	13	0	South east	
	East	0	271	432	410	287	132	44	44	41	38	28	13	0	East	
	South east	0	312	486	514	470	382	233	114	72	38	28	13	0	North east	
	South	0	158	233	297	344	366	379	366	344	297	233	158	0	North	
	South west	0	13	28	38	73	114	233	382	470	514	486	312	0	North west	
	West	0	13	28	38	41	44	44	132	287	410	432	271	0	West	
	North west	0	13	28	38	41	44	44	44	41	54	88	47	0	South west	
	Horizontal	0	44	208	378	527	609	637	608	527	379	208	44	0	Horizontal	

Table 17.9 (c) Solar heat gain through ordinary glass (W/m²)

20° North Latitude		a.m.												p.m.	20° South Latitude	
Time of yr.	Exposure	6	7	8	9	10	11	Noon	1	2	3	4	5	6	Exposure	Time of yr.
June 21	North	88	129	104	79	60	54	47	54	60	79	104	129	88	South	Dec. 22
	North east	256	486	454	385	262	120	47	44	44	44	38	28	9	South east	
	East	256	467	505	451	303	129	44	44	44	44	38	28	9	East	
	South east	88	126	230	208	138	66	44	44	44	44	38	28	9	North east	
	South	10	28	38	44	44	44	44	44	44	44	38	28	9	North	
	South west	10	28	38	44	44	44	44	66	139	208	230	196	88	North west	
	West	10	28	38	44	44	44	44	129	303	451	505	467	256	West	
	North west	10	28	38	44	44	44	47	50	262	385	454	486	256	South west	
	Horizontal	35	189	382	555	681	732	789	732	681	555	382	189	35	Horizontal	
July 23 & May 21	North	63	88	73	54	47	44	44	44	47	54	73	88	63	South	Jan. 21 & Nov. 21
	North east	224	415	435	350	230	98	44	44	44	41	38	25	9	South east	
	East	237	467	514	457	312	145	44	44	44	41	38	25	9	East	
	South east	98	221	268	249	180	91	44	44	44	41	38	25	9	North east	
	South	10	25	38	41	44	44	44	44	44	41	38	25	9	North	
	South west	10	25	38	41	44	44	44	91	180	249	268	221	98	North west	
	West	10	25	38	41	44	44	44	145	312	457	514	467	237	West	
	North west	10	25	38	41	44	44	44	98	23	350	435	416	224	South west	
	Horizontal	25	174	372	552	681	757	792	757	681	552	372	173	25	Horizontal	
Aug. 24 & April 20	North	19	32	25	41	44	44	44	44	44	41	35	32	19	South	Feb.20 & Oct. 23
	North east	142	350	372	281	158	57	44	44	44	41	35	22	6	South east	
	East	167	448	521	470	334	161	44	44	44	41	35	22	6	East	
	South east	92	281	357	349	309	173	63	44	44	41	35	22	6	North west	
	South	6	22	35	44	63	76	82	76	63	44	35	22	6	North	
	South west	6	22	35	41	44	44	63	173	309	341	356	281	91	North west	
	West	6	22	35	41	44	44	44	161	334	471	520	448	167	West	
	North west	6	22	35	41	44	44	44	41	157	281	372	350	142	South west	
	Horizontal	16	151	338	527	662	241	779	741	662	527	338	151	16	Horizontal	
	North	0	19	35	41	44	44	44	44	44	41	35	19	0	South	
	North east	0	362	274	186	69	44	44	44	44	41	35	19	0	South east	

(Contd)

Table 17.9c (*Contd*)

	East	0	410	514	470	328	142	44	44	44	41	35	19	0	East	
Sept. 22	South east	0	312	429	442	379	265	129	47	44	41	35	35	0	North east	March 22
&	South	0	25	69	120	164	199	205	199	164	120	69	69	0	North	&
March 22	South west	0	19	35	41	44	47	129	265	379	442	429	429	0	North west	Sept. 22
	West	0	19	35	41	44	44	44	142	328	470	514	410	0	West	
	North west	0	19	35	41	44	44	44	44	69	186	274	262	0	South west	
	Horizontal	0	95	293	483	626	710	735	710	625	483	293	95	0	Horizontal	
	North	0	13	28	38	41	44	44	44	41	38	28	13	0	South	
	North east	0	139	164	91	41	44	44	44	41	38	28	13	0	South east	
	East	0	312	464	445	315	155	44	44	41	38	28	13	0	East	
Oct. 23	South east	0	287	461	506	470	375	233	85	41	38	28	13	0	North east	April 20
&	South	0	66	158	240	293	334	350	334	293	298	156	66	0	North	&
Feb. 20	South west	0	13	28	38	41	85	233	375	470	513	461	287	0	North west	Aug. 24
	West	0	13	28	38	41	44	44	155	315	366	464	312	0	West	
	North west	0	13	28	38	41	44	44	44	41	91	164	139	0	South west	6:8
	Horizontal	0	57	215	401	539	618	656	618	539	401	215	57	0	Horizontal	
	North	0	10	25	35	41	41	41	41	41	35	25	9	0	South	
	North east	0	76	82	44	41	41	41	41	41	35	25	9	0	South east	
	East	0	224	404	401	287	136	41	41	41	35	25	9	0	East	
Nov. 21	South east	0	230	454	517	498	426	287	145	50	35	25	9	0	North east	
&	South	0	88	218	315	388	429	445	429	388	315	217	88	0	North	
Jan. 21	South west	0	10	25	35	50	145	287	426	498	517	454	230	0	North west	
	West	0	10	25	35	39	41	41	136	287	401	404	224	0	West	May. 21
	North west	0	10	25	35	39	41	41	41	41	35	82	76	0	South east	&
	Horizontal	0	16	151	319	461	542	568	460	460	319	151	16	0	Horizontal	July 23
	North	0	6	22	35	38	41	41	41	38	35	22	6	0	South	
	North east	0	44	57	38	38	41	41	41	38	35	22	6	0	South east	
	East	0	177	372	382	268	107	41	41	38	35	22	6	0	East	
	South east	0	186	438	527	502	423	306	189	63	35	22	6	0	North east	
Dec. 22	South	0	79	233	350	416	461	370	461	416	350	233	79	0	North	June 21
	South west	0	6	22	35	63	189	306	422	502	527	438	186	0	North west	
	West	0	6	22	35	38	41	41	107	268	382	372	177	0	West	
	North west	0	6	22	35	38	41	41	41	38	38	38	44	0	South west	
	Horizontal	0	13	44	290	426	508	536	508	426	290	290	12	0	Horizontal	

Table 17.9 (d) Solar heat gain through ordinary glass (W/m²)

30° North Latitude		a.m.												p.m.	30° South Latitude	
Time of yr.	*Exposure*	**6**	**7**	**8**	**9**	**10**	**11**	**Noon**	**1**	**2**	**3**	**4**	**5**	**6**	*Exposure*	*Time of yr.*
June 21	North	104	91	57	44	44	44	44	44	44	44	51	91	104	South	Dec. 22
	North east	331	438	410	305	173	60	44	44	44	44	38	32	16	South east	
	East	340	492	508	451	309	139	44	44	44	44	38	32	16	East	
	South east	132	237	284	284	230	139	54	44	44	44	38	32	16	North east	
	South	16	32	38	44	44	60	66	60	47	44	38	32	16	North	
	South west	16	32	38	44	44	44	54	138	230	284	284	237	132	North west	
	West	16	32	38	44	44	44	44	138	309	451	508	492	341	West	
	North west	16	32	38	44	44	44	44	60	173	310	410	438	331	South west	
	Horizontal	60	192	413	568	685	757	789	757	685	568	413	192	60	Horizontal	
July 23 & May 21	North	69	63	76	41	44	44	44	44	44	41	44	63	69	South	Jan. 21 & Nov. 21
	North east	293	413	388	281	145	50	44	44	44	41	38	28	13	South east	
	East	315	489	517	457	312	139	41	44	44	41	38	28	13	East	
	South east	132	259	315	315	262	167	69	44	44	41	38	28	13	North east	
	South	13	28	38	44	63	85	95	85	63	44	38	28	13	North	
	South west	13	28	38	41	44	44	44	167	262	315	315	259	132	North west	
	West	13	28	38	41	44	44	44	139	312	457	517	489	315	West	
	North west	13	28	38	41	44	44	44	50	145	281	388	413	293	South west	
	Horizontal	47	208	388	535	675	744	776	744	675	555	388	208	47	Horizontal	
Aug. 24 & April 20	North	19	25	34	41	41	44	44	44	41	41	35	25	19	South	Feb. 20 & Oct. 23
	North east	173	340	315	208	85	44	44	44	41	41	35	25	6	South east	
	East	208	148	520	467	322	145	44	44	41	41	35	25	6	East	
	South east	117	309	401	407	353	258	123	47	41	41	35	25	6	North east	
	South	6	25	41	85	148	183	199	183	148	85	41	25	6	North	
	South west	6	25	35	41	41	47	123	259	353	407	401	309	117	North west	
	West	6	25	35	41	41	44	44	145	322	467	520	464	208	West	
	North west	6	25	35	41	41	44	44	44	85	208	315	340	173	South west	
	Horizontal	19	148	338	508	631	710	741	710	631	508	338	464	19	Horizontal	
	North	0	16	32	38	41	44	44	44	41	38	32	16	0	South	
	North east	0	233	283	126	447	44	44	44	41	38	32	16	0	South east	

(Contd)

Table 17.9d (*Contd*)

Sept.22 & March 22	East	0	391	498	454	325	151	44	44	41	38	32	16	0	East	March 22 & Sept. 22
	South east	0	309	413	430	445	356	211	79	41	38	32	16	0	North east	
	South	0	28	57	189	259	309	311	309	259	189	57	28	0	North	
	South west	0	16	32	38	41	789	211	356	445	479	413	309	0	North west	
	West	0	16	32	38	41	44	44	151	325	454	498	391	0	West	
	North west	0	16	32	38	41	44	44	44	47	126	284	233	0	South west	
	Horizontal	0	70	571	426	565	637	668	637	564	426	256	79	0	Horizontal	
Oct. 23 & Feb. 20	North	0	9	25	35	38	41	44	41	38	35	25	9	0	South	April 20 & Aug. 24
	North east	0	104	123	57	38	41	44	41	38	35	25	9	0	South east	
	East	0	249	426	416	297	136	44	41	38	35	25	9	0	East	
	South east	0	230	448	515	502	429	290	148	47	35	28	9	0	North east	
	South	0	57	180	290	382	438	457	438	382	290	180	57	0	North	
	South west	0	9	25	35	47	148	290	429	502	514	448	230	0	North west	
	West	0	9	25	35	38	41	44	136	297	416	426	249	0	West	
	North west	0	9	25	35	38	41	44	41	38	57	123	104	0	South west	
	Horizontal	0	19	155	315	451	539	564	539	451	315	155	19	0	Horizontal	
Nov. 21 & Jan. 21	North	0	23	19	28	35	38	38	38	35	28	19	3	0	South	May 21 & July 23
	North east	0	25	50	28	35	38	38	38	35	28	19	3	0	South east	
	East	0	85	34	366	262	110	38	38	35	28	19	3	0	East	
	South east	0	88	401	508	511	451	328	202	73	28	19	3	0	North east	
	South	0	32	215	344	432	486	502	486	432	344	216	32	0	North	
	South west	0	3	19	28	73	202	328	451	511	508	401	88	0	North west	
	West	0	3	19	28	35	38	38	110	262	366	344	85	0	West	
	North west	0	33	19	28	35	38	38	38	35	28	50	25	0	South west	
	Horizontal	0	6	85	224	343	429	457	429	344	224	85	6	0	Horizontal	
Dec. 22	North	0	0	13	28	235	38	38	38	35	28	13	0	0	South	June 21
	North east	0	0	32	28	35	38	38	38	35	28	13	0	0	South east	
	East	0	0	290	331	252	101	38	38	35	28	13	0	0	East	
	South east	0	0	360	495	511	451	340	227	88	28	13	0	0	North east	
	South	0	0	202	356	448	502	514	502	448	356	202	0	0	North	
	South west	0	0	13	28	88	227	341	451	511	495	360	0	0	North west	
	West	0	0	13	28	35	38	38	101	88	331	290	0	0	West	
	North west	0	0	13	28	35	38	38	38	35	28	32	0	0	South west	
	Horizontal	0	0	55	189	306	385	413	385	306	189	60	0	0	Horizontal	

Table 17.9 (e) Solar heat gain through ordinary glass (W/m^2)

40° North Latitude		*a.m.*												*p.m.*	*40° South Latitude*	
Time of yr.	***Exposure***	**6**	**7**	**8**	**9**	**10**	**11**	**Noon**	**1**	**2**	**3**	**4**	**5**	**6**	***Exposure***	***Time of yr.***
	North east	101	63	38	41	44	44	44	44	44	41	38	63	101	South	
	North	372	420	353	230	495	44	44	44	44	41	38	32	19	South east	
	East	397	508	511	448	300	139	44	44	44	41	38	32	19	East	
	South east	161	278	344	350	312	224	107	44	44	41	38	32	19	North east	
June 21	South	19	32	38	60	110	139	170	139	110	60	38	32	19	North	Dec. 22
	South west	19	32	38	41	44	44	107	224	312	350	344	278	161	North west	
	West	19	32	38	41	44	44	44	139	300	448	511	508	397	West	
	North	19	32	38	41	44	44	44	44	95	230	353	420	372	South west	
	Horizontal	98	259	423	565	662	732	748	732	662	565	423	259	98	Horizontal	
	North	76	44	38	41	44	44	44	44	44	41	38	44	76	South	
	North east	334	401	331	208	82	44	44	44	44	41	38	32	16	South east	
	East	372	508	517	454	309	136	44	44	44	41	38	32	16	East	
July 23	South east	170	303	375	394	347	259	132	47	44	41	38	32	16	North east	Jan. 21
&	South	16	32	41	82	139	199	218	199	139	82	41	32	16	North	&
May 21	South west	16	32	38	41	44	47	132	259	347	394	375	303	170	North west	Nov. 21
	West	16	32	38	41	44	44	44	136	309	454	517	508	372	West	
	North west	16	32	38	41	44	44	44	44	44	208	331	401	334	South west	
	Horizontal	76	230	397	539	640	710	735	710	640	539	397	230	76	Horizontal	
	North	22	25	35	41	44	44	44	44	44	41	35	25	22	South	
	North east	214	322	259	145	50	44	44	44	44	41	35	25	9	South east	
	East	265	464	511	457	319	142	44	44	44	41	35	25	9	East	
Aug. 24	South east	151	331	435	461	438	338	208	79	44	41	35	25	9	North east	Feb. 20
&	South	9	25	76	161	281	306	322	306	281	161	76	25	9	North	&
April 20	South west	9	25	35	41	44	79	208	338	438	461	435	331	151	North west	Oct. 23
	West	9	25	35	41	44	44	44	142	319	457	511	464	265	West	
	North west	9	25	35	41	44	44	44	44	50	145	259	322	215	South west	
	Horizontal	23	148	315	473	584	647	675	647	584	473	315	148	28	Horizontal	
	North	0	16	28	38	41	41	44	41	41	38	28	16	0	South	
	North east	0	161	183	82	41	41	44	41	41	38	28	16	0	South east	
	East	0	366	470	438	312	142	44	41	41	38	28	16	0	East	
	South east	0	300	454	511	495	420	284	129	44	38	28	16	0	Noth east	March 22

(Contd)

Table 17.9e (*Contd*)

Sept. 22	South	0	38	139	256	347	385	442	385	347	256	139	38	0	North	&
&	South west	0	16	28	38	44	129	284	420	495	511	454	300	0	North west	Sept. 22
March 22	West	0	16	28	38	41	41	44	142	312	438	470	366	0	West	
	North west	0	16	28	38	41	41	44	41	41	82	183	161	0	South west	
	Horizontal	0	66	211	391	483	555	577	555	483	391	211	66	0	Horizontal	
	North	0	16	19	32	35	38	38	38	35	32	19	6	0	South	
	North east	0	110	104	38	35	38	38	38	35	32	19	6	0	South east	
	East	0	269	369	385	278	123	38	38	35	32	19	6	0	East	
Oct. 23	South east	0	266	416	508	514	454	338	199	63	32	19	6	0	North east	April 20
&	South	0	66	186	328	432	486	511	486	432	328	186	66	0	North	&
Feb. 20	South west	0	6	19	32	63	199	338	454	514	508	416	256	0	North west	Aug. 24
	West	0	6	19	32	35	38	38	123	278	385	369	268	0	West	
	North west	0	6	19	32	35	38	38	38	35	38	104	110	0	South west	
	Horizontal	0	25	91	202	319	388	407	388	319	202	91	25	0	Horizontal	
	North	0	0	9	22	28	32	35	32	28	22	9	0	0	South	
	North east	0	0	38	22	28	32	35	32	28	22	9	0	0	South east	
	East	0	0	287	315	233	104	35	32	28	22	9	0	0	East	
Nov. 21	South east	0	0	344	454	492	454	366	221	85	22	9	0	0	North east	May 21
&	South	0	0	186	328	438	498	524	498	438	328	186	0	0	North	&
Jan. 21	South west	0	0	9	22	85	221	366	454	492	454	344	9	9	North west	July 23
	West	0	0	9	22	28	32	35	104	233	315	287	0	0	West	
	North west	0	0	9	22	28	32	35	32	28	22	38	0	0	South west	
	Horizontal	0	0	50	136	230	290	325	290	230	136	50	0	0	Horizontal	
	North	0	0	6	19	28	32	32	32	28	19	6	0	0	South	
	North east	0	0	22	19	28	32	32	32	28	19	6	0	0	South east	
	East	0	0	227	271	215	98	32	32	28	19	6	0	0	East	
	South east	0	0	276	423	467	448	363	230	95	22	6	0	0	North east	
Dec. 22	South	0	0	161	312	423	498	520	498	423	312	161	0	0	North	June 21
	South west	0	0	6	22	95	230	363	448	467	423	278	0	0	North west	
	West	0	0	6	19	28	32	32	98	215	271	227	0	0	West	
	North west	0	0	6	19	28	32	32	32	28	19	22	0	0	South west	
	Horizontal	0	0	25	101	173	240	268	240	173	101	25	0	0	Horizontal	

Solution The peak solar heat gain through the glass areas will correspond to the time of the maximum solar radiation through the glass areas. Hence taking values from Table 17.9 for the south and west side glass areas on June 22 at different hours and adding, we get the average radiation heat gain per 2 m^2 glass area, as given in Table 17.10.

Table 17.10 Solar radiation gain through glass on June 22

Time	2 p.m.	3 p.m.	4 p.m.	5 p.m.
South glass, W/m^2	47	44	38	32
West glass, W/m^2	309	451	508	492
Total, W/m^2	356	495	546	524

It is seen that the peak load occurs at 4 p.m.

17.11 THE FLAT-PLATE SOLAR COLLECTOR

Solar radiation in buildings may be an unwanted quantity as far as summer cooling is concerned. But for winter heating it can be a very useful aid. Attempts have been made to trap this energy for heating water and air.

Utilization of solar energy requires a *solar collector* of which there are two types:

(i) *Absorption type* or the *flat-plate collector.*

(ii) *Reflecting type* or the *concentrating collector.*

The reflecting type has to be continuously rotated with the changing altitude of the sun. Also, its collecting area which is formed by a short width along the focal point/axis of the reflecting surface is very small. The flat-plate collector is quite popular. It is commonly employed for water or air heating.

A schematic diagram of a flat-plate collector is shown in Fig. 17.19.

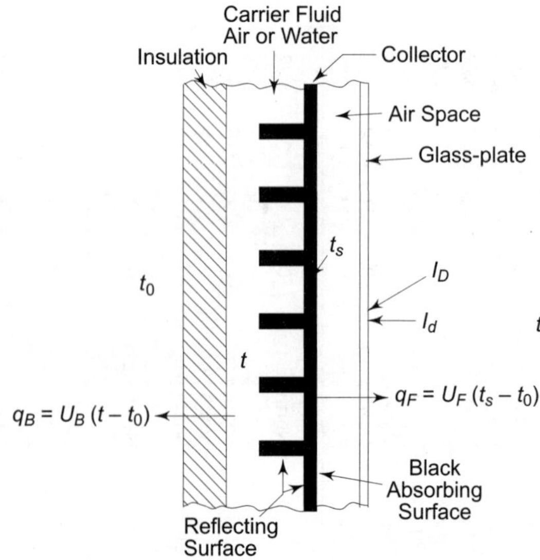

Fig. 17.19 Schematic diagram of a flat-plate collector

It essentially consists of a *flat surface of high absorptivity*. Generally, selectively coated black surfaces-with high absorptivity and low emissivity-of highly conducting materials like copper, aluminium and galvanized iron in that order are used. Heat is carried away by a *carrier fluid*, such as air or water, flowing on the back-side of the plate. Heat loss from the plate to the ambient from the front-side is reduced by covering it and sealing it with *transparent diathermaneous material*, such as glass or thin plastic film, which transmits high temperature short wavelength radiation from the sun to the plate, but is incapable of transmitting the low temperature long wavelength radiation from the plate to the ambient. Thus, sun's radiation is trapped in the collector. Further, the heat loss from the carrier fluid to the ambient from the back-side is eliminated to a great extent by providing insulation. And to improve heat transfer between the plate and the fluid, extended surface/fins are provided.

In the simplified analysis, by energy balance, we have for the heat $\dot{Q}$ received by the plate, and then transferred to the fluid and the ambient from the front side

$$\dot{Q} = (A_{sun}\, I_D\, \tau_D + A\, I_d\, \tau_d)\, \alpha_s = U_0\, A_0\, (t_s - t) + U_F\, A\, (t_s - t_0) \tag{17.30}$$

where A = Collector surface area

A_{sun} = Collector surface area receiving direct radiation I_D

I_D, I_d = Direct and diffuse components of solar radiation

τ_D, τ_d = Transmissivity of glass for direct and diffuse radiation

α_s = Absorptivity of plate for radiation

A_0 = Extended surface area on the side of fluid

U_0 = Overall heat-transfer coefficient from plate to fluid

U_F, U_B = Overall heat-transfer coefficients from plate to ambient from front side, and from fluid to ambient from back side.

t_0, t_s, t = Ambient, plate surface and fluid temperatures.

Considering an elemental plate surface area dA over which the temperature of the fluid rises by dt, we have:

$$\dot{m}Cdt = U_0\, \frac{A_0}{A}\, dA\, (t_s - t) - U_B\, dA\, (t - t_0) \tag{17.31}$$

where m is the mass flow rate of the carrier fluid, and C is its specific heat.

Now, eliminating t_s between Eqs (17.30) and (17.31), and integrating, we get

$$\int_{t_1}^{t_2} \frac{-K_2 \mathrm{d}t}{K_1 - K_2(t - t_0)} = -K_3 \tag{17.32}$$

where
$$K_1 = \frac{\left(\frac{A_{sun}}{A} I_D \tau_D + I_d \tau_d\right)\alpha_s}{1 + \frac{U_F}{U_0}\frac{A}{A_0}} = \frac{\dot{Q}/A}{1 + \frac{U_F}{U_0}\frac{A}{A_0}}$$

$$K_2 = \frac{U_F}{1 + \frac{U_F\, A}{U_0\, A_0}}$$

$$K_3 = \frac{A}{\dot{m}C} K_2$$

Equation (17.32) gives, for the rise in temperature of the carrier fluid

$$\Delta t = t_2 - t_1 = \left[\frac{K_1}{K_2} - (t_1 - t_0)\right](1 - e^{-K_3}) \tag{17.33}$$

The rate of heat collection is given by

$$\dot{Q} = \dot{m} C \,\Delta t$$

and the collector efficiency by

$$\eta_c = \frac{\text{Heat collected per unit area}}{\text{Solar intensity}} = \frac{\dot{Q}/A}{I_D + I_d}$$

It can be seen that the heat collection rate depends on a number of factors including the orientation of the collector. In general, the collector should face south in the northern hemisphere, and its surface should be normal to the sun's rays as far as possible. The *optimum angle* of tilt from the horizontal position for winter operation is 10 to 20 degrees greater than the latitude angle of the place. The optimum angle of tilt for summer operation is taken as equal to the latitude angle.

Equation (17.33) shows that for maximum t_2, K_3 should be ∞, so that $\dot{m} = 0$. Under these conditions, $\dot{Q} = 0$ and $t_2 = t_1 = t_s$. Then, if A_0 and A are equal, we have

$$(U_F + U_B)(t_s - t_0) = \frac{\dot{Q}}{A} \tag{17.34}$$

Equation (17.34) gives the maximum value of the surface temperature t_s. This temperature is called the *stagnation temperature* of the plate.

In case the collector is used for evaporation of a liquid as a boiler for power generation or as a generator for vapour absorption refrigeration system, the temperature of the carrier fluid is constant, viz., $t_2 = t_1$. Then, if h_{fg} is the latent heat of vaporization of the carrier fluid, Eq. (17.31) will give

$$\dot{m} h_{fg} = U_0 A_0 (t_s - t) - U_B A (t - t_0)$$

and solving this with Eq. (17.30), we get

$$\dot{m} = \frac{A}{h_{fg}}[K_1 - K_2 (t - t_0)] \tag{17.35}$$

since $\qquad K_3 = 0.$

Example 17.7 ***Solar Air Heater Design***

A flate plate solar collector at latitude of 34° for heating 300 kg/hr of outside air from –10°C to 0°C on a clear January 21st day. Design for solar noon time. Keep angle of tilt to the horizontal as 50°. Take: $A_{sun} = A$, $\tau_D = \tau_d = 0.8$, $\alpha = 0.9$, $U_F = 5.2$, $U_0 = 30$, $U_B = 0.6$ *W/m².K,* $A_0/A = 2$

Solution Here $\quad l = 34°;\ d = -19.9°,\ h = 0,\ \alpha = 0$

$$\phi = 90 - 50 = 40°$$

$$\sin\beta = \cos l \cos h \cos d + \sin l \sin d$$

$$= \cos 34° \cos 0° \cos 19.9° - \sin 34° \sin 19.9° = 0.5885$$

$$\Rightarrow \quad \beta = 36°$$

$$\cos\theta = \cos\beta\cos\alpha\cos\phi + \sin\beta\sin\phi$$
$$= 0.808 \times 1 \times 0.766 + 0.5885 \times 0.0428 = 0.9985$$
$$I_n = 1082\, e^{-0.182/\sin\beta} = 798 \text{ W/m}^2$$
$$I_D = I_n \cos\theta = 797 \text{ W/m}^2$$
$$F_{ss} = \frac{1}{2}(1 + \sin\phi) = \frac{1}{2}(1 + 0.6426) = 0.813$$
$$C = 0.058 \text{ on Jan 21}$$
$$I_d = CF_{ss}\, I_n = 0.058\,(0.813)\,(798) = 37.6 \text{ W/m}^2$$

Calculations for the collector

$$K_1 = \frac{(I_D\tau_D + I_d\tau_d)\alpha_s}{1+\dfrac{U_F}{U_0}\left(\dfrac{A}{A_0}\right)} = \frac{0.8\times 0.9(797+37.6)}{1+\dfrac{5.2}{30}\left(\dfrac{1}{2}\right)} = 552.8$$

$$K_2 = \frac{U_F}{1+\dfrac{U_F A}{U_0 A_0}} + U_B = \frac{5.2}{1+\dfrac{5.2}{30}\left(\dfrac{1}{2}\right)} + 0.6 = 5.384$$

The temperature of entering air is $t_1 = t_0$. Hence $t_1 - t_0 = 0$.

$$t_2 - t_1 = 10 = \left[\frac{K_1}{K_2} - (t_1 - t_0)\right](1 - e^{-K_3})$$

$$= \left[\frac{552.8}{5.384} - 0\right](1 - e^{-K_3}) = 102.7\,(1 - e^{-K_3})$$

$$\Rightarrow \qquad K_3 = 2.33 = \frac{A}{mC_p}\,K_2 = \frac{A(5.384)}{(300/3600)(1{,}005)}$$

$$\Rightarrow \qquad A = 36.3 \text{ m}^2$$

Note that t represents the temperature of air.

References

1. Bull L C, 'Solar radiation and air conditioning', *Heat. Vent. Engr.*, April 1961.
2. Carrier Air Conditioning Co., *Handbook of Air Conditioning System Design*, McGraw-Hill, New York, 1965.
3. Chawla O P, *Proceedings of the Q.I.P Short Term Course on Applications of Solar and Wind Energy*, IIT Delhi, 1977.
4. Eastop T D and J M Gasiorek, *Air Conditioning through Worked Examples, Longmans* Green, London, 1968.
5. Groundwater I S, *Solar Radiation in Air Conditioning*, Crosby Lockwood, 1957.
6. Jones W P, *Air Conditioning Engineering*, 2nd Ed., Edward Arnold, London 1973.

7. Moon P, 'Proposed standard solar radiation curves for engineering use', *J. Franklin Institute*, Vol. 230, p. 583, 1940.
8. Olgyay and Olgyay, *Solar Control and Shading Devices*, Princeton University Press, Princeton, 1957.
9. Robinson N, *Solar Radiation*, Elsevier, Amsterdam, 1966.
10. Stewart J P, 'Solar heat gain through walls and roofs for cooling load calculations', *Trans. ASHVE*, Vol. 54, pp. 361–68, 1948.
11. Thekaekara M P and A J Drummond, 'Standard values for the solar constant and its special components', *Nat. Phys. Sci.*, Vol. 229, p. 6, 1971.
12. Threlkeld J H, *Thermal Environmental Engineering*, Prentice Hall, Englewood Cliffs, N.J., 1962.

Revision Exercises

17.1 A factory in Delhi (latitude 30°N) has a sloping glass roof 30 m long and 6 m wide, the glass facing south-west lengthwise and making an angle of 30° with the horizontal. Calculate the heat gain through the glass in kW at solar noon on June 21. The following information is given:

Intensity of direct solar radiation on a plane normal to the sun's rays	910 W/m^2
Intensity of diffuse radiation	142 W/m^2
Transmissivity of glass	0.8
Reflectivity of glass	0.1
Inside heat-transfer coefficient	10 W/m^{2}K
Outside heat-transfer coefficient	28 W/m^{2}K
Outside air temperature	45°C
Inside air temperature	35°C

17.2 (a) Calculate the area of the shaded portion of a window that is recessed by 50 cm from the surface of the wall. The following information is given:

Altitude of the sun	43°
Solar azimuth	66° west of south
Orientation of window	Facing south-west
Dimensions of window	3 m × 3 m

(b) Also calculate the depth of recess that would result in the entire window being shaded.

17.3 (a) Figure 17.20 shows the orientation of buildings *A* and *B* located at a place at 20°N latitude. Make a table showing the dimensions of the area of the shadow produced on the face of building *A* by building *B* at noon on the 21st of every month.

(b) If the intensity of the direct solar radiation normal to sun's rays is 850 W/m^2, and the sun's declination is 23°, calculate the value of the direct solar radiation on the road-side face of building *A*.

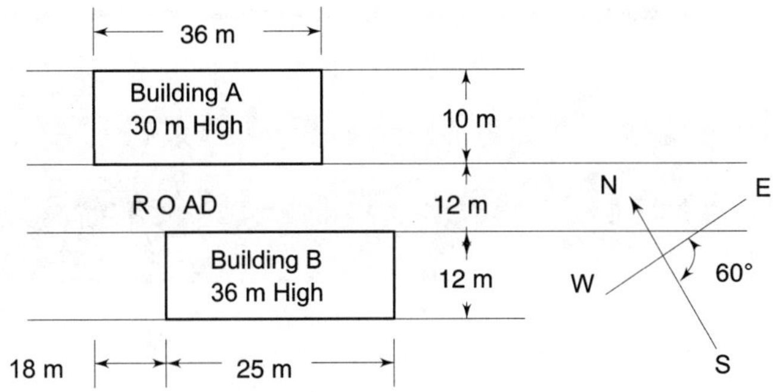

Fig. 17.20 Figure for Problem 17.3

17.4 Determine the following at 3 P.M. solar time at 30°N latitude.

(i) Altitude and solar azimuth angle on June 21.

(ii) Angle of incidence on June 8 on a vertical wall facing south.

(iii) Direct solar radiation on June 21.

(iv) Diffuse radiation on a roof facing SE with $\phi = 60°$ on July 21.

(v) Instantaneous heat gain through 3 m wide × 2.5 m high glass window facing SE recessed 0.4 m inside. Outside and inside temperatures are 40°C and 25°C.

17.5 A flat plate solar collector is to be designed for Delhi (30°N latitude) to evaporate 50 kg/h of liquid ammonia. The plane of the collector plate facing south may be tilted at 45° to the horizontal. Determine the required collector plate area. Assume suitable values for heat—transfer coefficients, etc.

Chapter

18

Heat Transfer through Building Structures

18.1 FABRIC HEAT GAIN

After the solar and transmission heat gain through glass, the most important heat gain or loss to be considered in the air conditioning of buildings is the heat transfer through walls, roof, ceiling, floor, etc., i.e., the building structure. The load due to such heat transfer is often referred to as the *fabric heat gain or loss*. In this connection, it is to be considered whether a particular wall or roof is exposed to the sun or not. In the case of a *sunlit* wall or roof, the heat gain of the room will be more in comparison to a *shaded* one, as the outside surface temperature of the wall or roof will increase above the outside air temperature due to the incident solar radiation.

The conduction heat transfer through the wall or roof will depend on the thickness and the thermal conductivity of the material used. In addition, there will be convection and radiation from both the outside and inside surfaces. Hence, the steady-state heat transfer is expressed in terms of an *overall heat-transfer coefficient* U and the overall temperature difference between the outside and inside $\Delta t = (t_0 - t_i)$ as given by Eq. (18.1). Also, the wall may consist of composite layers of different materials including insulation. In that case, U will incorporate the effect of all the materials. In the first instance, therefore, it is necessary to evaluate the value of the overall heat-transfer coefficient U.

Further, since the outside air temperature and solar radiation vary almost periodically over the 24 hours, it is required to establish a method to evaluate the transient heat transfer instead of using Eq. (18.1) which is only applicable to steady-state heat transfer.

18.2 OVERALL HEAT-TRANSMISSION COEFFICIENT

A wall may be composite, consisting of many sections of different construction and insulating materials. Also, the outside and inside wall surfaces may exchange heat by convection and radiation with the surrounding atmosphere. Thus, there will be more than one thermal resistance to heat transfer. Taking into account the number of layers of different materials with varying thickness Δx and thermal conductivity k, we have for the overall heat-transfer coefficient and overall thermal resistance R

$$q = U\Delta t = \frac{\Delta t}{R} \tag{18.1}$$

so that the overall heat-transfer coefficient may be calculated from the relation

$$\frac{1}{U} = R = \frac{1}{f_0} + \sum \frac{\Delta x}{k} + \sum \frac{1}{C} + \frac{1}{f_i} \tag{18.2}$$

In Eq. (18.2), *conductances* C have been included which are equivalent to the value of $k/\Delta x$ for a material. For some materials such as plaster, hollow tiles, etc., data are available in the form of conductance C instead of the thermal conductivity k. These values of C are only applicable to the prescribed thickness Δx. Also, f_0 and f_i in the equation represent the heat-transfer coefficients for combined convection and radiation from the outside and inside wall surfaces respectively. These terms are also referred to as *surface conductances*.

The properties of thermal conductivity, conductance, specific heat, density, etc., are referred to as *thermophysical properties*. The same for the common building and insulating materials are given in Table 18.1.

Note *Typical value of U for 22.5 cm brick wall with 1.25 cm plaster on both sides is 2.14 W/m^2°C. The same for 13.75 cm RCC slab roof with 1.9 cm plaster is 3.7 W/m^2°C. Note the need for insulating roofs, particularly, since they are exposed to sol. air temperature and not just DBT of outside air (see Sec. 18.3.1).*

18.2.1 Surface Conductance

The heat transfer between the wall surface and the surrounding air is mainly by convection and to some extent by radiation.

In still air, the convective heat transfer is by natural convection. The heat flux by natural convection can be approximated by a relation of the type

$$q_C = K\,(\Delta t)^{5/4} \tag{18.3}$$

Expressing the heat flux in terms of the convective coefficient f_C, we have

$$q_C = f_C\,(\Delta t) \tag{18.4}$$

From Eqs (18.3) and (18.4), the convective coefficient is

$$f_C = K\,(\Delta t)^{1/4} = 1.42\,(\Delta t/L)^{1/4} \tag{18.5}$$

where L is the vertical or horizontal dimension in *m*.

In actual cases, however, the air is in motion as a result of the wind velocity, and the convective coefficients are higher.

The radiation component of the heat flux is expressed by the relation

$$q_R = \varepsilon\sigma\,(T_w^4 - T_\infty^4) \tag{18.6}$$

Table 18.1 Thermophysical properties of selected building and insulating materials

Material	*Description*	*Specific Heat* kJ/kg.K	*Density* kg/m^3	*Thermal Conductivity* W/m.K	*Conductance* W/m^{2}K
Asphalt				0.74-0.76	
Bricks	Common	0.84	1600	0.77	
	Face brick	0.84	2000	1.32	
	Diatomaceous (fired)			0.24	
	Fire brick (500 to 1100)	0.96	2000	1.04 to 1.09	
	Magnesite (200 to 1200)	1.13			
Woods	Ply	—	544	0.1	
	Hard	2.39	720	0.158	
	Soft	2.72	512	0.1	
Masonary Materials	Concrete	0.88	1920	1.73	
	Plaster, cement	0.796	1885	8.65	
	Hollow clay tiles, 10 cm	—	—	—	5.23
	20 cm	—	—	—	3.14
	30 cm	—	—	—	2.33
	Hollow concrete blocks,	10 cm	—	—	8.14
		20 cm	—	—	5.23
		30 cm	—	—	4.54
	Foam concrete (Precast slabs for roof)		210–704	0.043–0.128	
Glass	Window	0.84	2700	0.78	
	Corosilicate		2200	1.09	
Insulating Materials	Mineral or glass wool	0.67	24–64	0.038	
	Rockwool	—	64	0.067	
	Fibreglass board	0.7	64-144	0.038	
	Cork board	1.884	104–128	0.038	
	Cork granulated	1.88	45–120	0.045	
	Expanded polysterene (Thermocole)	—	30	0.037	
	PUF (Polyurethane Foam)			0.0173	
	Diatomaceous earth	—	320	0.061	
	Felt	—	330	0.052	
	Insulex, dry	—	—	0.064	
	Kapok	—	—	0.035	
	Magnesia	—	270	0.067	
	Asbestos	0.816	470–570	0.154	

where ε is the emissivity of the wall surface, and T_w and T_∞ are the absolute temperatures of the wall and air respectively. Also, the Stefan-Boltzmann constant is

$$\sigma = 5.669 \times 10^{-8} \text{ W/m}^2 \text{ K}^4$$

Expressing in terms of a radiative coefficient f_R, we have

$$q_R = f_R\,(\Delta t) \tag{18.7}$$

so that

$$f_R = \frac{\varepsilon\sigma}{\Delta t}\,(T_w^4 - T_\infty^4) \tag{18.8}$$

The combined convection and radiation heat flux is given by

$$q = q_C + q_R = (f_C + f_R)\,\Delta t = f\Delta t$$

where $f = (f_C + f_R)$ represents the net surface conductance or heat-transfer coefficient. In turn, $1/f$ represents the thermal resistance of the surface. Practical values of the outside and inside wall-surface coefficients for various orientations, air velocities and surface emissivities are given in Table 18.2.

Table 18.2 Surface heat transfer coefficients f, W/m^2 K

Orientation of Surface	*Air Velocity*	*Direction of Heat Flow*	*Surface Emissivity* 0.9	0.7	0.5
Horizontal	Still air	Up	9.4	5.2	4.4
Horizontal	Still air	Down	6.3	2.2	1.3
Vertical	Still air	Horizontal	8.5	4.3	3.5
Any position	25 kmph	Any	35	—	—
Any position	12.5 kmph	Any	23.3	—	—

18.2.2 Air Spaces

The construction of many structures includes an *air space* between the wall materials. This air space may be narrow, such as a cavity of 5 cm or less in a 55 cm thick brick wall, or large such as the air space of a void between a flat ceiling and a pitched roof. The air space helps to reduce heat transfer and thus acts as an insulation. Heat transfer is greatly affected by surface emissivities and the width and orientation of the air space.

A greater natural convection results when the air space is vertical than when it is horizontal or sloping.

When the surfaces are close together, the natural air movement is suppressed and the heat transfer is by conduction alone. The thickness being very small, the resistance of the gap is very low and conductance is high.

As the width of the gap increases, the conduction component of heat transfer decreases rapidly to the extent that it becomes negligible, but convection comes into play as a result of which the overall conductance of the gap is not decreased very much. This is illustrated in Fig. 18.1

It is seen that for a gap of more than 2.5 cm, there is hardly any improvement in the insulation characteristic of the air space. This is due to an increase in internal convection. The conductive heat transfer is insignificant and the convective coefficient reaches a constant value. For a gap of 3 cm or more, the heat transfer may, therefore, be considered mainly by convection and radiation.

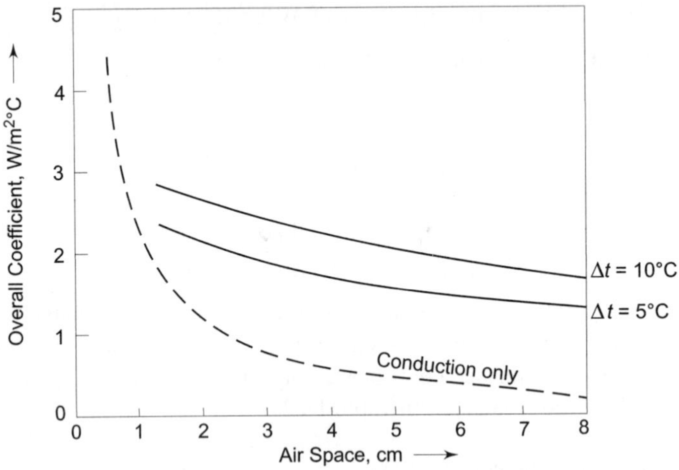

Fig. 18.1 Heat transfer across an air gap

Consider an air gap in a wall as shown in Fig. 18.2. The two surfaces of the wall are at temperatures t_1 and t_2. Let the convective and radiative components of the heat flux between the two surfaces be q_C and q_R respectively.

Now convection occurs on both the surfaces. Thus if f_C is the convective coefficient for one surface, the total convective resistance is given by

$$\frac{1}{f_C} + \frac{1}{f_C} = \frac{2}{f_C}$$

The reciprocal of this, viz., $f_C/2$ becomes the effective convective coefficient between the two surfaces, and the convective heat flux is given by

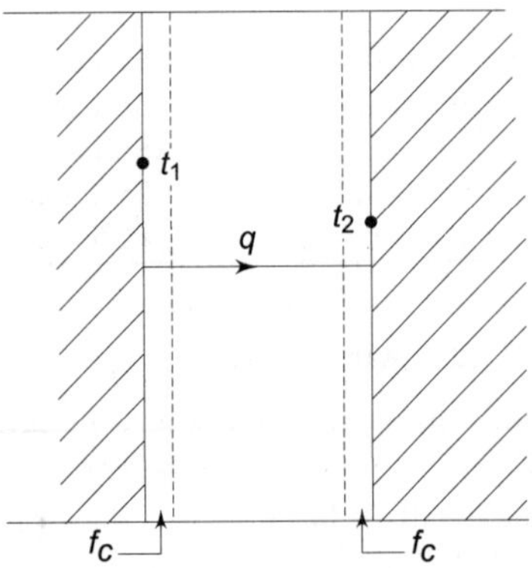

Fig. 18.2 Air gap in a wall

$$q_C = \frac{f_C}{2}\,(t_1 - t_2) \tag{18.9}$$

Considering the surfaces as comprising infinite parallel planes, we have for the radiative flux

$$q_R = F_{12}\sigma\,(T_1^4 - T_2^4) = f_R(t_1 - t_2) \tag{18.10}$$

where

$$F_{12} = \frac{1}{\dfrac{1}{\varepsilon_1} + \dfrac{1}{\varepsilon_2} - 1}$$

is the *shape factor* in which ε_1 and ε_2 are the emissivities of the two wall surfaces. With non-reflecting surfaces, as is usual in building construction, the emissivities can be taken as equal to 0.82. Equation (18.10) gives the value of the radiative coefficient f_R. Then, by adding Eqs (18.9) and (18.10), we obtain for the total heat flux between the wall surfaces

$$q = q_C + q_R$$
$$= \left(\frac{f_C}{2} + f_R\right)(t_1 - t_2) = C(t_1 - t_2) \tag{18.11}$$

where C represents the overall conductance between the two wall surfaces. Table 18.3 gives the thermal conductance for air spaces of three widths with different orientations and at two different mean temperatures.

Table 18.3 Thermal conductance C of air spaces

Position	*Direction of Heat Flow*	*Width* cm	*Conductance* W/m^2 °C
	Mean Temperature = 10°C		
Horizontal	Up	2.1	6.7
		11.6	6.2
	Down	2.1	5.7
		4.2	5.1
		11.6	4.8
Vertical	Horizontal	2.1	5.8
		11.6	5.8
	Mean Temperature = 32°C		
Horizontal	Up	2.1	7.7
		11.6	7.2
	Down	2.1	7.0
		4.2	6.2
		11.6	5.8
Vertical	Horizontal	2.1	7.0
		11.6	6.9

Example 18.1 ***Effect of Air Space and Insulation***

(a) Figure 18.3 shows the construction of a wall consisting of 1.5 cm of cement plaster, 10 cm each of hollow clay tile and air space, 20 cm of concrete block and 10 cm of face brick. Assume an outside velocity of 25 kmph. Calculate the heat-transmission coefficient of the wall with and without the air space.
(b) Calculate the same if the air space is filled with a bat of glass-fibre insulation.

Solution (a) From Tables 18.1, 18.2 and 18.3

$f_i = 8.5$ W/m^2 K (still air)

$f_0 = 35$ W/m^2 K (at 25 kmph)
$k_1 = 8.65$ W/mK

$C_2 = 5.23$ W/m^{2}K
$C_3 = 6.9$ W/m^{2}K
$C_4 = 5.23$ W/m^{2}K
$k_5 = 1.32$ W/mK

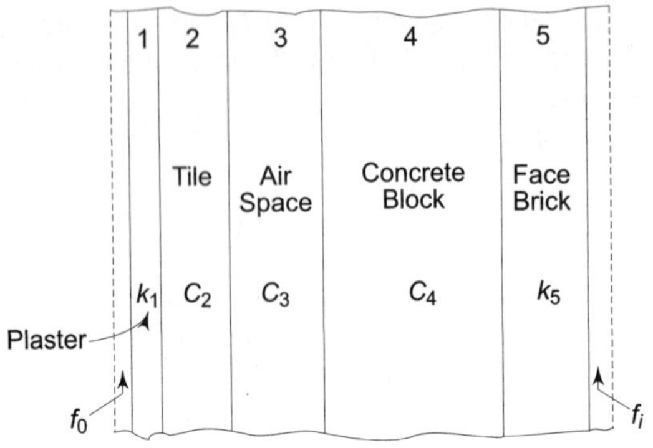

Fig. 18.3 Construction of wall for Example 18.1

With air space

Total thermal resistance

$$R = \frac{1}{f_0} + \frac{\Delta x_1}{k_1} + \frac{1}{C_2} + \frac{1}{C_3} + \frac{1}{C_4} + \frac{\Delta x_5}{k_5} + \frac{1}{f_i}$$

$$= \frac{1}{35} + \frac{0.015}{8.65} + \frac{1}{5.23} + \frac{1}{6.9} + \frac{1}{5.23} + \frac{0.1}{1.32} + \frac{1}{8.5} = 0.7609 \text{ m}^2 \text{ K/W}$$

Overall heat-transfer coefficient

$$U = \frac{1}{R} = \frac{1}{0.7609} = 1.33 \text{ W/m}^2 \text{ K}$$

Without air space

$$R = 0.7609 - \frac{1}{C_3} = 0.7609 - \frac{1}{6.9} = 0.63 \text{ m}^3 \text{ K/W}$$

$$U = \frac{1}{0.63} = 1.62 \text{ W/m}^2 \text{ K}$$

Note *The air space reduces heat transfer by (1.62–1.33)/1.62 or 19 per cent. The contribution of air space is small in this heavy construction. Air space is more effective in lighter constructions, and also in roofs exposed to sun (s*unlit roofs).

(b) *Air space filled with glass fibre* From Table 18.1 for glass fibre

$$k_3 = 0.038 \text{ W/mK}$$

Then

$$R = 0.63 + \frac{0.1}{0.038} = 2.5316 \text{ m}^2 \text{ K/W}$$

$$U = \frac{1}{2.5316} = 0.31 \text{ W/m}^2 \text{ K}$$

Note *Case(a) represents a typical construction of a wall for an air-conditioned building, whereas case (b) represents the wall construction for a cold storage with a minimum low temperature insulation thickness of 10 cm.*

Glass wool reduces heat transfer by (1.62 – 0.31)/1.62, viz., 80.8%.

18.3 PERIODIC HEAT TRANSFER THROUGH WALLS AND ROOFS

Heat transmission through the walls and roofs of building structures is not steady and is therefore, difficult to evaluate. The two principal factors causing this are:

(i) The variation of the outside air temperature-over a period of 24 hours.

(ii) The variation of the solar radiation intensity that is incident upon the surface over a period of 24 hours.

The phenomenon is further complicated by the fact that a wall has a thermal capacity due to which a certain amount of heat passing through it is stored and is transmitted to the outside and/or inside at some later time.

Figure 18.4 shows a typical variation of the outside air temperature and radiation on a hot summer day. The maximum temperatures usually occur just 2–3 hours after solar noon while the minimum temperatures occur just before sunrise. The outside air temperature t_0 follows nearly a harmonic variation. The mean line is shown at temperature t_{0m}.

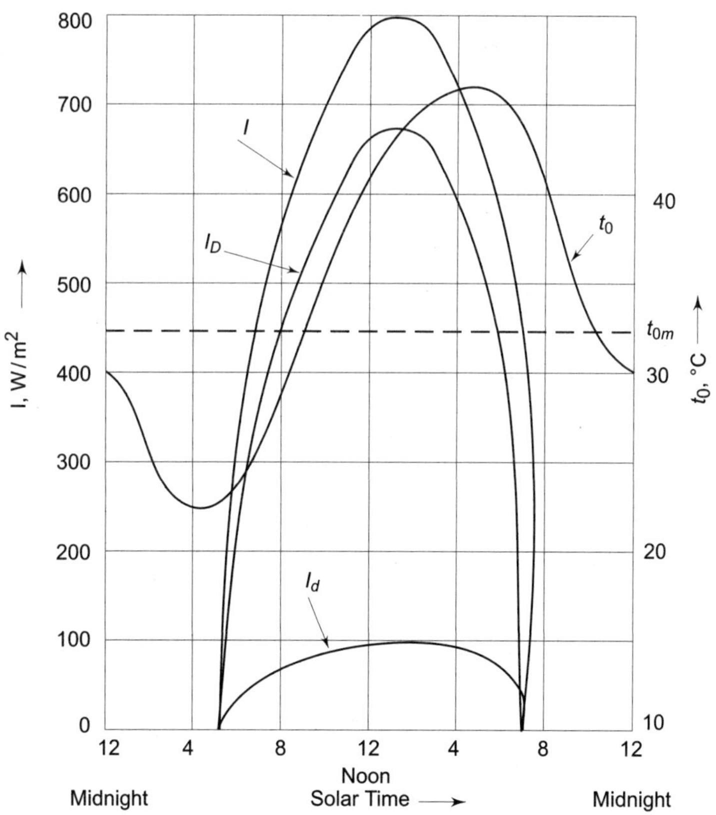

Fig. 18.4 Typical variation of solar radiation and outside air temperature during day

Whereas the combined effect of the solar radiation and outside air temperature can be incorporated into a single effective temperature as discussed in Sec. 18.3.1, the problem requires a solution of the governing equation for unsteady-state one-dimensional heat transfer, viz.,

$$\frac{\partial t}{\partial \tau} = \alpha \frac{\partial^2 t}{\partial x^2} \tag{18.12}$$

where t is the temperature at any section of the wall at a distance x from the surface at time τ, and α is the thermal diffusivity given by

$$\alpha = \frac{k}{\rho C}$$

where k is the thermal conductivity and ρC is the heat capacity of the wall, in which ρ and C are density and specific heat respectively. Equation (18.12) is to be solved with the boundary condition of periodic variation of the outside air temperature and solar radiation. The analytical solution of the problem requires many assumptions and is extremely cumbersome. In this chapter we shall lay stress on the numerical finite difference solution and the empirical methods invariably used by practicing engineers.

18.3.1 Sol-Air Temperature

For calculations of heat transfer through structures, it has been found convenient to combine the effect of the outside air temperature and incident solar radiation intensity into a single quantity as was introduced by Mackey and Wright.[7] For the purpose, an expression for the rate of heat transfer from the environment to the outside surface of the wall may be written as

$$q_0 = f_0 (t_0 - t_{s0}) + aI \tag{18.13}$$

where f_0 is the outside film-coefficient of heat transfer, t_{s0} is the temperature of the outside surface, a the absorptivity of the surface and I the total radiation intensity, as shown in Fig. 18.5.

Introducing an equivalent temperature t_e we may write for the heat-transfer rate

$$q_0 = f_0(t_e - t_{s0}) \tag{18.14}$$

Then, from Eqs (18.13) and (18.14)

$$t_e = t_0 + \frac{aI}{f_0} \tag{18.15}$$

This temperature t_e is called the *sol-air temperature* and can be considered as an equivalent outside air temperature such that the total heat transferred is the same as due to the combined effect of the incident solar radiation and outside air and the wall temperature difference.

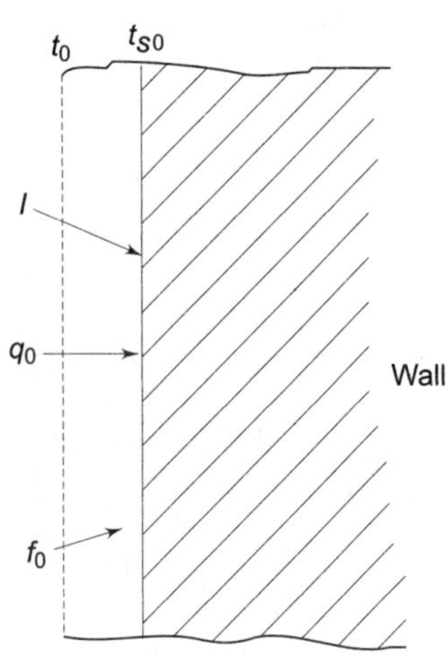

Fig. 18.5 Heat transfer to outside surface of a building wall

Example 18.2 *Calculate the instantaneous sol-air temperature for a wall with the following conditions:*

Total of direct and diffuse solar radiation	*260 W/m²*
Absorptivity of surface	*0.9*
Outside surface heat-transfer coefficient	*23 W/m² K*
Outside air temperature	*35°C*

Solution Sol-air temperature

$$t_e = t_0 + \frac{aI}{f_0} = 35 + \frac{0.9 \times 260}{23} = 45.2°\text{C}$$

Note *For all purposes of heat transfer through building structures, the sol-air temperature may be used instead of the dry bulb temperature of the outside air and the solar radiation separately.*

18.3.2 Analytical Solution to Periodic Heat-transfer Problem

In this section, we shall see the form of the analytical solution of Eq. (18.12) derived by Alford *et al.*[1]

Figure 18.6 shows an infinite wall of homogeneous construction for which the boundary conditions in terms of the heat fluxes q_0 and q_i on the outside and inside surface respectively are

$$q_0 = -k\left(\frac{\partial t}{\partial x}\right)_{x=0} = f_0(t_e - t_{s0}) \tag{18.16}$$

$$q_i = -k\left(\frac{\partial t}{\partial x}\right)_{x=L} = f_i(t_{si} - t_i) \tag{18.17}$$

where t_{s0} and t_{si} are the outside and inside wall-surface temperatures, and the outside sol-air temperature is given by the periodic equation[9]

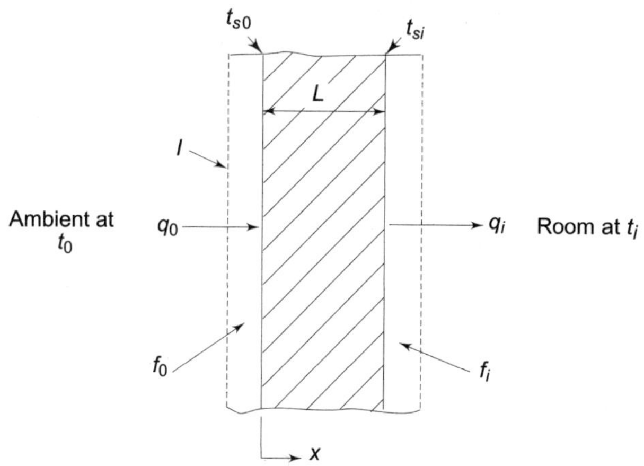

Fig. 18.6 One-dimensional heat transfer through a homogeneous wall

$$t_e = t_{e_m} + t_{e_1} \cos(\omega_1 \tau - \psi_1) + t_{e_2} \cos(\omega_2 \tau - \psi_2) + \ldots + \tag{18.18}$$

where t_{e_m} is the mean outside sol-air temperature t_e and ω_1, ψ_1, etc., are numerical values in the various harmonics. It is obvious that a solution for the inside wall-surface temperature t_{si} will also be periodic. Its solution along with Eq. (18.17) gives the following solution[9] for heat gain of the space maintained at a constant temperature t_i.

$$\begin{aligned} q = q_i &= f_i (t_{si} - t_i) \\ &= U(t_{e_m} - t_i) + U\lambda_1 t_{e_1} \cos(\omega_1 \tau - \psi_1 - \phi_1) \\ &\quad + U\lambda_2 t_{e_2} \cos(\omega_2 \tau - \psi_2 - \phi_2) + \ldots + \end{aligned} \tag{18.19}$$

The first term on the right-hand side in Eq. (18.19) represents the mean steady-state heat-transfer rate and the subsequent terms give the amplitudes of the harmonics with λ_n denoting the *decrement factor* for each harmonic, and ϕ_n the *time lag* between the harmonic of the sol-air temperature and the harmonic of the inside wall-surface temperature.

18.4 FINITE DIFFERENCE APPROXIMATION OF ONE-DIMENSIONAL HEAT TRANSFER THROUGH WALL

The drawback of the analytical solution is that it is cumbersome and only approximate due to the rejection of a large number of terms to make it possible to find a continuous mathematical solution.

For a numerical solution[5, 6] of Eq. (18.12) for the temperature distribution in space and time, the wall may be divided into a grid with space intervals of Δx as shown in Fig. 18.7. The results for the temperature will then be found for nodal points $(m - 1)$, m, $(m + 1)$, … , etc., which lie at the midpoints of these space elements, representing the average temperature of each element. Also, the temperature will be determined at time intervals of $\Delta\tau$.

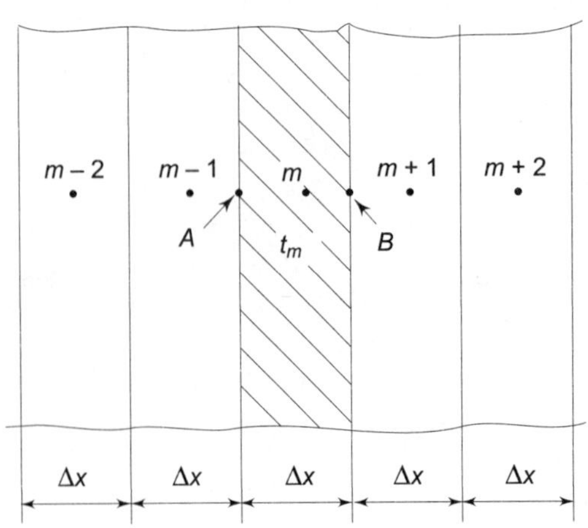

Fig. 18.7 Space grid in the conduction region

18.4.1 Node in the Conduction Region

Consider a section of the space element represented by any nodal point m in the conduction region within the wall for which the temperature is t_m. The temperature gradients at points A and B, to the left and right of m, at $\Delta x/2$ from m, can be written as follows:

$$\left(\frac{\partial t}{\partial x}\right)_B = \frac{t_{m+1} - t_m}{\Delta x}$$

$$\left(\frac{\partial t}{\partial x}\right)_A = \frac{t_m - t_{m-1}}{\Delta x}$$

Then the differential of the temperature gradient in Eq. (18.12) at m is expressed as

$$\left(\frac{\partial^2 t}{\partial x^2}\right)_m = \frac{\left(\frac{\partial t}{\partial x}\right)_B - \left(\frac{\partial t}{\partial x}\right)_A}{\Delta x}$$

$$= \frac{t_{m+1} + t_{m-1} - 2t_m}{(\Delta x)^2} \tag{18.20}$$

The time derivative in Eq. (18.12) is approximated by

$$\frac{\partial t^2}{\partial \tau} = \frac{t_m - t_m}{\Delta \tau} \tag{18.21}$$

where t_m^1 is the temperature at m after 1 $\Delta\tau$ time interval from the instant when the temperature at m was t_m.

Substitution of expressions from Eqs (18.20) and (18.21) into Eq. (18.12) gives for point m at any instant of time $(\tau + \Delta\tau)$

$$\frac{t_m^1 - t_m}{\Delta \tau} = \frac{\alpha}{(\Delta x)^2}\,(t_{m+1} + t_{m-1} - 2t_m)$$

whence
$$t_m^1 = \frac{\alpha \Delta \tau}{(\Delta x)^2}\,(t_{m+1} + t_{m-1}) + \left[1 - \frac{2\alpha\Delta\tau}{(\Delta x)^2}\right] t_m \tag{18.22}$$

or substituting a parameter M, such that

$$M = \frac{(\Delta x)^2}{\alpha \Delta \tau} \tag{18.23}$$

we have
$$t_m^1 = \frac{1}{M}\,(t_{m+1} + t_{m-1}) + \left[1 - \frac{2}{M}\right] t_m \tag{18.24}$$

Thus, the temperature t_m^1 at point m after a time interval 1 $\Delta\tau$ can be found in terms of temperatures t_{m+1} to the right, t_{m-1} to the left, and the local temperature t_m at any instant of time τ. Equations similar to (18.24) can be written for all points and solved simultaneously by a calculator or on a computer.

Some general remarks concerning the use of numerical method for the solution of transient heat conduction problem may now be made. It may be observed from Eq. (18.24) that if

$$\frac{2}{M} > 1$$

then the coefficient of t_m will be negative, so that for given values of t_{m+1} and t_{m-1}, the higher the value of t_m at anytime, the lower will be the temperature t_m^1 at time $(\tau + \Delta\tau)$. The result is incompatible and leads to instability in calculations. This mathematical anomaly can be avoided by choosing grid intervals in such a way that

$$\frac{2}{M} = \frac{2\alpha\,\Delta\tau}{(\Delta x)^2} \le 1$$

$$M = \frac{(\Delta x)^2}{\alpha\,\Delta\tau} \ge 2 \tag{18.25}$$

Thus, once the space interval Δx and the value of M are established, the time interval $\Delta\tau$ is fixed. If the minimum value of M is chosen, viz., $M = 2$, Eq. (18.24) becomes

$$t_m^1 = \frac{t_{m+1} + t_{m-1}}{2} \tag{18.26a}$$

or if $M = 3$, then

$$t_m^1 = \frac{t_{m+1} + t_m + t_{m-1}}{3} \tag{18.26b}$$

Accordingly, the temperature at any point x at time $(\tau + \Delta\tau)$ is equal to the mean value of the two temperatures at the adjacent points to the left and right of the point at time τ, in case $M = 2$, or equal to the mean value of the local temperature and the temperatures to the left and right of the point, in case $M = 3$.

The solutions for values of M equal to 2 and 3 can be physically interpreted from Fig. 18.8 which presents a graphical solution called the *Schmidt plot*.[8]

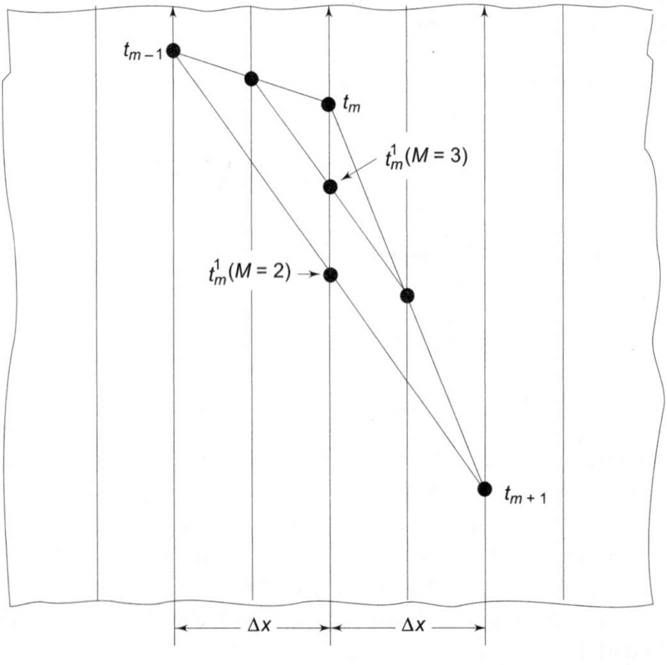

Fig. 18.8 Schmidt plot for graphical solution of nodal temperatures in the conduction region

It is seen that $M = 3$ gives a solutions t_m^1, nearer to the local value of the temperature t_m. Usually, a larger value of M gives a more accurate solution as a result of the smaller values of $\Delta\tau$ for a fixed Δx. But this, evidently, requires more computation time. On the other hand, a smaller value of M results in larger time intervals and hence less computational time and less accuracy in calculations. This restriction automatically limits our choice of $\Delta\tau$ and hence that of the value of the parameter M.

18.4.2 Nodes at the Surface and Next to the Surface

Consider a nodal point m located on the surface of the wall exposed to a convection and radiation environment as shown in Fig. 18.9. The equivalent temperature of the environment is equal to the sol-air temperature t_e which takes into account the effect of the convective film as well as radiation. The temperature at the surface of the wall is t_s which is to be determined. The energy balance for node S gives

$$f(t_e - t_s) + \frac{2k}{\Delta x}(t_m - t_s) = 0 \tag{18.27}$$

where f is the heat-transfer coefficient on the wall and k is the thermal conductivity of the material of the wall. Equation (18.27) simplifies to

$$\left(f + \frac{2k}{\Delta x}\right) t_s = f t_e + \frac{2k}{\Delta x} t_m$$

whence

$$t_s = \left[\frac{f}{f + \dfrac{2k}{\Delta x}}\right] t_e + \left[\frac{2k/\Delta x}{f + \dfrac{2k}{\Delta x}}\right] t_m$$

$$= \left(\frac{\text{Bi}}{\text{Bi}+2}\right) t_e + \left(\frac{2}{\text{Bi}+2}\right) t_m \tag{18.28}$$

where $\text{Bi} = \dfrac{f\Delta x}{k}$ is the *Biot number* which is a dimensionless number representing a convection boundary.

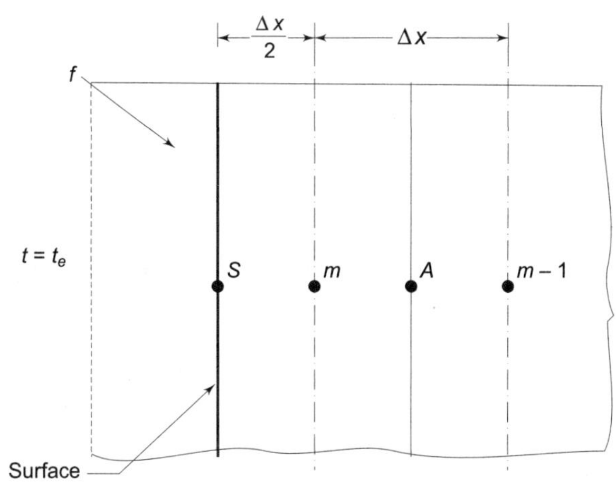

Fig. 18.9 Nodal point at the surface

For the nodal point m at a space interval of $\Delta x/2$ from the convection boundary, Eq. (18.12) can be written in the finite difference form. We first find the temperature gradient at A

$$\left(\frac{\partial t}{\partial x}\right)_A = \frac{t_m - t_{m-1}}{\Delta x}$$

and the temperature gradient at S

$$\left(\frac{\partial t}{\partial x}\right)_S = \frac{t_s - t_m}{\Delta x/2}$$

Hence we have at the mid point m

$$\left(\frac{\partial^2 t}{\partial x^2}\right)_m = \frac{\left(\frac{\partial t}{\partial x}\right)_S - \left(\frac{\partial t}{\partial x}\right)_A}{\Delta x}$$

$$= \frac{1}{(\Delta x)^2}\,[2t_s + t_{m-1} - 3t_m] \tag{18.29}$$

Substituting in Eq. (18.12) we get the finite difference approximation of the temperature at node m at time $(\tau + \Delta\tau)$ as

$$t_m^1 = \frac{1}{M}\,(2t_s + t_{m-1}) + \left(\frac{M-3}{M}\right) t_m \tag{18.30}$$

It can be seen that the stability criterion for this case requires that

$$M \geq 3$$

In the limiting case when $M = 3$, we have

$$t_m^1 = \frac{1}{3}\,(2t_s + t_{m-1}) \tag{18.31}$$

18.4.3 Procedure for Heat-transfer Calculations for a Homogeneous Wall

With periodic variation of the outside air temperature t_0 and solar radiation I, it is first necessary to evaluate the periodic value of the sol-air temperature t_e preferably on an hourly or half-hourly basis. These values are then imposed as the outside boundary conditions. In addition, as the initial condition, an approximate temperature distribution through the wall at a certain time $\tau = 0$ is to be assumed. Further, the inside air temperature t_i may be assumed constant, and also equal to that maintained by air-conditioning equipment. The calculations may now proceed for the temperatures at various nodes at time intervals $1\,\Delta\tau$, $2\Delta\tau$, $3\Delta\tau$, etc., until the total time of 24 hours is completed. The calculations may be stopped if at the end of 24 hours the same initial values are obtained as assumed, or may be repeated with new initial values until the two values are in agreement.

The following example illustrates the calculation procedure.

Example 18.3 ***Calculation of Heat Transfer through Wall by Finite Difference Method***

A 25 cm thick brick wall is exposed to the periodic temperature and incident radiation variation given in Table 18.4. Determine the average and peak loads on the air conditioner maintaining the room at 25°C per unit area of the wall.

Given for the wall:

Thermal conductivity, $k = 1.5$ *W/m°C*

Thermal diffusivity, $\alpha = 8 \times 10^{-7}$ m^2/s

Absorptivity of surface, $a = 0.8$

Outside wall coefficient, $f_0 = 23$ $W/m^2°C$

Inside wall coefficient, $f_i = 7$ $W/m^2°C$

Table 18.4 Values of outside air temperature and radiation for Example 18.3

Time	t_0 °C	I W/m²	*Time*	t_0 °C	I W/m²
12 (Midnight)	30	0	12(Noon)	38.5	1000
1 a.m.	29.5	0	1 p.m.	39.5	960
2	29	0	2	40.5	825
3	28.5	0	3	41.5	645
4	28.5	0	4	39.5	385
5	28	0	5	39	190
6	28	47	6	38	47
7	29	186	7	36	0
8	31.5	390	8	34.5	0
9	33.5	640	9	33.5	0
10	35.5	814	10	33	0
11 a.m.	37	954	11 p.m.	31.5	0

Solution To minimise the effort and time involved with hand calculations, it may be considered sufficient to divide the wall into two sections only, as shown in Fig. 18.10 so that $\Delta x = 12.5$ cm. The corresponding nodal points are *so* and *si* on the outside and inside surfaces, and 1 and 2 inside the wall in the conduction region. The stability criterion requires that

$$M \geq 3$$

$$\Delta\tau \leq \frac{(\Delta x)^2}{3\alpha} = \frac{(0.125)^2}{3(8 \times 10^{-7})} = 6510 \text{ s}$$

Choosing $\Delta\tau = 1$ h $= 3600$ s for the convenience of the outside air temperature and radiation data given, we have

$$M = \frac{(\Delta x)^2}{\alpha\Delta\tau} = \frac{(0.125)^2}{(8 \times 10^{-7})(3600)} = 5.43$$

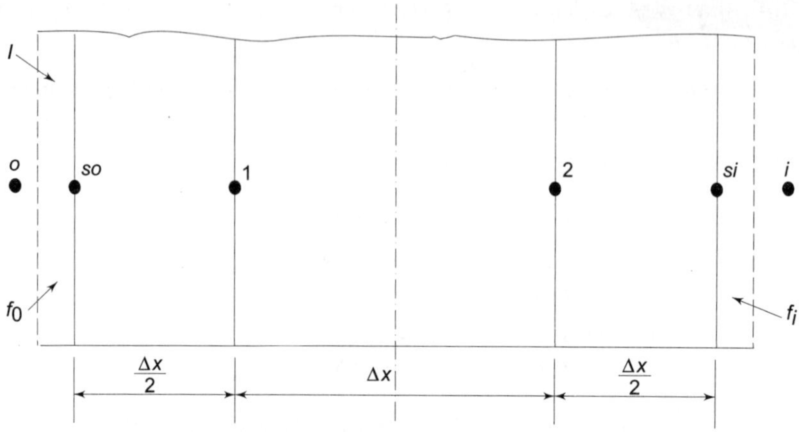

Fig. 18.10 Nodal points for wall in Example 18.3

Outside and inside surface Biot numbers

$$\text{Bi}_0 = \frac{f_0 \Delta x}{k} = \frac{23(0.125)}{1.5} = 1.917$$

$$\text{Bi}_i = \frac{f_i \Delta x}{k} = \frac{7(0.125)}{1.5} = 0.583$$

The finite difference relations for nodal temperature will now be established. Outside surface temperature

$$t_{so} = \frac{\text{Bi}_0}{\text{Bi}_0 + 2} t_e + \frac{2}{\text{Bi}_0 + 2}$$

$$t_1 = \frac{1.917}{3.917} t_e + \frac{2}{3.917} t_1 = 0.489\, t_e + 0.511\, t_1$$

Inside surface temperature

$$t_{si} = \frac{\text{Bi}_i}{\text{Bi}_i + 2} t_i + \frac{2}{\text{Bi}_i + 2}$$

$$t_2 = \frac{0.583}{2.583} t_i + \frac{2}{2.583} t_2 = 0.2257\, t_i + 0.7743\, t_2$$

Temperature at node 1

$$t_1^1 = \frac{2t_{so} + t_2}{M} + \frac{M-3}{M} t_1 = \frac{2}{5.43} t_{so} + \frac{1}{5.43} t_2 + \frac{2.43}{5.43} t_1$$

$$= 0.3683\, t_{so} + 0.1842\, t_2 + 0.4475\, t_1$$

Temperature at node 2

$$t_2^1 = \frac{2t_{si} + t_1}{M} + \frac{M-3}{M} t_2 = \frac{2}{5.43} t_{si} + \frac{1}{5.43} t_1 + \frac{2.43}{5.43} t_2$$

$$= 0.3683\, t_{si} + 0.1842\, t_1 + 0.4475\, t_2$$

Table 18.5 lists the coefficients of these equations in terms of left, right and local temperatures, called *influence coefficients*, for facilitating calculations.

Table 18.5 Influence coefficients

	Influence Coefficients		
Node	***Left***	***Local***	***Right***
so	0.489	0	0.511
1	0.3683	0.4475	0.1842
2	0.1842	0.4475	0.3683
si	0.7743	0	0.2257

Before proceeding further, the following steps are to be taken:
(a) Use of the excess temperature θ defined by

$$\theta = t - t_i$$

where t_i is equal to a constant room temperature of 25°C. The equations derived for temperature can also be used for excess temperatures.
(b) Sol-air temperature are evaluated first.
Thus for example at 12 noon

$$t_0 = 38.5°\text{C}$$
$$I = 1000 \text{ W/m}^2$$

The sol-air temperature is

$$t_e = t_0 + \frac{aI}{f_0} = 38.5 + \frac{0.8(1000)}{23} = 38.5 + 34.8 = 73.3°\text{C}$$

And the excess sol-air temperature is

$$\theta_e = 73.3 - 25 = 48.3°\text{C}$$

The calculated values of sol-air temperatures at different hours of the day are given in Table 18.6.
(c) Since the variation in temperature and radiation is periodic, the initial temperature at various nodal points are not known. Assumptions for the nodal temperatures at any instant of time can be made as equal to either the outside sol-air temperature or the room temperature. But this will make the calculations more tedious and time-consuming. Hence we assume the temperatures by a close guess of the expected temperatures as follows:

Time	θ_0	θ_e	θ_{so}	θ_1	θ_2	θ_{si}	θ_i
12 (Midnight)	5 (Data)	5	7.6	10.0	10.0	8.1	0 (Data)

We may now proceed with the calculations of finite difference approximations of the temperatures every hour. Thus at 1 a.m.

$$\theta_e = 4.5°\text{C}$$
$$\theta_{so} = (0.489)\,(4.5) + (0.511)\,(10) = 7.3°\text{C}$$

Note *For the calculation of* θ_{so} *at 1 a.m., the left-side temperature* θ_e *is taken at the same instant, viz., at 1 a.m. itself, whereas the right-side temperature,* θ_1 *is taken at a time,* $1\Delta\tau$ *earlier, viz., at 12 o'clock.*

$$\theta_1 = (0.3683)(7.6) + (0.4475)(10) + (0.1842)(10) = 9.1°C$$

$$\theta_2 = (0.1842)(107) + (0.4475)(10) + (0.3683)(8.1) = 9.3°C$$

$$\theta_{si} = (0.7743)(10) + (0.2257)(0) = 7.7$$

Proceeding likewise, the calculations are made for every hour and are given in Table 18.6.

Table 18.6 Finite difference approximations of excess temperatures

Time,	*Excess Temperature in*			°C	
Hours	***e***	***so***	**1**	**2**	***si***
12 (Midnight)	5	7.6	10.0	10.0	8.1
1 a.m.	4.5	7.3	9.1	9.3	7.7
2	4	6.6	8.5	8.7	7.2
3	3.5	6.1	7.8	8.1	6.7
4	3.5	5.7	7.2	7.5	6.3
5	3	5.1	6.7	7.0	5.8
6 a.m. (Radiation in)	4.6	5.7	6.2	6.3	5.4
7	10.5	8.3	6.1	6.0	5.0
8	20.1	12.9	6.9	5.7	4.6
9	30.8	18.6	8.9	5.5	4.4
10	38.8	23.5	11.8	5.7	4.3
11	45.2	28.1	15.0	6.3	4.4
12 (Noon)	48.3	31.3	18.2	7.2	4.9
1 p.m.	47.9	32.7	21.0	8.4	5.6
2	44.2	32.3	23.0	9.7	6.5
3	38.9	30.8	24.0	11.0	7.5
4	27.9	25.9	24.1	12.1	8.5
5	20.6	22.4	22.6	13.0	9.4
6	14.6	18.7	20.8	13.4	10.1
7 p.m. (Radiation out)	11.0	16.0	18.7	13.5	10.4
8	9.5	14.2	16.7	13.3	10.5
9	8.5	12.7	15.2	12.9	10.3
10	8.0	11.7	13.9	12.4	10.0
11 p.m.	6.5	10.3	12.8	11.8	9.6
12 (Midnight)	5.0	9.0	11.7	11.2	9.1

(Contd)

1 a.m.	4.5	8.2	10.6	10.5	8.7	
2	4.0	7.4	9.7	9.9	8.1	
3	3.5	6.7	8.9	9.2	7.6	
4	3.5	6.3	8.3	8.6	7.1	
5	3.0	5.7	7.6	8.0	6.6	
6 a.m. (Radiation in)	4.6	6.1	7.0	7.4	6.2	Final Solution
7	10.5	8.7	6.7	6.9	5.7	
8	20.1	13.3	7.5	6.4	5.3	
9	30.8	18.9	9.4	6.2	5.0	
10	38.8	23.8	12.3	6.3	4.8	
11 a.m.	45.2	28.4	15.4	6.9	4.9	
12 (Noon)	48.3	31.5	18.6	7.7	5.3	
1 p.m.	47.9	32.9	21.3	8.8	6.0	
2	44.2	32.5	23.3	10.1	6.8	
3	38.9	30.9	24.3	11.3	7.8	
4	27.9	26.1	24.3	12.4	8.8	
5	20.6	22.5	22.8	13.3	9.6	
6	14.6	18.8	20.9	13.7	10.3	
7 p.m. (Radiation out)	11.0	16.1	18.8	13.8	10.6	
8	9.5	14.3	16.9	13.5	10.7	
9	8.5	12.8	15.3	13.1	10.5	
10	8.0	11.7	14.0	12.5	10.1	
11 p.m.	6.5	10.3	12.9	11.9	9.7	
12 (Midnight)	5.0	9.0	11.8	11.3	9.2	
1 a.m.	4.5	8.2	10.7	10.6	8.7	
2	4.0	7.4	9.8	9.9	8.2	
3	3.5	6.7	8.9	9.3	7.7	
4	3.5	6.3	8.2	8.6	7.2	
5	3.0	5.7	7.6	8.0	6.7	
6 a.m. (Radiation in)	4.6	6.1	7.0	7.4	6.2	

While proceeding with calculations it is found that agreement occurs at 6 a.m. on the next day. The values shown between dashed lines in Table 18.6 are, therefore, repetitive and, hence, calculations may be discontinued. In case there is no agreement, the calculations may be continued until agreement occurs.

The average value of θ_{si} over a 24-hour period can be found by adding all the values and dividing by 24. Thus

$$(\theta_{si})_{\text{mean}} = 7.7°\text{C}$$

Also, the maximum value of θ_{si} is found from Table 18.6 at 8 p.m.

$$(\theta_{si})_{\max} = 10.7°\text{C}$$

The load on the air conditioner due to heat transfer through the wall per unit area of the wall can be evaluated by

$$q = f_i(t_{si} - t_i)$$

Average load on the air conditioner

$$q_{mean} = 7(7.7 - 0) = 53.9 \text{ W/m}^2$$

Maximum load on the air conditioner

$$q_{max} = 7(10.7 - 0) = 74.9 \text{ W/m}^2$$

which occurs at 8 p.m., five hours after the maximum outdoor temperature of 41.5°C is experienced.

Note *If the air conditioner has to do an effective job, it must meet the requirement of peak load. Since the average load is less, its running time will be partial, viz., (53.9/74.9) = 0.72, or 72 per cent.*

18.5 EMPIRICAL METHODS TO EVALUATE HEAT TRANSFER THROUGH WALLS AND ROOFS

There are two approaches to empirical calculations of heat transfer through walls and roofs. They are:

(i) The decrement factor and time lag method.
(ii) The equivalent temperature differential method.

Both the methods use analytical-experimental results for their formulations. The equivalent temperature differential method is more commonly used by the air-conditioning engineers as it is also applicable to sunlit walls and roofs.

18.5.1 Using Decrement Factor and Time Lag

If the thermal capacity of the wall is ignored, then the instantaneous rate of heat transfer through the wall at any time τ is given by

$$Q = UA(t_e - t_i) \tag{18.32}$$

and on an average basis, the mean heat flow is given by

$$Q_m = UA(t_{e_m} - t_i) \tag{18.33}$$

For the sake of simplicity, in this Chapter and the next the *dot* above Q has been dropped to indicate the *rate*.

But most building materials have a finite thermal capacity which is expressed as

$$mC = \rho CV = \rho C\,(A\Delta x)$$

where
m = Mass of wall
ρ, C = Density and specific heat of wall material
A = Cross-sectional area of wall
Δx = Wall thickness.

It has been seen that there is a two-fold effect of thermal capacity on heat transfer:

(i) There is a time lag between the heat transfer at the outside surface q_0 and the heat transfer at the inside surface q_i.
(ii) There is a decrement in the heat transfer due to the absorption of heat by the wall and subsequent transfer of a part of this heat back to the outside air when its temperature is lower.

The use of the rigorous analytical method to determine the time lag ϕ and decrement factor λ is quite complicated. The use of finite difference approximation for each wall, roof, etc., for each building is also time-consuming from the point of view of a practising engineer. Hence an empirical approach based on the determination of ϕ and λ for standard wall constructions, and their use for calculations can be employed.

It is observed that the specific heat of most materials is about 0.84 kJ/kg.K. The thermal capacity of most materials, therefore, essentially depends on their density and thickness. The IHVE Guide gives value of the time lag and decrement factor as a function of the wall thickness and density of construction materials. Figure 18.11 gives values of time lag for three different densities, while in Fig. 18.12, the effect of density on the decrement factor has been considered insignificant. In addition to these figures, Tables 18.7 and 18.8 give values for λ and ϕ for certain constructions taken from the ASHRAE Handbook.

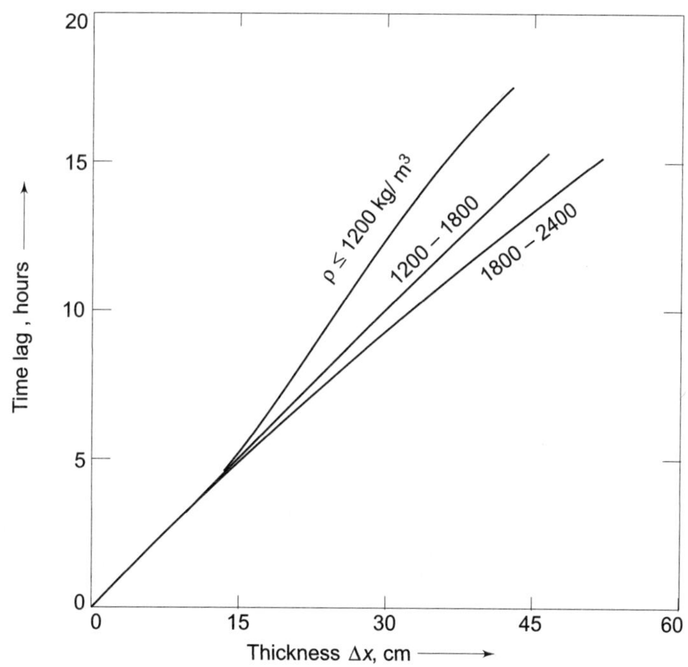

Fig. 18.11 Time lag of walls

Considering the effect of thermal capacity, the actual heat transfer at anytime τ is

$$Q_\tau = UA\,(t_{e_m} - t_i) + UA\lambda\,(t_{e_{\tau-\phi}} - t_{e_m}) \tag{18.34}$$

where $t_{e_{\tau-\phi}}$ is the sol-air temperature at time $\tau - \phi$, i.e., ϕ hours before the heat transfer is to be calculated.

A comparison of Eq. (18.34) with Eq. (18.33) shows that Q_τ can be greater or less than Q_{mean}, depending on whether $t_{e_{\tau-\phi}}$ is greater or less than t_{e_m}, ϕ hours before. The second term in Eq. (18.34), therefore, represents the periodic component which is equal to the sum of all such component harmonics in Eq. (18.19).

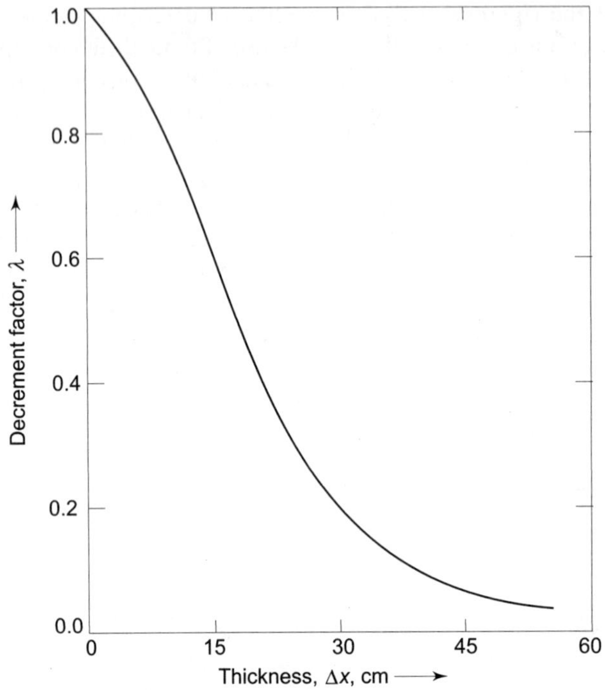

Fig. 18.12 Decrement factor of walls

It is evident that if the wall is thick, the decrement factor will be small as is also seen from Tables 18.7 and 18.8. For example, from Table 18.8, the decrement factor for a 15 cm concrete roof is 0.48 whereas for a 5 cm concrete roof, it is 0.83. Thus in the case of a very thick wall, the second term on the right-hand side in Eq. (18.34) can be ignored so that Eq. (18.33) holds

$$Q_\tau = UA\,(t_{e_m} - t_i) = Q_m$$

which implies that the heat transfer across the wall remains uniform at its mean value throughout the day. It is, therefore, advantageous to provide thicker walls in buildings that are not air conditioned. Such buildings will not become excessively hot in summer or excessively cold in winter.

Table 18.7 Amplitude decrement factor and time lag of shaded walls[2]

Construction	*Decrement Factor, λ*	*Time Lag, φ hours*
10 cm brick or stone veneer + frame	0.62	4
20 cm hollow tile or		
20 cm cinder block	0.48	5
10 cm brick or concrete	0.69	3
20 cm brick or concrete, or		
30 cm hollow tile or cinder block	0.39	5

Table 18.8 Amplitude decrement factor and time lag of roofs exposed to sun[2]

Construction	*Decrement Factor, λ*	*Time Lag, ϕ hours*
5 cm concrete	0.83	3
5 cm concrete + 5 cm insulating board	0.69	5
10 cm concrete	0.64	4
15 cm concrete	0.48	5
15 cm concrete + 5 cm insulating board	0.26	7

Opposite conditions prevail when the wall is too thin. In the limiting case, when the wall thickness approaches zero, the decrement factor λ tends to unity and the time lag ϕ tends to zero. In that case Eq. (18.32) is applicable

$$Q_\tau = UA\,(t_e - t_i)$$

i.e., the heat transfer through the wall is equal to its instantaneous value.

Accordingly, for thick wall, the heat gain does not vary much, whereas for thin walls, it varies considerably over 24 hours. The effect of the type of construction on heat gain is shown in Fig. 18.13. It is seen that a light wall with a low thermal capacity having a time lag of about 3 hours has a maximum heat gain at 3 p.m., and great variation in heat transfer over a 24-hour period. A heavier wall with high thermal capacity has a reduced and more uniform heat gain, and the peak occurs much later, say at 12 midnight, with a corresponding time lag of 12 hours.A still heavier construction may result in very small and uniform heat-transfer rate.

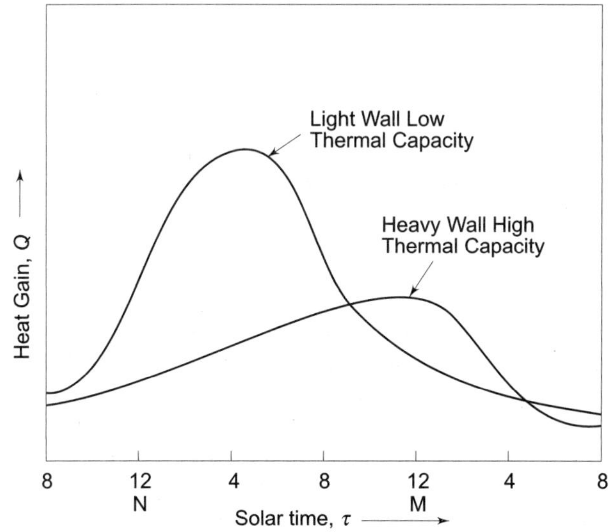

Fig. 18.13 Comparison of heat transfer through light and heavy construction

Thus in a locality where the daily range of variation of the outside air temperature is small, it is immaterial what thickness of wall is provided. But in a locality where the daily range of temperature is large, it is desirable to have thick walls so as to cut the cooling load in summer and the heating load in winter. Moreover, such walls will not allow the inside temperature to rise very much during the day and drop at night, and thus maintain a reasonably uniform and moderate inside temperature even without air conditioning.

Also, in buildings that are not conditioned, *night ventilation* helps to maintain them cooler during the day.

Example 18.4 *If the wall in Example 18.3 has a density of 2400 kg/m³, calculate the maximum heat gain of the room and the heat gain at 5 p.m. per unit area of wall, using the time lag and decrement factor approach.*

Solution From Figs 18.11 and 18.12 for 25 cm thickness, 2400 kg/m³ density, we have:

Time lag, $\phi = 6.7$ hours

Decrement factor, $\lambda = 0.455$

From Table 18.6, the mean value of the excess sol-air temperature is $\theta_{e_m} = 19.1°\text{C}$.

Mean sol-air temperature

$$t_{e_m} = \theta_{e_m} + t_i = 19.1 + 25 = 44.1°\text{C}$$

Maximum sol-air temperature is at 12 noon

$$t_{e_{\max}} = 48.3°\text{C}$$

Overall heat-transfer coefficient

$$\frac{1}{U} = \frac{1}{f_0} + \frac{\Delta x}{k} + \frac{1}{f_i} = \frac{1}{23} + \frac{0.25}{1.5} + \frac{1}{7}$$

$$U = 2.833 \text{ W/m}^2\text{°C}$$

Maximum heat gain of the room will occur 6.7 hours after the maximum sol-air temperature of 48.3°C is reached, viz., at 6.7 hours after noon.

Maximum heat gain

$$q_{\max} = U[(t_{e_m} - t_i) + \lambda(t_e - t_{e_m})]$$

$$= 2.833\ [19.1 + (0.455)\ (48.3 - 19.1)] = 91.8 \text{ W/m}^2$$

Note *A comparison can be made with the calculations by the finite difference method of Example 18.3. Here the time of peak load is obtained as 6.7 p.m. instead of 8 p.m. and the maximum heat gain is 91.8 W/m² instead of 74.9 W/m². Of course, the accuracy of the finite difference procedure can be improved by taking smaller time and space increments.*

For calculating the heat gain at 5 p.m., we take the sol-air temperature from the table about 6.7 hours before the time, or say at 11 a.m., to be on the safer side. Sol-air temperature at 11 a.m.

$$t_e = 45.2°\text{C}$$

Heat gain at 5 p.m.

$$q = 2.833\ [19.1 + (0.455)\ (45.2 - 19.1)] = 87.8 \text{ W/m}^2$$

18.5.2 Equivalent Temperature Differential (ETD) or Cooling Load Temperature Difference (CLTD) Method

Equation (18.34) for heat transfer through walls and roofs can also be expressed in terms of an equivalent temperature differential Δt_E defined by the equation

$$Q = UA\,(t_{e_m} - t_i) + UA\lambda\,(t_{e_{\tau-\phi}} - t_{e_m})$$

$$= UA\,\Delta t_E = f_i\,(t_{si} - t_i) \quad (18.35)$$

so that $$\Delta t_E = (t_{e_m} - t_i) + \lambda(t_{e_{\tau-\phi}} - t_{e_m}) \quad (18.36)$$

Thus Δt_E when multiplied by UA for the construction gives the heat transfer rate. We can see that Δt_E depends on:

(i) Decrement factor λ and time lag ϕ, which in turn depend on the thermo-physical properties of the construction.
(ii) The outside air temperature t_0 and solar radiation intensity I.
(iii) Room temperature t_i.

Thus, the equivalent temperature differential approach takes care of the exposure of the wall or roof to the sun. Tables of Δt_E can be prepared for fixed values of t_0 and t_i, for different types of constructions, and as a function of latitude and time for roofs, and latitude, time and orientation for walls. The Carrier Handbook[3] makes use of this approach and Tables 18.9 and 18.10 give the required values for walls and roofs respectively. It will be noticed on seeing these tables that the effect of density and wall thickness is incorporated by specifying the mass of the wall per unit area of its cross-section. Further, it must be pointed out here that these tables have been established from calculations made on an analogue computer using Schmidt's method based on the conditions given below:

(i) Latitude 40°N, but normally suitable for latitudes 0 to 50°N, for the hottest summer period
(ii) An outdoor daily range of a dry bulb temperature of 11.1°C (20°F)
(iii) An outside and inside design temperature difference of 8.3°C (15°F)
(iv) Dark colour walls and roofs with absorptivity of 0.9
(v) A specific heat of the construction material of 1.005 kJ/kg.K.

When there is a departure from these conditions, the following corrections may be applied.

(i) The values of t_0 and t_i are additive to Δt_E. Hence add or subtract the difference of $t_0 - t_i$ and 8.3°C.
(ii) If the daily range is different from 11.1°C, then apply effective corrections as follows:

(a) For each 1°C difference less than 11.1°C	Add 0.25° for medium construction. Add 0.5° for heavy construction.
(b) For each 1°C difference greater than 11.1°C	Subtract 0.25° for medium construction. Subtract 0.5° for heavy construction.
(c) Maximum correction	2° for medium and 3° for heavy construction.
(d) Light construction	No correction.

(iii) For south latitude use the following exposure values:

South Latitude	*Exposure Value*
North-east	South-east
East	East
South-east	North-east
South	North (shade)
South-west	North-west
West	West
North-west	South-west
North (shade)	South

Example 18.5 *If the wall in Example 18.4 is facing west, calculate the heat gain of the room per unit area of the wall, on an hourly basis between 1 p.m. and 6 p.m. The outdoor maximum and minimum temperatures are 40 and 22°C respectively. The outside and inside design temperatures are 40 and 25°C respectively. What is the time of maximum heat gain from the wall?*

Solution Thickness and density of wall

$$\Delta x = 0.25 \text{ m}$$
$$\rho = 2400 \text{ kg/m}^2$$

Mass of wall

$$m = \rho(\Delta x)A = (2400)\,(0.25)\,(1) = 600 \text{ kg/m}^2 \text{ area}$$

Overall heat-transfer coefficient

$$\frac{1}{U} = \frac{1}{f_i} + \frac{\Delta x}{k} + \frac{1}{f_0} = \frac{1}{7} + \frac{0.25}{1.5} + \frac{1}{23}$$
$$U = 2.83 \text{ Wm}^{-2}\text{ K}^{-1}$$

Outdoor daily range = $(t_0)_{max} - (t_0)_{min} = 40 - 22 = 18°C > 11.1°C$

Outside and inside temperature difference = $t_0 - t_i = 40 - 25 = 15°C > 8.3°C$

Correction for equivalent temperature difference

$$\Delta t'_e = (15 - 8.3) - (18 - 11.1)\,(0.25) = 4.98°C$$

The following table gives the values of equivalent temperature differentials from Table 18.9 and those obtained after correction by adding $\Delta t'_e$ to the values from the table. The table also gives the calculated values of the heat flux from the relation

$$q = U\Delta t_E$$

Time	$(\Delta t_E)_{table}$	$(\Delta t_E)_{corrected} = (\Delta t_E)_{table} + 4.98°C$	*Q*
p.m.	°C	°C	W/m^2
1	4.25	9.23	2.61
2	4.75	9.73	27.5
3	5.5	10.48	29.7
4	6.32	11.3	32.0
5	8.34	13.32	37.7
6	9.69	14.67	41.5

Note *It will be seen from Table 18.9 that the maximum equivalent temperature difference occurs at 8 p.m. Thus, the maximum heat gain from the wall is at 8 p.m.*

Example 18.6 ***Orientation of a Building***

A 20 m × 40 m × 3.5 m high building having a flat roof is located near Bombay (about 20°N latitude). For summer, based on June 21, show which of the orientations (A) or (B) is better (Fig. 18.14).

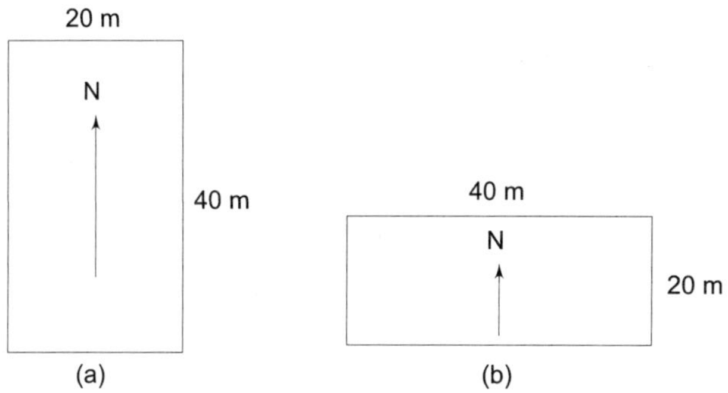

Fig. 18.14 The two orientations of a building for Example 18.6

Solution Since the roofs are horizontal flat, both the orientations will receive same solar radiation, and will have same equivalent temperature differential for roof. However, the walls will receive different radiations, and hence will have different temperature differentials.

Daily solar radiation from Table 17.9 (c):

$$\text{N-Wall} = 88 + 129 + 104 + 79 + 60 + 54 + 47 + 54 + 60 + 79 + 104 + 129 + 80 = 1,075\ \text{W/m}^2$$

$$\text{E-Wall} = 256 + 467 + 505 + 451 + 303 + 129 + 44(4) + 38 + 28 + 9 = 2,362\ \text{W/m}^2$$

$$\text{S-Wall} = 10 + 28 + 38 + 44\ (7) + 38 + 28 + 9 = 459\ \text{W/m}^2$$

$$\text{W-Wall} = \text{Same as E-wall} = 2,362\ \text{W/m}^2$$

Wall areas are $20 \times 3.5 = 70\ \text{m}^2$, and $40 \times 3.5 = 140\ \text{m}^2$

Total solar radiation received by the four walls are:

Orientation A

$$\Sigma IA = 70\ (1075) + 70(459) + 2 \times 140\ (2362) = 643{,}490\ \text{W}$$

Orientation B

$$\Sigma IA = 140\ (1075) + 140\ (459) + 2 \times 70\ (2362) = 545{,}440\ \text{W}$$

Orientation B receives less radiation than orientation A. Hence, B is better than A.

Note *Any orientation between A and B will receive more radiation than B. Hence, B is the best.*

Table 18.9 Equivalent temperature differential (°C) for walls^2 : 24-hour operation

Exposure	*Mass wall* kg/m^2	*a.m.*					*p.m.*							
		8	9	10	11	12	1	2	3	4	5	6	7	8
North-east	106	12.2	12.8	13.3	10.6	7.8	7.2	6.7	7.2	7.8	7.8	7.8	6.7	5.5
	319	–1.1	2.8	13.3	12.2	11.1	8.3	5.5	6.1	6.7	7.2	7.8	7.2	6.7
	532	2.2	2.2	2.2	5.5	8.9	8.3	7.8	6.7	5.5	6.1	6.7	6.7	6.7
	744	3.1	3.3	3.3	3.3	3.3	5.5	7.8	8.9	7.8	6.7	5.5	5.5	5.5
East	106	16.7	18.3	20	19.4	17.8	11.1	6.7	7.2	7.8	7.8	7.8	6.7	5.5
	319	0	11.7	16.7	17.2	17.2	10.6	7.8	7.2	6.7	7.2	7.8	7.2	6.7
	532	3.0	4.4	7.8	11.1	13.3	13.9	13.3	11.1	10	8.9	7.8	7.8	7.8
	744	5.5	5.0	4.1	5.0	5.5	8.3	10.0	10.6	10	9.4	8.9	7.8	6.7
South-east	106	7.2	10.6	14.4	15	15.6	14.4	13.3	10.6	8.9	8.3	7.8	6.7	5.5
	319	0	17.2	11.1	7.8	15.6	14.4	13.9	11.7	10	8.3	7.8	7.2	6.7
	532	3.3	3.3	3.3	6.1	8.8	9.4	10.0	10.6	10	8.9	7.8	7.2	6.7
	744	4.4	4.4	4.4	3.9	3.3	6.1	7.8	8.3	8.9	10.0	8.9	8.3	7.8
South	106	–2.2	0.55	2.2	7.8	12.2	15	16.7	15.6	14.4	11.1	8.9	6.7	5.5
	319	–2.2	–1.17	–1.1	3.9	6.7	11.1	13.3	13.9	14.4	12.8	11.1	8.3	6.7
	532	1.1	1.1	1.1	1.7	2.2	4.4	6.7	8.3	8.9	10.0	10.0	8.3	7.8
	744	3.3	2.8	2.2	2.2	2.2	2.2	2.2	3.8	5.5	7.3	7.8	8.3	8.9

(Contd.)

Table 18.9 (*Contd*)

Exposure	*Mass wall* kg/m²	*a.m.*					*p.m.*							
		8	9	10	11	12	1	2	3	4	5	6	7	8
	106	–2.2	–1.1	0	2.2	3.3	10.6	14.4	18.9	22.2	22.8	23.3	16.7	13.3
South-west	319	0	0	0	0.55	1.1	4.4	6.7	13.3	17.8	19.4	20.0	19.4	18.9
	532	3.3	2.8	2.2	2.8	3.3	3.9	4.4	6.7	7.8	10.6	12.2	12.8	13.3
	744	4.4	4.4	4.4	3.9	3.3	3.3	3.3	3.3	4.4	5.0	5.51	8.3	10.0
	106	–2.2	–1.1	0	1.7	3.3	7.7	11.1	17.8	22.2	25.0	26.7	18.9	12.2
	319	0	0	0	1.1	2.2	3.9	5.5	10.6	14.4	18.9	22.2	22.8	20.0
West	532	3.3	3.3	3.3	3.3	3.3	3.9	4.4	5.5	6.7	9.4	11.1	13.9	15.6
	744	5.5	5.0	4.4	4.4	4.4	5.0	5.5	5.5	5.5	6.1	6.7	7.8	8.9
	106	–2.2	–1.1	0	1.7	3.3	5.5	6.7	10.6	13.3	18.3	22.2	20.6	18.9
North-west	319	–2.2	–1.7	–1.1	0	1.1	3.3	4.4	5.5	6.7	11.7	16.7	17.2	17.8
	532	2.2	2.2	2.2	2.2	2.2	2.2	2.2	2.7	3.3	5.0	6.7	9.4	11.1
	744	3.3	3.3	3.3	3.3	3.3	3.3	3.3	3.3	3.3	3.9	4.4	5.0	5.5
	106	–2.2	–1.7	–1.1	0.55	2.2	4.4	5.5	6.7	7.8	7.2	6.7	5.5	4.4
North	319	–2.2	–1.7	–1.1	–0.55	0	1.7	3.3	4.4	5.5	6.1	6.7	6.7	6.7
	532	0	0	0	0	0	0.55	1.1	1.7	2.2	2.8	2.8	2.8	4.4
	744	0	0	0	0	0	0	0	0.55	1.1	1.7	2.2	2.8	3.3

(*Contd.*)

Table 18.10 Equivalent temperature differential (°C) for dark-coloured, sunlit and shaded roots[3]: 24-hour operation

Condition	*Mass of Roof* kg/m²	*a.m.*							*p.m.*				
		6	7	8	9	10	11	12	1	2	3	4	5
Exposed	53	–2.2	–3.3	–3.9	–2.8	–0.55	3.9	8.3	13.3	17.8	21.1	23.9	25.6
to	106	0	–0.55	–1.1	–0.55	1.1	5.0	8.8	12.8	16.7	20.0	22.8	23.9
sun	212	2.2	+1.7	–1.1	1.7	3.3	5.5	8.8	12.8	15.6	18.3	21.1	22.2
	319	5.0	4.4	3.3	3.9	4.4	6.1	8.8	12.2	15.0	17.2	19.4	21.2
	524	7.2	6.7	6.1	6.1	6.7	7.2	8.8	12.2	14.4	15.6	17.8	19.4
Covered	106	–2.8	–1.1	0	11.1	2.2	5.5	8.8	10.6	12.2	11.1	10.0	8.6
with	212	–1.7	–1.1	–0.55	–0.55	0	2.8	5.5	7.2	8.3	8.3	8.8	8.3
water	319	–0.55	–1.1	–1.1	–1.1	–1.1	–1.1	2.8	3.9	5.5	6.7	7.8	8.3
Sprayed	106	–2.2	–1.1	0	1.1	2.2	4.4	6.7	8.3	10.0	9.4	8.8	8.3
	212	–1.1	–1.1	–0.55	–0.55	0	1.1	2.8	5.0	7.2	7.8	7.8	7.8
	319	–0.55	–1.1	–1.1	–1.1	–1.1	0	1.1	2.8	4.4	5.5	6.7	7.2
	106	–2.8	–2.8	–2.2	–1.1	0	1.1	3.3	5.0	6.7	7.2	7.8	7.2
Shaded	212	–2.8	–2.8	–2.2	–1.7	–1.1	0	1.1	2.8	4.4	5.5	6.7	7.2
	319	–1.7	–1.7	–1.1	–1.1	–1.1	–0.55	0	1.1	2.2	3.3	4.4	5.0

(Contd)

Table 18.10 (*Contd*)

Condition	*Mass of Roof* kg/m^2	*p.m.*							*a.m.*				
		6	7	8	9	10	11	12	1	2	3	4	5
Exposed	53	25.0	22.8	19.4	15.6	12.2	8.9	5.5	3.9	1.7	0.55	–0.55	–1.7
to	106	23.9	22.2	19.4	16.7	13.9	11.1	8.3	6.7	4.4	3.3	2.2	1.1
sun	212	22.8	21.7	19.4	17.8	15.6	13.3	11.1	9.4	7.2	6.1	5.0	3.3
	319	21.7	21.1	20.0	18.9	17.2	15.6	13.9	12.2	10.0	8.9	7.2	6.1
	524	20.6	20.6	19.4	18.9	18.9	17.8	16.7	15.0	12.8	11.1	10.0	7.8
Covered	106	7.8	6.7	5.5	3.3	1.1	0.55	0.55	–0.55	–1.1	–1.7	–2.2	–2.8
with	212	8.3	7.8	6.7	5.5	3.9	2.8	1.7	0.55	–0.55	–1.1	–1.7	–1.7
water	319	8.9	8.3	7.8	6.7	5.5	4.4	3.3	2.2	1.7	1.1	0.55	0
	106	7.8	6.7	5.5	3.3	1.1	0.55	0	–0.55	–1.1	–1.1	–1.7	–1.7
Sprayed	212	7.8	7.2	6.7	5.0	3.9	2.2	1.7	0.55	0	0	–0.55	–0.55
	319	7.8	7.2	6.7	6.1	5.5	4.4	3.3	2.2	1.1	0.55	0	–0.55
	106	6.7	5.5	4.4	2.8	1.1	0.55	0	–0.55	–1.7	–2.2	–2.8	–2.8
Shaded	212	6.7	6.1	5.5	4.4	3.3	2.2	1.1	0	–0.55	–1.7	–2.2	–2.8
	319	5.5	5.5	5.5	5.0	4.4	3.3	2.2	1.1	0.55	0	–0.55	–1.1

18.6 NATURAL VENTILATION THROUGH INFILTRATION

Infiltration is the name given to the leakage of outside air through door openings, and through cracks and interstices around windows and doors into conditioned space. Even though the air inside is slightly pressurized, the leakage does take place which is principally due to the following factors:

(i) *Stack effect*, particularly in tall buildings
(ii) *Wind pressure*
(iii) Entry and exit of occupants effecting change of air due to door openings.

It must be pointed out here that corresponding to every infiltration there is an equivalent amount of *exfiltration*. In effect, infiltration involves an exchange between the outside and inside air.

Infiltration, as a result of stack effect, wind effect, and through doors and windows and other openings can be treated as contributing to *natural ventilation*.

Note *If natural ventilation is not enough to provide for desired level of comfort then recourse to mechanical ventilation is necessary.*

18.6.1 Stack Effect

Differences between temperatures and humidities produce differences in the densities of air between the outside and inside of buildings. As a result, pressure differences occur causing flow of air known as chimney or stack effect.

When the inside temperature is lower than the outside, the stack effect produces positive inside pressure at lower levels and negative inside pressure at high levels. Consequently, the outward flow of air takes place at lower levels and the inward flow at higher levels, with the neutral zone in the middle as shown in Fig. 18.15. The reverse is true when the inside is at a higher temperature than the outside. Thus, we have

In Summer Infiltration at the top and exfiltration at the bottom.

In Winter Infiltration at the bottom and exfiltration at the top.

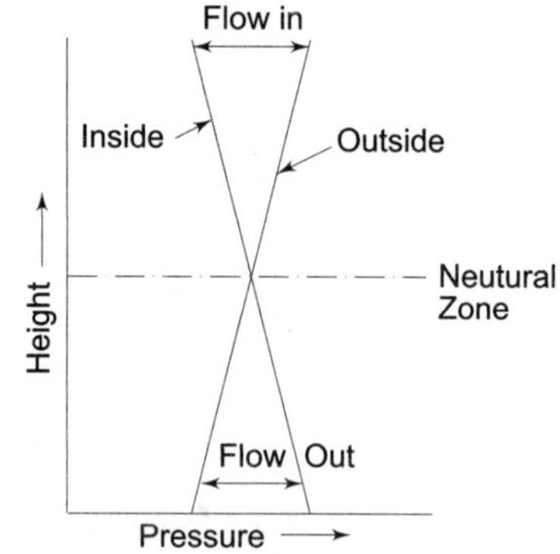

Fig. 18.15 Stack effect in summer air conditioning

The infiltration from the stack effect is generally small but is greatly influenced by the height of the building and the presence of staircases and elevators. The ASHRAE Guide gives the following formula to evaluate the infiltration rate Q_v in m^3/s due to stack effect

$$\dot{Q}_v = 0.172\, A\sqrt{H(t_i - t_0)} \tag{18.37}$$

where H is the building height in m, t_i and t_0 are the inside and outside temperatures in °C, and A is the area in m^2 available for the flow of air into or out of the building. The corresponding velocity of the air flow is

$$C = \frac{\dot{Q}_v}{A} = 0.172\sqrt{H(t_i - t_0)} \tag{18.38}$$

The pressure difference inducing this flow will then be given by

$$\frac{\Delta p}{\rho} = \frac{C^2}{2}$$

Assuming an effectiveness of 0.8, and a standard air density of 1.2 kg/m^3, we have

$$\Delta p = \frac{1.2}{2}\left(\frac{0.172}{0.8}\right)^2 H(t_i - t_0)$$

$$= 0.028\, H(t_i - t_0),\ \text{N/m}^2 \tag{18.39}$$

18.6.2 Wind Action

The flow of air due to wind over a building creates regions in which the static pressure is higher or lower than the static pressure in the undisturbed air stream. The pressure is positive on the windward side resulting in the infiltration of air, and negative on the leeward side resulting in exfiltration. This is illustrated in Fig. 18.16. In a tall building, the wind velocity is very high towards the top of the building and hence the leakage rate is also higher.

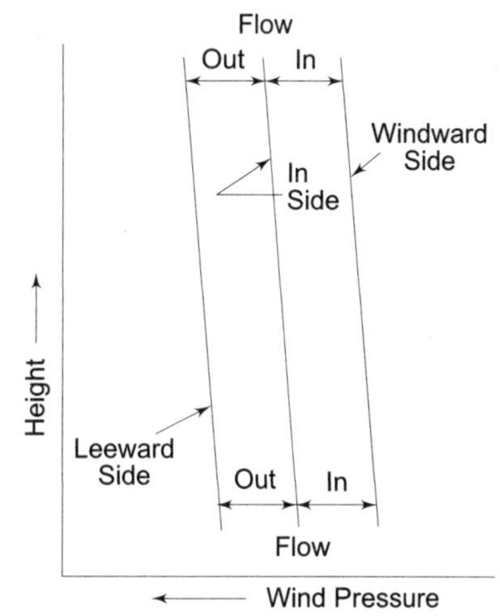

Fig. 18.16 Wind effect on leakage of air

There are two methods of estimating the infiltration of air into conditioned space due to wind action. They are:

(i) *Crack method,* and
(ii) *Air-change method.*

In the crack method, the estimate is based on measured leakage characteristics and the width and length of cracks around windows and doors. The air-change method assumes a certain number of air changes per hour for each space depending on its usage. The crack method is generally regarded as more accurate and is used in the case of windows. The air change method is more convenient to use in the case of doors.

The leakage of air in this case is a function of the wind pressure Δp which can be determined by knowing the wind velocity C using the equation (Sec. 21.2)

$$\Delta p = 0.00047\, C^2 \tag{18.40}$$

where Δp is in cm H_2O and C is in km/h. It is a common practice to take 0.64 Δp only as the pressure difference between the outside and inside air to evaluate the infiltration rate, which can be expressed by an equation of the type

$$\dot{Q}_v = C(\Delta p)^n \tag{18.41}$$

where C is a constant, and n lies between 0.5 and 1. For non-weather-stripped windows, the following relation may be used

$$\dot{Q}_v = 0.125\, (\Delta p)^{0.63} \tag{18.42}$$

where $\dot{Q}_v$ is in L/m of crack length, and Δp is in N/m^2. The leakage rates of most window cracks fall between the four classes of double-hung wood windows as given in Table 18.11. Similarly, Table 18.12 gives the leakage rates through cracks in doors on the windward side for different wind velocities and different door constructions.

Table 18.11 Infiltration through double-hung windows in m^3/h/m of crack[3]

Window type	*Pressure Difference, cm* H_2O				
	0.25	**0.50**	**0.75**	**1.00**	**1.25**
Non-weather-stripped, loose fit	7.1	11.3	14	18	21
Non-weather-stripped, average fit	2.5	4	5.3	6.4	7.4
Weather-stripped, loose fit	2.5	4	5.3	6.4	7.4
Weather-stripped, average fit	1.3	2.1	2.8	3.3	3.9

Table 18.12 Infiltration through doors-crack method[3]

Type of door	*cmm per Linear Metre of Crack*					
	Wind Velocity, kmph					
	8	**16**	**24**	**32**	**40**	**48**
Glass door						
Good installation						
3.2 mm crack	0.3	0.6	0.9	1.21	1.49	1.77
Average installation						
4.76 mm crack	0.45	0.93	1.3	1.86	2.23	2.7
Poor installation						
6.4 mm crack	0.6	1.21	1.77	2.42	2.42	3.53
Ordinary wood or metal door						
Well fitted						
W-stripped	0.04	0.06	0.08	0.12	0.16	0.2
Well fitted						
Now W-stripped	0.08	0.11	0.17	0.24	0.31	0.39
Poorly fitted						
Not W-stripped	0.08	0.21	0.34	0.48	0.61	0.78
Factory door						
3.2 mm crack	0.3	0.6	0.9	1.21	1.49	1.77

The ASHRAE Guide publishes data that indicates that wind pressure causes a leakage of air even through brick and concrete walls. Its numerical value is, however, very small. For example, for a 21.5 cm plastered brick wall at 24 kmph wind velocity, it is only 0.000356 cmm/m^2 of the wall area.

In the air-change method, it is required to use experience and judgement to estimate the overall value of the infiltration rate. In practice, the following values of air-changes per hour can be used with reasonable precision for rooms with the extent of windows and external doors given.

No windows or exterior doors	:	$\frac{1}{2}$
Exterior doors or windows on one side	:	1
Exterior doors or windows on two sides	:	$1\frac{1}{2}$

Exterior doors or windows on three sides : 2
Entrance halls : 2

These values include the exchange of air due to door openings which is discussed below.

18.6.3 Infiltration due to Door Openings

Infiltration through doors depends on the type of door, as well as its usage. Often the tables developed for doors give the infiltration rates through doors which include the leakage rates through cracks due to door openings. Table 18.13 through 18.15 from the Carrier Handbook give the infiltration rates through doors on the windward side for various door constructions and usage, and for a wind velocity of 12 kmph. For other doors, the values may be multiplied by 0.6. For other wind velocities, the values may be multiplied by the ratio of the velocities.

Table 18.13 Infiltration through doors on adjacent walls (wind velocity 12 kmph)[3]

Description	cmm/m^2 *Area*		*cmm*	
			Standing Open	
	No Use	*Average Use*	*No Vestibule*	*Vestibule*
Revolving Doors-				
Normal Operation	0.24	1.58	—	—
Panels Open	—	—	34	25
Glass Door-4.75 mm Crack	1.37	3.0	20	14
Wood Door-	0.3	1.98	20	14
Small Factory Door	0.23	1.98	—	—
Garage and Shipping				
Room Door	0.61	1.37	—	—
Ramp Garage Door	0.61	2.06	—	—

Table 18.14 Infiltration through swinging doors on opposite walls (wind velocity 12 kmph)[3]

% Time 2nd Door is Open	*cmm/Pair of Doors % Time 1st Door is Open*				
	10	**25**	**50**	**75**	**100**
10	2.8	7	14	21	28
25	7.1	18	35	53	71
50	14	35	71	106	142
75	21	53	106	159	210
100	28	71	142	210	280

It may be noted that vestibules may decrease the infiltration to the extent of 30 per cent when the door usage is light. However, when the door usage is heavy, vestibules are of little value.

18.6.4 Load due to Infiltration

Infiltration involves the heat gain or loss to the conditioned space due to the replacement of the conditioned inside air by the undesirable outside air. This load includes both sensible and latent, and is evaluated in the same manner as the ventilation load from the infiltration rate $(cmm)_1$. If ventilation air is greater than infiltration/exfiltration air then infiltration may not be considered seprately.

Table 18.15 Infiltration through doors based on usage (wind velocity 12 kmph[3])

Application	*cmm/Person in Room/Door*		
	182 cm	*91 cm Swinging Door*	
	Revolving Door	*No Vestibule*	*Vestibule*
Bank	0.184	0.227	0.17
Barber Shop	0.113	0.142	0.108
Soda Shop	0.156	0.198	0.15
Cigar Store	0.566	0.85	0.637
Department Store	0.182	0.227	0.17
Dress Shop	0.057	0.071	0.108
Drug Store	0.156	0.198	0.15
Hospital	—	0.099	0.637
Lunch Room	0.113	0.142	0.17
Men's Shop	0.076	0.105	0.054
Restaurant	0.057	0.071	0.15
Shoe Store	0.076	0.099	0.074

Example 18.7 *A room is 5.5 m × 3.5 m × 3 m high. The total perimeter of an average fit non-weatherstripped window crack in the room is 10 m. The wind velocity is 9 m/s. Calculate the air-change rate of infiltration:*

(i) Using Table 18.11.

(ii) Using the empirical relation for the frictional resistance to air flow

$$\dot{Q}_v = 0.125\,(0.64\,\Delta p)^{0.63}\ \text{L/m}$$

where Δp is in N/m^2

Solution Volume of room = (5.5) (3.5) (3) = 57.75 m^3

Wind pressure

$$\Delta p = \frac{\rho C^2}{2} = \frac{(1.2)(9)^2}{2} = 48.6\ \text{N/m}^2 = \frac{48.6}{98.1} = 0.5\ \text{cm H}_2\text{O}$$

(i) Infiltration rate for average fit non-weatherstripped windows from Table 18.11 at 0.5 cm H_2O pressure difference is 4.0 m^3/h/m crack. Total infiltration rate

$$\dot{Q}_v = (0.4)\,(10) = 40\ \text{m}^3/\text{h}$$

$$\text{Air-change rate} = \frac{40}{57.75} = 0.69$$

(ii) Total infiltration rate

$$\dot{Q}_v = \frac{0.125(48.6\times 0.64)(10)(3600)}{1000} = 39.2\ \text{m}^3/\text{h}$$

$$\text{Air-change rate} = \frac{39.2}{57.75} = 0.68$$

Note *It is, thus, seen that the infiltration rate from just one large non-weatherstripped window is more than half the air-change per hour.*

18.7 PASSIVE HEATING AND COOLING OF BUILDINGS

A *passive solar building* may be defined as that in which thermal energy flows by natural means, enabling the system to function without external power. The operation of a passive system involves the control of the flow of thermal energy, and includes the ability to completely stop energy from escaping or entering the building (e.g., by insulation) and the ability to vary the timing or location of energy flow inside the building (e.g., by opening or closing spaces to each other).

A space-heating system comprises the space, a solar collector and a thermal storage. A space-cooling system comprises the space, an environmental sink (sky and ground) and a thermal or cold storage. The heat exchange between the three elements falls into two categories, viz., forced exchange (using fans, pumps, etc.) and natural exchange (involving conduction, natural convection and radiation). If all the significant exchanges linking the three elements of a heating or cooling system involve forced flow, the system is classified as *active*, and if they involve purely natural flow, the system is classified as *passive*. A mixture of the two may be called a *hybrid* system.

Three basic passive solar concepts have been defined; each involving differential relationships relating the sun, storage mass and living space. They are *direct gain, indirect gain* and *isolated gain*. Each concept will now be dealt with, separately, from the viewpoints of principles, requirements, variations and controls.

It must be stated here that without the controls, the addition of the passive system with large glass areas and storage masses can cause much discomfort due to winter heat losses and summer overheating.

18.7.1 Direct Gain

This is the most common passive solar gain building. The living space is directly heated by the sun as shown in Fig. 18.17 and serves as a live-in collector. The basic requirements for direct-gain building are:

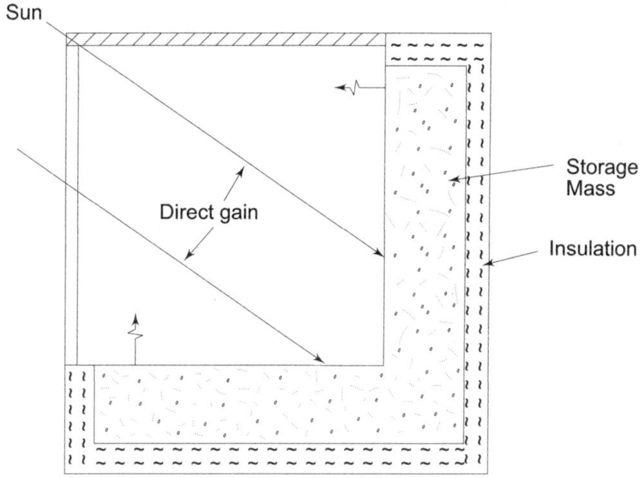

Fig. 18.17 Direct gain principle

(i) A large south-facing glazing (collector area) with the living space exposed directly behind.
(ii) A floor and/or wall storage mass of significant dimension for solar exposure and for thermal capacity.
(iii) A method for isolating the storage from the exterior climatic conditions by using insulation between the storage mass and outdoors.

The best location of the storage mass is often decided by the physical laws governing natural heat flow by radiation and convection. Typical variations of this mass are given below:

(a) External building walls
(b) Internal walls
(c) Floor
(d) Free-standing masses inside the space.

In addition to storage location, there may be significant variation in storage materials providing thermal capacities and different time-lag properties. Storage materials may vary from concrete, brick, sand and ceramics, to water and other liquids either singly or in various combinations.

To add to the efficiency and usefulness of direct gain, several controls must be considered. To prevent unwanted heat gain, sunshading is required for the large expanse of south-facing glass. Due to the high location of the southern summer sun, overhangs can provide adequate protection for vertical southern glazing, but some other solutions must be found for tilted glazing, or those with east and west orientations faced with low sun angles.

To prevent unwanted heat loss, for example on sunless winter days and nights, movable insulation panels, shutters, etc., may be used. These will also prevent heat gain on hot summer days. Other controls include exhaust and ventilation arrangements.

18.7.2 Indirect Gain

In an indirect gain building, a storage mass collects heat directly from the sun, stores it, and then transfers it to the living space. The storage wall is placed just behind the glass as shown in Fig. 18.18. The basic requirements of such a wall, called the *mass Trombe wall*, is a large glass area with a large storage mass directly behind it to intercept the sun's rays.

Again, variations are possible in the material used and the thickness of wall, depending on which the radiant distribution from the storage mass to the living space can be either instantaneous or delayed. It is possible to have distribution by natural convection through openings at the top of the wall as shown in Fig. 18.19 in addition to radiation and convection directly from the wall. If the openings are controllable, natural convection can be stopped and started as and when desired. Also, insulation placed behind the storage mass can eliminate radiant distribution.

The following controls are possible for the optimum operation of a mass Trombe wall building.

(i) In winter, when the sky is overcast and during the night, external insulation must be provided to prevent undesirable heat loss.
(ii) Similarly in summer, during the day, overhangs or external insulation will serve the purpose of preventing overheating of the living space.

(iii) The mass Trombe wall has the potential of providing induced ventilation during summer if exhaust vents are provided at the top. For this purpose, inlet openings for fresh air must be provided from a shaded or cooler area, as shown in Fig. 18.19. Thus sun's heat can be used to induce air movement to augment natural ventilation.

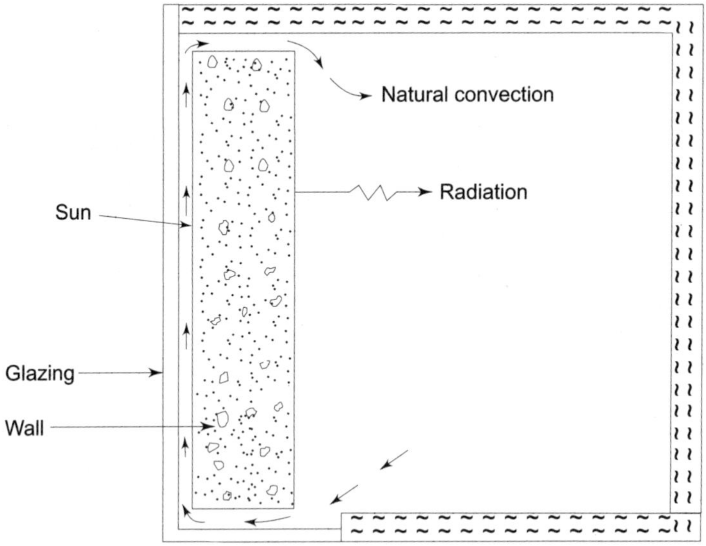

Fig. 18.18 Indirect gain principle illustrated by a mass trombe wall

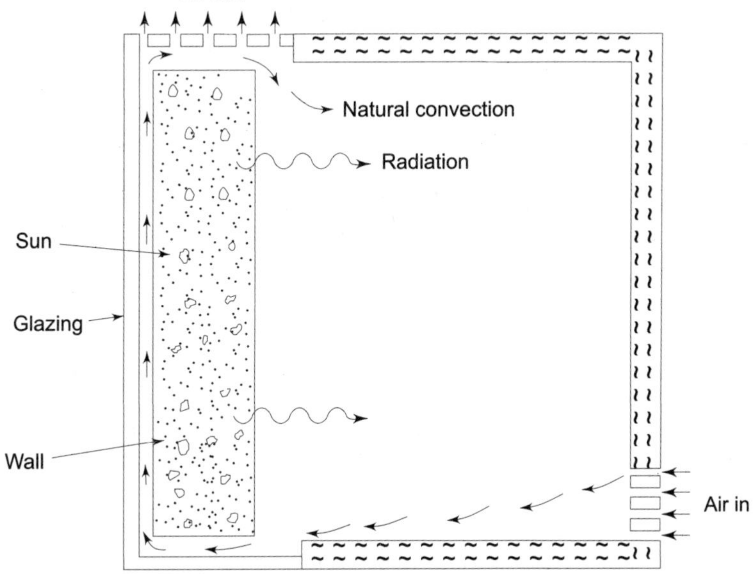

Fig. 18.19 Mass trombe wall with induced ventilation

The roof pond is another indirect gain-type construction in which water is the storage mass, and it is located at the roof instead of the floor or wall. The body of the water is directly exposed to the sun. The thermal storage is also provided by the ceiling mass. External insulation is necessary for control, as in the case of the mass Trombe wall, to cut off the day heat gain in summer and the night loss in winter.

A clear night sky acts as a large heat sink. Thus, the heat collected by water during the day and stored in its mass is dissipated to the sky during the night. During the day, therefore, the water mass acts as a *cold storage*, drawing the heat away from the living space. It is obvious that such a system will be effective in a climate with a large swing in day and night temperatures.

Direct evaporative cooling has already been discussed. There are other methods which use *Indirect* evaporative cooling. For example, a shaded roof pond also acts as an indirect evaporative cooler. The water is cooled to nearabout the wet bulb temperature of air. Thus, if the roof is completely shaded, there will be a loss of heat from the living space-which is at the dry bulb temperature of air—to the outside.

An underground construction provides another example of this type which is suitable both for passive solar cooling and heating. The ground temperature at a depth of about 10 m is uniform throughout the year, irrespective of the ambient conditions. This is due to the decrement in temperature oscillations as heat travels between the ground surface and the inside. Care must, however, be taken about the ground water level before deciding about the depth of construction. Also, a vapour barrier should be incorporated in the construction for preventing moisture transfer.

18.7.3 Isolated Gain

This is different from either direct or indirect gain concepts: In this, the collector and storage function somewhat independently from the building. An example is a solarium, in which solar radiation is collected in a secondary space and the living space is heated or cooled by convection. Also, evaporatively-cooled air can be used to cool large masses of rock beds at night, which act as cold storages. Room air can be circulated through these beds during the day for cooling.

18.8 WATER VAPOUR TRANSFER THROUGH STRUCTURES

As a result of vapour pressure difference existing on the two sides, water vapour flows across building walls and roofs, and across insulation over pipes; etc. This results in latent heat gain of rooms. The load is, however, quite insignificant, but there are other harmful effects as a result of condensation.

The process of vapour flow is similar to the flow of heat. Water vapour flows from high to low vapour pressure regions. As water vapour flows through the structure, its temperature also decreases. And, if at any section the local temperature falls below the dew point temperature, condensation occurs. This condensation is called *concealed condensation*.

Condensation not only causes deterioration of materials (such as steel rusting, wood decay, etc.) but also damages the insulation, and increases the thermal conductivity, thus, increasing cooling load in conditioned spaces, cold storages, frozen food storages, refrigerator cabinets, suction line and chilled water piping, supply air ducts, etc. Control of moisture transfer is, therefore, essential to ensure satisfactory operation in heating, ventilating and air-conditioning (HVAC), and refrigeration systems. In winter heating applications, condensation occurs with warm humid air inside the room and cold air outside, and in refrigeration applications, it is the outside air which is humid and inside air is at low temperature and low vapour pressure.

18.8.1 Vapour Transfer Equation

Vapour transfer rate $\dot{m}_v$ per unit area A is given by Fick's law Eq. (1.76), viz.,

$$\frac{\dot{m}_v}{A} = -\mu \frac{\mathrm{d}p_v}{\mathrm{d}x} \tag{18.43}$$

where μ is called *permeability*. It is like thermal conductivity in heat transfer. Integrating between outside and inside vapour pressures p_v and p_{v_0} and thickness L of construction, we get

$$\frac{\dot{m}_v}{A} = \mu \frac{p_{v_o} - p_{v_i}}{\Delta x} = \mu \frac{\Delta p_v}{\Delta x} \tag{18.43 a}$$

Sometimes, we may use the *permeance coefficient* designated as $M = \mu/\Delta x$. It is similar to heat-transfer coefficient. Thus, we have

$$\frac{\dot{m}_v}{A} = M \Delta p_v \tag{18.44}$$

The units of permeability μ will be

$$\left[\frac{\text{g}\cdot\text{cm}}{\text{hr}\cdot\text{m}^2\cdot\text{cm Hg pressure}}\right]$$

while the units of permeance coefficient M will be

$$\left[\frac{\text{g}}{\text{hr}\cdot\text{m}^2\cdot\text{cm Hg pressure}}\right]$$

The numerical values of μ and M for some common materials are given in Table 18.16.

Note *If condensation occurs, Fick's law and hence, the above relations are not applicable. After condensation, the water vapour pressure follows the saturation curve.*

Example 18.8 *A frozen food storage wall is exposed to the outdoor conditions of 39.5°C DBT and 28°C WBT in Chennai. The inside conditions are maintained at –18°C DBT and 80% RH. The wall has the following construction:*

Plaster (outside): 1.25 cm	*k = 8.65 W/mK, M = 4 g/hr.m^2.cmHg*
Brick : 22.5 cm	*k = 0.77 W/mK, μ = 2.2 g.cm/hr.m^2.cmHg*
Thermocole : 10 cm	*k = 0.037 W/mK, μ = 4.0 g.cm/hr.m^2 cmHg*

Check for possible condensation of moisture.

Solution See Fig. 18.20.

Table 18.16 Permeabilities and permeance coefficient of material

Material		*Permeability, μ* $\frac{\text{g}\cdot\text{cm}}{\text{hr}\cdot\text{m}^2\ \text{cm Hg}}$	*Permeance coefficient, M* $\frac{\text{g}}{\text{hr}\cdot\text{m}^2\cdot\text{cm Hg}}$
Construction materials			
Concrete		2.2	—
Brick (10 cm)		2.2	0.22
Asbestos cement board		—	0.15
Plaster (1.25 cm)		—	4.0
Gypsum		—	12.6
Wood		0.28–0.38	—
Insulating materials			
Still air		83	—
Mineral wool		80	—
Thermocole		1.38–4.0	—
Expanded			
Polyurethane		0.28–1.1	—
Foils			
Aluminium	(1 mil)	—	0
Polyethylene	(2 mil)	—	0.43
	(4 mil)	—	0.022
	(6 mil)	—	0.017

Psychrometric properties obtained by calculations are as follows:

	DB T	*WB T*	*RH*	p_s	p_v'	p_v
Condition	°C	°C	%	cm Hg	cm Hg	cm Hg
Outside	39.5	28	–	53.9	2.835	2.267 (Eq. 14.49)
Inside	–18	–	80	19.83	–	1.586

Overall heat-transfer coefficient assuming $f_o = 20$, and $f_i = 7$ W/m^2.K

$$\frac{1}{U} = \frac{1}{20} + \frac{0.0125}{8.65} + \frac{0.225}{0.77} + \frac{0.1}{0.037} + \frac{1}{7} = 3.189$$

$$\Rightarrow \quad U = 0.314 \text{ W/m}^2\text{.K}$$

Heat flux

$$q = U\Delta t = 0.314\,(39.5 + 18) = 18.055 \text{ W/m}^2$$

$$= f_o(t_o - t_1) = \frac{k_{12}}{\Delta x_{12}}(t_1 - t_2) = \frac{k_{23}}{\Delta x_{23}}(t_2 - t_3)$$

$$= \frac{k_{34}}{\Delta x_{34}}(t_3 - t_4) = f_i\,(t_4 - t_i)$$

Solving, we get the temperatures as shown in Fig. 18.20

$$t_1 = 38.6°\text{C},\ t_2 = 38.57°\text{C},\ t_3 = 33.3°\text{C},\ t_4 = -15.5°\text{C}$$

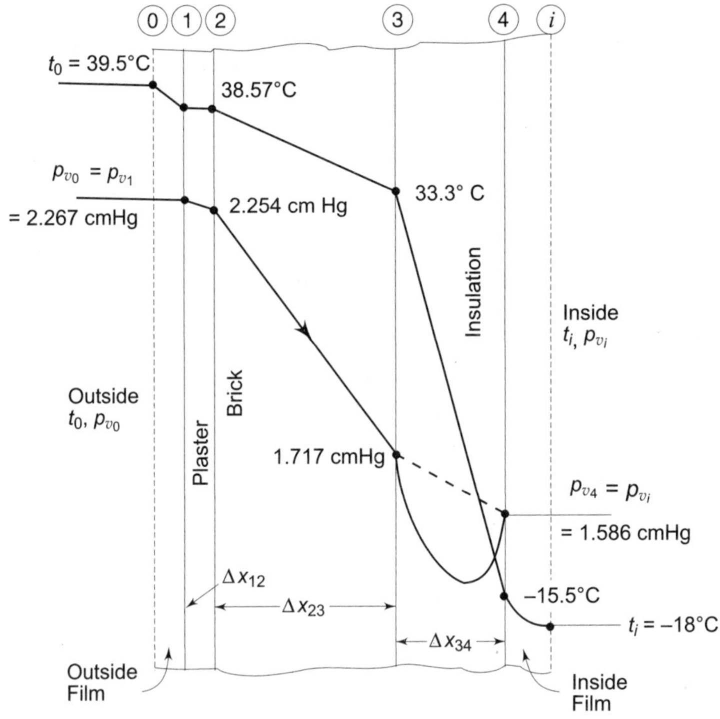

Fig. 18.20 Temperature and water vapour pressure distribution inside wall for Example 18.8

Similarly, we have for overall permeance coefficient

$$\frac{1}{M} = \frac{1}{M_{12}} + \frac{\Delta x_{23}}{\mu_{23}} + \frac{\Delta x_{34}}{\mu_{34}} = \frac{1}{4} + \frac{22.5}{2.2} + \frac{10}{4}$$

$$\Rightarrow \qquad M = 0.0771 \text{ g/hr.m}^2\text{.cm Hg}$$

Moisture transfer/mass flux (Given: $p_{v_1} = p_{v_0} = 2.267, p_{v_4} = p_{v_i} = 1.586$ cm Hg)

$$\frac{\dot{m}_v}{A} = M\,\Delta p_v = 0.0771\,(2.267 - 1.586) = 0.0525 \text{ g/hr.m}^2$$

$$= M_{12}\,(p_{v_1} - p_{v_2}) = \frac{\mu_{23}}{\Delta x_{23}}\,(p_{v_2} - p_{v_3}) = \frac{\mu_{34}}{\Delta x_{34}}\,(p_{v_3} - p_{v_4})$$

$$\Rightarrow \qquad p_{v_2} = 2.254 \text{ cm.Hg}, p_{v_3} = 1.717 \text{ cm.Hg}, p_{v_4} = 1.586 \text{ cm Hg} = p_{v_i}$$

The saturation pressures of water at t_3 and t_4 are:

$$(p_{v_s})_{t_3 = 33°\text{C}} = 3.84 \text{ cm Hg}, (p_{v_s})_{t_4 = -15.5°\text{C}} = 1.185 \text{ cm Hg}$$

We see that $p_{v_3} < (p_{v_s})_3$, but $p_{v_4} > (p_{v_s})_4$. Hence, condensation will occur between sections 3 and 4 inside insulation.

Note $p_{v_i} = p_{v_4} > (p_{v_s})_4$. *Hence, moisture will flow into insulation from inside room also and condense.*

18.8.2 Vapour Barriers

The migration of water vapour and its subsequent condensation poses a serious problem in summer, in cold storage structures where the outside vapour pressure is very high and inside vapour pressure is very low, and in winter, in buildings where the vapour pressure inside might be very high as compared to outside. *Water vapour barriers* are, therefore, applied to thermal insulation on cold pipes and refrigerated spaces.

The effectiveness of vapour barriers depends on its permeance which should be as low as possible. And it should be applied on the side of insulation exposed to higher vapour pressure.

The vapour barriers are of the following types:

(i) *Structural barriers*, e.g., plastics, aluminium sheets, etc.

(ii) *Membrane barriers*, e.g., aluminium foils, plastic films, etc. These are often bonded with insulation.

(iii) *Coating barriers* such as asphalts, resins, etc.

Note *The vapour barriers must be perfectly sealed, so that there is no leakage path for water vapour.*

In Example 18.8, vapour barrier must be applied at Section 3.

References

1. Alford J S, J E Ryan and F O, Urban 'Effect of heat storage and variation in outdoor temperature and solar intensity on heat transfer through walls', *ASHVE Trans.*, Vol. 45, 1939, pp. 393–397.
2. ASHRAE, *Handbook of Fundamentals*, 1972.
3. Carrier Air Conditioning Co., *Handbook of Air Conditioning System Design*, McGraw-Hill, New York, 1965.
4. Danter E, 'Periodic heat flow characteristics of simple walls and roofs', *J. IHVE*, Vol. 28, 1960, pp. 136–146.
5. Dusinberre G H, *Heat Transfer Calculations by Finite Differences*, International Text Book Co., Scranton, Pennsylvania, 1961.
6. Kadambi V and F W Hutchinson, *Refrigeration, Air Conditioning and Environmental Control in India*, Prentice-Hall of India, New Delhi, 1968.
7. Mackey C O and L T Sr Wright, 'Periodic heat flow-homogeneous walls or roofs', *ASHVE Trans.*, Vol. 50, 1944, p. 293.
8. Schmidt E, 'Uber die auwendung der differenzenrechung aug technische anheize and abkuhlungsprobleme', *Bietr. Tech. Mech. und Tech. Physik*, F'opple Festchrift, Berlin, 1924.
9. Threlkeld J L, *Thermal Environmental Engineering*, Prentice-Hall, Englewood Cliffs, N.J., 1962.

Revision Exercises

18.1 A composite wall is made up of 10 cm of common brick against 15 cm of concrete with 1.25 cm of plaster on the inside wall. Assume still air in room at 25°C. The outside air is at 40°C. The wind velocity is 25 kmph. Find:

(a) The thermal resistance of wall
(b) The overall heat-transfer coefficient
(c) The steady-state heat-transfer rate per unit area of wall
(d) The heat transfer if the outside and inside film resistances are disregarded.
Use Tables 18.1 and 18.2 for values of thermal conductivities and film coefficients.

18.2 A controlled temperature test room is held at 25°C. The outside design temperature is 40°C. The wall is constructed with 2.5 cm pine boards ($k = 0.12$ Wm^{-1} K^{-1}) on the inside and the outside with 10 cm mineral wool block insulation ($k = 0.05$ Wm^{-1} K^{-1}) in between. The pine boards are placed in position over the insulation by 1.875 cm steel through-bolts placed at 30 cm centres. Compute the overall heat-transfer coefficient:
(i) Not considering the presence of bolts.
(ii) Considering the presence of bolts.
The film coefficient on both sides of the wall is 9.4 Wm^{-2} K^{-1}.

18.3 A building wall consists of 25 cm concrete (k = 1.75 Wm^{-1} K^{-1}) and 1.9 cm plaster (k = 87 Wm^{-1} K^{-1}) on the inside surface. The outside and inside surface heat-transfer coefficients can be taken as 34 and 9.4 Wm^{-2}K^{-1} respectively. The outside temperature is –18°C. The room is held at 23.5°C DBT and 16.8°C WBT.
(a) What is the temperature on the inside wall surface?
(b) Will the moisture condense on the wall?
(c) How many layers of 1.25 cm thick fibre-board insulation (k = 0.048 Wm^{-1} K^{-1}) should be applied on the inside wall surface to prevent moisture to condense on it?

18.4 A 35 cm thick building wall is constructed from a material with a thermal conductivity of 0.87 Wm^{-1} K^{-1} and a thermal diffusivity of 0.564×10^{-6} m^2/s. The inside temperature of the building is maintained constant at 25°C, while the outside air temperature varies sinusoidally with time, the maximum being 42°C and the minimum 28°C. The period of this temperature variation is 24 hours. The film coefficients on the inside and outside surfaces are respectively 10 and 30 Wm^{-2} K^{-1}. Using the finite-difference method, compute the temperatures of the inside wall surface as a function of time, and find the time at which this temperature becomes a maximum. Also, find the maximum cooling load on the air conditioner per unit area of wall.

18.5 An air-conditioned room receives heat only from one wall which is 3 m wide and 3 m high and is facing west. The wall also has a glazed window 1.5 m × 1.5 m. The window is shaded internally by venetian blinds. Compute the heat gain of the room from the wall at 4 p.m. in June. The following information is given.

Outside state	: 42°C DBT,	26°C WBT
Inside state	: 25°C DBT,	50% RH
Solar intensity on glass	: 690 W/m^2	
U-value of wall	: 1.7 Wm^{-2} K^{-1}	

U-value of glass	: 5.7 $Wm^{-2} K^{-1}$
Transmissivity of glass	: 0.8
Absorptivity of glass	: 0.06
Outside wall surface	
Heat-transfer coefficient	: 30 $Wm^{-2}K^{-1}$
Absorptivity of wall surface	: 0.9
Time lag for wall	: 5.7 hours
Decrement factor for wall	: 0.62
Mean sol-air temperature over 24 hours	: 40°C

Variation of outside conditions:

Sun time, hours	9	10	11	Noon	13	14	15	16
Ambient temperature, °C	33	35	37	39	40	41	42	43
Solar intensity, W/m^2	75	75	75	80	240	340	420	420

18.6 A building has 28 double-hung 1.5 m × 1 m wood-sash non-weatherstripped windows of average construction. The windows are evenly distributed on all the four walls. In addition, there are four 2 m × 1 m poorly fitting doors. The wind velocity is 24 kmph NW. What are the sensible and latent heat gains of the building due to infiltration? The outside conditions are 40°C DBT, 27°C WBT, and the inside conditions are 24°C DBT, 50%RH. Also find the amount of moisture to be evaporated per hour to maintain the inside relative humidity at 50 per cent.

Chapter

19

Load Calculations and Applied Psychrometrics

19.1 PRELIMINARY CONSIDERATIONS

The importance of accurate load calculations for air-conditioning design and selection of equipment can never be overemphasized. In fact, it is on the precision and care exercised by the designer in the calculations of the cooling load for summer and the heating load for winter that a trouble-free successful operation of an air-conditioning plant, after installation, would depend.

An important consideration in this exercise is the date and time for which these calculations are made. The date would depend on the local climatic conditions. Although the longest day in summer is June 21, the hottest and most humid day may occur in July. Similarly, the coldest day may occur in January or even in February instead of December 21. Again, though the maximum temperature may occur outside at 1 or 2 p.m., the maximum heat gain of the room may occur at 3 or 4 p.m. due to the direct solar radiation through glass on the west side, or even later due to the time lag for the heat transfer through the structure. Further, the application for which the building is intended to be used would also govern the choice of time. For example, for an office building in winter that is not used at night, the time for load calculations may be taken during the early hours of the morning, although the maximum heating load may occur at night. Similarly, an office building in summer may have the maximum cooling load at 7 p.m. due to the time lag, but since no occupants would be present at that time, the time for load calculations may be taken as 4 or 5 p.m.

The major components of load in buildings are due to the direct solar radiation through the west glass, transmission through the building fabric or structure and fresh air for ventilation. In the case of applications such as theatres and auditoriums, the occupancy load is predominant.

A detailed discussion of the solar radiation incident on a surface and its transmission through glass has been given in Chapter 17. Further, in Chapter 18, we have studied the methods of calculating heat transmission and infiltration through structures. These form the components of load on the building from the external *environment*. The internal and system heat gains or losses also form the major components of other loads.

Thus, components that may cause colling loads include the followimg:
External : Walls, roof, windows, partitions, ceiling, and floor
Internal : Lights, people (occupancy), appliances, and equipment
Infiltration : Air leakage and moisture migration
System:Outside air (ventilation air), duct gain, reheat, fan and pump energy.

In this chapter, the methods for the evaluation of the above mentioned and other individual loads are presented, followed by a summary of all loads at the end along with an example and a calculation sheet illustrating the procedure that is followed by practising engineers. In the first instance here, the cooling load estimation is given followed by that of the heating load.

19.2 INTERNAL HEAT GAINS

The sensible and latent heat gains due to occupants, lights, appliances, machines, piping, etc., within the conditional space, form the components of the *internal heat gains*.

19.2.1 Occupancy Load

The occupants in a conditioned space give out heat at a metabolic rate that more or less depends on their rate of working. The relative proportion of the sensible and latent heats given out, however, depends on the ambient dry bulb temperature. The lower the dry bulb temperature, the greater the heat given out as sensible heat.

Typical values of heat given out are given in Table 19.1. The values for restaurants include the heat given out by food as well. It will be seen that the sensible heat (S) gain does not vary much with activity, more and more heat being liberated as latent heat (L) thus making up for total heat.

Table 19.1 Heat liberated due to occupancy

Activity	*Metabolic Rate* W	*Heat Liberated,* W *Room Dry Bulb Temperature,* °C							
		20		22		24		26	
		S	*L*	*S*	*L*	*S*	*L*	*S*	*L*
Seated at rest	115	90	25	80	35	75	40	65	50
Office work	140	100	40	90	50	80	60	70	70
Standing	150	105	45	95	55	82	68	72	78
Eating in restaurant	160	110	50	100	60	85	75	75	85
Light work in factory	235	130	105	115	120	100	135	80	155
Dancing	265	140	125	125	140	105	160	90	175

The usual problem in calculating the occupancy load lies in the estimation of the exact number of people present.

19.2.2 Lighting Load

Electric lights generate sensible heat equal to the amount of the electric power consumed. Most of the energy is liberated as heat, and the rest as light which also eventually becomes heat after multiple reflections.

Lighting manufacturers give some guidance as to the requirement of power for different fittings to produce varying standards of illumination. In connection with fluorescent tubes, it may be stated that the electric power absorbed at the fitting is about 25 per cent more than necessary to produce the required lighting. Thus, a 60 W tube will need 75 W at the fitting. The excess of 15 W is liberated at the control gear of the fitting.

As a rough calculation, one may use the lighting load equal to 33.5 W/m^2 to produce a lighting standard of 540 lumens/m^2 in an office space; 20 W/m^2 is minimum. After the wattage is known, the calculation of the heat gain is done as follows:

$$\text{Fluorescent: } Q = \text{Total watts} \times 1.25$$

$$\text{Incandescent: } Q = \text{Total watts}$$

19.2.3 Appliances Load

Most appliances contribute both sensible and latent heats. The latent heat produced depends on the function the appliances perform, such as drying, cooking, etc. Gas appliances produce additional moisture as product of combustion. Such loads can be considerably reduced by providing properly designed *hoods* with a positive exhaust system or suction over the appliances.

The appliances in the conditioned space are a common feature in cafetarias. Table 19.2 gives a general guidance about the sensible and latent heat loads of un-hooded common restaurant appliances.

Table 19.2 Appliance load, W

Appliance	*Sensible*	*Latent*	*Total*
Coffee brewer $\frac{1}{2}$ gal	265	65	329
Coffee warmer $\frac{1}{2}$ gal	71	27	98
Egg boiler	353	235	60
Food warmer/m^2 of plate	1150	1150	2300
Griddle frying with frying top of 46 cm × 36 cm	912	500	1412
Toaster, 360 slices/h	1500	382	1882

Electric motors contribute sensible heat to the conditioned space. A part of the power input is directly converted into heat due to the inefficiency of the motor and is dissipated through the frame of the motor. This power is

$$(\text{Input}) (1 - \text{Motor efficiency})$$

The rest of the power input is utilized by the driven mechanism for doing work which may or may not result in heat gain to the space. This depends on whether the energy input goes to the conditioned space or outside it.

19.2.4 Piping, Tanks, Evaporation of Water from a Free Surface and Steam

Heat is added to the conditioned space from running pipes carrying hot fluids due to heat transfer. On the other hand, cold pipes take away heat from the space. Open tanks containing warm water contribute both sensible heat and latent heat to the space due to evaporation. This can be calculated by knowing the rate of evaporation and energy balance.

In industrial air conditioning, products have often to be dried. This involves both the latent heat gain and the sensible heat gain to the space from the hot surfaces of the dryer depending upon the drying rate. For these calculations, a knowledge of the heat and mass transfer coefficients is essential.

When steam is entering the conditioned space, the sensible heat gain is very little. It is equal to only the difference in the enthalpy of steam at the steam temperature and the enthalpy of water vapour at the room dry-bulb temperature. The main load is in the form of the latent heat gain. Thus

$$\text{SHG} = (\text{kg/s})\,(t_{\text{steam}} - t_i)\,(1.88)\ \text{kW} \tag{19.1}$$

$$\text{LHG} = (\text{kg/s})\,(2500)\ \text{kW} \tag{19.2}$$

19.2.5 Product Load

In the case of cold storages, the enclosures are insulated with at least 10-15 cm of thermocole and are almost completely sealed. Thus, many of the loads present in buildings for comfort air conditioning are either absent or lessened in the case of cold storages. However, in addition to the heat which is removed from products at the time of initial loading, there is also the heat produced by the commodities during storage. This *heat of respiration* forms a sizable product load even at a storage temperature of 0°C. At higher temperatures, it is more. The approximate rate of evolution of heat by various products at different temperatures is given in Table 19.3.

Table 19.3 Heat of respiration of products in J/kg per 24 hours

Product	***Storage Temperature***		
	0°C	**4.4°C**	**15.6°C**
Apples	312–1560	625–2810	2390–8215
Bananas	—	—	—
Cabbage	1248	1770	4265
Carrots	2183	3640	8420
Cauliflower	—	4680	10500
Cherries	1352–1871	—	11440–13725
Cucumbers	—	—	2290–6860
Grape fruit	416–1040	730–1350	2290–4160
Grapes, American	624	1250	3640
Grapes, European	312–416	—	2290–2705
Lemons	520–936	625–1975	2390–5200
Melons	1350	2080	8840
Mushrooms	6446	—	—
Onions	728–1144	830	2495
Oranges	416–1040	1350–1665	3850–5405
Peaches	936–1456	1455–2080	7590–9670
Pears	728–936	—	9150–13,725

(Contd.)

Table 19.3 (*Contd*)

Product	***Storage Temperature***		
	0°C	**4.4°C**	**15.6°C**
Peas	8526–8733	13,520–16,635	40,860–46,265
Plums	416–728	935–1560	2495–2910
Potatoes, immature	—	2705	3015–7070
Potatoes, mature	—	1350–1870	1560–2705
Strawberries	2807–3950	3745–7070	16,220–21,105
Tomatoes, green	625	1145	6445
Tomatoes, ripe	1040	1350	5820
Turnips	1975	2290	5510

19.2.6 Process Load

The procedure of calculating the cooling and heating load for various industrial air-conditioning processes is specific for each process. The requirements for the process may involve the control of one or more of the following factors:

(i) Regain of moisture content by hygroscopic materials, such as cotton silk, tobacco, etc., and the accompanying heat liberated.
(ii) Drying load.
(iii) Rate of chemical and biochemical reactions.
(iv) Rate of crystallization, freezing, freeze-drying, etc.
(v) Sensible cooling load.

For details of these loads, one may refer to the ASHRAE Handbook.[1]

19.3 SYSTEM HEAT GAINS

The system heat gain is the heat gain (or loss) of an air-conditioning system comprising its components, viz., ducts, piping, air-conditioning fan, pumps, etc. This heat gain is to be initially estimated and included in the total heat load for the air-conditioning plant. The same should be checked after the whole plant has been designed.

19.3.1 Supply Air Duct Heat Gain and Leakage Loss

The supply air, normally, has a temperature of 10 to 15°C. The duct may pass through an unconditioned space having an ambient temperature of 40°C. This results in a significant heat gain till the air reaches the conditioned space even though the duct may be insulated.

The heat gain can be calculated using the following expression

$$Q = UA(t_a - t_s) \tag{19.3}$$

where U is the overall heat-transfer coefficient, A is the surface area of the duct system exposed to the ambient temperature t_a and t_s is the supply air temperature.

As a rough estimate, a value of the order of 5 per cent of the room sensible heat may be added to the total sensible heat if the whole supply duct is outside the conditioned space, and proportionately less if some of it is within the conditioned space.

It has been found that duct leakages are of the order of 5 to 30 per cent depending on the workmanship. Air leakages from supply ducts result in a serious loss of the cooling capacity unless the leakages take place within the conditioned space.

If all ducts are outside the conditioned space which, normally, is strctly avoided, a 10 per cent leakage is to be assumed which should be considered as a complete loss. When only a part of the supply duct is outside the conditioned space, then only the leakage loss of this portion is to be included. The fraction of 10 per cent, to be added in such a case, is equal to the ratio of the length outside the conditioned space to the total length of the supply duct.

19.3.2 Heat Gain from Air-Conditioning Fan

The heat equivalent of an air-conditioning fan horsepower is added as the sensible heat to the system. If the fan motor is outside the air stream, the energy lost due to the inefficiency of the motor is not added to the air. There are two types of air supply systems.

Draw-through System In the draw-through *system,* the fan is drawing air through the cooling coil and supplying it to the conditioned space. This is the most common system. In this system, the fan heat is in addition to the supply air heat gain. The heat should, therefore, be added to the room sensible heat.

Blow-through System In the blow-through *system*, fan blows air through the cooling coil before being supplied to the conditioned space. In this system, the fan heat is added after the room to the return air. Thus the fan heat is a load on the cooling coil. The heat should, therefore, be added to the grand total heat.

The fan efficiencies are of the order of 70 per cent for central air-conditioning plant fans and about 50 per cent for package air-conditioner fans.

The fan horsepower depends on the quantity of air supplied and the pressure rise, viz., the total pressure developed by the fan. The supply air quantity in turn depends on the dehumidified rise, which is of the order of 8 to 14°C. The fan total pressure depends on the system pressure loss which comprises the pressure drop through the duct-work, grilles, filters, cooling coil, etc. The approximate values of pressures to be developed by fans are given in Table 19.4.

Table 19.4 Fan pressure for different duct systems

Sl. No.	*Type of Ductwork*	*Fan Pressure, cm* H_2O
1.	No ductwork (package units)	1.25 to 2.5
2.	Moderate ductwork, low velocity systems	1.9 to 3.75
3.	Considerable ductwork, low velocity systems	3.0 to 5.0
4.	Moderate ductwork, high velocity systems	5.0 to 10.0
5.	Considerable dutwork, high velocity systems	7.5 to 15.0

Once the supply air-rate and pressure developed are known, the fan power can be calculated. But these cannot be known until the load calculations have been completed. Hence the procedure is to initially assume fan heat between 2.5 and 7.5 per cent of the room sensible heat and check the value after the design has been completed. Designers usually take 5% of RSH as fan heat.

19.3.3 Return Air Duct Heat and Leakage Gain

The calculation of the heat gain for return air ducts is done in exactly the same way as for supply air ducts. But the leakage in this case is that of the hot and humid outside air into the duct because of suction within the duct. If the ducts are outside the conditioned space, an inleakage up to 3 per cent may be assumed depending on the length of the duct. If there is only a short connection between the conditioning equipment and the space, this leakage may be neglected.

19.3.4 Heat Gain from Dehumidifier Pump and Piping

The horsepower required to pump water through the dehumidifier adds heat to the system and is to be considered like that of other electric motors.

For this purpose, pump efficiencies may be assumed as 50 per cent for small pumps and 70 per cent for large pumps.

The heat gain of dehumidifier piping may be calculated as a percentage of the grand total heat as follows:

(i) Very little external piping: 1% of GTH.
(ii) Average external piping: 2% of GTH.
(iii) Extensive external piping: 4% of GTH.

Note *It is to be noted that all heat gains after the room are not to be added to room heat gains, but to the grand total heat load that directly falls on the conditioning equipment. These include the return air duct heat and leakage gain, dehumidifier pump power, dehumidifier and piping losses, as outlined above and the fan sensible heat in the case of the blow-through system.*

19.3.5 Safety Factor

Safety factor is strictly a factor of probable error in the estimation of the load. For the purpose, additional 5 per cent heat should be added to the room sensible and latent heats.

19.4 BREAK-UP OF VENTILATION LOAD AND EFFECTIVE SENSIBLE HEAT FACTOR

The ventilation-air requirements, depending on individual applications are given in Table 16.2. The minimum requirement is taken as 0.2 cmm per person. This is based on a population density of 5 to 7.5 m^2 per person and a ceiling height of 2.4 m. When people are smoking, the minimum ventilation requirement is 0.4 to 0.7 cmm per person.

In Sec. 15.5.1, we have seen the calculation for the load due to ventilation air. It was also pointed out in Sec. 15.5.2, that the bypass factor of the cooling equipment affects the position of the grand sensible heat factor line. As a matter of fact, the effect of the bypass factor is such as to add (X) (m_{a_o}) amount of the outside air directly to the room, and to allow only $(1 - X)$ (m_{a_o}) to pass through the apparatus. Although the room air is also bypassed, this does not affect the break-up of the load as the room air going to the room does not change the load situation. Thus, we can say that a part of the ventilation load forms a component of the room load. This bypassed outside air load is proportional to the bypass factor X. It has both sensible and latent heat components. The other part-which is proportional to $1 - X$, both sensible and latent, which is bypassed around the apparatus—is added to the equipment load.

Thus the bypassed outside air loads on the room are:

$$\text{SH} = (\text{OASH})\,(\text{BPF}) \tag{19.4}$$

$$\text{LH} = (\text{OALH})\,(\text{BPF}) \tag{19.5}$$

These loads are imposed on the room in exactly the same manner as the infiltration load. Accordingly, the effective room loads are modified as follows:

Effective room sensible heat

$$\text{ERSH} = \text{RSH} + (\text{OASH})\,(\text{BPF}) \tag{19.6}$$

Effective room latent heat

$$\text{ERLH} = \text{RLH} + (\text{OALH})\,(\text{PBF}) \tag{19.7}$$

The *effective sensible heat factor* (ESHF) is the ratio of the effective room sensible heat to the effective room total heat

$$\text{ESHF} = \frac{\text{ERSH}}{\text{ERSH} + \text{ERLH}} = \frac{\text{ERSH}}{\text{ERTH}}$$

The term is devised to relate the bypass factor, apparatus dew point, room sensible and latent heats and bypassed outside air loads and to permit a simplified calculation of the dehumidified air quantity as in Sec. 19.7.

The remaining outside air loads which are proportional to 1 – BPF are accounted in the grand total heat which remains unaffected by the bypass factor.

19.5 COOLING LOAD ESTIMATE

The components of the cooling load for air-conditioning can now be summarized as follows. The load is classified as the room load that falls on the room directly, and the total load that falls on the apparatus.

Room Load A. Room Sensible Heat (RSH)

(i) Solar and transmission heat gain through walls, roof, etc.

(ii) Solar and transmission heat gain through glass.

(iii) Transmission gain through partition walls, ceiling, floor, etc.

(iv) Infiltration.

(v) Internal heat gain from people, power, lights, appliances, etc.

(vi) Additional heat gain not accounted above, safety factor, etc.

(vii) Supply duct heat gain, supply duct leakage loss and fan horsepower.

The sum of all the above gives the *room sensible heat* (RSH) load. For the purpose of psychrometric analysis, the following component is also included in the room sensible heat.

(viii) Bypassed outside air load.

The sum of items (i) to (viii) gives the *effective room sensible heat* (ERSH).

B. Room Latent Heat (RLH)

(i) Infiltration.
(ii) Internal heat gain from people, steam, appliances, etc.
(iii) Vapour transmission.
(iv) Additional heat gain not accounted above, safety factor, etc.
(v) Supply duct leakage loss.
The sum of these gives the *room latent heat* (RLH). The other latent heat gain considered for psychrometric analysis is:
(vi) Bypass outside air load.

The sum of items (i) to (vi) above gives the *effective room latent heat* (ERLH).

The sum of items A and B above gives the *effective room total heat* (ERTH). But the sum of items (i) to (vii) in A, and (i) to (v) in B gives the *room total heat* (RTH).

Grand Total Load on Air-Conditioning Apparatus A. Sensible Heat

(i) Effective room sensible heat (ERSH).
(ii) Sensible heat of the outside air that is not bypassed.
(iii) Return duct heat gain, return duct leakage gain, dehumidifier pump horse-power and dehumidifier and piping losses.

The sum of items (i) to (iii) gives the *total sensible heat* (TSH).

B. Latent Heat

(i) Effective room latent heat (ERLH).
(ii) Latent heat of outside air which is not bypassed.
(iii) Return duct leakage gain.

The sum of items (i) to (iii) above gives the *total latent heat* (TLH). Finally, the sum of A and B above gives the *grand total heat* (GTH).

19.5.1 Heat Balance (HB) Method

Heat balance method is a rigourous approch to cooling load calculations. It requires the use of computers. The method is described in Ashrae Transactions pp. 459–502.

The HB methos considers four distinct processes.

1. Outside-face heat balance.
2. Wall conduction process.
3. Inside-face heat balance.
4. Room-air heat balance.

The heat balance on outside surface considers solar radiation, convective exchange, and conductive heat flux. Numerical methods are used for wall conduction process. The heat balance on inside surface considers heat fluxes from wall to air. Air heat balance takes into account internal loads, infiltration and ventilation air, system loads. etc.

19.6 HEATING LOAD ESTIMATE

An estimate of the heating load is made on the basis of the *maximum probable heat* loss of the room or space to be heated. Thus the plant for the heating system is to be so designed that it has a capacity just sufficient to meet the heating load requirement which develops when most severe weather conditions occur. In this respect very brief periods of severe weather need not be taken into account.

Accordingly, the following points in heating-load calculations are noteworthy.

(i) Transmission Heat Loss: The transmission heat loss from walls, roof, etc., is calculated on the basis of just the design outside and inside air temperature difference.

$$Q = UA\,(t_i - t_o) \tag{19.8}$$

Thus, no allowances need be made for the walls and roof being exposed to sun. Also, the time lag and decrement factor are to be ignored.

(ii) Solar Radiation: There is generally no solar radiation present and hence, there is no solar heat gain at the time of the peak load which normally occurs in the early hours of the morning.

(iii) Internal Heat Gains: Internal heat gains from occupants, lights, motors and machinery, etc., diminish the heating requirement. These negative loads should be accounted for in applications, such as theatres, assembly halls, stores, office buildings, etc., where these loads are constantly present. But allowance for these loads must be made only after careful consideration. An important aspect to keep in mind is the use of the space at night, weekends or other unoccupied periods.

Also, the heating plant should have sufficient capacity to bring up the inside temperature to the design value before the occupants come in.

Example 19.1 *For the air conditioning of a 4 m high single-story office building located at 30°N latitude, the plan of which is shown in Fig. 19.1, the following data are given.*

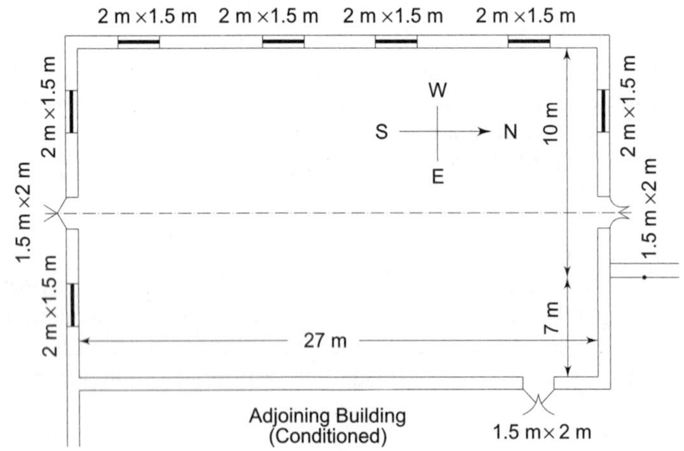

Fig. 19.1 Plan of building for Example 19.1

Plaster on inside wall	: $1\frac{1}{4}$ *cm*
Outside wall construction	: *20 cm concrete block*
	: *10 cm brick veneer*
Partition wall construction	: *33 cm brick*

Roof construction	*: 20 cm RCC slab with 4 cm asbestos cement board*
Floor construction	*: 20 cm concrete*
Densities, brick	*: 2000 kg/m³*
concrete	*: 1900 kg/m³*
plaster	*: 1885 kg/m³*
asbestos board	*: 520 kg/m³*
Fenestration (Weather-stripped, Loose fit)	*: $2 \text{ m} \times 1\frac{1}{2}$ m glass* $U = 5.9\ Wm^{-2}\ K^{-1}$
Doors	*: $1\frac{1}{2}$ m × 2 m wood panels* $U = 0.63\ Wm^{-2}\ K^{-1}$
Outdoor-design conditions	*: 43°C DBT, 27°C WBT*
Indoor-design conditions	*: 25°C DBT, 50% RH*
Daily range	*: 31°C to 43°C = 12°C*
Occupancy	*: 100*
Lights	*: 15,000 W fluorescent*

Assumed bypass factor of cooling coil 0.15

Find the room sensible and latent heat loads, and also the grand total heat load.

Solution Thermal conductivities from Table 18.1

$$k_{\text{glass}} = 0.78 \text{ W m}^{-1} \text{ K}^{-1}$$
$$k_{\text{concrete}} = 1.73 \text{ W m}^{-1} \text{ K}^{-1}$$
$$k_{\text{brick}} = 1.32 \text{ W m}^{-1} \text{ K}^{-1}$$
$$k_{\text{plaster}} = 8.65 \text{ W m}^{-1} \text{ K}^{-1}$$
$$k_{\text{asbestos}} = 0.154 \text{ W m}^{-1} \text{ K}^{-1}$$

Assumed film coefficients

$$f_0 = 23 \text{ W m}^{-2} \text{K}^{-1}$$
$$f_i = 7 \text{ W m}^{-2} \text{ K}^{-1}$$

Outside wall
$$\frac{1}{U} = \frac{1}{23} + \frac{0.1}{1.32} + \frac{0.2}{1.73} + \frac{1}{7} + \frac{0.0125}{8.65}$$
$$U = 3.5 \text{ W m}^{-2} \text{K}^{-1}$$

Partition wall
$$\frac{1}{U} = \frac{1}{7} + \frac{0.33}{1.32} + \frac{1}{7} + \frac{2(0.0125)}{8.65}$$
$$U = 1.86 \text{ W m}^{-2} \text{K}^{-1}$$

Roof
$$\frac{1}{U} = \frac{1}{23} + \frac{0.2}{9} + \frac{0.04}{0.154} + \frac{0.0125}{8.65} + \frac{1}{7}$$
$$U = 2.13 \text{ W m}^{-2} \text{K}^{-1}$$

Floor
$$\frac{1}{U} = \frac{1}{7} + \frac{0.2}{9}$$
$$U = 6.05 \text{ W m}^{-2} \text{ K}^{-1}$$

Area and volume of space $A = (27)(17) = 459 \text{ m}^2$

$V = (459)(4) = 1836 \text{ m}^3$

Ventilation rate for office

$$\dot{Q}_{v_o}/\text{person} = 0.28 \text{ cmm (from Table 16.2)}$$

$$\dot{Q}_{v_o} = 0.28\,(100) = 28 \text{ cmm}$$

$$\text{Number of air-changes of ventilation air} = \frac{(28)(60)}{1836} = 0.92 \text{ (Satisfactory)}$$

Mass of wall per unit area

$$\text{Outside wall: } 0.2\,(1900) + 0.1\,(2000) + 0.0125\,(1885) = 604 \text{ kg/m}^2$$

$$\text{Partition wall: } 0.33\,(2000) + 2(0.0125)\,(1885) = 707 \text{ kg/m}^2$$

$$\text{Roof: } 0.2\,(1900) + 0.04\,(520) = 401 \text{ kg/m}^2$$

Correction for equivalent temperature differentials

$$\text{For daily range of } 12°\text{C} = \frac{12 - 11.1}{2} = 0.45°\text{C}$$

$$\text{For } (t_o - t_i) \text{ of } 18°\text{C} = 18 - 8.3 = 9.7°\text{C}$$

$$\text{Total correction} = -\,0.45 + 9.7 = 9.25°\text{C}$$

Equivalent temperature differentials in °C, from Tables 18.9 and 18.10, and incorporating corrections:

	2 p.m.	**3 p.m.**	**4 p.m.**	**5 p.m.**	**6 p.m.**	**7 p.m.**
West wall	14.4	14.8	15.2	16.5	17.5	
North wall	9.6	10.2	9.6	11.3	11.7	
South wall	13.1	14.7	16.0	17.4	17.8	
Roof (exposed)	24.0	25.8	28.0	29.7	30.5	30.2

Rates of solar gains through glass on June 21 in W/m^2 from Table 17.8(d)

	2 p.m.	**3 p.m.**	**4 p.m.**	**5 p.m.**
West glass	309	451	508	492
North glass	44	44	51	91
South glass	47	44	38	32

$$\text{Door area} = 1\frac{1}{2} \times 2 = 3 \text{ m}^2$$

Glass areas

$$\text{West glass} = 4\left(2 \times 1\frac{1}{2}\right) = 12 \text{ m}^2$$

$$\text{North glass} = 2 \times 1\frac{1}{2} = 3 \text{ m}^2$$

$$\text{South glass} = 2\left(2 \times 1\frac{1}{2}\right) = 6 \text{ m}^2$$

Outside wall areas

$$\text{West wall} = (27)(4) - 12 = 96 \text{ m}^2$$

$$\text{North wall} = (10)(4) - 3 - 3 = 34 \text{ m}^2$$

$$\text{South wall} = (17)(4) - 3 - 6 = 59 \text{ m}^2$$

Partition wall areas

$$\text{East wall} = (27)(4) - 3 = 105 \text{ m}^2$$

$$\text{North wall} = (7)(4) = 28 \text{ m}^2$$

Estimated time of maximum cooling load:

From the above calculations, it is obvious that the major components of the variable cooling loads are solar and transmission heat gains through the west wall and glass, and the roof. Of these, glass and roof loads are the predominant loads. The roof load is maximum at 6 p.m. when the equivalent temperature differential is 30.5°C. The solar gain through the west glass has a maximum value of 508 W/m^2 at 4 p.m. Thus the time of maximum load is most likely to be near 5 p.m. Heat transfer through floor:

Assume a temperature difference of 2.5°C across the floor

Wind pressure

Assuming a wind velocity of 15 kmph, we have

$$\Delta p = 0.00047\ (15)^2 = 0.11 \text{ cm } H_2O$$

Infiltration rate for windows, from Table 18.11 for 0.11 cm wind pressure

$$= 2.5 \text{ m}^3/\text{h/m crack}$$

$$\text{Length of crack for 7 windows} = 7\left[2\left(2+1\frac{1}{2}\right)\right] = 49 \text{ m}$$

Occupancy load, from Table 19.1

$$\text{SHL} = 75 \text{ W/person}$$
$$\text{LHL} = 55 \text{ W/person}$$

Other assumptions

(i) Only 10% of the supply duct outside the conditioned space.
(ii) No return duct outside the conditioned space.
(iii) Fan horsepower, 5 per cent of RSH.

The details of cooling load calculations are given on the calculation sheet in Table 19.5.

Table 19.5 Calculation sheet for cooling load estimation for Example 19.1

SPACE USED FOR OFFICE SIZE 27 × 17 = 459 m^2 × 4 = 1836 m^2						
ESTIMATE FOR 5 p.m. LOCAL TIME HOURS OF OPERATION DAY TIME						
Conditions	DB	WB	%RH	DP	h, kJ/kg	ω, kg/kg
OUTDOOR	43	27	29	21.3	85.0	0.016
ROOM	25	18	50	15.7	50.85	0.01
DIFFERENCE	18				34.15	0.006
	100 PEOPLE × 0.28 cmm/PERSON = 28 cmm VENTILATION cmm = 28					
DOORS CRACK	3 DOORS × 3m^2 1.9813 cmm/m^2 = 17.8 cmm 49 m × 2.5/60 cmm/m = 2.0 cm INFILTRATION cmm = 17.8 + 2 = 19.8					

(Contd)

SENSIBLE HEAT				
		SOLAR GAIN-GLASS		
EAST GLASS	m^2	—	—	—
WEST GLASS	12 m^2	492	—	5,900
NORTH GLASS	3 m^2	91	—	270
SOUTH GLASS	6 m^2	32	—	190
SKY LIGHT	–m^2	—	—	—

LOAD CALCULATIONS

ITEM	AREA OR QUANTITY	SUN GAIN OR TEMP. DIFF. OR HUMIDITY DIFF.	FACTOR	W
	SOLAR TRANSMISSION GAIN-WALLS AND ROOF			
EAST WALL	–m^2	—	—	—
WEST WALL	96 m^2	16.5	3.5	5,540
NORTH WALL	34 m^2	11.3	3.5	1,345
SOUTH WALL	59 m^2	17.4	3.5	3,590
ROOF-SUN	459 m^2	29.7	2.13	29,035
ROOF-SHADED	–m^2	—	—	—
	TRANSMISSION GAIN-OTHERS			
DOORS	9 m^2	18	0.63	100
ALL GLASS	(12+3+6) m^2	18	5.9	2,230
PARTITION	(108+28) m^2	15.5	1.86	3,930
CEILING	–m^2	—	—	—
FLOOR	459 m^2	2.5	6.05	6,940
INFILTRATION	19.8 cmm	18	20.4	7,270
	INTERNAL HEAT GAIN			
PEOPLE	100	—	75	7,500
POWER	—	—	—	—
LIGHTS	15,000	—	1.25	18,750
APPLIANCES	—	—	—	—
ADDITIONAL	—	—	—	—
			SUB TOTAL	92,690
STORAGE (Neglected)	—	—	—	—
SAFETY FACTOR			5%	4,635
			ROOM SENSIBLE HEAT	103,090
		SUPPLY DUCT		
SUPPLY DUCT HEAT GAIN	0.5% + LEAKAGE	0.5% + Fan HP	5%	5,560
OUTDOOR AIR				

(Contd)

BYPASSED	28 cmm	18°C	20.4 × 0.15	1,540
	EFFECTIVE ROOM SENSIBLE HEAT			104,425
LATENT HEAT				
INFILTRATION	19.8 cmm	0.006	50,000	5,940
PEOPLE	100	—	55	5,500
STEAM	—	—	—	—
APPLIANCES	—	—	—	—
ADDITIONAL	—	—	—	—
VAPOUR TRANS	—	—	—	—
			SUB TOTAL	11,440
SAFETY FACTOR		5%		570
ROOM LATENT HEAT				12,010
SUPPLY DUCT LEAKAGE LOSS			0.5%	60
OUTDOOR AIR BYPASSED	28 cmm	0.006	50,000 × 0.15	1,260
	EFFECTIVE ROOM LATENT HEAT			13,330
	EFFECTIVE ROOM TOTAL HEAT			117,755
	OUTDOOR AIR TOTAL HEAT (On equipment)			
SENSIBLE	28 cmm	18	20.4 × (1 – 0.15)	8,740
LATENT	28 cmm	0.006	50,000 × (1 – 0.15)	7,140
RETURN DUCT HEAT GAIN	% + RETURN DUCT LEAKAGE GAIN	% + PUMP	% + DEUH. % PIPE GAIN	—
GRAND TOTAL HEAT				133,635 (38TR)

Note *Many designers do not take into account the infiltration load separately. It is considered to be taken care of by ventilation air if the ventilation cmm is greater than infiltration cmm. One such simplified load estimation calculation sheet for the ground, first and third floors of a television studio building, without considering infiltration load, is presented in Table 19.6. Note that in such a case, there is actually no infiltration as the room is under positive pressure. There is, however, exfiltration which is equivalent to exhaust of room air.*

19.7 PSYCHROMETRIC CALCULATIONS FOR COOLING

Figure 19.2 shows the condition of the mixture of the recirculated room air and ventilation air entering the apparatus at 1, and leaving the apparatus at 2 which is the same as the supply air state s, the effective surface of the apparatus being at S. The condition line 1–2 represents the psychrometric process in the air-conditioning apparatus, and hence the GSHF line. Further, the leaving air state 2 is governed by the BPF of the apparatus, although, at the same time, it must lie on the RSHF line $i - 2^*$.

Accordingly, the dehumidified air quantity can be calculated either from room sensible heat balance, viz., process $s - i$ in the room

$$(\text{cmm})_d = \frac{\text{RSH}}{0.0204(t_i - t_s)} \tag{19.8}$$

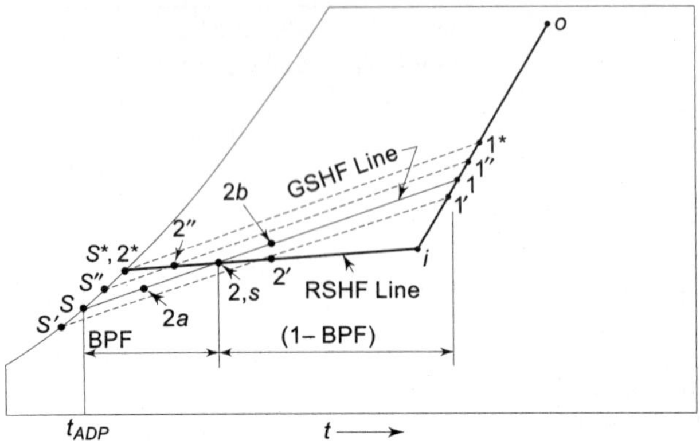

Fig. 19.2 Effect of bypass factor

or from the total sensible heat balance, viz., process 1–2 in the apparatus.

$$(\text{cmm})_d = \frac{\text{TSH}}{0.0204(t_1 - t_2)} \tag{19.9}$$

It can be seen that the two expressions give the same value of the cmm_d.

Note *Another simplified version of load calculation sheet for a studio building is given in Table 19.6 below.*

Table 19.6 Simplified load estimation sheet

FLOOR : GROUND, FIRST, THIRD
SIZE : 857.44 m^2 × 4 m
SEASON : SUMMER
PEAK LOAD : AT 8:00 P.M.

CONDITIONS	DBT	WBT	% RH	SP. HU., g/kg
OUTSIDE	43.3°	23.9°	20	11.0
ROOM	25°	—	50	10.0
DIFFERENCE	18.3°			1.0

DATA: U_{wall} = 1.78 W/m^2 K, U_{roof} = 1.316 W/m^{2}K
1. SOLAR & TRANSMISSION HEAT GAIN

DIRECTION	AREA (m^2)	ΔT_e (°C)	$Q_S = UA\Delta T_e$ (kW)
N E Wall	65.76	16.4	1.920
E Wall	105.16	17.4	3.257
SE Wall	31.72	16.4	0.926
S Wall	125.68	17.4	3.893
W Wall	137.4	25.7	6.285
N W Wall	31.72	21.5	1.214
N Wall	61.44	14.1	1.542
ROOF	—	—	—
Total			19.037

(Contd)

2. TRANS. GAINS THROUGH PARTITION/DOOR

ITEM	U (W/m^2K)	AREA (m^2)	ΔT (°C)	$Q = UA\,\Delta T$ (kW)
PARTITION-1	1.57	25.68	15.8	0.637
PARTITION-2	1.434	71.4	15.8	1.618
TOILET Doors	2.08	2.5	18.3	0.095
SWING Doors	2.08	2.5	18.3	0.095
F.C.C Doors	3.98	9.0	18.8	0.066
Total				2.511

3. OCCUPANCY HEAT GAIN

No. of PEOPLE	$FACTOR_S$	$FACTOR_L$	Q_S (kW)	Q_L (kW)
30	65	75	195	225

4. ELECTRICAL LOAD (kW)

NORMAL	8
FLUORESCENT	25

5. VENTILATION

VOLUME (m^3)	NO. OF AIR CHANGES	$(cmm)_0$
857.44×4	1.2	68.6

6. TOTAL ROOM HEAT GAINS

SAFETY FACTOR = 5%

RSH = 59.31 kW

RLH = 2.36 kW

RTH = 64.02 kW

7. EFFECTIVE ROOM HEAT GAINS

(a) FAN H.P. = 5% (b) DUCT LEAKAGE = 0.5%

(c) DUCT HEAT GAIN = 0.5%. (d) BPF = 0.15

OUTDOOR AIR BYPASSED LOADS

$Q_S = 0.0204 \times 18.3 \times 68.6 \times 0.15 = 3.84$ kW

$Q_L = 50 \times 1 \times 10^{-3} \times 68.6 \times 0.15 = 0.515$ kW

ERSH = 66.94 kW ERLH = 3.05 kW

ESHF = 0.956 ADP = 13.5°C

8. DEHUMIDIFIED AIR QUANTITY

$(cmm)_d = ERSH/0.0204 \times (t_i - ADP) \times (1 - BPF)$

$= 35.1$

(Contd)

9. RETURN AIR QUANTITY

$(\text{cmm})_i = (\text{cmm})_d - (\text{cmm})_0 = 282.4$

10. ENTERING AND LEAVING AIR TEMP.

$t_1 = 28.6°\text{C},\ t_2 = 15.8°\text{C}$

11. OUTDOOR AIR HEAT (PART OF GTH)

$Q_S = 0.0204 \times 18.3 \times 68.6 \times (1 - 0.15) = 21.76$ kW

$Q_L = 50 \times 1 \times 10^{-3} \times 68.6 \times (1 - 0.15) = 2.92$ kW

12. GRAND TOTAL HEAT (i.e., LOAD)

TSH = 88.7 kW

TLH = 4.97 kW

GTH = 93.67 kW (26.64 TR)

GSHF = 0.9469

19.7.1 Effect of Bypass Factor

The effect of the bypass factor of the apparatus may now be analysed in detail. It is to be noted that in the air conditioning of any building, the following two things are fixed, irrespective of the bypass factor of the apparatus.

(i) Room sensible and latent heat gains.

(ii) Ventilation air quantity.

As a result of these two, the total sensible and latent heat gains, and hence the grand sensible heat factor are also fixed. Hence the slope of the GSHF line cannot be changed. Thus to satisfy the given conditions, the GHHF line will float parallel to itself depending on the bypass factor of the apparatus, in such a manner that the leaving air states 2, 2′, 2″, etc., lie on the RSHF line as shown in Fig. 19.2. Another outcome of this is that the mixture state 1 must also correspondingly float in its position along the line $i - o$.

Thus, when the bypass factor increases, the following observations can be made.

(i) The supply air state 2 moves towards the room state i.

(ii) Dehumidified rise $(t_1 - t_2)$ is decreased.

(iii) Dehumidified air quantity $(\text{cmm})_d$ is, therefore, increased.

(iv) The fresh air quantity remaining fixed, the recirculated room air quantity $(\text{cmm})_i$ must increase.

(v) The apparatus dew point S is lowered and hence the running cost of the refrigeration equipment is increased.

(vi) The cost of the cooling apparatus, however, is decreased (Less Cost with Higher BPF).

(vii) The mixture state 1 moves towards i.

(viii) The recirculated room air quantity, therefore, increases

It is seen that when the BPF is zero, the ADP is at 2^* and the process line is in the limiting position $1^* - 2^*$. This corresponds to minimum dehumidified and supply air quantity.

19.7.2 Calculation by Iteration Procedure

An important point to note here is that point 2 or s in Fig. 19.2 is obtained simply by the intersection of GSHF and RSHF lines so that the supply state at s satisfies both the room sensible heat and total sensible heat requirements. In actual practice, however, the supply air state may be obtained at point 2, 2a or 2b, depending on the bypass factor of the apparatus for a fixed entering air state at 1.

Thus, for the leaving air state 2 to coincide with the supply air state s, it is possible to have two alternative conditions as follows:

(i) If point 1 is fixed, it is necessary to have an apparatus with a given bypass factor X as shown in Fig. 19.2 so that 2 coincides with s, and does not fall at 2a or 2b.

(ii) If the BPF is fixed, the GSHF line 1 – 2 will float parallel to the solid line as shown in Fig. 19.2. The GSHF line will assume a position depending on the value of the BPF, the extreme position being $1^* - 2^*$ corresponding to the case when the BPF is equal to zero. The RSHF line $i - 2 - 2^*$, however, is unchangeable.

Normally, the bypass factor of the apparatus is initially assumed depending on the nature of application. In that case, alternative (ii) is applicable. Thus the GSHF line is to be floated until point 2 is located, which divides $1 - S$ in the ratio of BPF and 1 – BPF, as shown in Fig. 19.4. In other words, the dehumidified air quantity cannot be calculated unless point 2, and therefore, point 1 is known, from where the GSHF line is to be drawn. At the same time, point 1 cannot be known unless the dehumidified air quantity is calculated, from which the recirculated room air quantity is determined by taking the difference

$$(\text{cmm})_i = (\text{cmm})_\text{d} - (\text{cmm})_o$$

and then applying the mixing rule to locate point 1.

Thus, the calculation procedure involves iteration, and is cumbersome and time-consuming. To simplify the procedure, the concept of the effective sensible heat factor as outlined in Sec. 19.5 is used.

19.7.3 Use of Effective Sensible Heat Factor to Determine Apparatus Dew Point and Dehumidified Air Quantity[3]

Figure 19.3 shows an equivalent system in which the bypassed outside air is shown separately. Thus, if the air leaving the apparatus did not contain the bypassed outside air, its state would be at 2′ as shown in Fig. 19.4. On mixing with the bypassed outside air at 0, one obtains the conditioned air at 2. Thus point 2′ lies on the extended line 0 – 2 – 2.′

The physical significance of the points 2 and 2′ may now be assessed. The conditioned air at 2 is able to balance RSH and RLH, and state 2 lies on the RSHF line. The conditioned air at 2′ is able to balance the bypassed OASH and OALH in addition to the room loads, and hence state 2′ lies on ESHF line.

To locate point 2′, first the ESHF line is to be drawn from i. Its intersection with the saturation curve also determines S, the coil ADP. Now, an approximation can be made by considering triangles S – 2 – 2′ and S – 1 – i as similar, so that

$$\text{BPF} = \frac{2-\text{S}}{1-\text{S}} = \frac{2'-\text{S}}{i-\text{S}}$$

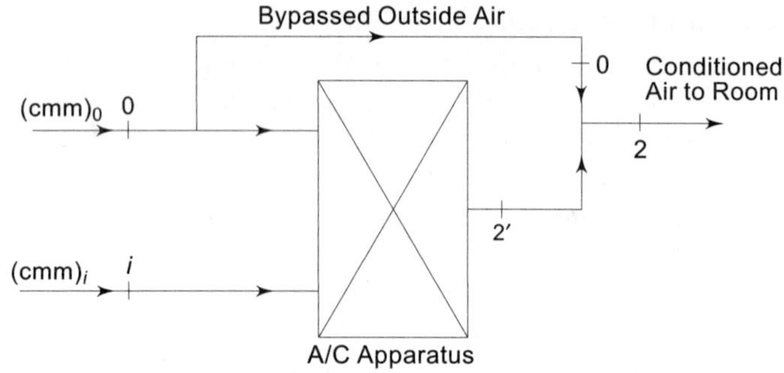

Fig. 19.3 Equivalent circuit for bypassed outside air

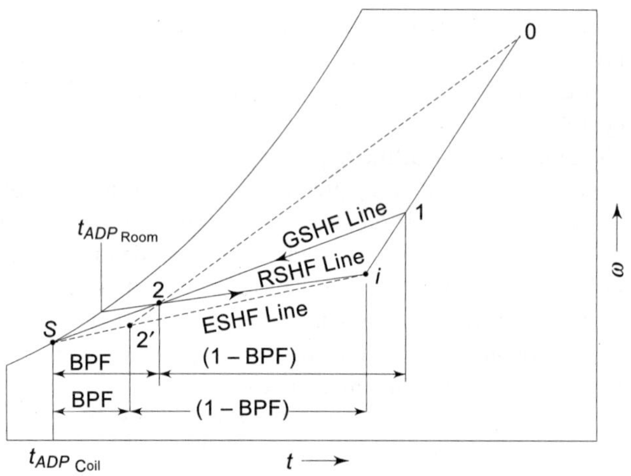

Fig. 19.4 Psychrometric process with bypassed outside air and effective sensible heat factor

The assumption is not exactly correct since $2 - 2'$ is not parallel to $o - i$. However, in practice, the points 2 and 2′ are so close to each other, and the point o which lies on both these lines is so far away, that for all practical purposes the approximation is valid. Accordingly, the dehumidified air quantity can be calculated using state 2′ and ERSH as follows

$$(\text{cmm})_d = \frac{\text{ERSH}}{0.0204(t_i - t_2')}$$

$$= \frac{\text{ERSH}}{0.0204(t_i - t_{\text{ADP}})(1 - \text{BPF})} \quad (19.10)$$

Equation (19.10) is normally used instead of Eqs (19.8) and (19.9).

19.8 SELECTION OF AIR-CONDITIONING APPARATUS FOR COOLING AND DEHUMIDIFICATION

After calculating the dehumidified air quantity, the air conditioning apparatus may be selected from the manufacturer's performance tables/charts. The necessary data required for this selection are the following:

(i) Grand total heat.
(ii) Dehumidified air quantity.
(iii) Apparatus dew point.

These will determine the size of the apparatus, and the refrigerant temperature.

The performance data also give the bypass factor of the apparatus.

If the BPF of the selected apparatus is not the same as the originally assumed BPF, the portion of the load estimation form involving BPF is to be adjusted accordingly.

Once the apparatus has been selected from ESHF, ADP, BPF and GTH, the entering and leaving air conditions are easily determined.

There are two types of cooling and dehumidifying apparatuses:

(i) Coil equipment.
(ii) Air washer or Spray equipment.

Normally, a coil consists of a series of tubes through which brine, chilled water or expanded refrigerant flows. Air is drawn over the tubes by a fan. The coil may also be in the form of rows of tubes in a number of circuits.

Thus Fig. 19.5 shows a 4-row coil with 7 rows of tubes in series in just one circuit.

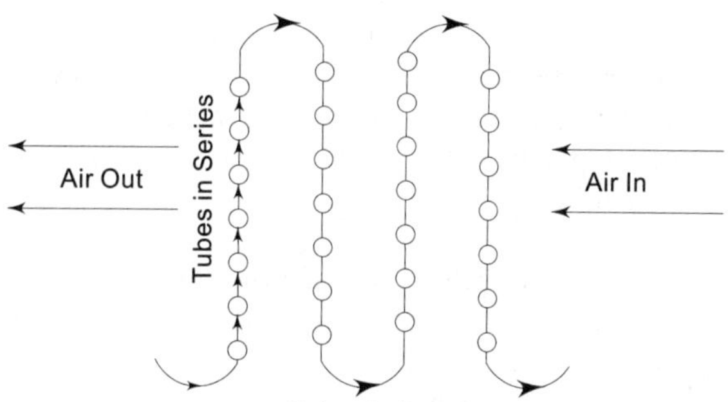

Fig. 19.5 Configuration of 4-row coil with 7 lengths of tubes

As the air passes over the coil, it is cooled and dehumidified or heated depending on the temperature of the fluid inside. The medium inside the coil is correspondingly required to be heated or cooled, external to the apparatus.

The amount of the coil surface not only affects the heat transfer but also the bypass factor. The BPF also depends on the number of rows. The larger the number of rows, the closer is the leaving air temperature to the cooling medium temperature.

Table 19.7 gives typical values of bypass factors for finned coils. These values apply to coils with 1.6 cm OD tubes spaced approximately at 3 cm centres. For other coils, data should be obtained from the manufacturer. Laboratory tests reveal that if the BPF of a single row coil is X, then the BPF of a coil n rows deep is X^n. Values in Table 19.7 conform to this result. Typical cooling coils in air conditioning have 6 fins/cm. Values of bypass factors for the same as a function of air velocity, also termed face velocity (FV), are presented in Fig. 19.6.

Air washers with chilled-water sprays can also be used for cooling and dehumidification in place of coils. Air washers are rather more versatile than coils in the sense that they can be used in addition for heating and humidifying as well as for evaporative cooling.

Table 19.7 Typical bypass factors for finned coils[3]

Rows Deep or Depth of Coils	*3 fins/cm*		*6 fins/cm*				
	Air Velocity (m/min)						
	90	*210*	*90*	*120*	*150*	*180*	*210*
1	.65	.74	.47				.61
2	.42	.55	.22	.274	.314	.346	.38
3	.27	.40	.10	.143	.176	.204	.23
4	.19	.30	.05	.076	.099	.12	.14
5	.12	.23	.02	.04	.056	.071	.09
6	.08	.18	.01	.022	.032	.042	.06
8	.03	.08					

Note *The importance of BPF lies in the fact that the outside air which gets bypassed becomes a part of the room load. The bypassed room air does not affect the load situation.*

The details of spray equipment are discussed in Chapter 20. Air washers with chilled-water sprays behave in much the same way as cooling coils, having an effective surface temperature equal to the required apparatus dew point.

The effect of the bypass factor on apparatus dew point has been illustrated in Fig. 19.2. The relationship between BPF and ADP and the considerations involved may be summarised as follows:

Smaller bypass factor:

(i) Higher ADP and higher refrigerant or brine or chilled-water temperature needed. Consequently a smaller refrigerating machine with a higher coefficient of performance is required.

(ii) More heat-transfer surface since more rows are necessary for a smaller BPF and also since the ADP is higher. Larger coil is, therefore, required.

Larger bypass factor:

(i) A lower ADP and lower refrigerant or brine or chilled-water temperature needed. Consequently a larger refrigerating machine with a lower coefficient of performance is required.

(ii) Less heat-transfer surface since less rows are necessary for a larger BPF and also since the ADP is lower.

Factors (i) and (ii) represent the operating and first costs respectively and are contradictory to each other. It is necessary to optimize for minimum total cost. As a guide, values of coil bypass factors from Table 19.8 for typical applications may be used.

Table 19.8 Typical bypass factors of finned coils for various applications

Coil BPF	*Type of Application*	*Example*
0.3 to 0.5	Small total load	Residence
0.2 to 0.3	Low SHF and large total load	Residence, small retail shop, factory
0.1 to 0.2	Typical comfort application	Department store, bank, factory
0.05 to 0.1	Higher latent loads or large outdoor-air load	Department store, restaurant
0 to 0.1	All outdoor-air applications	Hospital, operation rooms

Note *(i) In high SHF applications, the ADP is high. So we can use high BPF apparatus.*

(ii) In low SHF applications, the ADP is low. Hence we have to use low BPF apparatus.

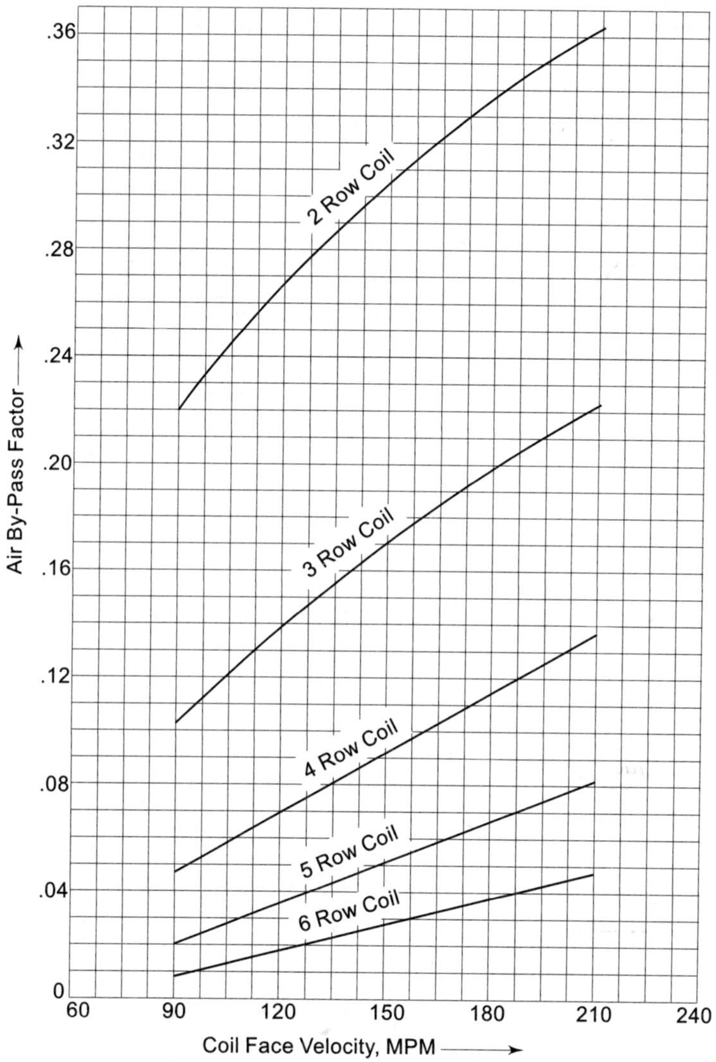

Fig. 19.6 Figure showing dependence of air bypass factor on face velocity and depth of rows for 6 fins/cm coils

Example 19.2 *A retail shop located in a city at 30°N latitude has the following loads.*

Room sensible heat:	*58.15 kW*
Room latent heat:	*14.54 kW*

The summer outside and inside design conditions are:

Outside: 40°C DB, 27°C WB

Inside: 25°C DB, 50% RH

70 cmm of ventilation air is used. Determine the following:

(i) Ventilation load. *(ii) Grand total heat.*

(iii) Effective sensible heat factor. *(iv) Apparatus dew point.*

(v) Dehumidified air quantity.

(vi) Condition of air entering and leaving apparatus.

Assume a suitable bypass factor of the cooling coil.

Solution

Table 19.9 Design conditions for Example 19.2

Condition	*DBT*	*WBT*	*RH*	ω	*h*	*v*
	°C	°C	%	kg/kg	kJ/kg d.a.	m^3/kg d.a.
Outside	40	27		0.0175	85.0	0.912
Inside	25		50	0.01	50.8	0.858

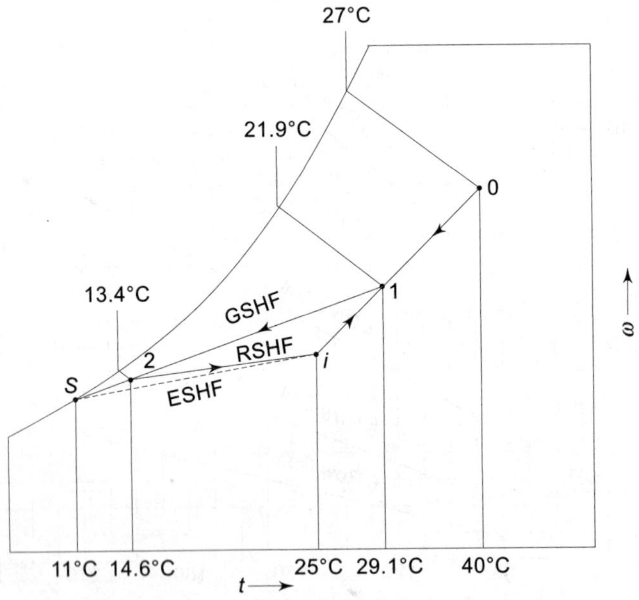

Fig. 19.7 Psychrometric processes for Example 19.2

Refer to Fig. 19.7.

(i) Outside air sensible heat

$$\text{OASH} = 0.0204(70)\,(40 - 25) = 21.42 \text{ kW}$$

Outside air latent heat

$$\text{OALH} = 50(70)\,(0.0175 - 0.01) = 26.25 \text{ kW}$$

Outside air total heat or ventilation load

$$\text{OATH} = \text{OASH} + \text{OALH} = 21.42 + 26.25 = 47.67 \text{ kW}$$

(ii) Total sensible heat

$$TSH = RSH + OASH = 58.15 + 21.42 = 79.57 \text{ kW}$$

Total latent heat

$$TLH = RLH + OALH = 14.54 + 26.25 = 40.79 \text{ kW}$$

Grand total heat

$$GTH = TSH + TLH = 79.57 + 40.79 = 120.36 \text{ kW}$$

(iii) Assume a bypass factor of 0.15 (Table 19.7)
Effective room sensible heat

$$ERSH = 58.15 + 0.15\,(21.42) = 61.36 \text{ kW}$$

Effective room latent heat

$$ERLH = 14.54 + 0.15\,(26.25) = 18.48 \text{ kW}$$

Effective sensible heat factor

$$ESHF = \frac{61.36}{61.36 + 18.48} = 0.77$$

(iv) Draw a 0.77 SHF line from inside conditions as shown in Fig. 19.7. The intersection with the saturation curve gives the ADP as 11°C.

(v) Dehumidified air quantity

$$(\text{cmm})_d = \frac{\text{ERSH}}{(0.0204)\,(t_i - t_{\text{ADP}})\,(1 - \text{BPF})}$$

$$= \frac{61.36}{(0.0204)\,(25 - 11)\,(1 - 0.15)} = 253 \text{ cmm}$$

(vi) Recirculated room air

$$(\text{cmm})_i = (\text{cmm})_d - (\text{cmm})_o = 253 - 70 = 183 \text{ cmm}$$

Basing calculations on cmm instead of mass flow rates as all the other calculations are based on the standard density of air, we obtain the following results.

Entering air dry bulb temperature

$$t_1 = \frac{183(25) + 70(40)}{253} = 29.1°\text{C}$$

Entering air wet-bulb temperature, on the line joining i to 0 and at the above temperature

$$t_1' = 21.9°\text{C}$$

Leaving air dry bulb temperature

$$t_2 = t_{\text{ADP}} + (t_1 - t_{\text{ADP}})\text{BPF} = 11 + (29.1 - 11)\,(0.15) = 13.7°\text{C}$$

Leaving air wet-bulb temperature is obtained by locating point 2 at the above temperature on the line joining 1 to S. Thus

$$t_2' = 13.2°\text{C}$$

19.8.1 High Latent Cooling Load Application

Sometimes, a situation may arise when the GSHF line does not intersect the saturation curve, or even when it does intersect, an absurdly low value of the apparatus dew point is obtained. In the former case it is not possible to design a cooling coil which would satisfy the SHF of the room and in the latter, an extremely uneconomical refrigerating machine would result with a very low suction pressure and high power consumption.

Such a situation arises when the latent heat load is very high in comparison with the total heat load due to the very humid outdoor design conditions or when very high internal latent loads are present.

In all such applications, an appropriate apparatus dew point t_S is selected and the air is reheated to the RSHF line after leaving the cooling coil as shown in Fig. 19.8. For the purpose, the following procedure is followed.

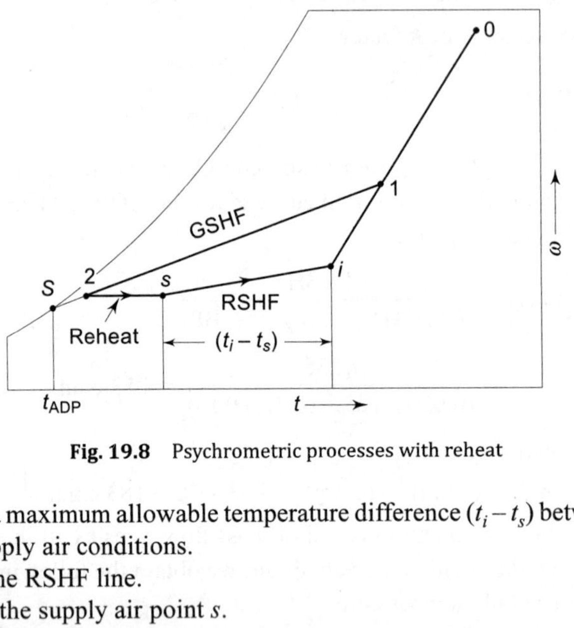

Fig. 19.8 Psychrometric processes with reheat

(i) Select a maximum allowable temperature difference $(t_i - t_s)$ between the room and supply air conditions.
(ii) Draw the RSHF line.
(iii) Locate the supply air point s.
(iv) Then the leaving air state 2 lies on the horizontal from s.
(v) Select a suitable coil with ADP so that 2 lies on this horizontal.
(vi) Calculate the reheat from t_2 to t_s.
(vii) Calculate the air quantity to take up the RSH from t_s to t_i or the sensible heat including reheat from t_2 to t_i.

There are two other alternatives to reheat. They are:

(i) Alter the room design condition to a condition of higher humidity. This will make the GSHF line intersect the saturation curve.
(ii) Use a coil of a lower bypass factor so that ESHF is increased and a higher ADP is obtained.

Even though these methods may not eliminate the need for reheat completely, they can at least reduce the reheat requirement. Thus the above methods can be used simultaneously with reheat.

Example 19.3 *A laboratory having an unusually large latent heat gain is required to be air conditioned. The design conditions and loads are as follows:*

Summer design conditions	:	*40°C DBT, 27°C WBT*
Inside design conditions	:	*25°C DBT, 50% RH*
Room sensible heat	:	*34.9 kW*
Room latent heat	:	*18.6 kW*

The ventilation air requirement is 85 cmm. Determine the following:

(i) *Ventilation load.*
(ii) *Room and effective sensible heat factors.*
(iii) *Apparatus dew point and amount of reheat for economical design.*
(iv) *Supply air quantity.*
(v) *Condition of air entering and leaving coil and supply air temperature.*
(vi) *Grand total heat.*

Assume a suitable bypass factor.

Solution The design conditions are the same as those given in Table 19.9. Refer to Fig. 19.9.

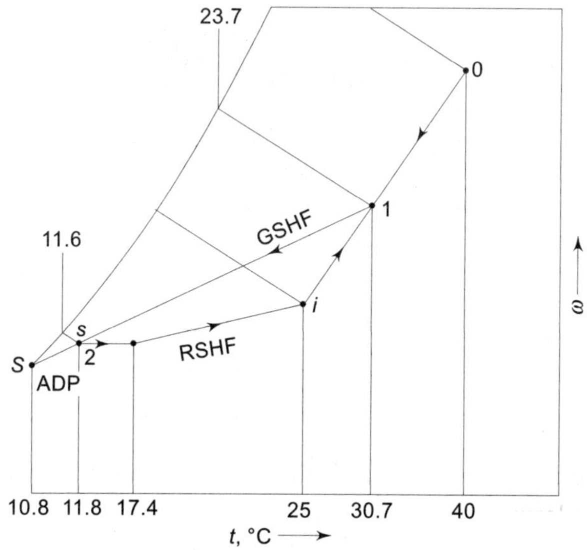

Fig. 19.9 Psychrometric processes for Example 19.3

(i) Outside air sensible heat

$$\text{OASH} = 0.0204\ (85)\ (40 - 25) = 26.0\ \text{kW}$$

Outside air latent heat

$$\text{OALH} = 50\ (85)\ (0.0175 - 0.01) = 30.6\ \text{kW}$$

Ventilation load or outside air total heat

$$\text{OATH} = 26.0 + 30.6 = 56.6\ \text{kW}$$

(ii) Room sensible heat factor

$$\text{RSHF} = \frac{34.9}{34.9 + 18.6} = 0.65$$

As it is a low sensible heat-factor application, we choose a coil with a low bypass factor of 0.05 for economical design. Effective room sensible heat

$$\text{ERSH} = 34.9 + (0.05)\,(26.0) = 36.2 \text{ kW}$$

Effective room latent heat

$$\text{ERLH} = 18.6 + (0.05)\,(30.6) = 20.63 \text{ kW}$$

Effective sensible heat factor

$$\text{ESHF} = \frac{36.2}{36.2 + 20.1} = 0.64$$

Note *The room sensible heat factor is very low. The effect of choosing a low BPF coil is to keep ESHF close to RSHF itself.*

(iii) The ESHF line intersected the saturation curve. But the ADP is as low as 3°C. If we had not selected a low BPF coil, the ESHF line would not have intersected at all. For economical design, let us select an ESHF of 0.75. Drawing this line from i, we get

$$\text{ADP} = 10.8°\text{C}$$

For calculating the required reheat, we write the expression for the modified ESHF

$$\text{Modified ESHF} = \frac{\text{ERSH} + \text{Reheat}}{\text{ERSH} + \text{ERLH} + \text{Reheat}}$$

$$0.75 = \frac{36.2 + \text{Reheat}}{36.2 + 20.1 + \text{Reheat}}$$

$$\Rightarrow \quad \text{Reheat} = 24.1 \text{ kW}$$

(iv) Dehumidified air quantity

$$(\text{cmm})_d = \frac{\text{ERSH} + \text{Reheat}}{0.0204(t_i - t_{\text{ADP}})\,(1 - \text{BPF})}$$

$$= \frac{36.2 + 24.1}{0.0204\,(25 - 10.8)\,(1 - 0.05)} = 220$$

Note *The supply air quantity is the same as the dehumidified air quantity.*

(v) Recirculated room air

$$(\text{cmm})_i = (\text{cmm})_s - (\text{cmm})_0 = 220 - 85 = 135$$

Entering air dry-bulb temperature

$$t_1 = \frac{139(25) + 85(40)}{220} = 30.7°\text{C}$$

Entering air wet-bulb temperature, from the psychrometric chart

$$t'_1 = 23.7°\text{C}$$

Leaving air dry-bulb temperature

$$t_2 = t_{\text{ADP}} + (t_1 - t_{\text{ADP}})\ \text{BPF}$$

$$= 10.8 - (30.7 - 10.8)\,(0.05) = 11.8°\text{C}$$

Leaving air wet-bulb temperature on the line 1-S from the psychrometric chart

$$t'_2 = 11.6°\text{C}$$

Supply air temperature

$$t_s = t_i - \frac{\text{RSH}}{0.0204\,(\text{cmm})_s} = 25 - \frac{34.9}{0.0204(220)} = 17.4°\text{C}$$

(vi) Grand total heat

$$\text{GTH} = \text{Room load} + \text{Ventilation load} + \text{Reheat}$$
$$= (34.9 + 18.6) + 56.6 + 24.1 = 134.2 \text{ kW}$$

19.8.2 All Outdoor Air Application

Most *all outdoor air* applications also imply high latent heat load applications. A typical application of this kind is a hospital operation theatre. After doing the necessary load calculations, and determining the apparatus dew point and dehumidified air quantity, we may come across the following three situations for which the procedures are also given.

Case I Dehumidified air quantity is equal to the ventilation air requirement.

The solution in this case is simple as shown in Fig. 19.10. The fresh air is directly taken through the air-conditioning apparatus and is then supplied to the room.

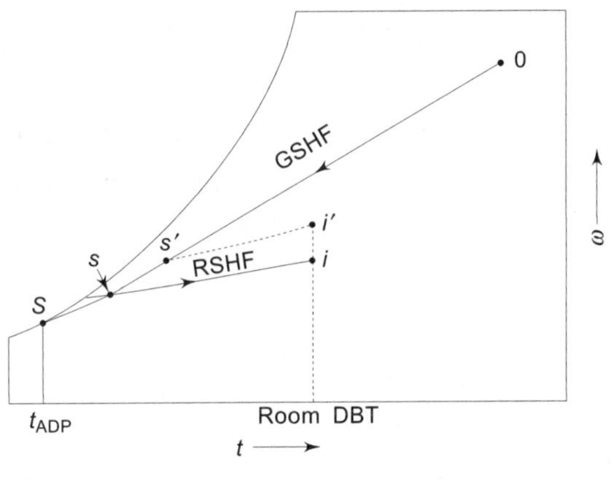

Fig. 19.10 Psychrometric processes for all outdoor air applications—dehumidified air quantity equal to ventilation air requirement

Case II Dehumidified air quantity is less than the ventilation air requirement.

There are two alternatives:

(a) If the difference in air quantities is very small, then a coil with a larger bypass factor may possibly be used. The effect of this will be to raise the temperature of the air leaving the coil and hence to increase the necessary supply air quantity to make it equal to ventilation air. However, leaving air state s' (Fig. 19.10) would not, then lie on RSHF line. Hence, inside condition would shift from i to i', viz., to a state of higher humidity if room DBT is to be maintained the same as before.

(b) If the difference in air quantities is large, it will be necessary to first condition the necessary outdoor air and then to reheat to artificially increase the room sensible heat load so that a larger quantity of supply air is required. The processes are illustrated in Fig. 19.11. It is to be seen that a suitable ADP and BPF have to be selected. Also, the method is wasteful of energy.

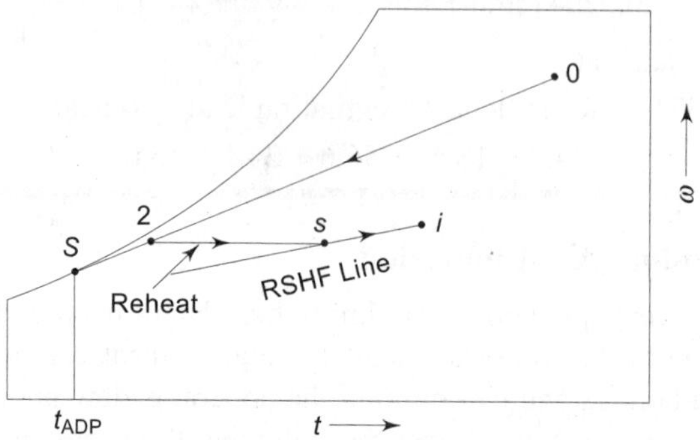

Fig. 19.11 Psychrometric processes for all outdoor air applications—dehumidified air quantity less than ventilation air requirement

Case III Dehumidified air quantity is more than the ventilation air requirement.

In this case, calculations have to be repeated by taking the outdoor air equal to the calculated dehumidified air quantity, as illustrated in Example 19.4.

Example 19.4 *A laboratory is to be air conditioned with outside and inside design conditions as in Example 19.2. It is specified that all outdoor air must be used. The room sensible and latent heat loads are 14.5 and 3.2 kW respectively. The minimum fresh air requirement is 50 cmm. Determine:*

(i) Ventilation load.
(ii) Effective sensible heat factor and apparatus dew point.
(iii) Dehumidified air quantity.
Assume a suitable bypass factor.

Solution As the ventilation load is large, the effective sensible heat factor is likely to be very low unless a very low bypass factor is chosen. Hence, choose a coil with a BPF of 0.05. Then

$$\text{OASH} = 0.0204\,(50)\,(40 - 25) = 15.3 \text{ kW}$$

$$\text{OALH} = 50\,(50)\,(0.0175 - 0.01) = 18.75 \text{ kW}$$

$$\text{OATH} = 15.3 + 18.75 = 34.05 \text{ kW}$$

$$\text{ERSH} = 14.5 + (0.05)\,(15.3) = 15.31 \text{ kW}$$

$$\text{ERLH} = 3.2 + (0.05)\,(18.75) = 4.14 \text{ kW}$$

$$\text{ESHF} = \frac{15.31}{15.31 + 4.14} = 0.784$$

$$\text{ADP} = 11.3° \text{ (from psychrometric chart)}$$

$$(\text{cmm})_d = \frac{15.31}{(0.0204)\,(25 - 11.3)\,(1 - 0.05)} = 57.5$$

Note *The dehumidified air quantity is more than the minimum fresh air requirement. The calculations are to be repeated with the ventilation air equal to 57.5 cmm. The results are as follows:*

(i) Ventilation load OATH = 39.16 kW

(ii) Effective sensible heat factor ESHF = 0.782
Apparatus dew point ADP = 11.3°C

(iii) Dehumidified air quantity $(\text{cmm})_d = 57.9$

The calculations are to be repeated, again with a ventilation air of 57.9 cmm until agreement is reached between assumed and calculated values.

19.9 EVAPORATIVE COOLING

As discussed in Chapter 15, evaporative cooling is obtained during the process of adiabatic saturation. It is a process of the removal of sensible heat from air and an equivalent addition of latent heat to it in the form of added water vapour.

Evaporative cooling is a process in which heat is neither added to nor it is removed from the water outside the air washer. Water is simply recirculated by a pump.

Evaporative cooling is commonly used when the outdoor conditions are very dry. This means that the wet-bulb depression of air is very large. In a dry climate, evaporative cooling can give some relief by removing the sensible heat from the room. But the humidity cannot be controlled.

Another defect of the evaporative cooling system is the large quantity of air that must be supplied to meet the room sensible heat load as the temperature difference between the room and supply air is generally small. Thus, whereas in air conditioning, the supply air quantity may be of the order of 8–10 air changes per hour, in the case of evaporative cooling, the same may be of the order of 20 air changes. This quantity increases rapidly as the humidifying efficiency of the air washer decreases as illustrated in Example 19.5.

Example 19.5 *In an industrial evaporative cooling application with outside conditions as in Example 19.2, the inside is to be maintained at a maximum relative humidity of 55 per cent. The room sensible heat is 581.5 kW. All outdoor air must be used. Find the room dry-bulb temperature and supply air quantity as a function of humidifying efficiencies of 80, 85, 90, 95 and 100 per cent.*

Solution (i) Room dry-bulb temperature:

To maintain 55 per cent RH in the room, it can be seen from Fig. 19.12 that the room DBT varies along 55 per cent RH line, depending on the humidifying efficiency η_H.

Also, as the room DBT is lowered with decreasing saturation efficiency, the supply air quantity increases.

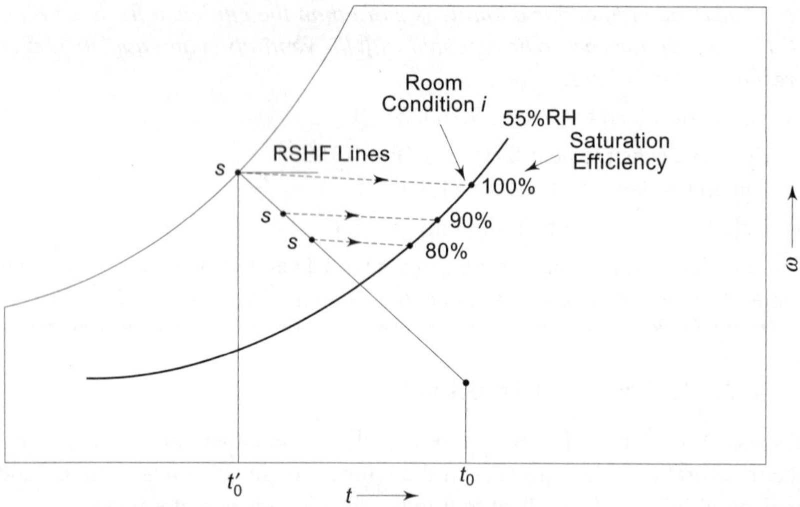

Fig. 19.12 Evaporative cooling—processes for Example 19.5

Thus to maintain an acceptable room DBT and supply air quantity, the following equation may be used.

$$t_s = t_0 - \eta_H (t_0 - t'_0) = 40 - \eta_H (40 - 27) = 40 - 13\ \eta_H$$

After finding t_s, the room DBT t_i can be found by moving along the horizontal from t_s up to the 55% RH line. The values are given in Table 19.10.

Table 19.10 Room DBT and supply air quantity variation with humidifying efficiency

η_H	t_s	t_i	$t_i - t_s$	$(cmm)_s$
100	27	37.8	10.8	2640
95	27.7	37.5	9.8	2910
90	28.3	37.1	8.6	3240
85	29.0	36.9	7.9	3610
80	30.0	36.7	6.7	4250

Table 19.10 also gives the supply air temperature rise $(t_i - t_s)$. It may be noted that the supply air temperature rise decreases more rapidly than the room DBT. Correspondingly, the supply air quantity increases rapidly.

(ii) Supply air quantity:

The supply air quantity can be calculated for various temperature rises using the equation

$$(\text{cmm})_s = \frac{581.5}{(0.0204)(t_i - t_s)}$$

The calculated values of $(\text{cmm})_s$ are also given in Table 19.10. The increase in the quantity of supply air with a slightly lower room DBT may create a feeling of discomfort due to the air blast. Also, the fan size and fan power will increase, though the size of the spray chamber will be reduced.

19.9.1 Limitations of Evaporative Cooling

These can now be listed as follows:

(i) The lowest possible DBT of air off the cooler is at 100% efficiency. It is then equal to WBT of ambient air. It is conceded that evaporative cooling is satisfactory only in areas where the DBT exceeds 32°C, and the WBT is below 21°C.

(ii) Evaporative cooling removes only sensible heat. Even though a satisfactory DBT may be obtained in the room, the relative humidity off the cooler and in the room is very high.

(iii) Supply air quantity is very large leading to conditions of draft.

(iv) The cooler is to be kept in good repair to obtain high efficiency, otherwise the DBT of air off the cooler will not be sufficiently low to cool the space.

19.10 BUILDING REQUIREMENTS AND ENERGY CONSERVATION IN AIR CONDITIONED BUILDINGS

The total amount of energy consumption in air conditioning is quite substantial. It is known that one ton of refrigeration in central air-conditioning plants requires 1.25 kW of power approximately. This is on the basis of roof not exposed to sun, and not too much glass areas in the walls. This one TR is sufficient for office space of 18–22.5 m^2, or 12–14 seats in a theatre. Thus, for an office of approximately 1850 m^2 or a theatre of 1250 seating capacity, the A/C load is 100 TR, requiring a power consumption of about 125 kW. This shows that the power consumption is sizable and there is need to minimize it. Further, the cost of air conditioning constitutes about 60% of the cost of building. Hence, the need to cut cooling loads to minimize the size of the plant.

Energy conservation in the air conditioning of buildings can be achieved by adopting the following measures:

(i) Minimisation of solar gain.

(ii) Other building design features and thermal properties of construction materials.

(iii) Minimizing infiltration and ventilation load.

(iv) Use of natural ventilation for cooling.

(v) Use of thermal storage.

(vi) Plant selection.

(vii) Plant maintenance.

(viii) Permitting drift in room design conditions.

19.10.1 Minimisation of Solar Gain

Solar radiation accounts for 40–70% of cooling load in many buildings. Factors affecting are; orientation of the building, fenestration, preventive measures to intercept solar heat, etc. It is conceded that maximum area of exposed walls and windows should be in North-South direction as against East-West direction. This means the longer side of the building to face N-S directions.

Further, the following design features will considerably decrease the cooling load due to fenestration:

(a) *Reduction in glass areas on the western side* The *W*-glass adds about 510 W/m^2 during the hot afternoon. Thus, 6.9 m^2 of unshaded *W*-glass contributes 1 TR. The same glass area on the N-side contributes only 1/15 TR.

(b) *Direct sunlight on W, N-W and S-W should be avoided* This can be done by suitable sunshades which will permit just enough light but limit direct solar radiation. The amount of overhang can be calculated from

$$\text{Overhang} = \text{Factor} \times \text{Shadow height}$$

where the factor recommended for the period April-September for various latitudes is given in Table 19.11.

Table 19.11 Shadow factors of overhangs

Latitude	*25°*	*30°*	*35°*	*40°*	*45°*	*50°*
Factor	0.37	0.48	0.59	0.71	0.85	1.02

(c) *Using curtains/venetian blinds on windows inside the space* However, external shading is more effective than internal shading.

(d) *Indium oxide (In_2O_3) coated on the external side reflecting glass surfaces can be used.*

(e) *Heat absorbing/tinted glass can be used.*

A comparison of various methods of cutting heat gain through glass is given in Table 19.12.

Table 19.12 Heat gain through glass with various devices

Device	*Heat Gain*
External shading	0.15–0.35
Reflecting glass	0.3–0.4
Internal shading	0.4–1.0
Heat absorbing glass	0.65 approx.

19.10.2 Other Building Design Features and Thermal Properties of Construction Materials

The measures include the following:

(a) *Air is a good insulator.* Hollow tiles with air trapped in them are most ideally suited for walls. Similarly, roofs and floors can be cast with hollow space full of air. A double-wall construction with hollow bricks and air-gap is commonly used in hot desert areas.

(b) *Roof exposed to sun must be insulated* with a minimum of 5 cm thick expanded polysterene or equivalent insulation. This will reduce the roof load from 3 TR to 0.1 TR in case of 100 m^2 of concrete slab.

(c) *Roof can be painted with aluminium paint* to reflect solar radiation.

(d) *Spraying the roofs with water* during sun periods.

(e) *Use of small surface to volume (A/V) ratio.* Division of a building like blocks *A* and *B* separated as shown in Fig. 19.13 increases the surface area and hence transmission gain. In the same way, it could be seen that buildings of larger aspect ratio will have more transmission heat gain.

(f) Wall transmission gains can also be minimised by applying insulation on exposed walls.

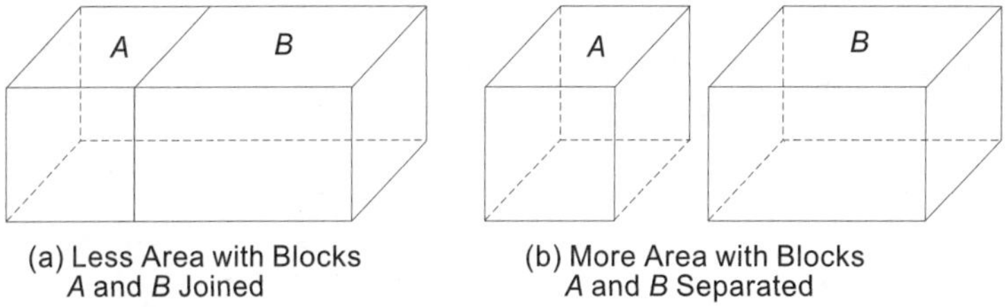

Fig. 19.13 Division of a building with blocks A and B separated thus increasing surface area

19.10.3 Minimizing Infiltration and Ventilation Load

Infiltration load is generally very heavy. If, however, infiltration cannot be reduced then no ventilation air need be taken separately with the recirculated room air before the A/C apparatus. It must be noted that every 5 cmm of fresh air requires about 1 TR.

It is recommended to keep the outside air dampers closed at all times unless the building is completely occupied. Infiltration is expected to take care of the fresh air requirement when the dampers are closed.

An interesting method to conserve energy is to cool the ventilation air regeneratively with the help of exhaust room air using a heat-pipe heat exchanger. Figure 19.14 shows the construction of heat pipes (a) without a wick and (b) with a wick. A number of such heat pipes are employed to form a regenerative heat exchanger as shown in Fig. 19.15.

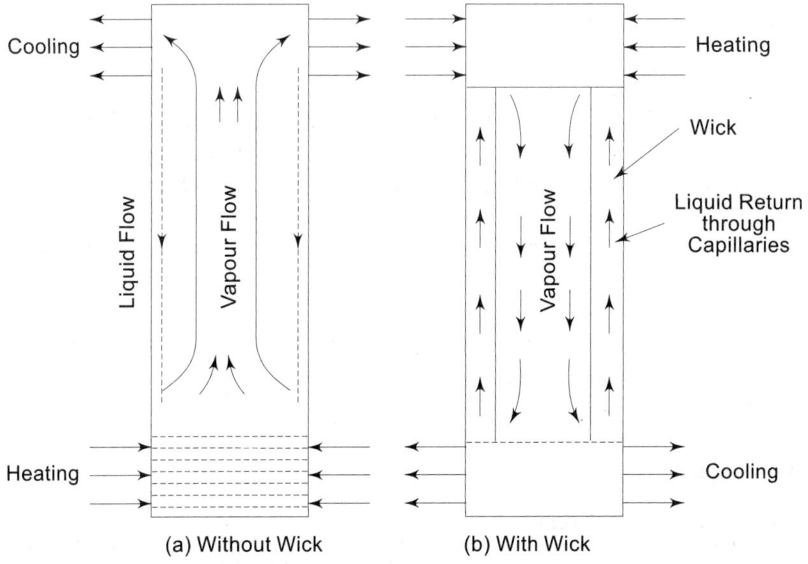

Fig. 19.14 Heat-pipe construction

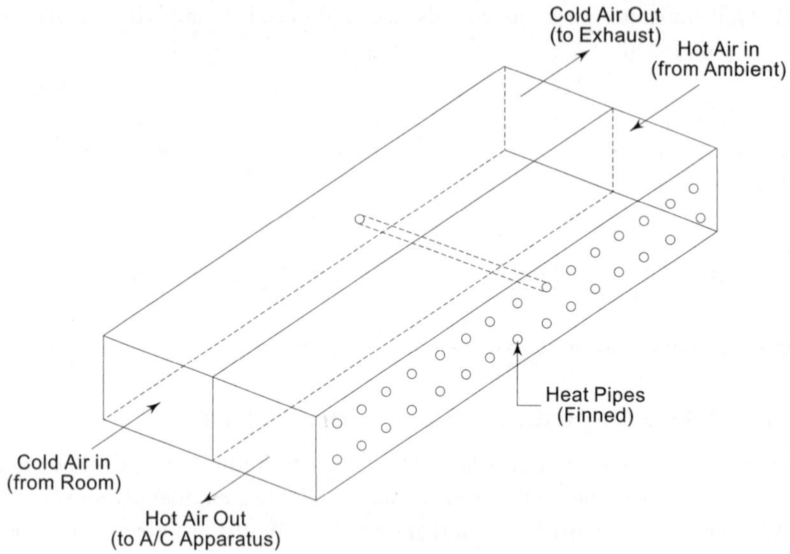

Fig. 19.15 Heat-pipe regenerative heat exchanger

19.10.4 Use of Natural Ventilation for Cooling

When the outside WBT is lower than room WBT of 18-20°C corresponding to room design conditions of 25°C DBT and 50 ± 5% RH, the air conditioning plant can be shut down, and only natural ventilation employed instead. For the purpose, larger fresh air intakes have to be provided at plant inlet.

Such a condition occurs for many days during March-April and September-October, and also at night. If it is resorted to, it will result in considerable power saving and longer life of plant.

19.10.5 Use of Thermal Storage

The method involves the use of properly sized and heavily insulated chilled water tanks which are charged at night storing surplus refrigeration when cooling load demand is low, and outside WBT is also low. This chilled water can be used during the day at the peak load period. The method mainly enables the plant to operate at higher COP thus decreasing power consumption. Although it also decreases the size of the plant, the capital cost is not reduced as the cost of the tanks and insulation is added.

This *sensible heat storage* is not really the best as it requires very large-sized chilled water tanks. *Latent heat storage* employing change of phase of materials is a better method. One such material is *sodium-sulphate decahydrate* (Na_2SO_4: 10 H_2O) which freezes at 7–10°C. This will require only 10% volume as compared to chilled water.

19.10.6 Plant Selection

(a) *Selecting high efficiency compressors*: Power consumption of reciprocating compressors in central A/C plants is in the range of 0.75–1.25 kW/TR. Hence, one with lowest kW/TR should be selected. This could mean a saving of as much as 40 kW for a 100 TR theatre A/C plant. Energy saving is thus quite enormous.

(b) *Capacity control of reciprocating compressors*: Suction pressure operated cylinder unloaders enable reduction in power proportional to capacity at part loads. External controls with solenoid valves and hot gas bypass do not offer any reduction in power consumption.

(c) *Multistage compression in centrifugal compressors*: Two-stage compression has high efficiency than single-stage compression. It consumes less kW/TR. For example, we find:

Single-stage requires: 0.8 kW/TR approx.

Two-stage requires: 0.71 kW/TR approx.

Thus, 11.5% saving in power can be achieved with 2-stage. Still higher efficiency can be achieved by providing an economiser/flash intercooler. In a single 500 TR centrifugal compressor, the saving during the life of the plant would be substantial.

(d) *Multi-evaporators with individual or multistage compressors*: In very large A/C installations, it is possible to consider cooling of air in two stages, first with chilled water at 15.5°C, and then with chilled water at 7°C; to achieve final supply air temperature of 14-15°C. Power requirement to cool water to 15.5°C is less than that for chilling to 7°C by about 30%. Hence, for a 500 TR air-conditioning plant, the saving is of the order of 55 kW.

(e) *Optimal design of refrigeration and air conditioning equipment*: The design of condenser and particularly the evaporator are crucial. The cost of equipment can be reduced by selecting tubes and augmentation devices to increase heat-transfer coefficients. However, these may result in increased refrigerant-side pressure drop, and hence higher running cost. The two cost factors have to be balanced to obtain total minimum cost.

Similarly, chillers should be operated at the highest possible *leaving water temperature* (LWT). Again, the overall cost has to be minimised.

(f) *Most designers oversize the equipment*: This results not only in more capital cost, but also more energy consumption.

(g) *Thermostat construction and adjustment*: Its influence is very great in room air conditioners and refrigerators. Thermostats with a differential of 5°C in the middle position have a rather short cut-in time. This often results in re-starting of the compressor even before the pressures have equalized. This causes frequent overload trippings. The result is—increase in energy consumption of the order of 4 kWH in one-ton air conditioners and 0.5 kWH in refrigerators in a day. This energy loss can be easily avoided by setting the thermostat saturated suction temperature differential for 'on'-'off' at an increased value of 8 ± 1°C.

19.10.7 Plant Maintenance

The need for proper maintenance of R and A/C equipment cannot be over emphasized.

(a) *Cooling tower maintenance*: One of the most vulnerable part is the cooling tower. The pipes and nozzles may easily get clogged due to algae, muck, leaves, dust, etc., resulting in poor cooling of water. If temperature of water returned to condenser is increased by 1°C, a 6% increase in power consumption is inevitable. And if the amount of water circulated is also reduced, a further increase in energy consumption will result. Proper maintenance of cooling towers, therefore, is an absolute necessity.

Performance of existing cooling towers can also be upgraded. A *cellular fill* is found to be better. PVC pipes could be used in place of GI ones. These will be non-corrosive. Similarly, ceramic non-corroding nozzles can be used.

Location of cooling towers is important. No high temperature flue gases, and exhausts from boiler, kitchen, laundry, etc., be permitted to enter cooling tower intake.

(b) *Removing fouling from condenser and chiller tubes*: This is another important source of deterioration of plant performance. Fouling results in high condenser and low evaporator pressures, and thus reduced capacity as well as high energy consumption. Dirty surfaces of air-cooled condenses also have the same effect. This overloading may also reduce the life of the plant. Hence, descaling of condensers and chillers must be done at intervals depending upon the hardness of water to reduce temperature gradient across the surface for heat transfer.

Descaling can keep down the energy consumption by approximately 20 per cent.

Water treatment may be necessary to prevent scaling and fouling of heat exchangers.

(c) *Cleaning of air filters*: Choking of air filters results in reduced flow of air through the cooling coil. At the same time, fan power consumption is increased.

Due to less air flowing over the coil, the evaporator pressure continually drops. Once again, the capacity of the air conditioner is reduced and the power consumption is increased. Even though the evaporator temperature and the temperature of air leaving coil are lowered, there is not enough cooling as the dehumidified air quantity is reduced. Very often, the cause for complaints in room air conditioners is simply unclean filters, and not the system which is mostly working satisfactorily. Air filters must, therefore, be frequently cleaned.

(d) *Overhauling of compressors*: Timely overhauling is necessary. Due to wearing out of valve reeds and piston rings, the volumetric efficiency is considerably reduced. There is, thus, a loss of refrigerating capacity. The friction losses remaining unchanged, there is simply a wastage of power.

19.10.8 Permitting Drift in Room Design Conditions

Most of the air conditioning equipment is governed by *bypass control* which maintains a constant DBT in room. The system and its psychrometry are illustrated in Figs 23.16 and 23.17 respectively. In this, the entering recirculated room air and outdoor air mixture is purposely diverted around the cooling coil. As against the inherent *coil bypass*, this is referred to as *system bypass*.

This method lacks humidity control as described in Sec. 23.6.2. In addition, therefore, one can provide controls allowing for the space temperature swing to 26°C in summer and 16°C in winter at the time of *peak cooling and heating loads*, thus reducing the size and energy consumption of the A/C system for the building. It has now been found that small drifts in temperature and humidity from the recommended steady-state conditions do not result in decline in comfort level. An acceptable level of change in temperature is $\leq$ 0.6°C per hour from a 25°C base, and in humidity is $\leq$ 16 torr with normal summer clothing and sedentary activity. The controls can be devised accordingly.

Note *Other energy conservation measures include the following:*

(i) Winter heating by heat pump. It will provide 3.4 kW of heating for 1 kW of power consumption.

(ii) Providing air locks in all major entrances to minimise infiltration.

(iii) Shading of the building by trees blooming in summer. The trees provide shade, and also act as evaporative coolers.

(iv) Replacing all resistance type regulators, dimmers, etc., with electronic ones.

(v) Installing timers, controls, etc., to automatically switch off lights, and air handling units (AHUs) when the space is not in use.

(vi) Preferring central A/C plants which employ water-cooled condensers in place of a number of window-type A/Cs which employ air-cooled condensers.

References

1. ASHRAE *Guide and Databook*, 'Applications Volume', 1964.
2. Carrier Air Conditioning Co., *Handbook of Air Conditioning System Design*, McGraw-Hill, New York, 1965.
3. Carrier W H, R E, Cherne, W A Grant and W H Roberts, *Modern Air Conditioning, Heating and Ventilating*, Pitman, New York, 1959.

Revision Exercises

19.1 Calculate the total heat gain of a restaurant at its peak occupancy load at about 1 p.m. when 100 diners and 15 employees are present. Given:

Heat gain through walls and roof	: 2500 kJ/h
Heat gain through glass areas	: 500 kJ/h
Number of fluorescent tube lights	: 60
Rating of each tube light	: 40 W
Rating of toasters inside space	: 2650 W
Sensible heat gain per diner	: 250 kJ/h
Latent heat gain per diner	: 260 kJ/h
Sensible heat gain per employee	: 305 kJ/h
Latent heat gain per employee	: 545 kJ/h
Inside design conditions	: 25°C DB, 19°C WB
Outside design conditions	: 40°C DB, 27°C WB
Ventilation requirement	: 0.4 cmm/person

19.2 A space to be conditioned has the following data.

Size of space	: 30 m × 30 m × 4 m high
West glass	: 15 m^2
South glass	: 15 m^2
Solar gain through west glass	: 508 W/m^2 at 4 p.m
Solar gain through south glass	: 38 W/m^2 at 4 p.m.
Overall heat-transfer coefficient of roof	: 2.5 W/m^2K
Overall heat-transfer coefficient of wall	: 3.5 W/m^2K

Overall heat-transfer coefficient of glass	: 6 W/m^{2}K
Door in E-wall	: 3 m × 2.5 m
Overall heat-transfer coefficient of door	: 1.5 W/m^{2}K
Equivalent temperature differentials at 4 p.m.	
E-wall	: 15°C
W-wall	: 10.5°C
N-wall	: 6.1°C
S-wall	: 10.5°C
Roof	: 17.8°C
Infiltration through window cracks	: 5.3 m^2/h/m
Infiltration through door openings	: 3 cmm/m^2
Occupancy	: 100
Sensible heat gain per occupant	: 75 W
Latent heat gain per occupant	: 55 W
Lighting	: 33.5 W/m^2 fluorescent
Outside design conditions	: 43°C DB, 27°C WB
Inside design conditions	: 25°C DB, 50% RH
Ventilation air	: 0.24 cmm/person.

Assume a suitable fan heat and bypass factor of the air-conditioning apparatus. Calculate:

(i) Room sensible heat gain.
(ii) Room latent heat gain.
(iii) Outside air sensible, latent and total heat gains.
(iv) Grand total cooling load on apparatus.
(v) Effective sensible heat factor and apparatus dew point.
(vi) Dehumidified and recirculated room air quantities.

19.3 A laboratory has the following heat gains:

Sensible heat	: 35 kW
Latent heat	: 20 kW

The design conditions are as follows:

Outside	: 40°C DB, 27°C WB
Inside	: 22°C DB, 50% RH

The ventilation air requirement is 80 cmm. A cooling coil with a bypass factor of 0.05 must be used. An apparatus dew point of 10°C must be maintained. Determine:

(i) Amount of reheat required.
(ii) Supply air quantity.
(iii) Dry bulb and wet bulb temperatures of air entering apparatus.
(iv) Dry bulb and wet bulb temperatures of air leaving apparatus.
(v) Supply air temperature.

19.4 The following are determined for an office building:

Outside design conditions	: 35°C DBT, 28°C WBT
Inside design conditions	: 25°C DBT, 50% RH
Room sensible heat gain	: 46.5 kW

Room latent heat gain : 8.1 kW
Ventilation air : 60 cmm

A 4-row D-X coil (BPF = 0.1) is to be used. Determine:

(i) Room ADP.
(ii) DBT of air off the coil.
(iii) Total quantity of supply air required.
(iv) Coil ADP and refrigerant temperature.
(v) Refrigerating capacity of the coil.
(vi) Effect of coil BPF.

19.5 A conditioned space with *partial recirculation* (Fig. 16.2) is to be maintained at 24°C DB, 17°CWB. The local outside conditions are 34.5°C DB, 25.6°C WB. The sensible heat gain of space is 44.5 kW. The latent heat gain from occupants and infiltration, but excluding ventilation, is 5.4 kW. Based on occupancy, 36 cmm of ventilation are required. Find:

(a) Temperature of supply air entering the conditioned space.
(b) Volume of air passing through the space.
(c) State of air entering the conditioner.
(d) Required ADP.
(e) Coil BPF.
(f) System bypass ratio.

Chapter

20

Design of Air-Conditioning Apparatus

20.1 AIR-CONDITIONING APPARATUS

Air-conditioning apparatuses have been generally classified into two major types, viz., coil equipment and washer or spray equipment.

In the *coil equipment*, air comes in contact with a surface and not the conditioning medium which may be either a refrigerant or water (chilled or heated), brine, steam, etc.

In the *washer equipment* or *spray equipment*, air comes in direct contact with the conditioning water (chilled, recirculated or heated) or hygroscopic fluids.

In either case the air comes in contact with a surface; which is wet in the case of the washer or spray equipment, and dry in the case of the coil equipment. However, in the case of the coil equipment also, if the temperature of the surface is below the dew point temperature of air, condensation of moisture from the air on to the surface will take place, and the surface eventually becomes wet.

20.2 HEAT AND MOISTURE TRANSFER IN AIR-CONDITIONING APPARATUS

When heat is transferred between unsaturated air and a *dry surface*, the driving force is the difference in the dry-bulb temperature of the air and the temperature of the surface. This driving force is also termed *temperature potential* and the heat transfer is termed *sensible heat transfer*.

But when heat is transferred between unsaturated air and a *wetted surface*, another driving force other than temperature difference results. This driving force is the difference in the vapour pressure in the unsaturated air and the saturation vapour pressure at the temperature of the wetted surface. The force causes a transfer of moisture resulting either in its condensation or evaporation. This driving force may be termed *vapour pressure potential* and the accompanying heat transfer as *latent heat transfer*.

When both the driving forces are present, we find that both sensible and latent heats are transferred resulting in *total heat transfer*, such that

$$dQ_T = dQ_S + dQ_L$$

The combined driving force in this case may be termed *enthalpy potential* as outlined below.

20.2.1 Enthalpy Potential[6]

Consider an elementary wetted surface at temperature t_S as shown in Fig. 20.1. Above the wetted surface, there exists a film of air through which temperature and vapour pressure gradients exist. In the immediate vicinity of the wetted surface, the air is saturated at t_S, ω_S and h_S. Under equilibrium conditions, the rate of diffusion of water vapour through the air film will equal the rate of condensation or evaporation of water on the wetted surface.

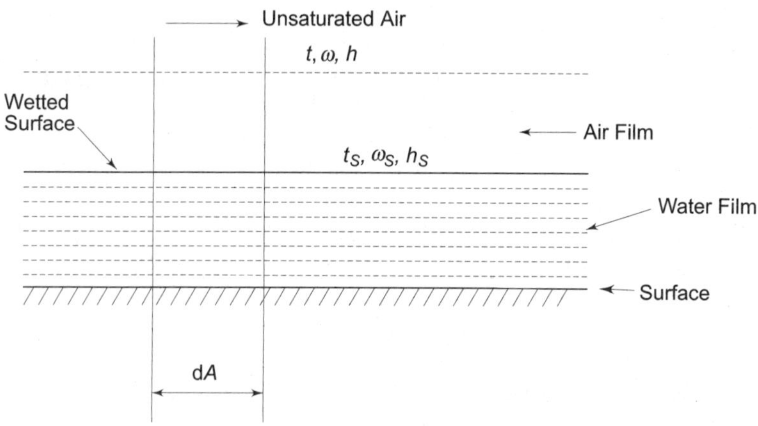

Fig. 20.1 Conditions with unsaturated air flowing over a wetted surface

Thus if f_g is the film coefficient of heat transfer through the air film, we have for the sensible heat transfer between the air and the wetted surface

$$\mathrm{d}\dot{Q}_S = f_g\,(t - t_S)\,\mathrm{d}A \tag{20.1}$$

Also, if k_ω is the diffusion coefficient of water vapour through the air film, and if h_{fg_S} is the latent heat at t_S, we have for latent heat transfer between unsaturated air and wetted surface

$$\mathrm{d}\dot{Q}_L = h_{fg_S}\,\mathrm{d}\dot{m}_v \tag{20.2}$$

where

$$\mathrm{d}\dot{m}_v = k_\omega\,(\omega - \omega_S)\mathrm{d}A \tag{20.3}$$

so that

$$\mathrm{d}\dot{Q}_L = k_\omega\,(\omega - \omega_S)\,h_{fg_S}\,\mathrm{d}A \tag{20.4}$$

As already stated in Chapter 15 according to Lewis, we have

$$k_\omega = \frac{f_g}{\mathrm{Le}\,C_p} \cong \frac{f_g}{C_p} \quad \text{for Le} = 1$$

Combining Eqs (20.1) and (20.2) we have for total heat transfer

$$\mathrm{d}\dot{Q}_T = f_g\,(t - t_S)\,\mathrm{d}A + \frac{f_g}{\mathrm{Le}\,C_p}\,h_{fg_S}\,(\omega - \omega_S)\,\mathrm{d}A$$

$$= \frac{f_g}{C_p} [C_p (t - t_S) + \frac{h_{fgS}}{\text{Le}} (\omega - \omega_S)] \, dA \tag{20.5}$$

where $\quad h_{fg_S} = h_v - h_{f_S}$.

Rearranging

$$d\dot{Q}_T = \frac{f_g}{C_p} [(C_p t + \frac{h_{fgS}}{\text{Le}} \omega) - (C_p t_S + \frac{h_{fgS}}{\text{Le}} \omega_S)] \, dA \tag{20.6}$$

Taking average values of the humid specific heat C_p and latent heat as h_{fg_o}, we may write

$$d\dot{Q}_T = \frac{f_g}{C_p} (h - h_S) \, dA + \frac{f_g}{C_p} \left(\frac{1}{\text{Le}} - 1\right) h_{fg_S} (\omega - \omega_S) \, dA$$

$$= \frac{f_g}{C_p} (h - h_S) \, dA, \quad \text{for} \quad \text{Le} = 1 \tag{20.7}$$

The quantity $(h - h_S)$ is the *enthalpy potential*. Thus, the driving force for the total heat transfer is the difference in the enthalpy h of unsaturated air and the enthalpy h_S of saturated air at the temperature of the wetted surface. And the coefficient determining this heat transfer is equal to $f_g/C_p = k_\omega$.

20.2.2 Surface Temperature and Direction of Process[6]

The following examples illustrate the usefulness of the concept of enthalpy potential. Depending on the wetted surface temperature, and the state of the unsaturated air in contact with the wetted surface, the direction of the process can be explained. Let the air being conditioned be at state 1 as shown in Fig. 20.2, with the dry bulb, wet bulb and dew point temperatures of t, t' and t_d respectively. Then the cases arise as follows:

Case I Cooling and Dehumidification (Process $1 - S_1$)
The surface temperature is lower than dew point temperature of air.

$t_{S_1} < t$	dQ_S is from air to surface
$\omega_{S_1} < \omega$	dQ_L is from air to surface
$h_{S_1} < h$	dQ_T is from air to surface

Case II Simple Cooling (Process $1 - S_2$)
The surface temperature is equal to the dew point temperature.

$t_{S_2} = t_d < t$	dQ_S is from air to surface
$\omega_{S_2} = \omega$	dQ_L is zero
$h_{S_2} < h$	dQ_T is from air to surface
	$dQ_T = dQ_S$

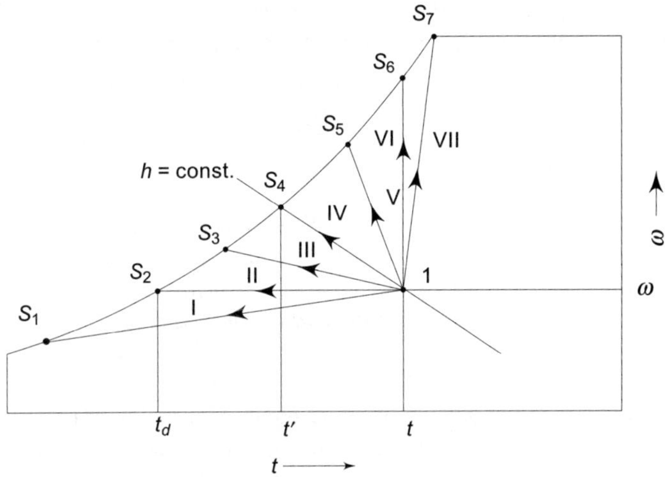

Fig. 20.2 Wetted surface temperatures and directions of processes

Case III Cooling and Humidification (Process $1 - S_3$)

The surface temperature is higher than the dew point temperature, but lower than the wet bulb temperature of air.

$t_d < t_{S_3} < t$	dQ_S is from air to surface
$\omega_{S_3} > \omega$	dQ_L is from surface to air
$h_{S_3} < h$	dQ_T is from air to surface

The air is, on the whole, losing energy to the surface.

Case IV Adiabatic Saturation (Process $1 - S_4$)

The surface temperature is equal to wet bulb temperature of air.

$t' = t_{S_4} < t$	dQ_S is from air to surface
$\omega' = \omega_{S_4} > \omega$	dQ_L is from surface to air
$h' = h_{S_4} = h$	dQ_T is zero
	$dQ_S = dQ_L$

Case V Cooling and Humidification (Process $1 - S_5$)

The surface temperature is higher than the wet bulb temperature, but lower than the dry bulb temperature of air.

$t' < t_{S_5} < t$	dQ_S is from air to surface
$\omega_{S_5} > \omega$	dQ_L is from surface to air
$h_{S_5} > h$	dQ_T is from surface to air

The air is, on the whole, gaining energy from the surface.

Case VI Simple Humidification (Process $1 - S_6$)

The surface temperature is equal to the dry bulb temperature of air.

$t_{S_6} = t$	dQ_S is zero
$\omega_{S_6} > \omega$	dQ_L is from surface to air
$h_{S_6} > h$	dQ_T is from surface to air
	$dQ_T = dQ_L$

Case VII: Heating and Humidification (Process 1 – S_7)
The surface temperature is higher than the dry bulb temperature of air.

$t_{S_7} > t$ — dQ_S is from surface to air

$\omega_{S_7} > \omega$ — dQ_L is from surface to air

$h_{S_7} > h$ — dQ_T is from surface to air

Note *It may be seen that in Cases III to V, there will be no latent heat transfer if the surface is dry. On the other hand, in Case I there will be condensation of moisture from the air on the surface, irrespective of the surface being either dry initially or wet. Latent heat transfer is always present in sprayed coils.*

20.2.3 Effective Surface Temperature[7]

The surface temperature in air-conditioning equipment varies from one end of the apparatus to the other end. Nevertheless, the *effective surface temperature* t_S can be considered as the uniform surface temperature that would produce the same leaving air state as the varying surface temperature. In the case of a cooling and dehumidifying equipment, it is equal to the apparatus dew point.

As an illustration Fig. 20.3 shows the entering and leaving states for air and chilled water in a counterflow cooling and dehumidifying apparatus.

The wet bulb temperature of air decreases from t'_1 to t'_2, whereas the chilled water temperature rises from t_{w_2} to t_{w_1}. The temperature of the surface changes from t_{S_1} to t_{S_2} from one end to the other end of the heat exchanger. The effective surface temperature t_S would lie between t_{S_1} and t_{S_2} and would correspond to the same entering and leaving states of both fluids.

The selection of the air-conditioning equipment is usually based on the concept of the effective surface temperature and bypass factor. However, the actual design of such an equipment would involve the basic principles of heat and mass transfer between the air and the heating or cooling media through the air-conditioning apparatus.

Note *In the case of a direct-expansion cooling coil, the chilled water is replaced by the evaporating refrigerant so that* $t_{w_1} = t_{w_2} = t_r$. *In the case of a spray equipment with chilled water which itself forms the surface, the temperature of water is the same as the surface temperature, so that* $t_w = t_S$, $t_{w_1} = t_{S_1}$, $t_{w_2} = t_{S_2}$, *and so on.*

20.2.4 Numerical Procedure for Heat and Mass Transfer Calculations between Unsaturated Air and Wetted Surface[2]

Consider a process of heat and mass transfer between unsaturated air and a wetted surface of area dA at temperature t_S. Then, referring to Fig. 20.1 we may write for the moisture transfer $d\dot{m}_v$

$$d\dot{m}_v = \dot{m}_a d\omega = k_\omega dA\,(\omega_s - \omega) \tag{20.8}$$

where $\dot{m}_a$ is the mass flow rate of dry air.
Equation (20.8) may be written as

$$\frac{d\omega}{\omega_S - \omega} = \frac{k_\omega dA}{\dot{m}_a} \tag{20.9}$$

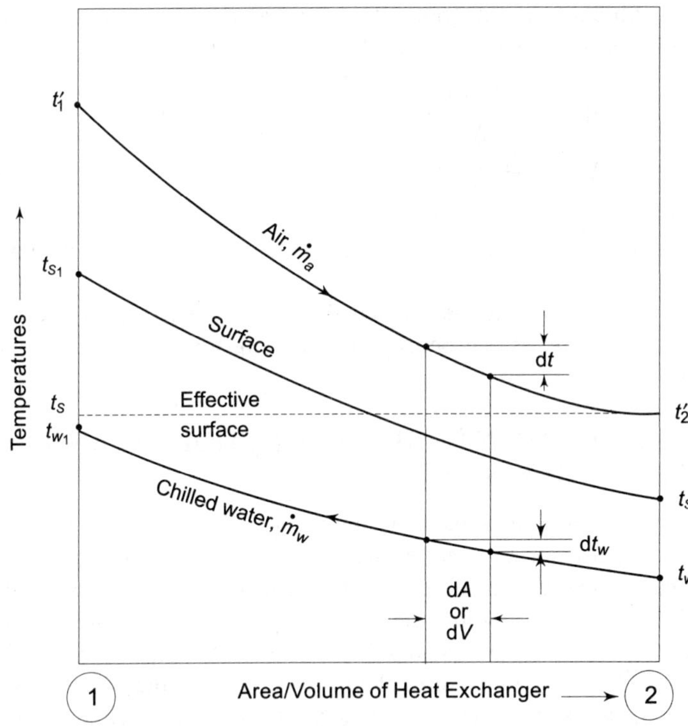

Fig. 20.3 Entering and leaving states of air and chilled water in a cooling and dehumidifying apparatus

The integration of Eq. (20.9) between the two ends of the exchanger gives

$$\int_1^2 \frac{d\omega}{\omega_S - \omega} = \frac{k_\omega A}{\dot{m}_a} \tag{20.10}$$

The dimensionless term $k_\omega A/\dot{m}_a$ is called the *number of transfer units* (NTU).

Again, writing for the total heat transfer between air and the wetted surface, we have

$$d\dot{Q}_T = \dot{m}_a dh = \frac{f_g}{C_p} dA\,(h_S - h)$$

$$= k_\omega\, dA\,(h_S - h) \tag{20.11}$$

The integration of Eq. (20.11) gives

$$\int_1^2 \frac{dh}{h_S - h} = \frac{k_\omega A}{\dot{m}_a} \tag{20.12}$$

Combining Eqs (20.9) and (20.11), we obtain

$$\frac{dh}{d\omega} = \frac{h_S - h}{\omega_S - \omega} \quad \text{or} \quad \frac{dh}{h_S - h} = \frac{d\omega}{\omega_S - \omega} = \frac{k_\omega dA}{\dot{m}_a} \tag{20.13}$$

Thus, for an actual apparatus, the value of $k_\omega A/\dot{m}_a$ may be experimentally obtained by plotting either $1/(h_S - h)$ against h or $1/(\omega_S - \omega)$ against ω along the process, and subsequently integrating the curve by numerical integration, e.g., by applying Simpson's rule according to which

$$\int_a^b y\mathrm{d}x = \frac{(b-a)}{3N}\,[y_0 + y_N + 4\,(y_1 + y_3 + \ldots + y_{N-1}) + 2\,(y_2 + y_4 + \ldots + y_{N-2})] \tag{20.14}$$

where a and b are the limits of integration.

The range $(b - a)$ is divided into N equal parts where N is an even number. The ordinates of the curve are $y_0, y_1, y_2, \ldots, y_{N-2}, y_{N-1}, y_N$.

Once the value of the number of transfer units $k_\omega\, A/\dot{m}_a$ is obtained, the heat and mass transfer process can be established along the exchanger for any given operating conditions.

20.3 COIL EQUIPMENT—DESIGN OF COOLING AND DEHUMIDIFYING COILS

The design of coils for only sensible heating and cooling is simple and follows the procedure prescribed for any heat exchanger. Coils for simultaneous cooling and dehumidification involve sensible as well as latent heat transfer and since they have major applications in air conditioning, their design procedure is outlined here.

A section of a cooling and dehumidifying coil with just one illustrative fin is shown in Fig. 20.4. The temperatures of air, wetted surface, outside and inside metal wall and refrigerant at any section are t, t_S, t_{m_o}, t_{m_i} and t_r respectively. The thermal resistances to heat transfer consist of those due to the air film, condensate water layer, metal wall and refrigerant film. In addition, there is a mass-transfer resistance between the air and the wetted surface.

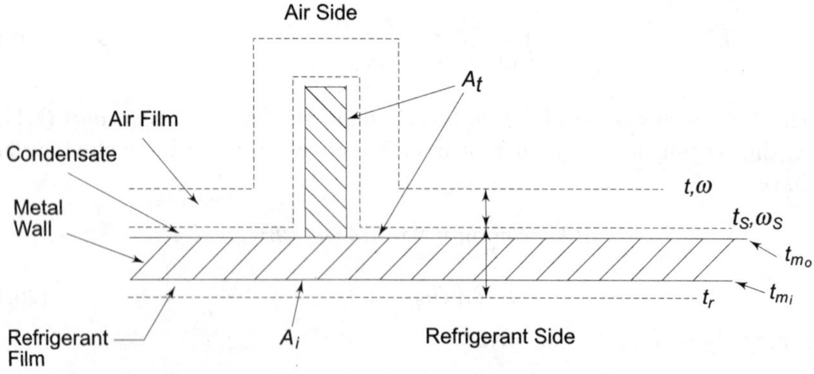

Fig. 20.4 Section of finned cooling and dehumidifying coil: thermal resistances

The resistances can be divided into two sections[5] which are given below.

Section I: This section is that of the air film. In this section both sensible and latent heats are transferred. The sensible heat Q_S is transferred from the air to the surface by virtue of the temperature difference $(t - t_S)$. Also, the latent heat Q_L is transferred by virtue of the partial pressure difference $(p_v - p_{v_S})$ or the specific humidity difference $(\omega - \omega_S)$, where p_{v_S} is the water vapour pressure and ω_S is the specific humidity at temperature t_S of the wetted surface.

Let us consider that the coil is a finned-type. The total heat transfer through section *I* with the inside tube surface area dA_i is

$$d\dot{Q}_T = d\dot{Q}_S + d\dot{Q}_L = d\dot{Q}_T$$

$$= \frac{f_g}{C_p}(h - h_s)\,\eta_f R\, dA_i \tag{20.15}$$

where η_f is the fin efficiency and

$$R = \frac{A_t}{A_i} = \text{Ratio of the total fin-side surface area } A_t \text{ to the inside surface area } A_i$$

Section II: This section comprises the three thermal resistances due to the condensate film, metal wall and refrigerant film. The heat is transferred as sensible heat only by virtue of the temperature difference $(t_S - t_r)$. Then, if U_i represents the overall heat-transfer coefficient for the three resistances based on the inside tube surface area, the heat transfer through the section is given by

$$d\dot{Q}_{II} = U_i(t_S - t_r)\, dA_i \tag{20.16}$$

By energy balance

$$d\dot{Q}_I = d\dot{Q}_{II}$$

so that we have from Eqs (20.15) and (20.16)

$$U_i\,(t_S - t_r)\, dA_i = \frac{f_g}{C_p}(h - h_S)\,\eta_f R\, dA_i$$

or

$$\frac{t_S - t_r}{h - h_s} = \frac{\eta_f f_g R}{U_i C_p} = \frac{k_\omega}{U_i}\,\eta_f R \tag{20.17}$$

The right-hand side is constant in case f_g and U_i are assumed to be constant. Equation (20.17) can be used for design by dividing the coil into a number of segments since t_S is varying.

Equation (20.17) can be written in another simplified manner. Writing the total and sensible heat transfers separately, we have for Section I

$$d\dot{Q}_T = d\dot{Q}_I = \frac{f_g}{C_p}(h - h_S)\,\eta_f R\, dA_i \tag{20.15}$$

$$d\dot{Q}_S = f_g\,(t - t_S)\,\eta_f R\, dA_i \tag{20.18}$$

Taking the ratio of the total heat to the sensible heat transfer we have

$$\frac{\text{TH}}{\text{SH}} = \frac{d\dot{Q}_T}{d\dot{Q}_S} = \frac{h - h_S}{C_p(t - t_S)} = \frac{1}{\text{GSHF}}$$

where GSHF is the required grand sensible heat factor of the apparatus. Thus

$$(h - h_S) = \frac{1}{\text{GSHF}}\, C_p\,(t - t_S) \tag{20.19}$$

Substituting from Eq. (20.19) into Eq. (20.15), we obtain for the total heat transfer

$$d\dot{Q}_T = \frac{f_g}{\text{GSHF}} (t - t_S)\, \eta_f R\, dA_i$$
$$= f'_g (t - t_S)\, \eta_f R\, dA_i \tag{20.20}$$

where

$$f'_g = \frac{f_g}{\text{GSHF}} \tag{20.21}$$

Thus f'_g can be taken as the *total heat-transfer coefficient* on the air side, when the sensible heat-transfer coefficient is f_g. Putting dQ_{II} in Eq. (20.16) equal to dQ_I in Eq. (20.20), we obtain the energy balance relation in the form

$$\frac{t_S - t_r}{t - t_S} = \frac{f'_g}{U_i} \eta_f R \tag{20.22}$$

The overall heat-transfer coefficient U_i can be calculated employing the usual relation for thermal resistances in series, viz.,

$$\frac{1}{U_i A_i} = \frac{1}{f_r A_i} + \frac{\Delta x}{k A_m} + \frac{1}{f_c \eta_f R A_i} \tag{20.23}$$

where

f_r = Refrigerant side heat-transfer coefficient

Δx = Thickness of metal wall

k = Thermal conductivity of metal wall

f_c = Conductance of condensate layer

A_i, A_m, A_t = Inside, mean and outside tube surface areas.

Table 20.1 gives the representative values of the air-side film heat-transfer coefficients f_g for different coil-face velocities. The following simple equation can be fitted into these values.

$$f_g = 38\,(FV)^{0.5}.$$

Table 20.2 similarly gives the approximate values of the refrigerant-side heat-transfer coefficients f_r for direct expansion R 22 coils as a function of tube diameter and refrigerant flow.

Table 20.1 Air side film conductance[2] (with half of face area as free area and staggard tubes)

Coil Face Velocity, **FV, m/s**	*Conductance*, f_g, $\mathbf{Wm^{-2}\,K^{-1}}$
0.508	23.3
1.016	35.8
1.524	45.4
2.032	54.5
2.54	62.5
3.048	69.8

Table 20.2 Boiling refrigerant conductance for R 22 D-X coil

Refrigerant Flow	***Tube OD,* cm**		
TR/Circuit	***1.27***	***1.5875***	***1.905***
0.5	1420		
0.7	2129		
0.8		1420	
1.2		2271	1420
1.5		2839	
1.7			2129

For a precise estimation of f_r, well-known boiling heat transfer correlations given in Chapter 9 may be used. For chilled-water coils, the water-side coefficient can be found by using the Dittus-Boelter equation.

Alternatively, we express the overall heat-transfer coefficient U_t' for Sections I and II together for which the overall temperature difference is $t - t_r$, and base it on the total fin-side surface area $A_t = RA_i$,

$$\begin{aligned} \mathrm{d}\dot{Q}_T &= U_t'(t - t_r)\,\mathrm{d}A_t \\ &= f_g'\,(t - t_S)\,\eta_f \mathrm{d}A_t \\ &= U_i\,(t_S - t_r)\,\mathrm{d}A_i \end{aligned} \tag{20.24}$$

Then the overall resistance based on the temperature potential $(t - t_r)$ is

$$\frac{1}{U_t'\mathrm{d}A_t} = \frac{1}{f_g'\eta_f \mathrm{d}A_t} + \frac{1}{U_i \mathrm{d}A_i}$$

$$\Rightarrow \quad \frac{1}{U_t'} = \frac{1}{f_g'\eta_f} + \frac{R}{U_i} \tag{20.25}$$

Substituting the complete expression for U_i, we have

$$\frac{1}{U_t'} = \frac{1}{f_g'\eta_f} + \frac{R}{f_r} + \frac{\Delta x R A_i}{k A_m} + \frac{1}{f_c\,\eta_f} \tag{20.26}$$

It may be noted that the design of a cooling coil involves the selection of number of circuits, face velocity, refrigerant or chilled-water temperature, number of rows along with parameters such as fin and tube spacing, arrangement of fins, etc.

Examples 20.1 and 20.2 illustrate the design procedure for a direct-expansion cooling coil. Whereas in Example 20.1, a stepped approach is used and Simpson's rule is employed, in Example 20.2, the coil is not divided into sections, but an overall design is given.

20.3.1 Air-Side Heat-Transfer Coefficient

Exact calculation of heat-transfer coefficient on the air side is based on the correlations of Elmahdy and Biggs[3] for flow of dry air over finned tubes as given below

$$f_g = \frac{JGC_p}{\mathrm{Pr}^{2/3}} \tag{20.27}$$

where J is the Colburn j-factor given by

$$J = C_1 \, \mathrm{Re}^{C_2} \tag{20.28}$$

in which C_1 and C_2 are constants which are functions of geometric parameters of the finned tubes. These can be expressed as

$$C_1 = 0.159 \left(\frac{F_T}{F_H}\right)^{0.141} \left(\frac{D_h}{F_T}\right)^{0.065} \tag{20.29}$$

$$C_2 = -\,0.323 \left(\frac{F_T}{F_H}\right)^{0.049} \left(\frac{F_s}{F_T}\right)^{0.077} \left(\frac{F_D}{S_2}\right)^{0.549} \tag{20.30}$$

wherein we have:

F_T = Fin thickness

F_H = Fin height = $r_2 - r_1 = r_2 - r_0 \cong \dfrac{1}{2}\left[\dfrac{S_1 + S_2}{2} - D_0\right]$

F_s = Fin spacing

F_D = Fin diameter = $2r_2 = 2\left(\dfrac{S_1 S_2}{4\pi}\right)^{0.5} \cong \dfrac{S_1 + S_2}{2}$

S_1 = Tube spacing in a row

S_2 = Row spacing

The hydraulic diameter D_h is given by

$$D_h = 4S_2 \, \frac{A_C}{A_t} \tag{20.31}$$

where A_c is the minimum/clear flow area and A_t is the total air-side area of the coil. The coefficient for wetted surface is then

$$f_g' = \frac{f_g}{\mathrm{GSHF}}$$

The parameter for fin efficiency becomes

$$m = \sqrt{\frac{2 f_g'}{k_{\mathrm{fin}} F_H}} \tag{20.32}$$

so that the fin efficiency can be evaluated from

$$\eta_f = \frac{\tanh m F_H}{m F_H} \tag{20.33}$$

and the air-side thermal resistance from the expression

$$\frac{1}{\eta_f f_g' A_t} \tag{20.34}$$

20.3.2 Diffusion Coefficient D and Mass-transfer Coefficient k_ω

For sensible heat transfer per unit area on the air side

$$\frac{\dot{Q}_S}{A} = f_g(t - t_S) = k\left(\frac{\partial t}{\partial y}\right)_{y=0,\ \text{at wetted surface}}$$

The solution of the above equation has the form

$$\text{Nu} = \frac{f_g L}{k} = f(\text{Re, Pr}) = a\left(\frac{Lu\rho}{\mu}\right)^b \left(\frac{C_p \mu}{k}\right)^c \tag{20.35}$$

For moisture transfer per unit area, we have in terms of diffusion coefficient D

$$\frac{\dot{m}_v}{A} = -D\rho\left(\frac{\partial \omega}{\partial y}\right)_{y=0}$$

We have also defined mass-transfer coefficient k_ω by

$$\frac{\dot{m}_v}{A} = k_\omega(\omega_S - \omega)$$

Thus, we have

$$k_\omega(\omega_S - \omega_{\text{air}}) = -D\rho\left(\frac{\partial \omega}{\partial y}\right)_{y=0}$$

The solution of the above equation has the form

$$\frac{k_\omega L}{\rho D} = f(\text{Re, Sc}) = a\left(\frac{Lu\rho}{\mu}\right)^b \left(\frac{\mu}{\rho D}\right)^c \tag{20.36}$$

where $\text{Sc} = \mu/\rho D$ is the Schmidt number.

Comparing Eqs (20.35) and (20.36), we get

$$\text{Le} = \frac{f_g}{C_p k_\omega} = \frac{k}{\rho C_p D} = \left(\frac{\alpha}{D}\right)^{1-C}$$

where $\alpha = k/\rho\ C_p$ is the thermal diffusivity. The following correlations are recommended for Lewis number.

For forced convection of air

$$\text{Le} = \left(\frac{\alpha}{D}\right)^{2/3} \tag{20.37}$$

For natural convection of air

$$\text{Le} = \left(\frac{\alpha}{D}\right)^{0.48} \tag{20.38}$$

For forced convection of air the values of α, D and Le have been compiled and are given in Table 20.3 for various temperatures. The evaluation of properties has been done at the wetted surface temperature:

Table 20.3 Thermal diffusivity, diffusion coefficient and Lewis number for forced convection of dry and saturated moist air

Temp. °C	*Degree of Saturation*	α m²/s	D m²/s	Le
10	0	1.987	2.325	0.901
	1	1.985		0.9
20	0	2.077	2.433	0.9
	1	2.072		0.898

(Contd)

Table 20.3 (*Contd.*)

Temp. °C	*Degree of Saturation*	α m²/s	D m²/s	Le
30	0	2.245	2.637	0.898
	1	2.234		0.895
40	0	2.418	2.849	0.896
	1	2.393		0.89
50	0	2.596	3.066	0.895
	1	2.531		0.88
60	0	2.694	3.187	0.894
	1	2.588		0.87

Note *It is seen that Le does not change very much. Average value of Le for dry air is 0.897, and that for saturated air is 0.889. The average of both is 0.893, which can be accepted as its value in general.*

Example 20.1 ***Design of Chilled-Water Coil-Overall LMTD Method***

Design a chilled-water cooling coil for each of the ground, first and third floors of the studio building calculations of which are given in the load estimation sheet in Table 19.6. The coil has to conform to some of the specifications given by the manufacturer which are as follows:

5/8 in Cu tubes : $OD = D_o = 15.8$ *mm*

$ID = D_i = 14.4$ *mm*

No. of tubes/row : 34

No. of circuits : 34

Frontal pitch (Tube spacing) : $S_1 = 38.1$ *mm*

Back pitch (Row spacing) : $S_2 = 43.2$ *mm*

Al fins :

Fin thickness : $F_T = 0.233$ *mm*

Fin density : 50/m

Fin spacing : $F_s = \dfrac{1}{50} = 0.002\ m = 2\ mm$

Thermal conductivity of fin : $k_{Al} = 201 W/m.K$

Fouling factor (inside) : $0.00009\ m^2\ K/W$

With these construction features, the area parameters are:

$$\frac{\text{Total outside finned surface area}}{\text{Inside tube surface area}} = \frac{A_t}{A_i} = 10.567$$

$$\frac{\text{Total outside finned surface area}}{\text{Face area}} = \frac{A_t}{FA} = 15.55$$

$$\frac{\text{Clear free flow area}}{\text{Face area}} = \frac{A_C}{FA} = 0.5$$

Estimate the size and rows of coil. Make suitable assumptions.

Solution *Coil Face Area* *Assume* coil face velocity, $FV = 2.54$ m/s

Coil face area $\quad FA = \dfrac{\text{cmm}_s}{FV} = \dfrac{351}{2.54 \times 6} = 2.303 \text{ m}^2$

Air-Side Heat-Transfer Coefficient Air enters at 28.6°C, and leaves at 15.8°C. The properties of air at the mean temperature of 22°C are:

$$\rho = 1.205 \text{ kg/m}^3$$
$$\mu = 1.82 \times 10^{-5} \text{ Ns/m}^2$$
$$C_p = 1.0057 \text{ kJ/kg.K} \quad \text{Pr} = 0.71$$

Hydraulic diameter

$$D_h = 4\,A_c\,\frac{S_2}{A_t} = 4\left(\frac{0.5}{15.55}\right)(43.2 \text{ mm}) = 5.564 \text{ mm}$$

Fin diameter and fin height

$$F_D \cong \frac{S_1 + S_2}{2} = \frac{38.1 + 43.2}{2} = 40.65 \text{ mm}$$

$$F_H \cong \frac{1}{2}\left(\frac{S_1 + S_2}{2} - D_0\right) = \frac{1}{2}(40.65 - 15.8) = 12.425 \text{ mm}$$

Constants of Eq. (20.28)

$$C_1 = 0.159\left(\frac{F_T}{F_H}\right)^{0.141}\left(\frac{D_h}{F_T}\right)^{0.065}$$

$$= 0.159\left(\frac{0.233}{12.425}\right)^{0.141}\left(\frac{5.564}{0.233}\right)^{0.065} = 0.11155$$

$$C_2 = -0.323\left(\frac{F_T}{F_H}\right)^{0.049}\left(\frac{F_s}{F_T}\right)^{0.077}\left(\frac{F_D}{S_2}\right)^{0.549}$$

$$= -0.323\left(\frac{0.233}{12.425}\right)^{0.049}\left(\frac{2}{0.233}\right)^{0.077}\left(\frac{40.65}{43.2}\right)^{0.549} = -0.3034$$

Mass velocity based on clear minimum flow area $A_c = 0.5$ (FA)

$$G_{\max} = \frac{\text{cmm}_s \rho}{60} \times \frac{1}{0.5(FA)} = \frac{351\,(1.205)}{60\,(0.5)\,(2.303)} = 6.096 \text{ kg/m}^2\text{.s}$$

Reynolds number

$$\text{Re} = \frac{G_{\max}\,D_o}{\mu} = \frac{6.096\,(0.0158)}{1.82 \times 10^{-5}} = 5292$$

Colburn J-factor

$$J = C_1 \,\text{Re}^{C_2} = 0.11155\,(5292)^{-0.3034} = 0.008274$$

Dry air-side heat-transfer coefficient

$$f_g = \text{J}\, G_{\max}\, C_p\, \text{Pr}^{-2/3}$$
$$= 0.008274\,(6.096)\,(1005.7)\,(0.71)^{-2/3} = 63.74\ \text{W/m}^2.\text{K}$$

Wet air-side heat-transfer coefficient

$$f'_g = \frac{f_g}{\text{GSHF}} = \frac{63.74}{0.9469} = 67.31\ \text{W/m}^2.\text{K}$$

Fin parameter

$$m = \sqrt{\frac{2\,f'_g}{k_{Al}\,F_H}} = \sqrt{\frac{67.31 \times 2}{201\,(0.233 \times 10^{-3})}} = 53.3$$

Fin efficiency

$$\eta_f = \frac{\tanh m\,F_H}{m\,F_H} = \frac{\tanh\,[(53.3)\,(12.425 \times 10^{-3})]}{53.3\,(12.425 \times 10^{-3})} = 0.92$$

Water-Side Heat-Transfer Coefficient Assume entering water temperature, EWT = 7.8°C, and leaving water temperature, LWT = 13.3°C. Thus, ΔT_W = LWT – EWT = 5.5°C.

Properties of water at the mean temperature of 10.6°C are:

$$k_w = 0.588\ \text{W/m.K},\ \mu_w = 1.31 \times 10^{-3}\ \text{Ns/m}^2,\ \text{Pr}_w = 9.12$$

Mass flow rate of chilled water and water velocity

$$\dot{m}_w = \frac{\dot{Q}}{C_p} = \frac{93.67}{4.1868\,(5.5)} = 4.068\ \text{kg/s}$$

$$u_w = \frac{\dot{m}_w}{\rho\,N\,\dfrac{\pi\,D_i^2}{4}} = \frac{4.068}{1000\,(34)\,\dfrac{\pi}{4}\,(0.0144)^2} = 0.735\ \text{m/s}$$

Note *This water velocity is low. It will result in low heat-transfer coefficient, and hence more rows and high capital cost. A very high water velocity will, however, cause erosion of tubes.*

Reynolds number

$$\text{Re}_w = \frac{D_i\,u_w\,\rho_w}{\mu_w} = \frac{(0.0144)\,(0.735)\,(1000)}{1.31 \times 10^{-3}} = 6814$$

Water-side heat-transfer coefficient

$$f_i = \left(\frac{k_w}{D_i}\right) 0.023\,\text{Re}^{0.8}\,\text{Pr}^{0.4}$$

$$= \left(\frac{0.588}{0.0144}\right) 0.023\,(6814)^{0.8}\,(9.12)^{0.4} = 2651\ \text{W/m}^2.\text{K}$$

Overall heat-transfer coefficient (neglecting Cu-tube resistance)

$$\frac{1}{U_t} = \frac{1}{\eta_f\, f_g'} + \frac{1}{f_i}\frac{A_t}{A_i} + \frac{1}{f_{\text{fouling}}}\frac{A_t}{A_i}$$

$$= \frac{1}{0.92\,(67.31)} + \frac{1}{2651}\,(10.567) + 0.00009\,(10.567)$$

$$\Rightarrow \qquad U_t = 47.4 \text{ W/m}^2\text{.K}$$

Log-mean temperature difference

$$\Delta T_m = \text{LMTD} = \frac{(28.6 - 13.3) - (15.8 - 7.8)}{\ln \dfrac{28.6 - 13.3}{15.8 - 7.8}} = 11.26°\text{C}$$

Total fin-side surface area

$$A_t = \frac{\dot{Q}}{U_t\, \Delta T_m} = \frac{98.67 \times 10^3}{47.4\,(11.26)} = 175.5 \text{ m}^2$$

Number of rows

$$n = \frac{A_t}{\left(\dfrac{A_t}{\text{FA}}\right)\text{FA}} = \frac{175.5}{(15.5)\,(2.303)} = 4.92 \ (\approx 5 \text{ rows})$$

Note *This is a high sensible heat factor application, (GSHF = 0.946). BPF assumed is 0.15. For such an application number of rows calculated, 5, is too high. Area calculated, 175.5 m^2, is very large. The coil will have high capital cost. The manufacturer's specification, and assumed design values need modifications. The following points emerge.*

(i) *From Table 19.7 and Fig. 19.6, it is seen that for a BPF of 0.14, one needs only a 4-row coil with 6 fins with a face velocity of 210 mpm (3.5 m/s).*

(ii) *The manufacturer is making coils with a fin density of only 50 fins per metre. This comes to 0.5 fin/cm. It has resulted in a large coil. Hence, the fin density must be increased to about 5-6 fins/cm.*

(iii) *The present face velocity assumed is 2.54 m/s. It is giving low air-side, and hence low overall heat-transfer coefficient. The air velocity could be increased to 3 to 3.5 m/s.*

(iv) *Water velocity assumed/calculated is 0.735 m/s. It can be raised to at least 1 m/s. This will increase water-side heat-transfer coefficient.*

(v) *The air-side coefficient is, however, the controlling coefficient. The overall coefficient U is lower than air-side coefficient. Hence, increase in water-side coefficient will not have much effect. Nevertheless, it will affect LWT.*

(vi) *Further, the effect of changing EWT, and hence the refrigerant evaporation temperature on total cost should also be studied. All this is best done on a digital computer.*

Ramachandran and Arora[4] have developed detailed computer programs for the simulation, design and optimization of both direct-expansion and chilled-water cooling coils.

Example 20.2 ***Design of Direct-Expansion Coil: Heat-Transfer Coefficient***

Given

4.717 m³/s of air at 32.2°C DBT and 50 per cent RH enters a finned-coil direct-expansion dehumidifier with the coil surface area per row of 42.02 m². The refrigerant temperature is 4.4°C. The air-side heat-transfer coefficient may be taken as 83 W/m²K. The overall heat-transfer coefficient from the air-water interface to the refrigerant bulk based on the fin-side surface area is 275 W/m²K. Calculate the required number of coil rows for the leaving air state at 9.7°C WBT.

Solution Refer to Fig. 20.5.

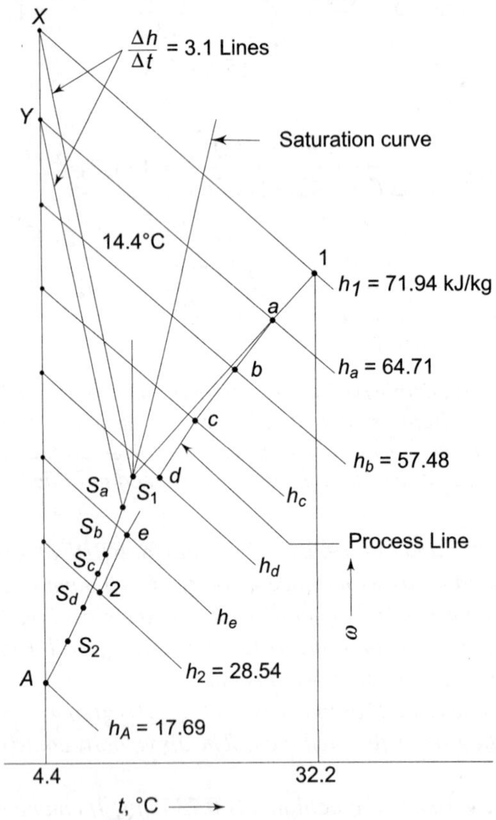

Fig. 20.5 Figure for Example 20.2

Air inlet conditions

$$t_1 = 32.2°\text{C};\ \text{RH}_1 = 50\ \%;\ h_1 = 71.94\ \text{kJ/kg};$$
$$v_1 = 0.901\ \text{m}^3/\text{kg}$$

Mass flow rate of air

$$\dot{m}_a = \frac{4.717}{0.901} = 5.235\ \text{kg/s}$$

At 4.4°C and 100 per cent RH

$$h_A = 17.69\ \text{kJ/kg}$$

Mass-transfer coefficient

$$k_\omega = f_g/C_p \text{ Le} = \frac{83}{(1021.6)\,(0.893)} = 0.09 \text{ kg/s.m}^2$$

Enthalpy of air at the exit at 9.7°C WBT

$$h_2 = 28.54 \text{ kJ/kg}$$

Enthalpy difference between inlet and outlet air conditions

$$h_1 - h_2 = 71.94 - 28.54 = 43.4 \text{ kJ/kg}$$

Let the difference be divided into six equal increments of 7.23 kJ/kg each. Now

$$U_t A_t = U_i A_i$$

so that

$$\frac{U_i}{R} = U_t = 275 \text{ W/m}^2\text{K (given)}$$

Hence, taking $\eta_f = 1$

$$\frac{h - h_S}{t_S - t_r} = \frac{U_i}{k_\omega R} = \frac{U_t}{k_\omega} = \frac{275/1000}{0.09} = 3.1 \text{ kJ/kg.K}$$

The above equation can also be solved graphically for each increment as shown in Fig. 20.5. The 4.4°C DBT line cuts the h_1 line at X. A line of slope $\Delta h/\Delta t = \frac{\Delta (h - h_S)}{\Delta (t_S - t_r)} = 3.1$ may now be drawn from point X to cut the saturation line at S_1. Point S_1 on the saturation line satisfies this condition of slope of 3.1. The interface temperature at this section t_{S_1} is then found to be 14.4°C. Join point 1 to point S_1 and where it cuts the $h_a = h_1 - 7.23 = 64.71$ kJ/kg line, gives point a on the process line.

Now, the enthalpy line h_a cuts the 4.4°C vertical at Y. Again, draw a line of slope 3.1 from Y, to cut the saturation line at S_a. Join a to S_a. This cuts the $h_b = 64.71 - 7.23 = 57.48$ kJ/kg line at b. In a similar manner, the remaining points c, d, e and 2 may be located and the process line completed.

Table 20.4 may thus be drawn up. The same may be established numerically as well. Using the above equation for calculation of h_S by trial and error for finding suitable values of t_S, we obtain Table 20.4.

Table 20.4 Stepped calculations for Example 20.2

h, kJ/kg	$h_1 = 71.94$	64.71	57.48	50.25	43.02	35.79	$h_2 = 28.54$
h_S, kJ/kg	$h_{S_1} = 39.58$	36.32	33.52	30.50	27.70	24.68	$h_{S_2} = 22.12$
$1/(h - h_S)$	0.0309	0.0352	0.0417	0.0506	0.0653	0.0900	0.158

Then from Simpson's rule

$$\int_{28.54}^{71.94} \frac{dh}{h - h_S} = \frac{71.94 - 28.54}{3 \times 6} [0.0309 + 0.1558 + 4(0.0352$$
$$+ 0.0506 + 0.09) + 2(0.0417 + 0.0653)] = 2.66$$
$$= \frac{k_\omega A}{\dot{m}_a}$$

$$\Rightarrow \qquad A = \frac{2.66\,(5.235)}{0.09} = 154.7 \text{ m}^2$$

Required number of rows

$$n = \frac{154.7}{42.02} = 3.7 = 4 \text{ (Select)}$$

Example 20.3 ***D – X Coil Design: Heat Transfer Coefficients Appropriately Selected***

0.472 m³/s of air are cooled from 26.7°C DBT and 21.1°C WBT to 12.8°C DBT and 12.2°C WBT. Design a suitable direct-expansion cooling coil.

Select 1.5875 cm OD tubes at 3.81 cm centres staggered and 0.71 mm wall thickness. The total fin-side surface area per row is 22 m² for each m² of face area. The inside surface area per row is 1.2 m² per m² of face area. Assume the resistance of the metal wall as 0.0044 W^{-1} m²K.

Solution Extended surface to inside tube surface area ratio

$$R = \frac{\text{Finned surface/row/m}^2\text{ FA}}{\text{Inside tube surface/row/m}^2\text{ FA}} = \frac{22}{1.2} = 18.3$$

Figure 20.6 shows the path of the conditioning process. If the entering and leaving air states are joined, and the line extended to intersect the saturation curve, we obtain the apparatus dew point of the coil

$$t_{ADP} = 10.4°\text{C}$$

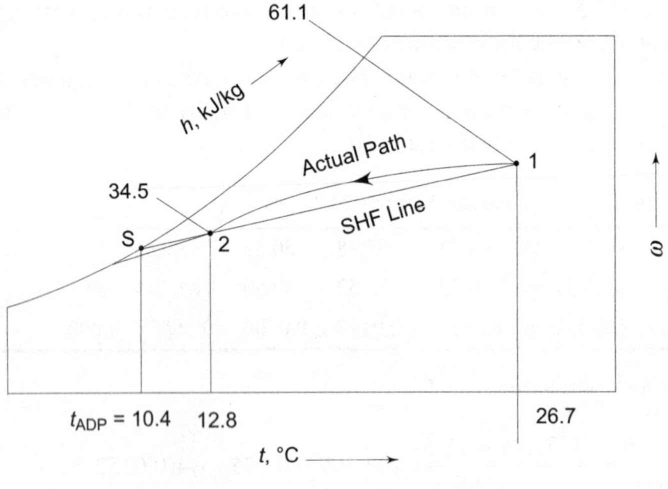

Fig. 20.6 Figure for Example 20.3

This represents the average temperature t_S of the coil. But the surface temperature t_S will be higher on the air-entering side and lower on the air-leaving side. The coil bypass factor can also be found.

$$X = \frac{t_2 - t_{ADP}}{t_1 - t_{ADP}} = \frac{12.8 - 10.4}{26.7 - 10.4} = 0.147$$

The total and sensible heat-transfer rates are as follows:

$$\text{TH} = \dot{Q}_T = \dot{Q}_v \, \rho \, (h_1 - h_2)$$
$$= (0.472)\,(1.2)\,(61.1 - 34.5) = 15.0 \text{ kW}$$
$$\text{SH} = \dot{Q}_S = \dot{Q}_v \, \rho \, C_p \, (t_1 - t_2)$$
$$= (0.472)\,(1.2)\,(1.0216)\,(26.7 - 12.8) = 8.04 \text{ kW}$$

$$\text{Ratio} = \frac{\dot{Q}_T}{\dot{Q}_S} = \frac{1}{\text{SHF}} = \frac{15}{8.04} = 1.866$$

Since the surface temperature is not the same throughout the coil, this ratio will actually not be the same. However, the assumption of a constant ratio leads to only a small error and results in a slight oversizing of the coil (see the actual path in Fig. 20.6).

Take a face velocity of 2.54 m/s.

Face area

$$\text{FA} = \frac{\dot{Q}_v}{\text{FV}} = \frac{0.472}{2.54} = 0.186 \text{ m}^2$$

Air-side film conductance (from Table 20.1)

$$f_g = 62.5 \text{ W m}^{-2} \text{ K}^{-1}$$

Total heat-transfer coefficient

$$f'_g = \frac{\text{TH}}{\text{SH}} f_g = 1.866\,(62.5) = 116.53 \text{ W m}^{-2} \text{ K}^{-1}$$

Assume a coil loading of 1 TR per circuit.

Boiling heat-transfer coefficient (from Table 20.2)

$$f_r = 1704 \text{ W m}^{-2} \text{ K}^{-1}$$

Assuming the fin efficiency to be equal to unity and neglecting the resistance of the thin condensate layer, we have for the overall heat-transfer coefficient

$$\frac{1}{U'_t} = \frac{1}{f_r}\frac{A_t}{A_i} + \frac{\Delta x}{k}\frac{A_t}{A_m} + \frac{1}{f'_g}$$
$$= \frac{18.3}{1704} + 0.0044 + \frac{1}{116.63}$$
$$U'_t = 42.17 \text{ W m}^{-2} \text{K}^{-1}$$

First approximation:

Assume a refrigerant temperature of $t_r = -1$°C. Then the temperature differentials at the two ends of the cooling coil are

$$\Delta t_1 = t_1 - t_r = 26.7 - (-1) = 27.7\text{°C}$$
$$\Delta t_2 = t_2 - t_r = 12.8 - (-1) = 13.8\text{°C}$$

Log mean temperature difference

$$\Delta t_m = \frac{27.7 - 13.8}{\ln \dfrac{27.7}{13.8}} = 19.95\text{°C}$$

Total fin side surface area

$$A_t = \frac{\dot{Q}}{U_t' \Delta t_m} = \frac{15 \times 10^3}{(42.17)(19.95)} = 17.83 \text{ m}^2$$

If n is the number of rows, then

$$A_t = n(\text{FA}) \text{ (surface area/row/m}^2 \text{ FA)}$$

$$17.83 = n(0.186)(22)$$

whence $n = 4.4$ rows

Surface temperatures, from Eq. (20.24)

$$t_{S_1} = t_1 - \frac{U_t}{f_g'}(t_1 - t_r) = 26.7 - \frac{42.17}{116.63}(26.7 + 1) = 16.7°\text{C}$$

$$t_{S_2} = t_2 - \frac{U_t'}{f_g'}(t_2 - t_r) = 12.8 - \frac{42.17}{116.63}(12.8 + 1) = 7.8°\text{C}$$

Note *As the number of rows is 4.4, it is not a whole number. Hence, further approximation is necessary.*

20.4 OPTIMAL DESIGN OF COOLING AND DEHUMIDIFYING COILS

It will be seen from Example 20.3 that for the given construction of the cooling coil, two choices are available, viz., to select a 6-row coil or a 4-row coil in lieu of the calculated result of 4.4 rows. Let us see the effect of selecting 6 or 4 rows on other parameters.

20.4.1 6-Row Coil

It can be seen that if a 6-row coil is selected, it will result in the following changes:

(i) Higher refrigerant temperature t_r can be used and hence a higher surface temperature t_S is maintained.

(ii) Leaving conditions at a lower temperature as a result of lower BPF: Back calculations show that the leaving air conditions are:

$$\text{DBT} = t_2 = 12.2°\text{C}$$

$$\text{WBT} = t_2' = 11.7°\text{C}$$

$$h_2 = 33.5 \text{ kJ/kg}$$

(iii) Increased total heat removal

$$\dot{Q}_T = \dot{Q}_v \, \rho \, \Delta h = (0.472)(1.2)(61.1 - 33.5) = 15.63 \text{ kW}$$

Required mean temperature difference

$$\Delta t_m = \frac{15.63}{(42.17)(6)(0.186)(22)} = 15.3°\text{C}$$

Refrigerant temperature required

$$t_r = 2.8°\text{C (As against} - 1°\text{C selected in Example 20.3)}$$

20.4.2 4-Row Coil

A smaller number of rows will result in the following:

(i) Lower refrigerant temperature

$$t_r = -2.8°\text{C}$$

and a lower mean surface temperature.

(ii) Leaving conditions at a higher temperature as a result of higher BPF.

(iii) Decreased heat removal

$$\dot{Q}_T = 14.95 \text{ kW}$$

A comparison will thus show that a 6-row coil will cost more. However, for the same condensing temperature, say $t_k = 37°$C although a 4-row coil would cost less but other costs will increase as follows:

(i) It will require 14 per cent more compressor displacement. As a result of lower saturated suction temperature.

(ii) It will require 10 per cent more power for a capacity of 14.95 kW at $t_r = -2.8°$C, as against a capacity of 15.63 kW at $t_r = +2.8°$C for a 6-row coil.

Obviously, for an optimal design, an analysis of costs is required. It may be pointed out here that other choices are also available. These include:

(i) *Lower face velocity:* This will result in a lower surface temperature and bring the outlet condition nearer to 2. Again, the face area and hence the cost of cooling coil will increase.

(ii) *Higher tons refrigeration per circuit:* This can be adopted only if it is practical without excessive pressure drop. This will give a higher conductance f_r on the refrigerant side, which will lower the surface temperature t_S and allow the use of a 4-row coil at the same refrigerant temperature of $t_r = -1°$C.

(iii) *Another type coil:* A coil of different construction may be used, i.e., a coil with a lower metal resistance and a lower or higher area ratio R.

All the alternatives can be compared for the optimal design of the cooling coil.

Note *Usually, a refrigerant temperature 3 to 5°C below the coil ADP provides an economical selection.*

20.5 SPRAY EQUIPMENT—DESIGN OF AIR WASHERS AND COOLING TOWERS

In a spray chamber, air is brought in direct contact with a dense spray of water. The schematic diagram of an air-spray equipment is shown in Fig. 20.7.

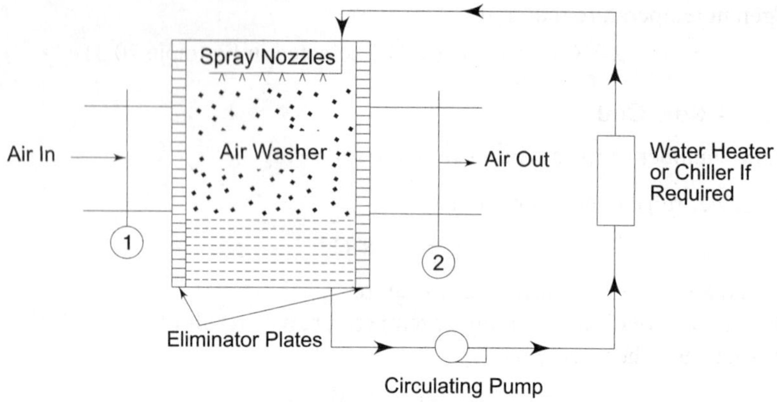

Fig. 20.7 Spray equipment

The air washer and cooling tower are two common types of spray equipment used in air conditioning. In spray equipment, there is direct contact between air and the sprayed water. Consider such a heat and mass exchanger in which the mass flow rates of the dry air and water are $\dot{m}_a$ and $\dot{m}_w$ respectively. The processes undergone by air and water are similar to those shown in Fig. 20.3. Equations derived in Sec. 20.2 can be applied to the process. However, in the case of spray equipment, the wetted-surface area is not clearly known as it is formed by the surfaces of individual droplets in the total volume of the equipment. We, therefore, denote the interfacial contact area of the surface of water droplets per unit volume of the equipment by a, so that the contact area in a differential volume dV is d$A = a$dV, and the total area $A = aV$. Also, the temperature of water t_w in the equations replaces the temperature t_S of the surface.

The energy balance equation for the process can be written using Eq. (20.7) as

$$d\dot{Q} = -\dot{m}_w C_w dt_w = \dot{m}_a dh = k_\omega (adV)(h_S - h) \tag{20.39}$$

where C_w is the specific heat of water. Equation (20.39) on integration gives

$$-\int_1^2 \frac{C_w dt_w}{h_S - h} = \frac{k_\omega aV}{\dot{m}_w} \tag{20.40}$$

Equation (20.40) can be used in conjunction with Eq. (20.12). Comparing the two, we have

$$\frac{dh}{C_w dt_w} = -\frac{\dot{m}_w}{\dot{m}_a} \tag{20.41}$$

Hence, if the properties of moist air are taken in the form of an enthalpy-temperature chart, which is another form of the psychrometric chart shown in Fig. 20.8, a line AB of constant slope given by $C_w \dot{m}_w / \dot{m}_a$ can be drawn for any process provided the initial state of water t_{w_1} and that of air at 1, and their mass flow rates are known. Such a line is called the *energy balance line*. The line AB is drawn from A, the point where the horizontal line corresponding to h_1 cuts the vertical line corresponding to t_{w_1}, upto the point B which is on the vertical corresponding to the water inlet temperature t_{w_2}. Thus the enthalpy h of air varies along the line AB at any section according to the water temperature t_w. The vertical distance between the saturation line and the line AB represents the enthalpy potential $(h_S - h)$ at any section at the water temperature t_w. See how the driving potential changes from $(h_{S_1} - h_1)$ at one end to $(h_{S_2} - h_2)$ at the other end.

To determine the value of the dimensionless performance coefficient termed NTU, viz., $k_\omega aV/\dot{m}_w$, for any equipment from the performance curve as in Fig. 20.3, one can also plot a graph between $1/(h_S - h)$ against t_w, and evaluate the area of the curve for substitution in Eq. (20.40).

In the case of pumped recirculation or evaporative cooling, ω_S is constant, and hence Eq. (20.9) can be integrated directly, so that

$$-\ln \frac{\omega_S - \omega_2}{\omega_S - \omega_1} = -k_\omega \frac{aV}{\dot{m}_a}$$

$$\frac{\omega_S - \omega_2}{\omega_S - \omega_1} = e^{-Z} \tag{20.42}$$

where

$$Z = \frac{k_\omega aV}{\dot{m}_a}$$

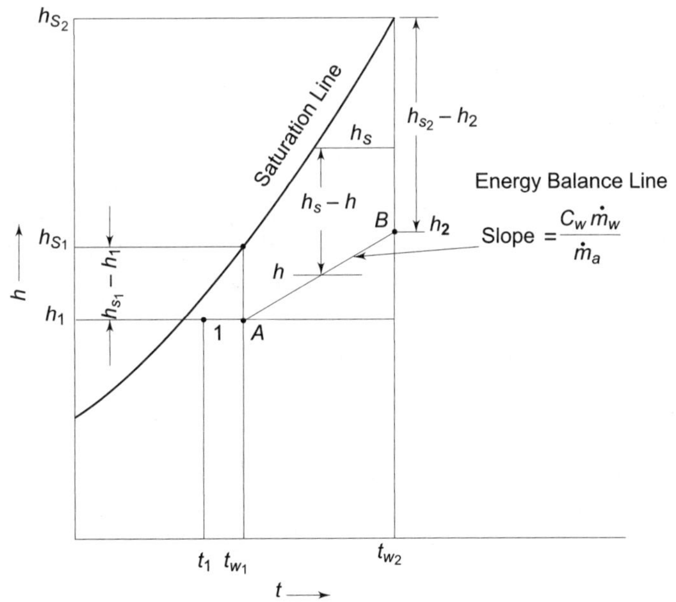

Fig. 20.8 Process line for air washer on *h-t* chart

The process is shown in Fig. 20.9. It will be seen that the air washer humidifying efficiency and bypass factor are given in terms of Z by the following expression

$$\eta_H = (1 - X) = \frac{\omega_2 - \omega_1}{\omega_S - \omega_1} = 1 - e^{-Z}$$

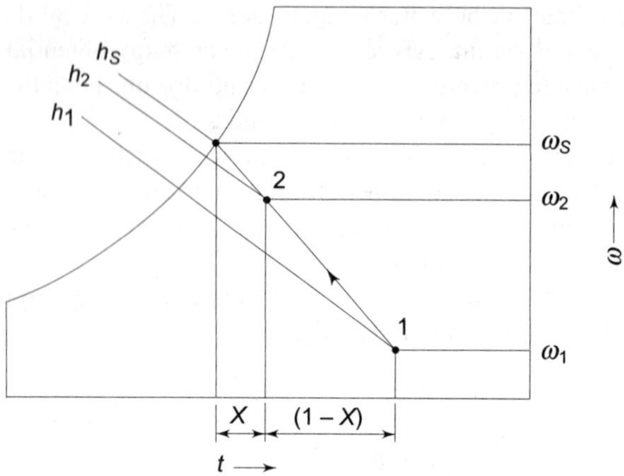

Fig. 20.9 Process line for air washer on psychrometric chart

Example 20.4 illustrates the design procedure for a spray dehumidifier. The same procedure is adopted in respect of air washers in general (humidifier and evaporative cooler) and cooling towers.

Example 20.4 *Sprayed Coil*

Air at 32.2°C DBT and 50 per cent RH, enters a spray-type dehumidifier at the rate of 4.717 m^3/s. Chilled water enters at 4.4°C and leaves at 11.2°C. The ratio of water to air mass flow rate is 1.2. The face velocity of air is 2.032 m/s. The value of the product $k_\omega a$ may be taken as 1.334 kg/sm^3. Calculate the length of the dehumidifier and the state of air at the exit assuming parallel flow.

Solution Refer to Figs 20.10 and 20.11

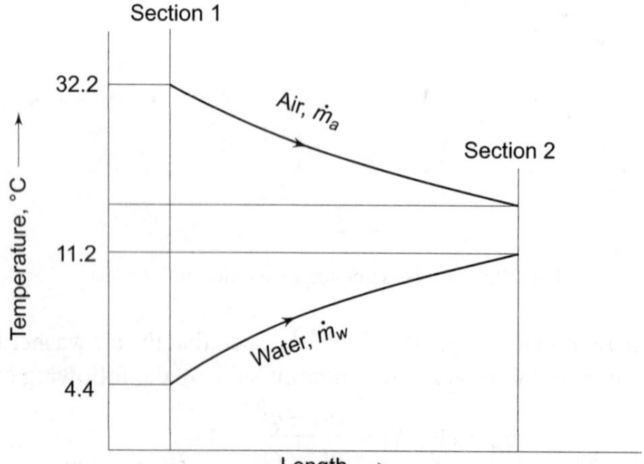

Fig. 20.10 Air and water temperatures for Example 20.4

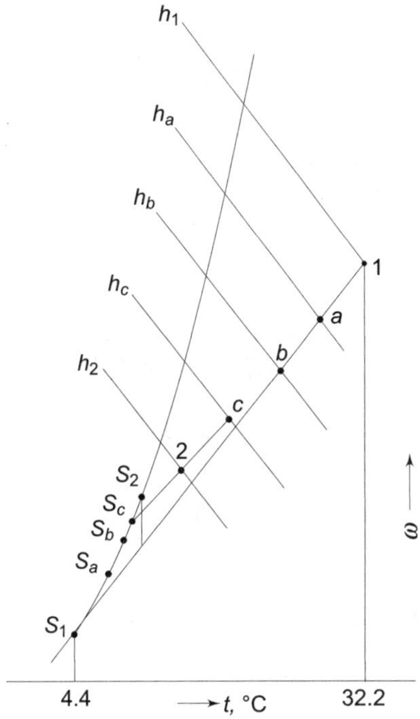

Fig. 20.11 State changes of air and water for Example 20.4

Air inlet conditions

$$t_1 = 32.2°\text{C, RH}_1 = 50\%,\ h_1 = 71.94 \text{ kJ/kg d.a.,}$$
$$v_1 = 0.901 \text{ m}^3\text{/kg d.a.}$$

Now

$$\frac{\mathrm{d}h}{C_w\,\mathrm{d}t_w} = -\frac{\dot{m}_w}{\dot{m}_a} = -1.2$$

Note *The negative sign is taken for parallel flow.*

Also, we have

$$-C_w \int_1^2 \frac{\mathrm{d}t_w}{h_S - h} = \frac{k_\omega aV}{\dot{m}_w} \quad \text{or} \quad C_w \int_{4.4}^{11.2} \frac{\mathrm{d}t_w}{h - h_S} = \frac{k_\omega aV}{\dot{m}_w}$$

Divide 6.8°C temperature interval into four equal parts with ordinates at 4.4, 6.1, 7.8, 9.5 and 11.2°C. Table 20.5 can be prepared by calculating h by energy balance for each increment, i.e.,

$$\Delta h = -1.2\, C_w\, \Delta t_w = -1.2\,(4.187)\,(1.7) = -8.54 \text{ kJ/kg d.a.}$$

Table 20.5 Stepped calculations for Example 20.4

Section		***1***	***a***	***b***	***c***	***2***
t_w	°C	4.4	6.1	7.8	9.5	11.2
h	kJ/kg	71.94	63.40	54.86	46.32	37.78
h_S	kJ/kg	17.72	21.07	24.58	28.31	32.24
$h - h_S$	kJ/kg	54.22	42.33	30.28	18.01	5.54
$1/(h - h_S)$		0.0184	0.0236	0.0330	0.0555	0.1805

Using Simpson's rule

$$\int_{4.4}^{11.2} \frac{dt_w}{h - h_S} = \frac{(11.2 - 4.4)}{3 \times 4} [0.0184 + 0.1805 + 4(0.0236 + 0.555) + 2(0.033)]$$

$$= 0.3294$$

Hence

$$\frac{k_\omega aV}{\dot{m}_w} = 0.3294\ (4.187) = 1.3792$$

Mass flow rate of dry air

$$\dot{m}_a = \frac{\dot{Q}_v}{v} = \frac{4.717}{0.901} = 5.235 \text{ kg/s}$$

Mass flow rate of water

$$\dot{m}_w = 5.235 \times 1.2 = 6.282 \text{ kg/s}$$

Voume of dehumidifier

$$V = \frac{1.3792\ \dot{m}_w}{k_\omega a} = \frac{6.282\ (1.3792)}{1.334} = 6.495 \text{ m}^3$$

Face area $\quad FA = \dfrac{4.717}{2.032} = 2.322 \text{ m}^2$

Length of dehumidifier

$$L = \frac{V}{FA} = \frac{6.495}{2.322} = 2.8 \text{ m}$$

In order to find the state of air at exit, the process can be plotted on the psychrometric chart as shown in Fig. 20.11. Joining 1 and S_1, we get a, joining a and S_a, we get *b*, and so on. Finally joining c and S_c we get the leaving air state 2, for which the dry bulb and wet bulb temperatures are

$$t_2 = 15°\text{C} \quad \text{and} \quad t_2' = 13.6°\text{C}$$

20.5.1 Significance of Performance Coefficient $k_\omega aV/m_w$ in Cooling Tower Selection

To obtain the value of $k_\omega\ aV/\dot{m}_w$, the integral on the left hand side of Eq. (20.40) must be evaluated.

The value of the integral cannot be obtained directly. For the purpose, the cooling tower is subdivided into "N" sections, and the integral is replaced by a summation of values prevailing in each section.

Thus, Eq. (20.40) takes the form

$$\frac{k_\omega aV}{\dot{m}_w} = C_w \sum_{i=1}^{N} \frac{(\Delta t_w)_i}{(h_s - h)_i} \tag{20.43}$$

This gives the value of the coefficient as a function of $\dot{m}_w/\dot{m}_a$.

If the summation is done for specified values of

(i) WBT of air t_1' at inlet

(ii) Water inlet temperature t_{w_1}

(iii) Water outlet temperature t_{w_2}

then the summation gives *required value* of the coefficient. A plot of the same as a function of $\dot{m}_w/\dot{m}_a$ represents the *thermal demand* as shown by curve A in Fig. 20.12.

The cooling tower must fulfil this condition. According to the curve, the larger the value of $\dot{m}_w/\dot{m}_a$, the larger is the coefficient and hence the demand on the cooling tower in terms of size/volume 'V' and the wetted surface area ratio 'a' which means smaller droplet size/atomization, number of stacks, packing, etc.

On the other hand, if the summation is done, by actual measurements in an experiment on the cooling tower, we get *available value* of the coefficient for that cooling tower. It is found that entering WBT of air t'_1 and inlet water temperature t_{w_1} do not affect the coefficient. However, the values of $\dot{m}_w$ and $\dot{m}_a$ do influence water cooling, and hence affect the coefficient. The variation of the coefficient for a cooling tower as a function of $\dot{m}_w/\dot{\boldsymbol{m}}_{\boldsymbol{a}}$ can be expressed by

$$\frac{k_\omega \, aV}{\dot{m}_w} = x \left[\frac{\dot{m}_w}{\dot{m}_a}\right]^{-y} \tag{20.44a}$$

where x and y are determined experimentally. Taking logarithms, this relation takes the form of a straight line

$$\ln\left(\frac{k_\omega \, aV}{\dot{m}_w}\right) = \ln x - y \ln\left(\frac{\dot{m}_w}{\dot{m}_a}\right) \tag{20.44b}$$

A typical relation for a cooling tower

$$\frac{k_\omega \, aV}{\dot{m}_w} = 0.79\left(\frac{\dot{m}_w}{\dot{m}_a}\right)^{-0.72} \tag{20.44 c}$$

is shown by line B in Fig. 20.12.

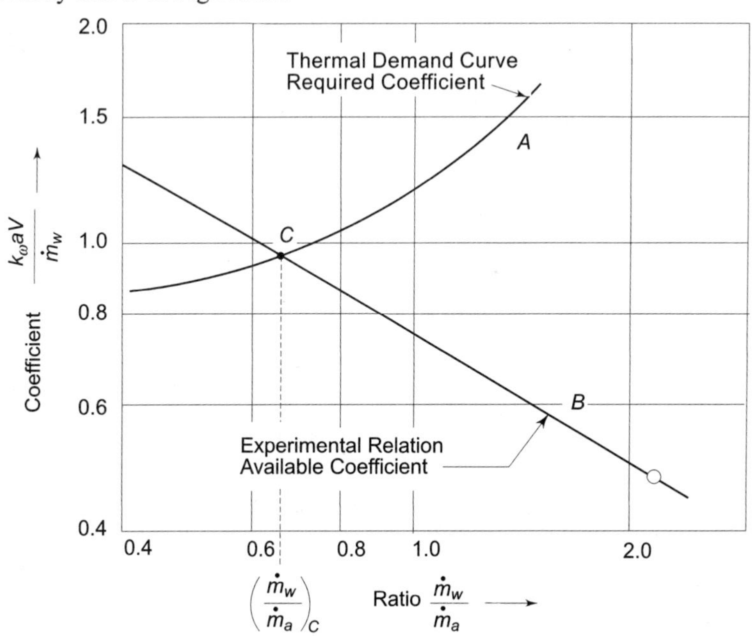

Fig. 20.12 $\frac{K_\omega aV}{\dot{m}_w}$ as a function of $\frac{\dot{m}_w}{\dot{m}_a}$: experimental and thermal demand curves

Note *$k_\omega = f_g/Le\ C_p$ depends mainly on f_g and hence on $\dot{m}_a$.*

The intersection of curve A and line B, viz., the operating point C, represents a complete match between the capacity of the cooling tower and the required cooling of water at the rate $(\dot{m}_w/\dot{m}_a)_C$ for given t'_1, t_{w_1} and t_{w_2}. At a lower value of $\dot{m}_w/\dot{m}_a$, the cooling capacity of the tower is greater than required, while at a higher value of $\dot{m}_w/\dot{m}_a$ the capacity is less than required.

Example 20.5 ***Experimental Determination of Performance Coefficient***

A test is performed on an induced draft counterflow cooling tower. The following observations are made:

Water flow rate	*: 12.62 kg/s*
Air flow rate	*: 11.9 kg/s*
Entering water temperature	*: 36.3°C*
Leaving water temperature	*: 32.1°C*
Ambient air conditions	*: 43.3°C DBT, 25.6°C WBT*

Determine the value of the performance coefficient $k_\omega aV/\dot{m}_w$. If the dimensions of the tower are length $L = 3.9624$ m, width $W = 2.616$ m and height $H = 2.438$ m, what is the value of $k_\omega a$?

Solution Let the cooling tower be considered as divided into 6 sections as shown in Fig. 20.13 with the water temperature dropping by 0.7°C in each section. The

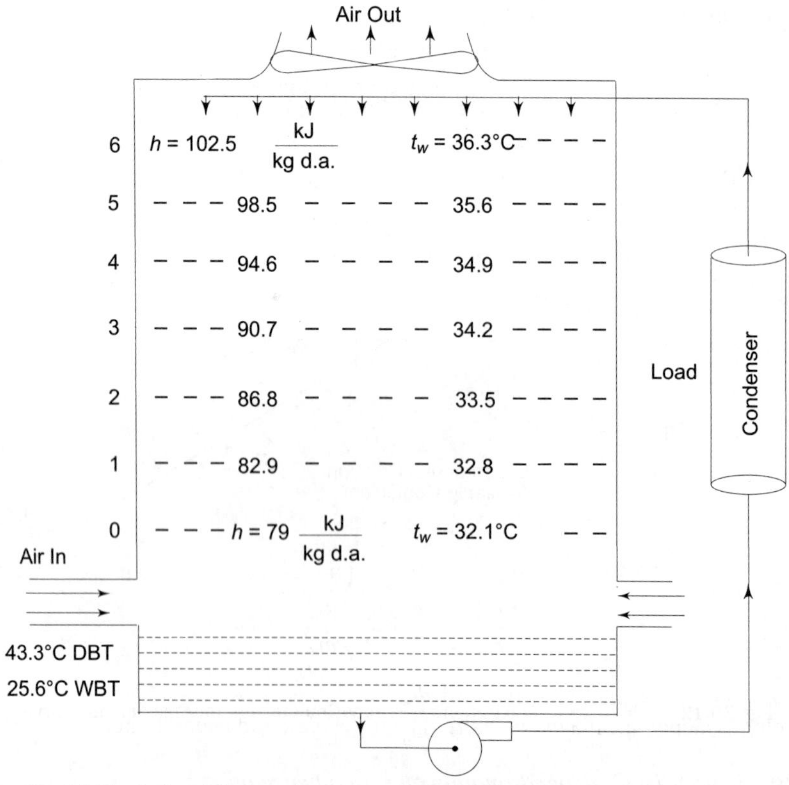

Fig. 20.13 Induced draft cooling tower: figure showing sections for Example 20.5

values of t_w and $h_S = f(t_S = t_w)$ are given in Table 20.6 for each section. The values of enthalpy of air for each section are calculated from the change in enthalpy corresponding to $\dot{m}_w/\dot{m}_a = 12.62/11.9 = 1.06$ given by

$$\Delta h = \frac{\dot{m}_w}{\dot{m}_a} C_{pw}\, \Delta t_w = (1.06)\,(4.1868)\,(0.1) = 3.9 \text{ kJ/kg}$$

The values are also given in Table 20.6.

Table 20.6 Values of properties at sections for Example 20.5

Section	0	1	2	3	4	5	6
$t_S = t_w$, °C	32.1	32.8	33.5	34.2	34.9	35.6	36.3
h_S, kJ/kg	111.2	115.1	119.5	123.6	128.4	133.0	137.8
h, kJ/kg	79.0	82.9	86.8	90.7	94.6	98.5	102.4
$\dfrac{1}{h_S - h}$	.03106	.03106	.03058	.0304	.02959	.02899	.02825
	y_0	y_1	y_2	y_3	y_4	y_5	y_6

Using Simpson's rule

$$\int \frac{dt_w}{h_S - h} = \frac{\Delta t_w}{3N}\,[y_0 + y_6 + 4\,(y_1 + y_3 + y_5) + 2\,(y_2 + y_4)]$$

$$= \frac{4.2}{3 \times 6}\,[.03106 + .02825 + 4\,(.03106 + .0304 + .02899) + 2\,(.03058 + .02959)]$$

$$= 0.12634$$

Performance coefficient

$$\frac{k_\omega a V}{\dot{m}_w} = C_w \int \frac{dt_w}{h_S - h} = 4.1868\,(0.12634) = 0.529$$

Cross-sectional area of flow of air and volume of cooling tower

$$A = \text{WL} = 2.616 \times 3.9624 = 10.366 \text{ m}^2$$

$$V = \text{AH} = 10.366 \times 2.438 = 25.27 \text{ m}^3$$

Value of coefficient $k_\omega a$

$$k_\omega a = 0.529\,\frac{\dot{m}_w}{V} = 0.529\,\frac{(12.62)}{25.27} = 0.2642 \text{ kg/s.m}^3$$

Note (i) *This value of performance coefficient is much below the value given by line B in Fig. 20.12.*

(ii) *More the number of sections taken, better is the accuracy.*

(iii) *Calculations can also be done by using Eq. (20.43) for summation, instead of Simpson's rule, taking mean values for sections 0–1, 1–2, etc.*

20.5.2 Crossflow Cooling Tower

Cooling towers without fans are known as *atmospheric cooling towers* in which the air velocity depends on *wind velocity*. The flow of air and water in them is *crossflow*. Some forced or induced draft cooling lowers are also designed as crossflow towers. This reduces the height of the tower. Such towers are convenient for installation on the roofs of high-rise buildings.

Calculations in crossflow towers can be done by dividing the tower in N number of horizontal and vertical sections. The volume of each section is thus reduced to V/N. Similarly, the area $A = aV$ as well as the performance coefficient are reduced accordingly. Example 20.6 illustrates the procedure by dividing an atmospheric cooling tower into 4 horizontal sections.

Example 20.6 *Atmospheric Cooling Tower*

Water enters an atmospheric cooling tower at 7.57 kg/s at 36.4°C. The dimensions of the tower are W = 1.853 m, L = 2.134 m, H = 3.886 km. The wind velocity can be taken as minimum/normal equal to 3 miles per hour. The performance coefficient for the cooling tower as a whole can be taken as $k_\omega aV/\dot{m}_w = 0.8$*. Find the temperature of water leaving cooling tower, and the heat removed, if the WBT of ambient air is 28.3°C.*

Solution Volume of the tower

$$V = \text{WLH} = 1.853\ (2.134)\ (3.886) = 15.366\ \text{m}^3$$

Face area of the tower

$$\text{FA} = \text{WH} = 1.853\ (3.886) = 7.2\ \text{m}^2$$

Wind velocity

$$C = \frac{3 \times 1760 \times 3 \times 0.3048}{3600} = 1.341\ \text{m/s}$$

Divide the tower into 4 sections horizontally as shown in Fig. 20.14.

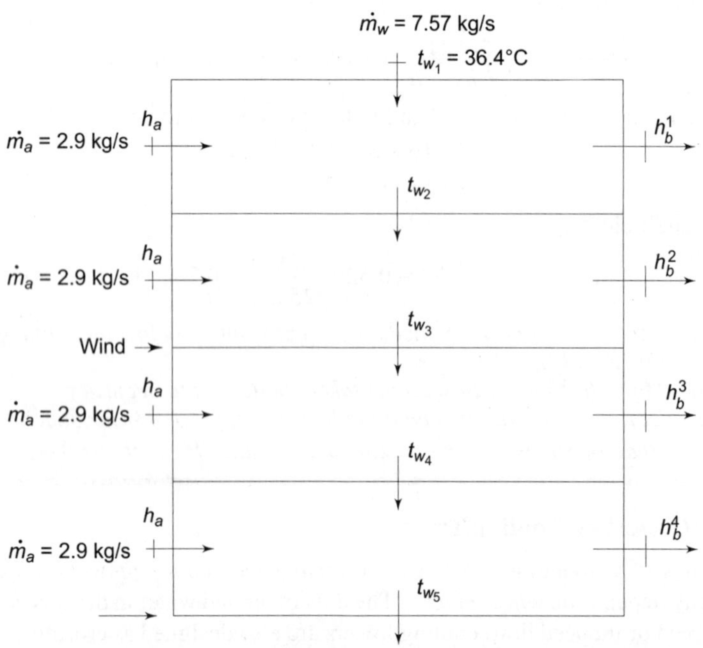

Fig. 20.14 Atmospheric cooling tower divided into 4 sections for Example 20.6

Flow rate of air through the whole tower

$$\dot{m}_a = (\text{FA})\,(\text{C})\,\rho = (7.2)\,(1.341)\,(1.2) = 11.59 \text{ kg/s}$$

Flow rate of air through each section

$$\dot{m}_a/4 = \frac{11.59}{4} = 2.9 \text{ kg/s}$$

Coefficient $k_\omega\, a$ for each section

$$k_\omega\, a = \left(\frac{k_\omega\, aV}{\dot{m}_w}\right)\frac{\dot{m}_w}{V} = \frac{0.8\,(7.57)}{15.366} = 0.394 \text{ kg/s.m}^3$$

Let the temperature of water leaving sections be t_{w_2}, t_{w_3}, t_{w_4} and t_{w_5}. The enthalpy of air entering is h_a = 92 kJ/kg d.a. at the given WBT of 28.3°C. Let the enthalpy of air leaving sections be h_b^1, h_b^2, h_b^3 and h_b^4. Volume of each section is 15.366/4 = 3.8415 m^3. Calculations for sections are then done as follows:

Section 1:

$$t_{w_\text{in}} = t_{w_1} = t_{S_1} = 36.4°\text{C}$$

$$h_S = h_{S_1} = 116.5 \text{ kJ/kg d.a.}$$

$$h_a = 92 \text{ kJ/kg d.a.}$$

$$\Delta\dot{Q} = (k_\omega\, a)\,(V)\,(h_S - h)$$

$$= (0.394)\,(3.8415)\,(116.5 - 92) = 37.08 \text{ kW}$$

$$\Delta h = \frac{\Delta Q^1}{\dot{m}_a^1} = \frac{37.08}{2.9} = 12.8 \text{ kJ/kg d.a.}$$

$$\Delta t_w = \frac{\Delta Q^1}{\dot{m}_w\, C_w} = \frac{37.08}{7.57\,(4.1868)} = 1.17°\text{C}$$

$$h_b^1 = 92 + 12.8 = 104.8 \text{ kJ/kg d.a.}$$

$$t_{w_\text{out}} = t_{w_2} = 36.4 - 1.17 = 35.23°\text{C}$$

Values similarly calculated for all section are given in Table 20.7.

Table 20.7 Section-wise calculations for crossflow atmospheric cooling tower in Example 20.6, Fig. 20.14

Section	$t_{w_\text{in}} = t_S$	h_{S_1}	h_a	$\Delta\dot{Q}$	Δh	Δt_w	h_b	t_{w_out}
	°C	**kJ/kg d.a**	**kJ/kg d.a.**	**kW**	**kJ/kg d.a.**	**°C**	**kJ/kg d.a.**	**°C**
1–2	36.4	116.5	92	37.08	12.8	1.17	104.8	35.23
2–3	35.23	115.0	92	34.8	13.54	1.09	105.5	34.15
3–4	34.15	114.0	92	33.3	12.96	1.05	105.0	33.1
4–5	33.1	113.0	92	31.77	12.36	1.0	104.4	32.09

Note that the wetted surface temperature for each section has been taken as equal to the temperature of water at inlet to the section. For example, for section 1-2, $t_S = t_{w_1}$. Results can be improved if we take

$$t_S = \frac{t_{w_1} + t_{w_2}}{2} = \frac{36.4 + 35.23}{2} = 35.82°\text{C}$$

for section 1–2, and similar average values for all the sections. Doing accordingly, in the second approximation, we find the heat-transfer rates, and the temperatures of water leaving sections are as follows:

$\Delta\dot{Q}$, kW	36.1	34.2	32.8	31.3
t_w, °C	35.26	34.18	33.14	32.2

Thus, the temperature drop of water is 36.4–32.2 = 4.2°C. Wet bulb approach is 32.2 – 28.3 = 3.9°C.

Note *The cooling tower efficiency is found as follows:*

$$\eta_{tower} = \frac{\text{Temperature drop of water}}{\text{Maximum possible temperature drop}}$$

$$= \frac{36.4 - 32.2}{36.4 - 28.3} = \frac{4.2}{8.1} = 0.52\ (52\%)$$

Total heat-transfer capacity of the tower is 35.26 + 34.18 + 33.14 + 32.2 = 134.8 kW = $\dot{Q}_k$, equal to the heat rejected in condenser.

Note *(i) If the COP of the refrigerating machine is taken as 3, this will correspond to a refrigerating capacity of [3/(1+3)] 134.8 = 101 kW (28.8 TR).*
(ii) Heat transfer can also be found as $\dot{Q}_k = \dot{m}_w C_w \Delta t_w$.

References

1. Arora C P, *Heat and Mass Transfer*, Khanna Publishers, Delhi, 1979.
2. Eastop, T D and J M Gasiorek, *Air Conditioning Through Worked Examples,* Longmans, London, 1968.
3. Elmahdy A H and R C Biggs, 'Performance simulation of multirow dry (and/or wet) heat exchangers', *Proc. Sixth International Heat Transfer Conf.*, Toronto, pp. 327–332, 1978.
4. Ramachandran P V and C P Arora, 'Computer aided design of cooling coils', *Proc. 16th International Congress of Refrigeration*, Paris, 1983.
5. Ramsey M A, *Design Problems in Air Conditioning and Refrigeration*, Industrial Press, New York, 1966.
6. Stoecker W F, *Refrigeration and Air Conditioning*, McGraw-Hill, New York, 1958.
7. Stoecker W F, *Principles for Air Conditioning Practice*, Industrial Press, New York, 1968.
8. Treybal R E, *Mass Transfer Operations*, McGraw-Hill, New York, 1955.

Revision Exercises

20.1 (a) 30 cmm of air is to be cooled from 25°C DBT and 50 per cent RH to 12°C DBT and 11°C WBT using a direct-expansion coil with 1.27 cm O.D. tubes with 0.71 mm wall thickness and staggered at 3.81 cm centres.

Determine the total fin-side surface area per row per m^2 of face area, and also per m^2 of the inside-tube surface area. The coil has 5 fins/cm.

(b) Assume the air-side heat-transfer coefficient as 62.5 W/m^2K and the refrigerant-side coefficient as 1500 W/m^2K. The thermal conductivity of copper is 385 W/m.K. Assuming the fin efficiency to be 95 per cent, and neglecting the thermal resistance of the condensate layer, find the value of the coefficient for total heat transfer based on the fin-side surface area.

(c) If the coil contains four rows, determine the evaporating temperature of the refrigerant.

20.2 Water at the rate of 1 kg/s is to be cooled in a cooling tower from 32°C to 28°C. The ambient dry bulb and wet bulb temperatures are 40°C and 26°C respectively. Taking the ratio of rates of the mass flow of water to mass flow of air as 0.75, calculate:

(a) Required tower volume

(b) Condition of air at exit

(c) Height of tower if the air velocity at the exit is 1.5 m/s.

The mean value of the product of the coefficient of diffusion of water vapour into air based on the specific humidity and surface area of water droplets per unit volume of equipment, viz., $k_w a$, may be taken as 0.5 kg/sm^3.

20.3 Air enters an evaporative cooler at the rate of 4.5 m^3/s at 40°C DBT and 26°C WBT. The required relative humidity at exit is limited to 60 per cent. Calculate:

(a) The dry bulb temperature of air at exit.

(b) The length of air washer if the face velocity at the inlet is 2 m/s.

(c) The required humidifying efficiency of the air washer.

Take the mean value of the product $k_w a$ for the apparatus as 1.3 kg/s m^3.

Chapter

21

Transmission and Distribution of Air

The air-handling system consists of:

(i) The air distribution system comprising various inlets for recirculated air and outlets for the supply air.

(ii) The duct system including the return duct, supply duct, and air-conditioning apparatus comprising of dampers, filters, coil/air washer, etc.

(iii) The fan which provides the necessary energy to move the air.

Figure 21.1 shows how air is handled in a simple air-conditioning system. It is seen that a closed loop is formed for the circulation of air. The reference point in this loop is the room itself which can be considered at atmospheric pressure. The air enters the return duct through the inlets from the room and continues to drop in pressure until it reaches the fan. The fan raises the pressure. Thereafter, the pressure starts dropping in the supply duct until the air is released to the space. Therefore, the pressure on the suction side of the fan is negative, and on the discharge side, positive. The pressure in the room is atmospheric, i.e., zero gauge.

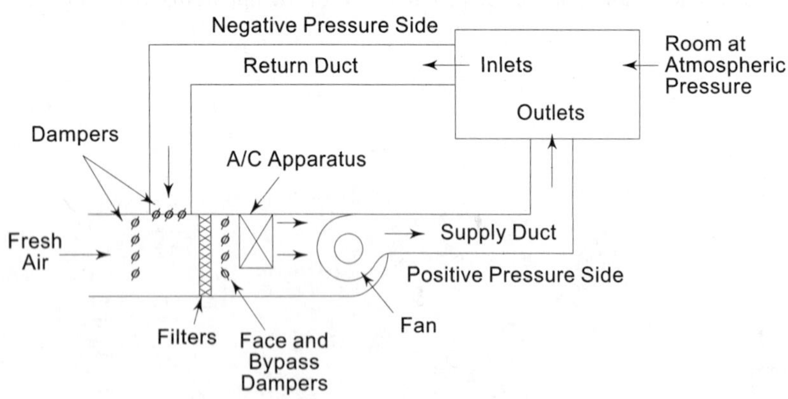

Fig. 21.1 Schematic air-flow diagram for an air-conditioning system

In this chapter, we shall concern ourselves with the distribution aspects of the system, design of ducts and the pressure required to be developed by the fan. The aspects pertaining to fans are discussed in Chapter 22.

21.1 ROOM AIR DISTRIBUTION

The requirement of good room air distribution is to create a proper combination of temperature, humidity and air motion in the occupied zone which is normally at 1.8 m above the floor level. The maximum variation in temperature in a single room should not be more than 1°C. The desirable air velocity is 9.1 mpm at the occupancy level.

Draft is defined as any localized feeling of coolness or warmth of any portion of the body due to both air movement and air temperature, with humidity and radiation considered constant. The warmth or coolness of a draft is measured above or below the controlled room condition of 24.4°C DBT at the centre of the room and air moving at approximately 9.1 mpm velocity. To define the difference Δt in effective temperature for comfort, Rydberg and Norback[3] use the following equation

$$\Delta t = (t - 24.4) - 0.1276\,(C - 9.1) \tag{21.1}$$

where t is the local temperature in °C and C is the local velocity in mpm. It will be seen that a 1°C difference in temperature is equivalent to 9.1 mpm difference in velocity. It may be noted that proper air distribution calls for entrainment of room air by supply air to bring the velocity to 9.1 mpm in the entire space.

The most desirable direction of cooled supply air is from the front side towards the face of a person. Cold air directed towards the feet or from the back is undesirable while warm air directed towards the feet is comfortable and directed towards the face is undesirable.

From the considerations of permissible limits of sound levels, the maximum recommended velocities from air outlets are given in Table 21.1.

Table 21.1 Recommended outlet velocities

Application	*Outlet velocity, mpm*
Broadcasting studios	90–150
Residences, Private offices, Theatres	150–225
Cinema theatres	300
General offices	330–375
Stores, upper floors	450
Main floor	600

The principles of air distribution involve the following factors.

Blow or Throw: Blow or throw L is the distance travelled by the supply air stream in the horizontal direction on leaving the air outlet and reaching a velocity of 15 mpm. The velocity is measured in the occupied zone at 1.8 m above the floor level. The localized condition just at the end-point of blow will not be comfortable due to lower temperature than 24.4°C, and higher velocity of 15 mpm compared to a velocity of 9 mpm after diffusion of supply air in room.

The desirable length of blow is up to three-fourths of the distance to the opposite side of wall.

Drop: Drop is the vertical distance the air moves after it leaves the outlet and reaches the end of the blow.

Induction Ratio or Entrainment Ratio: The air leaving the outlet is *primary* air. This supply air entrains some room air, called the *secondary air*. The sum of primary and secondary air is called *total air*. The induction ratio R is defined as the ratio of total air to primary air. The induction is governed by the momentum equation

$$m_1 C_1 + m_2 C_2 = (m_1 + m_2)\, C_3 \tag{21.2}$$

where the subscripts 1, 2, and 3 refer to primary, secondary and total air respectively.

Blow depends on the initial supply air velocity, temperature difference between the supply and room air and induction ratio. Induction, in turn, depends on the primary air velocity and perimeter of primary air stream.

In addition to the primary and secondary air streams, air motion also develops in the room as a result of natural convection due to the density difference between the room air and the colder or hotter supply air.

The induction ratio at distance x from the outlet is

$$R = \frac{m_x}{m_1} = K \frac{C_1}{C_x} \tag{21.3}$$

where C_x is the velocity at distance x, and K is the entrainment coefficient $\cong 2$ for round ducts.

Spread: Spread is the angle of divergence of an air stream after it leaves the outlet. The spread can be both horizontal and vertical.

Air outlets can have three types of vanes, viz., straight, converging or diverging vanes. Outlets with straight vanes produce a spread of 14 to 24° (approximately 19°) in both the vertical and horizontal planes. Outlets with converging vanes have approximately the same spread as straight vanes but the blow is about 15 per cent longer. Outlets with diverging vanes give a fanning effect and have a marked effect on the direction and blow. Figure 21.2 shows an outlet with vertical vanes. The end vanes are turned at an angle of 45° and intermediate vanes are turned by progressively smaller angles. Such an outlet has a horizontal spread of 60°. The blow is reduced by about 50 per cent as compared to that of a straight-vane outlet.

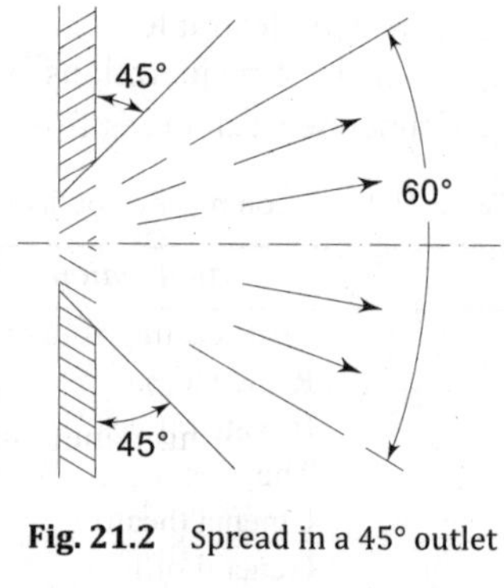

Fig. 21.2 Spread in a 45° outlet

21.1.1 Types of Supply Air Outlets

The outlets may be classified as high sidewall, low sidewall, ceiling and floor outlets according to their location. There are also numerous outlet designs which are described by their construction features. Accordingly, there are four basic types of outlets.

(i) Grille outlets.
(ii) Slot diffuser outlets.
(iii) Ceiling diffuser outlets.
(iv) Perforated ceiling panels.

Grille outlets may have *adjustable bar grilles* which are the most common types with vertical and horizontal vanes. They may have *fixed bar grilles* in which case vanes are not adjustable, but they are fixed either straight or at an angle. The application of grille outlets is in a high sidewall location. They are not acceptable in ceiling location for comfort application as they may cause draft conditions. The accessories to grille outlets are shown in Fig. 21.3. These help in the uniform distribution of air over the whole grille area. A combination of a grille and a damper is called a *register*.

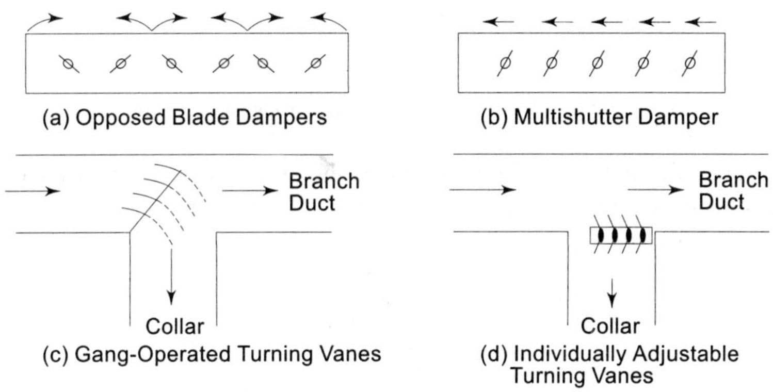

Fig. 21.3 Accessories to grille outlets

A slot diffuser is an elongated outlet with an aspect ratio of 25 : 1 and a maximum width of 7.5 cm. Their application is in high sidewall locations or perimeter installations in floors. They should not be installed in ceilings. Accessories to slot diffusers are dampers normally available as integral equipment.

Ceiling diffusers are mounted in the ceilings. Multi-passage round, square or rectangular are the most common types. They consist of a series of flaring rings or louvers, which may be of adjustable pattern or variable area type. They may also be fitted with dampers.

Perforated ceiling panels use the confined space above the ceiling as a supply *plenum*. Air is delivered to the room through holes or slots. The upper and lower limits of the plenum pressure are 3.5 mm and 0.025 mm H_2O respectively. The recommended supply air rates are 0.3 to 4.5 cmm/m^2 of the floor area. Perforated ceiling panels are suited to large zones of uniform room temperature.

21.1.2 Mechanism of Flow through Outlet

The mechanism of flow of air from the duct and through the outlet to the room is shown in Fig. 21.4. A_c is the *core area* or the area of grille opening in which the air flows with a velocity C_c. A_{fa} is the *free area* of the grille through which air can pass. The ratio A_{fa}/A_c is R_{fa} so that $C_{fa} = C_c/R_{fa}$. A_0 is the area at the vena *contracta* formed outside the grille. If C_d is the discharge coefficient of the outlet, and C_0 is the velocity at the vena contracta.

$$C_d = \frac{A_0}{A_{fa}} = \frac{A_0}{A_c R_{fa}} \tag{21.4}$$

$$C_0 = \frac{C_c A_c}{A_0} = \frac{C_c}{C_d R_{fa}} \tag{21.5}$$

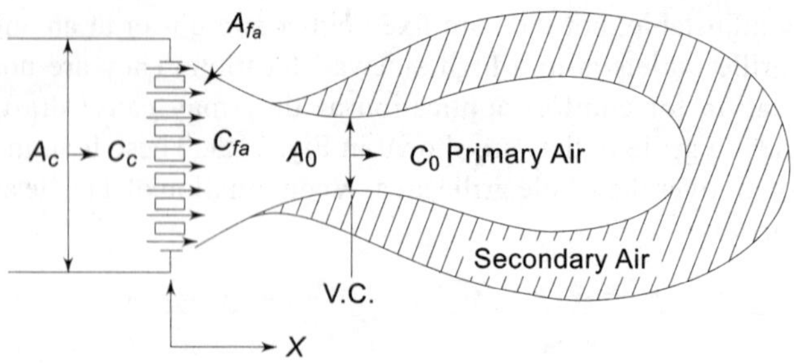

Fig. 21.4 Mechanism of flow of air through a duct outlet

The zone of interest is at 25 to 100 times the diameter or width of the outlet in the x direction. In this zone, the velocity at any x is given by

$$\frac{C_x}{C_0} = \frac{KD_0}{x} = \frac{K'\sqrt{A_0}}{x}$$

$$C_x = \frac{K'C_0\sqrt{A_0}}{x} = \frac{K'\dot{Q}_v}{x\sqrt{A_0}} = \frac{K'\dot{Q}_v}{x\sqrt{A_c C_d R_{fa}}} \tag{21.6}$$

where $\dot{Q}_v$ is the volume delivered by the outlet and $K' = 1.13\ K$.

The tested values of K are given in Table 21.2. Equation (21.6) can also be used to calculate the throw L by putting $C_x = 15$ mpm = 0.25 m/s.

$$L = x = \frac{K'}{0.25}\frac{\dot{Q}_v}{\sqrt{A_c C_d R_{fa}}} \tag{21.7}$$

As far as the entrainment ratio R is concerned, it is given by the following empirical relations in which $\dot{Q}_x$ represents the volume of the total air at any distance x from the outlet and $\dot{Q}_v$ is the volume of primary air.

For circular jets

$$R = \frac{\dot{Q}_x}{\dot{Q}_v} = \frac{2}{K'}\frac{x}{\sqrt{A_0}} = 2\frac{C_0}{C_x} \tag{21.8}$$

For long slots of width H_0

$$R = \frac{\dot{Q}_x}{\dot{Q}_v} = \sqrt{\frac{2}{K'}}\sqrt{\frac{x}{H_0}} = \sqrt{2}\frac{C_0}{C_x} \tag{21.9}$$

Table 21.2 Recommended values of K

Type of Outlet	*K*	
	C_0 = *25 to 50* m/s	C_0 = *100 to 500* m/s
Free openings,		
round or square	5.0	6.2
rectangular, $AR < 40$	4.3	5.3
Grilles, $R_{fa} \geq 0.4$	4.1	5.0
Perforated panels		
R_{fa} = 0.03 to 0.05	2.7	3.3
R_{fa} = 0.1 to 0.2	3.5	4.3

Example 21.1

A grille has a core area of 0.3 m × 0.5 m. The free flow area is 90 per cent. The discharge coefficient may be taken as 0.8. The recommended value of coefficient K′ is 5.0. Find the core velocity, and cmm of air delivered, so that the air velocity is 0.25 m/s for a throw of 15 m.

Solution Core area, and area at vena contracta

$$A_c = (0.3)\,(0.5) = 0.15 \text{ m}^2$$

$$A_0 = A_c\, R_{fa}\, C_d = (0.15)\,(0.9)\,(0.8) = 0.108 \text{ m}^2$$

Now

$$\frac{C_x}{C_0} = \frac{K'\sqrt{A_0}}{x} = \frac{5\sqrt{0.108}}{15} = 0.1095$$

$$\frac{C_x}{C_c} = \frac{C_x}{C_0 C_d R_{fa}} = \frac{0.1095}{(0.8)\,(0.9)} = 0.1521$$

Core velocity and volume delivered

$$C_c = \frac{C_x}{0.1521} = \frac{0.25}{0.1521} = 1.644 \text{ m/s}$$

$$\dot{Q}_v = C_c\, A_c = (1.644)\,(0.15) = 0.247 \text{ m}^3/\text{s}$$

21.1.3 Considerations for Selection and Location of Outlets

The selection and location of supply air outlets is governed by the following considerations:

(i) The amount of air to be delivered by the outlet should be proportional to the load of the part of the space for which it is installed.

(ii) The selection of the type of outlet is governed by the ceiling height, nature of room occupancy, etc.

(iii) The location of the outlet should be governed by the condition of uniform air distribution and *rapid temperature equalization.*

(iv) The selection of size of the outlet can be made from the manufacturer's catalogue data according to the air delivery, core velocity, distribution pattern, sound levels, throw, drop, spread, etc.

As a corollary to this, the outlets should be located so as to neutralize the *concentrated loads*, such as those that result from exterior windows, electronic equipment, etc. In buildings in which the lighting load is heavy, i.e., more than 55 W/m^2, and the ceiling height is more than 4.5 m, it is desirable to locate the outlets below the lighting load.

In such and similar cases of concentrated loads, the return grilles or inlets can be located adjacent to these loads so that warm air (in the case of cooling) is withdrawn from the source instead of being dissipated in the conditioned space. This arrangement is also suitable to remove the fumes, pollutants, etc., from their sources in the space.

In their distribution pattern, the outlets may have characteristics in between the behaviour at the two extremes.

At one extreme are ceiling diffusers with radial flow. As a result of the large perimeter area of the primary air, they will have high entrainment rate and rapid temperature equalization in room. The air will, however, quickly slow down and will have a short throw.

At the other extreme are slot diffusers. They have low entrainment rate and slow temperature equalization. But they have a long throw.

Thus, generally speaking, ceiling diffusers can deliver more air to a space than grilles and slot diffusers. Because of their high entrainment, ceiling diffusers may also be used in systems with low supply air temperatures. In spite of the low supply air temperature, induction will result in rapid temperature equalization.

The same cannot be done in the case of slot diffusers and grilles. In their case, this temperature difference may not exceed 11°C. They are used only when the throw required is very long.

21.1.4 Distribution Patterns of Outlets[1]

The general distribution patterns of various types of supply air outlets will now be described. The representation for primary air, total air, natural convection air and stagnant zone will be as shown in Fig. 21.5. It will be seen that the distribution patterns follow differently for cooling and heating. The best distribution pattern is one in which the whole room air is set in motion, and there are neither any stagnation zones nor zones of draft at the occupancy level.

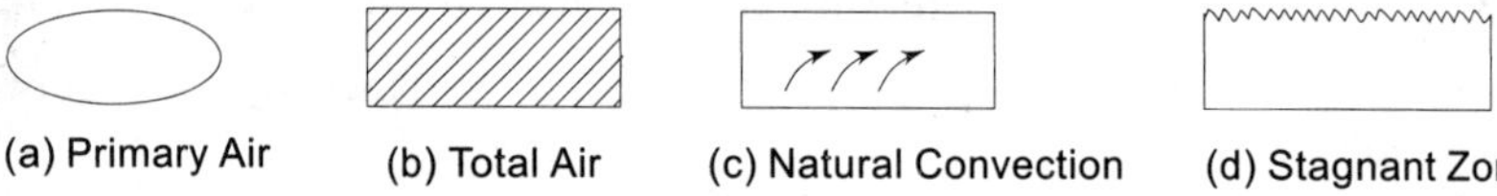

Fig. 21.5 Representation of primary, total and natural convection air and stagnation zone

A. High Sidewall Grilles/Ceiling Diffusers Discharging Air Horizon tally Figures 21.6 (a) and (b) shows the distribution patterns for cooling and heating respectively for a high sidewall grille. The variation of the vane setting may affect the flow to some extent but the general pattern will be the same.

It is seen that during cooling, the total air drops on the occupied zone at some distance from the outlet, depending on $\dot{Q}_v$, C_c, $(t_i - t_s)$, deflection setting, ceiling effect and type of loading in space. It may be noted that the throw is about three-fourths of the room width, and in no case should the air overthrow *otherwise* draft conditions will result.

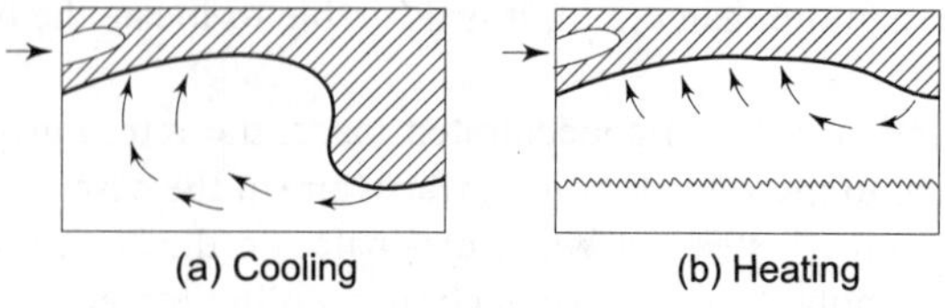

Fig. 21.6 Distribution patterns for high sidewall grille

During heating, the total warm air tends to rise. This results in a large stagnation zone. A degree of over-blow may be helpful in minimizing the stagnation zone.

The general pattern for ceiling diffusers projecting air horizontally is similar though symmetrical on the two sides as shown in Figs 21.7(a) and (b). There is hardly any stagnation zone for cooling application though the same cannot be said for the case of heating. During heating, cold air from the walls tends to drop but warm air tends to remain near the ceiling. A large stagnation zone results. An attempt must be made to direct the air towards the cold walls.

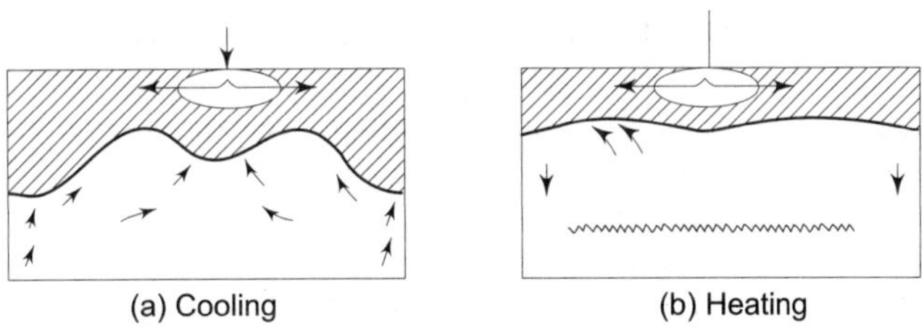

Fig. 21.7 Distribution patterns for ceiling diffusers projecting air horizontally

These outlets with high location are practically suited for cooling with a large dehumidified rise, of the order of 13 to 17°C. They are selected on the basis of drop and throw.

B. Floor Registers Discharging Air Vertically Floor registers normally discharge primary air in a straight vertical jet as shown in Fig. 21.8. Ultimately the total air, after reaching the ceiling, fans out. In the case of cooling, it falls out soon after travelling a short distance.

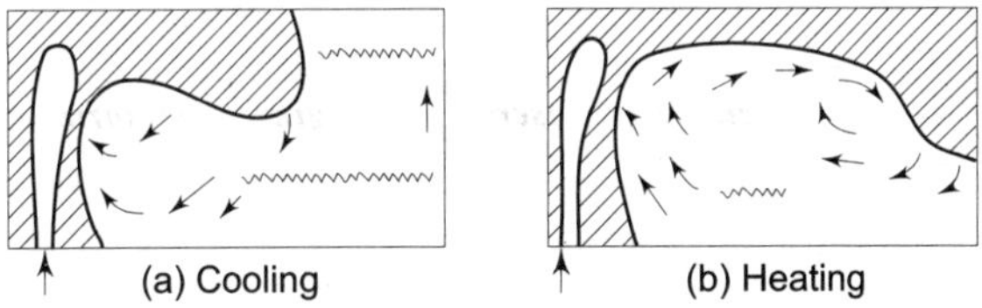

Fig. 21.8 Distribution patterns for flow registers

The cooling diagram shows stagnation region above the terminal point of the total air. In a large space, this stagnation zone may extend much farther and to a lower level. In the case of heating, the total air follows the ceiling and then descends down if flowing along the cold exterior walls.

There is a better temperature equalization for heating than for cooling. In these outlets, generally, an increase of the supply air velocity will improve the air distribution. These outlets are more suited for heating only.

C. Floor Diffusers Discharging Air in a Spreading Jet Floor diffusers are similar to floor registers. The only difference is in the nature of the jet which is spreading in this case, instead of being *nonspreading* as seen from Fig. 21.9. Although the characteristics are similar, the stagnation zone is much larger during cooling but smaller during heating.

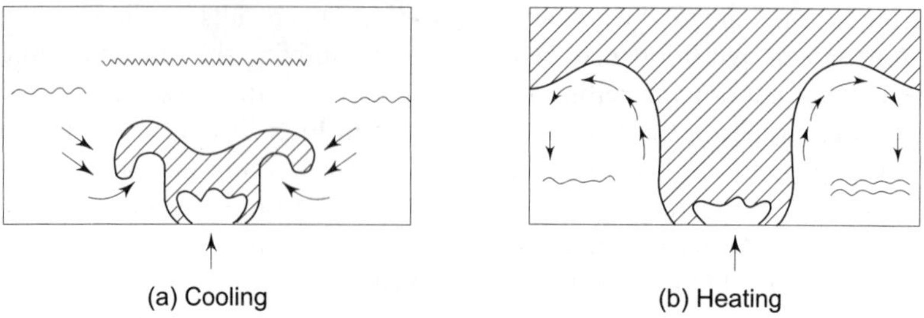

Fig. 21.9 Distribution patterns for floor diffusers

These outlets are suitable when the heating requirement is severe and primary, and the cooling requirement moderate and secondary.

Floor outlets are not permissible when people are seated such as in theatres. But where people are moving, as in stores, they are quite permissible. However, a very low dehumidified rise, say, not more than 8°C should be used. This will require a large volume flow. One disadvantage of floor outlets is that they become dust collectors.

D. Low Sidewall Outlets Discharging Air Horizontally As is seen from Fig. 21.10, the total air during cooling remains near the floor level resulting in low temperature in the occupied zone and a large stagnation zone above. During heating, the warm air rises and temperature equalization takes place except in the region of total air.

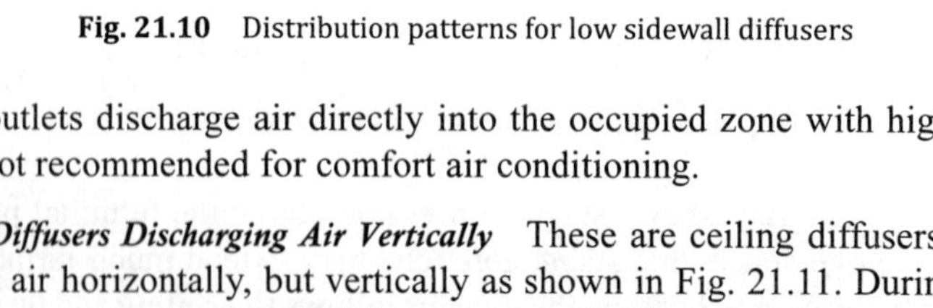

Fig. 21.10 Distribution patterns for low sidewall diffusers

These outlets discharge air directly into the occupied zone with high velocity. They are not recommended for comfort air conditioning.

E. Ceiling Diffusers Discharging Air Vertically These are ceiling diffusers which do not project air horizontally, but vertically as shown in Fig. 21.11. During cooling, the total air drops to the floor and then fans out, finally rising along the walls. The stagnation region is near the ceiling. During heating, the total air, after reaching the floor, returns back towards the ceiling. There is no stagnation zone.

These outlets have completely different distribution patterns for cooling and heating because of the different throws obtained. They are, therefore, used either for cooling or for heating, but seldom for both. For cooling, we require low values of supply air volume, velocity and temperature difference, whereas for heating, the same should be high to get proper throw.

Nevertheless, ceiling diffusers can be conveniently applied to ducts or plenums in the ceiling in large spaces and halls/auditoriums.

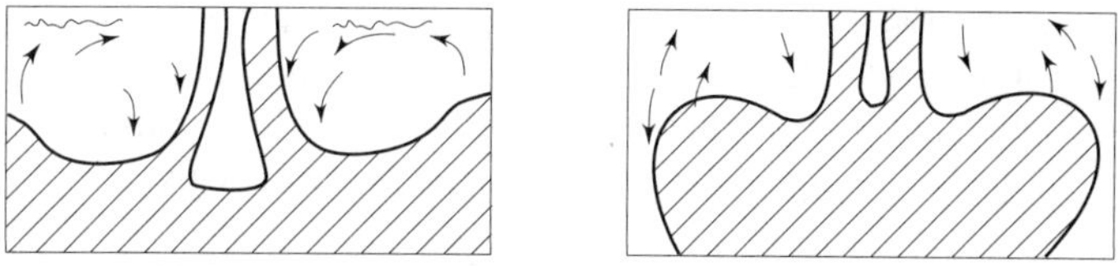

Fig. 21.11 Distribution patterns for ceiling diffusers projecting air vertically

21.1.5 Locating Return Air Openings

While a great deal has been written about supply air outlets, not enough is known about locating return air openings. The most important considerations involved in the location of inlets are the following:

(i) There should not be any short-circuiting of air between the supply outlet and return inlet. The return air opening should be complementary to the flow pattern generated by the supply air.

(ii) The return air which is either too cool in winter or too warm in summer or contains dirt, gases or odours, should be removed without causing any drafts or stratification in the conditioned space by making the undesirable product to move in its natural direction of flow. For example, to remove tobacco smoke, return air openings located high in the room will be most effective. But if the air contains cotton lint as in textile mills, returns located in the floor will do a better job.

(iii) The velocity of air decreases rapidly as one moves away from the inlet. Thus draft conditions near the inlets are rare. However, the recommended velocities are approximately 240 mpm for inlets above the occupied zone and 180 mpm within the occupied zone.

(iv) The consideration of noise is a little more severe in the case of return inlets as they are normally located near the ear-level. In that case the velocity in the return inlet should not exceed 75 per cent of the velocity in the supply air outlet.

From the point of view of location, a wall return near the floor is the best. A ceiling return may short-circuit the cold air. A wall return near the ceiling is as undesirable as a ceiling return.

Floor returns should be avoided because they act as dust collectors. Whenever floor returns are used, it is necessary to provide a low velocity settling chamber.

Also, as stated earlier, in the case of concentrated load, the return opening should be located near the load.

21.2 TOTAL, STATIC AND VELOCITY PRESSURES

The terminology used pertaining to the flow of air will now be described. For this purpose, we begin with the steady-flow energy equation for flow through a pipe or duct which can be written as

$$\mathrm{d}\left(\frac{C^2}{2}\right) + \mathrm{d}h + g\mathrm{d}z = \mathrm{d}q - \mathrm{d}w \tag{21.10}$$

where the various terms denote changes in kinetic energy, enthalpy and potential energy on the left-hand side and the heat added and work done on the right-hand side respectively. Since no external work is done and the heat transfer is negligible, and the enthalpy change, in a reversible adiabatic process, is given by

$$dh = v dp = \frac{1}{\rho} dp \tag{21.11}$$

Equation (21.10) becomes

$$d\left(\frac{C^2}{2}\right) + \frac{1}{\rho} dp + g dz = 0 \tag{21.12}$$

Further, in the case of the flow of air through ducts, the pressure changes are so small that the density change is negligible. Equation (21.12) can, then, be integrated to give the well-known *Bernoulli's equation* for the flow of incompressible fluids, viz.,

$$\frac{p}{\rho} + \frac{C^2}{2} + gz = \text{constant} \tag{21.13}$$

where ρ is the density of air. Equation (21.13) is in terms of the *head* of the fluid, commonly used in hydraulics. However, in air conditioning, Eq. (21.13) is rewritten in the following form

$$p + \frac{\rho C^2}{2} + \rho gz = \text{constant} \tag{21.14}$$

which is in terms of the *pressure* of the fluid. In Eq. (21.14), the three terms represent the *static pressure* (SP) p_s, *velocity pressure* (VP) p_v, and pressure p_z due to the datum head respectively. Equation (21.14) can, therefore, be written as

$$p_S + p_V + p_Z = \text{constant} = p_T \tag{21.15}$$

where p_T is termed as the *total pressure* (TP).

In the absence of changes of datum head, we can take

$$p_S + p_V = P_T = \text{constant} \tag{21.16}$$

Thus, if air flowing with a velocity C is brought to a state of rest or stagnation, there will be a conversion of velocity pressure p_V into static pressure. The static pressure p_S will then rise to a value equal to that of the total pressure p_T. The velocity pressure in turn is then given by

$$p_V = p_T - p_S \tag{21.17}$$

Both static and total pressures may be taken as gauge pressures. Equation (21.17) is the basic equation governing the operation of a *Pitot tube* used for the measurement of velocity of air by measuring the static and total pressures.

If the velocity pressure is measured in mm H_2O, and velocity in m/s, and ρ equal to 1.2 kg/m^3 is taken as the standard density of air, the velocity pressure is given by

$$p_V (9.81) = (1.2)\frac{C^2}{2}$$

$$\Rightarrow \qquad p_V = \left(\frac{C}{4.04}\right)^2 \text{ mm H}_2\text{O} \qquad (21.18)$$

If p_V is measured in N/m^2, and the velocity remains in m/s, then

$$p_V = \frac{1}{2}\rho C^2 = \frac{1}{2}(1.2)\, C^2 = 0.6\, C^2 \qquad (21.19)$$

and expressing the velocity in terms of the velocity pressure, we have

$$C = 1.291\sqrt{p_V} \qquad (21.20)$$

The expression used for wind pressure in Sec. 18.6.2 can also be derived. Thus if wind velocity is measured in kilometres per hour, and wind pressure, i.e., the corresponding velocity pressure, in centimetres of water, we have

$$p_V\ (98.1) = \frac{1.2}{2}\left(\frac{C \times 1000}{3600}\right)^2$$

$$\Rightarrow \qquad p_V = 0.00047\, C^2 \text{ cm H}_2\text{O} \qquad (21.21)$$

Note *1 mm H_2O = 9.81 Pa, and in inch units, 1 in H_2O = 249 Pa*

21.2.1 Flow through a Duct

Consider now the flow of air between two sections 1 and 2 of an insulated duct as shown in Fig. 21.12. By mass balance between the two sections, we have

$$\dot{m} = \rho_1 A_1 C_1 = \rho_2 A_2 C_2 \qquad (21.22)$$

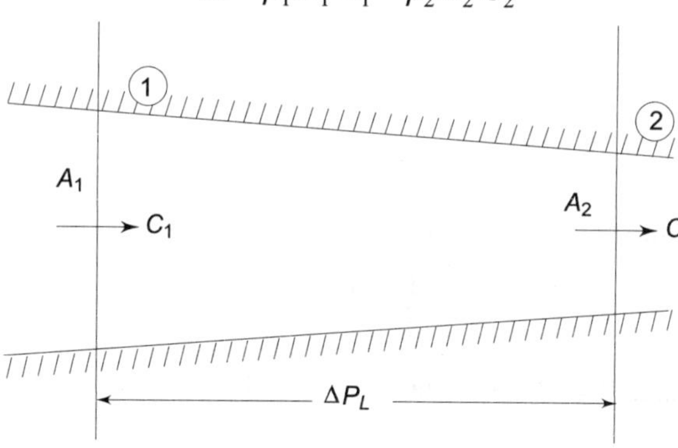

Fig. 21.12 Flow of air through a duct with drop in total pressure

Assuming constant density of air, we have for the volume flow rate

$$\dot{Q}_v = A_1 C_1 = A_2 C_2 \qquad (21.23)$$

Further, by energy balance, assuming frictionless flow and no pressure drop, we have from Eq. (21.16)

$$p_{S_1} + p_{V_1} = p_{S_2} + p_{V_2} = p_T \qquad (21.24)$$

i.e., the total pressures at the two sections are equal. However, pressure drop may take place along the duct due to friction and other causes such as sudden changes in area and direction. An increase in the datum head also results in a drop in static pressure, which is not accounted for in Eq. (21.24). As a result of pressure drop, the energy due to static pressure will be converted into heat and appear in the form of internal energy or temperature rise of air which is normally insignificantly small. However, to account for the drop in static and total pressures, Eq. (21.24) is modified to

$$p_{S_1} + p_{V_1} = p_{S_2} + p_{V_2} + \Delta p_L \tag{21.25}$$

where Δp_L represents the total pressure drop or loss between the two sections. The changes in pressure due to changes in datum may also be included in the numerical value of Δp_L.

Further, if between the two sections a fan or blower is introduced as shown in Fig. 21.13 which will do work on the gas and will have the effect of raising its total pressure, Eq. (21.25) is further modified to

$$p_{S_1} + p_{V_1} + \text{FTP} = p_{S_2} + p_{V_2} + \Delta p_L \tag{21.26}$$

where FTP is the pressure rise due to the fan work and is called the *fan total pressure*.

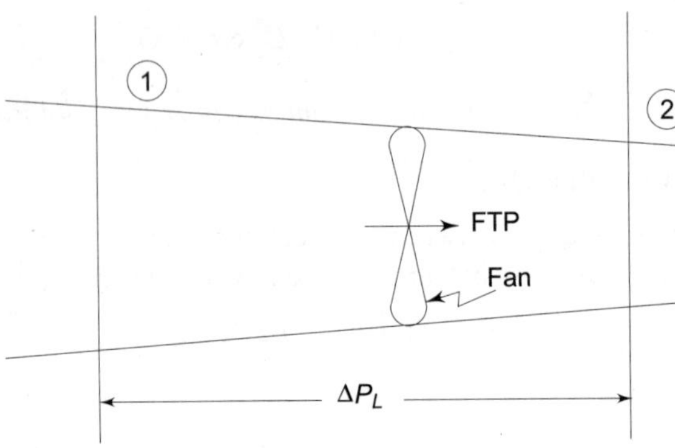

Fig. 21.13 Flow of air through a duct with fan

Writing in terms of energy, Eq. (21.26) becomes

$$\frac{p_{S_1}}{\rho_1} + \frac{C_1^2}{2} + \frac{\dot{W}}{\dot{m}} = \frac{p_{S_2}}{\rho_2} + \frac{C_2^2}{2} + \frac{\dot{L}}{\dot{m}} \tag{21.27}$$

where $\dot{W}$ represents the fan work and $\dot{L}$ represents the losses or increase in internal energy.

Example 21.2

The main supply air duct of an air-conditioning system is 100 cm × 90 cm in cross-section and carries 10 m³/s of air. It branches off into two ducts, one 80 cm × 80 cm and the other 80 cm × 60 cm. If the mean velocity in the larger branch is 9 m/s, find the mean velocities in the main duct and smaller branch.

Solution Refer to Fig. 21.14.

Duct areas

$$A_1 = (1.0)\,(0.9) = 0.9 \text{ m}^2$$
$$A_2 = (0.8)\,(0.8) = 0.64 \text{ m}^2$$
$$A_3 = (0.8)\,(0.6) = 0.48 \text{ m}^2$$

Flow rates $\dot{Q}_{v_1} = 10 \text{ m}^3/\text{s}$

$\dot{Q}_{v_2} = A_2 C_2 = (0.64)\,(9) = 5.76 \text{ m}^3/\text{s}$

$\dot{Q}_{v_3} = \dot{Q}_{v_1} - \dot{Q}_{v_2} = 10 - 5.76 = 4.24 \text{ m}^3/\text{s}$

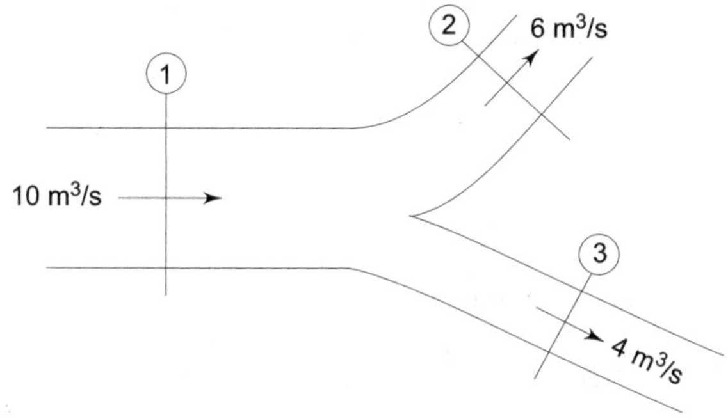

Fig. 21.14 Figure for Example 21.2

Air velocities $C_1 = \dfrac{\dot{Q}_{v_1}}{A_1} = \dfrac{10}{0.9} = 11.1 \text{ m/s}$

$C_3 = \dfrac{\dot{Q}_{v_2}}{A_2} = \dfrac{4.24}{0.48} = 8.83 \text{ m/s}$

21.2.2 Pressure Drop in Ducts

As flow continues in a duct, the static pressure of air drops. This drop in pressure takes place due to two factors:

(i) Duct friction, and
(ii) Change of direction and/or velocity.

While the former is called the frictional pressure drop or *friction loss*, the latter is termed the momentum pressure drop or *dynamic loss*. If the cross-sectional area of the duct remains constant, the mean velocity also remains constant. The changes in velocity are only due to area changes.

As a result of pressure drop, the gas actually expands. But the change in volume is negligible. Hence, for all practical purposes, the density of air can be considered as constant.

21.3 FRICTION LOSS IN DUCTS

Frictional losses are usually expressed by means of *D' Arcy's formula* or the Fanning equation.

$$\frac{\Delta p_f}{\rho} = gH_f = \frac{fLC^2}{2D_m} \tag{21.28}$$

where Δp_f = frictional pressure drop

H_f = friction head or loss in metres of air

ρ = Density of air

f = Friction factor

L = Length of duct

C = Mean duct velocity

D_m = Hydraulic mean diameter

$$= \frac{\text{Cross-sectional area}}{\text{Perimeter}} = \frac{A}{P}$$

For a circular duct,

$$A = \frac{\pi D^2}{4} \text{ and } P = \pi D$$

so that

$$D_m = \frac{D}{4}$$

Equation (21.28) for a circular duct is, therefore, written as

$$\Delta p_f = \rho g \, H_f = \frac{4fL}{D}\left(\frac{\rho C^2}{2}\right) = \frac{4fL}{D} p_V \tag{21.29}$$

Thus, the frictional pressure drop is proportional to the dynamic head or velocity pressure $\rho C^2/2$. It is directly proportional to the length L and inversely proportional to the diameter D of the duct. It is also proportional to the friction factor f, which in turn depends on Reynolds number. In ducts, the flow is normally turbulent for which the *Colebrook-White*[2] relation can be used for the evaluation of the friction factor, viz.,

$$\frac{1}{\sqrt{f}} = -4 \log_{10}\left(\frac{k_s}{3.7D} + \frac{1.255}{\text{Re}\sqrt{f}}\right) \tag{21.30}$$

where k_s = Absolute roughness of the duct wall in metres.

The value of k_s may be taken as 0.00015 m for ducts of sheet-metal construction. Rearranging of Fanning equation (21.29) with the help of Equation (21.30) and substitution of properties of air yields the following simplified relation for frictional pressure drop which is due to *Fritzsche*[2]

$$\Delta p_f = \frac{0.01422\, C^{1.852} L}{D^{1.269}} \tag{21.31}$$

in which Δp_f is in N/m^2 for standard air at 0°C and 1.01325 bar (specific volume equal to 0.778 m^3/kg), C is in m/s and L and D are in m.

Replacing the velocity by the volume flow rate, i.e.,

$$C = \frac{\dot{Q}_v}{A} = \frac{4\dot{Q}_v}{\pi D^2} \tag{21.32}$$

in Eq. (21.31), we can express Δp_f as a function of $\dot{Q}_v$ instead of C

$$\Delta p_f = \frac{0.022243\, \dot{Q}_v^{1.852}\, L}{D^{4.973}} \text{ N/m}^2$$

$$= \frac{0.002268\, \dot{Q}_v^{1.852}\, L}{D^{4.973}} \text{ mm H}_2\text{O} \tag{21.33a}$$

or in terms of C and $\dot{Q}_v$, by eliminating D,

$$\Delta p_f = \frac{0.012199\, C^{2.4865} L}{\dot{Q}_v^{0.6343}} \text{ N/m}^2 \tag{21.34}$$

where $\dot{Q}_v$ is in m^3/s.

For calculating D, Eq. (21.33a) can also be written as

$$D = \frac{0.4652\,\dot{Q}_v^{0.372}}{(\Delta p_f / L)^{0.2}}, \text{ if } \Delta p_f \text{ is in N/m}^2$$

$$= \frac{0.294\,\dot{Q}_v^{0.372}}{(\Delta p_f / L)^{0.2}}, \text{ if } \Delta p_f \text{ is in mm H}_2\text{O} \qquad (21.33\text{ b})$$

Equations (21.31), (21.33) and (21.34) are three different forms of Fritzsche's formula which give quite accurate results. The expressions are used to develop duct *friction charts* as drawn in Figs 21.15 and 21.16 for low and high velocities respectively. These charts are plotted for the volume flow rate $\dot{Q}_v$ in m^3/s as a function of the *friction rate*, i.e., the friction pressure drop per unit length $\Delta p_f / L$ in N/m^3. The other parameters are the duct diameter D, and mean duct velocity C. The charts are valid for air at 20°C and 1.01325 bar and clean galvanized iron (GI) ducts with joints and seams having good commercial practice for which $k_s = 0.00015$ m.

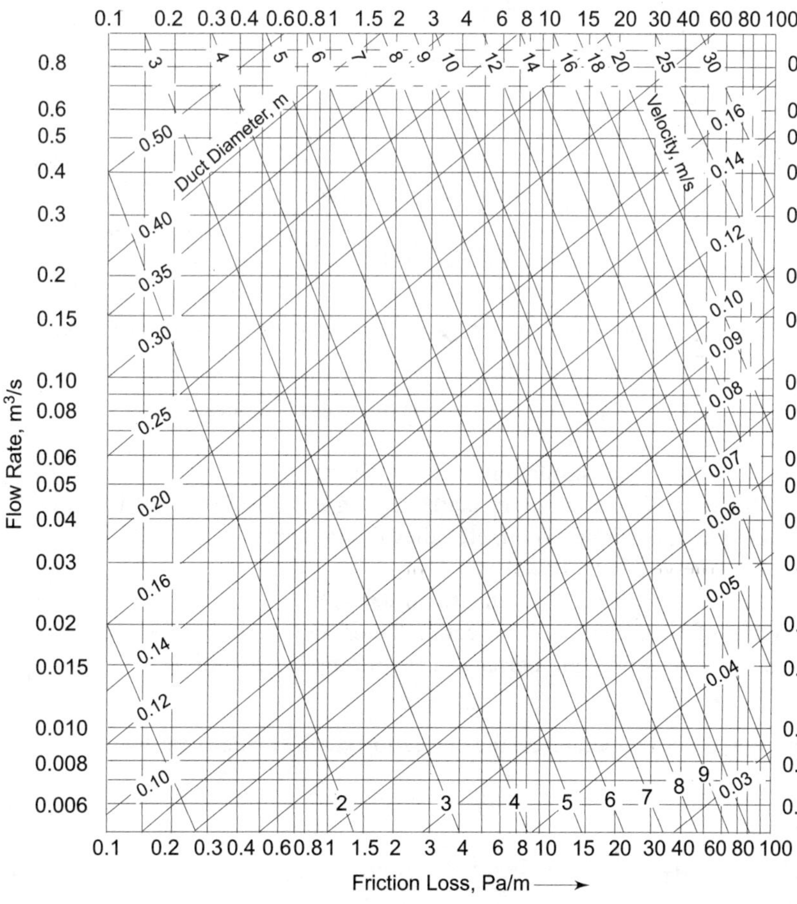

Fig. 21.15 Duct friction chart: low flow rates

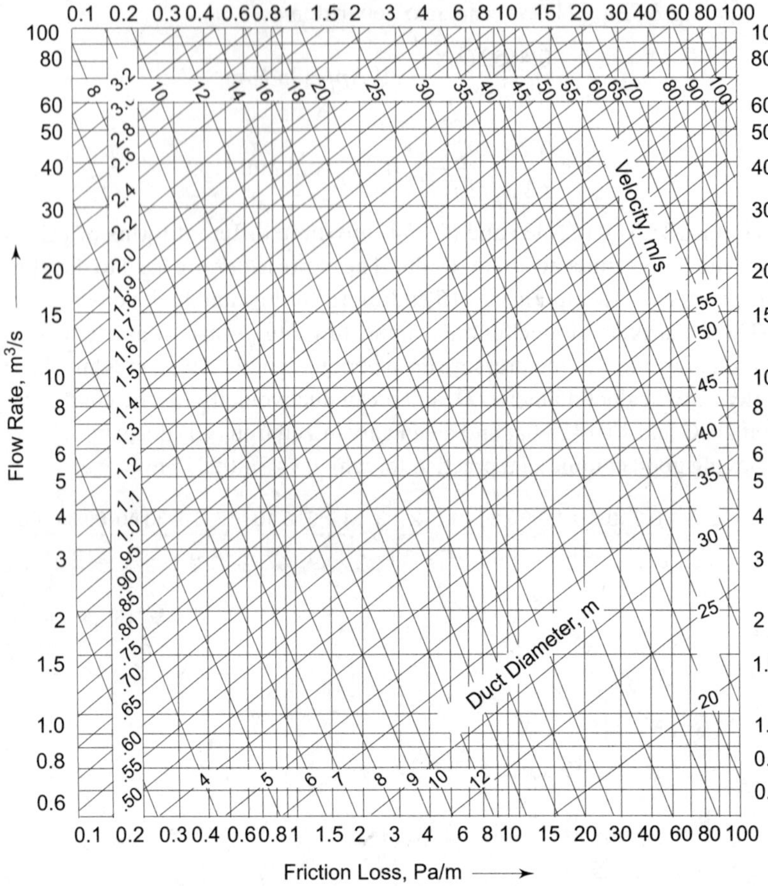

Fig. 21.16 Duct friction chart: high flow rates

The following are the limitations of these charts:

(i) If the duct is made of other materials, e.g., plastics, expanded polystyrene, fibre glass, concrete, wood, etc., they do not apply.

(ii) The charts apply to air only.

(iii) The charts are based on standard air density.

For ducts of other materials, correction factors have to be applied. For small differences in the density of air, correction can be applied according to

$$\Delta p_f \propto \rho$$

For air at other temperatures, the correction can be made according to

$$\Delta p_f \propto \frac{1}{T^{0.857}}$$

21.3.1 Rectangular Equivalents of Circular Ducts

Air ducts are usually sized first for round sections. Then if rectangular ducts are required, ducts are sized to provide the same flow rates and to have the same rate of pressure drop as for round ducts. From Eq. (21.28),

$$\frac{\Delta p_f}{L} = f\,\frac{\rho C^2}{2D_m} = f\left(\frac{\rho C^2}{2}\right)\frac{P}{A} \tag{21.35}$$

Substituting $\dot{Q}_v/A$ for C and transposing, we obtain from Eq. (21.35)

$$\dot{Q}_v = \sqrt{\frac{2}{\rho f}\left(\frac{\Delta p_f}{L}\right)}\sqrt{\frac{A^3}{P}} \tag{21.36}$$

Consider two ducts each handling the same $\dot{Q}_v$, one circular and one rectangular. The friction rate $\Delta p_f/L$ is the same and ρ and f are fixed. Hence, for both ducts A^3/P is the same.

Then if D is the diameter of the round duct, and a and b are the large and small dimensions of the rectangular duct, we have

$$\left(\frac{\pi D^2}{4}\right)^3 \times \frac{1}{\pi D} = \frac{(ab)^3}{2(a+b)}$$

whence
$$D = \left(\frac{32}{\pi^2} \times \frac{a^3 b^3}{a+b}\right)^{1/5}$$

$$= 1.265\left[\frac{(ab)^3}{(a+b)}\right]^{1/5} = 1.265\,\frac{(ab)^{0.6}}{(a+b)^{0.2}} \tag{21.37}$$

It may be noted that the velocities will not be the same in the two ducts, thus, affecting the value of the friction factor. The mean velocity in a rectangular duct will be less than that in its circular equivalent. The normal expression used for equivalence is, therefore, modified to

$$D = 1.3\,\frac{(ab)^{0.625}}{(a+b)^{0.25}} \tag{21.38}$$

Equation (21.38) or Table 21.3 may be used to determine one dimension of a rectangular duct, if the other is assumed, as equivalent to a circular duct whose diameter is known. Although round ducts require the least metal to carry a given quantity of air, rectangular ducts are used often because of the following reasons:

(i) Space considerations: as they fit easily in building construction and occupy less building space without being conspicuous.

(ii) Ease of fabrication.

Square ducts are closest to round ducts. They require less material than rectangular ducts. The material required for the rectangular duct increases with the *aspect ratio*, viz., a/b. Hence, the attempt is always made to keep the aspect ratio close to unity.

Table 21.3 Circular equivalents of rectangular ducts for equal friction and flow rate

Side Recta-ngular Duct	***4.0***	***4.5***	***5.0***	***5.5***	***6.0***	***6.5***	***7.0***	***7.5***	***8.0***	***9.0***	***10.0***	***11.0***	***12.0***	***13.0***	***14.0***	***15.0***	***16.0***				
3.0	3.8	4.0	4.2	4.4	4.6	4.8	4.9	5.1	5.2	5.5	5.7	6.0	6.2	6.4	6.6	6.8	7.0				
3.5	4.1	4.3	4.6	4.8	5.0	5.2	5.3	5.5	5.7	6.0	6.3	6.5	6.8	7.0	7.2	7.4	7.6				
4.0	4.4	4.6	4.9	5.1	5.3	5.5	5.7	5.9	6.1	6.4	6.8	7.1	7.3	7.6	7.8	8.1	8.3				
4.5	4.6	4.9	5.2	5.4	5.6	5.9	6.1	6.3	6.5	6.9	7.2	7.5	7.8	8.1	8.4	8.6	8.9				
5.0	4.9	5.2	5.5	5.7	6.0	6.2	6.4	6.7	6.9	7.3	7.6	8.0	8.3	8.6	8.9	9.1	9.4				
5.5	5.1	5.4	5.7	6.0	6.3	6.5	6.8	7.0	7.2	7.6	8.0	8.4	8.7	9.0	9.4	9.6	9.8				
Side Recta-ngular Duct	*6*	*7*	*8*	*9*	*10*	*11*	*12*	*13*	*14*	*15*	*16*	*17*	*18*	*19*	*20*	*22*	*24*	*26*	*28*	*30*	*Side Recta-ngular Duct*
6	6.6																				6
7	7.1	7.7																			7
8	7.4	8.2	8.8																		8
9	8.0	8.6	9.3	9.9																	9
10	8.4	9.1	9.8	10.4	10.9																10
11	8.8	9.5	10.2	10.8	11.4	12.0															11
12	9.1	9.9	10.7	11.3	11.9	12.5	13.1														12
13	9.5	10.3	11.1	11.8	12.4	13.0	13.6	14.2													13
14	9.8	10.7	11.5	12.2	12.9	13.5	14.2	14.7	15.3												14
15	10.1	11.0	11.8	12.6	13.3	14.0	14.6	15.3	15.8	16.4											15
16	10.4	11.4	12.2	13.0	13.7	14.4	15.1	15.7	16.3	16.9	17.5										16
17	10.7	11.7	12.5	13.4	14.1	14.9	14.5	16.1	16.8	17.4	18.0	18.6									17
18	11.0	11.9	12.9	13.7	14.5	15.3	16.0	16.6	17.3	17.9	18.5	19.1	19.7								18
19	11.2	12.2	13.2	14.1	14.9	15.6	16.4	17.1	17.8	18.4	19.0	19.6	20.2	20.8							19

(*Contd*)

Side Recta-ngular Duct	*6*	*7*	*8*	*9*	*10*	*11*	*12*	*13*	*14*	*15*	*16*	*17*	*18*	*19*	*20*	*22*	*24*	*26*	*28*	*30*	*Side Recta-ngular Duct*
20	11.5	12.5	13.5	14.4	15.2	15.9	16.8	17.5	18.2	18.8	19.5	20.1	20.7	21.3	21.9						20
22	12.0	13.1	14.1	15.0	15.9	16.7	17.6	18.3	19.1	19.7	20.4	21.0	21.7	22.3	22.9	24.1					22
24	12.4	13.6	14.6	15.6	16.6	17.5	18.3	19.1	19.8	20.6	21.3	21.9	22.6	23.2	23.9	25.1	26.2				24
26	12.8	14.1	15.2	16.2	17.2	18.1	19.0	19.8	20.6	21.4	22.1	22.8	23.5	24.1	24.8	26.1	27.2	28.4			26
28	13.2	14.5	15.6	16.7	17.7	18.7	19.6	20.5	21.3	22.1	22.9	23.6	24.4	25.0	25.7	27.1	28.2	29.5	30.6		28
30	13.6	14.9	16.1	17.2	18.3	19.3	20.2	21.1	22.0	22.9	23.7	24.4	25.2	25.9	26.7	28.0	29.3	30.5	31.6	32.8	30
32	14.0	15.3	16.5	17.7	18.8	19.8	20.8	21.8	22.7	23.6	24.4	25.2	26.0	26.7	27.5	28.9	30.1	31.4	32.6	33.8	32
34	14.4	15.7	17.0	18.2	19.3	20.4	21.4	22.4	23.3	24.2	25.1	25.9	26.7	27.5	28.3	29.7	31.0	32.3	33.6	34.8	34
36	14.7	16.1	17.4	18.6	19.8	20.9	21.9	23.0	23.9	24.8	25.8	26.6	27.4	28.3	29.0	30.5	32.0	33.0	34.6	35.8	36
38	15.0	16.4	17.8	19.0	20.3	21.4	22.5	23.5	24.5	25.4	26.4	27.3	28.1	29.0	29.8	31.4	32.8	34.2	35.5	36.7	38
40	15.3	16.8	18.2	19.4	20.7	21.9	23.0	24.0	25.1	26.0	27.0	27.9	28.8	29.7	30.5	32.1	33.6	35.1	36.4	37.6	40
42	15.6	17.1	18.5	19.8	21.1	22.3	23.4	24.5	25.6	26.6	27.6	28.5	29.4	30.4	31.2	32.8	34.4	35.9	37.3	38.6	42
44	15.9	17.5	18.9	20.2	21.5	22.7	23.9	25.0	26.1	27.2	28.2	29.1	30.0	31.0	31.9	33.5	35.2	36.7	38.1	39.5	44
46	16.4	17.8	19.2	20.6	21.9	23.2	24.3	25.5	26.7	27.7	28.7	29.7	30.6	31.6	32.5	34.2	35.9	37.4	38.9	40.3	46
48	16.5	18.1	19.6	20.9	22.3	23.6	24.8	26.0	27.2	28.2	29.2	30.2	31.2	32.2	33.1	34.9	36.6	38.2	39.7	41.2	48
50	16.8	18.4	19.9	21.3	22.7	24.0	25.2	26.4	27.6	28.7	29.8	30.8	31.8	32.8	33.7	35.5	37.3	38.9	40.4	42.0	50
52	17.0	18.7	20.2	21.6	23.1	24.4	25.6	26.8	28.1	29.2	30.3	31.4	32.4	33.4	34.3	36.2	38.0	39.6	41.2	42.8	52
54	17.3	19.0	20.5	22.0	23.4	24.8	26.1	27.3	28.5	29.7	30.8	31.9	32.9	33.9	34.9	36.8	38.7	40.3	42.0	43.6	54
56	17.6	19.3	20.9	22.4	23.8	25.2	26.5	27.7	28.9	30.1	31.2	32.4	33.4	34.5	35.5	37.4	39.3	41.0	42.7	44.3	56
58	17.8	19.5	21.1	22.7	24.2	25.5	26.9	28.2	29.3	30.5	31.7	32.9	33.9	35.0	36.0	38.0	39.8	41.7	43.4	45.0	58
60	18.1	19.8	21.4	23.0	24.5	25.8	27.3	28.7	29.8	31.0	32.2	33.4	34.5	35.5	36.5	38.6	40.4	42.3	44.0	45.8	60
62	18.3	20.1	21.7	23.3	24.8	26.2	27.6	29.0	30.2	31.4	32.6	33.8	35.0	36.0	37.1	39.2	41.0	42.9	44.7	46.5	62
64	18.6	20.3	22.0	23.6	25.2	26.5	27.9	29.3	30.6	31.8	33.1	34.2	35.5	36.5	37.6	39.7	41.6	43.5	45.4	47.2	64
66	18.8	20.6	22.3	23.9	25.5	26.9	28.3	29.7	31.0	32.2	33.5	34.7	35.9	37.0	38.1	40.2	42.2	44.1	46.0	47.8	66
68	19.0	20.8	22.5	24.2	25.8	27.3	28.7	30.1	31.4	32.6	33.9	35.1	36.3	37.5	38.6	40.7	42.8	44.7	46.6	48.4	68
70	19.2	21.0	22.8	24.5	26.1	27.6	29.1	30.4	31.8	33.1	34.3	35.6	36.8	37.9	39.1	41.3	43.3	45.3	47.2	49.0	70

(*Contd*)

Side Rectangular Duct	*6*	*7*	*8*	*9*	*10*	*11*	*12*	*13*	*14*	*15*	*16*	*17*	*18*	*19*	*20*	*22*	*24*	*26*	*28*	*30*	*Side Rectangular Duct*
72															39.6	41.8	43.8	45.9	47.8	49.7	72
74															40.0	42.3	44.4	46.4	48.4	50.3	74
76															40.5	42.8	44.9	47.0	49.0	50.8	76
78															40.9	43.3	45.5	47.5	49.5	51.5	78
80															41.3	43.8	46.0	48.0	50.1	52.0	80
82															41.8	44.2	46.4	48.6	50.6	52.6	82
84															42.2	44.6	46.9	49.2	51.1	53.2	84
86															42.6	45.0	47.4	49.6	51.6	53.7	86
88															43.0	45.4	47.9	50.1	52.2	54.3	88
90															43.4	45.9	48.3	50.6	52.8	54.8	90
92															43.8	46.3	48.7	51.1	53.4	55.4	92
96															44.6	47.2	49.5	52.0	54.4	56.3	96
Side Rectangular Duct	*32*	*34*	*36*	*38*	*40*	*42*	*44*	*46*	*48*	*50*	*52*	*56*	*60*	*64*	*68*	*72*	*76*	*80*	*84*	*88*	*Side Rectangular Duct*
32	35.0																				32
34	36.0	37.2																			34
36	37.0	39.4																			36
38	38.0	39.2	40.4	41.6																	38
40	39.0	40.2	41.4	42.6	43.8																40
42	39.9	41.1	42.4	43.6	44.8	45.9															42
44	40.8	42.0	43.4	44.6	45.8	46.9	48.1														44
46	41.7	43.0	44.3	45.6	46.8	47.9	49.1	50.3													46
48	42.6	43.9	45.2	46.5	47.8	48.9	50.2	51.3	52.6												48

(*Contd*)

Side Recta-ngular Duct	32	34	36	38	40	42	44	46	48	50	52	56	60	64	68	72	76	80	84	88	Side Recta-ngular Duct
50	43.5	44.8	46.1	47.4	48.8	49.8	51.2	52.3	53.6	54.7											50
52	44.3	45.7	47.1	48.3	49.7	50.8	52.2	53.3	54.6	55.8	56.9										52
54	45.0	46.5	48.0	49.2	50.6	51.8	53.2	54.3	55.6	56.8	57.9										54
56	45.8	47.3	48.8	50.1	51.5	52.7	54.1	55.3	56.5	57.8	58.9	61.3									56
58	46.6	48.1	49.6	51.0	52.4	53.7	55.0	56.2	57.5	58.8	60.0	62.3									58
60	47.3	48.9	50.4	51.8	53.3	54.6	55.9	57.1	58.5	59.8	61.0	63.3	65.7								60
62	48.0	49.7	51.2	52.6	54.2	55.5	56.8	58.0	59.4	60.7	62.0	64.3	66.7								62
64	48.7	50.4	52.0	53.4	55.0	56.4	57.7	59.0	60.3	61.6	62.9	65.3	67.7	70.0							64
66	49.5	51.1	52.8	54.2	55.8	57.2	58.6	59.9	61.2	62.5	63.9	66.3	68.7	71.1							66
68	50.2	51.8	53.5	55.0	56.6	58.0	59.5	60.8	62.1	63.4	64.8	67.3	69.7	72.1	74.4						68
70	50.9	52.5	54.2	55.8	57.3	58.8	60.3	61.7	63.0	64.3	65.7	68.3	70.7	73.1	75.4						70
72	51.5	53.2	54.9	56.5	58.0	59.6	61.1	62.6	63.9	65.2	66.6	69.2	71.7	74.1	76.4	78.8					72
74	52.1	53.9	55.6	57.2	58.8	60.4	61.9	63.3	64.8	66.1	67.5	70.1	72.7	75.1	77.4	79.9	83.2				74
76	52.7	54.6	56.3	57.9	59.5	61.2	62.7	64.1	65.6	67.0	68.4	71.0	73.6	76.1	78.4	80.9	84.2				76
78	53.3	55.2	57.0	58.6	60.3	62.0	63.4	64.9	66.4	67.9	69.3	71.8	74.5	77.1	79.4	81.8	84.2				78
80	53.9	55.8	57.6	59.3	61.0	62.7	64.1	65.7	67.2	68.7	70.1	72.7	75.4	78.1	80.4	82.6	85.2	87.5			80
82	54.5	56.4	58.2	60.0	61.7	63.4	64.9	66.5	68.0	69.5	71.0	73.6	76.3	79.0	81.4	83.8	86.2	88.6			82
84	55.1	57.0	58.9	60.7	62.4	64.1	65.7	67.3	68.8	70.3	71.3	74.5	77.2	79.9	82.4	84.8	87.2	89.6	91.9		84
86	55.7	57.6	59.5	61.3	63.0	64.8	66.4	68.0	69.5	71.1	72.6	75.4	78.1	80.8	83.3	85.8	88.2	90.6	92.9		86
88	56.3	58.2	60.1	62.0	63.7	65.4	67.0	68.7	70.3	71.8	73.4	76.3	79.0	81.6	84.2	86.8	89.2	91.6	93.9	96.3	88
90	56.9	58.8	60.7	62.6	64.4	66.0	67.8	69.4	71.1	72.6	74.2	77.1	79.9	82.5	85.1	87.8	90.2	92.6	94.9	97.3	90
92	57.4	59.4	61.3	63.2	65.0	66.8	68.5	70.1	71.8	73.3	74.9	77.9	80.8	83.4	86.0	88.7	91.2	93.6	95.9	98.3	92
94	57.9	60.0	61.9	63.8	65.6	67.5	69.2	70.8	72.5	74.1	75.6	78.7	81.7	84.3	86.9	89.6	92.1	94.6	96.9	99.3	94
96	58.4	60.5	62.4	64.4	66.2	68.2	69.8	71.5	73.2	74.3	76.3	79.4	82.6	85.2	87.8	90.5	93.0	95.6	97.9	100.3	96

21.4 DYNAMIC LOSSES IN DUCTS

In flow through ducts, whenever direction or velocity changes, the pressure loss is greater than if there had been uninterrupted flow. The additional loss, in excess of the *straight-duct friction loss*, is the dynamic one. The dynamic losses in ducts are caused by the following:

(i) Changes in direction, i.e., due to elbows, bends, etc.

(ii) Changes in area or velocity, i.e., due to enlargement, contraction, suction and discharge openings, dampers, etc.

Normally, the dynamic pressure loss Δp_d is proportional to the velocity pressure, and is, therefore, expressed as a product of the downstream velocity pressure p_V and a *dynamic loss coefficient* K found experimentally. Thus

$$\Delta p_d = K p_V = K\left(\frac{\rho C^2}{2}\right) \tag{21.39}$$

where C is normally the downstream velocity.

The losses in elbows, fittings, etc., are also expressed in terms of an *equivalent length* L_e of the duct, so that

$$\Delta p_d = K p_V = \frac{4 f L_e p_V}{D} \tag{21.40}$$

and the relationship between the dynamic loss coefficient and the equivalent length is

$$K = \frac{4 f L_e}{D} \tag{21.41}$$

21.4.1 Pressure Losses in Elbows, Bends and Tees

The values of the dynamic loss coefficient K and/or equivalent length L_e for different types of elbows bends and tees as shown in Fig. 21.17 (refer to the ASHRAE Guide and Data Book,1963 for further details) and are given here in Table 21.4. The

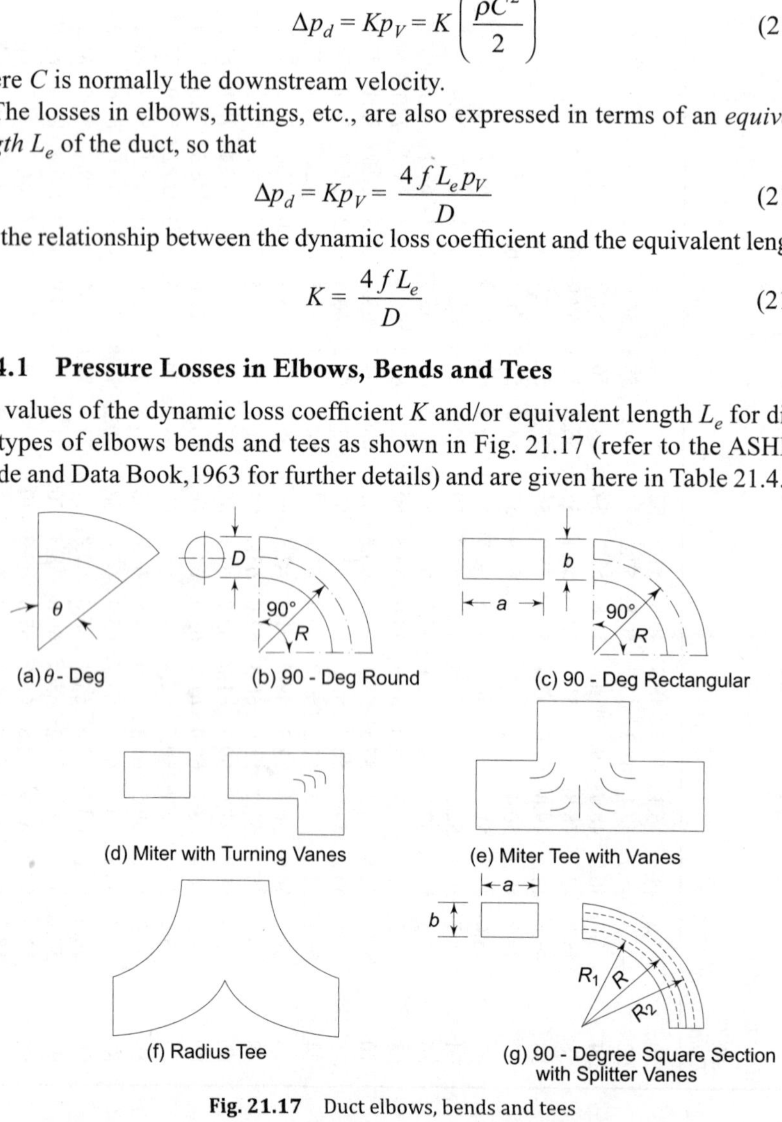

Fig. 21.17 Duct elbows, bends and tees

coefficients are found nearly independent of the air velocity and are affected by the geometry and roughness of duct walls only.

Table 21.4 Pressure losses due to elbows, bends and tees[1]

Reference in Fig. 21.17	*Type*	*Conditions*			*Pressure loss* K	L_e/D	L_e/b
(a)	θ-deg	Rectangular or round; with or without vanes			$\frac{\theta}{90}$ times value for similar 90-deg elbow		
(b)		Miter			1.3	65	
	90-deg round section	$R/D = 0.5$			0.9		
		1.0			0.33	17	
		1.5			0.24	12	
		2.0			0.19	10	
		a/b	R/b				
		0.25	Miter		1.25		25
			0.5		1.25		25
			0.75		0.6		12
(c)	90-deg rectangular section		1.0		0.37		7
			1.5		0.19		4
			Miter		1.47		49
		0.5	0.5		1.1		40
			0.75		0.5		16
			1.0		0.28		9
			1.5		0.13		4
		1.0	Miter		1.5		75
			0.5		1.0		50
			0.75		0.41		21
			1.0		0.22		11
			1.5		0.09		45
		4.0	Miter		1.38		110
			0.5		0.96		65
			0.75		0.37		43
			1.0		0.19		17
			1.51		0.07		6
(d)	Miter with turning vanes	$K = 0.1$ to 0.35 depending on manufacture					
(e)	Miter tee with vanes	Consider equal to a similar elbow, base loss on upstream velocity					
(f)	Radius tee	As above					
(g)	90-deg section with splitter vanes	R/b	R_1/b	R_2/b	K		L_e/b
		Miter	0.5				28
		0.5	0.4		0.7		19
		0.7	0.6				
		1.0	1.0		0.13		7.2
		1.5			0.12		
		Miter	0.3	0.5			22
		0.5	0.2	0.4	0.45		16
		0.75	0.4	0.7	0.12		
		1.0	0.7	1.0	0.1		
		1.5	1.3	1.6	0.15		

To minimise the pressure loss in bends, midfeathers or splitters are used when the aspect ratio is small. They are placed nearer to the throat where the flow is accelerating. Experiments show that it is the curve ratio (Fig. 21.18)

$$\mathrm{CR} = \frac{\text{Throat radius}}{\text{Heal radius}} = \frac{R_0}{R_1} = \frac{R_1}{R_2} = \cdots = \frac{R_n}{R_{n+1}}$$

which determines the pressure loss. Here

$$R_1 = \frac{R_0}{\mathrm{CR}}$$

$$R_2 = \frac{R_1}{\mathrm{CR}} = \frac{R_0}{(\mathrm{CR})^2}, \text{ and so on.}$$

$$R_{n+1} = \frac{R_0}{(\mathrm{CR})^{n+1}}$$

$$\Rightarrow \quad \mathrm{CR} = \left(\frac{R_0}{R_{n+1}}\right)^{\frac{1}{n+1}} = k^{\frac{1}{n+1}} \qquad (21.42)$$

where $k = \dfrac{R_0}{R_{n+1}}$ = Overall curve ratio

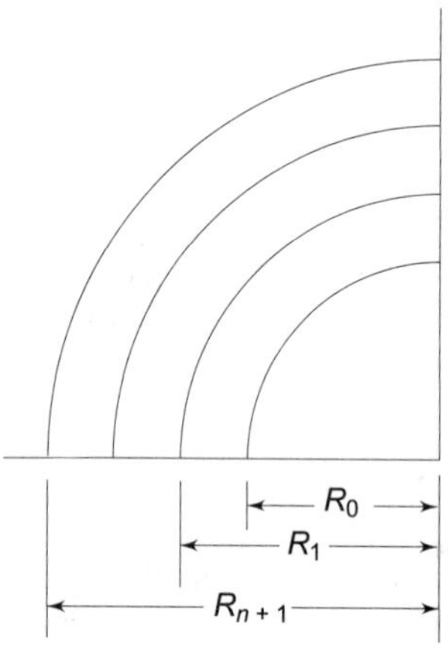

Fig. 21.18 Splitters in duct

In the case of larger aspect ratios, the use of turning vanes is recommended instead of splitters.

Example 21.3 *A bend for a duct is 1500 mm wide and 250 mm high. The centre line radius to width ratio is unity. Determine the best position for the insertion of one splitter.*

Solution Centre line radius, width and ratio

$$R_c = 0.5\,(R_0 + R_{n+1})$$

$$W = R_{n+1} - R_0$$

$$\frac{R_c}{W} = \frac{0.5\,(R_{n+1} + R_0)}{(R_{n+1} - R_0)} = \frac{0.5\,(1+k)}{(1-k)} = 1$$

$$\Rightarrow \quad k = \frac{1}{3}$$

Curve ratio for number of splitters $n = 1$

$$\mathrm{CR} = (k)^{\frac{1}{n+1}} = \left(\frac{1}{3}\right)^{\frac{1}{1+1}} = 0.577$$

Throat and heal radii

$$R_{n+1} - R_0 = 1500, \quad \frac{R_0}{R_{n+1}} = \frac{1}{3}$$

$$\Rightarrow \quad R_0 = 750 \text{ mm and } R_{n+1} = 2250 \text{ mm}$$

Radius of splitter

$$R_1 = \frac{R_0}{\text{CR}} = \frac{750}{0.577} = 130 \text{ mm}$$

21.4.2 Loss due to Enlargement and Static Regain

An enlargement in a duct-run results in a decrease in velocity, and hence in the conversion of the velocity pressure into static pressure. The increase in static pressure, as a result of the conversion from velocity pressure, is termed *static regain*, denoted by SR. If the enlargement is not accompanied with pressure loss, there is full conversion of the velocity pressure into static pressure. If, however, the enlargement is accompanied with pressure loss, the increase in the static pressure or static regain is reduced by the amount of the pressure loss.

Consider the case of a gradual enlargement as shown by continuous lines in Fig. 21.19 in which the velocity changes gradually from C_1 to C_2. The figure also shows a gradual reduction of the velocity pressure along the length with its simultaneous conversion into static pressure. Since there is no loss of pressure

$$p_{T_1} = p_{S_1} + p_{V_1} = p_{S_2} + p_{V_2} = p_{T_2}$$

and we have for maximum static regain

$$(\text{SR})_{\max} = p_{S_2} - p_{S_1} = p_{V_1} - p_{V_2} \tag{21.43}$$

Thus in the case of gradual enlargement, there is 100 per cent static regain.

A gradual enlargement represents an idealized case. In the case of an actual enlargement, the total pressure decreases from p_{T_1} to p_{T_2} because of friction and turbulence. The pressure loss is

$$\Delta p_L = p_{T_1} - p_{T_2} = (p_{S_1} + p_{V_1}) - (p_{S_2} + p_{V_2})$$

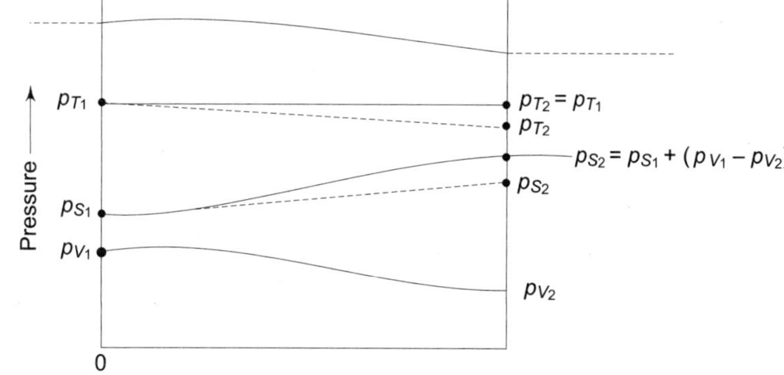

Fig. 21.19 Pressure variation in gradual and sudden enlargements

Hence, the static regain in the actual case is given by

$$\text{SR} = (p_{S_2} - p_{S_1}) = (p_{V_1} - p_{V_2}) - \Delta p_L$$
$$= R(p_{V_1} - p_{V_2}) \tag{21.44a}$$

where R is termed the static regain factor. Thus

$$\Delta p_L = (1 - R)\,(p_{V_1} - p_{V_2}) \tag{21.44b}$$

Now consider the case of a *sudden or abrupt enlargement* of a duct section as shown in Fig. 21.20 which corresponds to the case of maximum pressure drop. Applying momentum equations at sections 1 and 2, we obtain

$$(p_{S_1} - p_{S_2})\,A_2 = \dot{m}(C_2 - C_1) \tag{21.45}$$

where the mass flow rate $\dot{m}$ is given by the continuity equation

$$\dot{m} = \rho C_1 A_1 = \rho C_2 A_2 \tag{21.46}$$

Combining Eqs (21.45) and (21.46), we obtain for static regain

$$\text{SR} = p_{S_2} - p_{S_1} = \rho(C_1 C_2 - C_2^2) \tag{21.47}$$

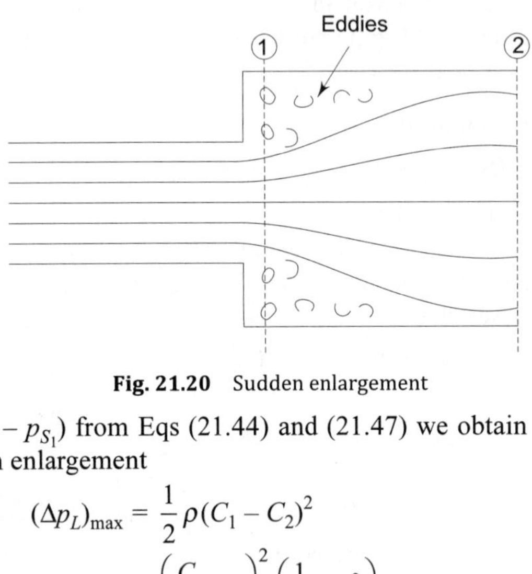

Fig. 21.20 Sudden enlargement

Equating $(p_{S_2} - p_{S_1})$ from Eqs (21.44) and (21.47) we obtain for pressure drop through a sudden enlargement

$$(\Delta p_L)_{\max} = \frac{1}{2}\rho(C_1 - C_2)^2$$
$$= \left(\frac{C_1}{C_2} - 1\right)^2 \left(\frac{1}{2}\rho C_2^2\right)$$
$$= (A_2/A_1 - 1)^2 \left(\frac{1}{2}\rho C_2^2\right) = (K_2)_{\max}\; p_{V_2} \tag{21.48}$$

where

$(K_2)_{\max} = (A_2/A_1 - 1)^2$ is the maximum value of the dynamic loss coefficient.

The above expression represents the maximum possible loss due to an enlargement, and hence corresponds to a minimum value of static regain, which is

$$(\text{SR})_{\min} = (p_{V_1} - p_{V_2}) - (\Delta p_L)_{\max}$$
$$= \frac{1}{2}\rho(C_1^2 - C_2^2) - \frac{1}{2}\rho(C_1 - C_2)^2$$
$$= 2\left(\frac{C_1}{C_2} - 1\right)\left(\frac{1}{2}\rho C_2^2\right)$$

$$= 2\left(\frac{A_2}{A_1} - 1\right)\left(\frac{1}{2}\rho C_2^2\right)$$

$$= 2\,(\lambda - 1)p_{V_2} \tag{21.49}$$

where λ is the *area ratio*. In Fig. 21.19 the broken lines show the pressure changes due to an abrupt expander.

For the case of actual enlargements, $\Delta p_L = K_2\, p_{V_2}$. Table 21.5 gives the values of a coefficient ratio K_r as a function of the included angle θ of the sides. Here K_r is the ratio of the actual loss to the maximum possible loss, viz.,

$$K_r = \frac{\Delta p_L}{(\Delta p_L)_{\max}} = \frac{K_2}{(K_2)_{\max}}$$

Table 21.5 Loss coefficients K_r for expansion and K_2 for contraction[1]

Case	*Conditions*	*Loss Coefficient*
Expansion	θ, deg	K_r
	5	0.17
	7	0.22
	10	0.28
	20	0.45
	30	0.59
	40	0.73
Contraction	θ, deg	K_2
	30	0.02
	45	0.04
	60	0.07

21.4.3 Loss due to Contraction

Consider now a sudden or *abrupt contraction* as shown in Fig. 21.21. It may be noted that turbulence occurs at two places, viz., at the shoulders of the contraction in the large section, and at a section shortly after the vena contracta. It may be observed that the major source of pressure loss is at the neck. Thus, the loss corresponds to that of sudden expansion from the velocity C_0 at the vena contracta to the downstream velocity C_2, i.e.,

$$\Delta p_L = \frac{1}{2}\rho(C_0 - C_2)^2 \tag{21.50}$$

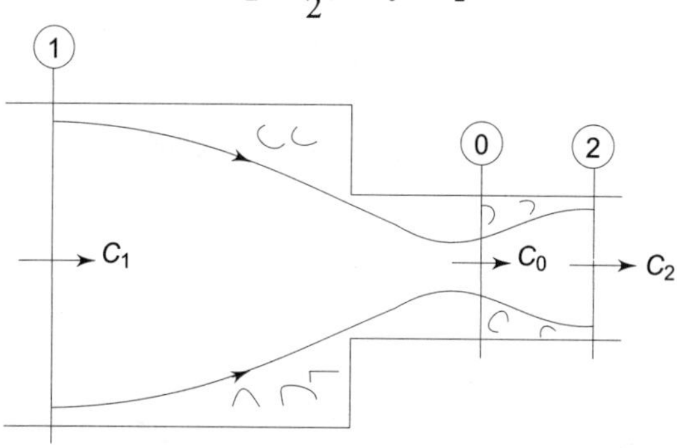

Fig. 21.21 Sudden contraction

Putting the area coefficient

$$C_c = \frac{A_0}{A_2} = \frac{C_2}{C_0}$$

we have

$$\Delta p_L \approx \frac{1}{2}\rho C_2^2 \left(\frac{1}{C_c} - 1\right)^2 \tag{21.51}$$

For circular orifices and for the flow of water, C_c is taken as 0.62. This gives

$$\Delta p_L = \frac{1}{2}\rho C_2^2 \left(\frac{1}{0.62} - 1\right)^2$$

$$= 0.375\left(\frac{1}{2}\rho C_2^2\right) = 0.375\, p_{V_2}$$

Thus the dynamic loss coefficient K_2 is equal to 0.375. Experimental results show that a more accurate value of the dynamic loss coefficient for air flow in ducts is 0.5 for an abruptly reducing duct section. Table 21.5 gives the values of K_2 for different included angles θ of the sides of the duct contraction.

Example 21.4

(a) In a duct enlargement, the velocity reduces from 20 to 5 m/s.

(i) If it is a 20°-angle gradual enlargement, the pressure loss is given by

$$\Delta p_L = 0.45\ (p_{V_1} - p_{V_2})$$

Calculate the static regain.

(ii) What will be the static regain in an abrupt enlargement?

(b) A 10 cm diameter duct converges gradually to 7.5 cm diameter. The static pressure and velocity just upstream of the reducer are 3 cm H_2O and 7.6 m/s respectively. The dynamic loss coefficient of the reducer is 0.1.

Calculate:

(i) The flow rate.

(ii) The pressure loss in the reducer.

(iii) The pressure indicated by a U-tube water manometer connected to pressure tappings at the upstream and downstream of the reducer.

Solution (a) (i) *Gradual enlargement*:

Pressure drop ($1 - R = 0.45$)

$$\Delta p_L = 0.45\ (p_{V_1} - p_{V_2})$$
$$= 0.45\ (0.6 \times 20^2 - 0.6 \times 5^2)$$
$$= 0.45\ (225) = 101.3 \text{ N/m}^2$$

Static regain

$$\text{SR} = (p_{V_1} - p_{V_2}) - \Delta p_L$$
$$= 225 - 101.3 = 123.7 \text{ N/m}^2$$

(ii) *Abrupt enlargement*:

Dynamic loss coefficient

$$(K_2)_{max} = \left(\frac{A_2}{A_1} - 1\right)^2 = \left(\frac{C_1}{C_2} - 1\right)^2 = \left(\frac{20}{5} - 1\right)^2 = 9$$

Pressure drop

$$(\Delta p_L)_{\max} = (K_2)_{\max}\, p_{V_2} = 9\,(0.6 \times 5^2) = 135 \text{ N/m}^2$$

Static regain

$$\text{SR} = (p_{V_1} - p_{V_2}) - \Delta p_L = 225 - 135 = 90 \text{ N/m}^2$$

(b) *Gradual reducer*:

(i) Flow rate

$$\dot{Q}_v = A_1 C_1 = \frac{\pi\,(0.1)^2}{4}\,(7.6) = 0.0597 \text{ m}^3\text{/s}$$

(ii) Downstream velocity

$$C_2 = \frac{\dot{Q}_v}{A_2} = \frac{(0.0597)^4}{\pi (0.075)^2} = 13.5 \text{ m/s}$$

Velocity pressures

$$p_{V_1} = \left(\frac{7.6}{4.04}\right)^2 = 3.54 \text{ mm } H_2O$$

$$p_{V_2} = \left(\frac{13.5}{4.04}\right)^2 = 11.17 \text{ mm } H_2O$$

Pressure loss

$$\Delta p_L = K_2 p_{V_2} = (0.1)\,(11.17) = 1.12 \text{ mm } H_2O$$

(iii) Upstream total pressure

$$p_{T_1} = p_{S_1} + p_{V_1} = 30 + 3.54 = 33.54 \text{ mm } H_2O$$

Downstream total pressure

$$p_{T_2} = p_{T_1} - \Delta p_L = 33.54 - 1.12 = 32.42 \text{ mm } H_2O$$

Downstream static pressure

$$p_{S_2} = p_{T_2} - p_{V_2} = 32.42 - 11.17 = 21.25 \text{ mm } H_2O$$

Pressure indicated by the manometer

$$\Delta p = p_{S_1} - p_{S_2} = 30 - 21.25 = 8.75 \text{ mm } H_2O$$

21.4.4 Losses at Suction and Discharge Openings

In an abrupt suction opening, the air is accelerated as it approaches the opening, forming a vena contracta inside the duct. The area changes from infinity to the duct area. In such a case the dynamic loss coefficient is 0.85. By making a formed entrance of bell-mouth shape, the loss coefficient can be reduced to 0.03 (Fig. 21.22).

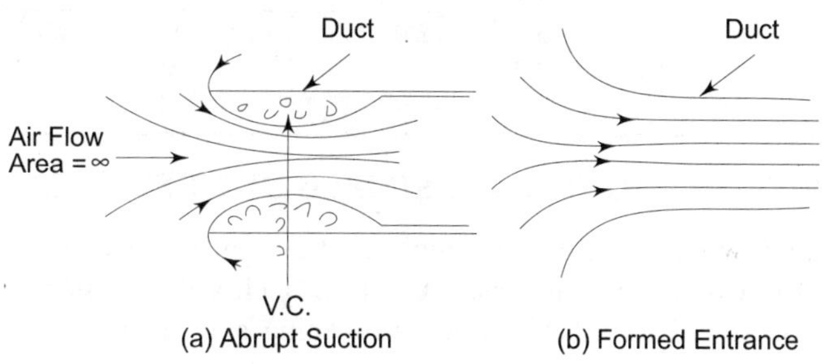

Fig. 21.22 Suction openings

At the discharge opening, the air is virtually at atmospheric pressure. After discharge, all the velocity pressure at the exit is completely dissipated.

21.4.5 Pressure Loss in Divided Flow Fittings

Whenever air is diverted to a branch as in Fig. 21.23, there is a velocity reduction in the *straight-through section*. If there is no loss, the change in the velocity pressure is completely converted into regain in static pressure. However, due to the dynamic loss, the actual static pressure regain is reduced by the static regain factor R. The value of R is 0.9 for well-designed and constructed round ducts with no reducing section immediately after take off. In less ideal conditions, e.g., in rectangular ducts of high-aspect ratio, or take-offs closely following an upstream disturbance, the regain coefficient can be as low as 0.5. The total pressure loss is given by Eq. (21.44 b)

$$\Delta p_L = (1 - R)\,(p_{V_1} - p_{V_2})$$

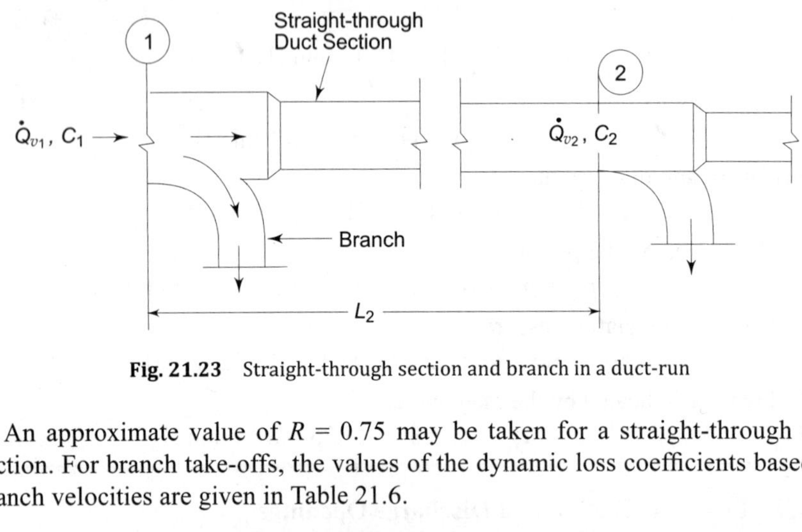

Fig. 21.23 Straight-through section and branch in a duct-run

An approximate value of R = 0.75 may be taken for a straight-through duct section. For branch take-offs, the values of the dynamic loss coefficients based on branch velocities are given in Table 21.6.

Table 21.6 Dynamic loss coefficients for branch take-offs

Take-off Angle, Deg.	*Ratio of Velocity in Branch to Velocity in Main Duct*						
	0.4	**0.6**	**0.8**	**1.0**	**1.5**	**2.0**	**3.0**
90	6.5	3.1	2.0	1.5	0.95	0.74	0.62
60	5.0	2.2	1.3	0.77	0.47	0.47	0.58
45	3.5	1.3	0.64	0.43	0.4	0.45	0.54

21.5 AIR FLOW THROUGH A SIMPLE DUCT SYSTEM[2]

Consider the flow of air through a simple duct system for example as shown in Fig. 21.26. The overall pressure distribution for such a duct system is shown in Fig. 21.24. Section 1 is just after suction opening and Section 4 is just before discharge opening. Air is sucked in from conditioned space at zero gauge pressure, through the suction-side duct 1–2, by a fan. The fan raises the total pressure from p_{T_2} to p_{T_3}, and drives the air in the discharge-side duct 3–4, before it enters the conditioned space, finally reaching zero gauge pressure again.

Figure 21.24 also shows the distribution of static, velocity and total pressures through the system. Let us consider the variation of these pressures.

Suction-side Let the pressure loss at entry of the suction opening and due to friction of the inlet grille be Δp_E. Then the total pressure at 1 is given by Bernoulli's equation

$$0 = p_{T_1} + \Delta p_E$$
$$p_{T_1} = -\Delta p_E$$

Static pressure at 1

$$p_{S_1} = p_{T_1} - p_{V_1} = -(\Delta p_E - p_{V_1})$$

Thus the static pressure at 1 is negative. We define *static depression* or vacuum if the pressure is negative as follows:

$$\text{Static depression} = 0 - p_S$$

Hence the static depression at 1 is equal to $(\Delta p_E + p_{V_1})$.

Let Δp_{1-2} be the pressure loss in the suction-side duct 1-2. Then the total pressure at 2 is given by

$$p_{T_2} = p_{T_1} - \Delta p_{1-2} = -(\Delta p_E + \Delta p_{1-2})$$

Static pressure at 2

$$p_{S_2} = p_{T_2} - p_{V_2} = -(\Delta p_E + \Delta p_{1-2} + p_{V_2})$$

Static depression at 2 is

$$\Delta p_E + \Delta p_{1-2} + p_{V_2}$$

Discharge-side Let the pressure loss at the outlet in the discharge grille be Δp_0. Then total pressure at 4

$$p_{T_4} = \Delta p_0 + p_{V_4}$$

Static pressure at 4

$$p_{S_4} = p_{T_4} - p_{V_4} = \Delta p_0$$

Let Δp_{3-4} be the pressure loss of the discharge side duct 3-4. Then the total pressure at 3 is given by

$$p_{T_3} = p_{T_4} + \Delta p_{3-4} = \Delta p_{3-4} + \Delta p_0 + p_{V_4}$$

Static pressure at 3

$$p_{S_3} = p_{T_3} - p_{V_3} = \Delta p_{3-4} + \Delta p_0 + p_{V_4} - p_{V_3}$$

Finally, the pressure developed by the fan, called the *fan total pressure*, FTP, is given by

$$\begin{aligned}\text{FTP} &= p_{T_3} - p_{T_2} \\ &= \Delta p_E + \Delta p_{1-2} + \Delta p_{3-4} + \Delta p_0 + p_{V_4}\end{aligned} \tag{21.52}$$

Thus the fan total pressure consists of the following:

(i) The pressure loss of suction opening and friction past the inlet grille.
(ii) The pressure loss of the return duct and air-conditioning apparatus.
(iii) The pressure loss of the supply duct.
(iv) The friction past the outlet grille.
(v) The velocity pressure at the outlet.

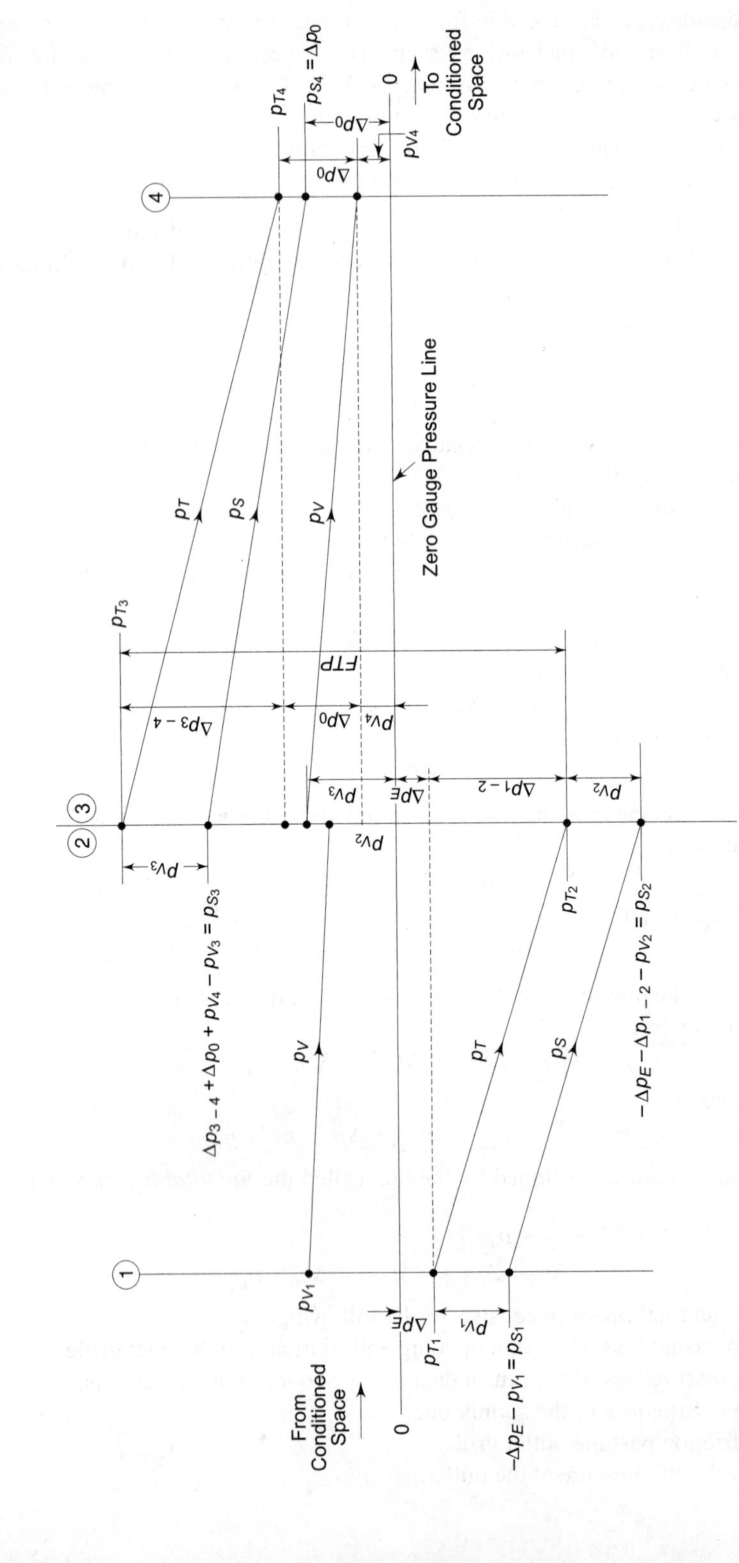

Fig. 21.24 Pressure variation in a duct system

In brief, it is equal to the total pressure drop plus the velocity pressure at the outlet. It may be pointed out that the pressure loss Δp_{1-2} of the return duct and air-conditioning apparatus may include pressure drops in air filters, dampers, cooling coil or air washer, heater, etc., which may form a part of the air-conditioning equipment.

We may also define *fan static pressure*, FSP, as the fan total pressure minus the velocity pressure at fan outlet, i.e.,

$$\text{FSP} = \text{FTP} - p_{V_3}$$

Also, the *fan velocity pressure*, FVP, is the dynamic head at the fan outlet

$$\text{FVP} = p_{V_3}$$

The theoretical power requirement of the fan, normally referred to as *air horsepower* is given by

$$\dot{W} = \dot{Q}_v \text{ (FTP)}$$

where $\dot{Q}_v$ is in m^3/s and FTP is in N/m^2, which can be converted into hp. W is in watts

21.6 AIR-DUCT DESIGN

The essential economics of an air transmission system is achieved by a proper balance between the initial or first cost and operating cost for the given flow rate of air. The first cost is determined by the cost of the duct system which depends on duct sizes. The operating cost is determined by the fan power consumption which depends on the pressure drop in the air-handling equipment and the duct system. The pressure drop can be reduced by increasing the size of the ducts but it will increase the first cost. Hence the need for a proper balance.

A few general rules are stated below which should be followed in the design of ducts.

(i) Air should be conveyed as directly as possible to economise on power, material and space.

(ii) Sudden changes in direction should be avoided. When bends are essential, turning vanes should be used to minimise the pressure loss.

(iii) Air velocities in ducts should be within permissible limits to minimise noise.

(iv) Diverging sections should be made gradual. The angle of divergence should not exceed 20°.

(v) Rectangular ducts should be made as nearly square as possible. This will ensure minimum duct surface, and hence cost, for the same air-carrying capacity. An aspect ratio of less than 4 : 1 should be maintained.

(vi) Ducts should be made of smooth materials such as galvanized iron (GI) or aluminium sheet metals. Whenever other materials are used, allowance should be made for the roughness of the material.

(vii) Dampers should be provided in each branch outlet for balancing the system.

(viii) Avoid duct obstructions.

21.6.1 Duct Construction

The most commonly used duct material is GI sheet metal. Aluminium sheet metal may be used where there are benefits to be derived due to its lighter weight and resistance to moisture. Heavy gauge black-steel may be used for a kitchen exhaust. Cement asbestos may be used for underground air distribution and for exhausting corrosive materials. Fibre glass ducts are in use in low velocity applications where thermal insulation or accoustic treatment is required.

Duct reinforcing is done by sheet metal itself or by steel or extruded aluminium angles.

21.6.2 Pressure Classifications

Air imposes two types of loads on a duct structure:

(i) The static load due to the mean static pressure differential across the duct wall. This is the dominating load.

(ii) The pulsating load due to turbulent air flow. This load is comparatively small.

Accordingly, the ducts are classified as follows:

(i) *Low-pressure system*
Velocities < 600 mpm and static pressure ≤ 5 cm H_2O gauge.

(ii) *Medium-pressure system*
Velocities < 600 mpm or static pressure up to 15 cm H_2O gauge.

(iii) *High-pressure system*
Velocities > 600 mpm or static pressure over 15 cm and up to 25 cm H_2O gauge.

The low-pressure system permits the use of simple forming methods. A majority of duct construction belongs to the low-pressure type. The gauge of the sheet metal depends on the dimension of the longest side as given in Table 21.7.

The recommended maximum duct velocities for the low-velocity system are given in Table 21.8.

Medium and high pressure rectangular ducts may be used to meet the space limitations. It is generally a more expensive construction. Tie rods as well as transverse reinforcements are used. Recommended GI gauges for the two systems are also given in Table 21.7.

Table 21.7 Recommended GI gauges

Longest side in cm			*GI Sheet Metal Gauge*
Low Pressure	***Medium Pressure***	***High Pressure***	
Upto 30	—	—	26
30–75	Upto 45	—	24
75–135	45–120	Upto 120	22
135–210	120–180	120–180	20
Above 210	Above 180	180–240	18
—	—	Above 240	16

21.6.3 Duct Design Procedures

There are three common methods for the sizing of ducts. They are:

(i) Equal friction method

(ii) Velocity reduction method

(iii) Static regain method

Equal Friction (EF) Method In the equal friction method, the frictional pressure drop per unit length of the duct is maintained constant throughout the duct system. The procedure is to select a suitable velocity in the main duct from sound level considerations as given in Table 21.8. Knowing the air flow rate and the velocity in the main duct, the size and friction loss are determined from the chart. The remaining ducts are then sized, maintaining the friction loss per unit length at this value for their respective air-flow rates.

This method of sizing ducts, automatically, reduces the air velocity in the direction of flow. The method is generally recommended because of its simplicity. Moreover, the use of a calculator, called the *ductulator*, speeds up the design work.

If an equal friction design has a mixture of short and long runs of duct, the shortest duct will need a considerable amount of dampering. This is a drawback of the equal friction design.

To determine the total friction loss in the duct system, it is necessary to calculate the loss in the duct run that has the highest resistance.

Velocity Reduction Method In this method, the main duct is designed in the same manner as in the equal friction method. Thereafter, arbitrary reductions are made in the air velocity as we go down the duct run. Equivalent diameters are found, as before, from the friction chart.

Though the method allows safe velocities, it is not normally adopted unless the person using it has considerable practical experience and knowledge to design within reasonable accuracy.

The starting velocities should not exceed those recommended in Table 21.8.

Note *High velocity duct/pipe system, with velocity approaching 1800 mpm, are used for air conditioning in ships and aircraft to economise on spaces.*

Table 21.8 Recommended maximum duct velocity for low-velocity system (mpm)

Application	*Controlling factor:Noise*	*Controlling Factor:Friction*			
		Main Ducts		*Branch Ducts*	
	Main Ducts	*Supply*	*Return*	*Supply*	*Return*
Residences Apartments. Hotel bed rooms	180	300	240	180	180
Hospital bed rooms	300	450	390	360	300
Offices. Libraries	360	600	450	480	360
Theatres Stores. Banks	240	390	330	300	240
Restaurants	450	600	450	480	360
Cafetarias	540	600	450	480	360

Static Regain (SR) Method The principle of the static regain method is to maintain a constant static pressure before each terminal and each branch. This is achieved by sizing the duct in such a manner that after each branch or outlet, the static pressure gain—due to reduction in velocity—exactly balances the pressure drop in the succeeding duct section.

Figure 21.23 shows the section of the duct under consideration for sizing between two branch take-offs. Let Δp_f and Δp_d represent the friction and dynamic pressure losses of this section. The velocity at 2 is less than the velocity at 1. The static pressure regain is given by Eq. (21.44a). Equating the losses to regain, we have

$$\Delta p_{f_2} + \Delta p_{d_2} = R(p_{V_1} - p_{V_2}) \tag{21.53}$$

The above equation can be solved for C_2, and thus the duct diameter may be found from the air flow and velocity. The procedure, however, involves iteration as the equation is not explicit in C_2.

The method meets the essential requirement of maintaining uniform static pressure at all branches and outlets. Thus it is a balanced system and the duct work designed accordingly does not require the use of dampers anywhere.

Note *The SP more or less remains constant throughout the duct system designed by SR method. The friction and dynamic pressure losses in total pressure, therefore, come from decrease in VP.*

Example 21.5 *An air duct system is provided as shown in Fig. 21.25.*

(i) Determine the dimensions of AB, BC, CD using the equal friction method. Choose a friction rate of 0.08 mm H_2O/m length of duct. Use the following formula for friction rate:

$$\Delta p_f/L = 2.268 \times 10^{-3}\ \dot{Q}_v^{1.852}/D^{4.973} \text{ mm } H_2O/m$$

where $\dot{Q}_v$ is in m^3/s and D is in m.

(ii) Determine the total and static pressures at point A. Assume free exit at each outlet.

Losses are given by:

For elbow: 0.25 p_{V_2}

For branch: 0.2 p_{V_2} + Elbow loss

For straight-through section: 0.25 × Difference of velocity pressures

(iii) Find the diameter of BE so that no dampering is required in the section.

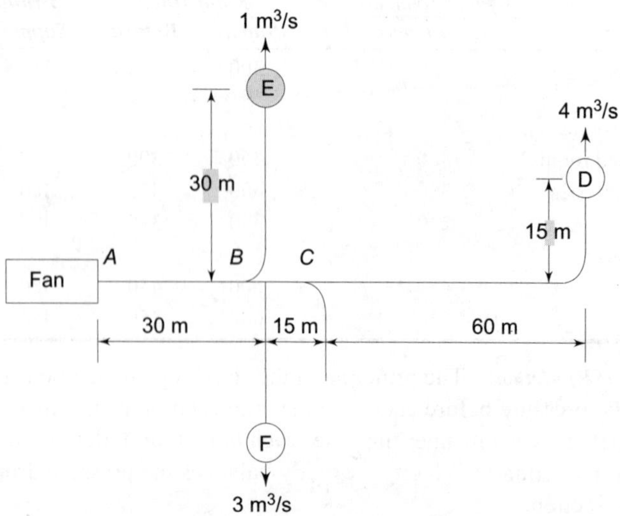

Fig. 21.25 Duct layout for Example 21.5

Solution (1) Air flow at A

$$\dot{Q}_{vA} = 1 + 4 + 3 = 8 \text{ m}^3/\text{s}$$

Rearranging the friction rate Eq. (21.33)

$$\frac{\Delta p_f}{L} = \frac{0.002268\, \dot{Q}_v^{1.852}}{D^{4.973}}$$

gives for the duct diameter

$$D = \frac{0.2948\, \dot{Q}_v^{0.372}}{(\Delta p_f/L)^{0.2}}$$

Substituting for the friction rate, we get

$$D = \frac{0.2948\, \dot{Q}_v^{0.372}}{(0.08)^{0.2}} = 0.4886\, \dot{Q}_v^{0.372}$$

Using the above equation, the diameters are calculated for different sections for the given air-flow rates, and are given in Table 21.9 along with the length, area, velocity, velocity pressure and frictional pressure drop of the section.

Table 21.9 Calculations for Example 21.5

Section	*L*	$\dot{Q}_v$	*D*	*A*	*C*	p_V	$\Delta p_f = 0.08\, L$
	m	m^3/s	m	m^2	m/s	mm H_2O	mm H_2O
AB	30	8	1.06	0.882	9.7	5.76	2.4
BC	15	7	1.01	0.8	8.75	4.72	1.2
CD	75	4	0.82	0.528	7.6	3.54	6.0
Frictional pressure drop from A to D							9.6

(ii) Dynamic losses between A and D

Loss in discharge opening

$$\Delta p = p_V = 3.54 \text{ mm } H_2O$$

Elbow loss

$$\Delta p = 0.25\,(p_V)_{CD}$$
$$= 0.25\,(3.54) = 0.88 \text{ mm } H_2O$$

Fitting losses

$$\Delta p = (1 - R)\,(p_{V_1} - p_{V_2})$$

$$\Delta p_B = 0.25\,(5.76 - 4.72) = 0.26 \text{ mm } H_2O$$

$$\Delta p_C = 0.25\,(4.72 - 3.54) = 0.3 \text{ mm } H_2O$$

Total dynamic losses, with Δp_O pressure drop at the outlet

$$\Delta p_d = 3.54 + 0.88 + 0.26 + 0.3 = 5 \text{ mm } H_2O + \Delta p_O$$

Total frictional and dynamic losses between A and D

$$\Delta p = \Delta p_f + \Delta p_d$$
$$= 9.6 + 5.76 + \Delta p_O = 15.4 \text{ mm } H_2O + \Delta p_O$$

Total pressure at A $\quad (p_T)_A = 15.4 \text{ mm } H_2O \text{ gauge} + \Delta p_O$

Static pressure at A $(p_S)_A = P_T - p_V$

$= 15.4 + \Delta p_O - 5.76 = 9.6$ mm H_2O gauge $+ \Delta p_O$

(iii) Total pressure at B $(p_T)_B = (p_T)_A - (\Delta p_L)_{AB}$

$= 15.4 + \Delta p_O - 2.4 = 13$ mm $H_2O + \Delta p_O$

Pressure loss in BE $(\Delta p)_{BE} = (0.25 + 0.2 + 1)\, p_V + \Delta p_f + \Delta p_O$

$= 1.45\, p_V + \Delta p_f + \Delta p_O$

The first term on the right-hand side represents the sum of the dynamic losses in the elbow at B, the area change in the branch and the velocity pressure at discharge. Equating it to the total pressure at B for complete balancing, we obtain

$$\Delta p_C + 1.45\, p_V + \Delta p_f = 13 + \Delta p_O$$

The above equation can only be solved by trial and error.

Assume $D = 0.4$ m for BE

Friction rate $$\frac{\Delta p_f}{L} = \frac{0.002268\,(1)^{1.825}}{(0.4)^{4.973}} = 0.216$$

Friction loss $$\Delta p_f = (0.216)\,(30) = 6.5 \text{ mm } H_2O$$

Velocity $$C = \frac{\dot{Q}_v}{A} = \frac{1}{\frac{\pi}{4}(0.4)^2} = 7.96 \text{ m/s}$$

Velocity pressure $$p_V = \left(\frac{7.96}{4.04}\right)^2 = 4 \text{ mm } H_2O$$

Thus, the pressure loss in BE

$$(\Delta p)_{BE} = (1.45)\,4 + 6.5 + \Delta p_O = 12.3 + \Delta p_O$$

which is approximately equal to the total pressure at B. Hence an assumed diameter of 0.4 m for section BE is satisfactory.

Note *The above example illustrates the design of ducts mainly by the equal friction method. However, the design of branch BE has been done on the basis of static regain method. This approach provides for the complete balancing of the flow in branches without introducing an additional resistance by way of dampering, and at the same time reduces the duct size. On the basis of the equal friction method, the diameter of section BE would have been 0.48 m instead of 0.4 m. If the diameter is further reduced to 0.39 m, the pressure drop will increase to 13.6 mm H_2O.*

Example 21.6 *A centrifugal fan with 90 cm × 70 cm outlet is moving standard air at a rate of 11.5 m^3/s through a system which consists of straight inlet and outlet ducts. The inlet duct is 90 cm in diameter and 15 m long. The outlet duct is 100 cm in diameter and 60 m long. There is a fan diffuser between the fan discharge and the 100 cm diameter duct for which the pressure loss is one-third the difference in velocity pressures. The pressure drop at the filter, damper and cooling coil in the inlet duct is 15 mm H_2O. The loss at the entry to the inlet is 0.5 × velocity pressure. The friction factor for the outlet duct is 0.0035, and that for the inlet duct is 0.004.*

(i) *Determine the fan total pressure.*
(ii) *Determine the static pressures at fan inlet and outlet.*
(iii) *Plot the variation of the total pressure and static pressure along the duct system.*

The air is sucked in by the inlet duct and delivered by the outlet at atmospheric pressure.

Solution Refer to Fig. 21.26

(i) Area, velocity and velocity pressure in the inlet duct

$$A_1 = A_2 = \frac{\pi}{4}(0.9)^2 = 0.636 \text{ m}^2$$

$$C_1 = C_2 = \frac{\dot{Q}_v}{A_1}$$

$$= \frac{11.5}{0.636} = 18.1 \text{ m/s}$$

$$p_{V_1} = p_{V_2} = (C_1/4.04)^2$$

$$= \left(\frac{18.1}{4.04}\right)^2 = 20.1 \text{ mm H}_2\text{O}$$

Area, velocity and velocity pressure at fan discharge

$$A_3 = (0.9)(0.7) = 0.63 \text{ m}^2$$

$$C_3 = \frac{11.5}{0.63} = 18.25 \text{ m/s}$$

$$p_{V_3} = \left(\frac{18.25}{4.04}\right)^2 = 20.4 \text{ mm H}_2\text{O}$$

Area, velocity and velocity pressure in the outlet duct

$$A'_3 = A_4 = \frac{\pi}{4}(1)^2 = 0.785 \text{ m}^2$$

$$C'_3 = C_4 = \frac{11.5}{0.785} = 14.64 \text{ m/s.}$$

$$p'_{V_3} = p_{V_4} = \left(\frac{14.65}{0.04}\right)^2 = 13.14 \text{ mm H}_2\text{O}$$

The various pressure losses will now be calculated.

Inlet loss

$$\Delta p_E = 0.5\, p_{V_1} = 0.5\,(20.1) = 10.05 \text{ mm H}_2\text{O}$$

Inlet duct friction loss

$$\Delta p_{f_{1-2}} = \frac{4 f_1 L_{1-2}}{D_1} p_{V_1}$$

$$= \frac{4(0.004)(15)(20.1)}{0.9} = 5.36 \text{ mm H}_2\text{O}$$

Loss at the filter, damper and cooling coil

$$\Delta p_{d_{1-2}} = 15 \text{ mm H}_2\text{O (given)}$$

Loss in the fan diffuser

$$\Delta p = \frac{1}{3}(p_{V_3} - p'_{V_3}) \text{ (say, given)}$$
$$= \frac{1}{3}(20.4 - 13.14) = 2.42 \text{ mm } H_2O$$

Outlet duct friction loss

$$\Delta p_{f_{3-4}} = \frac{4 f_3 L_{3-4}}{D_3} p_{V_3}$$
$$= \frac{4(0.0035)(60)(13.14)}{1} = 11.04 \text{ mm } H_2O$$

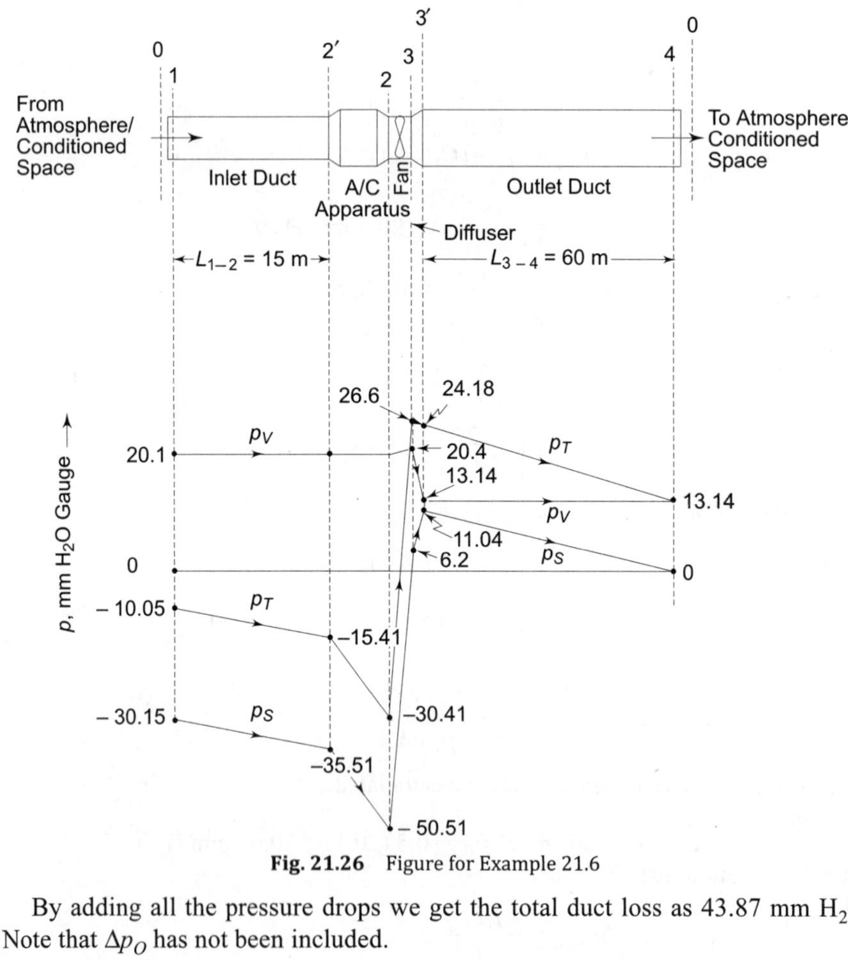

Fig. 21.26 Figure for Example 21.6

By adding all the pressure drops we get the total duct loss as 43.87 mm H_2O. Note that Δp_O has not been included.

Fan total pressure

$$\text{FTP} = \text{Pressure loss in duct system} + \text{outlet velocity pressure}$$
$$= 43.87 + 13.14 = 57.01 \text{ mm } H_2O$$

(ii) Equating total pressures in the atmosphere and fan inlet

$$0 = \Delta p_{0-2} + p_{V_2} + p_{S_2}$$

whence the static pressure at the fan inlet is

$$\begin{aligned} p_{S_2} &= -\Delta p_{0-2} - p_{V_2} \\ &= -(10.05 + 5.36 + 15) - 20.1 \\ &= -30.41 - 20.1 = -50.51 \text{ mm } H_2O \quad \text{(Negative)} \end{aligned}$$

Energy balance of fan from Eq. (21.52)

$$p_{S_3} + p_{V_3} = p_{S_2} + p_{V_2} + \text{FTP}$$

whence the static pressure at the fan outlet is

$$\begin{aligned} p_{S_3} &= p_{S_2} + p_{V_2} + \text{FTP} - p_{V_3} \\ &= -50.51 + 20.1 + 57.01 - 20.4 = 6.2 \text{ mm } H_2O \quad \text{(Positive)} \end{aligned}$$

(iii) The total pressures and static pressures at the other sections are as follows:

Section 1:

$$\begin{aligned} p_{T_1} &= p_{T0} - \text{Inlet loss} \\ &= 0 - 10.05 = -10.05 \text{ mm } H_2O \end{aligned}$$

$$\begin{aligned} p_{S_1} &= p_{T_1} - p_{V_1} \\ &= -10.05 - 20.1 = -30.15 \text{ mm } H_2O \end{aligned}$$

Section 2′:

$$\begin{aligned} p'_{T_2} &= p_{T_1} - \text{Inlet duct friction loss} \\ &= -10.05 - 5.36 = -15.41 \text{ mm } H_2O \end{aligned}$$

$$\begin{aligned} p'_{S_2} &= p'_{T_2} - p'_{V_2} \\ &= -15.41 - 20.1 = -35.51 \text{ m } H_2O \end{aligned}$$

Section 2:

$$\begin{aligned} p_{T_2} &= p'_{T_2} - \text{Loss in A/C apparatus} \\ &= -15.41 - 15 = -30.41 \text{ mm } H_2O \end{aligned}$$

$$p_{S_2} = -50.51 \text{ mm } H_2O$$

Section 3:

$$\begin{aligned} p_{T_3} &= p_{T_2} + \text{FTP} \\ &= -30.41 + 57.01 = 26.6 \text{ mm } H_2O \end{aligned}$$

$$p_{S_3} = 6.2 \text{ mm } H_2O$$

Section 3′:

$$\begin{aligned} p'_{T_3} &= p_{T_3} - \text{Loss in fan diffuser} \\ &= 26.6 - 2.42 = 24.18 \text{ mm } H_2O \end{aligned}$$

$$\begin{aligned} p'_{S_3} &= p'_{T_3} - p'_{V_3} \\ &= 24.18 - 13.14 = 11.04 \text{ mm } H_2O \end{aligned}$$

Section 4:

$$\begin{aligned} p_{T_4} &= p'_{T_3} - \text{Outlet duct friction loss} \\ &= 24.18 - 11.04 = 13.14 \text{ mm } H_2O \end{aligned}$$

$$\begin{aligned} p_{S_4} &= p_{T_4} - p_{V_4} \\ &= 13.14 - 13.14 = 0 \end{aligned}$$

Note *The values are plotted in Fig. 21.26. The variation within the A/C apparatus between sections 2′ and 2 has been taken on an aggregate basis.*

Example 21.7 *Determine the duct sizes of sections A to G of the duct system in Fig. 21.27, and total pressure at fan outlet for the following three designs:*

(i) Equal friction (EF) method.
(ii) Static regain (SR) method.
(iii) EF method in section B, and SR in the rest.

Assume velocity in main duct A as 450 mpm (≈ 1500 fpm). Also assume dynamic loss coefficient in elbow K = 0.22, and static regain factor in fitting R = 0.75 (1 – R = 0.25). Static pressure at each outlet is 3 mm H_2O.

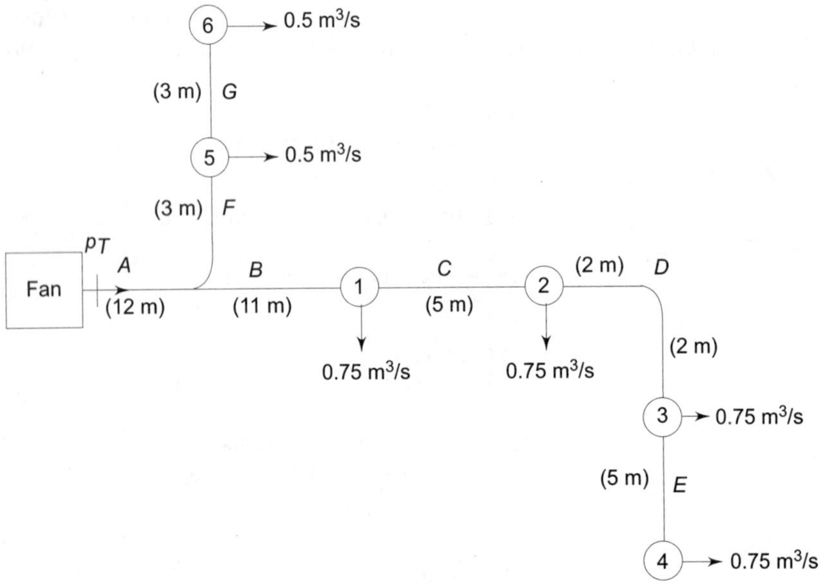

Fig. 21.27 Duct system for Example 21.7

Solution Volume flow rate, velocity, area, diameter, friction rate and VP in A

$$(\dot{Q}_v)_A = 4(0.75) + 2(0.5) = 4 \text{ m}^3/\text{s}$$

$$C_A = \frac{450}{60} = 7.5 \text{ m/s}$$

$$A_A = \frac{\dot{Q}_v}{C} = \frac{4}{7.5} = 0.5333 \text{ m}^2 \ (5333 \text{ cm}^2)$$

$$D_A = \sqrt{\frac{4A}{\pi}} = \sqrt{\frac{4(0.5333)}{\pi}} = 0.824 \text{ m (82.4 cm)}$$

$$\frac{\Delta p_f}{L} = \frac{0.002268\,\dot{Q}_v^{1.852}}{D^{4.973}} = \frac{0.002268\,(4)^{1.852}}{(0.824)^{4.973}} = 0.0774 \text{ mm H}_2\text{O/m}$$

$$p_{VA} = \left(\frac{C}{4.04}\right)^2 = \left(\frac{7.5}{4.04}\right)^2 = 3.45 \text{ mm H}_2\text{O}$$

(i) Equal Friction Method: EF method diameters for sections along with their lengths, flow rates and velocity pressures are given in Table 21.10.

Table 21.10 EF method duct diameters for Example 21.7

Section	*A*	*B*	*C*	*D*	*E*	*F*	*G*
$\dot{Q}_v$, m^3/s	4	3	2.25	1.5	0.75	1	0.5
L, m	12	11	5	4	5	3	5
D, cm	82.4	73.7	66	57	44	49	31
p_V, mm H_2O	3.45	2.98	2.65	2.12	1.49	1.72	2.69

Friction pressure drop in the longest duct-run A to E

$$\Delta p_f = \left(\frac{\Delta p_f}{L}\right) L = (0.0774)\,(37) = 2.864 \text{ mm } H_2O$$

Dynamic pressure drops in A to E

$$\begin{aligned}
{}_A(\text{Fitting})_B &= 0.25\,(3.45 - 2.8) &&= 0.118\\
{}_B(\text{Fitting})_C &= 0.25\,(2.8 - 2.65) &&= 0.083\\
{}_C(\text{Fitting})_D &= 0.25\,(2.65 - 2.12) &&= 0.133\\
{}_D(\text{Fitting})_E &= 0.25\,(2.12 - 1.49) &&= 0.158\\
\text{Elbow}_D &= 0.22\,(2.12) &&= 0.466\\
\hline
\text{Total} & &&= 0.958\\
+\text{ VP at outlet E} & &&= 1.49\\
+\text{ SP at outlet E} & &&= 3.0\\
\hline
\text{Total} & &&= 5.448 \text{ mm } H_2O
\end{aligned}$$

Total pressure at fan outlet

$$p_T = \Delta p_f + \Delta p_d = 2.864 + 5.448 = 8.312 \text{ mm } H_2O$$

(ii) *Static Regain Method:*

Sample calculations for section B

Assume $D_B = 0.78$ m. Then

$$A_B = 0.4776 \text{ m}^2,\ C_B = 6.282 \text{ m/s},\ p_{VB} = 2.417 \text{ mm } H_2O$$

$${}_A(\Delta p \text{ Fitting})_B = 0.25\,(3.45 - 2.417) = 0.258$$

$$\Delta p_f = \frac{0.002268\,(3)^{1.852}(11)}{(0.78)^{4.973}} = 0.781$$

Loss, $\Delta p_L = 0.258 + 0.781 = 1.039$

Regain, $\Delta p_R = 0.75\,(3.45 - 2.417) = 0.774 < \Delta p_L$

Hence, increase the diameter to increase regain. We find that at $D_B = 0.8$ m, $\Delta p_L = 0.895$, $\Delta p_R = 0.948$ just exceeds Δp_L. Therefore, take $D_B = 0.8$ m

Calculations are done similarly for all sections. Results are given in Table 21.11.

Table 21.11 SR method duct diameters for Example 21.7

Section	*A*	*B*	*C*	*D*	*E*	*F*	*G*
D, cm	82.4	80	72	67	52	48	39
p_V, mm H_2O	3.45	2.185	1.873	1.11	0.765	1.873	1.074
Δp_L, mm H_2O		0.895	0.13	0.558	0.258	1.072	0.539
Δp_R, mm H_2O		0.948	0.234	0.572	0.258	1.182	0.6

Friction pressure drop in A

$$(\Delta p_f)_A = \left(\frac{\Delta p_f}{L}\right) L = 0.0774\ (12) = 0.929\ \text{mm H}_2\text{O}$$

Friction and dynamic pressure loss in sections B to E (Table 21.11)

$$_B(\Delta p_L)_E = 0.895 + 0.13 + 0.558 + 0.258 + 1.072 + 0.539$$
$$= 3.452\ \text{mm H}_2\text{O}$$

Total pressure at fan outlet

$$p_T = (\Delta p_f)_A + {}_B(\Delta p_L)_E + (VP + SP)\ \text{at outlet}\ E$$
$$= 0.929 + 3.452 + 0.765 + 3.0 = 8.146\ \text{mm H}_2\text{O}$$

Note *Total pressure at fan outlet in SR method is nearly the same as in EF method. But, comparing the diameters in Tables 21.10 and 21.11, we find that SR diameters are much larger. Capital cost of ducts of SR design is, thus, higher. However, the operation of duct system is very satisfactory. Since no dampering is required at outlets, the volumes delivered are not affected by damper adjustments.*

(iii) *EF Method in Section B, SR Method in the Rest*: As seen above, SR method increases the overall cost of the duct system. A compromise can be made by extending EF method to the first section B in the longest duct-run. This will reduce dimensions of sections. However, provision will have to be made in the design of section F for additional drop in pressure equal to the friction loss in B. So, the diameter of B is taken as $D_B = 73.7$ cm from Table 21.10 with $p_{vB} = 2.98$ mm H_2O. The diameters of C, D and E that are found using the SR method are given in Table 21.12.

Table 21.12 Diameters: EF method in B, SR method in rest for Example 21.7

Section	*A*	*B*	*C*	*D*	*E*	*F*	*G*
D, cm	82.4	73.7	69	65	51	46	37
p_V, mm H_2O	3.45	2.98	2.221	1.253	0.827	2.221	1.326
Δp_L, mm H_2O			0.512	0.682	0.296		0.665
Δp_R, mm H_2O			0.577	0.726	0.32		0.672

Friction pressure drop in B, and pressure drop in fitting AB

$$(\Delta p_f)_B = (0.0774)11 = 0.851\ \text{mmH}_2\text{O}$$
$$_A(\Delta p\ \text{fitting})_B = 0.25\ (3.45 - 2.98) = 0.3\ \text{mm H}_2\text{O}$$

Total pressure at fan outlet

$$p_T = (\Delta p_f)_A + (\Delta p_f)_B + {}_A(\Delta p_{\text{fitting}})_B + {}_C(\Delta p_L)_E + (VP + SP)\ \text{at outlet}\ E$$
$$= 0.929 + 0.851 + 0.3 + (0.512 + 0.682 + 0.296) + 0.827 + 3.0$$
$$= 7.397\ \text{mm H}_2\text{O}$$

Static pressure available at inlet to section F

$$= p_T - (\Delta p_f)_A = 7.397 - 0.929 = 6.468\ \text{mm H}_2\text{O}$$

Hence, for section F, we must have

$$_A(\Delta p_{\text{fitting}})_F + \Delta p_{\text{elbow}} + (\Delta p_f)_F + (VP + SP)\ \text{at outlet}\ F = 6.468$$

By iteration, we find that $D_F = 0.46$ m, when $p_{VF} = 2.221$, and

$$(\Delta p_{\text{fitting}})_F = 0.25\,(3.45 - 2.221) = 0.307$$

$$\Delta p_{\text{elbow}} = 0.22\,(2.221) = 0.489$$

$$(\Delta p_{\text{friction}})_F = \frac{0.002268\,(1)^{1.852}}{(0.4)^{4.973}} = 0.324$$

Substituting values in the above equation

$$\text{LHS} = 0.307 + 0.489 + 0.324 + 2.221 + 3.0 = 6.341 \approx \text{RHS} = 6.468$$

Finally diameter of G, by SR method, is found to be 0.37 m.

21.7 PROCESSING, TRANSMISSION AND DISTRIBUTION OF AIR IN CLEAN ROOMS

In a clean room, particles within the room are removed and controlled by

(i) *preventing* the entry of particles by using HEPA filters,
(ii) *purging* the room of particles generated by changing the air in the room,
(iii) *prohibiting* the generation of particles by proper selection of construction materials, clothing, personnel, etc.,
(iv) *protecting the product* from direct impact and settling of particles, and
(v) *providing* support for cleaning of material and personnel before entry into the clean room.

Cleanliness levels have been defined in Chap. 16. The two major developments in clean room design are *HEPA filters*, and the concept of *laminar flow*.

21.7.1 Filters

Filters use the principles of straining, impingement, and interception.

There are three grades of filters, viz., *roughing filters*, *pre-filters* and *HEPA filters*.

Roughing filters use the principle of straining. Roughing filters used in clean rooms have a number of layers of nets, with a layer provided with resin impregnated synthetic media, all stitched together. Pre-filters use the method of impingement which depends on the inertia force of the filtered particles. HEPA filters use a combination of impingement and interception.

The filtration media in HEPA filters is made of submicronic glass fibres. It is a continuous sheet formed into accordion pleats. The glass fibres are heat resistant, retaining their strength upto 320°C. Aluminium separators are inserted between glass fibre sheet folds to form a *filter pack*. The filter pack is then sealed into an aluminium frame using suitable adhesive, and also provided with rubber gaskets on the sealing faces.

HEPA filters are tested both before installation, and once again, at site after installation to ensure that there are no leakages. The filter efficiency is computed as below:

$$\eta = \frac{\text{No. of particles before filter} \times \text{No. of particles after filter}}{\text{No. of particles before filter}}$$

The tests are conducted with a uniform velocity of 0.46 m/s (90 fpm). Disctyl phthalate (DOP) smoke is introduced upstream of the filters. The entire filter bank is then scanned using a properly sized aerosol photometer probe.

HEPA filters should have a filtration efficiency of 99.97% down to 0.3 micron size particles. They can handle their rated capacities at an initial pressure drop of 25 mm H_2O, and are normally operated to a final pressure drop of 75 mm H_2O *before being discarded.* The life of HEPA filters can be enhanced by using roughing filters, and microfine pre-filters, particularly in the *make-up/ventilation* air circuit, prior to them.

21.7.2 Conventional Flow Clean Rooms

Conditioned air in the clean rooms should be directed to obtain the *cleanest air at the most critical work areas*. Thereafter, as the particles are entrained, they are led to the less critical areas. These criteria require introduction of large quantities of air at low velocities.

Further, all rooms should be maintained at positive pressure, i.e.—at a static pressure sufficiently above atmospheric pressure—so as to prevent infiltration of contaminated outside air and permit only exfiltration.

Conventional flow patterns of air are described in Sec. 21.1.4. A conventional flow clean room A/C system (cooling and dehumidification) is illustrated in Fig. 21.28.

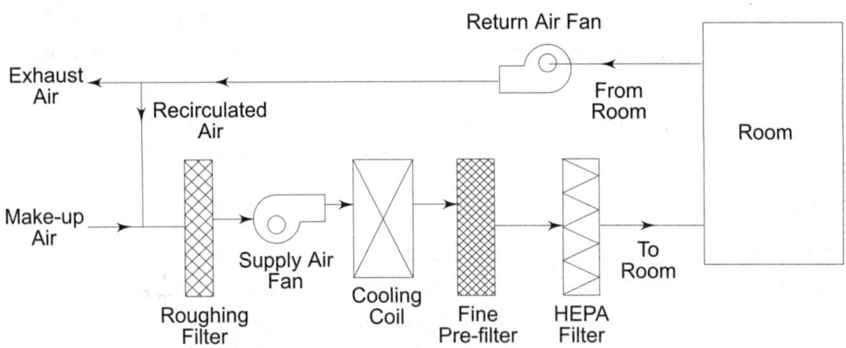

Fig. 21.28 Conventional flow clean room A/C systems

Major design features of the conventional flow clean rooms are 20 air changes per hour, positive pressure, final filters to be HEPA filters, installed only on discharge side of the fan, and to be preferably *terminally mounted*, people/particle generating activities near air exit, etc. The building materials used should not chip off or flake. All surfaces should be flush finished. Surface finish is done by apoxy paint, etc. Special lint free garments and clothing are to be used. Further, support areas are provided for *air showers* and *air locks*, restricting entry of personnel and introducing clean room discipline.

21.7.3 Laminar Flow Clean Rooms

It was found that even after taking all the measures described above, cleanliness levels could not be improved any further in conventional flow rooms. Major drawback is the build-up of smaller size particles. This method also does not prevent *cross-contamination* from one work area to another.

There was a major breakthrough in this respect with the application of *laminar flow* to clean rooms. It was realized that, instead of providing multiple outlets, if the entire ceiling or one wall were to be made of HEPA filters and used as supply panel, and the entire flooring or opposite wall were to be made up of return air grilles with prefilters installed, then the air distribution system would be the most effective. The advantages of laminar flow could be listed as follows:

(i) Air flows in parallel stream lines.
(ii) No cross-contamination of stream lines.
(iii) Predictable air path.
(iv) Independent of operation/activity/personnel.
(v) HEPA filters give lowest contamination.
(vi) No build-up of particles.
(vii) Isolating operations possible.

There are two versions of laminar flow; *downflow* in which the air is supplied from the ceiling and return is taken from the floor, and crossflow in which the air is supplied from one wall and return is taken from the opposite wall. The processing, transmission and distribution of the two is shown in Figs 21.29 and 21.30 respectively.

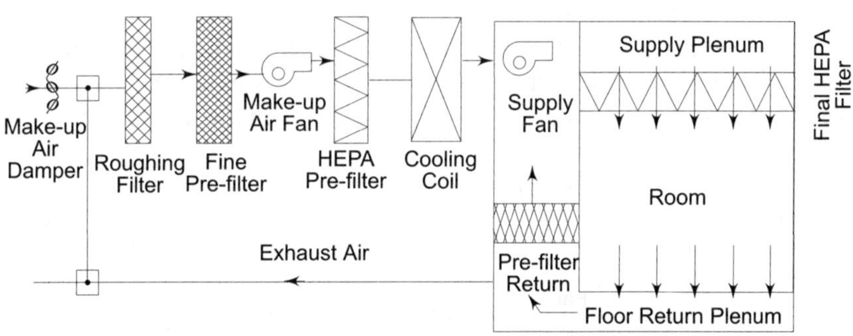

Fig. 21.29 Downflow laminar flow clean room A/C system

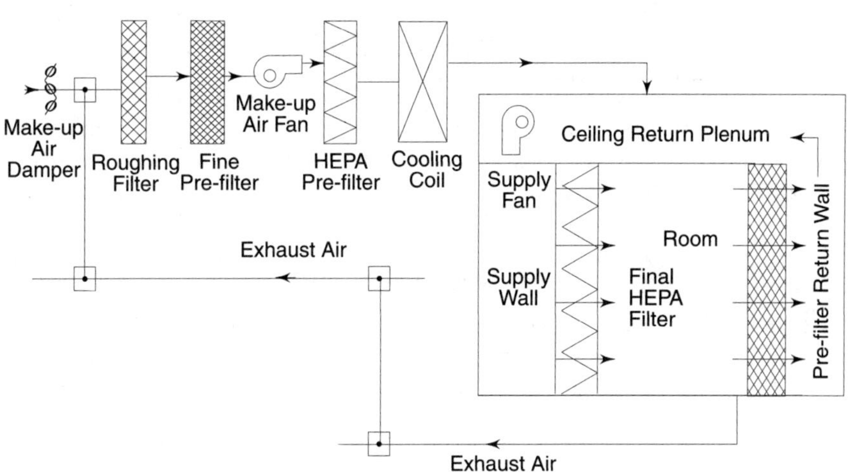

Fig. 21.30 Crossflow laminar flow clean room A/C system

In downflow, since the floor is made of open grilles, it is not possible to place heavy machinery on the floor. In crossflow, the disadvantage is that the downstream contamination is higher than upstream.

The clean room make-up/ventilation air dampers and exhaust air dampers are inter-connected. The exhaust air ducts should be carefully routed so that the laminar flow is maintained.

Figures 21.29 and 21.30 show that only make-up air is processed in the A/C central equipment. However, pre-filtered return air may also be recirculated depending upon SHF, ADP and other requirements.

21.8 AIR LOCKS, AIR CURTAINS AND AIR SHOWERS

Air locks are simply the interlocking spaces between conditioned and unconditioned/outside spaces. Provision of such air locks at the entrance to buildings minimises heat gain and infiltration of outdoor air directly into conditioned space.

Air curtains also serve the same purpose in more critical situations such as the entrances where the doors keep on opening too frequently. An air curtain is a layer of air which is blown across a door, parallel to the wall, in order to inhibit infiltration of unconditioned/outdoor air into conditioned space. This layer of air moves at such a velocity and angle that the air which tries to infiltrate/penetrate this curtain will be opposed by it. The air curtains have either *vertical downward flow* or *horizontal flow*. The flow may have either ducted or non-ducted return. The *effectiveness* of an air curtain to prevent heat flux or infiltration of air is to the extent of 60–80%. Note that the air curtain will maintain higher pressure in the layer as compared to the infiltrating air.

One of the major sources of particle contamination in clean room is personnel. Particles generated per person while carrying out various activities can range from 10^5 per minute for a standing/sitting person to 3×10^7 per minute for a person involved in vigorous activity. One of the steps that can be taken to control the contamination from personnel is to have *air shower* at entrance. Air shower can be installed in air locks. The air supplied in air shower can be separately processed.

References

1. 'Fundamentals and equipment', *ASHRAE Guide and Databook*, 1963.
2. Jones W P, *Air Conditioning Engineering*, Edward Arnold, London, 1973.
3. Rydberg J. and P. Norback, 'Air distribution and draft', *ASHVE Transactions*, Vol. 55, 1949, p. 225.

Revision Exercises

21.1 In the duct layout shown in Fig. 21.31, outlets 1 and 2 deliver 20 cmm each and outlet 3 delivers 28 cmm. Select a velocity of 8 m/s in section A. Size the duct system using the equal friction method and determine its static pressure requirement.

21.2 Size the duct system of Fig. 21.31, using the static regain method, and determine its static pressure requirement.

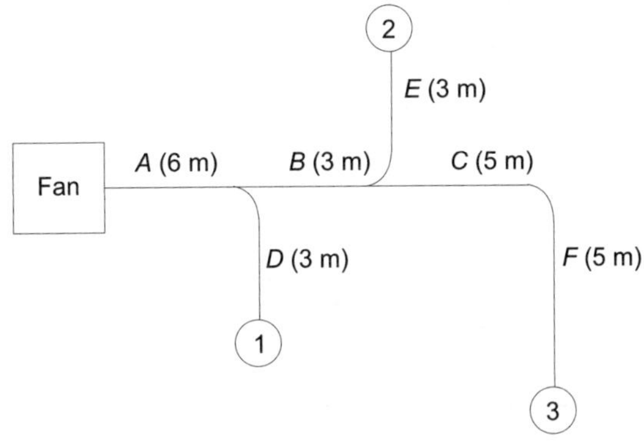

Fig. 21.31 Figure for Problems 21.1 and 21.2

21.3 (a) For the system shown in Fig. 21.32, size the ducts on a rate of pressure drop of 0.1 mm H_2O/m length.

(b) Modify the diameter of branch duct to outlet 1 so that no dampering is required at the outlet.

(c) Calculate the fan total and static pressures. The pressure drops in equipment are as follows:
Filter: 10 mm H_2O
Damper: 5 mm H_2O
Cooling coil: 15 mm H_2O
Mixing section: 5 mm H_2O

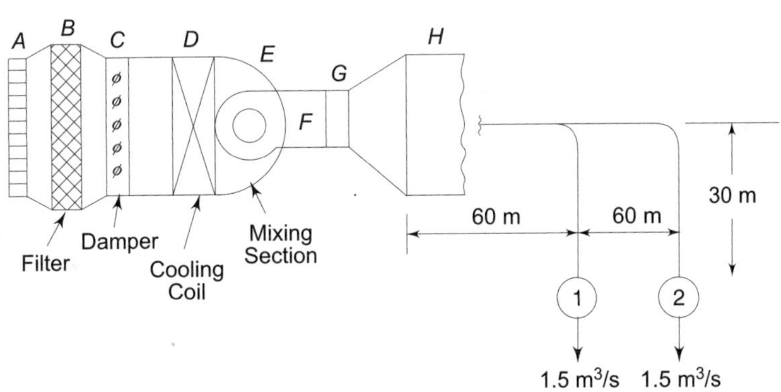

Fig. 21.32 Figure for Problem 21.3

The dynamic loss coefficients *K* for all expanders are to be taken as applying to the difference between the upstream and downstream velocity pressures, and for all reducers as applying to the downstream velocity pressure only. The values are given in the following table.

Section	*K*	*Condition*
Inlet	1.4	Mean face velocity = 4 m/s
Expander AB	0.35	Mean face velocity at filter = 1.5 m/s
Reducer BC	0.02	Mean face velocity at damper = 3 m/s.
Reducer EF to fan suction	0.02	
Reducer at fan discharge	0.3	
Straight-through duct suction	0.25	
Elbow	0.23	
Grille	0.5	

Chapter

22

Fans

22.1 TYPES OF FANS

Ultimately it is the fan which moves the air through the entire duct system and conditioned space. Two types of fans can be used for the transmission of air:

(i) Centrifugal fans

(ii) Axial-flow fans

Whenever a system has duct-work, centrifugal fans have to be used as the static pressure drop is considerable. But when there is no duct-work, propellers or axial-flow fans can be used. Nevertheless, in window-type and package units, simple drum-type centrifugal fans are used, whereas most exhaust fans are of the axial type, as they occupy less space, and can handle large volumes.

The centrifugal fan has the advantage of quiet and efficient operation at high pressures. Another advantage is the ease with which the centrifugal fan inlet can be connected to larger apparatus sections and its outlet to smaller supply duct sections.

Axial-flow fans are suitable for handling large air volumes and can be used where noise-level considerations are not important. They are, therefore, used in industrial ventilation and air-conditioning systems. The majority of fans are of the centrifugal or radial-flow type.

22.2 FAN CHARACTERISTICS

The required fan work can be calculated by knowing the flow rate and fan total pressure using Eq. (21.57) and including the fan efficiency in it. Thus

$$\dot{W} = \frac{\dot{Q}_v(\text{FTP})}{\eta} \tag{22.1}$$

Typical fan-characteristic curves are shown in Fig. 22.1. Point A on the fan total pressure-volume flow curve represents the condition of the open inlet and outlet. At this point

$$\text{FSP} = 0$$

$$\text{FTP} = \text{FVP}$$

It is seen that as $\dot{Q}_v$ decreases, FTP increases.

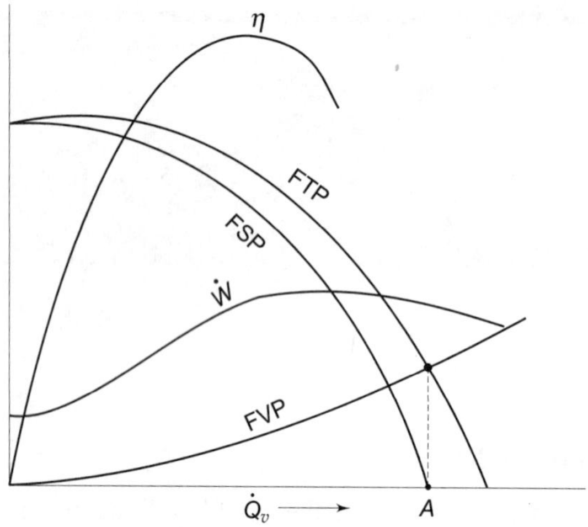

Fig. 22.1 Typical characteristic curves for fans

In terms of the diameter D and speed N of the fan, it is seen that the fan total pressure, FTP or Δp_T is proportional to the density and square of the velocity, which in turn is proportional to the product DN. Thus

$$\text{FTP} = \Delta p_T \propto \rho D^2 N^2 \tag{22.2}$$

Such a relation can also be obtained by dimensional analysis. The volume flow rate is proportional to the fan area and velocity, so that

$$\dot{Q}_v \propto D^2 (DN) \propto D^3 N \tag{22.3}$$

Thus, there are different $\Delta p_T - \dot{Q}_v$ characteristics, similar to the one shown in Fig. 22.1, for different speeds. Accordingly, the fan total pressure decreases as flow increases. Proceeding from above, we obtain a similarity relation for fan power consumption as follows:

$$\begin{aligned} \dot{W} &= (\dot{Q}_v)(\Delta p_T) \\ &\propto (D^3 N)(\rho D^2 N^2) \\ &\propto \rho D^5 N^3 \end{aligned} \tag{22.4}$$

22.3 CENTRIFUGAL FANS

A centrifugal or a radial-flow fan consists of an impeller running in a casing, normally of volute shape. The air enters axially and is discharged at the periphery. The work done on the air by the blades primarily depends on the tip speed (diameter and rpm) of the impeller and blade angles. The thermodynamic analysis and design of centrifugal fan is identical to that of a centrifugal compressor. The work done and the head and pressure developed are given by

$$\begin{aligned} \frac{\Delta p_T}{\rho} &= g\,H = w = C_{u_2} u_2 - C_{u_1} u_1 \\ &= \frac{u_2^2 - u_1^2}{2} + \frac{C_2^2 - C_1^2}{2} + \frac{C_{\text{rel}_1}^2 - C_{\text{rel}_2}^2}{2} \end{aligned} \tag{22.5}$$

$$H = \frac{w}{g} \tag{22.6}$$

$$\Delta p_T = \rho w$$
$$= \rho\,(C_{u_2}u_2 - C_{u_1}u_1) \tag{22.7}$$

The nomenclature used is the same as in Chapter 6. In the case of radial entry, $C_{u_1} = 0$. Then

$$\Delta p_T = \rho C_{u_2}u_2 \tag{22.8}$$

Three forms of blade designs and their respective outlet velocity triangles are shown in Fig. 22.2. They differ according to the outlet blade angle β_2.

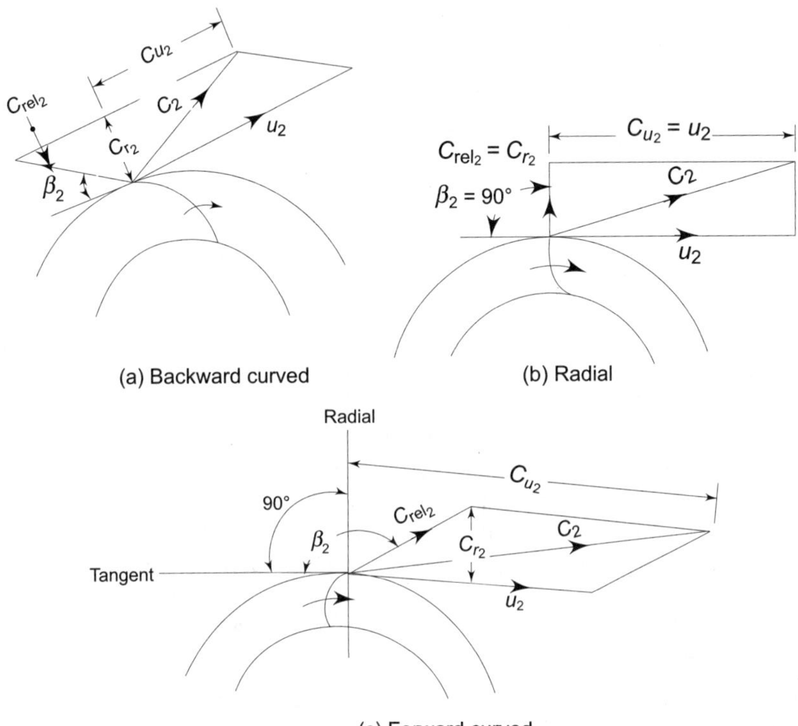

Fig. 22.2 Blade profiles and outlet velocity triangles for centrifugal fans

(i) Backward curved blades, $\beta_2 < 90°$

$$C_{u_2} = (u_2 - C_{r_2}\cot\beta_2) < u_2$$

(i) Radial blades, $\beta_2 = 90°$

$$C_{u_2} = u_2$$

(iii) Forward curved blades, $\beta_2 > 90°$

$$C_{u_2} = (u_2 + C_{r_2}\cot\beta_2) > u_2$$

It is thus seen that the pressure developed for a given *tip speed* u_2 is greatest for forward-curved fans.

The volume flow through the fan is given by

$$\dot{Q}_v = \pi D_2 b_2 C_{r_2} = \pi D_1 b_1 C_{r_1} \tag{22.9}$$

where D is the impeller diameter, b is the width of the shrouds and C_r is the radial velocity:

In practice, the radial velocities at the inlet and outlet of impeller are kept the same, and are of the order of

$$C_{r_1} = C_{r_2} \approx 0.2\, u_2$$

The relationship between the pressure developed and flow from Eqs (22.8) and (22.9) and the expression for C_{u_2} is obtained as follows:

$$\Delta p_T = \rho u_2^2 \pm \rho u_2 \frac{\dot{Q}_v}{\pi D_2 b_2} \cot \beta_2 \tag{22.10}$$

Putting, *Head coefficient* $\psi = \dfrac{\Delta p_T}{\frac{1}{2}\rho u_2^2}$ (22.11)

Flow coefficient $\phi = \dfrac{\dot{Q}_v}{\frac{1}{4}\pi D_2^2 u_2}$ (22.12)

we get

$$\psi = 2\left(1 \pm \phi \frac{D_2}{4b_2} \cot \beta_2\right) \tag{22.13}$$

For radial blades, $\cot \beta_2 = 0$

$$\Delta p_T = \rho u_2^2 \tag{22.14}$$

$$\psi = 2 \tag{22.15}$$

For backward-curved blades, the second term in Eq. (22.13) is negative, whereas for forward-curved blades, it is positive. Thus, forward-curved fans have a *rising characteristic* for Δp_T vs. $\dot{Q}_v$, backward-curved have a *droping characteristic* and radial fans develop constant pressure. The actual characteristics are different from these to the extent of losses as in the case of centrifugal compressors. In general, the pressure developed decreases with the volume flow for all the three types. Further, both the pressure developed and volume flow tend to increase with β_2.

22.3.1 Comparison of Characteristics of Backward and Forward Curved Blade Fans

Figure 22.3 compares the head developed by the three types of fans having the same diameter and speed. It may be noted that forward-curved fans develop highest pressure for a given diameter and speed. They are also available as high volume flow fans.

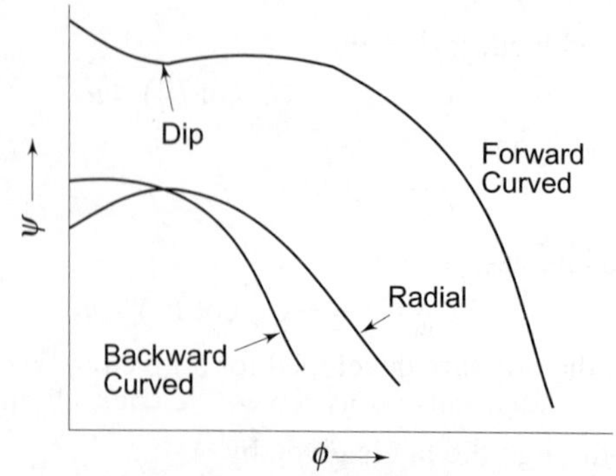

Fig. 22.3 Comparison of head-flow characteristics of forward-curved, radial and backward-curved blade fans

Figures 22.4 and 22.5 show the performance characteristics of the forward- and backward-curved blade fans respectively. The performance of the radial blade is similar to that of the forward-curved blade except that there is no *dip* in the static pressure curve.

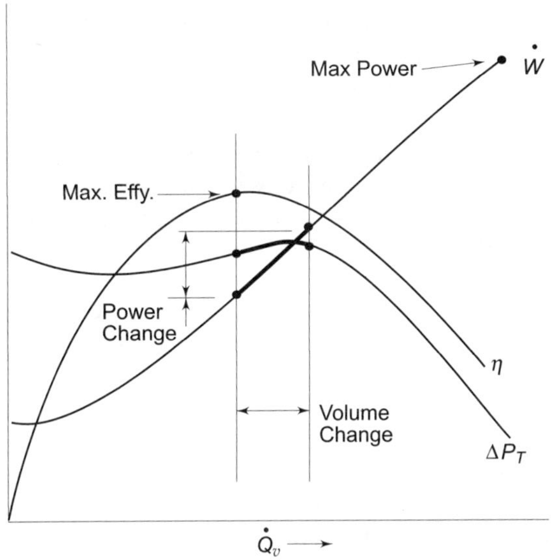

Fig. 22.4 Performance characteristics of forward-curved fans

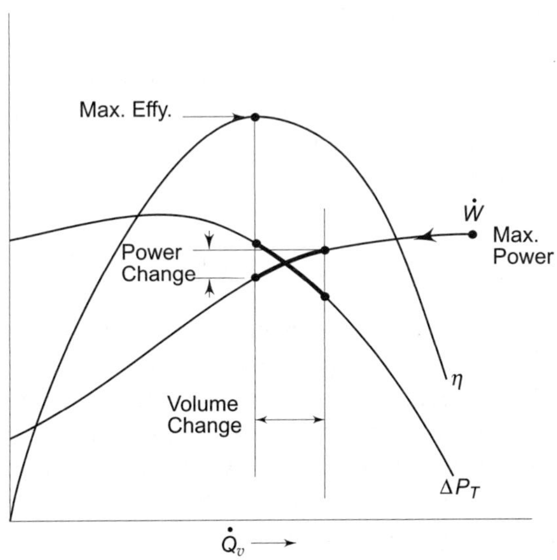

Fig. 22.5 Performance characteristics of backward-curved fans

Forward-curved and radial fans rarely have efficiencies higher than 75 per cent. Backward-curved fans of sheet metal construction have an efficiency up to 80 per cent and those with aerofoil section have even up to 90 per cent.

The important characteristic features of forward-curved and radial fans may be summarised as follows:

(i) The $\Delta p_T - \dot{Q}_v$ characteristic is flatter. Thus a small change in Δp_T, viz., the resistance of the duct system, filter, etc., greatly affects the volume delivered.

(ii) The point of maximum efficiency for forward-curved fans is at a volume flow rate corresponding to the very flat part of the $\Delta p_T - \dot{Q}_v$ curve. Thus, if a fan is selected to operate at maximum efficiency, as is desirable, a slight change in the system resistance not only changes the volume delivered, but also drops the efficiency correspondingly.

(iii) The $\dot{W} - \dot{Q}_v$ characteristic is steeper. Thus a small increase in the volume delivered causes a considerable increase in the power consumption. This may overload the electric motor since it is selected for design duty rather than maximum possible fan power. Thus, the motor has to be oversized by 25 to 30 per cent.

(iv) Finally, they have lower efficiencies.

As against the above, the characteristics of backward-curved fans are briefly as follows:

(i) The $\Delta p_T - \dot{Q}_v$ characteristic is steeper. Even with change in resistance of the duct system, therefore, the volume delivered remains fairly constant.

(ii) The η-$\dot{Q}_v$ characteristic is such that the peak efficiency occurs at a steeper part of the $\Delta p_T - \dot{Q}_v$ curve. Thus a change in the system resistance does not significantly affect $\dot{Q}_v$, and hence the efficiency.

(iii) The $\dot{W} - \dot{Q}_v$ characteristic is flatter. It is often referred to as the *non-overloading* characteristic. The maximum power required is only a little in excess of the design power consumption.

(iv) Efficiencies are generally higher.

For these reasons, backward-curved fans are commonly used in air conditioning. However, a backward-curved fan must run at a higher speed to develop the same pressure as a forward-curved fan. Accordingly, forward-curved fans are smaller and slower running. Thus they tend to be quieter and cheaper for FTP up to 750 N/m^2 (7.6 mm H_2O).

22.4 AXIAL-FLOW FANS

Figure 22.6 shows the schematic representation of an axial-flow fan. The flow of air is substantially parallel to the axis of the impeller. Blades are of aerofoil section. The tips run with as fine a clearance as practicable in a cylindrical casing.

Air enters in the axial direction and leaves with a rotational component due to the work done. The absolute velocity of the leaving air is higher than the axial velocity. Recent designs have guide vanes downstream which remove the rotational component, converting some excess velocity into more useful static pressure. Another method is to use pre-rotation vanes (upstream of the impeller). These rotate the air in a direction opposite to that of the impeller. The air, then, leaves axially. True axial discharge from either is only possible for a single-operating condition.

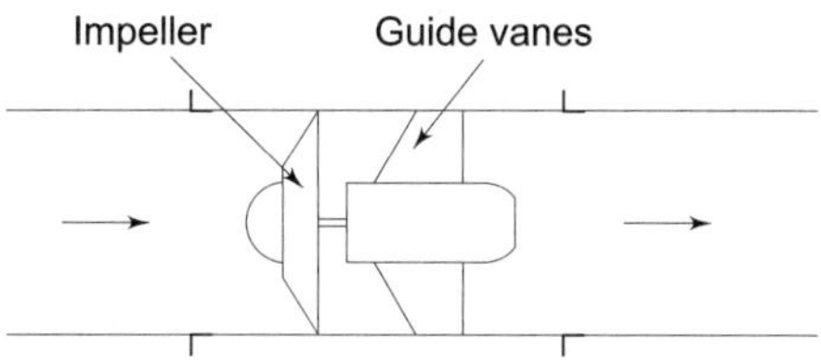

Fig. 22.6 Axial-flow fan

22.5 SYSTEM CHARACTERISTICS

A duct system may be designed for a certain air flow and the pressure drop of the system may then be calculated. The particular design flow and pressure drop represent one point of the *system characteristic* curve. If the volume flow is altered, the pressure drop will also be affected.

Consider a duct system in which the total pressure drop is calculated by adding the pressure drops of different sections, which are in turn proportional to their respective velocity pressures, as follows:

$$\begin{aligned}\Delta p &= K_1 \frac{1}{2} \rho C_1^2 + K_2 \tfrac{1}{2} \rho C_2^2 + \cdots \\ &= K_1 \frac{1}{2}\rho\left(\frac{\dot{Q}_v}{A_1}\right)^2 + K_2 \frac{1}{2}\rho\left(\frac{\dot{Q}_v}{A_2}\right)^2 + \cdots \\ &= \dot{Q}_v^2 \sum \frac{K\rho}{2A^2} \qquad (22.16)\end{aligned}$$

We thus obtain an expression similar to Eq. (22.16). However, once a system is designed and is in operation, the parameters K, ρ and A are constant. Then

$$\Delta p \propto \dot{Q}_v^2 = R\,\dot{Q}_v^2 \qquad (22.17)$$

Thus, the pressure drop has a parabolic variation with flow. The nature of system characteristic OA is shown in Fig. 22.7. The curve must pass through the origin, as the pressure drop through the system is zero when there is no flow.

By analogy with electricity, we can derive the concept of the resistance R of the duct system. Here the pressure drop Δp is equivalent to voltage. Comparing Eq. (22.17) to the corresponding equation in electricity,

$$V = R_e I$$

we have

$$R = \frac{\Delta p}{\dot{Q}_v^2} \qquad (22.18a)$$

Also,

$$\Delta p_2 = \Delta p_1 \,(\dot{Q}_{v_2}/\dot{Q}_{v_1})^2 \qquad (22.18b)$$

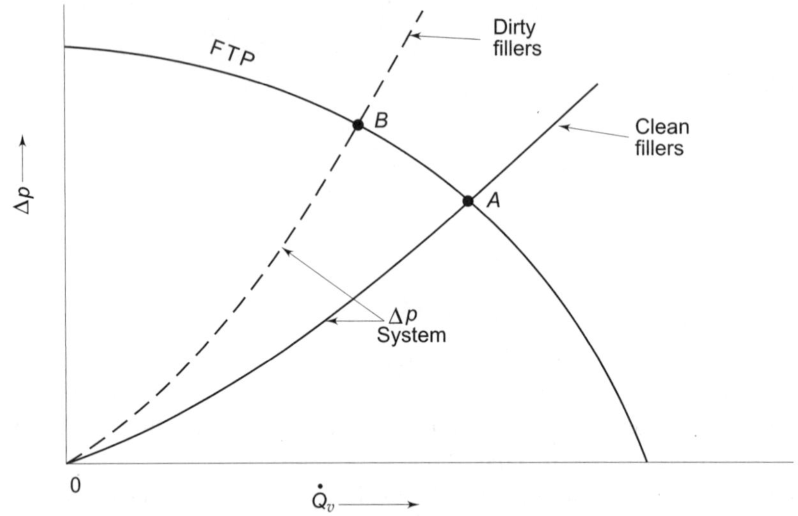

Fig. 22.7 Fan-system interaction

Accordingly, if one point is given, say, $\Delta p = 531$ N/m^2, $\dot{Q}_v = 3$ m^3/s then one can draw the table for system characteristic

$$\Delta p = 531 \left(\frac{\dot{Q}_v}{3}\right)^2 = 59\,\dot{Q}_v^2$$

as follows:

$\dot{Q}_v$, m^3/s	0.5	1	1.5	2	2.5	3
Δp, N/m^2	14.75	59	133	236	369	531

22.5.1 Systems in Series and in Parallel[3]

This concept of systems is useful in solving complex problems such as in mine ventilation, clean rooms, etc. The resistances may be in series as in Fig. 22.8(a), or in parallel as in Fig. 22.8(b). If the resistances are in series, the flow is the same through each resistance, but the pressure drop is different. The total pressure drop is obtained by adding the individual pressure drops. Thus:

$$\Delta p = \Delta p_1 + \Delta p_2 + \Delta p_3 + \cdots$$

Expressing the pressure drop from Eq. (22.17), we have

$$\dot{Q}_v^2 R = \dot{Q}_v^2 R_1 + \dot{Q}_v^2 R_2 + \dot{Q}_v^2 R_3 + \cdots$$

whence

$$R = R_1 + R_2 + R_3 + \cdots \tag{22.19}$$

Thus the overall resistance in series is equal to the sum of individual resistances. If the resistances are in parallel, the pressure drop through each is the same. The overall air flow is equal to the sum of individual flows. Accordingly:

$$\dot{Q}_v = \dot{Q}_{v_1} + \dot{Q}_{v_2} + \dot{Q}_{v_3} + \cdots$$

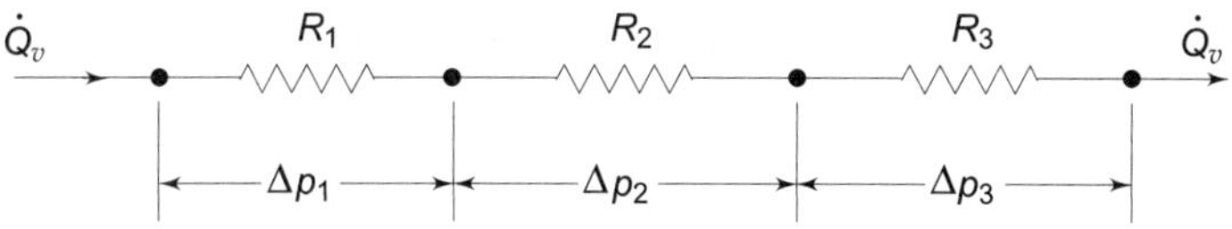

(a) Resistance in series

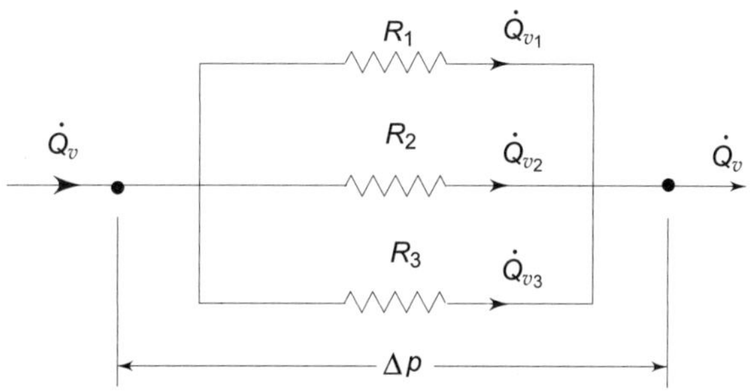

(b) Resistances in parallel

Fig. 22.8 Flow resistances in series and in parallel

Again, making use of Eq. (22.17) as before, we have

$$\sqrt{\frac{\Delta p}{R}} = \sqrt{\frac{\Delta p}{R_1}} + \sqrt{\frac{\Delta p}{R_2}} + \sqrt{\frac{\Delta p}{R_3}} + \cdots$$

whence

$$\frac{1}{\sqrt{R}} = \frac{1}{\sqrt{R_1}} + \frac{1}{\sqrt{R_2}} + \frac{1}{\sqrt{R_3}} + \cdots \tag{22.20}$$

Equation (22.20) can be used to evaluate the overall resistance of flow circuits in parallel.

22.5.2 Fan-System Interaction[2]

Figure 22.7 shows the fan and system characteristics superimposed on each other. For a fixed system, the system resistance is constant, and the system characteristic from Eq. (22.18) is of the parabolic form

$$\Delta p \propto \dot{Q}_v^2$$

Thus for a given system of duct-work, space, etc., the pressure drop is proportional to the square root of the volume handled. The intersection of the fan and system characteristics gives the point of operation A. This is the only single operating point for the combined fan-duct work system.

Now, if the system resistance changes, say due to unclean filters, the system characteristic will move to curve O–B. The operating point will then shift to B resulting in decreased flow.

22.5.3 Effect of Change in Fan Speed

The effect of change in the fan speed is shown in Fig. 22.9. For the same fan, i.e., for the same diameter, it is seen from Eqs (22.2) and (22.3) that

$$\Delta p_T \propto N^2 \quad \text{and} \quad \dot{Q}_v \propto N.$$

Note *Pressure increases more rapidly than* $\dot{Q}_v$ *with speed, as also shown in Fig. 22.9.*

Accordingly, the variation in power consumption is given by

$$\dot{W} \propto (\dot{Q}_v)(\Delta p_T)$$
$$\propto N^3$$

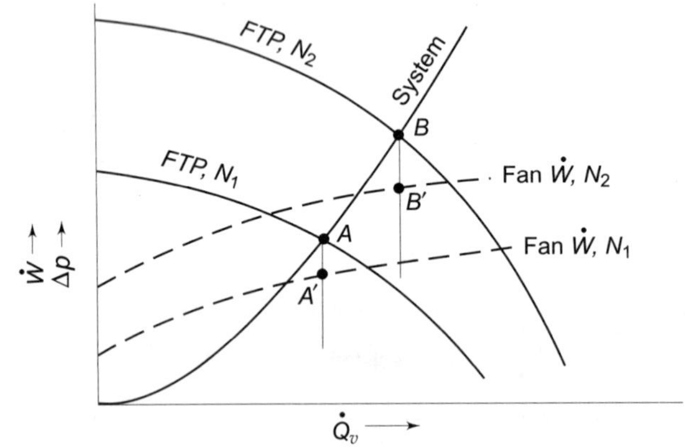

Fig. 22.9 Effect of change in fan speed

By increasing the fan speed from N_1 to N_2, the point of operation shifts from A to B for FTP and from A' to B' for fan power .

22.5.4 Effect of Change in Air Density

Both the fan total pressure and system pressure loss are directly proportional to density while the volume handled remains unchanged. The fan power is also proportional to density. Thus, as shown in Fig. 22.10, due to the increase in the density of air from ρ_1 to ρ_2, the operating point shifts from A to B for FTP and from A' to B' for $\dot{W}$. Accordingly, the fan pressure, system pressure drop and fan power, all increase with an increase in density. But whether the volume and mass handled would also increase or not has to be seen from the actual characteristics.

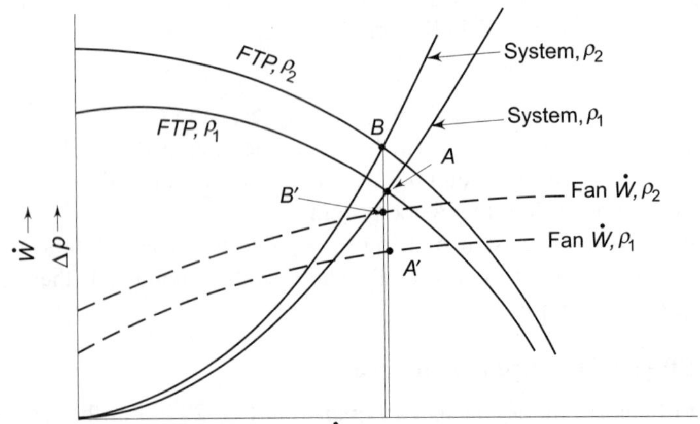

Fig. 22.10 Effect of change in gas density

22.5.5 Simple Fan and System Network

Conventional air-conditioning systems have air-transmission and distribution system as shown in Fig. 22.11.

Consider the parallel branches, B, C and D for which the required volume flows $\dot{Q}_B$, $\dot{Q}_C$ and $\dot{Q}_D$ are known. The pressure losses in them, Δp_B, Δp_C and Δp_D can be calculated. It is unlikely that all losses will be equal. But under operating conditions, they must be the same, say Δp. Then the flow rates will change to $\dot{Q}_B$, $\dot{Q}_C$ and $\dot{Q}_D$ according to Eq. (22.18), viz.,

$$\dot{Q}'_B = \dot{Q}_B \sqrt{\frac{\Delta p'}{\Delta p_B}}$$

$$\dot{Q}'_C = \dot{Q}_C \sqrt{\frac{\Delta p'}{\Delta p_C}}$$

and

$$\dot{Q}'_D = \dot{Q}_D \sqrt{\frac{\Delta p'}{\Delta p_D}}$$

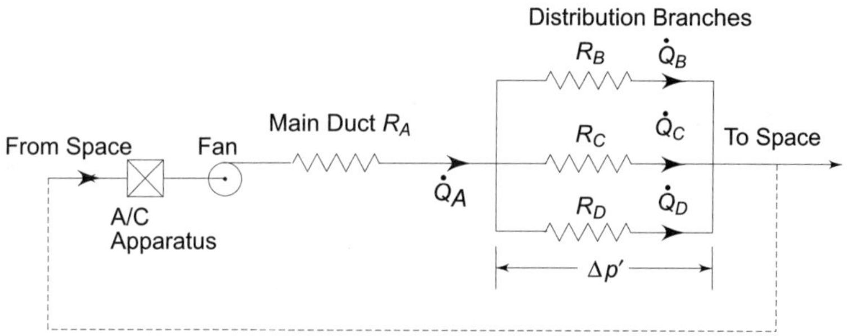

Fig. 22.11 Simple fan-system network

Example 22.1 *A system has a total pressure loss of 531 N/m² when handling 3 m³/s of standard air. The characteristics of the fan installed on the system when running at 480 rpm are given in Table 22.1.*

Table 22.1 Characteristics of fan in Example 22.1

Flow, m³/s	0	0.5	1.0	1.5	2.0	2.5	3.0	3.5
FTP, N/m²	350	385	410	427	433	424	400	343
Power, kW				1.185	1.354	1.537	1.715	
Efficiency, %				54	64	69	70	

(i) *Calculate the quantity of air handled, FTP developed, power consumption and efficiency of the fan when operating on the system.*

(ii) *If the output of the fan-duct system is reduced to 2 m³/s by partly closing a damper, calculate the power wasted across the damper.*

(iii) *State if this fan is suitable or not to deliver 3 m³/s of air through the system. Calculate the speed at which this fan must be run if it is to deliver 3 m³/s of air.*

Solution Refer to Fig. 22.12.

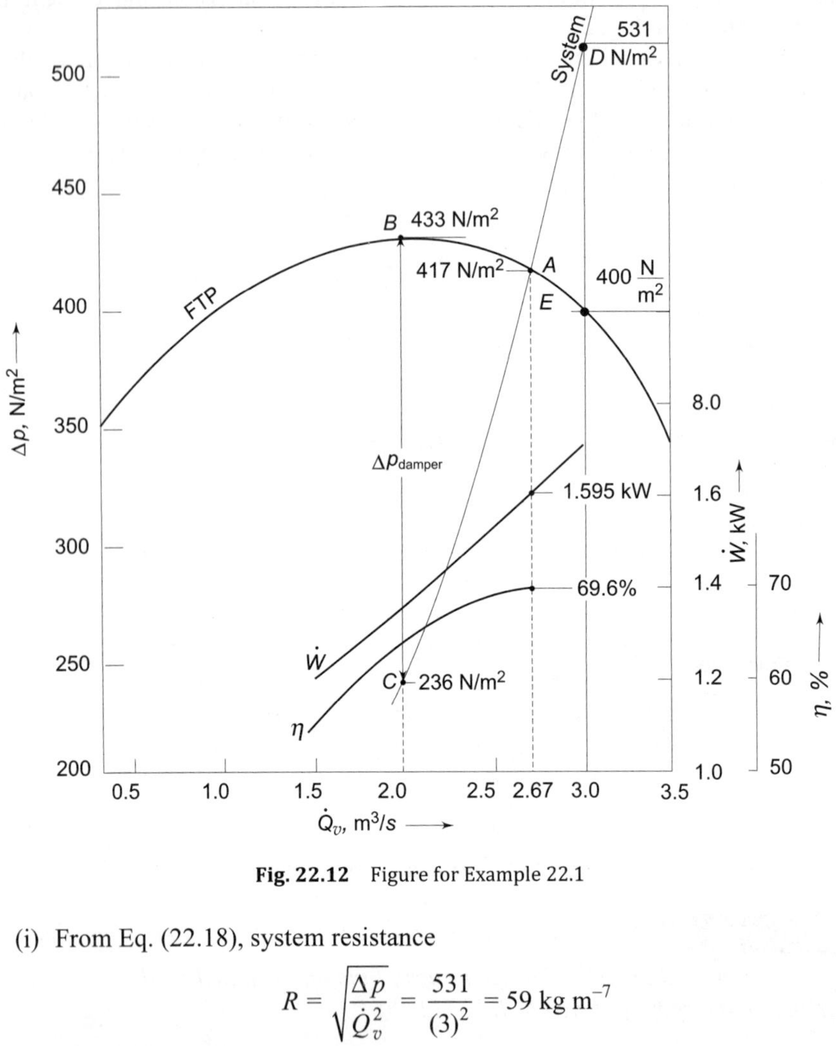

Fig. 22.12 Figure for Example 22.1

(i) From Eq. (22.18), system resistance

$$R = \sqrt{\frac{\Delta p}{\dot{Q}_v^2}} = \frac{531}{(3)^2} = 59 \text{ kg m}^{-7}$$

System pressure drop for various flow rates from $\Delta p = R\,\dot{Q}_v^2$

$\dot{Q}_v$, m^3/s	0	0.5	1.0	1.5	2.0	2.5	3.0	3.5
Δp, N/m^2	0	14.75	59	132.75	236	368.75	531	722.75

The fan and system Δp vs. $\dot{Q}_v$ characteristics are plotted in Fig. 22.12. The point of intersection is obtained at *A*. The following can be read from the graph.

Air handled, $\dot{Q}_v = 2.67$ m^3/s

FTP developed, $\Delta p_T = 417$ N/m^2

Fan characteristics for the power consumption $\dot{W}$ and efficiency are also plotted in Fig. 22.12. Their values at 2.67 m^3/s volume handled are as follows:

Power consumption, $\dot{W} = 1.595$ kW

Efficiency, $\eta = 69.6\%$

(ii) At 2 m^3/s volume handled, the point of intersection is at B. The pressure developed by the fan at this point is 433 N/m^2. Thus

Pressure loss of the system with the damper

$$\Delta p_B = 433 \text{ N/m}^2$$

Pressure loss of the system if the damper had not been there

$$\Delta p_C = 236 \text{ N/m}^2$$

Pressure drop in the damper

$$\Delta p_{\text{damper}} = \Delta p_B - \Delta p_C$$
$$= 433 - 236 = 197 \text{ N/m}^2$$

Power wasted across the damper

$$\Delta \dot{W} = \dot{Q}_v (\Delta p_{\text{damper}})$$
$$= 2 (197) = 394 \text{ W}$$

(iii) The fan is not suitable to deliver 3 m^3/s of air through the system since the pressure loss of the system at this flow rate at D is 531 N/m^2, whereas the total pressure developed by the fan at E is only 400 N/m^2. To develop the required pressure, the rpm must be increased according to

$$\frac{\Delta p_2}{\Delta p_1} = \frac{N_2^2}{N_1^2}$$

whence

$$N_2 = N_1 \sqrt{\frac{\Delta p_2}{\Delta p_1}}$$
$$= 480 \sqrt{\frac{531}{400}} = 553 \text{ rpm}$$

22.6 FAN ARRANGEMENTS

The fan arrangements are standardized for the drive, rotation, motor position, suction and discharge. Thus, there can be a belt or direct drive and bearings on one side with the wheel overhung or bearings on both sides. The rotation may be clockwise or counter-clockwise. The discharge may be top horizontal or bottom horizontal, upblast or downblast, top angular down or up, or bottom angular down or up. The suction is commonly from one side but may be from both sides also.

Further, a multiple number of fans may be used. The arrangement for the purpose will be either in series or in parallel. The processing, transmission and distribution of air in clean rooms using a number of filters as discussed in Sec. 21.7 is an illustration of the use of a multiple number of fans.

22.6.1 Fans in Series

When two fans are employed in series,

(i) The flow rate $\dot{Q}_v$ through each fan is the same, and

(ii) The overall fan total pressure Δp_T is equal to the sum of individual FTPs minus the losses in the connections.

The combined characteristic of two fans in series can, therefore, be drawn by adding the FTP of each fan for each $\dot{Q}_v$ as shown in Fig. 22.13. To obtain the combined characteristic, it is assumed that the characteristic of each fan is known for volumes greater than those which are achieved by fans when running with suction and discharge unconnected to the system. This information is rarely available. The characteristic in this region may be extrapolated as shown in Fig. 22.13.

Further, it is preferable to use identical fan units in series as it is unlikely that efficient operation would result otherwise.

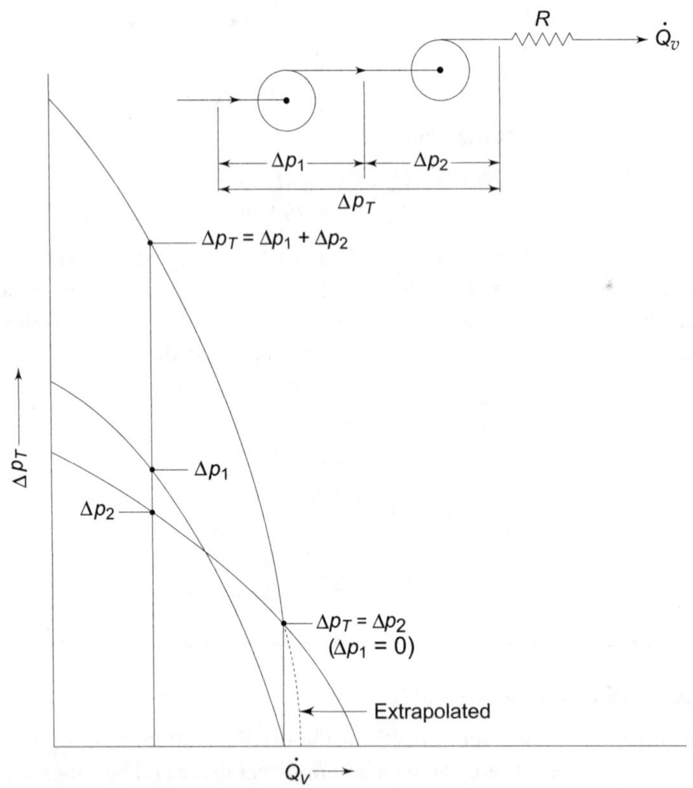

Fig. 22.13 Combined characteristic of fans in series

22.6.2 Fans in Parallel

When two fans are employed in parallel,

(i) The total pressure Δp_T across each fan is the same, and
(ii) The total volume handled $\dot{Q}_v$ is equal to the sum of the volumes handled by individual fans.

The combined characteristic of two fans in parallel can, therefore, be drawn by adding $\dot{Q}_v$ of each fan for the same Δp_T as shown in Fig. 22.14. Again, in order to draw this combined characteristic in full, it is necessary to know the *reverse-flow characteristic* of one of the fans with the impeller running in the normal direction which is normally not known.

As with series operation, identical fans should normally be used in parallel.

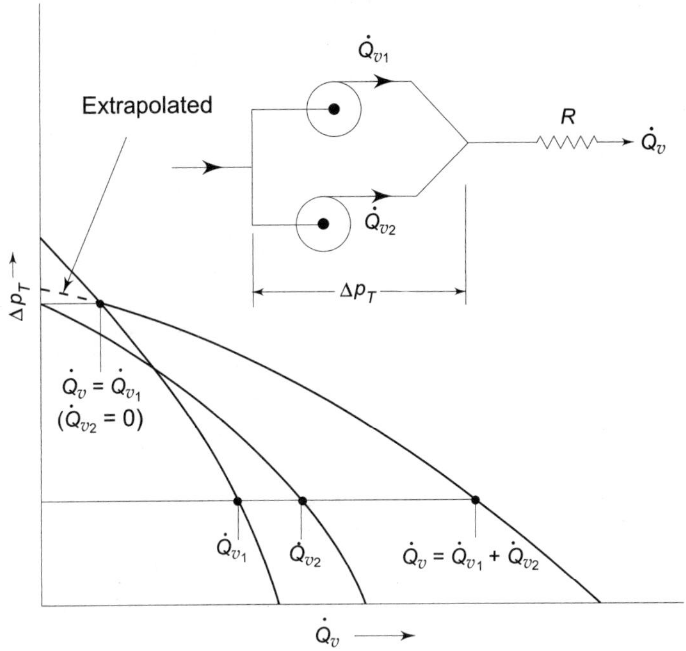

Fig. 22.14 Combined characteristic of fans in parallel

Example 22.2 *A large workshop is ventilated by a supply fan and duct system, the pressure loss through which is 30.5 mm H_2O when the flow is 3 m^3/s. There is also an exhaust fan with another duct system for which the pressure loss is 43.2 mm H_2O when the flow is 2.5 m^3/s.*

(a) Determine the pressure of air in the workshop and volume of ventilation air.
(b) If there is loss of air by leakage, determine the pressure in the space, supply air volume flow and exhaust air volume flow. The characteristic of the leakage system is such that the leakage rate is 0.5 m^3/s for a pressure difference of 12.5 mm H_2O.

The two fans are identical and their characteristics are given in Table 22.2.

Table 22.2 Characteristics of fans in Example 22.2

$\dot{Q}_v$, m^3/s	0	1	2	3	4
FTP, mm H_2O	76	77	74.2	59.9	27.9
HP	0.83	1.43	2.23	2.9	2.9

Solution Refer to Fig. 22.15

(a) Supply system resistance

$$R_1 = \frac{\Delta p_1}{\dot{Q}_{v_1}^2} = \frac{30.5}{(3)^2} = 3.39$$

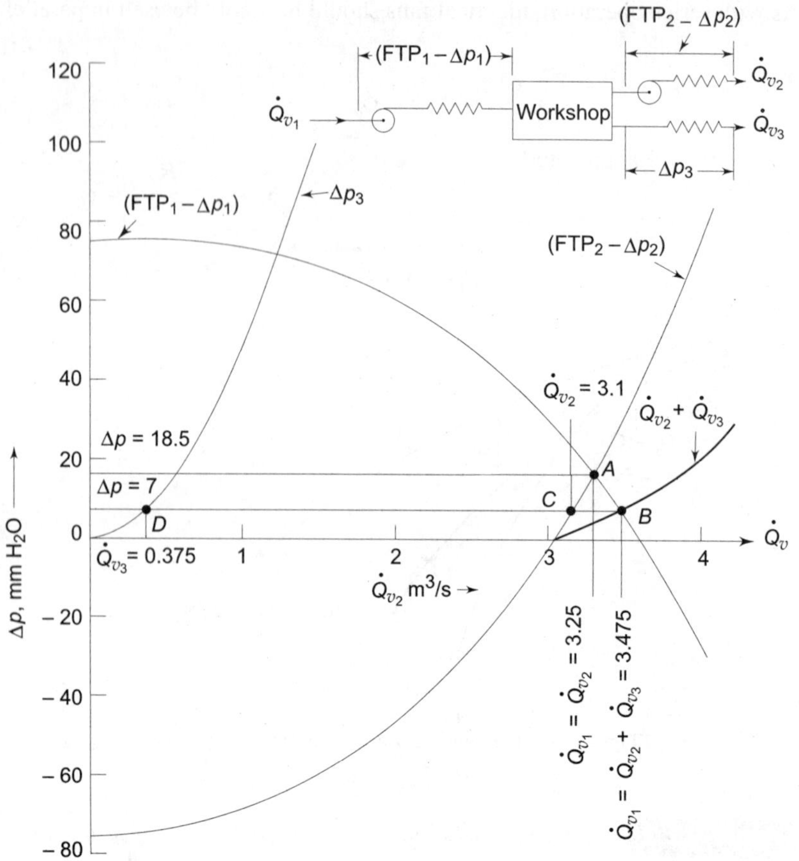

Fig. 22.15 Figure for Example 22.2

Supply system characteristic from $\Delta p_1 = R_1\, \dot{Q}_{v_1}^2$:

$\dot{Q}_{v_1}$, m³/s	0	1	2	3	4
Δp_1, mm H_2O	0	3.39	13.56	30.5	54.2

Resultant pressure in the room due to the supply fan-duct system is $\text{FTP}_1 - \Delta p_1$, as plotted in Fig. 22.15, which is positive:

$\dot{Q}_{v_1}$, m²/s	0	1	2	3	4
$(\text{FTP}_1 - \Delta p_1)$, mm H_2O	76	73.6	60.6	29.4	– 26.3

The pressure is plotted in Fig. 22.15.
Exhaust system resistance

$$R_2 = \frac{\Delta p_2}{\dot{Q}_{v_2}^2} = \frac{43.2}{(2.5)^2} = 6.912$$

Exhaust system characteristic, $\Delta p_2 = R_2 \dot{Q}_{v_2}^2$:

$\dot{Q}_{v_2}$, m³/s	0	1	2	3	4
Δp_2, mm H_2O	0	6.912	27.6	62.2	110.6

Resultant pressure in the room due to the exhaust fan-duct system is $(\text{FTP}_2 - \Delta p_2)$, which is negative.

$\dot{Q}_{v_2}$, m³/s	0	1	2	3	4
$(\text{FTP}_2 - \Delta p_2)$, mm H_2O	– 76	– 70.1	– 46.6	– 2.3	82.7

This pressure is also plotted in Fig. 22.15. The final resultant pressure in the room corresponds to the point *A* at the intersection of the two curves which gives the following:

Pressure developed by the supply system

= Pressure drop of the exhaust system
= Pressure in the workshop (gauge)
= 18.5 mm H_2O

Ventilation rate $\dot{Q}_{v_1} = \dot{Q}_{v_2} = 3.25\ \text{m}^3/\text{s}$

(b) Leakage system resistance

$$R_3 = \frac{\Delta p_3}{\dot{Q}_{v_3}^2} = \frac{12.5}{(0.5)^2} = 50$$

Leakage system characteristic, $\Delta p_3 = R_3 \dot{Q}_{v_3}^2$

$\dot{Q}v_3$, m³/s	0	0.5	1.0	1.5
Δp_3, mm H_2O	0	12.5	50.0	112.5

The Leakage system characteristic is also plotted in Fig. 22.15.

Note *The leakage system and exhaust duct plus exhaust fan system are in parallel. The resultant of the two is obtained by adding $\dot{Q}_{v_2}$ and $\dot{Q}_{v_3}$ for the same pressure drop, i.e., for*

$$(\text{FTP}_2 - \Delta p_2) = -\Delta p_3$$

The resultant of the two parallel systems of leakage and exhaust is also plotted in Fig. 22.15. Its intersection with the supply system characteristic at *B* gives the following:

Pressure in the workshop = 7 mm H_2O gauge

Ventilation rate $\dot{Q}_{v_1} = \dot{Q}_{v_2} + \dot{Q}_{v_3} = 3.475\ \text{m}^3/\text{s}$

Reading for the characteristic of the exhaust system and leakage system for the same pressure drop of 7 mm H_2O, we obtain the following:

Volume handled by exhaust system (Point C) $\dot{Q}_{v_2} = 3.1\ \text{m}^3/\text{s}$

Leakage volume (Point D) $\dot{Q}_{v_3} = 0.375\ \text{m}^3/\text{s}$.

References

1. Church, A H, *The Centrifugal Pumps and Blowers*, John Wiley, New York, 1994.
2. Jones, W P, *Air Conditioning Engineering*, Edward Arnold, London, 1973.
3. Osborne, W C, *Fans*, Pergamon Press, Oxford, 1966.

Revision Exercises

22.1 A fan with characteristics as in Table 22.2 supplies air to a system for which the total pressure loss is 61 mm H_2O for a volume flow rate of air of 2.5 m^3/s. Find the flow rate of air and fan total efficiency.

22.2 A fan with characteristics as given in Table 22.2 delivers air to a system at a temperature of 18°C and 760 mm Hg barometric pressure at a rate of 2.5 m^3/s. To obtain the same flow of air at 20°C and 610 mm Hg barometric pressure, two such fans are used in series. Show that the arrangement will supply the same mass flow rate of air.

22.3 The total pressure drop through a heat exchanger is 56 mm H_2O for a volume flow rate of air of 2.5 m^3/s at 18°C. A fan with characteristics as in Table 22.2 is used to supply air at 15°C which leaves the exchanger at 93°C to the atmosphere through an area of 0.09 m^2. Show where the fan should be placed to obtain a greater mass flow of air.

22.4 The performance figures for a centrifugal fan driven by a constant speed motor are tabulated below. Plot these and superimpose a shaft-power curve. From this determine the shaft power at the design output of 50 m^3/s and also the power if the output is reduced to 30 m^3/s by damper regulation.

If, instead of using damper regulation, the fan speed is reduced by a hydraulic coupling of constant torque, calculate the reduction in the power input to the fan shaft.

Output m^3/s	*SP* mm H_2O	*Efficiency* %
0	85	0
10	92.5	46
20	95	66
30	90	70
40	80	67
50	65	60
60	47.5	48
70	25	32

22.5 The characteristics of a supply fan installed on a system are given in Table 22.1. The system has a total pressure loss of 450 N/m^2 when handling 3 m^3/s of air.

(a) Calculate the quantity of air handled, total pressure developed, power consumption and efficiency of the fan.

(b) The filter in the above fan-duct system offers a pressure drop of 90 N/m^2 for an air flow of 3 m^3/s when it is clean. Find the fan power consumed in the clean filter.

(c) The fan delivers only 2.3 m^3/s of air when the filters have become dirty. Find the pressure drop in dirty filters and the fan power wasted due to the filters being dirty.

22.6 A fan supplies air to a space through a duct system in which the pressure drop is 30.5 mm H_2O for a flow rate of 3 m^3/s. Another fan also supplies air to the space through another duct system in which the pressure drop is 43.2 mm H_2O for a flow rate of 2.5 m^3/s. Both the fans are identical and have characteristics as given in Table 22.2. From the conditioned space, the air is sucked in by both the fans, (which are in parallel) through a return duct-and-apparatus system in which the pressure drop is 30.5 mm H_2O for a flow rate of 5.5 m^3/s. Determine the volume handled and power consumption of each fan.

Chapter

23 Refrigeration and Air-Conditioning Control

If the refrigeration load and capacity of a plant remained constant, there would not be any need to have any control. However, in practice, both do fluctuate. The cooling load changes due to input conditions and changes in the outside conditions. The plant capacity also changes due to the changes in the outside conditions. Also, since the plant is designed for peak load, its capacity is usually greater than the immediate demand, resulting in a trend to establish a suction pressure and temperature that are lower than those necessary to fulfil the peak demand.

Automatic control makes the plant independent of skilled supervision. It also achieves a higher degree of accuracy in maintaining the required temperatures, pressures and humidities. Further, it offers flexibility in the operation of complicated plants on various impulses from different parts of the system ensuring overall control and protection.

23.1 BASIC ELEMENTS OF CONTROL

In every control device, there is a *controlled variable* or controlled condition such as the temperature of the conditioned space. Figure 23.1 shows the basic elements of a control system in the form of a block diagram. The *detecting element* feeds a signal to the controlling unit, when any deviation from the desired value of the controlled condition, i.e., from the *set value*, occurs. The measuring element of the control unit then measures the extent of this deviation and triggers the *actuating element*. The *positioner* is responsible for setting the value of the controlled variable. In case, the force from the *measuring element* is not strong enough to actuate the regulating unit, servomotors or relays can be used. The *transmitting element* is at times necessary between the actuating and regulating elements. It is a part of the control unit proper. Often, the functions of the various elements are combined in one unit. The reader may refer to the construction of expansion valves given in Chapter 8 and try to identify the various elements.

It may be noted that the response of the control unit to an impulse is subject to *lags* and *retards* of different kinds which are due to friction between parts, heat capacities of walls, etc.

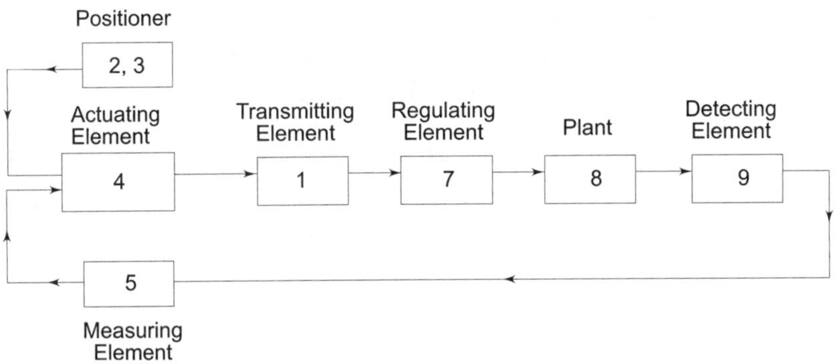

Fig. 23.1 Block diagram of basic elements of a control unit

The common items of control involved here are given below.

(i) Expansion devices:
   (a) Automatic expansion valve
   (b) Thermostatic expansion valve

   The capillary tube does not do any control function.

(ii) Cut-outs for refrigeration units:
   (a) High pressure cut-out
   (b) Low pressure cut-out

(iii) Back pressure valves

(iv) Liquid level regulating devices:
   (a) High-side float valve
   (b) Low-side float valve

(v) Flow-regulating devices:
   (a) Solenoid valves
   (b) Check valves
   (c) Water and brine valves

(vi) Thermostats:
   (a) Evaporator thermostat
   (b) Room thermostat

(vii) Humidity controls (Humidstats)

(viii) Motor controls

(ix) Modulating motors.

These are the general ones. There are many others. A detailed discussion of these and other controls is, however, outside the scope of this book. Only the salient features of some pertaining to refrigeration and air-conditioning control are given here.

23.2 DETECTING ELEMENTS[1]

These are sensing devices, often having the measuring device and positioner built into them. The common types of detecting elements are bimetallic, bulb-and-bellow, temperature sensitive electric resistance, electromagnetic and humidity sensitive.

23.2.1 Bimetallic Elements

A bimetallic element comprises of two strips of dissimilar materials with different coefficients of thermal expansion, joined together and fixed at one end. Any change in temperature results in non-uniform expansion of the two strips and consequent bending as shown in Fig. 23.2. These elements are used in thermostats. Two well-known materials used as a pair and their coefficients of expansion are:

Brass	:	17.5×10^{-6} per°C
Invar (36% Ni + Steel)	:	0.9×10^{-6} per°C

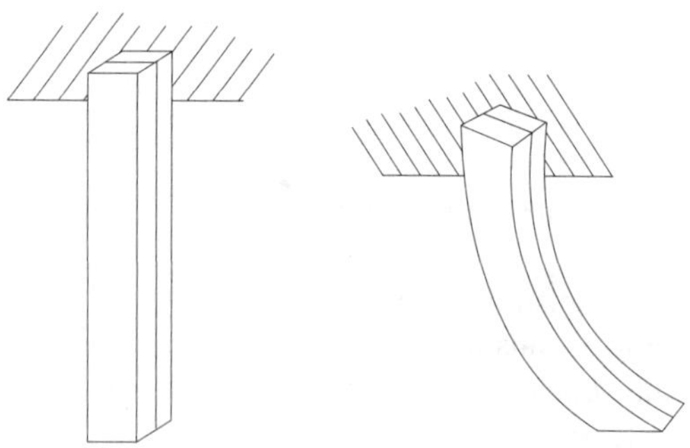

Fig. 23.2 Bimetallic element

It is seen that the magnitude of movement of the element depends on the length of the strip. Hence a greater length is required for increased sensitiveness. Often, the bimetallic elements are bent in U-form or wound in different shapes.

23.2.2 Bulb-and-Bellow Elements

The typical construction of a bulb-and-bellow element is shown in Fig. 23.3. The control spring provided acts as a positioner to preset the value of the control variable. It is seen that the control variable is the temperature, but with the bulb removed, the element can be connected to a system and this be used to control pressure. Its use has already been explained in the case of the two expansion valves, both as temperature-sensitive and pressure-sensitive elements. Simple bellows are used in low pressure cut-outs provided in all condensing units. By regulating the tension of the spring, one can select a vapour pressure (and hence temperature) at which the assembly will actuate the switch or valve.

The bellow and bulb can be either *liquid-charged* or *gas-charged.* In the liquid-charged type, the vapour pressure in the element will increase with increasing temperature of the bulb according to the vapour pressure curve ABC in Fig. 23.4. It may be noted that the vapours tend to condense on the coldest part. Thus if the bellows are cooler than the bulb, the liquid will accumulate in the bellows and no liquid may remain in the bulb which will effectively become gas-charged and, therefore, insensitive to temperature charge. Hence, it is required that the element should have sufficient charge so as to leave some liquid in the bulb even if the bellows and capillary tube are completely filled with the liquid.

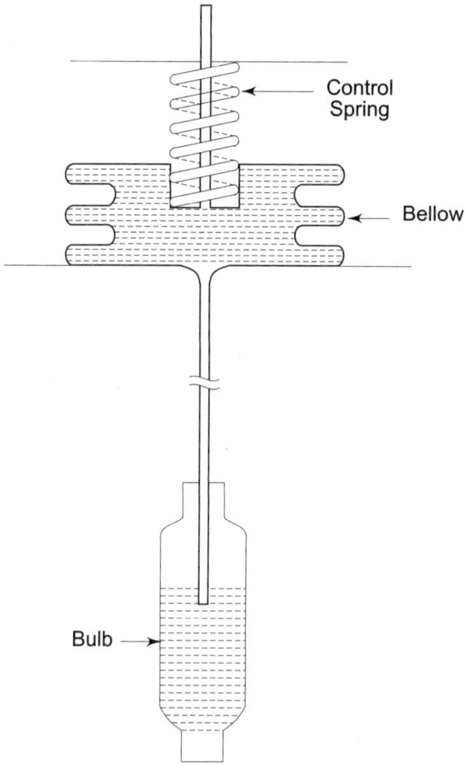

Fig. 23.3 Bulb-and-bellow element

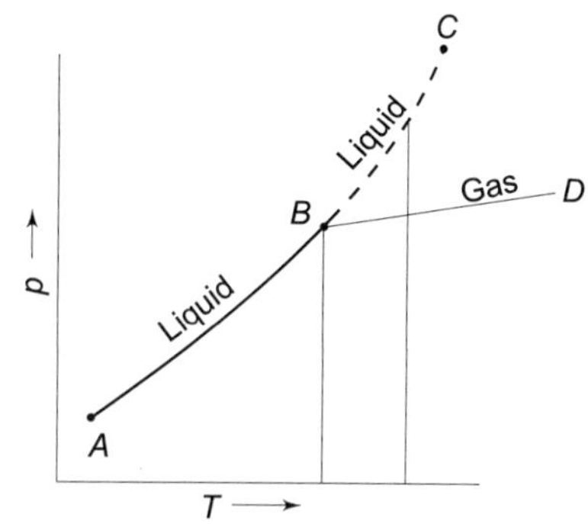

Fig. 23.4 Variation of pressure with temperature in liquid and gas charged bulbs

As stated above, the gas-charged types are the ones with a small liquid charge at low temperature of operation, so that, at higher temperatures, the element is completely filled with gas, and the pressure variation is along the constant-volume line BD in Fig. 23.4. Such elements are used for low-temperature operation, and when the bulbs are exposed to high temperatures, they go out of operation. They are called fade-out elements. The fade-out principle limits the maximum pressure of operation and prevents excessive pressures.

23.2.3 Electric Resistance Elements

The electric resistance of elements changes with temperature. Whereas copper has a positive temperature coefficient of 0.004 per°C and constantan (60% Cu + 40% Ni) has a resistance more or less independent of temperature, there are semiconductors which have negative coefficients of about 0.04 per °C. The latter are well-suited in automatic control because of their numerical high-temperature dependence. They are called *negative temperature coefficient* or NTC resistors, and are made by the ceramic process from mixtures of metal oxides, such as NiO, Mn_2O_3, CO_2O_3, etc. They can be made in small sizes to minimise time lag.

Figure 23.5 shows the detecting element forming a part of the Wheatstone bridge.

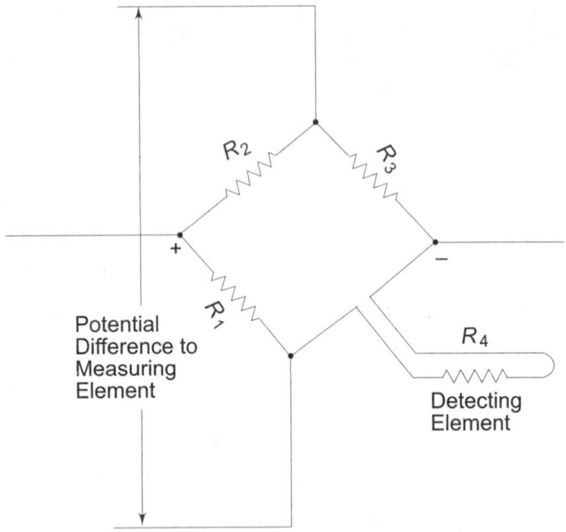

Fig. 23.5 Wheatstone bridge for electric resistance elements

23.2.4 Electromagnetic Elements

The magnetic field in a solenoid can be influenced by an iron armature in a coil. By moving the armature up or down actuated by a sensing element, a change in current will occur, and by means of an amplifier, this change can be made strong enough to actuate a control device, e.g., liquid level regulating devices. On the other hand, solenoid valves are used as two-position controls in the liquid lines of refrigerant circuits. A typical construction of a solenoid operating a valve is shown in Fig. 23.6.

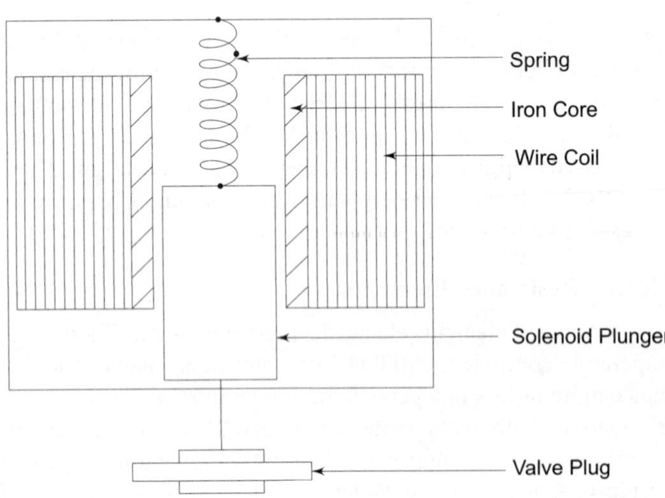

Fig. 23.6 Solenoid

Note *The solenoid falls down by its own weight when not energised. When it is energised, it is lifted up.*

23.2.5 Humidity Sensitive Elements

Two temperature sensitive elements may be used to measure the dry bulb and wet bulb temperatures and, hence, the humidity. Figure 23.7 shows one such arrangement using bellow elements. Two thermistors can also be similarly used, coupled into a Wheatstone bridge. Such a device will operate on the basis of increase/decrease in the difference (DBT – WBT) of air for the control of humidity.

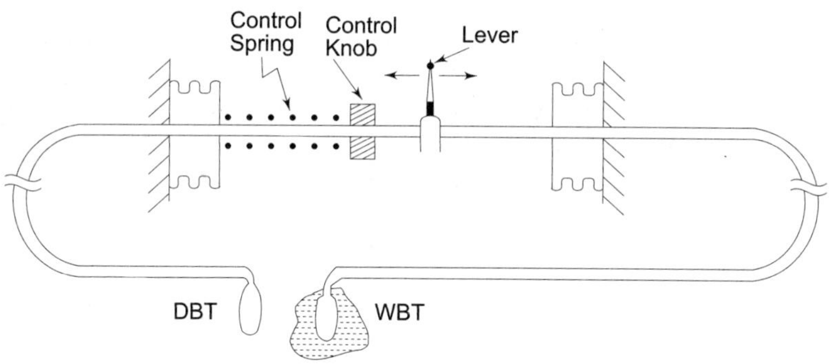

Fig. 23.7 Using bellow elements for humidity control

In addition to the above, hygroscopic elements, such as a strand of hair, can be used for humidity control. A simple arrangement is shown in Fig. 23.8. Elongation of the hair due to an increase in relative humidity results in a clockwise turning of the roller, moving lever to the left, which may operate a switch.

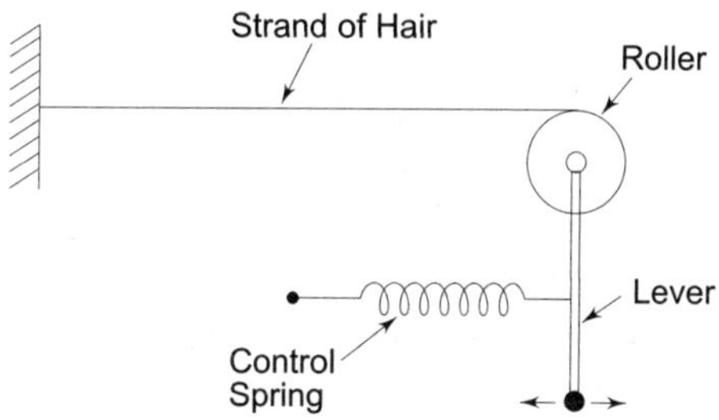

Fig. 23.8 Hair as a humidity-sensitive element

Recently, the hygroscopic nature of some salts, e.g., $CaCl_2$, $LiCl_2$, etc., has also been used in humidity measurement and control devices. It is found that their electric resistance changes with change in humidity.

23.2.6 Some Pressure Sensitive Elements

In addition to bellows, other pressure sensitive elements are diaphragms, bourdon tubes and piezo-electric crystals.

23.3 ACTUATING ELEMENTS[3]

In response to impulses from detecting elements, the actuating elements bring about adjustment in plants. They may operate directly or indirectly. In the latter case, additional transmitting elements are used to transfer the actuating impulse to the regulating unit.

The actuating elements can be categorised as follows:

(i) Electric switches
(ii) Relays
(iii) Solenoids
(iv) Electric or pneumatic motors.

There are a variety of electric switches with considerable difference in their mechanical and electrical construction.

A relay is a device which, by a weak pilot current from the detecting and measuring element assembly, actuates switches for heavy-duty circuits. An illustration of a single-phase relay is shown in Fig. 23.9 in combination with a solenoid.

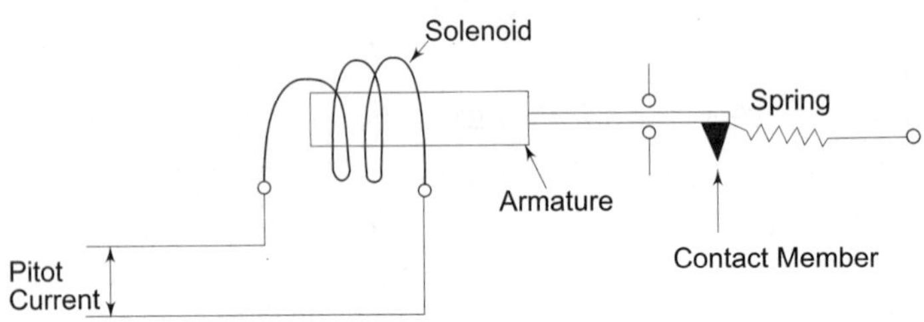

Fig. 23.9 A single phase relay

In another construction, the spring is omitted and it becomes a simple solenoid. The solenoid, if in vertical position, falls by its own weight when not energised. When it is energised, it is lifted up. Such a relay is used as *starting relay* for operating switches in hermetically sealed units for changing over from starting winding to running winding, after rated speed is attained.

Electric or pneumatic motors are used for operating dampers or valves which are required to open or close more slowly than a solenoid coil will permit. According to their mode of operation, motors are of the following two essential types:

(i) Two-position motors
(ii) Modulating motors.

A *spring-return motor*, shown in Fig. 23.10, is an illustration of a two-position motor. When the controller element closes the switch, the motor winding is energised from *A* to *B*. This starts the motor which drives a linkage to open or close a valve or damper. A cam, also mounted on the motor shaft, rotates and at the proper time throws the *limit switch* from *B* to *C*. The whole winding is now energised from *A* to *C*. This added coil resistance reduces the current to a *holding level*. The motor then stops and is held in this position. Later, when the switch opens, a spring returns the motor to its original position.

Modulating motors are used for *proportional* or *floating control*. They are reversible and capable of stopping and holding at any position in the cycle. They are either *reversible two-phase induction* type or *shaded pole* type.

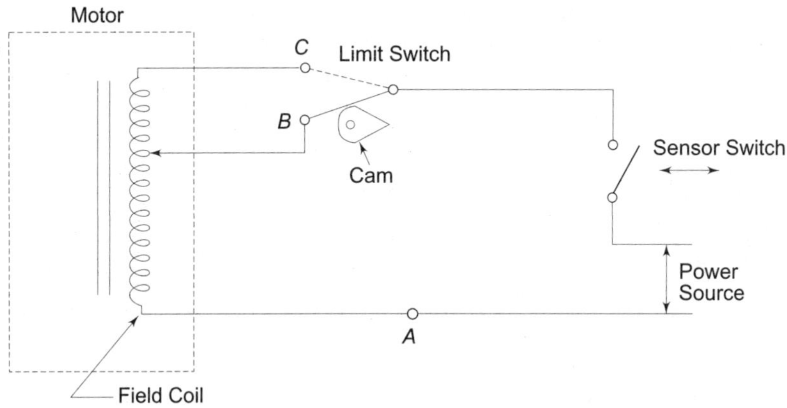

Fig. 23.10 Circuit diagram of a spring-return motor

The principle of operation of a reversible two-phase induction motor can be explained with the help of Fig. 23.11. In this motor, there are two field windings, I and II. Power may be supplied either at *A* or at *B*. The other power connection is made at *C*.

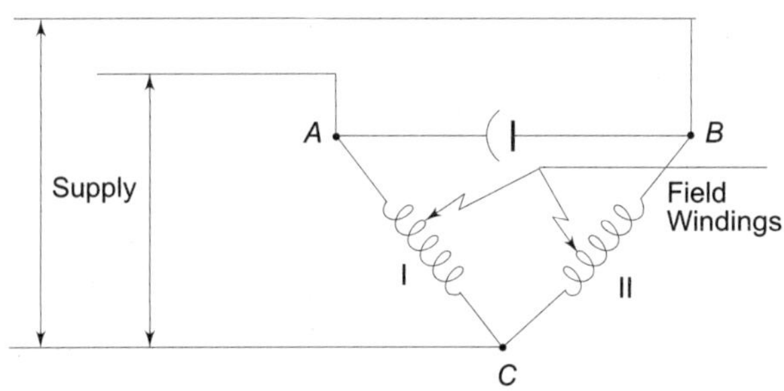

Fig. 23.11 Reversible two-phase motor

If *A* and *C* are connected, coil I is directly powered, while coil II is indirectly powered through a capacitor which introduces a phase difference. Thus a rotary motion is imparted to the motor armature to open or close dampers, or operate any other device.

On the other hand, if *B* and *C* are connected, the motion of the motor is in the reverse direction.

A fully-modulating control would, however, require negative feedback at the motor. The complete modulating system is shown in Fig. 23.12. The operating principle is explained in Fig. 23.13.

In the balanced position, the wiper arms of both the controller and feedback potentiometers *A* and *B*, and relay arm *C* are centred as in Fig. 23.13 (a). The current through the two coils of the balancing relay is equal.

With sensor action in response to pressure or temperature changes, the controller potentiometer arm is deflected, resulting in an unbalanced condition. The current difference in the two coils of the relay causes the relay arm to swing to one contact as shown in Fig. 23.13 (b). This starts the motor and causes it to actuate a valve or damper.

The movement of the motor also moves the feedback potentiometer wiper which brings in added resistance and offsets the effect of the controller. A new balanced condition for the relay is obtained as shown in Fig. 23.13 (c). Again, the relay coil currents are equal, the relay arm is centred, and the motor stops.

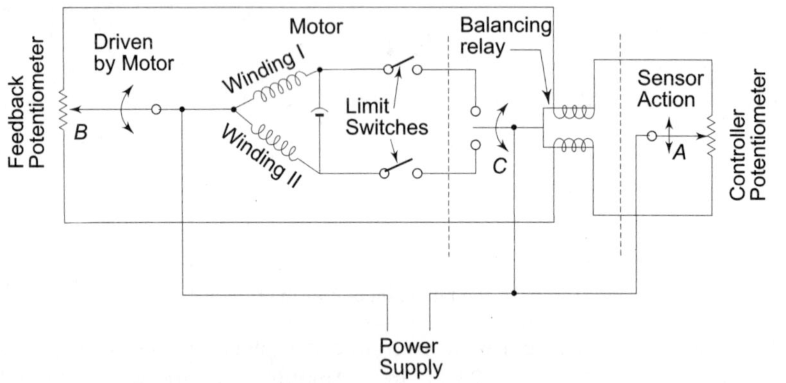

Fig. 23.12 Circuit diagram of a fully modulating control unit

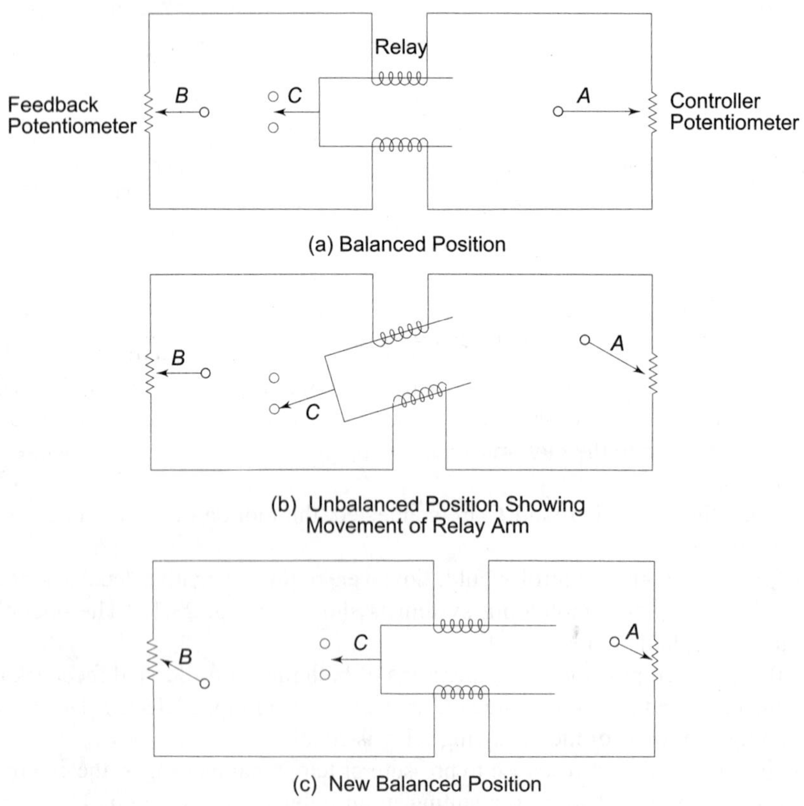

Fig. 23.13 Operation of a feedback modulating controller

Limit switches are used to stop the motor at the end of the desirable maximum stroke. Sensor action in the opposite direction reverses the motion.

23.4 ELECTRIC MOTORS AND CONTROLS[2]

Both single-phase and three-phase alternating current motors are used to drive pumps, fans and compressors. Single-phase motors range from 35 W to 7.5 kW. Three-phase motors are employed in sizes upwards of 1 kW. Motors are designed to operate at temperatures of about 40°C above the ambient temperature in open-type compressors, and about 55°C above the ambient in hermetically-sealed compressors. The motors are also designed to take a small percentage of the overload over a period of time. But continuous operation under overload conditions will result in the overheating of windings and consequent burn-out. It is, therefore, necessary to provide adequate safety control on all motors.

In refrigeration and air-conditioning practice, *induction motors* are commonly used as they can take up a high-starting torque. The name is derived from the fact that the magnetic field in the rotor is induced by the current flowing in the stator windings. Low starting torque (LST) motors are used in a compressor meant for use in conjunction with a capillary tube because the system is expected to equalise the pressures between the suction and discharge sides during the off-cycle period. Higher starting torque (HST) motors are meant for systems which are controlled by expansion valves.

23.4.1 Three-phase Induction Motors

There are two types of three-phase induction motors:

(i) Squirrel cage

(ii) Wound rotor.

In the squirrel-cage motor, the rotor is in the form of a simple squirrel cage. In the wound-rotor motor, the rotor also has a three-phase variable resistance windings.

Both the types have three-phase stator windings alternately distributed around the stators. There is a pole for each of the phases. A two-pole, three-phase motor in fact has six poles.

The rotor in an induction motor always lags behind the stator winding in speed. It rotates at a speed that is slightly less than the speed of the rotating stator field. The difference in speed is called the *magnetic* or *rotor slip*. The speed of the stator is determined by the frequency of the current and the number of poles as follows:

$$\text{rpm} = \frac{(\text{Frequency})\,120}{\text{Number of poles}}$$

A four-pole, 50-cycle A.C. motor will have a stator speed of

$$\text{rpm} = \frac{(50)\,(120)}{4} = 1500$$

This is *synchronous speed* of the rotor. However, for such a motor, the rotor speed is normally 1440 rpm. Thus the rotor slip, expressed as a percentage of the *synchronous speed* is

$$\frac{1500 - 1440}{1500} = 4\%$$

The slip also implies a loss of power of the motor due to the generation of heat. In this case, 4 per cent of the motor power input is converted into heat. Hence the motor efficiency is 96 per cent.

In the case of wound-rotor motors, the slip and hence the speed can be varied by varying the resistance of the rotor windings. But this would result in inefficient operation. By far, the simpler squirrel-cage motors are most commonly used.

23.4.2 Single-phase Induction Motors

In the case of a single-phase motor, when the stator winding is energised, there is only one stator pole and no rotating field. The magnetic field set-up in the rotor is in line with that of the stator. Therefore, no starting torque is developed and the rotor does not move.

If, however, the rotor is started somehow, the current induced in the rotor will lag behind the current in the stator, and the motor will continue to run.

All single-phase motors have squirrel-cage type rotors, but they differ only in the method employed to produce the necessary starting torque as discussed above. Accordingly, they are of the following types:

(i) Split phase
(ii) Capacitor start and induction run
(iii) Capacitor start and capacitor run
(iv) Permanent capacitor
(v) Shaded pole.

The difference in their construction is shown in Fig. 23.14. All the types have two stator windings, called the *starting* and *running windings*, connected in parallel.

In the *split-phase* type the starting winding, which is of thinner wire, has a high resistance and low inductance. The running winding, on the other hand, has a low resistance and high inductance. The relatively higher resistance in the starting winding can only withstand the main voltage for a very short period, otherwise it burns out. As a result, of the difference in inductances, when the two are energised, the current in the running winding lags behind the current in the starting winding by about 30°, thereby implying that the single phase is split. In effect a phase difference is created, a rotating field is formed and the motor starts. When the rotor reaches a speed of 70–75 per cent of the rated speed, a shaft mounted *centrifugal switch* or a starting relay in series with the starting winding as shown in Fig. 23.14 (a), opens and cuts off the starting winding. The motor then runs on the running winding only. These motors are available only in 35–350 W sizes.

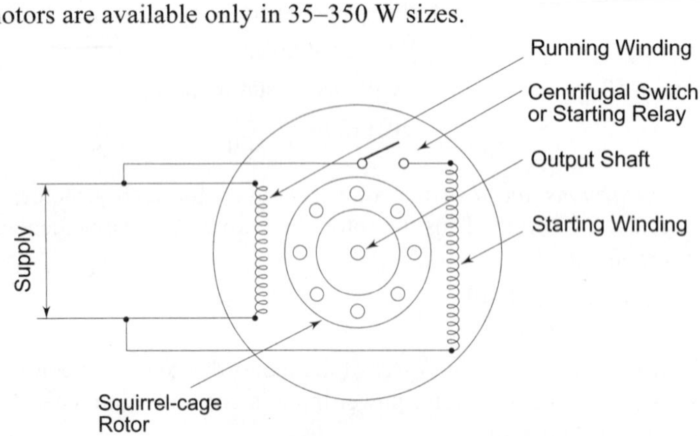

Fig. 23.14(a) Split-phase

The windings of the *capacitor-start-induction-run* motor are similar to those of the split-phase motor. This motor has a capacitor in addition to the centrifugal switch in series with the starting winding as shown in Fig. 23.14 (b). With this, the phase difference between the two windings can be increased to 90°. As a result, this motor can develop a higher-starting torque. They are available in sizes up to 560 W.

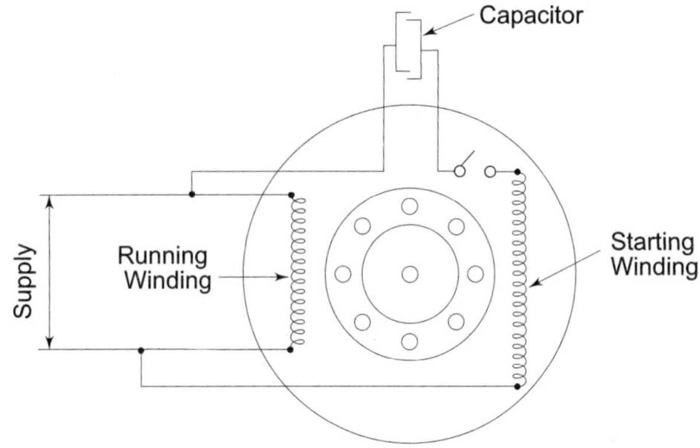

Fig. 23.14(b) Capacitor-start-Induction-run

Example 23.1 ***Conversion of Split-phase Motor into Capacitor-Start Motor***

A 5.48 cm^3 displacement volume R 12 compressor is used in a refrigerator. It has a 120 W split-phase motor. When used with R 134a and R 152a, it was found that the motor was tripping due to inadequate starting torque.

(a) A capacitor was introduced in the starting winding in order to increase the starting torque by increasing the phase difference between the starting and running windings from 33.3° to 90°. What is the ratio of new starting torque to original starting torque?

(b) Cheque if the new starting torque is adequate.

Solution (a) Original starting torque

$$\tau_0 = K\,(\sin 33.3°)$$

where K is the constant of motor depending on size, winding material, etc. New starting torque

$$\tau = K\,(\sin 90°)$$

Ratio

$$\frac{\tau}{\tau_0} = \frac{\text{Torque with capacitor}}{\text{Torque without capacitor}} = \frac{\sin 90°}{\sin 33.3°} = \frac{1}{0.55} = 1.82$$

(b) Pressures for standard cycle for refrigerator, in bar

	R 12	*R 134a*	*R 152a*
Condenser at 55°C	13.61	14.92	13.32
Evaporator at –25°C	1.24	1.0685	0.9765

Torque available in motor (from R 12 values)

$$\tau_{12} = \frac{1}{2} V_p (p_k - p_o) = \frac{1}{2}(5.48 \times 10^{-6})(13.61 - 1.24)10^5 = 3.38 \text{ Nm}$$

Torque required with R 134a and R 152a

$$\tau_{134a} = \frac{1}{2}(5.48 \times 10^{-6})(14.92 - 1.0685)10^5 = 3.8 \text{ Nm}$$

$$\tau_{152a} = \frac{1}{2}(5.48 \times 10^{-6})(13.32 - 0.9765)10^5 = 3.4 \text{ Nm}$$

Torque provided by introducing capacitor

$$\tau = 1.82\ \tau_{12} = 1.82\ (3.38) = 6.1 \text{ Nm}$$

It is greater than τ_{134a} and τ_{152a}. Hence, it is adequate.

Note *The phase difference of 90° was created by introducing capacitor of 40 μF in the starting winding.*

The *capacitor-start-capacitor-run* motor has two capacitors as shown in Fig. 23.14 (c). In this case, the centrifugal switch merely cuts off the starting capacitor from the circuit. The motor operates with both the windings in the circuit. The running capacitor is employed to correct the power factor. As a result, this motor has a higher starting torque as well as a higher efficiency. These motors are employed in sizes ranging from 375 W to 7.5 kW.

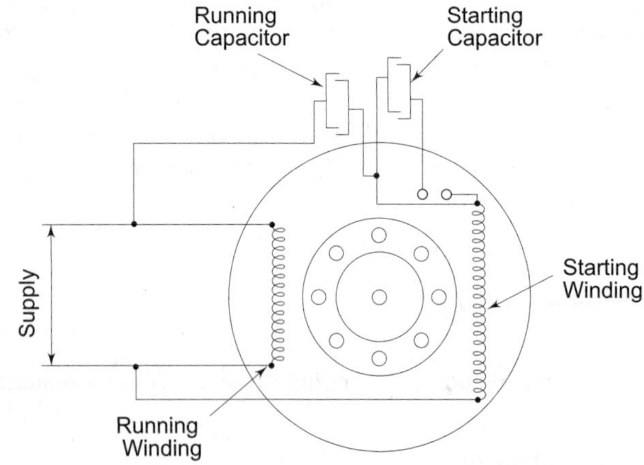

Fig. 23.14(c) Capacitor-start-capacitor-run

The *permanent-capacitor* motor shown in Fig. 23.14(d) does not have either a starting capacitor or a starting switch. A small capacitor permanently remains in the circuit creating the phase difference between the starting and running windings. It also serves the purpose of power-factor correction. The starting torque of these motors is very low. These are used in small fans only.

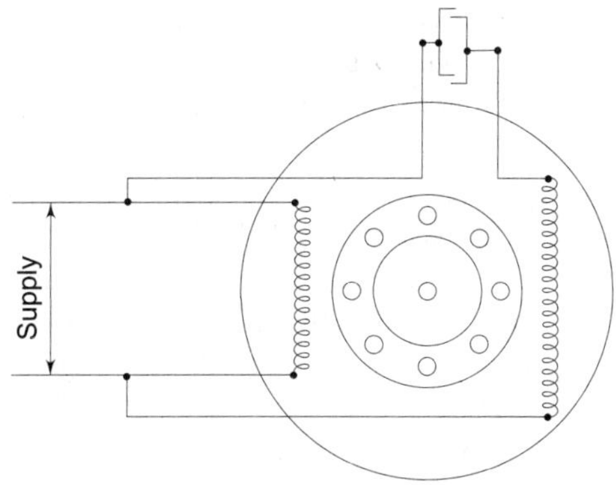

Fig. 23.14(d) Permanent capacitor

The construction of the *shaded-pole* motor shown in Fig. 23.14(e) is quite different. In this, the main stator winding is arranged to form two poles. Also, there is a *shading coil* covering a portion of each pole in the form of a few short-circuited coils of copper wire. The shading coil acts as a starting winding by distorting the magnetic field and thus producing a small starting torque. These motors are used in applications requiring extremely small-sized motors such as modulating motors for control.

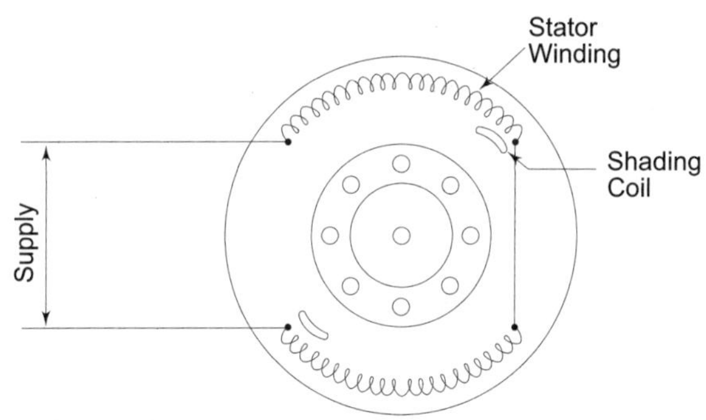

Fig. 23.14(e) Shaded Pole

23.4.3 Hermetically-sealed Units

A hermetically-sealed unit consists of a compressor and its motor coupled together and enclosed in a hermetically-sealed dome. The discharge line muffler, necessary for attenuating vibration and noise, is also enclosed within the dome. Also, the suction line is installed in such a way that cold suction vapour passes over the motor windings, thus enabling them to cool, before being led to the compressor-suction valve. A separate charging line is provided which communicates with the suction space in the dome.

The application range of a conventional motor is from idling to full-load torque because its winding temperature increases with increasing load. In the case of a hermetic motor, the temperature of the windings is reduced on increasing load, to an extent, as a result of greater cooling by the increased mass of the high-density suction vapour entering the compressor. The hermetic motor can, therefore, be utilized quite close to its breakdown torque without being thermally overloaded. However, if the load decreases, there is a risk of its overheating. This results from a decreased evaporator pressure at reduced load, and hence a lower density of the suction vapour which may seriously affect the dissipation of heat from the windings.

Another important feature of hermetic motors is the winding insulation. Due to recent developments of a new type of polyester-enamel material for winding wire, the compressor windings can withstand temperatures as high as 130°C. The life of such motors is more than 10,000 hours.

The presence of air is detrimental to the system. It increases the pressure ratio and discharge temperature. In the presence of moisture and oxygen, acids are formed in the system contributing to the dissolution of copper, i.e., copper crystals are formed on the sliding surfaces, such as bearings, pistons and crank shafts. From this it is clear that in a hermetic system a drier is essential since it ensures trouble-free operation during the entire life-span of the system.

23.4.4 Thermal Overload Protection for Hermetically-Sealed Units

Although refrigerant vapour and lubricating oil are used for the cooling of windings in hermetically-sealed units, it is absolutely essential that a protection is provided for the thermal overloading of these units. A *thermal overload* with a temperature sensitive element fastened directly to the motor parts is, therefore, installed in these units. The overload operates through a *starting relay* which is normally employed to replace the shaft-mounted centrifugal switch. In a compressor, although the starting winding is designed to take a relatively large current, but under locked rotor conditions—which could arise due to many factors, such as poor supply voltage, unbalanced pressure conditions, etc.—the starting winding temperature rise is very fast and may damage the insulation of starting windings eventually leading to *burn-out*. This is avoided by the use of the overload-cum-starting relay. The overload is a combination of a bi-metal and heater element. It is very precisely calibrated to operate under specific current and temperature conditions. The bimetallic strip must cut out at a particular fixed temperature and cut in again when the temperature is dropped from 70 to 84°C.

23.5 CONTROLS IN REFRIGERATION EQUIPMENT

A few common controls are described below.

23.5.1 High and Low Pressure Cut-outs

Refrigerant compressors are provided with *high pressure* (HP) and *low pressure* (LP) *cut-outs*. Their designs are similar. The high pressure cut-out is merely a safety control. When the head pressure increases beyond a set-point, the HP cut-out cycles off the compressor in order to ward off the possible damage to the compressor. When the head pressure subsequently drops, the circuit is once again closed and the compressor starts. Some cutouts, however, require manual resetting. Because of the possibility of scale formation in condenser tubes, and more important, the failure of water supply, high pressure cut-outs are essential in systems with water-cooled condensers.

A low pressure cut-out is used both as a safety control as well as temperature control. Since the suction pressure is governed by the evaporator temperature, a low pressure cut-out actuated by changes in suction pressure can be used to indirectly control the evaporator temperature. It is suitable as a temperature controller in remote installations in which the compressor is located quite some distance away from the evaporator.

23.5.2 Capacity Control of Reciprocating Compressors

The methods of capacity control of reciprocating compressors have been described in Chapter 6. The simplest method is that of the on-off control operated by the room or chilled-water thermostat.

A commonly used method known as the *pump-down cycle* method involves the use of both a thermostat and a low pressure cut-out as shown in Fig. 23.15. In this cycle, the room thermostat, instead of starting and stopping the compressor, operates a solenoid valve installed in the liquid line of the refrigerant. With a rise in the temperature of the room or chilled water, the thermostat closes the circuit, energising the solenoid valve which is lifted up, and allows the refrigerant to enter the evaporator. The flow of the refrigerant into the evaporator raises the suction pressure, closes the low pressure cut-out and starts the compressor. With a drop in the temperature of the controlled medium, the thermostat operates the solenoid valve and stops the flow of the liquid refrigerant to the evaporator. As the compressor continues to run, the suction pressure and evaporator temperature continue to drop until the low pressure cut-out opens and stops the compressor.

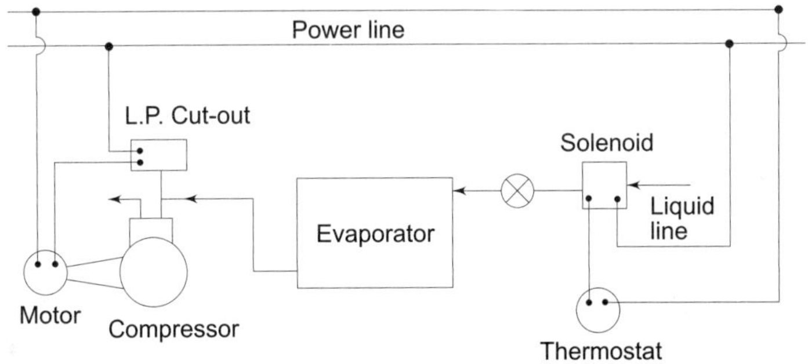

Fig. 23.15 Pump-down cycle

There are many advantages of the pump-down cycle. It provides a dual control and eliminates hunting of the compressor. An important advantage is the elimination of crank case oil dilution by the absorption of the refrigerant which results in inadequate lubrication, excessive carry-over of the oil into the discharge line due to foaming at low pressure and possible damage to the compressor due to the formation of slugs of refrigerant and oil.

Large reciprocating compressors have a stepped-capacity control, achieved by the loading and unloading of compressor cylinders by the lifting of suction valves controlled by a thermostat or low pressure control.

23.5.3 Electrical Disturbances

Some of the faults in refrigeration plants along with their symptoms, mainly appearing as electrical disturbances, are summarised below in Table 23.1.

Table 23.1 Electrical disturbances

	Projector trips during operation	*Compressor starts, but projector trips immediately after start*	*Compressor tries to start without success, projector trips*	*Compressor will not start, no projector trip*
	1	*2*	*3*	*4*
A. Faults in Electrical Components				
Wrong motor projector	×	×	×	
Defective motor projector				×
Wrong start relay	×	×	×	
Defective start relay			×	×
Wrong start capacitor			×	
Defective start capacitor			×	
Defective fan	×			
Defective thermostat				×
Open circuit in compressor, electrical equipment or leads				×
Earthing	×			
B. Faults in Electrical Circuit				
Extreme overvoltage	×	×		
Extreme under voltage or weak line	×	×	×	
Wrong rated voltage	×	×	×	
Mains disconnected (fuses blown)				×
C. Defective Compressor or Wrong Application of Compressor				
Mechanical defects			×	
Wrong application range (evaporating temp.)	×	×		
Wrong application (ambient temp).	×			
Wrong voltage range	×	×	×	
D. Defects in Refrigeration System				
Dirty condenser	×			
Capillary tube partly or completely blocked			×	
System overcharged	×	×		
System or Cabinet				

(Contd.)

Table 23.1 *(Contd.)*

E. Wrong Design of Refrigeration System or Cabinet				
Condenser too small	×			
Wrong condenser design			×	
Wrong evaporator design (too large volume)	×			
Too high equalizing pressure		×		
Wrong capillary tube			×	
Bad mounting conditions for compressor or condenser	×			
Wrong thermostat differential			×	
F. Other Conditions				
Very low compressor temperature		×		

23.6 CONTROLLING ROOM CONDITIONS AT PARTIAL LOAD

An air-conditioning plant is selected for design peak load conditions. But most of the time the plant operates under partial load which is either due to change in the outside conditions or due to withdrawal of load. The control of room conditions under part load is, therefore, very important. For this purpose, one or more of the following methods may be employed:

(i) On-off control of the air-handling unit or refrigeration unit.
(ii) Controlled bypass of the air-entering apparatus.
(iii) Varying water flow in the chilled-water coil.
(iv) Reheating the supply air.
(v) Controlling the volume of the supply air.

23.6.1 On-off Control

In the on-off control of the air-handling unit or fan-coil unit, the fan motor is operated intermittently by the room thermostat, the bulb of which is installed in the return air. In this, the refrigerant or chilled water continues to flow through the cooling coil. This results in fluctuating room temperature. In the on-off control of the refrigeration unit, the compressor operates intermittently. The air continues to flow over the cooling coil and in the room. This method also causes fluctuations in room conditions but ensures continuous air movement and is quite convenient in small air conditioners.

23.6.2 Bypass Control Using Face and Bypass Dampers

The bypass control maintains a constant DBT in the room by bypassing a part of the return air entering the cooling coil from the apparatus and thus increasing the supply air temperature. The schematic diagram of a bypass control with a typical D-X coil is shown in Fig. 23.16. The method is employed on chilled water coils as well. Its psychrometrics is illustrated in Fig. 23.17. As distinct from the inherent bypass of the air due to the bypass factor of the cooling coil, in bypass control the entering air is purposely diverted around the cooling coil. By far, this is the most common method of control adopted by the industry.

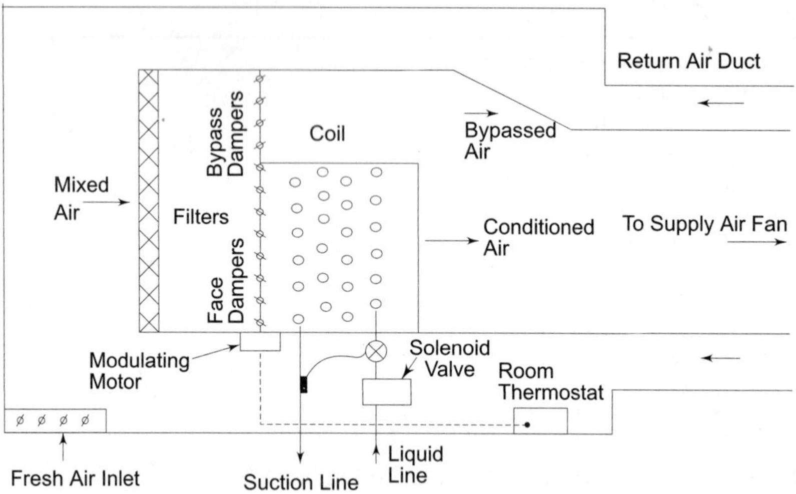

Fig. 23.16 Face and bypass control

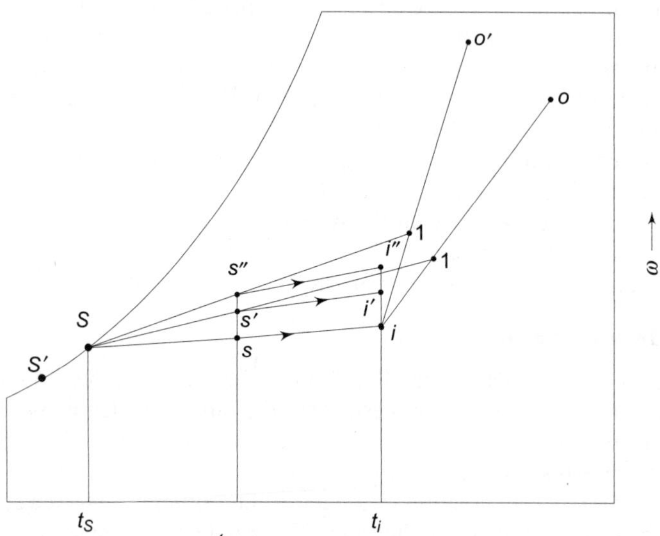

Fig. 23.17 Psychrometry of bypass control

The method of bypassing the mixture rather than the room air is inferior because in this case the outside air is also bypassed and is introduced into the room, thus causing an increase in relative humidity of the room. Why the return air alone should be bypassed in preference to the mixture or outside air, is explained below.

Referring to Fig. 23.17, let S at ADP of the coil represent the supply air state at full load, assuming the BPF of the coil as zero. Now consider a case when the RSH load decreases to 50 per cent and the RLH load remains the same. This may coincide with a reduction in the outside air dry bulb temperature. Assume, therefore, that the outside air condition changes from o to o'. At this reduced 50 per cent RSH load, the condition of the supply air should be at s as shown in Fig. 23.17, where

$$\frac{i - s}{i - S} = 0.5$$

Presuming that the ADP of the coil is unaffected, a bypass of the mixture at 1 around the coil would result in a condition of the supply air slightly above s, viz., at s' on the line joining 1 to S. Since the change in the condition of the supply air would occur along a new RSHF line, which is steeper due to the RLH load remaining the same, the condition of the room air achieved by this controlled bypass by a thermostat will be at i'. The space condition will thus be at a higher moisture content than state i.

A bypass of the outside air at o around the coil would result in the supply air condition little more above s, and at a still higher humidity in the room.

If the change in the outside air condition from o to o' is also considered, then the bypass of the mixture at $1'$ would result in a supply air condition at s'', and the room condition at a further higher humidity at i''.

Thus when any outside air is bypassed, the effect is to raise the room humidity. Hence the bypass should be normally that of recirculated room air and should be arranged as in Fig. 23.18 in preference to the arrangement in Fig. 23.16.

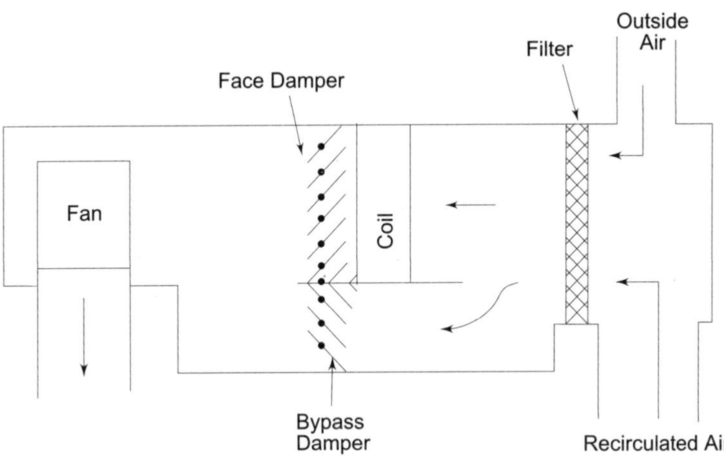

Fig. 23.18 Arrangement with bypass of return air only

Even if no outside air is bypassed, the entering air state will be above 1 and the room humidity will again be higher. This leads to a lack of humidity control. However, the decrease in the flow of air over the cooling apparatus decreases the load on the cooling coil. As a result, the ADP drops from S to S' and the air leaves at a lower temperature. The supply air state, therefore, is closer to s than to s' and s'' and the change in the room humidity is not to the extent as shown. Nevertheless, coil icing may result at higher bypass rates. This may be eliminated by using a pump-down cycle by installing a solenoid valve as shown in Fig. 23.15.

A bypass control can be used in chilled water coils also. The method is conveniently employed in single zone as well as multizone units for the air conditioning of large spaces.

Example 23.2

(a) Given for the air conditioning of a plant:

Inside design conditions	*: 24°C DB, 50% RH*
Outside desing conditions	*: 40°C DB, 25°C WB*
Room sensible heat	*: 40 kW*
Room latent heat	*: 10 kW*
Ventilation air	*: 25% of supply air.*

Calculate the state and rate of supply air, and the design duty of the cooling coil. Assume the bypass factor of the coil as zero.

(b) A thermostat is employed to maintain the room temperature in the above plant using face and bypass dampers. The outside air dry bulb temperature changes to 34°C, the wet bulb temperature remaining the same, and thereby the room sensible heat load is reduced to half while the latent heat load remains the same. Determine to the first approximation:

(i) The amount of air bypassed by the control.

(ii) The relative humidity maintained in the room.

Solution Refer to Fig. 23.19

(a) *At full load*: Room sensible heat factor

$$\text{RSHF} = \frac{40}{40+10} = 0.8$$

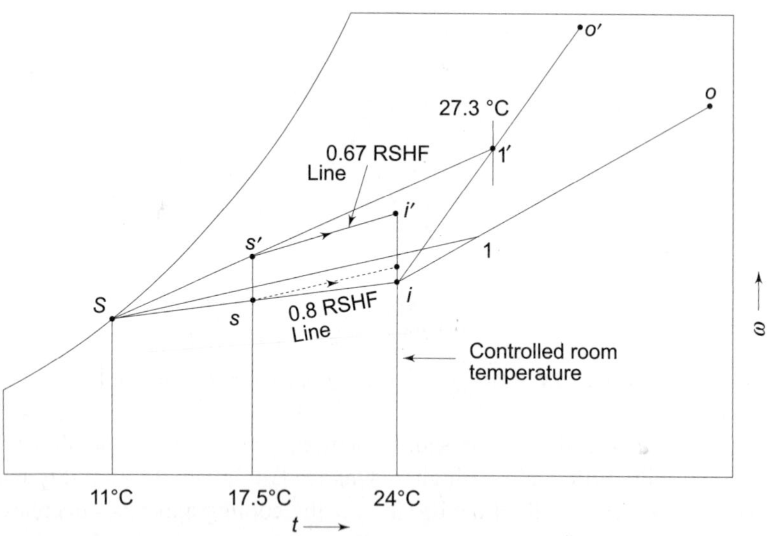

Fig. 23.19 Figure for Example 23.2

Apparatus dew point from psychrometric chart (Room ADP, BPF = 0)

$$t_S = 11°\text{C}$$

Dehumidified or supply air quantity

$$(\text{cmm})_d = \frac{\text{RSH}}{0.0204(t_i - t_S)} = \frac{40}{0.0204(24-11)} = 150.8$$

Ventilation air $(\text{cmm})_0 = 0.25\,(\text{cmm})_d = 0.25\,(150.8) = 37.7$
Recirculated room air $(\text{cmm})_i = (\text{cmm})_d - (\text{cmm})_0 = 150.8 - 37.7 = 113.1$
Ventilation load

$$\begin{aligned}\text{OATH} &= 0.02\,(\text{cmm})_0\,(h_o - h_i)\\ &= 0.02\,(37.7)\,(76.2 - 48.1) = 21.2 \text{ kW}\end{aligned}$$

Design duty of cooling coil

$$\begin{aligned}\text{GTH} &= \text{RTH} + \text{OATH}\\ &= (40 + 10) + 21.2 = 71.2 \text{ kW}\end{aligned}$$

(b) *At part load:*

Supply air temperature at 50 per cent RSH

$$\begin{aligned}t_s &= \frac{1}{2}(t_i + t_S)\\ &= \frac{1}{2}(24 + 11) = 17.5°\text{C}\end{aligned}$$

Bypassed room air

$$\begin{aligned}(\text{cmm})_{i,\ \text{bypassed}} &= \frac{t_i - t_s}{t_i - t_S}\,(\text{cmm})_d\\ &= \frac{1}{2}(\text{cmm})_d\\ &= \frac{1}{2}(150.8) = 75.4\end{aligned}$$

Recirculated air through apparatus

$$(\text{cmm})_{i,\ \text{apparatus}} = 150.8 - 75.4 = 75.4$$

The ventilation air through the coil remains the same.
Dry bulb temperature of the entering air through the coil

$$\begin{aligned}t'_2 &= \frac{(\text{cmm})_{i,\ \text{apparatus}}\, t_i + (\text{cmm})_0 t'_0}{(\text{cmm})_{i,\ \text{apparatus}} + (\text{cmm})_0}\\ &= \frac{75.4\,(24) + 37.7(34)}{75.4 + 37.7} = 27.3°\text{C}\end{aligned}$$

Join the entering air state 1′ to the apparatus dew point S. The new supply air condition lies at s' on the line $1' - S$ at a DBT of 17.5°C. New room sensible heat factor

$$\text{RSHF}' = \frac{20}{20 + 10} = 0.67$$

Draw a 0.67 RSHF line from s' to the controlled DBT line of 24°C. This locates the controlled room conditions at i'. The relative humidity at this point read from the psychrometric chart is 60%.

Note *A more elaborate calculation would show that the coil ADP is lowered and the actual relative humidity maintained in the room is not so high.*

23.6.3 Varying Flow of Water in Chilled-Water (C-W) Coil

In this case the room thermostat operates a three-way or straight-through valve, supplying chilled water to the cooling coil through a modulating motor or a two-position control. The valve throttles the water thereby varying the flow rate. The three way valve mixes fresh chilled water with return chilled water, and this varies the entering water temperature (EWT). The temperature of water does not remain constant in C-W coil. It is essential in the case of a chilled-water coil, therefore, to have a counterflow arrangement between air and water as shown in Fig. 23.20.

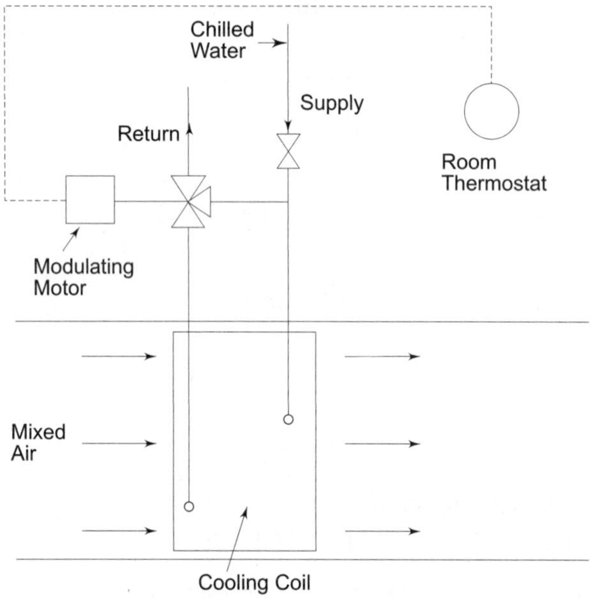

Fig. 23.20 Control by varying chilled water flow and/or temperature

Since the water temperature rise will be greater than the design rise due to smaller flow rate, or the mixed water temperature will be higher at part load, the surface temperature of the coil and hence the ADP will normally increase. Thus dehumidification of air will be reduced, and the supply air state will be above *s*′ or *s*′ in Fig. 23.17. Consequently, this method would maintain a higher humidity in the room than the method of controlled bypass of air. Nevertheless, the method is conveniently applied in the air conditioning of multistoreyed buildings with a large number of rooms, each having its individual fan-coil unit (FCU) and chilled-water supply. Varying C-W flow is another common FCU method of control, along with bypass control, preferred by the industry.

Note *If it is desired that the relative humidity does not alter, a return-air humidstat can control the water flow rate, while the thermostat can control the bypass of air. Such a combination offers the possibility of maintaining both temperature and humidity, simultaneously, in the conditioned space.*

23.6.4 Reheat Control

Any increase in the latent heat load and decrease in the sensible heat load of the room can be handled by employing reheat of the supply air. Reheat thus provides an artificial load on the room as illustrated in Fig. 19.8 and maintains a constant dry bulb temperature. But there is no control on the room relative humidity.

The psychrometrics of reheat control is explained in Fig. 23.21. The broken lines in the figure represent the processes at part load. The room thermostat senses the temperature of the room air. As soon as the temperature drops below the design value, the thermostat operates the reheat coil. The leaving air at 2 is reheated to *s* and then supplied to the room along the part-load sensible-heat-factor line *s*–*i*. The position of the point s can be located on the intersection of the horizontal from 2 with the line of minimum RSHF drawn from *i*.

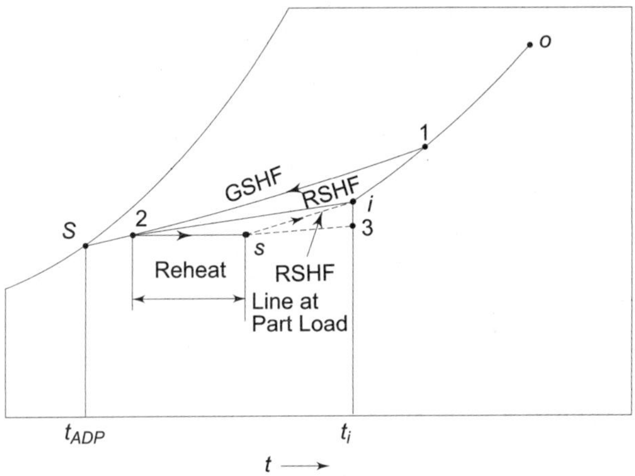

Fig. 23.21 Psychrometrics of reheat control

If the plant is operating under part-load conditions, the latent-heat load may also reduce along with the sensible heat load. Then, the resulting room condition will be at 3 and the RSHF line will be *s*–3.

As reheat adds to the requirement of refrigerating capacity, it is wasteful of energy, and is, therefore, not commonly used in air conditioning except for high latent-load applications.

23.6.5 Volume Control

The control of the quantity of the supply air to room essentially provides the same type of control as obtained by bypassing the recirculated room air. A problem encountered with volume control is the non-uniform distribution of air within the conditioned space. It is, therefore, desirable to see that the supply air quantity at partial load is sufficient for satisfactory distribution.

23.7 INDUCTION SYSTEM

In the constant-volume induction system, the central plant conditions only fresh air. The supply air is maintained at a constant temperature of about 10°C. A constant volume of this primary air is led to the room through an induction unit as shown in

Fig. 23.22. This *primary air* induces a stream of *secondary room air*. The mixture can be further treated in the coil installed in the unit.

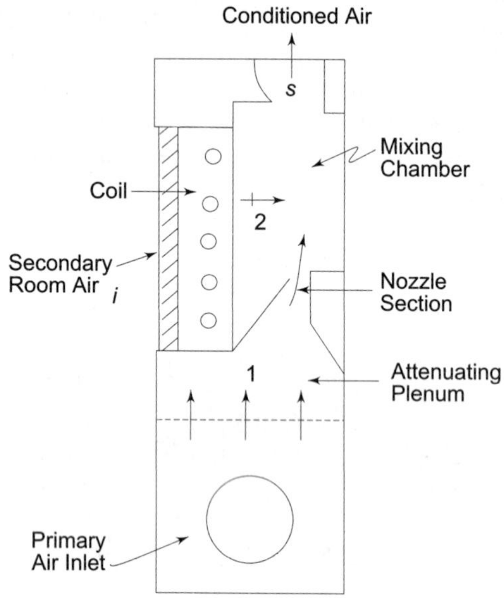

Fig. 23.22 Induction unit

The all-air-constant-volume induction system is particularly suitable for spaces having a large ratio of the floor area to height, requiring horizontal duct-work and piping. The treated fresh-air duct in this case is designed as a high-pressure system.

Example 23.3 *The air-conditioning system of a building has a central plant which processes fresh air from 43°C DB and 24°C WB to 12°C DB and 11°C WB. The rooms have induction units with individual thermostat control to maintain the room at 24°C by modulating the supply of chilled water through the room coil in which only sensible cooling takes place. The induction ratio is 4 by volume. The room sensible and latent heat gains are 5 and 1 kW respectively. The primary air delivered to the room is 8 cmm. Calculate:*

(a) The relative humidity in the room.
(b) The cooling loads on the central plant and room induction unit.
(c) The grand total heat.

Solution Refer to Fig. 23.23. The fresh air is treated from o to 1. The primary air is at 1. The secondary room air at *i* is sensibly cooled to 2 in the induction unit. The supply state s is obtained by the mixing of 2 and 1 in the given ratio.

Secondary room-air quantity

$$(\text{cmm})_2 = 4\ (\text{cmm})_1 = 4(8) = 32\ \text{cmm}$$

Supply air quantity

$$(\text{cmm})_s = 8 + 32 = 40\ \text{cmm}$$

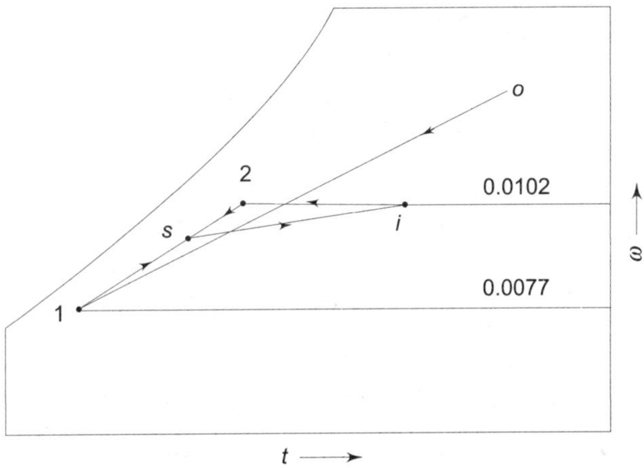

Fig. 23.23 Figure for Example 23.3

Supply air temperature

$$t_s = t_i - \frac{\text{RSH}}{0.0204\,(\text{cmm})_s}$$

$$= 24 - \frac{5}{0.0204\,(40)} = 17.9°\text{C}$$

(a) Room latent-heat balance:

$$\text{RLH} = 50(\text{cmm})_s\,(\omega_i - \omega_s)$$

$$= 50\,(\text{cmm})_s\left(\omega_i - \frac{4\omega_i + \omega_1}{5}\right)$$

$$1 = 50(40)\left(\omega_i - \frac{4\omega_i + 0.0077}{5}\right)$$

whence $\omega_i = 0.0102$ kg/kg

The room relative humidity from the psychrometric chart at t_i = 24°C and ω_i = 0.0102 kg/kg is found to be 53.5%.

(b) Specific humidity of air after cooling in the induction unit:

$$\omega_2 = \omega_i = 0.0102 \text{ kg/kg}$$

Dry bulb temperature at 2

$$t_2 = \frac{(\text{cmm})_s\,t_s - (\text{cmm})_1 t_1}{(\text{cmm})_2}$$

$$= \frac{40(17.9) - 8(12)}{32} = 19.4°\text{C}$$

Enthalpies at, o, i, 1 and 2 from the psychrometric chart

$$h_o = 68.0 \text{ kJ/kg}$$
$$h_i = 50.0 \text{ kJ/kg}$$
$$h_2 = 45.0 \text{ kJ/kg}$$
$$h_1 = 31.5 \text{ kJ/kg}$$

Cooling load on the room-induction unit

$$\dot{Q} = 0.02\,(\text{cmm})_2\,(h_i - h_2)$$
$$= 0.02(32)\,(50 - 40) = 6.4\text{ kW}$$

Cooling load on the central plant

$$\text{OATH} = 0.02\,(\text{cmm})_1\,(h_o - h_1)$$
$$= 0.02\,(8)\,(68 - 31.5) = 5.84\text{ kW}$$

(c) Grand total heat

$$\text{GTH} = 6.4 + 5.84 = 12.24\text{ kW}$$

Example 23.4 ***Year-round A/C System with Sprayed Coil and Induction Unit.***

Figure 23.24 shows the schematic arrangement of the central plant of an year-round A/C system with induction unit. The room units have individual thermostats controlling water flow in coils. The induction ratio (IR) is 4.

The system is designed for sprayed coil with C-W spray in summer, and pumped recirculation in winter.

(a) Summer design conditions are:

Outside air	*: 43.3°C DBT, 28.3°C WBT*
Air off cooling coil	*: 11.7°C DBT, 10.6°C WBT*
Primary air	*: 15.6°C DBT*
RSH gain	*: 4.5 kW*
RLH gain	*: 0.5 kW*

Design room temperature: 22.2°C

Calculate:

(i) Primary air quantity. Mixed supply air temperature is limited to 13.3°C.
(ii) Cooling load on room coil in induction unit, and condition of air leaving room coil.
(iii) Relative humidity in the room.

(b) Winter design conditions are:

Outside air	*: –1.1°C DBT, 80% RH*
Air off pre-heater	*: 26.7°C*
Humidifying efficiency	*: 80% (of sprayed coil)*
RSH loss	*: 3.3 kW*
RLH gain	*: 0.5 kW*

Assuming the unit handles the same supply air volume, and the room is still to be maintained at 22.2°C, calculate:

(i) Primary air temperature.
(ii) Temperature of mixed supply air.
(iii) Relative humidity in the room.

Solution (a) *Summer Air Conditioning*: (Fig. 23.25)

(i) Mixed supply air quantity

$$\dot{m}_{a_s} = \frac{\text{RSH}}{C_p(t_i - t_s)} = \frac{4.5}{1.0216\,(22.2 - 13.3)} = 0.5\text{ kg/s}$$

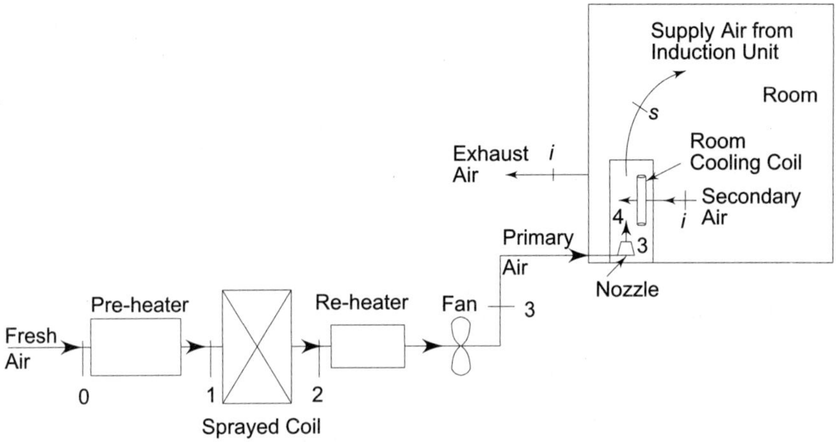

Fig. 23.24 Schematic arrangement of year-round A/C system for Example 23.4

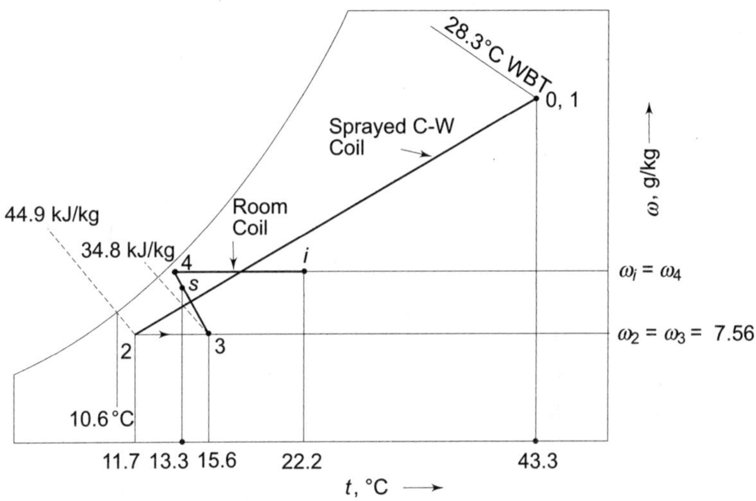

Fig. 23.25 Summer A/C psychrometrics for Example 23.4

Primary air quantity

$$\dot{m}_{a_3} = \frac{1}{5}\dot{m}_{a_s} = 0.1 \text{ kg/s}$$

Secondary air quantity

$$\dot{m}_{a_4} = \frac{4}{5}\dot{m}_{a_s} = 0.4 \text{ kg/s}$$

(ii) Condition of air off sprayed coil

$t_2 = 11.7°\text{C}$, $t'_2 = 10.6°\text{C}$, $\omega_2 = 7.56$ g/kg d.a., $h_2 = 44.9$ kJ/kg d.a.

Moisture balance of room gives

$$\omega_i = \omega_3 + \frac{\text{RLH}}{\dot{m}_{a_3}(h_{fg})_{22.2°\text{C}}}$$

$$= 7.56 + \frac{0.5}{0.1}\left(\frac{1000}{2561}\right) = 9.51 \text{ g w.v./kg d.a.}$$

$$\Rightarrow \quad h_i = 47.0 \text{ kJ/kg d.a. (As a function of } t_i, \omega_i)$$

Cooling load on room cooling coil, by energy balance of room

$$\dot{m}_{a_3}(h_i - h_3) = \text{RSH} + \text{RLH} - \dot{Q}_{\text{Coil}}$$

$$\Rightarrow \quad \dot{Q}_{\text{Coil}} = \text{RSH} + \text{RLH} - \dot{m}_{a_3}(h_i - h_3)$$

$$= 4.5 + 0.5 - 0.1\,(47 - 34.8) = 3.78 \text{ kW}$$

Condition of air leaving room coil

$$h_4 = h_i - \frac{\dot{Q}_{\text{Coil}}}{\dot{m}_{a_4}} = 47 - \frac{3.78}{0.4} = 37.6 \text{ kJ/kg}$$

$$\omega_4 = \omega_i = 9.51 \text{ g.w.v./kg d.a.}$$

$$\Rightarrow \quad t_4 = 15°\text{C}$$

(iii) Relative humidity in room

$$(\text{RH})_{\text{room}} = 58\% \text{ (As a function of } t_i, \omega_i)$$

(b) *Winter Air Conditioning*: (Fig. 23.26)

Volumes handled in summer ($v_3 \cong v_4 \cong v_s = 0.83$ m^3/kg d.a.)

$$\dot{Q}_{v_s} = 0.5\,(0.83) = 0.415 \text{ m}^3\text{/s}, \; \dot{Q}_{v_3} = 0.081 \text{ m}^3\text{/s},$$

$$\dot{Q}_{v_4} = 0.324 \text{ m}^3\text{/s}$$

Assume $v_s \cong v_3 \cong v_4 = 0.85$ m^3/kg d.a. for winter.

Mass flow rates in winter, with volume flow rates same as in summer

$$\dot{m}_{a_s} = \frac{\dot{Q}_{v_s}}{v_s} = \frac{0.415}{0.85} = 0.49 \text{ kg/s}, \; \dot{m}_{a_3} = 0.098 \text{ kg/s}, \; \dot{m}_{a_4} = 0.392 \text{ kg/s}$$

Humidity off air washer

$$\omega_3 = \omega_2 = \omega_1 + \eta_H(\omega_1' - \omega) = 2.8 + 0.8\,(8.8 - 2.8) = 7.6 \text{ g/kg d.a.}$$

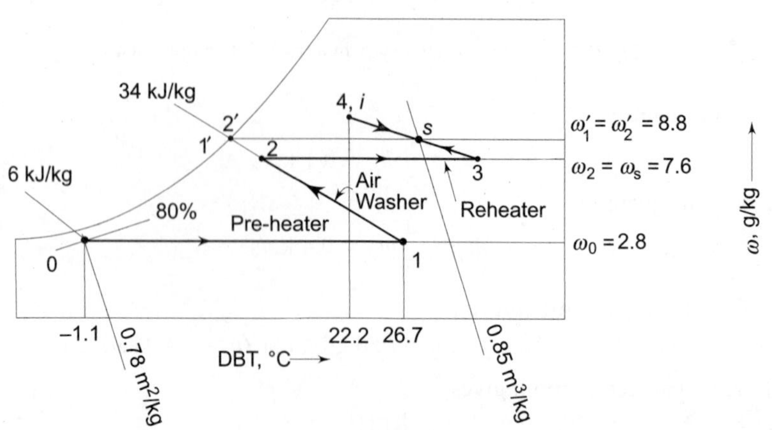

Fig. 23.26 Winter A/C psychrometrics for Example 23.4

Moisture balance of room gives

$$\omega_i = 7.6 + \frac{0.5}{0.098}\left(\frac{1000}{2561}\right) = 9.59 \text{ w.v./kg d.a.}$$

$$\Rightarrow \qquad h_i = 47.5 \text{ kJ/kg d.a. (As a function of } t_i, \omega_i)$$

Energy balance of room gives

$$h_3 = h_i + \frac{\text{RSH} + \text{RLH}}{\dot{m}_{a_s}} = 47.5 + \frac{3.3 + 0.5}{0.098} = 76.1 \text{ kJ/kg d.a.}$$

(i) Primary air temperature

$$t_3 = 56°\text{C (As a function of } \omega_3, h_3)$$

(ii) Room sensible heat balance gives

$$h_s = h_i + \frac{\text{RSH}}{\dot{m}_{a_s}} = 47.5 + \frac{3.3}{0.49} = 54.2 \text{ kJ/kg d.a.}$$

Temperature of mixed supply air

$$t_s = \frac{4t_i + t_3}{5} = \frac{4(22.2) + 56}{5} = 29°\text{C}$$

Note *Check v_s. The assumption is found to be satisfactory.*

References

1. Andersen, S A, *Automatic Refrigeration*, MacLaren, Nordborg, Denmark, 1959.
2. Dossat, R I, *Principles of refrigeration*, John Wiley, New York, 1961.
3. Haines R W, *Control Systems for Heating, Ventilating and Air Conditioning*, Van Nostrand Reinhold, New York, 1977.

Revision Exercises

23.1 (a) An air-conditioning plant with face and bypass control by a room thermostat is designed to maintain the conditioned space at 25°C DB and 50 per cent RH, when the outside design conditions are 43°C DB and 24°C WB, and the room sensible and latent heat gains are 80 and 20 kW respectively. The ventilation air is 75 cmm. Calculate the state and rate of the supply air and design duty of the cooling coil. Assume the bypass factor of the apparatus as zero.

(b) If the room total heat is reduced to half when the outside-design conditions change to 34°C DB and 24°C WB, determine the amount of mixture of the recirculated room air and ventilation air bypassed and the relative humidity maintained in the conditioned space.

23.2 (a) The inside design conditions in a space are 21°C DB and 0.0078 kg/kg moisture content. The outside design conditions are 30°C DB and 0.011 kg/kg moisture content. The room sensible and latent heat gains are 17.5 and 3.5 kW respectively. The design supply air temperature is 15.5°C. The ventilation air is 25% of supply air.

The air-conditioning plant comprises a mixing chamber, an air washer with chilled-water circulation and a reheater. A room thermostat controls the reheater and a room humidstat controls the chilled-water flow in the air washer. Calculate, under the design conditions, the state and rate of supply air, cooling load on the air washer and amount of reheat.

(b) If the room sensible and latent heat gains change to 20.5 and 5.5 kW respectively when the outside condition changes to 33°C DB and 22°C WB, determine approximately the conditions maintained in the room and the amount of reheat.

Assume that the supply air rate and cooling duty of the air washer remain the same.

23.3 The peak cooling load requirements for a large banquet hall are as follows:

Inside design conditions : 25°C DB, 50% RH
Outside design conditions : 40°C DB, 24°C WB
Space sensible heat : 380,000 kJ/hr
Space latent heat : 250,000 kJ/hr
Ventilation air : 190 cmm

Determine the space cooling load, the reheat load and the total refrigerating capacity. Make suitable assumptions.

If the space sensible heat load decreases by 10%, what will be the condition maintained in the room with coil bypass control?

Chapter

24 Applications in Food Refrigeration/Processing and Industrial Air Conditioning

Air conditioning for comfort is one of the most common application of refrigeration, and it has been dealt with at length in the preceding chapters. However, the most significant application of refrigeration is in food preservation, whether it is by way of processing or for storage. Processing is done by heating, heat drying, etc., and by refrigeration such as in chilling, freezing or freeze-drying. Storage may be of either chilled or frozen product. Some of the important products involved in processing are candy, beverages, meat, poultry, fish, bakery and dairy products, fruits and vegetables, fruit-juice concentrates, precooked foods, etc. The common products preserved by storage after chilling are fruits such as apples, pears, grapes, citrus fruits, etc., vegetables such as onions, potatoes, tomatoes, etc., and dry fruits, candies, milk, eggs and their products. Storage under frozen conditions is resorted to for preserving the food value as well as to store perishable products over a long period. The common items of frozen food are fish, meat, poultry, and some vegetables such as peas, beans, carrots, cauliflower, etc. It may be noted that all products are not amenable to freezing.

An interesting feature of the chilled and frozen food industry is the cold chain that must be maintained from the farm to the consumer. An important link in this chain is that of transport refrigeration.

In this chapter, a few problems in food refrigeration and dehydration have been studied from the point of view of thermodynamic and heat transfer principles. It must be stated here that processing by simple heat drying is a more economical way of food preservation. However, with many products, it may involve loss of quality in which case it is not considered suitable. The study of the drying process involves psychrometric principles.

24.1 TYPICAL EXAMPLES OF FOOD PROCESSING BY REFRIGERATION[1] AND STORAGE

A few typical examples of food processing by refrigeration and the operating conditions involved may be of interest to the readers.

24.1.1 Candy Manufacture

The proportion of various ingredients in candy manufacture may vary. The main constituents are, however, sucrose and glucose and in some cases dextrose. These and other ingredients used are sensitive to temperature and/or humidity. They change readily to the liquid form depending on temperature and humidity. Accordingly, temperatures must not be higher than 21°C and the relative humidity should be less than 60 per cent.

Chocolate is the main item of candy, and cocoa butter is its principal ingredient. For most practical purposes, 30°C may be considered as the freezing point of chocolate. At usually encountered ambient temperatures of 45°C, the chocolate will become soft and will deteriorate fast in quality. Its latent heat of fusion is approximately equal to 70–93 kJ/kg. The average values of specific heats before and after freezing are 2.34 and 1.26 kJ/kg.K respectively. Chocolate bars are cast in metal moulds which are pre-cooled to 15°C. The dew-point temperature of the cold room in which solidification takes place and the packing room should be below 15°C so that there is no condensation of moisture on the product. The usual temperature maintained is between 4–10°C. In the case when solidification is carried out in a tunnel, chilled air at 4°C is made to flow counter-current to the direction of movement of the product.

24.1.2 Beverage Processing

The main items of manufacture are beers, wines and carbonated drinks.

Beer is manufactured from malted barley. Barley is first seeped in water at a temperature of 10 to 15°C for 2–3 days until germination, after which it is spread over, by draining off the water, for further germination. The grain is then malted in slowly revolving or agitated drums in which the moisture content is reduced to 4 per cent in a kiln. The malted dried sprouts are separated from the grain and stored for future use. These contain starch and diastase.

This malt is then crushed and suspended into water and made into a mash. Another portion is cooked separately with corn or rice. The two streams are mixed so that the temperature is about 70°C. After a period of rest for the mash, the diastase of the malt splits the starch of the grain into sugar maltose ($C_{12}H_{22}O_{11}$) and to some extent glucose ($C_6H_{12}O_6$) by enzymatic action. The *sweet wort* thus produced is transferred to a kettle where *hops* are added. After boiling, it is quickly cooled to 7 to 10°C and after adding yeast, is transferred to a cellar for fermentation under refrigeration. Here maltose is converted into ethyl alcohol and carbon dioxide as follows:

$$C_{12}\,H_{22}\,O_{11} + H_2O \rightarrow 4\,C_2H_5OH + 4\,CO_2$$

This is an exothermic reaction. The heat of reaction is 652 kJ/kg of maltose which must be continuously removed. The temperature in the fermentation tank is maintained at about 7–12°C by chilled wear or brine coils.

Refrigeration is also required in the *stock cellar* where the beer is retained from 3 weeks to 3 months for aging, and in the *finishing cellar* where it is polished by filtration and carbonation. The carbon dioxide produced during fermentation is liquified for use for carbonation later. Further, hops are stored at a temperature of –2 to 0°C and at a relative humidity of 50 to 60 per cent.

In wine-making also, refrigeration is required in three stages, viz.,

(i) for the control of temperature during fermentation,

(ii) for the removal of excess potassium tartrate by cold precipitation, and

(iii) for storage.

The quality of wine is greatly affected by the temperature of fermentation which should not exceed 29°C for red wine and 15°C for white wine. The minimum cooling requirement for fermentation is 31 kJ/kg of *must*. For grapes of high-sugar content, it can be up to 52 kJ/kg. Often chilled-water cooling coils are used in concrete fermentors.

For the cooling of wine for precipitating excess potassium bitartrate, the lowest possible temperatures above the freezing point of wine are used. The freezing points vary between –13 and – 6°C.

The quality of wine also depends on the temperature and length of storage. Some wines require as low a temperature as 7°C for storage.

In carbonated-drink manufacturing plants, refrigeration is required to cool the water, syrup or finished product.

24.1.3 Bakery Products

Bread is the most important bakery product. During the preparation of the mix for the dough, the heat of hydration and heat of friction of the churner, equivalent to the electrical input power, are generated. As the temperature of dough should not exceed 24–25°C, the best way to keep the temperature low is to add chilled water at 2–4°C for the preparation of the mix. The mixer is jacketed with a direct-expansion coil at a temperature of 0°C.

The doughs are then placed in troughs for fermentation for a period of $3\frac{1}{2}$ to 5 hours. During the process, the temperature may rise by 4–6°C. It is, therefore, desirable to keep them in a room at a temperature of 26°C.

Refrigeration is needed in the manufacture of other bakery products also in order to maintain uniformity of the product, prevent spoilage and for storage.

24.1.4 Meat Products

Ammonia-refrigeration systems are commonly used for the chilling of carcass and for holding them in rooms. The cooling is by chilled-brine spray or sprayed coil.

In the case of brine, a 20 per cent solution of NaCl is chilled to a temperature –7 to –9°C. The brine absorbs heat and moisture from the air. The diluted brine leaving the room must be strengthened regularly. The majority of installations these days use a sprayed-coil system with air being cooled with brine sprayed over a direct-expansion coil.

The average specific heat of carcass is 3.0 kJ/kg.K. The carcass should be reduced to a temperature of 1.5°C in a short time after slaughter. The room temperature could be as low as – 3.5°C. Refrigeration is also required for chilling water for washing the product.

More and more of meat products are, however, being preserved by freezing. The processing steps before freezing involve cutting, chipping, grinding, forming, pressing and slicing. The recommended storage temperature is – 29°C for a storage life of 12 months or more for beef, lamb, pork, etc. At – 23°C, a reasonable length of storage between 8 to 18 months is possible. The product quality is seriously impaired if the temperature goes up to – 9°C at any stage.

24.1.5 Poultry Products

Poultry meat is preserved both by chilling and freezing. Chilling requires a lower capital investment whereas freezing offers quality and flexibility of operation.

A major fraction of poultry meat is processed and transported by liquid chilling by the use of flake ice. Air chilling results in a considerable amount of dehydration at the surface. The approximate time required for chilling to 7°C is about 115 minutes for dressed birds, 50 minutes for eviscerated birds and 25 minutes for cut-up ones.

As for freezing, the use of air-blast tunnels operating at air temperatures of –29 to – 40°C and air velocities of 2.5 m/s or more is recommended. This ensures rapid freezing which is essential to obtain good quality and appearance.

Cooling is also required in poultry farms. A temperature higher than 29°C results in reduced egg production and weight and thinner shells. A temperature of 24°C is recommended for poultry-keeping.

24.1.6 Fishery Products

Refrigeration is required in many ways in the processing, preservation and transport of fresh and frozen fish and their products. Care starts from the stage of catching in fishing boats and trawlers.

Icing of fish is first done to chill it to a temperature of 0°C. The melt-water also helps to wash off the slime and bacteria. Often the fish are stored in tanks using *refrigerated sea water* at –1°C. It is found that *shrimp* when stored with refrigerated sea water is superior to that stored with ice. The storage life of chilled fish is normally 10–15 days.

Again, more and more of fish are being frozen these days. The valued varieties of fish that are frozen include salmon, shrimp, shell fish, etc. *Individual quick freezing* (IQF) techniques are preferred for these products. Quick freezing has many advantages. It prevents bacterial spoilage and ensures rapid handling, packaging and good appearance of the product, and also makes optimum utilization of the freezer space. Quick-freezing is achieved by using low refrigeration temperatures.

The common methods of freezing are described in Sec. 24.4.1. *Contact plate freezers* are suitable with large size products. Blocks of fish are frozen in multilayered plates in contact with primary or secondary refrigerant. For obtaining individual freezing of small-sized fish, such as shrimp, two methods are employed. *Air-blast freezers* use air at – 29 to – 40°C flowing at 2.5 m/s or more. *Immersion freezers* make use of liquid freon or even liquid nitrogen at atmospheric pressure. The freezing time depends on the water content and thermophysical properties of fish. It also depends on the refrigeration temperature. Frozen fish is normally stored at a temperature of –23°C or lower.

24.1.7 Fruits and Vegetables

Storage requirements of most fruits and vegetables are given in Chapter 16. It may be stated here that rapid pre-cooling after harvesting increases the storage life considerably. Further, *controlled atmosphere* (CA) storage greatly reduces the respiration rate and retards the deterioration of the product, thus increasing its storage life. It may otherwise raise the requirement of the storage temperature. A controlled atmosphere aims at maintaining the oxygen level at a low value of 3–5 per cent and the carbon dioxide level at a high concentration of 3–10 per cent.

A few vegetables, such as peas, beans, carrots and cauliflower are also found suitable for freezing. These are partially cooked before freezing. Quick freezing is not so critical in the case of vegetables. Nevertheless, airblast freezers are commonly used for individual quick freezing of these products. Indirect immersion freezing is also being resorted to on a small scale for packaged products.

Dehydro-freezing is another process that has been found successful in the preservation of some products, such as apples, apricots, peas, etc. This involves partial dehydration, and then freezing. The process retains the advantages of both, without the disadvantages of either. One clear advantage is the reduced refrigeration load during freezing. Another is the reduced weight of the product for transport. In simple dehydration, there is a marked deterioration in quality during the last stages of drying. This is eliminated in dehydro-freezing.

24.1.8 Mushroom Cultivation and Storage: A Case Study

Mushroom cultivation, processing, transportation and storage has increased manifold in the last few years. It provides an interesting case study. The stages involved are as follows:

Stage	*Area of Treatment*	*Inside Conditions*	*Method*	*Remarks*
I	Growth period (1 Month)	15.5°C DBT, 90% RH	A/C by chilled-water plant and steam injection	
II	Pre-cooling (45 min)	3°C DBT, 90% RH	D-X plant with Sprayed-coil	Rapid pre-cooling after harvesting
	(5 min)	–1°C DBT, 90% RH		Slight Cool-down from 3°C to –1°C
III	Cold Storage (1-2 hours)	–1°C DBT, 90% RH	D-X plant and product cooler	To maintain room condition at 90% RH, –1°C, dehumidified rise is only 3.5°C approx. Air flow, is, therefore, very high
IV	Packing Hall (5 – 10 min)	Normal A/C		
V	Cold Room (1-2 days)	–1°C DBT, 90% RH	D-X plant and product coolers	As in Stage III
VI	Transportation and storage: Conditions of –1°C DBT and 90% RH will have to be maintained throughout.			

24.1.9 Dairy Products

Most dairy plants pre-cool milk to 2.5°C before processing for the separation of cream and blending for producing standardized quality. There are hot as well as cold milk separators. In the latter, the separation process takes place at a temperature of 4°C. After separation and blending, the milk is sent for *pasteurization* which is accomplished by heating followed by the chilling of milk. Heating is done either by hot water or steam and holding the milk at 62°C for about 30 minutes. The milk is then cooled first by water from a cooling tower and then by chilled water. The rate of flow of chilled water should be such that its rise in temperature is not more than 5.5°C. When a direct-expansion system is used for cooling, the refrigerant flow is controlled by an automatic expansion valve so that the temperature does not drop below –2°C lest the milk may freeze. A water-chilling equipment normally consists of an *ice-bank type* water chiller consisting of an insulated tank with water, direct-expansion coil and pump. This system ensures a constant temperature of chilled water of 0 – 1°C. Also, it takes care of load fluctuations by allowing excess refrigeration to be stored in the form of ice on immersed coils.

Refrigeration is also required for cooling the cream before churning and the mechanical separation of butter. For the storage of butter, a temperature of 0 to 4°C is satisfactory. But for a storage for 6 months or more, a temperature of – 23 to – 18°C must be used. This is the same as for some frozen products. Thus butter can be kept in freezer compartment of refrigerators for storing beyond a period of six months or more.

Ice-cream manufacture is another industry that makes great use of refrigeration. The ice-cream freezer has to whip in a certain amount of air in the mix, simultaneously with freezing. The product is 0 to 100 per cent frozen between the temperature range of – 2.5 to – 55°C approximately.

24.1.10 Cold Storage Design

Cold storage is a low temperature refrigeration application. To keep the size of the refrigeration unit small, and power consumption low, it is necessary to construct the storage such that the cooling load is minimized.

Accordingly, 10 to 15 cm thick insulation is applied on the walls and ceiling. The common insulating materials are expanded polysterene (thermocole) and polyurethane foam (PUF). Their respective thermal conductivities are as follows:

Thermal Conductivity,	Thermocole	PUF
W/m°C	0.037	0.0173

It is seen that PUF has half the value of thermal conductivity of thermocole. PUF is, therefore, preferable to thermocole. Manufacturers, these days, are making sandwich PUF panels also which are used to assemble pre-fabricated cold stores. These panels are in the form of a sandwich consisting of two steel plates with core of a rigid PUF in between. Foaming is performed in a hydraulic press. PUF is injected in a mould under high pressure. The foaming agent is a CFC-free substance say, propane. The value of overall heat-transfer coefficient of a 10 cm PUF panel is:

$$U = 0.2 \text{ W/m}^2\text{°C}$$

Note *The flooring can also be provided with a PUF slab, and the requisite water-proofing and RCC finish will have to be done.*

The load calculation for a cold store is separated into the following main sources of heat for a given 24-hour period.

(i) Transmission load.
(ii) Air change (infiltration) load.
(iii) Miscellaneous loads like electric motors, lighting, people, etc.
(iv) Product load, sensible.
(v) Product load, respiration.

The average air changes per 24 hours for storage rooms above 0°C due to the door openings and infiltration may range from 10 for small cold room of 90 m^3 volume to 1 for large cold storage of about 900 m^3 volume.

Although most of the load in refrigerated room is due to transmission, air changes and product, the loads due to lighting, motors and people working should not be overlooked.

The product load consists of three separate components:

(a) *Sensible*: This is the amount of heat that must be removed to bring the product to storage temperature. It depends on specific heat of the product.
(b) *Latent*: This is the amount of heat that must be removed to freeze the product. It depends on latent heat of fusion. And it is applicable in the case of frozen food storage. Example: 2500 kJ/kg of heat removal required for the hardening of ice-cream.
(c) *Respiration*: Fresh fruits and vegetables are alive. They undergo changes even when stored at low temperatures releasing heat of respiration. Typical values of this sensible heat are given in Table 19.3.

Note *(i) When product load is calculated on per hour basis then a correction factor for taking into account pull-down time other than 24-hours must be applied as follows:*

$$\frac{\text{24 hours}}{\text{Pull-down time}}$$

(ii) A safety factor of 10% is also added to allow for any omissions and inaccuracies.

When load in kW has been determined, the-condensing unit is selected first to give a capacity greater than the cooling/freezing load. The fan-coil unit is then selected to balance the capacity of the condensing unit.

The nature of the product governs the choice of relative humidity to be maintained in the room. The relative humidity, in turn, determines the choice of refrigeration temperature. *Higher the humidity, higher the refrigeration temperature.* The table below gives recommended temperature differences between the air in the storage room and the saturated suction temperature of the refrigerant in the forced convection evaporator, depending on room relative humidity.

Product type	*RH*	*Temp. Difference*	*Remarks*
Fruits, Vegetables, flowers, etc.	90%	4 – 5°C	Minimum moisture loss
General storage, packed products.	80–85%	6°C	Products require lower RH
Pharmaceuticals, beer, potatoes, onions, tough-skin fruits, etc.	65–80%	7–9°C	Products require moderate RH
Prep./cutting rooms, candy, etc.	50–65%	9–12°C	Products not affected by RH

Note *In cold storage applications, it is recommended that load be divided into multiple units to provide for some capacity control, since the refrigeration system has to be selected for the worst (maximum load) conditions which may occur even for one day in the year.*

Important considerations in the selection and location of fan-coil units are:

(i) Air should be directed to move where there is heat gain, e.g., towards walls and ceiling.

(ii) Total supply of air is around 40-80 air changes per hour. This is so because the dehumidified rise is very small.

Example 24.1 ***Cold Storage Psychrometrics***

A small cold storage has the dimensions 8.5 m × 5 m × 2.5 m height. The design conditions are:

Outside:	*43°C DBT,*	*27°C WBT*
Inside:	*2°C DBT,*	*90% RH*

The transmission load is estimated as 1.5 kW. The fan motor and lighting load is assumed as 0.3 kW. The product loading is 2000 kg of fruits/vegetables every day. Assume 6 air changes per day equivalent of infiltration due to door openings. Determine:

(i) Refrigerating capacity required.

(ii) Approximate saturated suction temperature of refrigerant.

(iii) Number of air changes of supply air.

Solution Refer to Fig. 24.1.

As the inside condition is very close to saturation curve, all psychrometric properties are found by calculation rather than by using psychrometric chart. The same are as follows:

	DBT,°C	WBT,°C	RH,%	p'_v, mmHg	p_s, mmHg	p_v, mmHg	DPT, °C	ω, g/kg	h, kJ/kg
Outside	43	27	29	26.74	64.8	18.79		15.77	83.91
Inside	2		90		4.93	4.437	–0.39	3.653	11.16

Volume of space $V = 8.5 \times 5 \times 2.5 = 106.25\ \text{m}^3$

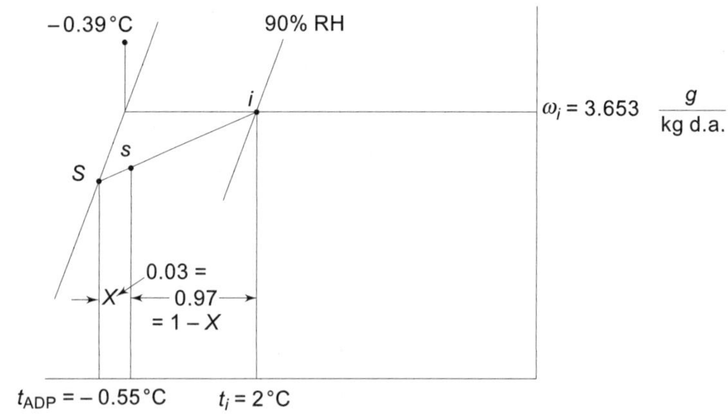

Fig. 24.1 Psychrometric points for Example 24.1

(i) Product load (specific heat of fruits/vegetables = 3.77 kJ/kg.K)

$$\dot{Q} = \frac{2000\,(3.77)\,(43-2)}{24 \times 3600} = 3.58 \text{ kW}$$

Infiltration air $\quad \dot{Q}_v = \dfrac{106.25 \times 6}{24 \times 60} = 0.44$ cmm

Air change load $\quad$ SH = 0.0204 (0.44) (43 – 2) = 0.37 kW

LH = 50 (0.44) (.01577 – .003653) = 0.267 kW

Total sensible and latent heats (with 10% safety factor)

TSH = (1.5 + 0.3 + 3.58 + 0.37) 1.1 = 6.33 kW

TLH = (0.267)1.1 = 0.29 kW

Grand total heat and sensible heat factor

GTH = 6.33 + 0.29 = 6.62 kW (1.9 TR)

$$\text{SHF} = \frac{6.33}{6.62} = 0.956$$

(ii) Apparatus dew point (by calculation)

$$\frac{.0204(2 - t_{\text{ADP}})}{.0204(2 - t_{\text{ADP}}) + 50(.003653 - \omega_{\text{ADP}})} = 0.956$$

By iteration

$$t_{\text{ADP}} = -0.55°\text{C}$$

Saturated suction temperature of refrigerant

$$t_0 = t_{\text{ADP}} - \Delta T \text{ for 90\% RH} \cong -0.55 - 5 = -5.55°\text{C}$$

(iii) The room DPT is – 0.39°C. And coil ADP is –0.55°C. The supply air condition lies between 2°C and –0.55°C. It has to be rather close to –0.55°C, otherwise the supply air quantity will be extremely large. Hence, one has to choose a coil of almost zero BPF. For the purpose, a large number of rows have to be taken, say, from 6 to 12. Let us choose a coil with BPF = 0.03.

Supply air temperature

$$\frac{t_i - t_s}{t_i - t_{ADP}} = 1 - X = 1 - 0.03 = 0.97$$

$$\Rightarrow \qquad t_s = 0.03\, t_i + 0.97\, t_{ADP}$$

$$= 0.03\,(2) + 0.97\,(-\,0.55) = -\,0.47°C$$

Supply air quantity $\quad \text{cmm}_s = \dfrac{\dot{Q}_s}{.0204\,(t_i - t_s)} = \dfrac{6.33}{.0204(2 + 0.47)} = 125.6$

Air change of supply air quantity per hour

$$= \frac{125.6 \times 60}{106.25} = 71$$

It is seen that number of air changes of supply air is very large in a cold storage application since the room condition at low temperature (2°C) and high humidity (90%) is very close to the saturation curve. It implies that dehumidified rise is very very small (2.47°C in this case) even after selecting a coil with large number of rows and very low BPF (0.03 in this case). Even then, the number of air changes of 71 is excessive. For vegetables/fruits, it should be between 30–60. Air changes in excess of 60 would lead to abnormal moisture loss and deterioration of the product. The supply air quantity, and hence the air changes can be reduced either by decreasing the load, or by decreasing the supply air temperature. The following measures are, therefore, recommended:

(i) Reduce transmission load by applying further insulation.
(ii) Reduce infiltration load by minimising door openings.
(iii) Limit daily loading of product, and hence cooling load.
(iv) Decrease coil BPF by increasing number of rows.

Another alternative is to have *jacketed cold store* if the supply air quantity is large. The supply air can be made to flow in a jacket all around the store along the walls. Thus, air velocity will not be large near the product to cause undesirable moisture loss.

Note *ADP is – 0.55°C, below zero. Hence, it is a frosted coil. Not only timely de-frosting is necessary, but the coil itself can not have more than 3 fins/cm, otherwise the flow will get blocked too soon due to frosting.*

24.2 TRANSPORT REFRIGERATION

Refrigerated transport is the main link of the cold food chain. Some salient features of it are discussed in this section.

24.2.1 Refrigerated Trucks and Trailers

These vehicles are refrigerated to maintain temperatures of either 1.5 to 4°C for cold foods or –18°C for frozen foods. The types of refrigeration systems used are given below.

Product Subcooling With the use of improved insulating materials, it is possible to drastically cut the transmission load of vehicles. In that case, use may be made of the storage capacity of the product itself for cold by subcooling it to as low a temperature as possible before transporting for short distances. Thus even after reaching the destination, the temperature of the product is below the temperature of the storage requirement. Examples are tankers for milk, orange juice, etc.

Using Water Ice The top of the product can be suitably iced. Again, it is a satisfactory method for short distances and for some products only. The refrigerating effect produced by the melting of ice is 335.4 kJ/kg.

Water Ice in Bunker with Forced Air Circulation Figure 24.2 shows the sketch of an ice bunker that is fitted in front of an insulated vehicle. Air is sucked over ice by the blower taking its drive from the engine. A $\frac{1}{2}$ HP blower will add a heat equivalent of 0.37 kW. A mixture of ice and salt can also be used for lower temperatures up to – 9°C.

Using Dry Ice Dry ice is used in many small retail trucks for the delivery of frozen food, such as ice-cream. The usual positioning of the dry-ice blocks is in the ceiling. The cooling is by natural convection. When forced convection is employed, dry ice may be placed in bunkers just like water ice. The refrigerating effect produced by the sublimation of dry ice, which takes place at a temperature of – 78.5°C, is 605.5 kJ/kg. In the use of dry ice, care must be taken to avoid burns due to low temperature, and suffocation due to lack of oxygen.

Using Liquid Nitrogen or Liquid Carbon Dioxide Spray In recent times, due to the expansion of the industry for the manufacture of oxygen, by liquefaction of atmospheric air, liquid nitrogen has become available almost as a by-product. It has, therefore, come in use for the transport of frozen food. It may, however, be noted that liquid nitrogen has a normal boiling point of –195.6°C. This is an extremely low temperature from the point of view of COP and refrigeration economy. Liquid nitrogen is, therefore, not recommended if it is not available as a by-product.

In both liquid nitrogen and liquid carbon dioxide cooling, the refrigerant is introduced into the vehicle through nozzles, the flow being controlled by thermostats and solenoid valves. The storage vessels for the refrigerants are kept within the refrigerated space. Whereas a liquid nitrogen vessel need not be under pressure, a liquid carbon dioxide vessel has to be under a pressure of about 5.2 bar as at atmospheric pressure it cannot exist as a liquid. Spray nozzles for liquid carbon dioxide should, therefore, be designed to avoid blockage due to snow formation during pressure drop.

Eutectic Plates with Station Charging Eutectic plates forming channels are placed all round the body of the vehicle. They contain an eutectic solution which can be frozen by a refrigerant flowing in a coil immersed in the solution. The coil is connected to a mechanical refrigeration system at the charging station. Cooling is produced by the evaporation of the primary refrigerant in the coil. When the solution is frozen, the truck loaded with the product for delivery is ready for departure.

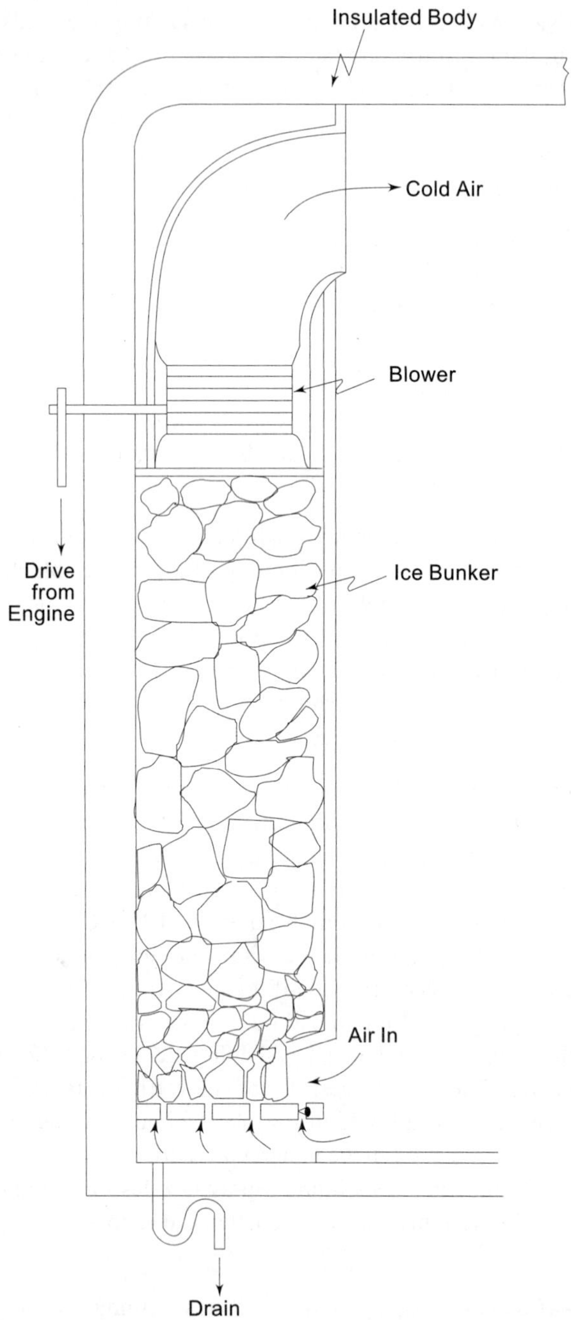

Fig. 24.2 Ice bunker in transport refrigeration

Eutectic plates are made in many sizes varying from 45 to 90 cm in width. A typical plate of 76 cm × 168 cm × 6.65 cm has a refrigerating capacity of 17935 kJ. Standard plates operate at temperatures of – 51 to – 3.5°C.

Eutectic Plates with Vehicle-Mounted Condensing Unit This system is becoming increasingly popular because of economy as well as reliability for the transport of frozen foods. Eutectic plates are used for maintaining the product at the required temperature. However, when the frozen eutectic has melted, the vehicle-mounted condensing unit can be started. Further, the condensing unit has an auxiliary drive mounted on the vehicle when it is in use and an electric motor drive for use at the charging station.

Mechanical Refrigeration with Independent Engine or Electric Motor The mechanical-refrigeration system mounted on the vehicle has an independent engine. Some are also equipped with an electric motor for stand-by operation at the charging station. These units are available in two ranges:

(i) 5.86 to 10.3 kW capacity at 1.5°C.

(ii) 1.76 to 5.28 kW capacity at – 18°C in a 40°C environment.

By far, this is one of the most common methods.

Mechanical Refrigeration Deriving Power from Vehicle Engine or Transmission There are many practical designs available. The refrigeration compressor may be either directly coupled to the engine, or may take off the power through transmission. In one design, the engine runs an alternator. The A.C. current is rectified and is used to run D.C. motors for compressor and fans. The refrigeration is, however, produced only when the vehicle engine is in operation.

24.2.2 Refrigerated Railway Cars

Most refrigerated railway cars use ice bunkers with water ice or ice-and-salt mixture. The recharging of ice is required at intermediate stations on the route. Nowadays, the mechanical-refrigeration system is being increasingly adopted. It is provided with an independent diesel-generator set so that refrigeration is independent of the car movement. The normal generator capacity is 20 kW.

24.2.3 Marine Refrigeration

A special feature of marine transport is the varying climate, ranging from extreme hot to extreme cold, through which the ship has to pass during the course of its journey. The insulation and fittings should be suitable both for warm and cold water routes. Thus a vapour barrier for moisture should be provided on both sides of the insulation. Corkboard continues to be favoured as an insulation because of its many desirable characteristics, such as structural strength, fire-resistant property, low permeability for moisture, etc.

Refrigeration is used in ships for cargo and stores of passenger ships. It is also used in trawlers for fishing.

In the case of cargo, the refrigeration system should be capable of providing any temperature between – 23.5 and 12.5°C. R 12 with reciprocating compressors is presently used as a refrigerant. R 22 is not recommended because of its problem of critical oil miscibility. Compound compression is employed. Parallel operation is not recommended with R 12 as there is a tendency for oil to migrate and flood one compressor while starving another. Accordingly, each evaporator has an independent compressor. This involves the use of a large number of compressors. The system can be made more economical by the use of parallel brine circuits in cooler coils and fewer condensing units. Capacity control is obtained by speed variation or cylinder cut-outs and cylinder unloading. Condensers are of the shell-and-tube type using sea water for cooling. Corrosion-resistant cupro-nickel material is, therefore, used for tubes and end covers. A receiver capable of holding 20 per cent more in addition to the whole charge is essential. The liquid line should emerge from both ends of the condenser, and later joining into one. This ensures continuous draining of the liquid during *roll* or *pitch* of the ship.

In addition to the cargo, ships must have their own stores. These stores have to be much bigger for passenger vessels. They also have a ventilated area for some of the items, such as onions and potatoes.

Fishing vessels use ice for short distances from the shore, but deep-sea fishing trawlers which remain away for months together must have mechanical refrigeration.

24.2.4 Refrigeration in Air Transport

Refrigerated air transport of some commodities can be justified on the basis of saving in the time and preservation of quality. In some passenger aircraft, the cargo compartments are cooled by the air-conditioning system itself. In cargo aircraft, the perishables are pre-cooled before shipment. For transit refrigeration, if necessary, refrigerant packages of water ice, dry ice or other substances are used.

24.3 COOLING AND HEATING OF FOODS

Several of the most interesting problems confronting the food technologist involve unsteady state heat-transfer. The analytical solutions to differential equation

$$\frac{\partial T}{\partial \tau} = \alpha \frac{\partial^2 T}{\partial x^2}$$

for certain unsteady state heat-transfer situations have been plotted in a series of graphs called *Gurnie-Lurie charts*.

Figure 24.3 represents this solution expressing the temperature as a function of time and position in an infinite slab being heated or cooled. Here:

T_∞ = Ambient temperature
T = Temperature at time τ
r = Distance from centre line to point under consideration
$r_m = \frac{1}{2}$ Thickness of infinite slab
f_g = Heat transfer coefficient at slab surface
k = Thermal conductivity of slab
ρ = Density of slab
C_p = Specific heat of slab
τ = Time of heating or cooling
$X = r/r_m$

The figure gives a plot of the dimensionless temperature θ against Fourier number Fo with Biot number Bi as parameter. The dimensionless temperature θ is defined by

$$\theta = \frac{T_\infty - T}{T - T_i}$$

where T_i is the initial uniform temperature of the slab. The dimensionless time, viz., the Fourier number Fo is defined as

$$\text{Fo} = \frac{k\tau}{\rho C_p r_m^2} = \frac{\alpha\tau}{r_m^2}$$

where $\alpha = k/\rho C_p$ is the thermal diffusivity of the material. The parameters X and Bi define respectively the dimensionless distance

$$\text{X} = \frac{r}{r_m}$$

and the ratio of the convective resistance to the conductive resistance, viz.,

$$\text{Bi} = \frac{f_g r_m}{k}$$

known as the Biot number.

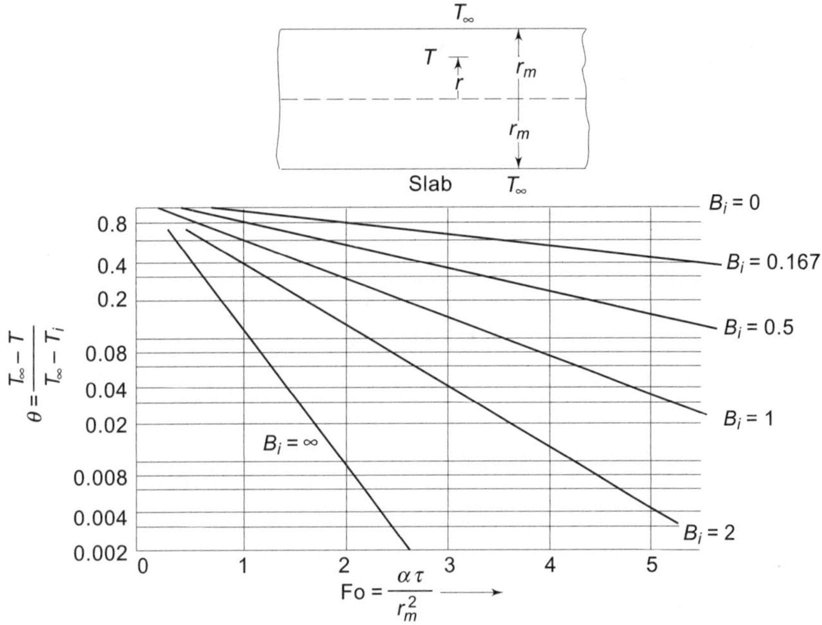

Fig. 24.3 Gurnie-lurie chart for cooling or heating of a slab

A numerical solution can also be found in the same manner as for heat transfer through a building wall. In this respect, a knowledge of the precise values of thermophysical properties of foods cannot be over-emphasized.

Example 24.2 *Determine the time required for the centre of a thin slab of meat to reach a temperature of 4°C. The temperature of the cooling medium is 2°C. The initial temperature of the product is 35°C. The thickness of the slab is 10 cm. The slab is cooled from both sides. The heat-transfer coefficient is 23 W/m²°C. The properties of meat are:*

$$k = 0.6 \text{ W/m°C}$$
$$C_p = 3.65 \text{ kJ/kg°C}$$
$$\rho = 1040 \text{ kg/m}^3$$

Solution We have

$$t_\infty = 2°\text{C}$$
$$t_i = 35°\text{C}$$
$$t = 4°\text{C}$$
$$r_m = 0.05 \text{ m}$$

At the centre

$$X = \frac{r}{r_m} = 0$$

$$\theta = (t_\infty - t)/(t_\infty - t_i) = \frac{2-4}{2-35} = 0.06$$

$$\frac{1}{\text{Bi}} = \frac{k}{f_g\, r_m} = \frac{0.6}{23\,(0.05)} = 0.52$$

Hence, from the Gurnie-Lurie chart

$$\text{Fo} = \frac{k\tau}{\rho C_p r_m^2} = 2.94$$

from which we get the required time as

$$\tau = \frac{(1040)\,(3.55\times 10^3)\,(0.05)^2\,(2.94)}{0.6} = 45227 \text{ s} = (12.6 \text{ hours})$$

Example 24.3 *A package of meat 1.83 m × 1.83 m × 7.8 cm, initially at 35°C, is brought in contact with a cooling medium at 3°C. Determine, by numerical method, the temperature distribution in the slab after an elapse of one hour. Given for meat:*

$$k = 0.606 \text{ W m}^{-1}\text{ K}^{-1}$$
$$\rho = 1200 \text{ kg m}^{-3}$$
$$C_p = 3.3 \text{ kJ kg}^{-1}\text{ K}^{-1}$$

Solution It is assumed that heat transfer through the faces with the 7.8 cm side is negligible compared to that transferred through the two large faces. Now the thermal diffusivity

$$\alpha = \frac{k}{\rho C_p}$$

$$= \frac{0.602}{(1200)\,(3.3\times 10^3)} = 0.15\times 10^{-6} \text{ m}^2\text{s}^{-1}$$

Choose a space element of Δx = 1.3 cm (0.013 m) and a value of M = 2, so that the time interval is

$$\Delta\tau = \frac{(\Delta x)^2}{\alpha M}$$

$$= \frac{(0.013)^2}{(0.15\times 10^{-6})\,(2)} = 563 \text{ s } (9.4 \text{ min or } 0.156 \text{ h})$$

The space grid is as shown in Fig. 24.4. Assuming a high heat-transfer coefficient at the surface, the temperature at the surface may be considered to reach 3°C instantaneously, or in two steps, viz., initially (35 + 3)/2 = 19°C and then 3°C after the first time interval of 563 s. The node 3 is at the centre line. The temperature at the various nodes after 1 $\Delta\tau$ are calculated as follows:

$$t'_1 = \frac{t_0 + t_2}{2} = \frac{19 + 35}{2} = 27°\text{C}$$

$$t'_2 = \frac{t_1 + t_3}{2} = \frac{35 + 35}{2} = 35°\text{C}$$

and so on. The complete set of calculations at various nodes and time intervals of 1 $\Delta\tau$, 2$\Delta\tau$, $\cdots$, etc., are presented in Table 24.1. The graphical procedure is also illustrated by broken lines in Fig. 24.4. Only half the thickness is shown in the figure because of symmetry. The results for temperature distribution are presented up to 7 $\Delta\tau$ as it is 341 s more than one hour.

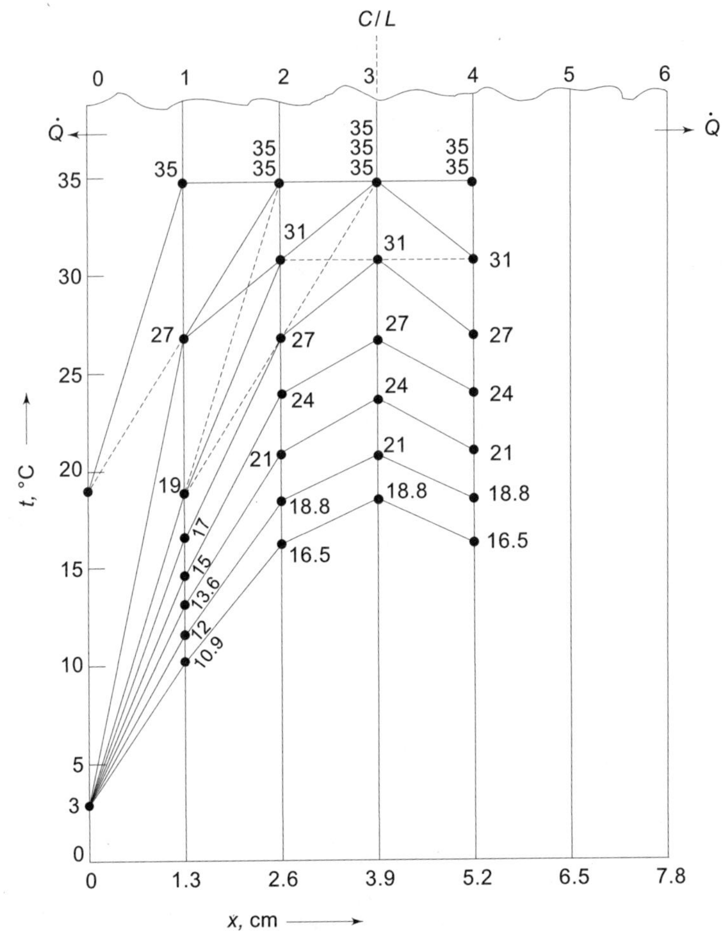

Fig. 24.4 Schmidt plot for Example 24.2

Table 24.1 Calculations of temperatures for Example 24.3

Number of $\Delta\tau$	*Time s*	t_0	t_1	t_2	t_3	t_4	t_5	t_6
0	0	19	35	35	35	35	35	19
1	563	3	27	35	35	35	27	3
2	1129	3	19	31	35	31	19	3
3	1689	3	17	27	31	27	17	3
4	2252	3	15	24	27	24	15	3
5	2815	3	13.5	21	24	21	13.5	3
6	3378	3	12	18.8	21	18.8	12	3
7	3941	3	10.9	16.5	18.8	16.5	10.9	3

24.4 FREEZING OF FOODS

Since properly frozen foods have a much longer storage life than cold-stored foods and also retain better quality, the emphasis is on the development of frozen-food storages and techniques of quick freezing. Among the common products that are preserved by freezing are fish, meat, vegetables, such as beans, sprouts, carrots, cauliflower, peas and spinach, various tissues and organs, etc. Frozen shrimp is one of the most valuable foreign-currency earners for India.

Freezing requires the removal of heat, and the product from which heat is removed falls in temperature in the manner shown in Fig. 24.5. The first stage involves sensible cooling. The temperature drops rapidly to the freezing temperature below 0°C. The second stage involves the extraction of latent heat of fusion during which the temperature changes very little. This stage is known as the period of *thermal arrest.* When about 75 per cent of the water is converted into ice, the temperature begins to fall again in the third stage in which the remaining water is frozen. The unfrozen water has a higher concentration of salts and other compounds, the effect of which is to depress the freezing point. Even at temperatures as low as – 30°C, a proportion of water in the product tissues remains in the unfrozen state.

In order to produce a good product, the freezing must be accomplished quickly so that no change takes place in the product salt concentrations. Changes take place during freezing as the salts tend to separate out. The product is altered from its *native* state. The process is termed *denaturation.* Slow freezing is, therefore, detrimental as it means a longer time being spent in the zone of denaturation activity which is in the region of –1 to – 2°C.

Table 24.2 gives the freezing rates for classification of the type of freezing. Thus, freezing rates vary from 2 to 1000 mm/h of the product thickness.

Table 24.2 Freezing rates[4]

Thickness Frozen mm/h	*Type*
2	Slow bulk freezing in a blast-room
5 to 30	Quick freezing in an air blast or plate freezer
50 to 100	Rapid freezing of small products
100 to 1000	Ultra rapid freezing in liquefied gases, such as liquid nitrogen, liquid carbon dioxide or liquid freon

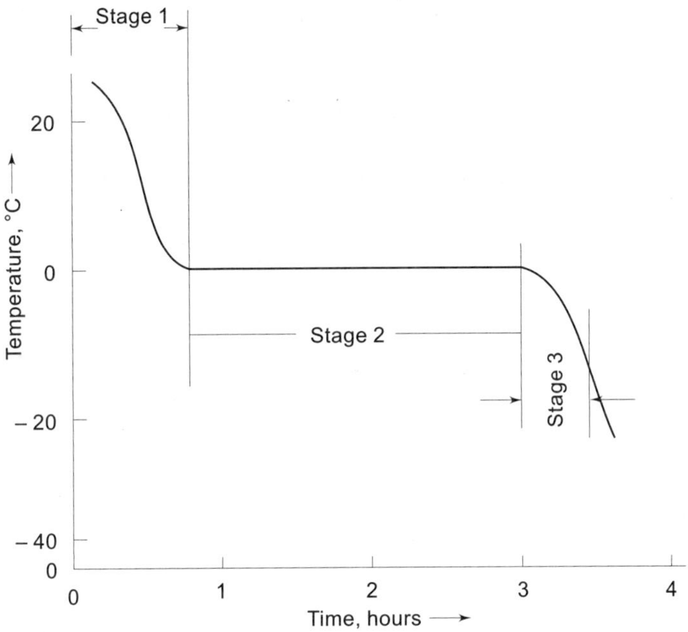

Fig 24.5 Temperature-time graph of a product during freezing

24.4.1 Types of Freezers

There are three basic types of freezers:

(i) *Air-blast freezers* in which cold air is blown over the product.

(ii) *Contact or plate freezers* in which the product is brought in direct contact with a refrigerated surface.

(iii) *Immersion or spray freezers* in which the product is immersed in or sprayed with a liquid refrigerant.

Air blast and contact-plate freezers are the two most common types used in industry. Contact-plate freezers do not have the versatility of the air-blast freezers. The former can only be used to freeze regularly shaped blocks. The latter can be used for a variety of irregular shaped as well as small sized products such as peas, french fries, shrimp, etc., which results in *individual quick freezing* (IQF) of each piece. Air blast freezers can be freezing tunnel type, belt freezers or fluidized bed type freezers. Immersion freezers use *liquid freezants* at atmospheric pressure, the corresponding temperatures being – 196°C with N_2 and –30°C with R 12.

To achieve quick freezing in air blast freezers, the air flow rates should be high. Figure 24.6. Shows the influence of air velocity on freezing time. A high air velocity will mean shorter freezing time but more fan power. An optimum air speed of 5 m/s is recommended for most purposes. In *continuous* air-blast freezers, even higher speeds, of the order of 10 to 15 m/s, could also be economically employed as that would result in smaller freezer length and lower cost due to reduced freezing time.

Figure 24.7 shows the schematic diagram of an air-blast freezer in which a moving belt carries the product. The belt can also be replaced by trolleys and the product stacked in multilayered trays. The air is handled by axial-flow fans developing a pressure of about 14 cm H_2O gauge. The air is in cross-flow to the product. Likewise, there may be three or more fan-coil units along the length of the belt. To minimise dehydration, the temperature of the air is kept close to that of the product. The temperature rise of air is thus limited to about 2°C only. Accordingly, the fan handles a large quantity of air. Also, the cooling coil is 12–16 rows deep to achieve BPF ≅ 0, in order to obtain ADP as high as possible for higher COP in this low refrigeration temperature application. As the frost formation is more in the front rows, their spacing must be larger so as to maintain a certain free-flow area with continuous operation without defrosting for at least 8 hours and without affecting the fan-air flow.

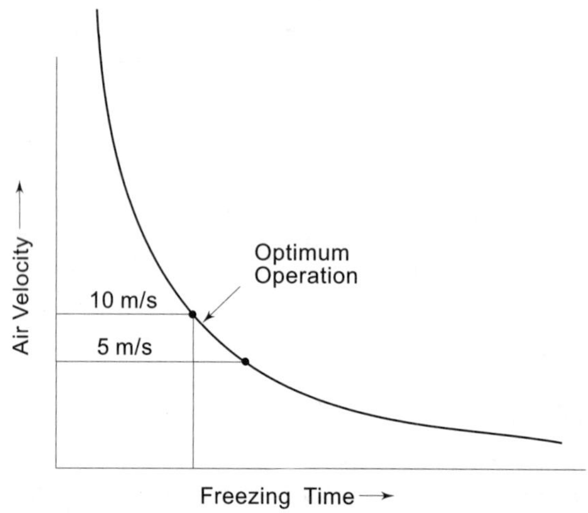

Fig. 24.6 Influence of air velocity on freezing time

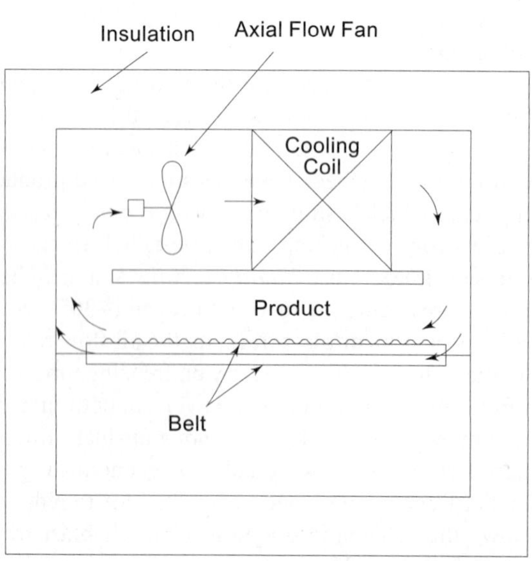

Fig. 24.7 A continuous belt air-blast freezer with cross-flow of air

In well-designed freezers, the fan power is of the order of 25 to 30 per cent of the refrigeration load. If the air-flow rate is very large, the fan power may be greater. As a result of small dehumidified rise and high air flow the temperature rise of air over the product will be very small. An average rise in air temperature of about 1.5–2°C is permissible.

Air-blast freezers with *batch operation* use trolleys or shelves for loading the product. Those with *continuous operation* make use of conveyors or belts. The latter are used only if the product can be frozen quickly, the freezing time not exceeding 30 minutes, otherwise the length of the freezer will become very large. To reduce the length, sometimes double, triple or spiral belts are used. The speed of the belt should be kept variable to increase its versatility to accommodate the varying freezing time requirements of various products and their thicknesses.

Example 24.4 *200 kg/h of fish are to be frozen, the product requiring a residence time of 20 minutes, in an air-blast freezer with a 1 m wide stainless steel belt. The possible belt loading density is 3 kg/m². Find the length of the freezer.*

Solution

$$\text{Belt loading} = 200 \times \frac{20}{60} = 66.6 \text{ kg}$$

$$\text{Belt loading per unit length of belt} = \frac{3}{1} = 3 \text{ kg/m}$$

$$\text{Length of belt required} = \frac{66.6}{3} = 22.2 \text{ m}$$

Note *To allow for the loading and unloading of fish outside the freezer, a total length of about 25 m would be required.*

24.4.2 Refrigeration Load in Freezers

Example 24.5 illustrates the procedure for calculating the refrigeration load in freezers. In addition to the product load, other loads that need to be considered are the fan heat, pump heat, heat leakage through insulation, heat load due to internal lighting, containers and conveyers, heat load due to defrosting, infiltration load, etc.

Example 24.5 *32,400 kg/day of cold fish is to be frozen to – 30°C in 10 cm thick blocks each weighing 45 kg. The secondary refrigerant temperature is – 40°C. The evaporating refrigerant temperature is – 47°C. The fish enters at 30°C. The freezing cycle time may be taken as 4 hours. Given:*

Specific heat of thawed fish = 3.77 kJ/kg°C

Latent heat of fusion of fish = 251.2 kJ/kg

Specific heat of frozen fish = 1.67 kJ/kg°C

Calculate the number of blocks frozen per cycle and refrigeration duty of the plant for 18 hours running time.

Solution Number of blocks/day $= \dfrac{32400}{45} = 720$ blocks/day

Number of freezing cycles/day $= \dfrac{24}{4} = 6$

Number of blocks frozen per cycle $= \dfrac{720}{6} = 120$ blocks/cycle

Fish loading $\dot{m} = \dfrac{32400}{(24)(3600)} = 0.375$ kg/s

Sensible cooling to a freezing temperature of 0°C

$$\dot{Q}_1 = (0.375)(3.77)(30) = 42.4 \text{ kW}$$

Latent heat of fusion

$$\dot{Q}_2 = (0.375)(251.2) = 94.2 \text{ kW}$$

Sensible cooling to – 30°C

$$\dot{Q}_3 = (0.375)(1.67)(30) = 18.8 \text{ kW}$$

Total product load

$$\dot{Q}_p = 42.4 + 94.2 + 18.8 = 155.4 \text{ kW}$$

Product load for 18 hours running time

$$Q = 155.4 \times \frac{24}{18} = 207.2 \text{ kW}$$

24.4.3 Calculation of Freezing Time

The time taken to lower the temperature of the product from its initial temperature to a given temperature at its thermal centre, is called the *freezing time*. The final temperature is generally the intended storage temperature of the product. For example, in the case of fish, the recommended storage temperature is – 30°C. To ensure quick freezing, the freezer temperature must be below this temperature. It is desired that after freezing, the temperature of the thermal centre should be reduced to at least – 20°C so that the average temperature of the fish is near the storage temperature of – 30°C. The freezing time will, therefore, be the time required to reduce the thermal centre from its initial temperature to – 20°C. The *residence time* of the product in the freezer is, therefore, equal to its freezing time.

A precise calculation of the freezing time for irregular-shaped product is difficult. But for uniformly-shaped products such as rectangular blocks, suitable relations have been proposed. However, they often do not take into account the pre-cooling from the initial temperature to the final temperature. They assume that the product has been chilled initially, and that all extraction of heat is at the freezing temperature. The presence of other factors such as packing, etc., may also give erroneous results. Nevertheless, calculations by the finite-difference method, using a computer, give very good results.

A solution for the calculation of temperature distribution throughout a mass, in which a change of state is occurring, has been proposed by Neuman[10] for freeze-drying. The same can be applied for freezing. The equations expressing the temperature as a function of the time and position in an infinite slab with a change of state are as follows:

$$t_1 = \frac{t_s}{\text{erf}\ \lambda}\ \text{erf}\ \frac{x}{2\,(\alpha_1\ \tau)^{1/2}} \tag{24.1}$$

$$t_2 = t_i - \frac{(t_i - t_s)}{\text{erfc}\left(\dfrac{\alpha_1}{\alpha_2}\right)^{1/2}}\ \text{erfc}\ \frac{x}{2\,(\alpha_2\ \tau)^{1/2}} \tag{24.2}$$

where

t_1 = Temperature in the frozen section

t_2 = Temperature in the thawed section

t_s = Temperature at which change of state occurs

t_i = Initial temperature of the thawed material

x = Distance from surface of slab

α_1 = Thermal diffusivity of the frozen material

α_2 = Thermal diffusivity of the thawed material

τ = Time

λ = Factor determined from the following equation

$$\frac{e^{-\lambda^2}}{\text{erf}\ \lambda} - \left(\frac{k_2}{k_1}\right)\left(\frac{\alpha_1}{\alpha_2}\right)^{1/2} \frac{(t_i - t_s)\exp[-(\alpha_1/\alpha_2)\,\lambda^2]}{t_s\ \text{erfc}\,[\lambda\,(\alpha_1/\alpha_2)^{1/2}]} = \frac{\lambda\,\Delta h \pi^{1/2}}{C_1\ t_s} \tag{24.3}$$

where

Δh = Latent heat of fusion

k_1, C_1 = Thermal conductivity and specific heat of the frozen material

k_2, C_2 = Thermal conductivity and specific heat of the thawed material.

The conditions assumed in these equations are as follows:

(i) All the temperatures are excess temperatures measured above the surface temperature.

(ii) Region $x > 0$ is initially at the constant temperature t_i with surface $x = 0$ maintained at zero for $\tau = 0$.

(iii) Also t_2 approaches t_i as x approaches infinity and $t_1 = 0$ at $x = 0$.

In other words, it is assumed that the surface comes to the temperature of the freezing medium immediately and the surface temperature is at 0°. If the surface temperature is other than zero on any temperature scale, an excess temperature scale must be employed so that on the excess-temperature scale the surface temperature is at 0°. This involves subtracting or adding a constant to each temperature involved, in order to obtain the excess temperature to be used in the foregoing equations.

Example 24.6 *A package of meat 5.1 cm thick is being frozen in a plate or contact freezer. The meat is initially at 4.4°C. It has a moisture content of 75 per cent and its freezing-point average is – 5°C. The meat is being frozen from both sides and the refrigerant is at – 29°C. Determine the time for the meat to pass through the freezing zone. The thermophysical properties of thawed (2) and frozen (1) materials are as follows:*

$$k_2 = 0.571 \text{ W m}^{-1}\,°\text{C}^{-1}$$
$$k_1 = 1.04 \text{ W m}^{-1}\,°\text{C}^{-1}$$
$$\rho_2 = 1057 \text{ kg m}^{-3}$$
$$\rho_1 = 961 \text{ kg m}^{-3}$$
$$C_2 = \text{specific heat of thawed meat} = 0.75\,(4.18) + 0.25\,(1.38) = 3.488 \text{ kJ kg}^{-1}\,°\text{C}^{-1}$$
$$C_1 = \text{specific heat of frozen meat} = 0.75\,(2.09) + 0.25\,(1.38) = 1.918 \text{ kJ kg}^{-1}\,°\text{C}^{-1}$$

Also calculate the time for the meat centre to reach – 17.8°C.

Note *The specific heat of dried material is 1.38. The specific heats of water and ice are taken as 4.18 and 2.08 respectively.*

Solution In this case, the surface temperature is at – 29°C. In order to make the surface-excess temperature zero, we have a temperature scale

$$t = t' - (-29) = t' + 29$$

where t' is the actual temperature and t is the excess temperature. Thus, on the excess-temperature scale

$$t_i = 4.4 + 29 = 33.4$$
$$t_s = -5 + 29 = 24$$

Also, taking the latent heat of fusion of water as 335.5 kJ/kg, we have for the latent heat of fusion of meat (with 75% water content)

$$\Delta h_{\text{meat}} = 0.75\,(335.5) = 252 \text{ kJ/kg}$$

Thermal diffusivities

$$\alpha_2 = \frac{0.571}{(1057)\,(3.488 \times 10^3)} = 1.549 \times 10^{-7} \text{ m}^2 \text{ s}^{-1}$$

$$\alpha_1 = \frac{1.04}{(961)\,(1.918 \times 10^3)} = 5.64 \times 10^{-7} \text{ m}^2 \text{ s}^{-1}$$

In order to calculate λ we solve Eq. (24.3) by trial and error as shown in Table 24.3.

Table 24.3 Solving Eq. (24.3) by Iteration

Assumed λ	*LHS*	*RHS*
0.1	24.1	0.975
0.5	3.78	4.85
0.4	5.28	3.88

By interpolation for LHS = RHS, we get

$$\lambda = 0.46$$

For the whole meat to pass through the excess freezing point temperature t_s at 24°C we must have at the centre line

$$t_1 = t_s$$

Hence from Eq. (24.1)

$$\lambda = \frac{x}{2\,(\alpha_1\,\tau)^{1/2}}$$

Since freezing takes place from both sides

$$x = 5.1/2 = 2.55 \text{ cm } (0.0255 \text{ m})$$

Hence

$$\alpha_1 \tau = \left(\frac{x}{2\lambda}\right)^2$$

$$= \left(\frac{0.0255}{2 \times 0.46}\right)^2 = 7.673 \times 10^{-4}$$

so that

$$\tau = \frac{7.673 \times 10^{-4}}{5.64 \times 10^{-7}} = 1360 \text{ s } (0.38 \text{ h})$$

The time for the meat centre to pass through –17.8°C is obtained by putting

$$t_1 = -17.8 + 29 = 11.2$$

in Eq. (24.1). Thus

$$11.2 = \frac{24}{\text{erf } 0.46} \text{erf } \frac{0.0255}{2(5.64 \times 10^{-7}\,\tau)^{1/2}}$$

whence

$$\tau = 6660 \text{ s } (1.85 \text{ h}).$$

Note *In practice, the contact plates would not assume the freezing-medium temperature. In addition, the thermal resistance of the metal wall is to be considered. Accordingly, the freezing time will actually be more than the calculated value.*

24.4.4 Freezing in Air

When freezing in contact-plate freezers, it may be assumed that the surface of the object immediately attains the freezing-medium temperature since the heat-transfer coefficient is high. However, attention is drawn to the note written above. But, when freezing in air, as in an air-blast freezer, the surface does not immediately attain the freezing-medium temperature. The film of air offers an additional convective resistance to the heat transfer. In order to calculate the freezing time under these conditions employing Eq. (24.1), it is necessary to convert the heat-transfer coefficient into an equivalent thickness of the starting material, x_e, approximated by

$$x_e = \frac{k_2}{f_g} \tag{24.4}$$

where k_2 is the thermal conductivity of the thawed material and f_g is the heat-transfer coefficient of air. If freezing occurs from both sides, x_e is added to the half-thickness of the material.

Example 24.7 *Considering the package of meat in the preceding example, how long would it take for the package to pass through the freezing zone, and for the centre to reach –17.8°C, if the package were frozen in an air-blast freezer rather than the contact-plate freezer. Air is at –28.9°C. The initial temperature is 4.4°C. Assume the heat-transfer coefficient in the freezer to be 17 W m^{-2} °C^{-1}.*

Solution Considering Eq. (24.3), λ will remain the same as in the preceding example, viz. $\lambda = 0.46$.

Additional equivalent thickness of thawed material

$$x_e = \frac{0.571}{17} = 0.0336 \text{ m}$$

Hence, the resulting half-thickness

$$x = 0.02544 + 0.0336 = 0.0591 \text{ m}$$

From Eq. (24.1), the time for the package to pass through the freezing zone is given by

$$\tau = \frac{1}{\alpha_1}\left(\frac{x}{2\lambda}\right)^2$$

$$= \frac{1}{5.64 \times 10^{-7}}\left(\frac{0.0591}{2 \times 0.46}\right) = 7317 \text{ s } (2.03 \text{ h})$$

The time for the centre to reach – 17.8°C is given by

$$\frac{11.2}{24} \operatorname{erf} 0.46 = \operatorname{erf} \frac{0.0336}{2(5.64 \times 10^{-7}\tau)^{1/2}}$$

whence

$$\tau = 12115 \text{ s } (3.37 \text{ h}).$$

24.4.5 Modified Plank's Equation for Calculation of Freezing Time

Plank's formula[1] for calculating the freezing time τ has been widely used. It is given below.

$$\tau = \frac{\Delta h}{v\,\Delta t}\left[P\frac{\Delta x}{f_g} + R\frac{(\Delta x)^2}{k}\right] \tag{24.5}$$

where

Δh = Heat extracted between the initial temperature of freezing surface t_s and the final temperature

v = Specific volume of fish

Δt = Temperature difference between the initial freezing point t_s and the temperature of the freezing medium t_f

Δx = Thickness of product

f_g = Heat-transfer coefficient of air

k = Thermal conductivity of the frozen material

$$P = \frac{1}{6} \text{ for a sphere, } \frac{1}{4} \text{ for a cylinder and } \frac{1}{2} \text{ for a slab}$$

$$R = \frac{1}{24} \text{ for a sphere, } \frac{1}{6} \text{ for a cylinder and } \frac{1}{6} \text{ for a slab.}$$

For a rectangular block of $(\Delta x) \times (\beta_1 \Delta x) \times (\beta_2 \Delta x)$, the values of P and R can be found from Fig. 24.8 which is due to Ede[3].

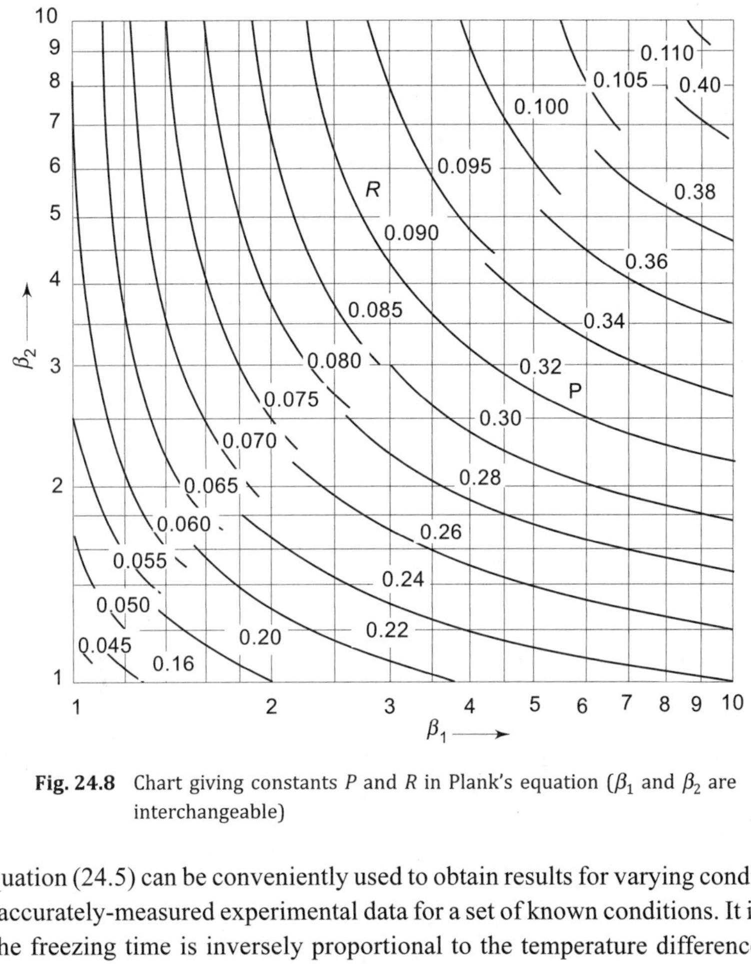

Fig. 24.8 Chart giving constants P and R in Plank's equation (β_1 and β_2 are interchangeable)

Equation (24.5) can be conveniently used to obtain results for varying conditions, from accurately-measured experimental data for a set of known conditions. It is seen that the freezing time is inversely proportional to the temperature difference, and nearly proportional to the square of the product thickness.

Nagaoka et al.[9] have modified Plank's equation to include the effect of difference in the initial and final temperatures as follows:

$$\tau = [1 + 0.008\,(t_i - t_s)] \left(\frac{\Delta h}{v\,\Delta t}\right) \left[P \frac{\Delta x}{f_g} + R \frac{(\Delta x)^2}{k}\right] \tag{24.6}$$

Here, t_s is the freezing surface temperature.

Example 24.8 *(a) The freezing time for a 100 mm thick block of fish is found to be 200 min (3 h 20 min) when the freezing-medium temperature is –35°C. What is the freezing time if the freezing medium temperature is changed to – 25°C? The initial freezing point of fish is –1°C.*

(b) What is the freezing time if the block thickness is reduced to 75 mm?

All other conditions remain same in both cases.

Solution (a) Effective temperature difference

$$\Delta t = t_s - t_f = -1 - (-35) = 34°C$$

Changed effective temperature

$$\Delta t' = -1 - (-25) = 24°C$$

Required freezing time

$$\tau' = \tau \left(\frac{\Delta t}{\Delta t'}\right)$$

$$= 200 \left(\frac{34}{24}\right) = 283 \text{ min (4 h 43 min)}$$

(b) Required freezing time with reduced thickness

$$\tau' = \tau \left(\frac{\Delta x'}{\Delta x}\right)^2$$

$$= 200 \left(\frac{75}{100}\right)^2 = 112.5 \text{ min (1 h 52.5 min)}$$

24.4.6 Ice Manufacture

The time required to freeze a given thickness of ice in cans immersed in a brine tank is determined firstly by the temperature of the ice-freezing surface, in this case, that of brine. Most raw waters, treated or untreated, cannot be frozen without cracking at a temperature below –12°C. The brine velocity also has a marked influence on the freezing time and should not exceed above 105 mpm. The brine movement is caused by the hydraulic gradient best designed for about 3.2 to 3.8 cm in the tank length. The brine level in the tank should permit the submergence of water in the ice cans below the brine level.

A uniform rate of harvesting exactly according to the time schedule is necessary for maximum yield. The rate of ice-freezing drops rapidly as the ice layer becomes thicker and the thermal resistance to heat transfer increases due to the lower thermal conductivity of ice as compared to that of water. A 28 cm × 56 cm can holding 145 kg of water in –11°C brine will make 127 kg of ice in 24 hours and consume an additional 14 hours to freezing the remaining 18 kg.

When brine agitation is moderate, 4.5 to 7.5 mpm, the total freezing time for US standard cans, 28 cm thick, may be expressed by Eqs (24.7) and (24.8)

$$\tau = \frac{0.6 b^2}{(-t)} \tag{24.7}$$

where

τ = Total freezing time for the block, hours

b = Thickness of ice block (short side), cm

t = Temperature of brine, °C

$$\tau = \frac{mn}{1000} \times 24 \text{ or } n = \frac{41.67\,\tau}{m} \tag{24.8}$$

where

n = Number of cans per metric ton of ice produced in 24 hours

m = Mass of ice block, kg

Equation (24.9) is obtained by combining Eqs (24.7) and (24.8) and sovling for the brine temperature

$$t = -\frac{25b^2}{mn} \tag{24.9}$$

Thus, for a given plant, the daily output is a function of the brine temperature only.

The number of cans per ton is the usual unit for rating tank capacity, which also establishes the necessary brine temperature for a given daily output.

For block thickness above 30 cm or below 25 cm, Eq. (24.9) should not be used, because of the following reasons:

(i) It assumes that all resistance to heat flow is dependent upon ice thickness, whereas the resistance actually consists of
   (a) ice resistance (thickness), and
   (b) surface resistance (from brine to ice), which is independent of the ice thickness.

(ii) It does not take into consideration the flow of heat across the narrow sides of the can, which becomes increasingly important as the shape of the can approaches a square.

The value of the heat-transfer coefficient f varies with the brine velocity. For very high brine velocities, $f = 285$ W.m^{-2} K^{-1}. For ordinary ice tanks with relatively slow-moving brine, $f = 140$ W.m^{-2} K^{-1}.

24.5 FREEZE DRYING

Freeze drying is a successful process of liquid separation from a product in a frozen state, achieved by sublimation under vacuum. The sublimation serves to obtain a product that retains even its volatile components and initial quality, and the vacuum is used to maintain the physical state as frozen and to direct the vapour flow. At present, the main problems to the application of the process in general are; (a) the relatively high cost of the freeze-dried product by sublimation-dehydration due to high vacuum, (b) very low temperature refrigeration to pre-freeze the product and then to condense the sublimated vapour in a freeze-condenser, (c) the complicated operational control and (d) the long duration of the freeze-drying time. Characteristically, however, the cost is not the determining factor for the manufacture of certain products including blood plasma, life-saving drugs such as gamma-globulin, and high value food products such as mushrooms, shrimp, prawns, etc.

Figure 24.9 shows a model of the freeze-drying process with radiant heating. The profile shows the temperature distribution.

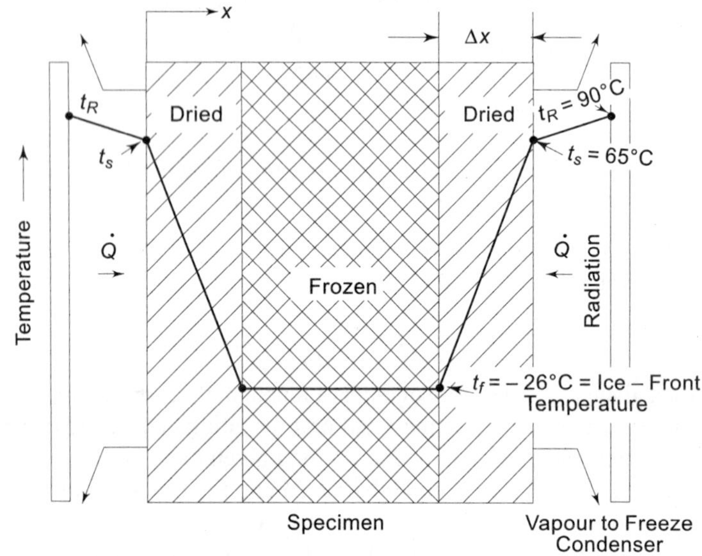

Fig. 24.9 Temperature distribution during freeze-drying of a specimen with radiant heating (Example 24.9)

The dehydration takes place in a vacuum chamber. The pressure in the chamber must be maintained below the triple-point pressure, normally below 5 mm Hg, otherwise the product will begin to thaw.

The rate of dehydration is governed by either one of the following two factors:

(i) The rate of heat transfer from the heat source to the ice front.

(ii) The rate of vapour diffusion from the ice front to the freeze condenser.

Usually, the rate of dehydration is controlled by the transfer of heat which may be supplied by either a radiant-heat source, or by contact heating platens sandwiching the product, but having pores to permit the sublimated vapours to escape. In such a case, there is an obvious similarity between freezing and freezing-drying. Neuman and Plank's solutions for freezing can, therefore, be used to calculate freeze-drying time as well. A simple heat and mass-transfer model is, however, presented below. Numerical methods can also be successfully used.

It must be mentioned that the heat input is limited by the maximum allowable temperature of the dried material at the surface so that the product is not damaged by overheating.

24.5.1 Heat and Mass Transfer through the Dried Material[2]

Let the surface temperature of the material be maintained at the maximum allowable value, say t_s, and the ice-front temperature of the frozen layer be t_f. We also assume that the partial pressure of the water vapour in the drying chamber is maintained constant at p_s. The thickness of the dried layer at any instant of time is Δx. Under these conditions, the heat-transfer rate is given by

$$\dot{Q} = \frac{k\,A}{\Delta x}\,(t_s - t_f) \tag{24.10}$$

where k is the thermal conductivity of the dried material.

The mass-transfer rate is given by

$$\dot{m} = \frac{k_d\,A}{\Delta x}\,(p_f - p_s) \tag{24.11}$$

where k_d is the diffusion coefficient or *permeability* of the dried material, and p_f is the saturation pressure at t_f.

Assuming that the rate of heat transfer is equal to the rate of mass transfer multiplied by the latent heat of sublimation Δh_s, we obtain by energy balance from Eqs (24.10) and (24.11).

$$p_f = p_s + \frac{k}{k_d\,\Delta h_s}\,(t_s - t_f) \tag{24.12}$$

If p_s, t_s, k, k_d and Δh_s are taken as constants, we have a linear relation between p_f and t_f as shown in Fig. 24.10. The figure also shows the thermodynamic pressure-temperature relationship for water. The point of intersection f gives the ice-front temperature t_f and pressure p_f which will be constant for the assumed conditions.

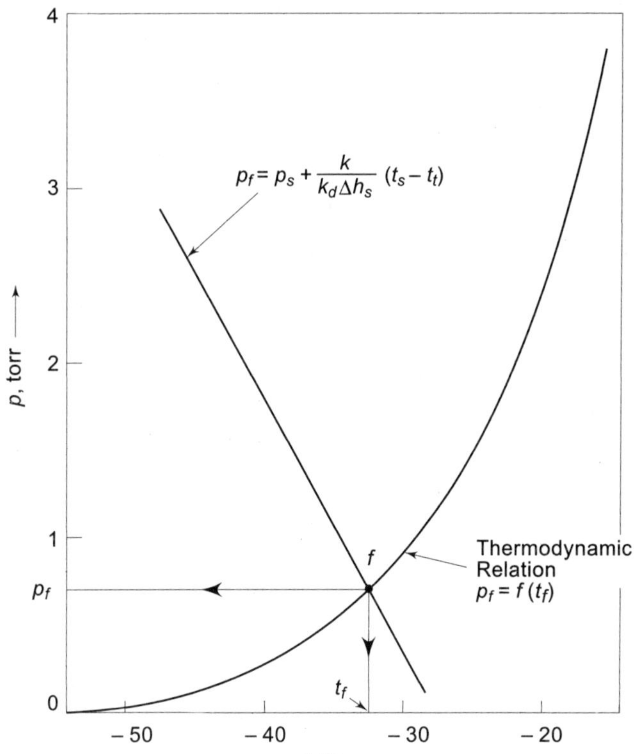

Fig. 24.10 Graphical determination of ice front temperature and vapour pressure during freeze drying

Under these conditions, the freeze-drying time can be determined by writing Eq. (24.10) in the differential form at any instant of time τ when the ice front is at a distance x from the surface, viz.,

$$\dot{Q}\,\mathrm{d}\tau = kA(t_s - t_f)\frac{\mathrm{d}\tau}{\mathrm{d}x} = \rho Ax\,(w_i - w_o)\,(\Delta\, h_s) \tag{24.13}$$

where

ρ = Density of solids in the material, viz., the dried material

w_i = Initial moisture content per unit mass of solids

w_o = Final moisture content per unit mass of solids.

The integration of Eq. (24.13) with $(t_s - t_f)$ constant, yields for the drying time

$$\tau = \frac{(\Delta x)^2\,\rho\,(w_i - w_o)\,\Delta h_s}{8k\,(t_s - t_f)} \tag{24.14}$$

It may be stated that the largest source of error in the above method is in generalizing the surface temperature and assuming it to be constant. In actual practice, as the dried-layer thickness grows, the surface temperature has to be increased to overcome the increased thermal resistance. Similarly, the ice-front temperature is also affected by the pressure drop through the dried layer. Numerical techniques were used by Trifonova, Arora and Sharma[12,17] to obtain a more accurate result for the drying time. As an approximate method, the Neuman solution for freezing can also be applied to freeze drying.

Example 24.9 *(a) A 1.125 cm layer of frozen shrimp is freeze dried by contact with heating platens. The chamber is maintained at 0.43 mm Hg pressure. The heating platens are raised linearly from the freezing temperature to the maximum allowable temperature of 65°C. The product contains 80 per cent water. The density and thermal conductivity of the dried product are respectively 335 kg/m³ and 0.015 W/mK. Determine the freeze-drying time. The heating is done from both sides.*

(b) If the heating is done by radiation, determine the maximum permissible temperature of the radiating surface.

Solution (a) The sublimation temperature from properties of water

$$t_f = -\,26°\text{C (at 0.43 mm Hg pressure)}$$

Heat of sublimation at –26°C

$$\Delta h_s = 2462 \text{ kJ/kg}$$

Average heating surface temperature

$$t_s = \frac{65 + (-26)}{2} = 19.5°\text{C}$$

Initial moisture constant

$$w_i = \frac{0.8}{0.2} = 4 \text{ kg water/kg solids}$$

Freeze-drying time

$$\tau = \frac{(\Delta x)^2 \rho (w_i - w_o) \Delta h_s}{8k (t_s - t_f)}$$

$$= \frac{(0.01125/2)^2(335)(4-0)(2462\times10^3)}{8(0.015)(19.5-(-26))}$$

$$= 19{,}118 \text{ s } (5.3 \text{ h})$$

(b) Let T_R be the absolute temperature of the radiating surface. Then by energy balance between heat transfer to the surface by radiation and heat transfer by conduction between the surface and the ice front, we have

$$\frac{k}{\Delta x}(t_s - t_f) = 5.669 \left[\left(\frac{T_R}{100}\right)^4 - \left(\frac{T_S}{100}\right)^4\right]$$

$$\frac{(0.015\,(65+26)}{(0.01125/2)} = 5.669 \left[\left(\frac{T_R}{100}\right)^4 - \left(\frac{273+65}{100}\right)^4\right]$$

whence

$$T_R = 363 \text{ K}$$
$$t_R = 90°\text{C}$$

24.5.2 Freeze Drying of Yoghurt: Influence of Concentration of Milk Solids

Yoghurt is a common ingredient of diet because of its nutritional and therapeutic value. Accordingly, efforts should be made to incorporate it in the manufacturing of products, such as yoghurt-flavoured baby foods, by freeze drying process.

The high cost of freeze drying can be reduced, particularly in the case of yoghurt, by reducing the initial moisture content of the product. Sharma, Arora and Mital[11] conducted a study to determine the effect of *milk solids* in yoghurt on its freeze drying rate, and the quality of the reconstituted product.

For yoghurt preparation, the milk was heated to 85°C for 30 min. The total solids in milk were increased by adding non-fat dry milk (NFDM) (moisture 5%, fat 2.7%, protein 35%) at 2, 5 and 8% (w/v) levels. The resultant solids in the yoghurt were 14.7% without addition of NFDM, and 16.3%, 18.8% and 20.4% with the addition of NFDM. The milk was then cooled and inoculated with the *cultures Streptococcus thermophilous*-YHS and *Lactobacillus delbrueckii* subsp. *bulgraicus*-YHL in the ratio of 1:1 at 3% *v*/*v*.

The inoculated milk was incubated at 42°C for 3.5 hours. The yoghurt thus prepared was kept at 4 ± 1°C until transferred to a *product tray* for freezing at –24°C.

The experimental set-up is shown in Fig. 24.11. The compressor maintained a temperature of – 54 ± 3°C in the freeze condenser. A temperature controller automatically adjusted the heater-platen temperature to 45 ± 3°C. The schematic of the product tray is shown in Fig. 24.12. The pressure in the vacuum chamber ranged from 1.3 m bar at the beginning to 0.52 m bar at the end of drying run. After drying, the samples were stored in glass-stoppered containers at 4°C.

There are two methods of heating:

(i) Contact-plate heating.
(ii) Radiant heating.

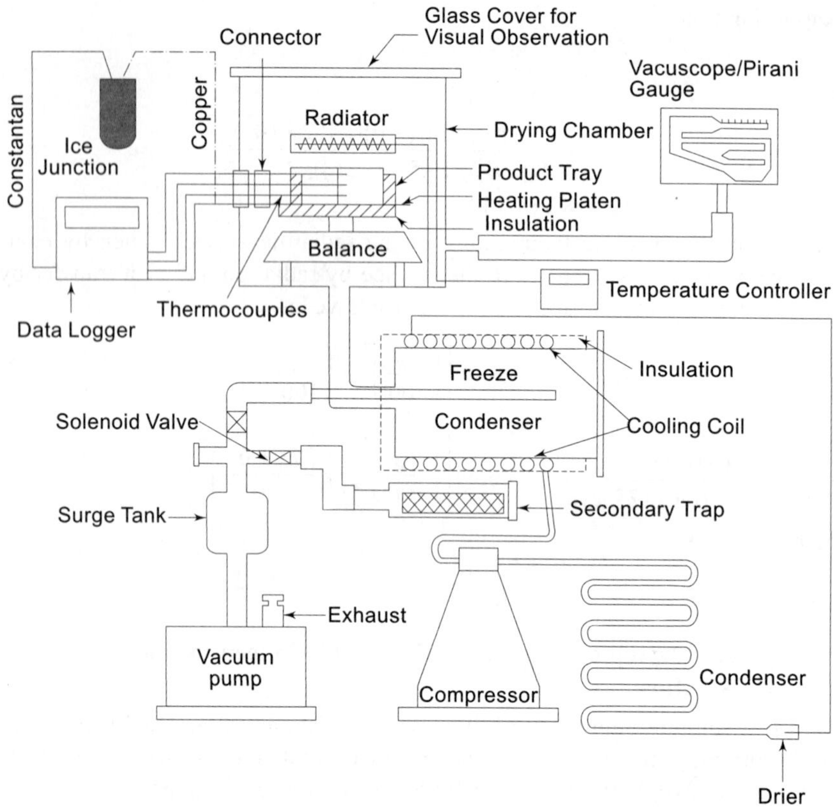

Fig. 24.11 Experimental set-up of freeze drying apparatus

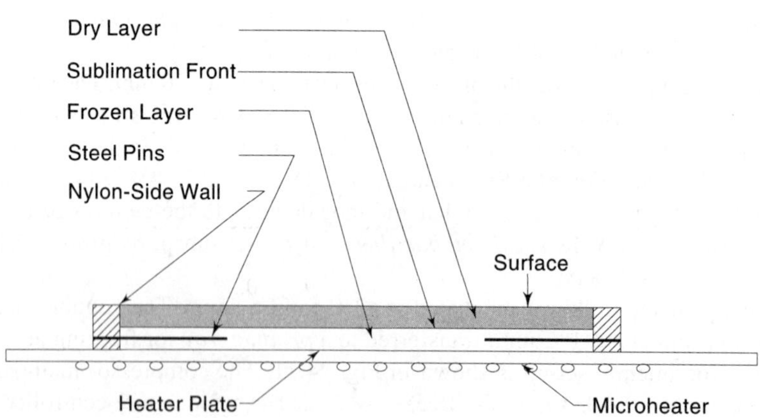

Fig. 24.12 Schematic of product tray

Sharma et al. used contact-plate heating in this work. Samples were placed symmetrically on the aluminium platen fitted with a microheater. To promote good contact between the product bottom and the tray, steel pins fixed radially through the side walls and extending into the product were used, as shown in Fig. 24.12.

The weight loss as a function of drying time for the yoghurt samples with different total solids (14.7–20.4%) is shown in Fig. 24.13. After drying for 4.5 h, moisture determinations showed that the moisture loss of the samples had become negligible.

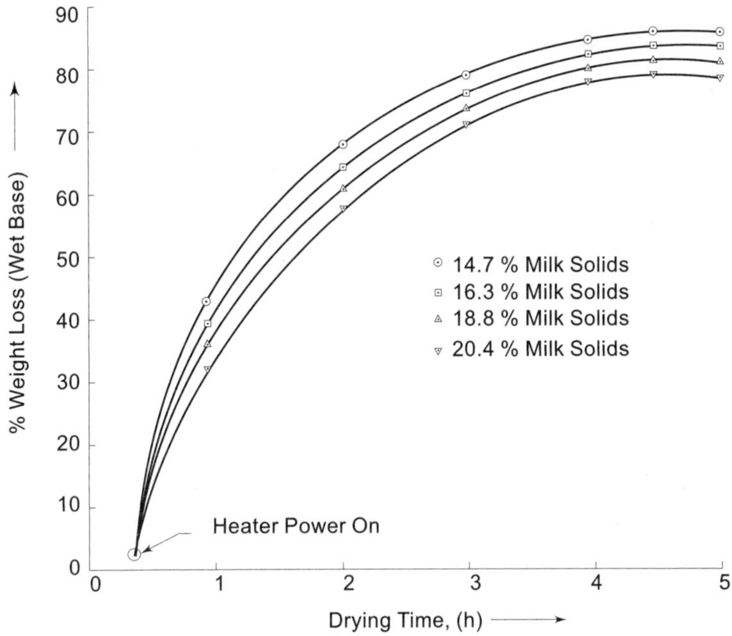

Fig. 24.13 Weight loss in different yoghurt samples during freeze-drying at constant heater-platen temperature

Therefore, the end point of drying was taken as 4.5 h. Due to the large thermal contact resistance between the heater-platen and the product-tray interface, the temperature at the product bottom-tray interface varied from – 19 to – 3°C during sublimation.

The drying rate curves are shown in Fig. 24.14. And the production rate, and the drying time per unit output of freeze-dried yoghurt are shown in Fig. 24.15. With the increase in solids concentration in yoghurt from 14.7 to 20.4%, the production rate of freeze-dried yoghurt increased from 0.2196 to 0.3067 kg/m^2 h, and the drying time per unit output decreased from 1.265 to 0.8658 h/g (Table 24.4). This shows an increase of 40.6% in the production rate, and a reduction of 28.9% in the drying time per unit output.

Therefore, productivity increases more than that accounted for by the lower amount of water needed to be evaporated. With the increase in solids concentration, the eutectic temperature decreases and partial melting of the frozen core during drying may occur. Secondly, yoghurt being a cultured product, the *sensory characteristics* of the freeze-dried product will be a limiting factor in deciding the solids concentration in yoghurt.

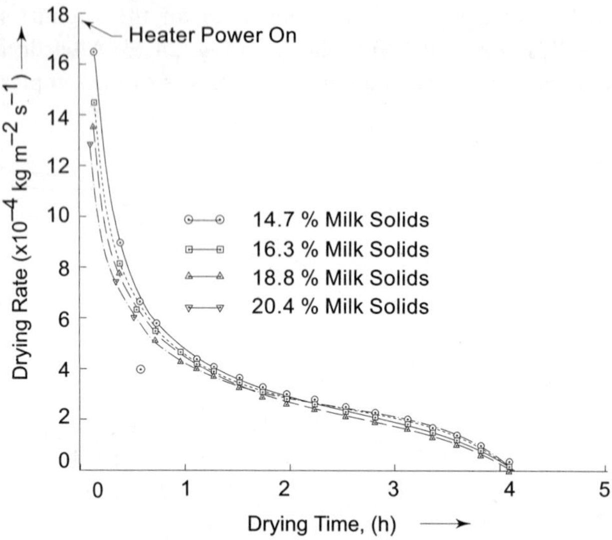

Fig. 24.14 Drying rate curves for yoghurt containing different concentration of milk solids

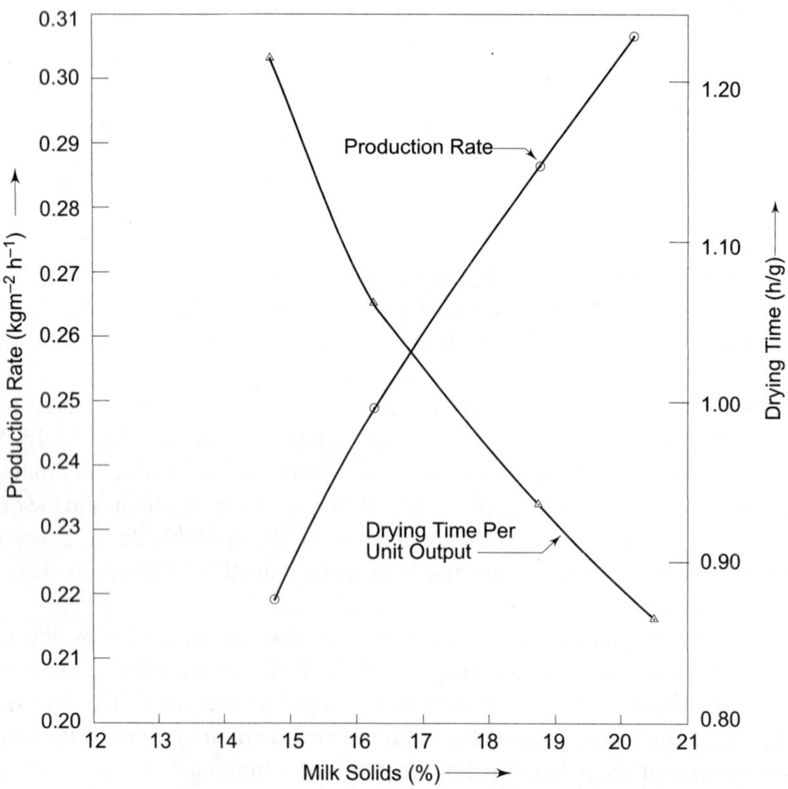

Fig. 24.15 Production rate and drying time per unit output of freeze dried yoghurt as influenced by concentration of milk solids

Table 24.4 Influence of concentration of milk solids in yoghurt on the production rate and the drying time

Concentration of milk solids	*Drying time/ unit output*	*Reduction in drying time*	*Production rate*
%	h/g	%	kg/m^2h
14.7	1.2650	–	0.2186
16.3	1.0718	15.27	0.2495
18.8	0.9380	25.85	0.2851
20.4	0.8658	31.56	0.3067

Table 24.5 shows the proximate composition of freeze-dried yoghurt samples prepared using different concentration of solids in milk. Freeze-dried yoghurt prepared from whole cow milk contained 26.3% protein, 29.8% fat and 5.02% ash. As expected, the addition of NFDM at 2, 5 and 8% levels enhanced protein and ash content of the products, whereas their fat contents decreased. The concentration of lactic acid in the freeze-dried product increased from 5.8 to 6.95, and the average bulk density increased from 0.29 to 0.42 g/cm^3.

Table 24.5 Proximate composition of freeze-dried yoghurt at different concentrations of milk solids

Conc. of solids in yoghurt (%)	*Freeze-dried yoghurt*					
	Moisture (%)	*Protein (%)*	*Fat (%)*	*Lactic acid (%)*	*Ash (%)*	*Bulk density* (g/cm^3)
14.7	4.3	26.3	29.8	5.80	5.02	0.29
16.3	4.5	26.8	28.0	6.52	5.38	0.32
18.8	5.3	27.6	23.1	6.61	5.51	0.40
20.4	5.9	30.5	20.2	6.95	6.33	0.42

The acceptability of freeze-dried yoghurt samples was determined after reconstitution. The acceptability scores of the samples prepared from yoghurt containing 14.7, 16.3, 18.8 and 20.4% total solids were 5.08, 6.5, 7.58 and 6.41 respectively. *Tukey's test analysis* also showed that the samples containing 16.3 to 20.4% solids were superior to the sample containing 14.7% solids.

It was also observed that increasing the level of solids in milk upto 18.8% improves the acceptability of the product. A further increase in concentration of solids lowers the acceptability.

Note *Computer simulation of freeze drying behaviour can be obtained using very simple equation/s, while simultaneously determining, experimentally, the parameters and properties for the process and the product as done by Sharma and Arora for freeze drying yoghurt and given in Table 24.6.*

Table 24.6 Parameter values for yoghurt freeze drying

	Parameter	*Value*
Specific heat, dried layer	C_{pd} (cal/g–K)	0.46
Specific heat, external space (pores)	C_{pe} (cal/g–K)	0.25
Specific heat, water vapour	C_{pwv} (cal/g–K)	0.45
Thickness of sample	L (cm)	1.125
Distance from radiator	Z (cm)	1.5
Porosity	σ	0.75
Enthalpy of sublimation	ΔH (cal/g)	670.0
Density, frozen layer	ρ_f (g/cm^3)	0.92
Density, dried layer	ρ_d (g/cm^3)	0.38
Density, external space	ρ_e (g/cm^3)	0.00005
Thermal Cond., dried layer	λ_d (cal/cm-s-K)	.00015
Thermal Cond., frozen layer	λ_f (cal/cm-s-K)	.005
Thermal, Cond., External space	λ_e (cal/cm-s-K)	.00005
Thermal diffusivity, dried layer	α_d (cm^2/s)	.00085812
Thermal diffusivity, external space	α_e (cm^2/s)	4.0
Emissivity, radiation	ε_R	0.9
Emissivity, surface	ε_S	0.8

24.5.3 Comparison of Radiant and Contact-plate Heating

Sharma and Arora[13] showed that the production rate is not affected significantly by product thickness in radiant heating. Hence large product thicknesses can be used to neutralize and reduce the *dead time* between loadings.

In contrast, the production rate improves with decreasing product thickness in contact heating through the bottom of frozen layer.

In radiant heating, the production rate improves slightly with an increase in chamber pressure from 0.01 to 0.5 mm Hg.

In contact heating through frozen region, and for simultaneous heat transfer through both frozen and dried region, low operating vacuum chamber pressures are essential to ensure that the temperature of the frozen core is as low as possible, and hence allow the use of higher platen temperature to shorten the drying time.

For all modes of heat transfer and thicknesses, it is only the *melt constraint* that needs to be controlled during sublimation, and *burn constraint* after drying.

24.6 HEAT DRYING OF FOODS[2]

When the moisture content of foods is reduced to 5 to 10 per cent, the microorganisms become sufficiently inactive. Dehydration by heating is thus the most economical method of food preservation. However, the delicate characteristics of foods require skillful operation of the process so as to preserve its nutrition value and flavour. Many industrial processes also require the drying of solids.

A few accepted terms in drying terminology may now be given. *Moisture content w*, as different from *W* the total moisture in the product, is the percentage of moisture by weight of the bone-dried solid. The *equilibrium-moisture content* is that to which a material can be dried under given conditions of air temperature and humidity. The process of drying essentially takes place in two stages as shown in Fig. 24.16. These are:

(i) Constant-rate period
(ii) Falling-rate period.

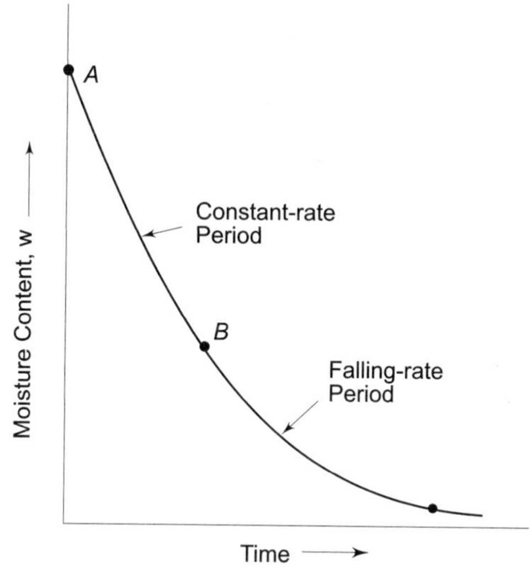

Fig. 24.16 Drying process

During the constant-rate period, the surface of the product remains wet and its temperature and evaporation rate remain constant. The primary resistance to heat and mass transfer is at the surface. When the surface appears to develop dry patches, the rate of drying begins to decrease, as the internal resistance of the material to *liquid diffusion* to the surface becomes significant. When the constant-rate period ends and the falling-rate period begins (at *B* in Fig. 24.16), the moisture content of the material is termed as the *critical moisture content.*

24.6.1 Drying During Constant-Rate Period

Drying during the constant-rate period is equivalent to the evaporation of moisture from a wetted surface. Thus, equations derived in Chapter 20 are applicable. In this case, the surface remains at a constant temperature. If heat transfer is only by convection from the air, the surface approaches the wet-bulb temperature of air. But if heat transfer takes place by conduction and radiation in addition to convection, the surface will attain a temperature in between the dry and wet bulb temperatures of air.

The equilibrium between heat and mass transfer can be expressed as follows

$$\frac{dW}{d\tau} = \frac{f_t\, A\,(t - t_S)}{\Delta h_S} = k_\omega A(\omega_S - \omega) \tag{24.15}$$

where W = Mass of moisture in product

$\frac{dW}{d\tau}$ = Drying rate

f_t = Total heat-transfer coefficient including conduction, convection and radiation

Δh_S = Latent heat of vaporization at the wetted surface temperature t_S

and the other terms have their usual meaning as given in Chapter 20. When $f_t = f_g$, i.e., equal to the convective coefficient, the surface temperature t_S approaches the wet bulb temperature of air.

The convective heat-transfer coefficient for air blowing parallel to a surface is given by

$$f_g = 14.327\ G^{0.8} \tag{24.16}$$

and for air blowing perpendicular to a surface is given by

$$f_g = 24.199\ G^{0.37}$$

where G is the mass velocity of air in kg/s m^2.

Example 24.10 *To control the temperature and drying rate of bread, a humidified baking oven is maintained at 166°C dry bulb and 60°C wet bulb temperatures. The walls of the oven radiate heat as black bodies. The air is circulated in the oven by natural convection so that the heat-transfer coefficient is given by*

$$f_g = 1.974\ (t - t_S)^{0.25} \tag{24.17}$$

The oven is 0.3 m cube and the bread-pan dimensions are 0.3 m × 0.1 m × 0.1 m. Determine the drying rate during the constant-rate drying period. Neglect drying from the base of the pan.

Solution The heat transfer is by radiation and natural convection. The heat-transfer coefficients depend on the surface temperature. The solution is, therefore, found by iteration.

Vapour pressure of water at 60°C

$$p_v' = 19910 \text{ N/m}^2$$

Vapour pressure of water in air steam, from the Carrier equation

$$p_v = p_v' - \frac{(p - p_v')(t - t')(1.8)}{2800 - 1.3\,(1.8t + 32)}$$

$$= 19910 - \frac{(101325 - 19910)(166 - 60)(1.8)}{2800 - 1.3\,(1.8 \times 166 + 32)}$$

$$= 13355 \text{ N/m}^2$$

Specific humidity of air

$$\omega = 0.622\ \frac{p_v}{p - p_v}$$

$$= 0.622\ \frac{13355}{(101325 - 13355)} = 0.094 \text{ kg/kg}$$

Assume the surface temperature of bread to be above wet bulb temperature, say 65°C

Convective coefficient

$$f_g = 1.974\,(166 - 65)^{0.25} = 7.09 \text{ W m}^{-2}\text{ K}^{-1}$$

Radiative coefficient

$$f_R = \frac{5.669 \times 10^{-8}\,(T^4 - T_S^4)}{(T - T_S)}$$

$$= \frac{5.669\,(4.39^4 - 3.38^4)}{(166 - 65)} = 13.52 \text{ W m}^{-2}\text{ K}^{-1}$$

Total heat-transfer coefficient

$$f_t = f_g + f_R = 7.09 + 13.52 = 20.61 \text{ W m}^{-2}\text{ K}^{-1}$$

$$k_\omega = \frac{f_g}{C_p} = \frac{7.09}{1.0216 \times 10^3} = 6.94 \times 10^{-3} \text{ kg s}^{-1}\text{ m}^{-2}$$

Vapour pressure at the surface temperature of 65°C

$$p_{v_S} = 25{,}000 \text{ N/m}^2$$

Specific humidity at surface

$$\omega_S = 0.622\,\frac{25{,}000}{101325 - 25{,}000} = 0.2037 \text{ kg/kg}$$

Latent heat of vaporization at 65°C

$$\Delta h_S = 2346 \text{ kJ/kg}$$

Checking the assumed value of t_S from Eq. (24.15)

$$\frac{f_t\,(t - t_S)}{\Delta h_S} = k_\omega\,(\omega_S - \omega)$$

$$\frac{20.61\,(166 - 65)}{2346 \times 10^3} \neq 6.94 \times 10^{-3}\,(0.2037 - 0.094)$$

$$\text{LHS} = 0.8873 \times 10^{-3} \neq 0.7613 \times 10^{-3} = \text{RHS}$$

Assume $t_S = 67°\text{C}$. Then,

$$f_g = 6.227 \text{ Wm}^{-2}\text{ K}^{-1}$$

$$f_R = 13.615 \text{ W m}^{-2}\text{ K}^{-1}$$

$$f_t = 19.842 \text{ Wm}^{-2}\text{ K}^{-1}$$

$$k_\omega = 6.095 \times 10^{-3} \text{ kg s}^{-1}\text{ m}^{-2}$$

$$p_{v_S} = 27{,}375 \text{ N/m}^2$$

$$\omega_S = 0.23$$

$$\Delta h_S = 2341 \text{ kJ/kg}$$

From Eq. (24.15), $$\frac{19.842\,(166 - 67)}{2341 \times 10^3} = 6.095 \times 10^{-3}\,(0.23 - 0.094)$$

$$\text{LHS} = 0.839 \times 10^{-3} \approx 0.829 \times 10^{-3} = \text{RHS}$$

The assumed value of a surface temperature of 67°C is quite satisfactory. Hence the initial drying rate

$$\frac{dW}{d\tau} = \frac{f_t\, A\,(t - t_S)}{\Delta h_S}$$

$$= \frac{(19.842)\,[0.3 \times 0.1 + 2 \times 0.3\,(0.3 + 0.1)]\,(166 - 67)}{2341 \times 10^3}$$

$$= 0.227 \times 10^{-3} \text{ kg/s}$$

24.6.2 Drying During Falling-rate Period

Drying during the falling-rate period begins when the product reaches a critical moisture-content level. At this stage, the rate of migration of water from the interior to the surface becomes less than the rate of evaporation of water from the surface. Accordingly, during this period, the rate of drying is controlled by the rate of migration of water, i.e., by the internal resistance of the product to the mass transfer. If the rate of heat transfer is high, the vaporization may take place within the material. In this case, the conductive heat-transfer resistance of the material will also be significant. The surface temperature, as also the temperature at the interface between the dried and wet material, will be controlled by the relative thermal and mass-transfer resistances.

There are different mechanisms of migration of water from the interior. Some are summarised by Görling[7] as follows:

(i) Liquid movement due to capillary forces,
(ii) Diffusion of liquid due to concentration difference,
(iii) Surface diffusion in the liquid absorbed at the boundary of the solid,
(iv) Water-vapour diffusion by partial-pressure difference.

It is possible that one or more of the above mechanisms, and/or other mechanisms of moisture migration are present simultaneously. In most food products, however, the moisture migration occurs by diffusion.

It is found that the diffusion coefficient depends on moisture content. It is constant down to about 15 per cent moisture content. At this stage, a second-diffusion coefficient, which is about one-fifth the first-diffusion coefficient controls the process.

The governing equation for the transfer of moisture by diffusion is analogous to the transient heat-conduction equation. Thus the methods of solution are also the same. Accordingly, the Gurnie-Lurie[1] charts may also be used for estimating the moisture transfer.

In the analogy, the liquid diffusivity D is analogous to the thermal diffusivity $\alpha = k/\rho\, C_p$ or $\alpha\rho\, C_p$ is analogous to the thermal conductivity k. The mass-transfer equivalence of the heat-transfer coefficient is determined from the mass-balance equation

$$f_g \frac{(t - t_S)}{\Delta h_S} = k_\omega\,(\omega_S - \omega) \tag{24.18}$$

whence

$$f_g = k_\omega \frac{\Delta h_S\,(\omega_S - \omega)}{(t - t_S)} \tag{24.19}$$

which is, therefore, equivalent to $k_\omega C_p$ since $\omega_S - \omega$ is dimensionless and $\Delta h_S/(t - t_S)$ is equivalent to C_p. Thus in Gurnie-Lurie charts, the surface-resistance ratio

$$\frac{k}{f_g r_m} = \frac{1}{\text{Bi}} \text{ is analogous to } \frac{D\rho C_p}{k_\omega C_p r_m} \text{ or to } \frac{D\rho}{k_\omega r_m}$$

in which k has been replaced by $D\rho C_p$ and f_g by $k_\omega C_p$. The density term represents the density of bone-dried material. The drying rate is then given by

$$\frac{\mathrm{d}W}{\mathrm{d}\tau} = k_\omega A(\omega_S - \omega) \tag{24.21}$$

where W is the moisture content of the material at any time τ and $\mathrm{d}W/\mathrm{d}\tau$ represents the moisture gradient.

In the use of Gurnie-Lurie charts, the Fourier number $\alpha\tau//r_\text{m}^2$ may be replaced by $D\tau/r_m^2$. As for the dimensionless temperature, it may be replaced by the dimensionless moisture content W/W_i. For this purpose, the initial-moisture content should be taken as equal to the critical-moisture content.

Example 24.11 *297 g of a food product in the form of a slab are dried. The weight vs. time readings are given in Table 24.7.*

Table 24.7 Experimental weight vs. time data for Example 24.11

Time *min*	*Weight* *g*	*Time* *min*	*Weight* *g*
0	297	535	149.9
25	283.5	637	132.2
55	273.5	777	107.2
195	233.5	883	93.1

The weight of the bone-dried material is 79 g. The dimensions of the material are 12.1 cm × 12.1 cm × 2 cm.

(a) Determine the critical-moisture content of the product and the drying rate during the constant-rate period.

(b) The dry and wet bulb temperatures of the air are 54.5 and 37.8°C respectively. If the heat transfer is only by convection, determine the heat-transfer coefficient of the air and diffusion coefficient of water vapour during the constant-rate period.

(c) Determine the liquid diffusivity during the falling-rate period.

Solution (a) Initial-moisture content

$$w_i = \frac{297 - 79}{79} = \frac{218}{79} = 2.76$$

The values of moisture content w, and dimensionless moisture content W/W_i are evaluated in Table 24.8.

Table 24.8 Moisture content vs. time for Example 24.11

Time min	*w*	*W/W$_i$*
0	2.76	1.0
25	$\frac{283.5-79}{79}=2.59$	0.935
55	$\frac{273.5-79}{79}=2.46$	0.892
195	$\frac{233.5-79}{79}=1.95$	0.709
535	$\frac{149.9-79}{79}=0.895$	0.325
635	$\frac{132.2-79}{79}=0.675$	0.245
777	$\frac{107.2-79}{79}=0.375$	0.129
883	$\frac{93.1-79}{79}=0.082$	0.082

The drying curve, w vs. τ is plotted in Fig. 24.17. It is seen that the plot is a straight line only up to $\tau = 70$ min and $W = 2.4$.

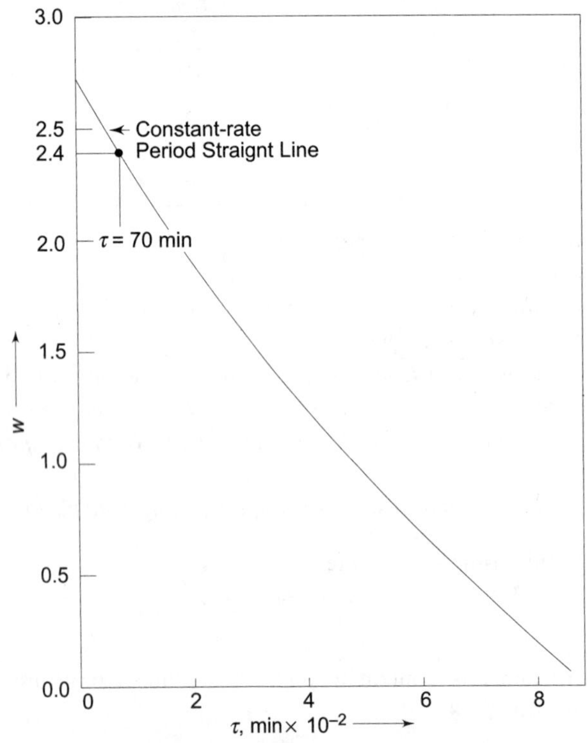

Fig. 24.17 W vs. τ plot for Example 24.11

Thus, constant-rate drying ends at this point so that the critical-moisture content is

$$w_c = 2.4$$

Rate of change of moisture content

$$\frac{\mathrm{d}w}{\mathrm{d}\tau} = \frac{2.76 - 2.4}{70/60} = 0.3 \text{ kg water/kg solid/hour}$$

Drying rate dring the constant rate period

$$\frac{\mathrm{d}W}{\mathrm{d}\tau} = \left(\frac{\mathrm{d}w}{\mathrm{d}\tau}\right) \text{(Mass of solid)}$$

$$= (0.3)\,(0.079) = 0.0237 \text{ kg/h}$$

(b) Surface area of the product

$$A = 2(12.1 \times 12.1) + 2(12.1 + 12.1) \times 2$$

$$= 390 \text{ cm}^2$$

By moisture balance during the constant-rate period

$$\frac{\mathrm{d}W}{\mathrm{d}\tau} = \frac{f_g\, A\,(t - t_S)}{\Delta h_S} = k_\omega\, A(\omega_S - \omega)$$

whence

$$f_g = \left(\frac{0.0237}{3600}\right) \frac{(2410 \times 10^3)}{(390 \times 10^{-4})\,(54.5 - 37.8)}$$

$$= 24.3 \text{ W m}^{-2} \text{ K}^{-1}$$

Now

$$p'_v = 6552 \text{ N/m}^2 \text{ (at 37.8°C)}$$

$$p_v = 6552 - \frac{(101325 - 6552)\,(54.5 - 37.8)}{2800 - 1.3\,(1.8 \times 54.5 + 32)}$$

$$= 5950 \text{ N/m}^2$$

$$\omega = 0.622 \left(\frac{5950}{101325 - 5950}\right) = 0.039$$

$$\omega_S = 0.622 \left(\frac{6552}{101325 - 6552}\right) = 0.043$$

Hence

$$k_\omega = \left(\frac{0.0237}{3600}\right) \frac{1}{(390 \times 10^{-4})\,(0.043 - 0.039)}$$

$$= 0.0422 \text{ kg s}^{-1} \text{ m}^{-2}$$

From the similarly relation

$$k_\omega = \frac{f_g}{C_p}$$

$$= \frac{24.3}{1.0216 \times 10^3} = 0.024 \text{ kg s}^{-1} \text{ m}^{-2}$$

(c) The value of $w_i = w_c = 2.4$ may be taken as the initial moisture-content for the falling-rate period. Considering 70 min as zero time, the values of W/W_c, beginning from the new zero time, are evaluated and given in Table 24.9.

Table 24.9 Dimensionless moisture content vs. time for falling-rate period for Example 24.11

Time ***h***	$W/W_i = W/W_c$
0	1.0
2.67	0.75
7.0	0.417
11.35	0.167
12.85	0.1
13.6	0.071

Density of bone-dried material

$$\rho = \frac{0.079}{(12.1 \times 12.1 \times 2) \times 10^{-6}} = 269.8 \text{ kg.m}^{-3}$$

The liquid diffusivity D will now be evaluated by trial and error.

Assume $D = 10^{-4}$ m^2 h^{-1}

Surface resistance ratio

$$\frac{D\rho}{k_\omega r_m} = \frac{(10^{-4}/3600)(269.8)}{(0.024)(0.01)} = 0.031 \quad \text{and} \quad \frac{k_\omega r_m}{D\rho} = \frac{1}{0.031} = 32$$

which is equivalent to Bi in Fig. 24.2.

From Table 24.6

$$\frac{W}{W_i} = 0.1 \text{ at } \tau = 12.85 \text{ h}$$

Diffusion term

$$\frac{D\tau}{r_m^2} = \frac{(10^{-4}/3600)(12.85 \times 3600)}{(0.01)^2} = 12.85$$

From the Gurnie-Lurie chart (Fig. 24.2).

$$\frac{W}{W_i} < 0.002 \neq 0.1$$

It is evident that the value of D is such that the equivalent of Bi, viz.,

$$\frac{k_\omega r_m}{D} = \infty$$

From Fig. 24.2, for this value and for $W/W_i = 0.1$, we have

$$\frac{D\tau}{r_m^2} = 1.02$$

so that

$$D = \frac{(1.02)(0.01)^2}{(12.85)(3600)} = 2.205 \times 10^{-9} \text{ m}^2 \text{ s}^{-1}$$

24.7 TUNNELS VENTILATION

All road tunnels require ventilation to remove pollutants produced from exhaust of vehicles. Natural ventilation can suffice if the tunnel length is 300 m or so. For longer tunnels, mechanical ventilation is necessary.

There are two type of ventilation requirements. These are 'normal ventilation' and 'emergency ventilation'.

Emergency ventilation is necessary in case, for example, if there is fire. Normal ventilation must be capable of handling emergency requirement.

There is the third type also which is 'temporary ventilation'. This is needed during construction. The purpose can be served by portable equipment, which can be removed after work is over.

There are two systems for tunnel ventilation: 'Longitudinal ventilation' and 'transverse ventilation'.

Longitudinal ventilation is by far very common. It introduces 'in' and removes air 'from' tunnel at one or more points. A typical layout is shown in Fig. 24.18. Fig. 24.18 (a) shows system with 'jet injection' of air at one point, while Fig. 24.18 (b) shows a module with one jet injection and one exhaust fan combination.

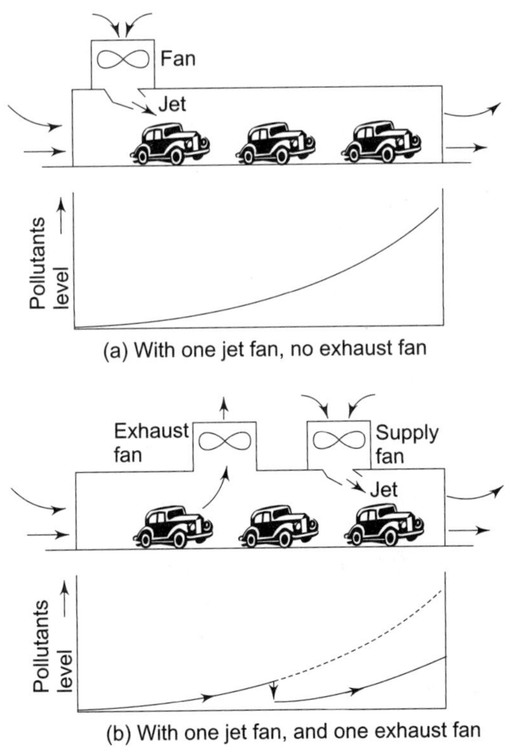

Fig. 24.18 Longitudinal ventilation

The figure, also shows increasing pollutants level towards the exhaust side. It is seen that the pollutants level towards the end of the tunnel can be decreased by employing exhaust fan/s.

Note that, in longitudinal ventilation, air flows longitudinally in the direction of the moving vehicles.

In transversve ventilation, air flows in a direction transverse to the movement of vehicles, Transverse ventilation requires supply and exhaust air fans and ducts as shown in Fig. 24.19.

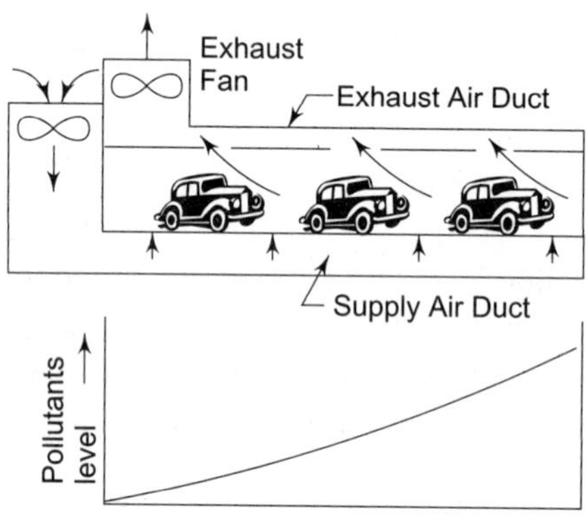

Fig. 24.19 Transverse ventilation

Such arrangement is used in long tunnels. This system takes care of the emergency ventilation requirement as well. During a fire the exhaust fans in the system start working at their maximum capacities.

Ventilation air quantities can be determined by estimation of poisonus CO, and CO_2 and NO_x emissions at maximum traffic level. Based on monitoring the CO levels, for example, centralized control is provided to regulate fan speeds and operate dampers in the system.

24.8 STATION AIR CONDITIONING

The net internal sensible heat gain for a typical 2-track metro station catering to, say, 40 trains per track traveling at 80 km/h speed may be of the order of 1.5 MW nearly.

To remove just this sensible heat only (excluding latent heat) simply by a ventilation system, allowing for 1.7°C rise in temperature of ouside air, will require about, say, 660 m^3/s of outside air. Such a ventilation system anyway will be very costly, not much less costly than the air-conditioning system. Further, air velocities on platform would reach intolerable propertions. And when one thinks of the letent heat load to cater to, one can imagine how unbearable the conditions on the platfoms will be. Hence air conditioning is necessary.

The same amount of sensible heat gain plus the latent heat gain and ventilation air loads for keeping the station air temperature 5°C below the ambient temperature could be handled by about 2.6 MW of refrigeration.

24.9 MINE AIR CONDITIONING AND VENTILATION

Air conditioning and ventilation of mines requires very special expertise. The description presented here is only a brief overview of the topic.

Mine ventilation is required to supply oxygen to underground facilities, to remove dangerous substances like hydrocarbon methane (CH_4), radon, strata gases, dust, blasting fumes, diesel emission, etc., and also to remove heat, and help control humidity. Use of ventilation is limited by the wet bulb temperature which should not exceed the prescribed range which is considered satisfactory between 25.5 to 29°C. Accordingly, ventilation can be done to a depth of 2500 m at the most. Below this depth, air temperatures in the intake shaft reach very high values, and air conditioning becomes absolutely necessary.

'Stope' in the mine is a production site where ore is actually mined. Actual cooling load is calculated at the average stope temperature, which is lower than the acceptable temperature for ventilation.

Heat enter the mines from wall rocks.

Other loads are from electric motors, lights, substation losses, calorific value of diesel burnt, and occupancy. Another important source of heat entering mine is due to adiabatic compression of air descending the shaft.

Finally, heat enters from ground water.

Wall rock heat flow is the main source of heat in deep rock mines. Note that the temperature at the earth's core is estimated about 5700°C. Heat flux from core to mine's internal surface may be as high as 50 W/m^2 which corresponds to thermal conductivity k = 5.5 W/mK, and thermal diffusivity α = 0.008 m^3/h of surrounding earth.

For estimation of wall rock heat flow, one may refer to ASHRAE Handbook Application, 2003.

Note that air descending a shaft increases in pressure.

This may be referred to as 'adibatic compression'. As a result, temperature of descending air increases. Load due to this adiabtic compression is given by

$$Q = Q_v\, \rho E\, \Delta L$$

where the load Q is in W, Q_v is air flow in shaft in m^3/s, ρ is air density about 1.12 kg /m^3, E is energy added per unit distance of elevation about 0.01 kJ/kg m, and ΔL is the change in elevation. For example, load of adiabatic compression of 150 m^3/s of air at ρ = 1.12, flowing down 1500 m deep shaft is

$$\dot{Q} = (150)(1.12)(0.01)(1500) = 2520 \text{ W}$$

Groundwater load is an important factor. For example, if 1.5 L/s of water at 53°C leaks into shaft sump which is at 29°C, the load is

$$\dot{Q} = \left(1.5\frac{L}{s}\right)\left(1\frac{\text{kg}}{L}\right)\left(4.1868\frac{\text{kJ}}{\text{kg K}}\right)(53-29)\text{K}$$

$$= 151 \text{ kW}$$

For mine air conditioning, we may have a surface plant or an underground plant. Underground plants do the same job as surface plants. They have the advantage of being closer to work areas. As a result, they give better efficiency and utilization.

But heat rejection is limited by the temperature of mine exhaust air which is used for cooling in condenser. Maintenance is also more difficult. Building spary chambers is more costly.

Mine A/C plant may either cool air or supply chilled water. Figure 24.20 shows a combination system. It can cool air and chill water also.

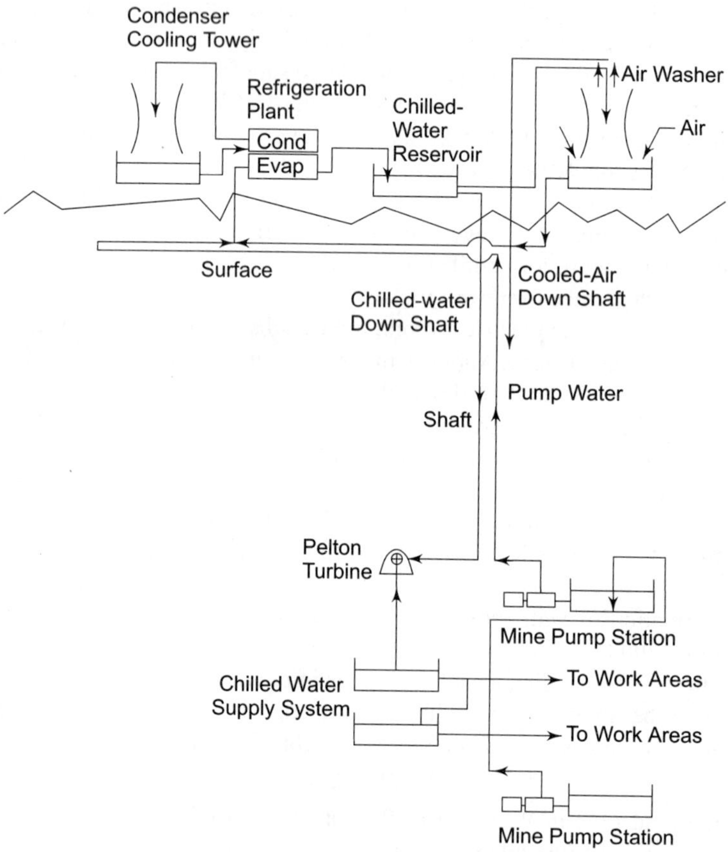

Fig. 24.20 Integrated cooling system

Such a surface plant provides a higher fraction of cooling capacity to cool intake air in bulk in summer. In winter, a higher fraction of cooling capacity is used to chill air conditioning water. This water is delivered underground.

Figure 24.20 shows that chilled water from chilled water reservoir is taken to the following:

(i) ***Air Washer*** From air washer, cooled-air is sent down shaft.

(ii) ***Underground Chilled Water Supply System*** Chilled water first goes to pelton turbine which produces power from the 'head' of water corresponding to depth below ground level. After doing work in turbine, water is available for chilled-water fan-coil unit (FCU), and other services.

A mine pump station returns the water to surface. This return water plus return water from air washer together enter the evaporator/chiller of the refrigeration plant. The plant uses centrifugal or screw compressoor. Most common refrigerant is HCFC 22. Ammonia is also commonly used in surface plants. Other method being developed includes air cycle refrigeration. Air is compressed on the surface. It is sent underground to a turbine where it turns a generator and exits at –40°C.

References

1. *ASHRAE, Handbook of Applications*, 1968.
2. Charm, S E, *The Fundamentals of Food Engineering*, AVI, Westport, Connecticut, 1963.
3. Ede, A J, 'The calculation of freezing and thawing of foodstuffs', *Modern Refrigeration*, Vol. 52, 1949, p. 52.
4. 'Freezing in fisheries', Food and Agricultural Organisation, *Fisheries Technical Paper,* No. 167, 1977.
5. Gilliland, E R and T K Sherwood, 'Drying of solids IV', *Ind. Engg. Chem.*, Vol. 25, p. 1134.
6. Goldblith, S A, L Rey and W W, Rothmayer, *Freeze Drying and Advanced Food Technology*, Academic Press, London, 1975.
7. Gorling, P, *Fundamental Aspects of the Dehydration of Foodstuffs*, Macmillan, New York, 1958.
8. Harper J C and A L, Tappel, *Advances in Food Research*, Vol. 7, Academic Press, New York, 1957.
9. Nagaoka, J, S Takagi and S, Hotani 'Experiments on the freezing of fish in an air-blast freezer', *Proc. IX International Congress of Refrigeration*, Paris, Vol. 2, 1953, p. 4.
10. Neuman, K, 'Advances in the method and application of freeze drying', *ChemiIngenieur Technik*, Jan. 1957, p. 267.
11. Sharma N K, C P Arora and B K Mital, 'Influence of concentration of milk solids on freeze-drying rate of yoghurt and its quality', *J. Food Process Engg.*, Vol. 15, No. 3, pp. 187–198, 1972.
12. Sharma N K and C P, Arora, 'Prediction of transient temperature distribution during freeze drying of yoghurt', *Drying Technology*, Vol. 11, No. 7, pp. 1863–1883, 1993.
13. Sharma N K and C P, Arora, 'Influence of product thickness, chamber pressure and heating conditions on production rate of freeze-dried yoghurt', *Int. J. Refrigeration*, Vol. 18, No. 5, pp. 297–307, 1995.
14. Sherwood, T K, 'Application of theoretical diffusion equations to the drying of solids', *Trans. American Institute of Chemical Engineers*, Vol. 27, 1931, p. 190.
15. Tressler D K and C F, Evers, *The Freezing Preservation of Foods*, Vols. I and II, AVI, Westport, Connecticut, 1957.
16. Trifonova L I, '*Theoretical and experimental investigation of freeze dehydration process*, Ph D. thesis, IIT, Delhi, 1975.

17. Trifonova L I and C.P. Arora, 'Symmetrical steady state freeze drying and optimization of radiator temperature, *XIV Int. Cong. Refrigeration, Moscow*, 1975.

Revision Exercises

24.1 (a) 10 cm thick meat slabs, initially at 35°C, are kept in a cold room at 1°C. Using numerical method, find the time required for the centre of the slab to reach a temperature of 4°C. The heat-transfer coefficient of air may be taken as 10 $Wm^{-2} K^{-1}$. The thermophysical properties of meat are as follows:

$$\text{Thermal conductivity} = 0.65 \text{ W m}^{-1} \text{ K}^{-1}$$
$$\text{Specific heat} = 3.5 \text{ kJ kg}^{-1} \text{ K}^{-1}$$
$$\text{Density} = 1020 \text{ kg m}^{-3}$$

(b) Compare your result with that obtained by using Gurnie-Lurie charts.

24.2 (a) A 10 cm thick meat slab is being frozen from both sides in a contact-plate freezer. The meat is initially at 5°C and has a moisture content of 75 per cent. The average freezing point is – 5°C. The plates are maintained at – 18°C. Employing Neuman solution, determine the time for the meat centre to pass through the freezing zone.

(b) Determine the time required if the freezing is carried out in an air-blast freezer with air supplied at – 18°C and the heat-transfer coefficient of air is 20 $Wm^{-2} K^{-1}$.

All other data remain the same.

The thermophysical properties of thawed and frozen meat are given below.

Property	*Thawed Meat*	*Frozen Meat*	*Units*
Thermal conductivity	0.57	1.04	$W m^{-1} K^{-1}$
Specific heat	3.48	1.91	$kJ kg^{-1} K^{-1}$
Density	1057	961	$kg m^{-3}$

24.3 Solve Problem 24.2 using modified Plank's equation.

24.4 Determine the freeze-drying time for a tray of 1.2 cm thick, frozen lean steaks in contact with heating platens on both sides at 90°C. The drying chamber is maintained at 1 mm Hg pressure. The thermophysical properties of dried and frozen meat are as follows:

Property	*Dried Meat*	*Frozen Meat*	*Units*
Thermal conductivity	0.017	1.0	$W m^{-1} K^{-1}$
Specific heat	1.34	8.3	$kJ kg^{-1} K^{-1}$
Density	320	960	$kg m^{-3}$

Use Neuman's solution.

24.5 1360 kg/h of magnesium hydroxide are to be dried from 82 to 4 per cent moisture content (based on a bone-dry basis) on a belt drier with air blown at a dry bulb temperature of 71°C and a wet bulb temperature of 37.5°C. The make-up air is at 40°C dry bulb and 25°C wet bulb. The material enters the drier at a temperature of 30°C. The test-drying time is 25 min. The drier bed is loaded with 34 kg/m^2 of bone-dry material. The depth of the bed is 10 cm. The air velocity is 1.27 m/s.

(a) If the commercial-drying time is 70 per cent more than the test-drying time, find the width and length of a perforated belt assuming a suitable free-flow area.

(b) Determine the evaporation rate in the drier, supply air rate and make-up air rate.

(c) Establish the heat balance of the system. The specific heat of wet material is 1.2 kJ/kg.K.

24.6 Trays of material cut into 1.25 cm cubes at an initial moisture content of 70% are being dried in a counter-current tunnel drier. The trays are 15.2 cm deep. The space between the trays is 2.5 cm. The temperature of cubes entering the drier is 30°C. The production rate is to be 45 kg of the product per hour. Assume the final product has 2 per cent moisture. The heat-transfer coefficient of air may be calculated from Eq. (24.16). The spacing between trays is 7.5 cm. The trays are mounted on trucks, the cross-section of which is 0.9 m × 0.8 m Calculate:

(a) The drying rate in the constant-rate period and the mass-transfer coefficient.

(b) The critical-moisture content.

(c) The length of the drier.

The diffusivity of the dried material is 1.8×10^{-9} m^2/s and its density is 320 kg/m^3.

Appendix

A

Thermodynamic Properties Correlations for Refrigerants

A.1 CORRELATIONS FOR THERMODYNAMIC PROPERTIES OF R 12

All the correlations given by Watson* as written below have been used:

Vapour Pressure Correlation

$$\ln P_s = P_1 + \frac{P_2}{T_s} + P_3 \ln T_s + P_4 T_s^5 \quad \text{(A.1.1)}$$

where P_s is in bar. And constants are as follows:

$$P_1 = 4.749214603 \times 10^1$$
$$P_2 = -3.783397029 \times 10^3$$
$$P_3 = -5.852561143$$
$$P_4 = 1.568931884 \times 10^{-8}$$
$$n = 3$$

From Eq. (A.1.1), we obtain the saturation pressure versus saturation temperature gradient

$$\left[\frac{\mathrm{d}P_s}{\mathrm{d}T_s}\right] = P_s\left[-\frac{P_2}{T_s^2} + \frac{P_3}{T_s} + nP_4 T_s^{n-1}\right] \quad \text{(A.1.2)}$$

for substitution in Clapeyron equation to determine the latent heat of vaporization of R 12.

Equation of State and Vapour Phase Enthalpy and Entropy

Perel' shtein's equation is used as given below

$$P = R\,T\rho \sum_{i=1}^{9} E_i \rho^{i-1} \quad \text{(A.1.3)}$$

* Watson J.T.R., *Thermophysical properties of refrigerant R 12*, Department of Industry/National Engineering Lab., Edinburgh, 1975.

where P is in bar, ρ is in kg/dm^3 and the constants are

$$E_1 = 1.0$$
$$E_2 = A_1 + A_2/\tau + A_3/\tau^2 + A_4/\tau^3$$
$$E_3 = A_5 + A_6/\tau + A_7/\tau^2$$
$$E_4 = A_8 + A_9/\tau + A_{10}/\tau^2$$
$$E_5 = A_{11} + A_{12}/\tau$$
$$E_6 = 0.0$$
$$E_7 = A_{13} + A_{14}/\tau$$
$$E_8 = 0.0$$
$$E_9 = A_{15}$$

$A_1 = 4.18883659$ $\quad A_2 = -11.4858003$ $\quad A_3 = 8.64725830$
$A_4 = -3.51695112$ $\quad A_5 = -9.66065502$ $\quad A_6 = 21.9447268$
$A_7 = -10.8428817$ $\quad A_8 = 17.9920989$ $\quad A_9 = -41.2886382$
$A_{10} = 23.7864658$ $\quad A_{11} = 0.0485266372$ $\quad A_{12} = -0.162237281$
$A_{13} = 4.63277362$ $\quad A_{14} = -6.16887911$ $\quad A_{15} = 1.29950293$
$R = 0.6875616$ bar dm^3/kg.K.

τ is reduced temperature defined as T/385.15, where 385.15 K is the critical temperature of R 12.

From Eq. (A.1.3), we obtain

$$\left[\frac{\mathrm{d}P}{\mathrm{d}T}\right]_\rho = \rho R + \rho^2 R\left[A_1 - \frac{A_3(385.15)^2}{T^2} - \frac{A_4(385.15)^3}{T^3}\right] + \rho^3 R\left[A_5 - \frac{A_7(385.15)^2}{T^2}\right] + \rho^4 R\left[A_8 - \frac{A_{10}(385.15)^2}{T^2}\right] + \rho^5 R A_{11} + \rho^7 R A_{13} + \rho^9 R A_{15} \quad \text{(A.1.4)}$$

Substituting from the above, we obtain the following expressions for the vapour phase enthalpy and entropy of R 12.

$$h = h_o + Pv - RT_o + \int_{T_o}^{T} (C_p^{\,o} - R)\mathrm{d}T + \rho R\,T\left(\left[A_1 - A_3\left(\frac{385.15}{T}\right)^2 - A_4\left(\frac{385.15}{T}\right)^3\right] - E_2\right) + \frac{\rho^2 RT}{2}\left(\left[A_5 - A_7\left(\frac{385.15}{T}\right)^2\right] - E_3\right) + \frac{\rho^3 RT}{3}\left(\left[A_8 - A_{10}\left(\frac{385.15}{T}\right)^2\right] - E_4\right)$$

$$+\frac{\rho^4 RT}{4}[A_{11}-E_5]+\frac{\rho^6 RT}{6}[A_{13}-E_7]$$

$$+\frac{\rho^8 RT}{8}[A_{15}-E_6] \qquad (A.1.5.)$$

$$s = s_o + \int_{T_o}^{T}(C_p^{\rho}-R)\frac{dT}{T} - R\ln\left[\frac{RT\rho}{P_o}\right]$$

$$-\rho R\left[A_1 - A_3\left[\frac{385.15}{T}\right]^2 - A_4\left[\frac{385.15}{T}\right]^3\right]$$

$$-\frac{\rho^2 R}{2}\left[A_5 - A_7\left[\frac{385.15}{T}\right]^2\right]$$

$$-\frac{\rho^3 R}{3}\left[A_8 - A_{10}\left[\frac{385.15}{T}\right]^2\right]-\frac{\rho^4 A_{11} R}{4}$$

$$-\frac{\rho^6 R A_{13}}{6}-\frac{\rho^8 R A_{15}}{8} \qquad (A.1.6)$$

Correlation for Saturated Liquid Density

$$\rho_L - \rho_c = \sum_{i=1}^{6} D_i (T_c - T)^{i/3} \qquad (A.1.7)$$

where ρ_L is in kg/dm^3, and the constants are as follows:

$D_1 = 0.2477199$ $\quad D_2 = -0.1480948$ $\quad D_3 = 0.008001550$

$D_4 = -0.01962269$ $\quad D_5 = 0.0023223$ $\quad D_6 = -0.0001057677$

Zero-Pressure Constant Volume Specific Heat

$$C_v^o = C_{v1} + C_{v2}\,T + C_{v3}\,T^2 + C_{v3}\,T^3 \qquad (A.1.8)$$

where the units of specific heat are in kJ/kg.K, and the constants are

$C_{v1} = 0.0479836$ $\qquad C_{v3} = -2.94985 \times 10^{-6}$

$C_{v2} = 0.00238154$ $\qquad C_{v4} = 1.37374 \times 10^{-9}$

A.2 CORRELATIONS FOR THERMODYNAMIC PROPERTIES OF R 134a

The correlations given by Wilson and Basu* have been used:

Vapour Pressure Correlation

$$\ln P_s = P_1 + \frac{P_2}{T_s} + P_3\,T_s + P_4\,T_s^2 + \frac{P_5(P_6 - T_s)}{T_s}\ln(P_6 - T_s) \qquad (A.2.1)$$

* Wilson D.P. and Basu R.S., 'Thermodynamic properties of a new statospherically safe working fluid-Refrigerant 134a', *ASHRAE Trans.*, Vol. 94, pp. 2095–2118, 1988.

where P_s is in kPa, and the constants are

$$P_1 = 24.8033988$$
$$P_2 = -0.3980408 \times 10^4$$
$$P_3 = -0.2405332 \times 10^{-1}$$
$$P_4 = 0.2245211 \times 10^{-4}$$
$$P_5 = 0.1995548$$
$$P_6 = 0.3748473 \times 10^3$$

Also, differentiating Eq. (A.2.1), the derivative is obtained as below:

$$\left[\frac{dP_s}{dT_s}\right] = P_s\left[-\frac{P_2}{T_s^2}+\frac{P_3}{T_s}+2P_4T_s-\frac{P_5P_6\ln(P_6-T_s)}{T_s^2}-\frac{P_5P_6}{T_s(P_6-T_s)}+\frac{P_5}{(P_6-T_s)}\right] \quad \text{(A.2.2)}$$

Equation of State and Vapour Phase Enthalpy and Entropy

A 13-constant Martin-Hou equation of state has been used.

$$P = \frac{RT}{(v-b)}+\frac{E_1+E_2T+E_3e^{-kT_r}}{(v-b)^2}+\frac{E_4+E_5T+E_6e^{-kT_r}}{(v-b)^3}+\frac{E_7}{(v-b)^4}+\frac{E_8+E_9T+E_{10}e^{-kT_r}}{(v-b)^5}$$
$$= \frac{RT}{(v-b)}+\sum_{i=2}^{5}\frac{A_i+B_iT+C_ie^{-kT_r}}{(v-b)^i} \quad \text{(A.2.3)}$$

where P is in kPa, v is in m^3/kg, and the constants are:

$$b = 0.3455467 \times 10^{-3} \qquad k = 5.475$$
$$E_1 = -0.1198051$$
$$E_2 = 0.1137590 \times 10^{-3}$$
$$E_3 = -3.531592$$
$$E_4 = 0.1447797 \times 10^{-3}$$
$$E_5 = -0.8942552 \times 10^{-7}$$
$$E_6 = 0.6469248 \times 10^{-2}$$
$$E_7 = -1.049005 \times 10^{-7}$$
$$E_8 = -6.953904 \times 10^{-12}$$
$$E_9 = 1.269806 \times 10^{-13}$$
$$E_{10} = -2.051369 \times 10^{-9}$$
$$R = 0.081488162 \text{ kJ/kg.K}$$

From Eq. (A.2.3), we obtain

$$\left[\frac{d\boldsymbol{P}}{d\boldsymbol{T}}\right]_\rho = \frac{R}{v-b}+\frac{E_2-E_3ke^{-kT_r}/T_c}{(v-b)^2}+\frac{E_5-E_6ke^{-kT_r}/T_c}{(v-b)^3}$$

$$+ \frac{E_9 - E_{10}\, k\, e^{-kT_r} / T_c}{(v - b)^5} \tag{A.2.4}$$

Substituting from the above, we obtain the following expressions for the vapour phase enthalpy and entropy of R 134a.

$$h = h_o + \int_{T_o}^{T} (C_p^0 - R)\, \mathrm{d}T + \sum_{i=2}^{5} \left[\frac{A_i}{(i-1)(v-b)^{(i-1)}} \right] \tag{A.2.5}$$

$$+ \sum_{i=2}^{5} \left[\frac{C_i\, e^{-kT_r} (1 + kT_r)}{(i-1)(v-b)^{(i-1)}} \right] + (Pv - RT_o)$$

$$s = s_o + \int_{T_o}^{T} \frac{C_p^o\, \mathrm{d}T}{T} - R \ln \left[\frac{RT}{P_o (v-b)} \right] + \sum_{i=2}^{5} \left[\frac{B_i}{(i-1)(v-b)^{(i-1)}} \right]$$

$$+ \sum_{i=2}^{5} \left[\frac{C_i (k/T_c)\, e^{-kT_r}}{(i-1)(v-b)^{(i-1)}} \right] \tag{A.2.6}$$

Correlation for Saturated Liquid Density

$$\rho_L = \rho_c + \sum_{N=1}^{4} D_N (1 - T_r)^{N/3} \tag{A.2.7}$$

where ρ_L is in kg/m^3 and the constants are:

$$D_1 = 819.6183$$
$$D_2 = 1023.582$$
$$D_3 = -1156.757$$
$$D_4 = 789.7191$$

Zero-Pressure Constant Pressure Specific Heat

$$C_p^o = C_{p1} + C_{p2}T + C_{p3}T^2 + C_{p4}T^3 + \frac{C_{p5}}{T} \tag{A.2.8}$$

where C_p^o is in kJ/kg.K and the constants are:

$$C_{p1} = -\,0.5257455 \times 10^{-2}$$
$$C_{p2} = 0.3296570 \times 10^{-2}$$
$$C_{p3} = -2.017321 \times 10^{-6} \qquad C_{p4} = 0.0$$
$$C_{p5} = 15.82170$$

A.3 CORRELATIONS FOR THERMODYNAMIC PROPERTIES OF R 152a

The correlations given by Kamei et al*. have been used:

---

* Kamei A., Piao C.C., Sato H. and Watanabe K. 'Thermodynamic charts and cycle performance of HFC-134a and HFC-152a', *ASHRAE Trans.*, Vol. 96, Part 1, pp. 141–149, 1990.

Vapour Pressure Correlation

$$\ln \frac{P_s}{P_c} = \left(\frac{T_c}{T_s}\right) [-7.46588\, T_{r1} + 1.89468\, T_{r1}^{\ 1.5} - 2.57557\, T_{r1}^{2.5}] \quad \text{(A.3.1)}$$

where P_s and P_c are in Pa, and

$$T_{r1} = 1 - \frac{T_s}{T_c}$$

From Eq. (A.3.1), we obtain

$$\left[\frac{dP_s}{dT_s}\right] = P_s \left[\frac{7.46588 T_c}{T_s^2} - \frac{1.26312 T_c^{2/3}}{T_s^{5/3}} + \frac{1.030228 T_c^{2/5}}{T_s^{7/5}}\right] \quad \text{(A.3.2)}$$

Equation of State and Vapour Phase Enthalpy and Entropy

Soave Redlich-Kwong type equation is used

$$P = \frac{RT}{v-b} - \frac{A_c\, a(T)}{v(v-b)} \quad \text{(A.3.3)}$$

where P is in Pa, v is in m^3/kg and

$$A_c = 0.42748\ \frac{R^2 T_c^2}{P_c}$$

$$a(T) = 1 + \left(1 - \frac{T}{T_c}\right)\left[m + n\left(\frac{T_c}{T}\right)\right]$$

$$b = \frac{0.08664\, RT_c}{P_c}$$

$$R = \frac{R_o}{M}$$

$R_o = 8.31451$ kJ/kmol·K, $M = 66.051 \times 10^{-3}$ kg/mol

$m = 0.45 \qquad n = 1.15$

From Eq. (A.3.3), we obtain

$$\left[\frac{dP}{dT}\right]_\rho = \frac{R}{(v-b)} + \frac{A_c\left[\dfrac{nT_c}{T^2} + \dfrac{m}{T_c}\right]}{v(v+b)} \quad \text{(A.3.4)}$$

Substituting from the above, we obtain the following expressions for the vapour phase enthalpy and entropy of R 152a.

$$h = h_o + \frac{A_c}{b}\left[1 + m - n + 2n\frac{T_c}{T}\right] \ln\left[\frac{v}{v+b}\right]$$

$$+ Pv - RT + \int_{T_o}^{T} C_P^o\, dT \quad \text{(A.3.5)}$$

$$s = s_o + \frac{A_c}{b}\left[\frac{m}{T_c} + \frac{nT_c}{T^2}\right] + R\ln\left[\frac{(v-b)}{RT}\right] + \int_{T_o}^{T} C_P^o \frac{dT}{T} \quad \text{(A.3.6)}$$

Correlation for Saturated Liquid Density

$$\frac{\rho_L}{\rho_c} = 1 + 1.86756\ T_{r1}^{0.326} + 0.99263\ T_{r1}^{0.836} \quad \text{(A.3.7)}$$

where ρ_L and ρ_c are in kg/m^3

Zero-Pressure Constant Pressure Specific Heat

$$C_p^o = 2.07225 + 5.72223 \times 10^{-2}\ T - 3.48012 \times 10^{-5}\ T^2 + 8.10711 \times 10^{-9}\ T^3 \quad \text{(A.3.8)}$$

where C_p^o is in kcal/kmol K.

A.4 CORRELATIONS FOR THERMODYNAMIC PROPERTIES OF R 22

Correlations given by Reynolds* have been used:

Vapour Pressure Correlation

$$\ln P_s = P_1 + \frac{P_2}{T_s} + P_3 \ln T_s + [P_4 \times T_s] + P_5\ \frac{(\gamma - T_s)}{T_s}\ \ln(\gamma - T_s) \quad \text{(A.4.1)}$$

where P_s is in Pa, and the constants are

$$P_1 = 7.1554148092$$
$$P_2 = -\,4.8189575050 \times 10^3$$
$$P_3 = -7.8610312200$$
$$P_4 = 9.0806824483 \times 10^{-3}$$
$$P_5 = 4.4574670300 \times 10^{-1}$$
$$\gamma = 381.17$$

From Eq. (A.4.1) we obtain

$$\left[\frac{dP_s}{dT_s}\right] = P_s\left[-\frac{P_2}{T_s^2} + \frac{P_3}{T_s} + P_4 - \frac{P_5\,\gamma\ln(\gamma - T_s)}{T_s^2} - \frac{P_5\,\gamma}{T_s(\gamma - T_s)} + \frac{P_5}{(\gamma - T_s)}\right] \quad \text{(A.4.2)}$$

* Reynolds C W, *Thermodynamic properties in S.I.: Graphs, tables and computational equations for forty substances*, Stanford University, Stanford, CA 94305, USA, 1979.

Equation of State and Vapour Phase Enthalpy and Entropy

Modified Martin-Hou equation is used.

$$P = \frac{RT}{v-b} + \sum_{i=2}^{5} \frac{(A_i + B_i T + C_i e^{-kT/T_c})}{(v-b)^i} + \frac{(A_6 + B_6 T + C_6 e^{-kT/T_c})}{e^{\alpha v}(1+ce^{\alpha v})} \tag{A.4.3}$$

where P is in Pa, v is in m^3/kg, and the constants are

$R = 96.1467$	$k = 4.2$
$b = 1.24855636 \times 10^{-4}$	$\alpha = 8781.3417$
$A_2 = -1.16981908 \times 10^2$	$A_3 = -2.92952588 \times 10^{-2}$
$A_4 = 2.41919261 \times 10^{-4}$	$A_5 = -2.43458381 \times 10^{-7}$
$A_6 = 9.40022615 \times 10^{11}$	$B_2 = 1.16431240 \times 10^{-1}$
$B_3 = 2.30319412 \times 10^{-4}$	$B_4 = -6.79667708 \times 10^{-7}$
$B_5 = 6.30201766 \times 10^{-10}$	$B_6 = -2.07580650 \times 10^9$
$C_2 = -1.18409710 \times 10^3$	$C_3 = 2.48896136$
$C_4 = 0.0$	$C_5 = -1.20619716 \times 10^{-6}$
$C_6 = 0.0$	

From Eq. (A.4.3), we obtain

$$\left[\frac{dP}{dT}\right]_\rho = \frac{R}{(v-b)} + \sum_{i=2}^{5} \frac{1}{(v-b)^i}\left[B_i - \frac{C_i k e^{-kT/T_c}}{T_c}\right] + \frac{B_6 - \dfrac{C_6 k e^{-kT/T_c}}{T_c}}{e^{\alpha v}(1+e^{\alpha v})} \tag{A.4.4}$$

Substituting from the above, we obtain the following expressions for the vapour phase enthalpy and entropy of R 22.

$$h = h_o + \int_{T_o}^{T} C_v^o \, dT - \sum_{i=2}^{5}\left[\frac{A_i + C_i e^{-kT/T_c}\left(1+\dfrac{kT}{T_c}\right)}{(1-i)(v-b)^{(i-1)}}\right] + \frac{A_6 + C_6 e^{-kT/T_c}\left(1+\dfrac{kT}{T_c}\right)}{\alpha e^{\alpha v}} + (Pv - RT_o) \tag{A.4.5}$$

$$s = s_o + \int_{T_o}^{T} C_v^o \frac{dT}{T} + \mathrm{R}\ln(v-\mathrm{b}) + \sum_{i=2}^{5} \frac{B_i - C_i \dfrac{k}{T_c} e^{-kT/T_c}}{(1-i)(v-b)^{(i-1)}} - \frac{B_6 - C_6 \dfrac{k}{T_c} e^{-kT/T_c}}{\alpha e^{\alpha v}} \tag{A.4.6}$$

Correlation for Saturated Liquid Density

$$\rho_L = \sum_{i=1}^{5} D_i\, T_{r1}^{(i-1)/3} + D_6\, T_{r1}^{1/2} + D_7\, T_{r1}^{2} \tag{A.4.7}$$

where ρ_L is in kg/m^3, and $T_{r1} = 1 - \dfrac{T}{T_c}$

and the constants are

$D_1 = 5.24766060 \times 10^2$
$D_2 = 8.75161285 \times 10^2$
$D_3 = 5.88662575 \times 10^2$
$D_4 = -\,3.57093464 \times 10^2$
$D_5 = 3.27951374 \times 10^2$

Zero-Pressure Constant Volume Specific Heat

$$C_v^o = \sum_{i=1}^{4} C_{vi}\, T^{i-1} + \frac{C_{v5}}{T^2} \tag{A.4.8}$$

where C_v^o is in J/kg.K, and the constants are

$C_{v1} = 1.17767818 \times 10^2$
$C_{v2} = 1.69972960$
$C_{v3} = -\,8.83043292 \times 10^{-4}$
$C_{v4} = 0.0$
$C_{v5} = 3.32541759 \times 10^5$

A.5 CORRELATIONS FOR THERMODYNAMIC PROPERTIES OF R 290 AND R 600a

Reynolds* also gives correlations for R 290 and R 600a which are as follows.

Vapour Pressure Correlation

$$\ln (P_s/P_c) = \left(\frac{T_c}{T_s} - 1\right) \sum_{i=1}^{8} P_i \left(\frac{T_s}{T_p} - 1\right)^{i-1} \tag{A.5.1}$$

The constants are given in the table below:

Constants	*Propane*	*Isobutane*
P_1	$-\,6.230993$	-6.3016457
P_2	$-\,4.4226860 \times 10^{-1}$	2.1880736×10^{-1}
P_3	$-\,1.8839624$	-1.1288158
P_4	3.6383362×10^{-1}	2.2391095
P_5	1.5177354×10^{1}	1.0653363
P_6	1.1216551×10^{2}	9.3322720
P_7	2.7635840×10^{2}	2.4836848×10^{1}
P_8	2.358535×10^{2}	3.7187854×10^{1}
T_p	300	300
P_c	4.2359300×10^{6}	3.6845470×10^{6}

From Eq. (A.5.1), we obtain

$$\left[\frac{\mathrm{d}P_s}{\mathrm{d}T_s}\right] = P_s\left[-\frac{T_c}{T_s^2}\sum_{i=1}^{8} P_i\left[\frac{T_s}{T_p}-1\right]^{i-1}\right.$$

$$\left.+\left[\frac{T_c}{T_s}-1\right]\frac{1}{T_p}\left[\frac{T_s}{T_p}-1\right]^{i-2}\sum_{i=1}^{8}(i-1)\,P_i\right] \tag{A.5.2}$$

Equation of State and Vapour Phase Enthalpy and Entropy

Reynolds has used the following equation of state for both R 290 and R 600a.

$$P = \rho RT + \left(E_1\,RT - E_2 - \frac{E_3}{T^2} + \frac{E_4}{T^3} - \frac{E_5}{T^4}\right)\rho^2$$

$$+\left(b\,RT - a - \frac{\mathrm{d}}{T}\right)\rho^3 + \alpha\left(a + \frac{\mathrm{d}}{T}\right)\rho^6 \tag{A.5.3}$$

$$+\,c\,\frac{\rho^3}{T^2}\,(1+\gamma\rho^2)\,e^{-\gamma\rho^2}$$

where P is in Pa, and ρ is in kg/m^3.

The constants for the two refrigerants are given in the table below:

Constants	*Propane*	*Isobutane*
R	1.887326×10^{2}	1.430797×10^{2}
E_1	1.366892×10^{-3}	2.018128×10^{-3}
E_2	2.579108×10^{2}	2.964140×10^{2}
E_3	3.401044×10^{7}	2.489763×10^{7}
E_4	1.076728×10^{9}	1.163672×10^{9}
E_5	3.375879×10^{10}	6.371519×10^{10}
b	1.096523×10^{-5}	9.906333×10^{-6}
a	7.856721×10^{-1}	4.100261×10^{-1}
d	1.639769×10^{2}	1.029360×10^{2}
c	1.661103×10^{5}	1.072632×10^{5}
α	5.728034×10^{-9}	5.253972×10^{-9}
γ	9.157270×10^{-6}	8.208362×10^{-6}

From Eq. (A.5.3), we obtain,

$$\left(\frac{\mathrm{d}P}{\mathrm{d}T}\right)_\rho = \rho R + \rho^2\left[E_1\,R + \frac{E_3}{T^3} - \frac{E_4}{T^4} + \frac{E_5}{T^5}\right] + \rho^3\left[bR + \frac{\mathrm{d}}{T^2}\right]$$

$$-\frac{\alpha\rho^6\,\mathrm{d}}{T^2} - \frac{2c\rho^3\,(1+\gamma\rho^2)e^{-\gamma\rho^2}}{T^3} \tag{A.5.4}$$

Substituting from the above, we obtain the following expressions for the vapour phase enthalpies and entropies of R 290 and R 600a.

$$h = h_o + Pv - R\,T_o + \int_{T_o}^{T}(C_p^o - R)\,\mathrm{d}T$$

$$+\rho\left[\left(E_1 R+\frac{E_3}{T^3}-\frac{E_4}{T^4}+\frac{E_5}{T^5}\right)-\left(E_1 RT-E_2-\frac{E_3}{T^2}-\frac{E_4}{T^3}+\frac{E_5}{T^4}\right)\right]$$

$$-\frac{\rho^2}{2}a-\frac{\rho^5}{5}\left[\frac{\alpha d}{T}+\frac{\alpha d}{T^2}\right]-\frac{3c\,e^{-\gamma\rho^2}}{T^2\gamma}(\rho-1)$$

$$-\frac{3c}{T^2}(1-e^{-\gamma\rho^2})-\frac{\rho^2 e^{-\gamma\rho^2}}{2\gamma} \tag{A.5.5}$$

$$s=s_o+\int_{T_o}^{T}\frac{dT}{T}+R\ln\left[\frac{RT\rho}{P_o}\right]-\rho\left[E_1 R+\frac{2E_3}{T^3}-\frac{3E_4}{T^4}+\frac{4E_5}{T^5}\right]$$

$$-\frac{\rho^2}{2}\left[bR+\frac{d}{T^2}\right]+\frac{\rho^5\alpha d}{5T^2}-\frac{2ce^{-\gamma\rho^2}}{T^3\gamma}(\rho+1) \tag{A.5.6}$$

$$+\frac{2c}{\gamma T^3}\left(1-e^{-\gamma\rho^2}\right)-\frac{c\rho^2 e^{-\gamma\rho^2}}{\gamma T^3}$$

Correlation for Saturated Liquid Density

$$\rho_L=\sum_{i=1}^{6}D_i T_{r1}^{(i-1)/3} \tag{A.5.7}$$

where ρ_L is in kg/m^3, and

$$T_{r1}=1-\frac{T}{T_c}$$

and the constants are given in the table below:

Constants	*Propane*	*Isobutane*
D_1	1.9738193×10^2	1.9450561×10^2
D_2	-2.1307184×10^1	-9.1725345×10^1
D_3	3.3522024×10^3	2.4446128×10^3
D_4	-7.7040243×10^3	-2.7219989×10^3
D_5	7.5224059×10^3	1.9324597×10^2
D_6	-2.5663363×10^3	8.7037158×10^2

Zero-Pressure Constant Volume Specific Heat

$$C_v^o=\sum_{i=1}^{6}C_{vi}T^{i-2} \tag{A.5.8}$$

where $(C_v)^o$ is in J/kg. K and the constants are as follows:

Constnats	*Propane*	*Isobutane*
C_{v1}	2.0582170×10^5	1.7563902×10^5
C_{v2}	-1.9109547×10^3	-1.7524300×10^3
C_{v3}	1.1622054×10^1	1.1642389×10^1
C_{v4}	-9.7951510×10^{-3}	-1.0197170×10^{-2}
C_{v5}	4.5167026×10^{-6}	4.9006615×10^{-6}
C_{v6}	-8.6345035×10^{-10}	-9.8234416×10^{-10}
u_o	4.2027216×10^5	3.9342075×10^5
s_o	2.1673997×10^3	1.8189390×10^3

Appendix

B

Tables

B.1 THERMOPHYSICAL PROPERTIES OF AIR AT ATMOSPHERIC PRESSURE*

The values of μ, k, C_p and Pr are not strongly pressure-dependent and may be used over a fairly wide range of pressures.

T (K)	ρ (kg/m^3)	C_p (kJ/kg K)	μ (kg/m s × 10^5)	v (m^2/s ×10^6)	k (W/mk)	α (m^2/s × 10^4)	Pr
100	3.9010	1.0266	0.6924	1.923	0.009246	0.02501	0.770
150	2.3675	1.0099	1.0283	4.343	0.013735	0.05745	0.753
200	1.7687	1.0061	1.3289	7.49	0.01809	0.10165	0.739
250	1.4128	1.0053	1.488	9.49	0.02227	0.13161	0.722
300	1.1774	1.0057	1.983	15.68	0.02624	0.2216	0.708
350	0.9980	1.0090	2.075	20.76	0.03003	0.2983	0.697
400	0.8826	1.0140	2.286	25.90	0.03365	0.3760	0.689
450	0.7833	1.0207	2.284	28.86	0.03707	0.4222	0.683
500	0.7048	1.0295	2.671	37.90	0.04038	0.5564	0.680
550	0.6423	1.0392	2.848	44.34	0.04360	0.6532	0.680
600	0.5879	1.0551	3.018	51.34	0.04649	0.7512	0.680
650	0.5430	1.0635	3.177	58.51	0.04953	0.8578	0.682
700	0.5030	1.0752	3.322	66.25	0.05230	0.9672	0.684
750	0.4709	1.0856	3.481	73.91	0.05509	1.0774	0.686
800	0.4405	1.0978	3.625	82.29	0.05779	1.1951	0.689
850	0.4149	1.1095	3.765	90.75	0.06028	1.3097	0.692
900	0.3925	1.1212	3.899	99.3	0.06269	1.4271	0.696
950	0.3716	1.1321	4.023	108.2	0.06525	1.5610	0.699
1000	0.3524	1.1417	4.152	117.8	0.06752	1.6779	0.702
1100	0.3204	1.160	4.44	136.6	0.0732	1.969	0.704
1200	0.2947	1.179	4.92	159.1	0.0782	2.251	0.707
1300	0.2707	1.197	4.93	182.1	0.0837	2.583	0.705
1400	0.2515	1.214	5.17	205.5	0.0891	2.920	0.705
1500	0.2355	1.230	5.40	229.1	0.0946	3.262	0.705
1600	0.2211	1.248	5.63	254.5	0.100	3.609	0.705

* From *National Bureau of Standards* (USA), Circ, 564, 1955.

B.2 THERMOPHYSICAL PROPERTIES OF SATURATED WATER AND STEAM*

Temperature (°C)	*Specific-Volume* (m^3/kg)		*Isobaric Specific Heat Capacity* (kJ/kg K)		*Thermal Conductivity* (W/m K)		*Dynamic Viscosity* (g/ms)		*Prandtl Number*	
t	v_f	v_g	C_{pf}	C_{pg}	k_f	k_g	μ_f	μ_g	Pr_f	Pr_g
Triple point										
0.01	0.00100	206.2	4.217	1.854	0.509	0.0173	1.755	0.0088	13.02	0.942
10	0.00100	106.4	4.193	1.860	0.587	0.0185	1.301	0.0091	9.29	0.915
20	0.00100	57.8	4.182	1.866	0.603	0.0191	1.002	0.0094	6.95	0.918
30	0.00100	32.9	4.179	1.875	0.618	0.0198	0.797	0.0097	5.39	0.923
40	0.00100	19.5	4.179	1.885	0.632	0.0204	0.651	0.0101	4.31	0.930
50	0.00100	12.05	4.181	1.899	0.643	0.0210	0.544	0.0104	3.53	0.939
60	0.00102	7.68	4.185	1.915	0.653	0.0217	0.462	0.0107	2.96	0.947
70	0.00102	5.05	4.190	1.936	0.662	0.0224	0.400	0.0111	2.53	0.956
80	0.00103	3.41	4.197	1.962	0.670	0.0231	0.350	0.0114	2.19	0.966
90	0.00104	2.36	4.205	1.992	0.676	0.0240	0.311	0.0117	1.93	0.976
100	0.00104	1.673	4.216	2.028	0.681	0.0249	0.278	0.0121	1.723	0.047
125	0.00107	0.770	4.254	2.147	0.687	0.9272	0.219	0.0133	1.358	1.047
150	0.00109	0.392	4.310	2.314	0.687	0.0300	0.180	0.0144	1.133	1.110
175	0.00112	0.217	4.389	2.542	0.679	0.0334	0.153	0.0156	0.990	1.185
200	0.00116	0.127	4.497	2.843	0.665	0.0775	0.133	0.0167	0.902	1.270
225	0.00120	0.0383	4.648	3.238	0.644	0.0427	0.1182	0.0179	0.153	1.36
250	0.00125	0.0500	4.867	3.772	0.616	0.0495	0.1065	0.0191	0.841	1.45
275	0.00132	0.0327	5.202	4.561	3.582	0.0587	0.0972	0.0202	0.869	1.57
300	0.00140	0.0216	5.762	5.863	0.541	0.0719	0.0897	0.0214	0.955	1.74
325	0.00153	0.0142	6.861	8.440	0.493	0.0929	0.0790	0.0230	1.100	2.09
350	0.00174	0.00880	10.10	17.15	0.437	0.1343	0.0648	0.0258	1.50	3.29
360	0.00190	0.00694	14.6	25.1	0.400	0.168	0.0582	0.0275	2.11	4.89
374.15	0.00317	0.00317	–	∞	0.24	0.94	0.045	0.045	∞	∞
Critical point	0.00317	0.00317	–	∞	0.24	0.94	0.045	0.045	∞	∞

*Haywood, R W, *Thermodynamic tables in S.I. units*, Cambridge University Press, 1968.

B.3 THERMODYNAMIC PROPERTIES OF R 744 (CARBON DIOXIDE)*

Table B.3.1 Properties below critical temperature

Saturated Temperature (°C)	*Saturated Pressure* (MPa)	*Saturated Liquid and Vapour* — *Specific volume* (m^3/kg)		*Specific enthalpy* (kJ/kg)		*Specific entropy* (kJ/kg K)		*Vapour* — *Superheated* — By 30°C		By 60°C	
t	p	v_f	v_g	h_f	h_g	s_f	s_g	h	s	h	s
– 56.6											Triple point
– 40	1.005	0.00090	0.0382	zero	321.1	zero	1.377	355.4	1.507	383.0	1.611
– 35	1.20	0.00091	0.0320	9.7	322.2	0.39	1.352	356.9	1.485	385.6	1.588
– 30	1.43	0.00093	0.0270	19.5	323.1	0.079	1.328	358.7	1.464	388.0	1.566
– 25	1.68	0.00095	0.0229	29.5	323.7	0.119	1.304	360.4	1.442	390.3	1.545
– 20	1.97	0.00097	0.0195	39.7	323.7	0.158	1.280	361.8	1.421	392.5	1.525
– 15	2.29	0.00099	0.0166	50.2	323.2	0.198	1.256	363.0	1.401	394.5	1.505
– 10	2.65	0.00102	0.0142	60.9	322.3	0.238	1.231	363.9	1.381	396.2	1.486
– 5	3.04	0.00105	0.0122	72.0	320.5	0.278	1.205	364.6	1.361	397.8	1.467
0	3.48	0.00108	0.0104	83.7	318.1	0.320	1.178	364.9	1.342	399.3	1.449
5	3.97	0.00111	0.00879	96.0	312.9	0.364	1.143	364.9	1.322	400.4	1.431
10	4.50	0.00116	0.00743	109.1	307.2	0.407	1.107	364.7	1.302	401.4	1.414
15	5.08	0.00121	0.00623	123.3	301.0	0.454	1.071	364.0	1.282	402.2	1.396
20	5.73	0.00129	0.00516	139.1	292.3	0.506	1.028	362.9	1.261	402.7	1.379
25	6.44	0.00140	0.00413	159.7	279.9	0.573	0.976	361.5	1.241	403.0	1.362
30	7.21	0.00169	0.00294	191.2	253.1	0.682	0.886	359.6	1.220	402.9	1.345
31.05	7.38	0.00214	0.00214	223.0	223.0	0.780	0.780	359.1	1.216	402.9	1.341
										Critical point	

*Haywood R W, *Thermodynamic Tables in S.I. Units*, Cambridge University Press, 1968

Table B.3.2 Thermodynamic properties of gaseous R 744 (carbon dioxde)

p, MPa (T_{sat}, K)		Sat.	300	400	500	600
1.0	v	.0384	.0538	.0742	.0938	.1130
(233)	h	321.1	384.3	477.9	577.5	683.2
	s	1.377	1.617	1.886	2.108	2.300
2.0	v	.0190	.0254	.0364	.0465	.0564
(253.6)	h	322.7	373.7	472.8	574.3	681.0
	s	1.275	1.460	1.746	1.972	2.166
5.0	v	.0064	.0078	.0137	.0182	.0224
(287.5)	h	304.1	331.3	456.6	564.7	674.7
	s	1.086	1.178	1.542	1.784	1.984
10.0	v			.0062	.0089	.0111
	h			427.5	549.0	664.5
	s			1.356	1.628	1.838
20.0	v			.0026	.0043	.0055
	h			367.3	519.6	646.2
	s			1.106	1.448	1.680

Temperature of Vapour/Gas, K (columns 300–600)

B.4 THERMODYNAMIC PROPERTIES OF R 290 (PROPANE)*

Table B.4.1 Saturation table of R 290

Temp.	p_{sat}	*Specific Liquid* $v_f \times 10^3$	*Volume Vapour* v_g	*Enthalpy* h_f	h_{fg}	h_g	*Entropy* s_f	s_{fg}	s_g
(°C)	(bar)	(m³/kg)	(m³/kg)	(kJ/kg)			(kJ/kg.K)		
– 50	.699	1.729	.5828	83.41	432.92	516.33	.5363	1.9400	2.4763
– 48	.769	1.735	.5338	87.90	430.79	518.69	.5562	1.9133	2.4695
– 46	.843	1.742	.4897	92.41	428.64	521.05	.5760	1.8870	2.4630
– 44	.923	1.748	.4500	96.93	426.48	523.40	.5957	1.8610	2.4567
– 42	1.009	1.754	.4142	101.46	424.29	525.75	.6153	1.8355	2.4508
– 40	1.101	1.761	.3818	106.01	422.09	528.10	.6347	1.8103	2.4450
– 38	1.200	1.768	.3525	110.57	419.87	530.44	.6540	1.7855	2.4395
– 36	1.305	1.774	.3259	115.14	417.63	532.77	.6732	1.7610	2.4342
– 34	1.417	1.781	.3017	119.73	415.38	535.10	.6923	1.7368	2.4291
– 32	1.537	1.788	.2797	124.32	413.10	537.42	.7112	1.7130	2.4242
– 30	1.664	1.795	.2596	128.93	410.81	539.74	.7300	1.6895	2.4195
– 28	1.799	1.802	.2413	133.55	408.50	542.05	.747	1.6662	2.4150
– 26	1.942	1.809	.2246	138.19	406.16	544.35	.7673	1.6433	2.4106
– 24	2.094	1.816	.2092	142.84	403.81	546.65	.7858	1.6207	2.4065
– 22	2.255	1.823	.1951	147.50	401.43	548.94	.8042	1.5983	2.4025
– 20	2.425	1.831	.1822	152.18	399.03	551.21	.8224	1.5762	2.3986
– 18	2.605	1.838	.1703	156.87	396.61	553.48	.8406	1.5544	2.3949
– 16	2.794	1.846	.1593	161.58	394.16	555.74	.8586	1.5327	2.3914

(Contd.)

Temp.	p_{sat}	*Specific Liquid* $v_f \times 10^3$	*Volume Vapour* v_g	*Enthalpy* h_f	h_{fg}	h_g	*Entropy* s_f	s_{fg}	s_g
(°C)	**(bar)**	**(m^3/kg)**	**(m^3/kg)**	**(kJ/kg)**			**(kJ/kg.K)**		
– 14	2.994	1.853	.1492	166.31	391.68	557.99	.8766	1.5114	2.3879
– 12	3.205	1.861	.1399	171.06	389.17	560.23	.8944	1.4902	2.3846
– 10	3.426	1.869	.1312	175.82	386.63	562.46	.9122	1.4692	2.3814
– 8	3.659	1.877	.1232	180.61	384.06	564.67	.9299	1.4484	2.3783
– 6	3.904	1.885	.1158	185.42	381.45	566.87	.9476	1.4278	2.3754
– 4	4.161	1.894	.1090	190.25	378.81	569.06	.9651	1.4074	2.3725
– 2	4.430	1.902	.1026	195.11	376.12	571.24	.9826	1.3871	2.3697
0	4.712	1.911	.0966	200.00	373.40	573.40	1.0000	1.3669	2.3669
2	5.007	1.919	.0911	204.91	370.63	575.54	1.0174	1.3469	2.3643
4	5.316	1.928	.0860	209.86	367.81	577.67	1.0347	1.3271	2.3617
6	5.639	1.937	.0812	214.83	364.95	579.77	1.0519	1.3073	2.3592
8	5.976	1.947	.0767	219.83	362.03	581.86	1.0691	1.2876	2.3567
10	6.329	1.956	.0725	224.87	359.07	583.93	1.0862	1.2681	2.3543
12	6.696	1.966	.0686	229.93	356.05	585.98	1.1033	1.2486	2.3519
14	7.079	1.975	.0649	235.02	352.99	588.01	1.1203	1.2292	2.3495
16	7.478	1.985	.0614	240.16	349.86	590.02	1.1372	1.2099	2.3472
18	7.893	1.996	.0582	245.32	346.67	592.00	1.1542	1.1907	2.3448
20	8.325	2.006	.0552	250.52	343.43	593.95	1.1710	1.1715	2.3425
22	8.774	2.017	.0523	255.75	340.12	595.87	1.1878	1.1523	2.3401
24	9.241	2.028	.0497	261.02	336.75	597.77	1.2045	1.1332	2.3378
26	9.726	2.039	.0471	266.32	333.32	599.63	1.2212	1.1142	2.3354
28	10.230	2.051	.0448	271.69	329.76	601.45	1.2379	1.0950	2.3329
30	10.752	2.063	.0425	277.07	326.18	603.24	1.2545	1.0759	2.3304
32	11.294	2.076	.0404	282.48	322.52	605.00	1.2709	1.0569	2.3278
34	11.855	2.088	.0384	287.93	318.78	606.71	1.2873	1.0378	2.3252
36	12.437	2.102	.0365	293.41	314.97	608.38	1.3036	1.0188	2.3224
38	13.040	2.115	.0347	298.93	311.07	610.00	1.3198	.9997	2.3195
40	13.664	2.130	.0330	304.48	307.09	611.57	1.3359	.9806	2.3166
42	14.309	2.144	.0314	310.07	303.01	613.08	1.3520	.9615	2.3134
44	14.977	2.160	.0299	315.70	298.84	614.54	1.3679	.9422	2.3101
46	15.667	2.176	.0285	321.37	294.56	615.93	1.3837	.9229	2.3066
48	16.381	2.192	.0271	327.08	290.18	617.26	1.3994	.9035	2.3029
50	17.119	2.210	.0258	332.82	285.68	618.51	1.4149	.8840	2.2989
52	17.881	2.228	.0246	338.61	281.07	619.68	1.4303	.8644	2.2947
54	18.668	2.247	.0234	344.44	276.33	620.77	1.4456	.8446	2.2902
56	19.480	2.268	.0222	350.31	271.46	621.77	1.4606	.8247	2.2854
58	20.319	2.289	.0212	356.21	266.46	622.67	1.4755	.8046	2.2802
60	21.184	2.312	.0202	362.15	261.34	623.48	1.4902	.7844	2.2746

Table B.4.2 Superheat table: R 290 vapour

p, bar (t_{sat}, °C)	sat.	t,°C –15	0	30	45	70	95	120	145	170
1.01	v 0.4143	0.4673	0.4970	0.5554	0.5840	0.6323	0.6800	0.7274	0.7746	0.8217
(–41.14)	h 526.98	566.21	589.79	639.62	665.97	712.07	760.93	812.56	866.93	923.98
	s 2.4553	2.6158	2.7046	2.8776	2.9627	3.1023	3.2398	3.3755	3.5096	3.6422
1.50	v 0.2860	0.3111	0.3318	0.3723	0.3922	0.4249	0.4574	0.4897	0.5219	0.5539
(– 32.69)	h 536.59	563.90	587.82	638.15	664.67	710.98	760.02	811.79	866.27	923.41
	s 2.4251	2.5349	2.6251	2.8001	2.8856	3.0258	3.1639	3.3000	3.4344	3.5671
2.50	v 0.1758	0.1808	0.1941	0.2196	0.2320	0.2522	0.2722	0.2918	2.3114	0.3308
(– 20.46)	h 550.06	558.92	583.61	635.01	661.92	708.75	758.16	810.19	864.92	922.22
	s 2.3887	2.4235	2.5168	2.6958	2.7826	2.9245	3.0636	3.2004	3.3354	3.4687
3.50	v 0.1296		0.1349	0.1541	0.1632	0.1782	0.1927	0.2021	0.2212	0.2352
(– 7.86)	h 565.73		579.16	631.76	659.09	706.45	756.25	508.60	863.51	921.02
	s 2.3902		2.4404	2.6239	2.7122	2.8558	2.9961	3.1338	3.2693	3.4030
4.50	v 0.1009		0.1019	0.1176	0.1250	0.1370	0.1486	0.1599	0.1711	0.1822
(– 1.64)	h 571.53		574.44	628.38	656.17	704.09	754.31	806.97	862.15	919.84
	s 2.3680		2.3786	2.5675	2.6574	2.8029	2.9445	3.0831	3.2194	3.3535
5.50	v 0.0831			0.0943	0.1006	0.1108	0.1205	0.1299	0.1392	0.1484
(5.12)	h 578.82			624.86	653.14	701.68	752.33	805.32	860.74	918.63
	s 2.3601			2.5201	2.6118	2.7594	2.9024	3.0419	3.1788	3.3135
6.50	v 0.0706			0.0781	0.0837	0.0926	0.1010	0.1092	0.1171	0.1250
(11.01)	h 585.02			621.18	650.01	699.21	750.32	803.64	559.31	917.40
	s 2.3536			2.4784	2.5722	2.7220	2.8664	3.0070	3.1446	3.2797
7.50	v 0.0613			0.0662	0.0713	0.0792	0.0867	0.0939	0.1009	0.1078
(16.21)	h 590.31			617.31	646.76	696.67	748.26	801.93	857.87	916.17
	s 2.3477			2.4404	2.5365	2.6888	2.8348	2.9764	3.1148	3.2505

(Contd)

p, bar (t_{sat}, °C)	sat.	t, °C – 15	0	30	45	70	95	120	145	170
8.50	v 0.0541			0.0571	0.0617	0.0690	0.0758	0.0823	0.0885	0.0947
(20.87)	h 594.87			613.30	643.37	694.05	746.16	800.20	856.41	914.92
	s 2.3421			2.4052	2.5037	2.6588	2.8064	2.9492	3.0883	3.2245
9.50	v 0.0483			0.0498	0.0542	0.0609	0.0671	0.0730	0.0788	0.0843
(25.10)	h 598.83			609.01	639.82	691.36	744.01	798.43	854.93	913.66
	s 2.3367			2.3714	2.4728	2.6311	2.7805	2.9244	3.0644	3.2012
10.50	v 0.0436			0.0438	0.0481	0.0543	0.0601	0.0656	0.0708	0.0759
(29.01)	h 602.32			604.44	636.20	688.58	741.82	796.64	853.43	912.38
	s 2.3313			2.3386	2.4437	2.6051	2.7565	2.9017	3.0425	3.1799
11.50	v 0.0396				0.0429	0.0489	0.0543	0.0594	0.0643	0.0690
(32.66)	h 605.46				632.33	685.71	739.58	794.82	851.92	911.09
	s 2.3262				2.4153	2.5805	2.7341	2.8806	3.0222	3.1603
12.50	v 0.0363				0.0386	0.0443	0.0494	0.0542	0.0588	0.0632
(36.11)	h 608.31				628.25	682.74	737.28	792.97	850.38	909.79
	s 2.3213				2.3874	2.5569	2.7128	2.8607	3.0033	3.1421
13.50	v 0.0334				0.0348	0.0404	0.0453	0.0498	0.0541	0.0582
(39.38)	h 610.92				623.92	679.76	734.93	791.08	848.82	908.48
	s 2.3165				2.3597	2.5343	2.6926	2.8420	2.9856	3.1250
14.50	v 0.0310				0.0316	0.0370	0.0417	0.0460	0.0500	0.0539
(42.51)	h 613.34				619.29	67659	732.51	789.16	847.24	907.15
	s 2.3119				2.3317	2.5122	2.6731	2.8242	2.9688	3.1089
15.50	v 0.0288					0.0340	0.0385	0.0427	0.0465	0.0502
(45.50)	h 615.56					673.29	730.03	787.21	845.64	905.80
	s 2.3073					2.4903	2.6543	2.8071	2.9528	3.0937
16.50	v 0.0269					0.0314	0.0358	0.0397	0.0434	0.0469
(48.38)	h 617.59					669.86	727.48	785.21	844.02	904.45
	s 2.3027					2.4687	2.6360	2.7907	2.9375	3.0792

* Ashok Babu. T P, *A Theo. and Experimental Investigations of Alternatives to CFC 12 in Refrigerators*, Ph. D. Thesis, IIT Delhi, 1997.

B.5 THERMODYNAMIC PROPERTIES OF R 22*

Table B.5.1 Saturation table of R 22

Temp.	p_{sat}	*Specific* *Liquid* $v_f \times 10^3$	*Volume* *Vapour* v_g	*Enthalpy* h_f	h_{fg}	h_g	*Entropy* s_f	s_{fg}	s_g
(°C)	(bar)	(m³/kg)	(m³/kg)	(kJ/kg)			(kJ/kg.K)		
– 50	0.644	.695	.3246	145.05	238.96	384.01	.7792	1.0708	1.8500
– 48	0.713	.698	.2952	147.12	237.84	384.96	.7884	1.0563	1.8447
– 46	0.787	.701	.2690	149.20	236.70	385.90	.7976	1.0420	1.8396
– 44	0.868	.704	.2456	151.29	235.55	386.84	.8067	1.0279	1.8346
– 42	0.955	.706	.2246	153.39	234.38	387.77	.8158	1.0139	1.8297
– 40	1.049	.709	.2057	155.51	233.19	388.70	.8249	1.0001	1.8250
– 38	1.151	.712	.1888	157.63	231.99	389.62	.8339	.9865	1.8204
– 36	1.259	.715	.1735	159.76	230.77	390.53	.8429	.9730	1.8160
– 34	1.376	.718	.1597	161.90	229.53	391.43	.8519	.9597	1.8116
– 32	1.501	.721	.1472	164.06	228.27	392.33	.8608	.9466	1.8074
– 30	1.635	.725	.1358	166.22	227.00	393.22	.8698	.9335	1.8033
– 28	1.778	.728	.1256	168.40	225.70	394.10	.8786	.9206	1.7993
– 26	1.930	.731	.1162	170.59	224.39	394.97	.8875	.9079	1.7953
– 24	2.092	.734	.1077	172.78	223.05	395.84	.8963	.8952	1.7915
– 22	2.265	.738	.0999	174.99	221.70	396.69	.9051	.8827	1.7878
– 20	2.448	.741	.0928	177.21	220.32	397.53	.9139	.8703	1.7841
– 18	2.643	.744	.0864	179.44	218.93	398.37	.9226	.8580	1.7806
– 16	2.849	.748	.0804	181.68	217.51	399.19	.9313	.8458	1.7771
– 14	3.068	.751	.0750	183.93	216.07	400.00	.9400	.8337	1.7737
– 12	3.299	.755	.0699	186.20	214.61	400.81	.9486	.8218	1.7704
– 10	3.543	.759	.0653	188.47	213.13	401.60	.9572	.8099	1.7670
– 8	3.801	.763	.0611	190.75	211.62	402.37	.9658	.7981	1.7639
– 6	4.072	.766	.0572	193.05	210.09	403.14	.9744	.7864	1.7608
– 4	4.358	.770	.0536	195.36	208.54	403.89	.9830	.7748	1.7577
– 2	4.659	.774	.0502	197.67	206.96	404.63	.9915	.7632	1.7547
0	4.976	.778	.0471	200.00	205.36	405.36	1.000	.7518	1.7518
2	5.308	.782	.0443	202.34	203.73	406.07	1.0085	.7404	1.7489
4	5.657	.787	.0416	204.69	202.08	406.77	1.0169	.7291	1.7460
6	6.023	.791	.0391	207.05	200.40	407.45	1.0254	.7179	1.7432
8	6.406	.795	.0368	209.42	198.69	408.11	1.0338	.7067	1.7405
10	6.807	.800	.0347	211.81	196.95	408.76	1.0422	.6956	1.7377
12	7.226	.805	.0327	214.20	195.19	409.39	1.0506	.6845	1.7351
14	7.665	.809	.0309	216.61	193.40	410.01	1.0589	.6735	1.7324
16	8.123	.814	.0291	219.03	191.57	410.60	1.0673	.6625	1.7298
18	8.601	.819	.0275	221.46	189.72	411.18	1.0756	.6516	1.7272
20	9.099	.824	.0260	223.90	187.83	411.73	1.0839	.6407	1.7246
22	9.619	.830	.0246	226.36	185.91	412.27	1.0922	.6299	1.7221
24	10.160	.835	.0233	228.83	183.95	412.78	1.1005	.6190	1.7195
26	10.723	.840	.0220	231.31	181.96	413.27	1.1088	.6082	1.7170

(Contd)

Temp.	p_{sat}	Specific Liquid $v_f \times 10^3$	Volume Vapour v_g	Enthalpy h_f	h_{fg}	h_g	Entropy s_f	s_{fg}	s_g
(°C)	(bar)	(m³/kg)	(m³/kg)	(kJ/kg)			(kJ/kg.K)		
28	11.309	.846	.0208	233.81	179.93	413.74	1.1770	.5975	1.7145
30	11.919	.852	.0197	236.31	177.86	414.18	1.1253	.5867	1.7120
32	12.552	.858	.0187	238.84	175.75	414.59	1.1336	.5759	1.7095
34	13.210	.864	.0177	241.38	173.60	414.98	1.1418	.5652	1.7070
36	13.892	.870	.0168	243.93	171.41	415.33	1.1501	.5544	1.7045
38	14.601	.877	.0160	246.50	169.16	415.66	1.1583	.5437	1.7020
40	15.335	.884	.0151	249.08	166.87	415.95	1.1666	.5329	1.6995
42	16.096	.891	.0144	251.68	164.53	416.21	1.1749	.5220	1.6969
44	16.885	.898	.0136	254.30	162.13	416.43	1.1831	.5112	1.6943
46	17.702	.906	.0129	256.94	159.67	416.61	1.1914	.5003	1.6917
48	18.548	.914	.0123	259.59	157.15	416.75	1.1998	.4893	1.6891
50	19.423	.922	.0117	262.27	154.57	416.84	1.2081	.4783	1.6864
52	20.328	.930	.0111	264.97	151.92	416.89	1.2165	.4672	1.6837
54	21.265	.939	.0105	267.69	149.19	416.88	1.2249	.4560	1.6809
56	22.232	.949	.010	270.43	146.38	416.81	1.2333	.4447	1.6780
58	23.232	.958	.0095	273.20	143.48	416.68	1.2418	.4333	1.6751
60	24.266	.969	.0090	276.00	140.49	416.49	1.2504	.4217	1.6721

Table B.5.2 Superheat table: R 22 vapour

t(°C)	v,(m³/kg)	h,(kJ/kg)	s, (kJ/kg. K)	t(°C)	v,(m³/kg)	h,(kJ/kg)	s, (kJ/kg. K)
Saturation temperature, –20°C				**Saturation temperature, – 10°C**			
– 20	.0928	397.5	1.7841				
– 15	.0951	400.7	1.7969				
– 10	.0974	404.0	1.8095	– 10	.0653	401.5	1.7671
– 5	.0997	407.3	1.8219	– 5	.0670	404.9	1.7800
0	.1019	4106.	1.8341	0	.0687	408.4	1.7927
5	.1041	413.9	1.8461	5	.0703	411.8	1.8052
10	.1063	417.3	1.8580	10	0719	415.2	1.8174
15	.1085	420.6	1.8697	15	.0735	418.7	1.8295
20	.1107	423.9	1.8813	20	.0750	422.1	1.8414
25	.1128	426.3	1.8928	25	.0766	425.6	1.8531
Saturation temperature, 0°C				**Saturation temperature, 5°C**			
0	.0471	405.3	1.7518	5	.0404	407.1	1.7446
5	.0484	408.9	1.7649	10	.0415	410.8	1.7578
10	.0496	412.5	1.7777	15	.0425	414.5	1.7708
15	.0508	416.1	1.7903	20	.0436	418.2	1.7834
20	.0520	419.6	1.8026	25	.0446	421.8	1.7958
25	.0532	423.3	1.8148	30	.0456	425.5	1.8080
				35	.0467	429.2	1.8200
				40	.0477	432.8	1.8319
				45	.0487	436.5	1.8435
				50	.0496	440.2	1.8550

(Contd)

t(°C)	v,(m^3/kg)	h,(kJ/kg)	s, (kJ/kg. K)	t(°C)	v,(m^3/kg)	h,(kJ/kg)	s, (kJ/kg. K)
Saturation temperature, 10°C				**Saturation temperature, 15°C**			
10	.0347	408.6	1.7377	10			
15	.0357	412.4	1.7511	15	.0300	410.2	1.7311
20	.0366	416.2	1.7642	20	.0308	414.0	1.7556
25	.0376	420.0	1.7769	25	.0317	417.8	1.7578
30	.0385	423.4	1.7894	30	.0325	421.5	1.7707
35	.0394	427.1	1.8017	35	.0334	425.2	1.7833
40	.0403	431.0	1.8137	40	.0342	429.0	1.7956
45	.0412	434.4	1.8256	45	.0349	432.8	1.8078
50	.0420	437.9	1.8373	50	.0357	436.5	1.8197
Saturation temperature, 20°C				**Saturation temperature, 25°C**			
20	.0260	411.5	1.7246	25	.0226	413.0	1.7183
25	.0267	415.4	1.7383	30	.0233	417.1	1.7322
30	.0278	419.3	1.7517	35	.0240	421.1	1.7458
35	.0286	423.3	1.7646	40	.0247	425.1	1.7590
40	.0290	427.1	1.7774	45	.0254	429.0	1.7718
45	.0297	431.1	1.7899	50	.0260	433.0	1.7844
50	.0304	434.8	1.8021	55	.0266	437.0	1.7967
55	.0311	438.8	1.8141	60	.0272	441.0	1.8087
60	.0318	442.5	1.8258	65	.0278	444.8	1.8206
Saturation temperature, 30°				**Saturation temperature, 32°C**			
30	.0197	414.0	1.7120	35	.0191	417.1	1.7182
35	.0204	418.3	1.7262	40	.0197	421.4	1.7322
40	.0210	422.4	1.7400	45	.0203	425.5	1.7458
45	.0216	426.6	1.7534	50	.0209	429.7	1.7591
50	.0222	430.5	1.7664	55	.0214	433.8	1.7719
55	.0228	434.6	1.7791	60	.0220	437.9	1.7845
60	.0234	438.7	1.7915	65	.0225	442.0	1.7968
65	.0239	442.6	1.8036	70	.0231	446.0	1.8089
				75	.0236	450.0	1.8207
				80	.0241	454.0	1.8323
Saturation temperature, 34°C				**Saturation temperature, 36°C**			
35	.0179	415.7	1.7099				
40	.0185	420.0	1.7243	40	.0173	418.7	1.7162
45	.0191	424.4	1.7382	45	.0179	423.1	1.7304
50	.0196	428.5	1.7517	50	.0185	427.4	1.7442
55	.0202	432.8	1.7647	55	.0190	431.7	1.7575
60	.0207	436.9	1.7775	60	.0195	436.0	1.7704
65	.0212	441.0	1.7899	65	.0200	440.0	1.7830
70	.0217	445.0	1.8021	70	.0205	444.3	1.7954
75	.0222	449.0	1.8141	75	.0210	448.2	1.8074
80	.0227	453.0	1.8258	80	.0214	452.1	1.8193

(Contd)

t(°C)	v,(m³/kg)	h,(kJ/kg)	s, (kJ/kg. K)	t(°C)	v,(m³/kg)	h,(kJ/kg)	s, (kJ/kg. K)
Saturation temperature, 38°C				**Saturation temperature, 40°C**			
40	.0162	417.3	1.7080	40	.0151	415.9	1.6995
45	.0168	421.9	1.7225	45	.0157	420.4	1.7144
50	.0173	426.2	1.7365	50	.0162	424.9	1.7287
55	.0178	430.6	1.7501	55	.0168	429.3	1.7426
60	.0183	434.8	1.7632	60	.0172	433.6	1.7560
65	.0188	439.0	1.7760	65	.0177	438.0	1.7690
70	.0193	443.4	1.7885	70	.0182	442.1	1.7817
75	.0198	447.3	1.8008	75	.0187	446.2	1.7940
80	.0202	451.2	1.8127	80	.0191	450.5	1.8061
				85	.0195	454.8	1.8180
Saturation temperature, 42°C				**Saturation temperature, 45°C**			
45	.0147	419.0	1.7061	45	.0133	416.5	1.6931
50	.0152	423.5	1.7208	50	.0138	421.3	1.7084
55	.0157	428.0	1.7349	55	.0143	426.0	1.7231
60	.0162	432.4	1.7486	60	.0148	430.5	1.7372
65	.0167	436.8	1.7618	65	.0153	435.1	1.7509
70	.0172	441.2	1.7747	70	.0157	439.4	1.7641
75	.0176	445.4	1.7872	75	.0161	443.6	1.7769
80	.0180	449.5	1.7995	80	.0165	448.0	1.7895
85	.0185	451.7	1.8115	85	.0170	452.4	1.8017
				90	.0174	456.6	1.8137

* Ashok Babu T P, *A Theoretical and Experimental Investigation of Alternatives to CFC 12 in Refrigerators,* Ph. D. Thesis, IIT Delhi, 1997.

B.6 THERMODYNAMIC PROPERTIES OF R 717 (AMMONIA)

Table B.6.1 Saturation table of R 717 (ammonia)

Temp.	***Pressure***	***Specific Volume* (m³/kg)**		***Enthalpy* (kJ/kg)**		***Entropy* kJ/kg.K)**	
(°C)	**(bar)**	$v_f \times 10^3$	v_g	h_f	h_g	s_f	s_g
–60	.2199	1.40	4.685	–69.5	1373.2	–0.1095	6.6592
–55	.3029	1.41	3.474	–47.5	1382.0	–0.0071	6.5454
–50	.4103	1.42	2.617	–25.4	1390.6	0.0926	6.4382
–45	.5474	1.43	2.000	–3.3	1399.0	0.1904	6.3369
–40	.7201	1.45	1.547	18.9	1407.2	0.2865	6.2410
–35	.9349	1.46	1.212	41.2	1415.2	0.3808	6.1501
–30	1.1990	1.48	.961	63.6	1422.8	0.4735	6.0636
–28	1.3202	1.48	.878	72.5	1425.8	0.5101	6.0302
–26	1.4511	1.48	.809	81.5	1428.7	0.5465	5.9974
–25	1.5216	1.49	.770	86.0	1430.2	0.5646	5.9813
–24	1.5922	1.49	.737	90.5	1431.6	0.5827	5.9652
–22	1.7441	1.49	.677	99.6	1434.4	0.6186	5.9336

(Contd)

Temp.	*Pressure*	*Specific Volume* (m^3/kg)		*Enthalpy* (kJ/kg)		*Entropy* kJ/kg.K)	
(°C)	(bar)	$v_f \times 10^3$	v_g	h_f	h_g	s_f	s_g
–20	1.9074	1.50	.622	108.6	1437.2	0.6543	5.9025
–18	2.0826	1.50	.573	117.7	1439.9	0.6898	5.8720
–16	2.2704	1.51	.528	126.7	1442.6	0.7251	5.8420
–15	2.3709	1.52	.508	131.3	1443.9	0.7426	5.8223
–14	2.4714	1.52	.488	135.8	1445.2	0.7601	5.8125
–12	2.6863	1.53	.451	144.9	1447.7	0.7950	5.7835
–10	2.9157	1.53	.417	154.2	1450.2	0.8296	5.7550
–9	3.0360	1.53	.402	158.6	1451.4	0.8469	5.7409
–8	3.1602	1.54	.387	163.2	1452.6	0.8641	5.7269
–7	3.2884	1.54	.373	167.8	1453.8	0.8812	5.7131
–6	3.4207	1.54	.359	172.4	1455.0	0.8983	5.6993
–5	3.5571	1.55	.346	176.9	1456.1	0.9154	5.6856
–4	3.6977	1.55	.334	181.6	1457.2	0.9324	5.6721
–3	3.8426	1.55	.322	186.2	1458.4	0.9493	5.6586
–2	3.9920	1.56	.310	190.8	1459.5	0.9663	5.6453
–1	4.1458	1.56	.299	195.4	1460.6	0.9831	5.6320
0	4.3043	1.57	.289	200.0	1461.7	1.0000	5.6189
1	4.4674	1.57	.279	204.6	1462.7	1.0167	5.6058
2	4.6353	1.57	.269	209.3	1463.8	1.0335	5.5929
3	4.8081	1.57	.260	213.9	1464.8	1.0502	5.5800
4	4.9859	1.58	.251	218.5	1465.8	1.0669	5.5672
5	5.1687	1.58	.243	223.2	1466.8	1.0835	5.5545
6	5.3567	1.59	.235	227.8	1467.8	1.1001	5.5419
7	5.5500	1.59	.227	232.5	1468.8	1.1167	5.5294
8	5.7487	1.59	.219	237.1	1459.7	1.1332	5.5170
9	5.9528	1.60	.212	241.8	1470.7	1.1496	5.5046
10	6.1625	1.60	.205	246.5	1471.5	1.1661	5.4924
11	6.3778	1.60	.198	251.2	1472.5	1.1825	5.4802
12	6.5989	1.61	.192	255.9	1473.3	1.1988	5.4681
13	6.8259	1.61	.186	260.6	1474.2	1.2152	5.4561
14	7.0588	1.61	.180	265.3	1475.4	1.2314	5.4441
15	7.2979	1.62	.174	270.0	1475.9	1.2477	5.4322
16	7.5431	1.62	.169	274.8	1476.2	1.2639	5.4204
17	7.7946	1.62	.164	279.5	1477.5	1.2801	5.4087
18	8.0525	1.63	.158	284.8	1478.3	1.2963	5.3971
19	8.3169	1.63	.154	289.0	1479.0	1.3124	5.3855
20	8.5879	1.64	.149	293.8	1479.8	1.3285	5.3740
21	8.8657	1.64	.144	298.5	1480.5	1.3445	5.3626
22	9.1503	1.64	.140	303.3	1481.2	1.3606	5.3512
23	9.4418	1.65	.136	308.4	1481.9	1.3765	5.3399
24	9.7403	1.65	.132	312.9	1482.5	1.3925	5.3286

(Contd)

Temp.	Pressure	Specific Volume (m^3/kg)		Enthalpy (kJ/kg)		Entropy kJ/kg.K)	
(°C)	(bar)	$v_f \times 10^3$	v_g	h_f	h_g	s_f	s_g
25	10.046	1.66	.128	317.7	1483.2	1.4084	5.3175
26	10.359	1.66	.124	322.5	1483.8	1.4243	5.3063
27	10.680	1.67	.128	327.3	1484.4	1.4402	5.2953
28	11.007	1.67	.117	332.1	1485.0	1.4560	5.2843
29	11.343	1.67	.114	336.9	1485.8	1.4718	5.2733
30	11.686	1.68	.110	341.8	1486.1	1.4876	5.2624
31	12.037	1.68	.107	346.6	1486.7	1.5033	5.2516
32	12.396	1.69	.104	351.5	1487.2	1.5191	5.2408
33	12.763	1.69	.101	356.3	1487.7	1.5348	5.2300
34	13.139	1.70	.098	361.2	1488.1	1.5504	5.2193
35	13.522	1.70	.096	366.1	1488.6	1.5660	5.2086
36	13.915	1.71	.092	370.9	1489.0	1.5816	5.1980
37	14.314	1.71	.090	375.9	1489.4	1.5972	5.1874
38	14.724	1.72	.088	380.8	1489.8	1.6128	5.1768
39	15.143	1.72	.085	385.7	1490.1	1.6283	5.1663
40	15.570	1.72	.083	390.6	1490.4	1.6437	5.1558
41	16.006	1.73	.080	395.5	1490.7	1.6592	5.1453
42	16.451	1.73	.078	400.4	1490.9	1.6747	5.1349
43	16.906	1.74	.076	405.4	1491.2	1.6901	5.1244
44	17.370	1.74	.074	410.4	1491.4	1.7055	5.1140
45	17.843	1.75	.072	415.4	1491.5	1.7209	5.1036
46	18.326	1.75	.070	420.4	1491.7	1.7363	5.0932
47	18.819	1.76	.068	425.4	1491.8	1.7517	5.0827
48	19.322	1.76	.066	430.4	1491.8	1.7671	5.0723
49	19.835	1.77	.065	435.4	1491.9	1.7825	5.0618
50	20.359	1.77	.063	440.5	1491.8	1.7979	5.0514
51	20.892	1.78	.061	445.6	1491.8	1.8134	5.0409
52	21.436	1.78	.060	450.7	1491.7	1.8289	5.0303
53	21.991	1.79	.058	455.9	1491.5	1.8444	5.0198
54	22.556	1.79	.056	461.1	1491.3	1.8600	5.0092
55	23.132	1.80	.055	466.3	1491.1	1.8757	4.9985

Table B.6.2 Superheat table: R 717 (ammonia) vapour

t_{sat}	p_{sat}	Degree of Superheat of Vapour 50°C			100°C		
		v	h	s	v	h	s
(°C)	(bar)	(m^3/kg)	(kJ/kg)	(kJ/kg.K)	(m^3/kg)	(kJ/kg)	(kJ/kg.K)
–40	0.718	1.82	1517	6.667	2.08	1624	7.016
–35	0.932	1.45	1526	6.572	1.76	1634	6.919
–30	1.196	1.24	1535	6.483	1.45	1644	6.827

(Contd)

t_{sat}	p_{sat}	Degree of Superheat of Vapour 50°C			100°C		
		v	h	s	v	h	s
(°C)	(bar)	(m³/kg)	(kJ/kg)	(kJ/kg.K)	(m³/kg)	(kJ/kg)	(kJ/kg.K)
–25	1.516	.96	1544	6.399	1.15	1654	6.741
–20	1.9	.78	1553	6.319	.90	1664	6.659
–15	2.36	.61	1561	6.243	.73	1674	6.581
–10	2.91	.53	1570	6.171	.59	1683	6.507
–5	3.55	.42	1578	6.102	.49	1693	6.437
0	4.29	.36	1586	6.036	.42	1702	6.370
5	5.16	.30	1594	5.974	.35	1711	6.307
10	6.15	.25	1601	5.914	.285	1720	6.247
15	7.28	.22	1608	5.856	.25	1729	6.189
20	8.57	.185	1615	5.801	.215	1737	6.133
25	10.01	.165	1622	5.748	.18	1746	6.080
30	11.67	.137	1628	5.697	.16	1754	6.030
35	13.5	.118	1634	5.648	.14	1762	5.982
40	15.54	.110	1640	6.601	.12	1770	5.935
45	17.82	.090	1646	5.555	.105	1778	5.890
50	20.33	.070	1651	5.510	.085	1785	5.847

B.7 THERMODYNAMIC PROPERTIES OF R 12*

Saturation Temp.	*Saturation Pressure*	*Saturated Liquid and Vapour*						*Vapour Superheated*			
								By 20		By 40°C	
t (°C)	p (bar)	v_f (kJ/kg)	v_g (m^3/kg)	h_f (kJ/kg)	h_g (kJ/k)	s_f (kJ/kg.K)	s_g (kJ/kg.K)	h (kJ/kg)	s (kJ/kg.K)	h (kJ/kg)	s (kJ/kg.K)
–40	0.6417	0.66	0.2421	0	169.0	0	0.7274	180.8	0.7737	192.4	0.8178
–35	0.8069	0.67	0.1950	4.4	171.9	0.0187	0.7220	183.3	0.7681	195.1	0.8120
–30	1.0038	0.67	0.1595	8.9	174.2	0.0371	0.7171	185.8	0.7631	197.8	0.8068
–25	1.2368	0.68	0.1313	13.3	176.5	0.0552	0.7127	188.3	0.7586	200.4	0.8021
–20	1.5089	0.69	0.1089	17.8	178.7	0.0731	0.7088	190.8	0.7546	203.1	0.7979
–15	1.8256	0.69	0.0911	22.3	181.0	0.0906	0.7052	193.2	0.7510	205.7	0.7942
–10	2.1912	0.70	0.0767	26.9	183.2	0.1080	0.7020	195.7	0.7477	208.3	0.7909
–5	2.610	0.71	0.0650	31.4	185.4	0.1251	0.6991	198.1	0.7449	210.9	0.7879
0	3.086	0.72	0.0554	36.1	187.5	0.1420	0.6966	200.5	0.7423	213.5	0.7853
5	3.626	0.72	0.0475	40.7	189.7	0.1587	0.6942	202.9	0.7401	216.1	0.7830
10	4.233	0.73	0.0409	45.4	191.7	0.1752	0.6921	205.2	0.7381	218.6	0.7810
15	4.914	0.74	0.0354	50.1	193.8	0.1915	0.6902	207.5	0.7363	221.2	0.7792
20	5.673	0.75	0.0308	54.9	195.8	0.2078	0.6885	209.8	0.7348	223.7	0.7777
25	6.516	0.76	0.0269	59.7	197.7	0.2239	0.6869	212.1	0.7334	226.1	0.7763
30	7.450	0.77	0.0235	64.6	199.6	0.2399	0.6854	214.3	0.7321	228.6	0.7751
35	8.477	0.79	0.0206	69.5	201.5	0.2559	0.6839	216.4	0.7310	231.0	0.7741
40	9.607	0.80	0.0182	74.6	203.2	0.2718	0.6825	218.5	0.7300	233.4	0.7732
45	10.843	0.81	0.0160	79.7	204.9	0.2877	0.6812	220.6	0.7291	235.7	0.7724
50	12.193	0.83	0.0142	84.9	206.5	0.3037	0.6797	222.6	0.7282	238.0	0.7718
60	15.259	0.86	0.0111	95.7	209.3	0.3358	0.6777	226.4	0.7265	242.4	0.7706
70	18.859	0.90	0.0087	107.1	211.5	0.3686	0.6738	230.2	0.7240	246.2	0.7650

*Haywood R W, *Thermodynamics Tables in S.I. Units*, Cambridge University Press, 1968, p.22.

B.8 THERMODYNAMIC PROPERTIES OF R 152a*

Table B.8.1 Saturation table of R 152a

Temp.	p_{sat}	*Specific Volume* Liquid $v_f \times 10^3$	*Specific Volume* Vapour v_g	*Enthalpy* h_f	h_{fg}	h_g	*Entropy* s_f	s_{fg}	s_g
(°C)	(bar)	(m³/kg)	(m³/kg)	(kJ/kg)			(kJ/kg.K)		
–40	0.479	0.957	0.5989	139.78	338.59	478.38	0.7552	1.4522	2.2074
–38	0.530	0.961	0.5453	142.52	337.36	479.88	0.7673	1.4346	2.2019
–36	0.585	0.964	0.4973	145.28	336.11	481.39	0.7794	1.4172	2.1966
–34	0.644	0.968	0.4543	148.80	334.81	482.89	0.7916	1.4000	2.1915
–32	0.709	0.972	0.4156	150.90	333.49	484.39	0.8037	1.3829	2.1866
–30	0.778	0.976	0.3808	153.75	332.13	485.88	0.8158	1.3659	2.1818
–28	0.853	0.980	0.3495	156.63	330.74	487.37	0.8280	1.3491	2.1771
–26	0.933	0.984	0.3212	159.54	329.31	488.85	0.8402	1.3324	2.1726
–24	1.020	0.988	0.2956	162.47	327.85	490.33	0.8524	1.3158	2.1682
–22	1.112	0.992	0.2724	165.45	326.35	491.80	0.8646	1.2994	2.1640
–20	1.211	0.996	0.2514	168.43	324.83	493.26	0.8768	1.2831	2.1599
–18	1.317	1.001	0.2323	171.47	323.25	494.72	0.8891	1.2669	2.1560
–16	1.431	1.005	0.2149	174.52	321.65	496.17	0.9014	1.2508	2.1522
–14	1.552	1.009	0.1991	177.60	320.01	497.61	0.9136	1.2348	2.1484
–12	1.681	1.014	0.1846	180.71	319.33	499.05	0.9259	1.2189	2.1448
–10	1.818	1.018	0.1714	183.87	316.61	500.47	0.9383	1.2031	2.1414
–8	1.963	1.023	0.1593	187.02	314.87	501.89	0.9505	1.1875	2.1380
–6	2.118	1.028	0.1482	190.24	313.05	503.30	0.9629	1.1718	2.1347
–4	2.282	1.032	0.1380	193.47	311.23	504.70	0.9752	1.1563	2.1315
–2	2.456	1.037	0.1287	196.73	309.35	506.09	0.9876	1.1409	2.1285
0	2.640	1.042	0.1201	200.02	307.44	507.46	1.000	1.1255	2.1255
2	2.835	1.047	0.1122	203.35	305.48	508.83	1.0124	1.1102	2.1226
4	3.040	1.052	0.1049	206.71	303.48	510.18	1.0248	1.0949	2.1197
6	3.257	1.057	0.0981	210.09	301.44	511.53	1.0372	1.0798	2.1170
8	3.486	1.063	0.0919	213.48	299.38	512.86	1.0496	1.0648	2.1144
10	3.726	1.068	0.0862	216.93	297.25	514.18	1.0620	1.0498	2.1118
12	3.980	1.074	0.0808	220.39	295.09	515.48	1.0745	1.0348	2.1093
14	4.246	1.079	0.0759	223.87	292.90	516.77	1.0869	1.0200	2.1069
16	4.526	1.085	0.0713	227.41	290.63	518.04	1.0994	1.0051	2.1045
18	4.819	1.091	0.0670	230.97	288.33	519.30	1.1119	0.9903	2.1022
20	5.127	1.097	0.0631	234.54	286.00	520.54	1.1244	0.9756	2.1000
22	5.450	1.103	0.0594	238.13	283.64	521.77	1.1369	0.9610	2.0978
24	5.787	1.109	0.0559	241.77	281.20	522.98	1.1494	0.9463	2.0957
26	6.141	1.116	0.0527	245.44	278.73	524.16	1.1620	0.9317	2.0937
28	6.510	1.122	0.0497	249.12	276.22	525.34	1.1745	0.9172	2.0917
30	6.896	1.129	0.0469	252.86	273.62	526.48	1.1872	0.9026	2.0897
32	7.300	1.136	0.0443	256.57	271.05	527.61	1.1996	0.8882	2.0879
34	7.720	1.143	0.0419	260.34	268.38	528.72	1.2123	0.8737	2.0860
36	8.159	1.150	0.0396	264.19	265.61	529.80	1.2251	0.8591	2.0842
38	8.616	1.157	0.0374	267.99	262.87	530.86	1.2377	0.8448	2.0825

(Contd)

Temp.	p_{sat}	*Specific* *Liquid* $v_f \times 10^3$	*Volume* *Vapour* v_g	*Enthalpy* h_f	h_{fg}	h_g	*Entropy* s_f	s_{fg}	s_g
(°C)	**(bar)**	**(m³/kg)**	**(m³/kg)**	**(kJ/kg)**			**(kJ/kg.K)**		
40	9.092	1.165	0.0354	271.86	260.03	531.89	1.2505	0.8303	2.0808
42	9.587	1.172	0.0335	275.68	257.22	532.91	1.2630	0.8162	2.0792
44	10.103	1.180	0.0317	279.66	254.22	533.88	1.2760	0.8016	2.0776
46	10.638	1.188	0.0300	283.57	251.26	534.83	1.2888	0.7873	2.0760
48	11.195	1.197	0.0285	287.49	248.27	535.76	1.3015	0.7730	2.0745
50	11.774	1.206	0.0270	291.50	245.14	536.64	1.3145	0.7586	2.0731
52	12.374	1.214	0.0256	295.52	241.97	537.49	1.3275	0.7442	2.0717
54	12.997	1.224	0.0242	299.55	238.76	538.32	1.3405	0.7298	2.0703
56	13.643	1.233	0.0230	303.59	235.51	539.10	1.3536	0.7155	2.0690
58	14.313	1.243	0.0218	307.74	232.10	539.84	1.3669	0.7009	2.0678
60	15.008	1.253	0.0207	311.90	228.63	540.53	1.3803	0.6863	2.0666
62	15.727	1.264	0.0196	316.07	225.12	541.18	1.3937	0.6717	2.0654
64	16.471	1.275	0.0186	320.25	221.54	541.79	1.4073	0.6571	2.0644
66	17.242	1.286	0.0177	324.44	217.92	542.36	1.4208	0.6425	2.0634

Table B.8.2 Superheat table: R 152a vapour

p, bar (t_{sat}, °C)		sat.	–15	0	30	45	t,°C 70	95	120	145	170
1.01	v	0.2973	0.3097	0.3299	0.3695	0.3891	0.4214	0.4535	0.4854	0.5172	0.5488
(–24.13)	h	489.13	498.31	513.55	544.97	561.27	589.42	618.85	649.60	681.64	714.96
	s	2.1858	2.2215	2.2782	2.3857	2.4369	2.5188	2.5969	2.6710	2.7413	2.8076
1.50	v	0.2055		0.2195	0.2471	0.2607	0.2829	0.3050	0.3268	0.3485	0.3700
(–14.84)	h	495.91		511.48	543.54	560.05	588.46	618.08	648.96	681.10	714.49
	s	2.1672		2.2245	2.3335	2.3852	2.4678	2.5465	2.6207	2.6912	2.7576
2.50	v	0.1265		0.1275	0.1451	0.1537	0.1676	0.1812	0.1947	0.2080	0.2211
(–1.51)	h	505.32		507.01	540.52	557.49	586.47	616.47	647.63	679.98	713.54
	s	2.1449		2.1509	2.2634	2.3161	2.4000	2.4793	2.5544	2.6253	2.6921
3.50	v	0.0916			0.1013	0.1077	0.1181	0.1282	0.1380	0.1477	0.1573
(8.18)	h	511.84			537.36	554.85	584.42	614.884	646.28	678.85	712.57
	s	2.1314			2.2149	2.2689	2.3541	2.4344	2.5101	2.5815	2.6486
4.50	v	0.0717			0.0769	0.0822	0.0906	0.0987	0.1066	0.1143	0.1219
(15.82)	h	516.82			534.06	552.11	582.33	613.18	644.93	677.72	711.61
	s	2.1219			2.1770	2.2323	2.3190	2.4001	2.4765	2.5483	2.6158
5.50	v	0.0588			0.0613	0.0658	0.0731	0.0799	0.0865	0.0930	0.0993
(22.31)	h	520.86			530.58	549.26	580.19	611.49	643.56	676.57	710.64
	s	2.1147			2.1452	2.2018	2.2901	2.3722	2.4492	2.5216	2.5894
6.50	v	0.0498			0.0504	0.0545	0.0609	0.0669	0.0727	0.0782	0.0837
(27.96)	h	524.22			526.90	546.29	577.98	609.77	642.17	675.42	709.66
	s	2.1090			2.1172	2.1755	2.2654	2.3486	2.4263	2.4991	2.5672
7.50	v	0.0431				0.0461	0.0519	0.0573	0.0625	0.0674	0.0722
(32.96)	h	527.04				543.18	575.71	608.02	640.76	674.26	708.68
	s	2.1041				2.1519	2.2437	2.3279	2.4063	2.4796	2.5480
8.50	v	0.0379				0.0397	0.0451	0.0500	0.0547	0.0591	0.0634
(37.50)	h	529.50				539.91	573.38	606.24	639.34	673.10	707.70
	s	2.1001				2.1304	2.2242	2.3095	2.3886	2.4624	2.5312

(*Contd*)

p, bar (t_{sat}, °C)		sat.	−15	0	30	45	t, °C 70	95	120	145	170
9.50	v	0.0338				0.0345	0.0397	0.0443	0.0485	0.0526	0.0565
(41.65)	h	531.63				536.45	570.97	604.42	637.90	671.92	706.71
	s	2.0366				2.1104	2.2063	2.2928	2.3727	2.4469	2.5161
10.50	v	0.0304					0.0353	0.396	0.0435	0.0473	0.0509
(45.49)	h	533.49					568.48	602.57	636.45	670.73	705.72
	s	2.0936					2.1898	2.2776	2.3582	2.4329	2.5024
11.50	v	0.0277					0.0316	0.0357	0.0394	0.0429	0.0463
(49.06)	h	535.13					565.89	600.67	634.97	669.54	704.72
	s	2.0910					2.1743	2.2634	2.3448	2.4200	2.4899
12.50	v	0.0253					0.0285	0.0324	0.0360	0.0393	0.0424
(52.41)	h	536.56					563.20	598.74	633.48	668.33	703.72
	s	2.0886					2.1598	2.2503	2.3324	2.4082	2.4783
13.50	v	0.0233					0.0258	0.0296	0.0330	0.0361	0.0391
(55.57)	h	537.84					560.38	596.75	631.96	667.12	702.72
	s	2.0865					2.1459	2.2379	2.3208	2.3971	2.4676
14.50	v	0.0215					0.0235	0.0272	0.0305	0.0334	0.0363
(58.56)	h	538.95					557.44	594.72	630.42	665.89	701.71
	s	2.0847					2.1326	2.2262	2.3100	2.3867	2.4576
15.50	v	0.0199					0.0214	0.0251	0.0282	0.0311	0.0338
(61.38)	h	539.89					554.33	592.63	628.87	664.66	700.69
	s	2.0830					2.1198	2.1250	2.2997	2.3770	2.4483
16.50	v	0.0186					0.0196	0.0232	0.0263	0.0290	0.0316
(64.06)	h	540.69					551.03	590.49	627.29	663.41	699.67
	s	2.0874					2.1073	2.2045	2.2900	2.3678	2.4394
17.50	v	0.0174					0.0180	0.0216	0.0245	0.0272	0.0296
(66.62)	h	541.35					547.49	588.28	625.67	662.16	698.65
	s	2.0801					2.0951	2.1943	2.2808	2.3591	2.4311

*Ashok Babu T P, A *Theo. & Expt. Investigation of Alternatives to CFC 12 in Refrigerators*, Ph.D. Thesis, IIT Delhi, 1997.

B.9 THERMODYNAMIC PROPERTIES OF R 600a (ISOBUTANE)*

Table B.9.1 Saturation table of R 600a (isobutane)

Temp.	p_{sat}	*Specific Liquid* $v_f \times 10^3$	*Volume Vapour* v_g	*Enthalpy* h_f	h_{fg}	h_g	*Entropy* s_f	s_{fg}	s_g
(°C)	(bar)	(m³/kg)	(m³/kg)	(kJ/kg)			(kJ/kg.K)		
–40	0.287	1.604	1.1461	109.64	393.42	503.06	0.6446	1.6873	2.3320
–38	0.318	1.609	1.0432	114.14	391.52	505.66	0.6639	1.6649	2.3288
–36	0.351	1.614	0.9514	118.64	389.63	508.27	0.6829	1.6429	2.3258
–34	0.387	1.619	0.8692	123.13	387.76	510.89	0.7017	1.6214	2.3231
–32	0.425	1.624	0.7955	127.61	385.91	513.52	0.7203	1.6002	2.3206
–30	0.467	1.630	0.7292	132.09	384.06	516.16	0.7388	1.5795	2.3183
–28	0.512	1.635	0.6696	136.57	382.23	518.80	0.7571	1.5591	2.3162
–26	0.561	1.641	0.6158	141.05	380.40	521.45	0.7752	1.5391	2.3143
–24	0.612	1.646	0.5673	145.53	378.58	524.12	0.7932	1.5194	2.3126
–22	0.668	1.652	0.5233	150.02	376.77	526.79	0.8110	1.5001	2.3111
–20	0.728	1.657	0.4834	154.51	374.95	529.46	0.8287	1.4811	2.3098
–18	0.791	1.663	0.4472	159.00	373.14	532.15	0.8463	1.4624	2.3087
–16	0.859	1.669	0.4143	163.51	371.33	534.84	0.8638	1.4440	2.3077
–14	0.931	1.674	0.3843	168.02	369.51	537.54	0.8811	1.4258	2.3069
–12	1.008	1.680	0.3569	172.55	367.69	540.24	0.8984	1.4079	2.3063
–10	1.090	1.686	0.3319	177.09	365.87	542.95	0.9155	1.3903	2.3058
–8	1.176	1.692	0.3089	181.64	364.03	545.67	0.9326	1.3729	2.3055
–6	1.269	1.698	0.2880	186.20	362.19	548.39	0.9495	1.3557	2.3053
–4	1.366	1.704	0.2687	190.78	360.34	551.12	0.9664	1.3388	2.3052
–2	1.469	1.710	0.2510	195.38	358.47	553.85	0.9832	1.3220	2.3052
0	1.578	1.717	0.2347	200.00	356.59	556.59	1.0000	1.3054	2.3054
2	1.693	1.723	0.2196	204.64	354.70	559.33	1.0167	1.2891	2.3057
4	1.815	1.730	0.2058	209.30	352.78	562.08	1.0333	1.2728	2.3062
6	1.943	1.736	0.1930	213.98	350.85	564.83	1.0499	1.2568	2.3067
8	2.078	1.743	0.1811	218.69	348.90	567.58	1.0664	1.2409	2.3073
10	2.220	1.750	0.1701	223.42	346.92	570.34	1.0829	1.2252	2.3080
12	2.369	1.757	0.1600	228.18	344.92	573.10	1.0993	1.2096	2.3089
14	2.526	1.764	0.1505	232.96	342.90	575.86	1.1157	1.1941	2.3098
16	2.690	1.771	0.1417	237.78	340.84	578.62	1.1320	1.1787	2.3108
18	2.863	1.778	0.1336	242.62	338.76	581.38	1.1484	1.1635	2.3119
20	3.043	1.786	0.1260	247.50	336.65	584.15	1.1646	1.1484	2.3130
22	3.233	1.793	0.1189	252.40	334.51	586.91	1.1809	1.1333	2.3142
24	3.431	1.801	0.1123	257.34	332.33	589.68	1.1972	1.1184	2.3155
26	3.638	1.809	0.1061	262.32	330.12	592.44	1.2134	1.1035	2.3169
28	3.854	1.817	0.1004	267.33	327.87	595.20	1.2296	1.0887	2.3183
30	4.080	1.826	0.0950	272.37	325.59	597.96	1.2458	1.0740	2.3198
32	4.315	1.834	0.0899	277.46	323.26	600.72	1.2620	1.0593	2.3213

(*Contd*)

Temp.	p_{sat}	*Specific Liquid* $v_f \times 10^3$	*Volume Vapour* v_g	*Enthalpy* h_f	h_{fg}	h_g	*Entropy* s_f	s_{fg}	s_g
(°C)	**(bar)**	**(m^3/kg)**	**(m^3/kg)**	**(kJ/kg)**			**(kJ/kg.K)**		
34	4.561	1.843	0.0852	282.59	320.89	603.48	1.2781	1.0447	2.3228
36	4.817	1.852	0.0807	287.75	318.48	606.23	1.2943	1.0302	2.3244
38	5.083	1.861	0.0766	292.95	316.03	608.98	1.3104	1.0156	2.3261
40	5.361	1.870	0.0726	298.20	313.52	611.72	1.3266	1.0012	2.3278
42	5.649	1.879	0.0690	303.48	310.97	614.45	1.3427	0.9867	2.3294
44	5.949	1.889	0.0655	308.81	308.37	617.18	1.3589	0.9723	2.3312
46	6.261	1.899	0.0622	314.18	305.72	619.91	1.3750	0.9579	2.3329
48	6.585	1.910	0.0591	319.61	303.01	622.62	1.3911	0.9435	2.3346
50	6.920	1.920	0.0562	325.07	300.26	625.33	1.4073	0.9291	2.3364
52	7.269	1.931	0.0535	330.58	297.44	628.02	1.4234	0.9147	2.3381
54	7.630	1.943	0.0509	336.14	294.57	630.70	1.4395	0.9004	2.3399
56	8.004	1.954	0.0485	341.75	291.63	633.37	1.4557	0.8860	2.3416
58	8.392	1.966	0.0462	347.41	288.62	636.03	1.4718	0.8716	2.3434
60	8.793	1.979	0.0440	353.12	285.55	638.67	1.4879	0.8571	2.3451
62	9.209	1.991	0.0419	358.88	282.42	641.30	1.5041	0.8426	2.3467
64	9.638	2.005	0.0399	364.69	279.21	643.90	1.5202	0.8281	2.3483
66	10.083	2.019	0.0381	370.56	275.93	646.49	1.5364	0.8136	2.3499

Table B.9.2 Superheat table: R 600a (isobutane) vapour

p, bar (t_{sat}, °C)		sat.	−5	0	30	45	t, °C 70	95	120	145	170
0.25	v	1.3024	1.5216	1.5506	1.7242	1.8108	1.9552	2.0992	2.2432	2.3870	2.5307
(−42.68)	h	499.59	554.13	561.85	610.61	636.59	682.24	730.82	782.28	836.55	893.56
	s	2.3365	2.5556	2.5841	2.7535	2.8372	2.9754	3.1121	3.2473	3.3812	3.5137
0.50	v	0.6804	0.7543	0.7690	0.8568	0.9006	0.9733	1.0459	1.1183	1.1906	1.2628
(−30.04)	h	515.93	553.14	560.90	609.86	635.92	681.67	730.34	781.86	836.19	893.25
	s	2.3078	2.4536	2.4823	2.6524	2.7364	2.8749	3.0113	3.1473	3.2813	3.4138
0.75	v	0.4865	0.4984	0.5084	0.5677	0.5971	0.6460	0.6943	0.7433	0.7918	0.8401
(−19.00)	h	542.90	552.12	559.93	609.10	635.24	681.10	729.85	781.45	835.83	892.93
	s	2.3578	2.3926	2.4215	2.5924	2.6767	2.8155	2.9527	3.0884	3.2226	3.3553
1.01	v	0.3551	0.3654	0.3729	0.4173	0.4394	0.4759	0.5123	0.5485	0.5845	0.6205
(−11.91)	h	540.36	551.03	558.80	608.29	634.51	680.50	729.34	781.00	835.44	892.59
	s	2.3060	2.3464	2.3755	2.5473	2.6318	2.7711	2.9086	3.0444	3.1788	3.3116
1.50	v	0.2461		0.2476	0.2784	0.2936	0.3187	0.3436	0.3684	0.3930	0.4175
(−1.40)	h	554.68		556.92	606.77	633.16	679.36	728.37	780.17	934.72	891.96
	s	2.3054		2.3136	2.4871	2.5723	2.7123	2.8503	2.9866	3.1212	3.2542
2.50	v	0.1520			0.1625	0.1720	0.1877	0.2031	0.2183	0.2334	0.2484
(13.74)	h	575.52			603.51	630.27	676.97	726.35	778.45	833.23	890.66
	s	2.3100			2.4053	2.4918	2.6334	2.7727	2.9098	3.0451	3.1786
3.50	v	0.1101			0.1127	0.1199	0.1315	0.1428	0.1540	0.1650	0.1759
(24.50)	h	590.30			600.06	627.25	674.49	724.28	776.68	831.71	889.34
	s	2.3149			2.3476	2.4357	2.5792	2.7193	2.8578	2.9938	3.1279
4.50	v	0.0863				0.0908	0.1002	0.1093	0.1183	0.1270	0.1357
(33.41)	h	602.61				624.06	671.91	722.14	744.88	830.16	887.99
	s	2.3218				2.3911	2.5368	2.6783	2.8179	2.9546	3.0893

(*Contd*)

p, bar (t_{sat}, °C)		sat.	–5	0	30	45	t,°C 70	95	120	145	170
5.50	v	0.0708				0.0722	0.0802	0.0880	0.955	0.1028	0.1101
(40.91)	h	612.93				620.69	669.22	719.93	773.03	828.59	886.63
	s	2.3282				2.3530	2.5012	2.6448	2.7851	2.9226	3.0579
6.50	v	0.0599					0.0664	0.0732	0.0797	0.0861	0.0923
(47.49)	h	621.94					666.41	717.65	771.13	826.98	885.25
	s	2.3342					2.4700	2.6154	2.7569	2.8953	3.0312
7.50	v	0.0518					0.0561	0.0623	0.0681	0.0738	0.0793
(53.34)	h	629.86					663.46	715.29	769.18	825.34	883.85
	s	2.3396					2.4416	2.5891	2.7319	2.8713	3.0078
8.50	v	0.0456					0.0483	0.0539	0.0592	0.644	0.0693
(58.61)	h	636.88					660.34	712.84	767.18	823.66	882.42
	s	2.3442					2.4152	2.5660	2.7093	2.8497	2.9869
9.50	v	0.0405					0.0420	0.0473	0.0522	0.059	0.0615
(63.41)	h	643.17					657.04	710.29	765.12	821.95	880.97
	s	2.3481					2.3900	2.5426	2.6885	2.8300	2.9680
10.50	v	0.0364					0.0369	0.0419	0.0465	0.0509	0.0551
(67.83)	h	648.85					653.52	707.62	762.99	820.20	879.50
	s	2.3514					2.3655	2.5213	2.6691	2.8117	2.9506

* Ashok Babu T P, A *Theo. & Expt. Investigation of Alternatives to CFC 12 in Refrigerators*, Ph.D. Thesis, IIT Delhi, 1997.

B.10 THERMODYNAMIC PROPERTIES OF SATURATED R 11

Temp.	*Pressure*	*Specific Volume* (m^3/kg)		*Enthalpy* (kJ/kg)		*Entropy* (kJ/kg.K)	
(°C)	(bar)	$v_f \times 10^3$	v_g	h_f	h_g	s_f	s_g
0	.4018	.652	.4031	200.0	388.9	1.0000	1.6915
1	.4192	.653	.3875	200.9	389.4	1.0031	1.6908
2	.4373	.654	.3726	201.7	389.9	1.0063	1.6902
3	.4560	.655	.3584	202.6	390.4	1.0094	1.6895
4	.4754	.656	.3448	203.5	390.9	1.0125	1.6889
5	.4953	.657	.3319	204.3	391.4	1.0156	1.6883
6	.5160	.658	.3195	205.2	392.0	1.0186	1.6877
7	.5373	.659	.3077	206.1	392.5	1.0217	1.6871
8	.5593	.660	.2964	206.9	393.0	1.0248	1.6865
9	.5821	.661	.2856	207.8	393.5	1.0278	1.6859
10	.6055	.662	.2753	208.7	394.0	1.0309	1.6854
11	.6297	.663	.2655	209.5	394.5	1.0339	1.6849
12	.6547	.663	.2561	210.4	395.0	1.0369	1.6844
13	.6804	.665	.2470	211.3	395.5	1.0399	1.6839
14	.7070	.666	.2384	212.1	396.0	1.0429	1.6834
15	.7343	.667	.2301	213.0	396.5	1.0459	1.6829
16	.7625	.668	.2222	213.3	397.0	1.0489	1.6824
17	.7915	.669	.2146	214.7	397.5	1.0519	1.6819
18	.8214	.670	.2073	215.6	398.0	1.0549	1.6815
19	.8521	.671	.2004	216.9	398.6	1.0579	1.6811
20	.8838	.672	.1937	217.4	399.1	1.0608	1.6806
21	.9164	.673	.1872	218.2	399.6	1.0638	1.6802
22	.9499	.674	.1811	219.1	400.1	1.0667	1.6798
23	.9844	.675	.1752	220.0	400.6	1.0696	1.6794
24	1.0198	.676	.1695	220.8	401.1	1.0725	1.6791
25	1.0562	.677	.1640	221.7	401.6	1.0755	1.6787
26	1.0937	.678	.1588	222.6	402.1	1.0784	1.6793
27	1.1321	.679	.1538	223.5	402.6	1.0813	1.6780
28	1.1716	.680	.1489	224.3	403.1	1.0841	1.6776
29	1.2122	.681	.1442	225.2	403.6	1.0870	1.6773
30	1.2538	.683	.1398	226.1	404.1	1.0899	1.6770
32	1.3405	.685	.1313	227.9	405.1	1.0956	1.6764
34	1.4318	.687	.1235	229.6	406.1	1.1013	1.6758
36	1.5278	.690	.1161	231.4	407.1	1.1070	1.6753
38	1.6287	.692	.1094	233.1	408.1	1.1126	1.6748
40	1.7346	.694	.1032	234.9	409.1	1.1182	1.6743
45	2.0228	.700	.0892	239.3	411.5	1.1321	1.6732
50	2.3464	.707	.07764	243.8	413.9	1.1458	1.6724
55	2.7083	.713	.06780	248.2	416.3	1.1593	1.6717
60	3.1110	.720	.05945	252.7	418.7	1.1727	1.6711

B.11 THERMODYNAMIC PROPERTIES OF R 290/ R 600a MIXTURE*

Table B.11.1 Thermodynamic properties of saturated mixture of R 290/R 600a
Composition: Propane = 50%/Isobutane = 50% (by mass)

Bubble t	*Bubble p*	$v_f \times 10^3$	h_f	s_f	$v_g \times 10^3$	h_g	s_g
(°C)	(bar)	(m³/kg)	(kJ/kg)	(kJ/kg. K)	(m³.kg)	(kJ/kg)	(kJ/kg. K)
–40	0.750	1.682	107.82	0.7531	520.6454	510.88	2.3998
–38	0.819	1.688	112.35	0.7724	479.6430	513.64	2.3954
–36	0.893	1.694	116.89	0.7915	442.3487	516.42	2.3913
–34	0.973	1.700	121.42	0.8104	408.5479	519.22	2.3874
–32	1.057	1.706	125.97	0.8292	377.9978	522.04	2.3836
–30	1.148	1.712	130.51	0.8478	350.1872	524.87	2.3801
–28	1.244	1.719	135.06	0.8663	324.8254	527.72	2.3768
–26	1.346	1.725	139.62	0.8846	301.9677	530.59	2.3737
–24	1.455	1.731	144.18	0.9029	280.7593	533.47	2.3708
–22	1.570	1.737	148.76	0.9210	261.3344	536.37	2.3681
–20	1.693	1.744	153.34	0.9389	243.5231	539.29	2.3655
–18	1.822	1.751	157.94	0.9568	227.1760	542.23	2.3631
–16	1.959	1.757	162.55	0.9746	212.3951	545.18	2.3608
–14	2.104	1.764	167.17	0.9922	198.5801	548.16	2.3587
–12	2.257	1.771	171.80	1.0098	185.8725	551.15	2.3567
–10	2.418	1.778	176.45	1.0272	173.9780	554.16	2.3549
–8	2.588	1.785	181.12	1.0446	163.2164	557.18	2.3531
–6	2.767	1.792	185.81	1.0619	153.2965	560.23	2.3515
–4	2.955	1.799	190.52	1.0791	143.9803	563.29	2.3501
–2	3.153	1.806	195.25	1.0963	135.3818	566.37	2.3487
0	3.360	1.814	200.00	1.1134	127.2927	569.47	2.3474
2	3.578	1.821	204.78	1.1304	119.9372	572.59	2.3463
4	3.806	1.829	209.58	1.1473	113.0286	575.73	2.3452
6	4.045	1.837	214.40	1.1642	106.5303	578.89	2.3442
8	4.295	1.845	219.26	1.1811	100.6241	582.06	2.3433
10	4.556	1.853	224.14	1.1979	94.9280	585.25	2.3424
12	4.829	1.861	229.05	1.2147	89.7230	588.46	2.3416
14	5.115	1.870	233.99	1.2313	84.8589	591.69	2.3409
16	5.413	1.878	238.97	1.2480	80.3081	594.94	2.3402
18	5.723	1.887	243.07	1.2646	75.9500	598.21	2.3396
20	6.047	1.896	249.01	1.2812	71.9572	601.49	2.3390
22	6.384	1.905	254.08	1.2977	68.2095	604.80	2.3384
24	6.735	1.915	259.18	1.3142	64.6884	608.12	2.3379
26	7.100	1.924	264.32	1.3306	61.3771	611.46	2.3374
28	7.479	1.934	269.51	1.3471	58.2603	614.82	2.3368
30	7.874	1.944	274.72	1.3635	55.3241	618.20	2.3363

(Contd)

Bubble t	***Bubble p***	$v_f \times 10^3$	h_f	s_f	$v_g \times 10^3$	h_g	s_g
(°C)	(bar)	(m³/kg)	(kJ/kg)	(kJ/kg. K)	(m³.kg)	(kJ/kg)	(kJ/kg. K)
32	8.284	1.955	279.97	1.3798	52.4820	621.60	2.3358
34	8.709	1.966	285.26	1.3961	49.8727	625.01	2.3352
36	9.150	1.977	290.58	1.4123	47.4090	628.45	2.3347
38	9.608	1.988	295.94	1.4285	45.0814	631.90	2.3341
40	10.082	2.000	301.34	1.4446	42.8811	635.37	2.3334
42	10.574	2.012	306.78	1.4607	40.7998	638.87	2.3327
44	11.083	2.024	312.26	1.4767	38.8300	642.38	2.3319
46	11.610	2.037	317.78	1.4927	36.9647	645.91	2.3310
48	12.155	2.051	323.34	1.5086	35.1974	649.45	2.3300
50	12.720	2.065	328.95	1.5244	33.4648	653.02	2.3289
52	13.303	2.080	334.60	1.5402	31.8771	656.60	2.3277
54	13.907	2.095	340.29	1.5559	30.3701	660.21	2.3 263
56	14.530	2.111	346.03	1.5715	28.9389	663.83	2.3247
58	15.174	2.128	351.81	1.5870	27.5787	667.47	2.3230
60	15.840	2.145	357.63	1.6024	26.2849	671.13	2.3211
62	16.526	2.164	363.49	1.6177	25.0531	674.81	2.3189
64	17.235	2.184	369.39	1.6328	23.8792	678.51	2.3166
66	17.967	2.205	375.31	1.6478	22.7593	682.23	2.3140

Table B.11.2 Superheat table: R 290/R 600a vapour mixture composition: R 290 = 50%/R 600a = 50% (by mass)

p, bar (*Bubble t*, °C)		*Values at Dew t*	(*Temp. of Superheated Vapour*, °C) −15	0	30	45	70	95	120	145	170
1.01	*v*	0.3806	0.4121	0.4379	0.4891	0.5146	0.5569	0.5989	0.6409	0.6827	0.7244
(−33.15)	*h*	520.25	545.20	568.13	617.06	643.08	688.79	737.39	788.84	843.10	900.08
	s	2.4659	2.5726	2.6593	2.8298	2.9139	3.0524	3.1894	3.3248	3.4588	3.5913
1.50	*v*	0.2647	0.2748	0.2927	0.3280	0.3455	0.3745	0.4033	0.4320	0.4605	0.4890
(−23.40)	*h*	534.33	544.46	567.46	616.49	642.55	688.32	736.97	788.47	842.76	899.77
	s	2.4553	2.5047	2.5919	2.7631	2.8473	2.9863	3.1235	3.2591	3.3932	3.5258
2.50	*v*	0.1637		0.1706	0.1927	0.2036	0.2215	0.2392	0.2567	0.2741	0.2914
(−9.24)	*h*	555.31		566.02	615.29	641.45	687.34	736.10	787.69	842.06	899.14
	s	2.4471		2.5012	2.6738	2.7587	2.8983	3.0360	3.1721	3.3065	3.4394
3.50	*v*	0.1182			0.1339	0.1419	0.1550	0.1679	0.1806	0.1932	0.2057
(1.24)	*h*	571.41			614.04	640.30	686.34	735.22	786.90	841.35	898.49
	s	2.4454			2.6124	2.6978	2.8382	2.9766	3.1131	3.2479	3.3810
4.50	*v*	0.0933			0.1025	0.1090	0.1196	0.1299	0.1400	0.1500	0.1600
(9.31)	*h*	584.15			612.74	639.12	685.32	734.32	786.10	840.63	897.85
	s	2.4460			2.5669	2.6530	2.7943	2.9332	3.0702	3.2053	3.3388
5.50	*v*	0.0766			0.0818	0.0874	0.0963	0.1049	0.1134	0.1218	0.1300
(16.30)	*h*	595.43			611.38	637.89	684.26	733.40	785.29	839.90	897.19
	s	2.4477			2.5283	2.6152	2.7574	2.8970	3.0345	3.1700	3.3038
6.50	*v*	0.0649			0.0674	0.0723	0.0801	0.0877	0.0950	0.1022	0.1092
(22.41)	*h*	605.47			609.95	636.62	683.18	732.45	784.46	839.17	896.53
	s	2.4499			2.4950	2.5828	2.7260	2.8663	3.0043	3.1402	3.2743
7.50	*v*	0.0562			0.0569	0.0613	0.0683	0.0750	0.0815	0.0879	0.0941
(27.81)	*h*	614.50			608.44	635.28	682.06	731.49	783.62	838.42	895.86
	s	2.4523			2.4654	2.5542	2.6986	2.8397	2.9783	3.1146	3.2490

(*Contd*)

p, bar (Bubble t, °C)		Values at Dewt	– 15	0	30	(Temp. of Superheated Vapour, °C) 45	70	95	120	145	170
8.50	v	0.0496				0.0529	0.0594	0.0655	0.0713	0.0770	0.0826
(32.66)	h	622.72				633.88	680.91	730.51	782.76	837.66	895.19
	s	2.4547				2.5285	2.6742	2.8162	2.9553	3.0921	3.2268
9.50	v	0.0439				0.0459	0.0518	0.0574	0.0628	0.0680	0.0730
(37.37)	h	630.82				632.41	679.71	729.50	781.89	836.90	894.51
	s	2.4573				2.5034	2.6506	2.7935	2.9333	3.0706	3.2056
10.50	v	0.0396				0.0405	0.0461	0.0514	0.0563	0.0611	0.0658
(41.45)	h	637.90				630.85	678.47	728.47	781.00	836.12	893.82
	s	2.4595				2.4813	2.6302	2.7741	2.9145	3.0523	3.1877
11.50	v	0.0357					0.0410	0.0459	0.9506	0.0550	0.0593
(45.55)	h	645.12					677.18	727.41	780.10	835.34	893.13
	s	2.4619					2.6094	2.7545	2.8957	3.0340	3.1698
12.50	v	0.0327					0.0370	0.0417	0.0461	0.0502	0.0542
(49.14)	h	651.48					675.84	726.31	779.18	834.54	892.43
	s	2.4638					2.5911	2.7375	2.8794	3.0182	3.1544
13.50	v	0.0300					0.0335	0.0380	0.0422	0.0461	0.0499
(52.54)	h	657.58					674.44	725.19	778.24	833.73	891.72
	s	2.4657					2.5735	2.7212	2.8640	3.0033	3.1399
14.50	v	0.0277					0.0305	0.0348	0.0388	0.0425	0.0461
(55.79)	h	663.45					672.96	724.03	777.28	832.91	891.00
	s	2.4673					2.5563	2.7055	2.8492	2.9892	3.1262

(Contd)

p, bar (Bubble t, °C)		Values at Dew t	Temp. of Superheated Vapour, °C)								
			– 15	0	30	45	70	95	120	145	170
15.50	v	0.0256					0.0277	0.0320	0.0358	0.0394	0.0428
(58.90)	h	665.11					671.40	722.84	776.30	832.08	890.27
	s	2.4688					2.5393	2.6904	2.8351	2.9757	3.1131
16.50	v	0.0238					0.0253	0.0295	0.0332	0.0366	0.0399
(61.87)	h	674.58					669.74	721.60	775.29	831.23	889.54
	s	2.4701					2.5226	2.6758	2.8216	2.9628	3.1007
17.50	v	0.0221					0.0231	0.0273	0.0309	0.0342	0.0373
(64.73)	h	679.86					667.97	720.32	774.27	830.37	888.80
	s	2.4711					2.5059	2.6615	2.8085	2.9504	3.0888
18.50	v	0.0208					0.0214	0.0255	0.0290	0.0322	0.0352
(67.13)	h	684.34					666.05	718.99	773.22	829.50	888.05
	s	2.4718					2.4911	2.6493	2.7974	2.9400	3.0788
19.50	v	0.0195					0.0195	0.0237	0.0272	0.0303	0.0332
(69.74)	h	689.23					663.95	717.59	772.14	828.61	887.29
	s	2.4723					2.4741	2.6357	2.7853	2.9287	3.0680

* Ashok Babu T P, A *Theo. & Expt. Investigation of Alternatives to CFC 12 in Refrigerators*, Ph.D. Thesis, IIT Delhi, 1997.

B.12 THERMODYNAMIC PROPERTIES OF R 718 (WATER)*

Table B.12.1 Thermodynamic properties of saturated water and steam, temperature from triple point to 50°C

t_{sat} (°C)	p_{sat} (kPa)	v_f (L/kg)	v_g (m^3/kg)	h_f (kJ/kg)	h_{fg} (kJ/kg)	h_g (kJ/kg)	s_f (kJ/kg. K)	s_g (kJ/kg. K)
0.01	0.611	1.000	206.2	0.0	2501.6	2501.6	0	9.157
2	0.705	1.000	179.9	8.4	2496.8	2505.2	0.031	9.105
4	0.813	1.000	157.3	16.8	2402.1	2508.9	0.061	9.053
6	0.935	1.000	137.8	25.2	2487.4	2512.6	0.091	9.001
8	1.072	1.000	121.0	33.6	2482.6	2516.2	0.121	8.951
10	1.2276	1.000	106.4	42.0	2477.9	2519.9	0.151	8.902
12	1.401	1.000	93.8	50.4	2473.2	2522.6	0.180	8.854
14	1.597	1.001	82.9	58.8	2468.5	2527.2	0.210	0.806
16	1.817	1.001	73.4	67.1	2463.8	2530.9	0.239	8.759
18	2.062	1.001	65.1	75.5	2459.0	2534.5	0.268	8.713
20	2.34	1.002	57.8	83.9	2454.3	2538.2	80.296	8.668
22	2.64	1.002	51.5	92.2	2449.6	2541.8	0.325	8.624
24	2.98	1.003	45.9	100.6	2444.9	2545.5	0.353	8.581

(Contd)

t_{sat} (°C)	p_{sat} (kPa)	v_f (L/kg)	v_g (m^3/kg)	h_f (kJ/kg)	h_{fg} (kJ/kg)	h_g (kJ/kg)	s_f (kJ/kg. K)	s (kJ/kg. K)
25	3.17	1.003	43.4	104.8	2442.5	2547.3	0.367	8.559
26	3.36	1.003	41.4	108.9	2440.2	2349.1	0.381	8.538
28	3.78	1.004	36.7	117.3	2435.4	2552.7	0.409	8.496
30	4.246	1.004	32.9	125.7	2430.7	2556.4	0.437	8.455
32	4.75	1.005	29.6	134.0	2425.9	2560.0	0.464	8.414
34	5.32	1.006	26.6	142.4	2421.2	2563.6	0.491	8.374
36	5.94	1.006	24.0	150.7	2416.4	2567.2	0.518	8.335
38	6.62	1.007	21.6	159.1	2411.7	2570.8	0.545	8.296
40	7.384	1.008	19.55	167.5	2406.9	2574.4	0.572	8.258
42	8.20	1.009	17.69	175.8	2402.1	2577.9	0.599	8.221
44	9.10	1.009	16.04	184.2	2397.3	2581.5	0.625	8.184
46	10.09	1.010	14.56	192.5	2392.5	2585.1	0.651	8.148
48	11.16	1.011	13.23	200.9	2387.7	2588.6	0.678	8.113
50	12.349	1.012	12.05	209.3	2382.9	2592.2	0.704	8.078

*Haywood RW, *Thermodynamic Tables in SI Units*, Combridge University Press, 1960, p. 9.

Table B.12.2 Thermodynamic properties of saturated water and steam, pressures from triple point to 20 bar*

p (kPa)	t (°C)	v_f (L/kg)	v_g (m^3/kg)	h_f (kJ/kg)	h_{fg} (kJ/kg)	h_g (kJ/kg)	s_f (kJ/kg. K)	s_g (kJ/kg. K)
1	2	3	4	5	6	7	8	9
0.611	0.01	1.000	206.2	+0.0	2501.6	2501.6	0.0	9.157
0.8	3.8	1.000	159.7	15.8	2492.6	2508.6	0.058	9.058
1.0	7.0	1.000	129.2	29.3	2485.0	2514.4	0.106	8.977
1.2	9.7	1.000	108.7	40.6	2478.7	2519.3	0.146	8.910
1.4	1.20	1.000	93.9	50.3	2473.2	2523.5	0.180	8.854
1.6	14.0	1.001	82.8	58.9	2468.4	2527.3	0.210	8.805
1.8	15.9	1.001	74.0	66.5	2464.1	2530.6	0.237	8.763
2.0	17.5	1.001	67.0	73.5	2460.2	2533.6	0.261	8.725
2.2	19.0	1.002	61.2	79.8	2456.6	2536.4	0.282	8.690
2.4	20.4	1.002	56.4	85.7	2453.3	2539.0	0.302	8.659
2.6	21.7	1.002	52.3	91.1	2450.2	2541.3	0.321	8.630
2.8	23.0	1.002	48.7	96.2	2447.3	2543.6	0.338	8.603
3.0	24.1	1.003	45.7	101.0	2444.6	2545.6	0.354	8.578
3.5	26.7	1.003	39.5	111.8	2438.5	2550.4	0.391	8.523
4.0	29.0	1.004	34.8	121.4	2433.1	2554.5	0.422	8.475
4.5	31.0	1.005	31.1	130.0	2428.2	2558.2	0.451	8.433
5.0	32.9	1.005	28.2	137.8	2423.8	2561.6	0.476	8.396
6	36.2	1.006	23.74	151.5	2416.0	2567.5	0.521	8.331
7	39.0	1.007	20.53	163.4	2409.2	2572.6	0.559	8.277
8	41.5	1.008	18.10	173.9	2403.2	2577.1	0.593	8.230

(Contd)

p (kPa)	t (°C)	v_f (L/kg)	v_g (m^3/kg)	h_f (kJ/kg)	h_{fg} (kJ/kg)	h_g (kJ/kg)	s_f (kJ/kg. K)	s_g (kJ/kg. K)
1	2	3	4	5	6	7	8	9
9	43.8	1.009	16.20	183.3	2397.9	2581.1	0.622	8.188
10	45.8	1.010	14.67	191.8	2392.9	2584.8	0.649	8.151
11	47.7	1.011	13.42	199.7	2388.4	2588.1	0.674	8.118
12	49.4	1.012	12.36	206.9	2284.3	2591.2	0.696	8.087
13	51.1	1.013	11.47	213.7	2380.4	2594.0	0.717	8.087
14	52.6	1.013	10.69	220.0	2376.7	2596.7	0.737	8.033
15	54.0	1.014	10.02	226.0	2373.2	2599.2	0.755	8.009
16	55.3	1.015	9.43	231.6	2370.0	2601.6	0.772	7.987
17	56.6	1.015	8.91	236.9	2366.9	2603.8	0.788	7.966
18	57.8	1.016	8.45	242.0	2363.9	2605.9	0.804	7.946
19	59.0	1.017	8.03	246.8	2361.1	2607.9	0.818	7.927
20	60.1	1.017	7.65	251.5	2358.4	2609.9	0.832	7.909
22	62.2	1.018	7.00	260.1	2353.3	2613.5	0.858	7.876
24	64.1	1.019	6.45	268.2	2348.6	2616.8	0.882	7.846
26	65.9	1.020	5.98	275.7	2344.2	2619.9	0.904	7.819
28	67.5	1.021	5.58	282.7	2340.0	2622.7	0.925	7.793

(Contd)

p (kPa)	t (°C)	v_f (L/kg)	v_g (m^3/kg)	h_f (kJ/kg)	h_{fg} (kJ/kg)	h_g (kJ/kg)	s_f (kJ/kg. K)	s_g (kJ/kg. K)
1	2	3	4	5	6	7	8	9
30	69.1	1.022	5.23	289.3	2336.1	2625.4	0.944	7.770
35	72.7	1.025	4.53	304.3	2327.2	2631.5	0.988	7.717
40	75.9	1.027	3.99	317.7	2319.2	2636.9	1.026	7.671
45	78.7	1.028	3.58	329.6	2312.0	2641.7	1.060	7.631
50	81.3	1.030	3.24	340.6	2305.4	2646.0	1.091	7.595
55	83.7	1.032	2.96	350.6	2299.3	2649.9	1.119	7.562
60	86.0	1.033	2.73	359.9	2293.6	2653.6	1.145	7.533
65	88.0	1.035	2.53	368.6	2288.3	2656.9	1.170	7.506
70	90.0	1.036	2.36	376.8	2283.3	2660.1	1.192	7.480
75	91.8	1.037	2.22	384.5	2278.6	2663.0	1.213	7.457
80	93.5	1.039	2.087	391.7	2274.1	2665.8	1.233	7.435
85	95.2	1.040	1.972	398.6	2269.8	2668.4	1.252	7.415
90	96.7	1.041	1.869	405.2	2265.6	2670.9	1.270	7.395
95	98.2	1.042	1.777	411.5	2261.7	2673.2	1.287	7.377
100	99.6	1.043	1.694	417.5	2257.9	2675.4	1.303	7.360
101	100.0	1.044	1.673	419.1	2256.9	2676.0	1.307	7.355
110	102.3	1.046	1.549	428.8	2250.8	2679.6	1.333	7.328
120	104.8	1.048	1.428	439.4	2244.1	2683.4	1.361	7.298

(Contd)

p (kPa)	t (°C)	v_f (L/kg)	v_g (m^3/kg)	h_f (kJ/kg)	h_{fg} (kJ/kg)	h_g (kJ/kg)	s_f (kJ/kg. K)	s_g (kJ/kg. K)
1	2	3	4	5	6	7	8	9
140	109.3	1.051	0.236	458.4	2231.9	2690.3	1.411	7.247
150	111.4	1.053	1.159	467.1	2226.2	2693.4	1.434	7.223
160	113.3	1.055	1.091	475.4	2220.9	2696.2	1.455	7.202
170	115.2	1.056	1.031	483.2	2215.7	2699.0	1.475	7.181
180	116.9	1.058	0.977	490.7	2210.8	2701.5	1.494	7.162
190	118.6	1.059	0.929	497.8	2206.1	2704.0	1.513	7.144
200	120.2	1.061	0.885	504.7	2201.6	2706.3	1.530	7.127
220	123.3	1.064	0.810	517.6	2193.0	2710.6	1.563	7.095
240	126.1	1.066	0.746	529.6	2184.9	2714.5	1.593	7.066
260	128.1	1.069	0.693	540.9	2177.3	2718.2	1.621	7.039
280	131.2	1.071	0.646	551.4	2170.1	2721.5	1.647	7.014
300	133.5	1.074	0.606	561.4	2163.2	2724.7	1.672	6.991
320	135.8	1.076	0.570	570.9	2156.7	2727.6	1.695	6.969
340	137.9	1.078	0.538	579.9	2150.4	2730.3	1.717	6.949
360	139.9	1.080	0.510	588.5	2144.4	2732.4	1.738	6.930
380	141.8	1.082	0.485	596.8	2138.6	2753.4	1.757	6.912
400	143.6	1.084	0.462	604.7	2133.0	2737.6	1.776	6.894

(*Contd*)

p (kPa)	t (°C)	v_f (L/kg)	v_g (m^3/kg)	h_f (kJ/kg)	h_{fg} (kJ/kg)	h_g (kJ/kg)	s_f (kJ/kg. K)	s_g (kJ/kg. K)
1	2	3	4	5	6	7	8	9
420	145.4	1.086	0.442	612.3	2127.5	2739.8	1.795	6.878
440	147.1	1.088	0.423	619.6	2122.3	2741.9	1.812	6.862
460	148.7	1.089	0.405	626.7	2117.2	2743.9	1.829	6.847
480	150.3	1.091	0.389	633.5	2112.2	2745.7	1.845	6.833
500	151.8	1.093	0.375	640.1	2107.4	2747.5	1.860	6.819
550	155.5	1.097	0.342	655.8	2095.9	2751.7	1.897	6.787
600	158.8	1.101	0.315	670.4	2085.0	2755.5	1.931	6.758
650	162.0	1.105	0.292	684.1	2074.7	2750.9	1.962	6.730
700	165.0	1.108	0.273	697.1	2064.9	2762.0	1.902	6.705
750	167.8	1.112	0.2554	709.3	2055.5	2764.8	2.020	6.682
800	170.4	1.1115	0.2403	720.9	2046.5	2767.5	2.046	6.660
850	172.9	1.1118	0.2268	732.0	2037.9	2769.9	2.071	6.639
900	175.4	1.1121	0.2148	742.6	2029.5	2772.1	2.094	6.619
950	177.7	1.1124	0.2040	752.8	2021.4	2774.2	2.117	6.601
1000	179.9	1.1127	0.1943	762.6	2013.6	2776.2	2.138	6.583

(Contd)

p (kPa)	t (°C)	v_f (L/kg)	v_g (m^3/kg)	h_f (kJ/kg)	h_{fg} (kJ/kg)	h_g (kJ/kg)	s_f (kJ/kg. K)	s_g (kJ/kg. K)
1	2	3	4	5	6	7	8	9
1050	182.01	1.1130	0.1855	772.0	2005.9	2778.0	2.159	6.566
1100	184.1	1.1133	0.1774	781.1	1998.5	2779.7	2.179	6.550
1150	186.0	1.1136	0.1700	789.9	1991.3	2781.3	21.98	6.534
1200	188.0	1.1139	0.1632	798.4	1984.3	2782.7	2.216	6.519
1250	189.8	1.1141	0.1569	806.7	1977.4	2784.1	2.234	6.505
1300	191.6	1.1144	0.1511	814.7	1970.7	2785.4	2.251	6.491
1400	195.0	1.1149	0.1407	830.1	1957.7	2787.8	2.284	6.465
1500	198.3	1.1154	0.1317	844.7	1945.2	2789.9	2.314	6.441
1600	201.4	1.1159	0.1237	858.6	1933.2	2791.7	2.344	6.418
1700	204.3	1.1163	0.1166	871.8	1921.5	2793.4	2.371	6.396
1800	207.1	1.1168	0.1103	884.6	1910.3	2794.8	2.398	6.375
1900	209.8	1.1175	0.1047	896.8	1899.3	2796.1	2.423	6.355
2000	212.4	1.1177	0.0995	908.6	1888.6	2797.2	2.447	6.337

* Haywood R W, *Thermodynamic Tables in S.I. Units*, Cambridge University Press, pp. 10–13.

Table B.12.3 Enthalpy (in kJ/kg) and entropy (in kJ/kg. K) of superheated steam*

Temperature	*Pressure, bar*													
	0		*0.1*		*0.5*		*1*		*5*		*10*		*20*	
	h	*s*	*h*	*s*	*h*	*s*	*h*	*s*	*h*	*s*	*h*	*s*	*h*	*s*
50	2595	Infinite	2593	8.176										
75	2642		2640	8.317										
100	2689		2688	8.449	2683	7.695	2676	7.362						
125	2736		2735	8.572	2731	7.822	2726	7.492						
150	2784		2783	8.689	2780	7.941	2776	7.614						
175	2832		2831	8.799	2829	8.053	2826	7.720	2800	6.947				
200	2880		2880	8.905	2878	8.159	2875	7.835	2855	7.059	2827	6.692		
225	2929		2928	9.005	2927	8.260	2925	7.937	2909	7.169	2886	6.815	2834	6.412
250	2978		2977	9.101	2976	8.356	2975	8.034	2961	7.272	2943	6.926	2902	6.545
275	3027		3027	9.193	3026	8.449	3024	8.127	3013	7.369	2998	7.029	2965	6.663
300	3077		3077	9.282	3076	8.538	3074	8.217	3065	7.461	3052	7.125	3025	6.770

* Haywood R W, *Thermodynamic Tables in S.I. Units*, Cambridge University Press, 1968, pp. 16–17.

Table B.12.4 Water vapour pressure (1 mmHg = 133 Pa)

t_{sat} (°C)	p_{sat} (mm Hg)	t_{sat} (°C)	p_{sat} (mm Hg)
–50	0.02955	0	4.579
–48	0.0378	1	5.29
–46	0.0481	2	4.93
–44	0.0609	3	5.69
–42	0.0768	4	6.10
–40	0.0966	5	6.54
–38	0.1209	6	7.01
–36	0.1507	7	7.51
–34	0.1873	8	8.05
–32	0.23188	9	8.61
–30	0.2859	10	9.21
–29	0.317	11	9.84
–28	0.351	12	10.52
–27	0.389	13	11.23
–26	0.430	14	11.99
–25	0.476	15	12.79
–24	0.526	16	13.63
–23	0.580	17	14.53
–22	0.640	18	15.48
–21	0.705	19	16.48
–20	0.747	20	17.54
–19	0.852	21	18.65
–18	0.937	22	19.83
–17	1.029	23	24.07
–16	1.130	24	22.38
–15	1.239	25	23.76
–14	1.359	26	25.21
–13	1.488	27	26.74
–12	1.629	28	28.35
–11	1.783	29	30.40
–10	1.948	30	31.82
–9	2.128	31	33.70
–8	2.329	32	35.66
–7	2.535	33	37.73
–6	2.763	34	39.90

(Contd)

t_{sat} (°C)	p_{sat} (mm Hg)	t_{sat} (°C)	p_{sat} (mm Hg)
−5	3.011	35	42.18
−4	3.278	36	44.56
−3	3.567	37	47.07
−2	3.380	38	49.69
−1	4.217	39	52.44
0	4.579	40	55.32
41	58.34	71	243.9
42	61.50	72	254.6
43	64.80	73	265.7
44	68.26	74	277.2
45	71.88	75	289.1
46	75.65	76	301.4
47	79.60	77	314.1
48	83.71	78	327.3
49	88.02	79	341.0
50	92.51	80	355.1
51	97.20	81	396.7
52	102.1	82	384.9
53	107.2	83	400.6
54	112.5	84	416.8
55	118.0	85	433.6
56	123.8	86	450.9
57	129.8	87	468.7
58	136.1	88	487.1
59	142.6	89	506.1
60	149.4	90	525.8
61	156.4	91	546.1
62	163.8	92	567.0
63	171.4	93	588.6
64	179.3	94	610.9
65	187.5	95	633.9
66	196.1	96	657.6
67	205.0	97	682.1
68	214.2	98	707.3
70	233.7	100	760.0

B.13 OUTDOORS DESIGN DATA*

Place		*Summer*			*Monsoon*			*Winter*		*Latitude*
	DBT	*WBT*	*RH*	*DBT*	*WBT*	*RH*	*DBT*	*WBT*	*RH*	
Bombay	35	28.3	60	29.4	27.8	88	18.3	14.4	65	18.54
Calcutta	37.8	28.3	49	32.2	30.0	85	13.3	89	55	22.32
Delhi	43.3	23.9	20	35.0	28.3	60	7.2	5.0	70	28.35
Guwahati	32.3	25.7	59	31.7	27.8	78	11.1	8.3	69	26.11
Kanpur	42.8	25.0	23	36.1	28.9	58	7.2	5.6	80	26.26
Madras	39.4	27.8	41	28.3	26.7	88	18.3	13.9	60	13.04

*Temperatures in °C, relative humidity in per cent and latitude in degrees.

B.14 THE ERROR FUNCTION*

η	erf η	η	erf η	η	erf η
0.00	0.00000	0.76	0.71754	1.52	0.96841
0.02	0.02256	0.78	0.73001	1.54	0.97059
0.04	0.04511	0.80	0.74210	1.56	0.97263
0.06	0.06762	0.82	0.75381	1.58	0.97455
0.08	0.09008	0.84	0.76514	1.60	0.97635
0.10	0.11246	0.86	0.77610	1.62	0.97804
0.12	0.13476	0.88	0.78669	1.64	0.97962
0.14	0.15695	0.90	0.79691	1.66	0.98110
0.16	0.17901	0.92	0.80677	1.68	0.98249
0.18	0.20094	0.94	0.81627	1.70	0.98379
0.20	0.22270	0.96	0.82542	1.72	0.98500
0.22	0.24430	0.98	0.83423	1.74	0.98613
0.24	0.26570	1.00	0.84270	1.76	0.98719
0.26	0.28690	1.02	0.85084	1.78	0.98817
0.28	0.30788	1.04	0.85865	1.80	0.98909
0.30	0.32863	1.06	0.86614	1.82	0.98994
0.32	0.34913	1.08	0.87333	1.84	0.99074
0.34	0.36936	1.10	0.88020	1.86	0.99147
0.36	0.38933	1.12	0.88079	1.88	0.99216
0.38	0.40901	1.14	0.89308	1.90	0.99279
0.40	0.42809	1.16	0.89910	1.92	0.99338
0.42	0.44749	1.18	0.90484	1.94	0.99392
0.44	0.46622	1.20	0.91031	1.96	0.99443
0.46	0.48466	1.22	0.91553	1.98	0.99489
0.48	0.50272	1.24	0.92050	2.00	0.995322
0.50	0.52050	1.26	0.92524	2.10	0.997020
0.52	0.53790	1.28	0.92973	2.20	0.98137
0.54	0.55494	1.30	0.93401	2.30	0.998857
0.56	0.57162	1.32	0.93806	2.40	0.999311
0.58	0.58792	1.34	0.94191	2.50	0.999593
0.60	0.60386	1.36	0.94656	2.60	0.999764
0.62	0.61941	1.38	0.94902	2.70	0.999866
0.64	0.63459	1.40	0.95228	2.80	0.999925
0.66	0.64938	1.42	0.95638	2.90	0.999959
0.68	0.66278	1.44	0.95830	3.00	0.999978
0.70	0.67780	1.46	0.96105	3.20	0.999994
0.72	0.69143	1.48	0.96365	3.40	0.999998
0.74	0.70468	1.50	0.96610	3.60	1.000000

* Boelter L M K, V H Cherry, H A Johnson, and R C Martinelli, *Heat-transfer Notes*, Univ. of California Press, Berkeley, California, 1948.

B.15 CONVERSION TABLES

Lengths

(m)	(ft)
1	3.2808
0.3048	1

Masses

(kg)	(lb)
1	2.2046226
0.45359237	1

Areas/Surfaces

(m^2)	(ft^2)
1	10.764
0.092903	1

Densities

(kg/m^3)	(lb/ft^3)
1	0.062428
16.02	1

Volumes

(L)	(m^3)	(ft^3)	(US Gallons)
1000	1	35.3	264.2
	0.0283	1	7.484
3.785	0.003785	0.1336	1

Forces

(N)	(kg_f)	(lb_f)
1	0.102	0.22481
9.80665	1	2.20462
4.448	0.4536	1

Pressures

(bar)	(ata) $\left(\frac{kgf}{cm^2}\right)$	$\left(\frac{lb_f}{in^2}\right)$	(atm)	(torr) or (mm Hg)	(m H_2O)
1	1.02	14.5	0.987	750	10.2
0.9807	1	14.22	0.968	736	10.0
0.0689	0.0703	1	0.068	51.7	0.703
0.01325	1.033	14.7	1	760	10.33
0.00133	0.00136	0.01934	0.001316	1	0.0136

Velocities

(m/s)	(ft/min)
1	196.9
0.00508	1

Flow Rates

(L/s)	(m^3/s)	$\frac{m^3}{min}$, cmm	$\frac{ft^3}{min}$, cfm	$\frac{Gal}{min}$, gpm
1		0.06	2.1	
16.67		1	35.3	
0.472		0.0283	1	
1000	1	60	2119	15,850

Dynamic Viscosities

(Pa s) or $\left(\frac{Ns}{m^2}\right)$ or $\left(\frac{kg}{ms}\right)$	$\left(\frac{lb_f\ h}{ft^2}\right)$	$\left(\frac{lbm}{fts}\right)$	(cP)
1	5.8016×10^{-6}	0.672	1,000
1.488		1	

Kinematic Viscosities and Diffusivities

$\left(\frac{m^2}{s}\right)$	$\left(\frac{ft^2}{s}\right)$	(cS)
1	10.76	10^6
0.0929	1	

(Contd)

$$1\text{P (poise)} = 1\left(\frac{\text{dynes}}{\text{cm}^2}\right) = \left(1\frac{\text{g}}{\text{cm s}}\right) = \left(0.1\frac{\text{kg}}{\text{m s}}\right)$$

$$1\text{ St (stoke)} = 1\left(\frac{\text{cm}^2}{\text{s}}\right) = \left(10^{-4}\frac{\text{m}^2}{\text{s}}\right)$$

Power

(kW)	(Btu/h)	(hp)
1	3,412	1.341

Energies

(J) or (N m) or (W s)	(kgf m)	(ft lbf)	(kW h)	Horsepower hours Metric	Horsepower hours Imperial	(kcal)	(Btu)
1	0.102	0.738	0.0^5278	0.0^6378	0.0^6372	0.0^3239	0.0^3948
9.80665	1	7.23	0.0^5272	0.0^537	0.0^5365	0.0^234	0.0^2929
1.356	0.1383	1	0.0^6377	0.0^6512	0.0^6505	0.0^3324	0.0^21285
3.6×10^6	0.67×10^5	2.66×10^6	1	1.36	1.341	860	3412
2.65×10^5	2.7×10^5	1.953×10^6	0.736	1	9.986	632	2510
2.68×10^6	2.74×10^5	1.98×10^6	0.746	10.14	1	641	2540
41	427	3090	0.0^21163	0.0^21581	0.0^2156	1	3.97
1055	107.6	778	0.0^3293	0.0398	0.0^393	0.252	1

Enthalpies

$\left(\frac{\text{kJ}}{\text{kg}}\right)$	$\left(\frac{\text{kcal}}{\text{kg}}\right)$	$\left(\frac{\text{Btu}}{\text{lb}}\right)$
1	0.239	0.42993
4.1868	1	1.8
2.326	0.556	1

Heat Transfer Coefficients

$\left(\frac{\text{W}}{\text{m}^2\text{ K}}\right)$	$\left(\frac{\text{k cal}}{\text{m}^2\text{ h°C}}\right)$	$\left(\frac{\text{Btu}}{\text{ft}^2\text{ hr°F}}\right)$
1	0.86	0.17612
1.163	1	0.205
5.68	4.88	1

Entropies and Specific Heats

$\left(\frac{\text{kJ}}{\text{kg K}}\right)$	$\left(\frac{\text{kcal}}{\text{kg °C}}\right)$	$\left(\frac{\text{Btu}}{\text{lb °F}}\right)$
1	0.2389	0.2389
4.1868	1	1
4.1868	1	1

Heat Radiation

$\left(\frac{\text{W}}{\text{m}^2\text{ K}^4}\right)$	$\left(\frac{\text{k cal}}{\text{m}^2\text{ h K}^4}\right)$	$\left(\frac{\text{Btu}}{\text{ft}^2\text{ hr°F}}\right)$
1	0.86	0.0302
1.163	1	0.0350
33.1	28.5	1

Thermal Conductivities

$\left(\frac{\text{W}}{\text{m K}^-}\right)$	$\left(\frac{\text{k cal}}{\text{m h °C}}\right)$	$\left(\frac{\text{Btu}}{\text{ft hr°F}}\right)$	$\left(\frac{\text{Btu-in}}{\text{ft}^2\text{ hr°F}}\right)$
1	0.86	0.57	6.934
1.163	1	0.671	
1.73	1.488	1	

Index

A

B

C

F

G

H

O

P

Q

R

S

T

U

V

W

Y

Z